Observing and Cataloguing Nebulae and Star Clusters
From Herschel to Dreyer's New General Catalogue

The *New General Catalogue*, created in 1888, is *the* source for referencing bright nebulae and star clusters, both in professional and amateur astronomy. With 7840 entries, it is the most-used historical catalogue of observational astronomy, and NGC numbers are commonly referred to today. However, the fascinating history of the discovery, observation, description and cataloguing of nebulae and star clusters in the nineteenth century has largely gone untold, until now.

This well-researched book is the first comprehensive historical study of the NGC, and is an important resource to all those with an interest in the history of modern astronomy and visual deep-sky observing. It covers the people, observatories, instruments and methods involved in nineteenth-century visual deep-sky observing, as well as prominent deep-sky objects. The book also compares the NGC with modern object data, demonstrating how important the NGC is in observational astronomy today.

Dr WOLFGANG STEINICKE, FRAS, is a committee member of the Webb Deep-Sky Society and Director of its Nebulae and Clusters section, a core team member of the international NGC/IC Project, Head of the History Section of the VdS, Germany's largest national association of amateur astronomers, and a member of the Working Group for the History of Astronomy of the Astronomische Gesellschaft. He frequently gives conference talks and courses, and contributes to astronomical magazines. This is his fourth book.

John Louis Emil Dreyer (1852–1926)

Observing and Cataloguing Nebulae and Star Clusters
From Herschel to Dreyer's New General Catalogue

Wolfgang Steinicke, FRAS

Webb Deep-Sky Society

CAMBRIDGE UNIVERSITY PRESS
Cambridge, New York, Melbourne, Madrid, Cape Town, Singapore,
São Paulo, Delhi, Dubai, Tokyo, Mexico City

Cambridge University Press
The Edinburgh Building, Cambridge CB2 8RU, UK

Published in the United States of America by Cambridge University Press, New York

www.cambridge.org
Information on this title: www.cambridge.org/9780521192675

© W. Steinicke 2010

This publication is in copyright. Subject to statutory exception
and to the provisions of relevant collective licensing agreements,
no reproduction of any part may take place without the written
permission of Cambridge University Press.

First published 2010

Printed in the United Kingdom at the University Press, Cambridge

A catalogue record for this publication is available from the British Library

Library of Congress Cataloguing in Publication data
Steinicke, Wolfgang.
  Observing and cataloguing nebulae and star clusters : from Herschel to Dreyer's new general
  catalogue / Wolfgang Steinicke.
    p. cm.
  Includes bibliographical references and index.
  ISBN 978-0-521-19267-5 (hardback)
  1. Nebulae–Catalogs. 2. Stars–Clusters–Catalogs. 3. Nebulae–Charts, diagrams, etc.
  4. Stars–Clusters–Charts, diagrams, etc. I. Title.
  QB853.S736 2010
  523.1′135–dc22   2010022752

ISBN 978-0-521-19267-5 Hardback

Cambridge University Press has no responsibility for the persistence or
accuracy of URLs for external or third-party internet websites referred to in
this publication, and does not guarantee that any content on such websites is,
or will remain, accurate or appropriate.

To my wife Gisela.

Contents

Preface		*page* xi
1	**Introduction**	**1**
	1.1 The significance of the New General Catalogue	1
	1.2 Motivation and method	4
	1.3 Milestone catalogues of non-stellar objects and major topics	6
	1.4 Structure, presentation and conventions	11
2	**William Herschel's observations and parallel activities**	**14**
	2.1 Objects discovered prior to Herschel	15
	2.2 Structure and content of the Herschel catalogues	18
	2.3 Caroline Herschel and other discoverers	28
	2.4 Herschel's eight classes and modern object types	30
	2.5 Brightness of the objects	34
	2.6 Herschel's class IV: planetary nebulae	35
	2.7 Von Hahn's observations of planetary nebulae	41
	2.8 Special objects	44
	2.9 Additions by John Herschel and Dreyer	47
	2.10 Later publications and revisions of Herschel's catalogues	49
3	**John Herschel's Slough observations**	**52**
	3.1 Structure and content of the Slough catalogue	53
	3.2 Identification of the catalogue objects	54
	3.3 John Herschel's new objects	55
	3.4 Additions and drawings	59
	3.5 Olbers' review of the Slough catalogue	61
4	**Discoveries made in parallel with John Herschel's Slough observations**	**63**
	4.1 Harding's list of new nebulae	63
	4.2 Wilhelm Struve: nebulae in the *Catalogus Novus*	66
	4.3 Cacciatore and his nebula	69
	4.4 Dunlop and the first survey of the southern sky	71
5	**John Herschel at the Cape of Good Hope**	**77**
	5.1 Structure and content of the Cape catalogue	77
	5.2 Identification of catalogue objects	79
	5.3 John Herschel's new objects	80
	5.4 Classification, supplements and drawings	83
6	**The time after Herschel's observations until Auwers' list of new nebulae**	**88**
	6.1 Lamont and the nebulae	88
	6.2 The short career of Ebenezer Porter Mason	94

6.3	Two 'new' nebulae of Bianchi	96
6.4	Lord Rosse: first observations at Birr Castle	98
6.5	Admiral Smyth and his Bedford Catalogue	119
6.6	Hind – discoverer of some remarkable objects	121
6.7	Observations at Harvard College Observatory	124
6.8	Lassell's first nebula	131
6.9	Ernest Laugier and the first catalogue of accurate positions of non-stellar objects	134
6.10	Cooper and his Markree catalogue	135
6.11	Secchi and de Vico: observations at the Collegio Romano	139
6.12	Winnecke's observations in Göttingen, Berlin, Bonn and Pulkovo	144
6.13	Auwers' first discoveries in Göttingen	146
6.14	D'Arrest: the Leipzig 'Erste Reihe' and early observations in Copenhagen	148
6.15	Schönfeld and the nebulae in the Bonner Durchmusterung	161
6.16	Brorsen and Bruhns: comet discoverers on the wrong track	166
6.17	Tempel's observations in Venice, Marseille and Milan	170
6.18	Schmidt's first discoveries in Athens	174
6.19	Auwers' work 'William Herschel's Verzeichnisse von Nebelflecken und Sternhaufen'	178

7 Compiling the General Catalogue — 188

7.1	Lord Rosse's publication of 1861	188
7.2	Considering the Harvard observations	203
7.3	Chacornac and his 'variable nebula' in Taurus	205
7.4	D'Arrest's contribution to the General Catalogue	207
7.5	Lassell and his 48″ reflector on Malta	211
7.6	Content and structure of the General Catalogue	217

8 Dreyer's first catalogue: the supplement to Herschel's General Catalogue — 225

8.1	Dreyer's biography	225
8.2	Origins and intention of the GCS	230
8.3	Harvard objects	231
8.4	Schweizer's new 'nebula'	232
8.5	Schönfeld's Mannheim observations	233
8.6	D'Arrest's masterpiece: *Siderum Nebulosorum Observationes Havnienses*	237
8.7	Marth on Malta: 600 new nebulae	251
8.8	Schmidt's positional measurements and discovery of new nebulae	258
8.9	Winnecke's observations in Karlsruhe and Straßburg	261
8.10	Tempel in Arcetri	263
8.11	Rümker and the 'circumpolar nebulae'	267
8.12	Ferrari – in the shadow of Secchi	269
8.13	Observations in Marseille: Voigt, Stephan, Borrelly and Coggia	269
8.14	Vogel's observations in Leipzig and Vienna	277
8.15	In the footsteps of his father: Otto Struve	281
8.16	Nordic combination: Schultz, Dunér and Pechüle	284
8.17	Holden, Tuttle and a possible case of imposture	294
8.18	Further Birr Castle observations and the publication of 1880	295
8.19	The structure of the GCS	318

9	**Compilation of the New General Catalogue**	**323**
	9.1 Dreyer's unpublished 'second supplement'	323
	9.2 Star charts and nebulae of Peters	324
	9.3 Tempel's new nebulae and a controversial treatise	327
	9.4 Harvard's new guard: Austin, Langley, Peirce, Searle and Winlock	335
	9.5 Warner Observatory: Lewis Swift and his son Edward	337
	9.6 Dearborn Observatory: Safford, Skinner, Burnham and Hough	357
	9.7 Todd and the search for the trans-Neptunian planet	369
	9.8 Stephan's nebulae in the NGC	372
	9.9 The Reverend Webb and his planetary nebula NGC 7027	378
	9.10 New nebulae discovered by Pechüle	383
	9.11 Baxendell's 'unphotographable nebula'	384
	9.12 Common's discoveries with the 36-inch reflector	385
	9.13 Pickering's spectroscopic search for planetary nebulae	389
	9.14 Copeland: on Pickering's trail	392
	9.15 Barnard: the best visual observer	395
	9.16 Holden at Washburn Observatory	403
	9.17 Harrington's galaxy	408
	9.18 Hall, the Martian moons and a galaxy	409
	9.19 Palisa, Oppenheim and the new Vienna University Observatory	410
	9.20 Hartwig's observations in Dorpat and Straßburg	412
	9.21 Ellery, Le Sueur, MacGeorge, Tuner, Baracchi and the Great Melbourne Telescope	415
	9.22 Bigourdan, master of the NGC	422
	9.23 Young's discovery in Princeton	427
	9.24 Lohse at the Wigglesworth Observatory	428
	9.25 The Leander McCormick Observatory: Stone, Leavenworth and Muller	431
	9.26 The first photographic discovery: NGC 1432	437
10	**The New General Catalogue: publication, analysis and effects**	**439**
	10.1 Dreyer's publication of 1888	439
	10.2 Content of the NGC and statistical analysis	443
	10.3 Corrections and additions to the NGC	459
	10.4 Revisions of the NGC	467
11	**Special topics**	**472**
	11.1 Positional measurements	472
	11.2 Drawings of nebulae: facts and fiction	473
	11.3 M 51 and the spiral structure of nebulae	482
	11.4 Hind's Variable Nebula (NGC 1555) and its vicinity	498
	11.5 D'Arrest, Dreyer and the variable nebulae	513
	11.6 The Pleiades nebulae	521
12	**Summary**	**562**
	12.1 The subject and line of questioning	562
	12.2 The importance of the New General Catalogue and the motivation of the work	562
	12.3 Objects, observers and methods	563
	12.4 Milestones of the cataloguing of nebulae and star clusters	564
	12.5 Statistical analysis and the way ahead	565

Appendix		**567**
Timeline		567
Abbreviations and units		571
Telescope data		573
References		**583**
Internet and image sources		**619**
Name index		620
Site index		629
Object index		631
Subject index		647

Preface

My enthusiasm for nebulae and star clusters goes back a long way – they were the targets of my first telescopic explorations of the night sky. This book treats the history of their discovery, visual observation and cataloguing. It is naturally focused on the nineteenth century – the fascinating epoch of classical astronomy, characterised by precious achromatic refractors and massive metal-mirror reflectors. Only a few astronomers searched for non-stellar objects systematically – foremost among them William Herschel and his son John. We owe to both of them the first comprehensive catalogues. The development reached its climax with the New General Catalogue (NGC) by John Louis Emil Dreyer – which is still a standard source for both amateur and professional astronomers.

Initially this immense work appeared to me as a mysterious treasure, arousing my scientific curiosity. What was behind the 7840 objects and who were the discoverers? By using the NGC, I gradually became familiar with its content. However, due to erroneous and incomplete data, it was often difficult to match the entries with the real sky. Over the years, due to my research on the historical sources and visual observing, the catalogue became a close companion. Many secrets could be disclosed – and, of course, my admiration for Dreyer increased.

The many years of investigation resulted in the 'Revised New General and Index Catalogue', which connects the original data with modern ones. In a second step, the historical background (discoverers, dates, instruments, sites) was revealed, which eventually led to the 'Historic NGC'. Both catalogues have seen many updates and are an essential basis of my recent German Ph.D. thesis at Hamburg University (Steinicke 2009). Actually, this book is an enhanced version of it and the first comprehensive popular presentation of the subject.

I want to thank the many supporters of my work – first of all my wife Gisela, who was a great help from the beginning. She worked on the indices, corrected the manuscript and was a valuable companion on many astronomical tours. Of course, special thanks must go to my friends of the NGC/IC Project, especially Harold Corwin and Malcolm Thomson. This international organisation deals with the correct identification of all objects in Dreyer's catalogues – and is, without any shadow of a doubt, very successful in this task!

During my research, I was supported by many famous institutions – I may just mention a few: Armagh Observatory, the Royal Astronomical Society, the Royal Observatory Edinburgh, the Institute of Astronomy (Cambridge), Arcetri Observatory, Birr Castle Museum and the Webb Deep-Sky Society. Special thanks go to the librarians Peter Hingley (RAS), John McFarland (Armagh Observatory), Françoise Le Guet Tully (Nice Observatory), Bertil Dorch (Danish National Library), Volker Mandel (Heidelberg-Königstuhl Observatory) and Anke Vollersen (Hamburg-Bergedorf Observatory).

Moreover, the list of important contributors is long; among them are Brent Archinal, Bob Argyle, Simone Bianchi, Wilhelm Brüggenthies, Ron Buta, Lutz Clausnitzer, Steve Coe, Glen Cozens, David Dewhirst, Wolfgang Dick, William Dreyer, Hilmar Duerbeck, Sue French, Hartmut Frommert, Steve Gottlieb, Michael Hoskin, Arndt Latußeck, John McConnell, Stewart Moore, Yann Pothier, Peter Schliebeck, William Tobin and Gudrun Wolfschmidt.

Of course, special thanks go to Steven Holt, Vince Higgs, Jonathan Ratcliffe and Claire Poole of Cambridge University Press. It was a pleasure to work with them. Last but not least, I am proud to have been supported by the late Mary Brück, who sadly died in 2008.

Wolfgang Steinicke
Umkirch, May 2010

1 • Introduction

1.1 THE SIGNIFICANCE OF THE NEW GENERAL CATALOGUE

Besides the point-like stars, the sky offers a large number of objects showing an extended structure. Except for a few, they are not visible without the aid of a telescope. In terms of their optical appearance, there are star clusters (resolvable objects) and nebulae (unresolvable objects). In 1862 Eduard Schönfeld, an astronomer at Mannheim Observatory, gave the following definition:[1] *'Nebulae or nebulous patches are celestial objects, which do not contrast with the sky background as shining points, like individual stars, but present the impression of a more or less extended and diffuse area of light.'*[2]

Long before the invention of the telescope, the open clusters of the Pleiades and Praesepe and the diffuse spot of the Andromeda Nebula were known. Later the telescopic exploration of the sky brought many more cases to light. Soon it became evident that some nebulae are disguised clusters of stars; the best examples are globular clusters, the compact and star-rich versions of open clusters. Other objects, such as the bright nebulae in Orion and Andromeda, could not be resolved, even with the largest telescopes. However, in 1864 the new astrophysical method of spectroscopy revealed that the Orion Nebula is a mass of gas (mainly hydrogen and helium). On the other hand, the Andromeda Nebula is a galaxy, consisting of many hundreds of billions of stars, which was eventually proved in the twentieth century.[3]

Nebulae and star clusters are 'non-stellar' objects.[4] In terms of the criteria of form, individuality, physical relation and existence, the following types are meant by this term (Fig. 1.1 shows examples):

- open clusters and globular clusters (here often subsumed as 'star clusters')
- emission nebulae, reflection nebulae and dark nebulae (commonly known as galactic nebulae, which includes remnants of novae and supernovae)
- planetary nebulae
- galaxies (including quasars)

Galaxies are by far the dominating non-stellar objects (see Table 10.12). Their forms and types are manifold.[5] Star clusters, galactic nebulae and planetary nebulae are Milky Way objects.[6]

This definition is quite helpful to rate the success of a discoverer. The measure is the percentage of non-stellar objects. This is relevant, because often visual observation could not decide whether a nebula is real or whether the 'nebulous' impression was only simulated by a pair or small group of stars; the latter is a common phenomenon with a small telescope. Sometimes a subsequent observation shows a blank field; the object could have been a comet or the position was wrong. Thus the following cases must be determined in the discoverer's balance:

- stellar object: single star, star pair, star pattern (asterism)
- part of an object (e.g. galaxy)

---

[1] Schönfeld (1862b: 48).
[2] The terms 'nebula' and 'nebulous patch' (in German: *Nebel* and *Nebelfleck*) were mostly used synonymously; occasionally 'nebula' describes a spacious, diffuse object (such as the Orion Nebula) and 'nebulous patch' a small, confined object (such as a faint galaxy).
[3] A comprehensive review was given by Wolfschmidt (1995).
[4] In amateur astronomy these are 'deep-sky objects', i.e. targets beyond the solar system; see Steinicke (2004a).
[5] Here the ordinary Hubble classification is used, see Sandage (1961) and Sandage and Bedke (1994); the later scheme of de Vaucouleurs is explained in Buta *et al.* (2007). For fine images of galaxies see also Ferris (1982).
[6] For star clusters see Archinal and Hynes (2003); for galactic nebulae see Coe (2006); for planetary nebulae see Hynes (1991).

Figure 1.1. The globular cluster M 13 in Hercules, the open cluster M 44 (Praesepe) in Cancer and the emission nebula M 42 (Orion Nebula).[7]

- comet
- lost object (existence unknown); declared here as 'not found' (NF)

The best-known discoverers of nebulae and clusters are Charles Messier, William Herschel and his son John. Many others have fallen into oblivion. The same is true for published object lists; only the Messier catalogue (M), the New General Catalogue (NGC) and the Index Catalogue (IC) are still in use.

Unrivalled in general use – in both amateur and professional astronomy – is the M-number, designating bright, large non-stellar objects. The standard reference for smaller, fainter objects is still the NGC. The transition from Messier's catalogue to the NGC is a quantum leap: from 103 to 7840 objects! Attempts to establish an intermediate step or alternatives failed; being restricted to amateur astronomy, they had no influence. Examples are Patrick Moore's Caldwell catalogue of the 'best' 103 non-Messier objects (Moore 1995) and the 'Herschel 400' list with William Herschel's 'best' objects.[8]

Today M–NGC–IC is the primary sequence used to designate non-stellar objects (beyond it, the realm of special catalogues begins). The Andromeda Nebula is designated M 31; its NGC-number (NGC 224) being secondary. The North America Nebula in Cygnus, which is not contained in the Messier catalogue, is known as NGC 7000 (see Fig. 2.17). The

Figure 1.2. John Louis Emil Dreyer (1852–1926); about 1874.[9]

nearby Pelican Nebula bears no NGC-number and is designated IC 5070.

The New General Catalogue, published in 1888 in the *Memoirs of the Royal Astronomical Society*,[10] is inseparably connected with the name of John Louis Emil Dreyer[11] (frontispiece and Fig. 1.2) – the central person

---

[7] Images of non-stellar objects presented without mentioning a source are from the author's archive. A nice collection was presented by Vehrenberg (1983).

[8] There is a 'Herschel 2500' list of all objects discovered by him. Because the original catalogues were used, the data are not reliable.

[9] The late Dreyer can be seen on the frontispiece.

[10] Dreyer (1888b).

[11] This is the English version, which will be used here; the Danish is Johan Ludvik Emil Dreyer and the German Johann Louis Emil Dreyer.

1.1 The significance of the NGC

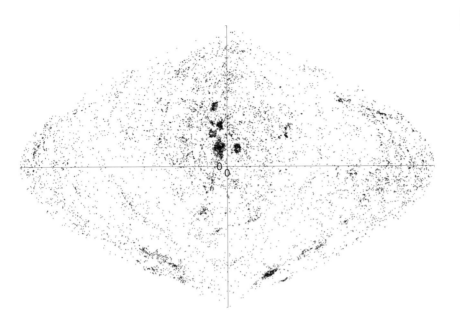

Figure 1.3. A plot of all 13 226 NGC/IC objects. The 'clusters' above centre ($\alpha = 12^h$, $\delta = 0°$) are mainly due to selection effects from photographic IC II surveys; the largest contains Virgo Cluster galaxies. The oval spot below right ($\alpha = 5^h$, $\delta = -70°$) represents objects in the Large Magellanic Cloud.

of the present work. Dreyer might be much less well known than his predecessor Charles Messier. This is due to the strong connection of name and catalogue: while the Messier catalogue is commonly known, there never was a 'Dreyer catalogue'. But Dreyer's merit for astronomy is much larger: he listed all of the non-stellar objects known up to the end of 1887, with all data necessary for their identification (position, description, source). The NGC is a standard work, which had (and still has) an enormous influence on observational astronomy.

Studying the Messier catalogue, with 103 objects and a moderate number of discoverers (23), is a manageable task – but the NGC with 7840 entries and more than 100 discoverers, is not![12] This is the reason why there have been many publications on the history of the Messier catalogue, but hitherto none about the NGC claiming to be comprehensive. The present work is the first.

Owing to the large number of contributing observers, instruments and sites, the NGC seems to be pretty inhomogeneous, but it has a common basis: all objects (except one) were found visually. This is different for

its two supplements with altogether 5386 entries. The first, the Index Catalogue (IC I), appeared in 1895 and the Second Index Catalogue (IC II) came out in 1908.[13] Already the IC I contains objects that had been found by photography, but in the IC II this was the dominating method. The photographic surveys (e.g. by Max Wolf) focused on certain areas of the sky. Thus the object distribution in the IC is very inhomogeneous (Fig. 1.3).

Modern catalogues, resulting from digital ground-based or orbital surveys, differ very much from the NGC/IC – especially in size: the latest contains more than 100 million records! Individual objects have no value, being lost in the statistical analysis. There is a large range of catalogues based on special selections:

- object type (e.g. galaxy, planetary nebula, star cluster)
- sky area (e.g. Milky Way region, constellation)
- spectral range (e.g. blue, visual, infrared)

The NGC is much different. The total number of entries is large – but not too large. Thus it is manageable, which has important consequences for current observations: Dreyer's catalogue still offers

---

[12] For catalogues it is better to speak about entries rather than objects. There are, for instance, many NGC numbers for which no object exists (at any rate, not at the given place).

[13] Dreyer (1895, 1908). The combined catalogue is often abbreviated NGC/IC.

No.	G. C.	J. H.	W. H.	Other Observers.	Right Ascension, 1860·0.	Annual Precession, 1880.	North Polar Distance, 1860·0.	Annual Precession, 1880.	Summary Description.	Notes.
					h m s	s	° ′	″		
1	1	...	...	d'A	0 0 4	+3·07	63 4·3	−20·1	F, S, R, bet ★11 and ★14	
2	6246	...	...	Ld R★	0 0 6	3·07	63 6·0	20·1	vF, S, s of G.C. 1	
3	5080	...	...	m 1	0 0 6	3·07	82 28	20·1	F, vS, R, alm stell	
4	5081	...	...	m 2	0 0 16	3·07	82 23	20·1	eF	
5	...	...	...	St XII	0 0 37	3·08	55 25·0	20·1	vF, vS, N = ★13, 14	
6	...	...	...	Sw II	0 1 5	3·08	58 15·6	20·1	eF, vS, eE	
7	2	4014	...	...	0 1 14	3·07	120 41·2	20·1	eF, eL, mE, vgvlbM	
8	5082	...	...	O Struve	0 1 17	3·08	66 59	20·1	vF, N in n end	
9	5083	...	...	O Struve	0 1 27	3·08	67 0	20·1	F, R, ★9, 10 sf	
10	3	4015	...	...	0 1 28	3·06	124 38·9	20·1	F, cL, vlE, glbM	
6999	5981	...	...	m 432	20 53 38	3·60	118 36	13·8	eeF, vS	
7000	4621	2096	V 37?	...	20 53 48	2·14	46 13·1	13·8	F, eeL, dif nebulosity	
7001	4622	2095	...	...	20 53 55	3·08	90 44·6	13·9	eF, S, E 0°	

Figure 1.4. The first ten entries of the New General Catalogue, to which NGC 6999–7001 are appended (Dreyer 1888b).

the primary targets (mainly galaxies). Their moderate brightness allows astrophysical studies with medium-sized telescopes; and with the biggest, like the Very Large Telescope (VLT) or the Hubble Space Telescope (HST), extremely detailed observations are possible. Thus NGC-numbers are part of the astronomer's daily routine. The catalogue might be the most used in modern observational astronomy. It therefore has both historical and astrophysical importance. The New General Catalogue marks the transition from (old) astronomy to (new) astrophysics, represented by spectroscopy, photography and photometry.[14] Dreyer has created the last 'visual' catalogue containing all types of non-stellar objects in the whole sky.

1.2 MOTIVATION AND METHOD

The New General Catalogue and both Index Catalogues were published as a book by the Royal Astronomical Society (RAS) in 1953.[15] Enthusiasm for the printed NGC – if there is any – might not result from its physical appearance. At first glance the work has the brittle charm of a phone book (Fig. 1.4). Without previous astronomical knowledge it will soon be shelved.

Present-day amateur astronomers interested in observing nebulae and star clusters are familiar with the term 'NGC', but its objects rate as 'faint' and thus difficult to observe. Consequently, the visual observer dealing with them does not rank as a 'beginner'. This implies that NGC objects do not possess the same popularity as Messier objects. While M 1, M 13, M 31, M 42 and M 57 belong to the standard repertoire of amateurs, naturally only a few NGC-numbers circulate. Table 1.1 presents a (subjective) sample of popular objects – most of them are better known through their common names. About 95 NGC objects bear a (more or less official) proper name. Unfortunately, since the late twentieth century there has been a certain inflation of new names, mostly created by American observers and based on photographic images.

The majority of the NGC objects are anonymous; but, of course, the unknown makes the catalogue interesting – and motivates investigations. If one takes, for instance, NGC 7000 (the North America Nebula in Cygnus; see Fig. 2.17), the following question arises: what is hidden behind the preceding and following entries, NGC 6999 and NGC 7001? Dreyer gives only the bare minimum of information (Fig. 1.4): cross references, position and coded description. For NGC 7001 one reads 'eF, S, E 0°', meaning 'excessively faint, small, extended 0°', which describes an extremely faint and small object, elongated north–south. To find its place on a modern star chart, the coordinates must be precessed to the epoch 2000 (declination results from 'North Pole Distance'). Thus it is not easy to get on the right track. Fortunately the necessary work has

---

[14] Concerning the instrumental aspects, see Staubermann (2007).
[15] This publication is used here as the NGC/IC standard reference, cited as 'Dreyer (1953)'. Unchanged editions were printed in 1962 and 1971.

1.2 Motivation and method

Table 1.1. *Examples of popular NGC objects without Messier-numbers*[16]

NGC	Discoverer	Date	Type	V	Con.	Remarks
104	Lacaille	1751	GC	4.0	Tuc	47 Tucanae
253	C. Herschel	23.9.1783	Gx	7.3	Scl	Silver Dollar Galaxy; Figs. 2.14 and 7.12 left
292	Vespucci	1501	Gx	2.2	Tuc	Small Magellanic Cloud
869	Hipparch	−130	OC	5.3	Per	Double Cluster (with NGC 884)
891	W. Herschel	6.10.1784	Gx	10.1	And	Edge-on galaxy with absorption lane; Fig. 2.15
1435	Tempel	19.10.1859	RN		Tau	Merope Nebula; Fig. 11.31
1499	Barnard	3.11.1885	EN		Per	California Nebula; Fig. 9.58
1555	Hind	11.10.1852	RN		Tau	Hind's Variable Nebula; Fig. 11.17
2237	Swift	1865	EN		Mon	Rosette Nebula (with NGC 2238/39/46); Fig. 9.14
2261	W. Herschel	26.12.1783	RN		Mon	Hubble's Variable Nebula; Fig. 6.56
2392	W. Herschel	17.1.1787	PN	9.1	Gem	Eskimo Nebula; Fig. 6.16
3242	W. Herschel	7.2.1785	PN	7.7	Hya	Ghost of Jupiter
3372	Lacaille	1751	EN		Car	Eta Carinae Nebula; Fig. 11.26
4038	W. Herschel	7.2.1785	Gx	10.3	Crv	The Antennae (with NGC 4039); Fig. 2.30 left
4565	W. Herschel	6.4.1785	Gx	9.5	Com	Edge-on galaxy; Fig. 7.12 right
4755	Lacaille	1751	OC	4.2	Cru	Jewel Box
5128	Dunlop	29.4.1826	Gx	6.6	Cen	Centaurus A; Fig. 4.12
6543	W. Herschel	15.2.1786	PN	8.1	Dra	Cat's Eye Nebula; Fig. 6.6 right
6822	Barnard	17.8.1884	Gx	8.7	Sgr	Barnard's Galaxy; Fig. 9.56
6888	W. Herschel	15.9.1792	EN		Cyg	Crescent Nebula; Fig. 8.56 centre
6992	W. Herschel	5.9.1784	EN		Cyg	Veil Nebula (with NGC 6960/95)
7000	W. Herschel	24.10.1786	EN		Cyg	North America Nebula; Fig. 2.17
7009	W. Herschel	7.9.1782	PN	8.0	Aqr	Saturn Nebula; Figs. 2.3, 6.3 and 8.40 right
7293	Harding	Sept.? 1823	PN	7.3	Aqr	Helix Nebula; Fig. 4.3
7331	W. Herschel	5.9.1784	Gx	9.5	Peg	
7662	W. Herschel	6.10.1784	PN	8.3	And	Blue Snowball; Fig. 2.21
7789	C. Herschel	30.10.1783	OC	6.7	Cas	

already been done: the author's 'Revised New General and Index Catalogue'.[17] This work shows that NGC 6999 and NGC 7001 are galaxies in the constellations Microscopium and Aquarius with 14.0 mag and 13.5 mag, respectively (Fig. 1.5).

The fascination of the NGC is thus partly due to its mysterious, almost cryptic data. Each entry offers an object, whose discovery story and physical nature must be revealed. There are cases, where one literally grasps at nothing. Cautiously noting 'not found' (NF), the term leaves open whether the object is non-existent or perhaps real at another place. Anyway, whomsoever wants to uncover the secrets of the NGC must consult the real sky by visual observing or – which is much easier – inspecting a photographic sky map, such as the Digitized Sky Survey (DSS).

---

[16] The abbreviations are explained in the appendix.
[17] See the author's website: www.klima-luft.de/steinicke. The modern data are also used in many 'planetarium programs' showing a digital image of the sky.

Figure 1.5. The galaxy NGC 7001 in Aquarius (DSS).[18]

However, for a definite identification of an NGC object, this is not sufficient. In many cases the historical sources must be taken into account, i.e. the original notes of the observers. Here the catalogue holds secrets too: who was 'Mr. Wigglesworth', owner of an observatory where 'J. G. Lohse' discovered 18 objects? Who were the Harvard astronomers Austin, Langley, Peirce, Searle, Wendell and Winlock? What is meant by the 'Melbourne observations' or by a source called 'Greenwich IX yr C', noted for NGC 2392?[19] Of course, Dreyer could count on the knowledge of the nineteenth-century observers, but today these names and terms say very little. Thus modern astronomers are pragmatic and mainly interested in astrophysical data. Nevertheless, it is fascinating to fathom out the stories behind the NGC entries.

Many non-stellar objects were catalogued prior to the NGC, e.g. in John Herschel's General Catalogue (GC) of 1864. Therefore the present work must check the cross references to the GC and other catalogues. The deeper one digs, the more Dreyer's achievement in creating a homogeneous catalogue from a large number of different observations and sources becomes clear. Never having been able to overview the relevant sky areas, this was like a 'blind puzzle'. The compilation of the NGC at the desktop was a hard and error-prone task, especially concerning the different qualities of observations and records. Today there are (digital) photographic maps to verify the identity of the objects, but, even with the aid of computers and the Internet, this is not straightforward!

For the modern analysis of the NGC it was useful to divide the catalogue into subsets corresponding to the individual observers and their different data qualities. To track the cross references, an analogous procedure was applied for the earlier catalogues. For Messier's catalogue this analysis has already been made, but in the case of the NGC new ground was broken. The method was as follows: cut the NGC and its forerunners into pieces, sort them by applying various filters and join the results. This leads to new insights about the catalogues involved, concerning their substance and historical evolution. One of the many results is that some of Dreyer's data must be revised. There are errors concerning discoverers, sources and identifications of objects or identities between them. The same is true for William and John Herschel – but both had the benefit that they discovered or observed most of their catalogued objects themselves. Thus the Herschel data are more homogeneous than the records Dreyer had to deal with.

The goal of this work is extracting primary structures from the various sources to picture the motivation and importance of the observations of nebulae and star clusters in the nineteenth century.[20] In Dreyer's New General Catalogue the consideration has had to be focused. Individual nebulae and star clusters play an important role, but, in the face of their large number, they can be presented only as examples – nevertheless, more than 2000 NGC objects are mentioned.

1.3 MILESTONE CATALOGUES OF NON-STELLAR OBJECTS AND MAJOR TOPICS

This work is focused on the nineteenth century, but the origins date earlier. The most important persons were

---

[18] Most images of deep-sky objects are from the Digitized Sky Survey (DSS); see http://archive.eso.org/dss/dss.

[19] See Section 10.1.1 for the answer.

[20] A contemporary and comprehensive outline of nineteenth century astronomy is due to Agnes Mary Clerke (Clerke 1893).

Table 1.2. *Milestone catalogues leading to the NGC*

Author	Milestone	Abbr.	Reference	Entries	New	Suppl.
Messier	Messier catalogue	M	Messier (1781)	103	103	7 (1921–66)
W. Herschel	Three catalogues	H	Herschel W. (1786, 1789, 1802)	2500	2427	8 (1847)
J. Herschel	Slough catalogue	SC (h)	Herschel J. (1833a)	2307	473	6 (1847)
J. Herschel	Cape catalogue	CC (h)	Herschel J. (1847)	1714	1421	
W. Parsons	Birr Castle (1861 publ.)	LdR	Parsons W. (1861a)	989	295	
Auwers	List of new nebulae	Au	Auwers (1862a)	50	46	
J. Herschel	General Catalogue	GC	Herschel J. (1864)	5057	419	22 (1864)
d'Arrest	Siderum Nebulosorum	SN	d'Arrest (1867a)	1942	307	
Dreyer	GC Supplement	GCS	Dreyer (1878a)	1166	1149	6 (1878)
Dreyer	Birr Castle (1880 publ.)		Parsons L. (1880)	1840	94	
Dreyer	New General Catalogue	NGC	Dreyer (1888b)	7840	1700	49 (1888)

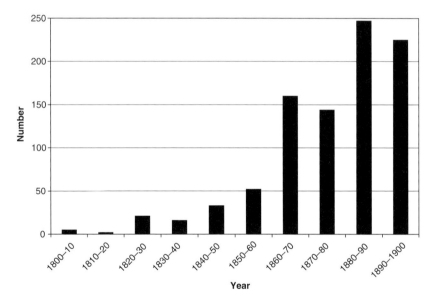

Figure 1.6. Numbers of articles on visual observations of non-stellar objects from 1800 to 1900.

undoubtedly Charles Messier and William Herschel. The three catalogues of the latter defined the decisive basis of John Herschel's observations at Slough and Feldhausen (Cape of Good Hope). The resulting Slough and Cape catalogues are among the milestones which form the backbone of this work (Table 1.2). In the column 'Abbr.' the usual catalogue abbreviation is given; 'New' shows the number of new (independent) objects, compared with earlier works (see particular sections); 'Suppl.' gives the number and year for objects added later.

Besides these major catalogues there exists a considerable number of other publications related to occasional visual observations and discoveries of non-stellar objects. Figure 1.6 shows the increase in number of articles during the nineteenth century.

The period 1860–70, during which new, ambitious observers entered the scene, such as d'Arrest, Auwers,

Schönfeld, Schmidt and Winnecke, is remarkable. Additionally, John Herschel's General Catalogue was an impetus. Another climax, with numerous observations by Stephan, Swift, Tempel and Barnard, came in 1880–90. Particularly productive years were 1886 with 54 publications (e.g. by Barnard and Swift), 1862 with 39 (e.g. on variable nebulae), 1885 (34) and 1883 (27). The growth culminated with Dreyer's New General Catalogue, initiating many activities. The aftermath, leading to amendments of the NGC (from the Index Catalogues to the modern revisions), is treated here too.

Besides the milestones, some important topics that cannot be timed are treated. They are essential parts of the development and concern objects, observers, methods and instruments. The topics are

(1) **discovery**: visual, photographic, spectroscopic
(2) **cataloguing**: observation, data processing
(3) **description and condition**: brightness, form, neighbourhood
(4) **nature and evolution**: resolvability, classification, change
(5) **telescopes and observers**: reflector–refractor, amateur–professional
(6) **astrometry**: position, reference system, proper motion, double nebulae, satellites
(7) **astrophysics**: spectroscopy, photography, photometry

(1) **Discovery**. This is fundamental, since the NGC and its forerunners were created to list not only known objects but also newly discovered ones. The most successful discoverers (see Fig. 10.3) were William Herschel (2416 objects), John Herschel (1691), Albert Marth (582), Lewis Swift (466), Edouard Stephan (420) and Ludwig d'Arrest (321).

Nearly all NGC objects were found by visual observation (7817). The year 1877 marks the beginning of new methods: 22 objects were discovered with the aid of a (visual) spectroscope or objective prism; 15 by Edward Pickering and 7 by Ralph Copeland. Photography of nebulae was still not established at that time.[21] Only one object was found: the Maia Nebula (NGC 1432) in the Pleiades, by the brothers Henry. Later, astrophysical methods massively affected the Index Catalogue.

Many discoveries were made while observing other types of objects, e.g. single, double and variable stars, minor planets and comets (important cases are mentioned in the text). It is remarkable that many visual observers were 'lone fighters'.

(2) **Cataloguing**. Generally, there are two types of catalogue, in which the objects are those discovered either by a single observer or by different observers. Examples for the first type are William Herschel's three catalogues, the lists of Dunlop, Marth, Swift and Stephan and the Birr Castle observations. The Messier catalogue, GC, GCS (Dreyer's supplement to the GC), NGC and IC are examples of the second type, which is usually claimed by its author to be complete up to a certain date.

Catalogues of nebulae and star clusters differ in structure, arrangement, epoch and many other aspects. Messier and William Herschel sorted the entries by discovery date; Dunlop arranged the objects found in the southern hemisphere by 'South Pole Distance' (SPD).[22] John Herschel's Slough catalogue introduced right ascension (AR) as the ordering element, which became the standard. A mixed type is constituted by 'zone catalogues' listing the objects in declination zones (ordered by AR within zones). Examples are Caroline Herschel's reduction of her brother's data and the work of Johann Georg Hagen.[23] Early catalogues of non-stellar objects did not use a standard epoch (the twentieth century established 1900, 1950 and 2000). Smaller lists were often referenced to the epoch of observation. John Herschel (SC, CC) and Auwers used 1830; the epoch of the GC (1860) was adopted by the GCS, NGC and IC. Most catalogues give absolute positions (coordinates), but others only relative ones, e.g. those of William Herschel, Herman Schultz and Guillaume Bigourdan. John Herschel and Dreyer spent much time to standardise the case. However, the situation for star catalogues was even worse: the nineteenth century saw a large number of position lists, differing by limiting magnitude or sky area. Some NGC discoverers, such

---

[21] The photography of the Sun, Moon, planets and bright stars was already advanced.

[22] Most classic catalogues (up to the IC II of 1908) use 'North Pole Distance' (NPD) instead of declination ($\delta = 90° - $ NPD). One modern list is sorted by NDP: Patrick Moore's Caldwell catalogue (Moore 1995).

[23] The 'Deep sky field guide' (DSFG) of *Uranometria 2000.0* is a modern zone catalogue, listing about 30 000 non-stellar objects (Cragin and Bonanno 2001).

as Harding, Chacornac, Hind, Cooper and Peters, created their own star catalogues. With the advent of the Bonner Durchmusterung (epoch 1855), consisting of a catalogue and an atlas, a certain standard was defined.[24] The first star catalogue with astrophysical data was the Henry Draper Catalogue (HD) of 1918.

(3) **Description and condition**. A continuous cause for discussion was the description of objects. Basically, the astronomers could gauge the sky only visually. To record their impressions concerning the brightness and structure of an object, and share the results with others, texts and sketches were the only media – a subjective matter, depending on ability, experience and talent. To objectify it, William Herschel developed a standardised (coded) description. Anyway, due to the uncertainty of the data, many objects could not be identified correctly. This led to a high error rate in the historical catalogues. Not until the twentieth century did a transition from qualitative to quantitative data for nebulae and star clusters (e.g. integrated/surface brightness, size, type) take place.

When a nebula or cluster is observed visually, the vicinity is relevant. This concerns not only the star field (e.g. for orientation) but also other non-stellar objects, which may be associated with the nebula or cluster. There are double and multiple nebulae. In these cases, William Herschel assumed an analogy with double (multiple) stars that was based on gravitational interaction. Concerning clusters of nebulae, it is interesting to ask whether such agglomerations were recognised by the nineteenth-century observers or even interpreted as hierarchical structures. A fascinating case is the rich galaxy cluster in Coma Berenices.

(4) **Nature and evolution**. On the basis of their observations, William and John Herschel thought about the nature and evolution of nebulae and star clusters. Helpful tools were classification and standardised description. William Herschel defined eight classes, which are barely related to modern object types. Key issues about the cosmogony of nebulae were resolvability and change. The question was whether all nebulae are star clusters, in which case a sufficient aperture would eventually unmask them. Otherwise, true (unresolvable) nebulae should exist, supposed to consist of a luminous 'gas' or 'fluid'. Such matter would naturally condense into stars by virtue of gravitational attraction. Thus changes of brightness and shape of nebulae should be detectable over a sufficient period of time. A popular idea was the 'nebular hypothesis', which had both enthusiastic advocates and strong opponents. The key object was the Orion Nebula, where William Herschel believed already to have seen change – a controversial matter. Other dubious cases were Hind's Variable Nebula (NGC 1555) and the Merope Nebula (NGC 1435); for the latter, even its existence was doubted.[25] Moreover, observers were confronted with strange things like planetary and spiral nebulae (such as M 51), whose features were not understood. The structure of the latter was interpreted – in the sense of the nebular hypothesis – as revolving nebulous matter.

(5) **Telescopes and observers**. The 'reflector–refractor' relation was a permanent issue in the nineteenth century. These two optical systems normally lived in peaceful harmony and had their typical users: amateur and professional astronomers, respectively,[26] but occasionally there was heated discussion about the pros and cons of each system.

According to George Biddell Airy, the reflector was *'almost exclusively the instrument of amateurs'* – the owners were wealthy, independent amateur astronomers.[27] Examples are William and John Herschel, William Lassell, John Ramage (see Fig. 1.7 left), William Parsons (Lord Rosse) and his son Lawrence. They had the freedom to observe nebulae and to deal with questions about their physical nature – omitting accurate measurements. A large reflector, such as Lord Rosse's 72-inch, was ideally suited. However, in professional astronomy such instruments were rare. Though two big reflectors were erected in Marseille and Melbourne in the 1860s, the breakthrough did not come until the early twentieth century; a trendsetter was Ritchey's 60-inch, which was installed in 1908 on Mt Wilson.

Through Fraunhofer's inventions, the refractor became the privileged instrument of professional

---

[24] For the history of star catalogues and charts see Tirion *et al.* (1987: xv–xlii).

[25] See Chapter 11 for details. The text contains other interesting examples (see the table of contents), e.g. 55 And, NGC 1333, BD –0° 2436, GC 80, NGC 1988, II 48, NGC 7027, NGC 6677/79 and the trapezium in M 42.

[26] See the interesting list of private British observatories (Anon 1866b).

[27] Airy (1849). Concerning the Victorian epoch, Allan Chapman created the term 'grand amateur' (Chapman A. 1998); see also Ashbrook (1984: 32–37).

Figure 1.7. Large private telescopes. Left: the 38-cm reflector of John Ramage (erected in 1820 at Greenwich);[28] right: James Buckingham's 54-cm refractor.

astronomy in the nineteenth century. One of the first to benefit was Wilhelm Struve with the 9.6-inch in Dorpat.[29] With a refractor mounted equatorially and equipped with precise setting-circles and micrometers, accurate positional measurements could be made. Owing to their classical education, most observers at university, royal and government observatories worked in the field of astrometry. Their assistants had to concentrate on astrometry – time-consuming routine work producing large amounts of data. Practical skills and patience were needed, to ensure their careers. The primary targets were minor planets and comets (measuring relative positions at the refractor) as well as single or double stars (measuring absolute positions with the meridian-circle).[30] As a by-product, new nebulae were occasionally detected.

Only a few amateurs used large refractors; among the four front-runners, three were British. In 1856 the Italian Ignazio Porro had erected a 52-cm refractor in Paris. In 1862 James Buckingham built a 54-cm refractor (with optics by Wray) on Walworth Common (Fig. 1.7 right). Even larger, but optically defective and only short-lived, was John Craig's instrument with 61-cm aperture (built by Chance/Gravatt), which was installed in 1852 in Wandsworth. The largest was owned by Robert Newall in Gateshead: a 63-cm refractor, made by Cooke in 1869. It strongly suffered from the bad weather at the site.

(6) Astrometry. For instrumental and personal reasons, the precise measurement of nebular positions was a slowly growing matter. It depended on the will and authority of the director to interrupt the routine observations and turn the refractor onto nebulae. By measuring relative positions between the object and a nearby reference star it was hoped to determine its proper motion. Such data could yield information about the cosmic order (distances) of the nebulae. Frequent observations over a long period of time were necessary. However, the diffuse appearance of a nebula limited the precision.[31] Much work was done by Laugier, d'Arrest, Auwers, Schönfeld, Vogel, Rümker, Schmidt and Schultz. Related fields were the investigation of double nebulae (claiming a similarity to double

---

[28] In 1823 Ramage cast a 21" mirror with focal length 25 ft in Aberdeen, but the appropriate telescope was never built (Anon 1836; Dick T. 1845: 308–311).

[29] A duplicate was the Berlin refractor (erected in 1835), which was used by Galle and d'Arrest for the discovery of Neptune in 1846. For telescope data see the appendix.

[30] The relative position gives the coordinate differences between object and reference star. From the known star position for a certain epoch the absolute position (right ascension, declination) of the object can be determined; this calculation is called 'reduction'.

[31] The determination of a parallax was therefore impossible.

stars), 'satellites' of planetary nebulae (supposed orbiting stars) and the construction of a celestial reference system based on nebular positions (e.g. Stephan at Marseille Observatory).

(7) **Astrophysics**. Finally, it must be stated that visual observations made to reveal the nature and evolution of nebulae and star clusters did not lead to any substantial progress. Starting from William Herschel, the basic ideas (such as the nebular hypothesis) were only slightly modified during the nineteenth century. Classical observation methods were improper as a means to get reliable statements about the physics of non-stellar objects. Neither large telescopes, with their enormous light-gathering power, nor extensive measurement campaigns to determine precise positions were able to terminate the various speculations. But the situation was changed by a simple ingenious stroke: spectroscopy.

Shortly after John Herschel's GC, the new method was systematically applied to bright, unresolved nebulae.[32] William Huggins' revolutionary studies brought astonishing results: some objects, such as planetary nebulae and the Orion Nebula, have discrete spectra and are thus gaseous. In other cases, such as the Andromeda Nebula, a continuous spectrum was detected, implying a starry nature.[33] Major contributions were also due to d'Arrest, Pickering and Copeland. Photography of nebulae was not relevant in the 1860s and 1870s. Most objects were not bright enough to be detected on the insensitive plates of the time (which had many defects). Photographic identification of nebulae (which could have been very useful for the compilation of the NGC) was out of reach. Another field, the photometry of nebulae, was still undeveloped; thus magnitudes had to be estimated visually. Useful techniques were eventually introduced in the twentieth century.

1.4 STRUCTURE, PRESENTATION AND CONVENTIONS

This work is characterised by its large amount of information and data. Altogther 2154 of the 7840 NGC objects are mentioned (see the appendix).[34] To get the necessary overview, a systematic presentation is needed.

Though chronology should be a central element, it is not a sufficient guide per se. There are many interconnected aspects to the story, which must be treated in a parallel manner. These are

- observers (discoverers) and their biographies
- discovery, description and cataloguing of objects
- instruments and sites (observatories)
- observing methods and the development of astronomy during the nineteenth century

The great amount of facts can be handled only by enriching the text with a considerable number of tables, graphics and figures. There are 239 tables, presenting objects, discoveries, observers and instruments, 35 graphics[35] about statistics, historical development and magnitude distribution and 324 figures[36] showing object images, drawings/sketches, portraits, instruments and observatories.

It would certainly be useful to present the 'Historic NGC', first published by the author in 2006.[37] It contains modern data, identifications, discoverers, dates and instruments for all NGC objects. However, due to its size, it was impossible to incorporate it. Nevertheless, in connection with the original NGC, it is an essential basis of this work. Of course, both catalogues reflect the historical development and the above-mentioned major topics only in a very condensed form, so a great amount of additional information is needed. Therefore, the present work surpasses the original and the modern 'Historic NGC', using their data only as examples.

Take, for instance, the question of 'priority', i.e. who discovered an object first. Often the observers were unaware of the existing catalogues and publications, but, even if earlier data were known, some problems remained. Owing to the incomplete or erroneous nature of the data, observers were unable to discern the correct priority. To clarify the situation, a standard procedure is presented here. First, all new objects of an observer are listed in a table (if the number is not too large) showing the relevant data. Next the status of the objects must be determined. Actually, two kinds of identities are possible: (a) 'catalogue identity' – the object appears more than once in the observer's list; and (b) 'NGC identity' – the object (entry) is identical with other ones (normally

---

[32] For the development of astrophysics, see Leverington (1995).
[33] These objects were called 'white nebulae'.
[34] Additionally 107 IC objects are mentioned.
[35] All graphics were made by the author.
[36] Most of them are from the author's archive.
[37] See the author's website: www.klima-luft.de/steinicke.

associated with different observers). Afterwards it must be checked that the objects had not been found earlier by other observers. The result is a list of NGC objects for which the observer possesses the priority. It is now easy to calculate an individual's success rate, i.e. the number of non-stellar objects discovered.

Generally, the structure of the tables is uniform. The catalogue number (priority objects are in bold print), discoverer, date, type and constellation are listed.[38] A visual (integrated) magnitude is not given for pairs/groups of stars and emission/reflection nebulae (it is not an adequate measure for these types). Moreover, there are remarks, e.g. pointing to identities, other observers or names.

For observers with large numbers of discoveries, only special objects are listed: brightest/faintest, most northern/southern and first/last discovered. The distributions of objects' types and magnitudes are given both in a table and graphically. Additional notes concern the discovery of galaxy groups, clusters and special appearances (e.g. ring or edge-on galaxy). If relevant, information about further persons, relations, organisations, observations, measurements or publications is presented.

The graphics present results of statistical analyses, concerning brightness distribution, temporal development, number of discoveries, publications and instruments. The photographs of persons, telescopes and observatories are from the author's archive, the RAS and other sources. Most object images were taken from the Digitized Sky Survey (DSS), processed and labelled by the author. The orientation resembles the real sky: north up, east left; the scale is given in arcmin ('). Sketches and drawings have been copied from the original publications (sometimes the orientation is changed).

Additional structure elements are quotations, notes and references. Unfortunately, only 90 sources are given by Dreyer in the NGC. The present work contains more than 1600; mainly references to the original and secondary literature.[39] For most catalogues, lists or papers a cited object can be easily found by its designation. Many articles, especially those from the *Astronomische Nachrichten* (*AN*), are pretty short, so an exact page (or column) for the quotation is not needed.

Of course, for longer quotations the page is given. All foreign-language texts were translated by the author. For those from previous centuries it was decided to keep the original structure and style to a large extent. However, often sentences were lengthy and dodgy, e.g. those by Tempel, written in German. In these cases the translation was a challenge. Titles of books and articles have been translated too; translations are given either in brackets, following the text, or in a footnote.

The RAS archive is a valuable source. This concerns the Herschel family (Bennett 1978) and letters and manuscripts of astronomers, such as Lord Rosse and Lassell. Other archives (if available) were consulted for those astronomers or observatories with significant contributions to the NGC. For Great Britain and Ireland some were built up by Hoskin and others: Birr Castle (Lord Rosse), Markree Castle (Cooper), Dunsink Observatory and Armagh Observatory.[40] Sadly, Dreyer's estate could not be located and it is doubted by the author that there is anything left.[41] The available information about other eminent astronomers, such as Tempel and d'Arrest, is sparse too.

The sequence of chapters generally follows the milestones defined in Table 1.2. Special sections are devoted to the discoverers contributing to the respective catalogue; one contains a detailed statistical analysis based on modern data. Looking at the table of contents, it is apparent that the number of sections (and subsections) increases with the years. This is due to the growing number of persons involved: William Herschel's three catalogues were supported only by his sister Caroline. Four discoverers contributed to the Slough and Cape catalogues (John Herschel, James Dunlop, Wilhelm Struve and Niccolò Cacciatore). In Auwers' list of new nebulae 14 persons are involved; in the GC there are 13. In the cases of the GCS and NGC we have 30 and 38 additional discoverers, respectively. Moreover, many other persons, who contributed by making measurements, corrections etc., are mentioned in the text.

---

[38] The international names and abbreviations of the constellations are used.

[39] Bigourdan gave some useful references (Bigourdan 1917b).

[40] Bennett and Hoskin (1981) (Birr Castle); McKenna-Lawlor and Hoskin (1984) (Markree); Hoskin (1982b) (Markree und Dunsink); Buttler and Hoskin (1987) (Armagh).

[41] Dreyer's estate seems to be lost (see Section 8.1.5). Some of his documents could be inspected by the author at Armagh Observatory and are considered here. Important letters of Dreyer to Hagen were found at the Vatican Observatory (see Section 10.1.1).

All discoverers are introduced by a short biography;[42] published obituaries and other biographical sources are given. If several persons were active at an observatory, the presentation is bundled. Examples are Birr Castle, Marseille, Cambridge (Harvard), Uppsala, Chicago (Dearborn), Vienna, Melbourne and Charlottesville (Leander McCormick). If an observer contributed to more than one catalogue, separate subsections are introduced; d'Arrest, Tempel, Stephan and the observers at Harvard and Birr Castle are examples.

Chapter 10 is central, describing the structure and content of the New General Catalogue. Here all information is combined. A complete review of the NGC discoverers and their success rates is presented in Table 10.6. The problem of missing data is treated too. Finally, supplements, corrections and revisions of the NGC are critically discussed.

Chapter 11 is reserved for special themes and important objects. It starts with a comprehensive summary of nineteenth-century campaigns to determine precise positions of non-stellar objects. Next the history of nebular drawings is presented, with special emphasis on the problem of objectivity. The remaining sections deal with special objects, demonstrating the controversial character of visual observations. The popular galaxy M 51 in Canes Venatici is representative for the discussion about the reality of spiral structure, since it was first detected by Lord Rosse in April 1845. Much excitement was caused by variable nebulae, like NGC 1555 in Taurus, which was discovered in 1852 by Hind. Another case is the Merope Nebula (NGC 1435) in the Pleiades, which was found in 1859 by Tempel in Venice.

The extensive appendix contains a timeline (with 152 major events concerning the history of the NGC), abbreviations/units and data about telescopes. References and Internet sources are given. Moreover, indices concerning persons, sites (observatories) and designated objects are presented. The work is closed with a comprehensive subject index.

---

[42] An exception is Dreyer, who is represented by a longer biography in Section 8.1.

2 • William Herschel's observations and parallel activities

When William Herschel (Fig. 2.1) started his systematic search for non-stellar objects in autumn 1783, only about 100 were known (Messier's final catalogue had appeared two years earlier). Hence the opportunity to discover new ones with his superior telescopes was great. Herschel took advantage of it and in the end his catalogues grew to about 2500 entries – enough to analyse the nature and evolution of nebulae and star clusters by standardised methods developed by him. Generally Herschel's work is characterised by his great interest in the nature of celestial objects – perhaps a consequence of the lack of precise measurements. With his equipment, positions were determined relative to reference stars, often lying at a considerable distance. As regretted 50 years later by Auwers and d'Arrest, the precision was not high enough for using the data to determine proper motion. Thus a valuable past reference point was lost. Another crucial innovation was his coded textual description, which he introduced not only to identify the objects, but also to classify them. Herschel defined eight classes (I to VIII), which became a powerful tool with which to derive an evolutionary scenario about the nature of nebulae and clusters.

Herschel had not received a scientific education; he was self-taught and had both the limitations and the sturdy independence of an autodidact.[1] Nevertheless, he undertook systematic studies of the heavens. Herschel had therefore already dealt with 'astrophysics' (much more than with astrometry), which was strongly influenced by Newton's theory of gravity. His theories did not rest on (objective) physical experiments, but on the interpretation of (subjective) visual observations. No doubt due to his superior equipment and revolutionary mind, he was quite solitary. There was no way in which the astronomical community could repeat his observations or assess his 'natural history' vision of

Figure 2.1. William Herschel (1738–1822).[2]

the heavens against accepted professional models.[3] Herschel tried to build a single series that linked 'true nebulosity' (appearing 'milky') with 'resolved' clusters. To reach this goal, a large sample of objects had to be accumulated. He introduced a 'morphological method' into astronomy.[4] Key objects (like specimens) were the Orion Nebula, 'resolved' star clusters and the planetary nebulae, the latter being primary targets of contemporary observers like Friedrich von Hahn too. However, he repeatedly changed his ideas about nebulae. Finally the objects were arranged by age, starting from much diffused nebulosity, which contracts into stars,

---

[1] Hoskin (1982c: 143).

[2] For a comprehensive collection of Herschel portraits, see Turner A. (1988).

[3] Schaffer (1980), Hoskin (1982c).

[4] Later this method was repeated in the work of Fritz Zwicky; see his *Morphological Astronomy* (Zwicky 1957).

building highly condensed star clusters. Herschel's method is similar to that of a naturalist, defining the life-history of a plant by pointing out specimens at successive stages of evolution. Since all stages are present at once, a natural sequence can be drawn by watching, classifying and ordering – one does not have to wait for the (slow) ageing of an individual object.

Because much has already been written about the life of William Herschel, a biography is omitted here.[5] The focus is on his observations leading to discoveries of new objects and their cataloguing. First the structure and content of his three published catalogues is presented, followed by their statistical analysis. Central aspects are the appearance of Herschel objects in the NGC and their identification. With the exception of Michael Hoskin's famous research, these issues had hitherto not been treated in the literature. William Herschel gave his name to the British 4.5-m reflector on La Palma and lately he was honoured by the European 'Herschel Space Observatory', a 3.5-m infrared telescope launched on 15 May 2009.

2.1 OBJECTS DISCOVERED PRIOR TO HERSCHEL

There are already many publications about early discoveries of nebulae and star clusters.[6] A pretty much uncovered aspect is their relevance for the NGC, which is described here. A central role is undoubtedly played by the Messier catalogue of 1781 with its 103 entries.[7] Charles Messier arranged it by the date of discovery or position measurement of the objects. It was Auwers who published in 1862 the first reduced version, sorted by right ascension for 1830 (Auwers 1862a). He included measurements by d'Arrest for 43 objects. Auwers made Messier's catalogue usable for the professional astronomer and laid the foundations, together with William Smyth, for its later popularity. The latter had already presented a sample of 65 objects in his Bedford Catalogue of 1844, with notes about history and visual observation (see Section 6.5). In 1877 Edward Holden published a list that identifies the Messier objects in John Herschel's General Catalogue (Holden 1877a). At the beginning of the twentieth century, John Ellart Gore presented comprehensive textual descriptions (Gore 1902) based on his visual observations and photographs of Isaac Roberts made in the 1890s. In 1917 Harlow Shapley and Helen Davis undertook a (successful) attempt to introduce the Messier catalogue to professional astronomers (Shapley and Davis 1917). The M- and NGC-number, equatorial coordinates (1900), galactic coordinates and type (name) are listed. However, M 91 is not identified as an NGC object; M 25 is IC 4725 (see Table 2.2).

The NGC contains 138 objects found prior to William Herschel; Table 2.1 shows their discoverers.[8] In the cases of Cassini and Pigott, Giovanni Domenico Cassini and Edward Pigott are meant. The latter saw M 64 on 23 March 1779 (Pigott 1781), i.e. a bit earlier than Bode (4 April 1779) and Messier (1 March 1780). The instruments ('Instr.') used are E = naked eye and Rr = refractor. For some discoveries the date is uncertain. 'Remarks' show the most important Messier objects and independent discoverers.[9] Amerigo Vespucci was the first northern-hemisphere observer to see the Small Magellanic Cloud (SMC), as was revealed recently (Dekker 1990). He saw it in 1501, i.e. 20 years earlier than Magellan. According to Humboldt, the Large Magellanic Cloud (LMC) was described by As-Sufi as a 'white ox'.[10] M 44 (Praesepe) and M 7 (Ptolemy's Cluster) are the second- and third-nearest NGC objects (see Table 10.17).

Actually the NGC lists 140 early objects, but two are double entries: the open cluster M 47 in Puppis and the planetary nebula M 76 in Perseus. M 47 was found

---

[5] See e.g. Holden (1881), Ball R. S. (1895: 200–218), Dreyer (1912a), MacPherson (1919), Lubbock (1933), Buttmann (1961), Hamel (1988), Gärtner (1996) and Hoskin (1959, 2003b, 2007).

[6] See e.g. Schultz (1866b), Wolf R. (1890, vol. 2: 600–609), Sawyer-Hogg (1947a–c), Duncan (1949), Gingerich (1953a, b, 1954, 1960, 1967, 1987), Glyn Jones (1967a, b, 1975, 1991), Nilson (1973: 449–455), Stoyan et al. (2008) and the website of Frommert: www.seds.org/messier/xtra/history/deepskyd.html.

[7] Messier published three catalogues with 45, 70 and 103 objects, respectively (Messier 1774, 1780, 1781).

[8] Biographical information on most of them can also be found in Johann Elert Bode's compilation in the *Berliner Jahrbuch* (Bode 1813).

[9] For Ihle's discovery of M 22, see Lynn (1886a). It is remarkable that M 31 was not noticed by Tycho Brahe and Galileo; for its history and Marius' rediscovery in 1612, see Webb (1864c) and Lynn (1886b). The origin of the name Praesepe is explained by Lynn (1905a).

[10] Humboldt (1850: 599); undoubtedly the SMC was seen much earlier too.

Table 2.1. *Early discoverers of NGC objects*

Discoverer	Number	Instr.	Date	Remarks
As-Sufi	1	E	905?	M 31 (Andromeda Nebula); Marius 1612[11]
Aratos	1	E	−260?	M 44 (Praesepe); Hipparch −130
Aristotle	2	E	−325?	M 39, M 41
Bevis	1	3" Rr	1731	M 1 (Crab Nebula)
Bode	4	Rr	1774–77	M 53, M 93; M 81 and M 82 (Bode's Nebulae)
Cassini	1	5" Rr	1711	M 50
Darquier	1	3.5" Rr	1779	M 57 (Ring Nebula)
de Chéseaux	6	Rr	1745?	Including M 4, M 16, M 17 (Omega Nebula), M 35, M 71
Flamsteed	1	9.7" Rr	1690	NGC 2244
Halley	2	Rr	1677–1714	M 13, ω Centauri
Hipparch	2	A	−130	Double Cluster NGC 869/884 (χ Per)[12]
Hodierna	11	Rr	1654?	Including M 6, M 33, M 34, M 36–38, M 47, M 71
Ihle	1	Rr	1665	M 22; Halley 1715
Kirch	2	Rr	1681–1702	M 5, M 11 (Wild Duck Cluster)
Koehler	3	Rr	1779	M 59, M 60, M 67
Lacaille	23	0.5" Rr	1751	Including M 55, M 83
Legentil	2	Rr	1749	M 8 (Lagoon Nebula), M 32
Mairan	1	Rr	1731	M 43
Maraldi	2	Rr	1746	M 2, M 15
Méchain	26	3" Rr	1779–82	Last object: NGC 6171 (M 107), April 1782
Messier	40	3.3" Rr	1771–81	Last object: M 80, 4.1.1781
Oriani	1	3.6" Rr	1779	M 61
Peiresc	1	Rr	1610	M 42 (Orion Nebula); Cysat 1611
Pigott	1	5" Rr	1779	M 64 (Black Eye); Bode 1779
Ptolemy	1	E	−138?	M 7 (Ptolemy's Cluster); Halley 1678
Vespucci	1	E	1501	NGC 292 (SMC); Magellan 1521

first by Hodierna and is catalogued as NGC 2422; while Messier's discovery was listed by Dreyer as NGC 2478. The case of M 76 is different: this is a bipolar nebula, whose brighter part (NGC 650) is credited to Méchain. Later William Herschel saw both 'components'; the fainter (I 193) is catalogued as NGC 651.[13] Lord Rosse even saw a 'spiral' structure.

Four Messier objects are missing in the NGC: M 24, M 25, M 40 and M 45 (Table 2.2). About M 25 Dreyer writes as follows in the introduction of the IC I: *'Two clusters in Messier's catalogue do not appear in the New General Catalogue, and may perhaps be mentioned here.'*[14] The second one is M 48; unfortunately wrong coordinates are given. This might be the reason why Dreyer did not notice that the object had already been catalogued as NGC 2548. For M 25 he changed

---

[11] The Andromeda Nebula was the first non-stellar object to be found with a telescope.

[12] The double cluster is usually referred as 'h + χ Persei' (with h = NGC 869 and χ = NGC 884), but, as shown quite recently, it should be only χ Per (O'Meara and Green 2003). Actually, Bayer assigned this 'star' to represent the combined light of the double cluster. Another mystery is the fact that Messier did not include the object in his catalogue (Burnham R. 1966: 1440).

[13] NGC 650/51 is known as the Little Dumbbell Nebula, a creation of Leland Copeland (Copeland L. 1960).

[14] Dreyer (1953: 243).

Table 2.2. *Missing, double and added Messier objects in the NGC*

M	NGC	Discoverer	Date	Type	V	Con.	Remarks and References
24		Messier	20.6.1764	Star cloud	2.5	Sgr	IC 4715, Barnard Aug. 1905
25		de Chéseaux	1745?	OC	4.6	Sgr	IC 4725, Bailey 1896
40		Messier	24.10.1764	2 stars	9.0	UMa	Winnecke 12.10.1863
45				OC	1.5	Tau	Pleiades
47	2422	Hodierna	1654?	OC	4.4	Pup	Messier 19.2.1771; W. Herschel 4.2.1785 (VIII 38)
	2478	Messier	19.2.1771				
48	2548	Messier	19.2.1771	OC	5.8	Hya	C. Herschel 8.3.1783 (VI 22)
76	650	Méchain	9.5.1780	PN	10.1	Per	
	651	W. Herschel	12.11.1787				I 193
91	4548	Messier	18.3.1781	Gx	10.1	Com	W. Herschel 8.4.1784 (II 120); Dreyer: NGC 4571?
102	5866	Méchain	27.3.1781	Gx	9.9	Dra	W. Herschel 5.5.1788 (I 215); Dreyer: NGC 5928
104	4594	Méchain	11.5.1781	Gx	8.3	Vir	W. Herschel 9.5.1784 (I 43); Flammarion (1917)
105	3379	Méchain	24.3.1781	Gx	9.5	Leo	W. Herschel 11.3.1784 (I 17); Sawyer-Hogg (1947c)
106	4258	Méchain	July 1781	Gx	8.3	CVn	W. Herschel 9.3.1788 (V 43); Sawyer-Hogg (1947c)
107	6171	Méchain	April 1782	GC	7.8	Oph	W. Herschel 12.5.1793 (VI 40); Sawyer-Hogg (1947c)
108	3556	Méchain	16.2.1781	Gx	9.9	UMa	W. Herschel 17.4.1789 (V 46); Gingerich (1954)
109	3992	Méchain	12.3.1781	Gx	9.8	UMa	W. Herschel 12.4.1789 (IV 51); Gingerich (1954)
110	205	Messier	10.8.1773	Gx	7.9	And	C. Herschel 27.8.1783 (V 18); Glyn Jones (1967a)

his mind in the IC II, because the cluster appears on plates taken in 1896 by Solon Bailey in Arequipa, Peru (Bailey 1908). It was now catalogued as IC 4725. M 45 (and the Hyades too) was ignored by Dreyer, though it was among the 13 non-NGC objects on Bailey's plates. He wrote, in the IC II introduction, '*the Pleiades and the Hyades I have not inserted*'.[15]

The identity of the large Sagittarius Star Cloud M 24 with IC 4715, which was photographed by Barnard in the summer of 1905 (Barnard 1908a), was not recognised by Dreyer. M 40 in Ursa Major was found independently by Winnecke in 1863. It is no. 4 in a list of new double stars published in 1869 (Winnecke 1869); see Section 6.12.2. A problematic case is M 91 (NGC 4548). For NGC 4571 Dreyer writes '*M 91??*', adding in the notes that '*M 91 must have been a comet*'. Obviously he follows Flammarion, assuming an identity of Messier's objects with the comet of 1779 (Flammarion 1917). The most controversial object is M 102, which has often been identified with the galaxy NGC 5866 in Draco, which originates from Solon Bailey.[16] Others believe the object to be a double

---

[15] Dreyer (1953: 287).

[16] Shapley and Davis (1917: 179); supported by Frommert (2006).

sighting of the bright galaxy M 101 in Ursa Major.[17] In the IC I notes, Dreyer gives a very different view. If Méchain's reference star is a typo, M 102 could be the galaxy NGC 5928 in Serpens: '*I assume that ι Draconis is an error for ι Serpentis.*'[18]

It is well known that the Messier-numbers M 104 to M 110 were added in the twentieth century by Camille Flammarion, Helen Sawyer-Hogg, Owen Gingerich and Kenneth Glyn Jones. All these objects are included in the NGC.

For 11 Messier objects Dreyer does not give the true discoverer. M 4 (NGC 6121), M 6 (NGC 6405) and M 8 (NGC 6523) are all credited to Lacaille, but they were found by de Chéseaux, Hodierna and Legentil, respectively. In the case of M 7 he does not mention Ptolemy, mentioning the 'modern' discoverers Halley and Lacaille. For M 36 (NGC 1960), Legentil instead of Hodierna is credited. Dreyer mentions Flamsteed and Legentil in the case of M 41 (NGC 2287), which he erroneously called 'M 14' (corrected in the IC I notes). This open cluster was first described by Aristotle. Hipparch's M 44 (Praesepe, NGC 2623) had already been seen (and named) by Aratos. Dreyer credits the galaxies M 49 (NGC 4472) and M 67 (NGC 2682) to Oriani. M 49 was found earlier by Messier; M 67, discovered by Koehler, is not at all an Oriani object. This error was made by Bigourdan too.[19] M 71 was found by de Chéseaux, not by Méchain, as Dreyer claims.

In the case of M 42 (NGC 1976) Dreyer mentions Cysat's observation of 1611, obviously following Rudolf Wolf,[20] but Peiresc had seen the Orion Nebula in 1610, as Bigourdan has shown in his paper 'La découverte de la Nébuleuse d'Orion (N.G.C. 1976) par Peiresc'.[21] Given these early observations and the fact that M 42 (3.7 mag) can even be glimpsed as a nebulous spot by the naked eye on a dark night, it is remarkable that Galileo had not found the nebula with his telescope. His 1610 map of the Sword of Orion does not show it (Gingerich 1987). This was noticed by Humboldt, who wrote '*How could the large nebula in the sword escape his attention?*'[22] Webb's idea was that '*We can only suppose that he may have mistaken it for the effect of moisture upon his eye-glass*' or '*engravers and copyists may have been in fault*' (Webb 1864c). In 1617 Galileo made another, even closer sketch of the area, which, however, shows only the principal stars with great accuracy (Graney 2007). As another reason for his failure, it was suggested recently that the nebula could have been rendered temporarily invisible by a flaring up of illumination from FU Orionis-type stars and reappeared later.[23] This, however, contradicts the observations of Cysat and Peiresc.

The pre-Herschel time ends with the observations of Méchain. Only five months after his last discovery (the globular cluster in Ophiuchus later named M 107), William Herschel took over, discovering the planetary nebula NGC 7009 in Aquarius on 7 September 1782.

2.2 STRUCTURE AND CONTENT OF THE HERSCHEL CATALOGUES

2.2.1 Herschel's sweeps and publication of the results

In Bath, Herschel had observed Messier objects with his 6.2" reflector,[24] not knowing the French catalogue at that time: the Orion Nebula (M 42) in 1774, the globular cluster M 13 in Hercules in 1779 and the Andromeda Nebula (M 31) in August 1780. He got Messier's second version (containing 70 objects) in December 1780.[25] Herschel later wrote '*As soon as the first of these volumes came into my hands, I applied my former 20-feet reflector of 12 inches aperture to them.*'[26] At that time, his largest telescope was the 'small 20 ft', a 12" reflector built in 1776 (Fig. 2.2 left).[27] By the end

---

[17] Proclaimed e.g. by O'Meara (2006).
[18] Dreyer (1953: 286).
[19] Bigourdan (1917b: E140).
[20] Wolf R. (1854); see also Webb (1864c: 258–266) and Lynn (1887).
[21] 'The discovery of the Orion Nebula (NGC 1976) by Peiresc' (Bigourdan 1916).
[22] Humboldt (1850: 506).
[23] This idea is due to Harrison (1984); see other views by Gingerich (1987) and Herczog (1998).
[24] Herschel had used this '7-foot' since 1778 in Bath. In August 1779 he started making with it a survey of all stars down to 8 mag to isolate as many double stars as he could discover, using them to determine stellar parallax. During this search, Uranus was found on 13 March 1781 (Schaffer 1981).
[25] Messier (1780); Herschel wrote about it four years later (Herschel W. 1784: 439–441).
[26] Herschel W. (1784: 439). See also Section 6.4.8.
[27] Telescope data are listed in the appendix. On Herschel's see Steavenson (1924), Maurer (1971, 1996) and Bennett (1976a).

2.2 Structure and content

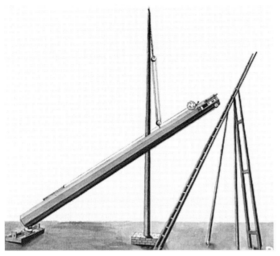

Figure 2.2. Herschel's reflectors in Datchet. Left: small 20-ft (12"); right: large 20-ft (18.7").

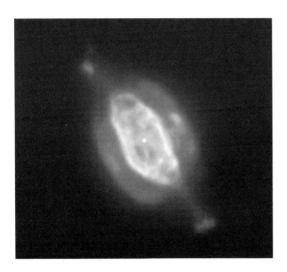

Figure 2.3. The Saturn Nebula NGC 7009 in Aquarius.[28]

of 1781 he had observed 24 Messier objects with it. However, being still absorbed with non-astronomical matter, he saw no reason for a systematic search for nebulae. In December 1781 his friend William Watson delivered him an exemplar of Messier's final catalogue with 103 objects.[29]

By the end of July 1782 Herschel moved to Datchet, where he observed Messier objects and double stars with the 12-inch. On 7 September 1782 he accidentally discovered his first nebula: NGC 7009 in Aquarius (Fig. 2.3); this is the very first non-stellar object to have been found with a reflector. However, the systematic search started a year later – motivated by observations by his sister Caroline.[30] With her small refractor, made by Herschel, she had found some non-stellar objects between August 1782 and October 1783 (see Section 2.3). Herschel was truly impressed and considered making his own observations. In spring 1783 he tested a 3.5" refractor for this task and subsequently tried the 'small 20 ft'. Since both instruments were not successful, he decided to build a larger telescope. On October 23, 1783 the 'large 20 ft', equipped with an 18.7" mirror, was ready – this was Herschel's standard telescope for his search for nebulae.[31]

Herschel effected his observations at three different sites, all near Windsor Castle.[32] From 2 August 1782 to early June 1785 he worked in Datchet (Berkshire), then moving to Clay Hall, Old Windsor. From 3 April 1786 he lived in the Observatory House on Windsor Road, Slough (Fig. 2.4). Table 2.3 gives the first and last objects discovered at the three sites. During the whole period Herschel observed on 401 nights. The numbers of objects found in Datchet and Slough are nearly equal.

---

[28] See also the sketches of Lamont (Fig. 6.3) and Vogel (Fig. 8.40).
[29] RAS Herschel 1/13.W.11.
[30] Hoskin (1979).
[31] Bennett (1976b), Ashbrook (1984: 127–132).
[32] See Dreyer (1912a: xxxvii).

20 William Herschel's observations

Table 2.3. *Herschel's observing sites near Windsor and his first and last objects discovered at these sites (see the text)*

Site	Number	Date	Sweep	Object	No.	C	NGC	Type	V	Con.
Datchet	1079 (43%)	7.9.1782	–	IV 1	1	1	7009	PN	8.0	Aqr
		5.5.1785	409	II 425	1079	2	5990	Gx	12.3	Ser
Clay Hall	345 (14%)	17.7.1785	415	VII 18	1080	2	6823	OC	7.1	Vul
		28.3.1786	550	II 567	1424	2	5101	Gx	10.5	Hya
Slough	1076 (43%)	17.4.1786	553	II 568	1425	2	4270	Gx	12.1	Vir
		26.9.1802	1111	III 978	2500	3	3057	Gx	12.9	Dra

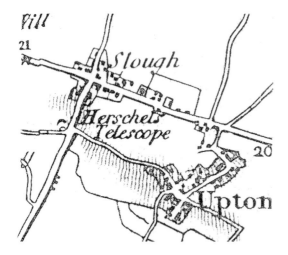

Figure 2.4. The site of Herschel's 'Observatory House' in Slough (Maurer 1996).

The 12-inch yielded only the first nebula (NGC 7009) – all later discoveries were made with the 18.7-inch.

The systematic observations ('sweeps'), were started on 28 October 1783. This night already brought a success, the first 18.7-inch discovery: II 1 (NGC 7184; Fig. 2.5), a galaxy in Aquarius.

All sweeps (1 to 1112) and discovered objects (1 to 2500) were numbered. However, these numbers are given only in Herschel's unpublished observing journals (not in his three catalogues 'C'), which are now at the RAS archive.[33] In the first 41 sweeps (up to 13 December 1783) Herschel used the telescope in 'front-view' mode,[34] standing on a platform in front of the tube. It pointed to the south (meridian) and could be moved horizontally by 30° to each side. He performed slow oscillations of 12° to 14° and made notes every 5 minutes.[35] Then the reflector was raised or lowered by 8' or 10' to repeat the procedure; 10 to 20 of them defining a sweep. Afterwards the telescope was reset to a slightly different declination (normally 2° to 3°) for another sweep. At the beginning, one sweep per night was executed, later three or four. The method, which was carried out without any assistance, turned out to be ineffective and yielded only eight nebulae, among them the galaxy NGC 253 (V 1, 30 October 1783) in Sculptor, which Caroline Herschel had already discovered on 23 September 1783 (William Herschel: '*It is Carolina's*'[36]). The main problem was caused by the frequent illumination needed to record the observation on the platform, which meant that the eye could not adapt properly.

Herschel tested alternative procedures and with sweep 46 (18 December 1783) a new standard was established. Now the south-looking telescope was moved vertically only (with the aid of an assistant); objects and reference stars passed the eye-piece due to the Earth's rotation. In every sweep three to five standard stars were observed. Herschel now used the Newtonian focus, sitting on a chair, which was fixed on a ladder at the side of the reflector (Fig. 2.2 right).[37]

---

[33] The RAS offers a digital version (DVD) of the archive. The content is explained by Bennett (1978) (his quotation form is used here).

[34] In this mode, the eye-piece points directly, i.e. without a secondary mirror, to the main mirror. The obstruction by the observer's head was negligible. The 'front-view' form is also called 'Herschelian' or 'Le Mairean' and was invented in 1728 (Mitchell O. 1851: 227–229). William Herschel tested it for his 12" and 18.7" reflectors, see his description in the notes of the first catalogue (Herschel W. 1786: 499).

[35] Herschel W. (1786: 458–459); see also Hoskin (2005c) and Dreyer (1912a: xxxix).

[36] RAS Herschel W. 2/1.7.

[37] The platform was re-installed in September 1786, occasionally using the reflector in 'front-view' mode (Herschel

2.2 Structure and content

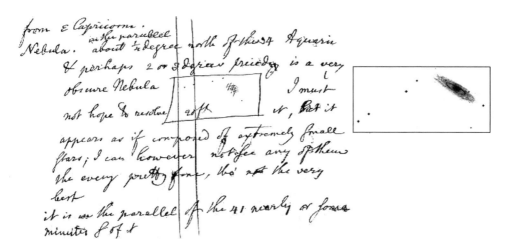

Figure 2.5. Left: Herschel's note and sketch for II 1 = NGC 7184 (RAS Herschel W. 2/1.7); right: DSS image showing the galaxy and the prominent star chain (north is up).

The observational results were shouted to Caroline Herschel, who was sitting at the window in the nearby house to record them.[38] The data were now entered in separate 'sweep-books'.[39] The first discovery with this new, effective method was II 5 = NGC 1032, a galaxy in Cetus (sweep 47, 18 December 1783).

Between 1786 and 1802 William Herschel published three catalogues in volumes 76, 79 and 92 of the *Philosophical Transactions of the Royal Society*[40] (Table 2.4), listing altogether 2500 objects. They were later revised by Dreyer and included in the *Scientific Papers of Sir William Herschel* (Dreyer 1912a); some new objects were added there (see Section 2.9).

William Herschel categorised his objects into eight classes (Table 2.5). The first five contain 'nebulae', i.e. unresolved objects, differentiated by brightness (I–III), type (IV) and extent (V). The last three classes describe 'clusters', i.e. resolved star clusters, where he distinguishes in terms of concentration and richness. As shown later, the Herschel classes are only weakly correlated with modern object types (especially for nebulae).

Table 2.6 gives the first and last objects from the three catalogues. Herschel put exactly 1000 objects in his first two catalogues (the publication of no. 1 roughly coincides with the end of the Datchet observations). His sense for numerical harmony seems to be stronger than the observational facts: it did not bother him that the discoveries made on the respective last nights landed in different catalogues. Actually, the objects III 376 and II 403 (found on 26 April 1785) were published with a time gap of three years; and even 13 years in the case of III 747 and I 216 (3 February 1788).

The title of Herschel's last catalogue promises another 500 objects. Actually his manuscript of 29 June 1802, prepared for the Royal Society, contains only 497. To get a round number, he made a last observation. The neglected near-pole regions were most promising for this task. On 26 September 1802 (sweep 1111) Herschel discovered three nebulae at a declination of +80°, added to his list as I 288, III 977 and III 978. These are the galaxies NGC 2655 (Camelopardalis), NGC 2908 and NGC 3057 (both in Draco). The three objects were communicated to the Royal Society by Caroline Herschel, who wrote that '*The reason for the addition is that, on casting up, the number of Nebulae was found 3 less than 500.*'[41]

W. 1786: 499); see also Dreyer (1912a: xlii). Herschel tried binocular vision a few times too (Dreyer 1912a: xliii).

[38] For the role of Caroline Herschel, see Hoskin (2002b) and Ashworth (2003).

[39] They were presented to the Royal Society by John Herschel in 1863 and are now in the RAS archive. The general appearance of the sweep record was explained by Dreyer (1912a: xlii).

[40] Herschel published from 1780 to 1818 in the *Philosophical Transactions* (except 1813 and 1816).

[41] Hoskin (2005c: 317).

Table 2.4. William Herschel's three catalogues

C	Title	*Phil. Trans.*	Date	Objects
1	Catalogue of one thousand new nebulae and clusters of stars	76, 457–499 (1786)	27.4.1786	1000
2	Catalogue of a second thousand new nebulae and clusters of stars	79, 212–255 (1789)	11.6.1789	1000
3	Catalogue of 500 new nebulae, nebulous stars, planetary nebulae, and clusters of stars	92, 477–528 (1802)	1.7.1802	500

Table 2.5. William Herschel's classification and object numbers in his three catalogues

Class	Description	C1	C2	C3	Sum
I	Bright nebulae	1–93 (93)	94–215 (122)	216–188 (73)	288
II	Faint nebulae	1–402 (402)	403–768 (366)	769–907 (139)	907
III	Very faint nebulae	1–376 (376)	377–747 (371)	748–978 (231)	978
IV	Planetary nebulae	1–29 (29)	30–58 (29)	59–78 (20)	78
V	Very large nebulae	1–24 (24)	25–44 (20)	45–52 (8)	52
VI	Very condensed and rich clusters of stars	1–19 (19)	20–35 (16)	36–42 (7)	42
VII	Compressed clusters of small stars and large stars[42]	1–17 (17)	18–55 (38)	56–67 (12)	67
VIII	Coarsely scattered clusters of stars	1–40 (40)	41–78 (38)	79–88 (10)	88
	Sum	1000	1000	500	2500

Table 2.6. First and last discovered objects in the Herschel catalogues

C	Date	Site	Sweep	Object	No.	NGC	Type	V	Con.
1	7.9.1782	Datchet	–	IV 1	1	7009	PN	8.0	Aqr
	26.4.1785	Datchet	402	III 376	1000	3821	Gx	12.8	Leo
2	26.4.1785	Datchet	402	III 377	1001	3837	Gx	12.7	Leo
	3.12.1788	Slough	889	III 747	2000	1961	Gx	10.9	Cam
3	3.12.1788	Slough	889	III 748	2001	2366	Gx	10.9	Cam
	26.9.1802	Slough	1111	III 978	2500	3057	Gx	12.9	Dra

However, Herschel discovered eight more objects in Draco and Ursa Major: three on 9 September 1802 and five during his last sweep (1112) on the 30th.[43] The final object was NGC 3063 (II 909 = no. 2508; see Table 2.22), which is only a pair of stars in Ursa Major, 2′ west of the galaxies NGC 3065 (II 333) and NGC 3066 (II 334); the object had already been found on 10 April 1785 in Datchet (Fig. 2.6). Because William Herschel did not want to exceed the magical number 500, he flatly omitted them.[44] His son John later published them in the Cape catalogue (see Section 2.9).

---

[42] Instead of 'faint' or 'bright' Herschel sometimes used 'small' or 'large'.

[43] For the last sweep, see RAS Herschel W. 2/3.8 (report of Caroline Herschel).

[44] He never thought about reaching another 1000 objects, for which about 50 additional sweeps would have been necessary

Figure 2.6. Herschel's final discovery: the star pair NGC 3063, found on 30 September 1802 (DSS).

Figure 2.7 shows the annual numbers of objects discovered. After a phase of orientation (1782–83), the years 1784 and 1785 were the most productive. The following decrease is due to the time-consuming construction of the 40-ft reflector.[45] After 1790 the number remained, except for 1793 (77 objects), below 50. Herschel was now married and had many social duties (his sister had moved to the neighbouring 'cottage'). Between 18 October 1794 and 22 November 1797 there was no sweep. At that time he concentrated on the determination of the relative brightness of stars and the secular variation of their light, measuring the magnitudes of nearly 3000 stars with astonishing accuracy compared with later photometric catalogues.

While observing the moons of Uranus, Herschel made an accidental find on 4 March 1796: I 272 = NGC 3332 (see Section 9.7.2).[46] The planet was used as a 'reference star'. This was the case too for III 934 (NGC 3080, 1 April 1894) and II 898 (NGC 3107, 22 March 1894). All of these objects are galaxies in Leo. From 22 November to 20 December 1797 Herschel searched at high declinations (74° to 80°) in the constellations Ursa Minor, Draco and Camelopardalis, discovering altogether 25 galaxies. Among them are bright objects such as NGC 4589 (I 273) with 10.7 mag.

Figure 2.8 shows the average declinations searched by Herschel. He started to the south, increasing to about 10° in 1785 (completion of the first catalogue), then going up to reach +45° in 1788 (the second catalogue). Then he moved to lower declinations. Finally, from 1897 to 1802, he focused on the near-pole regions, which was a much more difficult and time-consuming task.

It is interesting to look at Herschel's favourite constellations and observing seasons. The constellations Virgo, Ursa Major, Coma Berenices and Leo clearly dominate (Fig. 2.9). The most productive was spring, which confirms the last table: 46% of all objects were found in March/April (Fig. 2.10). The summer months, with late darkness, brought only a few. The most successful night was on 11 April 1785 in Datchet. Herschel discovered 74 objects, most of them in Coma Berenices (47). All but one (NGC 4209 = II 375, a star) are galaxies.

Table 2.7 lists Herschel's most southern and most northern discoveries. It is truly remarkable that he was able to detect the galaxies NGC 3621 (Fig. 2.11), NGC 6569 and NGC 5253.[47] Though pretty bright, they reach elevations of only 6.8°, 6.5° and 8°, respectively, in Slough!

As mentioned already, Herschel neglected the region around the northern pole. This was due to the construction of his telescope. For high declinations it had to be moved backwards beyond the zenith to the north. This was a problematic matter and sometimes the mechanism failed. Thus only a few such observations were executed, as can be seen from Caroline Herschel's

---

(Hoskin 2005c). Actually, the 64-year-old Herschel had not enough energy to do this. After 1802 he returned to the double stars, but, however, on 31 May 1813 he tried another sweep (1113) with a new mirror. Stopped by clouds, it lasted only half an hour (Dreyer 1912a: xliii).

[45] For Herschel's largest telescope, see Dreyer (1912a: xlv–lvi), Bennett (1976a) and Hoskin (2003c).

[46] The object is identical to NGC 3342 (III 5), which was found on 18 January 1784. Note the large class difference (brightness).

[47] NGC 5253 was the second galaxy in which a supernova was detected (after S And in M 31 in 1885; see Section 9.20.2). On 12 December 1895 Williamina Fleming noticed a 'new star' of 8 mag on a plate exposed in March at Arequipa (Pickering E. 1895). The object was later designated Z Cen. However, in 1925 Max Wolf discovered a 12.5-mag 'nova' in NGC 4424 on a plate taken on 15 April 1895 in Heidelberg (Wolf M. 1925). The supernova was named VW Vir. The host galaxy in Virgo was found by d'Arrest (27.2.1865).

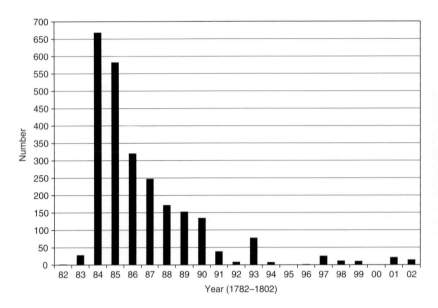

Figure 2.7. William Herschel's annual discoveries.

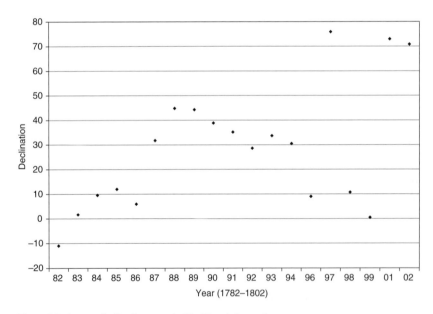

Figure 2.8. Average declinations searched by Herschel over the years.

compilation of the reference stars sorted by North Pole Distance (NPD).[48] There is not one nearer than 5° to the pole and only 12 between 5° and 10°. John Herschel later filled the gaps, discovering the most northern NGC object: Polarissima Borealis (NGC 3172).

2.2.2 Structure of the catalogues

All three Herschel catalogues have the same structure. Each class (I–VIII) has its own object table, sorted by discovery date.[49] Table 2.8 shows the meanings of the

---

[48] RAS Herschel C. 3/3.2; see also Hoskin (2005c).

[49] Here Herschel follows Messier, who, by the way, published his compilations in three steps too.

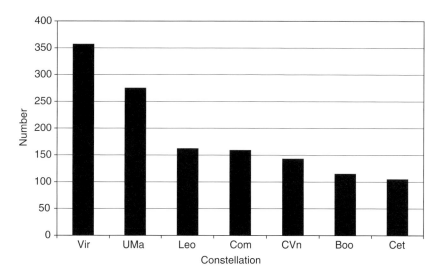

Figure 2.9. Discoveries by constellation.

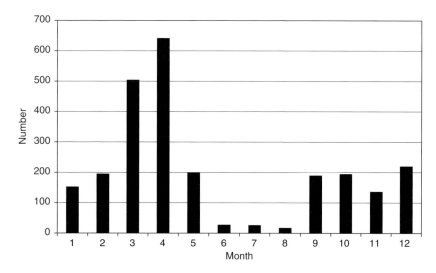

Figure 2.10. Discoveries by month.

columns. The table entries carry a running number, which is continued in the subsequent catalogue(s). II 426 is, for instance, the entry no. 426 in the table of class II objects (contained in the second catalogue in this case).

The data for distance and direction relate the object to the individual reference star. Mainly Flamsteed's *Catalogus Britannicus* of 1725 was used for this task; additional stars were taken from the catalogues of Bode, Lacaille and Wollaston.[50] Occasionally another, already catalogued nebula was used; for instance, I 100 (NGC 584) was used in the case of III 431 (NGC 586) on 10 September 1785.[51] Often the reference star is located quite far from the object; even 12.5° in the case of 85 Geminorum and II 48 (NGC 2672) in Cancer. Since Herschel did not use a micrometer, this sometimes caused wrong positions. Distances were simply determined with the aid of the (much smaller) diameter of the field of view. Herschel's standard eye-piece (focal length 39 mm) at the 18.7" reflector gave a power of 157

---

[50] For star catalogues see the list of Chambers (1890: 487–495).

[51] Even the 'Georgian Planet' Uranus was taken as a 'reference star' (see Section 2.2.1).

Table 2.7. *Most southern and most northern objects (site: D = Datchet, S = Slough)*

Object	NGC	Decl.	Date	C	Site	Type	V	Con.
II 638	5253	−31 38	15.3.1787	2	S	Gx	10.1	Cen
II 201	6569	−31 49	13.7.1784	1	D	GC	8.4	Sgr
I 241	3621	−32 48	17.2.1790	3	S	Gx	9.4	Hya
II 704	1184	+80 47	16.9.1787	2	S	Gx	12.5	Cep
III 974	6251	+82 32	1.1.1802	3	S	Gx	12.9	UMi
III 975	6252	+82 34	1.1.1802	3	S	Gx	14.8	UMi

Figure 2.11. Herschel's most southern discovery, the galaxy NGC 3621 in Hydra (DSS).

with a field of view of 15′ 4″. Sometimes other magnifications were applied (240, 300 or 320).

Herschel's code (the last column) was the basis of nearly all textual descriptions of non-stellar objects in the nineteenth century, particularly in the catalogues of his son (SC, CC, GC) and in Dreyer's GCS and NGC. The features are characterised by simple abbreviations.[52] Examples are brightness (F = faint, B = bright), size (S = small, L = large) and form (E = elongated, R = round). Additional letters, including e (exceedingly), v (very), p (pretty) and c (considerably), give further differentiation.[53] The column also contains remarks and further information, e.g. on magnification or Caroline Herschel's observations. In the case of class IV objects (planetary nebulae) remarks can be lengthy. A few notes can be found below the tables of the first two catalogues (e.g. corrections to I 54, II 1 and II 239).

As an example, the entry for the 'faint nebula' II 5 (Table 2.9) is presented here; the galaxy NGC 1032 in Cetus (13.8 mag) was observed eight times. According to the (first) catalogue, it is located 5^s west and 46′ north of δ Ceti, which is correct. The description means pretty bright, small, little elongated, brighter middle.

Owing to identities, the number of entries in Herschel's catalogues can be reduced (Table 2.10). In 34 cases there is an identity between entries in the three catalogues. A typical example is II 57 = II 546 (both listed in the second catalogue). The object was discovered on 15 March 1784 in Datchet (II 57) and was found again on 3 March 1786 in Clay Hall (II 546). It is NGC 2872, a galaxy in Leo (11.9 mag).

One object was even listed four times: the Trifid Nebula M 20 (NGC 6514) in Sagittarius (Fig. 2.12). On 12 July 1784 (Datchet) Herschel found three individual nebulae (actually separated by dark lanes), catalogued as V 10, V 11 and V 12: '*three nebulae, faintly joined, form a triangle*'. Unfortunately his position was 30′ too far south. This error made it possible to discover the nebula a second time (26 May 1786), now catalogued as IV 41 at the correct position. Auwers was the first to notice the identity.[54] John Herschel created the popular name 'Trifid' (meaning 'threefold') in his note about the observation on 1 July 1828 in Slough (h 1991): '*very large, trifid, three nebulae with a vacuity in the midst*'.[55] Interestingly, the object was not identified with Messier's M 20 by

---

[52] Attempts to introduce new codes, e.g. by John Herschel and by Schultz (see Section 8.16.5), were unsuccessful.

[53] Dreyer (1953: 12–13). The most important abbreviations are collected in the appendix.

[54] Auwers (1862a: 57).

[55] Citations related to objects from the catalogues of William and John Herschel or Dreyer are referred to simply by their catalogue number.

Table 2.8. *Meanings of the columns*

Column	Content	Remarks
1	Object number	Number in the particular class
2	Discovery date	Sorting order
3	Reference star	Mainly from Flamsteed's catalogue
4	Direction (AR)	p (preceding = west), f (following = east)
5	Distance (AR)	M. S. = minute, second
6	Direction (Decl)	n (north), s (south)
7	Distance (Decl)	D. M. = degree, minute
8	Observations	Number of observations (1–8)
9	Description	Herschel code; remarks

Table 2.9. *The fifth entry of Herschel's class II*

II.	1783	Stars.		M. S.		D. M.	Ob.	Description
5	Dec. 18	82 (δ) Ceti	p	0 5	n	0 46	8	pB. S. lE. bM.

Table 2.10. *Independent objects in Herschel's catalogues*

	C1	C2	C3	Sum
Entries	1000	1000	500	2500
Catalogue identity	20	10	4	34
NGC identity	13	8	7	28
Balance	967	982	489	2438

William Herschel, John Herschel, Mason and Auwers.[56] The possible reason (besides the positional confusion) is that Messier described it as a 'star cluster'. The identification of the four Herschel nebulae with M 20 was first presented in the General Catalogue (GC 4355).

In 28 cases there is an identity with another NGC object (also contained in the Herschel catalogues). A typical example: NGC 3611 = II 521, a galaxy in Leo (11.9 mag), discovered on 27 January 1786 in Clay Hall. It is identical with NGC 3604 = II 626, found on 30 December 1786 in Slough. Another case is NGC 4664 = II 39 (Datchet, 23 February 1784), a galaxy of 10.3 mag in Virgo. Herschel saw it once again on 30 April 1786 in Slough, but now as a class I

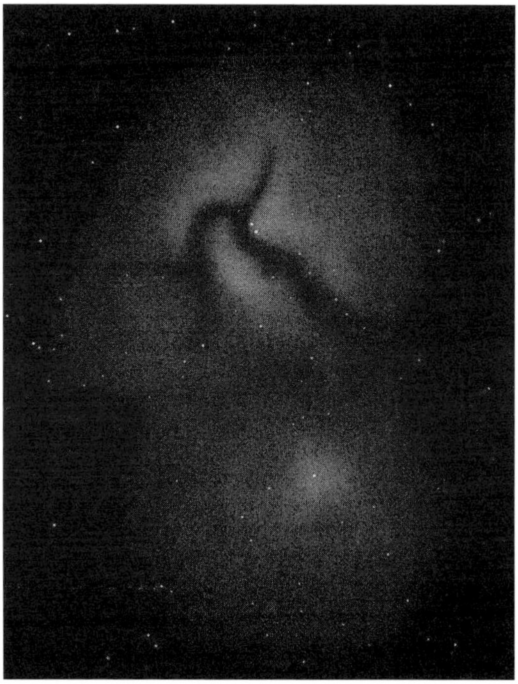

Figure 2.12. The Trifid Nebula (M 20) in Sagittarius; drawing by Trouvelot, Harvard College Observatory (Winlock 1876).

object (I 142). Dreyer, not recognising the identity, catalogued it as NGC 4665. But there is still a third

---
[56] M 20 is listed in the Cape catalogue as h 3718.

NGC-number: NGC 4624. Dreyer now refers to John Herschel's 'new' object h 1390, observed on 9 April 1828 in Slough.

Three cases show a 'combined' identity (already counted in Table 2.10): NGC 4124 = II 33 = II 60 is equal to NGC 4119 = II 14; NGC 4470 = II 18 = II 498 is the same as NGC 4610 = II 19; and finally NGC 4526 = I 31 = I 38 is equal to NGC 4560 = I 119. All of these objects are Virgo galaxies.

The balance gives 2438 independent Herschel objects, which is, compared with other catalogues, a high value (98%). Obviously William Herschel made a very good job of his cataloguing. Subsequent observers were less successful, despite using much better equipment.

2.3 CAROLINE HERSCHEL AND OTHER DISCOVERERS

Not all 2438 independent objects can be credited to this outstanding observer. As Table 2.11 shows, 32 had been found earlier by others; mainly Méchain, Caroline Herschel and Messier.

Eight objects from the Herschel catalogues were discovered by Caroline Herschel (Fig. 2.13).[57] For her early observations in Datchet (28 August 1782 to 4 July 1783) she used a small refractor with magnification 14.5 and 3° field of view, which had been made by her brother. William Herschel told her how to use it and suggested that she observe double stars, nebulae and star clusters. Caroline was very successful, despite being not much interested in the theoretical background.

On 30 September 1782 she independently found M 27, which had remained unobserved by her brother until that time.[58] William Herschel was impressed and eventually started his own search for nebulae. Later he built two larger telescopes for his sister.[59] The first was an azimuthal 4.5" reflector with a power of 24. Caroline

---

[57] For Caroline Herschel's life and work, see Lubbock (1933), Kemps (1955), Kerner C. (2004), Hoskin (2002b, 2003a, b, 2007) and Wilson B. (2007).

[58] M 27 was later called the Dumbbell Nebula by John Herschel. On 24 August 1827 he wrote (h 2070, the Tarantula Nebula) that 'The central mass may be compared to a vertebra or a dumb-bell.'

[59] For Caroline Herschel's telescopes, see Hoskin and Warner (1981) and Hoskin (2005a, b).

Table 2.11. *Discoverers of independent objects in the Herschel catalogues*

Discoverer	C1	C2	C3	Sum
W. Herschel	955	967	484	2406
C. Herschel	2	5	1	8
de Chéseaux		1		1
Flamsteed	1			1
Hipparch	2			2
Hodierna	3	1		4
Mairan	1			1
Méchain	2	4	3	9
Messier	4	1		5
Oriani		1		1
Sum	970	980	488	2438

Figure 2.13. Caroline Herschel (1750–1848).

used this 'small sweeper' from 8 July 1783 onwards, discovering some nebulae and star clusters. From 17 March 1791 (now in Slough) she owned the 'large sweeper', a 9.2" reflector, which brought to light no new objects.[60]

---

[60] In Slough she observed from the roof of a small detached building to the north of the dwelling house, which was used as library.

2.3 Caroline Herschel and other discoverers

Table 2.12. *Caroline Herschel's objects (sorted by date; see the text)*

P	H	h	NGC	Date	Date (H)	C	Note (H)	Type	V	Con.	Remarks
1	VII 12	440	2360	26.2.1783	4.2.1785	1	CH	OC	7.2	CMa	Refractor
1	VII 27	436	2349	4.3.1783	24.4.1786	2	CH 1783	OC		Mon	Refractor
0	VI 22	496	2548	8.3.1783	1.2.1786	2	CH 1783	OC	5.8	Hya	Refractor; M48, Messier 19.2.1771
1	VII 59	2066	6866	23.7.1783	11.9.1790	3		OC	7.6	Cyg	
0	VIII 72		6633	31.7.1783	30.7.1788	2	CH 1783	OC	4.6	Oph	de Chéseaux 1745?
2			IC 4665	31.7.1783				OC	4.2	Oph	Bailey 1896
0	V 18	44	205	27.8.1783	5.10.1784	1	CH 23.9.1783	Gx	7.9	And	M 110, Messier 10.8.1773
1	V 1	61	253	23.9.1783	30.10.1783	1	CH	Gx	7.3	Scl	'It is Carolina's'
2		36	189	27.9.1783				OC	8.8	Cas	J. Herschel 27.10.1829
1	VIII 78	25	225	27.9.1783	26.2.1788	2	CH 1784	OC	7.0	Cas	
1	VIII 65		659	27.9.1783	3.11.1787	2	CH 1783	OC	7.9	Cas	
0	VII 32	174	752	29.9.1783	21.9.1786	2		OC	5.7	And	Hodierna 1654?
1	VI 30	2284	7789	30.10.1783	18.10.1787	2	CH 1783	OC	6.7	Cas	
2		2048	6819	12.5.1784				OC	7.3	Cyg	Harding Sept.? 1823; J. Herschel 31.7.1831
1	VIII 77	2182	7380	7.8.1787	1.11.1788	2	CH 1787	OC	7.2	Cep	Slough

Table 2.12 lists Caroline Herschel's discoveries, sorted by date. Except for the last (NGC 7380), all were made in Datchet. The sources are the three Herschel catalogues and the observing journals of William and Caroline Herschel.[61]

Caroline Herschel found 11 objects. The eight with priority P = 1 are contained in the Herschel catalogues; the three with P = 2 bear no Herschel designation. The remaining objects (P = 0) were found earlier by other observers. The column 'Note (H)' gives William Herschel's (incomplete) notes. All objects are listed in the NGC, except IC 4665. This large open cluster was credited by Dreyer to the Harvard astronomer Solon Bailey, who photographed it 1896 at Arequipa Observatory. It is remarkable that all 11 of Caroline Herschel's objects exist. All are open clusters, except NGC 253, the bright galaxy in Sculptor (Fig. 2.14), which comes only 15° above the horizon in Datchet.

William Herschel erroneously credits his sister for three other objects. VII 13 = NGC 2204 (CH 26 February 1783) should read VII 12 (NGC 2360); and VIII 64 = NGC 381 (CH 1783) should be NGC 189 (not observed by William Herschel). The most curious case is the prominent edge-on galaxy NGC 891 = V 19 (Fig. 2.15), which was discovered by William Herschel on 6 October 1784. In the notes following the first

---
[61] See also Hoskin (2005b).

Figure 2.14. NGC 253 in Sculptor, discovered by Caroline Herschel on 23 September 1783 (DSS).[62]

Figure 2.15. Herschel's sketch of the edge-on galaxy NGC 891 with its 'black division' (Herschel W. (1811), Fig. 12).

catalogue one reads that Caroline found the nebula on 27 August 1783. This is, however, a typo: V 18 = NGC 205 = M 110 is meant, which was seen by her on the given date (but had already been discovered by Messier on 10 July 1773). It is probable that she made a clerical error. In the catalogue one correctly reads the remark 'CH' for the entry V 18.

In Table 2.13 the other discoverers of Herschel objects are listed. Though the astronomer has looked up nearly all of the nebulae and star clusters of Messier's catalogue, it sometimes happened that an object was thought to be new. In the appendix of the *Scientific Papers of Sir William Herschel*, Dreyer compiled 'Unpublished observations of Messier's nebulae and clusters'.[63] They are based on Herschel's notes in his observing journals (his manuscripts contain a Messier list[64]). Most objects were observed with the 18.7-inch; for M 2, M 5, M 42, M 72 and M 74 the 48-inch ('40 ft reflector') was used – a rare matter.[65]

Not observed were M 61, M 91 and M 102, which obviously could not be identified by Herschel. Additionally, it is not astonishing that there are no reports for M 44 (Praesepe) and M 45 (the Pleiades).

Only 2 of the 2500 catalogued objects are missing from the NGC. The emission nebula V 35 in Orion, which was discovered on 1 February 1786 in Clay Hall, was later listed by Dreyer as IC 434. For VI 8 there is no appropriate object. Dreyer has ignored Herschel's observation made in Datchet on 25 April 1784. Thus the number of independent NGC objects in the three catalogues is 2437, of which 2405 must be credited to Herschel.

2.4 HERSCHEL'S EIGHT CLASSES AND MODERN OBJECT TYPES

Next the relation between Herschel's classes and modern types is treated. Table 2.14 shows that, as expected, the classes I to III strongly correlate with galaxies (96%). Herschel could not resolve 28 globular clusters (GC) and classified them as 'nebulae' of classes I to III. In the case of very remote objects, this is comprehensible. The top scorers are the Intergalactic Wanderer NGC 2149 (I 218) in Lynx, found on 31 December 1788, and NGC 7006 (I 52) in Delphinus, found on 21 August 1784; the distances of these globular clusters are 182 000 ly and 135 000 ly, respectively.[66]

Most interesting is class IV ('planetary nebulae'), which is very inhomogeneous (see Section 2.6). In class V ('very large nebulae'), 63% of the objects are galaxies and 20% are emission nebulae. Obviously, the form did not play an essential role, since there are edge-on galaxies in this class, e.g. NGC 891 (Fig. 2.15), NGC 253 (Fig. 2.14), NGC 4565, NGC 4631 and NGC 5907, and also face-on galaxies, such as M 33, M 106, NGC 2403 and NGC 2997. The emission nebulae in class V are mostly irregular; the North America Nebula (NGC 7000, V 37), Veil Nebula (NGC 6992, V 15)

---

[62] See also Lassell's sketch, Fig. 7.12 left.
[63] Dreyer (1912a, vol. 2: 651–660).
[64] RAS Herschel W. 4/33.1.
[65] M 42 was the first object looked up using the '40-foot', while it was still under construction (19 February 1787). However, it is remarkable that it was so little used for observations of nebulae. The reason might be that the mirror soon tarnished and the image consequently became bad. Later, Proctor remarked that it was a matter of public notoriety in England that the 48" mirror '*bunched a star into a cocked hat*' (Wolf C. 1886b: 199). Some objects (e.g. M 31) were observed by Herschel with his 'X-feet', a chunky 24" reflector of focal length 10 ft.
[66] See Steinicke (2003e).

Table 2.13. Objects discovered prior to Herschel (without C. Herschel; sorted by H-number; site: D = Datchet, C = Clay Hall, S = Slough)

H	M	NGC	Date (H)	C	Site	Discoverer	Date	Type	V	Con.	Remarks
I 7	49	4472	23.1.1784	1	D	Messier	19.2.1771	Gx	8.3	Vir	
I 17	105	3379	11.3.1784	1	D	Méchain	24.3.1781	Gx	9.5	Leo	
I 43	104	4594	9.5.1784	1	D	Méchain	11.5.1781	Gx	8.3	Vir	Sombrero Galaxy
I 139	61	4303	17.4.1786	2	S	Oriani	5.5.1779	Gx	9.3	Vir	
I 186		5195	12.5.1787	2	S	Méchain	21.3.1781	Gx	9.6	CVn	Companion of M 51
I 193	76	651	12.11.1787	2	S	Méchain	5.9.1780	PN	10.1	Per	Component of M 76
I 215	102	5866	5.5.1788	2	S	Méchain?	April 1781	Gx	9.9	Dra	
II 120	91	4548	8.4.1784	1	D	Messier	18.3.1781	Gx	10.1	Com	
III 1	43	1982	3.11.1783	1	D	Mairan	1731	EN	6.8	Ori	
IV 61	109	3992	12.4.1789	3	S	Méchain	3.12.1781	Gx	9.8	UMa	
V 10–12	20	6514	12.7.1784	1	D	Messier	5.6.1764	EN+OC	8.5	Sgr	IV 41, Trifid Nebula
V 17	33	598	11.9.1784	1	D	Hodierna	1654?	Gx	5.5	Tri	
V 18	110	205	5.10.1784	1	D	Messier	10.8.1773	Gx	7.9	And	C. Herschel 27.8.1783
V 43	106	4258	9.3.1788	2	S	Méchain	July 1781	Gx	8.3	CVn	
V 46	108	3556	17.4.1789	3	S	Méchain	16.2.1781	Gx	9.9	UMa	
VI 22	48	2548	1.2.1786	2	C	Messier	19.2.1771	OC	5.8	Hya	C. Herschel 8.3.1783
VI 33		869	1.11.1788	2	S	Hipparch	−130	OC	5.3	Per	Double Cluster (χ Per)
VI 34		884	1.11.1788	2	S	Hipparch	−130	OC	6.1	Per	Double Cluster (χ Per)
VI 40	107	6171	12.5.1793	3	S	Méchain	April 1782	GC	7.8	Oph	
VII 2		2244	24.1.1784	1	D	Flamsteed	17.2.1690	OC	4.8	Mon	12 Mon
VII 17		2362	6.3.1783	1	D	Hodierna	1654?	OC	3.8	CMa	
VII 32		752	21.9.1786	2	S	Hodierna	1654?	OC	5.7	And	C. Herschel 29.9.1783
VIII 38	47	2422	4.2.1785	1	D	Hodierna	1654?	OC	4.4	Pup	NGC 2478, Messier 19.2.1771
VIII 72		6633	30.7.1788	2	S	de Chéseaux	1745?	OC	4.6	Oph	C. Herschel 31.7.1783

Table 2.14. *Herschel's classes and modern types*

Class	Description	Gx	EN	RN	PN	OC	GC	GxP	Star	Stars	NF	Sum
I	Bright nebulae	254	1	1	3	1	16	1		1		278
II	Faint nebulae	853	2	1	4	1	10		1		2	874
III	Very faint nebulae	934	3	1	5	2	2	4	4	8	2	965
IV	Planetary nebulae	39	7	5	20	2	2		2			77
V	Very large nebulae	31	12	3	1			1			1	49
VI	Very condensed and rich clusters of stars	2			1	28	8			2	1	42
VII	Compressed clusters of small stars and large stars					61				4	1	66
VIII	Coarsely scattered clusters of stars					69				17	1	87
	Sum	2113	25	11	34	164	38	6	7	32	8	2438

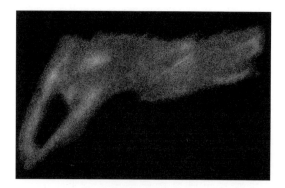

Figure 2.16. Herschel's illustration of an 'extensive diffused nebulosity' (Herschel W. (1811), Fig. 1).

Figure 2.17. The North America Nebula in Cygnus (NGC 7000), photographed by Max Wolf in 1902.

and Flame Nebula (NGC 2024, V 28) are prominent examples, though Herschel had seen only their brightest parts. His publication of 1811 contains 42 sketches demonstrating the different forms of nebulae (Herschel W. 1811). The largest object shown there was intended to illustrate one of his 52 regions with 'extensive diffused nebulosities' (Fig. 2.16), which later caused some confusion (see Section 11.6.15).

Herschel's class V and his 52 obscure regions are related, as shown by the prominent North America Nebula NGC 7000 (Fig. 2.17), located about 3° east of Deneb in Cygnus. The story is worth telling here, because it illustrates the historical development of observations. Such background information will be presented for several interesting objects in the book.

Herschel discovered the nebula during sweep 620 on 24 October 1786 in Slough; the reference star was 57 Cygni, about 1° to the west (the overexposed star at the middle right edge of Fig. 2.17). In his observation journal one reads of a '*very large diffused nebulosity, plainly visible, between 7 or 8' l, 6' b and losing itself gradually*'. The object was entered in the second catalogue as V 37 and the description matches that given in the journal. The position given there is near the centre of the nebula. The next journal note (same sweep) is interesting: '*All this time suspected diffuse nebulosity through the whole breadth of the sweep*'. Herschel gives two positions, one at the 'west coast of Florida' the other at the 'Californian coast'. These observations were later entered as nos. 44 and 46 in the list of 52 regions with 'extensive diffused nebulosity' (Herschel W. 1811). Number 44 is described as '*faint milky nebulosity scattered over this space, in some places pretty bright*' (diameter 2.8°) and no. 46 as '*suspected nebulosity joining to plainly visible diffused nebulosity*' (diameter 3.7°). The following text additionally gives that '*In No. 44 we have an instance of faint nebulosity which, though pretty bright in some places, was completely lost from faintness in others; and No. 46 confirms the same remark.*' On 11 September 1790 Herschel made a second observation using the same reference star (sweep 959). In the journal is noted the following: '*Faint milky nebulosity scattered over their space; in some places pretty bright. The brightest part of it about the place of my V 37.*' Again two positions are given (matching the former). There is no doubt that Herschel has discovered NGC 7000. He has seen not merely the brightest spots, but a large fraction of the whole nebula. This is astonishing, because normally this needs a filter (as would be used today).

On 21 August 1829 John Herschel looked up the object and catalogued his observation as h 2096 in the Slough catalogue. However, he was not sure about the identification, noting 'V 37?'. The position is that of his father and the description reads as follows: '*An immense nebula all around this place, but ill defined to fix the limits. RA that of V 37, from working list, not settled by the observation.*' On the basis of the three observations John Herschel listed the nebula as GC 4621 in the General Catalogue (noting 'V 37?' again) with the description '*F, eeL, diff. neb*'. This was adopted by Dreyer in the NGC (including 'V 37?' and William Herschel's position). There were only two further visual observations of NGC 7000 during the nineteenth century, which were made by Bigourdan in Paris with his 12" refractor. He could confirm the nebula on 16 August 1884 and 25 September 1889.

Now the German astronomer Max Wolf enters the scene. He photographed the nebula on 1 June 1891 with the 5" Kranz portrait lens at his private observatory in Heidelberg.[67] He reported his observation in a paper 'Ueber grosse Nebelmassen im Sternbild des Schwans'.[68] The 3-hour exposure showed '*a large and bright, exceedingly subtle plotted, fan-shaped nebula, whose brightest part hitherto was known as G.C. 4621.*' However, the term 'America' is not mentioned! Another plate was taken on 12 and 13 July 1901 with the 16" Bruce refractor on Königstuhl (exposure time 4.75 hours). It is the frontispiece of the first volume of the *Publikationen des Astrophysikalischen Observatoriums Königstuhl-Heidelberg*. The subtitle reads 'Der Amerika-Nebel im Cygnus' ['The America Nebula in Cygnus']. The same volume contains a work of Wolf's young assistant August Kopff, titled 'Die Vertheilung der Fixsterne um den grossen Orion-Nebel und den America-Nebel'.[69] Wolf himself mentions the object once more in his paper 'Über eine Eigenschaft der großen Nebel',[70] which treats the connection of nebulae and 'vacancies'. Obviously, the name occurred to Wolf while he was looking at the first (really good) image of July 1901.

In January 1903 Barnard wrote a paper on 'Diffused nebulosities in the heavens' (Barnard 1903). There one reads that the object "*was first photographed by Dr. Max Wolf some twelve years ago* [1891!] *and has lately been called by him 'America Nebula' from its striking resemblance to North America as shown on maps and globes.*" In a footnote Barnard added that "*The 'North America Nebula' would perhaps be more definite, for it is North America to which Dr. Max Wolf intends the compliment.*" This is the origin of the popular name for NGC 7000. Going back to William Herschel, it is surprising that he saw nebulosity in his eye-piece with only 15′ field of view. This is only a small fraction of the field presented in Fig. 2.17, which has a width of more than 6°. It is extremely difficult to notice any contrast in the eye-piece when the nebulosity completely covers the field of view. There is nothing known about whether Herschel had switched between different areas (with and without nebulosity) for comparison.

Classes VI to VIII are naturally dominated by open clusters (81%), followed by random star groups (12%). However, there are two galaxies in class VI: NGC 3055 (VI 4) in Sextans (VI 4, 24 January 1784) and NGC 6412 in Draco (VI 41, 12 December 1797). Herschel's 1814 publication contains another 17 sketches, among

---

[67] Often a wrong discovery date is given: 12 December 1890, see e.g. Vehrenberg (1983: 222). During that night Wolf took a plate of the region around ζ Orionis (Wolf M. 1891a), showing the Flame Nebula NGC 2024 and a new one, later catalogued as IC 448. Undoubtedly there was no time to turn to the Cygnus region (in a winter night!). Regarding Wolf's discoveries in Orion and Cygnus, see also Clerke (1891).

[68] 'On the large nebulous masses in the constellation Cygnus' (Wolf M. 1891c).
[69] 'The distribution of the fixed stars around the Great Orion Nebula and the America Nebula' (Kopff 1902).
[70] 'On a feature of the large nebulae' (Wolf M. 1903).

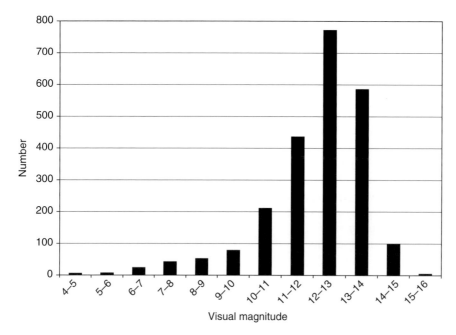

Figure 2.18. The brightness distribution of the Herschel objects.

them a few star clusters (Herschel W. 1814). Some single stars are in classes II, III and IV. In these cases, Herschel supposed there to be a nebula around the star – an erroneous perception, occasionally supported by other observers. The brightest star is NGC 5856 (IV 71) = BD +19° 2924 in Bootes with 6.0 mag. Herschel wrote that '*A star 7.6m. enveloped in extensive milky nebulosity. Another star 7m. [BD +19° 2935?] is perfectly free from such appearance.*' Finally, the low number of missing objects ('not found') is remarkable. As has already been stated, Herschel was a very diligent observer, correctly reporting the data with the support of his sister Caroline.

2.5 BRIGHTNESS OF THE OBJECTS

Using modern data, the brightness statistic of Herschel's objects can be derived (Fig. 2.18).[71] The mean is 12.0 mag, which is pretty bright, compared with values reported by subsequent observers, who were often observing with smaller telescopes.

The distribution reflects the situation of a widely unexplored sky prior to Herschel. The likelihood of encountering bright objects was high – certainly some fainter ones were overlooked at that time. Concerning the spatial distribution, one must consider that Herschel's sweeps primarily depended on bright reference stars (e.g. from Flamsteed's catalogue). A complete survey of the sky visible from southern England was not executed (and not planned). This can be shown by examination of the list of reference stars compiled later by Caroline Herschel.[72] There are gaps around the celestial pole (δ = 85° to 90°) and in the declination zone 42° to 52°, with no observations between right ascension 17^h to 17.5^h and 19.5^h to 21.25^h. This explains why many bright objects remained for coming discoverers.

Herschel constructed metal mirrors using an alloy of copper and tin ('speculum metal'). Modern measurements reveal a reflectivity of 63% for red light (450 nm) and even 75% for blue light (650 nm) for this material. Owing to tarnishing of the surface, the values decreased by 10% within 6 months. Therefore the mirror had to be polished frequently. Herschel (and later Lord Rosse and Lassell) used several mirrors, to allow continuous observing. Herschel's 18.7" reflector might be equal to a modern 10-inch with an aluminised glass mirror. Considering the

---

[71] An analogous graphic is presented for all observers with a large number of discoveries.

[72] RAS Herschel C. 3/2.3; see also Hoskin (2005c).

Table 2.15. *William Herschel's brightest and faintest objects (site: D = Datchet, S = Slough)*

Object	NGC	Date	C	Site	Type	V	Con.
VII 17	2362	6.3.1785	1	D	OC	3.8	CMa
VIII 5	2264	18.1.1784	1	D	OC	4.1	Mon
VIII 25	2232	16.10.1784	1	D	OC	4.2	Mon
III 735	6241	29.4.1788	2	S	Gx	15.2	Her
III 807	4549	24.4.1789	3	S	Gx	15.2	UMa
III 64	2843	21.3.1784	1	D	Gx	15.5	Cnc

unknown sky and the unfavourable site, it is truly astonishing that Herschel discovered nebulae fainter than 15th magnitude. Comparing historical and modern refractors, the difference is much smaller: the excellent instruments of d'Arrest or Tempel with 11″ aperture are hardly outperformed by current achromats of equal size. Thus the discovery of so many nebulae with refractors in the nineteenth century is not surprising.

Table 2.15 lists the three brightest and three faintest objects from Herschel's catalogues. VII 17, the compact open cluster NGC 2362 around τ CMa (4.4 mag), was found by Hodierna in about 1654. NGC 2264 is the cluster around 15 Monocerotis, which is covered by faint nebulosity (Herschel saw parts as V 27 on 26 December 1785). At the south end is the famous cometary Conus Nebula.

Figure 2.19 shows the magnitude distributions for objects in classes I to III. This analysis makes sense, insofar as they are very homogeneous (98% are galaxies). The graphic suggests what Herschel meant by 'bright', 'faint' or 'very faint' in the case of nebulae: the resulting averages are 10.6 mag, 12.1 mag and 13.0 mag, respectively. However, the statistical variance is pretty high (Table 2.16). It is peculiar that the brightest objects in each class are globular clusters. Obviously, Herschel had – compared with galaxies – a different perception in these cases. Moreover, he coded equally the extreme magnitudes in classes I and II ('considerably bright' and 'pretty bright'); only class III differs ('faint', 'excessively faint').

The above-mentioned facts lead to the following conclusion: Herschel's qualitative brightness measure, as given by the class or in the object description, is only weakly correlated with modern visual magnitude. The estimation is too much influenced by the structure (and surface brightness) of the object and the individual's perception.

2.6 HERSCHEL'S CLASS IV: PLANETARY NEBULAE

For Herschel this class was a depository for objects that did not fit into other classes. It, therefore, is pretty inhomogeneous: besides true (physical) planetary nebulae (here abbreviated 'PN') there are many 'foreign bodies'. The question of how Herschel's term 'planetary nebula' came into being is interesting.

2.6.1 The origin of the term 'planetary nebula'

It is undisputed that the visual appearance of planets inspired William Herschel to call similar-looking nebulae (those collected in class IV) 'planetary'. On the other hand, Uranus was not explicitly mentioned in this context. However, the popular literature does indeed stress a connection Herschel–Uranus–planetary nebulae, quoting the similarity in colour of the planet and some PN.[73] Actually, Herschel did not report any colour for planetaries – but saw Uranus as being '*of the colour of Jupiter*' (22 October 1781), adding, on 2 October 1782, '*Planet unexpectedly appeared blueish*'.[74]

On 29 August 1782 Herschel observed M 57, the Ring Nebula in Lyra. The famous object was discovered on 31 January 1779 by Antoine Darquier. Messier, quoting his observation, wrote the following: '*pretty dull, but perfectly outlined; it is as large as Jupiter and resembles a fading planet*' (Messier 1781). For Herschel the nebulae looked '*extremely curious*' in the 6.2″ reflector and his sketch shows a '*perforated nebula or ring of stars*' (Fig. 2.20).[75] The term 'planetary' is not being used.

---

[73] Steinicke (2007b).
[74] Herschel W. (1783: 7–8).
[75] A very similar sketch was published by Bode (1785a, Fig. 6.) Herschel's description probably gave rise to the popular name

Table 2.16. *Extreme magnitudes in classes I to III*

Class	H	NGC	Type	V	Con.	Code
I	I 44	6401	GC	7.4	Oph	cB
	I 113	2830	Gx	13.9	Lyn	cB
II	II 197	6544	GC	7.5	Sgr	pB
	II 26	4453	Gx	14.9	Vir	pB
III	III 143	6717	GC	8.4	Sgr	F
	III 64	2843	Gx	15.5	Cnc	eF

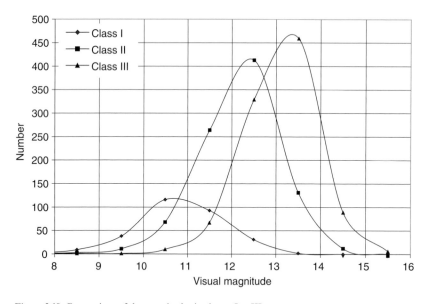

Figure 2.19. Comparison of the magnitudes in classes I to III.

Shortly thereafter, on 7 September 1782, Herschel discovered his first nebula: NGC 7009 in Aquarius[76] (named the Saturn Nebula by Lord Rosse in 1849). This object became the first one of his class IV, being published in 1786 in catalogue no. 1. In his observing journal one reads '*A curious Nebula or what else to call it I do not know. It is of a shape somewhat oval, nearly circular.*'[77] The essential sentence reads '*The brightness in all the powers does not differ so much as if it were of a planetary nature, but seems to be of the starry kind.*'

Concerning its behaviour with respect to magnification, the object behaved like a planet – however, there is no hint about its blueish colour.

On 30 September 1782 Caroline Herschel independently discovered M 27 (the Dumbbell Nebula) with her small refractor. Her brother noted that it was '*very curious with a compound piece; when comparing its place with Messier's nebulae, we find it is his 27*'. M 27 is not assigned as a 'planetary nebula'. But the term appears once again on 6 October 1784, when Herschel discovered NGC 7662 in Andromeda: '*wonderful bright, round planetary, pretty well defined disc, a little elliptical*'. Once again, he did not notice the striking colour of the PN, now known as the Blue Snowball.[78]

'Ring Nebula'. On 15 July 1847 William Mitchell saw '*many stars within the compass of the ring*' with the new 15" Merz refractor at Harvard College Observatory (Bond W. C. 1847).

[76] The name was created by Lord Rosse, who noted 'Saturn neb.' (h 2048) on 16 September 1849 (Parsons L. 1880: 159).

[77] RAS Herschel W. 4/1.13, 231.

[78] The name was created by Leland Copeland in 1960 (Copeland L. 1960).

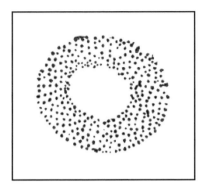

Figure 2.20. Herschel's sketch of M 57 (Herschel W. (1785), Fig. 5).

Figure 2.21. Herschel's sketch of IV 18 = NGC 7662 (Herschel W. (1811), Fig. 36).

The crucial term was first explained by Herschel in his paper 'On the construction of the heavens', written in late 1784.[79] At the beginning of the chapter 'Planetary nebulae' he notes '*a few heavenly bodies, that from their singular appearance leave me almost in doubt where to class them*'. He presents three examples: NGC 7009 = IV 1 ('*has much of a planetary appearance, uniform brightness*'), NGC 7662 = IV 18 ('*round, bright, pretty well defined planetary disc*'; Fig. 2.21) and NGC 1535 = IV 26 in Eridanus, discovered on 1 February 1784 ('*very bright, elliptical planetary, ill defined disc*'). Obviously, Herschel characterised objects as 'planetary' if they showed a round or oval disc with clearly defined edge and uniform surface brightness.

In Herschel's catalogues, class IV is titled 'Stars with burs, with milky chevelure, with short rays, remarkable shapes, &c', enlarging his definition of 1784. Perhaps he initially had the intention of listing objects with planetary discs only, but the variety of shapes forced him to use class IV for all peculiar cases. The term 'planetary' appears explicitly for 15 objects: 10 of them are PN, the rest are galaxies. Therefore, one must distinguish between 'planetary nebulae' (as members of class IV) and objects described as 'planetary', showing a smooth disc. Only the latter are related to PN. As d'Arrest pointed out: '*It is erroneous, when all 78 numbers of this fourth class IV, as is still done in new textbooks from time to time, are considered as planetary nebulae.*'[80]

In 1785 Johann Elert Bode published a short note on 'planetary-like nebulae' in the *Berliner Jahrbuch*,[81] presenting eight objects (Bode 1785b). William Herschel is not explicitly mentioned,[82] but Bode mentions observations with a telescope described as '*20-foot of 18.7-inch aperture*' and presents a sketch of M 57, which is very similar to Fig. 2.20. Moreover, part of the text looks much like Herschel's in his publication of 1785.[83] Bode wrote that '*Here I report some celestial objects, which, due to their peculiar appearance, let me strongly doubt in which class I should place them [...] The planetary-like shape of the first two [NGC 7009, NGC 7662] is so strange that we hardly consider them to be nebulae, their light being so smooth and vivid, their diameter so small and definite, it is therefore very improbable that they belong to these kinds of bodies.*'[84]

The connection of Herschel and PN-colour is a strange issue. There are only two cases where colour is mentioned for class IV objects.[85] The first is IV 22 (NGC 2467) in Puppis, which was discovered on 9 December 1784: '*faint red color visible*'. Actually, this is not a PN, but a mix of star cluster and emission nebula! For IV 27 (NGC 3242), discovered on 7 February 1785, Herschel notes '*planetary disc ill defined, but uniformly bright, the light of the colour of Jupiter*'. Today

---

[79] Herschel W. (1785); this and further papers on the 'Construction of the heavens' have been reprinted and analysed by Michael Hoskin (Hoskin 1963).

[80] d'Arrest (1856a: 359).

[81] Often called the *Astronomisches Jahrbuch*.

[82] Occasionally Bode published unauthorized versions of Herschel's papers in the *Berliner Jahrbuch*.

[83] Herschel W. (1785: 265–266).

[84] The other six PN are NGC 6572, NGC 6886, NGC 6894, NGC 1535 and NGC 3242.

[85] Strangely, Herschel saw a 'faint red color' in the brightest part of M 31 (Herschel W. 1785: 262).

this bright PN in Hydra is known as the Ghost of Jupiter.[86] However, in comparison with the cases with distinctive blue or blue–green colour, NGC 3242 is a rather pale example.[87]

William Herschel is not the originator of the often stressed relation Uranus–PN. It sounds plausible but it is a mere myth. The explicit connection is due to his son, John. On 3 April 1834 he discovered a remarkable object with his 18¼" reflector in Feldhausen (Cape of Good Hope), which is catalogued as h 3365. This is NGC 3918, a PN with 8.1 mag in Centaurus (Fig. 2.22).

In the Cape catalogue he wrote that it was '*perfectly round; very planetary; colour fine blue; [...] very like Uranus, only about half as large again and blue*'.[88] All fits now: 'planetary', blue colour, appearance of Uranus. John Herschel discovered some other objects of this kind. In the case of NGC 2867 (h 3163) in Carina he believed on 1 April 1834 to have found a new planet: '*just like a small planet*'. The following day it became obvious that the object '*has not moved perceptibly and is therefore not a planet*'. The case was mentioned in a letter to William Rowan Hamilton, dated 13 June 1835,[89] in which Herschel wrote '*Indeed, the first on which I fell was so perfectly planetary in its appearance, that it was not until several observations of it at the Royal Observatory* [at the Cape], *by Mr. Mclear* [Thomas Maclear], *had annihilated all suppositions of its motion, that I could relinquish the exciting idea that I had really found a new member of our own system, revolving in an orbit more inclined than Pallas.*'

In his popular textbook *A Treatise on Astronomy* John Herschel wrote about 'planetary nebulae': '*They have, as their name imports, exactly the appearance of planets*'.[90] He was also the first to report the colour of PN. He saw NGC 7009 as being '*light blue*', NGC

Figure 2.22. John Herschel's 'Uranus': the planetary nebula NGC 3918 in Centaurus (DSS).

7662 '*blueish white*' (both in Slough) and NGC 3242 '*sky-blue*' (in Feldhausen).[91] Nowadays it is easy to see such colours. Therefore the question of why William Herschel did not report them arises. A rather extreme possibility is that he was colour-blind in the case of faint light. For older visual observers, the colour perception can be reduced. At the beginning of William Herschel's career, when he discovered his first PN (NGC 7009) in 1782, he was already 44 years old – his son was 11 years younger when looking at this object for the first time. But another, more plausible reason is possible: mirror quality. Though of the same construction, John Herschel's telescope had a much better mirror than that used by his father, both in figure and in reflectivity. This was partly due to his skill as a chemist. Since the mirror rapidly tarnished at the Cape, Herschel got frequent practice at polishing. Therefore his reflector could have shown colour better than his father's.

2.6.2 Herschel's key object NGC 1514 and the content of class IV

William Herschel revised his ideas on the nature of nebulae several times (see also Section 6.4.8). His early observations of the Orion Nebula ('the most beautiful object in the heavens') with the 6.2" reflector in Bath

---

[86] Though Herschel mentions the resemblance to Jupiter, the popular name is due to Captain William Nobel, who wrote in 1886 '*a pale blue disk, looking just like the ghost of Jupiter*' (Nobel 1886). It is interesting that M 51 was described by Smyth as a '*ghost of Saturn, with his ring in vertical position*' (Smyth 1844: 302).

[87] The physical reason for PN colour is in most cases the strong O III emission line of oxygen.

[88] NGC 3918 is sometimes called the Blue Planetary Nebula.

[89] It is reprinted in Hoskin (1984); see also Jahn (1844: 79), where the year is erroneously given as 1836.

[90] Herschel J. (1833b: 378–379).

[91] Later Lord Rosse saw colour too; e.g. Struve's PN NGC 6210 (h 1970) showed an '*intense blue centre*' in the 72".

2.6 Herschel's class IV

Figure 2.23. The Dumbbell Nebula (M 27) in Vulpecula. Top: William Herschel's sketch of 1784 (RAS W. Herschel 4/1.7); bottom: a modern image.

convinced him that true nebulous matter exists, which he termed *'nebulosity of the milky kind'*.[92] This was based on supposed changes in form and brightness of parts of M 42.[93] On the other hand, Herschel later could resolve some nebulae with his 12-inch. An example is the globular cluster M 30 in Capricorn, which was observed on 21 August 1783: *'Plainly resolved into very small stars. It is a difficult step, i.e. if we divide the transition from the Pleiades [M 45] down to the Nebula of the Orion [M 42] into six steps this is perhaps the 4th towards the real nebulae.'*[94] Both objects were essential species of his 'natural history' of the heavens.

On the basis of observations of the Omega Nebula (M 17) and Dumbbell Nebula (M 27) with the 18.7-inch, Herschel changed his point of view. About M 17 he wrote on 22 June 1784 that *'the milky nebulosity seems to degenerate into the resolvable kind [...] this nebula is a stupendous Stratum of immensely distant fixed stars'*.[95] M 27 was described on 19 July 1784 as a *'double stratum of stars of a very great extent'* (Fig. 2.23 top). One further reads that *'The ends next to us are not only resolvable nebulosity but I really do see very many of the stars mixt with the resolvable nebulosity.'*

Now Herschel arrived at the conclusion that all nebulae must be star clusters – their resolution would be a question of distance and aperture only. In 1785 he developed an evolutionary scenario: the Universe started with widely distributed stars, which slowly condensed into larger agglomerations ('stratum', 'Milky Way') by virtue of gravitational forces, eventually fragmenting into many smaller clusters.[96] The density reaches its highest degree in globular clusters, ending as planetary nebulae, which *'may be looked upon as very aged* [globular clusters] *drawing on towards a period of change, or dissolution'*.[97]

However, an observation made on 13 November 1790 led Herschel to change his theory a second (and last) time. He discovered IV 69 (NGC 1514) in Taurus: *'A most singular phenomenon! A star of about 8th magnitude, with a faint luminous atmosphere'*.[98] He did not interpret the object as 'planetary', but as a 'star with atmosphere' (Fig. 2.24). What follows was a revision of his idea that all nebulae should be clusters. The dominant central star seems to be strongly correlated with the surrounding nebula and must therefore be formed by gravitational contraction. As explained in his paper 'On nebulous stars properly so called', Herschel now was

---

[92] Herschel W. (1784: 443).
[93] Hoskin (1979), Schaffer (1980).
[94] RAS Herschel W. 4/1.5.
[95] RAS Herschel W. 4/1.7: 643.
[96] Herschel W. (1785).
[97] Herschel W. (1785: 225).
[98] Herschel W. (1791: 82).

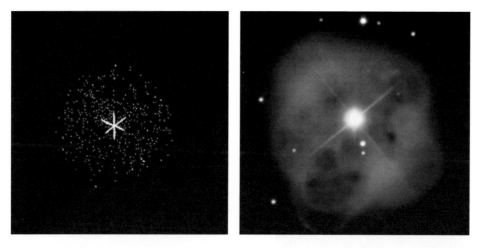

Figure 2.24. Left: Herschel's sketch of IV 69 = NGC 1514 (Herschel W. (1814), Fig. 8); right: a modern image.

convinced that at least some of the unresolved nebulae consist of a 'luminous fluid' (Herschel W. 1791). This is thought to be like an 'interstellar aether' and should not be confused with a 'gas'.[99] According to his final hypothesis, this true nebulosity would gradually condense into stars (clusters). His former picture (based on observations of M 17 and M 27) was partly reversed and his first idea (based on supposed changes in M 42) was eventually reactivated.

It is interesting to compare NGC 1514 and NGC 2392 (IV 45), the Eskimo Nebula in Gemini, found by Herschel on 17 January 1787. Because the latter holds a striking central star in a round nebulous envelope, it is astonishing that this object had not already changed his mind. However, he was surprised by the *'curious phenomenon'* and at first he could not believe it to be real: '*I suspected the glass* [eye-piece] *to be covered with damp, or my eye not yet to be in order*'. Herschel described the object as '*A star with a pretty strong milky nebulosity equally dispersed all around.*'[100] For Lord Rosse it was the prototype of a 'nebulous star' (see Section 6.4.11). He thought Herschel's NGC 1514 (h 311) to be a '*new spiral of an annular form round the star, which is central; spirality is very faint*'.[101] Thus both objects were interpreted quite differently by both observers.

Herschel accentuated his hypothesis in the publication of 1814 (Herschel W. 1814), presenting further examples of 'nebulous stars' (Table 2.17). As a connective link between them and his class IV objects, he introduced IV 73 (NGC 6826), a bright PN in Cygnus. Herschel wrote in his third catalogue that '*It is of a middle species, between the planetary nebulae and nebulous stars, and is a beautiful phenomenon.*' The reflection nebulae NGC 2167 and NGC 2170 in Monoceros are treated in Section 3.2.

Figure 2.25 shows that William Herschel's class IV contains only 20 true planetaries (PN). Most objects are galaxies (41); 13 are emission and reflection nebulae.

Examples of 'foreign' types in class IV are the emission nebula IV 41 (M 20, Trifid Nebula) in Sagittarius, globular cluster IV 50 (NGC 6229) in Hercules, galaxy IV 61 (M 109) in Ursa Major and double galaxy IV 28 (NGC 4038/39), The Antennae, in Corvus.[102] Also remarkable is IV 2, the reflection nebula NGC 2261 around the variable star R Mon in Monoceros. It is the prototype of a cometary nebula. Herschel discovered the object on 26 December 1783, describing it as 'fan-shaped'. Schmidt noticed the variability of the star in 1861; that of the nebula was first noticed by Hubble. NGC 2261 is known as Hubble's Variable Nebula (see

---

[99] Schaffer (1980: 90).
[100] RAS Herschel W. 2/3.6.
[101] Parsons W. 1861a: 148 and Fig. 7 (see Section 7.1).

[102] See Section 2.9. In the same night (7 February 1785) Herschel found the bright PN IV 27 = NGC 3242 in Hydra (Ghost of Jupiter). He also discovered the PN NGC 4361 in Corvus, but catalogued it as 'bright nebula' I 65.

2.7 Von Hahn's observations

Table 2.17. *Herschel's 'nebulous stars' (site: D = Datchet, C = Clay Hall, S = Slough)*

IV	NGC	Date	C	Site	Type	V	V*	Con.	Remarks
19	2170	16.10.1784	1	D	RN		10.6	Mon	Near NGC 2167, NGC 2182
25	2327	31.1.1785	1	D	EN		9.5	CMa	
36	2071	1.1.1786	2	C	RN		10.1	Ori	In M 78 complex
38	2182	24.2.1786	2	C	RN		9.3	Mon	Near NGC 2167, NGC 2170
44	2167	28.11.1786	2	S	EN		9.3	Mon	Near NGC 2170, NGC 2182
45	2392	17.1.1787	2	S	PN	9.1	10.5	Gem	Eskimo Nebula
52	7635	3.11.1787	2	S	EN		8.7	Cas	Bubble Nebula, centre of Sh2–162
57	6301	11.6.1788	2	S	Gx	13.5		Her	
58	40	25.11.1788	2	S	PN	12.3	11.5	Cep	
65	2346	5.3.1790	3	S	PN	11.6	11.6	Mon	
69	1514	13.11.1790	3	S	PN	10.9	9.5	Tau	'A most singular phenomenon'
71	5856	24.5.1791	3	S	star		6.0	Boo	See Section 2.4
74	7023	18.10.1794	3	S	EN+OC		7.4	Cep	

Section 6.18.2). This clearly shows how problematic the assignment PN–class IV–planetary is.

Vice versa, true planetaries (PN) appear in Herschel classes I, II, III, V and VI. Examples are NGC 7008 (I 192, 'bright nebula') in Cygnus, NGC 246 (V 25, 'large nebula') in Cetus and NGC 6804 (VI 38, 'rich cluster') in Aquila.

2.7 VON HAHN'S OBSERVATIONS OF PLANETARY NEBULAE

Friedrich von Hahn was among the few contemporaries of William Herschel observing nebulae. His favourite targets were objects with planetary or annular shape. He owned a castle near Remplin in Mecklenburg, which was equipped with a considerable observatory. The main telescope was the third largest outside Great Britain.[103]

2.7.1 Short biography: Friedrich von Hahn

Friedrich von Hahn was born on 27 July 1742 in Neuhaus, Holstein, where he spent his youth. Soon his interest in philosophy and science arose. At the age of 18 he began to study mathematics and astronomy at the University of Kiel. Later Hahn, being handicapped, was largely occupied by his manor. Therefore he had to wait until the age of 50 to practise astronomy. In about 1792 he erected a private observatory. The largest instrument, an 18.7" reflector with focal length 20 ft, first saw light in 1800 (Fig. 2.26). Herschel manufactured the metal mirror; the mechanical parts were constructed by Hahn.[104] The telescope had a wooden tube and no secondary mirror ('front-view'). Together with two smaller Herschel reflectors, with apertures of 12" (1794) and 8" (1793), respectively, it was used unshielded in the garden.[105] To measure positions, Hahn purchased a circle, equipped with a 2" refractor, from Cary. From 1801 it was located in a small dome on top of a four-storeyed tower. Hahn observed the Sun, planets, variable stars and nebulae, being focused on planetaries and the Orion Nebula (Hahn 1796). His main discovery was the

---

[103] The two largest were used by Schroeter and Schrader.

[104] Maurer (1996: 10); optically the telescope was a duplicate of Herschel's standard reflector. It was used without a secondary mirror ('front-view'); see Bode (1808: 204).

[105] See Bode (1794: 242).

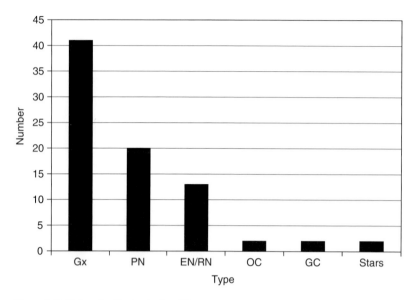

Figure 2.25. Object distribution in class IV.

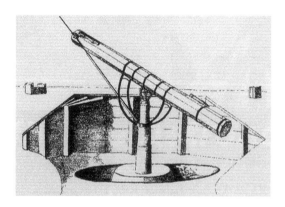

Figure 2.26. The 18.7″ reflector of Friedrich von Hahn in Remplin (Fürst and Hamel 1999).

central star of the Ring Nebula M 57. He was in close contact with Johann Elert Bode, Director of the Berlin Observatory. Friedrich von Hahn died on 9 October 1805 in Remplin at the age of 63.[106]

2.7.2 Observations of planetary nebulae and the discovery of the central star in M 57

Among Hahn's primary targets were Herschel's class IV objects. The observations were published in Bode's *Berliner Jahrbuch*, titled 'Ueber den planetarischen Nebelfleck bey μ Wasserschlange'.[107] The object is IV 27 (NGC 3242), which was named the Ghost of Jupiter in the twentieth century. Hahn asked whether planetary nebulae '*would show traces of proper motion, and thus being not clusters, but singular cosmic bodies.*' He adopted Herschel's idea that planetary nebulae are the final state of globular clusters.

For the necessary positional measurements, Hahn used the Cary circle. Unfortunately, in the small instrument the objects appeared '*with very pale light* [...], *standing only very faint illumination of the wires*', which made the reading pretty difficult.[108] For finding the objects, Francis Wollaston's star catalogue of 1789 was used, which '*contains all nebulae known at its making*'. Hahn observed NGC 3242 too, using the 12-inch with powers of 240 and up. The nebula appeared '*more brilliant than the outer planets, only at higher magnification does its light decrease*'. About the appearance he notes that it was '*quite round and circular, only on one side not complete, having the shape of the moon a few days before opposition* [full moon]'. Hahn's interpretation: '*One is tempted to assume that the supposed nebula*

---

[106] Obituary: Bode (1806); see also Fürst and Hamel (1983, 1999).

[107] 'About the planetary nebula near μ Hydrae' (Hahn 1799).
[108] The illumination of the cross-wires in the micrometer eyepiece is meant here.

is actually a sphere with a brilliant and a dark side, where only a small part of the latter is visible.'

A year later Hahn supposed that the shape and position of NGC 3242 had changed, writing '*this nebula does not show the form which it had at the time of its discovery by Dr. Herschel [7.2.1785]*' (Hahn 1800). One further reads that '*This astronomer describes it as quite round, which is now obviously no longer the case, it resembling [...] the moon a few days before opposition. It seems to have waned even more and the place is different from yesteryear.*' Among Hahn's targets was the variable star Mira (o Ceti), which is mentioned in his paper too. Curiously, he suspected the object to be a planetary nebula: '*It cannot be magnified like those* [planetary nebulae], *but it appears as a disc with little dazzling, and brighter than other stars.*'

In about 1795 Hahn examined the Ring Nebula in Lyra (M 57), which had been discovered in 1779 by Darquier in Toulouse with a 9.5-cm Dollond refractor. As early as in 1785 Bode had called attention to the object in his paper 'Ein Sternring oder ein Nebelfleck mit einer Oeffnung'.[109] He added Herschel's sketch. While observing the planetary nebula with the 12" reflector, Hahn noticed the central star. He bequeathed no date, but wrote, in 1800, '*In the famous star-ring near β Lyrae I find distinct changes. A few years ago the interior of the ring was so clear that I could distinguish in its centre a telescopic star with my 20 ft reflector. Now this telescope shows only faint fine clouds and the small star is no longer visible. A change has certainly happened*' (Hahn 1800). Hahn supposed '*It could be possible too that the transparent ring has changed its position and, relative to the sky background, infinitely beyond the ring, appears different now.*' It is remarkable that the star was seen in a 12-inch – even for today that is an extremely difficult task for such a small aperture. Nothing is said about M 57 observations with the 18-inch, which was built in 1800.[110]

It is remarkable that William Herschel did not notice the central star, despite using a reflector of similar size. It might have been visibile in the 40-ft, but his largest telescope was never pointed to M 57 (NGC 6720, h 2023). After Hahn the central star was seen by only a few observers (John Herschel was not among them).[111] The problem is the low contrast between the faint star (14.8 mag) and the interior of the ring, which is not black, but has a higher surface brightness than the sky background. To see the star, a large telescope with high magnification (enhancing the contrast) is needed. Lord Rosse always found the star 'pretty bright' in his 72" reflector (it was seen first on 5 August 1848); Angelo Secchi (about 1855) and Herman Schultz (13 August 1865) could see the central star in their 9.5" refractors.[112] However, Hermann Vogel's observation with the 27" refractor in 1884 was unsuccessful: '*The Vienna refractor shows the interior of the ring quite uniformly filled with faint nebulosity.*'[113] James Keeler could easily see the star in the 36" Lick refractor: '*with this instrument the central star was always easily visible, although it was too faint for observation with the spectroscope*' (Keeler 1892).

The first photography of the central star was achieved by Eugen v. Gothard on 1 September 1886 with a 10.25" Browning reflector in Herény, Hungary (Gothard 1886a). He noted that '*in the middle a round (possibly annular) core is visible.*'[114] Thereupon Rudolph Spitaler observed M 57 in autumn 1886 with the great Vienna refractor – and was disappointed: '*A small star near the centre was, however, not seen*' (Spitaler 1887). But on 25 July 1887, accompanied by his visitor Charles Young from Princeton, he proudly noted that '*at first glance, almost in the middle of the interior ring area, a bit northwest of the very centre, a small star was visible, just like it appears on Gothard's photography*' (Fig. 2.27). Julius Scheiner, who had imaged the planetary nebulae NGC 7009 and NGC 7662, claimed in 1892 that the "*central 'stars' are by no means stars in the usual sense of the word, but only irregularly shaped nebulous condensations*" (Scheiner 1892). This supported William Herschel's idea of 'nebulous stars'. In 1842 Arago, Director of the Paris Observatory, had already surmised that the central star (in most cases too remote to be visible) illuminates the planetary nebula, which explain its uniform light. Thus PN would be mere reflection nebulae, which is incorrect.[115]

---

[109] 'A star-ring or a nebula with an opening' (Bode 1785a).

[110] The common claim that the central star was discovered with the 18-inch is wrong; see e.g. Stoyan *et al.* (2008).

[111] Therefore the star remained unknown; it is, for instance, not mentioned in Smyth's observing guide (Smyth 1844).

[112] Parsons L. (1880: 152), Secchi (1856a), Schultz (1874: 99).

[113] Vogel (1884: 35).

[114] See Section 9.15.3 (Fig. 9.58).

[115] The PN gas is highly excited by the hot central star. See article 877 in *Outlines of Astronomy* (Herschel J. 1869).

Figure 2.27. Spitaler's fine drawing of M 57 (Spitaler 1891a).

2.8 SPECIAL OBJECTS

In Herschel's papers on 'Astronomical observations', which were published in 1811 and 1814, many objects are presented.[116] He focuses on morphology and classification; individual objects were treated to demonstrate the systematic features. An interesting sample is consitituted by the double (multiple) nebulae, which were studied for the first time by Herschel and categorised by distance (Table 2.18).[117] As discoverer of physical double stars, he supposed that there was a certain similarity between these two phenomena. Therefore, double nebulae should show orbital motion too. Later John Herschel revisited the issue.

The 15 'double nebulae with joined nebulosity' (defining the closest pairs) are listed in Table 2.19. The first discovered case is NGC 5194/95 (M 51), which was seen by Méchain on 21 March 1781 (M 76 is not a true example). In nine cases we have true double galaxies, which are included in modern catalogues (Vorontsov–Velyaminov, Holmberg, Arp).[118] It is, however, remarkable that such a prominent example as NGC 4676 (II 326, 13 May 1785) in Coma Berenices, now called The Mice (Arp 242, VV 244), is not included. Herschel simply did not noticed that it is a double object, whereas

Table 2.18. *Herschel's double and multiple nebulae*

Category	Number
Double nebulae with joined nebulosity	15
Double nebulae that are not more than two minutes from each other	23
Double nebulae at a greater distance than 2′ from each other	101
Treble, quadruple and sextuple nebulae	20/5/1

his son noted on 9 April 1831 '*query if bicentral*' (the nuclei are only 37″ apart).[119]

NGC 2905 is a distinctive H II region in the galaxy NGC 2903. The list also contains the planetary nebulae M 27 and M 76, interpreted by Herschel as 'double nebulae'. The smaller 'component' of M 76 was catalogued as I 193; Dreyer had considered the object as NGC 650/51. A similar case is II 316/17 (NGC 2371/72) in Gemini (Fig. 2.28 left). III 644 (NGC 5522) is nothing but a single galaxy and the 'companion' of III 45 (NGC 5174) is only a star (14.4 mag), 45″ south of the centre. The curious case of II 48, II 80 and NGC 2672/73 is treated in Section 8.16.3. Table 2.20 shows the six largest of Herschel's groups of nebulae.

It is interesting that Herschel has not included a 'quadruple', which was sketched earlier (Fig. 2.28 right): the galaxy NGC 4449 (9.4 mag) of the 'magellanic' type Sm in Canes Venatici. Unlike NGC 2371/72 (Fig. 2.28 left), for which Herschel used two entries (II 316/17), the object is listed as I 213. His sketch shows four condensations, corresponding to bright H II regions.

The quartets Arp 318 and HCG 61 are noteworthy.[120] The latter is one of Paul Hickson's 'compact groups', which was nicknamed in the twentieth century The Box (Fig. 2.29).[121] The third of Herschel's

---

[116] Herschel W. (1811, 1814).
[117] Herschel W. (1811: 285–289).
[118] For these catalogues see Steinicke (2004a, Section 2) and Steinicke and Jakiel (2006, Section I.3). On the *Arp Atlas of Peculiar Galaxies*, see also Kanipe and Webb (2006).
[119] The galaxy pair was eventually seen by Spitaler on 20 March 1892 with the Vienna 27″ refractor (Spitaler 1893). His 'Novae' nos. 51 and 52 were catalogued as IC 819 and IC 820 by Dreyer. The identity with NGC 4676 was later detected by Carlson (1940). The name The Mice is a twentieth-century product.
[120] For these catalogues, see Steinicke and Jakiel (2006).
[121] See Steinicke (2001a).

Table 2.19. *Herschel's 15 'double nebulae with joined nebulosity'*

Pair	Object	NGC	Date	Type	V	Con.	Catalogue	Remarks
(1)	I 56	2903	16.11.1784	Gx	8.8	Leo		
	I 57	2905	16.11.1784	GxP		Leo		H II region in NGC 2903
2	I 176	4656	20.3.1787	Gx	10.1	CVn	KPG 350	
	I 177	4657	20.3.1787	Gx	12.4	CVn	KPG 350	
3	I 178 = I 179	4618	9.4.1787	Gx	10.6	CVn	Arp 23, VV 73	
	I 178 = I 179	4618	9.4.1787	Gx	10.6	CVn	Arp 23, VV 73	
(4)	I 193	651	12.11.1787	PN	10.1	Per		M 76, Méchain 5.9.1780; bipolar
		650		PN	10.1	Per		M 76, Méchain 5.9.1780; bipolar
5	I 186	5195	12.5.1787	Gx	9.6	CVn	Arp 85, VV 1	Méchain 21.3.1781
		5194	17.9.1783	Gx	8.1	CVn	Arp 85, VV 1	M 51, Messier 13.10.1773
(6)	II 80 = II 48	2672	14.3.1784	Gx	11.6	Cnc	Arp 167	With NGC 2673, J. Stoney 19.12.1848
7	II 271	741	13.12.1784	Gx	11.3	Psc	VV 175	
	II 272	742	13.12.1784	Gx	14.3	Psc	VV 175	
8	II 309	5427	5.3.1785	Gx	11.4	Vir	Arp 271, VV 21	
	II 310	5426	5.3.1785	Gx	12.1	Vir	Arp 271, VV 21	
(9)	II 316	2371	12.3.1785	PN	11.2	Gem		Bipolar
	II 317	2372	12.3.1785	PN	11.2	Gem		Bipolar
10	II 832	3895	18.3.1790	Gx	13.1	UMa	Holm 294	
	I 248	3894	18.3.1790	Gx	11.6	UMa	Holm 294	
(11)	III 45	5174	15.3.1784	Gx	12.5	Vir		
	III 46	5175	15.3.1784	Star	14.4	Vir		45″ south of galaxy centre
(12)	III 644	5522	19.3.1787	Gx	13.4	Boo		Only single galaxy
13	IV 8	4567	15.3.1784	Gx	11.3	Vir	VV 219, Holm 427	
	IV 9	4568	15.3.1784	Gx	10.9	Vir	VV 219, Holm 427	
14	IV 28.1	4038	7.2.1785	Gx	10.3	Crv	Arp 244, VV 245	The Antennae (see Table 2.22)
	IV 28.2	4039	7.2.1785	Gx	10.4	Crv	Arp 244, VV 245	The Antennae
(15)		6853	30.9.1782	PN	7.4	Vul		M 27, Messier 12.7.1764

46 William Herschel's observations

Table 2.20. *Herschel's five quadruple nebulae and one sextuple nebula*

Objects	Date	NGC	Con.	Remarks
II 482–485	28.11.1785	833, 835, 838, 839	Cet	Arp 318
II 568–571	17.4.1786	4270, 4273, 4277, 4281	Vir	In Virgo Cluster
II 372, III 358–360	11.4.1785	4173, 4169, 4174, 4175	Com	HCG 61 (The Box)
II 371, III 356/57	11.4.1785	4134, 4131, 4132	Com	Only three galaxies
III 562–565	21.9.1786	703, 704, 705, 708	And	In Abell 262
III 391–396	27.4.1785	4070, 4069, 4074, 4061, 4065, 4076	Com	In CGCG 1202.0+2028

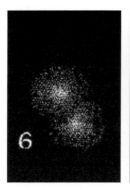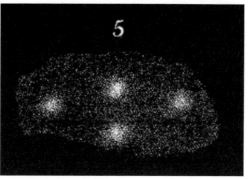

Figure 2.28. Left: Herschel's sketch of the 'double nebula' II 316/17 = NGC 2371/72; right: I 213 = NGC 4449 (I 213) shows four condensations (Herschel W. (1811), Figs. 6 and 5, respectively).

Figure 2.29. The Box in Coma Berenices: four galaxies within 8′, discovered by Herschel; the two stars NGC 4170/71 were found in 1864 by d'Arrest (DSS).

galaxy clusters (the sextuple is in a cluster, which was later catalogued by Zwicky).[122]

In addition to III 562–565, Herschel discovered two other members in Abell 262: NGC 679 (III 175, 13 September 1784) and NGC 687 (III 561, 21 September 1786). The remaining five NGC galaxies of the cluster are credited to d'Arrest (three) and the Birr Castle observers Lawrence Parsons and Mitchell. The objects in the quadruple II 568–571 (Herschel's first discovery in Slough) were difficult to identify; the problem was eventually solved by Schönfeld in 1862 (see Section 8.5.4).

Herschel was successful in other Abell clusters too. He found 23 galaxies in the Coma Cluster (Abell 1656), but did not notice its very structure (see Section 8.6.4). Seven of the 21 NGC galaxies in the Leo Cluster (Abell 1367) were seen on 26/27 April 1786 (Datchet); among them was the brightest member, NGC 3842

'quadruple nebulae' is only a chain of three galaxies, with NGC 4134 (12.9 mag) as the brightest (Herschel wrote '*a 4th suspected*'). The other groups are parts of

[122] Interestingly the term 'galaxy cluster' had already been used by Webb – albeit, to describe a galactic cluster, e.g. NGC 2301 (VI 27) in Monoceros (Webb 1859: 215).

Table 2.21. *Herschel's NGC galaxies in the Leo Cluster (Abell 1367)*

Object	NGC	Date	V
III 376	3821	26.4.1785	12.8
III 377	3837	26.4.1785	13.3
III 378	3842	26.4.1785	11.8
III 386	3860	27.4.1785	13.4
III 385	3862	27.4.1785	12.7
III 387	3875	27.4.1785	13.7
III 388	3884	27.4.1785	12.6

(Table 2.21). The remaining ones are due to d'Arrest (five), Stephan (five) and John Herschel (four).

Herschel succeeded at making a similar find in the NGC 507 group in Pisces. On 12 September 1784 he discovered the primary member (III 159, 11.3 mag) and three others: NGC 495 (III 156), NGC 496 (III 157) and NGC 508 (III 160). The remaining NGC galaxies of the group were found by John Herschel (two), Mitchell (two) and d'Arrest (one). On 26 September 1785 Herschel discovered three NGC galaxies in the Pegasus I Cluster, among them the brightest, NGC 7619 (II 439, 11.1 mag); the others are NGC 7623 (II 435) and NGC 7626 (II 440). This cluster, containing eight NGC galaxies, is not rich enough to be listed in the Abell catalogue; the others were found by Marth (two), John Herschel (one), d'Arrest (one) and Copeland (one).

On 17 March 1787 William Herschel discovered the first ring galaxy: NGC 4774 (III 618, VV 784) in Canes Venatici, described as 'eF, vF'. The object of 14.3 mag was nicknamed the Kidney Bean Galaxy by Zwicky. Other galaxies of this rare type were found by Marth, Stephan, Tempel and Leavenworth – though none of these visual observers could detect their peculiar structure.

2.9 ADDITIONS BY JOHN HERSCHEL AND DREYER

John Herschel and Dreyer added 15 objects to the catalogues (Table 2.22). John Herschel described the first two in his Slough catalogue, others followed in the Cape catalogue and GC, assigned as 'HON' ('Herschel omitted nebula'). Eventually Dreyer added two objects. In some cases an existing H-number was split; in other cases entries were appended to the particular class (the discovery date is thus later than that of William Herschel's final entry). Most additional objects were listed by Auwers (Au) in his revision of the three Herschel catalogues (see Section 6.19).

I 28 is the double galaxy NGC 4435/38 in the centre of the Virgo Cluster.[123] William Herschel noted '*One of two, at 4 or 5″ dist. B. cL.*' John Herschel entered these two objects as h 1274 and h 1275 in the Slough catalogue; h 1275, which is the brighter component NGC 4438 (10.0 mag), was identified as I 28. But h 1274 was equated with M 86 – a serious error, which was first noticed by Auwers[124] and eventually corrected by John Herschel in the GC, which gives GC 2991 = h 1274 = I 28,1 and GC 2994 = h 1275 = I 28,2. M 86 is now correctly identified as h 1253 = GC 2961.

Already William Herschel suspected IV 28 to be a double nebula '*opening with a branch or two nebulae very faintly joined*'. It is NGC 4038/39 in Corvus, known as The Antennae (Fig. 2.30 left); the components are only 1.5′ apart (Steinicke 2003d). It is remarkable that the pair was entered in class IV ('planetary nebulae'). In the Slough catalogue John Herschel separated it into IV 28.1 (h 1052 = NGC 4038) and IV 28.2 (h 1053 = NGC 4039).

V 29.1/2 is in the Slough catalogue too, where a common h-number was used (h 1252). V 29.1 (NGC 4395) is a galaxy of 10.0 mag in Canes Venatici, which was discovered on 2 January 1786. V 29.2 (NGC 4401) is a bright HII region 2′ southeast of the centre, which was found by John Herschel on 29 July 1827 (Fig. 2.30 right). He noted (h 1252) '*Two nebulae running into one another*.'[125] Two other HII regions, NGC 4399 and NGC 4400, were contributed by Bindon Stoney at Birr Castle (13 April 1850).

The designation VIII 1B is explained in the General Catalogue (note to GC 1480): "*This nebula is entered by C.H. as VIII. 1. B, with the remark 'not in print'.*" Caroline Herschel included the object in her zone catalogue (described in the next section), because it was erroneously missing from the first published Herschel catalogue. Since the discovery date (18 December 1783) follows that of VIII 1 = NGC 2509 (3 December), John Herschel introduced the suffix 'B'. It is, however, confusing that he assigns NGC 2319 in his entry h 423 as 'VIII. 1'.

---

[123] The pair is a member of Markarian's Chain, starting with M 84 and M 86.

[124] Auwers (1862a: 77).

[125] The Slough catalogue contains a sketch (Fig. 68).

Table 2.22. *Objects added by John Herschel and Dreyer (site: D = Datchet, S = Slough)*

Object	Source	h	Au	GC	NGC	Date	Site	Type	V	Con.	Remarks
I 28, 2	SC	1275		2994	4438	8.4.1784	D	Gx	10.0	Vir	I 28, 1 (h 1274) = NGC 4435
II 908	HON 3		*	1690	2650	30.9.1802	S	Gx	13.3	UMa	
II 909	HON 5		*	1972	3063	30.9.1802	S	2 stars		UMa	Dreyer: 'HON 5'; W. Herschel's final object
II 910	GC	1407	*	3179	4646	24.3.1791	S	Gx	13.4	UMa	II 794,2 (order changed); II 794,1 = NGC 4644
III 979	HON 6		*	2077	3210	26.9.1802	S	2 stars		Dra	Dreyer: 'HON'
III 980	HON 7		*	2078	3212	26.9.1802	S	Gx	13.7	Dra	Dreyer: 'HON'
III 981	HON 8		*	2079	3215	26.9.1802	S	Gx	13.2	Dra	Dreyer: 'HON'
III 982	HON 1		*	1679	2629	30.9.1802	S	Gx	12.3	UMa	Dreyer: 'HON'
III 983	HON 2		*	1682	2641	30.9.1802	S	Gx	14.0	UMa	Dreyer: 'HON'
III 984	SC	2296	*	5044	7810	17.11.1784	D	Gx	13.1	Peg	H.MS. (order changed)
III 985	D	1435		3224	4695	24.3.1791	S	Gx	13.5	UMa	'II 796' (order changed)
IV 28.2	SC	1053		2671	4039	7.2.1785	D	Gx	10.4	Crv	IV 28.1 (h 1052) = NGC 4038, The Antennae
IV 79	HON 4		*	1950	3034	30.9.1802	S	Gx	8.6	UMa	M 82, Bode 31.12.1774; Dreyer: '4 HON'
V 29.2	SC	1252		2962	4401	29.7.1827		GxP		CVn	H II region in NGC 4395 = V 29.1 (h 1252)
VIII 1B	SC	423		1480	2319	18.12.1783	D	Star group		Mon	VIII 1 = NGC 2509

III 984 is not in the Herschel catalogues. John Herschel recognised the object in an observing journal of his father. The Slough catalogue remarks for h 2296 'H.MS.' ('Herschel manuscript'). Dreyer introduced the designation III 984 in the *Scientific Papers of Sir William Herschel*.[126]

The eight 'HON' objects are listed in the appendix of John Herschel's Cape catalogue.[127] The abbreviation was used first in the GC and stands for 'Herschel omitted nebulae'.[128] All objects were discovered by William Herschel on 26 and 30 September 1802 in Slough. Since he had decided that his last catalogue should contain exactly 500 entries, they were omitted (see Section 2.2.2). Most of them are galaxies in the constellations Ursa Major and Draco. The story of IV 79 is remarkable. This 'planetary nebula' is identical with the bright

---

[126] Dreyer (1912a); see Dreyer's notes to the Herschel objects.
[127] Herschel J. (1847: 128).

[128] The GC uses 'H.O.N.', whereas Dreyer notes 'HON' in the NGC.

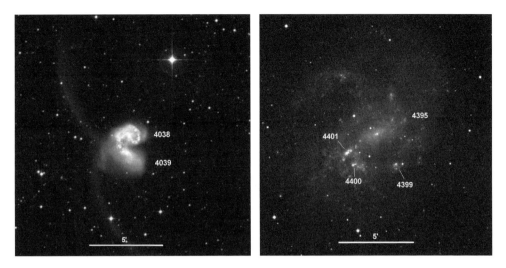

Figure 2.30. Additional Herschel objects. Left: The Antennae NGC 4038/39 (IV 28.1/2); right: NGC 4395 with the H II regions NGC 4399, NGC 4400 and NGC 4401 = V 29.1/2 (DSS).

galaxy M 82 in Ursa Major! It was discovered by Bode on 31 December 1774. The whole Herschel family is involved in this curious story (Steinicke 2007a).

For II 794 William Herschel noted two observations: 14 April 1789 (sweep 921) and 24 March 1791 (sweep 1001). The first was used for his entry in the third catalogue, described there as 'F, S'. In the GC John Herschel now claims that these observations concern different objects. He thus introduced two new designations: II 794,1 = h 1406 = GC 3177 (NGC 4644) and II 794,2 = h 1407 = GC 3179 (NGC 4646). The first object is II 794 of sweep 921. This was not recognised by him in the Slough catalogue, where we face a confusing situation: he identifies h 1407 with II 794, but takes the object from sweep 1001. On the other hand, h 1406 is assigned as 'Nova', but this is actually the object of sweep 921, named II 794 by his father. In the NGC Dreyer used the correct GC version. In the *Scientific Papers* the case was investigated once again. Now Dreyer introduces the designation II 910, instead of II 794,2, and II 794,1 gets back its old name II 794.

III 985 is a similar case. William Herschel noted two observations from his sweeps 921 and 1001 (the same as for II 794, see above). Once again two different objects are involved, which was not noticed by him. The first, found on 14 April 1789, was named II 796 (NGC 4686) in the third catalogue; the second, observed on 24 March 1791 was not listed. It first appears in Dreyer's *Scientific Papers* as III 985 (NGC 4695). Both objects are contained in the Slough catalogue and GC too. But there is confusion concerning the identification with H-numbers: h 1428 = GC 3216 (NGC 4686) is called 'II 795' and h 1435 = GC 3224 (NGC 4695) is called 'II 796'. The NGC gives the correct version. With the designations II 910, III 984 and III 985, introduced by Dreyer in the *Scientific Papers*, the usual order (by date) was given up. This makes his version different from William Herschel's original catalogues.

2.10 LATER PUBLICATIONS AND REVISIONS OF HERSCHEL'S CATALOGUES

Johann Elert Bode published Herschel's catalogues (in German translation) with a delay of two years in the *Berliner Jahrbuch* (Bode 1788, 1791, 1804b).[129] Right ascension and declination for 1786, 1790 and 1801, respectively, are given. However, Bode kept the original order. A flaw was the lack of precise star positions used to determine absolute positions. Bode's aim was not a new reduction, but to present Herschel's discoveries to a wide (German-speaking) audience. As d'Arrest and

---

[129] Herschel's first catalogue was also reprinted in Francis Wollaston's book *A Specimen of a General Astronomical Catalogue*, published in 1789 in London (Wollaston 1789). Bode translated other works of Herschel too; see e.g. Bode (1804a).

Auwers later pointed out, Bode's treatment is full of errors (d'Arrest 1856a; Auwers 1862a). D'Arrest wrote that, *'even if one ignores the inaccurate star positions used to determine the coordinates of the nebulae, the work is distorted by many errors'*. It therefore was hardly used.

Another treatment of the original data is due to Caroline Herschel. After her brother's death and her return to Hannover, she reduced the nebular positions for the epoch 1800.[130] The resulting catalogue was a folio-volume of 104 handwritten pages (Herschel C. 1827).[131] It was based on Flamsteed's star catalogue, which had been revised by her earlier. The objects are arranged in declination zones, starting with the circumpolar nebulae to 9° north polar distance, followed by the zones 10° to 14° and 15° to 16°, after which the zones up to the final one (121°) had a constant width of 1°. Inside a zone, the objects are ordered by right ascension. The sweep-number is given for each entry. The last two pages of the manuscript contain, among other things, errata to the three Herschel catalogues.

On 8 February 1828 Caroline Herschel received the gold medal of the Astronomical Society of London[132] *'for her recent reduction, to January 1800, of the Nebulae discovered by her illustrious brother, which may be considered as the completion of a series of exertions, probably unparalleled, either in magnitude or importance, in the annals of astronomical labour'* (South 1830). The work on the 'zone catalogue' began in Hannover in April 1824 and might have been finished in late 1827. Unfortunately it was never published and can have been seen by only a few people.[133] The catalogue was primarily intended for the use of her nephew, John Herschel. From it, 'working lists' for the Slough observations were compiled. Herschel would have had the opportunity to publish the zone catalogue, but Dreyer assumed that *'he shared the universal opinion at the time, that very few of his father's nebulae could be seen, or at least, usefully observed with any but the largest telescopes; but chiefly because he always intended to bring out a General Catalogue of all known Nebulae and Clusters, a task which the vast amount of valuable work he carried out did not allow him to complete till 1864'*.[134]

In 1826 another German edition of Herschel's catalogues was published by Wilhelm Pfaff in Leipzig (Pfaff 1826). Unfortunately, the number of errors was even larger than in Bode's version. Therefore d'Arrest criticised this work too, describing it as a *'very erroneous reprint of the unreduced nebulae catalogues'*.[135]

Herschel's catalogues, though receiving praise, were hardly used for subsequent observations. It was John Herschel's work that kept things going. His observations in Slough of a large fraction of the objects were published in 1833. The main value of the Slough catalogue is based on its reliable absolute positions and a homogeneous numbering (h). This made the old Herschel catalogues more or less obsolete. Why use the inconvenient original now that an updated version was available? Lord Rosse, for instance, referred to h-numbers exclusively (which he curiously wrote 'H').

This was partly changed by the work of Auwers, who published the first complete revision of the Herschel catalogues in 1862, titled 'William Herschel's Verzeichnisse von Nebelflecken und Sternhaufen'.[136] Auwers, of course, used the results of John Herschel. Thus he corrected not only the original catalogues, but also the Slough catalogue. Most errors were due to incorrect identifications or problems with reference stars. For instance, Auwers detected that William Herschel's Flamsteed-numbers in Lynx must be reduced by 1 for those greater than 38 (e.g. '39 Lyncis' must read '38 Lyncis'), which led to wrong positions for some objects (see Section 6.19).

The last step was made by Dreyer. Whilst working on the NGC (and later on the IC) he dealt intensively with Herschel's catalogues. He used the original notes of William and Caroline Herschel, which had been put at his disposal by the Royal Society. A most valuable source was the unpublished zone catalogue of the latter. The result of his revision appeared in the *Scientific Papers of Sir William Herschel*, edited by him in 1912 (Dreyer 1912a). This monumental work contains the original catalogues, enlarged by

---

[130] Dreyer (1912a: lxiii–lxiv).
[131] The manuscript is in the possession of the Royal Society.
[132] On 15 December 1830 the name was changed to 'Royal Astronomical Society'.
[133] Herschel C. (1827); compare with RAS Herschel C. 3/3.
[134] Dreyer (1912a: lxiv).
[135] d'Arrest (1856a: 360). Further volumes planned by Pfaff failed to appear.
[136] 'William Herschel's lists of nebulae and star clusters' (Auwers 1862a).

Table 2.23. *Different notation used for Herschel objects*

Author	Notation
William & John Herschel	VIII. 61
Smyth	61 H. VIII.
d'Arrest, O. Struve	H. VIII. 61
Auwers, Schönfeld, Schulz	VIII. 61
Winnecke, Rümker	H. VIII, 61
Tempel	VIII 61
Dreyer	VIII. 61, VIII 61

additional objects, cross references to the NGC and many notes.

Finally, a remark concerning the designation of Herschel objects, which appears not to be uniform in the nineteenth century.[137] This is demonstrated in Table 2.23 for object no. 61 in class VIII; the open cluster NGC 1778 in Auriga, which was discovered on 17 January 1787 in Slough.

---

[137] In the modern literature there is confusion too; a recent example is O'Meara (2007).

3 • John Herschel's Slough observations

Apart from William Herschel, his son John was the greatest discoverer of nebulae and star clusters (Fig. 3.1).[1] About his motivation he wrote in 1826 in Slough that *'The nature of nebulae, it is obvious, can never become more known to us than at present; except in two ways, – either by the direct observation of changes in the form or physical condition of some one or more among them, or from the comparison of a great number, so as to establish a kind of scale or graduation from the most ambiguous, to objects of whose nature there can be no doubt.'*[2] The first way had already been realised through his detailed observations of the Orion and Andromeda Nebulae[3] (Herschel J. 1826a, b). The second – the study of a large number of objects – was mastered in the years 1825–33, reproducing and extending the observations of his father.

John Herschel published the results as 'Observations of nebulae and clusters of stars, made at Slough, with a twenty-feet reflector, between the years 1825 and 1833' in the *Philosophical Transactions of the Royal Society* (Herschel J. 1833a).[4] The professional Slough catalogue (SC) appeared in the same year as his first astronomical textbook *A Treatise on Astronomy*, which had been written for the general public. The latter volume had great influence and was later enlarged to become the *Outlines of Astronomy*, first appearing in 1849. In 1836 Herschel

Figure 3.1. John Herschel (1792–1871).

received the RAS gold medal for this work.[5] During the laudation held on 12 February (Herschel was absent), George Biddell Airy, Astronomer Royal and President of the RAS, delivered an overview on the status of nebular research (Airy 1836).[6] Because of the wide popularisation of John Herschel's writing and his many duties, he even surpasses the eminence of his father.[7]

For his observations John Herschel used a reflector with an aperture of 18¼", which was completed in 1820. It used two mirrors, one made by his father alone and another one cast and ground under his father's supervision (Fig. 3.2). The telescope resembles William Herschel's famous

---

[1] A biography is omitted here; see e.g. Ball R. (1895: 247–271), Buttmann (1970), King-Hele (1992), Ring (1992) and Chapman A. (1993).

[2] Herschel J. (1826a: 487).

[3] On the Orion Nebula see Herschel J. (1826a); it contains a drawing made between February 1824 and March 1826. His subsequent paper treats the Andromeda Nebula (Herschel J. 1826b).

[4] The acknowledgment reads *'Received July 1, – Read November 21, 1833.'* John Herschel became a fellow of the Royal Society in 1813.

[5] John Herschel was a founder member of the Astronomical Society of London (which later became the RAS); see Dreyer and Turner (1923) and Whitrow (1970).

[6] At the same time this was done by Joseph v. Littrow in his book *Sterngruppen und Nebelmassen des Himmels* (Littrow J. 1835).

[7] His popularity remained in the twentieth century, as can be seen, for instance, in the forenames of the American astronaut Glenn, born in 1921, which are 'John Herschel'. By the way, the son of Thomas Maclear, Astronomer Royal at the Cape, was given the forenames 'George William Herschel'.

which was due to remarkable new features: absolute positions (for 1830), order by right ascension and new designation (h). The great homogeneity rests on the fact that all objects were observed and measured by John Herschel with the same telescope. The h-number got the new standard designation (e.g. h 50 = M 31).

3.1 STRUCTURE AND CONTENT OF THE SLOUGH CATALOGUE

The text covers 147 pages; followed by 8 tables with 91 figures (drawings, sketches). The main part consists of an introduction (6 pages), the catalogue including 'Errata and addenda' (117 pages) and an appendix, which also contains the explanations to the figures (24 pages).

The catalogue lists 2306 numbered entries ('No.' is equal to h-number). Table 3.1 shows the meanings of the columns. Following the last entry (h 2306 = NGC 7827), the first one is repeated (h 1 = III 868 = NGC 12). The column 'Synonym' primarily gives William Herschel's designation. Additionally, Messier objects and those found by Wilhelm Struve are mentioned; examples are 'M 27' (h 2060) and 'Σ 885' (h 385), respectively. John Herschel's own discoveries are listed as 'Nova'. Also a few stars are referred to here, marking the centre of the nebulous object: '15 Monoc.' (h 401), '50 Cassiop.' (h 179), '55 Androm.' (h 162) and 'θ Orionis' (h 360). The latter defines the prominent trapezium in the Orion Nebula.[10]

Usually, several observations are given for an object. The coordinates are mostly noted to 0.1^s (AR) and $1''$ (NPD); a lower accuracy is indicated. By the way, there is no NPD for the 'Nova' h 1039 ('*no PD taken*'). The descriptions rely on William Herschel's scheme. Occasionally, distance and position angle to nearby objects (nebulae, stars) or the air quality is noted. For some observations guests are mentioned: Mr Knorre (h 749), Lord Ardare and Mr Hamilton (h 1357), Mr Baily (h 1558), Capt. Smyth (h 1663) and Mr Struve (h 2081).

The last column gives the number of the sweep. Unfortunately there is no register listing the dates. They were eventually presented in 1847 as 'Synoptic table of the dates of the sweeps' in the Cape catalogue.[11]

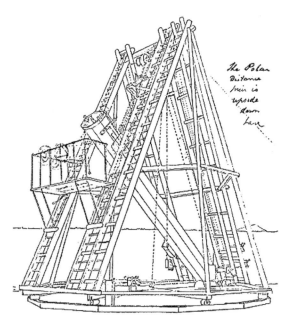

Figure 3.2. John Herschel's 18¼" reflector of focal length 20 ft that was completed in 1820 in Slough (Warner 1979).[8]

'large 20 ft' (Warner 1979). When John Herschel started observing, his intention was not so much the discovery of new nebulae and star clusters but rather he wanted to re-examine the three catalogues of his father. The main goals were identification and determination of exact positions. To realise this ambitious project, he compiled 'working lists' to direct his sweeps. They were based on Caroline Herschel's unpublished zone catalogue.[9]

John Herschel could observe a large fraction of his father's objects. Their data were partly confirmed, supplemented and corrected. Moreover he discovered many new ones. The resulting Slough catalogue contains

- known objects of William Herschel
- new objects of John Herschel
- objects that had been found earlier but not listed by William Herschel (e.g. from Messier's catalogue)
- objects discovered by others (mainly by Wilhelm Struve)

John Herschel compiled a (pretty complete) catalogue of all non-stellar objects known up to 1833. The work meant real progress and became a great success,

---

[8] See also Hoskin (1987); Fig. 3.
[9] RAS Herschel J. 1/5.1–4.
[10] Here John Herschel had discovered a sixth star in 1830; see Section 9.6.10.
[11] Herschel J. (1847: 129–131); for some sweeps the date gives no day (there is no information in his original reports).

Table 3.1. *Meanings of the columns*

Col.	Name	Explanation
1	No.	Running number (1 to 2306); later used 'h'
2	Synonym	Other designations (e.g. H, M, Σ) or 'Nova'
3	AR 1830.0	Right Ascension
4	N.P.D. 1830.0	North Polar Distance ($= 90° - \delta$)
5	Description and remarks	
6	Sweep	Number of observation (1–428)

Table 3.2. *John Herschel's view about the origin of the 2306 Slough objects*

Origin	Number	Remarks
W. Herschel	1697 + 2	803 objects missing
J. Herschel (Nova)	521	Some objects assigned 'Nova?'
Messier catalogue (M)	78	25 objects missing (of 103)
Struve (Σ)	6	
Stars	2	55 And, 50 Cas

Sweep 1 was made on 2 November 1823 and the final sweep 428 on 23 May 1832, but, due to a numbering error, no. 1 was not the first observation. Sweeps 53 and 54 were earlier (29 and 30 March 1821); moreover, sweeps 43 to 49, made from 8 May to 10 September 1823, lie between sweeps 1 (2 November 1823) and 2 (10 April 1825). About this issue Herschel noted in the Cape catalogue that the particular records '*have been mislaid, having been written on loose paper, and not found until after No. 42* [5.1.1827]'. The Slough observations thus cover a period of 11 years. The numbering of the sweeps was later continued at the Cape (sweeps 429 to 810).

Unfortunately the Slough catalogue gives no list of those William Herschel objects which were not observed or not found by his son. John Herschel refers to 1697 objects of his father, i.e. 803 of 2500 are missing (32%). Two observed by him, but not earlier catalogued, were included in the SC (see Table 2.22): 'HMS'(h 2296) and 'IV 28.2' (h 1053). For some objects, assigned 'Nova?', John Herschel was not sure whether they were new. Seventy-eight objects are referenced to Messier; thus 25 of 103 are missing. However, this result (shown in Table 3.2 and based on the information given by John Herschel) must be revised: there is a whole string of wrong identifications (see below).

3.2 IDENTIFICATION OF THE CATALOGUE OBJECTS

First of all, a modern analysis must reveal the number of independent Slough catalogue objects, i.e. which entries belong to the same object. As shown in Table 3.3, the net result is 2257.

In 18 cases ('SC identity') two h-numbers refer to the same object, bearing a single NGC-number. The identity was therefore, at the latest, recognised by Dreyer. Here are some examples. The first is h 4 = h 5, the galaxy NGC 16 in Pegasus. John Herschel assigns h 4 as 'Nova?' (the description already gives '*? if not IV. 15*') and h 5 is identified with IV 15. The problem was eventually solved in the GC. In contrast, he did not see the identity h 63 = h 64 (NGC 259, galaxy in Cetus); h 63 is called a 'Nova' (found on 18 October 1827) and h 64 = II 621 (16 October). The error probably occurred because the object was '*barely seen through fog*' on both nights. An interesting case is h 857 = h 875. For John Herschel h 875 is a 'Nova', described as a bright, extended nebula. This is actually the galaxy M 66 in Leo, however, catalogued at a wrong position. In the GC he notes '*h 857 = h875?*' (GC 2377). The final example is a double nebula, discovered on 2 May 1829 ('Nova'); the components are catalogued as h 1358 and h 1359. On 29 April 1830 he observed the double

Table 3.3. *Independent objects in the Slough catalogue*

	Number
Entries	2306
SC identity	18
NGC identity	31
Independent	2257

nebula IV 8/9, entered as h 1363 in the SC. But this object turned out to be identical with his earlier find and it is h 1363 = h 1358/59. This is the prominent galaxy pair NGC 4567/68 in Virgo, known as the Siamese Twins.[12] However, the pair is only 'optical', i.e. there is, due to their different distances away from us, no physical connection.

Additionally, a modern analysis revealed 31 cases with 'NGC identity'. Now both the h- and NGC-numbers are different. An extreme example is a galaxy in Triangulum: h 135 = NGC 614 = h 136 = NGC 618 = h 141 = NGC 627. It is the only case for which three different catalogue numbers are involved.

In contrast to Table 3.2, Table 3.4 gives the balance from a modern point of view. This concerns NGC identities too, considering objects that are not in the SC. For instance, the open cluster h 392 (NGC 2239) centred on 12 Monocerotis and discovered by John Herschel in May 1830 is identical to Flamsteed's NGC 2244. Seventy-seven Messier objects (not 78) are included in the SC, thus 26 are actually missing; 13 of them appear in the Cape catalogue. Though 'Messier' is entered in the first column, there are different discoverers (not resolved here). The additional Messier objects M 104 to M 110 can be found in the Slough catalogue too (except M 107); they are identified by H-number. A curiosity is the open cluster h 126. John Herschel mentions Wilhelm Struve (Σ 131) but overlooks that this is actually Méchain's M 103. All other Struve objects are correctly entered: h 385 = Σ 885, h 1970 = Σ 5 N., h 2000 = Σ Neb. 6, h 2067 = Σ 2630, h 2068 = Σ 2631 (see Section 4.2.3). The Slough catalogue contains nine discoveries of Caroline Herschel, but she is not mentioned explicitly; only the H-number is given.

One of the objects discovered at Slough was not due to John Herschel. For h 793 (NGC 3457), a galaxy of 12.5 mag in Leo found on 25 March 1827 (Fig. 3.3 left), he notes '*observed by Mr. Baily*'. This is Francis Baily (Fig. 3.3 right),[13] at that time President of the Astronomical Society of London and, as an intimate friend, a regular guest in Slough. Baily is well known through his work on star catalogues, e.g. the revision of Lalande's *Histoire Céleste* and the *British Association Catalogue* (BAC), which was published in 1845.

The five objects of Table 3.5 are not catalogued in the NGC. 'Synonym' gives John Herschel's (mostly incorrect) identification. h 162 is Flamsteed's star 55 Andromedae, characterised as '*nebulosa*' by Giuseppe Piazzi (see Section 6.14.6). John Herschel erroneously identifies h 305 ('*the first of 3*', observed on 17 October 1827) with III 569 (NGC 1367). The sky indeed offers three galaxies, forming a chain from west to east: IC 344, NGC 1417 and NGC 1418 (Fig. 3.4). Both NGC objects were found by William Herschel on 5 October 1785 and IC 344 by Lewis Swift on 23 December 1889 (no. 13 in his list IX). Dreyer investigated the case, but at first could not solve it (Dreyer 1891b). Eventually he presented the correct version in the IC I: '*h. 305 = Sw. IX*'. Thus John Herschel actually discovered one IC object (see Section 11.5.4).

For h 378 Herschel notes '*A star 7 m with a pretty strong nebulous atmosphere*', wrongly referring to IV 44 (Fig. 3.5), which is not his father's 'nebulous star' NGC 2167 (see Table 2.17). There is no NPD for h 1039; possibly this is h 1038 (NGC 4008), a galaxy in Leo. For h 1367 he suspected an identity with M 91, but probably h 1367 = h 1362 = NGC 4571. The galaxy M 91 in Coma Berenices is actually h 1345 = II 120 (NGC 4548).

3.3 JOHN HERSCHEL'S NEW OBJECTS

Herschel discovered 466 objects in Slough (Table 3.6); 385 of them are non-stellar (83%), mainly galaxies. A considerable number are stars (14%), among them a few bright ones (see Tables 6.20 and 10.14). The brightest is h 179 = 50 Cas (4.0 mag); he noted the following on 29 October 1831: '*I suspect this star to be nebulous.*' Dreyer catalogued it as NGC 771. Even more prominent is h 1332 = β CVn = NGC 4530 (4.3 mag); in early May 1828 Herschel noted '*8 Canum. Involved in*

---

[12] This is only an optical pair; the identification story is told in Steinicke (2003c); see also Steinicke (2001a). The popular name was created by the American amateur astronomer Leland Copeland in 1955 (Copeland L. 1955).

[13] Obituary: Anon (1844b) (written by John Herschel?).

Table 3.4. *Discoverers of independent objects in the Slough catalogue*

Discoverer	Number	Remarks
W. Herschel	1686	814 objects missing
J. Herschel	466	
C. Herschel	9	See Table 2.12
Messier (catalogue)	77	Different discoverers; 26 objects missing (13 in Cape catalogue)
W. Struve	5	See Table 4.2
Méchain	4	h 757 = M 105, h 831 = M 108, h 1030 = M 109, h 1175 = M 106
Messier	2	h 1376 = M 104, h 44 = M 110
Hipparch	2	h 207/212 = χ Per
Hodierna	2	h 174 = NGC 752, h 441 = NGC 2362
Flamsteed	2	h 162 = 55 And, h 392 = NGC 2239 = NGC 2244 (12 Mon)
Baily	1	h 793 = NGC 3457
Dunlop	1	h 1900 = NGC 5834
Sum	2257	

Table 3.5. *Slough objects not catalogued in the NGC*

h	GC	Date (h)	Synonym	Remarks
162	428	1.10.1828	55 And	Flamsteed 17.10.1691
305	756	16.10.1827	II 569	IC 344, Gx in Eri, 14.2 mag
378	1359	23.1.1832	IV 44	Star of 7 mag in Mon near NGC 2167
1039	2650	24.12.1827	Nova	NPD missing; perhaps h 1038 = NGC 4008
1367	3120	24.3.1830	M 91??	Possibly h 1362 = NGC 4571

Figure 3.3. Discovered by Francis Baily (right): the galaxy h 793 (NGC 3457) in Leo (DSS).

3.3 John Herschel's new objects   57

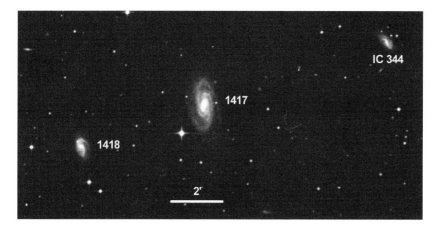

Figure 3.4. John Herschel's h 305 (IC 344) in Eridanus and its companions NGC 1417/18 (DSS).

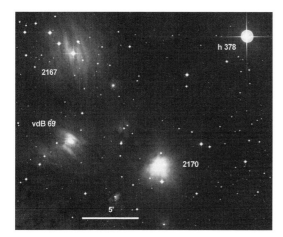

Figure 3.5. John Herschel's 'nebulous star' h 378 and the reflection nebulae NGC 2167 (IV 44) and NGC 2170 (IV 19) discovered by his father (DSS).

a considerable nebula.' For h 373 = 3 Mon = NGC 2142 (5.0 mag) he wrote '*I am sure this star has a faint nebulous atmosphere.*' In the SC appendix these cases are described as follows '*With regard to nebulous stars generally, I ought to mention that it has frequently occurred to me to notice a peculiar state of the atmosphere in which all large stars (above 7th magnitude) have appeared surrounded with photospheres of 2' or 3' or more in diameter, precisely resembling that about of the finer specimens of nebulous stars.*'[14] Herschel supports his father's view that 'nebulous stars' (like NGC 1514) are key objects to prove the existence of true nebulosity (see Section 2.6.2). Actually, these bright stars do not show any nebulosity (see Section 6.14.5). The star group NGC 6724 is the only object found with the '7-feet equatorial', a fine 5" Tully refractor purchased in 1828 from James South (normally used for precision measurements of double stars).

Twenty-three of Herschel's 433 objects are termed 'Nova'. Wrongly, 21 were thought to be in his father's catalogues and two in Messier's. The latter are h 1367, '*M 91??*' (see Table 3.5), and the open cluster h 2004 (NGC 6603), called M 24. But the true Messier object (Sagittarius Star Cloud) is listed in Dreyer's IC II as IC 4715 (see Table 2.2). Table 3.7 shows special objects and observational data.[15] In Slough Herschel discovered the most northern NGC object, the galaxy NGC 3172 in Ursa Minor, termed 'polarissima'[16] (h 250); the distance to Polaris is only 1.5° (Fig. 3.6).

John Herschel had less difficulty in observing near-pole objects than his father. However, the position measurement was far from easy: '*the AR cannot be determined exactly, and the PD is open to correction*'. For NGC 7010 it should be noticed that sweep 53 was the earliest; thus already Herschel's first observation brought a new object. On the other hand, his last one was discovered in sweep 419; the remaining nine sweeps (up to 428) brought nothing new.

---

[14] Herschel J. (1833a: 499).

[15] A similar table is given for all observers with a large number of discoveries.

[16] Later called Polarissima Borealis to distinguish it from Polarissima Australis (NGC 2573), which was found by John Herschel in 1837.

Table 3.6. *Types of the 466 objects discovered by John Herschel*

Type	Number	Remarks
Galaxy	329	Brightest: h 114 = NGC 29 (9.4 mag, Cas)
Open cluster	50	Brightest: see Table 3.7
Globular cluster	1	h 2029 = NGC 6749, 15.7.1827 (Aql, 12.4 mag)
Emission nebula	4	
Planetary nebula	1	h 2038 = NGC 6785, 21.5.1825 (Aql, 12.3 mag)
Single star	9	Brightest: h 179 = 50 Cas (NGC 771), 4.0 mag
Stars (pair to quartet)	16	
Star group	42	h 2024 = NGC 6724, found with 5" refractor
Not found	14	

Table 3.7. *Special objects and observational data of John Herschel's Slough catalogue*

Category	Object	Data	Date	Remarks
Brightest object	NGC 1981 = h 362	4.2 mag, OC, Ori	4.1.1827	
Faintest object	NGC 5441 = h 1740	15.6 mag, Gx, CVn	11.3.1828	
Most northern object	NGC 3172 = h 250	+89°, Gx, 14.4 mag, CMi	4.10.1831	Polarissima Borealis
Most southern object	NGC 823 = h 196	−25°, Gx, 12.8 mag, For	14.10.1830	
First object	NGC 7010 = h 2100	Gx, 13.0 mag, Aqr	6.8.1823	Sweep 53
Last object	NGC 3996 = h 1032	Gx, 13.9 mag, Leo	23.4.1832	Sweep 419; with h 1044 = NGC 4019 (Gx, Com)
Best night			22.11.1827	16 objects
Best constellation	Virgo			50 objects (48 galaxies)

Figure 3.6. Polarissima Borealis NGC 3172 (h 250); 'A' is the galaxy of 15.5 mag MCG 15–1–10 (DSS).

Figure 3.7 shows the brightness distribution of John Herschel's objects. All were observed with his 18¼" reflector, which offered a lowest magnification of 180. The average is 13.0 mag, a magnitude fainter than his father's (12.0 mag). Most of the northern-sky objects had already been found by the latter. The value is comparable to that obtained by Tempel (13.1 mag), who used an 11" refractor. It should be noted that John Herschel's magnitudes (mostly concerning stars) do not correspond to modern values.[17] An example is the globular cluster M 53 (h 1558), where he speaks about 'Stars very small, 12…20 m' (his magnitude '20' is roughly equal to 14 mag).

[17] See the comparison between the scales of John Herschel and Argelander (W. Struve) in Bigourdan (1917b: E144).

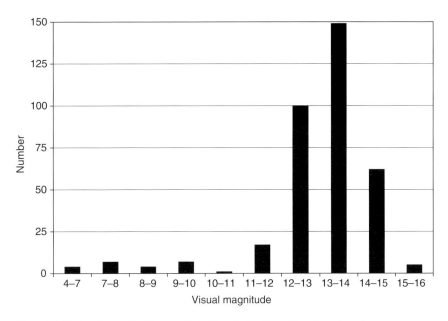

Figure 3.7. The brightness distribution of John Herschel's objects (Slough catalogue).

3.4 ADDITIONS AND DRAWINGS

Following the catalogue part is a list of errors added in proof. A 'new' object is presented too, which John Herschel forgot to enter: II 799 (NGC 5443), a galaxy in Ursa Major, which was discovered by William Herschel on 14 April 1789 in Slough. However, 'sweep 546' (noted for John Herschel's observation) cannot be true: the corresponding date (21 February 1835) is two years later than the publication of the SC. It probably must read 'sweep 346' (2 May 1831), because during that night he observed the nearby objects h 1736 = I 230 (NGC 5422) and h 1748 = I 231 (NGC 5473). In the General Catalogue the added object appears as GC 3758, assigned 'HON' ('Herschel omitted nebula'). This is confusing, because John Herschel has used this abbreviation in the GC also for those objects of his father which were added in the Cape catalogue. Dreyer distinguishes between 'HON' and 'hon'; thus in the New General Catalogue one reads for NGC 5443 the following: GC 3578 = h 1743A = II 799 = hon. His designation 'h 1753A' refers to the fact that the object (due to its right ascension) must follow h 1743 in the Slough catalogue. It is interesting that the SC ends with entry no. 2306 and the Cape catalogue (its successor) starts with no. 2308. Herschel has reserved no. 2307 for the missing object, writing in the Cape catalogue that the numbers were *"continued onward from No. 2307, the 'omitted' nebula added after the 2306 regularly entered in the order of R.A., in the Northern [Slough] Catalogue."*[18]

The SC appendix starts with a description of the sweeping method and the measurement of positions, but the main part is occupied by John Herschel's drawings plus explanations. The first table (Plate IX) contains 24 sketches, which demonstrate the meaning of 'brightness', 'central brightness', 'gradations of brightness from inwards' and 'elongation of form'. Instead of real objects, an illustrating sequence is shown; e.g. for 'brightness' (Figs. 1 to 6) eF–vF–F–pB–B–vB.

The remaining eight tables (Plates IX to XVI) show altogether 67 objects (Figs. 25 to 91), drawn in black on white. As was later noted by Chambers, all of the figures are mirror-inverted.[19] The associated text explains selected objects or classes (Table 3.8). Here one finds, for instance, a remark about the origin of the popular name Black Eye for the galaxy M 64 in Coma Berenices (Fig. 27): *"It was however seen by my Father, and shown to the late Sir Charles Blagden, who*

---

[18] Herschel J. (1847: 2).
[19] Chambers (1867: xi); Chambers presents a selection of John Herschel's drawings (in normal orientation and white on black).

Table 3.8. *Descriptions of figures concerning individual objects and classes*

Fig.	h	Object	NGC	Date	Type	Remarks
25	1622	M 51	5194/5	26.4.1830	Gx	Whirlpool Nebula
26	2060	M 27	6853	7.8.1828	PN	Dumbbell Nebula
27	1486	M 64	4826	21.4.1833	Gx	Black Eye
28	218	V 19	891	18.9.1828	Gx	
29	2023	M 57	6720	31.7.1829	PN	Ring Nebula
33	2088	V 15	6960	5.8.1829	EN	Veil Nebula
35	2008	M 17	6618	19.7.1828	EN	Omega Nebula
37	1357	V 24	4565	28.3.1832	Gx	
43	2050	IV 73	6826	8.9.1829	PN	Not h 2051
44–47					PN	'Planetary nebulae'
50–67					Gx, PN	'Long nebulae'
68–79					Gx, PN	'Double nebulae'
80, 82–85					Gx, EN	'Peculiarity'
81, 86–91					GC, OC, Gx, EN	'Clusters of stars'

likened it to the appearance of a 'black eye', an odd, but not inapt comparison."[20] William Herschel even gives the expected connection with fighting (sweep 699, 13 February 1787): '*contains one lucid spot like a star with a small black arch under it, so that it gives one the idea of what is called a black eye, arising from fighting.*'[21] John Herschel's idea about the nucleus of M 64 is interesting: '*I am much mistaken also if the nucleus of Messier's 64th nebula (which is a very large and bright object) be not a close double star; and what renders this (if verified by further observations) particularly remarkable, is the immediate neighbourhood of the nucleus, of a considerable vacuity or space comparatively free from nebulous matter, as if that portion had been absorbed into the star.*'[22]

John Herschel created the name Omega Nebula for M 17 in Sagittarius (Fig. 35): '*The figure of this nebula is nearly that of a Greek capital omega Ω, somewhat distorted and very unequally in light.*'[23] In the cases of the edge-on galaxies NGC 4565 in Coma Berenices and NGC 4594 in Virgo (M 104; Fig. 50) he saw a '*parallel appendage*'. The detached appearance is due to a strong absorption lane. For NGC 6826 in Cygnus (Fig. 43) he supposed (following his father) there to be a link between planetary nebulae and nebulous stars: '*This remarkable object, as my father rightly observes, appears to constitute a connecting link between the planetary nebulae and nebulous stars.*'[24] In John Herschel's 'List of figured nebulae' one reads 'h 2051', which should be 'h 2050'.

Herschel gives examples for five object classes. In the case of 'planetaries', NGC 7009, NGC 7662, NGC 6818 and NGC 6905 are shown. For a few he mentions 'satellites', which are faint stars in the vicinity. Already his father had mentioned them. William Herschel reported about a visit of Jean Dominique Cassini to Slough[25] during which the planetary nebula IV 18 (NGC 7662) was observed on 27 November 1787: '*Mr. Cassini observed that a very small fixt star nf the nebula appeared not unlike a satellite to it*'.[26] John Herschel's examples are h 2047 (NGC 6818) '*exactly like a planet and two satellites*', h 2075 (NGC 6905) '*it has four small stars near it like satellites*', and h 2241 (NGC 7662), which contains a double star.

---

[20] Herschel J. (1833a: 498); Blagden was Secretary of the Royal Society.
[21] Dreyer (1912a, vol. 2: 658).
[22] Hoskin (1987: 15).
[23] Herschel J. (1833a: 498).
[24] Herschel J. (1833a: 499).
[25] A few months earlier Lalande had visited Slough (8 August 1787); Herschel showed him the PN NGC 6818 (IV 51).
[26] RAS Herschel W. 2/3.7; 'nf' means 'north following'.

Table 3.9. *Drawings of double nebulae in the Slough catalogue*

Fig.	H	NGC	Date	Remarks
68	1252	4395, 4401	29.4.1827	V 29.1+2; NGC 4401 = H II region in NGC 4395 (Table 2.22)
69	1202	4303 (M 61)	7.4.1828	Single galaxy; Herschel: 'bicentral'
70	604	2903, 2905	24.2.1827	NGC 2905, H II region in NGC 2903; Herschel: 'double nebula'
71	1146	4214	11.3.1831	= galaxy NGC 4228 (h 1157); Herschel: 'double nucleus'
72	444, 445	2371, 2372	23.12.1827	Bipolar PN
73	2197, 2198	7443, 7444	9.9.1825	Double galaxy (separation 1.5′)
74	1405, 1408	4647, 4649 (M 60)	10.3.1826	Double galaxy (2.5′); Arp 116
75	1414, 1415	4656, 4657	10.4.1831	Double galaxy (3′)
76	1391, 1397	4627, 4631	29.4.1827	Double galaxy (2.2′); Arp 281
77	1905	5857, 5859	23.3.1827	Double galaxy (1.9′)
78	1358, 1359	4567, 4568	2.5.1829	= h 1363 (see Section 3.2); Siamese Twins (1.2′)
79	934, 936	3799, 3800	3.4.1826	Double galaxy (1.5′); Arp 83

John Herschel expected an 'orbital motion', reminding himself that *'these satellites of planetary nebulae ought to be especially attended to'*. The issue was later treated by Olbers and d'Arrest (see Section 6.14.2). Of course, no motion could be detected. In all cases the 'satellites' are mere stars in the line of sight, not associated with the nebula.

Seventeen of the 18 'long nebulae' are galaxies, among them M 65, M 66, M 104 and M 106. The remaining object is NGC 2261, the cometary reflection nebula in Monoceros, now known as Hubble's Variable Nebula. John Herschel shows six drawings of *'nebulae which offer some remarkable peculiarity of situation with respect to stars'*. Two are emission nebulae (M 20, NGC 6995), the rest are galaxies. The seven 'clusters of stars' are the globulars M 2, M 5, M 13 and M 30, the open cluster NGC 2304, the emission nebula M 1 (Crab Nebula) and the galaxy NGC 5964.

The 12 'double nebulae' are listed in Table 3.9. Herschel suspected (following his father) an analogy to double stars; of course, orbital motion was never detected. Mädler later counted the multiple nebulae contained in the Slough catalogue: 146 are double, 25 triple, 10 quadruple, 1 quintuple and 2 sextuple.[27]

3.5 OLBERS' REVIEW OF THE SLOUGH CATALOGUE

Olbers wrote a review, which was published in May 1834 in the *Astronomische Nachrichten* (Olbers 1834). It starts with the statement *'This is the long announced and expected treatise of the worthy son of his great father on nebulae and clusters; certainly one of the most important writings about these interesting but partly mysterious objects of the starry sky.'* About Herschel's original goal of observing all of his father's objects at least twice, Olbers remarks that, because *'the want for a comprehensive list of nebulae, ordered by right ascension, facing the improvement of achromats and the increase of astronomical activity to find and observe comets, has become so strong, he [John Herschel] has decided to publish his observations already now'*. One further reads that, *'If, however, we cannot get all [...], the gift presented now is already very rich.'* After a short review of the content of the catalogue, he praises Herschel's drawings, which are *'of such excellence as one is able to desire'*.

The *'extremely irregular distribution of the nebulae and clusters on the sphere'* (already seen by William Herschel) is remarkable. He adds that *'Eagerly one awaits, how Sir John's Durchmusterung of the southern sky will confirm or modify this peculiar behaviour.'* In this connection he mentions Dunlop's work, especially

---
[27] Mädler (1849: 466).

his drawings. Concerning those made by William and John Herschel, Olbers remarks that *'Only a few [sketched objects] are common; and there are cases where the son has noticed some peculiarities missed by the father.'* About the nature of the nebulae, he agrees with William Herschel *'That the largest part of those nebulae, previously not resolved into stars by telescopes, are but remote clusters is certain. Regardless, there are indeed many indications for true, partly huge light-nebulae, not containing stars but probably associated with them.'*

Olbers was strongly interested in John Herschel's 'satellites' of planetary nebulae: *'the figures show that some have faint stars in the vicinity, supporting the idea of orbiting companions.'* Olbers, as the discoverer of the second and fourth of the minor planets, was obviously fascinated by this idea, quoting Herschel: *'Such they may possibly be. The enormous magnitude of these bodies, and consequent probable mass (if they be not hollow shells), may give them a gravitional energy, which, however rare we may conceive them to be, may yet be capable of retaining in orbits, three or four times their own diameter, and in periods of great length, small bodies of a stellar character.'*[28] Olbers speculates that, by virtue of the large mass of a planetary nebula needed to keep the satellites, and the assumption that *'according to Newton the light has something material'*, the *'velocity of the emitted light* [might be] *strongly reduced.'* This would lead to visible effects: planetary nebulae *'must show also a much larger aberration than the surrounding stars, which should be easily detected by a few observations.'*

For Olbers the considerable amount of *'double nebulae'* is remarkable. Like Herschel he supposes there to be parallels to double stars: *'All the differences in separation, position and brightness found for double stars are also present for nebulous stars* [double nebulae]*.'* He concludes that a connection between the two different classes of objects is highly probable. Olbers laments that John Herschel did not publish his list of 'missing nebulous stars', i.e. objects of his father, which were not found. Olbers supposes that some short-period comets (like Encke's) could be among them. Such early observations would be of great value to improve their orbital parameters. Moreover, he criticises the fact that Herschel has not published the dates of his 427 sweeps. Once again, the comet expert is speaking, when he guesses that *'among the 500 new* [nebulae] *there might have been some comets'*.[29] Finally, Olbers wishes Herschel all the best for his observations at the Cape of Good Hope: *'May this praiseworthy expedition, made with sacrificing ardency, be fully crowned with the intended success, and may this astronomer of outstanding merit return healthy and lucky to perfect and finish his work that has even now for ever saved the name Herschel the most grateful admiration by present and future generations.'*

---

[28] Olbers has translated Herschel's text from the Slough catalogue; here the original is given (Herschel J. 1833a: 500).

[29] Probably NGC 1170, found in 1869 by Peirce, was a comet tail (see Section 9.4.6). Moreover, three objects in Swift's lists of nebulae might be comets (not entered in the NGC; see Section 9.5.4). Finally, IC 2120, discovered in Auriga by Bigourdan in 1890, most probably was comet 113P/Spitaler (Skiff 1996). A counter-example: in the Cape catalogue John Herschel described NGC 1325 (h 2534) as a *'miniature of Halley's* [comet]*'* and added a sketch (Plate VI, Fig. 17). However, this is an elongated galaxy with a superimposed star on one side, appearing like the head and tail of a comet.

4 · Discoveries made in parallel with John Herschel's Slough observations

Like his father, John Herschel dominated, with respect to the number and quality of his observations of nebulae and star clusters, a whole epoch. While he was working in Slough there was hardly any competition in this field – at any rate, concerning the northern sky. The yield of other observers was poor: 466 new objects of John Herschel stand against 9 discovered by others. Those are connected with the names of Karl Ludwig Harding, Friedrich Georg Wilhelm Struve and Niccolò Cacciatore. Fortunately there were objects not searched for by Herschel, such as minor planets and comets – here the chance for honour was considerably greater.

However, the situation for the southern hemisphere was different. Around 1826 James Dunlop, observing in the small Australian town of Paramatta, catalogued no less than 629 nebulae and star clusters. John Herschel, whose observations at the Cape of Good Hope did not start until 1834, was beaten by eight years.

4.1 HARDING'S LIST OF NEW NEBULAE

In 1824 Harding published a list of eight new nebulae, hardly attracting any interest – even John Herschel was not aware of it. The identity of these objects was not clarified until Winnecke's analysis in 1857. This also showed the difficulty of comparing different observations made during Harding's time: only a few astronomers noticed the few publications about non-stellar objects and therefore observations to verify the discoveries were seldom made.

4.1.1 Short biography: Karl Ludwig Harding

Karl Ludwig Harding, of English descent, was born on 29 September 1765 in Lauenburg, Germany (Fig. 4.1). As a theologian, he taught the son of Johann Hyronimus Schröter in Lilienthal and came into contact with

Figure 4.1. Karl Ludwig Harding (1765–1834).

astronomy there.[1] Schröter, possessing large telescopes made by William Herschel, primarily observed the Moon.[2] From 1800 until 1809 Harding was superintendent at the observatory. On 1 August 1804 he discovered the third minor planet, Juno, winning great fame. A year later he became Professor at Göttingen University (second to the famous Gauß). There Harding discovered the variable star R Virginis in 1809; three comets are due to him too.

---

[1] For Lilienthal Observatory and its astronomers see Drews and Schwier (1984) and Witt (2007a).
[2] The largest was a reflector with focal length 8.5 m, which was delivered in 1793.

Table 4.1. *Harding's eight nebulae (bold = discovered NGC object)*

No.	Date	Discoverer	Identity	NGC	Type	V	Con.	Remarks
1	Feb.? 1824	Herschel 1791	VI 37	2506	OC	7.6	Mon	
2	July? 1823	Méchain 1782	VI 40 = M 107	6171	GC	7.8	Oph	W. Herschel 1793, Capocci 1826
3	July? 1823	Messier 1764	M 14	6402	GC	7.6	Oph	
4	Sept.? 1823	Harding		**7293**	PN	7.3	Aqr	Au 48; Helix Nebula, Capocci 1824, C. H. F. Peters 1849?
5	Feb.? 1824	Herschel 1784	VI 6	2355	OC	9.7	Gem	
6	Sept.? 1823	Herschel 1784	VIII 20	6885	OC	8.1	Vul	NGC 6882 (W. Herschel VIII 22)
7	Sept.? 1823	Harding	h 2048	6819	OC	7.3	Cyg	C. Herschel 12.5.1784, J. Herschel 31.7.1831
8	Sept.? 1823	Herschel 1790	VII 59	6866	OC	7.6	Cyg	

In 1822 Harding published his *Atlas Novus Coelestis*, containing accurate positions of more than 40 000 stars; a second edition appeared in 1856 in Leipzig (published by Gustav Adolph Jahn). The charts cover the region around the ecliptic, where most minor planets were found. Karl Ludwig Harding lived in Göttingen until his death on 31 August 1834 at the age of 68.[3]

4.1.2 Harding's list of nebulae

On 28 April 1824 Harding sent a list of eight nebulae (Table 4.1) to Bode in Berlin, who published it in the *Astronomisches Jahrbuch*[4] (Harding 1824). The author briefly noted that *'During my surveys of the sky I found the following nebulae, which seem to be still unknown'*.

Harding does not give any details about his telescope or the dates of his observations. At that time he was working at Göttingen University Observatory and probably used an 8.5″ reflector. This telescope was built by William Herschel and installed at the (old) observatory by the master himself in July 1786 (Fig. 4.2).[5] Until 1822 Harding mainly determined star

Figure 4.2. Harding's 8.5″ reflector erected in 1786 by Herschel (Grosser 1998).

positions for his sky atlas with the Reichenbach meridian-circle with aperture 10 cm. Particularly his discovery of NGC 7293, a large planetary nebula with low surface brightness, makes it likely that he used the 'Herschel' for his observations, instead of the 'Reichenbach'.

Only one of Harding's objects was new: no. 4 = NGC 7293, the famous Helix Nebula in Aquarius

---

[3] Obituary: Anon (1835); see also Wittmann (2004).
[4] Also often called the *Berliner Jahrbuch*.
[5] Grosser (1998: 44).

4.1 Harding's list of new nebulae 65

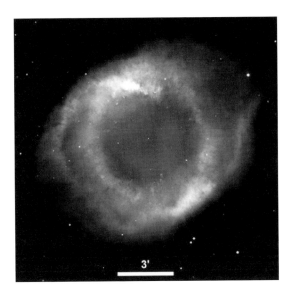

Figure 4.3. The large Helix Nebula NGC 7293 in Aquarius.

(Fig. 4.3), which he found in autumn 1823.[6] NGC 6819 (no. 7) had been discovered on 12 May 1784 by Caroline Herschel (see Table 2.12). Independently, John Herschel found this open cluster on 31 July 1831, so he entered it in the SC as 'Nova' h 2048. But he missed NGC 7293. Considering a size of almost 18′, the nebula might have been too large for the 15′ field of view of his reflector. Herschel did not know Harding's list. He also missed a note by Ernesto Capocci,[7] Director of Capodimonte Observatory in Naples, in the *Astronomische Nachrichten* (Capocci 1827). Capocci describes an object in Aquarius, found in late 1824: '*Very large nebula, faintly visible since 1824*'. He used the 17.6-cm Fraunhofer refractor (mounting by Reichenbach), which had been erected in 1812. However, Capocci's position is wrong; therefore d'Arrest could not find the object during his Leipzig observations.[8] D'Arrest erroneously suspected an identity with II 1 (NGC 7184).

---

[6] The name was created by Heber Curtis, writing in September 1912: "*I would suggest this interesting object be refered to as 'The Helical Nebula in Aquarius'.*" (Curtis 1912b). However, he classified NGC 7293 as a 'spiral nebula'. NGC 6543 was later described as a 'helical nebula' too (Curtis 1918b).
[7] Obituary: Peters C. H. F. (1864).
[8] d' Arrest (1856a: 352).
[9] Owing to this fact and Herschel's nescience, the Helix Nebula is not mentioned in Smyth's popular observing guide (Smyth 1844).

It took over 30 years until Harding's list was checked by Winnecke (Winnecke 1857).[9] He used coordinates for William Herschel's objects that had been reduced by his friend Arthur Auwers and made available to him in 1856 in Göttingen. Winnecke wrote that Auwers '*for some time past has been dealing with the cataloguing of Herschel's sky surveys, available in a pretty unusable form only, and has nearly finished the task*' (see Section 6.19). He states that no. 3 is identical to M 14 and that nos. 1, 2, 5, 6 and 8 had already been catalogued by William Herschel. By the way, Auwers overlooked the identity of no. 6 (NGC 6885) with Herschel's VIII 20, noting only '*no nebula; only 3 faint stars close together*'.[10] But this fits well with Herschel's description of a '*cluster of course scattered stars, not rich*'. VIII 20 is identical with VIII 22 (NGC 6882), which was discovered only one day later. In the case of no. 2 (NGC 6171) Winnecke was not aware that the object had been found by Méchain in 1782 (it was later called M 107); Capocci had seen it, too, in June 1826 (Capocci 1827). Object no. 7 (NGC 6819) was identified by Winnecke with John Herschel's 'Nova' h 2048, but this did not clear up the correct priority.

Only no. 4 (NGC 7293) was left to Harding by Winnecke (again not mentioning Capocci). His observation with the 7.8-cm Fraunhofer comet-seeker at Bonn Observatory showed it as a '*very large, but easily visible nebulous mass, lying between several stars*'. Auwers saw the nebula on 28 September 1858, using the 6″ heliometer at Königsberg Observatory: '*with a magnification of 45, and very near the horizon, appearing incredibly large (diameter 12′), similar to the nebula h. 2060 [M 27]; both heads are considerably stretched in PA about 115°. A fair number of stars are projected particularly on the preceding part of the southern head.*'[11] Further observations were made on 16 August and 12 November 1860. During the latter night, Auwers determined the size with the micrometer '*by placing it in the middle of the ring [...] the larger axis measures around 14′, the smaller 11–12′*'.[12]

For Auwers the Helix Nebula was an example to demonstrate that '*large, diffuse, faint objects can be detected more easily with small instruments than with big ones*'.[13] He saw

---

[10] Auwers (1862a: 76).
[11] Auwers (1865: 211).
[12] Auwers (1865: 218).
[13] Auwers (1862b); see also the discussion on NGC 1333 in Section 6.15.2.

it '*in a 2ft comet-seeker with 15-times magnification bright and conspicuous [...] but direct after that* [he could] *perceive it in a 6ft Fraunhofer with a power of 42 only by moving the telescope*'. This method is known as 'field sweeping'.[14] In 1862 Auwers entered the object as no. 48 in his list of new nebulae. It was due to this publication that John Herschel learned about the nebula – but he never observed it. The object was entered in the General Catalogue of 1864 as GC 4795, but is not assigned as 'planetary'.

The fact that NGC 7293 was not found by both Herschels, led to two other, curious discovery reports. The first is due to C. H. F. Peters of Dudley Observatory (Albany). He had sent a list of seven nebulae to Benjamin Gould, editor of the *Astronomical Journal*.[15] In the publication Gould noted these as nebulae '*observed by him* [Peters] *at Capodimonte in Naples, and which do not appear in the books*'. The last object ('*very large, preceding υ Aquarii*') is NGC 7293 – of all things, found at Capodimonte! The second report is from Jermaine Porter, of Cincinnati Observatory, who found the object on 18 October 1884: '*Although easily seen with a four-inch glass, it is not given by the Herschels*' (Porter 1884) – ignoring the GC. During his investigation he eventually noticed the publications of Harding, Capocci and Winnecke. Owing to the fact that d'Arrest in Leipzig had looked for the object in vain, Porter guessed that he '*looked for the nebula with too large power. This may also account for the absence from the catalogues of the Herschels.*' In a letter of 19 December 1884 Lewis Swift wrote '*The large bright nebula to which Professor Porter calls attention [...] I have been acquainted with for many years*' (Swift 1884a). He independently discovered the object in about 1858 with his 4.5" refractor in Rochester: '*I ran upon this object, and, finding it was not a comet, I supposed, of course, that an object so conspicuous was well known. The first intimation I had to the contrary, was in a communication from Dr. Peters.*'

4.2 WILHELM STRUVE: NEBULAE IN THE *CATALOGUS NOVUS*

In 1827 Wilhelm Struve published a catalogue of double stars, which has a remarkable appendix. Here nine nebulae found by him in Dorpat are listed; however, only two are new. John Herschel noticed the list with much interest. He observed the objects in Slough and listed the two new ones in his catalogue – with confusing designations.

4.2.1 Short biography: Friedrich Georg Wilhelm Struve

Wilhelm Struve was born on 15 April 1793 in Altona (near Hamburg) as son of the Rector of the Christianeum Acadamy, where he later attended school (Fig. 4.4). From 1808 he studied in Dorpat (Tartu), Estonia; first philology, then changing in 1811 to astronomy. Struve's teacher was Johann Sigismund Huth, Director of the University Observatory.[16] As early as in August 1811 he verified the orbital motion of the double star Castor, which had been suspected by William Herschel. Two years later, Struve wrote his thesis on the geographical location of the observatory, which had been erected in 1808 (Oestmann 2007). Afterwards he became Professor of Mathematics and Astronomy and in 1818, after Huth's death, he was appointed Director of Dorpat Observatory.

Struve was a protagonist in the race for the first stellar parallax, together with Bessel and Henderson (Hirshfeld 2001). In the years 1835–38 he measured the position of Vega with high precision, leading to a parallax of 0.261″. Parallel to this work, he realised his plan of a complete measurement of all known northern double stars. He eventually published the results in 1837 in St Petersburg, titled *Mensurae Micrometricae Stellarum Duplicium*. For his first series of observations (*Catalogus Novus*), he had already received the gold medal of the Astronomical Society of London in 1826.[17]

In spring 1839 Struve left Dorpat (his successor became Johann Heinrich Mädler), having been appointed Director of the new Pulkovo Observatory near St Petersburg, which was built according to his plans. There he erected the famous 38-cm Merz refractor in the same year.[18] Owing to ill heath he had

---

[14] Steinicke (2004a: 174).

[15] Peters C. H. F. (1856); see also Section 9.2.2.

[16] Huth's only contribution on non-stellar objects was his note 'An Nebelflecken' ['On nebulae'] in the *Berliner Jahrbuch* (Huth 1804).

[17] He received the award together with two other eminent double-star observers, John Herschel and James South (Dreyer and Turner 1923).

[18] It remained the largest until 1847; then it was beaten by the Merz refractor at Harvard College Observatory, with a lens 1.5 cm larger.

4.2 Wilhelm Struve: the *Catalogus Novus*

Figure 4.4. Wilhelm Struve (1793–1864).

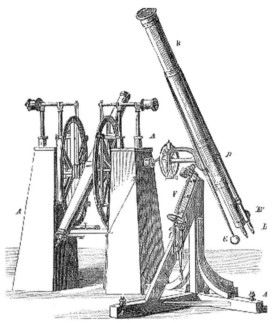

Figure 4.5. Struve's 24.4-cm Fraunhofer refractor and the Reichenbach & Ertel meridian-circle.

to retire in 1862. Wilhelm Struve died on 23 November 1864 in Pulkovo at the age of 71. He was married twice, with altogether 18 children – the origin of a dynasty of astronomers (Batten 1977). His son Otto succeeded him as Director of Pulkovo Observatory.[19]

4.2.2 The *Catalogus Novus*

Struve's main focus was the observation of double stars. In 1822 he published his first catalogue with 795 objects. He was in competition with John Herschel and James South.[20] In December 1824 Struve got the 24.4-cm Fraunhofer refractor (Fig. 4.5).[21] The superior instrument, characterised by John Herschel as '*probably the very best refracting telescope, ever made*' (Herschel J.

---

[19] Obituaries: Pritchard (1864) and Winnecke (1865); see also Ashbrook (1984: 20–26) and Batten (1988).
[20] In 1816 John Herschel began to observe double stars, mainly studying objects discovered by his father. During 1821–23 Herschel and South made measurements with Dollond refractors of aperture 9.5 cm and 12.7 cm, which they published in a common paper (Herschel J. and South 1824).
[21] Repsold J. (1908, vol. I: 108–109).

1826c), became the largest in the world, surpassing the 'Fraunhofer' at Capodimonte.

After Struve had received the telescope, packed in 22 boxes, he erected it single-handed. On Christmas Eve 1824 the refractor was operational (Struve F. G. W. 1826a, b). It offered magnifications from 175 to 750 and was equipped with an outstanding Fraunhofer micrometer. The humble Dorpat Observatory became a first-ranked scientific institution – an astonishing development for a one-man business. It was not until May 1827 that Struve got an assistant, Ernst Wilhelm Preuß, who mainly observed with the meridian-circle by Reichenbach & Ertel, which had been erected in 1821.

As we know from a letter written by Struve on 20 March 1825, he immediately planned a '*new survey of the sky relating to double stars*' (Struve F. G. W. 1826c, d). Using the refractor at a standard power of 214, all known pairs (with distances below 32″) between the pole and –15° declination should be measured. Struve was a scrupulous worker with great endurance. Once he observed for eight hours without any break, withstanding a temperature of –25 °C. Only two years later Struve could publish his *Catalogus Novus Generalis Stellarum Duplicium et Multiplicium* (Struve F. G. W.

Table 4.2. *Objects found by Wilhelm Struve in Dorpat*

No.	Discoverer	Id	NGC	h	Au	Type	V	Con.	Remarks
1	Le Gentil 1749	M 32	221	51		Gx	7.6	And	
2	W. Struve 1825		**629**		16	Star group		Cas	
3	W. Herschel 1801	I 286	3077	658		Gx	10.0	UMa	
4	Bode 1775	M 53	5024	1558		GC	7.7	Com	
5	W. Struve 1825		**6210**	1970		PN	8.8	Her	J. Herschel: Struve
6	W. Struve 18.7.1825		**6572**	2000		PN	8.1	Oph	Bessel 1822; J. Herschel: Struve
7	W. Struve 1825		**6648**		41	2 stars		Dra	Σ 2332 (sketch)
8	W. Herschel 1782	IV 1	7009	2098		PN	8.0	Aqr	Saturn Nebula
9	Maraldi 1746	M 15	7078	2120		GC	6.3	Peg	
Σ 131	Méchain 1781	M 103	581	126		OC	7.4	Cas	J. Herschel: Struve
Σ 885	W. Struve 1825		**2202**	385		OC		Ori	J. Herschel: Struve
Σ 2630	W. Struve 1825		**6871**	2067		OC	5.2	Cyg	J. Herschel: Struve
Σ 2631	W. Struve 1825		**6873**	2068		Star group		Sge	J. Herschel: Struve

1827); due to his philological passion it was written in Latin. The catalogue lists 3063 double/multiple stars, designated 'Σ' (2343 were new).[22]

4.2.3 The appendix: 'Nebulae detectae'

For those interested in nebulae, the appendix was the most interesting part. The list 'Nebulae detectae' contains nine objects with positions for 1826, numbered 1 to 9 (Table 4.2). Struve found them during his regular double-star observations. Alas, he does not give the dates, which might lie between 1825 and 1826. Struve discovered seven NGC objects (marked bold); four are non-stellar (a moderate rate).

Being an experienced double-star observer, John Herschel delved into Struve's *Catalogus Novus*. Of course, the appendix attracted his attention too, leading him promptly to check the nebulae with his reflector in Slough. He entered two of them in his catalogue: no. 5 (h 1970) and no. 6 (h 2000), both planetary nebulae. Unfortunately he called them 'Σ 5' and 'Σ 6', respectively. The designation 'Σ' should have been used exclusively for Struve's double stars, listed in the *Catalogus Novus*. Catalogue and appendix must be treated independently. Actually, there the catalogue contains a no. 5 and no. 6 too – the very same objects as Σ 5 and Σ 6. Especially treacherous is that the 'nebula' no. 7 of the appendix again appears in the main catalogue as double star Σ 2332! Vice versa, some catalogue objects are connected with nebulosity (the last four in Table 4.2).

Both SC objects, Struve's nos. 5 and 6, look stellar in small telescopes. This is the reason why they appear in early star catalogues, albeit without mention of their nebulosity. Object no. 5 (NGC 6210) is listed in Lalande's *Histoire Céleste* of 1801 as 8-mag star LL 30510 and no. 6 (NGC 6572) was measured on 26 June 1822 by Bessel in his zone-observations.[23] The non-stellar character was first announced by Struve and Herschel. Object no. 6 was later observed by Auwers, who measured the distance (N–*) between the nebula and a nearby Lalande star (Auwers 1862c). This was recorded by Schmidt too, who observed the object in Athens in 1860 (Schmidt 1861a). Auwers determined

---

[22] See also Struve's summary of 1828 in the *Edinburgh Journal of Science* (Struve F. G. W. 1828).

[23] NGC 6572 appears in zone 92 of Bessel's '8. Abtheilung' of the Königsberg observations (Bessel 1823: 107–108).

the diameter with the Königsberg heliometer, getting 14.6″ for no. 5 (5 August 1859) and 6.24″ for no. 6 (25 September 1859).[24]

What about the other Struve objects? Two (nos. 2 and 7) were ignored by John Herschel, being a star group and a star pair, respectively. Both were catalogued by Auwers, who was unable to see them with the Königsberg heliometer: NGC 629 (no. 2) was observed on 19 and 24 February 1861; NGC 6648 (no. 7) on 19, 21 and 24 February. Auwers suspected that Struve's positions were wrong.[25] Strangely, Struve did not notice that such prominent objects as M 32 (the companion of the Andromeda Nebula) and the globular clusters M 15 and M 53 are in his list. Obviously he was concentrating then on double stars – nebulae were mere by-products. Herschel might have been aware of their identity, but did not mention it in the Slough catalogue.

Four other Struve-numbers are in the SC, designated 'Σ' with some right. The objects are clusters or groups of stars. Why do they not appear in the appendix? The reason is simple: it is not the cluster itself which is meant, but a double star inside, which was entered by Struve in the *Catalogus Novus*. On the other hand, the 'Nebulae detectae' in the appendix describe unknown nebulous objects accidently discovered in the vicinity of double stars. However, no. 7 is a mixed case, sketched by Struve: a double star (Σ 2332) with a nebulous envelope (Fig. 4.6). The nebulosity does not exist; such an appearance was often suggested around groups or pairs of stars (and even single stars).

By the way, Struve visited John Herschel on 19 April 1830, both travelling from London to nearby Slough. There, in the evening dawn of that day, Herschel showed him the planetary nebula no. 6 = h 2000 in the 18¼″ reflector. He had first observed the object on 9 May 1828.

4.3 CACCIATORE AND HIS NEBULA

In 1826 Niccolò Cacciatore found a nebula in Corona Australis: the 6.3-mag globular cluster NGC 6541. The story of its discovery is curious, demonstrating some typical problems of the time.

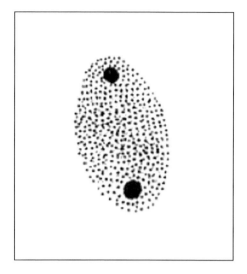

Figure 4.6. Struve's sketch of his object no. 7 = NGC 6648 (Struve F. G. W. 1827).

4.3.1 Short biography: Niccolò Cacciatore

Niccolò Cacciatore (Fig. 4.7) was born on 26 January 1780 in Casteltermini, Sicily. At the age of 17 he went to Palermo, first studying Greek, later mathematics and physics. Giuseppe Piazzi, Director of the local observatory, noticed him and Cacciatore was his assistant from 1800 to 1826 (with Piazzi calling him 'Don Nicola'). In 1811 Cacciatore became Professor of Astronomy and Geodesy. His main task was the determination of star positions for the Palermo catalogue, which was published in 1814. After 1807 Piazzi's eyesight deteriorated due to illness and Cacciatore did the bulk of the work. When Piazzi changed to Naples (Capodimonte) in 1817, Cacciatore became Director of Palermo Observatory. When it was partly destroyed during the revolution of 1820, he managed the rebuilding. He is also known for the controversial 'lost planet', a moving object found by him in May 1835 (Olbers 1836).[26] He married Emmanuela Martini in 1819. Five children were born, among them Gaetano Cacciatore, who succeeded him as Observatory Director after his death. Niccolò Cacciatore died in Palermo, weakened by cholera, on 28 January 1841 at the age of 61.[27]

---

[24] Auwers (1865: 194–195).
[25] Auwers (1865: 227).

[26] The object was thought to be the minor planet (7) Iris, which was later discovered by Hind, but Hind would disprove this in 1847.
[27] Anon (1844a).

Figure 4.7. Niccolò Cacciatore (1780–1841).

Figure 4.8. The 7.5-cm Ramsden refractor in Palermo (Witt 2007a).

4.3.2 The nebula in Corona Australis

The first to announce the discovery of a nebula by Cacciatore in Palermo was Baron von Zach in his journal *Correspondance Astronomique* (Zach 1826). Cacciatore's own report, written in Italian, was printed too. It starts with the sentence *'The nebula, which I found in the morning of 20 March deserves some attention.'* It must be noted that 20 March 1826 is mentioned here. The object was *'visible in the telescope in spite of moonlight and distinguishable in the cross-wire eye-piece'*. This suggests that Cacciatore found the nebula with the azimuthal 7.5-cm refractor by Ramsden used for the Palermo catalogue (Fig. 4.8).[28]

In a letter to the editor of the *Astronomische Nachrichten* (Schumacher), written on 29 July 1826, Olbers presented the discovery too, but dated it to 19 March (Olbers 1827). He judged Cacciatore's report (which is not added) not very convincing. Olbers suspected a comet, asking whether the nebula actually was at the given place before. He also missed a statement about its brightness compared with *'the magnitude of other nebulous stars noticed by Lacaille.'* Moreover, he asked whether *'this nebulous star remains visible when the field is illuminated, as normally applied by Piazzi and Cacciatore while measuring stars'*. This question is important in connection with comet observations, because the coma and a nebula can look much different if measured in an illuminated micrometer eye-piece. In most cases the coma appears more compact at discovery time; it is more stellar and thus less sensitive to extraneous light. Later Schmidt mentions a correspondence between Olbers and Bessel concerning the disputed place of Cacciatore's nebula (Schmidt 1868b).

Because of Olbers' questions, Schumacher decided to reprint Cacciatore's report (Cacciatore 1827). The quoted Italian text is not identical to that already published by Baron von Zach, since the date 19 March is given. Schumacher wrote that *'from the whole context*

---

[28] With this instrument, Piazzi discovered Ceres in 1801; on this event and Palermo Observatory see Witt (2007b).

[…] *one must assume that this nebula can be seen in the illuminated field*', adding, about the possibility of it being a comet, '*it would have been better had Mr Cacciatore noted whether he has seen it again later*'. The reason for the confusion lies in Cacciatore's text: he mentions that Lacaille and Piazzi saw a star at the position of the nebula, which he had observed in 1809 and 1810 too; but no nebula was seen at that time. While searching for the comet of 1826 Cacciatore eventually came upon the nebula ('*I was impressed by this wonderful nebula*'). Therefore the critical questions of Olbers about the existence and brightness of the object were justified.

Wilhelm von Biela took part in the discussion too, writing a letter to Schumacher (Biela 1827). Biela had visited Ernesto Capocci at Capodimonte Observatory in early 1827, getting news about the controversial object. Appended to his letter is an observation made by Capocci in June 1826, which shows that '*the nebula does not stand a strong illumination of the field*'; the given position of the '*Nebulosa di Cacciatore*' fits well. By the way, the appendix lists the places of another two objects, which had already been observed by Harding (see Table 4.1): M 107 and NGC 7293. Capocci's observations were made with a 17.6-cm Fraunhofer refractor.

Finally, two other persons enter the scene: James Dunlop and John Herschel. It was due to the former that Olbers absolved Cacciatore. In a letter of 17 November 1828 he wrote that, on the '*supposed new nebula, a decisive judgement can now be passed*' (Olbers 1828). He had detected the object as no. 473 in Dunlop's '*highly valuable*' catalogue of southern nebulae and star clusters. The description reads as follows: '*A very bright round highly condensed nebula, about 3′ in diameter. I can resolve a considerable portion round the margin, but the compression is so great near the centre, that it would require a very high power, as well as light to separate the stars; the stars are rather dusky*' (Dunlop 1828). Olbers states that '*There is no doubt that this nebula was always there, but merely overlooked by Lacaille due to the small size of his telescope, and also by Piazzi and earlier by Cacciatore* [1809–10] *probably because of the low altitude and the illuminated field of view.*'

John Herschel observed the object on 1 and 3 June 1834 in Feldhausen; it is listed as h 3726 in the Cape catalogue and classified as a globular cluster. He mentions Dunlop, giving the correct identification Δ 473 (see the next section). However, Herschel was not aware

Figure 4.9. The bright globular cluster NGC 6541 in Corona Australis (DSS).

of Cacciatore at that time. In the General Catalogue he treats the issue – this is obviously the result of research done for this work. The notes to GC 4372 contain an annotated translation of Cacciatore's report of 1827. Herschel writes that Cacciatore '*observed it as a nebula, he says on the 12th of March, 1826 (of course, therefore, Dunlop has the priority of date).*' Then he solves the problem concerning the star seen by Lacaille, Piazzi and Cacciatore at the location of the nebula: the positions of star and nebula differ by 18′ in declination, thus the nebula could not be visible in the eye-piece while observing the star! About the priority it must be said that Dunlop's observation was actually on 3 July 1826, i.e. later than Cacciatore's. Thus the latter was the true discoverer of the bright globular cluster in Corona Australis. Dreyer listed it as NGC 6541, correctly noting '*Cacciatore, Δ 473*' (Fig. 4.9).

4.4 DUNLOP AND THE FIRST SURVEY OF THE SOUTHERN SKY

Because of the few nebulae and star clusters discovered by Halley and Lacaille, the knowledge of the southern sky in the early nineteenth century was comparable to that of the northern before Messier and Herschel. It was James Dunlop who carried out the first systematic observations in 1826. At Paramatta Observatory, 25 km west of Sydney, he surveyed a quarter of the unexplored

Figure 4.10. James Dunlop (1793–1848), in about 1835.

terrain and published a catalogue of 629 objects. Though the quality of his observations was a matter of controversy, they nevertheless were an important contribution to Australian astronomy.[29] The observatory was the site with second lowest southern latitude (−34°, the same as Feldhausen) contributing to the NGC, beaten only by the Great Melbourne Telescope at −38°.

4.4.1 Short biography: James Dunlop

James Dunlop was born on 31 October 1793 in Dalry, Ayrshire, Scotland (Fig. 4.10). At an early age he became interested in astronomy, built his first telescope in 1810 and was an autodidact. Later he got in contact with Thomas Macdougall Brisbane, General of the Scottish Brigade and fellow amateur astronomer.[30] In 1821 Brisbane was sent to Australia as the sixth Governor of New South Wales.[31] He wanted to erect an observatory there. For this project he hired two competent companions: James Dunlop and Christian Carl Rümker (father of George Rümker) from Mecklenburg. In May 1822 the private observatory, which was located in Paramatta Park, near the Governor's house, became operational. Together with the South African Cape Observatory (founded in 1820), it was the first important observatory in the southern hemisphere (Anon 1880b). The primary goal was the cataloguing of all stars brighter than 8th magnitude and south of −33° declination. For this task a Troughton meridian-circle with aperture 8.9 cm was used.

By 1823 Rümker had left the observatory after a quarrel.[32] The reason for the tensions was that, due to his astronomical and mathematical abilities, he felt more qualified than Dunlop and thus required the priority for the observational results. For instance, Rümker had calculated the position for the first return of comet Encke and would actually discover it. But Brisbane reported that Dunlop had seen the comet first. Actually, this was rather accidental, since Dunlop was first in the queue to look through the eye-piece! Undoubtedly, the honour of the rediscovery of a comet (the second after Halley's) goes to Rümker.

Brisbane returned to England in early 1826 (a year earlier the observatory had been purchased by the state). Dunlop, now thrown back on his own resources, could finish the observations of altogether 7385 stars in early March 1826.[33] He was hopeful of becoming the official head of the observatory – but this did not happen. Ironically, the job was offered to Rümker, who returned to Paramatta in May 1826 as first Government Astronomer. He stayed until October 1829; from 1833 until 1857 Rümker was Director of Hamburg Observatory.

Disappointed, Dunlop left the observatory in March 1826, but decided to stay in Paramatta for private research on double stars, nebulae and star clusters. By 1824 he had observed non-stellar objects, compiling a small, unpublished catalogue. Dunlop's new site was the garden of his house in Hunter Street; the largest instrument was a self-made 22.8-cm reflector.[34] About his task he later wrote the following: '*Finding myself in possession of reflecting telescopes, which I considered capable of adding considerably to our knowledge of the*

---

[29] Haynes *et al.* (1996).
[30] Obituary: Anon (1860); see also Weitzenhoffer (1992).
[31] The Australian city of Brisbane bears his name.
[32] Schramm (1996: 94), Haynes *et al.* (1996: 44–46).
[33] The Brisbane catalogue was not published before 1835. The 8-mag star B 895 appears in John Herschel's Cape catalogue as h 2767 (NGC 1838) with the description '*chief of a very loose clustering mass*'.
[34] Dunlop was the owner of the first Australian reflectors, which were equipped with metal mirrors made by himself.

nebulae and double stars in that portion of the heavens, I resolved to remain behind to prosecute my favourite pursuits, in collecting materials towards the formation of a catalogue of nebulae and double stars in that hemisphere, and any other object which might have attracted my attention.'[35] Between 27 April and 24 November 1826 Dunlop discovered the considerable number of 629 nebulae and star clusters, writing '*The nebulae being a primary object with me, I devoted the whole of the favourable weather in the absence of the moon to that department.*' When the Moon was too bright or the visibility too bad, he observed double stars, cataloguing altogether 254 (Dunlop 1829a, b).

In 1827 Dunlop returned to London. The results of his private search for southern non-stellar objects were published in 1828 as 'A catalogue of nebulae and clusters of stars in the southern hemisphere, observed at Paramatta in New South Wales' (Dunlop 1828). The appendix presents 33 sketches and drawings. In February 1828 Dunlop and Brisbane received the gold medal of the Astronomical Society of London for their work. In his laudation President John Herschel called Paramatta the '*Greenwich of the southern hemisphere*' (Herschel J. 1830).

Thus honoured, Dunlop eventually was appointed Government Astronomer, succeeding Rümker. In November 1831 he went back to Paramatta, finding the observatory in a bad condition. During the refurbishment, a new home for Dunlop was built too. There he lived until August 1847; afterwards the observatory was closed. It was not until 1858 that Australian astronomy could be continued with the foundation of Sydney Observatory.[36] James Dunlop died on 22 September 1848 in Boora Boora, New South Wales, at the age of 54.[37]

4.4.2 Importance and content of Dunlop's catalogue

In 1829 Olbers described the result as the '*highly regarded catalogue of nebulae and clusters of stars in the southern hemisphere by Mr Dunlop*' (Olbers 1828). Dunlop was mentioned again in his review of the Slough catalogue, comparing John Herschel's drawings with those '*Mr Dunlop has added to his estimable list of more than 600 southern nebulae*' (Olbers 1834). In contrast to Olbers, John Herschel was not enthusiastic about Dunlop's work, as can be seen from his remarks in the Cape catalogue. His search for objects marked 'Δ' must have been frustrating: with great effort he could find only one third of them. Herschel's words show a certain despair, lamenting the wasted time: '*The rest of the 629 objects, comprised in that catalogue, have escaped my observations; and I am not conscious of any such negligence in the act of sweeping as could give rise to so large a defalcation, but, on the contrary, by entering them in my working lists (at least, until the general inutility of doing so, and loss of valuable time in fruitless search, thereby caused it to become apparent), took the usual precautions to ensure their rediscovery; and as I am, moreover, of opinion that my examination of the southern circumpolar region will be found, on the whole, to have been an effective one, I cannot help concluding that, at least in the majority of those cases, a want of sufficient light or defining power in the instrument used by Mr. Dunlop, has been the cause of his setting down objects as nebulae where none really exist. That this is the case, in many instances, I have convinced myself by careful and persevering search over and around the places indicated in his catalogue.*'[38]

Once praised by Herschel in 1828 for his star observations, Dunlop was now confronted with a negative judgement about his nebular work by the eminent astronomer, on the basis of Cape observations. Dunlop's bad name can still be traced in the modern literature. For instance, Buttmann wrote that the observations of Lacaille, Brisbane and Dunlop "*had been only humble beginnings, undertaken sporadically, often with inadequate means, and containing appreciable errors. For example, of the 600 nebulous objects listed by Dunlop in his catalogue published in 'Philosophical Transactions' in 1826 [1828], Herschel was able to find only about a third, because the positions were so unreliable and descriptions so inaccurate.*"[39]

Anyway, Dunlop's honour is saved by the modern analysis of his original data. The remarkable result is that, actually, 355 objects exist (56%), which is considerably more than John Herschel's 'third'![40] Thus

---

[35] Dunlop (1829a).
[36] Bhathal (2009).
[37] See also Cozens and White (2001).

[38] Herschel J. (1847: 4); this contains one of Herschel's typical 'never-ending' sentences.
[39] Buttmann (1970: 90).
[40] The analysis was supported by the work of Glen Cozens (private communication), who has access to Dunlop's original notes.

criticism of him appears unjustified. However, one must consider that, with his means (i.e. only by visual observations), Herschel was not able to pass a positive judgement. For him, like Dunlop before, the southern sky was 'terra incognita'. Especially in the regions of the Milky Way and the Magellanic Clouds, where he discovered many objects, the orientation is extremely difficult. With modern charts and data the matter is much easier. Moreover, Herschel possessed only Dunlop's catalogue, published in the *Philosophical Transactions*, not the original notes, containing valuable additional information. When looking at his Cape catalogue, the most identification problems appeared in the LMC and SMC. In all other regions the match between Dunlop and Herschel is satisfactory.

Dunlop's catalogue shows five columns. The first contains the running number (nos. 1 to 629) and the next two the coordinates for 1827: right ascension (AR) and 'South Pole Distance' (SPD). The values are nominally given to 1^s and $1'$, respectively, but they turn out to be very precise – an impressive result, achieved with a self-made telescope. Column four gives the object description, which is rather detailed, not using any (Herschel-type) abbreviations. In the last column, the number of observations is listed (up to eight). Dunlop has sorted the entries by SPD; Δ 1 is the most southern object (SPD 12° 14′ = –77° 46′ Decl.) and Δ 629 the most northern (SPD 61° 55′ = –28° 05′ Decl.). Unfortunately, both are missing (no. 629 could be the globular cluster NGC 6316 in Ophiuchus). No dates of the observations are given; fortunately, these can be found in Dunlop's original notes.

Dunlop was the discoverer of 225 NGC objects (Table 4.3), of which 223 are non-stellar – a surprisingly high rate! One of his emission nebulae is the supernova remnant NGC 1918 (Δ 88) in Mensa, which he found on 27 September 1826. Fourteen open clusters and five emission nebulae are actually a mix of open cluster and emission nebula. NGC 2818 (Δ 564, 28 May 1826) in Compass is a rare association of open cluster and planetary nebula (Fig. 4.11).[41] Dunlop describes the object as a *'pretty large faint nebula* [star cluster]. *There is a small nebula* [PN] *in the north preceding edge* […] *the large nebula is resolvable into stars with small nebula remaining.'* Concerning the conflict with John Herschel

---

[41] Occasionally the PN is designated (lacking a historical reason) as NGC 2818A.

Table 4.3. *Dunlop's 225 NGC objects, divided by type*

Type	Number	Remarks
Open cluster	120	
Galaxy	46	
Globular cluster	28	
Emission nebula	25	SNR: NGC 1918 (Men), Δ 88
Planetary nebula	4	
Star group	1	NGC 6415 (Sco), Δ 595+596
Not found	1	NGC 6529 (Sgr), Δ 569?

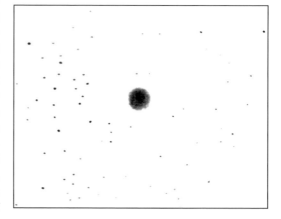

Figure 4.11. John Herschel's sketch of NGC 2818 (Δ 564, h 3154) in Compass; a rare combination of planetary nebula and open cluster (Herschel J. (1847), Plate V, Fig. 8).

about the quality of his data, it should be noted that 72 of his 225 NGC objects were later independently found by Herschel (i.e. there is no identification with a Dunlop object in the Cape catalogue).

The four individual PN are NGC 5189 (Δ 252, Mus, 1 July 1826), NGC 6302 (Δ 567, Sco, 5 June 1826), NGC 6326 (Δ 381, Ara, 26 August 1826) and NGC 6563 (Δ 606, Sgr, 3 September 1826). NGC 6302 is the famous Bug Nebula in Scorpius, which was independently found by Barnard in 1880 (see Section 9.15.2).

Table 4.4 lists special Dunlop objects and observational data. The work started on 27 April 1826 and ended on 24 November. During the first night

4.4 Dunlop and the southern sky

Table 4.4. *Special objects and observational data of Dunlop*

Category	Object	Data	Date	Remarks
Brightest object	NGC 3114 = Δ 297	OC, 4.2 mag, Car	8.5.1826	
Faintest object	NGC 1751 = Δ 78	OC, 14.5 mag, Dor	24.9.1826	
Most northern object	NGC 2243 = Δ 616	−31°, OC, 9.4 mag, CMa	24.5.1826	
Most southern object	NGC 602 = Δ 17	−73°, EN+OC, Hyi	1.8.1826	
First object	NGC 3766 = Δ 289	OC, 6.6 mag, Cen	27.4.1826	Lacaille 1751
Last object	NGC 1317 = Δ 547	Gx, 11.1 mag, For	24.11.1826	Fornax Galaxy Cluster
Best night			24.9.1826	32 objects
Best constellation	Dorado		28.5.–6.11.1826	85 objects (LMC)

Table 4.5. *Dunlop's discoveries in the Fornax Cluster*

NGC	Δ	Date	V	Remarks
1316	548	2.9.1826	8.4	Brightest member
1317	547	24.11.1826	11.1	NGC 1318, Schmidt 19.1.1865
1350	591	24.11.1826	10.4	At northern edge
1365	562	2.9.1826	9.5	Barred spiral
1380	574	2.9.1826	9.9	
1399	332	1.5.1826	9.4	

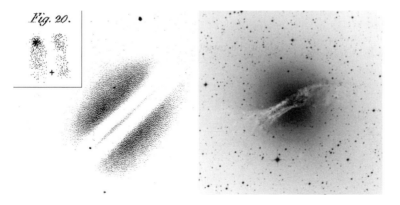

Figure 4.12. Three views of the peculiar galaxy NGC 5128 (Δ 482, h 3501) in Centaurus; centre: John Herschel's drawing (Herschel J. (1847), Plate IV, Fig. 2), upper left inset: Dunlop's sketch (Dunlop (1828), Fig. 20); right inset: DSS image (north at the top).

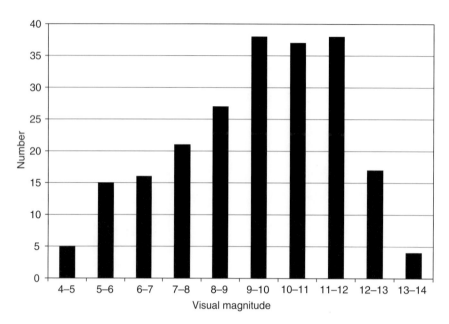

Figure 4.13. The brightness distribution of Dunlop's objects.

Dunlop found NGC 3766 (Δ 289); the open cluster in Centaurus had already been discovered by Lacaille in 1751. Dunlop's own initial object was NGC 5128 (Δ 482, 29 April 1826) in Centaurus (Fig. 4.12). This prominent galaxy not only is known for its chaotic structure (Arp 153; the brightest peculiar galaxy), but also is a strong radio source (Centaurus A).[42] The 85 objects in Dorado belong to the Large Magellanic Cloud (LMC).

Dunlop discovered six members of the Fornax Cluster Abell S373 (Table 4.5), among them NGC 1316, the brightest cluster galaxy. Most of the NGC galaxies were found by John Herschel (15); seven others are due to Julius Schmidt in Athens. However, none of the observers recognised the structure of the cluster. It is interesting to read John Herschel's description of the barred spiral galaxy NGC 1365 in the Cape catalogue (h 2552, observation of 28 November 1837): '*A very remarkable nebula. A decided link between the nebulae M 51 and M 27*' (see Section 11.3.1).

The brightness distribution of Dunlop's objects shows a mean of 9.3 mag (Fig. 4.13). Facing the small reflector and the unknown southern sky, the high value is plausible. The bar 13–14 includes the faintest object (NGC 1751; 14.5 mag).

---

[42] Steinicke (2002a).

5 · John Herschel at the Cape of Good Hope

From 1834 to 1838 John Herschel continued his observations of nebulae and star clusters in Feldhausen at the Cape of Good Hope with the 18¼" reflector, which had been built in Slough (Fig. 5.1).[1] Additionally he used his 5" Tully refractor for double stars and a comet-seeker that his father had made for Caroline Herschel. A central task was the exploration of the southern sky, which led to the discovery of a large number of non-stellar objects. It should be mentioned that Feldhausen was the site with the second lowest southern latitude (−34°) contributing to the NGC, beaten only by the Great Melbourne Telescope at −38°. Eight years earlier, James Dunlop had done important preliminary work at Paramatta (at −34° too). Herschel did not appreciate it very much, having big problems with the identification of Dunlop's objects. However, he also discovered a considerable number of northern objects, which are actually visible from Slough.

John Herschel was particularly interested in the Large and Small Magellanic Clouds (LMC and SMC), calling them Nubecula Major and Nubecula Minor, respectively, and the striking nebulous complex around the star η Carinae (formerly η Argûs). He also observed objects of Lacaille and southern nebulae and star clusters from the catalogues of his father, which were left behind in Slough. The Cape catalogue (CC) therefore contains

- discoveries of John Herschel
- discoveries of Dunlop
- southern objects of William Herschel
- objects discovered earlier by others and not catalogued by William Herschel, e.g. from Messier and Lacaille

Like the Slough catalogue, the Cape catalogue (published in 1847) was a great success, and John Herschel's scientific reputation reached another climax. The new work was much more than a mere list of southern deep-sky objects. With the integration of the southern sky, he could complete and finish his theory of cosmic structure, the nature of nebulae and star clusters and their evolution.[2] Actually, John Herschel was the only astronomer in history to examine visually the entire heavens with a large telescope. Moreover, at the Cape he made careful delineations of southern nebulae. A central goal, following the ideas of his father, was to demonstrate the transition from one type of object (nebula) to another (cluster). Owing to his comprehensive observations and theoretical work, he definitely became the dignified successor of the great William Herschel.

5.1 STRUCTURE AND CONTENT OF THE CAPE CATALOGUE

Herschel performed 382 sweeps in Feldhausen, continuing the Slough catalogue numbering: the first (no. 429) was made on 5 March 1834, the last (no. 810) on 22 January 1838.[3] Thus the systematic observations cover a period of almost four years. The observations were assisted by his mechanic John Stone; though the process was less efficient than that available to his father (who was supported by the industrious Caroline Herschel). Therefore, Herschel's wish to repeat all of the sweeps could not be realised: only 43% of the objects were observed more than once.

The observational results of the mission were not published until 1847 in London. The large-format book is titled *Results of Astronomical Observations Made during the Year 1834, 5, 6, 7, 8 at the Cape of Good Hope*

---

[1] Herschel J. (1835), Ashbrook (1984: 37–41). At the Cape Herschel produced even better mirrors than his father's. Sometimes the aperture was reduced to 12", e.g. when observing the trapezium in M 42 (Herschel J. 1847: 30).

[2] Hoskin (1987).

[3] The first observation with the 20-ft was made on 22 February 1834.

Figure 5.1. Herschel's observing site at Feldhausen with the 18¼" reflector (Herschel J. 1847).

(Herschel J. 1847).[4] The subtitle reads 'Being the completion of a telescopic survey of the whole surface of the visible heavens. Commenced in 1825.' The Cape catalogue finishes John Herschel's observations started in 1823 in Slough. In sum, now the first comprehensive survey of nebulae and star clusters covering the whole sky was available. Because no northern objects were found between 1833 and 1838, the two Herschel catalogues represent the state of nebular discoveries for the year 1838.

The text part covers 177 pages.[5] It contains an introduction (16 pages), a description of the observations and drawings in 90 articles (41 pages), a list of stars in the η Argûs nebula (6 pages), the main catalogue (78 pages), an appendix with additional objects of William Herschel and lists of sweeps and figures (3 pages) and a description of the objects' distribution on the sky and their classification, with a special focus on the Magellanic clouds (18 pages). In the following 14 pages, Herschel gives the positions of stars, nebulae and star clusters in the Large and Small Magellanic Clouds (919 and 244 objects, respectively). A few non-stellar objects listed here were later (in the GC and NGC) designated by numbers in brackets (LMC: ( ), SMC: [ ]). Examples are (147) = GC 1002 = NGC 1785 in the LMC and [162] = GC 231 = NGC 422 in the SMC. Finally, known errors and added objects are given (1 page), followed by the 10 tables with altogether 59 drawings. The appendix contains an addition to William Herschel's last catalogue. There his final objects (discovered on 26 and 30 September 1802) were not included in order to hold to the magical number of 500 (see Section 2.2.1). They were first listed by John Herschel in the Cape catalogue and later designated 'HON' ('Herschel omitted nebulae') in the General Catalogue (see Section 2.9).

The Cape catalogue lists 1708 regular entries, continuing the numbers of the SC: h 2308 to 4015. The structure is similar to the Slough catalogue (see Table 3.1); the epoch 1830 is adopted too. Mostly, more than one observation for a single object is given. The column 'Synon.' contains William Herschel, John Herschel, Dunlop, Messier and two stars from the Astronomical Society Catalogue (ASC) of 1827 and Brisbane's 'Catalogue of 7385 southern stars' of 1835.[6] For new objects this column is empty (i.e. there

---

[4] The delay was caused by Herschel's many duties after his return to England. It also took much time to collect all his results into one volume (perhaps two volumes would have been a better choice).

[5] This includes only that part of the publication dealing with the observation of nebulae and clusters. In other parts double stars, planets and the Sun are treated.

[6] See Chambers' list of star catalogues (Chambers 1890: 487–495).

Table 5.1. *The origins of the 1708 Cape catalogue objects as given by John Herschel*

Origin	Number	Remarks
W. Herschel	211	Objects from William Herschel's catalogues
J. Herschel (h)	6	Slough catalogue objects with no other designation
J. Herschel (new)	1274	New objects
Dunlop (Δ)	200	
Messier (M)	15	
Brisbane (B)	1	h 2767 = B 895 = NGC 1838 (OC in Dor)
Astronomical Society Catalogue (ASC)	1	h 3115 = ASC 1001 = 19 Pup = NGC 2542

Table 5.2. *Independent objects in the Cape catalogue*

	Number
Entries	1708
Catalogue identity	5
NGC identity	18
Independent	1685

is no designation 'Nova'). Table 5.1 shows the origin of the catalogue objects from John Herschel's point of view. As for the SC earlier, this is subject to revision. As shown in the following, there are a lot of wrong identifications.

5.2 IDENTIFICATION OF CATALOGUE OBJECTS

The total number of 1708 is first reduced by five double entries ('catalogue identity'; Table 5.2). These are h 2399 = h 2398, h 3231 = h 3229, h 3275 = h 3274, h 3967 = h 3966 and h 3972 = h 3971. Additionally 18 cases that turned out to be identical due to a modern analysis of the NGC ('NGC identity') must be considered. Here are three examples. The first is h 2446 (NGC 729), a galaxy in Fornax, found by Herschel on 30 November 1837; it was already contained in the catalogue as h 2445 (NGC 727, 1 September 1834). On 2 November 1834 Herschel discovered h 2783 (NGC 1855), a globular cluster in the LMC. He saw it again on 23 November catalogued as h 2782 (NGC 1824). But the real discoverer was Dunlop (Δ 119, 2 August 1826),

which was not noticed by Herschel. The third case is h 3950, a 13.6-mag galaxy in Grus, discovered on 30 August 1834 and confirmed two days later. However, on 23 October 1835 the object was seen again, but, due to coordinate problems, Herschel doubted the identity, noting that '*it is not impossible that this may be a different nebula*'. In the General Catalogue he entered his observations as two objects: GC 4812 and GC 4822; the latter is assigned as 'h 3950 (no. 2)?'. Dreyer catalogued them as NGC 7322 and NGC 7334 (noting '? = G.C. 4812'). The supposed identity was confirmed by a modern analysis. As Table 5.2 shows, the net result is 1685 independent CC objects.

Table 5.3 shows the result of a modern analysis, considering NGC identities. Eighty-nine objects are already in the SC (not influencing their independence). The CC contains 1207 new objects of John Herschel. Halley's find is the naked-eye globular cluster ω Centauri. Méchain's globular cluster was designated M 107 in the twentieth century. Looking at both the SC and CC, 13 Messier objects are missing in the end: M 4, M 24, M 25, M 38, M 40, M 43, M 45, M 73, M 76, M 79, M 82, M 90 and M 92. Only one CC object is not in the NGC: h 2420 in Sculptor, which was found on 23 October 1835; Herschel supposed it to be a nebulous star, but ignored it in the GC. It is a star of only 9.1 mag (GSC 7541–7).[7] Of the 213 objects discovered by Dunlop, 72 were independently seen by Herschel. Thus in the column 'Synon.' of the Cape catalogue a Δ-object is missing.

---

[7] GSC refers to the Guide Star Catalogue, which was compiled for the Hubble Space Telescope.

Table 5.3. *Discoverers of the independent objects in the Cape catalogue*

Discoverer	Number	Remarks
W. Herschel	209	71 objects already in SC
J. Herschel (new)	1207	
J. Herschel (SC)	6	Objects already in SC
C. Herschel	2	h 2345 = h 61 = NGC 253 (Gx, Scl), h 3079 = h 440 = NGC 2360 (OC, CMa)
Messier catalogue	22	Various discoverers; 9 objects already in SC
Cacciatore	1	h 3726 = NGC 6541 (GC, Cra)
Dunlop	213	72 independently found by J. Herschel
Halley	1	h 3504 = NGC 5139 = ω Centauri (CC, Cen)
Hodierna	3	One object already in SC: h 3077 = h 441 = NGC 2362 (OC, CMa)
Lacaille	19	
Méchain	1	h 3637 = NGC 6171 = M 107 (GC, Oph)
Vespucci	1	h 2356 = NGC 292 = SMC (Gx, Tuc); brightest NGC object
Sum	1685	

5.3 JOHN HERSCHEL'S NEW OBJECTS

The Cape catalogue contains 1207 objects found by Herschel. Concerning the types listed in Table 5.4, 1168 of them are non-stellar, mostly galaxies. This gives a very high rate of 97%. The brightest planetary nebula is remarkable: NGC 3918 (h 3365), found on 3 April 1834. On a later observation (5 April 1837) Herschel notes that it was '*very like Uranus, only about half as large again and blue*'. This was the first concrete hint that a PN looks like the planet Uranus (see Section 2.6.1). The brightest single star is ASC 1001 (19 Pup, see Table 5.1); on 11 December 1836 Herschel noted '*A fine nebulous star 6 m, in the following part of the cluster VII. 11* [h 3114 = NGC 2359], *and almost unconnected with it. The nebula is faint, but I feel confident that it is not the nebulous haze.*' In a subsequent 'Notandum' one reads that there is '*Nothing more difficult than to prove a nebulous star of the 6th m and above.*' A nebula around 19 Puppis could not be found, like in other cases too (see Tables 6.20 and 10.14).

Table 5.5 shows special objects and observational data. Possibly the open cluster NGC 2451 (h 3099) had already been seen by Hodierna in about 1754. The large bright object appeared in Herschel's reflector '*too loose to be a fit object for ordinary magnification*'. Five days after his initial sweep (no. 429), Herschel discovered his first southern object, the bright galaxy NGC 2887. In the same night he observed Dunlop's globular cluster Δ 265 (h 3152, NGC 2808) in Carina. In his last sweep (no. 810), Herschel found the open cluster NGC 2849 (Fig. 5.2 right) and the galaxy NGC 3210. For five entries, no sweep is mentioned (thus there is no date): h 2928, h 2929, h 2942, h 2951, h 2957 and h 2962. All belong to the LMC, where Herschel discovered 198 objects.

Concerning John Herschel's scanned region, Hoskin is not right in saying that he '*searched the southern skies that were invisible from England*'.[8] Actually he looked in the northern sky too: on 18 March 1836 four objects in Leo were found. Besides NGC 3447 (h 3300), these are the missing NGC 2953 (h 3182) and the galaxies NGC 2954 (h 3181) and NGC 3075 (h 3207).[9] His next southern declination is −2° (Eridanus), where he discovered the galaxy NGC 1322 (h 2532, 5 October 1836). Herschel found altogether 101 objects (15%) lying north of his most southern Slough object. This is NGC 823 in Fornax (δ = −25° 26′), catalogued as

---

[8] Hoskin (1990: 331).
[9] In Feldhausen they nevertheless reached an altitude of 40°.

5.3 John Herschel's new objects

Table 5.4. *Types of the 1207 objects discovered by John Herschel at the Cape*

Type	Number	Remarks
Galaxy	818	Brightest: NGC 1553 = h 2630 (9.0 mag, Dor)
Open cluster	275	Brightest: see Table 5.5
Globular cluster	14	Brightest: NGC 6558 = h 3731 (8.6 mag, Sgr)
Emission nebula	44	
Planetary nebula	17	Brightest: NGC 3918 = h 3365 (8.1 mag, Cen)
Reflection nebula	1	h 3548 = NGC 5367 (Cen)
Single star	5	Brightest: NGC 2542 = h 3115 = 19 Pup (4.7 mag)
Star pair to quartet	5	
Star group	22	
Not found	6	

Table 5.5. *Special objects and observational data from John Herschel's Cape catalogue*

Category	Object	Data	Date	Remarks
Brightest object	NGC 2451 = h 3099	2.8 mag, OC, Pup	1.2.1835	Hodierna 1754?
Faintest object	NGC 1135 = h 2498	15.5 mag, Gx, Hor	11.9.1836	
Most northern object	NGC 3447 = h 3300	+16°, Gx, 13.3 mag, Leo	18.3.1836	Plus three other objects in Leo
Most southern object	NGC 2573 = h 3176	−89°, Gx, 13.4 mag, Oct	29.3.1837	Polarissima Australis
First object	NGC 2887 = h 3168	Gx, 11.7 mag, Car	8.3.1834	Sweep 430
Last object	NGC 2849 = h 3160	OC, 12.5 mag, Vel	22.1.1838	Sweep 810; with h 3225 = NGC 3120 (Gx, Ant)
Best night			23.12.1834	38 objects (34 in LMC)
Best constellation	Dorado			199 objects (168 in LMC)

h 196 and also observed at the Cape (h 2460); one of six cases catalogued both in the SC and in the CC (see Table 6.3). It is interesting that among the 101 objects there are four that had already been discovered by William Herschel in Slough: NGC 3505 (h 3312) = NGC 3508 (II 507) in Crater, NGC 3528 (h 3316) = NGC 3495 (II 498) in Leo, NGC 4776 (h 3437) = NGC 4759 (II 559) in Virgo and NGC 4994 (h 3471) = NGC 4993 (III 766) in Hydra.

The southern counterpart of Herschel's 'polarissima' (NGC 3172) is Polarissima Australis NGC 2573 (h 3176) in Octans (Fig. 5.2 left). The galaxy lies only 1° from the southern pole. As in the case of NGC 3172, the position was difficult to determine: '*Being so near the pole, the RA may be in error many minutes of time.*'

John Herschel discovered the most NGC members in the rich galaxy clusters in Fornax (Abell S373), Hydra (Abell 1060) and Centaurus (Abell 3526). He must be credited for 15 of 28 NGC galaxies in the Fornax Cluster, lying partly in Eridanus (Table 5.6). The bulk was found on 28 and 29 November 1837. The remaining members are due to Schmidt in Athens (seven) and Dunlop in Paramatta (six), among them the brightest (NGC 1316, 8.4 mag). Another object, h 2582 = NGC 1437 (28 November 1837), is identical to NGC 1436. Four galaxies where classified by Herschel as

Table 5.6. *Members of the Fornax Galaxy Cluster (Abell S373) discovered by John Herschel*

NGC	h	Date	V	Con.	Remarks
1310	2524	22.10.1835	12.1	For	Herschel: globular cluster
1315	2526	13.11.1835	12.6	Eri	
1326	2535	29.11.1837	10.6	For	
1336	2537	22.10.1835	12.1	For	
1341	2540	29.11.1837	11.5	For	
1351	2544	19.10.1835	11.5	For	
1373	2556	29.11.1837	12.4	For	Galaxy trio
1374	2557	29.11.1837	11.0	For	Galaxy trio
1375	2558	29.11.1837	12.2	For	Galaxy trio
1379	2561	25.12.1835	10.7	For	Herschel: globular cluster
1387	2564	25.12.1835	10.8	For	Herschel: globular cluster
1419	2576	22.10.1835	12.5	Eri	
1427	2579	28.11.1837	10.8	For	
1436	2581	9.1.1836	11.7	Eri	= NGC 1437; Herschel: globular cluster
1460	2587	28.11.1837	12.5	Eri	

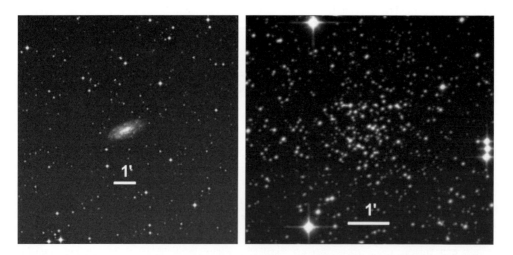

Figure 5.2. Left: Herschel's Polarissima Australis, NGC 2573 (h 3176) in Octans; right: his last object, the open cluster NGC 2849 (h 3160) in Vela (DSS).

globular clusters. For two (NGC 1379 and NGC 1387) this is explainable by their roundish shape (type E), but the others are barred spirals (type SB).[10]

Table 5.7 lists Herschel's 10 discoveries in the Hydra Cluster. Apart from a straggler (NGC 3307),

all of these galaxies were first observed in late March 1835, among them the two brightest, NGC 3309 and NGC 3311. These elliptical galaxies form a close pair (separation 1.5′), as was noticed by Herschel ('double nebula'). He also recognised the group character. There are altogether 12 NGC galaxies in the cluster; the remaining two were found by Austin. In May 1834

---

[10] Here Hubble's classification is meant.

Table 5.7. *John Herschel's galaxies in the Hydra Cluster (Abell 1060)*

NGC	h	Date	V
3285	3270	24.3.1835	12.1
3305	3277	24.3.1835	12.8
3307	3278	22.3.1836	13.7
3308	3279	24.3.1835	12.0
3309	3280	24.3.1835	11.6
3311	3281	30.3.1835	11.6
3312	3282	26.3.1835	11.9
3314	3283	24.3.1835	12.5
3316	3284	26.3.1835	12.7
3336	3289	24.3.1835	12.3

Table 5.8. *John Herschel's galaxies in the Centaurus Cluster (Abell 3526)*

NGC	h	Date	V
4575	3403	8.5.1834	12.3
4601	3405	8.5.1834	13.6
4603	3406	8.5.1834	12.4
4616	3408	6.5.1834	13.3
4622	3409	6.5.1834	12.5
4645	3412	8.5.1834	11.8
4650	3413	26.5.1834	11.8
4661	3415	6.5.1834	13.5
4672	3416	8.5.1834	13.1
4677	3418	8.5.1834	12.9
4683	3422	8.5.1834	12.8
4706	3427	6.5.1834	12.8
4729	3430	8.5.1834	12.4
4730	3431	8.5.1834	12.7
4743	3429	8.5.1834	12.4
4744	3433	8.5.1834	12.7

Herschel discovered 16 of 17 NGC galaxies in the Centaurus Cluster (Table 5.8). The brightest member, NGC 4696 (10.2 mag), was first seen by Dunlop on 7 May 1826.

Figure 5.3 shows the brightness distribution of John Herschel's Cape objects. The mean is 11.1 mag, which is a magnitude brighter than that found by his father (12.0 mag). The Slough catalogue gives an average of 13.0 mag. The difference is due to the mostly unknown southern sky, where John Herschel could discover (like Dunlop some years earlier) many bright objects.

5.4 CLASSIFICATION, SUPPLEMENTS AND DRAWINGS

5.4.1 John Herschel's classification of nebulae and star clusters

In the Cape catalogue John Herschel developed a classification scheme for nebulae and star clusters that was intended to unify the classes and codes of his father (Table 5.9).[11] William Herschel had already expanded his original scheme in the publication of 1802 to 12 classes, without this receiving much attention, however.[12] D'Arrest speaks about the '*philosophical classification of the nebulae*'.[13] John Herschel's classification is based on three classes (Table 5.9). The most important (I) contains the 'regular nebulae', which include globular clusters; the rest are either 'irregular nebulae' (II) or 'irregular clusters' (III). Members of class I, containing at least 90% of all objects, are categorised according to five criteria: 'magnitude', 'brightness', 'roundness', 'condensation' and 'resolvability'. Each is split into five grades. In the case of 'magnitude' the grade ranges from 'great' (1) to 'minute' (5). Interesting is the 'resolvability' with its grades 'discrete' (1), 'resolvable' (2), 'granulate' (3), 'mottled' (4) and 'milky' (5). Thus the extremes are star clusters and (true) nebulae.

In 1856 d'Arrest critically remarked about this scheme that '*The application of a third, recently proposed classification of the nebulae, must precede at least a repeated checking of all objects on both hemispheres, which might take many years, would need special instruments to be executed, and therefore will not happen in the near future.*' John Herschel had even tried to 'digitise' his scheme, which was later copied by Schultz (see Section 8.6.5). However, the new method was not accepted in the end, the classic description of William Herschel remaining more or less unchanged. From a

---

[11] Herschel J. (1847: 140–141). Another classification of non-stellar objects was introduced by Stephen Alexander in 1852 (Alexander S. 1852). He defined seven classes with various subclasses, illustrated by examples and drawings.
[12] W. Herschel introduced the new classes in 12 chapters (Herschel W. 1802: 478–502).
[13] d'Arrest (1856a: 360).

Table 5.9. *John Herschel's classification of nebulae and star clusters*

Class	Designation	Explanation	Examples
I	Regular nebulae	Distinct central condensation or regular symmetry around a central point (in general: elliptical, planetary, annular, globular)	Round/elliptical galaxies, NGC 2392, M 57, M 13
II	Irregular nebulae	Irregular shape, usually large	Orion Nebula, η Carinae Nebula, M 27, M 8, M 51, NGC 2261
III	Irregular clusters	Star clusters with no symmetry	Open clusters, W. Herschel's class VII & VIII

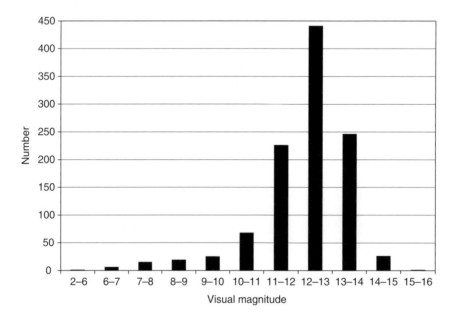

Figure 5.3. The brightness distribution of Herschel's discoveries in the Cape catalogue.

modern point of view, John Herschel's classification hardly corresponds to the physical nature of non-stellar objects.

5.4.2 Objects without h-number

Curiously the Cape catalogue contains 18 entries which received no h-number; they are listed in Table 5.10 (the preceding h-number is given there for orientation).

Mostly these are regions in the Milky Way (MW) or the Magellanic Clouds (LMC and SMC). With Herschel's coordinates and descriptions, identification is possible in most cases. The series of clusters mentioned after h 2797 obviously consists of NGC 1869, NGC 1871 and NGC 1873 in Dorado. The trio of nebulae (after h 2843) is NGC 1929 (h 2840), NGC 1934 (h 2842) and NGC 1937 (h 2845). The first object listed after h 3713 is NGC 6480, a star group in Scorpius. John Herschel designates the object in the General Catalogue (GC 4342) as h 3713' (h 3701' and h 3702' can be found there too). Unfortunately Dreyer forgot the mark and therefore h 3713 appears twice in the NGC (the same applies for h 3702). There are drawings for three of Herschel's objects without h-number; in the 'List of figured nebulae' they are called h 3684½, h 3702½ and h 3713½.

5.4 Classification, supplements and drawings 85

Table 5.10. *Herschel's entries without h-number*

After h	Date	Identification	Con.	Herschel's description (extract)
2335	11.4.1834	Western SMC edge	Tuc	Preceding edge of the Nubecula Minor, seen as a mere nebula
2368	5.11.1836	Northern SMC edge	Tuc	Upper edge of the Nubecula Minor – resolvable
2668	12.12.1836	LMC	Men	Nubecula Major
2794	16.12.1835		Dor	Southern part of the field is here illuminated by the faint light of the Nubecula Major
2797	2.1.1837	(NGC 1869/71/73)	Dor	θ Doradus marks the northern limit of a great irregular series of clusters
2825	23.12.1834	LMC	Men	Edge of nebula light of the Nubecula Major
2843	13.12.1835	(NGC 1929/34/37)	Dor	Middle of a group of three nebulae
2854	12.12.1836	Southern LMC edge	Men	Southern edge of Nubecula Major
2868	34.11.1834	Δ 136 = NGC 1966	Dor	Chief star, 9 m of a large irregular cluster
2980		LMC	Dor	Here follows a region of very small scattered stars
3684	1.6.1834	MW	Sco	The edge of the Milky Way is here quite sharp and definite (Fig. V.3)
3701'	7.6.1837	NGC 6415	Sco	A great nebulous projection of the Milky Way
3702'	8.6.1837	NGC 6421	Sco	A most remarkable, well-insulated, semi-nebulous Milky Way patch (Fig. V.1)
3707	15.7.1836		Sgr	Here begins an enormous region of stars in the Milky Way
3713	27.6.1837	MW	Sco	Milky Way sharply terminated
3713'	3.9.1834	NGC 6480	Sco	An extraordinarily bright nebulous part of the Milky Way (Fig. V.2)
3716	3.9.1834	MW	Sgr	Southern extremity of a great nebulous promontory of the Milky Way
3739	3.9.1834	MW	Sgr	Milky Way superb

The appendix of the CC contains a table 'Omitted observations of Nebulae, &c., and supplementary nebulae', listing 13 objects, ordered by right ascension. For seven, already catalogued, only observations are added. The rest are objects that Herschel simply forgot to list (Table 5.11). Since the main catalogue ends with no. 4015, the numbering was continued – breaking the right-ascension order. Dreyer called the added objects 'hon' ('Herschel omitted nebulae'); the lower-case letters were to assign that John Herschel is meant (for William Herschel 'HON' was used).

Five 'supplementary nebulae' must be credited to John Herschel, raising the total number of discoveries in Feldhausen to 1212. The open cluster NGC 2658 (h 4017) had already been found by Dunlop and correctly identified by Herschel as Δ 609.

5.4.3 John Herschel's drawings and his cosmology

Like in the Slough catalogue, Herschel presents a collection of his drawings. In 9 tables there are altogether 59 figures, some showing more than one object (giving a total of 100). For identification there is a 'List of figured nebulae', sorted by h-number.[14] Table 5.12 lists only the larger drawings. The column 'Object' lists Herschel's designation. Unfortunately a date can be given only for some drawings. Numbers in brackets in the column 'h' refer to the Slough catalogue.

Herschel's Table X shows the distribution of all objects from the two catalogues for the northern and

[14] Herschel J. (1847: 143).

Table 5.11. *Herschel's 'supplementary nebulae' listed in the appendix*

h	NGC	Date	Type	V	Con.	Remarks
4016	1905	2.1.1837	OC	13.2	Dor	Dreyer: (hon)
4017	2658	16.1.1836	OC	9.2	Pyx	Δ 609, Dunlop 28.5.1826
4018	2973	5.2.1837	3 stars		Ant	Dreyer: hon
4019	3244	22.4.1835	Gx	12.3	Ant	Dreyer: hon
4020	6404	27.6.1837	OC	10.6	Sco	
4021	6708	9.6.1836	Gx	12.0	Tel	Dreyer: hon

Table 5.12. *Large drawings of John Herschel*

Fig.	Object	h	NGC	Date	Type	Con.	Remarks
I.1	M 8	3722	6523	27.6.1837	EN	Sgr	
I.2	κ Crucis	3435	4755	14.3.1834	OC	Cru	Jewel Box[16]
II.1	M 17	(2008)	6618		EN	Sgr	Omega Nebula
II.2	IV 41, V 1, V 2, V 3	3718	6514	24.5.1835	EN	Sgr	M 20, Trifid Nebula
II.3	c Orionis		1977		EN	Ori	GC 1188[17]
II.4	30 Doradus	2941	2070	16.12.1835	EN	Dor	Tarantula Nebula
VII.1	Nubecula Minor	2356	292		Gx	Tuc	SMC
VII.2	Nubecula Major				Gx	Dor	LMC
VIII	θ Orionis	(360)	1976		EN	Ori	M 42, Orion Nebula
IX	η Argûs	3295	3372		EN	Car	η Carinae Nebula

southern hemispheres. The sky is divided into areas of 1^h in right ascension and 15° in declination. In each area only the number of objects is entered. In the chapter 'Of the law of distribution of nebulae and clusters of stars over the surface of the heavens' (pages 133–137) he discusses the results of his visualization. Obviously, the distribution resembles the Milky Way. Moreover, about a third of all objects are lying in a wide, irregular band centred on '*the nebulous region of Virgo [...] situated almost precisely in one pole of the Milky Way*' and '*within this area there are several local centres of accumulation, where the nebulae are exceedingly crowded*'.[15] Additionally there are larger regions with low density. The southern hemisphere shows, apart from the Magellanic Clouds, a more homogeneous distribution.

Concerning the theoretical sections in the Cape catalogue, they are, as Hoskins has stressed, ill-organised and unsystematic.[18] Though inspired by his father, John Herschel's cosmology, which he first presented in his popular textbook *A Treatise on Astronomy* (1833), was not an evident theory. He often changed his arguments and ideas. Once more, the basic question was the relation of nebulae and star clusters, focused on the terms 'resolvability' and 'change'.[19] John Herschel claimed to have attacked the issue by deploying observational facts, whereas his father, in his opinion, had

---

[15] Herschel J. (1847: 135–136); this is the Virgo Galaxy Cluster.
[16] This name is due to Herschel, who spoke in his Cape catalogue of '*a superb piece of fancy jewelery*' (Herschel J. 1847: 17).
[17] The nebula was photographed by the brothers Henry on 27 January 1887 (Henry 1887).
[18] Hoskin (1987: 3).
[19] Airy (1836).

relied on speculation. Unfortunately, the delay between Herschel's return to England and the publication of the results of almost nine years was not utilised for a thorough elaboration of his cosmological ideas. Moreover, the Cape catalogue was published in an already changing era of nebular research, marked by the success of Lord Rosse's superior telescopes. The results had already been circulated and led to influential books, such as those written by the Glasgow astronomer John Pringle Nichol.[20] Anyway, it more and more became clear that even the largest telescopes were unable to decide whether the nebulae were disguised star clusters or whether true nebulosity existed. Thus observation did not lead to any decision about the nature and evolution of these cosmic objects.

After 1847, Herschel would never again have the occasion to present a theory of nebulae. With the publication of the Cape catalogue he declared '*I have made up my mind to consider my astronomical career as terminated.*'[21] This was not quite correct, since his final great work, the General Catalogue, appeared in 1864. Though this work was a 'mere' compilation of data, devoid of theory, it paved the way for Dreyer.

---

[20] Nichol (1837, 1846).

[21] Hoskins (1987: 23).

6 • The time after Herschel's observations until Auwers' list of new nebulae

In 1841 Wilhelm Struve wrote that *'The study of the nebular heaven seems to be the exclusive dominion of the two Herschels.'*[1] Thus the following question arises: was there still anything to discover after the intensive observations of William and John Herschel? The clear answer is yes – for several reasons. First, the two Herschels had not completely surveyed the sky. Their sweeps left remarkable gaps, especially at high declinations and at greater distances from their reference stars. Moreover, new and better telescopes were available in the post-Herschel era, revealing much fainter objects. Finally, new campaigns were started, in which many visual observers (normally professionals) participated. Though their targets were mostly comets and minor planets, many new nebulae were found accidentally. For the search small refractors with great light-gathering power were used: the typical comet-seeker was optimal for observing large, faint nebulae too. Larger telescopes with apertures of 10–30 cm were needed to trace a comet or to create 'ecliptical charts' showing stars down to 13 mag – such charts were important tools by means of which to identify minor planets. Using them, occasionally faint non-stellar objects could be detected.

By virtue of these observations, the number of new nebulae slowly increased from 1845. Some discoverers published lists, but there were at first no new compilations. John Herschel's Slough catalogue was the primary reference to check the identity of objects. The two central questions were the following.

- Is the object a nebula (not moving) or a comet (moving)?
- Is the nebula new or already known?

Much more time was spent on known objects. The central issue was the determination of precise positions. The first relevant work in this field was done by Johann von Lamont in Munich, Laugier in Paris and d'Arrest in Leipzig. A few new nebulae were found during the measurements.

Other distinguished astronomers interested in nebulae were the observers at Harvard College Observatory, Hind at Bishop's Observatory in London, Cooper at Markree Castle, Secchi at the Collegio Romano, Winnecke in Göttingen, Berlin, Bonn and Pulkovo and Schönfeld, who discovered a few nebulae during his observations for the Bonner Durchmusterung. Additionally Lassell, Brorsen, Bruhns and Tempel, who found the Merope Nebula in 1859, must be mentioned. Beside the observations, new ideas about the nature of the nebulae were rare. Among the contributors were Lamont in Munich, Mason at Yale College and Bond at Harvard College Observatory. Their publications contain important drawings.

It was Arthur Auwers who compiled a first list of new nebulae, published in 1862 and containing 50 objects. It appears in the appendix of his important work 'William Herschel's Verzeichnisse von Nebelflecken und Sternhaufen' – the first modern revision of the three Herschel catalogues. This work initiated Auwers' remarkable career as founder of fundamental astronomy. He was an utterly exact observer, as his many publications show.

6.1 LAMONT AND THE NEBULAE

For a while Lamont dealt intensively with William Herschel's ideas about the nature and evolution of nebulae and star clusters, but his work does not cease with mere theory; he also planned systematic observations. At the observatory in Bogenhausen, a suburb of Munich, he used a fine 28.5-cm Merz refractor. Unfortunately, his proposal did not lead to appreciable action. Lamont's observations remained fragments, as was already being lamented by contemporary observers. Nevertheless, he made some interesting drawings, e.g. of his favourite object, the Orion Nebula.

---

[1] Struve W. (1847: 48).

Figure 6.1. Johann von Lamont (1805–1879).

6.1.1 Short biography: Johann von Lamont

Johann von Lamont was born on 13 December 1805 in Corriemulzie near Braemar, Scotland (Fig. 6.1). At the age of 11 he was sent to Regensburg, Germany, to attend the Schottenstift St Jakob. He learned German, Latin and Greek, turning later to mathematics and science. In the summer of 1827 he visited the Munich University Observatory, at Bogenhausen. Its Director, Johann Georg von Soldner, was impressed by young Lamont and he became an assistant in March 1828. Lamont graduated two years later and in 1833, after Soldner's death, he took over Soldner's duties and became the official Director in 1835. His main task was positional astronomy, fully in the tradition of Soldner.

For a short period (1835–1839) Lamont was interested in nebulae, observing with the new 28.5-cm Merz refractor. Afterwards he returned to classical astronomy, publishing the 'Erstes Münchener Sternverzeichnis' ['First Munich star catalogue'] with 34 674 stars. Other fields of activity were geophysics, especially measuring the magnetic field of the Earth, and meteorological studies. In 1845 and 1846 Lamont had seen Neptune twice, before the discovery by Galle and d'Arrest in Berlin – albeit cataloguing the planet as a star of 9th and 8th magnitude.[2] In 1853 he became Ordinary of the astronomical faculty. Johann von Lamont died, never having married, on 6 August 1879 in Bogenhausen at the age of 73.[3]

6.1.2 Lamont's public lecture *Ueber die Nebelflecken*

In his late years, Soldner, impressed by the '*strange discoveries, made by* [John] *Herschel in the remote southern sky with his large telescopes*',[4] had suggested observing nebulae in Munich. Lamont picked up this idea and was focused on it for some years. Concerning the nature and evolution of nebulae, he criticised some of William Herschel's ideas, focused on the question of 'resolvability' and 'change'. Lamont was aware of the different aspects of the term 'nebulae': diffuse masses of (true) nebulosity, unresolvable even with the largest telescopes and subject to change, and disguised star clusters, appearing 'nebulous' by virtue of their distance. To determine whether there are actually two classes or only one (the latter implying that all objects are resolvable), he designed an observing programme including measurements and drawings. On his target list were his favourite object, the Orion Nebula,[5] the open cluster M 11 in Scutum, the Omega Nebula M 17 in Sagittarius and a selection of other interesting objects (e.g. planetary nebulae).

Lamont explained the basics of his observing programme in a public talk *Ueber die Nebelflecken*, which he gave on 25 August 1837 in a meeting of the Bavarian Academy of Science in Munich.[6] The printed version contains an appendix, which was intended to show the first results. Unfortunately Lamont's promising start was followed by an early end: his last observation is dated October 1837 (see the next section).

In his talk, Lamont criticised Herschel's idea of a permanent evolution of nebulae – from a diffuse state (true nebula) to a spherical end (star) – and hence that the whole sequence should be visible in the sky, represented by individual objects. In his opinion, the previous considerations were only qualitative, being based on the descriptions of a large sample of objects. He laments the lack of quantitative studies, i.e. measurements, like

---

[2] This was noticed by Hind in 1850 (Hind 1850a).

[3] Obituaries: Dunkin (1880) and Anon (1880a).
[4] Lamont (1838: 144).
[5] See also Lamont (1843).
[6] 'About the nebulae' (Lamont 1837).

those Wilhelm Struve had made for double stars: '*Half a century ago, and nearly coinciding with the double stars, the observation of nebulae began; but the successes, like the method, were different. Whereas the result of the former belongs to the strangest discoveries of astronomy, all efforts made on the latter brought not much more than the mere knowledge of their existence and celestial positions.*'[7]

According to Lamont, yet another idea of Herschel's is not supported by observations: since time and distance are related by the finite velocity of light, the near, bright (and thus old) objects should already have converted into spherical bodies (stars), whereas the distant, faint (young) objects should appear incomplete, i.e. as 'disrupted masses' – with the two extremes being connected by intermediate steps. But Lamont states that obviously there are near (bright) nebulae, such as M 42, which are in a diffuse early state, and, at the same time, there are distant (faint) star clusters showing great regularity: the globular clusters.

Lamont concludes that there is no difference between old and young objects, thus direct evidence for evolution is lacking and the transformation of matter into stars must be completed everywhere. All objects are in (gravitational) equilibrium, i.e. they are composed of stars. That some appear resolved, but others diffuse, is only a matter of distance. With increasing aperture, the diffuse nebulae must show first substructures (individual nebulous areas) and finally single stars. In a large telescope the nearby Orion Nebula, for instance, must appear as a star cluster. Lamont believed that he could confirm this for his favourite object: he had measured a large number of substructures, which should finally (with sufficient power) turn out to be stars.

6.1.3 The observing programme

With the modest telescopes of the observatory which were available when Lamont became Director, systematic observations of nebulae could not be realised. He wrote that, '*If the Munich institution is to be successful in this field, it must be equipped with the necessary tools.*'[8] Lamont's advice was heeded: '*After Fraunhofer had finished the Dorpat refractor, a first proof of a high, and*

Figure 6.2. The 28.5-cm Merz refractor at Munich Observatory, Bogenhausen (Witt 1999).

up to that time unrivalled art, he began to build a much larger telescope of 12 Paris inches aperture for the Munich Observatory.' After Fraunhofer's death, his successor, Georg Merz, could deliver only a refractor of aperture 10.5 Paris inches (28.5 cm), offering magnifications from 110 to 1200 (Fig. 6.2). In 1835 the telescope was erected in a new building.[9]

Lamont regarded his refractor as superior to Herschel's 18.7" reflector for several reasons: its optical quality, precise equatorial mounting and equipment (e.g. the excellent Fraunhofer micrometer). This allowed exact measurements. His observations of nebulae were focused on the following points:

- resolution: into stars or at least into nebulous parts (as an indication for single stars, analogous to the existence of smaller subgroups in star clusters); measurement of the stars in the vicinity, proving a possible association with the nebula
- shape: from sharply defined (e.g. planetary nebula) to diffuse (e.g. the Orion Nebula); check of variability by comparison with Herschel's descriptions

---

[7] Lamont (1837: 13).
[8] Lamont (1838: 134).

[9] During 1835–39 the Munich refractor was the world's most powerful (though not the largest); it was soon surpassed by the 15-inch at Pulkovo (Brachner 1988; Häfner 1990; Witt 1999).

Table 6.1. *Objects observed by Lamont; 'N1' and 'N2' are the numbers of drawings in Ueber die Nebelflecken (Lamont 1837) and 'Refractor-Beobachtungen …' (Lamont 1869), respectively*

No.	NGC	Date	M	H	h	Type	V	Con.	N1	N2
1	224	13.10.1836	31		50	Gx	3.5	And		2
2	2415	1.5.1837		II 821	456	Gx	12.2	Gem		
3	2681	1.5.1837		I 242	530	Gx	10.2	UMa		
4	2683	1.5.1837		I 200	532	Gx	9.7	Lyn		
5	2695	2.4.1837		II 280	536	Gx	11.9	Hya	8	10
6	2775	1.5.1837		I 2	564	Gx	10.4	Cnc		
7	2903	3.4.1837		I 56	604.1	Gx	8.8	Leo		
8	3423	1.5.1837		IV 6 = II 131	777	Gx	10.9	Sex		
9	3623	7.5.1837	65		854	Gx	9.2	Leo	6	7
10	3627	7.5.1837	66		857 = 875	Gx	8.9	Leo		
11	4565	20.6.1837		V 24	1357	Gx	9.5	Com		
12	4594	20.6.1837	104	I 43	1376	Gx	8.3	Vir		
13	4762	5.6.1837		II 75	1466	Gx	10.1	Vir		13
14	4900	5.6.1837		I 143	1509	Gx	11.3	Vir		
15	5033	5.6.1837		I 97	1564	Gx	10.0	CVn		
16	5377	5.6.1837		I 187	1712	Gx	11.3	CVn		
17	5775	28.6.1837		III 554	1885	Gx	11.8	Vir		
18	5813	28.6.1837		I 127	1896	Gx	10.5	Vir		
19	5838	28.6.1837		II 542		Gx	10.9	Vir		
20	5907	25.8.1837		II 759	1917	Gx	10.4	Dra		
21	6210	7.7.1837			1970	PN	8.8	Her	1	
22	6572	7.7.1837			2000	PN	8.1	Oph		
23	6720	21.10.1837	57		2023	PN	8.8	Lyr		14
24	6781	7.7.1837		III 743	2037	PN	11.4	Aql	7	3
25	6818	26.7.1837		IV 51	2047	PN	9.3	Sgr	2	11
26	6853	10.7.1837	27		2060	PN	7.4	Vul		5,6
27	6905	10.7.1837		IV 16	2075	PN	11.1	Del	5	12
28	7009	7.10.1836		IV 1	2098	PN	8.0	Aqr	4	4
29	7331	2.8.1837		I 53	2172	Gx	9.5	Peg		
30	7332	2.8.1837		II 233	2173	Gx	11.0	Peg		4
31	7479	2.8.1837		I 55	2205	Gx	10.9	Peg		
32	7662	27.7.1837		IV 18	2241	PN	8.3	And	3	9

- rotation or oscillation (relative to fixed points): strongly elliptical or low-mass objects should show a large 'spring force'
- proper motion: determination of an 'advancement in space' for sufficiently near objects

From October 1836 to October 1837 Lamont observed 32 objects (Table 6.1) that were compact enough to allow precise measurements (galaxies, planetary nebulae). Using the micrometer, he determined their positions relative to reference stars (distance and position angle). For some objects drawings were made. Unfortunately the results were not published until 1869 (Lamont 1869).

Lamont was interested in possible changes in position of nebulae. For h 536 (NGC 2695 = II 208) and h 2075 (NGC 6905 = IV 16) he claimed that the positions had varied, since William Herschel's descriptions

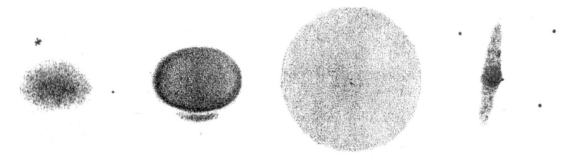

Figure 6.3. Examples of Lamont's drawings (left to right): NGC 2695, NGC 7009, NGC 6905 and M 65 (Lamont 1837).

no longer fitted. The positions of both objects (a galaxy in Hydra and a planetary nebula in Delphinus) were not measured. D'Arrest stated in 1854 that the position of IV 16 had remained constant (d'Arrest 1856b), which was confirmed by Auwers with the Königsberg heliometer (Auwers 1862c). Lamont further believed that the position angle of h 854 (M 65) had changed significantly. Another case was the planetary nebula h 2098 (NGC 7009, the Saturn Nebula), which had been observed by both Herschels several times, for which they gave discrepant descriptions: round, oval, one time larger, then smaller, in one case a bit displaced. D'Arrest could not confirm this in 1855 (d'Arrest 1856b); again he was supported by Auwers, who precisely measured the size of the nebula (23.5″ × 18.0″) and the position angle (72.8°) with the Königsberg heliometer (Auwers 1862c). Concerning the contradictory statements, Lamont correctly concluded '*I could mention many examples of this kind* [NGC 7009], *but I believe to have shown sufficiently how different the nebulae appear, depending on air quality and telescope.*' Some of the 'changing' nebulae were sketched by Lamont (Fig. 6.3).

The nebula no. 19, which was found on 28 June 1837 near no. 18 = h 1896 (galaxy NGC 5813 in Virgo), was new to Lamont since it is not listed in the Slough catalogue. Though his position is very rough, the description and position relative to two nearby stars clearly indicate the galaxy NGC 5838 (Fig. 6.4). The object was discovered by William Herschel in 1786 and listed as II 542. Why did Lamont not notice this? John Herschel's catalogue of 1833 was the standard source for the identifications of non-stellar objects. Lamont should have known that Herschel had not observed and catalogued all of the objects his father did. The nebula no. 19 is among the missing 814!

Figure 6.4. Lamont's 'new' nebula no. 19: the galaxy NGC 5838 (II 542) in Virgo (DSS).

Perhaps it was too time-consuming for him to dig into William Herschel's catalogues, which contain no coordinates.

6.1.4 Problems with the interpretation and presentation of the results

About a year after his Munich talk, Lamont changed his mind. Obviously due to the poor observational results, he doubted that all nebulae are mere star clusters and took the opposite position (Lamont 1839) – like William Herschel before, who had even revised his ideas about nebulae and star clusters twice.

Lamont now reflects that, if one observes a cluster in a small telescope or defocuses the image in a larger one, it appears nebulous, showing an irregular or diffuse boundary. Since a few nebulae do indeed look like this (e.g. M 31), Lamont states that *'this can all be imitated by an artificial transformation of star clusters'*.[10] He rhetorically states *'If we thus assume that all nebulae consist of stars, it would lead to the conclusion that they must be irregularly bounded.'*[11] But *'The experience firmly shows the opposite. Perfectly regular bodies are by no means rare, and sharp boundaries, paired with round or oval shape, can often be found in the odd class of planetary nebulae.'* This causes a contradiction: *'Either the induction is not allowed, or the assumption that all nebulae consist of stars is unfounded.'* For Lamont only the latter was acceptable, but with a bit of resignation he wrote that *'Many years, probably centuries, will pass to come to a decision.'*[12] Realising that he was not able to find out the truth with his (small) instrument, Lamont turned away from the nebulae in 1839. Thirty years later Lamont wrote *'I soon abandoned the study of nebulae, because I was convinced that all are remote clusters of stars and noticeable changes might be expected only in large periods of time.'*[13] Of course, he was right and, fortunately, the time gap was not as large as expected. As early as in 1864 Huggins could show with the aid of spectroscopy that the Andromeda Nebula (showing a blurred border) is indeed a kind of 'defocused star cluster', whereas planetary nebulae (showing sharp boundaries) do not consist of stars (Huggins W. 1864).

Looking critically at Lamont's observing programme, the published data show (compared with later works of others) no great substance, neither in quality nor in quantity. Nevertheless, it was a first and, above all, theoretically based attempt to step out of the long shadow of William Herschel. In the end, the instrumental resources were not sufficient to achieve any progress.

Johann Heinrich Mädler, Director of Dorpat Observatory, comments on Lamont's programme in the chapter 'Die Nebelflecke und die ihnen ähnlichen Bildungen'[14] of his book *Populäre Astronomie* [*'Popular astronomy'*]: *'On the basis of the work of the two Herschels, a few years ago Lamont started a study of all nebulae with the 18ft telescope, which raises the observatory at Bogenhausen to be among the first in the world. He not only wants to determine precise positions, but, moreover, plans to give a detailed physical description of those masses.'*[15] However, Lamont did not intend to observe 'all nebulae'. About the current state, Mädler noted in 1849 that *'Up to now, we have got only a few, but highly important, results of this work.'*

Mädler was a supporter of William Herschel's ideas that both resolvable and true nebulae exist. He first states *'that most nebulae are resolvable, i.e. true star clusters'*.[16] Owing to their irregular form, they must consist of stars rather than of nebulous matter, since only *'a body composed of separate, independent masses (stars) is able to take every shape'*. However, true nebulae are different: *'For the more regular nebulae, though not just sharply bounded like planetaries, the assumption is acceptable that they do not at all consist of stars, but of diluted luminous matter, quasi-star-like, related to the denser bodies of actual stars like comets to planets.'*[17] Concerning the planetary nebulae, he writes *'For the strange planetary nebulae this is actually the only explanation, because a sharp boundary of a cluster consisting of many distant stars would be at least a remarkable coincidence, which, given the innumerable equally possible forms, would surely not repeat itself 78 times out of 2500.'* Here Mädler refers to the 78 objects of class IV ('planetary nebulae'). He also agrees with Herschel's idea about the creation of stars and clusters by means of gravity: *'What is now still a nebula will shine as a star cluster one day.'*[18] The chapter ends with a description of 40 nebulae and 23 clusters, arranged by h-number. Wherever possible, he cites Lamont's observations.

Also Schönfeld complained that Lamont had made no progress. In his talk 'Ueber die Nebelflecke' of 1861 he states that *'Lamont's work seems to be held up, as only a few results are published, but for years nothing new is announced.'*[19] In 1866 Auwers was still waiting for results, mentioning in his review of the General Catalogue Lamont's *'until 1863 published or at least announced works'* (Auwers 1866).

---

[10] Lamont (1839: 182).
[11] Lamont (1839: 183).
[12] Lamont (1839: 180).
[13] Lamont (1869: 305).
[14] 'The nebulae and similar bodies'.
[15] Mädler (1849: 449).
[16] Mädler (1849: 451).
[17] Mädler (1849: 452).
[18] Mädler (1849: 453).
[19] Schönfeld (1862b: 66–67).

6.2 THE SHORT CAREER OF EBENEZER PORTER MASON

In 1838 Mason built the largest telescope in the United States, making accurate observations and drawings of selected nebulae. From John Herschel he received much praise for this work. Moreover, in his homeland, Mason was regarded as one of the greatest scientific talents. Sadly he died at the early age of 21.

6.2.1 Short biography: Ebenezer Porter Mason

Ebenezer Porter Mason was born on 7 December 1819 in Washington, Litchfield County (Connecticut), where his father, Stephen Mason, was a pastor. His mother, Elizabeth Brown, died when he was three years old. At the age of eight he moved to his aunt Harriet Brown Turner in Richmond, Virginia. Supported by her, Mason's interest in astronomy grew, and he showed a great talent for drawing too. With the aid of a sky-globe he became familiar with the constellations. In 1830 he moved to Nantucket, Massachusetts, where his father had settled down. This was a famous place, since it was the home of the well-known astronomer William Mitchell and his daughter Maria Mitchell (who was of nearly the same age as Mason).[20]

At the age of 13 Mason attended the school in Hartford, learning Latin and French. In 1835, now 16, he entered Yale College in New Haven (Connecticut). Denison Olmstedt, Professor of Astronomy, soon recognised his great scientific talent.

With a lent 6" reflector of Herschel-design, Mason observed Jupiter and Saturn, but soon focused on double stars and nebulae. He was also interested in constructing telescopes. Together with Hamilton Smith, a fellow student of the same age, he built a Herschel reflector with a 12" metal mirror.[21] The wooden mounting was a simplified version of the 15-inch built by the Scot John Ramage. Smith was responsible for the financing and construction, whereas Mason concentrated on the scientific goals. However, he was not interested in mere visual observing or the search for new objects, but planned accurate studies of selected nebulae.

At Yale the first signs of tuberculosis appeared. A year before his examination in 1839, Mason became Olmstedt's assistant. Afterwards he was asked to write an appendix on 'Practical astronomy' for the new edition of Olmsted's *Introduction to Astronomy*. It should contain astronomical calculations and inform the reader about the use of telescopes.

At the same time Mason and Smith published the results of their observations of nebulae in the distinguished *Transactions of the American Philosophical Society*. The manuscript, titled 'Observations on nebulae with a fourteen feet reflector', was received on 17 April 1840. Mason's health became worse and he hoped that open-air work would improve it. In the summer of 1840 he helped to determine the border between Maine and Canada. After his return to Yale, he felt better and could finish his 'Practical astronomy', but soon the disease became dramatic. Mason was brought to his aunt in Richmond, where he died on 26 December 1840 at the age of 21. His publication on nebulae appeared posthumously (Mason and Smith 1841).

Olmstedt, shocked by the early death of his favourite student, wrote a book about Mason (Olmstedt 1842). There one reads '[Mason's] *in early youth developed talents for Practical Astronomy* [are] *so extraordinary, as to leave no doubt, that, had his life been spared, he would have risen to a rank among the first astronomers of the age* […] *Science will long mourn the loss of a youth signally endowed with that rare assemblage of qualities essential to the structure of a great astronomer in which are united the refined artist and the profound mathematician.*'[22]

Smith graduated in 1839 at Yale and went afterwards to Ohio City. The 12" reflector was re-erected there on a more stable mounting. From 1868 to 1900 he was Professor of Astronomy at Hobart College (Geneva, NY). Hamilton Smith died in 1903.

6.2.2 The observations of Mason and Smith

In his publication, Mason wrote that "*The main object of this paper is to inquire how far that minute accuracy which has achieved such signal discoveries in the allied department of 'the double stars' may be introduced into the observations of nebulae, by modes of examination and description more peculiarly adapted to this end than such

---

[20] On Maria Mitchell, see Kerner C. (2004).

[21] It became the largest American reflector, surpassing the 8.5-inch of Amasa Holcombe, in Southwick (MA).

[22] Obituary: Anon (1845); see also Treadwell (1943) and Ashbrook (1984: 388–392).

Table 6.2. *The nebulae studied by Mason*

Object	NGC	M	H	h	Con.	Type
Trifid Nebula	6514	20	IV 41, V 10–12	1991 = 3718	Sgr	EN+DN
Omega Nebula	6618	17		2008	Sgr	EN
Veil Nebula	6992		V 14	2092	Cyg	SNR
Veil Nebula	6995			2093	Cyg	SNR

as can be employed in general reviews of the heavens." Mason's quantitative method should produce data '*by which future changes, if there be any, can be recognised and detected in at least a few of these wonderful sidereal systems*'. This purpose matched that of Lamont in Munich, formulated two years earlier. The difference was that Lamont could hardly implement his programme, whereas Mason (not knowing Lamont) produced eminent results in a very short time! John Herschel was visibly impressed, writing in the Cape catalogue '*Mr. Mason, a young and ardent astronomer, a native of the United States of America, whose premature death is the more to be regretted, as he was (as far as I am aware) the only other recent observer who has given himself, with the assiduity which the subject requires, the exact delineations of nebulae, and whose figures I find at all satisfactory.*'[23]

In the summer of 1839, Mason effected an intensive observing programme for three large, bright nebulae: M 17, M 20 (both in Sagittarius) and the eastern arc of the Veil Nebula in Cygnus, with its two parts NGC 6992 and NGC 6995 (Table 6.2). For each object the procedure was identical: first, the positions of the brighter stars around the nebula were measured with the 5″ Dolland refractor of the Yale College Observatory, equipped with a filar-micrometer. Then fainter stars were added to the map by visual observation with the self-made 12″ reflector (the standard eye-piece had 30′ field of view). Finally, Mason drew the nebulae with maximum precision relative to the star pattern.

Mason observed the Omega Nebula M 17 six times, from 1 to 9 August 1839. Altogether 37 stars were measured with the Dolland refractor. For the Trifid Nebula M 20 there are six observations too, made from 12 June to 14 August 1839, measuring 29 stars. The object, a mix of emission nebula and dark nebula, shows considerable brightness differences.[24] To visualise them, Mason developed an isophotic chart, described as a '*method by lines of equal brightness*' (Mason and Smith 1841). The contours render the visual impression '*imagined to surround all those portions of the nebula which are of uniform brightness*' (Fig. 6.5). Mason was one of the first to see the fainter northern part of M 20, a pure reflection nebula. About 40 years later, Swift announced another 'nebula' in the vicinity: '*To my extreme surprise I detected still another very large nebula close following the trifid, which, strange to say, is also trifid in character, having a branch or prong extending to, and mingling with, Mason & Smith's nebula.*' (Swift 1885d). However, the object could not be verified.

The Veil Nebula in Cygnus was first seen by William Herschel on 5 September 1784 as a 'very large nebula', V 14. Another part was found by his son on 7 September 1825 (h 2093). Mason revealed that the two objects are connected: '*Mr. Smith and myself were both able to trace the nebula continuously from one to the other, and the reverse, so that these are now satisfactorily ascertained one immense nebula, stretching through several fields 30′ in diameter.*' With the Dollond refractor 196 stars were measured. Owing to Mason's observation, John Herschel corrected his own result: '*In conformity with Mr. Mason's remarks on my obs. of this neb. and with his elaborate and excellent monograph of which it forms a part, I have diminished the P.D. [north polar distance] of cat. of 1833 by 1°. It is evident that the index reading must have been mistaken, 1° or 0°, sweep 8 examined; the writing is clear and the reduction correct, but the conclusion from Mr. Mason's obs. is irresistible.*'[25] NGC 6992 and NGC 6995 form the eastern arc of the impressive supernova remnant. The western arc (near

---

[23] Herschel J. (1847: 7–8).

[24] William Herschel catalogued M 20 four times (see Section 2.2.2).

[25] Note to NGC 6995 in the New General Catalogue.

96   The time after Herschel

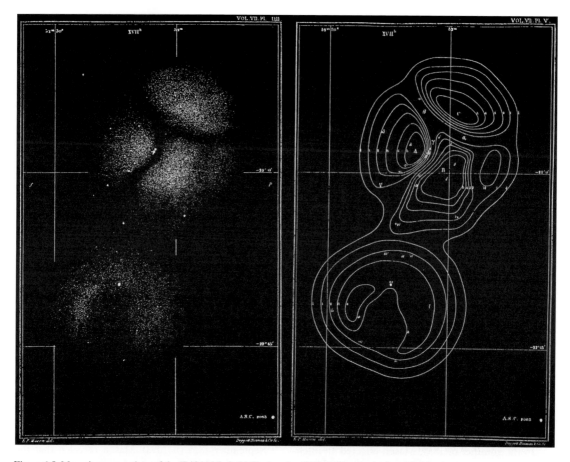

Figure 6.5. Mason's presentations of the Trifid Nebula M 20: drawing (left) and isophotic chart (right).

52 Cygni), which was discovered by William Herschel on 7 September 1784 (V 15 = h 2088), was catalogued by Dreyer as NGC 6960.

6.3 TWO 'NEW' NEBULAE OF BIANCHI

Similarly to Lamont's 'new' nebula, the two objects found by Bianchi are listed in William Herschel's catalogues. This is another instance of the fact that observers were too much fixed on the Slough catalogue. Though John Herschel's compilation is more detailed, it does not contain all of the objects catalogued by his father.

6.3.1 Short biography: Giuseppe Bianchi

Giuseppe Bianchi was born on 13 October 1791 in Modena, Italy. He studied mathematics and astronomy in Padua and Milan. In 1819 he became Professor of Astronomy at the University of Modena, where in 1826 he erected a small observatory equipped with instruments by Reichenbach and Amici. Bianchi observed there until 1859, when he was removed from office. Later he could use the private observatory of Montecuccoli. Giuseppe Binachi died on 25 December 1866 in Modena at the age of 75.[26]

6.3.2 Bianchi's discoveries

In 1839 Bianchi found two nebulae with the 5-ft meridian-circle by Reichenbach (Table 6.3). His observation was published in the *Astronomische Nachrichten* as a letter to the editor (Schumacher), dated 6 July 1839. He wrote that '*The first priority is to announce two small*

---

[26] Obituary: Ragona (1868).

Table 6.3. *Bianchi's two nebulae*

Date	NGC	H	GC	Type	V	Con.	Remarks
11.6.1839	6229	IV 50	4244	GC	9.4	Her	W. Herschel 12.5.1787
16.6.1839	6543	IV 37	4373	PN	8.1	Dra	W. Herschel 15.2.1786; Lalande: 9-mag star

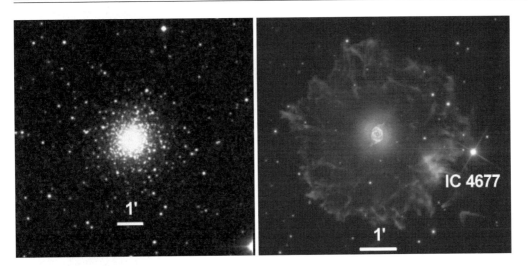

Figure 6.6. Bianchi's discoveries. Left: the globular cluster NGC 6229 (DSS); right: the planetary nebula NGC 6543 with the bright knot IC 4677, which was discovered by Barnard.

nebulae. I have found them during my observations of stars and there is no word or hint about them in the charts and catalogues which I could consult.' (Bianchi G. 1839). The first nebula, discovered on 11 June 1839, is the globular cluster NGC 6229 in Hercules (Fig. 6.6 left). The second was seen five days later: the bright planetary nebula NGC 6543 in Draco, located at the northern pole of the ecliptic (Fig. 6.6 right). Bianchi measured precise positions but could not find any suitable objects in John Herschel's Slough catalogue.

The first to react to Bianchi's note was Frederick Kaiser, Director of Leiden Observatory. In a letter to the *Astronomische Nachrichten* of 21 December 1839 (Kaiser 1839), he wrote about the second object (NGC 6543), on the basis of his observations with the 16.6-cm Fraunhofer refractor: '*The object in Draco is very strange. It is not round, as stated by Mr Bianchi, but elliptic. [...] This beautiful planetary nebula is neither new nor has it appeared in the sky recently.*' Curiously, Kaiser did not identify it with Herschel's IV 37, but states that '*it is listed on page 361 of the* Histoire Céleste *as a star of 9th magnitude*'. Indeed, the object is listed as a star in Lalande's catalogue *Histoire Céleste Française* of 1801 and was first measured on 26 July 1790 (LL 38303). On Bianchi's first, more extended nebula Kaiser wrote '*The nebula in Hercules is not listed in the* H.C. [*Histoire Céleste*]*, which indeed is less probable, as only the object in Draco appears exactly like a star of magnitude 8.9 in a low-magnifying telescope.*'

Eighteen years later Winnecke observed both objects, using the 6" heliometer at Bonn Observatory. He was the first to refer to William Herschel's catalogues (Winnecke 1857). As already in the case of Harding, he used Auwers' revision, which he got prior to its publication (1862). With the coordinates given there, Winnecke could easily see that Bianchi's nebulae had already been discovered in 1787 and 1786, and catalogued as IV 50 and IV 37, respectively. Thus the matter was similar to the case of Lamont's object no. 19 (II 542).

The objects are listed in John Herschel's General Catalogue as GC 4244 and GC 4373. Strangely the former (NGC 6229) was entered by William Herschel in class IV ('planetary nebula') too, though he could resolve the outer regions of the globular cluster. Otherwise, NGC 6543 (IV 37) is a true PN, now called the Cat's Eye Nebula. In 1872 d'Arrest reviewed the existing observations of this remarkable nebula to prove that it had a proper motion (d'Arrest 1872a). The result was negative. His spectroscopic investigation showed that '*The line spectrum of IV. 37 is striking and bright.*'

6.4 LORD ROSSE: FIRST OBSERVATIONS AT BIRR CASTLE

In the nineteenth century Birr Castle was among the most important centres for visual observations of nebulae.[27] Its great reflectors were used for this task from 1839 to 1878. The first date marks the erection of the 36-inch, which was surpassed in 1845 by the famous 72-inch. This section covers the first period until 1850, which was perhaps the most active time of William Parsons, Third Earl of Rosse. Lord Rosse, as he is generally called,[28] was a multi-talented man like William Herschel. He not only constructed the optics and mechanics of his telescopes, but also used them for visual observations and interpreted the results with scientific rigour. He was supported by able assistants, some of whom had considerable careers later.

Owing to the private character of the observatory, the main task was not classic astronomy, but the detailed observation and drawing of nebulae to reveal their nature – following the tradition of both Herschels. No doubt, the early Birr Castle observations – described in three papers of 1840, 1844 and 1850 in the *Philosophical Transactions* – brought many fresh ideas and much inspiration. Parallel to Lord Rosse, Romney Robinson, Director of Armagh Observatory and a frequent user of the great reflectors, published three reports, which appeared in 1840, 1845 and 1848. He was an eloquent opponent of the popular nebular hypothesis which claimed the existence of true nebulosity. Robinson's central theme was the resolvability of nebulae. However, due to Lord Rosse's discovery of the spiral structure of M 51 in April 1848, the interest shifted from 'resolution' to 'structure'. It is interesting to compare the different attitudes of these two eminent Irish astronomers.

6.4.1 Short biography: William Parsons (Lord Rosse)

William Parsons was born on 17 June 1800 in York, England (Fig. 6.7). As the eldest son of the Second Earl of Rosse, he received in accord with tradition the title Lord Oxmantown (Murphy 1965). He grew up at Birr Castle, which lies at the western border of the village of Parsonstown (now Birr) and was educated privately. In 1818 he spent a year at Trinity College, Dublin, then changing to Magdalen College, Oxford. There Parsons studied mathematics, physics, chemistry and engineering. In 1822 he got his degree in mathematics. Early in life he had to take on official duties, being, for instance, a Member of Parliament for his native King's County (now Offaly) from 1821 to 1834.

William Parsons' knowledge of astronomy grew and in 1824 he joined the young Astronomical Society of London. All his life he was an independent mind, acting freely on his own expense, equipped with a broad spectrum of scientific knowledge. As an engineer with great skill, he logically was inspired by William Herschel and the reflecting telescope. For William Parsons, being not a representative of classic (positional) astronomy, this was the ideal tool with which to step into 'astrophysics' – following his master.

Figure 6.7. William Parsons, Third Earl of Rosse (1800–1867).

---

[27] McKenna (1967: 287–289).
[28] Occasionally the name 'Lord Rosse' was later used for his son too; to avoid any confusion he is called Lawrence Parsons here.

Facing the high performance of Fraunhofer's refractors, William Parsons nevertheless was convinced that the reflector had not become obsolete. He recognised that refractors *'were limited to a very small scale, owing the impossibility of producing [...] glass lenses of apertures at all approaching the late Sir William Herschel's reflectors'*.[29] On the other hand, *'many practical men whom I have spoken to, seem to think that since Fraunhofer's discoveries, the refractor has entirely superseded the reflector and that all attempts to improve the latter instrument are useless'*. This was motivation for him to construct large reflectors. That he was right was acknowledged later by one of the most distinguished representatives of classic 'refractor' astronomy: the Astronomer Royal, George Airy. After visiting Birr Castle in summer 1848 he remarked *'it is easy to see that the vision with a reflecting telescope may be much more perfect than with a refractor'*.[30]

William Herschel had published only rough descriptions of his mighty telescopes, leaving out the essential details of their construction.[31] Therefore Lord Rosse had to acquire the whole process: casting and grinding of the mirror, the design of the optical system and the construction of the mounting. The mirror, usually called the 'speculum',[32] was made of an alloy of two parts copper and one part tin (sometimes adding a bit of arsenic or zinc). The production was difficult: some blanks had cracks or even broke.

In 1826 William Parsons built his first telescope, a reflector of aperture six inches and focal length 2 ft. Two years later he published a detailed report on its construction in the *Edinburgh Journal of Science*, titled 'Account of a new reflecting telescope' (Parsons W. 1828). Never being cagey about the details of his constructions, he was lauded for publishing his methods. In the same year he also developed a machine to polish metal mirrors. Then, in 1830, he built a 15" reflector (later mounted equatorially) and soon afterwards another one with aperture 24", both on Herschel-style mountings. Both mirrors were made of segments – anticipating a common technique for the great reflectors of today. The astonishing success created a stir and led to his admission to the Royal Society in 1831.

The same year, William Parsons took up office as Lord Lieutenant of King's County, responsible for the supply and safety of the people. The times were troubled, characterised by poverty and violence; a weapon was useful in the night. Parsons was a man of law and order – well respected by many, but occasionally treated with hostility. In 1836 he married Mary Wilmer-Field from Heaton-Hall, Yorkshire. The later Countess of Rosse became an ambitious photographer, developing the plates in her own darkroom.[33] Altogether eleven children were born, but only four reached adulthood: Lawrence (1840), later Fourth Earl of Rosse and Lord Rosse's scientific heir, Randal (1848), later Reverend, Richard (1851) and Charles (1854).[34] The latter became famous as inventor of the steam-turbine and co-owner of the telescope company Grubb-Parsons (Dublin).[35]

In September 1839 a 36" reflector, the '3-foot', was erected[36] – having twice the aperture of William Herschel's standard instrument. The first mirror was made of 16 segments, but it was replaced by a monolithic one in November 1840.[37] Having already speculated about a reflector of double size that year, it was not long before William Parsons started its construction. After some failures a first mirror was cast in April 1842. Though the 72" reflector and its environment (tube and mounting, held by two massive flanking walls) was soon

---

[29] Parsons W. (1828). Fraunhofer was jokingly referred to as 'Frauendevil in Munich' by Thomas de Quincy, a contemporary of Lord Rosse (de Quincey 1846: 575).

[30] Schaffer (1998a: 211).

[31] William Herschel treated the construction of large metal mirrors as a 'trade secret'; see Dreyer (1912a: xlviii).

[32] This is the Latin word for mirror. In the early period of telescope making, the term was used as a synonym for 'metal mirror', but later silvered glass mirrors were also called 'speculum' or 'glass speculum'.

[33] Her work is now on display at the Birr Castle Museum; some photographs, showing telescopes too, were published by the Birr Scientific Heritage Foundation; see Davison (1989).

[34] Four children (born in 1842, 1843, 1845 and 1850) died before christening; three others, Alice (1839), William (1844) and John (1846), died young (Davison 1989).

[35] Chant (1931).

[36] Claridge (1907). A similar construction was John Ramage's 15" reflector (Fig. 1.7 left), which was used during 1820–36 at the Royal Observatory Greenwich (Dick T. 1845: 308–311; Chapman A. 1989a).

[37] In the early 1860s Lord Rosse had replaced the round solid tube by a lighter truss-tube with quadratic cross-section; see pictures in Hoskin (1982c) and Davison (1989).

Figure 6.8. Lord Rosse at the 72" reflector.[38]

built,[39] it was not used regularly until March 1845. The famous telescope of focal length 54 ft (Fig. 6.8) – called the '6-foot' or 'Leviathan of Parsonstown' – surpassed even Herschel's legendary '40 ft' (aperture 48") of 1789. A larger telescope was not be erected until 1917, when the 100" Hooker reflector was made ready on Mt Wilson, California.

When William Parsons' father died in 1841, he became the Third Earl of Rosse. In 1843 Lord Rosse was elected President of the British Association for the Advancement of Science. In 1845 he was named as the representative of the Irish Republic in the London House of Lords, which office he could not really fulfil for several reasons. The great reflector stood by and he was excited about observing with it. After spectacular early successes, such as revealing the spiral structure of M 51 in April 1845, Lord Rosse was forced to stop all astronomical activity in 1846. The reason was the disastrous Irish potato famine, which lasted until 1848.[40] As Lord Lieutenant he was strongly involved, doing everything in his power to help the people. His social engagement brought him much sympathy.

From 1848 to 1854 Lord Rosse was President of the Royal Society. In 1851 he received their highest award, the Royal Medal, for his achievements in telescope making and his observations of nebulae, especially the discovery of spiral structure. Edward Sabine said in his laudation '*You have reopened a field of investigation which seemed almost exhausted by two of the most illustrious observers* [William and John Herschel] *at this or any age.*' (Sabine 1853). It is remarkable that Lord Rosse was never awarded the RAS gold medal, '*a somewhat curious omission which it is difficult to justify*' as Edmond Grove-Hills later commented.[41] From 1862 Lord Rosse was a member of the Royal Society commission for the 'Great Melbourne Telescope'. The 48-inch, built by Grubb, was the last large metal mirror reflector – however, it was found to be a complete failure (see Section 9.21).

All of Lord Rosse's duties resulting from his status and reputation were perfectly carried out – though they

---

[38] This drawing, showing the Earl at the 72-inch, is one of several made by Piazzi Smyth in 1850 (Brück M. 1988).
[39] The final shape was already visible in autumn 1843. A year later, the enormous instrument inspired Thomas Woods to write a 'travel guide' to the *Monster Telescopes Erected by the Earl of Rosse*, which was printed in Parsonstown (Woods 1844). Birr Castle became a site of pilgrimage.

[40] Moreover, his first child, Alice, died in 1847.
[41] Grove-Hills (1923: 114).

Table 6.4. *The scientific assistants of William and Lawrence Parsons (* labels those who were tutors of Lord Rosse's sons)*

Name	Life dates	Period of duty	Assistant of
William Hautenville Rambaut	13.4.1822–26.8.1893	Jan.–Jun. 1848	William Parsons
George Johnstone Stoney	15.2.1826–5.7.1911	Jul. 1848–Jun. 1850, Aug.–Dec. 1852, Aug. 1854, Apr. 1855, Mar. 1856	William Parsons
Bindon Blood Stoney	13.6.1828–5.5.1909	Jun. 1850–Sep. 1852	William Parsons
R. J. Mitchell*		Dec. 1853–May 1858	William Parsons
Samuel Hunter*		Feb. 1860–May 1864	William Parsons
Robert Stawell Ball*	1.7.1840–25.11.1913	24. Nov. 1865–Aug. 1867	Lawrence Parsons
Charles Edward Burton*	16.9.1846–9.7.1882	Feb. 1868–Mar. 1869	Lawrence Parsons
Ralph Copeland	3.9.1837–27.10.1905	Jan. 1871–May 1874	Lawrence Parsons
John Louis Emil Dreyer	13.2.1852–14.9.1926	Aug. 1874–Aug. 1878	Lawrence Parsons
Otto Boeddicker	17.11.1853–31.8.1937	1880–Feb. 1916	Lawrence Parsons

often prevented him from practising his astronomical interests. In the 1850s the nightly tasks were left more and more to his scientific assistants. Despite his great astronomical success, Lord Rosse all his life felt himself to be more like an engineer. His complete work was published in 1926 by his youngest son Charles: *The Scientific Papers of William Parsons, Third Earl of Rosse, 1800–1867* (Parsons C. 1926). He was pictured by his contemporaries as a modest, highly educated man of noble character. Robert Ball remarked that '*His character is in every respect a very amiable one; a mild gentle disposition, most careful never to utter a word that could give offence or cause pain to those with whom he had intercourse.*' (Ball V. 1940.)

From 1862 until his death, Lord Rosse was chancellor of Trinity College, Dublin. He died after a long illness (caused by a tumour in his knee) on 31 October 1867 in Monkstown near Dublin at the age of 67. His grave is at St Brendan's Church in Birr.[42]

6.4.2 The scientific assistants

William Parsons, and later his son Lawrence too, employed scientific assistants, who were responsible for observing and documentation (Table 6.4). There were gaps (e.g. in 1853) but hardly any overlaps. Johnstone Stoney worked for Lord Rosse several times. Some had duties as tutor of Lord Rosse's sons. Additionally, there were workers responsible for the technical aspects and maintenance of the telescopes, especially the positioning and guidance of the 72-inch. Not much is known about them. Robert Ball wrote that '*Four men had to be summoned to assist the observer. One stood at the winch to raise or lower, another at the lower end of the instrument to give it an eastward or westward motion, as directed by the astronomer, while the third had to be ready to move the gallery in and out, in order to keep the observer conveniently placed with regard to the eye-piece. It was the duty of the fourth to look after the lamps and attend to minor matters.*'[43] In the following, Lord Rosse's five scientific assistants are introduced (for those of Lawrence Parsons see Section 8.18).

6.4.3 Short biography: William Hautenville Rambaut

Not much is known about William Rambaut (Fig. 6.9 left), Lord Rosse's first scientific assistant. He was born on 13 April 1822 in Ireland. His grandfather, Jean Rambaut, came from Bordeaux and was married to Marie Hautenville (whence the second

---

[42] Obituaries: Robinson (1867b) and Challis (1869); see also Ball R. S. (1895: 272–288) and Chapman A. (1998: 96–100).

[43] Ball V. (1915: 64).

Figure 6.9. The first three scientific assistants at Birr Castle (from left to right): William Rambaut (1822–1893), Johnstone Stoney (1826–1911) and Bindon Stoney (1828–1909).

forename). William Rambaut was the nephew of the first wife (Elizabeth Rambaut) of Romney Robinson, long-time Director of Armagh Observatory. Rambaut's own nephew was Arthur Allcock Rambaut, assistant of Robert Ball at Dunsink Observatory (as successor of Dreyer) and later Astronomer Royal of Ireland.

In 1839 Rambaut entered Trinity College, Dublin, to study mathematics and physics. He was not awarded his degree until 1848; the delay was as a consequence of his clerical activities. Robinson convinced Lord Rosse to accept the aid of a scientific assistant. He recommended Rambaut and trained him for this task. Rambaut worked at Birr Castle from January to June 1848, mainly observing and sketching nebulae.[44] A few fine chalk drawings on black cardboard have remained, showing M 42, M 51, M 64, M 97 and M 99.[45] In August 1850 Rambaut became the second assistant of Robinson in Armagh. There he measured star positions for the First Armagh Catalogue with the meridian-circle, which were published in 1859. He married in 1862 and was ordained as a priest (from 1864 as the Reverend Rambaut). In July 1864 he was promoted to first assistant in Armagh, succeeding Neil Edmondson. He stayed at the observatory until September 1868; his successor was the Reverend Charles Faris.[46] Later he became the principal in Drumreilly, Kilmore. William Rambaut died on 26 August 1893 in Ireland at the age of 71 and is buried in Dublin.

6.4.4 Short biography: George Johnstone Stoney

Johnstone Stoney was born on 25 February 1826 in Oakley Park, Ireland (Fig. 6.9 centre). He studied mathematics and science at Trinity College, Dublin. In July 1848 he succeeded William Rambaut as Lord Rosse's assistant. Stoney was a keen observer, competently controlling the great reflector, which was in a very good condition at that time. He contributed micrometric measurements of M 51 and made a fine drawing of M 66 in 1849. He first stayed until June 1850 and returned to Birr Castle in August 1852, observing until the end of the year. A close friendship arose with Lord Rosse, who appreciated Stoney's advice all his life. When the new Queen's College, Galway, sought a Professor of Natural Philosophy, Lord Rosse, who always supported Stoney's scientific career, recommended him for the post. Stoney stayed there during the years 1853–57. During that time he often visited Birr Castle (observations in August 1854, April 1855 and March 1856 are documented).

In 1857 Stoney became secretary of Queen's University, Dublin and a member of the Royal Dublin Society. He became affiliated to the Royal Astronomical Society in 1860 and a year later to the Royal Society. Stoney had great knowledge about physics, astronomy and mathematics. He worked on electrodynamics and in 1874, convinced of the existence of an elementary electrical charge, he created the term 'electron'. In 1882 Queen's University closed and Stoney lost his position. Until 1893 he was Vice-President of the Royal Dublin Society, then he moved to London. There Johnstone Stoney died after a long

---

[44] A first person had been tried in autumn 1847, but it was not a success (Bennett 1990: 108). Moore incorrectly writes that Rambaut had been Lord Rosse's assistant for two years (Moore 1971).

[45] The drawings are stored in the cellar of Armagh Observatory.

[46] Dreyer (1883), Bennett (1990).

illness on 5 July 1911 at the age of 84, two years after his younger brother Bindon.[47]

6.4.5 Short biography: Bindon Blood Stoney

Bindon Stoney was born on 13 June 1828 in Oakley Park, Ireland (Fig. 6.9 right). He was two years younger than his brother Johnstone. At Trinity College, Dublin, he studied art, mathematics and engineering, being always among the best students. After university he could not practise a profession as engineer due to family affairs. He eventually decided to follow his brother to Birr Castle, succeeding him as Lord Rosse's assistant in June 1850. Lord Rosse wanted to employ a skilled drawer to render the Orion Nebula – and Bindon Stoney was the perfect choice. His observations and drawings at the 72-inch are excellent, as can be seen, for instance, from the beautiful illustrations of the Dumbbell Nebula (M 27) and the central part of the Orion Nebula (M 42), which he finished in 1852. Lawrence Parsons later wrote that he was a '*highly educated civil engineer [...] well accustomed to use his pencil*'.[48] Bindon Stoney is considered the discoverer of the spiral structure of the Andromeda Nebula.[49] His micrometric measurements of stars in M 27 and M 51 made during 1850–51 were recognised too.

In late September 1852 Bindon Stoney left Birr Castle to take up a job as engineer for the Spanish railway. Only a year later he returned to Ireland, where he became well known for some important constructions, such as the bridge over the river Boyne. He married in 1879 and was elected a Fellow of the Royal Society in 1881. He retired as chief-engineer of Dublin harbour in 1898. Bindon Stoney spent his last years in Australia, where he died on 5 May 1909 at the age of 80.[50]

6.4.6 Short biography: R. J. Mitchell

Not much is known about Mitchell.[51] He studied at Queens College, Galway. Probably after having been advised to do so by Johnstone Stoney, who was working there as a professor. Mitchell got a job as a scientific assistant of Lord Rosse in December 1853, succeeding Bindon Stoney (after a gap of more than a year). He was also responsible for the education of Lord Rosse's son Lawrence, who was 13 years old at that time (additionally he taught the younger sons Willmer, John and Randall[52]). Mitchell was a talented observer and drawer, making many important discoveries at the 72-inch. When the education of Lawrence Parsons needed a more profound knowledge in mathematics, he decided to leave Birr Castle in May 1858.[53] He was the scientific assistant with the longest stay (three years). Mitchell changed to a career in the civil service, as controller of the Irish Government's schools.

6.4.7 Short biography: Samuel Hunter

There is not much biographical information about Hunter, who may have been born in the mid 1830s. After Mitchell's departure, Lord Rosse inquired at the Dublin School of Design for a talented drawer, mainly to update and enlarge Bindon Stoney's drawing of the Orion Nebula. Hunter, who was then attending the well-known institution, was recommended. Owing to his additional astronomical knowledge, he was the ideal person. In February 1860 his service at Birr Castle began (after a gap of 1½ years with no assistant). Like his predecessor, Hunter showed great observing and drawing abilities. Lawrence Parsons praised his '*considerable amount of training as an artist*'.[54] Robert Ball noted later that '*He was a very enthusiastic astronomer; I often heard from the men* [technical helpers] *how he used to rout them all out of bed, sometimes in the middle of the night, and how sometimes after having closed up they reopened.*'[55] For the required drawing of M 42 (see Fig. 8.61) Hunter occasionally used the 72-inch in 'front-view' mode, i.e. without a secondary mirror. It was finished in 1864. To his duties belonged the education of Lord Rosse's sons Randall, Richard and Charles. Unfortunately Hunter had to leave Birr Castle

---

[47] Obituaries: Ball R. (1911), Jeans (1911) and Anon (1912).
[48] Parsons L. (1868: 57).
[49] Roberts later confirmed this feature by photography (de Vaucouleurs 1987).
[50] Obituary: Grubb (1910).
[51] Ball V. (1940: 199).

[52] Willmer and John died in 1855 and 1857, respectively; Lord Rosse's first child (and only daughter), Alice, had died in 1847.
[53] Gray took over Mitchell's job as tutor (responsible for Randall and Richard). After leaving Birr Castle a year later, he was followed by John Purser, who worked as a teacher for the next four years. Both had nothing to do with observations.
[54] Parsons L. (1868: 66).
[55] Ball V. (1940: 200–201).

in May 1864 due to ill-health, which was probably the consequence of exhaustion. Ball became his successor as scientific assistant.[56]

6.4.8 36-Inch observations: resolvability of nebulae and the nebular hypothesis

Regardless of the technical challenge, the motivation of Lord Rosse (at that time still Lord Oxmantown) for building large reflectors was simple: the question of the resolvability of nebulae. The issue goes back to William Herschel, who had changed his opinion several times (see Section 2.6.2). First he believed that some objects are star clusters, while others, like the Orion Nebula, might be true 'nebulae properly so-called'. On the basis of many observations (started already in Bath) he believed that he had seen changes in M 42, which he explained by invoking a deformable 'luminous fluid'. He again attacked the problem in 1784 in Datchet, by examining 29 'nebulae without stars' from Messier's (second) catalogue with his 12" reflector. He found out that all *'have either plainly appeared to be nothing but stars, or at least to contain stars, and to shew every other indication of consisting of them entirely'.*[57] He added another nine cases *'which in my 7, 10, and 20-feet reflectors shewed a mottled kind of nebulosity which I call resolvable'.*[58]

Then, having partly 'resolved' the Omega Nebula and the Dumbbell Nebula (M 17 and M 27) into stars with his new 18.7-inch, he was convinced that with sufficient telescope power all nebulae would eventually turn out to be mere star clusters – especially the Orion Nebula. Perhaps this was a reason for building his largest telescope, the legendary '40 ft', which was finished in August 1789. However, M 42 was observed only twice with the bulky reflector, yielding no spectacular news. Herschel's final swing came in 1790, when he discovered the planetary nebula IV 69 (NGC 1514), his famous 'star with an atmosphere'. The central star is surrounded by a diffuse envelope. He concluded that it must have been formed by gravity out of the 'luminous fluid'. Thus it was a true nebula, so he was right. For the rest of his life, William Herschel was convinced that there are two kinds of nebulae: the 'physical' (and starting point of cosmic evolution) and the 'optical' (unresolved star clusters as end-products of gravitational condensation).

John Herschel shared this view about the existence of true nebulosity – the Orion Nebula being his poster child too. He had observed the object during 1824–26 at Slough with the 18¼" reflector, partly assisted by the Scot John Ramage (Herschel J. 1826a). A fine drawing was made in January and February 1824. Herschel wrote *'Among all the theories which may be imagined respecting it, that which regards it as a self-luminous or phosphorescent material substance in a highly dilated or gaseous state, but gradually subsiding by the mutual gravitation of its molecules into stars and sidereal systems, must certainly in the present state of our knowledge be looked upon as the most probable.'* M 42 shows *'no appearance of being composed of small stars, and their aspect is altogether different from that of resolvable nebulae'.*

This was true for the Andromeda Nebula (M 31) too: *'evidently not stellar, but only nebula in a high state of condensation'.*[59] Airy wrote about both objects *'No one, I think, who has seen these in a telescope of great light [...] can persuade himself that these can be any thing but masses of nebulous matter.'*[60] However, John Herschel was critical about his M 42 drawing, noting that *'there can be no doubt that a telescope of greater power will exhibit many more of its phaenomena, and possibly discover other branches, or exhibit nebulous atmospheres about other small stars near, not noticed by me'.*[61] On the other hand, both Herschels had resolved many nebulae (mainly globular clusters). Therefore the concept of a coexistence of (true) nebulae and star clusters became the official interpretation among the contemporary astronomers. Consequently, John Herschel developed a new classification scheme of non-stellar objects, which he published in the Slough catalogue (see Section 5.4.1). It differentiates, for instance, 'resolvability' into five grades: 'discrete', 'resolvable', 'granulate', 'mottled' and 'milky'. The grades describe a (reversed) evolutionary sequence from star cluster to (true) nebula.

---

[56] The new tutors were Wittey, who stayed only a few months, and Lamprey (until November 1865).
[57] Herschel W. (1784: 440). Strangely, the galaxies M 31, M 51, M 65, M 66 and M 74 are in the sample.
[58] Among them were the supernova remnant M 1, the planetary nebulae M 27 and M 57 and the galaxies M 33, M 81, M 82 and M 101.

[59] Herschel J. (1833a: 377); see also Herschel J. (1826b).
[60] Airy (1836: 304–305).
[61] Herschel J. (1826a: 495); he even suggested that it would be desirable to *'re-accommodate the 40-feet reflector (now almost decayed by age and unavoidable exposure to weather)'*.

However, John Herschel was not always consistent in his 'cosmological' ideas.[62]

Besides these observational results, another (more theoretical) issue entered the scene: a cosmic version of the nebular hypothesis, which was brought into life in the remarkable year 1833[63] by the Cambridge lecturer William Whewell and John Pringle Nichol, who was from 1836 Director of Glasgow Observatory.[64] It was based on ideas of Pierre-Simon Laplace and William Herschel about the formation of the solar system. Nichol claimed the existence of true nebulosity in the form of a (spinning) cloud of luminous fluid, which gradually condenses into stars by virtue of gravity. Stars, which are born collectively, appear as clusters.

To popularise his view, Nichol published two books, which appeared in many editions: *The Architecture of the Heavens* (Nichol 1837) and *Thoughts on Some Important Points Relating to the System of the World* (Nichol 1846). Drawings of nebulae, as presented in these books, were intended to establish his reformist account of the law of progress at work in nature.[65] He saw evidence for true nebulosity in the observations of William and John Herschel. Both were unable to resolve the Orion Nebula into stars with their great telescopes and the former had reported a certain variability. Now, according to rumours, the latter had detected changes in the shape of the nebula too. This, however, was a false report. The truth came out in 1847 when the Cape catalogue appeared (though with much delay). There John Herschel compared his drawings made in Slough (1824) and Feldhausen (1837).[66] Indeed, article (67) is titled 'Of evidences of change in the nebula', but in the text he warns the reader not to believe in '*great and rapid changes undergone by the nebula itself*', explicitly stating '*I am far from participating in any such impression*'. The differences in the drawings are due to his '*inexperience in such delineations (which are really difficult) at an early period* [1824]'. He added that the '*supposed changes have originated partly from the difficulty of correct drawing, and, still more, engraving such objects*'. Additionally, an essential point might be the better viewing conditions in South Africa: the Orion Nebula rises much higher there. Therefore, John Herschel was sceptical about the nebular hypothesis (or parts of it), lamenting that visible evidence of condensation was still lacking. Though not a pure opponent, he was unable to accept the popular idea in its local, potentially atheistic form.[67]

Indeed, a fundamental conflict between (materialistic) science and religion appeared. The main protagonists were Nichol, advocating the 'atheistic' position, and the Reverend Romney Robinson[68] (Fig. 6.11 right), Director of Armagh Observatory. Robinson, an aggressive Ulster Episcopalian, preached against materialism, unbelief and evolution theory. He was an eloquent opponent of the nebular hypothesis, brushing off Nichol's *Architecture of the Heavens* as a catchpenny book designed for mass consumption.[69] He was convinced that true nebulosity does not exist and, therefore, all nebulae are mere star clusters. Resolution of nebulae into stars was supposed to remove the evidence for nebulous fluids in space. Consequently, all his observations were designed to convince the scientific community. Later Robinson was proud that the 'confirmation' of his view was exclusively based on Irish results.[70] His aversion to the nebular hypothesis (and its English and French advocates) sometimes led him to wishful thinking – against empirical evidence.

Being in close contact with both Nichol and Robinson, Lord Rosse often had to serve as a mediator. Though the protestant nobleman was an exponent of the Church of Ireland, he nevertheless appeared as an independent, free-thinking scientist. His goal was to prove whether a large reflector could reveal a starry nature of such 'hard' targets as the Orion Nebula or whether there

---

[62] Hoskin (1987).

[63] Obviously it was an 'annus mirabilis' of nebular astronomy, insofar as John Herschel's Slough catalogue and his famous textbook *A Treatise on Astronomy* appeared too.

[64] On the nebular hypothesis see also Williams (1852: 111–123); for a historical review see Clerke (1905); and for a modern presentation (including the effect on Birr Castle) see Schaffer (1989, 1998a). About the Glasgow Horselethill Observatory, see Roy (1993) and Hutchins (2008).

[65] Schaffer (1998b).

[66] Herschel J. (1826a), Plate VII; Herschel J. (1847), Plate VIII. The history of 'changes' in M 42 was later compiled by Holden (1882b).

[67] Schaffer (1998a: 201).

[68] About Robinson see e.g. Crowe (1971) or Bennett (1990).

[69] Schaffer (1998a: 208).

[70] Robinson was not even afraid of attacking the Englishman John Herschel. He wanted to send '*a Reflector on Irish principles*' to the Cape to '*put an extinguisher on Herschel's labours*' (Schaffer 1998b: 458).

are actually irresolvable nebulae. This was the controversial theoretical background for Lord Rosse's work – it was, indeed, a dangerous conglomerate of ideas.

In September 1839 Lord Rosse erected his first large telescope, the 36" reflector with 27-ft focal length. It had a closed metal tube on an azimutal mounting, which had been invented by William Herschel and modified by John Ramage (Fig. 6.10).[71] Of course, Robinson urged for quick results with which to attack Nichol. Until the end of the year, selected objects from John Herschel's catalogues of nebulae (SC) and double stars were observed. In June 1840 Lord Rosse wrote a first report for the Royal Society, which was published in the *Philosophical Transactions*. There he wrote about the prospects of his work: '*the present instrument will add something to the very little that is known respecting these wonderful bodies*' (Parsons W. 1840).

In contrast to John Herschel, Lord Rosse believed that he had seen signs of resolvability for some bright objects in the 36-inch: '*27 Messier, the annular nebula in Lyra* [M 57] *and, what is perhaps more curious, the edge of the great nebula in Andromeda* [M 31], *have shown evident symptoms of resolvability. No such appearance, however, was observed till the power reached six hundred, and sometimes it was more decisive with powers of eight hundred and one thousand.*' This issue was dominating, thus a search for new nebulae was not on the agenda: '*It is evident, therefore, that, except for the discovery of very faint nebulae, an object, perhaps, of but little interest, nothing would be effected by constructing a telescope of the greatest dimensions, unless it was at the same time proportionally perfect; that a mere light grasper would do nothing.*'

Having been invited by Lord Rosse, Romney Robinson and James South (Fig. 6.11 left) visited Birr Castle from 29 October to 8 November 1840 to test the new reflector.[72] The experienced visual observers came from Markree Castle, where they had used Cooper's 14" refractor. South, the celebrated double-star expert, ran a private observatory in Kensington. His fate was to be stuck with a refractor with an 11¾" Cauchoix lens, which had been erected in 1832, which produced a lot

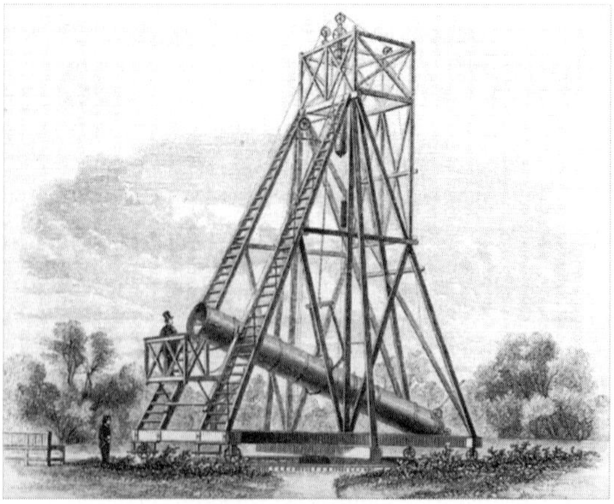

Figure 6.10. The 36" reflector in its first configuration with closed tube.

Figure 6.11. Regular visitors at Birr Castle. Left: James South; right: Romney Robinson.

of trouble.[73] For Robinson, who had been working at Armagh with an equatorial 15" Cassegrain (built by Grubb in 1835), the Birr Castle 36-inch was a quantum leap. He called it '*the most powerful telescope that has ever been constructed*', even superior to William Herschel's 48-inch.[74] During the following years he regarded the instrument as his personal property. For Robinson, Lord Rosse was the ideal partner: a man who

---

[71] William Rowan Hamilton was the first to point the 36-inch towards a celestial object. In the 1860s it was equipped with a truss tube of quadratic cross-section.

[72] The list of prominent Birr Castle visitors is long, containing, for instance, Lassell, Stokes, Airy, Sabine, Hamilton, Brünnow, Otto Struve, Piazzi Smyth and John Herschel.

[73] On the curious story of this refractor and the hard dispute with the Reverend Sheepshanks, see Bagdasarian (1985), Hoskin (1989) and McConnell (1994). South died on 19 October 1867, only 12 days prior to Lord Rosse. Robinson wrote an obituary (Robinson 1867a).

[74] Robinson (1840: 11). However, William Herschel had built two 24" reflectors (Bennett 1976a): one in 1797 for the King

6.4 Lord Rosse: Birr Castle

built him great telescopes, able to create the necessary evidence to support his ideas about nature.

Robinson, having the nebular hypothesis firmly in sights, had no doubt that all nebulae are star clusters. About his observation of the Ring Nebula M 57 he noted that '*it was resolved at its minor axis*'.[75] However, '*the great nebula in Orion and that of Andromeda shewed no appearance of resolution, but the small nebula near the latter is clearly resolvable*'.[76] Concerning resolution, South often was of a quite opposite view to him, always trying to slow down his enthusiasm. For instance, he corrected Robinson's report, replacing 'decidedly' by 'suspiciously'. His scepticism was based on his experience as a double-star observer using powerful refractors, which offer sharp images of stars. On the other hand, a large reflector could not convince him on this point. In 1845 (having already used Lord Rosse's 72-inch) he wrote '*During Sir W. Herschel's lifetime, with the 20-feet reflector at Slough I saw, amongst others, 3 Messier, 5 Messier, 13 Messier, 92 Messier, the annular nebula in Lyra, and the great nebula in Andromeda. No telescope of its size probably ever showed them better; yet on the same night the same instrument, when directed to α Lyrae, broke down under a power of about 300.*'[77] Thus nebulae and stars were quite different things for a reflector – and, therefore, their combination (resolved nebulae) was quite a delicate matter.

About the observation of M 1 (h 357, NGC 1952) in Taurus, Robinson remarked that '*it is ragged, bifurcated at the top, and has streamers running out like claws in every direction*'.[78] This might be the textual origin of the popular name Crab Nebula; obviously, the graphical one was the drawing of 1844 (see Fig. 6.12). Some weeks later he urged Lord Rosse to observe more '*nice round nebulae*', because '*I am anxious to know whether they all have tails and claws*'. Concerning the resolution of M 1, Robinson claimed that '*It is still more decidedly resolvable [as M 57], indeed I think it is resolved towards the centre.*' He added '*This telescope, however, shews the stars, as in figure 89* [Herschel's Fig. 81 in SC], *and some more plainly, while the general outline, besides being irregular and fringed with appendages, has a deep bifurcation to the south.*'[79] It is interesting that John Herschel too had noted '*barely resolvable*' in the Slough catalogue.[80] Robinson was '*confident that either a finer night or a few more inches aperture would actually resolve it beyond dispute*'.

At his request, the segmented mirror of the 36-inch was changed for the solid one on 5 November. However, the observations were terminated three days later due to strong moonlight (half moon had already occurred on the 2nd). South and Robinson departed for Dublin, where the latter reported the results of their Birr Castle observations to the Royal Irish Academy on 9 November (Robinson 1840).

In a one-hour talk, Robinson enthusiastically celebrated his key subject: the resolution of nebulae. However, the inevitable attack on the nebular hypothesis, on the basis of his own evidence, led to a dispute with South. He ignored the (sometimes contrary) impressions of his English friend, though not forbearing to comment on them. For instance, about the observation on 31 October 1840 Robinson reported '*Ring nebula and 1 Messier were examined by Sir J. South, who thought the appendages might be optical illusions produced by defective adjustment but there is no doubt of their reality and of the inaccuracy of the published drawings.*'[81] Like South, Lord Rosse was cautious about resolvability, writing in his report the following caveat: '*In describing the appearance of these bodies, I am anxious to guard myself from being supposed to consider it certain that they are actually resolvable, in the absence of that complete resolution which leaves no room for error; nothing but the concurring opinion of several observers could in any degree impart to an inference the character of an astronomical fact.*'[82] Fortunately, some objects resisted all attempts: the Orion Nebula ('*not the least suspicion of resolvability*') and M 31 ('*no appearance of resolvability*').

---

of Spain with focal length 25 ft (the 'Spanish') and a second in 1799 with a focal length of only 10 ft (the 'X-feet'). Whereas with the former only a few nebulae were observed, the latter was frequently used for this task.

[75] Robinson (1840: 10).
[76] The 'small nebula' could have been either M 31 or NGC 205.
[77] South (1846).
[78] Hoskin (1990: 334).

[79] Robinson (1840: 10); the object in Herschel's Fig. 89 is the galaxy NGC 5964.
[80] Slough catalogue, Fig. 81 (Herschel J. 1833a).
[81] Hoskin (1990: 335).
[82] Parsons W. (1840: 525 (footnote)).

Figure 6.12. Two of Lord Rosse's early drawings. Left: the Dumbbell Nebula M 27; right: the Crab Nebula M 1 (Parsons W. 1844a).

Four years after his first report, Lord Rosse presented another to the Royal Society, titled 'Observations of some nebulae' (Parsons W. 1844a).[83] It contains observations of *'two-thirds of the figured nebulae, and a few others in the general* [Slough] *catalogue'*. About Herschel he remarked *'that all we have seen strongly confirms the accuracy of Sir John Herschel's judgement in selecting the nebulae which he places in the class designated as resolvable'*.[84] He warned his readers that *'it still would be very unsafe to conclude that such will be always the case, and hence to draw the obvious inference that all nebulosity is but a glare of stars too remote to be separated by the utmost power of our instrument'*.[85] He further wrote that *'It is important from its bearing on future researchers; for the power of our instruments is sufficient to do more than to bring to light distinctly the peculiar characteristics of resolvability, these once observed with due caution and their reality ascertained beyond doubt, we shall conclude with little danger of error, that the object is really a cluster.'* However, Lord Rosse concedes that there might be nebulae, like M 42 or M 31 that will be unresolvable even with the largest telescopes, and therefore that they cannot be star clusters.

For the first time Lord Rosse presents drawings of nebulae, writing *'I believe they will be found tolerably correct.'*[86] No doubt, they were intended to demonstrate the power of his large reflector. Shown are the supernova remnant M 1, the globular cluster M 15 and three planetary nebulae: M 27 (drawn on 21 September 1843; Fig. 6.12 left), M 57 and NGC 6905. Neither in this nor in later publications is a date given for these drawings (with the exception of M 27).[87]

In accordance with Robinson's earlier description, the figure of M 1 in Taurus (Fig. 6.12 right) shows a 'crab'. Lord Rosse noted *'a cluster [...] we see resolvable filaments singularly disposed, springing from its southern extremity, and not, as is usual in clusters, irregularly in all directions [...] it is studded with stars, mixed however with nebulosity probably consisting of stars too minute to be recognized'*.[88] Dreyer later commented as follows on the curious 'crab': "The only published drawing which is a complete failure, is that of M. 1, the 'Crab Nebula', which has unfortunately been reproduced in many popular books. It was made with the 3-foot, and the long 'feelers' were never again seen with the 3-foot nor with the 6-foot." (Dreyer 1914). Obviously, Lord Rosse doubted his own drawing, since in 1855 he made a new one, showing M 1 very differently and much closer to reality. It is odd that Secchi's drawing of 1852 looks amazingly similar to Lord Rosse's 'crab' (Secchi 1856b).[89]

The observations of Lord Rosse and Robinson show a general problem: even nebulae that cannot be resolved with modern technology looked so. These

---

[83] A summary appeared in the *Proceedings of the Royal Society* (Parsons W. 1844b).
[84] Parsons W. (1844a: 323).
[85] Parsons W. (1844: 324).
[86] Parsons W. (1844a: 321). The figure number given refers to the Slough catalogue.
[87] Parsons L. (1880: 154).
[88] Parsons W. (1844a: 322).
[89] The different drawings of the Crab Nebula are treated in Section 11.2.

objects are either huge masses of gas and dust, i.e. true nebulae like M 42 or galaxies like M 31, which are too remote for individual stars to be seen visually. Why did such targets appear as 'star clusters' anyway? Perhaps two factors are important. First, the physical nature of these objects was unknown. When gazing intensively at faint diffuse structures, especially with high power, some kind of 'noise' arises on the retina, producing a certain 'mottled' structure. This is exacerbated by an instrument that does not yield perfect images. Obviously, the 36-inch (and the later 72-inch too) was of this kind. The second factor is expectation. Especially Robinson, decidedly rejecting the possibility of the existence of true nebulosity, suffered from the problem of wish-fulfilment: he simply saw what he wanted to see. Today the brain – biased by knowing the true nature – is alerted and the observer correctly perceives an unresolved nebula. Unfortunately, the issue of structural expectation is not dead for present-day visual observers – nowadays it is caused by the bad influence of 'pretty pictures'!

6.4.9 First observations with the 72-inch: the spiral structure of nebulae

Shortly after the 36-inch had been finished, Lord Rosse was already speculating about a Newtonian reflector of double the size. At the end of the 1840 publication one reads '*I think that a speculum of six feet aperture could be made to bear a magnifying power more than sufficient to render the whole pencil of light, and that in favourable states of the atmosphere it would act efficiently, without having recourse to the expedient which Newton pointed out as the last resort, that of observing from the summit of a high mountain. […] an instrument even of the gigantic dimensions I have proposed might, I think, be commenced and completed within one year.*'[90] Obviously, he was not satisfied by the results of the 36-inch concerning his major goal. After many attempts Lord Rosse was eventually successful on 13 April 1842 (at 9 p.m.) at casting a metal mirror of 72" diameter; it was 5.5" thick and had a mass of 3 tons. However, it took until September 1844 for the Leviathan of Parsonstown to see provisional first light.[91] South reported the event to *The Times*.[92] '*The leviathan telescope on which the Earl of Rosse has been toiling upwards of two years, although not absolutely finished, was on Wednesday last directed to the Sidereal Heavens. The letter which I have this morning received from its noble maker, in his usual unassuming style, merely states, that the metal only just polished, was of a pretty good figure, and that with a power of 500, the nebula known as No. 2., of Messier's catalogue, was even more magnificent than the nebula, No. 13 of Messier, when seen with his Lordship's telescope of 3 feet diameter, and 27 feet focus. Cloudy weather prevented him from turning the leviathan on any other nebulous object.*' The two objects mentioned are the globular clusters M 2 in Aquarius and M 13 in Hercules.

Having been invited by Lord Rosse, Robinson and South visited Birr Castle once again, staying from February until the end of March. On 11 February 1845 the trio celebrated the 'official' first light of the finished reflector (Fig. 6.13).[93] The targets were Sirius and 'some nameless clusters'. A main issue of interest to the visitors was the optical quality of the 72-inch. As their later reports show (see below), both were impressed by its 'space-penetrating power'. This must concern the great light collection of the mirror, whereas the sharpness of the image was probably not very high. Later, many praised the quality of the telescope,[94] but others were critical. For instance, Proctor remarked that '*we hear so little of any discoveries effected within the range of our own system by means of the great Parsonstown reflector*'.[95] In fact, there are no remarkable drawings of planets (when compared with those derived from use of contemporary refractors). Even more dramatic sounds the remark of an anonymous visitor, quoted by Charles Wolf: '*They showed me something which they told me*

---

[90] Parsons W. (1840: 527); concerning the hint of Newton about high-altitude observations see Holden (1896: 1). Robinson too announced in his 1840 talk that '*Lord Oxmantown is about to construct a telescope of unequalled dimensions. He intends it to be six feet aperture, and fifty feet focus, mounted in the meridian, but with a range of half an hour on each side of it.*' (Robinson 1940: 11).

[91] Robinson, visiting Birr Castle in February 1843 to inspect the progress, had advocated an equatorial mount. But Lord Rosse favoured a 'transit' telescope, which was much easier to build (Bennett 1990: 106).

[92] Dick T. (1845: 554).

[93] Hoskins (2002a).

[94] See, for instance, Airy (1849).

[95] Proctor (1869b: 755).

Figure 6.13. Lord Rosse's 72" reflector at Birr Castle.

was Saturn, and I believed them!'[96] The comment of Foucault (who built an 80-cm reflector with silvered glass mirror in 1862) was hard too: '*Lord Rosse's telescope is a joke*'.[97]

Owing to bad weather, no observations were possible after 11 February. Lord Rosse used the time to re-polish the metal mirror. Exposed to the open air, it soon tarnished, which led to a noticeable reduction of reflectivity. Johnstone Stoney later wrote that even when '*the speculum was quite fresh from the polisher* [the] *effect was lost in a very short time*'.[98] The sky cleared up on 4 March and the weather was stable until the 13th (new moon was on the 10th). The period was used intensively by the three astronomers. South summarised the results in a letter to *The Times* of 16 April 1845 (which was reprinted a year later in the *Astronomische Nachrichten*): '*The night of the 5th of March* [1845] *was, I think, one of the finest I ever saw in Ireland. Many Nebulae were observed by Lord Rosse, Dr. Robinson and myself. Most of them were, for the first time since their creation, seen by us as groups or clusters of stars; whilst some, at least to my eyes, showed no such resolution.*' (South 1846). For Robinson the 72-inch was the definitive tool with which to destroy the nebular hypothesis. Therefore he noted that '*of the 43 nebulae which have been examined all have been resolved*'.[99] Obviously, once again the two astronomers did not agree concerning the resolution of some of the nebulae.

For a second time Robinson informed the Royal Irish Academy about the results, in a talk given on 14 April 1845 (Robinson 1845).[100] His presentation matches with South's on many points, but is more detailed. Robinson mentions, for instance, the unusually low temperatures of less than −8 °C. He reports that 40 objects from the Slough catalogue were observed, which were divided into three classes:

- class 1: 'round and of nearly uniform brightness' (e.g. M 65 and NGC 5964)
- class 2: 'round, but appear to have one or more nuclei' (e.g. M 51 and M 96)
- class 3: 'extended in one direction, sometimes so much as to become long stripes or rays' (e.g. NGC 3628 and NGC 4565)

Concerning the first class, Robinson states that all objects could easily be 'resolved'; even in the 46-mm finding eye-piece (see Table 6.5). Most interesting is the second class, containing the spiral galaxy M 51. The objects of the third class were correctly interpreted as

---

[96] Wolf C. (1886b: 200).
[97] Tobin (2003: 204).
[98] Schaffer (1998b: 465).
[99] Hoskin (1990: 339).
[100] Two weeks after the casting of the 72-inch mirror (13 April 1842), Robinson had given an intermediate report to the Academy.

being of the same type as those of class 2, but seen more or less from the side (edge-on galaxies): *'they proved much more difficult for resolution, probably from greater optical condensation, and yielded most easily towards their minor axis'*. Except for 'resolution', his interpretation matches reality pretty well.

Robinson proudly concludes that *'no REAL nebula seemed to exist among so many of these objects chosen without bias; all appeared to be clusters of stars, and every additional one which shall be resolved will be an additional argument against the existence of such'*.[101] However, there might be many objects *'which from mere distance will be irresolvable in any instrument'*. For him, the 72-inch confirmed John Herschel's statement that *'a nebula, at least in the generality of cases, is nothing more than a cluster of distant stars'*. However, Herschel had actually told the British Association that it was a 'general law' that resolvability was limited to 'spherical nebulae' (i.e. globular clusters).[102] This sounds like defending Nichol's nebular hypothesis, though not accepting all parts of it. Facing Robinson's claims, which were based on observations and drawings made at Birr Castle, he asked *'is this really the appearance in the telescope, or has the artist intended to express his conception of its solid form by this shading?'* Herschel was convinced that true nebulae exist, e.g. the Orion Nebula. Even Lord Rosse had failed to resolve the object with the 36-inch in the winter of 1844–45.

Therefore, an attack on this key target with the 72-inch was overdue. However, M 42 could not be observed until Christmas 1845, when Lord Rosse hosted a distinguished astronomer, John Pringle Nichol from Glasgow. For Nichol, who was anxious about the result, the nebula looked impressive, though not resolved: *'not yet the veriest trace of a star'*.[103] The nebular hypothesis was rescued – for the moment. In February Lord Rosse noted that *'We are still in doubt as to the resolvability of the Nebula in Orion. The great instrument has shown us an immense number of stars in it, dense groups in the immediate vicinity of the Trapezium, but further evidence is I think wanting.'*[104] Of course,

for Robinson the case was already decided: *'The great Nebula in Orion was completely resolved in those places which presented the mottled appearance [...]. On February 20 a power of 470 showed the stars quite distinct there on a resolvable ground; and this clearly separated into smaller stars with 830, which the instrument bore with complete distinctness.'*[105] Especially Lord Rosse's comment on *'dense groups in the immediate vicinity of the Traprzium'* indicates that the Birr Castle observers might have been dazzled by the impressive mix of stars and nebulosity in the central region of M 42.[106] Actually, the next month Lord Rosse was convinced too, writing to Nichol on 19 March 1846 that *'we could plainly see that all about the trapezium is a mass of stars; the rest of the nebula also abounding with stars and exhibiting the characteristics of resolvability strongly marked'*.[107] Obviously, Nichol was both fascinated and shocked: *'Every shred of that evidence which induced us to accept as a reality, accumulations in the heavens of matter not stellar, is for ever and hopelessly destroyed.'* Was this the end of the celebrated nebular hypothesis?

Independently William Cranch Bond at Harvard College Observatory got the same result in September 1847, by observing M 42 with the new 15" Merz refractor: *'You will rejoice with me that the Great Orion Nebula has yielded to the powers of our incomparable telescope.'*[108] The object appeared to be 'sprinkled with stars'.[109] Not having heard about Lord Rosse's success of the same month, he claimed that he was the first to resolve the nebula. The alarmed Robinson, though happy about the unexpected confirmation, qualified the American observation as being biased by the Birr Castle results: *'this success must be in a great measure due to that precise knowledge of the phenomenon, and of the points where it might be looked for, which is afforded by Dr Nichol's work'*.[110] Obviously he damned his colleagues with the sins of his Scottish enemy.

---

[101] Robinson (1845: 17).
[102] Schaffer (1998a: 214).
[103] Hoskin (1982c: 145).
[104] Letter of Lord Rosse to James Challis, dated 14 February 1846 (Hoskin 1990: 341); see also Parsons L. (1868: 69).

[105] Robinson (1848: 6).
[106] On the trapezium, see Section 9.6.10.
[107] Nichol (1846: 55).
[108] Jones and Boyd (1971: 67). Bond also claimed the resolution of M 27 and M 57 on 15 July 1847. For the Ring Nebula he noted *'My friend Mr. William Mitchell of Nantucket, who was observing with us, was confident that he saw many stars within the compass of the ring.'* (Bond W. 1847.)
[109] Schaffer (1998a: 218).
[110] Robinson (1848: 7); see also Schaffer (1998b: 466).

However, the astronomical community was not convinced. For instance, Otto Struve critically wrote in 1853 that '*the alleged miracles of resolution are nothing but illusions*'.[111] Airy too rejected the inference that all nebulae could be resolved into stars.[112] Later Lawrence Parsons remarked about his observations of M 42 with the 36-inch (in 'front-view' mode) that '*when we consider this object carefully we shall see that this* [resolution] *is far from being the case*'.[113] Up to the 1860s the 'resolvability' of the Orion Nebula was an important issue, but later it was discreetly ignored. Lawrence Parsons put it tersely: '*The observations made before 1860 are of very little interest.*'[114] The reason was simple: in 1864 William Huggins' visual spectroscopy revealed the object to be a gaseous nebula! (Huggins W. 1864). The idea of changes in its shape did not live for much longer. One of the last 'confirmations' was due to Lawrence Parsons, when comparing the drawings of Bindon Stoney (1852) and Hunter (1864).[115]

6.4.10 Robinson's report of 1848 and Lord Rosse's publication of 1850

After his reports of 1840 and 1845, Robinson presented a last one, written on 16 March 1848. It is based on a talk given at the Royal Irish Academy (Robinson 1848). He immediately stresses the central point: '*Above fifty nebulae selected from Sir John Herschel's catalogue, without any limitation of choice but their brightness, were all resolved without exception.*'[116] The objects, mainly galaxies, '*consist of a central cluster, mostly globular, of comparatively large stars, surrounded by an exterior mass of much smaller and fainter stars, whose arrangement is often circular and thin like a disc*'. He provocatively asks whether '*nebulous matter has real existence?*'. No doubt, the nebular hypothesis had been damaged. However, after Lord Rosse's discovery in April 1845 that M 51 is a spiral nebula,[117] the observational (and theoretical) focus significantly changed from 'resolution' to 'structure'. Interestingly, both issues were later treated by Tempel in his treatise 'Über Nebelflecken' (see Section 9.3.2). The Arcetri astronomer doubted the existence both of true nebulosity and of spiral structure. Robinson, who died in 1882, would have been happy to read this.

In Lord Rosse's third publication of 1850 the observations were summarised from his point of view (Parsons W. 1850a).[118] At the beginning, the work is described as '*an account of the progress which has been made up to the present date in the re-examination of Sir John Herschel's Catalogue of Nebulae published in the Philosophical Transactions for 1833*'. Sadly, after the successful first phase of 72-inch observations in 1845, the disastrous Irish potato famine (1846–48) caused a cessation of astronomical activity. As Lord Lieutenant of King's County, Lord Rosse tried to relieve suffering with all his power. There was hardly any time for the new reflector. Only in spring 1846 could he use it for some observations, which at once brought spectacular news: the spiral structure of NGC 2903 (h 604) and M 99 (h 1173).

Robinson, the *eminence grise* of Birr Castle, was the driving force to resume the observations in 1848. Fearing an idleness of 'his' reflector, he urged Lord Rosse to engage a scientific assistant. He recommended for the job William Rambaut, who stayed until June and was soon followed by Johnstone Stoney. Especially the first half of 1848 was a period of intensive observing. First, a few technical aspects of the 72" reflector and its equipment will be presented.

Lord Rosse employed several technical helpers, who were responsible for the maintenance of the telescope. Normally three men were needed to move the heavy instrument. The observation of specific targets was a difficult task: the meridian had to be crossed suitably at night (being quite short in summer) and the sky had to be clear (a rare case). However, the region within 25° of the celestial pole was out of reach (Dreyer 1914).[119] Since a horizontal shift of 15° was possible, an

---

[111] Hoskin (1982c: 150). In 1852 Struve had proposed combined labours at Pulkovo and Parsonstown to clarify the issue, which, however, never happened (Schaffer 1998b: 462–463).
[112] Hoskin (1982c: 150).
[113] Parsons L. (1868: 72).
[114] Parsons L. (1880: 48); altogether 43 observations were made by 1861.
[115] Parsons L. (1868).
[116] Robinson (1848: 1).
[117] This issue is presented in detail in Section 11.3.1.

[118] A summary appeared in the *Proceedings of the Royal Society* (Parsons W. 1850b).
[119] Since Birr is at a northern latitude of 53° (see the site table in the appendix), this implies that the telescope could even reach 12° beyond the zenith. Therefore, near-zenith observations were possible.

Table 6.5. *Eye-pieces used at the 72" reflector (M = magnification, f = focal length, p = exit pupil); the 46-mm and 76-mm are finding eye-pieces*[120]

M	f (mm)	p (mm)
216	76	8.5
360	46	5.1
400	41	4.6
470	35	4.0
500	33	3.7
560	29	3.3
700	24	2.6
750	22	2.4
830	20	2.2
854	19	2.1
900	18	2.0
940	17	1.9
1000	16	1.8
1280	13	1.4
1550	11	1.2
1833	9	1.0
1928	8	0.9

object could be tracked for up to one hour (depending on its elevation). Though the telescope was able to perform Herschel-like sweeps, a systematic search for new objects was not on Lord Rosse's agenda. Anyway, if such a task was ever planned, it would have failed due to lack of time. Nevertheless, some 70 nebulae were accidentally discovered by June 1850, when the third report was submitted. They are not listed there, but appear in his comprehensive publication of 1861 (see Section 7.1).

From the reports of Robinson (1848) and Lord Rosse (1850) it can be derived which eye-pieces were used at the 72-inch (Table 6.5). Three could be mounted at once on a slide to allow a quick change; a finding eye-piece was always among them. Two were available, with focal lengths of 76 mm and 46 mm (three lens systems; the former was of 6" diameter). They offered powers of 216 and 360 with fields of view of 31′ and 13.7′, respectively. Robinson wrote that '*a nebula is easily found in this wide field; and bringing it into the centre, the eye-slide is shifted, and it is viewed with higher powers*'.[121] However, the 76-mm '*will only bring into action forty-three inches of the mirror, yet even though its optical power is very great*'. This is due to the ratio of the exit pupil (p) to the entrance pupil (P); the former is the diameter of the ray-bundle on leaving the eye-piece. Because a well-adapted eye has an entrance pupil of about 7 mm (iris diameter), some light is lost in the case of the 76-mm eye-piece ($p = 8.5$ mm); the telescope aperture cannot be used completely (vignetting). All other eye-pieces give $p < P$ and the full beam can enter the observer's eye.

Additionally a special micrometer eye-piece was available. Robinson wrote that '*any illumination of the wires extinguishes many of the faint nebulae [therefore] Lord Rosse used a very ingenious substitute, a screw whose threads were rubbed with phosphorus*'.[122] However, the tool was not often used during the first period (1839–48). Lord Rosse wrote that '*much fewer objects have been examined than would otherwise have been practicable. […] Without accurate micrometrical measurements any sketch can be of comparatively little value as an astronomical record.*'[123]

Concerning the observations, 3 March 1846 brought a surprising result: for h 604 (the galaxy NGC 2903 in Leo) Lord Rosse noticed '*a tendency to an annular or spiral arrangement*'.[124] This was confirmed two years later by Robinson: '*h 604 is also spiral, but without any other peculiarity*'.[125] In April 1846 the galaxy M 99 in Coma Berenices was observed by Lord Rosse: '*In the following spring an arrangement, also spiral but of a different character* [from that of M 51], *was detected in 99 Messier […] This object is easily seen, and probably a smaller instrument, under favourable circumstances, would show everything in the sketch.*'[126] Robinson wrote that '*the centre is a globular cluster, surrounded by spirals, in the brighter parts of which stars are seen with 470: these have the same direction as in Messier 51, namely from east to west, in receding from*

---

[120] The focal length of the 72-inch was 16.5 m. Later Dreyer even used an eye-piece with a power of 130 and a 26′ field of view (see Section 11.6.5); it probably had a focal length of $f = 127$ mm.

[121] Robinson (1848: 4).
[122] Robinson (1848: 4–5).
[123] Parsons W. (1844a: 321–322).
[124] Parsons W. (1850a: 511).
[125] Robinson (1848: 9).
[126] Parsons W. (1850a: 505); see also Steinicke (2004c).

Table 6.6. *Objects drawn by Lord Rosse and Johnstone Stoney, published in 1850*

Fig.	H	NGC	M	Drawer	Date	Con.	Type	V	Remarks
1	1622	5194	51	Lord Rosse	31.3.1848	CVn	Gx	8.1	With NGC 5195
2	1173	4254	99	Lord Rosse	11.3.1848	Com	Gx	9.7	
3	604	2903		Lord Rosse	5.3.1848	Leo	Gx	8.8	With NGC 2905
4	2205	7479		Lord Rosse	10.9.1849	Peg	Gx	10.9	
5	131	598	33	Lord Rosse	16.9.1849	Tri	Gx	5.5	'New spiral'
6	444	2371		Lord Rosse	22.12.1848	Gem	PN	11.2	With NGC 2372
7	854	3623	65	Lord Rosse	31.3.1848	Leo	Gx	9.2	
8	1909	5866	102	Lord Rosse	27.4.1848	Dra	Gx	9.9	
9	1397	4631		J. Stoney	19.4.1849	CVn	Gx	9.0	With NGC 4627
10	399	2261		Lord Rosse	16.1.1850	Mon	RN		
11	838	3587	97	Lord Rosse	11.3.1848	UMa	PN	9.9	Annular
12	464	2438		Lord Rosse	22.12.1848	Pup	PN	10.8	Annular; in M 46
13	2241	7662		J. Stoney	16.12.1848	And	PN	8.3	Annular
14	2098	7009		Lord Rosse	23.10.1848	Aqr	PN	8.0	
15	450	2392		J. Stoney	20.2.1849	Gem	PN	9.1	
16	361	1980		Lord Rosse	28.1.1849	Ori	EN+OC		Around ι Orionis
17	2060	6853	27	J. Stoney	9./16.9.1849	Vul	PN	7.4	

the centre'.[127] In h 731 (NGC 3310) in Ursa Major stars were seen too by Robinson, *'but no central cluster'*. The object was *'looking like a turbinated shell [...] seen edgewise'* (the galaxy is not edge-on).

Lord Rosse's publication contains 17 drawings, with detailed descriptions, made by Johnstone Stoney and himself (Table 6.6). They represent a selection of interesting types: spiral nebulae, planetary nebulae (especially annular ones) and very elongated objects. Lord Rosse reported about the making of the drawings that *'we necessarily employ the smallest amount of light possible, very feeble lamp-light, especially where the objects or their details are of the least degree of faintness'*.[128] In the case of a faint structure *'it is often necessary to mark it too strongly'*. But this method causes some problems (see Section 11.2): *'The most important error to guard against is that of supposing that the well-marked confines of the*

---

[127] Robinson (1848: 9).

[128] Parsons W. (1850a: 509).

6.4 Lord Rosse: Birr Castle

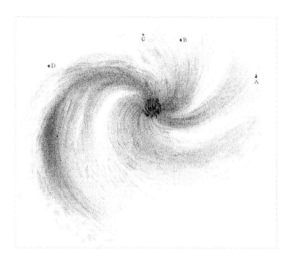

Figure 6.14. Lord Rosse's drawing of the spiral nebula M 99 in Coma Berenices (Parsons W. 1850a).

nebula really represent the boundaries of the object in space in all cases.' Moreover, '*the form would alter if additional optical power could be brought to bear upon it*'.

The illustrations of M 51 and M 99 (Fig. 6.14) are outstanding. Both show spiral structure, but differ in the form of the arms. The former was re-examined in early 1848 and Robinson '*fully verified* [the] *spiral arrangement*'.[129] But he could not, however, '*satisfy himself that it was to be traced in the three-feet*'. Lord Rosse's drawing of M 51 made on 31 March 1848 is considerably different from his first one of April 1845 (see Section 11.3.1). Obviously, due to the lack of measurements, only the main features could be drawn: '*the sketches are sufficient to convey a pretty accurate idea of the peculiarities of structure* [...] *many micrometrical measures are still wanting, and there are many matters of detail to be worked in before they will be entitled to rank as astronomical records*'.[130] This was eventually done by Johnstone Stoney: in the spring of 1849 and 1850 he determined the relative positions of 15 (foreground) stars in M 51, out to a distance of 5′ from the centre; for M 99 four stars were measured in April 1849. He used a bar-micrometer, whose threads were thick enough to be seen in the eye-piece without illumination.

At the end of his publication, Lord Rosse lists 14 cases with 'spiral or curvilinear' structures (Table 6.7).

Strangely, the most impressive ones, M 51 and M 99, are missing.[131] NGC 5195, the peculiar companion of M 51, is not mentioned either. Lord Rosse '*saw also spiral arrangement in the smaller nucleus*'.[132] However, in this case he was wrong. Contrariwise, he did not list M 33 (h 131, GC 352), though the first observation (16 September 1849) yielded '*New Spiral, s branch the brightest*' (Parsons L. 1880). Lord Rosse's sketch clearly shows four spiral arms.[133] All of the objects in Table 6.7, except one, are actually spiral galaxies. The exception is NGC 5557, an elliptical galaxy.

On the physical nature of spiral arms Lord Rosse commented that '*in many of the nebulae they are very remarkable, and seem even to indicate the presence of dynamical laws we may perhaps fancy to be almost within our grasp*'.[134] Robinson wrote about the feature that '*Their resemblance to bodies floating on a whirlpool is, of course, likely to set imagination at work, though the conditions of such a state are impossible there. A still more tempting hypothesis might rise from considering orbit motion in a resisting medium; but all such guesses are but blind.*'[135] Here for the first time the term 'whirlpool' arises in connection with spiral nebulae – it is now the popular name of M 51.[136]

Lord Rosse presents four objects (M 5, M 10, M 13 and NGC 2506) that show '*a tendency to an arrangement in curved branches, which cannot well be unreal, or accidental*'.[137] Actually these are globular clusters! Already John Herschel had noted that M 13 (h 1968) '*has hairy-looking curvilinear branches*'. For Lord Rosse the association of spiral nebulae with globular clusters was convincing, insofar as some spiral nebulae (e.g. M 99) were 'resolved' into stars – at least by Robinson.

A much different class is constituted by the rare annular nebulae (actually ordinary PN). Lord Rosse

---

[129] Robinson (1848: 9).
[130] Parsons W. (1850a: 503).
[131] See the analysis of Dewhirst and Hoskin (1991).
[132] Parsons W. (1850a: 510).
[133] See also Webb (1864b: 351–352, Webb 1864c: 452).
[134] Parsons W. (1850a: 503).
[135] Robinson (1848: 10).
[136] See William Tobin's research on the origin of the name 'whirlpool' (Tobin 2008); it is interesting that some authors (advocates of the nebular hypothesis) even anticipated the name, e.g. Nichol in 1833 writing '*supposing the condensations of one of these portions of nebulous matter to commence* [...] *motion like that of a whirlpool results*'.
[137] Parsons W. (1850a: 506).

Table 6.7. *Lord Rosse's spiral nebulae, mentioned in his 1850 publication (sorted by discovery date of the structure)*

NGC	M	h	Date	Con.	V	Remarks
5194	51	1622	April 1845	CVn	8.1	Not in list
2903		604	24.3.1846	Leo	8.8	Not in list
4254	99	1173	April 1846	Com	9.7	Not in list
3726		910	26.3.1848	UMa	10.0	
5195		1623	26.4.1848	CVn	9.6	Not in list
5557		1776	26.4.1848	Boo	11.0	
628	74	142	13./14.12.1848	Psc	9.1	
1637		327	19.12.1848	Eri	11.0	
1068	77	262	22.12.1848	Cet	8.9	Seyfert galaxy
3938		1002	17.3.1849	UMa	10.0	
7331		2172	12.9.1849	Peg	9.5	
3198		695	3.3.1850	UMa	10.0	
3368	96	749	3.3.1850	Leo	9.3	
4321	100	1211	9.3.1850	Com	9.3	
4501	88	1312	9.3.1850	Com	9.4	
4725		1451	9.3.1850	Com	9.3	
5055	63	1570	9.3.1850	CVn	8.5	
4579	58	1368		Vir	9.6	No date

mentions two northern objects. The first is the Ring Nebula M 57 in Lyra, but the second is not identified. From consulting John Herschel's textbook *Outlines of Astronomy*, which lists six annular nebulae from both hemispheres,[138] the second object must be NGC 6894 (h 2072) in Cygnus, which was first observed on 1 August 1848. Lord Rosse further wrote that '*we have found, that five of the planetary nebulae are really annular*'.[139] Three of them are sketched: M 97, NGC 2438 and NGC 7662 (see Table 6.6); the remaining ones are NGC 6905 (h 2075) in Delphinus and NGC 6818 (h 2047) in Sagittarius.

About the nebula in Ursa Major Robinson wrote '*97 Messier is a strange object. With the finding eye-piece it looks like the figure of 8 with a star at the intersection* [the central star], *but with 470 it is spiral with two centres. The principal one looks like a star, but with 830 gives the suspicion that it is a small cluster. The spirals related to this have the same direction as the former; but the other centre, which also looks like a minute star, has a smaller set in the opposite direction.*'[140] The drawing of M 97 in Lord Rosse's publication is remarkable: the planetary nebula looks like the head of an owl (Fig. 6.15 left).[141] Though the popular name Owl Nebula was not explicitly used by Lord Rosse and his assistants in any of his papers, it might nevertheless originate from Birr Castle. William Darby wrote in 1864 that M 97 was "*familiarly known in the Parsonstown Observatory as 'the owl nebula' from its resemblance of an owl*".[142] Robinson described the object as '*looking like the visage of a monkey*' (Robinson 1853). Another remarkable drawing of the planetary nebula was made by Rambaut

---

[138] Herschel J. (1869, vol. 2: 792).
[139] Parsons W. (1850a: 506).
[140] Robinson (1848: 9–10).
[141] John Herschel drew M 97 on 10 February 1831 (SC, Fig. 32), but saw only a '*large uniform nebulous disc*'. On the basis of the Birr Castle observations the object was later described by Nichol as a '*most fantastic figure, something like an 8*', and '*The odd configuration of the cactus tribe in the vegetable world, would thus seem repeated among the stars!*' (Nichol 1846: 23, added footnote).
[142] Darby (1864: 94). Later Roberts wrote that Lord Rosse '*figured the nebula as a circle filled in with details somewhat resembling the face of an owl*' (Roberts 1896b). In 1907 Barnard

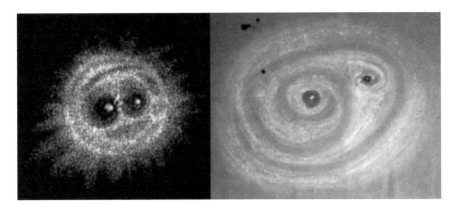

Figure 6.15. Drawings of the Owl Nebula, M 97, made in 1848 at Birr Castle. Left: Lord Rosse (Parsons W. (1850), Fig. 11); right: Rambaut.

in spring 1848. Here M 97 shows spiral structure (Fig. 6.15 right). Lord Rosse, again claiming a connection to spiral nebulae, noted that '*a double perforation appears to partake of the structure both of the annular and spiral nebulae*'. Concerning the stars in the owl's 'eyes', which appear in both drawings: only the left one is real (the second was only 'seen' twice by Lord Rosse, Rambaut and Robinson).

The connection between planetary and spiral nebulae was strengthened in 1852 by Robinson. In his talk at the 22nd meeting of the British Association in Belfast, titled 'Drawings to illustrate recent observations on nebulae', he claimed that '*the class of planetary nebulae might now be fairly assumed to have no existence, as all of them which I have examined prove to be either annular or of a spiral character*' (Robinson 1853).[143] M 97 is given as an example, showing '*in the six-feet a most intricate group of spiral arms*'. Here Robinson mentions other, unpublished drawings, which he had seen while visiting Birr Castle – obviously Rambaut's of M 97 was among them.

Though Lord Rosse could 'resolve' annular nebulae, he detected (in some cases) stars inside and outside the ring or even on it. His elegant interpretation was '*That a faint nebula should be more easily resolvable than a bright one is not unusual, neither is it contrary to probability.*'[144] He further wrote that '*in a faint nebula they* [these stars] *might be seen separate with an instrument of great aperture, while in the brighter and more closely packed nebula* [like M 27] *they were blended together, owing to imperfect definition, arising out of the state of the air, or instrument*'. For Lord Rosse h 2205 (the galaxy NGC 7479 in Pegasus) was a limiting case: '*Spiral, but query whether this is not more properly an annular than a spiral nebula.*'[145] This object was later selected by James Keeler as an example of the subjectivity of drawings (see Section 11.2.4). Another critic was Tempel, fighting against the idea of Lord Rosse and Robinson that spirality is a common phenomenon, appearing in various classes of objects (see Section 11.3).

Lord Rosse also treated the 'long elliptic or lenticular nebulae' M 65, M 102 (h 1909, NGC 5866) and NGC 4631 (h 1397).[146] For M 65 he supposed an '*elliptic*

---

published a paper on M 97, where Lord Rosse's drawing is compared with his own, which he made at the 40" Yerkes refractor (Barnard 1907c).

[143] It is surprising that this was later 'confirmed' by photography. In 1902/03 Martin Schaeberle took plates of M 57 and M 27 with a 13" reflector (f/1.5) at his private observatory in Ann Arbor. The appearance of the planetary nebulae was interpreted as 'great spirals'. He wrote about M 57 that '*The various exterior nebulosities, including Barnard's small nebula* [IC 1296], *and practically all the neighbouring stars, are apparently part of the same great* [two-branched] *spiral of which the well-known Ring nebula forms but a comparatively small central part.*' (Schaeberle 1903a, b). Already John Herschel (without,

however, knowing about its spiral structure) had called the barred spiral NGC 1265 (h 2552) in the Fornax Cluster a '*decided link between the nebulae M 51 and M 27*' (Herschel J. 1847: 61).

[144] Parsons W. (1850a: 506–507).
[145] Parsons W. (1850a: 511).
[146] Parsons W. (1850a: 508).

annular system seen very obliquely'. The absorption lane ('*dark chink*') of M 102 '*might indicate either a real opening, the system being an elliptical ring, or merely a line of comparative darkness*'. With the last guess he approached the truth pretty well. Lord Rosse also mentions the discovery of '*several groups of nebulae*', but '*new objects have not been as yet sought for*'. For a publication it was too early: '*we have not proceeded sufficiently far with it to make it worth while to enter upon in the present paper*'. At the end of his paper, Lord Rosse lists not only the 14 spiral nebulae, but also 13 nebulae 'with dark spaces', eight 'knotted nebulae' and three showing a 'ray with split'.

What was the impact of spiral structure on the nebular hypothesis? Of course, the situation became more complicated: the supposed starry nature of the nebulae was now paired with spirality. Robinson tried to rescue his view against the nebular hypothesis by picturing an object consisting of 'bodies' (stars), orbiting around the centre – which suggestion is not far from reality. For Nichol, its strongest advocate, a spiralling mass like that in M 51 fits surprisingly well the old Laplacian idea of a spinning primordial nebula from which the solar system formed. Thus Lord Rosse's spirals were even better evidence for true nebulosity than was the Orion Nebula against it – evolution matters more than resolution.[147] In the end, both Nichol and Robinson could live with the new situation. Diplomatically, Lord Rosse did not define the quality of the spiralling matter, only stating that it followed a 'dynamical law': '*That such a system should exist, without internal movement, seems to be in the highest degree improbable.*' Newton's gravitation, which was obviously acting in spiral nebulae, was treated again by Nasmyth in a paper written on 7 June 1855 (Nasmyth 1855). However, the Manchester astronomer speaks only about '*particles of a nebulous mass*', which spiral towards the centre. He compares the action with water in a basin, when it is escaping through a hole in the bottom. Later Hubble discussed the problem of internal motions of spiral nebulae (Hetherington 1974).

6.4.11 The Eskimo Nebula as a 'nebulous star', observed by Lord Rosse, Key and d'Arrest

That true nebulosity (in addition to the resolvable case) exists had already been demonstrated by the 'nebulous stars' of William Herschel. Of course, Lord Rosse was highly interested in these objects (though Robinson was very sceptical). As an example, NGC 1980 (h 361), a mix of open cluster and emission nebula around the star ι Orionis, is mentioned in his publication of 1850. He correctly suggests a connection between cluster and nebula, which '*is in some way connected with these bright stars*'.[148] However, his prototype of a 'nebulous star' was NGC 2392 (IV 45, h 450) in Gemini, which had been characterised by William Herschel as a '*very remarkable phenomenon*'. In 1849 the PN, now known as the Eskimo Nebula, was sketched by Johnstone Stoney (Fig. 6.16 left). Lord Rosse confessed that '*we have not seen the slightest indication of resolvability*'. He further reported that '*The annular form of this object was detected by Mr. Johnstone Stoney, my assistant, when observing alone.*' Besides the disc-like structure the published drawing shows a remarkable dark spot west of the central star.

NGC 2392 was later studied by Henry Cooper Key with an 18¼" reflector. The telescope was self-made (except for the equatorial Berthon mounting). It stood in Stretton Sugwas, only 6 km away from Hereford, the domicile of his friend, the Reverend Webb (see Section 9.9).[149] In the report of 26 February 1868, Key compares his observations of NGC 2392 (using powers of 150 to 666) with those of William and John Herschel, Lord Rosse and Lassell (Key 1868c). He wrote that '*The present appearance of this object, as seen in my instrument, is that of a bright, but somewhat nebulous star closely surrounded by a dark ring; this again by a luminous ring; then an interval much less luminous, and, finally, at some distance, an exterior luminous ring.*' Obviously, his reflector showed much more than had those of the earlier observers: '*While in my own instrument, which, although somewhat more powerful than the 20-foot reflectors (front view) of the Herschels, is of incomparably inferior power to the instruments of the Earl of Rosse and Mr. Lassell, the interior bright ring was visible at once during the first night's observation, and with a power of 510 is quite obvious.*' Key made a drawing (Fig. 6.16 second from right), adding on 23 March 1868 '*I have

---

[147] Schaffer (1989: 141–142).

[148] Parsons W. (1850a: 508).

[149] He was the first in Great Britain to make silvered glass mirrors (Key 1863). The Reverend Key was born in 1819 in London. In 1860 he became a member of the RAS. He died on 25 December 1879 in Stretton Sugwas at the age of 60. Obituary: Anon (1880c).

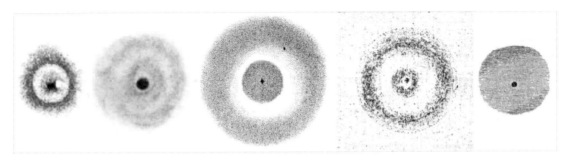

Figure 6.16. Drawings of the Eskimo Nebula NGC 2392 in Gemini (from left to right): J. Stoney, Birr Castle (Parsons W. 1850a); Secchi, Rome (Secchi 1856b); Lassell, Malta (Lassell 1867a); Key, Stretton Sugwas (Key 1868c); and d'Arrest, Copenhagen (d'Arrest 1867a).

seen a dark patch situated in the inner ring close to the central star.'

D'Arrest later observed NGC 2392 in Copenhagen (d'Arrest 1872a): *'concerning the shape and appearance of the ring nebula'*, he thought Key's sketch to be *'too systematic and regular'*. He further wrote that *'Between the outer ring and the dark space, the contrast is represented too stark, because the gap between centre and border is undoubtedly filled with nebulosity.'* D'Arrest presented his own sketch of the planetary nebula made on 19 February 1865 in the *Siderum Nebulosorum* (Fig. 6.16 right).[150] In 1872 he added *"I see the body still as described in my 'Obs. Havn.' [SN]: with eccentric core and bright outer border of unequal width and intensity. Following Rosse and Key this figure, however, could be called anyhow a ring."* (d'Arrest 1872a). The main topic of d'Arrest's report was a spectroscopic investigation of the object, finding that *'IV. 45 clearly and undoubtedly belongs to the gaseous nebulae'*. This was first noticed by Winlock and Peirce on 7 January 1869 at Harvard College Observatory (Pickering E. 1882a). With the aid of a visual spectroscope they detected three bright lines on a continuum. The object was also drawn by Secchi in Rome, using a 9.5" Merz refractor (in about 1854) and by Lassell on Malta in March 1862 with his 48" reflector (Fig. 6.16, second from left, centre).

6.5 ADMIRAL SMYTH AND HIS BEDFORD CATALOGUE

The year 1844 saw the publication of a famous guide for amateur astronomical observations: the two-volume *Cycle of Celestial Objects* by William Smyth. Volume 1 is an introduction to practical astronomy. Volume 2 is the famous Bedford Catalogue, a compendium of the author's observations with the 5.9" Tully refractor at his private Bedford Observatory. It treats a selection of nebulae and star clusters from the catalogues of William and John Herschel, Messier and Wilhelm Struve. The work was both praised and criticised. Undisputed is its great influence on amateur astronomers, among them Thomas William Webb, who published his *Celestial Objects for Common Telescopes* in 1859. No doubt Smyth had spotted a gap in the market, since professional astronomers usually did not popularise their observations.[151]

6.5.1 Short biography: William Henry Smyth

William Smyth was born on 21 January 1788 in Westminster (London); his father came from New Jersey (Fig. 6.17). In 1805 the young Smyth joined the Royal Navy. His last years of service (1813–24) were mostly spent in the Mediterranean area. In 1815, at the age of 27, he was appointed Captain and married Eliza Anne Warrington. Smyth's interest for astronomy was kindled in 1813 when he visited Giuseppe Piazzi, the discoverer of Ceres, at Palermo Observatory. The admiration for this astronomer was so great that his first son, who was born in 1819 in Palermo, was given the forenames Charles Piazzi. Smyth had three sons and two daughters.

---

[150] d'Arrest (1867a: 92).

[151] A remarkable exception was Joseph v. Littrow, Director of the Vienna Observatory, who described 316 non-stellar objects in his book *Sterngruppen und Nebelmassen des Himmels* ['Star groups and nebulosities in the heavens'] (Littrow J. J. 1835).

Figure 6.17. William Smyth (1788–1865) and his 5.9" Tully refractor at Bedford Observatory.

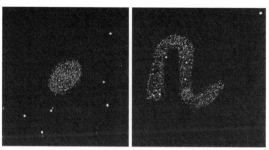

Figure 6.18. Smyth's sketches of M 1 in Taurus (left) and M 17 in Sagittarius (Smyth W. 1844).

In 1825, now 37 years old, Smyth returned to England and settled a year later in Bedford, 60 km north of London. In 1830 his private Bedford Observatory was ready to use (Smyth W. H. 1830). The main instrument was a 5.9" Tully refractor equipped with a mechanical drive constructed by the Reverend Sheepshanks, perhaps the first in England. The observations for the catalogue were finished in 1839; the observatory had served its purpose and was dismantled. John Lee, from nearby Aylesbury, purchased it together with all its instruments to erect his Hartwell Observatory in 1839. Smyth first moved to Cardiff and then came to London in 1842. There he finished the manuscript of the *Cycle of Celestial Objects*. The two-volume work eventually appeared in 1844 in London (Smyth W. 1844). For this he received the gold medal of the RAS (he was a member from 1821). The laudation was given by President Airy (Airy 1845). Smyth became his successor (1845–47).

Smyth now moved to Aylesbury, as John Lee's neighbour. Until 1859 he used his old refractor for a supplement to the Bedford Catalogue ('Speculum Hartwellianum'), which was published in 1860.[152] In 1858 he was appointed vice-admiral, becoming admiral four years later. William Henry Smyth died on 8 September 1865 in Aylesbury at the age of 77.[153] His son, Charles Piazzi Smyth, was Astronomer Royal of Scotland (1746–88) at the old Edinburgh Observatory on Calton Hill.

6.5.2 Structure and content of the Bedford Catalogue

According to Smyth, the Bedford Catalogue lists 850 objects for the epoch 1840. Nebulae, star clusters, single, double and multiple stars were described in the text, including notes about their discovery and previous observations by others. The book contains 38 sketches by Smyth (25 show Messier objects; Fig. 6.18).

The author states that 98 'nebulae' and 72 'clusters' are presented, which is incorrect (Table 6.8). Most non-stellar objects are from the catalogues of Messier and William Herschel. Smyth's notation is a bit strange, for instance 39 M. Cygni (M 39) for Messier objects (except 31 Messier = M 31) or 61 H. VIII. Aurigae (VIII 61 = NGC 1778) for Herschel objects. Six 'clusters' were taken from the Slough catalogue: h 22 = NGC 110, h 28 = NGC 146, h 124 = NGC 559, h 146 = NGC 657, h 227 = NGC 957 and h 2035 = N 6775. The last object is a star group in Aquila (called 2035 H. Aquilae). Two objects are planetary nebulae that

---

[152] Anon (1864), Anon (1865a). A second edition of the 'Cycle' was published in 1881 by Chambers, containing nearly twice as many objects (1604); see Chambers (1881).

[153] Obituaries: Fletcher (1866), Anon (1865b) and Anon (1866a); see also Sugden (1982).

Table 6.8. *Non-stellar objects in the Bedford Catalogue*

	H	M	h	Σ	Sum
Nebulae	61	31		2	94
Clusters	29	34	6		69
Sum	90	65	6	2	163

had been found by Wilhelm Struve; the designation varies: Σ. 5 N. Herculis (NGC 6210) and 6 Σ. N. Taurii Poniatowskii (NGC 6572). The latter is also named 6 Σ. N. Tauri Poniavii in the text;[154] the meaning of 'N' is unclear (perhaps 'nebula' or 'nova').

Besides the 'nebulae' there is a 'nebulous star': 69 H. IV. Tauri (counted as a nebula here). This is IV 69 (NGC 1514) in Taurus, Herschel's famous 'star with an atmosphere'. 45 H. IV Geminorum, the planetary nebula NGC 2392, is listed in the category 'stars and comets' (and counted as a 'nebula' too). The star 241 P. XIX. Aquilae from Piazzi's catalogue is erroneously listed as a 'nebula'. The object 22 M. (?) Draconis is interesting. Smyth suggests a nebula of Méchain. Obviously M 102 is meant, but the given position indicates William Herschel's II 757 (NGC 5879), a galaxy in Draco. The inclusion of 147 H. III. Andromedae (NGC 23) is strange too. The 11.9-mag galaxy in Pegasus is described as a '*double star in a course cluster*'. Since this is the only Herschel nebula of class III ('very faint nebulae'), it is improbable that Smyth saw the object. This shows that not all of his observations can be taken seriously.

The lack of rigour in Smyth's observations eventually led to the 'Sadler–Smyth scandal'.[155] The English amateur astronomer Herbert Sadler had checked the double-stars data,[156] publishing a critical assessment in the *Monthly Notices* (Sadler 1879). Smyth was blamed for having simply copied results of earlier observers, including their errors. Sadler's reproach not only damaged Smyth's reputation, but also attacked the RAS, which had awarded him a gold medal for the 'Cycle'. Consequently, the society forced Sadler to retract his claim. But he had prominent supporters: no less a person than Sherburne Wesley Burnham confirmed that Smyth's double-star data must be viewed with caution, writing in the popular magazine *English Mechanic* that '*No publication of original observations, in this or in any other language, can be named which contains so many serious errors.*'[157] Because of Sadler's paper, Burnham had observed all of the Bedford double stars in 1879–80, using the 18.5" refractor at Dearborn Observatory, Chicago. His result was that, for known pairs with moderate separation, Smyth's data were fairly acceptable, but for widely separated ones they turned out to be wrong in many cases. Exactly the latter data were due to Smyth's own micrometrical measurements, whereas the former had been taken from such observers as Wilhelm Struve. Edward Knobel later stated that not only Smyth's measurements but also his calculations were incorrect.[158] The scandal persuaded George Chambers to publish a corrected and enhanced version of Smyth's Bedford Catalogue in 1881 (Chambers 1881).

6.6 HIND – DISCOVERER OF SOME REMARKABLE OBJECTS

Around 1850 Hind was a very successful observer. As the scientific Director of the private Bishop's Observatory, he discovered some non-stellar objects, among them Hind's Variable Nebula (NGC 1555), the celebrated object in Taurus.

6.6.1 Short biography: John Russell Hind

John Hind was born on 12 May 1823 in Nottingham (Fig. 6.19 left). After having worked as an engineer, he got a job at the Magnetical and Meteorological Department of the Royal Greenwich Observatory in 1840. Afterwards, having been recommended by Airy, he was appointed as scientific director at Bishop's Observatory, London. Hind married in 1846 and became the father of six children. At that time it was not easy for an astronomer to earn money. Fortunately, besides the governmental or university observatories, like Greenwich, Cambridge and Oxford, there were a few private ones that had been established by wealthy amateurs.[159] One of them was George Bishop in London.

---

[154] 'Taurus Poniatowksi' is now classified as a part of Ophiuchus.
[155] See e.g. Ashbrook (1984: 51–55).
[156] Sadler checked William Herschel's double-star observations too, finding some errors.
[157] Sadler (1879: 184).
[158] Ashbrook (1984: 54).
[159] Hutchins (2008), Chapman A. (1998).

Figure 6.19. Left: John Russell Hind (1823–1895); right: bust of George Bishop (1785–1861) at the Royal Astronomical Society, London.

At his observatory Hind discovered 10 minor planets, 3 comets, the Nova Ophiuchi 1848 and the variable star U Geminorum (December 1855). For the minor planets (no. 7) Iris and (no. 8) Flora found in 1847, Hind and Bishop (as owner of the observatory) were honoured with a 'testimonial' by the RAS.[160] The laudation was given by the President, John Herschel (Herschel J. 1848). In 1853 Hind was awarded a gold medal for the discovery of eight minor planets, which was handed out by John Couch Adams (Adams 1853).

From 1853 to 1891 Hind was Superintendent of the *Nautical Almanac*, the highest officer for the production of the annual British ephemeris. This took much time and only occasional observations were possible. In 1863 he moved to Twickenham, the new site of Bishop's Observatory. There John Russell Hind died on 23 December 1895 at the age of 72.[161]

6.6.2 George Bishop and his private observatory in London

George Bishop was born on 21 August 1785 in Leicester (Fig. 6.19 right). He was a brewer and the largest British producer of wines. His private interest was astronomy. In 1836 he completed an observatory beside his 'South Villa' in London's Regent's Park; the place is now occupied by Regent's College (Howard-Duff 1985). The main instrument was a 7" Dollond refractor, offering magnifications from 45 to 1200 (Fig. 6.20). For his well-equipped observatory, Bishop hired some of the best astronomers. The first was the famous double-star observer William Rutter Dawes (1839–44). He was followed by John Russell Hind (1844–55), Norman Pogson (1850–51), who became known for his observations of variable stars, Eduard Vogel (1851–53), Albert Marth (1853–55), the discoverer of many nebulae, and finally Charles George Talmage (1860–61). Dawes later critically characterised Bishop's astronomical and human 'qualities': '*Mr. B. never did and never could observe at all, not even a transit; but after I left his observatory he put his own name to all my observations!!*'[162]

Bishop's support of astronomy yielded considerable fruits, above all in the field of minor planets and comets[163] (Bishop 1852). First, the *Berliner Akademische Sternkarten*, which had been initiated by Bessel, were used for the search.[164] The charts (epoch 1800), being limited to an area of 15° to both sides of the celestial equator, do not fully cover the ecliptic; the limiting magnitude is 9 to 10 mag. Bishop decided to create new 'ecliptical charts' for the epoch 1825, showing all stars of up to 11 mag in a strip of 6° width along the ecliptic (as references, stars from Lalande's *Histoire Céleste* and Bessel's 'Zonenbeobachtungen' were used). In the mid nineteenth century similar projects were realised by Edward Cooper (Markree), Jean Chacornac (Paris) and C. H. F. Peters (Hamilton College). Single charts appeared in a loose series.[165] The main work (observation, data reduction) was done by Hind. At Bishop's Observatory 11 minor planets were found (10 by Hind, 1 by Marth) – the 'ecliptical charts' fully served their purpose.[166]

---

[160] A 'testimonial' (as substitute for a gold medal) was presented only in 1848 – though for 12 persons at once! This was due to a dilemma of the RAS: the council was unable to decide who should be awarded a gold medal for the discovery of Neptune, Adams or Leverrier. After a controversial debate, the issue was adjourned; no medal was given and instead a 'testimonial' was created (Sampson 1923: 98–99).

[161] Obituaries: Plummer (1896) and Anon (1896).

[162] Letter from Dawes to George Knott of 19 December 1862 (Turner H. 1910: 393); see also Dawes (1865).

[163] Bishop (1852); see also Gould (1853).

[164] The charts appeared in irregular series (Bessel 1834); the work was finished in February 1854. With the aid of the chart 'Hora XXI', which had been made by Carl Bremiker in December 1844, the planet Neptune was discovered on 23 September 1846 in Berlin; see Hamel (1989).

[165] The charts are now at the RAS archive.

[166] Sometimes, however, plotted stars got lost. A curious example was Hind's 'four small stars' (Hind 1853b, c); a similar case later happened to Chacornac.

Figure 6.20. Bishop's Observatory with its 7" Dollond refractor in London's Regent's Park.

During 1840–57 Bishop was treasurer of the RAS and afterwards its president for two years. He died on 14 June 1861 in London at the age of 75.[167] Up to this time, 17 of 24 planned charts had been published. After Bishop's death the Dollond refractor and its dome moved to the home of his son, George Bishop Jr, at Meadowbank, Twickenham. There, in the summer of 1863, the new Meadowbank Observatory was established. It was closed in 1877 and the instruments were given to Capodimonte, Naples.

6.6.3 Hind's discoveries

On 30 March 1845, a year after Hind had been employed at Bishop's Observatory, he discovered his first non-stellar object, the globular cluster NGC 6760 in Aquila (Fig. 6.21). Half a year later, on 3 October he wrote to the editor of the *Astronomische Nachrichten* (Schumacher) that '*On the night of 1845 March 30, I found a faint nebula of a circular form [...] There does not appear to be any previous notice of this nebula, and Sir James South informs me that he can find no registered nebula within two degrees of this place. This is somewhat singular as the present object was found while searching for comets with our 11 feet refractor.*' (Hind 1845). Independently d'Arrest found the object in spring 1852 with the 11.7-cm Fraunhofer refractor of the Pleißenburg Observatory, Leipzig (d'Arrest 1852a). Inspecting Herschel's catalogues yielded no result;

Figure 6.21. Hind's first nebular discovery: the globular cluster NGC 6760 in Aquila (DSS).

Hind's discovery was obviously not known to him. D'Arrest describes the object (actually his first nebular find) thus: '*The nebula has a diameter of three minutes and its brightness is equal to that of a class I Herschel nebula.*' Auwers observed the object (Au 44) on 16 August 1860 with the Königsberg heliometer, finding it '*astonishingly faint, at most II. class, very gradually brighter middle, round, 2..2,5′ diam., frequently seen 1854–1858* [with the 6.3-cm Fraunhofer refractor], *and always of class I*'.[168]

---

[167] Obituaries: Anon (1862a) and Anon (1862b); see also Chapman A. (1998: 47–48).

[168] Auwers (1865: 216).

On 4 January 1850 Hind discovered his second nebula, the galaxy NGC 4125 in Draco. A short remark in his report of 28 March reads that it was *'tolerably bright, but small'*.[169] In the following issue of the *AN* he gives a bit more detail, though he now dates the observation on 5 January (Hind 1850b). The object is described as being *'of an elliptical form with a strong nuclear condensation'*. The paper also contains the first note on Hind's Crimson Star, the red shining variable carbon star R Leporis: *'I may mention also a remarkable crimson star in Lepus of about the 7th mag. the most curious coloured object I have seen [...] I found this star in October 1845 and have kept a close watch upon it since.'*

In contrast to other early observers, whose discoveries were often overlooked due to deficient publication, Hind acted more cleverly. He reported a new result in diverse magazines: the *Astronomische Nachrichten*, *Monthly Notices* (Hind 1850c) and *Comptes Rendus* (Hind 1850d). The last two give 4 January 1850 as the discovery date for NGC 4125.[170]

A third nebula, the globular cluster NGC 6535, was found in April 1852 (Hind 1852). At that time the location had been allocated to Ophiuchus; nowadays it is accorded to Serpens. Hind wrote that *'It is very small and rather faint, perhaps 1' in diameter, and is preceded by a very minute hazy-looking star.'* He mentions the exact date only in a letter to John Herschel of 12 May 1852: *'I detected a faint nebula on April 26th which I do not find in any Catalogue.'*[171]. Auwers observed the object (Au 38) on 10 October 1860 with the Königsberg heliometer: *'pretty faint, 2' large, round, gradually not much brighter middle'*.[172]

The first three nebulae are listed in a letter to John Herschel written on 7 February 1862.[173] Interesting is a handwritten note by Herschel on the letter: *'19. 3. 89 13 was also obs. by Petersen in 1849. It is also on Argelander's list'*. The digits give the position of NGC 6760 (AR 19^h 3^m, NPD 89° 13'). In 1850 Adolph Cornelius Petersen had reported a new nebula of Hind,[174] but an observation by him at Altona Observatory is not mentioned (Petersen 1850). The meaning of 'Argelander's list' is obscure; the place matches that of BD +0° 4122 (7.7 mag).

Hind made his most famous discovery on 11 October 1852 in Taurus: NGC 1555 (Hind 1853a). This is Hind's Variable Nebula, which is associated with the variable star T Tauri. Much had been reported about changing nebulae, but there was no convincing case – except Hind's object. In the context of the ideas of the two Herschels about the evolution of nebulae, the excitement among astronomers was great. Until then the visual observations of NGC 1555 had only fragmentarily been documented. A new investigation yielded more than 100 between 1852 and 1919; nearly all of the eminent observers were involved. The reports show a broad spectrum of conflicting and even curious results. The story of Hind's Variable Nebula and the nearby NGC 1554 (Struve's Lost Nebula) is the subject of Section 11.4. Table 6.9 shows Hind's discoveries; all four objects are non-stellar.

6.7 OBSERVATIONS AT HARVARD COLLEGE OBSERVATORY

Harvard College Observatory in Cambridge (MA) was founded in 1839; the first director was William Cranch Bond (Fig. 6.22 left).[175] Harvard was to receive a duplicate of the 15" Merz refractor which had been erected in 1839 at Pulkovo Observatory (it was at that time the largest in the world).[176] In late June 1847 the instrument became operational and some test objects were observed, among them the Andromeda Nebula (M 31) and M 57, the Ring Nebula in Lyra (Bond W. 1847, 1848). During 1847–48 a detailed study of the Orion Nebula (M 42) was made (Bond W. 1848; Bond G. 1867). Bond's son, George Phillips, was the most important visual observer of the early period. He had the privilege of being the first in the United States to discover a 'nebula': NGC 7150 on 10 February 1848, which is only a star group. By 1863

---

[169] Hind (1850a). The main content of the letter is Hind's detection that Lamont had seen Neptune twice as a star – before its 'official' discovery in 1846.

[170] In the *CR* paper, the nebula curiously does not lie in Draco, but in the '*grande Ourse*' (Ursa Major).

[171] RAS Herschel J. 12/1.6; see Bennett (1978).

[172] Auwers (1865: 217).

[173] RAS Herschel J. 12/1.6.

[174] Petersen refers to NGC 4125, but this should be NGC 6760.

[175] Bailey (1931), Jones and Boyd (1971), Rothenberg (1990), Stephens (1990).

[176] The lens of the Harvard refractor was 1.5 cm larger. It was surpassed by the 24" refractor of John Craig, which was erected in Wandsworth, England (Steel 1982).

Table 6.9. *Hind's four objects (sorted by date)*

Date	NGC	Au	GC	Type	V	Con.	Remarks
30.3.1845	**6760**	44	4473	GC	9.0	Aql	d'Arrest 1852
4.1.1850	**4125**	28	2735	Gx	9.6	Dra	
26.4.1852	**6535**	38	4369	GC	9.3	Ser	
11.10.1852	**1555**	20	839	RN		Tau	Breen, summer 1856

Figure 6.22. Harvard astronomers (from left to right): William Cranch Bond (1789–1859), Sidney Coolidge (1830–1863) and Horace Tuttle (1837–1923).

Coolidge, Safford and Tuttle (Table 6.10) had found new objects too. Safford, who later discovered many nebulae at Dearborn Observatory, will be treated in Section 9.6.

6.7.1 Short biography: George Phillips Bond

George Phillips Bond was born on 20 May 1825 in Dorchester, MA, and was led to astronomy by his father. In 1845 he graduated from Harvard, and he was employed as first assistant at the observatory a year later. Three years later a first study on the Andromeda Nebula, made with the new 15″ Merz refractor, appeared (Bond G. 1848). After the death of William Cranch Bond (1859) he became director and Professor of Astronomy. In November 1863 he was selected as the first American Associate of the Royal Astronomical Society. With Bond's directorship a new era of American astronomy began.[177]

On 7 February 1865 Bond was awarded the RAS gold medal. The laudation, given by Warren de la Rue, lists a remarkable number of efforts (de la Rue 1865): a comprehensive study of comet Donati 1858 (published in Volume 2 of the *Harvard Annals*), brilliant drawings of the Orion Nebula and of the Andromeda Nebula[178] (Bond G. 1867), the discovery of Hyperion, the eighth moon of Saturn (together with Lassell), and its inner ring, zone observations of stars near the celestial equator and the first astronomical photographs. George Phillips Bond could not receive the prize personally because he died on 17 February 1865 in Cambridge, MA, at the early age of 39 after a chronic illness of the lungs – a result of heavy smoking.[179]

6.7.2 Short biography: Phillip Sidney Coolidge

Sidney Coolidge, born on 22 August 1830 in Boston, MA, was the grandson of Thomas Jefferson (Fig. 6.22

---

[177] For the development of American astronomy and the scientific community in the United States during the period 1859–1940, see Lankford (1997).

[178] Bond was able to show that the 'changes' in M 31, which had been reported first by Boulliau in 1667 and later by Kirch (in 1676) and Legentil (in 1759), were by no means proved (Bond G. 1867).

[179] Obituary: Anon (1865c); see also Holden (1897a).

Table 6.10. *Observers at Harvard College Observatory during the years 1846–65*

Observer	Function
George Phillips Bond	First assistant 1846–1859, director 1859–1865
Phillip Sidney Coolidge	Research associate 1853–60
Truman Henry Safford	Assistant 1854–65; acting director 1865
Horace Parnell Tuttle	Assistant 1857–62 (younger brother of Charles Wesley Tuttle, assistant 1850–54)

centre). From 1839 to 1850 he lived in Europe. In Dresden he joined the Royal Military Academy. Back in America he worked for railway projects and the *Nautical Almanac* at Harvard College. Later he was an observer at the observatory, though without payment. His drawings of Saturn observed with the 15-inch showed unknown ring structures. His astronomical work was interrupted by military activities, e.g. in the American–Mexican War (1858). After his time at Harvard, Coolidge became a colonel of the Union Army infantry in the Civil War. He died on 19 September 1863 at the battle of Chickamauga, being only 29 years old.[180]

6.7.3 Short biography: Horace Parnell Tuttle

Horace Tuttle was born on 24 March 1837 in Newfield, ME (Fig. 6.22 right). Having been recommended by his older brother Charles Wesley Tuttle, who was already working at Harvard College Observatory, he joined the institution at the age of 20. He immediately discovered his first comet (in April 1857). For three others, found in the following year, Tuttle received the Prix d'Astronomie ('Prix Lalande') of the Académie des Sciences in Paris for 1858.[181] In 1862 he discovered the well-known periodic comet Swift–Tuttle (the source of the Perseid meteor shower). The relevant orbit computations were done together with Asaph Hall, who was active at Harvard College Observatory from 1857 to 1862.

In September 1862 Tuttle entered the Union Army, participating in the Civil War for nine months; then he served as an acting Navy paymaster. On 5 January 1866 he discovered the parent comet of the Leonids, the comet Tempel–Tuttle. In March 1875 Tuttle was dismissed from the Union Navy due to illegal money transactions and worked afterwards for the US Geographical and Geological Survey. Later he lived in Georgetown and from 1884 in Washington, where he observed at the US Naval Observatory (meeting Hall once again). In his later years he was employed by the US Mail Service and wrote articles for popular science magazines. Horace Tuttle died in August 1923 in Falls Church, VA, at the age of 96.[182]

6.7.4 Discoveries of nebulae

From 1847 the 15" Merz refractor (Fig. 6.23), which at that time was the largest telescope in America, was intensively used for visual observations of planets, comets, stars and selected nebulae (such as M 31 and M 42). Initially the Bonds, father and son, were concerned, who studied, for instance, the nebulae h 44, h 45 and h 51 (M 110, NGC 206 and M 32) associated with M 31.[183] Later Bond,[184] Coolidge, Safford and Tuttle found several new nebulae, '*met with by accident, while sweeping for comets or in the passage of Zones of stars*' (Bond G. 1864a).

In his 'List of new nebulae seen at the observatory of Harvard College' (published on 18 December 1863) Bond presented all of the objects found from 1848 onwards (Table 6.11). Later Dreyer used the source 'Bond' in different ways in the NGC (see the 'Remarks'). Bond's list has 33 numbered entries ('No.'). It consists of two parts: part I contains the verified objects (nos. 1 to 26) and part II '*objects supposed to be nebulae but requiring verification*' (nos. 27 to 33). Below

---

[180] Sheehan and O'Meara (1998).
[181] The argument of Jones and Boyd that the prize was given also '*for his additions to the catalogue of known nebulae visible in smaller telescopes*' is wrong (Jones and Boyd 1971: 117).
[182] Yeomans (1991: 238–239).
[183] Published by the President of Harvard University, Edward Everett (Everett 1850).
[184] In the following always George Phillips Bond is meant.

6.7 Harvard College Observatory

Figure 6.23. The 15" Merz refractor, which was installed in 1847.

Bond mentions two objects of Tuttle: NGC 1333 and NGC 6791. They were not included, since he knew from Auwers' list of new nebulae published in 1862 that they had been discovered earlier by Schönfeld and Winnecke, respectively (Auwers 1862a). 'Rr' shows the discovery refractor. The 4" is the comet-seeker by Bowditch, which was mainly used by Bond and Tuttle. The smaller one (2.5") was made by Quincy; both instruments were purchased in 1845.[185]

The column 'Z' indicates whether the object was found during the Harvard zone observations (marked 'Harvard Zones' in Bond's list). The first campaign (I) was executed by Bond and Charles Wesley Tuttle during 1852–53; 5500 stars between 0° and +0° 20' declination were measured with the 15-inch (Bond W. 1855). They determined the positions (epoch 1852) with a micrometer at magnification 141. New nebulae were listed by Bond in the 'Remarks' for the nearest zone star. The third campaign (III), which was performed during 1859–60 by Bond, Coolidge and Horace Tuttle, brought fourth some new objects (Winlock 1872). Coolidge's object no. 13 (NGC 3123) is not listed in the zone observations. The final zone (IV) brought Safford's NGC 2189 and NGC 2198 (Bond incorrectly gives 'J. H. Safford' in his list). These observations were not published; due to the great delay, their worth obviously became too small.

In 1866 Bond published his 'list of new nebulae' once again in the *Proceedings of the American Academy of Arts and Sciences* (Bond G. 1866). It appears in slightly modified form. For no. 2 (NGC 223 in Pisces) is added '*The nebulae h. 39–41 and 43 precede the above nearly in the same declination.*' The mentioned objects are the galaxies NGC 192 (h 39), NGC 196 (h 41) and NGC 201 (h 43), lying 40' west of NGC 223 (and NGC 219). The text implies h 40 (NGC 194) too, but this galaxy is 2.2° northwest. Actually, the group has a fourth member, NGC 197, but this was discovered by Marth only on 16 October 1863 (and hence was unknown to Bond). In Bond's modified list, two objects are deleted: no. 6 (NGC 1924) and no. 12 (NGC 2655), both of which had been discovered by William Herschel as III 447 and I 288, respectively. This is mentioned by him in an insert.

In 1862 Auwers ('Au') knew only of the nebulae from Bond's first zone observations, which had been published in 1855 (Auwers 1862a), and an object found by Tuttle (NGC 6643). Bond mentions it in a letter of 2 November 1861 to the *Astronomische Nachrichten*, which primarily treats Encke's comet (Bond G. 1862). The remaining numbers (Au 17 and Au 45) refer to objects of which Auwers had been told by their discoverers (Schönfeld and Winnecke). For Au 17 (NGC 1333) he also refers to Tuttle's independent discovery, published in 1859 in the *Monthly Notices* (Bond G. 1859). Auwers gives 1858 as the date, but 1855 is correct.[186] He was not aware that Tuttle's NGC 6643 (Au 40) had also been found by Schönfeld in about 1858 (the same applies for John Herschel and Dreyer). He always speaks about 'Tuttle's nebula'. There are two observations with the Königsberg heliometer (19 and 21 February 1862). About the first he wrote '*quite bright, gradually brighter to the middle, elongated in PA 50°, 2.5' long, 1.5' wide*'.[187] Later Borrelly reported

---

[185] Information on the telescopes can be found in Volume 8 of the *Harvard Annals*.

[186] John Herschel perpetuated this error in the GC.
[187] Auwers (1865: 227).

Table 6.11. Objects discovered at Harvard (sorted by date; see the text)

Date	Observer	Z	NGC	No.	Au	GC	HN	Type	V	Con.	Rr	Remarks
10.2.1848	Bond		7150	25		5077	1	Star group		Cyg	15"	First US object; Dreyer: 'G. P. Bond'
23.10.1848	Bond		7692	26		5079	2	Gx	14.8	Aqr	15"	
27.2.1850	Bond			27				NF		Cas	4"	
6.10.1850	Bond		2054	29		5354	3	4 stars		Ori	4"	Dreyer: 'G. P. Bond'
7.11.1850	Bond		7793	33		6233	4	Gx	9.0	Scl	4"	Dunlop 14.6.1826 (Δ 335)
30.12.1850	Bond			31				NF		Boo	4"	
18.2.1851	Bond		1624	28		879		OC+EN	11.8	Per	4"	W. Herschel 28.12.1790 (V 49)
11.9.1852	Bond		2515	11		5066	5	2 stars		Cnc	15"	Not found by d'Arrest
24.11.1852	Bond	I	6859	24	46	4538	6	3 stars		Aql	15"	
5.1.1853	Bond	I	223	2	4	119	7	Gx	13.4	Cet	15"	Not NGC 219; d'Arrest 8; L. Swift 21.11.1886
8.1.1853	Bond	I	391	3	9	211	8	Gx	13.5	Cet	15"	Dreyer: 'Bond, 1853'
26.2.1853	Bond	I	2399	9	24	1537	9	3 stars		CMi	15"	
26.2.1853	Bond	I	2400	10	25	1538	10	3 stars		CMi	15"	
9.5.1853	Bond	I	5632	20	33	3898	11	Star		Vir	15"	
9.5.1853	Bond	I	5651	21	34	3912	12	Star		Vir	15"	
9.5.1853	Bond	I	5658	22	35	3918	13	Star		Vir	15"	
1853	Bond	I				6000		Star	9.0	Aqr	15"	'Planetary?', BD +0° 4741
8.6.1855	Bond		5366	18		5074	14	Gx	13.8	Vir	15"	
5.2.1859	Tuttle	III	1333		17	710		RN		Per	2.5"	Schönfeld 31.12.1855 (BD +30° 548)

31.3.1859	Coolidge	?	3123	13		5067	15	NF		Sex	15"	
31.3.1859	Coolidge	III	3229	14		5068	16	3 stars		Sex	15"	
8.4.1859	Tuttle		2655	12		1691		Gx	10.1	Cam	4"	W. Herschel 26.9.1802 (I 288 = h 520)
29.4.1859	Coolidge	III	5404	19		5075	17	3 stars		Vir	15"	
30.4.1859	Coolidge	III	5200	16		5072	18	2 stars		Vir	15"	
30.4.1859	Coolidge	III	5310	17		5073	19	Star	12.9	Vir	15"	
3.5.1859	Coolidge	III	4582	15		5071	20	Star	13.4	Vir	15"	
3.5.1859	Coolidge	III		30				NF		Vir	15"	
17.7.1859	Tuttle		6791		45	4492		OC	9.5	Lyr	4"	Winnecke December 1853
1.9.1859	Tuttle		6643	23	40	4415	21	Gx	11.0	Dra	4"	Schönfeld 1858? (BD +74° 766)
15.11.1859	Coolidge	III	7403	32		6092	22	Star	13.9	Psc	15"	
16.12.1859	Coolidge	III	1312	5		5061	23	2 stars		Tau	15"	
25.1.1860	Coolidge	III	1251	4		5060	24	2 stars		Cet	15"	
7.2.1863	Bond		1924	6		1130		Gx	12.5	Ori	15"	W. Herschel 5.10.1785 (III 447)
19.3.1863	Safford	IV	2189	7a/b		5064	25/26	NF		Ori	15"	
19.3.1863	Safford	IV	2198	8		5065	27	NF		Ori	15"	
16.9.1863	Bond		219	1		5058	28	Gx	14.5	Cet	15"	Dreyer: 'G. P. Bond'

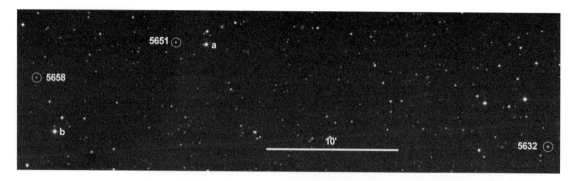

Figure 6.24. Bond's NGC 5632, NGC 5651 and NGC 5658 (nos. 20–22) in Virgo are merely faint stars (DSS).

about 'Changements observés dans la nébuleuse de Tuttle N.G.C. 6643'.[188] Here a letter from d'Arrest to John Herschel, written on 8 May 1863, is mentioned, in which d'Arrest describes an observation on 24 September 1862 (which is contained in the *Siderum Nebulosorum* too). Borrelly compares the available results and concludes that NGC 6643 is variable, which is not the case.

Equally, Auwers did not know of the independent discovery of Winnecke's NGC 6791 by Tuttle, because it was first published by Bond in 1863. Bond's nebulae NGC 2399 and NGC 2400 in Canis Major and NGC 5632, NGC 5651 and NGC 5658 in Virgo could not be found by Auwers on 15 April 1861 with the Königsberg heliometer.[189] For the latter trio (Fig. 6.24) he used the wrong reference star, applying a star termed 'a' (11.0 mag) instead of 'b' (10.5 mag).

Later d'Arrest investigated NGC 2399/2400, writing '*by the way, both star clusters are easily resolvable*' (d'Arrest 1865c). Actually both objects are merely star trios. The three 'nebulae' in Virgo turned out to be mere stars. NGC 6859 (Au 46) in Aquila was seen by Auwers on 28 September 1861 as a '*faint nebulous star 11–12^m, very small*' – this is a star trio too.[190]

The story of NGC 219 and NGC 223 (Fig. 6.25) is interesting. Bond discovered NGC 219 during the night on which he checked NGC 223, which had already been found in 1853. NGC 219 is 4' north preceding. The case of NGC 223 (Au 4), described by Auwers,

Figure 6.25. Bond's nebulae NGC 219 and NGC 223 (nos. 1 and 2) in Cetus (DSS).

is curious.[191] In Bond's zone observations three other nebulae are mentioned, though with '*ambiguous places*'. Auwers' observations on 28 and 30 September 1861 with the Königsberg heliometer showed that these must be h 39 (NGC 192), h 41 (NGC 196) and h 43 (NGC 201),[192] which had been discovered in 1790 by William Herschel. Auwers, who was generally a paragon of correctness, was not error-free either: the declination difference relative to the reference star (BD +0° 96) for h 43 has the wrong sign and he later called the nebula 'h 42'.[193]

---

[188] 'Changes observed in Tuttle's nebula NGC 6643' (Borrelly 1913).
[189] Auwers (1865: 221).
[190] Auwers (1865: 222).
[191] Auwers (1862a: 73).

[192] Auwers (1865: 222).
[193] Auwers (1862a: 73; 1865: 222).

About Bond's object NGC 2515 (GC 5066) d'Arrest wrote the following: '*Often and for a long time unsuccessfully I have searched also for Bond's nebula G.C. 5066. The polar distance is set too small by 4 degrees.*' (d'Arrest 1867b). The inclusion of further Harvard objects in the General Catalogue and Dreyer's Supplement (GCS) will be treated in Sections 8.2 and 9.3, respectively.

In 1908 Edward Pickering compiled a list of all 'Nebulae discovered at the Harvard College Observatory' (Pickering E. 1908). Analogously to the 'Harvard variable', he created the designation 'Harvard nebula' (HN). As column 'HN' shows, Pickering ignored doubtful objects (nos. 27, 30 and 31); and also those which had already been discovered by others. An exception is NGC 7793, for which he didn't know its identity with Dunlop's Δ 335 (found in 1826). Dreyer notes in the NGC that it was '*Like a comet (1850)*'. In Pickering's list NGC 219 appears twice (as HN 7 and HN 28); actually HN 7 is NGC 223.

What about the rating of the four Harvard observers? Bond thought he had discovered 19 objects, of which 3 must be credited to others (not known to him). The identity of no. 6 with William Herschel's III 447 was not noticed until the making of the HN list (the object was not observed by John Herschel). The identifications of nos. 28 and 33 result from a modern analysis. Bond's no. 2 (NGC 223) was found independently on 1 January 1862 by d'Arrest.[194] Lewis Swift had observed this galaxy twice (and thought it to be new): on 21 November 1886 and again on 12 November 1890 (curiously, the latter find was catalogued by Dreyer as IC 44). Of Bond's own 14 NGC objects only 5 are non-stellar (all galaxies), the rest are stars or star groups or are lost (i.e. cannot be found at the given position). Thus Bond's rating is pretty bad.

Coolidge contributed 10 objects; 9 are listed in the NGC (credited to him by Dreyer). But not a single one is real: all turned out to be stars or are missing. Safford is no better. His objects nos. 7a and 7b were described as '*two clusters* [...] *near two stars of 10.11th magnitude*' and for no. 8 one reads '*cluster* [...] *between two stars*' (Bond G. 1864a). In all cases only the reference stars can be seen – but no trace of any clusters! The case of Tuttle, who was an excellent observer, looks quite different. All four objects are real, but, unfortunately, had been discovered earlier by others; among them are two Schönfeld saw while measuring stars for the Bonner Durchmusterung (1852–59). However, Tuttle was the true discoverer of NGC 7581, a galaxy in Pisces (see Section 8.17).

6.8 LASSELL'S FIRST NEBULA

After William Herschel and Lord Rosse, William Lassell was the most important constructor of large reflecting telescopes.[195] Essential developments are due to him. All three were infected by the idea of 'doubling' the size:[196] Herschel increased the focal length from 20 ft to 40 ft (the aperture was even increased from 18.7" to 48"), Lord Rosse doubled the mirror diameter from 36" to 72" and Lassell replaced his 24" reflector by the famous 48-inch, both (in this way differing from Herschel and Lord Rosse) with equatorial mountings! He used the instruments to observe planets (discovering some moons), nebulae and star clusters. Lassell was one of the first to escape from the growing urban light and air pollution, making excellent observations and drawings during his two Malta expeditions.

6.8.1 Short biography: William Lassell

William Lassell was born on 18 June 1799 in Moor Lane, Bolton, Lancashire (Fig. 6.26). He first attended a day school and then changed at the age of nine to the Rochdale Academy. When the family moved to Liverpool in 1815 (the father had died in 1810) Lassell entered a merchant's office. In 1824 he commenced business as a beer brewer.[197] Owing to the rise of industry in the Liverpool area, this was quite profitable.

In his leisure time Lassell dealt with mechanics and astronomy, which combination inevitably led him to the construction of telescopes. As early as in 1820 he

---

[194] This refers to object 'no. 8' in d'Arrest's paper in the *Astronomische Nachrichten* No. 1500 (d'Arrest 1865a).

[195] See e.g. Ashbrook (1984: 136–141).

[196] The principle was applied by Foucault too, who built an 80-cm reflector (1862) shortly after his 40-cm (1859), both equipped with silvered glass mirrors; see Tobin (2003). George Ellery Hale continued working on it in the twentieth century: shortly after erecting the 100" Hooker reflector on Mt Wilson, he started his last project, the 200" on Mt Palomar, which bears his name.

[197] George Bishop and Richard Carrington were active in this trade too.

Figure 6.26. William Lassell (1799–1880).

Figure 6.27. A model of Lassell's equatorially mounted 24" Newtonian reflector (National Museums & Galleries on Merseyside, Liverpool).

built two 7-inch reflectors of Newtonian and Gregorian type; as usual, the mirrors were made of speculum metal. His astronomical interests led him to Maria King (her brothers Albert and Joseph were enthusiastic observers), whom he married in 1827. In the summer of 1830, Lassell constructed a 9" Newtonian reflector, which was erected on the roof of his home in Milton Street. With it he independently found the sixth star in the trapezium of the Orion Nebula (John Herschel saw it in 1830).[198] Only five miles away, in Ormskirk, lived another eminent amateur astronomer: William Rutter Dawes, with whom Lassell was closely connected throughout his life.[199]

In about 1837 Lassell moved to the rural Liverpool suburb of West Derby, naming his new domicile 'Starfield'. In summer 1840 an observatory with the 9-inch as main instrument was ready. The reflector was mounted on an equatorial fork; the tube and mounting were made of iron.[200] A first highlight was the construction of an equatorial 24" Newtonian reflector in 1845 (Fig. 6.27), an enlarged version of the 9-inch. Two years earlier, Lassell had visited Lord Rosse at Birr Castle, where he inspected his workshop and instruments with great interest; the 72-inch was then nearly complete. On the basis of this experience, Lassell built a steam-driven grinding machine. He also profited from the mechanical skill of another friend and amateur astronomer, James Nasmyth, who was living in Patricroft near Manchester.[201]

Lassell's primary targets for the 24-inch were the planets, concerning which he achieved his greatest success. On 10 October 1846 he discovered one of Neptune's moons, Triton (the planet had been known for just 2½ weeks). In September 1848 he found Hyperion, the eighth moon of Saturn (which was independently seen by William Cranch Bond at Harvard College Observatory). In October 1851 the third and fourth moons of Uranus – Ariel and Umbriel – followed.

In February 1849 Lassell was awarded the RAS gold medal for his achievements on reflectors and the discovery of Triton and Hyperion. John Herschel gave the laudation (Herschel J. 1849). Already in November 1848 George Biddell Airy had praised the telescopes of Lassell and Lord Rosse, proudly stating that '*the reflecting telescope is exclusively a British instrument in its invention and improvement, and almost exclusively in its use*' (Airy 1849).

In 1850 Otto Struve visited Starfield. His primary interest was the power of Lassell's 24" reflector in comparison with his 15" refractor at Pulkovo (which was

---

[198] See Section 9.6.10.
[199] Ashbrook (1984: 360–365).
[200] Lassell (1841, 1842). A comparably massive instrument was built in 1835 by Grubb: the 15" Cassegrain reflector for Romney Robinson at Armagh Observatory.

[201] Ashbrook (1984: 141–147).

at that time the best in the world, together with the Harvard refractor). A key object for the competition was the Orion Nebula, which fascinated both observers. The theme 'refractor vs. reflector' was a permanent issue in the nineteenth century (see Section 7.5.2).

Unfortunately, due to increasing air and light pollution, the Liverpool sky got worse ('Starfield' was often referred to jokingly as 'Cloudfield' by Dawes). Therefore, in autumn 1852, Lassell moved the 24-inch to Malta (the island was under British rule), where he observed on a hill near the coast in a suburb of the capital La Valetta until mid 1853. Not only the better seeing but also the southern latitude (36°) were favourable for many objects, such as the Orion Nebula.[202] On the other hand, Lassell's expectation concerning the planets could not be realised.

In January 1854 Lassell returned to Starfield. A year later he moved to Bradstones outside Liverpool. There in 1857 he began to build his masterpiece, the 48-inch, an equatorially mounted Newtonian reflector destined for Malta. Only a year later he was awarded a medal for the second time by the RAS. After having finished its construction, Lassell tested the great telescope in January 1860 at Sandfield Park near Liverpool. On 6 September 1861 it was transported by train to Marseille and shipped to Malta, where it arrived on the 26th. A comprehensive description of the reflector can be found in a letter of Lassell to the *Astronomische Nachrichten*, written in 1865, which contains a fine drawing (Lassell 1865a). The main mirror (made from an alloy of copper and tin) had a diameter of 1.22 m; actually there were two examples with different focal lengths (11.22 m and 11.38 m). A new feature of the telescope was that it had a truss-tube, which is much lighter than a solid one, allows rapid heat transfer and performs better in the wind.[203] It was erected in the open air, because its great dimensions did not allow the building of any shelter.

On Malta, Lassell was largely supported by his assistant Albert Marth. During his four-year visit, both made many observations, which were eventually published in 1867 in the *Memoirs of the Royal Astronomical Society*. Later Margaret Huggins wrote that, '*Had Mr. Lassell continued to observe at Malta, the discovery of the satellites of Mars might not have been left to America.*' (Huggins M. 1880).[204] Similarly to the case of John Herschel's stay at the Cape of Good Hope, Lassell had travelled with the whole family, i.e. his wife and daughters. One of them, Caroline, was a great artist, who made a fine drawing of the Orion Nebula in 1862, which was based on Lassell's observations.

After his return to England, Lassell bought an estate in Maidenhead, called 'Ray Lodge', where he erected his old 24-inch. He had offered his 48-inch for the planned Melbourne observatory, but the board, which was committed to the Great Melbourne Telescope (Lassell, Lord Rosse and Romney Robinson were members, among others), favoured a new instrument.[205] Angry about the decision, Lassell never erected his large reflector again. Both of the main mirrors were destroyed and only a secondary mirror remained (Bennett 1981).

However, Lassell did not remain idle. He constructed, for instance, an optimised grinding machine. In his last years an eye disease forced him to abandon observing. During 1870–72 he was President of the RAS. William Lassell died on 5 October 1880, at the age of 81 in Maidenhead. The 24-inch was given to the Greenwich Observatory by his daughter Caroline. In 1921 Dreyer catalogued Lassell's estate for the RAS.[206]

6.8.2 Lassell's observations of nebulae

Lassell's first discovery was accidental: on 31 March 1848 he found NGC 3121, a galaxy in Leo, with his 24-inch (Fig. 6.28). He casually mentions the observation in a letter to the *Astronomische Nachrichten* of 8 April 1848: '*a faint nebula […] was almost in the field at the same time as the Comet* [Mauvais 1847 IV].

---

[202] For a few months in 1856, Piazzi Smyth tested the island of Tenerife for astronomical observations.

[203] A bit earlier, Lord Rosse's 36-inch was equipped with a truss-tube too.

[204] Asaph Hall discovered the Martian moons in 1877 at the US Naval Observatory, Washington.

[205] The result was a 48″ Cassegrain with metal mirror constructed by Grubb; see Section 9.21.6.

[206] Obituaries: Huggins M. (1880), Huggins W. (1881) (copied in *Proc. Roy. Soc.* **31**, vii–x, 1881), Peters C. F. W. (1881) and Anon (1880e). For details about Lassell's work, see Chapman A. (1988); see also Chapman A. (1998: 100–102).

Figure 6.28. Lassell's first object, the galaxy NGC 3121 in Leo, which he discovered in 1848 in Liverpool with the 24″. The companion NGC 3119 was found by Marth on 14 December 1863 with the 48″ on Malta (DSS).

I know not whether it has previously been observed, but I estimated the Comet at about half its brightness' (Lassell 1848). Later Auwers included the object as no. 26 in his list of new nebulae (Auwers 1862a). With the Königsberg heliometer he saw it on 9 April 1861 as '*very faint, diameter 1.5'; star 9.10^m 4' np 14..15s*' and determined the position relative to BD +15° 2165, lying 4' to the northwest.[207]

During his first stay on Malta, Lassell did not find any non-stellar objects, but he used the 24″ reflector to observe and draw selected nebulae and star clusters, mainly in the southern sky. Like Lord Rosse, he tried to resolve some objects, especially the Orion Nebula (Lassell 1854a). Between 10 December 1852 and 12 March 1853 he made drawings of NGC 1535, NGC 1952 (M 1), NGC 1980, NGC 2022, NGC 2261, NGC 2371, NGC 2392 (see Fig. 6.16), NGC 2438, NGC 2440, NGC 3132 and NGC 3242 (Lassell 1854b). The publication gives object descriptions too.[208] During his second stay, Lassell discovered three nebulae with the 48-inch (all in 1863), which were listed by Herschel in the General Catalogue.[209]

6.9 ERNEST LAUGIER AND THE FIRST CATALOGUE OF ACCURATE POSITIONS OF NON-STELLAR OBJECTS

Laugier was the first to measure accurate positions of nebulae and star clusters, comparing them with earlier data of Messier and William Herschel. Although he had been expecting evidence for proper motions, he had to confess that the earlier data were not good enough for this task. As we now know, even the most precise measurements would not reveal any positional change. Anyway, Laugier's work greatly influenced this new branch of nebular research. Prominent followers were d'Arrest, Auwers and Schönfeld.

6.9.1 Short biography: Paul Auguste Ernest Laugier

Ernest Laugier was born on 22 December 1812 in Paris. At the age of 19 he entered the Ecole Polytechnique. As little as two years later he joined the Bureau des Longitudes as a 'membre adjoint'. Arago, Director of Paris Observatory, noticed his abilities and hired him as an 'astronome-adjoint' (later Laugier married Arago's niece). There he was one of the most active observers.[210] In 1842 he discovered a comet and calculated its orbit. For this achievement Laugier was awarded the 'Prix Lalande' of the Académie des Sciences for 1843. Later he made positional measurements of stars and nebulae. After the death of his patron and subsequent personnel changes by the new, authoritarian Director Leverrier, Laugier left the observatory in early 1854. In 1862 he became a nominal member of the Bureau des Longitudes and was responsible for the well-established almanac *Connaissance du Temps*. Ernest Laugier died on 5 April 1872 in Paris.[211]

6.9.2 Laugier's measurements

Laugier first dealt with nebulae and star clusters in 1847. His aim was the detection of proper motion by comparing past and present locations. He selected

---

[207] Auwers (1865: 221).
[208] See also Lassell (1854c).
[209] This is described in Section 7.5.

[210] Until 1 January 1854 Paris Observatory was the official observing station of the Bureau des Longitudes.
[211] Obituary: Dunkin (1873).

three compact-looking objects, whose positions had already been determined by Messier in 1764 with a meridian-circle under favourable circumstances: the open cluster M 11 (Scutum) with a dominant star in its centre and the globular clusters M 3 (Canes Venatici) and M 28 (Sagittarius). With Laugier's new measurements a period of 83 years was calculated.

The results were published in June 1847 in the *Comptes Rendus* (Laugier 1847). The data showed positional differences of up to 1.5′, which Laugier interpreted as evidence for proper motion. He ruled out observational errors, fully trusting Messier's data because of his great reputation as an observer. Encouraged, he planned a larger project: '*I hope to present shortly a catalogue of nebulae, whose positions are determined with great care using the equatorial of Gambey.*' This instrument is the 9.7-cm refractor made by Gambey in 1826 (lens by Lerebours), which had been erected on the roof of the Paris Observatory.

Laugier made his observations during the winter 1848/49. Afterwards he wrote a first report about methods and results (omitting the data) for the Académie des Sciences, titled 'Sur l'utilité d'un Catalogue de nébuleuses' (Laugier 1849). He now doubted Messier's positions; his own measurements with a filar-micrometer, using reference stars from the *British Association Catalogue* (BAC) of 1845, were much better. The resulting absolute positions were reduced to epoch 1850. Owing to this difference in quality, the detection of proper motions was thought to be impossible.

6.9.3 The catalogue of 53 nebulae and star clusters

Laugier's catalogue was eventually published in December 1853 (Laugier 1853). It lists 53 objects, though without identifications. Given are positions for 1850 plus calculated differences from those of Messier and William Herschel. Measuring extended objects without a compact centre was difficult: '*For nebulae with large diameter, I always measured the position of the brightest point.*' Whereas the differences from Herschel were generally small, those from Messier were as large as 50′. In Table 6.12 Laugier's objects (L) are identified with the main catalogues. Concerning the types, we have 33 galaxies, 8 planetary nebulae, 8 globular clusters, 3 open clusters, 2 reflection nebulae and 1 supernova remnant (M 1).

D'Arrest (dA) reported having observed 31 Laugier objects in the years 1855–56 (actually there are three more).[212] Generally he refers to the L-number, except for NGC 221, NGC 224 and NGC 1023 (marked '?'). Of course, the period between his and Laugier's measurements was much too short to reveal any proper motion, thus he could estimate only the accuracy of Laugier's positions. D'Arrest concluded that the probable error is about 6″. In one case (L 38 = NGC 4203) the difference in right ascension is considerable and he notes '*AR is obviously errant*'. In his opinion, the deviations noticed are due not to Laugier's measurements, but to the inaccurate positions of the reference stars.

For objects catalogued by William Herschel, Auwers (Au) analysed Laugier's positions too (Auwers 1862a). Three were omitted (marked '?'). He did not make observations, but merely reduced Herschel's relative positions. For some cases, Auwers could detect considerable differences between Laugier and Herschel.

A modern analysis of Laugier's positions shows that they are pretty good. The mean differences are 27″ in right ascension and 44″ in declination, which gives a positional accuracy of about 1′. In three cases, however, there are large deviations (Table 6.13); e.g. Laugier's position for M 82 (L 19) is more than 20′ off! Interestingly, the difference for the neighbouring galaxy M 81 (L 18) is only 1.7′. For M 78 (L 10) and M 27 (L 49) we have 1.3′; both are extended objects (RN, PN), whose centres are not clearly defined in a telescope of aperture 10 cm. The error in right ascension for NGC 4203 had already been noticed by d'Arrest. Laugier's position of NGC 7009 (L 50) is very good. This planetary nebula is an ideal measuring target: it is small and round with a central star of 12.1 mag.

6.10 COOPER AND HIS MARKREE CATALOGUE

In 1830 Edward Cooper founded the Irish Markree Observatory, which was characterised by George Biddell Airy as '*undoubtedly the most richly furnished private observatory known*' (Airy 1851). In contrast to other amateur astronomers, who favoured large reflectors, for example Lord Rosse and Lassell, Cooper decided to erect a refractor with a considerable aperture

---

[212] d'Arrest (1856a: 305).

Table 6.12. Laugier's catalogue (dA = d'Arrest, Au = Auwers)

L	NGC	Type	Con.	M	H	h	dA	Au
1	221	Gx	And	32		51	?	
2	224	Gx	And	31		50	?	
3	650/51	PN	Per	76	I 193			?
4	1068	Gx	Cet	77		262	*	
5	1023	Gx	Per		I 156	242	?	*
6	1052	Gx	Cet		I 63	254	*	*
7	1904	GC	Lep	79			*	
8	1931	OC	Aur		I 261	355		*
9	1952	SNR	Tau	1		357	*	
10	2068	RN	Ori	78		368	*	
11	2099	OC	Aur	37		369		
12	2261	RN	Mon		IV 2	399	*	*
13	2655	Gx	Cam		I 288	520	*	*
14	2683	Gx	Lyn		I 200	532	*	*
15	2841	Gx	UMa		I 205	584	*	*
16	2859	Gx	LMi		I 137	593	*	?
17	2903	Gx	Leo		I 56/57	604	*	*
18	3031	Gx	UMa	81		649		
19	3034	Gx	UMa	82	IV 79			*
20	3351	Gx	Leo	95		743	*	
21	3368	Gx	Leo	96		749		
22	3377	Gx	Leo		II 99	754	*	*
23	3379	Gx	Leo	105	I 17	757	*	*
24	3384	Gx	Leo		I 18	758	*	*
25	3414	Gx	LMi		II 362	773	*	*
26	3486	Gx	LMi		I 87	805	*	*
27	3489	Gx	Leo		II 101	806	*	*
28	3504	Gx	LMi		I 88	810	*	*
29	3521	Gx	Leo		I 13	818	*	*
30	3587	PN	UMa	97		838		
31	3623	Gx	Leo	65		854	*	
32	3627	Gx	Leo	66		857	*	
33	3665	Gx	UMa		I 219	881		*
34	3675	Gx	UMa		I 194	887		*
35	3941	Gx	UMa		I 173	1005		*
36	3992	Gx	UMa	109	IV 61	1030		*
37	4111	Gx	CVn		I 195	1088	*	*
38	4203	Gx	Com		I 175	1140	*	*
39	4216	Gx	Vir		I 35	1148	*	*
40	4361	PN	Crv		I 65	1231	*	*
41	4736	Gx	CVn	94		1456	*	
42	5272	GC	CVn	3		1663	*	
43	5904	GC	Ser	5		1916		
44	6205	GC	Her	13		1968	*	
45	6254	GC	Oph	10		1972	*	
46	6626	GC	Sgr	28		2010	*	
47	6705	OC	Sct	11		2019	*	
48	6720	PN	Lyr	57		2023	*	
49	6853	PN	Vul	27		2060	*	
50	7009	PN	Aqr		IV 1	2098	*	?
51	7078	GC	Peg	15		2120	*	
52	7089	GC	Aqr	2		2125	*	
53	7331	Gx	Peg		I 53	2172	*	*

Table 6.13. *Laugier objects with large deviations (Δα and Δδ are differences in right ascension and declination, respectively)*

L	Object	Actual position (2000)	Laugier's position (2000)	Δα (′)	Δδ (′)
13	NGC 2655	08 55 37.7 +78 13 25	08 55 29.6 +78 05 02	−2.03	−8.38
19	NGC 3034 (M 82)	09 55 54.0 +69 40 59	09 55 57.2 +70 01 19	+0.80	+20.33
38	NGC 4203	12 15 05.0 +33 11 51	12 14 23.5 +33 11 47	−10.37	−0.07

of 14″. His greatest achievement was the Markree catalogue, listing more than 60 000 stars along the ecliptic. During his observations with the great refractor, Cooper discovered some 'nebulous' objects and clusters. Except for one – an already known galaxy – there is nothing spectacular in his list.

6.10.1 Short biography: Edward Joshua Cooper

Edward Cooper was born on 1 May 1798 in St Stephen's Green, Dublin (Fig. 6.29 left). During his childhood, his mother encouraged his interest in astronomy. Being a pupil in Armagh, he came into contact with James Archibald Hamilton, the first Director of Armagh Observatory. Cooper studied in Eton and Oxford, leaving the university without any degree. Afterwards he travelled around Europe for nearly 10 years. After the death of his father (in 1830) he inherited Markree Castle near Colloney, Sligo. He soon planned an observatory and in 1831 he was lucky to get a 14″ lens from Cauchoix, Paris, for 1200 pounds sterling. At that time, this was the largest one in the world.[213] The corresponding telescope had a wooden tube on a Herschel mounting (Fig. 6.29 right). In 1834 it was fitted with an iron tube and an equatorial mounting made by Thomas Grubb (Dublin).[214] The heavy instrument rested on a pyramidal stone socket about 4 m high (this is partly visible in Fig. 6.30). There was no dome; the only shelter against wind was a surrounding wall of the same height. A regular visitor was Romney Robinson; the Director of nearby Armagh Observatory was impressed by the large telescope, which was superior to his 15″ Cassegrain with metal mirror. During 1843–45 Cooper spent some time in Nice and Naples. At Capodimonte he observed with his 3″ comet-seeker, making some drawings of nebulae.

Edward Cooper died on 28 April 1863 at the age of 64, a few months after he had lost his second wife; he left five daughters.[215] The heritage was taken over by his nephew Col. Edward Henry Cooper, who was interested in meteorology and astronomy.[216] Until 1874 Markree Observatory was left without supervision, then Col. Cooper hired the Dane William Doberck as director (Fig. 6.30).[217] The great refractor was in a bad condition (Doberck 1876). After a restoration, Doberck mainly observed double stars. In 1884 he published a short history of the observatory (Doberck 1884).[218] He was succeeded in 1883 by Albert Marth, who held the position until 1897. The last director was Frederick William Henkel. After Col. Cooper's death in 1902 the observatory was closed and the 14″ refractor was moved to Hongkong Observatory, where Doberck was Government Astronomer from 1883 until 1907. After an air raid in 1941 the damaged telescope was evacuated to Manila. There the Cauchoix lens was used until 1989 in a Littrow spectrograph for observations of the Sun.

6.10.2 Cooper's *Catalogue of Stars Near the Ecliptic*

Cooper's most important astronomical contribution was the *Catalogue of Stars Near the Ecliptic*, which he

---

[213] The telescope surpassed the 25-cm Lerebours refractor in Paris; in 1839 the 37.9-cm Merz refractor at Pulkovo became the largest.
[214] Glass (1997: 13–16).
[215] Obituary: Anon (1863a); for Cooper see also Coffey (1998) and Chapman A. (1998: 48–50).
[216] For the Markree Observatory archive see Hoskin (1982b).
[217] Doberck and Dreyer were friends with surprisingly parallel careers; see the biography of Doberck by MacKeown (2009). The photograph (Fig. 6.30), which was probably taken in about 1880, was found in one of Dreyer's albums.
[218] See also Fitzsimons (1980, 1982) and McKenna (1967: 289–290).

Figure 6.29. Left: Edward Cooper (1798–1863), from Coffey (1998). Right: the 14″ refractor on its first mounting.

Figure 6.30. William Doberck at the Markree refractor with its unique pyramidal stone socket.

made at Markree Observatory between 1848 and 1856. The data were needed in the search for minor planets. The existing standard, the *Berliner Akademische Sternkarten*, did not have a sufficient limiting magnitude and did not cover the whole area of the ecliptic, reaching only 15° to both sides of the equator (Cooper 1848). Cooper's catalogue (and the associated charts) contains stars to 12 mag, lying 6° around the ecliptic. Owing to their huge number, only hitherto uncatalogued stars were observed. In the middle of the nineteenth century comparable ecliptical charts were made by Hind (Bishop's Observatory, London), Chacornac (Paris Observatory) and C.H.F. Peters (Hamilton College, Clinton).

All observations were made with the 14″ refractor, which was fitted with a wide-angle eye-piece (magnification 80, field of view 30′) and a square bar-micrometer built by Cooper's assistant Andrew Graham (Cooper 1852). Owing to the faintness of the stars, the field was not illuminated. Graham did most of the work. On 26 April 1848 he discovered the minor planet (9) Metis with the 3″ comet-seeker by Ertel.[219] He left Markree in 1860 and went to Cambridge. Cooper used the comet-seeker to draw nebulae (e.g. the Orion Nebula). In February 1845 he discovered a comet.

The Markree catalogue lists positions of 60 066 stars for the epoch 1850. It was published in four volumes in Dublin (Cooper 1851–56).[220] Unfortunately the accompanying atlas could not be printed; Cooper's heirs bequeathed the hand-drawn charts to Cambridge Observatory, where they are still in the archive.

The catalogue contains eight objects, marked as 'nebulous' or 'cluster' (Table 6.14). The column 'Markree' shows volume, page and description. They

---

[219] Obituary: Anon (1909).
[220] The volumes appeared as follows: I, in 1851; II, in 1853; III, in 1854; and IV, in 1856.

Table 6.14. *Cooper's eight discoveries (sorted by date)*

Date	NGC	Markree	Au	GC	Type (V)	Con.	Remarks
9.4.1852	4989	2, 210 Neb.	31	3421 = 3426	Gx (13.3)	Vir	Auwers: not clearly visible (15.4.1861); W. Herschel 17.4.1784 (II 185)
22.10.1852	**46**	3, 20 Neb.	2	24	Star (11.7)	Psc	Auwers: a quite sharp star 11^m with no nebulosity (28./30.9.1861)
13.1.1853	**2218**	3, 53 Faint cluster	22	1402	NF	Gem	Auwers: not visible (15.4.1861)
23.12.1853	**2248**	3, 155 Small cluster	23	1426	2 stars	Gem	Auwers: just resolved very faint patch of 2–3′ diameter, the brightest star is 12^m (15.4.1861)
24.11.1854	1488	4, 55 Nebulous	19	791	Star group	Tau	Auwers: not observed
24.11.1854	**7122**	4, 38 Nebulous	47	4695	2 stars	Cap	Auwers: a star 11^m, surrounded by some faint stars 12–13^m, perhaps nebulous (28.9.1861)
17.1.1855	**5268**	4, 96 Neb.	32	3633	Star (11.3)	Vir	Auwers: not clearly visible (15.4.1861)
8.10.1855	**7447**	4, 183 Nebulous	49	4878	NF	Aqr	Auwers: not visible (28.9.1861); today visible (30.9.1861)

were entered by Auwers (Au) in his list of new nebulae.[221] Later it became clear that the first object (NGC 4989) is identical with William Herschel's II 185. It is also the only non-stellar object; five are stars or star groups that appeared 'nebulous' in the refractor and two are 'not found'. Auwers observed most of these objects with the Königsberg heliometer (his description is given in the last column).[222] Cooper's rating is inconceivably bad.

Three objects were also observed at Birr Castle (Parsons L. 1880). Dreyer observed NGC 1488 (GC 791; Fig. 6.31 left) on 1 December 1874: '*no nebulosity seen*'. Charles Burton noted on 5 January 1869 for NGC 2248 (GC 1426; Fig. 6.31 centre): '*numerous stars, nothing remarkable seen*'. NGC 7122 (GC 4695; Fig. 6.31 right) was observed by Robert Ball on 28 October 1866 ('*the only object found was a double star*') and again by Dreyer on 5 October 1875 ('*no nebulous star 10–11m found, only a double star*'). Tempel could not find NGC 7447 (Au 49) at several attempts in Arcetri (Tempel 1880b). The same applies for Burnham at the 36″ Lick refractor in 1891 (Burnham S. 1892b).

6.11 SECCHI AND DE VICO: OBSERVATIONS AT THE COLLEGIO ROMANO

Angelo Secchi was not only a pioneer of astrophysics (because of his innovative spectral analysis), but was interested in star clusters and nebulae too. He contributed a considerable number of observations and drawings. The same applies to his predecessor, Francesco de Vico. With the 6″ Cauchoix and 9.5″ Merz refractors, the Collegio Romano Observatory had two fine instruments. Secchi thought the latter to be even comparable to John Herschel's 18¼″ reflector.

6.11.1 Short biography: Angelo Secchi

Angelo Secchi was born on 29 June 1818 in the Italian town of Reggio Emilia (Fig. 6.32). He attended the

---
[221] Auwers (1862a).
[222] Auwers (1865: 221–222).

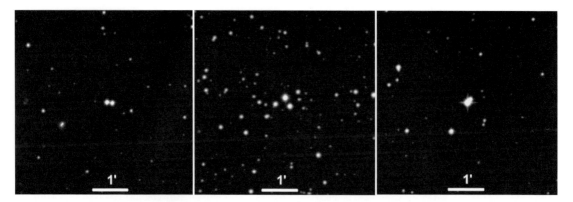

Figure 6.31. Three Cooper objects (from left to right): NGC 1488, NGC 2248 and NGC 7122 (DSS).

Figure 6.32. Angelo Secchi (1818–1878).

local Jesuit school, learning Greek and Latin. In 1833 he went to Rome, entering the Jesuit order at the minimum age of 15. He got his further education at the Collegio Romano, where his liking for astronomy and physics appeared. During 1841–44 he studied science in Loretto and went back to Rome to be ordained as a priest in 1847. In the meantime he worked as an assistant of Francesco de Vico at the Collegio Romano Observatory, which was located on the highest tower of the main building at that time. From 1825 it possessed a 15-cm refractor by Cauchoix. The political climate in Rome was turbulent and, when in 1848 a revolt broke out, Secchi and de Vico left Italy. After a short stay at Stoneyhurst, England, they moved to Georgetown University near Washington. De Vico died suddenly in 1849 and a year later Secchi became the new Director of the Collegio Romano Observatory at the age of 32.

Back in Rome he erected a new observatory on the roof of the Church of St Ignazio which was located on the northern side of the Collegio Romano. The planned dome of the church with a diameter of 17 m was never built and thus the walls were strong enough to hold the observatory buildings. It was ready in 1853, equipped with a fine 24-cm Merz refractor of focal length 4.35 m (Fig. 6.33 right). Secchi first observed double stars with it. The old Cauchoix was still used for solar observations (Fig. 6.33 left). Both instruments were housed in cylindrical shelters. During 1852–53 Secchi discovered three comets. Besides astronomy he was interested in seismology and meteorology. In 1853 he became a Fellow of the RAS and three years later was elected to the Royal Society. In the 1860s he entered the new field of spectroscopy, constructing several attachments for both telescopes. With the Cauchoix refractor, equipped with an objective prism, Secchi made a spectroscopic survey of 4000 stars. His studies led to the first classification of stellar spectra in 1863. In 1867 a catalogue of spectral types for 300 stars was published and in the following year he established the four 'Secchi classes'. This made him the 'father of astrophysics'.

In 1870 Italian troops occupied the Vatican and banished the Jesuits. Secchi protested and he was allowed to continue his work at the observatory in a kind of voluntary confinement. Angelo Secchi died on 26 February 1878 in Rome at the age of 59.[223]

---

[223] Obituary: Anon (1879a); see also Rigge (1918), Abietti (1960), Brück H. (1979) and Maffeo (2002).

Figure 6.33. Left: the 15-cm Cauchoix refractor (de Vico 1839); right: the 24-cm Merz refractor.

6.11.2 De Vico's observations and drawings

Francesco de Vico was born on 19 May 1805 in Macerata. He entered the Jesuit order in 1823 and became Director of the Collegio Romano Observatory in 1839, following Etienne Dumouchel. He discovered six comets. When in 1848 the Jesuits had to flee from Rome, he went to Georgetown University near Washington and became director of the local observatory. While visiting England during the same year de Vico suddenly died from typhus on 15 November 1848 in Liverpool.[224]

During 1839–43 de Vico made some fine observations of nebulae with the Cauchoix refractor. The published drawings were due to his assistant Francesco Rondoni (who was not a Jesuit) and are of high quality. The first article, 'La nebulosa d'Orione', appeared in 1839 and shows an image of the refractor (de Vico 1839).[225] The next and most important one is titled 'Nebulose' (de Vico 1841b). It contains a new version of the M 42 drawing (see Fig. 6.34 right). Further, a table of micrometrical positions for 24 stars in the nebula is given. Finally a drawing of the Andromeda Nebula is shown.

In 1843 a last article on observations of nebulae was published by de Vico, titled 'Nebulose' too (de Vico 1843). It contains drawings of three nebulae. The first is the globular cluster M 13 in Hercules, called '*la nebulosa sul piede di Ercole*' ['*the nebula at the foot of Hercules*'], observed with magnifications of 60 and 420. The remaining objects are the galaxy M 81 in '*la coda dell'Orsa maggiore*' ['*the tail of the Great Bear*'] and its neighbour M 82, described as '*altra nebulosa vicina alla precedente*' ['*another nebula, near to the former*'].

6.11.3 Secchi's observations of nebulae

Secchi used the 24-cm Merz refractor for observations and drawings of nebulae and star clusters. On 6 March 1853 he found a double nebula near γ Leonis, which result he published in the *AN* as 'Entdeckung eines neuen Nebelflecks'.[226] He states that the object is "*not listed in Herschel's 'Observations of nebulae and clusters of stars'* [Slough catalogue]". About this issue, he wrote a letter to Petersen in Altona, who mentions Secchi's find in a separate article (Petersen 1853). According to Secchi the components are separated by 3^s in right ascension and 3′ in declination. D'Arrest was the first to notice that the object '*is much too near to the place of the double nebula II. 28, 29* [NGC 3226/27 in Leo], *observed 70 years earlier by W. Herschel, to be not identical with it, in view of the rareness of such twofold nebulae*' (d'Arrest 1855). He was right: William Herschel had discovered it on 15 February 1784.[227] Secchi failed, like many others at that time, to avoid identifying 'new' objects that had merely not been catalogued by John Herschel.

---

[224] Obituary: Anon (1849).
[225] The work was also published in *Comptes Rendus* (de Vico 1841a).
[226] 'Discovery of a new nebula' (Secchi 1853).
[227] See Steinicke (2001a).

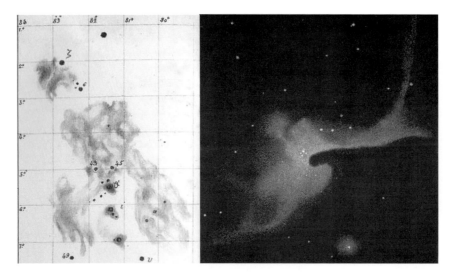

Figure 6.34. Images of the Orion Nebula. Left: Secchi's drawing of 4.3.1855, showing the nebulous region with NGC 2024 to the upper left (Secchi 1856b); right: de Vico's observation, rendered by Rondoni (de Vico 1841b).

In 1856 Secchi published the results of his observations of nebulae (Secchi 1856b). Starting in 1853, he made detailed drawings of 21 objects, which he presented in two tables (IV and V). Table 6.15 shows that most of them are planetary nebulae. His drawing of the region around the Orion Nebula is shown in Fig. 6.34 (left).

Some objects (marked *) are also described in another publication (Secchi 1856a). Three additional objects are mentioned there: the planetary nebulae M 57 in Lyra (NGC 6720, h 2023) and NGC 6905 (IV 16 = h 2075) in Delphinus and the globular cluster NGC 6943 in the same constellation. Secchi determined some diameters. For some objects, powers of up to 1000 were used.

Secchi divided the nebulae into three classes: planetary, irregular and elliptical. For the latter he suggested the existence of 'obscuring matter'. In the Andromeda Nebula, for instance, he saw two dark channels, interpreting them as absorbing masses projected onto the nebulous background.[228] In this connection, a published letter to Richard Carrington is interesting (Secchi 1857a). It mentions an observation of the open cluster NGC 6520 (VII 7, h 3721) in Sagittarius, lying in a crowded region of the Milky Way. About 8′ to the northwest, Secchi discovered a conspicuous dark object (Fig. 6.35): '*a perfectly dark spot of the shape of a pear, about 4^m large*'. He further wrote that '*This spot, by its contrast, shows that the galaxy* [Milky Way] *in that region is quite strewed with stars, which give a white aspect to the firmament.*' It is possible that Secchi has already realised the physical nature of such dark nebulae as absorbing matter – contrary to the idea of a 'hole' in the Milky Way.[229] It was conjectured that William Herschel had already seen this remarkable object, because he noted '*Hier ist wahrhaftig ein Loch im Himmel*'.[230] However, this is doubtful, since his observation indicates the ϱ Ophiuchi dark nebula.[231] Anyway, Secchi's object was independently found by Barnard in July 1883: '*small triangular hole in the Milky Way. Perfectly black, some 2′ diameter, much like a jet black nebula.*' It is no. 86 in his later catalogue of dark regions (see Section 9.15.2).

---

[228] See also Abietti (1960).

[229] This opinion was advanced by Hermann Brück (Brück H. 1979). For Webb, Secchi's discovery was just as important as the detection of the spiral structure of M 51 by Lord Rosse (Webb 1859: 231). In the case of the Omega Nebula (M 17) John Herschel had suggested absorption of nebulous matter (SC, h 2008, 24 September 1826).

[230] '*Here is truly a hole in the heavens*'; see Chambers (1890: 111) and Houghton (1942).

[231] Ashbrook (1984: 392–396). John Herschel detected many dark regions in the Milky Way; Chambers lists five and mentions as many as 49 others (Chambers 1890: 111).

Table 6.15. *Objects drawn by Secchi*

Fig.	M	NGC	H	h	Type	V	Con.	Remarks
VI-1		6818	VI 51	2047	PN	9.3	Sgr	*
VI-2		7009	VI 1	2098	PN	8.0	Aqr	*
VI-3		6572		2000	PN	8.1	Oph	*, Discovered by W. Struve
VI-4		7662	IV 18	2241	PN	8.3	And	*
VI-5		3242	IV 27	3248	PN	7.7	Hya	
VI-6		2261	IV 2	399	RN		Mon	Fig. 6.56
VI-7		6826	IV 73	2050	PN	8.8	Cyg	*, Blinking Planetary
VI-8	1	1952		357	SNR	8.4	Tau	Crab Nebula
VI-9	30	7099		2128	GC	6.9	Cap	
VI-10	27	6853		2060	PN	7.4	Vul	Dumbbell Nebula
VI-11		2438	IV 39	464	PN	10.8	Pup	
VI-12		2022	IV 34	365	PN	11.6	Ori	
VI-13		2392	IV 45	450	PN	9.1	Gem	
VI-14		2440	IV 64	3095	PN	9.4	Pup	
VI-15		2371/2	II 316/7	444/5	PN	11.2	Gem	Wrong declination sign
VI-16		3132		3228	PN	9.2	Vel	
V-1	9?	6333		1979	GC	7.8	Oph	Identification not safe
V-2	10?	6254		1972	GC	6.6	Oph	Identification not safe
V-3	11	6705		2019	OC	5.8	Sct	*, Identification probable
V-4	42	1976		369	EN	4.0	Ori	Orion Nebula; see Secchi (1865)
V-5	13	6205		1968	GC	5.8	Her	

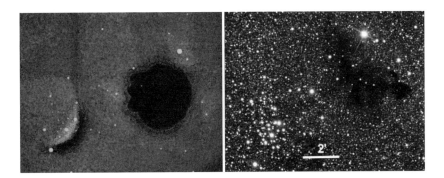

Figure 6.35. The remarkable dark nebula B 86 near the open cluster NGC 6520 in Sagittarius. Left: a drawing by Trouvelot made in August 1876 in Washington (Trouvelot 1884); right: a DSS image.

In his letter Secchi also compares the power of the Merz refractor with that of John Herschel's reflector: *'From repeated researches, indeed, I am persuaded that there is no great difference of penetrating power between my 9-inch Munich refractor and the [18¼"] telescope of Sir John Herschel.'* He repeats this in 1866 in his publication on the discovery of nebulae by his assistant Gaspare Ferrari (see Section 8.12.2).

Around 1865 Secchi also used a visual spectroscope at his Merz refractor to observe nebulae. He analysed M 8, NGC 6818 and NGC 6826 (Secchi 1865). Further results for M 57, NGC 6818, NGC 6572 and NGC 7662 can be found in a short paper titled 'Spectral studies on some of the planetary nebulae' (Secchi 1867). Therein, Secchi confirms the observations of Huggins: '*The planetary nebulae, instead, give the very fine lines indicated for the first time by Mr. Huggins.*' However, in the case of M 42 he could not detect any lines: '*the image cannot be linear, and the distinct monochromatic spectral lines are waiting*'. As for many other observers, the Orion Nebula was Secchi's favourite target.[232]

Figure 6.36. August Winnecke (1835–1897).

6.12 WINNECKE'S OBSERVATIONS IN GÖTTINGEN, BERLIN, BONN AND PULKOVO

Winnecke was one of the most important visual observers of nebulae in the nineteenth century. The stations of his remarkable career were Göttingen, Berlin, Bonn and Pulkovo. There he used various refractors, discovered objects and solved several identity problems. Winnecke always corresponded about his discoveries with his close friend Arthur Auwers, who confirmed many of his observations.

6.12.1 Short biography: Friedrich August Theodor Winnecke

August Winnecke was born on 5 February 1835 in Groß-Heere near Hildesheim, Germany (Fig. 6.36). His mother died a few days after his birth. Winnecke was brought up by his father, who was a priest, and his two sisters. From 1850 he attended the Lyceum in Hannover, where he was already scanning the sky with small refractors. On 29 September 1853 he entered Göttingen University to study astronomy. There Winnecke met Arthur Auwers and Carl-Ferdinand Pape; all three observed with Winnecke's 7.4-cm comet-seeker by Merz. Carl Friedrich Gauß, though he had for some time been an emeritus professor, supported the keen student. After two terms he went to Johann Franz Encke at Berlin Observatory in autumn 1854, where he discovered his first comet (of eight) on 15 January 1855. Winnecke graduated in August 1856 with a work on the double star η Coronae. In November 1856 he became an assistant of Argelander in Bonn, where he mainly observed with the 6″ heliometer. On 3 March 1858 he found his second comet (Pons–Winnecke).

When Wilhelm Struve visited Bonn Observatory in autumn 1857, he managed to acquire Winnecke for Pulkovo. He started in July 1858 as an 'adjoint astronomer' and measured star positions for the first Pulkovo star catalogue with the Repsold meridian-circle. In May 1864 Winnecke married Hedwig Döllen, daughter of the Pulkovo astronomer Wilhelm Döllen (who was married to Wilhelm Struve's oldest daughter, Charlotte). When Otto Struve, who became from 1862 the successor of his father, fell ill in 1864, Winnecke (who was already the vice-director) was assigned the management of the observatory. Dutifully he took over the new office alongside his immense observing programme – a heavy burden. The stress became too great and in December 1865 Winnecke had to leave Pulkovo for Bonn for medical treatment.

In 1867 he moved to Karlsruhe, to recover in its good climate. The Grand Duke offered him a room in the Amalien-Schlösschen of the Nymphengarten.[233] In the park, Winnecke observed with a 12.2-cm refractor by Reinfelder & Hertel. During 1867–71 he discovered five comets; the first on 27 September 1867 (Baeker–Winnecke). In 1869 he became secretary of the Astronomische Gesellschaft.

When in 1872 a new university was founded in Straßburg (which belonged to Germany at that time),

---

[232] A drawing can be found in Secchi (1857b).

[233] Hirsch (1932: 171).

which was to have a large observatory, Winnecke was assigned to it as its Director and Professor of Astronomy. During 1874–80 he could again practise his favourite task, the observation of nebulae. From 1875 he used a 16.3-cm orbit-sweeper by Reinfelder & Hertel on a Repsold mounting. With this unique instrument he discovered his last three comets between 1874 and 1877. In November 1880 the 48.7-cm Merz refractor was ready, which instrument was later mainly used by Kobold and Wirtz to continue Winnecke's observations of nebulae.[234] In early 1881 Winnecke's oldest son died. A year later, shortly after his appointment as President of Straßburg University, he fell ill again and had to resign all his offices. Finally, the doctors in Bonn could not help him and August Winnecke died there on 2 December 1897 at the age of 62. Auwers prefaced his obituary with the following words: '*On December 2nd a smooth death has cleared the darkness in which one of the most noble minds of German science has languished for many years.*'[235]

6.12.2 Winnecke's first discoveries

In December 1853 Winnecke discovered his first deep-sky object, the open cluster NGC 6791 in Lyra (Fig. 6.37). He owned a small comet-seeker by Merz with aperture 7.4 cm, equipped with a ring-micrometer. The observing site – the 'Hinterthürsche Garten' southeast of Göttingen – was used together with his friend Auwers, who was still attending secondary school (Winnecke had just matriculated).

Auwers later entered the object as no. 45 in his list of new nebulae (Auwers 1862a). He copied Winnecke's description, '*very faint; a miniature image of M. 74*', adding '*Already visible in a 30-inch. Fraunhofer* [9.3-cm comet-seeker], *but in an 8ft. heliometer still very faint.*' Auwers' observations with the Königsberg 6" heliometer were made on 25 September 1859 and 9 and 12 October 1860. About the first he wrote '*with magnification 45 a faint, apparently large object; with magnification 91 I saw only the densest part and supposed it to be a cluster of very faint stars*', and about the second observation '*it seems to be a faint small star cluster of about 1.5' diameter, with numerous stars of 12^m scattered around*'.[236]

Figure 6.37. Winnecke's first object, the rich open cluster NGC 6791 in Lyra (DSS).

On 23 July 1854 Winnecke discovered the open cluster NGC 6704 (Au 43) in Scutum with his comet-seeker (at the same site). He reported the find three years later in Bonn (28 February 1857) in his 'Notiz über Nebelflecke': '*In July 1854 I discovered a faint nebula in Scutum with the comet-seeker. In the Berlin refractor* [24.4-cm Fraunhofer[237]] *it appears as a faint star cluster, but the place might be interesting, because it looks cometary in a small telescope.*'[238] Auwers later added that '*With the Berlin refractor Winnecke saw at magnification 214 about 60 stars 13^m, calling the object a cluster of class VI.*' (Auwers 1862a). Here William Herschel's class VI ('very condensed and rich clusters of stars') is meant. According to Auwers, the declination given by Winnecke is wrong (Auwers 1862c). He saw the cluster on 25 September 1859 and 7 September 1861 with the Königsberg heliometer, writing about the first observation that '*With magnification 45 the nebula appears as a slightly north–south-elongated, faint, resolvable spot; with 115 it is not quite resolved, since most of the individual stars are too faint, only 15…20 starting from 11^m are visible; diameter 2..3'.*'[239]

Winnecke stayed at the Berlin Observatory from autumn 1854 to November 1856. In March 1855 he discovered the galaxy NGC 3222 (Au 27) in Leo with

---

[234] Holden (1891b).
[235] Obituaries: Auwers (1898) and Hartwig (1898b, c).
[236] Auwers (1865: 217).

[237] With this refractor, Galle discovered Neptune in 1846. It is a duplicate of Struve's Dorpat refractor (see Fig. 4.5).
[238] 'Note on nebulae' (Winnecke 1857).
[239] Auwers (1865: 194).

Table 6.16. *Winnecke's objects found until 1863 (sorted by date)*

Date	NGC	Au	GC	Type	V	Con.	Rr (cm)	Remarks
Dec. 1853	6791	45	4492	OC	9.5	Lyr	7.4	Tuttle 17.7.1859
23.7.1854	6704	43	4435	OC	9.2	Sct	7.4	'Nova Scuti'
March 1855	3222	27	2084	Gx	13.0	Leo	24.4	Schultz 1.4.1865
June 1855	6655	42	4423	2 stars		Sct	24.4	
12.4.1860	6366	36	4301	GC	9.5	Oph	7.4	
12.10.1863				2 stars		UMa	38.0	Winnecke 4 = M 40

the 24.4-cm Fraunhofer refractor, '*while observing the double nebula H. II. 28, 29* [NGC 3226/27]' (Winnecke 1857). He wrote further that '*It is much fainter than both components, round and slightly brighter to the middle.*' Auwers observed the nebula on 9 April 1861 with the Königsberg heliometer, noting that it was '*round, very faint (a bit brighter than Lassell's nebula)* [NGC 3121], *appears to have a star-like centre of 12–13^m; diameter 1".*'[240] With the Berlin refractor Winnecke found another object in June 1855: NGC 6655 (Au 42) in Scutum, cited by Auwers as '*small, pretty faint, elongated; 10" long, 3" wide*' (Auwers 1862a). Auwers was not able to find it with the Königsberg heliometer on two occasions (25 September 1859 and 15 August 1860).[241] However, d'Arrest was successful with the Copenhagen 11" refractor, but saw only a '*small, inconspicuous, easily resolvable star cluster*' (d'Arrest 1865c). Winnecke had probably observed a pair of 13.3-mag stars (distance 10"), about 5′ northwest of his rough position cited by Auwers.

During his work at Pulkovo (autumn 1857 to December 1865), Winnecke found the globular cluster NGC 6366 (Au 36) on 12 April 1860. According to Auwers he used the 7.4-cm comet-seeker, his own first instrument from Göttingen (Auwers 1862a). Table 6.16 shows Winnecke's early discoveries.

On 12 December 1863 Winnecke discovered a double star in Ursa Major at Pulkovo, using the 38-cm Merz refractor with a power of 320. Being the fourth of seven pairs published in 1869 (Winnecke 1869), it was later named Winnecke 4. He noted it as follows: '*Nova. Groombridge 1878. (9^m), (9.3^m). 1863,777 49"16 (1) 88°02 (3) 320 Pulk. R.*' The object had been catalogued by Stephen Groombridge as a single star (no. 1783).[242] John Mallas revealed in 1966 that Winnecke 4 is identical with the missing object M 40, which had been found by Messier in 1764 (Mallas 1966). Auwers had already noticed that there is a pair of stars at the position: '*Two close stars, found when searching for a nebula mentioned by Hevel, who obviously has seen the pair as a single star.*'[243]

6.13 AUWERS' FIRST DISCOVERIES IN GÖTTINGEN

Even prior to his famous astronomical career, Arthur Auwers discovered two galaxies in Göttingen. His early enthusiasm for nebulae, which was strongly influenced by his friend Winnecke, later yielded important results. Auwers was not only a keen observer but also impressed on account of the accuracy of his work, especially concerning the description of nebulae and their measured positions. He eventually became, together with Airy, the main exponent of fundamental astronomy. Both were feared for their professional and personal strength.[244]

6.13.1 Short biography: Julius Georg Friedrich Arthur Auwers

Arthur Auwers was born on 12 September 1838 in Göttingen (Fig. 6.38). His parents died young. Even as

---

[240] Auwers (1865: 221); see also Nugent (2002).
[241] Auwers (1865: 216).
[242] The other data concern the brightness of the components, observation date, distance, position angle and magnification.
[243] Auwers (1862a: 69); Hevelius, the true discoverer, catalogued the object as 'nebulous' in his great star catalogue Prodromus Astronomiae of 1690 (entry no. 1496).
[244] About Airy's reign at Greenwich, see Ashbrook (1984: 41–47) and Ball R. (1895: 289–302).

6.13 Auwers' first discoveries

Figure 6.38. Arthur Auwers (1838–1915).

Figure 6.39. Auwers' first discovery, the bright galaxy NGC 6503 in Draco (DSS).

a pupil he observed the sky. From 1847 he attended secondary school and changed to Pforta in 1854, where he took his examination three years later. Auwers owned a small 6.3-cm Fraunhofer refractor. He observed together with August Winnecke; his life-long friend was three years older and was studying in Göttingen at that time. Their targets were planets, comets, variable stars and nebulae. In autumn 1857 Auwers began his study of astronomy in Göttingen. Two years later he went to Königsberg Observatory as an assistant of Eduard Luther, and he graduated in 1862. There he made extensive measurements of nebular positions with the heliometer.

Auwers married the daughter of a former teacher. In summer 1862 he moved to Gotha Observatory to work for Peter Andreas Hansen. He determined the parallax of the star Groombridge 34 with the local 11.3-cm Steinheil refractor. In August 1866 Auwers succeeded Encke as Director of Berlin Observatory. At the same time, only 28 years old, he was affiliated to the well-known Preußische Akademie der Wissenschaften. During 1865–75 he was secretary of the Astronomische Gesellschaft, and he served as its President from 1881 until 1890.

Auwers was the leading representative of fundamental astronomy, which deals with the building of a standard reference system for the determination of positions and motions of stars. In 1888 he was awarded the RAS gold medal (Glaisher 1888) and in 1899 he got the Bruce medal of the Astronomical Society of the Pacific (Aitken 1899). Auwers died on 24 January 1915 in Berlin-Lichterfelde at the age of 76. For Simon Newcomb and others, Auwers was the most important astronomer of his time. Concerning his strength, Johann Adolf Repsold described him as a man of *'hardly flexible character'*.[245]

6.13.2 Auwers' discoveries of nebulae

On 22 July 1854 Auwers, at that time on vacation in Göttingen, discovered the galaxy NGC 6503 in Draco with his 6.3-cm refractor (Fig. 6.39). He showed the object to his friend Winnecke, who later wrote that *'In July 1854 Mr Auwers in Göttingen discovered a nebula in Draco with his Fraunhofer refractor of 29 Lines aperture, which is not listed in the catalogues of the two Herschels. After repeated attempts to find the object with the Berlin refractor* [24.4-cm Fraunhofer], *I eventually saw it on 1 April 1856, though considerably distant from the place once estimated. [...] It is a beautiful object; very bright, slightly brighter in the middle and extraordinarily extended form north preceding to south following* [northwest–southeast], *3–4' long, 50" wide.'* (Winnecke 1857).

---

[245] Obituaries: Seeliger (1915), Dyson (1915) (published also in *Proc. Roy. Soc.* **92**, xvi–xx, 1915), Eddington (1916) and Repsold J. (1918).

Auwers later added that *'In July 1851 Winnecke and I clearly revealed it as a double nebula in a comet-seeker of 34^L and magnification 50. But in 1855 Winnecke could not confirm this appearance at the Berlin refractor.'*[246] Having a visual magnitude of 10.2 mag and being at a high declination, the nebula was an easy object – even for Auwers' small refractor.

As so often, the question of why William Herschel had not already discovered this bright object arises. The main reason is that his sweeps did not cover the whole northern sky; in particular the polar region was barely searched (see Section 2.5). The field near ψ Draconis at declination +70° was observed in May–July 1788. There Herschel discovered only NGC 6434 – about 3° apart from NGC 6503. Auwers inspected the object on 19 February 1862 with the Königsberg heliometer, noting as follows: *'bright, 3–4' long, at most 1' wide; gradually brighter in the middle. There appears to be a faint star in the northern part, which makes the nebula nearly look like a double nebula.'*[247]

On 21 April 1860 Auwers made a special find at Königsberg Observatory: the Nova T Scorpii in the bright globular cluster M 80.[248] Norman Pogson wrote the following on 8 October 1860 about the phenomenon: *'the cluster known as 80 Messier changed, apparently, from a pale cometary looking object, to a well-defined star, fully of the seventh magnitude, and then returned to its usual and original appearance'* (Pogson 1860). He meant that the assumption of change in the nebula itself is absurd and correctly interpreted the case as a *'new variable star'*. Nevertheless, for a long time M 80 was considered as an example of a 'variable nebula' (see Section 11.5.1). In Athens Schmidt tried in vain to see the nova with his 15.7-cm refractor from 21 July 1860 to 31 January 1861 (Schmidt 1861b, 1868a).

On 5 March 1862, while observing M 86 in Virgo with the 6" Fraunhofer heliometer, Auwers discovered a second nebula: the galaxy NGC 4402 (Auwers 1862b).[249] It is located 10′ south of M 86, one of the central galaxies in the Virgo Cluster. He noted that it was *'faint, gradually brighter in the middle, much elongated east–west, 3′ long and 1′5 wide'* (see Fig. 6.60). Auwers observed NGC 4402 again on 2 and 30 April 1862.[250] Table 6.17 lists Auwers' discoveries sorted by date.

6.14 D'ARREST: THE LEIPZIG 'ERSTE REIHE' AND EARLY OBSERVATIONS IN COPENHAGEN

Heinrich Ludwig d'Arrest was one of the most important observers of non-stellar objects in the nineteenth century. He became well known for his assistance in the discovery of Neptune at the Berlin Observatory in 1846. His systematic observations of nebulae and star clusters with a small refractor at Leipzig Observatory set new standards. In parallel to Auwers, he dealt with William Herschel's catalogues and contributed to their revision. D'Arrest investigated interesting phenomena, such as nebulous stars, 'satellites' of planetary nebulae and variable nebulae. The climax of his work was reached in Copenhagen, using an excellent 11" Merz refractor. In September 1861 he started an extensive programme that must have strongly influenced other observers, especially Dreyer.

6.14.1 Short biography: Heinrich Ludwig d'Arrest

Heinrich Ludwig d'Arrest was born on 13 July 1822 in Berlin (Fig. 6.40). His ancestors were French Huguenots (thus his middle name is often written 'Louis'), who came from the small village of Arrest, in the Départment La Somme. Even as young boy, d'Arrest was interested in astronomy, especially visual observations. After finishing secondary school he studied mathematics in Berlin. When he was only 22 years old, he was allowed to use the telescopes at Berlin Observatory, directed by Johann Franz Encke. He independently found the comet Mauvais in 1844, two days after its Paris discovery.

In 1845 d'Arrest was appointed second assistant. A first highlight happened just a year later. D'Arrest was involved in the discovery of Neptune with the 24.4-cm

---

[246] Auwers (1862a: 75). The dates 1851 and 1855 are incorrect and should read 1854 and 1856, respectively. 34^L stands for an aperture of '34 Lines' (see the appendix on 'Telescope data'); here the 7.4-cm Merz refractor is meant.

[247] Auwers (1865: 227).

[248] The discovery was communicated by Eduard Luther, Director of Königsberg Observatory (Luther 1860); see also Auwers' later measurements with the heliometer (Auwers 1862c).

[249] See also Section 6.19.2.

[250] Auwers (1865: 229, 239).

6.14 D'Arrest: Leipzig and Copenhagen

Table 6.17. *The two objects discovered by Auwers*

Date	NGC	Au	GC	Type	V	Con.	Rr (cm)	Remarks
22.7.1854	**6503**	37	4351	Gx	10.2	Dra	6.3	
5.3.1862	**4402**	30	2965	Gx	11.8	Vir	15.0	Virgo Cluster

Figure 6.40. Heinrich Ludwig d'Arrest (1822–1875).

Fraunhofer refractor on 23 September 1846. He had the decisive idea of using the chart 'Hora XXI' of the *Berliner Akademische Sternkarten* for the search. It had been finished by Carl Bremiker in December 1844, and it was found in a drawer. The first observer, Johann Gottfried Galle, immediately marked the position, as calculated by Leverrier. Assisted by d'Arrest, he discovered the new planet only 30 minutes later.

On the suggestion of Encke, d'Arrest was engaged as an observer at Leipzig University Observatory on 1 May 1848. At that time it was still located on the Pleißenburg.[251] In 1850 he became an honorary doctor of the philosophical faculty and was awarded his *Habilitation* a year later. In November 1851 d'Arrest married Auguste Emilie Möbius, the daughter of the observatory director and well-known mathematician August Ferdinand Möbius.[252] They had two children (Doris Sophie and Louis). In spring 1852 d'Arrest was appointed senior lecturer, perhaps to prevent a possible departure to Washington.

From 1831 the main instrument of the observatory was an 11.7-cm refractor by Fraunhofer. On 28 June 1851, d'Arrest discovered with it the famous periodic comet which bears his name.[253] He soon was attracted by non-stellar objects and began systematic observations, which were published in 1856. When in the same year d'Arrest got a call from Russia, he could only be kept in Leipzig by the promise of a new observatory. Initially he promoted its erection at Johannistal (elevation 120 m), but all courting was in vain: d'Arrest changed to Copenhagen in early 1857. Though the government in Dresden had been willing to provide a comparable salary, it did not agree with his claim that it should pay a pension of equal amount for his father-in-law Möbius. The new observatory was eventually opened in 1861 under the directorship of Carl Christian Bruhns.

In Copenhagen d'Arrest became Professor of Astronomy and director of the observatory, which at that time was still located on top of the old 'Rundetaarn' ['round tower']. His predecessor, Christian Friis Rottbøl Olufsen, had headed it from 1832 until his death. However, astronomical observations were barely effective at that site, 34.8 m above the city centre. Therefore d'Arrest planned a new observatory, which was erected between 1859 and 1861 on the 'Østervold' ['eastern field'], outside Copenhagen. He quickly learned the Danish language and was interested in the culture and history of the country. Like his student Dreyer, d'Arrest was a recognised expert on Tycho Brahe. In mid 1861 the main instrument was erected: a 28.5-cm Merz refractor on a mounting constructed by Emil Jünger, a student of Ertel. D'Arrest was assisted by Hans Carl Frederik Christian Schjellerup, who worked as an observer in Copenhagen from 1851 but was ignored in the succession of Olufsen.[254]

On 21 October 1862 d'Arrest discovered the minor planet (76) Freia. During the Mars oppositions of 1862

---

[251] Möbius (1848). About the various observatories in Leipzig, see Ilgauds and Münzel (1995).
[252] He is well known for the 'Möbius strip'.
[253] Lynn (1909a).
[254] Dreyer (1888b), Thiele (1888); on Copenhagen Observatory see also Beekman (1983b) and Witt (2007c).

and 1864 he searched for moons, but the Merz refractor turned out not to be powerful enough. From the beginning he used the telescope for systematic observations of nebulae with a ring-micrometer. In 1867 he published the results in his magnum opus *Siderum Nebulosorum*, which was completely written in Latin (d'Arrest 1867a). For this work, d'Arrest was awarded the RAS gold medal in February 1875; the laudation was given by President John Couch Adams (Adams 1875). D'Arrest published some 140 articles on nebulae, comets, minor planets and celestial mechanics. In his later years he worked on the new field of spectroscopy, detecting in 1872 that line-spectrum objects (gaseous nebulae) are concentrated towards the Milky Way. In the same year he published a monograph about nebular spectroscopy, which contains a fine drawing of the Orion Nebula.[255] At the early age of 53 Heinrich Ludwig d'Arrest died of a heart attack on 14 June 1875 in Copenhagen.[256] His successor was the astronomer and mathematician Thorvald Nicolai Thiele.

Figure 6.41. D'Arrest's 11.7-cm Fraunhofer refractor at Leipzig Observatory (Bruhns 1872).

6.14.2 The Leipzig observations

Encouraged by Laugier's systematic nebular observations published in 1853, d'Arrest decided to bring about their significant enlargement. With the 11.7-cm Fraunhofer refractor[257] at Leipzig Observatory (Fig. 6.41) he had only a modest telescope at his disposal. Nevertheless, he wrote that *'For me it was surprising to learn that in a telescope of only 4½ inches aperture, being one of the smallest used today, according to pretty reliable estimate, almost a thousand nebulae are perceivable, i.e. about one third of those which have been made known at our latitudes with the largest reflectors.'*[258] Like Laugier, d'Arrest purposed *'to get a basis on which it could later be possible to determine a motion, either of individual nebulae or of the starry sky against the nebulae'*.[259] D'Arrest wondered at the little interest of astronomers in measuring positions of nebulae: *'The lower accuracy, however, one always will have to content oneself with in this part of sidereal astronomy, compared with observations of fixed stars, cannot at all be a sufficient reason for the sparse attention positional measurements of nebulae have hitherto received.'*[260] He planned to determine not only well-known objects, but also faint ones *'so far picked up only accidentally, but hardly ever actually observed'*.

D'Arrest started his observations on 18 May 1855. With the applied magnification of 42, most nebulae appeared small and round/oval; thus they could be measured well. The refractor was equipped with a Fraunhofer micrometer having a double ring. Besides the mean location (determined with the aid of reference stars from Bessel's and Argelander's zone observations), the brightness, form and shape were recorded too. Already on 15 June 1855 d'Arrest published a first note (d'Arrest 1855). There he lists errors in the catalogues of William and John Herschel, which he had noticed while preparing the observing programme and hitherto had overlooked. He also mentions a double nebula found on 6 March 1853 by Secchi (NGC 3226/27).

The selection of appropriate nebulae from William Herschel's catalogues required a reduction of the original relative positions to d'Arrest's epoch 1850. On

---

[255] Published in Denmark as *Undersøgelser over de nebulose Stjerner i Henseende til deres spektralanalytiske Egenskaber* (d'Arrest 1872b).
[256] Obituaries: Tromholdt (1875), Dreyer (1876a) and Anon (1876a); see also Steinicke (2003f).
[257] It is known as the 'sechsfüssiger Refractor'; in 1865 it was used by Vogel for his observations (see Section 8.14.2).
[258] d'Arrest (1856a: 296).
[259] d'Arrest (1856a: 295).

[260] d'Arrest (1856a: 296).

6.14 D'Arrest: Leipzig and Copenhagen

the basis of the current locations of the reference stars being used, absolute positions (right ascension and declination) had to be determined for each object. In November 1855 d'Arrest published his 'Verzeichnis von fünfzig Messier'schen und Herschel'schen Nebelflecken, aus Beobachtungen auf der Leipziger Sternwarte abgeleitet'.[261] There he compared his positions with those of Messier, John Herschel and Laugier.

In this article, d'Arrest mentions an issue of great interest: the change of position, brightness or form of nebulae. Lamont reported in 1837 that the planetary nebulae IV 16 (NGC 6905) and IV 18 (NGC 7662) had changed their positions significantly. William Herschel had earlier suggested the same for IV 1 (NGC 7009), the Saturn Nebula in Aquarius. D'Arrest investigated M 49, M 22, NGC 6905, NGC 7009 and NGC 7662, but could not find any differences. On the other hand, he also mentions some cases '*which really seem to show hints of motion*'. Nevertheless, he was cautious: '*due to the deficiency of most of the hitherto observed places, the decision about this most important issue still has to wait for some years*'. D'Arrest's candidates were II 4 (NGC 596),[262] II 282 (NGC 615) and M 55. The last case is the most significant, insofar as the right ascension showed a decrease of 2′ on comparing his 1855 position with Lacaille's (considered accurate) and those of Messier and John Herschel. D'Arrest gives examples for brightness variations too: h 3576 (NGC 5694) and I 1 (NGC 1055). On the basis of seven observations, William Herschel had described the latter as 'considerably bright', but '*nowadays [it] is among the faintest nebulae to be visible in my instrument*'. The report ends with a list of errors in John Herschel's Slough and Cape catalogues, noting respectfully, however, '*that the fifty corrections hitherto indicated* [by d'Arrest] *are but an infinitesimal number compared with the immense work Sir John's two catalogues is based on*'.

By 9 June 1856 d'Arrest had determined the positions of 230 objects; about 110 of them had not been observed since the two Herschels. Since each one was observed more than once (on different nights), d'Arrest made altogether more than 1000 measurements. They were published as *Resultate aus Beobachtungen von Nebelflecken und Sternhaufen – Erste Reihe*.[263] The addition 'Erste Reihe' ['first set'] suggests that he planned further campaigns. Indeed, shortly before the work appeared d'Arrest wrote '*I persistently continue the positional measurements*' (d'Arrest 1856c).

D'Arrest's observations proved to be more important than their forerunner (Laugier) and inspired a whole string of measuring campaigns, e.g. by Auwers, Schönfeld, Vogel and Schultz. Argelander enthusiastically wrote that '*D'Arrest's publication on nebulae has really pleased me; after all, this once more is a sterling work, initiated with just as much prudence and awareness as needed, then realised with skill and patience; our descendants will one day reap the rich benefits of this work.*' (Argelander 1856). The work was highly regarded (Anon 1856).

One chapter treats the precision of earlier measurements. For the question of proper motion of nebulae, an integrative assessment of these data was essential. According to the contemporary opinion, an effect could be detected only if the time interval was large and all contributing measurements were accurate. Owing to the lack of a comprehensive, systematic observing programme, an objective statement was impossible. It was the merit of d'Arrest that he changed this situation. He compared his results with those of Lacaille, Messier, William Herschel, John Herschel and Laugier. Lamont's Munich observations of 1836–37 could not be considered since they were not published until 1869.

Since Lacaille's catalogue contains only five of d'Arrest's objects, a reliable statement could not be given. About Messier he critically notes that '*The data were respected for a long time. But meanwhile the deficiency of Messier's bar-micrometer has become known from his many observations of comets. It is to be regretted too that, due to obvious mistakes, the positions of nebulae are often wrong by several arcminutes.*'[264] To make a long story short, Messier's positions were unusable – in view of the large time difference, this was a sad matter. This squares with the conclusion of Laugier, who at first was optimistic about Messier. Méchain was seen

---

[261] 'List of 50 Messier and Herschel nebulae, derived from observations at Leipzig Observatory' (d'Arrest 1856b).
[262] Also investigated by Auwers during 1860–61 (Auwers 1862c).
[263] 'Results from observations of nebulae and star clusters – first set' (d'Arrest 1856a).
[264] d'Arrest (1856a: 300).

differently, being praised by d'Arrest for his *'always careful observations'*. Using these important data, he could state that *'for the brighter nebulae, strong proper motions were not present'*.

About William Herschel's positions d'Arrest notes that, *'Given the known mounting of Herschel's large telescopes, it surely is respectable that with such an imperfect instrument, using ropes, rollers, clockfaces, and hands, the comparisons with Flamsteed stars achieved such a degree of accuracy with but a few exceptions; nevertheless it is to be regretted that this treasure too, sampled with unparalleled perseverance, will barely allow conclusions about motions'*.[265] Herschel himself, however, had noticed that the quality of his measurements, made during 1783–1802, had been increasing from 10–15′ at the beginning to about 1–2′ at the end.

The catalogues of John Herschel (SC and CC) were qualified by d'Arrest as *'the most complete of all works on nebulae [...] leaving far behind all earlier observations concerning number and accuracy'*.[266] Herschel himself estimated the average positional error to be 30–45″. D'Arrest actually got 15″ in right ascension and 19.5″ in declination for the Slough catalogue. The values for the Cape catalogue were even better: 12.7″ and 19.3″, respectively. Owing to the southern declination, the number of common objects was, of course, smaller in this case.

Laugier delivered the *'currently most accurate known positions'*, because *'they were specially measured with the very intention of getting a basis for future investigations about possible proper motions'*.[267] D'Arrest deduced a probable error of 5.83″ in right ascension and 5.7″ in declination, ignoring a few runaway cases like M 82.[268]

D'Arrest brought up the issue of 'brightness changes' in his publication too. On the basis of his observations, he recognised that some objects appeared brighter or fainter than in Herschel's observations made 70 years earlier: *'Here are some cases in which the present view differs, mostly in brightness, from the according descriptions given in Herschel's catalogues.'*[269] (Table 6.18). Of course, he was aware that *'the appearance of nebulae strongly depends on the particular condition of the atmosphere'*. Except for two objects (planetary nebulae), all are galaxies. NGC 821 (h 193), which was too faint in Leipzig, could later be seen (d'Arrest 1863b).

D'Arrest listed the *'barely relevant, mostly negative results'*.[270] He complained about the *'state of nearly complete ignorance we have about nebulae, if things other than their appearance are concerned'*. Concerning proper motions, *'we cannot mention a single reliable case at present'*. About the few, strong positional differences, he wrote that it is *'still uncertain, whether errors in the earlier comparisons are involved'*. D'Arrest comes to the conclusion that the *'most probable amount of the annual motion of nebulae relative to the system of fixed stars is less than 0″.411'*.[271]

The negative result was particularly evident for planetary nebulae, which cannot be distinguished from mere stars in meridian-circle observations. Therefore they had been subjected to very precise measurements in the past. Some appear in Bessel's zones or are listed as stars in Lalande's *Histoire Céleste* of 1801. Examples from the latter catalogue are NGC 3242 (LL 20204),[272] NGC 6210 (LL 30510), NGC 6543 (LL 38303) and NGC 7009 (LL 40762). D'Arrest stated their *'total indiscernibility of a proper motion over the last 60 years'*.[273] Parallax measurements (to determine the distance) were unsuccessful in these cases too. Winnecke had first tried in 1857 to measure the parallax of a planetary nebula, using the 6″ heliometer at Bonn Observatory; his target was NGC 7662 (h 2241) in Andromeda. D'Arrest mentions his own unpublished attempts for the planetary nebula IV 64 (NGC 2240). Some others followed – all getting negative results; e.g. Franz Brünnow at Dunsink Observatory in 1872 (Brünnow 1872) and Wilsing, observing NGC 6543 (IV 37) in Draco with the photographic 32-cm refractor of Potsdam Observatory (Wilsing 1893, 1894).

Another issue was 'satellites' of planetary nebulae. John Herschel had detected faint stars in the immediate neighbourhood of some nebulae. He wrote that they *'suggest the idea of accompanying satellites'*.[274] About the physical conditions he conjectured that *'The enormous*

---

[265] d'Arrest (1856a: 301).
[266] d'Arrest (1856a: 303).
[267] d'Arrest (1856a: 305).
[268] The galaxy M 82 was a problem not only for the Herschel family but also for Laugier too (Steinicke 2007a).
[269] d'Arrest (1856a: 297).
[270] d'Arrest (1856a: 308).
[271] d'Arrest (1856a: 356).
[272] The Ghost of Jupiter; Lalande observed the object on 21 April 1798.
[273] d'Arrest (1856a: 308).
[274] Herschel J. (1833a: 500).

6.14 D'Arrest: Leipzig and Copenhagen

Table 6.18. *Nebulae observed by William and John Herschel (H. and h., respectively) appearing brighter or fainter*

H	NGC	App.	Type	V	Con.	Remarks (d'Arrest)
II 99	3377	Brighter	Gx	10.2	Leo	First class
II 101	3489	Brighter	Gx	10.2	Leo	Very bright, certainly first class
III 44	4647	Brighter	Gx	11.4	Vir	First class, very faint for H. and h.
III 743	6781	Brighter	PN	11.4	Aql	Pretty faint for H., faint for h.; pretty bright, easily visible as planetary nebula
IV 69	1514	Brighter	PN	10.9	Tau	Faint for H. and h.; striking nebulous atmosphere
I 1	1055	Fainter	Gx	10.6	Cet	Second class
I 23	4452	Fainter	Gx	11.9	Vir	Second class
I 55	7479	Fainter	Gx	10.9	Peg	Pretty bright for H.; barely second class
I 62	701	Fainter	Gx	12.2	Cet	Pretty bright for H.; not visible
I 102	1022	Fainter	Gx	11.3	Cet	Not first class
I 104	7606	Fainter	Gx	10.8	Aqr	Second class
I 119	4560	Fainter	Gx	9.6	Vir	Very bright for H., bright for h.; at present extremely faint
I 131	3672	Fainter	Gx	11.4	Crt	Pretty bright for H.; barely second class
I 152	821	Fainter	Gx	10.8	Ari	Very bright for H.; too faint to be seen

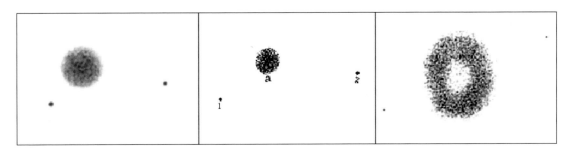

Figure 6.42. Drawings of the planetary nebula NGC 6818 in Sagittarius, showing two 'satellites'. From left to right: Herschel 1831, Lamont 1837 and d'Arrest 1861.[275]

magnitude of these bodies [PN], *and consequent probable mass (if they be not hollow shells), may give them* [the satellites] *a gravitating energy, which, however rare we may conceive them to be, may yet be capable of retaining in orbits, three or four times their own diameter, and in periods of great length, small bodies of a stellar nature.*' Herschel presented a few examples. For h 2047 = NGC 6818 (Fig. 6.42) he noted that it was '*exactly like a planet and two satellites*';[276] and for h 2075 = NGC 6905, '*it has four small stars near it like satellites*'. In the case of h 2241 = NGC 7662 there is even a double star near the nebula and Herschel urged that '*these satellites of planetary nebulae ought to be especially attended to*'. Shortly thereafter, Olbers noted the importance of the issue too.[277] D'Arrest could perceive only a few companions in his small telescope: '*those recovered were still at their places, which were so accurately determined by Sir John Herschel*'.[278] Auwers too had measured two

---

[275] Sources: Herschel J. (1833a), Fig. 46; Lamont (1869), Fig. 11; and d'Arrest (1861), Fig. 3.

[276] For NGC 6818 see also Steinicke and Stoyan (2005a).

[277] Olbers (1834); see Section 3.5.

[278] He published a sketch in the *Siderum Nebulosorum* (d'Arrest 1867a: 336).

cases with the Königsberg heliometer: NGC 6818 and NGC 6905.[279]

A related matter was double nebulae, which were investigated by d'Arrest for possible orbital motions. However, he found *'no change of relative positions, neither in position angle nor in distance, though the comparisons were made for 25 or even 70 years'*.[280] Using astrometric methods, no evidence that they were double stars could be found for double nebulae. The same applied for proper motion and parallax.

In the appendix d'Arrest gives absolute positions for all Herschel nebulae of classes I ('bright nebulae') and IV ('planetary nebulae'), reduced to epoch 1850. These amount to 288 and 78 objects, respectively – more than he used for his observations. The Herschel catalogues had two great disadvantages. First, they were arranged by discovery date (following Messier). Secondly, the position of an object was only given relative to a reference star, thus there were no coordinates. This made the identification of an object, especially a newly discovered one, pretty difficult. Therefore John Herschel's catalogues, sorted by right ascension for 1830, were consulted in these cases. However, this could be troublesome too. D'Arrest noted that *'one could mention at least four cases, appearing within a short period of time, in which the indication of a supposed new nebula or comet could be traced back to a nebula of* [William] *Herschel'*.[281]

About choosing the classes I and IV, d'Arrest wrote that *'Under these circumstances it appeared not useless, if, facing the lack of a strongly wanted general catalogue of nebulae, for the time being at least the brighter or, for their various appearance, peculiar nebulae were to be arranged.'*[282] D'Arrest urged that one should investigate class IV objects: *'it is errant that all 78 numbers of the fourth class, as is still done in the present-day textbooks, were suspected to be planetary nebulae'*.[283] He estimated that there were about 30 planetary nebulae in the whole sky, *'of which 20 can be observed in the northern hemisphere'*. D'Arrest guessed that class IV contains *'very diverse objects, among them fixed stars with nebulous envelopes and atmospheres, stars with fan-like plumes, nebulous streamers etc.'*.

For his reductions, d'Arrest used star positions from the *British Association Catalogue* (BAC) and Lalande's *Histoire Céleste*, both of which were contemporary to Herschel's observations. Unfortunately, d'Arrest was not aware of Arthur Auwers, who revised the Herschel catalogues as well.

6.14.3 D'Arrest's own discoveries

D'Arrest started his Leipzig observations with the 11.7-cm Fraunhofer refractor in April 1852. The first objects to be measured were M 49 (h 1294) and NGC 4526 (h 1329). Both were revisited by Auwers in 1861–62 with the Königsberg heliometer.[284] D'Arrest's first find was NGC 6760 in Aquila. He saw the globular cluster in May 1852 while searching for comets: *'The nebula has three minutes diameter and resembles in brightness Herschel's nebulae of the first class.'* (d'Arrest 1852a, b). Later he noticed that the object had been discovered by Hind in 1845, writing *'I use this occasion to correct an error made four years ago'*.[285]

On 23 August 1855 d'Arrest discovered NGC 607 in Cetus and NGC 7005 in Aquarius.[286] The former ('Nova?') was described as a nebulous star of 11 mag. Auwers confirmed this appearance with the Königsberg heliometer.[287] He entered the object as no. 15 in his list of new nebulae. On 15 February 1861 he termed d'Arrest's position of the 'Nova Ceti' *'totally errant'*, the mistake having been caused by a problem with the reference star.[288] Actually NGC 607 is a pair of stars (11.7 mag and 13.9 mag; distance 14″), 2′ off d'Arrest's position – which is not so bad at all. Then, 13′ northeast, lies the galaxy NGC 615 (II 282), which was also observed by Auwers. On 3 November 1861 he noted (on the observation of h 132 = NGC 596) that *'d'Arrest's nebulous star immediately appeared to me as strongly nebulous; it is a star 11ᵐ in a pretty dense nebula of small size; looking like a blurred star'*. It is interesting that the famous double-star observer Burnham could not find the pair NGC 607 while inspecting the

---

[279] Auwers (1865: 198–199).
[280] d'Arrest (1856a: 308).
[281] d'Arrest (1856a: 360).
[282] d'Arrest (1856a: 361).
[283] d'Arrest (1856a: 359).

[284] Auwers (1862c, 1865: 213).
[285] d'Arrest (1856a: 299).
[286] d'Arrest (1856a: 311, 350).
[287] Auwers (1862c, 1865: 218, 266).
[288] Auwers (1862c: 373).

6.14 D'Arrest: Leipzig and Copenhagen

Figure 6.43. The star trio NGC 7005 in Aquarius (DSS).

field with the 36″ Lick refractor in 1891. He suspected NGC 607 to be identical with the galaxy NGC 615 (Burnham 1892b).

NGC 7005 was seen by d'Arrest as a 'nebulous star cluster', appearing similar to M 73, only 40′ to the northwest (M 73 is a star quartet). In the Königsberg heliometer, Auwers correctly saw on 9 September 1861 'not a nebula, but only a triangle of three stars'.[289] The three stars are of 11–12 mag, surrounded by a few fainter ones (Fig. 6.43). Auwers' observation was confirmed by Schönfeld on 30 September.[290] The same applies to an object, termed 'Nova?' by d'Arrest, described on 17 October 1855 as 'a fairly bright nebula, 30″ diameter; condensed to a star in the middle'.[291] Schönfeld (13 August 1861) and Auwers (9 September 1861) agreed that this is not a nebula, but a mere star trio (11.6 to 14.2 mag).[292] John Herschel later catalogued the object as GC 4941, obviously convinced by d'Arrest's four, nearly consistent observations. It often happens that a compact group of stars with comparable brightnesses produces a nebulous appearance. This optical illusion is not easy to unmask, since there are cases with real nebulosity too. In Table 6.19 all d'Arrest's objects found in Leipzig are listed.

6.14.4 A 'lost' nebula in Coma Berenices

The two objects in Coma Berenices observed by d'Arrest on 29 March 1856 and termed 'Nova' triggered a vivid debate. These are the galaxies NGC 4473 and NGC 4477 (Fig. 6.44). The position of the first is correct; the second is but 5′ too far east ('*place not quite reliable due to the lack of appropriate reference stars*'[293]). John Herschel mentioned the case in an open letter to Hind on 4 April 1862, titled 'On the disappearance of a nebula in Coma Berenices' (Herschel J. 1862).

Herschel, calling the nebulae (A) and (B), unmistakably states that '*Both M. d'Arrest's Nebulae, and one more, of an order of brightness little inferior to the latter (B), occur in Sir W. Herschel's catalogues, (A) being identified with H. II. 114, and (B) with one of the two, H. II. 115 and II. 116.*' He wonders why d'Arrest has not seen the third, which is II 116 = NGC 4479, located 5′ southeast of (B) = II 115 = NGC 4477. Herschel wrote that "*M. d'Arrest's telescope has amply sufficient power to have shown the missing Nebula (the fainter of the two in all probability, or H. II. 116), which had sufficient illumination to be characterised as 'resolvable' and as 'pretty bright' or a full second-class Nebula.*" He has not observed this '*very rich region of the heavens*' (part of the Virgo Cluster). For Herschel, the 'missing nebula' indicates a possible change of heavenly bodies, writing: '*I strongly recommend the point for re-examination*'.

On 20 May 1862 d'Arrest reacted to Herschel's letter, starting with the sentence '*Sir John's authority is too great, and the matter too important that I could refrain from saying with certainty: no nebula has disappeared in Coma Berenices.*' (d'Arrest 1862c). About the fact that he could not identify his two nebulae in William Herschel's catalogues, he states that Herschel's data are '*not exact enough concerning the positions, so that, in areas where nebulae are crowded, it is nearly impossible to identify W. Herschel's nebulae with the objects being observed in the sky*'. With a jaundiced eye he wrote '*what an extraordinary treasure of supplemental and more detailed observations of former times must be in the hands of Sir John Herschel in the form of manuscripts*'. Indeed, it was only by using unpublished observational notes of his father that John Herschel was able to identify d'Arrest's objects. About the third nebula (II 116 = NGC 4479), d'Arrest wrote that it was '*too faint for the 6ft refractor of Leipzig Observatory*', but he owns that it '*is not much fainter than some other nebulae of second class. Moreover, the large number of nebulae in

---

[289] Auwers (1865: 222).
[290] Schönfeld (1862d: 115).
[291] d'Arrest (1856a: 354).
[292] Auwers (1865: 222), Schönfeld (1862d: 115).

[293] d'Arrest (1856a: 334).

156   The time after Herschel

Table 6.19. *D'Arrest's objects found in Leipzig (sorted by date); 'Source' = page in 'Erste Reihe' (* = d'Arrest 1852a)*

Date	NGC	Au	GC	Type	V	Con.	Source	Remarks
May 1852	6760	44	4473	GC	9.0	Aql	*	Hind 30.3.1845
23.8.1855	**607**	15	358	2 stars		Cet	311	
23.8.1855	**7005**	*	5983	Star group		Aqr	350	Near M 73
17.10.1855			*	4941	3 stars		Peg	354	
29.3.1856	4473	*	3030	Gx	10.2	Com	334	II 114, W. Herschel 8.4.1784
29.3.1856	4477	*	3025	Gx	10.4	Com	334	II 115, W. Herschel 8.4.1784

Figure 6.44. The galaxies NGC 4473, NGC 4477 and NGC 4479 in Coma Berenices (DSS).

using the Copenhagen 11" refractor.²⁹⁴ Once this had been confirmed by Auwers it was evident that no object had disappeared.²⁹⁵ However, the case was not closed elsewhere.

Jean Chacornac at Paris Observatory felt obliged too to react to John Herschel's letter; his reply is dated 22 May 1862 (Chacornac 1862c). On 19 April 1862, he had accidently observed the field in Coma Berenices with the new 80-cm Foucault reflector, finding a double nebula (NGC 4473/77). When he read about Herschel's 'missing nebula' in the *Monthly Notices* a little later, he tried again and could see the third nebula (NGC 4479). The faint object was confirmed also by Schönfeld in Mannheim with the 16.5-cm refractor. He at first had '*overlooked it, like d'Arrest, due to its faintness*' but, '*after John Herschel's recent request*', was eventually successful on 21 May 1862.²⁹⁶

Lassell and Hind mention the case in letters to Herschel (both dated 27 May 1862). Lassell observed II 115/116 (NGC 4477/79) on 19 and 21 May with the 48-inch on Malta, '*which I find are still in the position you have indicated [...] there is a very striking inferiority of brightness in II 116, such, that possibly D'Arrest might have overlooked it in his telescope*'.²⁹⁷ He added a sketch. Hind wrote about Chacornac's observation as follows: '*M. Chacornac saw the nebula in Coma Berenices [NGC 4479] which you suppose had disappeared, but this region makes the observer hurry up.*' Lately, d'Arrest had become a victim (like others before) of the '*chronological order of the old catalogues, which makes finding so very time-consuming, strenuous and difficult*'. On 5 May 1862 he was eventually able to observe all three objects,

---

²⁹⁴ d'Arrest (1867a: 229–231).
²⁹⁵ Auwers (1862a: 77).
²⁹⁶ Schönfeld (1862d: 113).
²⁹⁷ RAS Herschel J. 12/1.7.

Table 6.20. *Individual stars supposedly surrounded by 'nebulosity'; [ ] = page in d'Arrest (1856a).*

Star	Discoverer	NGC	h	GC	V	Description of d'Arrest	Remarks
55 And	Piazzi		162	428	5.4	No trace of nebulosity [311]	Piazzi's 'nebulosa' (Hora I. 190); Bode 6
50 Cas	J. Herschel	771	179	462	4.0		
BD +36° 567	Bessel			614	9.8	No nebula visible; comet? [313]	Bessel's star in B.Z. 527 (Perseus)
3 Mon	J. Herschel	2142	373	1337	5.0	Atmosphere not detectable [318]	
19 Pup	J. Herschel	2542	3115	1632	4.7		ASC 1001 = BAC 3073
β CVn	J. Herschel	4530	1332	3079	4.2	Not safely seen [335]	8 Canum (8 CVn)

with a brightness of only one tenth of that of B [NGC 4477] on May 21; sufficiently proving that it is variable to a great extent, whether it becomes invisible in our telescope or not.'[298]

Possibly Schmidt saw NGC 4479 on 29 April 1862 in Athens with the 6.2" Plössl refractor: *'In the middle between 2 Novae* [of d'Arrest = NGC 4473/77] *I repeatedly saw a faint nebula and another small nebula following it'* (Schmidt 1865c). There is no nebula between NGC 4473 and 4477, but NGC 4479 follows NGC 4477. Cooper at Markree apologised for being unable to search for II 115 (NGC 4477), writing on 3 May to John Herschel *'I have not your Catalogue with me, so I cannot get at the position of H II W. 116 to see if it be or not in the part of the heavens.'*[299]

6.14.5 The phenomenon of 'nebulous stars'

D'Arrest's observations include individual stars supposedly surrounded by 'nebulosity'. In Table 6.20 six objects are listed; four were observed in Leipzig. The case originates from William Herschel. Encouraged by the peculiar object NGC 1514, he was fascinated by 'nebulous stars' (see Tables 2.17 and 10.14). This passed to John Herschel (see Section 3.3) and Lord Rosse, whose prototype was NGC 2392 (see Section 6.4.11).

Whereas NGC 1514 and NGC 2392 are ordinary planetary nebulae with a central star, John Herschel's objects are curious. In the appendix of the Slough catalogue one reads *'With regard to nebulous stars generally, I ought to mention that it has frequently occurred to me to notice a peculiar state of atmosphere in which all large stars (above the 7th magnitude) have appeared surrounded with photospheres of 2' or 3' or more diameter, precisely resembling that about some of the finer specimens of nebulous stars.'*[300] For Herschel β CVn is a striking example (Fig. 6.45 left): *'Not the least doubt of a considerable atmosphere around this star.'* The star became popular by virtue of Smyth's Bedford Catalogue: it is listed as '8 Canum Venaticorum' in the category 'stars and comets'. Smyth wrote that *'the nebulous atmosphere is no further apparent in my instrument, than in giving the object an apparent derangement of focal definition'*.[301] John Herschel also called 3 Mon and 19 Pup 'nebulous stars'; 50 Cas was a 'suspected' case only.

The Birr Castle observations are interesting. Lord Rosse could confirm the case of 3 Mon on 30 November 1850: *'same appearance as ε Orionis* [V 34 = NGC 1990], *but v. much fainter'*;[302] his son Lawrence Parsons detected on 6 December 1872 that 19 Pup is surrounded by *'a faint nebulous atmosphere'*.[303] Both failed in the case of β CVn, noting *'I cannot decide about its nebulosity'* (19 April 1857) and *'I could not detect any neby about it'* (1 May 1867), respectively.[304] Lassell too was unable to see nebulosity with his 48-inch on Malta, as John Herschel mentions in the General Catalogue:

---

[298] RAS Herschel J. 12/1.6.
[299] RAS Herschel J. 12/1.5.
[300] Herschel J. (1833a: 499).
[301] Smyth W. (1844: 272).
[302] Parsons W. (1861a: 715).
[303] Parsons L. (1880: 66).
[304] Parsons L. (1880: 117).

158 The time after Herschel

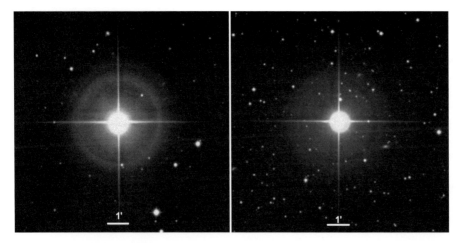

Figure 6.45. Left: β in the constellation Canes Venatici, catalogued as 'nebulous star' h 1332 (NGC 4530); right: 55 Andromedae, Piazzi's 'nebulosa' (DSS).

'*Yet Mr. Lassell saw no nebulosity about 8 Canum*' (note to GC 3079). This dates back from a letter of Lassell to Herschel (23 May 1862), in which he wrote that the star '*does not appear to me to have any nebula or appendage about it. It is a fine brilliant, full yellow star without anything further remarkable about it. It was viewed with powers 185 + 466*'.[305]

Despite some negative results, John Herschel took these cases as being undecided, for he eventually entered all of them in the GC. Dreyer was more careful, cataloguing four in the NGC '*for future occasional observations*' (note for NGC 771); among them was β CVn, though he had previously noted in the GCS that '*Is probably to be struck out*'.[306] However, in the NGC one learns that '*Tempel was uncertain whether the star was not after all slightly nebulous*' (note to NGC 4530).

'Bessel's star' was measured on 8 November 1832 by Bessel in zone 527 of his Königsberg meridian-circle observations and marked as a 'nebulous star'.[307] In the Bonner Durchmusterung Argelander could see only a star of 9.3 mag, which was confirmed by d'Arrest and Auwers. However, the former suggested that the object was '*probably a comet*'.[308] After observations on 28 September and 4 October 1861 with the Königsberg heliometer Auwers stated that '*a star marked in B. Z. 527 as nebulous appears free of nebulosity*'.[309] The same was reported by Schultz in Uppsala: '*Bessel's star (B.Z. 527) has definitely no nebulosity*' (Schultz 1865b). John Herschel catalogued the object as GC 614.

6.14.6 The case '55 Andromedae'

This 5.4-mag star in Andromeda was much discussed (Fig. 6.45 right). It appears as a 'nebulous star' in Bode's catalogue of 1777, listed as no. 6 (Bode 1777). Also Giuseppe Piazzi marked it 'nebulosa' in his catalogue *Præcipuarum Stellarum inerrantium Positiones Mediæ* of 1803.[310] John Herschel observed the object on 1 October 1828, entering it as h 162 in the Slough catalogue with the identification '55 Androm.'. The description reads as follows: "*A fine nebulous star with a strong atmosphere losing itself imperceptibly; diam 90″. It is also a double star h 1094; called 'nebulosa' by Piazzi.*" The term 'h 1094' refers to his fourth catalogue of double stars. In Herschel's textbook *A Treatise of Astronomy* 55 And is the key object in the chapter on 'nebulous stars'.[311]

William Smyth wrote in the Bedford Catalogue that the 'nebulosity' is '*perhaps a consequence of some small stars near it*'.[312] His own observations of 55 And

---

[305] RAS Herschel J. 12/1.7.
[306] Dreyer (1878a: 395).
[307] Bessel (1835: 96).
[308] d'Arrest (1856a: 313).
[309] Auwers (1862a: 77, 1865: 222).
[310] On the star catalogues mentioned, see Chambers (1890: 487–495).
[311] Herschel J. (1833a: 377).
[312] Smyth W. (1844: 43).

with the 5.9″ Tully refractor at Bedford Observatory yielded '*It sometimes had a blurred aspect to my gaze.*' The nebular appearance could not be confirmed by Lord Rosse, who wrote in his 1861 publication that he had '*Looked for 8 times. Dec. 18, 1848. Found star 7th or 8th mag. in place but saw no nebulous atmosphere.*' The object appears in his 'List of nebulae not found' in the appendix (see Section 7.1.5). At Birr Castle 55 And was observed 11 times altogether – a nebula was never detected.[313]

Herschel listed 55 And as GC 428 in the General Catalogue. In the notes one reads that, "*Although this star has been eight times examined by Lord Rosse without perceiving any nebulous atmosphere, yet as my observations are corroborative of Piazzi's designation of it as 'Nebulous', it is retained for occasional future examinations.*" D'Arrest noticed in 1864 that the star '*is at present free of all nebulosity*' (d'Arrest 1865a). He added that, '*In accordance with this, Mr Huggins has recently found the spectrum of 55 Andr. to be similar to that of an ordinary star. This case is one of a few that deserve special attention in the future.*'[314]

In 1867 Schjellerup tried to reveal the reason for Piazzi's description. In Lalande's *Histoire Céleste* of 1801 (a catalogue of Piazzi's time) the case is not mentioned. However, he found out "*that Piazzi, hitherto considered as the first observer of the star's nebulosity, only copied a note of Flamsteed, who had added 'Neb' to the star in his Catal. Brit. (Hist. Coel. T. III. p. 39)*" (Schjellerup 1867). Schjellerup refers to Flamsteed's *Catalogus Britannicus* of 1725 with the Historia Cœlestis,[315] containing 3310 stars. There '55 Andromedae' is listed as no. 217. Since Flamsteed had observed the star on 17 October 1691 '*soon after the great Andromeda nebula in the same night*', the note 'Neb' might have erroneously been transcribed from the Andromeda Nebula to the star. He therefore concludes that '*The assumption that the star has lost its nebulosity since Piazzi's time is certainly not allowed, and it was probably never surrounded by nebulosity.*' Dreyer supports Schjellerup in the GCS: '*Has probably never been nebulous*'.[316] Consequently the object was ignored in the NGC.[317]

A 'nebulos star' of another kind was found by Stephan. It is object no. 13 in his list X (Stephan 1880): '*This is the star 4811 Lalande, surrounded by a faint round nebulosity, a bit extended to the southeast.*' Actually, this is the 11-mag galaxy NGC 988 in Cetus, upon which a star of 7.1 mag is superimposed. A similar example is NGC 676, in Pisces, which was found by William Herschel on 30 September 1786 (IV 42). It is described as '*A star about 8 or 9 m. with very faint branches.*' The 9.5-mag star (BD +5 244) is only 9″ from the centre of the edge-on galaxy (the lowest value of all cases in the NGC). For John Herschel it was a '*most curious object*' (h 151), which he sketched on 24 September 1830; it is shown in Fig. 58 of the Slough catalogue.

6.14.7 First observations in Copenhagen

Shortly after his arrival in Copenhagen in early 1857, d'Arrest began to observe nebulae. He could use only a small Fraunhofer refractor with focal length 1.5 m on top of the 'round tower' in the city. Between 20 February 1857 and 13 March 1860 he observed 32 objects from the Slough catalogue. However, this cannot be seen as a follow-up to his Leipzig observations ('Erste Reihe').

The real task started on 1 September 1861, when the 11″ Merz refractor was ready to use at the new observatory on the Østervold (see Fig. 8.9 in Section 8.6.1). It turned out to be much better than the Leipzig telescope. D'Arrest started with the bright objects IV 1 (NGC 7009), IV 51 (NGC 6818) and I 55 (NGC 7479), then continuing to fainter ones (below 13 mag). He observed two nebulae from Herschel's class III ('very faint nebulae'), III 221 (NGC 7559) and III 222 (NGC 7563), and was even successful with h 2176 (NGC 7347), which had been noted as 'exceedingly faint' by John Herschel.

A detailed presentation of the refractor and the first observations (including those on the 'round tower') can be found in d'Arrest's report 'Instrumentum magnum aequatoreum in Specula Universitatis Havniensis super erectum', published in 1861.[318] From 1 September to 9 October 1861 he observed 90 objects, of which 12 are assigned as new ('Nova mihi'). A modern analysis even gives 14, but 3 can be identified with objects that had been discovered earlier (Table 6.21).

---

[313] Parsons L. (1880: 23).
[314] Huggins (1864: 442).
[315] Not to be confused with the catalogue of Lalande.
[316] Dreyer (1878a: 388).
[317] About 55 And, see also Wenzel (2004).

[318] d'Arrest (1861); see also d'Arrest (1862a).

Table 6.21. D'Arrest's first objects discovered in Copenhagen (sorted by date; P = page in d'Arrest (1861))

Date	NGC	dA	Au	GC	Type	V	Con.	P	Remarks
30.9.1861	1	1		1	Gx	12.8	Peg	28	≠ h 4 (= IV 15 = h 5 = NGC 16)
30.9.1861	7548	120	50	4914	Gx	13.4	Peg	38	
2.10.1861	588	25	13	348	GxP	13.5	Tri	31	H II region in M 33
2.10.1861	592	26	14	349	GxP	13.0	Tri	31	H II region in M 33
4.10.1861	4205	87	29	2798	Gx	13.0	Dra	34	
6.10.1861	5819				Gx	13.5	UMi	35	NGC 5808, W. Herschel (III 311) 16.3.1785
7.10.1861	370	7	5	197	3 stars		Psc	29	NGC 372, Dreyer 12.12.1876
7.10.1861	374	8	6	199	Gx	13.5	Psc	29	
7.10.1861	384	9	7	207	Gx	13.0	Psc	29	W. Parsons 4.11.1850
7.10.1861	385	10	8	208	Gx	13.0	Psc	30	W. Parsons 4.11.1850
8.10.1861	109	2	3	54	Gx	14.1	And	28	
8.10.1861	443	13	10	249	Gx	13.7	Psc	30	IC 1653, Javelle 17.10.1903
8.10.1861	447	14	11	250	Gx	13.6	Psc	30	IC 1656, Barnard 1888?
8.10.1861	504	17	12	291 = 292	Gx	13.2	Psc	30	J. Herschel (h 107) 22.11.1827

With the exception of two objects, Auwers recorded all in his list of new nebulae (Au), writing '*As these nebulae are very faint and reliably determined, I have not looked them up.*'[319] NGC 1 and NGC 5819 are missing, because d'Arrest did not mark them as 'Nova mihi'. The former case is remarkable (Fig. 6.46). The object is called 'h. 4' by d'Arrest, but, as noticed by him in 1862, John Herschel's h 4 (GC 8) is identical to his father's IV 15 (h 5, GC 12, NGC 16).[320] D'Arrest's report of 1861 describes an extended, faint, round object between two stars showing no central condensation. This cannot be NGC 16, but clearly fits a new nebula: NGC 1. John Herschel catalogued the 12.8-mag galaxy in Pegasus as GC 1. He refers to 'D'Arrest, 1', despite not knowing of d'Arrest's report. The reason is that on 8 May 1863 d'Arrest had sent a list of 125 new nebulae to Herschel for inclusion in the General Catalogue[321] (see column 'dA'). NGC 1 forms a double galaxy with NGC 2 (14.2 mag); the fainter companion was discovered by Lawrence Parsons on 20 August 1873. The second case is NGC 5819; the galaxy in Ursa Minor first appears in

Figure 6.46. The galaxy pair NGC 1 and NGC 2 in Pegasus (DSS).

---

[319] Auwers (1862a: 73).
[320] The reason was a wrong right ascension (d'Arrest 1867a: 1).
[321] RAS Herschel J. 12/1.8.

the NGC. D'Arrest lists it as 'H. III. 311 (?)'.[322] Indeed, its position and description indicate III 311 (NGC 5808); according to modern data, William Herschel's original position was wrong.

According to Dreyer's New General Catalogue, NGC 381 and NGC 385 had already been discovered in 1850 by Lord Rosse at Birr Castle. D'Arrest's last object, catalogued as GC 291 by Herschel, is identical with h 107 (GC 292 = NGC 504), as Dreyer states in the notes to NGC 504. Herschel has given an incorrect declination for h 107. NGC 588 and NGC 592 are H II regions in M 33. Though d'Arrest had already noticed the proximity to this large nebula, he did not mention a physical connection. Another H II region (NGC 595) was found by him on 1 October 1864 (it was independently seen by Safford on 1 November 1866). NGC 370 is a trio of stars of 13–14 mag (d'Arrest's position is 3′ southwest). Dreyer too found this object, in 1876 at Birr Castle (NGC 372), without noticing the identity.

D'Arrest's report 'Instrumentum magnum' closes with 10 drawings (Figs. 1 to 10). They were made at the Merz refractor and show mainly planetary nebulae. The figures show NGC 7009 (PN), NGC 2022 (PN), NGC 6818 (PN), NGC 1952 (M 1, SNR), NGC 6720 (M 57, PN), NGC 7479 (Gx), NGC 936 (Gx), NGC 6853 (M 27, PN), NGC 1535 (PN) and NGC 2023 (EN).

6.15 SCHÖNFELD AND THE NEBULAE IN THE BONNER DURCHMUSTERUNG

In the middle of the nineteenth century, Schönfeld was among the keenest observers of nebulae, comparable to Auwers, d'Arrest, Schmidt and Winnecke. His astronomical mentor was Argelander at Bonn Observatory. There he was responsible for the Bonner Durchmusterung (BD) – a monumental project realised with a humble comet-seeker. During the observations many (mostly known) nebulae were detected, some causing considerable confusion.

6.15.1 Short biography: Eduard Schönfeld

Eduard Schönfeld was born on 22 December 1828 in Hildburghausen, Southern Thuringia (Fig. 6.47).

Figure 6.47. Eduard Schönfeld (1828–1891).

He attended the local secondary school. Because his father did not allow him to study astronomy, he learned machine-building at academies in Hannover and Kassel. In 1849 Schönfeld changed to Marburg University, where he came into contact with astronomy. After visiting Argelander in 1851, he decided to study astronomy in Bonn. In 1853 he became an assistant at the local observatory (following Julius Schmidt, who moved to Olmütz) and a year later he received a doctorate. Schönfeld's Bonn years were filled by the work on the survey of the northern sky, in which he was assisted by Adalbert Krüger. The resulting catalogue contains 324 198 stars between +90° and −2° declination; the positions are reduced to epoch 1855. During nights with strong moonlight he observed variable stars, applying Argelander's methods.

In 1859 Schönfeld became Director of Mannheim Observatory, which was located in a tower near the castle and found to be in bad condition.[323] On the flat roof, a 16.5-cm refractor by Steinheil was erected in a small dome. Schönfeld soon began observing and measuring nebulae; the results were published in 1862 and 1875. As early as in 1861 he wrote a popular article about nebulae. Initially he also worked on manuscripts for the Bonner Durchmusterung (which appeared in 1861); later he published catalogues of variable stars (in 1866 and 1874).

Schönfeld and his friends Wilhelm Foerster and Christian Bruhns initiated the foundation of the Astronomische Gesellschaft (AG), which took place

---

[322] d'Arrest (1867a: 314).

[323] See e.g. Klare (1970) and Budde (2006).

on 28 August 1863.[324] From 1863 to 1869 he was a board member. After Argelander's death Schönfeld was appointed his successor in 1875. Back in Bonn he started the 'Southern Durchmusterung', which he made (only occasionally assisted) with the new 15.9-cm Schröder refractor. The observations were finished in 1881; the result, which was published in 1887, contains 133 695 stars between −2° and −23° (epoch 1855). At the instigation of Auwers, Schönfeld was elected a Corresponding Member of the Preußische Akademie der Wissenschaften in 1887. During the same year he became President of Bonn University. After a long heart disease, Eduard Schönfeld died on 1 May 1891 in Bonn at the age of only 62.[325]

6.15.2 NGC 1333 – a variable nebula?

The work on the Bonner Durchmusterung started on 25 February 1852. A comet-seeker by Fraunhofer with aperture 7.8 cm and only tenfold magnification was used (Fig. 6.48 right). The first observers were Julius Schmidt and Friedrich Thormann, who were replaced by Schönfeld and Krüger as little as a year later. The survey was executed in zones, starting at the southern limit of −2° declination. Each star was observed at least twice.[326] Doubtful cases were measured with the meridian-circle (mostly by Argelander) or verified at the 6" heliometer. The huge project was finished on 27 March 1859. Afterwards the catalogue and the atlas (charts) were produced.

The BD is not only interesting by virtue of its stellar content but also offers some nebulous objects. Though most of them were already known, there was no rigorous analysis by the Bonn astronomers. One of the few nebulae actually discovered in Bonn is NGC 1333 in Perseus (Fig. 6.48 left). The circumstances are remarkable. Schönfeld saw the reflection nebula on 31 December 1855 with the comet-seeker, at that time equipped with an orthoscopic eye-piece by Kellner. Though the new object was catalogued as BD +30° 548,[327] Schönfeld did not publish his find until October 1862.

Figure 6.48. Discovered with the Bonn 7.8-cm comet-seeker (right): the bright reflection nebula NGC 1333 in Perseus (DSS).

In the meantime it remained unnoticed, which explains the following message claiming a success by George Phillips Bond, Director of Harvard College Observatory. In a letter to Richard Carrington, dated 4 April 1859, he wrote '*A nebula, which is to be found neither in Sir J. Herschel's Catalogue [SC] nor D'Arrest's ['Erste Reihe'], was detected by Mr. Tuttle on the 5th of February in R.A. 3^h 21^m, Decl. +30° 55'. It follows a star of 9·10 mag. by 6 seconds, and is 2' north of it. It is barely visible in a telescope of 3 inches aperture.*' (Bond G. 1859). The instrument mentioned is the 3" comet-seeker by Quincy. Auwers noticed that Horace Tuttle's description was erroneous. Obviously the directions of the star's relative position had been mixed up and it should have read '*It precedes a star of 9·10 mag. by 6 seconds, and is 2' south of it.*'[328]

In September 1862 d'Arrest announced the 'Auffindung eines dritten variablen Nebelflecks'[329] – Tuttle's new nebula. On 22 August 1862 (a '*very favourable night*') he had observed the object with the 11" Merz refractor in Copenhagen, noting '*but at present it is as faint as a Herschel nebula of third class*'. Since Tuttle had seen the nebula in a tiny telescope, d'Arrest supposed that it must have become fainter – a case of variability. He felt supported in this conjecture when noticing the Bonn observation (BD +30° 548), which had been made with a small refractor too: '*this*

---

[324] Haffner (1963).
[325] Obituaries: Krüger (1891a, b), Foerster (1891) and Anon (1891a); see also Steiner (1990) and Schmidt-Kaler (1983).
[326] See e.g. Ashbrook (1984: 427–436).
[327] It appears in Vol. 4 on page 172. In the twentieth century NGC 1333 was catalogued by Sven Cederblad (Ced 16), who

gives the correct discoverer ('Schönfeld in 1855'); see notes in Cederblad (1946).
[328] Auwers (1862a: 76).
[329] 'Localization of a third variable nebula' (d'Arrest 1862f).

is Argelander's nebula in the Bonner Durchmusterung Bd. IV, pag. 172'. D'Arrest concluded that *'Now the change occurring is beyond doubt.'* The two other variable nebulae mentioned in his report are Hind's near T Tauri (NGC 1555) and Tempel's Merope Nebula (NGC 1435). It was striking that all three are *'separated by only a few degrees, 9° and 8°, respectively, in a region without any other nebulae, and that they have disappeared or are about to disappear nearly simultaneously'*. Nonetheless, d'Arrest was very cautious in his statements: one should *'take good care not to conclude that the previous view about the permanence of the nebulae will in general be shocked. The number of suspicious nebulae is, at least for me, nearly exhausted.'*

Being the true discoverer, Schönfeld was forced to reply. His comprehensive report, dated 3 October 1862, is titled 'Über den Nebelfleck Zone +30° No. 548 des Bonner Sternverzeichnisses, mit einigen Bemerkungen über die Nebelbeobachtungen in der Bonner Sternwarte überhaupt'.[330] He cites two of his own observations and another one by Krüger. The Bonn records give for Schönfeld's observations *'nebulous […] air changing but in general clear'* (31 December 1855) and *'nebula [air] very clear and transparent'* (3 February 1856). Krüger does not mention any nebula on 4 October 1856 in *'good air'*; he might have seen only a star. Thus Schönfeld supposed the object to be a nebulous star.

While visiting Bonn in 1857, Winnecke took the view (as told by Schönfeld in his report) that the nebula *'must be a new one'*. Since early March 1856 he had owned a copy of Auwers' (still unpublished) work on the Herschel nebulae; the object was listed neither there nor in the Slough catalogue. Winnecke's check on 15 November 1857 with the Bonn heliometer revealed a *'Very large faint nebula and star 9·10^m following 3^s, 2′5 north.'* The mentioned star of 9–10 mag, lying 3^s east and 2.5′ north of the nebula, is BD +30° 549. Later, however, Winnecke could not find his observation in the Bonn records: *'perhaps they were written on a loose sheet of paper, which I later handed out to Krüger'* (Winnecke 1863), but he refers to another one, made on 7 February 1858 with the heliometer: *'I saw the nebula with the divided lens […] the observed object is a star (a) 9·10^m near a very faint diffuse nebulous mass […] the added sketch shows star a exactly on the northern edge of the roundish nebula with 6′–7′ diameter.'* Later he saw the nebula at Pulkovo Observatory with an 11.8-cm refractor: *'Sept. 29, 1858 I looked at the nebula with the excellent telescope by Steinheil with a Gauß lens of 54′′′ aperture.'* He further wrote that *'I easily detected it with magnification 23 […] in the same night a comet-seeker of 34′′′ aperture [7.4-cm Plössl] and magnification 15 showed the nebula very clearly too. It was less easily visible with the heliometer of 72′′′ aperture [15.7-cm Merz] at magnification 65.'*

Schönfeld now compared the available sightings, especially regarding the apertures of the telescopes involved. About the observations with the 7.8-cm comet-seeker at Bonn Observatory he wrote that *'the nebula undoubtedly was perceptible […] but it could be seen distinctly only once, when the air was very clear and transparent [3.2.1856]'*. Tuttle as well had described observing the nebula in an even smaller refractor (6.4 cm) as 'barely visible'. In a mid-sized telescope of aperture about 15 cm the object appeared hardly better. Schönfeld referred to Winnecke's observation and his own, made with the 16.5-cm Steinheil refractor at Mannheim Observatory. In January/February 1861 and again on 20 September 1862 he had problems with seeing the nebula.[331]

The same happened to Auwers, who observed the 'Bonn nebula' with the 6″ heliometer at Königsberg Observatory on 11 September 1861: *'quite diffuse dull gleam of some minutes diameter […] star involved near the following edge (around 7^s f 2′ n of the brightest part). A precise position cannot be given.'*[332] A year later he described the object as *'blurred and of very dim light, but conspicuous in a small instrument due to its size'* (Auwers 1862b). About d'Arrest's statement he noted that *'Prof. d'Arrest concludes a variability due to the fact that the nebula was discovered (1858) with a 2ft comet-seeker and can currently be seen in the Copenhagen refractor only with difficulty, but being quite invisible in a 4ft telescope. The latter instrument might not have a sufficiently low eye-piece.'* About his own observations, he wrote that *'with the 42 times magnification of a 6ft telescope [15.7-cm Fraunhofer refractor] I saw the nebula on Sept. 24 [1862] very distinctly and also without*

---

[330] 'On the nebula zone +30° No. 548 of the Bonn star catalogue, with some general remarks about observations of nebulae at Bonn Observatory' (Schönfeld 1862a).

[331] Schönfeld (1862d: 110).

[332] Auwers (1865: 222).

difficulty in a 2ft comet-seeker, which has by no means a great light-gathering power'. Auwers entered the object as no. 17 in his list of new nebulae, but giving a wrong discovery date: '*Schönfeld 1858*'. Schönfeld had privately communicated his discovery to Auwers; perhaps the date of the letter was published erroneously.

Schönfeld comes to a similar conclusion: '*The perceptions in the three equally sized telescopes in Bonn, Mannheim and Königsberg, between 1857 Nov. 15 and 1862 Sept. 20, are very congruent. Also d'Arrest's estimation that the nebula is comparable to those of Herschel's third class fits well to the observations of 1857 and 1861–1862.*' He states that a small telescope with low power, such as a comet-seeker, can show a '*large dim area of light*' much better than a large one, if the air is very clear.[333] He stressed the exceptional character of the Bonn observations: '*Krüger and I often observed exceedingly faint objects.*' In his obituary, Krüger wrote about Schönfeld that he had '*a persistent visual acuity and the ability to fix certain light impressions exactly in his mind*'.[334]

Schönfeld's result is unambiguous – and does not support d'Arrest: '*After all mentioned here, I can by no means consider the differences in the observations of the nebula zone +30° No. 548 as clear evidence of its variability, but only as a striking example of how the visibility of very faint, large diffuse nebulae depends on magnification, air transparency and adaptation to the dark of the eye, so that, compared with ordinary fixed stars, aperture takes a back seat.*' In 1914 the NGC 1333 region was photographed by Barnard with the 10″ Bruce astrograph at Yerkes Observatory. The nebula appears '*roundish and not symmetrical with respect to the star [BD +30° 548] – its centre seems to be several minutes to the south*' (Barnard 1915).

6.15.3 BD −0° 2436, NGC 3662 and Argelander's mistake

Another remarkable case is BD −0° 2436, marked on chart 6 as a 'nebula', which turned out to be wrong (Fig. 6.49). On 29 March 1862 Julius Schmidt called attention to the object in *AN* 1360[335] (Schmidt 1862b).

---

[333] Compare this with the discussion on Harding's Nebula (NGC 7293) in Section 4.1.2.
[334] Krüger (1891a: 175).
[335] The title of his note was 'Über veränderliche Nebelgestirne' ['On variable nebulous stars'].

Figure 6.49. The galaxy NGC 3662 and the faint star x = BD −0° 2436 (DSS).

In the 15.7-cm Plössl refractor at Athens Observatory there appeared an '*exceedingly faint object with slight central condensation*'. Relative to a reference star (BD −0° 2440, about 30′ east), he determined the position of the 'Bonn nebula' and a few stars in the vicinity (tagged x, y, p and q). He now detected that BD −0° 2436 '*coincides exactly with my star x = 12ᵐ 13* [12–13 mag]*, which can be seen neither in the finder nor with the meridian-circle*', concluding that '*this nebula, once identified with the Bonn comet-seeker, now has reached the limit of visibility*'. Schmidt further remarks that '*even the* [Bonn] *meridian-circle did show it distinctly*'. In the catalogue BD −0° 2436 is marked 'B', which refers to a control observation with the meridian-circle (such observations were normally executed by Argelander).[336]

Schmidt therefore believed that he had found a new 'variable nebulous star' (in his note Hind's Variable Nebula is mentioned too). We are now faced with two different 'nebulae': Schmidt's nebula and the one shown on the BD chart, which appeared to him as a faint star (x). Figure 6.49 shows that the former is located about 4′ northwest of BD −0° 2436 (x). No doubt this new object, described as '*Neb., central brightness = 13ᵐ*', is William Herschel's IV 4 = h 879 = NGC 3662. The following question arises: what was seen once in the Bonn comet-seeker and called a 'nebula' – the faint star x or Schmidt's NGC 3662?

If Schmidt's object was seen, what was the reason for the position error of 4′? D'Arrest picked up the

---

[336] Bonner Durchmusterung Vol. 3, p. 24.

case, writing on 20 May 1862 that *'What was seen at the place in the years 1784, 1828 and 1862* [by W. Herschel, J. Herschel and Schmidt] *fits so well that we must be careful with our judgement until Professor Argelander enlightens us about which object was observed by him at the mentioned place with the meridian-circle.'* (d'Arrest 1862c). He added that *'The case stressed by Schmidt is peculiar, but not the only one of its kind. Two other nebulae were, for instance, observed with the comet-seeker, where I was unable to find any older observation.'* The first obviously is NGC 1333 and the second NGC 1555. A similar remark was made by Hind in a letter written on 27 May 1862 to Herschel: *'I suppose the nebula referred to by M. Schmidt* Astr. Nachr. *1360, must be regarded as appending a third instance of variability in these wonderful objects.'*[337]

In his report, d'Arrest mentions another issue. According to him, there are 46 nebulae in the first two volumes of the Bonner Durchmusterung, covering the zones −2° to +41°. In September 1862 he specifies his statement: 31 BD nebulae are in Messier's catalogue, 4 are planetary nebulae (discovered by William Herschel and Wilhelm Struve), 9 are Herschel nebulae of classes I–VI and 2 are new (d'Arrest 1862f).

Hermann Goldschmidt of Paris Observatory contributed an observation of Schmidt's nebula (NGC 3662), which he made in March and communicated in a letter of 20 October 1862 (Goldschmidt 1863a). While searching for the minor planet Hygiea, he detected a nebula on the chart: *'This nebula […] is on Berlin Academy chart Hora XI., made by Boguslawski, and was also seen with the Bonn comet-seeker.'*[338] Goldschmidt's observation was unsuccessful, since he *'missed the nebula, which was therefore invisible for a lens of 48 lines* [10.5-cm refractor] *and magnification 36'*. He further wrote: *'A nice group of stars in an irregular pattern, near to which the nebula should be located, proved its definite invisibility.'*

However, the story of BD −0° 2436 does not end here. On 18 April 1868 C. H. F. Peters discovered the minor planet (98) Ianthe with the 13.5" Spencer refractor at Hamilton College Observatory, Clinton. The next days he traced it, noting "*Not far from the field where I found the planet, the Bonner 'Durchmusterung'* [chart] *shows a nebula at No. −0° 2440*" (Peters 1868). He added that *'Actually the nebula precedes this star* [BD −0° 2436] *by 6^s, 2'30" north.'* Thus Peters' result is equal to Schmidt's (without mentioning the Athens astronomer). He assumes that "*It is, however, possible that in a smaller telescope, like that used for the 'Durchmusterung', the nebula covers a larger area.*"

As requested by d'Arrest in 1862, Argelander gave his view on 29 May 1868: *'The object was not seen with the comet-seeker'* (Argelander 1868b). This terminated d'Arrest's idea of variability – there was no sighting with a small telescope, whereupon it could be rated as bright. Argelander cleared up the confusion too. While checking BD fields with the meridian-circle he measured BD +13° 2386 (M 65) on 23 March 1856, recorded the data in his observing book and marked the object as a nebula. He then checked the region around the Herschel nebula IV 4 (NGC 3662), since *'there was a nebula on Boguslawski's Berlin chart […] whose position nearly coincides with the nebula h. 879 = H. IV. 4'.* Argelander *'observed an object on two well-matching* [micrometer] *bars, but it was doubtful at the declination microscope'*. Despite this uncertainty, he entered the object in the next row of his book, putting a horizontal line below.

During the BD compilation a mistake happened: *'I interpreted this line as a repetition of the nebula entered in the upper row* [BD +13° 2386], *and thus entered the object in question* [BD −0° 2436] *as a nebula in the catalogue.'* Argelander added that, *'Being alerted by Schmidt's and d'Arrest's remarks in the* Astronomische Nachrichten *Vol. 57, pp. 245 and 345 about the faintness of the nebula, Krüger investigated the field on 1862 May 18 and could not spot the nebula with full certainty as the air was not quite transparent.'* Krüger's observation might have been triggered by Schmidt's report (15 May) alone, since d'Arrest's article was published later (26 June). Argelander concludes as follows: *'Therefore I suppose, and after Mr Peters' remark I'm even sure now, that I had not observed the nebula, but Schmidt's star x.'* Thus BD −0° 2436 is not NGC 3662, which due to its faintness has no right to be in the BD (as a 'star'). In later editions, the object is marked 'D' (deleted). Concerning the nebulae in the BD, Argelander refers to Schönfeld (Schönfeld 1862b).

In this report, Schönfeld gives a selection of 27 nebulae and corrects the common assumption that *'all objects marked as nebula in the Bonn star-catalogue*

---

[337] RAS Herschel J. 12/1.6.
[338] Heinrich von Boguslawski observed at Breslau Observatory with a 9.7-cm refractor.

were really recognised as such during the zone observations with the comet-seeker [...] a large fraction of them was but noted as 9^m or 9·10^m, and later verified in a larger telescope, if a comparison with other charts or catalogues revealed nebulae instead of a star at the very places'. He states that 'On the other hand, the extensive comparison with existing catalogues of nebulae, and generally, the complete inclusion of all nebulae, visible under good circumstances with the comet-seeker, was beyond our goals.' Thus many nebulae are shown as stars on the charts: 'Consequently there are many more nebulae on the Bonn charts, as indicated by d'Arrest in Astr. Nachr. No. 1379.'

6.15.4 Nebulae catalogued in the BD

Which nebulae were actually found with the comet-seeker? Argelander had already addressed the issue and counted 62 objects, according to Karl von Littrow in his report 'Zählung der nördlichen Sterne im Bonner Verzeichnis nach Größe'[339] of 1 April 1862. In the complete BD catalogue (northern and southern parts) a nebula is marked 'Neb' in the column 'Helligkeit' (brightness). An analysis gives 99 cases. It is remarkable that all but one are non-stellar objects, which shows the skill of the observers, especially Schönfeld.

Of the 99 BD 'nebulae' the majority is listed in the catalogues of Messier and William Herschel, with almost equal numbers (Table 6.22). Only two objects are new: NGC 1333 (BD +30° 548) and NGC 6643 (BD +74° 766, the most northern object), both of which were discovered by Schönfeld. There is no publication for the latter. The galaxy in Draco was probably found in 1858, since zones of high declination were treated last. The survey was finished in March 1859, thus Tuttle's observation, dated 1 September 1859, came later (Bond G. 1862). It is interesting that the bright star Betelgeuse in Orion is marked as 'Neb'.

Table 6.23 shows the 97 non-stellar objects marked 'Neb' in the Bonner Durchmusterung; 69 are in the northern BD catalogue (7 more than von Littrow has counted) and 28 in the southern. The brightest object is M 31, the faintest NGC 4800. As explained by Schönfeld, the BD catalogue contains further nebulae

---

[339] 'Count of the northern stars in the Bonn catalogue regarding brightness' (Littrow K. 1869).

Table 6.22. *Identifications of the 99 BD 'nebulae' in earlier catalogues*

Catalogue	Number	Remarks
Messier	48	Including M 103
W. Herschel	43	Two found by C. Herschel
Others	4	Harding, W. Struve, Hind, Auwers
New	2	Schönfeld: NGC 1333, NGC 6643
Star	2	Betelgeuse (BD +7° 1055); 'NGC 3662' (BD −0° 2436)

entered as stars. Some were identified with the Bonn 6" heliometer; others resulted from a modern study.

6.16 BRORSEN AND BRUHNS: COMET DISCOVERERS ON THE WRONG TRACK

Brorsen and Bruhns are well known for their discoveries of comets, but they also found three non-stellar objects with small telescopes. One of them, the Flame Nebula NGC 2024 in Orion, was already known: six years after Brorsen, Albert Marth revealed that it had been discovered by William Herschel – another example of a sloppy investigation by an observer.

6.16.1 Short biography: Theodor Johann Christian Brorsen

Theodor Brorsen was born on 29 July 1819 in Norburg (on the island of Alsen), Denmark (Fig. 6.50 left). After attending schools in Christiansfeld and Flensburg he studied at Kiel University during the years 1842–45. His astronomical interest grew early in life and in 1846 he discovered two comets. Heinrich Schumacher offered him a job at Altona Observatory, where he found a third comet in 1847. In the same year Brorsen obtained an employment at the new observatory at Žamberk Castle in Senftenberg, in Bohemia of John Parish (the Baron of Senftenberg). In 1851 he discovered two more comets. Three years later he published his observations of the *Gegenschein*, which was first seen on the island of Alsen and later confirmed

Table 6.23. *The 97 non-stellar BD objects (sorted by NGC–number); objects in the southern extension are in zones −03 to −23*

BD	M	NGC	Discoverer	Type	V	Con.	BD	M	NGC	Discoverer	Type	V	Con.
−09 108		157	W. Herschel	Gx	10.4	Cet	+12 2502	59	4621	Koehler	Gx	9.7	Vir
+40 141	110	205	Messier	Gx	7.9	And	+12 2508	60	4649	Koehler	Gx	8.8	Vir
+40 147	32	221	Legentil	Gx	8.1	And	−05 3572		4697	W. Herschel	Gx	9.2	Vir
+40 148	31	224	As-Sufi	Gx	3.5	And	+26 2394		4725	W. Herschel	Gx	9.3	Com
+29 261	33	598	Hodierna	Gx	5.5	Tri	+41 2333	94	4736	Méchain	Gx	8.1	CVn
+15 238	74	628	Méchain	Gx	9.1	Psc	+47 2002		4800	W. Herschel	Gx	11.6	CVn
−00 412	77	1068	Méchain	Gx	8.9	Cet	+22 2526	64	4826	Pigott	Gx	8.5	Com
+30 548		1333	Schönfeld	RN		Per	+18 2701	53	5024	Bode	GC	7.7	Com
−19 716		1400	W. Herschel	Gx	10.9	Eri	+47 2063	51	5194	Messier	Gx	8.1	CVn
−19 720		1407	W. Herschel	Gx	9.7	Eri	+29 2450	3	5272	Messier	GC	6.3	CVn
−13 842		1535	W. Herschel	PN	9.6	Eri	+60 1493		5322	W. Herschel	Gx	10.1	UMa
+21 891	1	1952	Bevis	SNR	8.4	Tau	+55 1651	101	5457	Méchain	Gx	7.5	UMa
+00 1177	78	2068	Méchain	RN	8.0	Ori	+29 2489		5466	W. Herschel	GC	9.2	Boo
−06 1440		2185	W. Herschel	RN		Mon	−05 3890		5634	W. Herschel	GC	9.5	Vir
+12 1066		2194	W. Herschel	OC	8.5	Ori	+56 1783	102	5866	Méchain	Gx	9.9	Dra
−15 1743		2360	C. Herschel	OC	7.2	CMa	−20 4193		5897	W. Herschel	GC	8.4	Lib
+65 580		2403	W. Herschel	Gx	8.2	Cam	+02 2943	5	5904	Kirch	GC	5.7	Ser
−14 2129		2438	W. Herschel	PN	10.8	Pup	−22 4135	80	6093	Méchain	GC	7.3	Sco
−17 2105		2440	W. Herschel	PN	9.4	Pup	−12 4537	107	6171	Méchain	GC	7.8	Oph
+12 1926	67	2682	Koehler	OC	6.9	Cnc	+36 2768	13	6205	Halley	GC	5.8	Her
+51 1498		2841	W. Herschel	Gx	9.3	UMa	−01 3245	12	6218	Messier	GC	6.1	Oph
+69 543	81	3031	Bode	Gx	7.0	UMa	+47 2384		6229	W. Herschel	GC	9.4	Her
+70 588	82	3034	Bode	Gx	8.6	UMa	−21 4439		6235	W. Herschel	GC	8.9	Oph
−07 2951		3115	W. Herschel	Gx	9.1	Sex	−03 4031	10	6254	Messier	GC	6.6	Oph
−17 3140		3242	W. Herschel	PN	7.7	Hya	−18 4488	9	6333	Messier	GC	7.8	Oph
+12 2249	95	3351	Méchain	Gx	9.8	Leo	+43 2711	92	6341	Bode	GC	6.5	Her
+12 2253	96	3368	Méchain	Gx	9.3	Leo	−17 4794		6356	W. Herschel	GC	8.2	Oph

Table 6.23. (Cont.)

BD	M	NGC	Discoverer	Type	V	Con.
+00 2736		3521	W. Herschel	Gx	9.2	Leo
+55 1449	97	3587	Méchain	PN	9.9	UMa
+13 2386	65	3623	Messier	Gx	9.2	Leo
+13 2390	66	3627	Messier	Gx	8.9	Leo
+04 2451		3640	W. Herschel	Gx	10.3	Leo
−00 2436		3662	W. Herschel	Gx	13.0	Leo
+12 2364		3810	W. Herschel	Gx	10.6	Leo
+49 2087		3893	W. Herschel	Gx	10.2	UMa
−00 2518		4030	W. Herschel	Gx	10.6	Vir
+65 868		4125	Hind	Gx	9.6	Dra
+15 2429	98	4192	Méchain	Gx	10.1	Com
+48 2012	106	4258	Méchain	Gx	8.3	CVn
+13 2531	84	4374	Messier	Gx	9.2	Vir
+13 2533	86	4406	Messier	Gx	8.9	Vir
+08 2607	49	4472	Messier	Gx	8.3	Vir
+13 2546	87	4486	Messier	Gx	8.6	Vir
+28 2136		4559	W. Herschel	Gx	9.6	Com
+26 2371		4565	W. Herschel	Gx	9.5	Com
+13 2560	90	4569	Messier	Gx	9.4	Vir
+12 2496	58	4579	Messier	Gx	9.6	Vir
−10 3525	104	4594	Messier	Gx	8.3	Vir
+62 1245		4605	W. Herschel	Gx	10.1	UMa

BD	M	NGC	Discoverer	Type	V	Con.
−03 4142	14	6402	Messier	GC	7.6	Oph
+70 957		6503	Auwers	Gx	10.2	Dra
+66 1066		6543	W. Herschel	PN	8.1	Dra
+06 3649		6572	W. Struve	PN	8.1	Oph
−16 4820	17	6618	de Chéseaux	EN	6.0	Sgr
+74 766		6643	Schönfeld	Gx	11.0	Dra
−08 4745		6712	W. Herschel	GC	8.1	Sct
+29 3538	56	6779	Messier	GC	8.4	Lyr
−14 5523		6818	W. Herschel	PN	9.3	Sgr
+18 4290	71	6838	de Chéseaux	GC	8.4	Sge
+22 3878	27	6853	Messier	PN	7.4	Vul
−22 5331	75	6864	Méchain	GC	8.6	Sgr
−13 5783	72	6981	Méchain	GC	9.2	Aqr
−11 5511		7009	W. Herschel	PN	8.0	Aqr
+11 4577	15	7078	Maraldi	GC	6.3	Peg
−01 4175	2	7089	Maraldi	GC	6.6	Aqr
−21 6239		7293	Harding	PN	7.3	Aqr
+33 4550		7331	W. Herschel	Gx	9.5	Peg
+60 2535	52	7654	Messier	OC	6.9	Cas
+41 4773		7662	W. Herschel	PN	8.3	And
+55 3056		7789	C. Herschel	OC	6.7	Cas

Figure 6.50. Left: Theodor Brorsen (1819–1895); right: Christian Bruhns (1830–1881).

in Senftenberg.[340] After Parish's death the observatory was closed; the instruments went to Vienna and Spain. Brorsen left Senftenberg, returning in 1879 to his hometown Norburg, where he died on 31 March 1895 at the age of 75.[341]

6.16.2 Short biography: Carl Christian Bruhns

Carl Bruhns was born on 22 November 1830 in Ploen, Holstein (Fig. 6.50 right). In 1851 he went to Berlin, where he was employed as an assistant locksmith by Siemens & Halske. The company worked for Berlin Observatory. Bruhns, having great mathematical talent, asked the director Johann Encke for problems in celestial mechanics, which he solved during the night. Encke was impressed by Bruhns' abilities and engaged him in 1852 as second assistant. Two years later he became first assistant and studied astronomy. In Berlin he dealt with orbit calculations and discovered four comets: in 1853, 1855, 1857 (rediscovery of comet Brorsen) and 1858. In 1858 he became Adjoint Professor. When in 1860 d'Arrest left the old Pleißenburg Observatory in Leipzig, Bruhns (supported by Encke) succeeded him as observer and Associate Professor at the university. The administration entrusted him with the planning of the new Leipzig Observatory in Johannistal.[342] On 8 November 1861 Bruhns became its director, following August Möbius. There he discovered two more comets in 1862 and 1864. He was a co-founder of the Astronomische Gesellschaft in 1863. His popular *Atlas der Astronomie* appeared in 1872 (the star charts were, for instance, used by Tempel). Carl Bruhns died on 25 July 1881 in Leipzig at the age of only 50.[343]

6.16.3 Discoveries of Brorsen and Bruhns

In September 1856 Brorsen discovered the globular cluster NGC 6539 in Ophiuchus. The observation was made at Žamberk Castle (Senftenberg), probably using the 9.4-cm comet-seeker by Merz, which had been purchased in 1842. He published his find in the popular weekly magazine *Unterhaltungen im Gebiete der Astronomie, Geographie und Meteorologie*[344] (Brorsen 1856). It was edited from 1847 by Gustav Adolph Jahn in Leipzig.[345] Auwers included the object as no. 39 in his list of new nebulae. In his observation on 10 October 1860 with the Königsberg 6" heliometer the nebula '*looked faint, but pretty well at magnification 65; it appears to be a faint star group of about 3' diameter, centrally surrounded by numerous stars 12ᵐ*'. On 9 September 1861 he looked at Brorsen's nebula again.[346]

In 1850 Brorsen had found the emission nebula NGC 2024 in Orion at Žamberk Castle (Fig. 6.51 left). On 17 January 1851 Brorson wrote to the *Astronomische Nachrichten* that '*The mostly pretty covered sky in that period has not really favoured comet-seeking, I have, however, found a very faint, very extended, pretty irregular nebula, located about 15 minutes east of ζ Orionis, which is listed neither in the catalogue of the younger Herschel nor in Messier's.*' (Brorsen 1851). However, in 1856 Albert Marth found out that '*The nebula mentioned by Mr Brorsen in A.N. 32, 105 is H. V. 28.*' (Marth 1856). William Herschel had discovered the class V object on 1 January 1786 at Clay Hall (Old Windsor), writing of a '*wonderful black space included in remarkable much nebulosity, divided in 3 or 4 large patches; cannot take up less than ½ degree, but I suppose it to be much more extensive*'. Owing to its branched form (best seen on photographs), the object near the eastern belt-star Alnitak is

---

[340] The *Gegenschein* is sunlight reflected by interplanetary dust near the ecliptic.
[341] Obituary: Weyer (1895).
[342] Bruhns (1862), Ilgauds and Münzel (1995).
[343] Obituaries: Anon (1881a), Anon (1881b), Anon (1881c), Anon (1882) and Foerster (1881).
[344] The title translates as 'Conversations in the fields of astronomy, geography and meteorology'.
[345] Münzel (2000).
[346] Auwers (1865: 217, 222).

Table 6.24. *Objects found by Brorsen and Bruhns (sorted by date)*

NGC	Discoverer	Au	GC	Type	V	Con.	Remarks
2024	Brorsen 1850		1227	EN		Ori	W. Herschel 1.1.1786 (V 28)
6539	Brorsen 1856	39	4370	GC	8.9	Oph	
2175	Bruhns 1857	21	1366	EN		Ori	

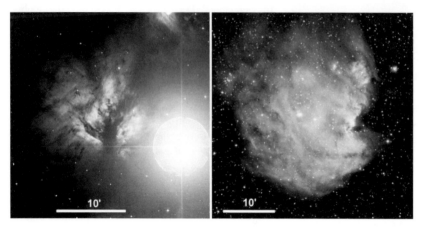

Figure 6.51. Emission nebulae in Orion: NGC 2024 east of ζ Orionis (left) and NGC 2175 (right) were discovered by Brorsen and Bruhns, respectively (DSS).

nicknamed the Flame Nebula. It was not observed by John Herschel, who eventually entered it in the General Catalogue as GC 1227. This triggered altogether 13 observations at Birr Castle (during 1873–78), made by Lawrence Parsons, Copeland and Dreyer. The former drew the nebula on 18 February 1874 with the 36-inch as 'Herschelian'.[347]

In 1857 Bruhns discovered the emission nebula NGC 2175 in Orion, using a comet-seeker at Berlin Observatory (Fig. 6.55 right). Auwers entered it (following a private communication by Bruhns) as no. 21 in his list of new nebulae, copying his description: '*star 8ᵐ with nebulosity*' (the star is BD +20° 1284, 7.6 mag).[348] He observed the object on 24 February 1861 with the Königsberg heliometer, seeing '*a considerable ray of milky, faint light, extended about 8′ north–south and 25′ east–west*'. He identified the star at the northeast edge as LL 11668 from Lalande's *Histoire Céleste*.[349] D'Arrest observed the nebula with the Copenhagen 11″ refractor: '*The extraordinarily large, faint nebula is one of those objects which are difficult to see with higher magnification. It took a long time to find it.*' (d'Arrest 1865c).

Brorsen and Bruhns found three objects, but only two were new (Table 6.24). Concerning the brightness of NGC 6539, it is astonishing that the globular cluster was not discovered earlier.

6.17 TEMPEL'S OBSERVATIONS IN VENICE, MARSEILLE AND MILAN

Wilhelm Tempel was one of the most important, but also controversial, visual observers in the second half of the nineteenth century. Through his numerous discoveries of comets, among them three periodic ones, and his excellent drawings of nebulae he won great fame.[350] Tempel acted first in Venice, Marseille and Milan (Brera), but the majority of his observing was done later

---

[347] Parsons L. (1880), Plate II. NGC 2024 was drawn by Tempel too, see Radrizzani (1989). See the study of the ζ Orionis region at Mt Wilson by Duncan (1921).
[348] Auwers (1862a: 74).
[349] Auwers (1865: 217, 227).
[350] Radrizzani (1989).

as Director of the Arcetri Observatory in Florence. He found a great number of non-stellar objects and published numerous papers. His first discovery was perhaps the most important: the Merope Nebula in the Pleiades – a very popular and much debated object.

6.17.1 Short biography: Wilhelm Leberecht Tempel

Wilhelm Tempel was born on 4 December 1821 in Niedercunnersdorf, Upper Lusetia (Fig. 6.52). Since he came from a poor parental home, he attended the village school for eight years. Owing to his abilities, and against his father's will, his teacher gave him private lessons in drawing and astronomy. Tempel was an autodidact with a great thirst for knowledge, who was often characterised as a melancholic, self-willed person. After having finished school, he left his home and went to Meißen to learn the trade of lithography. From 1841 he lived in Copenhagen for three years, working as a lithographer. Afterwards he went to Christiana (which is nowadays called Oslo).

In 1852 we find him in Venice, where he converted to the catholic faith and met his later wife Marianna Gambini, daughter of the gatekeeper of the Doge's Palace, but Tempel further travelled to Marseille and Bologna (1857). Back in Venice he bought a 10.8-cm refractor from Steinheil, of Munich, in autumn 1858 (Fig. 6.52). He observed from the staircase ('Escalier Lombard') of the Palazzo Contarini del Bovolo. Tempel repeatedly applied for an observatory job, but without any astronomical degree he got only refusals. In 1859 he made two important discoveries at Venice: his first comet and the Merope Nebula in the Pleiades. The supposed variability (and even existence) of the nebula was controversial for some decades.

Having become well known through his comet discoveries, Tempel tried his luck once again and asked for an appointment at Paris Observatory. Frustrated by the unkind reaction of director Leverrier, he moved to Marseille in March 1860. There he eventually got a job at the observatory, which was being managed by Benjamin Valz. However, the director resigned in September 1861 and Tempel had to leave, ending up working once again as a lithographer (his home was at Rue Pythagore 26). In Marseille he discovered five asteroids and eight comets with his small refractor, among them the famous periodic comet Tempel–Tuttle,

Figure 6.52. Wilhelm Tempel (1821–1889) with his 10.8-cm Steinheil refractor, in 1868 at his home in Marseille (Clausnitzer 1989).

which causes the Leonids.[351] Owing to the Franco-Prussian War he was expelled from France in January 1871. Fortunately, Giovanni Schiaparelli, Director of Milan Observatory (Brera), incorporated him into his staff.[352] There he discovered three comets within the next four years.

In 1875 Tempel became assistant astronomer at Arcetri Observatory in Florence. Soon he succeeded as director Giovan Battista Donati, who had died two years earlier. During the years 1869–72 Donati had erected the observatory, installing two Amici refractors with apertures of 9.4" and 11".[353] Tempel found both in a very bad condition. The lack of money affected not only the observatory – there was very little in the way of accessories and literature – but also his standard of living. Despite these adverse conditions, borne with stoicism by Tempel, he made many distinguished

---

[351] The spectacular Leonid shower of 13 November 1866 was described by Ball, who watched it at Birr Castle (Ball V. 1915: 70–71).

[352] About Tempel's time at Brera, see Radrizzani (1989).

[353] At that time the 11-inch was the largest telescope in Italy.

observations and drawings. With a few exceptions, his approximately 200 drawings of nebulae remained unpublished. The learned lithographer Tempel was often too highly demanding regarding their reproduction. The originals are now stored at Arcetri Observatory.

In 1881 Tempel was elected an Associate of the RAS. For his many discoveries of comets (he found his last comet in 1877 at Arcetri) he received many honours and prizes. He is the only one to have discovered two short-periodic comets: Tempel 1 (1867) and Tempel 2 (1873). From late 1886 his observing was terminated by ill health. After a long illness Wilhelm Tempel died on 16 March 1889 at the age of 67 in Arcetri.[354] Afterwards the observatory became dilapidated.[355] There is a crater named after Tempel on the Moon (of diameter 45 km).

6.17.2 The Merope Nebula and two objects in 'Eridanus'

In Venice Tempel discovered the reflection nebula near Merope in the Pleiades on 19 October 1859. He observed the star cluster with his Steinheil refractor at a power of 45 (field of view of 2°).[356] It was not until 23 December 1860, now in Marseille, that he reported the discovery in a letter to the editor of the *Astronomische Nachrichten* (C. A. F. Peters): '*Perhaps it is of interest that last year at Venice, having not observed the Pleiades for half a year, I found a large bright nebula on Oct. 19, which I first thought to be a great comet; convincing myself on the following evening the 20th that it has not moved.*' (Tempel 1861).

Auwers entered the object as no. 18 in his list of new nebulae. It appears as GC 768 in the General Catalogue and was listed by Dreyer as NGC 1435. The story of its discovery and subsequent visual and photographic observations is quite long (see Section 11.5).

Figure 6.53. Tempel's first galaxy: NGC 1398 in Fornax (DSS).

Questions regarding the existence, shape and variability of the Merope Nebula were controversial for over half a century.

In Marseille, using his private refractor, Tempel discovered NGC 1360 and NGC 1398 on 9 October 1861 (Fig. 6.53). Many years later, in May 1882, he published his find (Tempel 1882), obviously triggered by the reports of August Winnecke and Eugen Block, which were placed in the *Astronomische Nachrichten* in 1879 and 1880. These two authors had seen the objects independently in 1868 and 1879, respectively. Tempel wrote that, '*On my often-used small star chart of Bruhns, where I once marked the most conspicuous nebulae, seen while comet seeking – my telescope had no equatorial mounting and I had no catalogues – there are three nebulae in Eridanus, observed on Oct. 9, 1861.*'[357] The identity of the third object has not been confirmed, since Winnecke and Block noted only two. It could be the galaxy NGC 1371 (William Herschel's II 262 of 17 November 1784). Nowadays the planetary nebula NGC 1360 and the galaxy NGC 1398 are classified as being located in Fornax (until 1930 the northern part of Fornax belonged to Eridanus).

In the GCS Dreyer mentions for NGC 1360 (GC 5315) only Winnecke, who obviously informed him

---

[354] Obituaries: Schiaparelli (1889a, b), Anon (1889a) and Dreyer (1890); see also Hagen (1912), Eichhorn (1963) and Clausnitzer (1989).

[355] For Tempel's poor widow a subscription was opened amongst astronomers (Gautier E. 1889).

[356] The altazimuthal refractor had focal length 1.62 m and offered powers of 24 to 240. A 2.7-cm finder with focal length 24.4 cm was used. Today Tempel's telescope is installed as guider to the 36-cm Amici equatorial telescope at Arcetri Observatory; see Bianchi et al. (2010).

[357] The 'star chart' means Bruhns' *Atlas der Astronomie* of 1872.

Table 6.25. *Tempel's first discoveries*

NGC	GC	Date	Type	V	Con.	Remarks
1360	5315	9.10.1861	PN	9.4	For	L. Swift 1859, Winnecke Jan. 1868, Block 18.10.1879
1398		9.10.1861	Gx	9.8	For	Winnecke 17.12.1868, Block 18.10.1879
1435	768	19.10.1859	RN		Tau	Merope Nebula (Pleiades), Au 18

by letter. The NGC gives 'Swift 1857, Winnecke'. Indeed, it is correct that Lewis Swift was the first to see the planetary nebula, using his 4.5″ comet-seeker. However, this was not in 1857 but in 1859. The observation was not published until 1885 [358] – eight years after the GCS. Oddly, Dreyer does not mention Block in his catalogue. NGC 1360 is among the three NGC objects with the most independent discoveries (see Table 10.7). NGC 1398 does not appear in the GCS. It might have been omitted from Winneke's letter for some unknown reason. The NGC gives 'Winnecke, Block'. Tempel's early discoveries are listed in Table 6.25.

6.17.3 Tempel's publications in Marseille and Milan

In 1862, while he was living in Marseille, Tempel published in the *Astronomische Nachrichten* a drawing of the Orion Nebula engraved by himself[359] (Tempel 1862). It was based on observations made a year earlier with his small refractor. The editor, C. A. F. Peters, noted that *'It is undoubtedly of interest to see how an observer having a great drawing talent and a keen eye perceives this nebula with small magnification (24 to 40) in a refractor of only 4 inches aperture.'* In 1878 another drawing of M 42 appeared in the magazine *Sirius*, this one based on observations with the 11″ Amici refractor at Arcetri (Tempel 1878e). The editor, Hermann Klein, commented that *'Those who have viewed the Orion nebula in a telescope with great light-gathering power might be astonished about the wonderful accuracy with which the Arcetri astronomer has rendered the complex structure of the great nebula. Seldom is the ability of keen perception and the art of true geometrical reproduction combined in such a high degree as in the case of Mr Tempel.'*

In a paper written on 3 March 1864 in Marseille Tempel treated the issue of six 'components of Sirius' which Goldschmidt claimed to have detected at Paris in 1863 (Tempel 1864). He unmasks them as *'wrong stars'* due to *'ghost images in a bright eye-piece with great light-gathering power'*.[360] He was always critical concerning observations or drawings of other persons – his own were intransigently defended.

After his expulsion from France, Tempel moved to Milan (Brera), where he discovered a comet on 14 June 1871. In his report it is mentioned that he *'for some time past is busy with entries in charts and drawings of nebulae'* (Tempel 1871). The study of nebulae, noted here for the first time, was strongly related to his search for comets. No doubt the work at Brera marks the origin of his later passion at Arcetri. In contrast to Marseille, the Milan Observatory offered Herschel's catalogues and the Bonner Durchmusterung, according to a letter written on 8 July 1872 (Tempel 1873). In 1874 a collection of Tempel's observations was published as 'Osservazioni astronomiche diverse fatte nella specola di Milano'.[361] The report mainly deals with comets. However, in the appendix a fine drawing of the Pleiades with the Merope Nebula is shown.

The report gives a mysterious note about the possible discovery of a comet on 29 December 1871 in Cygnus. Tempel remarks that the Herschel nebula II 202 is located 1° south of its position. He suspects that this could be the 'comet', if Herschel has made a declination error of 1°.[362] On the following days,

---

[358] Swift (1885c: 39).
[359] See also Clausnitzer (1989: 26).

[360] On 31 January 1862 Clark had discovered Sirius B with the 18.5″ refractor built for Dearborn Observatory. Tempel made a sketch of Sirius and its surroundings; see Radrizzani (1989: 30).
[361] 'Miscellaneous astronomical observations made at Milan Observatory' (Tempel 1874).
[362] Such an error is possibly due to Caroline Herschel, who transcribed the original observation data.

Tempel was unable to detect the object, partly due to bad weather. Its position is occupied by the H II region Sh2–97 (LBN 151). The brightest portion could be seen by Herschel on 17 July 1784; his description of II 202 reads '*A resolvable patch; there are great numbers of them in this neighbourhood like forming nebulae, but this is the strongest of them; they are evidently congeries of small stars.*' Dreyer catalogued II 202 as NGC 6847 at Herschel's position, i.e. 1° south of Sh2–97. About 15' to the west lies the planetary nebula NGC 6852, which was discovered by Marth and mentioned by Dreyer as a candidate for II 202. However, Tempel's observation fits well with Herschel's, thus both of them possibly perceived a part of the extended emission nebula.

6.18 SCHMIDT'S FIRST DISCOVERIES IN ATHENS

Julius Schmidt was a keen visual observer. At his last and most important station, the Athens Observatory, he explored large fractions of the southern sky and encountered many new nebulae. Schmidt was an industrious writer, but not always in due time – perhaps as a result of the bad working conditions at the observatory. This caused some confusion. Schmidt was also interested in variable stars and – above all – the Moon. He has left an excellent lunar atlas.

6.18.1 Short biography: Johann Friedrich Julius Schmidt

Julius Schmidt was born on 26 October 1825 in Eutin, Oldenburg (Fig. 6.54). Already as a pupil at a Hamburg secondary school he was interested in astronomy. At the age of 14 he purchased Schroeter's work 'Selenotopographische Fragmente', on which he based his great affinity to the Moon. His father had built him a small telescope. His benefits were an extraordinarily keen eye and a great drawing talent. From 1841 to 1845 he had the opportunity to join the observatories in Altona (A. C. Petersen) and at the Hamburg Millerntor (Charles Rümker). In 1842 he observed many of the Virgo nebulae in Altona.

In 1845 Schmidt moved from Hamburg to the private observatory of Johann Friedrich Benzenberg in Düsseldorf-Bilk. The owner was occupied with the search for the proposed intra-Mercurial planet; therefore Schmidt could use only a small instrument for

Figure 6.54. Julius Schmidt (1825–1884).

his own observations. He stayed just a year, changing after Benzenberg's death to nearby Bonn Observatory. There, under Argelander, he got a thorough astronomical education. He especially learned how to observe variable stars and the determination of positions of nebulae and star clusters with the ring-micrometer, which had been made in 1846. For two years Schmidt was responsible for the meridian-circle, working for 'Hora V' of the *Berliner Akademische Sternkarten* (which was finished in February 1854). From 1852 he was involved in the Bonner Durchmusterung (together with Friedrich Thorman), but was ill at times. When he left Bonn after seven years of duty in 1853, Schönfeld took his place (Krüger followed Thorman). Schmidt's decision was due to Baron Eduard von Unkhrechtsberg, who visited Bonn to look for a director of his well-equipped private observatory in Olmütz, Moravia. With Argelander's recommendation Schmidt got the job. During 1853–54 he continued the observations of nebulae with the 13.6-cm Merz refractor (Schmidt 1854). Additionally he made micrometric determinations of the heights of lunar mountains. Schmidt stayed in Olmütz until 1858.

On 16 December 1858 Schmidt became Director of Athens Observatory, succeeding Georg Constantin Bouris. After an urgent refurbishment he started to observe the Sun, Moon, planets, comets and variable stars. Schmidt made many drawings (780 of Saturn alone). A special matter was the observation and documentation of lunar eclipses; there are detailed reports on 28 events from 1842 to 1879 (Ashbrook 1977). Also fascinating are his drawings of the Milky Way from naked-eye observations made during 1864–79;

6.18 Schmidt's first discoveries in Athens

Figure 6.55. Athens Observatory with the 15.7-cm Plössl refractor.

they were eventually published in 1923 by Antonie Pannekoek at Leiden Observatory.[363] But Schmidt's masterpiece was undoubtedly his monumental lunar map, published in 1878 as *Charte der Gebirge des Mondes nach eigenen Beobachtungen in den Jahren 1840–1874*.[364] John Birmingham wrote about it in 1879 that '*It is, in all truth, a performance highly creditable to the age in which we live, and to Teutonic intellect and perseverance*.'[365]

Using the 15.7-cm Plössl refractor, Schmidt could continue his observations of nebulae. In 1868 he was elected an honorary doctor of Bonn University, and he became an Associate of the RAS in 1874. On 24 November 1876 he discovered the 3-mag Nova Q Cygni (Schmidt 1877). Despite his humble equipment in Athens, Schmidt made numerous observations (among them about 85 000 of variable stars) and published nearly 340 articles, most of them in the *Astronomische Nachrichten*. Julius Schmidt died of a heart attack on 7 February 1884 at the age of 58. After having participated in a ceremony in the German embassy the evening before, he was found dead the next morning in his bed at the observatory home.[366]

6.18.2 Observations of nebulae and star clusters in Athens

When Schmidt came to Athens in late 1858, the observatory was in bad condition (Schmidt 1859). For the time being, systematic observations were impossible. The 15.7-cm Plössl refractor had to be urgently refurbished (Fig. 6.55). By 1860 the instrument had been sufficiently repaired to allow observations. A ring-micrometer was added to measure positions.

In spring 1860 Schmidt started observations of nebulae and star clusters (Schmidt 1861a). He admired d'Arrest's 'Beobachtungen der Nebelflecken und Sternhaufen', which had been published in 1856 in Leipzig. Among his first targets were M 13, M 30, M 31/32, M 51, M 77, NGC 6115 and the planetary nebulae NGC 6210, NGC 6572, NGC 6818 and NGC 7009. Schmidt wrote about his goals as follows: '*The Plössl refractor in Athens, in its present state only a middling instrument, gives but hope that the larger part of the nebulae listed in the catalogues of W. and J. Herschel could be re-observed and I believe that hardly another measuring campaign could be performed with better success than under the Greek sky.*' Despite good transparency, the seeing at the site was gauged badly due to the high temperatures: '*even at midnight the air inside the dome is not sufficiently balanced with the night air*'. Therefore, high-resolution observations (e.g. of double stars) were rare.

Variable stars were one of Schmidt's favourite targets; his knowledge was due to the seven years spent in Bonn as Argelander's assistant. This explains the

---

[363] 'Chart of the lunar mountains from my own observations in the years 1840–1874', published by Pannekoek in 1923; see also Ashbrook (1984: 373–379).
[364] Ashbrook (1984: 251–258).
[365] Birmingham (1879: 16).
[366] Obituaries: Krüger (1884), Lynn (1884) and Dunkin (1885).

Figure 6.56. The cometary reflection nebula NGC 2261 (Hubble's Variable Nebula) originating from the variable star R Monocerotis; drawing by Secchi (Secchi 1856a).

great interest in 'variable nebulae'. His publication 'Über einen neuen veränderlichen Nebelstern'[367] treats h 399 (NGC 2261), the popular cometary nebula in Monoceros discovered by William Herschel in 1783. Schmidt wrote that, *'When I observed the nebula h No. 399 on Jan. 24, 1861 in the great telescope during bright moonlight, I found the star south preceding the nebula to be missing.'* But on 31 January it was back again: *'visible at the first glance in the small eye-piece'*. The object is the variable star R Mon (Fig. 6.56). In 1916 Hubble revealed that the associated reflection nebula is variable too (Hubble 1916); this is the origin of the name Hubble's Variable Nebula.

In his paper, Schmidt also mentions the peculiar case of the Nova T Scorpii in the globular cluster M 80, which had been discovered by Auwers on 21 May 1859. He wrote *'I now can report a counterpart to this odd star.'* In 1862 Schmidt picked up the subject again. In his article 'Über veränderliche Nebelgestirne' he investigated NGC 1555 (Hind's Variable Nebula) and NGC 3662.[368] He was strongly interested in the case of the Merope Nebula too (Schmidt 1862c).[369]

---

[367] 'About a new variable nebulous star' (Schmidt 1861b).
[368] 'On variable nebulous stars' (Schmidt 1862b). See Sections 11.4 and 6.15.3.
[369] See Section 11.6.

Figure 6.57. Discovered by Schmidt: the remarkable trio of reflection nebulae NGC 6726, NGC 6727 and NGC 6729, associated with the variable star R Coronae Australis (DSS).

6.18.3 Schmidt's newly discovered nebulae

Schmidt's observations of nebulae naturally concentrated on the southern sky, using John Herschel's Cape catalogue as a reference. In early October 1861 he discovered his first object: NGC 6519 in Sagittarius – a star group only. He encountered it north of the globular cluster NGC 6522, which had been discovered by William Herschel in 1784 (I 49, h 3720). Schmidt's observation appeared, with some delay, in his report 'Beobachtungen der Nebelgestirne und des Faye'schen Cometen'[370] of 1865.

This report contains Schmidt's next find too: the remarkable trio NGC 6726, NGC 6727 and NGC 6729 in Corona Australis (Fig. 6.57), which is not listed in the Cape catalogue. NGC 6726/27 is a pair of reflection nebulae around stars of 7.3 mag and 9.5 mag, respectively (called A and B by Schmidt). The cometary nebula NGC 6729 (n) is associated with the variable star R CrA (x) at its northwest end (the duo was called nx). Schmidt's positions are pretty rough. He discovered the trio while observing the globular cluster NGC 6723 (h 3770 = Δ 573, Dunlop 2 June 1826), writing *'AB covered by strong nebula, whose diameter is 15^s [...] nx is a fine, but well-*

---

[370] 'Observations of the nebulous stars and Faye's comet' (Schmidt 1865c).

visible core-nebula'. The term 'core-nebula' describes an object with a compact centre (often a star). Schmidt detected another nebula near the bright double star C, 15′ southeast of A: *'Around C, whose components are of equal brightness, is a fainter nebula.'* Indeed, the double star is the centre of a large nebula discovered photographically on 4 August 1899 by Delisle Stewart with the Bruce astrograph in Arequipa (Peru). Stewart wrote *'not given in N.G.C., is involved in same nebula as N.G.C. 6726–7 and 6729. Whole region covered extending 2.2ᵐ in R.A., and 20′ in Dec.'*[371] The extended object, later catalogued by Dreyer as IC 4812 (*'star inv in eL neb'*), is much fainter than NGC 6726/27. Often pairs or compact groups of stars seem to have a nebulous envelope, which is an optical illusion. Therefore it is very doubtful that Schmidt really saw a bright part of IC 4812.

In mid 1866 Schmidt wrote a letter to the *Astronomische Nachrichten* about his discoveries in Corona Australis, which unfortunately got lost. He refers to it as *'letter of 1866 Sept. 1'* in a paper dated 5 October 1866 (Schmidt 1867a), but did not notice the loss until 1868. Immediately he put together the earlier results in a new report (Schmidt 1868a). There he noted the following about the nebula enclosing the stars A and B (NGC 6726/27): *'Both are surrounded by strong nebulosity, having a diameter of 14ˢ. 1861 June 15 I first detected the large conspicuous nebula around A and B, since then I never missed it.'* However, this discovery date of NGC 6726/27 differs from that in his earlier publication *AN* 1553 (Schmidt 1865c), which gives 8 February 1861. About the nebula around the double star C (IC 4812), whose components are now called c and c′, he wrote in his new report *'it has always remained improbable that there is a faint nebula around c c′'*. Schmidt also announced the variability of nx (NGC 6729 + R CrA): *'1865 Sept. 12 the variability of nx was detected, i.e. of both the nebula and the star. The nebula can totally disappear, and x can change from 11ᵐ to invisibility.'* He added that *'The strange group is a side piece to h. 3624 [M 80].'* A year earlier he had written (not knowing that his report of 1 September 1866 had been lost) that *'The nebula [n] and the star x west of it continue to be variable, so it seems, independently of each other.'* (Schmidt 1867a). The variability of NGC 6729 was later confirmed visually by Robert Innes (Reynolds 1916).

Further observations confirmed this behaviour (Schmidt 1869c, 1873). In mid 1876 Schmidt recognised that the star 5′ southeast of NGC 6729 is variable too; this is T CrA (called x′ by him). In his 1876 article, titled 'Ueber die Lage eines neuen veränderlichen Sterns …',[372] the situation is described in detail. The positions of the nebulae and stars in the field are now correct. However, the paper curiously presents a third variant of the discovery date for NGC 6726/27: October 1860. Anyway, 15 June 1861 is taken here as the most reliable one. The two variable stars x and x′ (R and T CrA) are again mentioned in Schmidt's last publication, which was written on 22 January 1884, only 16 days before his sudden death (Schmidt 1884). Incidentally, the remarkable trio NGC 6726/27/29 in Corona Borealis was found independently by Marth on 2 July 1864 (see Section 8.7.3).

Schmidt's 1865 paper 'Ueber die Beobachtung der Nebelgestirne'[373] lists four new nebulae near the bright galaxy M 49 in Virgo (Fig. 6.58), called n^1, n^2, n^3 and n^4 and described as *'exceedingly faint and small objects'*. As d'Arrest found out promptly, three of them are identical with objects of William Herschel: n^1 = II 18 (NGC 4470), n^3 = II 499 (NGC 4492) and n^4 = III 483 (NGC 4464). He observed them with the 11″ refractor, noting that they were *'bright and partially pretty large […] But at present n^2 is certainly invisible; there are only two very faint stars at its place.'* (d'Arrest 1865b). On the other hand, he was able to see a nebula (NGC 4467), which had been discovered in 1851 by Otto Struve, directly east of M 49. D'Arrest wrote that, because it *'was not noticed by Schmidt, the current invisibility of Schmidt's n^2 is evident.'* He also considers an identity with II 498, being aware that John Herschel had recently identified II 498 with II 18 (NGC 4470) in the General Catalogue. He formulates still another idea: *"It is not unlikely that all members of this beautiful group are merely light-knots of a large nebula [M 49], as this has been detected recently for similar objects. A temporary light variation of an individual 'knot' seems to be not unthinkable."* Schmidt does not give a clear discovery date for n^2 in his publication; probably it was 29 June 1861. Later the object, which is only a 14-mag star, was catalogued by Dreyer as GC 5655 (GCS) and NGC 4471.

---

[371] The nebula is HN 547 in Pickering E.C. (1908: 166).

[372] 'About the place of a new variable star…' (Schmidt 1876b).

[373] 'About the observation of nebulous stars' (Schmidt 1865a).

Figure 6.58. Galaxies around M 49 in Virgo (see the text; DSS).

On 10 October 1861 Schmidt discovered another object, NGC 32 in Pegasus. It is mentioned in connection with observations of comet Encke in his report 'Beobachtungen auf der Sternwarte zu Athen',[374] dated 1 March 1862. He determined the position of the supposed comet (called N) relative to a nearby star. The following day he noticed '*that N had been only a faint nebula, not the comet*'. Actually the object is a pair of stars. Auwers entered it as no. 1 in his list of new nebulae; the aperture of Schmidt's Plössl refractor is, however, incorrectly given as 7.3^z (19 cm). Table 6.26 lists all discoveries up to the end of 1861.

It is curious that Auwers knew of only Schmidt's last object (NGC 32). The reason is that it was published first! *AN* 1355 (Schmidt 1862a) appeared on 3 April 1862, not too late to be considered by Auwers for his list. Additionally, among Schmidt's first objects, NGC 32 is the only one appearing in the General Catalogue (GC 16); all of the others had to wait for Dreyer's GCS. This proved the deficient organisation at Athens Observatory – Schmidt's problem with trying to publish the Corona Australis discoveries is only one of many.

6.19 AUWERS' WORK 'WILLIAM HERSCHEL'S VERZEICHNISSE VON NEBELFLECKEN UND STERNHAUFEN'

Auwers' publication[375] of 1862 had a great influence on contemporary visual observers of nebulae, such as d'Arrest, Winnecke and Schönfeld. He was the first to present a reliable and comprehensive revision of William Herschel's three catalogues. It was, however, not based on his own observations. Auwers detected many errors in the catalogues. John Herschel, who was occupied with the compilation of the General Catalogue at that time, paid great respect to him too. Auwers was not only a man of theory, but a great observer too; his measurements of nebulae and star clusters, made in Göttingen and Königsberg, are among the most precise in the nineteenth century.

6.19.1 Origin and content

Auwers started in 1854, at the age of only 16, with the revision of Herschel's catalogues of 1786, 1789 and 1802. Including most of the nebulae later added by John

---

[374] 'Observations at Athens Observatory' (Schmidt 1862a).

[375] 'William Herschel's catalogues of nebulae and star clusters'.

Table 6.26. *Schmidt's first objects, those discovered by the end of 1861 (sorted by date)*

NGC	Date	Publ.	GC	Type	Con.	Remarks
6519	Oct. 1860	1865	5890	Star group	Sgr	
6726	15.6.1861	1865	5935	RN	CrA	Double nebula with NGC 6727, Marth 2.7.1864
6727	15.6.1861	1865	5936	RN	CrA	Double nebula with NGC 6726, Marth 2.7.1864
4471	29.7.1861	1865	5655	Star	Vir	Schmidt's object n^2 (14.0 mag); date doubtful
6729	15.6.1861	1865	5937	EN+RN	CrA	Associated with R CrA and T CrA, Marth 2.7.1864
32	10.10.1861	1862	16	2 stars	Peg	Au 1

Herschel (see Table 2.22), the total number of objects is 2510. The ambitious project was supported by the three-years-older Winnecke. Auwers' best friend later told about a '*just tackled excerpt of the Herschel catalogues contained in the* Phil. Trans. *and the Cape sojourn for private use*' (Winnecke 1863). A key event was Auwers' observation on 21 January 1854 of the large, faint galaxy NGC 6946 (IV 76) in Cygnus. Owing to its surprising visibility in a 7.4-cm comet-seeker by Merz, he decided '*to include also those faint nebulae estimated by Herschel to have diameters of several arcminutes [...], later experience has taught me that I have added with it nothing superfluous to the excerpt*'.

Undoubtedly, Herschel's catalogues were fundamental works about nebulae and star clusters, but hardly usable for practical purpose. Following Messier's precedent, the objects are sorted by discovery date. For each object only the coordinate differences relative to a reference star are given. This kind of presentation was just suitable for azimuthally mounted reflectors, as were used by both Herschels and by Lord Rosse, but relative positions were impractical for equatorial refractors equipped with setting circles.

The first to calculate coordinates for all Herschel objects was Bode, who published the results in the *Berliner Jahrbuch* (Bode 1788, 1791, 1804b). Beside Herschel's description, the right ascension and declination are given. However, the original order is preserved (the objects of each class are grouped in quintets). The next step was due to Caroline Herschel: in about 1827 she finished the 'zone catalogue', which gives the reduced positions of all objects for 1800. Though it was much better than Bode's work, it remained unpublished. John Herschel used the data for his 'working lists' to perform his systematic observations. The resulting Slough catalogue (SC) contains absolute positions (epoch 1830) for about two-thirds of his father's objects.

Not only because many Herschel nebulae are not listed in the SC, but also to get a unitary, up-to-date version of the three catalogues, Auwers wanted a complete revision. The result offered the chance to determine proper motions by comparing modern locations with those measured half a century ago. Another important aspect was identification. Owing to the lack of coordinates in the Herschel catalogues, this was a difficult task. Thus many observers eschewed such cumbersome work and identified objects with John Herschel's catalogues. No doubt many believed that they contained all known non-stellar objects. This often had the consequence that nebulae announced as new later turned out to be discoveries of William Herschel.

Auwers performed his work during 1854–56 in Göttingen. About the adverse conditions he wrote that, '*due to the nearly complete lack of literature, I could not use the* Phil. Trans. *for my reductions*'.[376] Thus the sources (*Philosophical Transactions*) were not at his disposal. He could use only the second edition of Wilhelm Pfaff's translation of Herschel's work, which appeared in 1850.[377] Auwers remarked about Pfaff that

---

[376] Auwers (1862a: 3).

[377] The first edition was published in 1826 in Leipzig (Pfaff 1826).

'The situation was insofar awkward, as the publication was defaced by a large number of errors – the position data contain around 350.'[378] When he eventually had access to the original publications, the results of his reductions could be checked. He was quite sure that the data *were brought into complete accordance with the* Philosophical Transactions *by repeated comparison and correction*'. For Auwers John Herschel's catalogues were of great value too, though he '*often detected large differences, whose reason I cannot explain*'. Since the Slough observations used Caroline Herschel's unpublished zone catalogue, Auwers suggests that '*perhaps Miss Caroline's reduction is based in many cases on other observations than those published*'. Concerning William Herschel's reference stars, Auwers (detecting a number of wrong designations) used the *British Association Catalogue* (BAC), the only star catalogue available in Göttingen. He determined precise coordinates, choosing the epoch of the Slough catalogue (1830). Afterwards the Herschel objects were (for the first time) sorted by right ascension. Early in March 1856, Auwers presented a first copy of the manuscript to his friend Winnecke in Berlin.[379] Later (in the publication of 1862) he proffered for his '*active support* [the] *most vivid thanks*'.[380]

At the same time, d'Arrest in Leipzig treated parts of the Herschel catalogues. He too was interested in the question of proper motion and criticised the useless publications of Bode and Pfaff. However, in contrast to Auwers, d'Arrest revised only the 366 objects of classes I ('bright nebulae') and IV ('planetary nebulae'). The reduced positions were published in 1856 in his 'Resultate aus Beobachtungen von Nebelflecken und Sternhaufen' (d'Arrest 1856a). D'Arrest was not aware of Auwers' work at that time (the same applies vice versa). Auwers later remarked that, '*When d'Arrest's work appeared, the complete reduction done by him was just finished, I had seen the need of such a task already in 1854; his wish, stated in his publication and supported by others, and the growing interest about nebulae during the last years, now induces me to use the opportunity, offered by Prof. Luther, to publish the reduced catalogue in the XXXIV. Abtheilung of the Königsberg Observations, to give to the astronomers this work which was primarily made for private use.*'[381]

Auwers' publication is dated 15 December 1861. Since autumn 1859 he had been working for Eduard Luther at Königsberg Observatory, who wrote in the preface that '*Mr A. Auwers, currently assistant at this observatory, already feeling the need for a complete reduction of William Herschel's observations of nebulae in 1854, initially realised this task for private use only. As now this highly meritorious work is suitable to remove the difficulties caused by the arrangement of the nebulae and star clusters in W. Herschel's three catalogues of the years 1786, 1789 and 1802 and hampering their practical use, an issue not being essentially improved by the existing revisions, I will present this editing of Herschel's observations of nebulae, meeting the wishes of several colleagues, as an appendix to the Königsberg Observations, together with the necessary preliminary remarks by Mr Auwers.*'[382] Certainly Winnecke was one of these 'colleagues'.

Auwers' work appeared in 1862 first as an appendix to the '34. Abtheilung der Königsberger Beobachtungen' and, later that year, as a book published by the Universitätsbuchhandlung Dalkowski, Königsberg. Auwers was unhappy about the latter edition, later writing that '*much to my regret and despite all due diligence in correcting the Königsberg catalogue of nebulae, a large number of defective types of the figure 4 is left during print, which, however, can always be distinguished from 1, but being aware of this issue, it needs special attention at some positions*'.[383] Like d'Arrest, Auwers estimated the precision of William Herschel's positions. About the instrumental conditions he remarked that '*The imperfect device and the whole mounting of the large telescope made it impossible to achieve a sufficient accuracy for the positions.*'[384] Herschel determined a position (while the object was centred in the field of view) by differential observations in the meridian, using reference stars. In the course of time he significantly improved his method. Thus Auwers distinguished three phases by allocating them different weights. The comparison with John Herschel's data (using d'Arrest's results) led to estimated positional errors (Table 6.27).

---

[378] Auwers (1862a: 3).
[379] Dreyer (1876a: 5).
[380] Auwers (1862a: 4).
[381] Auwers (1862a: 3).
[382] Auwers (1862a: preface (no pagination)).
[383] Auwers (1866: 178).
[384] Auwers (1862a: 4).

Table 6.27. *Estimated errors of William Herschels positions, according to Auwers*

Weight	Phase	Δα	Δδ
I	End of 1783 to end of 1784	8.3^s	1.45′
II	End of 1784 to autumn 1785	5.9^s	1.25′
III	Remaining time until 1802	4.8^s	1.05′

William Herschel's own values for phases I and II were even larger. Auwers lamented that there were no reliable data for objects observed by Herschel alone: '*Most of all, the publication of the original observations is desired for the sake of the remaining 650 nebulae; a large fraction of these objects are faint, thus our knowledge about their places might be limited for a long time to Herschel's determinations.*'[385]

The three Herschel catalogues contain many identification errors and typos (John Herschel is even worse). Auwers presents those found by himself, d'Arrest and Marth: the table lists 32 for William Herschel and 115 for John Herschel. In his revised catalogue, Auwers copied Herschel's coded English description. Moreover, he thought it to be '*far more comfortable than the cumbersome transcription by Pfaff*'.[386] Separate columns present the difference 'H. – h.' and 'synonyms', i.e. cross-identifications with the catalogues of d'Arrest, Laugier, Messier, Lacaille and Dunlop. Additionally there are notes for individual objects, a table of discovery dates and an object index, arranged by class and running number. Auwers completed his work by including Herschel's list of 52 regions with 'extensive diffused nebulosity'.[387] For the first time, coordinates (epoch 1830) and descriptions of these mysterious objects were made accessible (see Section 11.6.14).

Of special importance are the appendices containing the revisions of the catalogues of Messier and Lacaille and Auwers' list of new nebulae. The Messier catalogue with its 103 entries can be seen as the 'little brother' of Herschel's. There are certain similarities: both appeared in a series of three and entries are sorted by discovery date. Messier's coordinates refer to different epochs between 1758 and 1781. Auwers reduced them to 1830 and sorted the objects by right ascension. The two last entries are missing for an unknown reason: the controversial object M 102 (the galaxy NGC 5866) and the open cluster M 103. John Herschel had trouble with the latter: in the SC the object (h 126) is identified only with Wilhelm Struve's Σ 131.

Auwers compared the calculated positions with those of John Herschel, d'Arrest and Laugier. About Messier, Auwers remarks that his '*work is nothing but of historical interest, presenting the first systematic observations of nebulae*'. Because of the low instrumental and observational standard, he comes to the conclusion that '*Messier's places are not usable for the determination of proper motions*'.[388] Thus Auwers agrees with d'Arrest and Laugier. Argelander had criticised Messier's measurements too (Argelander 1868a). By the way, it is unknown whether Messier gave mean or apparent places; however, due to his low accuracy this is unimportant.

About Méchain's positions Auwers wrote that '*though Méchain's AR. might be better than Messier's, these data, based on so few comparisons, cannot be seen as reliable*'.[389] Since the data had been rated somewhat better by d'Arrest and Laugier, he added that '*One cannot even agree with d'Arrest that the mean values of Messier's and Méchain's positions give the places at that time to better than an arcminute.*' Auwers translated Messier's French descriptions into German and added some remarks. He could not have anticipated that it was precisely the Messier catalogue which would become one of the most popular – nevertheless he unintentionally supported its success.

Auwers similarly treated Lacaille's catalogue of 1755; all 42 southern objects are listed. Thomas Henderson had already reduced the positions during his time at the Cape Observatory (1831–33). Using

---

[385] Auwers (1862a: 5). Actually there are as many as 803 objects that were not re-observed by his son (see Section 3.2).
[386] Auwers (1862a: 7).
[387] Herschel W. (1811); see also Latußeck (2003a, b, 2004).
[388] Auwers (1862a: 70).
[389] Auwers (1862a: 71).

these data, Auwers calculated new coordinates for 1830 and sorted the objects by right ascension. Additionally, Lacaille's French descriptions were translated into German. In this case too it is Auwers' merit that this historical catalogue became widely known. Concerning the positions, he could compare them only with those in Herschel's Cape catalogue. Auwers concluded that, 'Considering that Lacaille has determined the places with imperfect micrometers at a telescope of only 6 lines aperture [13.6-mm refractor!] and 8-times magnification, one can easily agree with his opinion judging the observations as accurate as the instruments allow.'[390] It is surprising that Lacaille's errors are below 1′ – distinctly less than those of Messier and Méchain.

In 1863 a short review of Auwers' work appeared in the *Monthly Notices* (Anon 1863b). The writer stated that '*The absolute positions have been previously calculated by Bode and Caroline Herschel, and more recently as to two of the eight classes by D'Arrest, but a new and complete reduction of the observations appeared very desirable, and Mr Auwers has now availed himself of the opportunity given to him by Prof. Luther for the publication in the Königsberg* Memoirs *of the results calculated by him originally for his own use, about seven years ago.*'

6.19.2 Auwers' remarkable list of new nebulae

The most interesting appendix is Auwers' 'Verzeichniss neuer Nebelflecke' ['List of new nebulae'] with 50 entries. Except for two, it presents all objects discovered from 1824 to 1862 and not already contained in the catalogues of William and John Herschel. The coordinates are given for the epoch of the Slough catalogue (1830). Table 6.28 shows Auwers' list, enriched with modern data.

The object found last (Au 30 = NGC 4402) is due to Auwers himself. Because the introduction of his publication gives '*Königsberg, 1861 December 15*', it is evident that he compiled the list later and added it shortly before printing. Auwers could even include Schmidt's nebula NGC 32 (Au 1), published in *AN* 1355, which appeared on 3 April 1862 (Schmidt 1862a). On the other hand, Chacornac's nebula NGC 1988 (discovered on 19 October 1855) is missing: the discovery was not announced until 28 April 1863 (see Section 7.3.2).

In his publication Auwers speaks about 80–90 nebulae that had been indicated as being new by various observers (except John Herschel and Dunlop): '*Among them there are many that were erroneously thought to be new, namely those already observed by W. Herschel, and, of the nebulae which were really new to their discoverers, some were later listed in John Herschel's catalogues, so that in a compilation of all nebulae found since 1802 and not mentioned by h. or Δ* [John Herschel or Dunlop], *only 50 objects remain, of which 47 were discovered after 1845.*'[391] He presents the non-included objects in a separate account, distinguishing three different categories: (a) '*those nebulae, thought to be new by their discoverers, but later observed by h.*', (b) '*all other reported nebulae*' and (c) '*some nebulae noticed with meridian-circles but now missing*'.[392] This should be 30–40 objects. Table 6.29 lists 36 of them; Auwers' categories are separated by bold lines. Except for the objects of Petersen, Rümker and Maskelyne, all others are presented with references in the section of the particular observer.

Adolph Cornelius Petersen had found the open cluster NGC 2194 in 1849 with the 18-cm Repsold refractor at Altona Observatory (Petersen 1850). From 1827 he was employed as observer, and after the death of Heinrich Christian Schumacher in 1850 he succeeded him as director.[393] Auwers wrote the following about Petersen's nebula: '*At the position I have searched in vain, and the nebula is undoubtedly identical with H. IV. 5, thus Petersen's place is wrong by ⅓°.*'[394] Petersen was involved in the case of Secchi's double nebula NGC 3226/27 in Leo (see Section 6.11.3). In 1853 Secchi wrote him a letter reporting the discovery (Petersen 1853).

Auwers knew the two objects found by Charles Rümker from the Hamburg star catalogue (1843–42), where they are marked as 'nebula'.[395] Auwers could not see any trace of them in the Königsberg heliometer (Fig. 6.59) on 15 November 1861.[396] Rümker's son George gave his view on the issue in a letter to the *Astronomische Nachrichten* of 7 June 1865 (Rümker 1865a). Slightly incensed, he wrote '*It would be,*

---

[390] Auwers (1862a: 73).

[391] Auwers (1862a: 73).
[392] Auwers (1862a: 76–77).
[393] Obituary: Gould (1854).
[394] Auwers (1862a: 50).
[395] About this star catalogue, see Chambers (1890).
[396] Auwers (1865: 218).

Table 6.28. *The 50 objects of Auwers' list of new nebulae*

Au	NGC	GC	Observer	Date	Tel.	Type	V	Con.	Remarks
1	32	16	Schmidt	10.10.1861	6.2"	2 stars		Peg	
2	46	24	Cooper	22.10.1852	14"	Star	11.7	Psc	
3	109	54	d'Arrest	8.10.1861	11"	Gx	14.1	And	
4	223	119	Bond	5.1.1853	15"	Gx	13.4	Cet	IC 44, Swift 1890
5	370	197	d'Arrest	7.10.1861	11"	3 stars		Psc	NGC 372, Dreyer 12.12.1876
6	374	199	d'Arrest	7.10.1861	11"	Gx	13.5	Psc	
7	384	207	d'Arrest	7.10.1861	72"	Gx	13.0	Psc	W. Parsons 4.11.1850
8	385	208	d'Arrest	7.10.1861	72"	Gx	13.0	Psc	W. Parsons 4.11.1850
9	391	211	Bond	8.1.1853	15"	Gx	13.5	Cet	
10	443	249	d'Arrest	8.10.1861	11"	Gx	13.7	Psc	IC 1653, Javelle 17.10.1903
11	447	250	d'Arrest	8.10.1861	11"	Gx	13.6	Psc	IC 1656, Barnard 1888?
12	504	291 = 292	d'Arrest	8.10.1861	11"	Gx	13.0	Psc	J. Herschel (h 107) 22.11.1827
13	588	348	d'Arrest	2.10.1861	11"	GxP	13.5	Tri	H II region in M 33
14	592	349	d'Arrest	2.10.1861	11"	GxP	13.0	Tri	H II region in M 33
15	607	358	d'Arrest	23.8.1855	4.6"	2 stars		Cet	
16	629	373	W. Struve	1825	9.6"	Star group		Cas	
17	1333	710	Schönfeld	31.12.1855	3"	RN		Per	Tuttle 5.2.1859
18	1435	768	Tempel	19.10.1859	4"	RN		Tau	
19	1488	791	Cooper	24.11.1854	14"	2 stars		Tau	
20	1555	839	Hind	11.10.1852	7"	RN		Tau	Breen, summer 1856
21	2175	1366	Bruhns	1857		EN		Ori	
22	2218	1402	Cooper	13.1.1853	14"	Star group		Gem	
23	2248	1426	Cooper	23.12.1853	14"	Star group		Gem	
24	2399	1537	Bond	26.2.1853	15"	3 stars		CMi	
25	2400	1538	Bond	26.2.1853	15"	3 stars		CMi	
26	3121	2010	Lassell	31.3.1848	24"	Gx	12.9	Leo	
27	3222	2084	Winnecke	March 1855	9.6"	Gx	13.0	Leo	
28	4125	2735	Hind	5.1.1850	7"	Gx	9.6	Dra	
29	4205	2798	d'Arrest	4.10.1861	11"	Gx	13.0	Dra	
30	4402	2965	Auwers	5.3.1862	6"	Gx	11.8	Vir	
31	4989	3421 = 3426	Cooper	9.4.1852	14"	Gx	13.3	Vir	W. Herschel (II 185) 17.4.1784

Table 6.28. (*Cont.*)

Au	NGC	GC	Observer	Date	Tel.	Type	V	Con.	Remarks
32	5268	3633	Cooper	17.1.1855	14″	NF		Vir	
33	5632	3898	Bond	9.5.1853	15″	Star		Vir	
34	5651	3912	Bond	9.5.1853	15″	Star		Vir	
35	5658	3918	Bond	9.5.1853	15″	Star		Vir	
36	6366	4301	Winnecke	12.4.1860	3″	GC	9.5	Oph	
37	6503	4351	Auwers	22.7.1854	2.5″	Gx	10.2	Dra	
38	6535	4369	Hind	26.4.1852	7″	GC	9.3	Ser	
39	6539	4370	Brorsen	1856	3.8″	GC	8.9	Oph	
40	6643	4415	Schönfeld	1858?	3″	Gx	11.0	Dra	Tuttle 1.9.1859
41	6648	4419	W. Struve	1825	9.6″	2 stars		Dra	
42	6655	4423	Winnecke	June 1854	9.6″	2 stars		Sct	
43	6704	4435	Winnecke	23.7.1854	3″	OC	9.2	Sct	
44	6760	4473	Hind	30.3.1845	7″	GC	9.0	Aql	d'Arrest 1852
45	6791	4492	Winnecke	Dec. 1853	3″	OC	9.5	Lyr	Tuttle 17.7.1859
46	6859	4538	Bond	24.11.1852	15″	3 stars		Aql	
47	7122	4695	Cooper	24.11.1854	14″	2 stars		Cap	
48	7293	4795	Harding	Sept.? 1823	4″	PN	7.3	Aqr	Capocci 1824, Peters 1856
49	7447	4878	Cooper	8.10.1855	14″	NF		Aqr	
50	7548	4914	d'Arrest	30.9.1861	11″	Gx	13.4	Peg	

moreover, interesting to learn, if the remark of Dr Auwers merely aims to say that at the places given by my father no nebula exists, or, as I well-nigh guess, that at any time there was nothing to see at all.' He also criticised John Herschel, who '*has omitted* [both objects] *in his General-Catalogue, though he has listed a Bessel star* [of zone 527], *which is missing too, as No. 614*' (see Section 6.14.5).

George Rümker's attention was called to the two 'nebulae' by d'Arrest, who had searched for them in vain several times (during 1861–64) with the Copenhagen refractor.[397] Thereon Rümker reviewed his father's notes: observation no. 4542 was marked 'nebula'. The object is listed as an 8.8-mag star in the Bonner Durchmusterung (BD +45° 2144). For no. 8532 (a 9.6-mag star) he found the note 'Seq. Neb.' ['nebula following']. George Rümker suspected a typo: his father could have meant 'sequens Nord' ['north following']. This is due to a second observation by Charles Rümker on 28 October 1839, when the object was seen as a star '8″ *ad Bor. von 8532*' ['8 arcsec north of 8532']. George Rümker could confirm this on 8 October 1863 with the meridian-circle, finding the object in the micrometer eye-piece '*likewise as a star, but too faint to stand even a weak illumination of the field*'.

Nevil Maskelyne, Astronomer Royal in Greenwich from 1775 until his death in 1811, had detected a nebula while measuring stars on 14 February 1793. Auwers quoted him as follows: '*very plain notwithstanding the moonlight*'.[398] He noted that the nebula should '*precede a star by 9ᵐ 3ᵐ7ˢ,15 and 9′ 11″ south according to five comparisons*'. The reference star could be either BD −5° 536 (7.2 mag) or BD −5° 541 (7.1 mag); both had already been drawn in the *Berliner Akademische Sternkarte*. At the place in Eridanus no nebula was seen on 28 September 1861 with the Königsberg heliometer.[399] D'Arrest also mentions the case in an article titled 'Über einen angeblich von Maskelyne beobachteten,

---

[397] d'Arrest (1867a: 299, 341).

[398] Auwers (1862a: 77).

[399] Auwers (1865: 222).

Table 6.29. Objects not in Auwers' list (Disc. = discoverer)

Observer	Date	Disc.	Date	Identity	NGC	h	Type	V	Con.	Remarks
Harding (5)	<1824	Herschel	8.3.1784	VI 6	2355	439	OC	9.7	Gem	J. Herschel 25.5.1830
Struve (5)	1825				6210	1970	PN	8.8	Her	Bessel 1822; J. Herschel 9.5.1828
Struve (6)	18.7.1825				6572	2000	PN	8.1	Oph	
Harding (7)	<1824				6819	2048	OC	7.3	Cyg	C. Herschel 12.5.1784
Cacciatore	19.3.1826				6541	3726	GC	6.3	Cra	Dunlop (Δ 473) 2.6.1826; J. Herschel 1.6.1834
Harding (1)	<1824	Herschel	23.2.1791	VI 37	2506	480	OC	7.6	Mon	
Harding (2)	<1824	Méchain	April 1782	M 107	6171	3637	GC	7.8	Oph	W. Herschel (VI 40) 1793, Capocci 1826
Harding (3)	<1824	Messier	1.6.1764	M 14	6402	1983	GC	7.6	Oph	
Harding (6)	<1824	Herschel	9.9.1784	VIII 20	6885	2071	OC	8.1	Vul	NGC 6882, W. Herschel (VIII 22) 10.9.1784
Harding (8)	<1824	Herschel	11.9.1790	VII 59	6866	2066	OC	7.6	Cyg	
Capocci	1826	Méchain	April 1782	M 107	6171	3637	GC	7.8	Oph	W. Herschel (VI 40) 1793; Harding <1824
Struve (1)	1825	Legentil	29.10.1749	M 32	221	51	Gx	7.6	And	
Struve (3)	1825	Herschel	8.11.1801	I 286	3077	658	Gx	10.0	UMa	
Struve (4)	1825	Bode	3.2.1775	M 53	5024	1558	GC	7.7	Com	
Struve (8)	1825	Herschel	7.9.1782	IV 1	7009	2098	PN	8.0	Aqr	
Struve (9)	1825	Maraldi	7.9.1746	M 15	7078	2120	GC	6.3	Peg	
Bianchi	11.6.1839	Herschel	12.5.1787	IV 50	6229		GC	9.4	Her	
Bianchi	16.6.1839	Herschel	15.2.1786	IV 37	6543		PN	8.1	Dra	
Peters	1849?	Dunlop	10.5.1826	Δ 552	5986	3611	GC	7.6	Lup	J. Herschel 28.6.1834
Peters	1849?	Bode	17.12.1777	M 92	6341		GC	6.5	Her	
Peters	1849?	Dunlop	13.5.1826	Δ 557	6441	3705	GC	7.2	Sco	J. Herschel 28.6.1834
Peters	1849?	Herschel	22.5.1884	IV 12	6553	3730	GC	8.3	Sgr	
Peters	1849?	Messier	31.8.1780	M 69	6637	3747	GC	7.6	Sgr	Dunlop (Δ 613)
Peters	1849?	Dunlop	2.6.1826	Δ 573	6723	3770	GC	6.8	Sgr	J. Herschel 31.8.1834
Petersen	1849	Herschel	11.2.1784	V 15	2194		OC	8.5	Ori	d'Arrest 18.9.1862
Brorsen	1850	Herschel	1.1.1786	V 28	2024		EN		Ori	
d'Arrest	29.3.1856	Herschel	8.4.1784	II 114	4473		Gx	10.2	Com	
d'Arrest	29.3.1856	Herschel	8.4.1784	II 115	4477		Gx	10.4	Com	
d'Arrest	23.8.1855				7005				Aqr	Star group
d'Arrest	17.10.1855								Peg	3 stars
Secchi	6.3.1853	Herschel	15.2.1784	II 28	3226		Gx	11.4	Leo	Double nebula
Secchi	6.3.1853	Herschel	15.2.1784	II 29	3227		Gx	10.4	Leo	Double nebula
C. Rümker	27.5.1841			4542					CVn	BD +45° 2144
C. Rümker	15.8.1837			(8532)					Vul	Follows a star
Bessel	8.11.1832								Per	In B.Z. 527, BD +36° 567, GC 614
Maskelyne	14.2.1793								Eri	Comet 1792 II

185

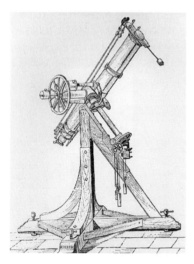

Figure 6.59. Auwers' main instrument, the 15-cm Fraunhofer heliometer at Königsberg Observatory.

Figure 6.60. M 86 and Olbers' nebula NGC 4402 in Virgo (DSS).

gegenwärtig unsichtbaren Nebelfleck',[400] giving a surprising explanation: *'one of the suspected nebulae from Auwers' list should be deleted, because it can evidently be shown that it was the second comet of 1792.'*

In a paper, he wrote on 21 March 1863 in Gotha, Auwers treated a nebula in Virgo (Auwers 1863). It was found in 1802 by Olbers and described as *'just as conspicuous as No. 143 Bode Virginis'*[401] (Olbers 1802). D'Arrest thought it to be identical with Auwers' discovery, listed as no. 30 (NGC 4402, Fig. 6.60). Auwers had indicated the find in *AN* 1391 (Auwers 1862b). For d'Arrest the case was a typical identification problem: *'How difficult it is, especially in this matter, to mind falsities, shows for instance the 9th nebula found in Königsberg, indicated in Astr. Nachr. No. 1391. Its place matches within a few arcminutes that of an object already discovered by Olbers, which is, however, overlooked in Auwers' list of new nebulae.'*[402] Auwers rejected d'Arrest's claim, stating that Olbers' nebula is nothing but M 86 (NGC 4406). Thus it has no right to be in his list. But things are different for his own discovery: *'the nebula listed as No. 30 [NGC 4402] is not identical with it [M 86] and, as far as I know, has been not observed previously. It is, by the way, very faint and was hardly visible in Olbers' instruments.'*

The identity of Harding's no. 5 (h 439, NGC 2355) with Herschel's VI 6 is only suspected by Auwers: *'I found this nebula 1858 February 10. Near the place of h.'s [h 439]; against it, at H.'s [VI 6] there was nothing to see in a 6f. Fraunhofer, so that VI. 6. is probably = h. 439, and H.'s AR. is wrong.'*[403] Auwers lists the globular cluster NGC 6171 twice (Harding no. 2, Capocci 1826), correctly identifying it as Herschel's VI 40. About Capocci's find in 1826, he merely remarks that the object had already been seen by Méchain (it is known as M 107 today).

However, Auwers did not include a nebula in Aquila that had been found by Charles Dien on 31 March 1855; probably the publication was not known to him (Dien 1855). The French astronomer reported that *'In the night of March 31 I observed a large nebula near 16h 02m [right ascension] north of ζ and ε Aquilae,*

---

[400] 'About a nebula ostensibly observed by Maskelyne, which is currently invisible' (d'Arrest 1863c).

[401] A star of Bode's *Catalogue de l'ascension droite et de la déclinaison de 5505 étoiles*, published in 1805 in Berlin, is meant; see Chambers (1890).

[402] d'Arrest (1863a); it is unknown what is meant by the '9th nebula'.

[403] Auwers (1862a: 51).

below the stars of the tail feather, which I had not noticed before and which is not contained in the charts and catalogues.' Actually, there is no nebula at that place.

6.19.3 A list of places for 40 nebulae

Auwers also tried to determine proper motions of nebulae in Königsberg, using the 6″ Fraunhofer heliometer (Auwers 1862c). The work was based on precise positions of reference stars measured by Johann Sievers[404] with the Reichenbach meridian-circle. Auwers determined the relative positions either directly with the heliometer (magnifications of 155 and 179) or by recording transits with a ring-micrometer (magnification 65). In relation to earlier results he stated '*that there are differences between earlier observers, which are considerably larger than that expected from the probable error of a single observation*'. Facing such '*systematic errors of the previous campaigns*' he pleaded for a limitation of the circle of observers. About his own task, he believes that it would be '*not to do anything in vain, if he spend the time left by regular observations at the Königsberg heliometer for most accurate positional measurements of a couple of nebulae*'. Auwers selected 40 objects (galaxies, planetary nebulae, open and globular clusters) and made 160 position measurements.

The accuracy of the reduced coordinates (epoch 1860) was calculated by Auwers to be 0.46″ to 2.16″, depending on the object's appearance (best are 'core-nebulae' showing a sharply defined centre). Compared with earlier campaigns, his data are extremely exact. Auwers set new standards, founding his reputation as a leading exponent of astrometry.

Many objects had already been measured by Lamont, Laugier and d'Arrest. Auwers derived the differences from Laugier and d'Arrest (18 and 34 common objects, respectively). His result gives the probable error of a nebula position:

- Laugier: right ascension = ±7.19″ sec δ and declination = ±4.47″
- d'Arrest: right ascension = ±4.75″ sec δ and declination = ±7.47″

Given the '*poorness of the instrument*' used by Laugier (a 9.7-cm Gambey refractor), Auwers certifies the French data as considerably accurate. Concerning d'Arrest's astonishing declination error, Auwers suggests that improper positions of the reference stars are to blame. Since they were not given, Auwers tried to guess these stars in critical cases. An example is the measurement of M 65 and M 66: '*d'Arrest has probably used W. 11,103, whose AR is found to be 0.70 smaller by new observations*'.[405] He therefore criticised d'Arrest: '*A catalogue of nebulae must contain the positional data of the reference stars to attain its maximum value.*'

Auwers published the complete data of the 40 nebulae which he had measured during 1859–62 with the heliometer in the '35. Abtheilung der Königsberger Beobachtungen' (Auwers 1865). Some objects were observed up to eight times. Additionally, reports about observations of 'new nebulae' contained in his list of 1862 are given. The work shows Auwers' enormous diligence. Afterwards he turned away from nebulae. His new task became 'fundamental astronomy', which deals with the creation of a highly accurate reference system based on measured star positions.

---

[404] From 1859 to 1865 Sievers was observer at Königsberg Observatory.

[405] Here the zone observations of Maximilian von Weisse (at Krakau, now called Kraków), which were published in 1846, are meant.

7 • Compiling the General Catalogue

John Herschel's 'Catalogue of nebulae and clusters of stars' – commonly known as the General Catalogue (GC) – was published in 1864 in Vol. 154, Part I, of the *Philosophical Transactions of the Royal Society* (Herschel J. 1864). At that time Herschel was already 71. After years of low astronomical activity, he had decided to collect all known nebulae and star clusters in a common catalogue. The work started in about 1859 and was finished on 23 June 1863.[1] This date refers to the main catalogue; the added 'Supplementary list' was not made until October 1863. Later in 1864 a hardcover version was published by Taylor and Francis, London. The GC differs from Herschel's earlier Slough and Cape catalogues (SC and CC), which mainly contain data based on his own observations. In the case of the SC (1833) his goal was a re-examination of the objects contained in the catalogues of his father, using Caroline Herschel's reduction. The CC (1847) lists, with the exception of Dunlop's objects, mainly his own southern-sky discoveries. On the other hand, the GC presents all non-stellar objects found by 1863 for both hemispheres – therefore, it is a 'database' in the modern sense. As shown in the following sections, the General Catalogue lists many new objects. They are mainly due to observations at Birr Castle (Lord Rosse and his assistants), Harvard (Bond, Coolidge, Safford), Paris (Chacornac), Copenhagen (d'Arrest) and Malta (Lassell). The last section describes the content and structure of the General Catalogue.

7.1 LORD ROSSE'S PUBLICATION OF 1861

Lord Rosse's final and most important publication (following those of 1840, 1844 and 1850) appeared in late 1861 in the *Philosophical Transactions*, titled 'On the construction of specula of six-feet aperture; and a selection from the observations of nebulae made with them' (Parsons W. 1861a).[2] It was received by the Royal Society on 5 June 1861; at that time Lord Rosse was nearly 61 years old. A short summary of the work was published during the same year as 'Further observations upon the nebulae, with practical details relating to the construction of large telescopes' (Parsons W. 1861b). There is also a comment by Thomas Backhouse on 'Lord Rosse on the nebulae', which was aimed at amateur astronomers (Backhouse 1863). Therein he complains that Lord Rosse's work '*has hardly received its attention as it deserves, and probably may have escaped some of our amateur friends altogether*'.

7.1.1 Structure and content

Lord Rosse's earlier publications were focused on telescopes and their construction. Though observations and drawings of nebulae were presented too, they are often mere examples to demonstrate the instrumental power. His final work is quite different. Nevertheless, the first 20 pages are allocated to a technical description of the 72-inch (illustrated by 11 figures; Fig. 7.1 shows an example). It is interesting to learn that occasionally a '*silvered speculum [...] for special purposes*' was used.[3] The metal secondary mirror was replaced by a silvered glass one. It was applied, for instance, on 15 November 1857 when observing h 69/70 (NGC 274/5): '*the silvered mirror shows the object brighter than before, but no new details*'.[4]

The new publication collects all of the objects which had been selected from John Herschel's Slough catalogue and observed by 1858. The observations are

---

[1] Buttmann (1970: 158).

[2] It is often cited as '*PT* 1861'.
[3] Parsons W. (1861a: 188).
[4] Citations referring to a certain h-number can be localised under that number in Lord Rosse's publication.

7.1 Lord Rosse's publication of 1861

Figure 7.1. Lord Rosse with employees at the 72″ reflector (Parsons W. (1861a), Fig. 5).

sorted by h-number.[5] Lord Rosse noted about the sample that '*we have examined all the brighter known nebulae except a few in the neighbourhood of the pole, and a great portion of the fainter nebulae*'.[6] He was mainly interested in peculiarities, such as '*the convolution of a spiral, dark lines, or dark spaces*' and the case of core nebulae: '*A star may have been mistaken for a condensed nucleus, or the reverse; and it is impossible to say which of the two suppositions is the more probable*'.[7] Such objects were studied in detail and drawings were produced: '*the more remarkable objects were selected for examination on favourable nights, when the details were carefully filled in, sometimes with the aid of the micrometer.*'[8]

For each observation a 'working list' was created, as later described by Robert Ball: "*The 'working list', as it is called, contains a list of all the nebulae which we want to observe. A glance at the book and the chronometer shows which of these is coming into the best position at the time. The necessary instructions are immediately given to the attendants. The observer, standing at the eye-piece, awaits the appointed moment, and the object comes before him. He carefully scrutinises it to see whether the great telescope can reveal anything which was not discovered by instruments of inferior capacity. A hasty sketch is made in order to record the distinctive features as accurately as possible. One beautiful object having been observed, the telescope is moved back to the meridian to be ready for the next vision of delight.*"[9]

Though the 72-inch was able to perform Herschel-like sweeps, there was no explicit search for new nebulae, but '*very many, however, have been found accidentally in the immediate neighbourhood of known nebulae, but for the most part they were faint objects presenting no features of interest*'.[10] One further reads that '*their places have been unusually entered roughly in the observing book, and a slight diagram made in the margin, so as to ensure their being easily found again*'.[11] Lord Rosse added that '*we have only entered observations when the micrometer was employed*'. Unfortunately, relative positions are missing for many discoveries, thus later identification (e.g. by John Herschel) was not always easy and led to several errors. Like the less interesting, brighter objects, the astronomers observed new ones only on one good night. Time was simply too valuable at a site with such problematic weather.

The site conditions are particularly mentioned: '*Here in winter the finest definition we have, and the blackest sky, is usually before eleven o'clock, after which the sky becomes luminous, and the fainter details of nebulae disappear. In spring and autumn the change is neither so early nor so decided; but the nights are shorter.*'[12] Later Robert Ball wrote the following about the issue: '*Birr Castle is not an ideal place for an observatory. It is near the Bog of Allen. Consequently, the skies were frequently overhung with clouds, to the distraction of the astronomer.*'[13] A year thus offered only a few good nights – a reason for Lord Rosse not to protract his gathering of results any longer: '*the progress made is necessary very slow, that I think it would be inexpedient longer to keep back this paper in the distant hope of making it in some respect more complete*'.[14] After all, it took three years from the last reported observation (May 1858) to the publication (June 1861). The delay was due to Lord Rosse's efforts at finding a successor for

---

[5] In his text, Lord Rosse uses 'H' instead of 'h'.
[6] Parsons W. (1861a: 125).
[7] Parsons W. (1861a: 147).
[8] Parsons W. (1861a: 125).
[9] Ball V. (1915: 67–68).
[10] Parsons W. (1861a: 125).
[11] Parsons W. (1861a: 146).
[12] Parsons W. (1861a: 125).
[13] Ball V. (1915: 66).
[14] Parsons W. (1861a: 126).

Mitchell. He eventually got a new assistant in February 1860: Samuel Hunter. The idea was that he should update and enlarge Stoney's drawing of the central part of the Orion Nebula of 1852. Hunter's drawing (see Fig. 8.61) was eventually published, with some additions by Lawrence Parsons, in 1868 (Parsons L. 1868). His other observations of 1860–61 did not appear until 1880 (Parsons L. 1880).

Unfortunately the quality of the 72″ mirror deteriorated over the years, never again reaching that of 1845: '*In the early observations with the 6-feet telescope we had the advantage of a very fine speculum.* [...] *In the mean time the speculum, which had been frequently dewed and occasionally cleaned, had lost its fine edge* [and it] *has lost much of its power*'.[15] This had consequences for practical observing: the goal had changed from 'resolving' nebulae, which needed the best conditions, to observing their shape: '*The question of resolvability, therefore, I think, must remain to be taken separately, when the finest instrumental means are available.*'

On 27 pages observational data, textual descriptions and small sketches are presented (in tabular form). The sketches serve for orientation, identification or presentation of special features, e.g. spiral structure. In nearly all of these sketches the orientation (N–S) is inserted. The main part is followed by 7 pages with 43 drawings of individual objects. Descriptions and sketches were taken from the original observing notes located '*in books, in which they were entered each night*'.[16] Lord Rosse adds that '*from time to time they were copied into a folio in the order of right ascension; and of the folio a copy was made for ordinary use in the Observatory*'. Absolute positions are not given (for the bulk of known objects there was no need for it). The work contains 989 different objects from the Slough catalogue. Many were observed several times: the record holder is the Orion Nebula (M 42, h 360) with 43 observations, closely followed by the Owl Nebula (M 97, h 838) with 42. The text presents only a selection ('*only a few good observations embodying the whole information we had been able to obtain*'[17]); besides the first observation, mainly those in which new results or certain changes appeared are given.

Unfortunately, the observer's name is rarely mentioned; sometimes even a date is missing. The first observation is dated 5 March 1848 (NGC 3310, h 731), the last 12 May 1858 (NGC 5775, h 1885). The collection covers a period of more than 10 years; Lord Rosse merely speaks of '*about seven years*'.[18] It starts with his first scientific assistant, Rambaut, and ends with the departure of Mitchell, who worked at Birr Castle from December 1853 to March 1858 (Table 6.4).

Many observations (not to say most of the later ones) were made by Lord Rosse's assistants; the head was strongly engaged in social duties and became too old for the arduous nightly task. He remarks that '*Though so many of the observations were made in my absence, they are not the less to be relied on: nothing was done by an unpractised hand, and no pains were spared to ensure accuracy.*'[19] This explains, too, why there were no observations in 1853: there was simply no scientific assistant. Lord Rosse attached great value to precise observing and exact documentation. Obviously, he could fully trust his assistants for this task: '*I refer with as much confidence to the observations of the two Mr. Stoneys and Mr. Mitchell as if I had on every occasion been present myself, because I know that they had thoroughly mastered the instrument and the methods of observing before they recorded a single independent observation; they were, besides, eminently cautious and painstaking.*' Only occasionally are there hints of team work, e.g. on 17 April 1855, when Lord Rosse noted about the observation of h 1357 (NGC 4565) that '*Mr. Stoney was with me*'. In April 1855 Johnstone Stoney had returned to Birr Castle for a short time.

7.1.2 Birr Castle drawings

The published drawings were made between 1849 and 1861 (Table 7.1). Besides single objects, pairs of galaxies and the galaxy group around NGC 70 in Andromeda are shown too. Some drawings were later analysed by Holden, who was trying to determine the true geometrical form of 17 'helical nebulae' (Holden 1889b); the objects are marked *.

Lord Rosse remarked about the drawings that '*they usually represent the objects a little stronger than they appear on an ordinary night, but not stronger than*

---

[15] Parsons W. (1861a: 147).
[16] Parsons W. (1861a: 149); see also Bennett and Hoskin (1981).
[17] Parsons W. (1861a: 146).
[18] Parsons W. (1861a: 125).
[19] Parsons W. (1861a: 148).

7.1 Lord Rosse's publication of 1861

Table 7.1. *Drawings in the 1861 publication*

Pl.	Fig.	NGC	H	M	Date	Drawer	Remarks
25	1	67, 68, 69, 70, 71, 72, 74	(15)		7.10.1855	Mitchell	
25	2	693	156		24.11.1851	B. Stoney	
25	3	972	232		12.10.1855	Mitchell	
25	4	1012	241		23.11.1857	Mitchell	
25	5	1023	242		27.12.1850	B. Stoney	
25	6	1068	262	77	24.11.1851	B. Stoney	Plate 1, Fig. 4 in Parsons L. (1880)
25	7	1514	311		9.1.1858	Mitchell	Hunter
25	8	1579	315		13.1.1858	Mitchell	
25	9	1637	327		28.12.1856	Mitchell	*
26	10	598	131	33	18.12.1857	Mitchell	
27	11	2245	393		28.2.1850	J. Stoney	B. Stoney
27	12	2316, 2317	421		23.11.1851	B. Stoney	
27	13	3184	688		1.2.1856	Mitchell	
27	14	3190	692		22.3.1857	Mitchell	
27	15	3395, 3396	765, 766		9.2.1855	Mitchell	*
27	16	3627	857	66	17.4.1849	J. Stoney	h 875 *
27	17	3953	1011		19.4.1857	Mitchell	
27	18	4038, 4039	1052, 1053		14.4.1852	B. Stoney	*
27	19	4051	1061		3.5.1851	B. Stoney	*
27	20	4151, 4156	1111, 1113		15.4.1851	B. Stoney	
27	21	4303	1202	61	7.3.1851	B. Stoney	*
27	22	4389	1245		30.3.1856	Mitchell	
27	23	4485, 4490	1306, 1308		27.3.1856	Mitchell	*
28	24	4536	1337		29.5.1856	Mitchell	*
28	25	4618	1385		27.3.1856	Mitchell	*
28	26	4656, 4657	1414, 1415		26.4.1851	B. Stoney	*
28	27	4710	1441		17.2.1855	Mitchell	
28	28	5112	1589		19.4.1857	Mitchell	*
28	29	5248	1650		29.3.1856	Mitchell	*
28	30	5378	1713		17.4.1855	Mitchell	
28	31	5857, 5859	1905		8.5.1861	Hunter	
28	32	6058	1946		9.5.1861	Hunter	*
28	33	6205	1968	13	26.5.1851	B. Stoney	
28	34	6905	2075		12.8.1855	Mitchell	
29	35	5457	1744	101	29.4.1861	Hunter	

Table 7.1. (Cont.)

Pl.	Fig.	NGC	H	M	Date	Drawer	Remarks
30	36	6946	2084		8.9.1850	B. Stoney	*
30	37	7008	2099		19.8.1855	Mitchell	
30	38	7177	2139		31.8.1854	Mitchell	
30	39	7331	2172		16.9.1854	Mitchell	
30	40	7662	2241		16.9.1852	B. Stoney	
30	41	7678	2245		17.10.1854	B. Stoney	Mitchell *
30	42	7814	2297		12.10.1855	Mitchell	Plate 5, Fig. 11 in Parsons L. (1880)
30	43	6853	2060	27	24.8.1851	B. Stoney	

on a fine night, when the air is clear and the sky black'.[20] He further wrote that *'Most of them have been repeatedly compared with the objects by different persons, and some have been several times sketched independently; so that I trust they are upon the whole accurate.'* He was not satisfied by the reproductions (*'many of the principal stars are too large'*), but did not see the reason in the gravure.

Most of the 43 drawings were made by Mitchell (24), followed by B. Stoney (14), Hunter (3) and J. Stoney (2). Lord Rosse lists the drawers in a table;[21] three assignments are incorrect (see 'Remarks'). For J. Stoney's drawing of M 66 (h 857) one reads 'h 875' – the error is due to John Herschel (see Section 3.2). In some cases the date can be derived only from Lawrence Parsons' publication of 1880. Owing to the good weather conditions in spring, one third of all drawings were made in March.

Three drawings were not made until April and May 1861 by Hunter. Shortly after, Lord Rosse sent the manuscript to the Royal Society, which received it on 5 June. About Hunter's drawings he notes (considering the fact that the objects had already been drawn by other assistants) *'We preferred Mr. Hunter's sketches, thinking they were upon the whole the most accurate, containing some additional details.'*[22] Hunter had graduated from the Dublin School of Design, and had been particularly engaged in February 1860 for a detailed drawing of the Orion Nebula. As early as in 1852 Bindon Stoney had drawn the inner part. Though Hunter's drawing was finished in 1864 (see Fig. 8.61), shortly before he left Birr Castle in May, it was published after Lord Rosse's death by his son Lawrence (Parsons L. 1868).

7.1.3 Lord Rosse's spiral nebulae

For 76 objects spiral structure was detected or supposed (Table 7.2); the terms 'spiral', 'branches' and 'arms' are used.

For some objects there is a sketch or even a drawing (the number is given under 'Fig.'); '1850' refers to Lord Rosse's publication of 1850 and 'JH' to the list of special Birr Castle objects in the General Catalogue.[23] Obviously, Herschel has entered only a few cases, according to Lord Rosse's descriptions. For GC 3751 he mixed up the galaxies h 1734 and h 1735, which was corrected in the NGC. The GC list contains three other 'spirals', that were not assigned as such by Lord Rosse: h 1196, h 1368 and h 4815. NGC 5985 (II 766) is listed under 'h 1934'. It is interesting to compare Table 7.2 with an analysis by Dewhirst and Hoskin, who left out a number of objects (Dewhirst and Hoskin 1991). Another list, which differs too, was compiled by Thomson (Thomson 2001).

The high hit rate is surprising: in most cases the object is indeed a spiral galaxy (type 'S'). For NGC

---

[20] Parsons W. (1861a: 148).
[21] Parsons W. (1861a: 149).
[22] Parsons W. (1861a: 149).

[23] Herschel J. (1864: 44).

Table 7.2. *Objects with spiral structure in the 1861 publication*

NGC	M	h	GC	Con.	Type	Sketch	Fig.	1850	JH
108		21	53	And	S				
205	110	44	105	And	E				
210		46	107	Cet	S				
278		71	158	Cas	S				
337		80	185	Cet	S	*	10		
598	33	131	352	Tri	S	*			
628	74	142	372	Psc	S			*	*
772		181	463	Ari	S				
1032		246	581	Cet	S				
1068	77	262	600	Cet	S		6	*	*
1514		311	810	Tau	PN		7		
1637		327	888	Eri	S		9	*	*
2068	78	368	1267	Ori	RN	*			
2537		491	1629	Lyn	S				
2608		512	1670	Cnc	S				
2776		563	1772	Lyn	S				
2903		604	1861	Leo	S				
3162		682	2034	Leo	S				
3184		688	2052	UMa	S	*	13		
3198		695	2066	UMa	S			*	*
3310		731	2158	UMa	S				
3344		739	2178	LMi	S				
3351	95	743	2184	Leo	S				
3367		748	2193	Leo	S				
3368	96	749	2194	Leo	S			*	*
3395		765	2216	LMi	S		15		
3485		804	2273	Leo	S				
3504		810	2287	LMi	S				
3510		813	2293	LMi	S				
3596		841	2350	Leo	S				
3623	65	854	2373	Leo	S				*
3631		858	2379	UMa	S	*			*
3646		866	2389	Leo	S				
3726		910	2445	UMa	S	*		*	*
3810		943	2499	Leo	S				*
3893		982	2559	UMa	S	*			*
3938		1002	2597	UMa	S	*		*	*
4017		1043	2655	Com	S	*			
4038/9		1052/3	2670/1	Crv	S		18		*
4051		1061	2680	UMa	S	*	19		*
4102		1085	2717	UMa	S	*			*
4123		1092	2733	Vir	S	*			*
4151		1111	2756	CVn	S	*	20		
4254	99	1173	2838	Com	S			*	

Table 7.2. (Cont.)

NGC	M	h	GC	Con.	Type	Sketch	Fig.	1850	JH
4303	61	1202	2878	Vir	S		21		
4321	100	1211	2890	Com	S			*	*
4414		1258	2972	Com	S	*			
4485		1306	3041	CVn	I		23		
4501	88	1312	3049	Com	S			*	*
4536		1337	3085	Vir	S		24		
4618		1385	3151	CVn	S		25		
4625		1392	3160	CVn	S		25		
4639		1403	3173	Vir	S				
4689		1431	3219	Com	S				
4725		1451	3249	Com	S	*		*	*
4736	94	1456	3258	CVn	S				*
5020		1556	3450	Vir	S				
5055	63	1570	3474	CVn	S			*	*
5112		1589	3511	CVn	S		28		
5194	51	1622	3572	CVn	S	*		*	*
5195		1623	3574	CVn	S	*		*	
5248		1650	3615	Boo	S	*	29		
5427		1735	3751	Vir	S	*			(*)
5457	101	1744	3770	UMa	S	*	35		
5468		1745	3777	Vir	S				
5557		1776	3843	Boo	E			*	*
5713		1857	3964	Vir	S				
5921		1919	4097	Ser	S				
5985			4132	Dra	S	*			
6218	12	1971	4238	Oph	GC				
6781		2037	4487	Aql	PN				
6905		2075	4572	Del	PN	*	34		*
6946		2084	4594	Cyg	S	*	36		
7662		2241	4964	And	PN		40		
7678		2245	4971	Peg	S		41		*

4151, the prototype of a Seyfert galaxy, the spiral arms are difficult to see, because the compact core dominates.[24] Most interesting are the exceptions: NGC 205 (elliptical galaxy, 'E') *'spirality suspected'*; NGC 1514 (planetary nebula) *'new spiral of an annular form round the star, which is central; spirality is very faint'*; NGC 2068 (reflection nebula) *'spiral arrangement sufficiently seen'*; NGC 4485 (irregular galaxy, 'I') *'suspected spirality'*; NGC 6218 (globular cluster) *'in the finder eyepiece the branches have a slight spiral appearance'*; NGC 6781 (planetary nebula) *'annular or perhaps spiral'*; NGC 6905 (planetary nebula) *'this planetary neb. is a beautiful little spiral'* – Schultz later wrote (observation of 14 September 1864) that *'the appearance gave no reason to think about such a form'* (Schultz 1865b); and NGC 7662 (planetary nebula) *'the outlying portions in the published sketch are parts of spiral branches'*. Lord Rosse's note on the elliptical galaxy NGC 5557 merely gives *'frequently observed; nothing certain'*. It is mentioned here

---

[24] Another example is NGC 1068 (M 77), which had already been revealed as 'spiral' in the 1850 publication (see Table 6.7).

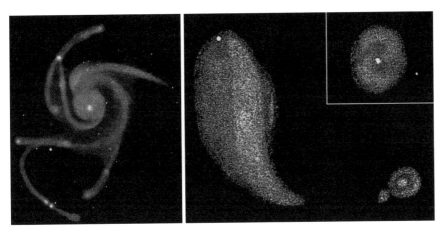

Figure 7.2. Lord Rosse's 'spiral nebulae'. Left: M 101, drawn by Hunter; right: the galaxy pair NGC 4485 and NGC 4490 (the larger object); inset: the planetary nebula NGC 1514.

because the 1850 publication states 'spiral or curvilinear' (see Table 6.7). NGC 2537 is called the Bear Paw Galaxy; it is a peculiar galaxy (Arp 6) in Lynx, showing no significant spiral structure.[25] Figure 7.2 shows examples of true and supposed spiral nebulae; NGC 1514 is William Herschel's 'star with an atmosphere' (see Section 2.6.2).

Two objects, indicated as being 'spiral or curvilinear' in 1850, are missing from Table 7.2: NGC 4578 (M 58, h 1368, GC 3121) and NGC 7331 (h 2172, GC 4815) – both are spiral galaxies. It is not known why the structure was not mentioned by Lord Rosse in 1861.

In 1860 George Phillips Bond published an article with the curious title 'On the spiral structure of the Great Orion Nebula' (Bond G. 1860). Like his father William Cranch Bond before, he had intensively observed M 42 with the 15″ Merz refractor of Harvard College Observatory. He wrote that "*It may, in fact, be properly classed among the 'spiral nebulae' under the definition given by their first discoverer, Lord Rosse; including in the term all objects in which curvilinear arrangement, not consisting of regular re-entering curves, may be detected.*" One further reads that '*The change from the previous notion of its configuration is not more considerable than that which took place with reference to the celebrated nebula 51 Messier.*' Bond could confirm the 'spiral structure' of the globular cluster M 13 too, which had been described by Lord Rosse in his 1850 publication as '*an unquestionable curvilinear sweep in the disposition of its exterior stars*'. Later Harrington treated this strange case too (see Section 9.17).

7.1.4 Micrometric measurements

A few objects, such as large bright nebulae, were measured with the micrometer at Birr Castle, accurately determining stars and knot structures. Lord Rosse reported that '*The powers used are very low, the ordinary working-eyepiece: with high powers the faint details vanish.*'[26] This was the reason to refrain from illuminating the field, using bars instead, which are reasonably visible in the dark. Lord Rosse presents Bindon Stoney's data for M 51 and M 27 (the Dumbbell Nebula) from 1850–51 and compares them with those of Johnstone Stoney (for M 51 only, first published in 1850) and Otto Struve's measurements with the 15″ refractor at Pulkovo. Owing to the lack of precise tracking, Lord Rosse concedes that '*Our measures of stars cannot therefore in accuracy compete with Struve's.*'[27] However, the much larger aperture had some advantages: '*we may the better see the faint details of the outlying portions of the nebulae*'. For M 51 and M 27 he noticed satisfying agreement between Struve and Bindon Stoney.

Later, no further detailed measurements were performed; the micrometer was used only to determine

---

[25] Steinicke (2002b); the name is due to Ron Buta (Glyn Jones 1981: 84).

[26] Parsons W. (1861a: 149).
[27] Parsons W. (1861a: 187).

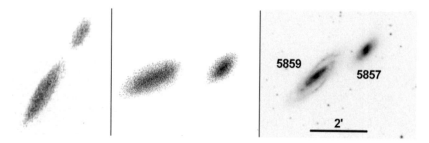

Figure 7.3. The double galaxy NGC 5857/59 in Bootes as seen by John Herschel (left) and Hunter (centre); right: DSS image.

position angles and distances between nebulae or to nearby stars. Lord Rosse wrote that *'There are no micrometer observations by Mr. Mitchell: I now regret it, as several cases of suspected change have recently been brought to light in arranging the materials of this paper. The fault, however, was mine. It appeared to me so highly improbable that any change would be detected, that I requested Mr. Mitchell to press on and not spend time on the micrometer.'*[28] As an example, he mentions h 1905 (Fig. 7.3 left), the double galaxy NGC 5857/59 in Bootes. Lord Rosse wrote the following: *'Herschel gives a drawing of it, the axes of the two nebulae in line.*[29] *On April 11, 1850, Mr. Johnstone Stoney remarks the two nebulae not in a line. April 17, 1855, Mr. Mitchell remarks the two nebulae are not in line, but the axes are parallel, and gives a diagram* [sketch 14 May 1855]. *At the present time they are neither in a line nor parallel, but inclined at an angle of sixteen degrees.'* There is a sketch by Hunter too, made on 8 May 1861 (Fig. 31) and reproduced in Fig. 7.3 (centre). The supposed displacement gave rise to speculations about an orbiting motion, which is not real.[30] Actually both galaxies are a bit displaced, but their axes are parallel (thus Mitchell was right). Later Dreyer could confirm this in Armagh by micrometrical measurements (Dreyer 1887d).

7.1.5 Lord Rosse's 'List of nebulae not found' and d'Arrest's reaction

At the end of the publication 35 missing John Herschel objects are listed (Table 7.3). To absolve him, Lord Rosse remarked that *'This is not to be considered as a list of missing nebulae, but merely of objects which were not found in the ordinary course of observing, and to which therefore it is desirable that attention should be directed.'*[31] Some objects were searched for in vain up to nine times (e.g. h 401 = NGC 2264). Actually there should be 36 cases, since Lord Rosse noted for h 136 (NGC 618) 'Sought for four times; not found.' The object was obviously overlooked when compiling the list.

D'Arrest, who was also intensively concerned with observing Herschel objects in Copenhagen, felt himself challenged by Lord Rosse's list. During his observations with the 11″ Merz refractor in autumn 1861, he proved the existence of some objects, but there were others he was unable to locate too. In 1865 the results were published as 'Ueber einige am Kopenhagener Refractor beobachtete Objekte aus Lord Rosse's 'List of nebulae not found''.[32] He wrote that he had not become aware of Lord Rosse's publication until October 1864, when it was cited by Huggins (Huggins W. 1864). He qualified it as *'undoubtedly the most important work on nebulae since 1833* [the Slough catalogue]*'*.

About the missing objects, d'Arrest wrote that *'12 objects were observed, but only for 4 others, not knowing the Parsonstown observations, do I have repeated notes about their invisibility'* (the four are marked * in column 'd'Arrest'). The remaining 20 had still not been observed at that time.[33] A modern analysis of Lord Rosse's list gives that three objects are actually missing ('not found'), three are stars (among them the case of

---

[28] Parsons W. (1861a: 148).
[29] Herschel J. (1833a), Fig. 77.
[30] Darby (1864 : 99). The author mentions another example: the galaxy pair M 60 and NGC 4647 in Virgo (see Tables 3.9 and 6.18).
[31] Parsons W. (1861a: 189).
[32] 'About some objects from Lord Rosse's 'List of nebulae not found' observed with the Copenhagen refractor' (d'Arrest 1865a).
[33] He was not aware of the additional case h 136.

Table 7.3. Lord Rosse's 'List of nebulae not found' (with h 136 added)

h	H	GC	NGC	d'Arrest	Type	Con.
57	V 20	132	247		Gx	Cet
136		364	618	*	Gx	Tri
162	(55 And)	428		*	Star	And
184	III 583	472	780	26.9.1864	Gx	Tri
206	III 457	510	864	1.1.1864	Gx	Cet
281	IV 43	639	1186	*	Gx	Per
284	III 578	646	1207		Gx	Per
314	III 587	851	1576		Gx	Eri
333	II 457	908	1665	3.10.1864	Gx	Eri
343		975	1757		NF	Eri
356		1133	1927		NF	Ori
401	VIII 5, V 27	1440	2264	*	OC+EN	Mon
468	III 479	1578	2459		OC	CMi
546		1743	2727		Gx	Hya
577		1791	2804	4.3.1862	Gx	Cnc
578		1794	2809	6.3.1862	Gx	Cnc
590	III 630	1832	2853		Gx	Lyn
669	III 65	2014	3129		2 stars	Leo
672		2019	3135		Gx	UMa
706		2094	3234		Gx	LMi
745	V 52	2189	3359		Gx	UMa
828	II 42	2315	3547		Gx	Leo
1307	I 83	3043	4494	5.2.1864	Gx	Com
1485	II 384	3319	4827	10.5.1863	Gx	Com
1535		3415	4978	29.12.1861	Gx	Com
1832	II 695	3920	5660		Gx	Boo
1948	III 74	4167	6073		Gx	Her
1974		4259	6257		Gx	Her
2062	III 144	4536	6857	6.12.1863	EN	Cyg
2073		4570	6903	11.8.1864	Gx	Cap
2113		4654	7051	26.8.1864	Gx	Aqr
2133		4710	7143		2 stars	Cyg
2137	III 930	4723	7165		Gx	Aqr
2148		4756	7210		NF	Peg
2250	III 213	4980	7691	12.8.1864	Gx	Peg
2302		5051	7822		EN	Cep

'55 Andromedae', see Section 6.14.6) and the rest are galaxies, open clusters and emission nebulae.

As a case of suspected variability, Lord Rosse had included h 401 (V 27 = VIII 5, GC 1440, NGC 2264) in his list. This is the coarse open cluster around 15 Monocerotis.[34] John Herschel had noted a '*nebulous haze*' (hence the second entry V 27 by his father), but

---

[34] The object was named the Christmas Tree Cluster by Leland Copeland. South of it lies the remarkable Conus Nebula.

Lord Rosse's observation yielded '*No neby. found, and only a few stars arranged in pairs; no cl. Has there been a change here?*' William and John Herschel had correctly seen a star cluster embedded within a faint nebula (which is now catalogued as Sh2–273).

In his article, d'Arrest added two problematic cases: h 63 and h 141. About h 63 (II 621) he wrote that '*it is most probably identical with h. 64*', which is correct (galaxy NGC 259 in Cetus). On John Herschel's h 141 he noted that it '*Is not in the heavens. Anyway, the place, assigned* [by Herschel] *as doubtful, is errant.*' The object (NGC 627) is identical with NGC 614 and NGC 618 – this is a complicated case, which was later investigated by Burnham too (Burnham S. 1892b). It is a galaxy in Triangulum.

7.1.6 New nebulae

Lord Rosse's publication of 1861 contains 295 new nebulae (mostly termed 'nova') found in the vicinity of Herschel nebulae. Unfortunately, there is no separate list; thus it is a laborious task to extract them from the textual descriptions. The objects came too late for Auwers' list of new nebulae and it was up to John Herschel to catalogue them in the GC, which entailed some problems.

Unfortunately, Lord Rosse's publication does not identify the true discoverers. Two things are essential to tackle the problem: the careful study of all available reports and knowledge of the period of duty for each of the Birr Castle assistants. One must consider that Lord Rosse, according to his own statements in publications and letters, was seldom an observer in later times, leaving the great reflector exclusively to his assistants. The investigation yields the most probable match between objects and persons – as long as further records (as yet unknown) do not give anything else.

According to this analysis, Lord Rosse (William Parsons) discovered only three nebulae – the rest must be credited to his assistants. This is supported by Dreyer, whose claim is particularly authorised by his important work at Birr Castle, writing in the NGC's introduction that '*The new nebulae found before 1861 (chiefly by G. J. Stoney, B. Stoney, and R. J. Mitchell) have been marked Ld. R.*'[35] Unfortunately, Dreyer's term 'Ld. R' (appearing in the column 'Other observers' of the NGC) was later taken in the literal sense. Thus Lord Rosse is usually credited for about 225 objects in the literature: as usual, the director wins the fame, leaving the assistants (and true discoverers) without recognition. Certainly, this was not wanted by Lord Rosse.

When Lord Rosse's work appeared in 1861, John Herschel immediately evaluated it. The moment was right, insofar as he was about to compile the General Catalogue. Herschel had to extract the new objects from the large table containing the descriptions. This was not an easy task, leading to various errors. Herschel overlooked objects or interpreted some as having already been catalogued. The result was a handwritten list, which was subjected to many corrections and additions.[36] It contained not only the new objects but also interesting data found in Lord Rosse's publication (e.g. about annular nebulae and missing objects).

In 1862–63 there was a lively correspondence between Herschel and Lord Rosse on nebulae.[37] Since most of the observations had been made by his assistants, Lord Rosse was unable to answer all of the questions. He thus needed their help. A letter of Hunter (28 July 1862) and a 'Memorandum' of Johnstone Stoney (24 April 1863) are interesting in this regard.[38] Lord Rosse forwarded them to Herschel, who entered the data in his table of Birr Castle observations. Surprisingly, seven objects appear in the GC that are not listed there. Among them are two that are first mentioned in Hunter's letter (h 688,a/b = GC 2049/50). Obviously, Herschel noticed some problems quite shortly before finishing the GC.

Since Lord Rosse gives no coordinates for the objects observed, Herschel was challenged. It was not simple to derive reliable positions from the often poor data. At best there were distances, position angles or even a sketch, but in most cases the descriptions merely give a direction (p, f, n, s = preceding, following, north, south). Accordingly, Herschel's positions for objects found at Birr Castle are often inaccurate. This later led to some confusion, as in the curious case of 'GC 80' (see the next section). It was increasingly often Dreyer

---

[35] Dreyer (1953: 9).
[36] RAS Herschel J. 12/1.2.
[37] Five letters from this time are in the RAS archive (Bennett 1978) and three at Birr Castle (Bennett and Hoskin 1981).
[38] RAS Herschel J. 12/1.2.

7.1 Lord Rosse's publication of 1861

Table 7.4. *Birr Castle objects in the GC (see the text)*

Observer	New objects	Listed in GC	Other discoverers	Rest
Bindon Stoney	114	110	1 (d'Arrest) + 1 (W. Herschel)	2
R. J. Mitchell	104	100	1 (d'Arrest)	3
Johnstone Stoney	74	71		3
William Parsons	3	3		
Total	295	284	3	8

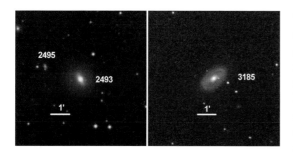

Figure 7.4. Birr Castle objects. Left: Mitchell's NGC 2495 = h 476,a near to NGC 2493 = h 476; right: J. Stoney's NGC 3185 = h 692,a (DSS).

who cleared up the identity of such objects. He had the advantage of a four-year stay at Birr Castle, with access to the original notes.

How many Birr Castle objects were entered in the General Catalogue? Table 7.4 gives the answer. Of the 295 objects found there, 284 are listed in the GC, mostly assigned by Herschel as 'R. nova' in his column 'Other authorities'. Two of them appear in the 'Supplementary list' (GC 5059 and GC 5069, both found by B. Stoney). For better identification he introduced an extended h-designation, presented in his column 'Sir J. H.'s catalogues of nebulae'. An example is h 476,a = GC 1605. The designation indicates that the object (the galaxy NGC 2495 in Lynx) was found near to h 476 (GC 1604, NGC 2493; see Fig. 7.4 left). Lord Rosse noted for h 476 '*Jan. 12, 1855. A F. star p. and a nebulous knot f.*'; the 'nebulous knot' is the new object (h 476,a) discovered by Mitchell. Bindon Stoney was the most successful observer, closely followed by Mitchell.

After treating the Birr Castle discoveries, Herschel turned to the 215 new objects of d'Arrest, sent to him by letter, which arrived on 8 April 1863 (see Section 7.4.4).

There were overlaps between the two lists – but also misunderstandings. For GC 4844 (NGC 7383) Herschel mentions d'Arrest and Lord Rosse ('R. nova'). Two other objects are credited to d'Arrest, but he assigns a special h-number too, indicating Lord Rosse: GC 878 = h 320,a (NGC 1622) in Eridanus and GC 2054 = h 692,a in Leo (NGC 3185; Fig. 7.4 right). Both galaxies were discovered by J. Stoney in January 1850; the latter was independently found by Schönfeld on 15 January 1861.

Besides these overlaps, there are identities (Table 7.5). Three nebulae, found at Birr Castle and Copenhagen, are listed twice in the GC, as an unrecognised matter. Of course, all priorities belong to the Birr Castle observers, since d'Arrest started his campaign in 1861. The column 'Near h' gives the Herschel nebula near where the new one was found (this reference is often the only way to localise a new object in Lord Rosse's publication); 'd'A' (in the column 'Remarks') refers to d'Arrest's list.

What about the 11 Birr Castle objects in Table 7.4, which were obviously ignored by Herschel? They are collected in Table 7.6.

Three are actually in the GC, but are not credited to Lord Rosse. For one (GC 442 = NGC 733), Herschel wrongly sees his father as the discoverer, but this is a new object found by Bindon Stoney on 11 October 1850. The two remaining ones are credited to d'Arrest. GC 82 is involved in the case of 'GC 80' (see the next section). GC 1616 is strange: a sketch in Lord Rosse's publication shows two nebulae near to h 483 = GC 1617 = NGC 2513, called β and γ. Obviously, in an addition to his Birr Castle table, Herschel credited β = GC 1616 to d'Arrest (d'A 151) and called γ = GC 1615 'R. nova'; but it should have been known to him that both nebulae had been discovered on 31 January 1851 at Birr Castle, long before d'Arrest had examined the field.

Table 7.5. *Identities and overlaps in the GC*

GC	NGC	Discoverer	Type	V	Con.	Near h	Remarks
213	394	Mitchell	Gx	14.0	Psc	87	= GC 215 (d'A 11)
464	770	Mitchell	Gx	13.0	Ari	181	= GC 461 (d'A 32)
878	1622	J. Stoney	Gx	13.1	Eri	320	h 320,a = d'A 44
2054	3185	J. Stoney	Gx	12.0	Leo	692	h 692,a = d'A 69; Schönfeld 15.1.1861
4844	7383	B. Stoney	Gx	14.1	Peg	2183	'R. nova', d'A 117; in group[39]
4942	7631	B. Stoney	Gx	13.1	Peg	2230	= GC 4943 (d'A 122)

Table 7.6. *The 11 Birr Castle objects ignored by John Herschel; some appear in the GCS (GC > 5000)*

GC(S)	NGC	Discoverer	Type	V	Con.	Near h	Remarks
82	169	Mitchell	Gx	12.4	And	32	d'A 3 (see the next section)
5129	316	B. Stoney	Star		Psc	78	GCS
442	733	B. Stoney	Star		Tri	169	Wrongly W. Herschel
5380	2373	J. Stoney	Gx	14.0	Gem	446	GCS
5383	2375	J. Stoney	Gx	13.8	Gem	446	GCS
1616	2511	B. Stoney	Gx	14.3	CMi	483	d'A 51
5435	2694	J. Stoney	Gx	14.7	UMa	535	GCS
5507	3016	Mitchell	Gx	12.9	Leo	642	GCS
5618	4109	B. Stoney	Gx	14.1	CVn	1088	GCS
		Mitchell	Gx	13.9	Vir	1880	NGC 5765B
6226	7752	Mitchell	Gx	14.0	Peg	2268	GCS

The remaining eight objects were simply overlooked by Herschel in Lord Rosse's work. According to Dreyer's analysis, while compiling the Birr Castle observations, all but one were listed in the GCS. The nebula near h 1880 (NGC 5765) was missed by him too, whereas Lord Rosse clearly noted '*2 neb, with 3 B. stars in the neighbourhood. Both F. and E.*' North of the 13.9-mag galaxy NGC 5765 (GC 3997) there is indeed a fainter one (14.6 mag), which is now called NGC 5765B (Fig. 7.5).

The modern analysis comes to the conclusion that only 264 objects found at Birr Castle by 1858 are actually new. As Table 7.7 shows, 24 must be credited to William Herschel and 2 to his son; 5 are identical.

Figure 7.5. The galaxy NGC 5765 and its close companion NGC 5765B, which were discovered by Mitchell (DSS).

[39] In a group with five other NGC galaxies; see Steinicke (2001c).

Table 7.7. *Only 264 of the 295 Birr Castle objects are actually new*

Observer	Objects	WH	JH	Identities	New
Bindon Stoney	114	10	1		103
R. J. Mitchell	104	7		2	95
Johnstone Stoney	74	6	1	3	64
William Parsons	3	1			2
Total	295	24	2	5	264

Table 7.8. *Additional Birr Castle objects not contained in the publication of 1861*

NGC	Discoverer	Date	GC(S)	Credited to
384	B. Stoney	4.11.1850	207	d'Arrest (JH)
385	B. Stoney	4.11.1850	208	d'Arrest (JH)
1699	Hunter	13.2.1860		L. Swift (D)
2321	J. Stoney	18.12.1849	6248	Lord Rosse (D)
4118	B. Stoney	21.4.1851	5618	Lord Rosse (D)
7387	Mitchell	9.9.1856	4847	d'Arrest (JH)

On looking at Lawrence Parsons' publication of the Birr Castle observations (1880) compiled by Dreyer, it is evident that six additional nebulae were found there before the GC appeared (Table 7.8). They are not in Lord Rosse's work of 1861. Three are nevertheless in the GC, but credited to d'Arrest, and two are listed in the GCS (GC 5618 and GC 6248). Hunter's nebula (his only one) is catalogued in the NGC, but Lewis Swift is mentioned (an observation of 22 December 1886). Added up, 301 objects had been found by 1861; however, some of them were not new: 247 were taken over into the NGC and 1 into the IC (see the next section). Finally, J. Stoney's GC 1807, the galaxy NGC 2826 in Lynx (again credited to d'Arrest by Herschel as d'A 63), is his own GC 1809 (h 581,d).

7.1.7 The Peculiar case of 'GC 80'

One of the objects ignored by Dreyer is GC 80 – a remarkable case. Three non-stellar objects are at this position in Andromeda (Fig. 7.6): the bright galaxy NGC 160 and, 11′ to the east, a double system consisting of the elongated galaxy NGC 169 and its faint oval companion IC 1559 (20′ south). Additionally, two bright stars are in the field: a (6.2 mag), 5′ north of NGC 162, and b (7.2 mag), 4′ northeast of NGC 169.

NGC 160 was discovered by William Herschel on 5 December 1785 and listed as III 476; John Herschel catalogued it as h 32 and GC 79. On 18 September 1857 Mitchell made an interesting observation of 'h 32', noting '*S; d. neb.; the n. one is E, sp, nf; bM.*' He saw a double nebula, the northern component of which appeared elongated with a bright centre. Obviously, this does not fit NGC 160, but corresponds to the pair NGC 169/IC 1559. There is an observation by d'Arrest, made on 22 August 1862 and reported in his list of 125 new nebulae sent to John Herschel. For object no. 3 one reads '*F. pL. (40″) R. star 6* [mag] *near; Dist = 3½′.*' This and the given position indicate NGC 169 too, though a companion is not mentioned.

How did Herschel use the data? The GC lists three objects in the field: GC 79 = h 32, GC 80 = h 32,a ('R. nova') and GC 82 ('D'Arrest, 3'). GC 80 is placed near to GC 79 (however, the coordinates are marked as inaccurate); the description reads '*Makes Dneb with h 32.*' Thus Herschel has interpreted Mitchell's new nebula (h 32,a) as a companion of NGC 160 (h 32); d'Arrest's nebula (GC 82) is seen as an independent object, placed (using his position) 11′ to the east. Herschel's data are the very reason for the trouble.

In the *Siderum Nebulosorum* (*SN*) d'Arrest presents his observations of NGC 160 and NGC 169 with

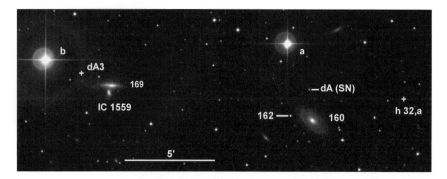

Figure 7.6. The galaxies NGC 160, NGC 162, NGC 169 and IC 1559 (DSS); for the meanings of the labels, see the text.

correct positions. The first took place on 22 August 1862, with him noting for NGC 160 (III 476, h 32) 'It is small and pretty faint, but not difficult to observe. Another nebula follows 2′ north, each one is obviously third class.'[40] The new nebula, 2′ north of NGC 160 and similar in brightness, is uncatalogued. The observation of NGC 169 is that from his list sent to Herschel, but d'Arrest added that 'If the Rosse nebula was always near to h 32, is unknown.' He questioned the Birr Castle observation. A second observation is dated 14 September 1865 (the GC had already appeared). On NGC 160 he wrote that 'The Rosse nebula is seen closely following, but not sufficiently exact.' This can only mean that d'Arrest has observed 'GC 80' – obviously the object near to NGC 160 that had already been noticed in 1862. His note on NGC 169 reads 'Pretty faint, large; double: nebula follows 25″ south.' Now d'Arrest has seen Mitchell's companion of NGC 169 for the first time. For NGC 169 he added 'Has the GC-number 82, one may be the Rosse nebula GC 80 or is different from it, this is still open.' He was not aware whether 'GC 80' is the companion of NGC 160 or NGC 169.

Now Herman Schultz enters the scene, observing the field on 5 September, 1867 with his 9.6″ refractor in Uppsala. For GC 82 (NGC 169) he suspected a double nebula: 'is probably S globular, and seems sometimes to be divided into two separate objects'.[41] Following Herschel, he searched for GC 80 in the vicinity of NGC 160 (h 32). Schultz in fact found an object hardly 1′ northeast of it, which he described as 'exceedingly faint' – and thus unwittingly confirms d'Arrest's observation

(the SN was still unknown to him). In his later publication, Schultz gives positions (epoch 1865) for the three objects (Schultz 1875). Those for NGC 160 (GC 79) and NGC 169 (GC 82) fit precisely; his 'GC 80' is only a 15-mag star 1.2′ northwest of NGC 160. The ominous subject – originally seen by Mitchell near NGC 169, then associated with GC 79 by Herschel (d'Arrest was unsure) – is now reborn in a new figure!

In the Birr Castle observations of 1880, Dreyer tries to solve the puzzle. About GC 82 he wrote that *"The observer [Mitchell] mistook this for h. 32, in consequence of which G.C. 80 was entered as a 'Nova R.'. There is a small diagram exactly like the one made on Oct. 22, 1857, and the position of the B. star shows clearly that 82 and not 79 was observed on both nights."* The attached sketch (Fig. 7.7) shows GC 82 as a double nebula, the following 6.2-mag star (estimated '7m') is correct. The publication contains Mitchell's second observation (22 October 1857) and later ones by Lawrence Parsons (1866) and Dreyer (1874/75). They all confirm that GC 82 (not GC 79 = h 32) is a double nebula. A third observation of Mitchell, made only two days later, is mentioned there, in which he notes about h 32 (NGC 160) 'See no comp.' obviously he now saw the right nebula.

There is an interesting observation of h 32 by Lawrence Parsons on 16 October 1866, where he noted 'f. neb follows 1½′.' Dreyer later explained that *"This 'f. neb'. is evidently the same as seen in 1874."* He refers to his own observation at Birr Castle on 5 November 1874 ('N. f. the neb. is a v. F. star or nebs. knot in Pos 74.0°, Dist. 69.9″.') When compiling the data for the 1880 publication, he later added 'Seen by Schultz, and taken for G.C. 80.' (Schultz' catalogue was published in April 1874.) Dreyer was able

---

[40] d'Arrest (1867a: 7).
[41] Schultz (1874: 74).

7.2 Considering the Harvard observations 203

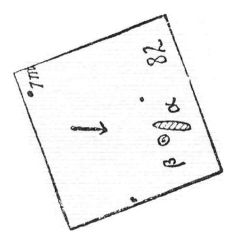

Figure 7.7. Mitchell correctly saw GC 82 as double nebula: NGC 169 and IC 1559 (sketch rotated).

to confirm his first observation a day after (6th): '*An e S, F neb. point, or probably a F. star n f 79 in Pos. 78.0°. Dist. 71.8″.*' This is exactly Schultz' 'GC 80', i.e. the 15-mag star northwest of NGC 160. Thus the object seen by him near to NGC 160 and interpreted as 'GC 80' had already been noticed by Lawrence Parsons (1866) and later confirmed by Dreyer (1874). In the GCS of 1878, Dreyer once more mentioned the case, writing about GC 80 in the notes that '*R. nova. Does not exist. 82 was undoubtedly observed instead of 79, which latter Nebula is not double. The description in Phil. Trans., 1861, agrees perfectly with the appearance of 82.*' He further wrote that "*Schultz's 'G.C. 80' has not been seen in Birr before 1874: I have therefore entered it in the Catalogue as a nova.*" For Dreyer, the designation 'GC 80' was suspect and he wanted to avoid it in the future. He uses the opportunity in the GCS, calling Schultz' object GC 5107. It is incomprehensible that he does not mention Lawrence Parsons (1866) as the discoverer of GC 79. The same applies to the fact that Mitchell's component of GC 82 was not catalogued, even though it had been confirmed by him and the Fourth Earl.

Dreyer, obviously fascinated by the case, revived it in the NGC. The catalogue gives NGC 160 = GC 79 = h 32 = III 476, NGC 162 = GC 5107 ('Schultz') and NGC 169 = GC 82 ('d'A, Ld R'). So the designation 'GC 80' is omitted and Schultz' star is NGC 162. Unfortunately, the description once again contains an error, since 'h 39 sp' should read 'h 32 sp' (NGC 160 is south preceding). In the notes for NGC 160 he states that '*This is not a double nebula as stated in the G.C. (No. 79, 80), as the observer at Birr Castle mistook G.C. 82 for h 32. There is, however, a vF neb. in Pos. 78°, Dist 72″, first seen by Schultz (No. 162 in this Cat.).*' Once again Mitchell's nebula near NGC 169 is missing.

The object eventually appears in the Second Index Catalogue as IC 1559, its inclusion having been forced by observations of Bigourdan and Javelle, who discovered it independently of one another. Dreyer mentions '*Ld. R., B. 245, J. 819*', giving the description '*vF, 0.5 ssf 169*'. Anyhow, Birr Castle is now mentioned ('Lord Rosse'). Javelle found IC 1559 on 20 November 1897 and Bigourdan's observation was even earlier (7 October 1885). The latter saw NGC 169 as a nebula with two cores: 'a' (NGC 169) and 'b' = '245 Big.' (IC 1559), writing "*The two cores differ little. Core a, 13.2–13.3 mag and 20′ diameter; it has a small stellar centre. Core b, 13.0 mag and 35′ diameter; it is brighter than 'a' and shows a slight central condensation, which appears barely stellar. The borders of the two condensations are touching; 'a' appears more distinct than 'b'.*" (Bigourdan 1904c). He correctly recognised Schultz' NGC 162 as a stellar object.

The conclusion is that, if one designates the object near NGC 160 (Schultz' 'GC 80') as NGC 162 (GC 5107), it was first noticed by d'Arrest in 1862 (before the GC), then seen by Lawrence Parsons in 1866 and again by Schultz a year later (after the GC). It is strange that for all three the 15-mag star (GSC 1738–1668) appeared as a 'nebula'. The 'true GC 80' – discovered by Mitchell near NGC 169, but wrongly placed by Herschel – now bears the designation IC 1559. The galaxy was independently found by Bigourdan and Javelle. Two errors came together in this case: (a) the Birr Castle publication of 1861 is wrong, in that not NGC 160 (h 32) was observed, but NGC 169 (GC 82) with its companion; and (b) John Herschel did not notice the error (being too old for an observation), placing GC 80 near to NGC 160 in the General Catalogue.

7.2 CONSIDERING THE HARVARD OBSERVATIONS

The objects discovered by Bond, Coolidge and Safford and included in the GC (main catalogue and 'Supplementary list') are listed in Table 7.9; compare this with Bond's published list in Section 6.7.4 ('No.' refers to it).

Table 7.9. *Harvard objects in the GC (from GC 5058 in the 'Supplementary list')*

GC	NGC	No.	Observer	Au	Date	Type	V	Con.	Remarks
119	223	2	Bond	4	5.1.1853	Gx	13.4	Cet	Not NGC 219; d'Arrest no. 6
211	391	3	Bond	9	8.1.1853	Gx	13.5	Cet	
879	1624	28	Bond		18.2.1851	OC+EN	11.8	Per	W. Herschel 28.12.1790 (V 49)
1130	1924	6	Bond		7.2.1863	Gx	12.5	Ori	W. Herschel 5.10.1785 (III 447)
1537	2399	9	Bond	24	26.2.1853	3 stars		CMi	
1538	2400	10	Bond	25	26.2.1853	3 stars		CMi	
1691	2655	12	Tuttle		8.4.1859	Gx	10.1	Cam	W. Herschel 26.9.1802 (I 288 = h 520)
3898	5632	20	Bond	33	9.5.1853	Star		Vir	
3912	5651	21	Bond	34	9.5.1853	Star		Vir	
3918	5658	22	Bond	35	9.5.1853	Star		Vir	
4415	6643	23	Tuttle	40	1.9.1859	Gx	11.0	Dra	Schönfeld 1858? (BD +74° 766)
4538	6859	24	Bond	46	24.11.1852	3 stars		Aql	
5058	219	1	Bond		16.9.1863	Gx	14.5	Cet	Last GC object found
5060	1251	4	Coolidge		25.1.1860	2 stars		Cet	
5061	1312	5	Coolidge		16.12.1859	2 stars		Tau	
5064	2189	7a/b	Safford		19.3.1863	NF		Ori	
5065	2198	8	Safford		19.3.1863	NF		Ori	
5066	2515	11	Bond		11.9.1852	2 stars		Cnc	
5067	3123	13	Coolidge		31.3.1859	NF		Sex	
5068	3229	14	Coolidge		31.3.1859	3 stars		Sex	
5071	4582	15	Coolidge		3.5.1859	Star		Vir	
5072	5200	16	Coolidge		30.4.1859	2 stars		Vir	
5073	5310	17	Coolidge		30.4.1859	Star	12.9	Vir	
5074	5366	18	Bond		8.6.1855	Gx	13.8	Vir	
5075	5404	19	Coolidge		29.4.1859	3 stars		Vir	
5077	7150	25	Bond		10.2.1848	Star group		Cyg	First American object
5079	7692	26	Bond		23.10.1848	Gx	14.8	Aqr	

Objects appearing in the main catalogue (GC-number less than 5058) were taken from Auwers' list of new nebulae (of 1862) by Herschel. This explains the wrong assignment of Tuttle (instead of Schönfeld) as the discoverer of GC 4415 (NGC 6643). The objects in the 'Supplementary list' (GC 5058 and higher) are from Bond's 'List of new nebulae and star-clusters seen at the observatory of Harvard College', which was sent to John Herschel in October 1863.[42] The addressee noted that '*the objects* [...] *were communicated to me by Professor Bond* [...], *too late for insertion in the main body of the catalogue*'.[43] Bond's list has two parts: the first contains confirmed objects (nos. 1 to 26) and the

---

[42] RAS Herschel J. 12/1.9.
[43] Herschel J. (1864: 137).

second '*objects supposed to be nebulae but requiring verification*' (nos. 27 to 33). Herschel made a mistake for GC 5058 (NGC 219): in the date '*Sept. 28, 1862*' it should read '1863'. This was the last object observed by Bond and the last GC object (concerning discovery date) too.

Bond's nos. 6 and 12 were ignored by Herschel, having been discovered already by his father (III 447 and I 288, respectively). Since nos. 27 to 33 are missing, Herschel obviously omitted the doubtful cases (part II). Anyway, no. 28 can be found in the GC (unnoticed by Herschel): it is GC 879 = V 49 (NGC 1624).

In December 1863, Bond's list appeared in the *Astronomische Nachrichten* (Bond G. 1864a). Two additional objects were included: NGC 1333 and NGC 6791 (see Table 6.11). Both came too late for Herschel, since his main catalogue was closed on 23 June 1863. In a note published shortly thereafter, Bond reported the identity of nos. 6 and 12 with William Herschel's III 447 and I 288, respectively (Bond G. 1864b).

7.3 CHACORNAC AND HIS 'VARIABLE NEBULA' IN TAURUS

Chacornac's announcement regarding the discovery of a 'variable nebula' in Taurus caused some confusion. He had initially seen a star at the place, which he entered in his 'Atlas ecliptique'. Later he detected a nebula there – but the object soon disappeared. Some of the best observers investigated the case, but could find only an 11-mag star. In the middle of the nineteenth century, Chacornac's observation landed on fertile soil: the chase after 'variable nebulae' started and their nature elicited considerable discussion.

7.3.1 Short biography: Jean Chacornac

Jean Chacornac was born on 21 June 1823 in Lyon. From 1851 he worked at Marseille Observatory (where he used a 4″ Steinheil refractor), and then moved to Paris in March 1854. There his main task was the making of the 'Atlas ecliptique', which partly appeared during 1856–63.[44] It shows stars of 7 to 13 mag along the ecliptic, with a scale four times larger than that of

the *Berliner Akademische Sternkarten*. Similar 'ecliptical charts' were made in the middle of the nineteenth century by Hind (Bishop's Observatory) and Cooper (Markree). All shared the problem of 'missing stars' (Chacornac 1855). With the aid of his 'Atlas ecliptique' Chacornac discovered five minor planets and some comets, the first on 13 May 1852. At Paris Observatory he had the privilege of being able to use the revolutionary 80-cm reflector by Foucault with a silver-on-glass mirror, which saw its first light on 17 January 1862. He immediately observed prominent objects, such as the Ring Nebula M 57 in Lyra (Chacornac 1862a) and the spiral nebula M 51 in Canes Venatici (Chacornac 1862b). In June 1863 he had to leave Paris due to ill health and moved to Villeurbanne near Lyon. There he built up a private observatory, mainly observing sunspots with a small telescope. Jean Chacornac died on 6 September 1873 in St Jean en Royans at the early age of 50.[45]

7.3.2 The discovery of NGC 1988

On 19 October 1855 Chacornac saw a 'nebula' in Taurus with the 25-cm Lerebours refractor of Paris Observatory.[46] He announced his discovery but years later in the *Bulletin quotidien de l'Observatoire impérial de Paris* (28 and 29 April 1863). In the 'Note de M. Chacornac sur une nébuleuse variable' one reads his claim to '*have found a new nebula very near to* ζ *Tauri*' (Chacornac 1863a, b). It was the 'disappearance' of the object on 20 November 1862 that caused the publication. The late note was the reason why Auwers did not enter the object in his list of new nebulae (which had already been printed at that time). In a subsequent report, titled 'Nébuleuse variable de ζ Taureau', Chacornac describes the object as an 11-mag star in a nebula: '*a small nebula, projected on a star*' (Chacornac 1863c). The star lies about 5″ northwest of ζ Tauri (3 mag) – thus the title is misleading. Dreyer catalogued it as NGC 1988 (Fig. 7.8).

Here is the story according to Chacornac's report: when he was making chart no. 17 of the 'Atlas ecliptique' (during 1853–54 in Marseille),

---

[44] Michaelis (1856). The work was completed by the brothers Henry with the aid of photography.

[45] Obituary: Fraissinet (1873).

[46] During the years from 1823–31 it was the largest refractor in the world. South gives a slightly different lens diameter: 9.2″, of which only 8.4″ are used (South 1826).

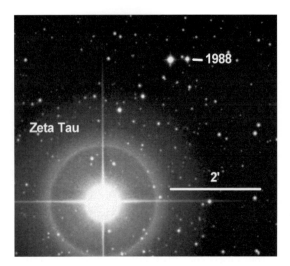

Figure 7.8. Chacornac's 'nebula' NGC 1988 near ζ Tauri (DSS).

nothing striking was at the place – only a faint star. Observations on 1 January and 17 December 1854 in Paris confirmed this. However, on 19 November 1855 Chacornac detected a 'nebula' at the place, which was clearly seen on 27 January 1856 when he checked the star positions of the chart. He wrote that *'it is remarkable that Hind has not seen it with his 7"-refractor'*. At that time Hind was working on similar maps at Bishop's Observatory, London. Chacornac described the object as a transparent cloud, reflecting the star's light, which appears *'quite different from the nebula 357 (Herschel II), as it does not give the impression of stellar points spread over the area'*. Obviously he has seen the Crab Nebula M 1 (h 357) in Taurus as a star cluster (*'this nebula proves indeed to be a star cluster'*). The same was claimed by Lord Rosse in his 1844 publication (see Section 6.4.8).

When observing the field on 20 November 1862, Chacornac was surprised that the object was missing – and supposed the rare case of a 'variable nebula'. He was already familiar with a similar one: Hind's nebula (NGC 1555), which was located in Taurus too. Hind later summarised Chacornac's discovery in his paper 'Chacornac's variable nebula near ζ Tauri' (Hind 1876). There he mentions d'Arrest's unsuccessful attempts to see the object in the Copenhagen refractor (12 September 1863 and 25 January 1865). The star is also listed in the Markree catalogue (observed on 16 January 1850), but *'without mention of surrounding nebula'*.

Tempel, working at Marseille Observatory, was informed by director Benjamin Valz about Chacornac's discovery and observed the field in early 1861 with his private 4″ Steinheil refractor. He was convinced that the 'nebula' must be a reflection in the eye-piece due to the bright star ζ Tauri, demonstrating his find immediately to Valz. In a letter to *Le Monde*, Tempel reacted to Chacornac's announcement, which had been placed a bit earlier in the French magazine too (Tempel 1863).[47] He doubted the object's variability, believing it to result from differences in air quality, but he conceded the following: *'I think that true variability is possible for very small and simply structured nebulae, which, however, do not deserve closer attention. But the new variable nebula, announced by Mr Chacornac, ranks among the larger ones and I really doubt it to be variable.'*

Tempel also criticised Chacornac's observing method: *'The charts of Mr Chacornac, the only ones reaching to the faintest stars, yield comprehensive matter to support my remarks.'* He added that, *'Of course, it is easier to draw stars than nebulae; Mr Chacornac's telescope has at least 10 inches, whereas my own one has only 4 inches aperture; nevertheless I sometimes could add a large number of faint stars to his charts; for instance, when checking his chart No. 26, I was able to enter 237 new stars. I believe that the atmosphere is much clearer in Marseille than in Paris.'* Tempel also mentions nebulae that had not been seen by the French observer: *'In his chart No. 34, Mr Chacornac shows a star of 12 mag near 11h 13m 32s and 3° 58'; at this place I often saw a pretty bright nebula, which is also on Argelander's maps* [BD]. *14' south (11h 12m 40s and 3° 44') there is another very faint nebula, missed by Mr Chacornac. Both nebulae are in* [John] *Herschel's catalogue.'* These are galaxies in Leo that had been discovered by William Herschel: NGC 3640 (II 33, h 864) and NGC 3645 (II 32) = NGC 3630 (h 861).

John Herschel, not knowing Tempel's result, catalogued 'Chacornac's recently discovered nebula' as GC 1191. Dreyer was the first to mention Tempel's observation in the notes to NGC 1988. In 1862 Winnecke noted the following about his observation with the 15″ Merz refractor at Pulkovo: *'of Chacornac's nebula near ζ Tauri, no trace was visible on Oct. 8'* (Winnecke 1866). d'Arrest entered the discussion on 15 September 1863 (d'Arrest 1863c). He correctly remarks that Chacornac's

---

[47] See Section 11.6.1.

nebula has still not been mentioned in the *Astronomische Nachrichten*. About his observation with the 11-inch in Copenhagen he noted that '*At the local refractor it is currently invisible.*' D'Arrest states that '*the number of safely recorded variables among the nebulae is no longer limited to a single case*' – therefore, Hind's nebula seemed not to be an exceptional object.

On 5 December 1876 Holden was unsuccessful in an attempt to find Chacornac's nebula with the 26″ refractor at the US Naval Observatory, Washington: '*Variable nebula; sky too bright to see it.*'[48] Four years later there was an anonymous call for observations in the magazine *Nature* (Anon 1880f). To make things palatable and because a summary of the background was lacking, the writer told the story covering 1853–64, '*though there has been no mention of late observations of the vicinity of the nebula*'. However, the resonance was poor. Only Burnham observed the object in 1891 with the 36″ refractor at Lick Observatory, noting that '*there is not the least trace of any nebulosity in this place, and there is no doubt of the correctness of Tempel's explanation*' (Burnham S. 1892b). He added that '*Too much time has been wasted in looking for this object, and particularly since there was no reason whatever for believing in its existence after Chacornac himself failed to see it in 1862.*' Chacornac's 'variable nebula' in Taurus (NGC 1988) is nothing but a single star of 11.0 mag (GC 1310–809), 5′ northwest of ζ Tauri.

7.4 D'ARREST'S CONTRIBUTION TO THE GENERAL CATALOGUE

This section treats d'Arrest's Copenhagen observations from the appearance of his monograph 'Instrumentum magnum' in late 1861 (see Section 6.14.7) until mid 1863, when John Herschel finished the General Catalogue. The period encompasses many discoveries by d'Arrest – important contributions to the GC. There was a crucial correspondence with Herschel on the subject.

7.4.1 The report of 1862

D'Arrest's Copenhagen observations of nebulae started on 30 September 1861. The first results were mentioned in his monograph 'Instrumentum magnum'. By May 1862 he had discovered more than 70 nebulae; a selection of 24 was presented in a report of 20 May, titled 'Vorläufige Mittheilungen, betreffend eine auf der Kopenhagener Sternwarte begonnene Revision des Himmels in Bezug auf die Nebelflecken'.[49] This is the first of many on the progress of his work, '*which will certainly take many more years*'.

D'Arrest judged the 11″ Merz refractor as outstanding for observing nebulae, ranking the instrument '*exactly between Herschel's twenty-foot reflectors in their most perfect condition, and the excellent telescope, so successfully used by Lassell, for instance in the years 1852–54*'. These metal mirror reflectors, with apertures of 18″ and 24″, respectively, were clearly larger. D'Arrest justifies his statement thus: '*Our telescope particularly shows not only all of the Herschel nebulae, including the most difficult, but also new ones in such numbers that my observations of 776 objects made during the last eight months contain more than 100 nebulae that are definitely unknown.*' A critical analysis shows that actually 50 objects were new.

For practical reasons d'Arrest did not determine absolute positions: '*Since I do all observations alone, it soon showed that the steady change between adjusting and observing the mostly faint objects and the microscopic reading of the setting-circles was unworkable, a disadvantage both for the eye and for the task itself.*' He thus limited the matter to getting only differences to nearby reference stars using bar- and ring-micrometers (at magnification 123). D'Arrest further explained as follows: '*I also do not see why objects that are observed only for their own sake and not to determine places of planets or comets should all be referred to the vernal equinox.*'

In his report d'Arrest mentions proper motions of nebulae too – a topic that had already occupied him in Leipzig. Referring to h 322 (NGC 1625), h 3248 (NGC 3242) and h 1779 (NGC 5566), he obtained a negative result: '*Even in the latter case I do not at all believe that a proper motion is recorded, I even take it for unlikely.*'

A main interest concerned 'variable nebulae'. D'Arrest mentions Hind's object in Taurus (see Section 11.4), John Herschel's missing nebulae in Coma Berenices (see Section 6.14.4) and the case of IV 4

---

[48] Newcomb (1879: 366).

[49] 'Preliminary announcement about a revision of the heavens concerning nebulae started at Copenhagen Observatory' (d'Arrest 1862c). The descriptions are, typically for d'Arrest, written in Latin.

= NGC 3662, which had been investigated by Schmidt (see Section 6.15.3). In this connection, 'lost nebulae' were treated too, namely II 48 (NGC 2672), I 155 (NGC 1453) and I 26 (NGC 3345). He had searched in vain for these objects (the first two are galaxies, the third is a pair of stars). The strange case of II 48 was later studied by Schultz (see Section 8.16.3).

7.4.2 D'Arrest's catalogue of double nebulae

A separate chapter in d'Arrest's report treats double nebulae. Already in Leipzig he had made his first studies, publishing the results in the 'Erste Reihe' (1856).[50] He was concerned about relative motions. According to John Herschel's ideas on the physical connection of the components, double nebulae could be early stages of double stars (see below).[51] For Herschel the phenomenon was pretty rare; a few objects were sketched in the Slough catalogue.[52] By the way, Keeler later took images of them with the Crossley reflector at Lick Observatory, finding hardly any congruence: '*The actual nebulae, as photographed, have almost no resemblance to the figures. They are, in fact, spirals, sometimes of very beautiful and complex structure*' (Keeler 1900c).

D'Arrest was unable to find anything about double nebulae in William Herschel's work. This is surprising, because the topic was presented in detail in the 1811 publication.[53] In his report of 1862, d'Arrest concludes '*that the number of probably physically connected nebulae is unexpectedly large compared with the rate of double stars among ordinary stars*'. On the basis of his own observations, 50 pairs up to a separation of 5′ are mentioned; the total number is estimated to be 200 to 300 (which is about 10% of all known nebulae). D'Arrest describes '*some new double nebulae*': NGC 2511/13, NGC 2698/99 (Fig. 7.9), NGC 2852/53 and NGC 5813/14. The galaxy NGC 2699 in Hydra was found by him, the rest are discoveries of William and John Herschel and Lord Rosse. D'Arrest treats still another case, for which measurements of 1785, 1827 and 1862 should reveal an 'orbital motion': NGC 2371/72 (II 316/17) in Gemini. Actually, this is a

Figure 7.9. D'Arrest's double nebula no. 18 = NGC 2698/99 (distance 4.5′) in Hydra (DSS).

bipolar planetary nebula, whose components show no relative motion.

On 2 June 1862 d'Arrest wrote a paper on double nebulae: 'Verzeichnis von 50, theilweise neuen, Doppelnebeln für den Anfang des Jahres 1861'.[54] There he presents the announced 50 cases in detail. Insofar as William Herschel had only mentioned some examples in a footnote of his 1811 publication, this is the first catalogue of double nebulae (four entries are triple systems). It is completely based on d'Arrest's observations made in Copenhagen. The separation of the components is limited to 5′; the given position is for the 'main nebula'. D'Arrest discovered 13 companions. Remarkable is the high hit rate: 48 cases are actually pairs (or trios) of galaxies. Occasionally there are even more galaxies involved, which were not detected by d'Arrest. One such object is the above-mentioned planetary nebula NGC 2371/72 (no. 14 in his list). Pair no. 26 consists of NGC 3928 (II 740) and a companion (NGC 3932 = GC 2593), which was probably found by d'Arrest on 4 December 1861. About its position he noted '*companion precedes 8s2 and is 2′ north*'. There is indeed a galaxy in the vicinity (MCG 8–22–23), but considerably more distant. D'Arrest's 'companion' is only a star of 13.2 mag. Prominent galaxy pairs in d'Arrest's list are NGC 3226/27 (no. 21), which had been observed by Secchi; NGC 4435/38 and NGC 4458/61 (nos. 31 and 32), both

---

[50] d'Arrest (1856a: 308).
[51] Herschel J. (1833a: 502).
[52] Herschel J. (1833a), Plate XV, Figs. 68–79.
[53] Herschel W. (1811: 285–289).

[54] 'List of 50, partly new double nebulae for the beginning of the year 1861' (d'Arrest 1862d).

members of Markarian's Chain in the Virgo Cluster; NGC 4567/68 (no. 36), the Siamese Twins ('*an elegant pair of nebulae*'); M 60/NGC 4647 (no. 38);[55] M 51/NGC 5195 (no. 40; see Section 11.3); NGC 7331/35 (no. 48) and NGC 7332/39 (no. 48).

In 1879 Camille Flammarion revived John Herschel's idea that such systems are progenitors of double stars (Flammarion 1879a, b). For verification he analysed the data on about 5000 known nebulae (obviously using the General Catalogue). The resulting sample of 13 double nebulae believed to show orbital motion was compared with data on double stars. However, Flammarion could not decide whether the supposed motion of a pair is a true revolution or due to different proper motions of the components. Nine systems in Flammarion's list are indeed double galaxies; the rest are galactic or planetary nebulae, among them M 17, M 20 and NGC 2371/72 (d'Arrest's no. 14). Another object is no. 49 in d'Arrest's list (NGC 7463/64/65; see Table 8.3).

7.4.3 Further publications

Until the General Catalogue was finished, d'Arrest published further articles in the *Astronomische Nachrichten*. In the first three, appearing in *AN* 1378 (23 August 1862), *AN* 1379 (1 September 1862) and *AN* 1393 (12 November 1862), the variability of nebulae was treated (Merope Nebula, NGC 1333).[56]

In a paper of 6 March 1863, d'Arrest gives corrections for eight Slough catalogue objects: '*Perhaps they can be useful for the upcoming edition of the General Catalogue, which Sir John has compiled from older and newer observations*' (d'Arrest 1863b). One case concerns possible brightness variations of h 193 (NGC 821), which are first mentioned in his 'Erste Reihe'. He again treats the Siamese Twins[57] NGC 4567/68 (no. 36 in his list of double nebulae). John Herschel had listed three objects: h 1363, h 1358 and h 1359. D'Arrest now states that there are identities: '*all these observations belong to H. IV. 8, 9, and there exists no other pair of nebulae in this region of the sky*'. He also had found a new double

---

[55] An orbiting motion of the pair was discussed; see Darby (1864: 99).
[56] These articles were treated in other contexts.
[57] The popular name is due to the American amateur astronomer Leland Copeland (Copeland L. 1955).

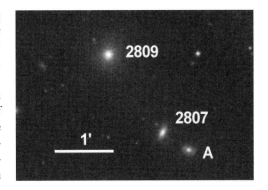

Figure 7.10. D'Arrest's double nebula NGC 2807/09 in Cancer (distance 2.5′); the third galaxy, NGC 2807A, was not observed (DSS).

nebula: '*h. 578. [NGC 2809] 'Nova' is a double nebula; the companion [NGC 2807, d'Arrest 17 February 1863] precedes 7.5s and is 118′ south.*' (Fig. 7.10).

7.4.4 A list of 125 new nebulae

In parallel with d'Arrest's publications, there was a growing correspondence with John Herschel about new nebulae that should be included in the upcoming General Catalogue. The RAS archive stores six letters in various languages; some are annotated by Herschel (Table 7.10).[58] In the first two, d'Arrest asked when he should send his list of new nebulae. He hesitated, because '*a part were observed only once and I will bring not a single wrong nebula into Herschel's catalogue*'. He also formulated an idea concerning an appendix to the GC, which should contain all his corrections to William Herschel's catalogues. However, Herschel did not follow d'Arrest's suggestion.

In the first letter d'Arrest gives Herschel a taste of his work, in terms of the observations on 4 March 1862, during which he discovered three objects: NGC 4269, NGC 4270 and NGC 4324. The second letter treats the supposed variability of the Merope Nebula. As another example, NGC 2071 (near M 78) is mentioned. In the third letter, of 1 February 1863, d'Arrest sent a list of 38 new nebulae with positions for 1830 (the SC epoch). This was consciously chosen to lighten Herschel's work. The fourth letter contains the 125 objects which Herschel eventually entered in the GC. At Herschel's

---

[58] RAS Herschel J. 12/1.8.

Table 7.10. *Letters of d'Arrest to John Herschel (1862–63)*

Letter	Date	Content	Language
1	12.03.1862	3 new objects: 4.3.1862	German
2	29.09.1862	Merope Nebula; NGC 2071	German
3	01.02.1863	38 new objects	French
4	08.05.1863	125 new objects (epoch 1860)	French
5	17.05.1863	Corrections (in GC)	French
6	26.05.1863	Corrections (in GC)	English

Table 7.11. *Objects of d'Arrest that are identical to ones in the NGC that had been discovered earlier*

NGC	GC	dA	Date	Type	V	Con.	NGC	Discoverer
523	306	21	23.8.1862	Gx	12.7	And	537	W. Herschel, 13.9.1784 (III 170)
674	398	27	2.12.1861	Gx	12.0	Ari	697	W. Herschel, 15.8.1784 (III 179)
1550	835	42	29.12.1861	Gx	12.1	Tau	1551	W. Herschel, 8.10.1785 (II 464)
3235	2095	72	29.12.1861	Gx	13.5	LMi	3234	J. Herschel, 24.12.1827 (h 706)
3760	2464	78	21.2.1863	Gx	11.4	Leo	3301	W. Herschel, 12.3.1784 (II 46)
5868	4060	100	27.4.1862	Gx	14.1	Vir	5865	W. Herschel, 4.11.1787 (II 684)

request, d'Arrest now used a different epoch: 1860 (that of the GC). The remaining letters contain corrections, which were considered by Herschel for inclusion in the General Catalogue.

Herschel mentions d'Arrest as the discoverer of 125 objects in the GC (i.e. his name is exclusively given in the column 'Other authorities'). Curiously, they are only nominally identical with the 125 of d'Arrest's list. Three objects (nos. 48, 76 and 77) are missing from the GC; for the first two, however, there is actually an entry. Herschel noticed the identity of no. 48 with his h 341 (GC 953, NGC 1726). No. 76 is William Herschel's II 493 (GC 2232, NGC 3413). For no. 77 d'Arrest communicated a corrected right ascension in his fifth letter. The object was not entered in the GC. There is, admittedly, no nebula at the position, but Herschel had not proved it (there were no observations at all for the GC). Herschel probably overlooked no. 77, since a similar correction for no. 78 was considered. Thirteen of the objects in d'Arrest's list were already to be found in his earlier monograph 'Instrumentum magnum' (see Table 6.21).

As shown, only 122 of d'Arrest's 125 objects are explicitly included in the GC. However, at the same time, there are three further discoveries, which are not in his list – thus balancing the total number. The first is GC 358 (NGC 607), for which Herschel wrote '*D'Arr. = Auw. N. 15*'. It is mentioned in d'Arrest's Leipzig observations (discovered on 23 August 1855) and no. 15 in Auwers' list of new nebulae. The second object (GC 4941) was found by d'Arrest on 17 October 1855 and entered in his 'Erste Reihe' too; it is not catalogued in the NGC.[59] The third (GC 5070, found on 4 March 1862) is mentioned in the first letter and was entered in Herschel's 'Supplementary list' (see also '433' in d'Arrest (1862c)). This is NGC 4270, a neighbouring galaxy of NGC 4269 (GC 2849) in Virgo. In the notes to GC 2849 Herschel indicates this companion. As Table 7.11 shows, some d'Arrest objects are identical with others that had been found earlier.

Generally d'Arrest's data are very accurate. Therefore it sounds curious that he holds two error records. In the case of NGC 3760 (Fig. 7.11) his right ascension is wrong by 1^h. The galaxy also shows the greatest number difference of all NGC identities: NGC 3301 – NGC 3760. Beyond that, d'Arrest is the only observer to have made even a second a

---

[59] Both objects are listed in Table 6.19.

Figure 7.11. For the galaxy NGC 3760 = NGC 3301 in Leo, d'Arrest made an AR error of 1^h (DSS).

1^h error: NGC 3575 (GC 5552), which was found on 21 February 1863, is William Herschel's NGC 3162 (II 43), a galaxy in Leo.

7.5 LASSELL AND HIS 48″ REFLECTOR ON MALTA

From September 1861 until mid 1865 Lassell stayed for a second time on Malta. Under the Mediterranean sky he erected his new masterpiece, the 48″ reflector. The site was offered him by the British government and was near to Fort Tigné on the north side of Quarantine Harbour (Lassell 1862c). The observations started in spring 1862. Lassell discovered some nebulae and was visited by Otto Struve, who carefully examined the 48-inch for comparison with his 15″ Merz refractor at Pulkovo. The result heated up the enduring discussion on 'reflector vs. refractor'.

7.5.1 The work on Malta

Lassell wrote two letters to the Royal Society, reporting his work on Malta. The first, addressed to President Edward Sabine, is dated 13 May 1862 (Lassell 1862a). After some technical problems and a severe winter, the 48″ reflector was ready to use. One of the first targets was the Orion Nebula. Lassell's observations led to an excellent drawing made by his daughter Caroline, a skilled artist. He did not publish it, but it was later engraved in Holden's monograph.[60] There are four letters of Lassell to John Herschel, written in 1862 and stored in the RAS archive (Table 7.12).[61]

Lassell investigated Lord Rosse's 'spiral nebulae', whose structure was clearly visible in the 48-inch. In his first letter to Herschel he wrote '*I am examining several of the nebulae Lord Rosse has figured. I confirm most fully the spiral character attributed to them by his Lordship.*' Added are drawings of M 100, M 99, NGC 5247 and M 83. Unfortunately the weather on Malta was even worse than during his first stay: '*the climate is altered* [...] *I have consequently done much less in observing than I had expected*'. Since only Messier's catalogue and Herschel's SC and CC were at his disposal, Lassell was eagerly awaiting Herschel's General Catalogue: '*I am delighted to hear that you are about to publish in one catalogue all the known nebulae* [...] *I have much felt the want of such a catalogue.*'

Another letter, of 26 September 1862, is addressed to the Secretary of the Royal Society, George Stokes (Lassell 1863); a copy went to Leverrier in Paris (Lassell 1862a). Lassell reported his observations of the planetary nebula NGC 7009 in Aquarius, using powers of 285 to 1480 (see Table 7.14): '*the nebula resembles in form the planet Saturn when the ring is seen nearly edgewise*' (a sketch was added). However, the name Saturn Nebula is due to Lord Rosse, who noted the following on 16 September 1849: '"*Saturn neb.*" *Pos. of ring 81°.*"[62]

Lassell's second letter to Herschel contains an invitation '*to see the nebula of Orion through this telescope*'. Concerning the observations he was on his own. There were only two workmen to maintain and operate the telescope, as mentioned in his letter to Edward Sabine (13 May 1862): '*For the luxury of observing two assistants are necessary, when the observer has really nothing to do but keep his eye at the telescope*' (Lassell 1862a). To minimise the effort required to turn the instrument to a new position, most observation were made near the meridian. To deal with the scientific work, Lassell strongly wanted a scientific assistant, as mentioned in his second letter: '*I much regret that too much splendid sky passes unused by me, and I am beginning to make*

---

[60] Holden (1882b: 76).
[61] RAS Herschel J. 12/1.7.
[62] Parsons L. (1880: 159).

Table 7.12. *Letters from Lassell to John Herschel (1862)*

Letter	Date	Content
1	23.5.1862	NGC 4477/79 (Coma Berenices); β CVn; spiral nebulae (with drawings of M 100, M 99, NGC 5347 and M 83)
2	26.9.1862	Search for an assistant; invitation to John Herschel
3	22.10.1862	Search for an assistant
4	1.11.1862	Marth's application; NGC 7662; appendix: various observations of nebulae (among them the discovery of NGC 7285)

Table 7.13. *Lassell's observations of southern nebulae (letter to Herschel, 1 November 1862)*

NGC	h	Type	V	Con.	Remarks
7135	3891	Gx	11.7	PsA	
7154	3900	Gx	12.3	PsA	
7507	3974	Gx	10.4	Scl	II 2
7582	3978	Gx	10.5	Gru	Dunlop 7.7.1826
7284	3943	Gx	12.2	Aqr	II 469
7285		Gx	11.9	Aqr	Lassell, late October 1862
7153	3898	Gx	13.5	PsA	
7152	3897	Gx	13.6	PsA	Not found by Lassell
7167	3905	Gx	12.2	Aqr	
7172	3908	Gx	11.8	PsA	
7173	3909	Gx	11.9	PsA	
7174	3910	Gx	13.1	PsA	
7258	3935	Gx	13.1	PsA	
7259	3936	Gx	13.1	PsA	

Table 7.14. *Lassell's eye-pieces used at the 48" reflector (M = magnification, f = approximate focal length, FoV = field of view)*

M	f (mm)	FoV (')
210	54	
228	50	13.7
231	49	15.5
285	40	10.5
466	24	
474	23	5.6
539	21	
576	20	
760	15	4
1060	11	
1480	8	

enquiry for some assistant on whom indeed at no distant period the main duty will probably fall.'

The last letter contains some observations of southern nebulae made in late October 1862. As Table 7.13 shows, these are galaxies in Pisces Austrinus, Aquarius, Sculptor and Grus – constellations that could easily be seen from Malta (latitude 36°). Except for one object (NGC 7285, discovered by Lassell), all are from Herschel's Cape catalogue.

Lassell wrote that Albert Marth had applied for the job as his scientific assistant: '*I have received an application from Dr. Marth of Durham to fill the position of observer with the four feet telescope.*' He added that '*from his long experience of the work and his general qualifications, he seems the most eligible of all candidates who have applied*'. Marth was eventually engaged in 1863 to search for new nebulae. However, Lassell's

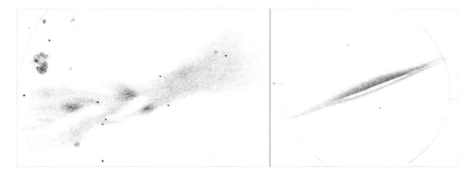

Figure 7.12. Drawings of Lassell (Lassell 1867a). Left: the galaxy NGC 253 in Sculptor (Fig. 1); right: the extreme edge-on galaxy NGC 4565 in Coma Berenices (Fig. 21^A).

own observations had priority: '*All these objects having first been satisfied, Mr. Marth has been very industriously scrutinizing the heavens when the moon was absent for nebulae.*' Despite these limitations, Marth was very successful: '*A good many new Nebulae however have been found, some description of which I hope to be able to give at a future time.*' Between 1863 and 1865 he discovered almost 600 objects – only William and John Herschel had found more to date (see Section 8.7). Marth catalogued them and Lassell published the result in 1867, titled 'A catalogue of new nebulae discovered on Malta with the four-foot equatoreal in 1863 to 1865' (Lassell 1867b). In the introduction of the 'Malta catalogue' Lassell explicitly appreciates Marth's effort. For the observations various eye-pieces were used; Table 7.14 collects the data given by Lassell.

In 1862 Lassell observed planets and their moons and drew galaxies, planetary and galactic nebulae. The high-quality figures are contained in Vol. 36 of the *Memoirs of the Royal Astronomical Society* (Lassell 1867a). Table 7.15 lists Lassell's drawings (that of M 1 was made in 1864). He identified the objects by means of John Herschel's catalogues (SC, CC, GC). For Fig. 15 Lassell noted the wrong object: not M 65 is shown, but M 66 (this error was later noticed by Vogel[63]). Some nebulae were drawn more than once; there are separate sketches showing the field stars too. An earlier publication contains another drawing of the Trifid Nebula (M 20), which was made in 1864 (Lassell 1865b). In 1889 Holden analysed 11 drawings of 'helical nebulae'

(marked *) to derive their true three-dimensional shape (Holden 1889b).

7.5.2 Otto Struve's visit and the 'refractor vs. reflector' rivalry

Unfortunately, John Herschel could not come to Malta, being too much concerned with the making of the General Catalogue at that time, but Otto Struve visited Lassell from 7 to 11 October 1863. Back at Pulkovo he wrote a detailed report about his stay (Struve O. 1866a). He had already visited Lassell in Liverpool to examine the 24-inch. Struve had inspected most of the big European telescopes and thus was able to judge their power.

When Struve arrived, Lassell was busy with erecting his old equatorial 9″ reflector. Of course, his main interest was focused on the large telescope. At the beginning, the observations, which were taking place 15 m above the ground, caused some problems: '*I will not hide that during the first night staying on the platform I felt uncomfortable, but the feeling disappeared in the following nights.*'[64] He added that '*Mr Marth tells too that in his first observing nights he had problems to master the fear; he even once got such a dizziness that he could rescue himself against falling only by jumping backwards into the tower with extraordinary willpower.*' By character, Marth was a fearful person, which didn't make things any easier.

For Struve the wind was a big problem, '*because it shakes both the platform and the telescope*'. He also

---

[63] Vogel (1888). Even John Herschel made an embarrassing mistake for M 66 in the SC (see Section 3.2).

[64] Struve O. (1866a: 525).

Table 7.15. *Lassell's drawings of nebulae (Malta 1862–64)*

Fig.	NGC	M	h	GC	Date	Remarks
1	253		61	138	18.10.1862	Fig. 7.12 left (see also Fig. 2.14)
2	1068	77	262	600	17.11.1862	*
3	1084		264	604	17.11.1862	*
4	1535		2618	826	22.11.1862	
6	1952	1	357	1157	1.1.1864	
8	2022		365	1225	3.2.1862	
9	2359		3075	1511	March 1862	*
10	2371		444	1519	24.3.1862	With NGC 2372
11	2392		450	1532	March 1862	2 drawings[65]
12	2903		604	1861	25.2.1862	With NGC 2905, 3 drawings, *
13	3132		3228	2017	16.4.1862	2 drawings
14	3242		3248	2102	24.3.1862	
15	3627	66	857	2377	21.4.1862	Not M 65 (h 854, GC 2373), *
16	4254	99	1173	2838	31.3.1862	*
17	4321	100	1211	2890	24.4.1862	2 drawings, *
18	4476		1296	3028	1862	With NGC 4478, 2 drawings
19	4477			3025	1862	With NGC 4479
20	4501	88	1312	3040	21.5.1862	
21	4565		1357	3106	29.4.1862	2 drawings; Fig. 7.12 right
22	4594	104	1376	3132	28.4.1862	
23	4621	59	1386	3155	29.4.1862	With NGC 4606
24	4634		1397	3167	27.3.1862	
25	4736	94	1456	3258	20.5.1862	
26	4826	64	1486	3321	22.4.1862	
27	5194	51	1622	3572	27.6.1862	With NGC 5195, 2 drawings, *
28	5236		3523	3606	20.5.1862	*
29	5247		1649	3614	20.5.1862	*
30	6337		3680	4290	21.6.1862	
31	6428		1989	4343	27.6.1862	
32	6514	20	1991	4355	27.6.1862	With stars
33	6618	17	2008	4403	1862	With stars, *
34	6781		2037	4487	29.8.1862	
35	6853	27	2060	4532	1862	With stars
36	6905		2075	4572	19.8.1862	
37	7009		2098	4628	23.8.1862	
38	7662		2241	4964	23.10.1862	

criticised the telescope's small field of view. Thus the observation of reference stars was difficult. There were, however, not enough catalogued stars for southern declinations. The northern Bonner Durchmusterung (published in 1861) reaches only to −2°, but on Malta it was possible to observe down to −40°. Thus appropriate reference stars, which should be visible in the eye-piece, were rare for these regions. About the finder Struve remarked that '*Alas, the telescope's finder is not*

---

[65] One is shown in Fig. 6.15; in the other, the nearby star (BD +21° 1610, 8.2 mag, 1.6′ north) appears nebulous too.

powerful enough for such purpose.'[66] This was the reason why Lassell had put his old 9-inch next to the large reflector; this arrangement should allow better orientation. The visitor criticised Lassell's micrometer too, describing it as '*not adequate to the optical power of the instrument*'.[67] Exact positional measurements, one of Struve's main tasks, could hardly be realised with Lassell's equipment. In comparison with Lassell's equipment, Struve's 15″ refractor at Pulkovo was judged to be a true precision instrument.

Struve observed Uranus and Neptune and their moons, double stars and nebulae. Special targets were the Ring Nebula in Lyra (M 57), Hind's Variable Nebula in Taurus (NGC 1555), the Saturn Nebula NGC 7009 in Aquarius and the Orion Nebula (M 42). About the latter he remarked '*Most of all, I was curious about inspecting the Orion nebula. However, my expectation of seeing all kinds of new forms and details not visible at Pulkovo did not come true.*'[68] Struve had planned parallel observations on Malta and at Pulkovo, where Winnecke was charged with using the 15″ refractor. In a report for 1863, Winnecke wrote that the '*Main intention of O. Struve's scientific sojourn was to compare the power of the new reflector with that of the Pulkovo refractor.*' (Winnecke 1866). Concerning the optical quality of the 48-inch, Struve amicably concluded that, '*Though I cannot estimate the sharpness of the images to be perfectly equal to those giving the Pulkovo refractor such a high benefit, it is but a pleasure to testify that Lassell's new reflector comes, to a considerable higher degree than any other large reflector known to me, close to our telescope in every way, and that it keeps the same image quality in all elevations against the horizon.*'[69]

Anyway, years later Struve's report led to a controversial debate about the power of Lassell's telescopes and the theme 'reflector vs. refractor' in general. In an article by Romney Robinson, Director of Armagh Observatory, titled 'On the relative power of achromatic and reflecting telescopes', one reads the following: '*Otto Struve thought that the Pulkovo Achromatic of 15 inches was equal to Lassell's 24-inch Newtonian*' (Robinson 1876). Struve urgently clarified that his Malta report actually refers to the 48-inch: '*the space-penetrating power of Lassell's new (4-feet) instrument can hardly be estimated superior to that of ours (the Pulkovo 15-inch) refractor*' (Struve O. 1877). This caused Lassell to react: '*There appears to me to be something so erroneous in this conclusion, that although very averse to controversy, and especially to enter the lists with my very distinguished friend, M. Struve, I crave permission to state my greatly differing opinion, and the grounds on which I have formed it.*' (Lassell 1877). On the basis of his observations of the Uranian moons Umbriel and Ariel, which had not been seen at Pulkovo, he believed that his 24″ reflector might even be superior to the 15″ refractor. As proof he quoted George Phillips Bond, past Director of Harvard College Observatory: '[Bond] *expressed verbally to myself, that my 2-foot Reflector [24″] was sensibly more efficient than the Harvard Telescope, which may be called a facsimile of that at Pulkovo*'. For Lassell the drawings of nebulae made with 24″ and 48″ reflectors clearly show that the larger telescope was superior – a statement that let the Pulkovo refractor appear in even worse light. For Struve's opinion, he had only one explanation: '[his] *unfavourable impression may perhaps in some degree have arisen from his visit to Malta having been made at a time of the year when the atmosphere is occasionally much disturbed by storms, which are of course greatly to the disadvantage of a very large telescope.*'

Later the subject 'reflector versus refractor' was treated in the magazine *Astronomical Register*, with contributions by Henry Cooper Key, Thomas William Webb and Charles Burton.[70] Key, the proud owner of a self-made 18¼″ reflector with silvered glass mirror, criticised an article by Norman Lockyer, who spoke about an '*enormous superiority possessed by refractors*'. This originated from observations of Saturn by Lassell (with a 48″ reflector) and Struve (with a 15″ refractor). Burton compared a 7.5″ Clark refractor with a self-made 9″ reflector (silver-on-glass). Tests on planets and double stars led to the result that '*a refractor of given aperture will probably be equal to a reflector of one-seventh the greater aperture in illuminating power, but falls perceptibly below the defining power*'. Tempel later continued the debate with every emphasis – especially against Dreyer. He claimed his 11″ refractor at Arcetri to be better than the Herschel reflectors, not hesitating

---

[66] Struve O. (1866a: 527).
[67] Struve O. (1866a: 529).
[68] Struve O. (1866a: 539).
[69] Struve O. (1866a: 530–531).

[70] Key (1868a, b), Webb (1869) and Burton (1872a, b).

Table 7.16. *Objects discovered by Lassell (sorted by date)*

Date	NGC	Au	GC(S)	m	Type	V	Con.	Remarks
31.3.1848	**3121**	26	2010		Gx	12.9	Leo	24″, Liverpool
Oct. 1862	**7285**		5078		Gx	11.9	Aqr	Pair with NGC 7284; GC Supplementary List
5.5.1863	**2620**		5426	125	Gx	13.7	Cnc	GCS
14.9.1863	**7489**		6124	523	Gx	13.4	Peg	GCS

even to claim that his equipment rivalled Lord Rosse's 72-inch (see Sections 8.10.2 and 11.3.4). As early as in 1857 Secchi had compared John Herschel's 18¼″ reflector with his 9.5″ Merz refractor, finding the two devices equivalent (see Section 6.11.2).

The dispute goes even back to Joseph von Fraunhofer and John Herschel. On the occasion of completing the famous 24.4-cm refractor for Dorpat Observatory, Fraunhofer said in a speech delivered on 10 July 1824 that *'The largest telescopes currently existing are reflectors with metal mirrors. As even a perfect metal mirror only reflects a small part of the incoming light, absorbing the larger part, reflectors must be tremendously large to achieve a large effect, and thus the intensity of the light received by the observer's eye is always very small.'* (Fraunhofer 1826). Consequently, *'reflectors could not be advantageously used for mathematical-astronomical observations, and, for instance, not every meridian-instrument was equipped with them.'* Therefore, *'currently nearly all astronomical observations are made with achromatic telescopes.'*

John Herschel could not accept this, in view of the success of his father and his own positive experience with reflectors. His letter to the *AN*, dated 15 August 1825, should serve *'to obviate an erroneous impression which may arise in the minds of those who read Fraunhofer's memoir, as to the great inferiority of reflecting telescopes in point of optical power, to achromatics in general, and more especially to those constructed with such delicacy as his own doubtless are'* (Herschel J. 1826c). He further wrote that *'M. Fraunhofer's expressions when speaking of the loss of light by metallic reflexion are, I think, somewhat too strong. […] A metallic mirror however reflects 0.673 [67.3%] of the incident light, or more than two thirds, and absorbs less than one third of the whole.'* Herschel believed that Fraunhofer has meant the double reflection in a Newtonian system (with two metal mirrors); in this case the value is only 45.2%. But, even then, he saw no reason to doubt the power of a large reflector: *'No one who has been half blinded by the entrance of Sirius or Lyra [Vega] into one of my Father's 20-feet reflectors, will say that the intensity of its light is small, nor, to take a less extreme case, will any one who uses one of M. Amici's Newtonian reflectors of 12 inches aperture (a perfectly convenient and manageable size, and of which he has constructed several) be disposed to complain of its want of light.'* Herschel stated that *'A reflector of 18 inches aperture would be equivalent to an achromatic of 15½, and one of 48 inches to an achromatic of 41½, a size we cannot suppose'* (the 48-inch was William Herschel's largest reflector).

7.5.3 Lassell's discoveries

As described in Section 6.8, Lassell's first find was the galaxy NGC 3121 in Leo, which he discovered in 1848 with the 24-inch in Liverpool. During his second stay on Malta, he discovered three galaxies with the 48-inch (Table 7.16).

Lassell's first discovery on Malta was NGC 7285, a galaxy in Aquarius, which forms an interacting pair (Arp 93) with NGC 7284 (Fig. 7.13). Herschel entered it as GC 5078 in the 'Supplementary list of nebulae and clusters'. Owing to its proximity to h 3943 (NGC 7284), he designates the nebula as 'h 3943,a' too, but gives no source (the same applies for the NGC). The object is mentioned in Lassell's fourth letter to Herschel, of 1 November 1862 (see Table 7.12). While observing h 3943 a *'nebulous star distant about a minute'* was noticed. Because his third letter is dated 22 October 1862, Lassell must have found NGC 7285 later in that month. The close pair NGC 7284/85 was observed by Howe at Chamberlin Observatory, Denver, in 1897 (Howe 1898a). About Dreyer's description of NGC

7.6 Content and structure

Figure 7.13. The double galaxy NGC 7284/85 in Aquarius (DSS).

7284 and NGC 7285 ('*cF, cS, lE, r, D star inv*' and '*Nebs. star 1′ dist. from 7284*') he remarked '*I judge 7285 to be simply one of the components of 7284. Both seemed to be nebulous stars.*'

Lassell's last discoveries were not published until 1867 in the Malta catalogue (see Section 8.7.3): the galaxies NGC 2620 (m 125) in Cancer and NGC 7285 (m 523) in Pegasus. The former was found on 5 May 1863, about a month before Marth started his systematic search. NGC 7285 was discovered by Lassell on 14 September 1863. At that time the constellation Pegasus was surveyed by Marth, who found 10 galaxies between the 13th and the 17th. Lassell's two objects came too late for the General Catalogue; they are listed in Dreyer's Supplement as GC 5426 and GC 6124.

7.6 CONTENT AND STRUCTURE OF THE GENERAL CATALOGUE

7.6.1 The making of the catalogue

In the introduction of his work, Herschel writes about the case history. The development of telescope technique, both for reflectors (Lord Rosse, Lassell, Nasmyth) and for refractors (Fraunhofer, Merz, Cooke), had a great influence on the observations of nebulae and star clusters. Precise positions of large samples were measured (Herschel marked d'Arrest's Leipzig observations) and, for selected objects, detailed studies were made, e.g. by Secchi, Bond and Mason. Facing this growing activity, Herschel realised the need for a comprehensive, up-to-date catalogue: '*it* [is] *extremely desirable to have presented in one work, without the necessity of turning over many volumes, a general catalogue of all the nebulae and clusters of stars actually known, both northern and southern, arranged in order of right ascension and reduced to a common and sufficiently advanced epoch which may serve as a general index to them, and enable an observer at once to turn his instrument on any one of them, as well as to put it in his power immediately to ascertain whether any object of this nature which he may encounter in his observation is new, or should be set down as one previously observed.*'[71]

Herschel addresses a general problem: many discoveries turned out to be known objects. Owing to the many publications and the large number of non-stellar objects, the necessary identity check was arduous, often leading to misinterpretations (e.g. a comet was thought to be a nebula and vice versa). He mentions further reasons for a new catalogue. Caroline Herschel reduced all of the objects found by her brother to the epoch 1800. She then sorted the data by right ascension and arranged them in declination zones of 1°. The resulting zone catalogue was the basis of John Herschel's four 'working lists'. Unfortunately, not all of the objects catalogued by William Herschel could be observed in Slough and at the Cape. Many were thus missing from the SC (1833) and CC (1847). For him, a 'general catalogue' was the opportunity to include all these objects '*which had escaped my own observations (a very numerous list)*'.

Supported by his sons, Herschel first calculated new positions for all of the objects of Caroline Herschel's zone catalogue. They were reduced to 1830, the epoch of the SC and CC. Sorting the data by right ascension should allow a direct comparison with his earlier catalogues to reveal errors and misidentifications. After a detailed analysis, Herschel precessed the coordinates to the final epoch 1860. For this task he got help from an 'experienced computer' deputed by Airy from the Royal Observatory, Greenwich. The work was finished on 6 February 1863.

On the 23rd Herschel received a surprising stimulus: "*I received, by the kindness of the Astronomer Royal*

---

[71] Herschel J. (1864: 2).

[Airy], *a copy of the important work of M. Auwers before alluded to, entitled 'William Herschel's Verzeichnisse von Nebelflecken und Sternhaufen, bearbeitet von Arthur Auwers, Königsberg, 1862', of whose existence this was my first notice.*"[72] Enthusiastically he added *'It may be readily supposed that I lost no time in comparing my own previous work with this of M. Auwers.'* Unfortunately, Auwers had no access to the valuable work of Caroline Herschel, since it was not published. Therefore he could use only William Herschel's catalogues, which give only one measurement for each object (position relative to the reference star). Auwers' calculated absolute positions were thus based on poor data. Caroline Herschel was in a much better position. She could use the original data of her brother, which normally present several measurements for each object, on the basis of different sweeps and reference stars. Therefore, the resulting coordinates were much more reliable (*'any suspiciously large deviation from the mean of all may be at once noticed and traced to its origin in the sweeping books'*[73]). This was the information Herschel could rely upon. He greatly admired the work of his aunt: *'I learned fully to appreciate the skill, diligence, and accuracy which that indefatigable lady brought to bear on a task which only the most boundless devotion could have induced her to undertake or enabled her to accomplish.'*[74] But Auwers' work was rated highly too: *'I must not omit to add that the comparison so instituted with M. Auwers's results has led me to the detection of several grave errors in my own work which would certainly otherwise have escaped notice (and in some cases have caused the loss of future observations by missetting the telescope), and whose rectification has added materially to its value.'* Of course, he detected errors in Auwers' data too: *'On the other hand, as no human work is perfect, I have been led to notice some errors in M. Auwers's work itself.'* These are listed in the General Catalogue. Auwers' publication appeared just in time and it was an important contribution, especially concerning the quality of Herschel's new catalogue.

The inclusion of new objects was problematic. The draft catalogue was a huge handwritten table, spread over many single pages. Inserting new rows was a challenging task, insofar as shifts and new numberings could cause errors. When the written rows became denser, Herschel thought about copying the whole table to get a better overview. But he eventually decided against this time-consuming work: *'I believe it is impossible to copy so voluminous a mass of figures and abbreviated writing without numerous errors.'*

When he was compiling the General Catalogue, Herschel's active time as an observer was past history. New objects were thus not checked at night. This can be seen by examining the last column giving the 'Total no. of times of obs. by h and H.': it shows a '0' for all objects not contained in his earlier catalogues. The only thing he could do was to be extremely careful at his desk work. Nevertheless, he made several errors – no wonder, considering the enormous amount of handwritten data.

7.6.2 The structure of John Herschel's publication

At the end of the introduction, Herschel lists the standard abbreviations (introduced by his father) to describe the objects. There follows a larger part with 'Notes on the catalogue', ordered by GC-number (for safety the common H- or h-designations are given too). In the case of a new object, the particular list (e.g. Auwers, d'Arrest, Lord Rosse) is mentioned. The notes contain 286 entries; many include remarks on several objects.

The GC is the first work to present 'References to published figures of nebulae'; 298 objects are listed. Most of the drawings are from John Herschel's catalogues and Lord Rosse's publications. Interesting is a list of 54 Birr Castle objects with peculiar structure: 'spiral' (S), 'dark space' (D), 'knotted' (K), 'rays with splits or clefts' (R). Obviously Herschel has not included all cases; perhaps some were overlooked (see Table 7.2). Finally there are corrections to William Herschel's catalogues and Auwers' publication.

The catalogue itself covers 92 pages. It consists of the main list with 5057 entries (nos. 1 to 5057) and the 'Supplementary list of nebulae and clusters' with an additional 22 entries (nos. 5058 to 5079). In both the objects are sorted by right ascension for 1860. Surprisingly, one reads in the section 'Explanation and arrangement of the catalogue' that *'the first* [column] *contains the general or current number in order from 1 up to 5063, the total number of objects comprised, including*

---

[72] Herschel J. (1864: 4).
[73] Herschel J. (1864: 5).
[74] Herschel J. (1864: 3).

7.6 Content and structure   219

Table 7.17. *The columns of the General Catalogue*

Column	Meaning
No. of Catalogue	GC-number (1–5079)
Sir J. H.'s Catalogue of Nebulae	h-number; Birr Castle objects with added ',a' etc.
Sir W. H.'s Classes and Nos.	H-designation (classes I to VIII)
Other Authorities	Other observers/discoverers; alternative designations
Right Ascension for 1860, Jan. 0	Right ascension; precision mostly 0.1 s (otherwise ± or digits omitted)
Annual Precession in right ascension for 1880	Precession (only for precise right ascension)
No. of Obs. used	Number of observation for right ascension; ( ) = doubtful observation, [ ] = data by d'Arrest
North Polar Distance for 1860, Jan. 0	N.P.D. = angle to north pole (90° – δ); precision mostly 0.1′ (otherwise ± or digits omitted)
Annual Precession in N.P.D. for 1880	Precession (only for precise NPD)
No. of Obs. used	Number of observation for NPD; ( ) = doubtful observation, [ ] = data by d'Arrest
Summary Description from a Comparison of all the Observations, Remarks, &c.	Description, remarks
Total No. of times of Obs. by h. and H.	Number of observations by William and John Herschel (0 for new objects); † or * = sign for figure or notes

six supplementary ones.[75] The 22 additional objects could not be included, because they came '*too late for insertion in the main body of the Catalogue*'. But this is incorrect, for Herschel has inserted all 22 (not only 6!) in the main catalogue at their appropriate places (according to right ascension), though not changing the existing numbering: '*the insertion in their proper order in R.A. would have involved altering all the numbering both of the catalogue and the annotations, &c., and would have proved a source of confusion and unavoidable error*'. The GC columns are given in Table 7.17.

7.6.3 The content of the GC

The GC is mainly based on John Herschel's two earlier catalogues (SC and CC), which include 89 common objects in an overlapping region of the northern and southern sky. Altogether 15 objects bear no h-number; they belong to the Large or Small Magellanic Cloud (abbreviated LMC and SMC) and were first listed in an appendix of the Cape catalogue. Of these, 13 were entered in the main catalogue and the remaining two in the 'Supplementary list'. These are GC 5062 = (123) and GC 5063 = (374), which appear in the Cape catalogue as numbers in round brackets. A large part is occupied by William Herschel objects that were not observed by his son. This covers nearly 90% of all GC entries. A further 438 objects were included from various lists. Curiously, no. 3398 is missing from the GC – there is not even an appropriate candidate! Table 7.18 shows all of the components.

Among the 'new' objects are some from Messier's catalogue, which is only partly contained in Herschel's SC and CC (see Tables 3.2 and 5.3). Surprisingly, the GC does not give all 103 numbers too – omitted are the problematic cases like M 48, M 91 and M 102. Nevertheless, the three objects (now related to these numbers) can be found in the General Catalogue: GC 1637 = NGC 2548 (Caroline Herschel), GC 3093 = NGC 4548 and GC 4058 = NGC 5866 (both William Herschel). A special case is the open cluster M 47, assigned as GC 1594 by Herschel, who noted '*Place*

---
[75] Herschel J. (1864: 7).

Table 7.18. *Components of John Herschel's General Catalogue (including the 'Supplementary list')*

Objects	In GC	Thereof in Suppl. List	Remarks
W. Herschel	2543	1	
J. Herschel (Slough)	587		
J. Herschel (Cape)	1507	2	Including 14 from the LMC, 1 from the SMC
Newly entered	419	19	
Missing entry	1		GC 3398

from Wollaston's Cat'.[76] Unfortunately this is the wrong place. Dreyer catalogued the object as NGC 2478, which is finally identical with the true M 47: GC 1551 = NGC 2422. Incorrect too is Herschel's identification of the open cluster GC 4397 (h 2004 = NGC 6603) with M 24.

GC 1953 in Ursa Major, which was discovered by William Herschel, is curious. In the General Catalogue it is called 'W. H. nova?', with the addition 'M. 81 ??'. In the notes John Herschel wrote that *'It would certainly be very extraordinary should three nebulae so extremely remarkable as M. 81 and 82 and this be found to lie so near together.'* But William Herschel's position is not that of the bright galaxy M 81. The correct Messier object M 81 (GC 1949 = h 649) bears no H-designation. He probably did indeed observe M 81, but noted a wrong position.[77] Definitely not listed in the GC are M 24, M 25 (both in Sagittarius), M 40 (Ursa Major) and M 45 (Taurus). For Herschel the Sagittarius Star Cloud M 24 appeared too large and the open cluster M 25 was not exactly localised. Both were first entered in Dreyer's Second Index Catalogue of 1908 as IC 4715 and IC 4725. M 40 was not found by William Herschel and, according to Auwers, the object is only a pair of stars (see Section 6.12.2). M 45 is the Pleiades, which were obviously too large for Herschel (and later for Dreyer) to be catalogued as a non-stellar object. The GC contains the seven objects added to the Messier catalogue in the twentieth century (M 104 to 110; see Table 2.2). All were observed by William Herschel.

John Herschel included two Lacaille objects from the 1755 list: Lac I.11 = GC 5076 (NGC 6634) and Lac I.13 = GC 4484 (NGC 6777). The former is incorrectly called 'M 69' in some catalogues; the true M 69, which was discovered by Messier, is identical with Δ 613 = GC 4411 = NGC 6637. Table 7.19 shows the discoverers of the 438 newly entered GC objects.

About the number of new objects from Auwers, d'Arrest and Lord Rosse, Herschel wrote of *'those of M. Auwers's catalogue of 'novae', those communicated to me for insertion by M. d'Arrest, and those noticed by Lord Rosse in his memoir of 1861, altogether 433 objects'*.[78] He included 36 of the 50 objects from Auwers' list of new nebulae (Au; see Table 6.28). The remaining 14 are assigned to d'Arrest (Au 3–8, 10–15, 29 and 50). Except for one (Au 15 = GC 358), all are contained in the list of 125 nebulae d'Arrest had sent to Herschel on 8 May 1863. Herschel gives no reference to Auwers in these cases (except for d'Arrest no. 6 = Au 4 = GC 119). Thus we have 126 d'Arrest objects (125 plus Au 15) and 36 from Auwers' list.

Herschel includes 271 Birr Castle discoveries (according to his own counts), called 'Rosse nova'. On adding all objects from d'Arrest, Auwers and Lord Rosse, we end up with Herschel's number of 433 entries. But this does not match the truth, as analysed in the sections on d'Arrest and Lord Rosse.

Referring to Lord Rosse's publication of 1861, Dreyer later remarked in the GCS introduction that "*It is to be regretted that the condensed form of this publication has made Herschel often make mistakes in the identification of the so-called 'novae', many of which are to be found among Sir William Herschel's Nebulae, as only Sir John Herschel's Slough-Catalogue was then used as a working-list by the observers at Birr.*"[79] He added that *'This has in many cases been noticed by D'Arrest, whose*

---

[76] Here Francis Wollaston's star catalogue of 1789 is meant.
[77] Steinicke (2007a).
[78] Herschel J. (1864: 6).
[79] Dreyer (1878a: 383–384).

Table 7.19. *Discoverers of the 438 objects newly entered into the GC (main catalogue and supplementary list)*

Discoverer	Main catalogue	Suppl. list	Source	Remarks
Auwers	2		Auwers	
Bessel	1		Zone observations	See Table 6.20
Bode	2		Messier	M 81, M 92
Bond	8	5	Auwers, Bond list	GC 5058 = last GC object found
Brorsen	1		Auwers	
Bruhns	1		Auwers	
Chacornac	1		*Paris Bulletin*	
Coolidge		8	Bond list	
Cooper	7		Auwers	
d'Arrest	105		List, Auwers	
de Chéseaux	1		Messier	M 35 (BAC 5455)
Harding	1		Auwers	
Hind	4		Auwers	
Hodierna	2		Messier	M 38, M 47 (Wollaston)
Lacaille	1	1	Messier, list	M 4
Lassell	1	1	Auwers, letter	
Méchain	2		Messier	M 76, M 79
Messier	3		Messier	M 69, M 73, M 90
Mitchell	98		PT 1861	
Parsons W.	2		PT 1861	
Safford		2	Bond list	
Schmidt	1		Auwers	
Schönfeld	2		Auwers	Tuttle
Stoney B.	100	2	PT 1861	
Stoney J.	65		PT 1861	
Struve W.	2		Auwers	
Tempel	1		Auwers	
Winnecke	5		Auwers	

suggestions with regard to these objects I of course was more able to confirm or reject than many others.' When compiling the GC, Herschel was not fully aware of the issue. The reason might be the difference of 17 years between the publication dates of the Cape catalogue and the General Catalogue – and perhaps his advanced age too. Thus Dreyer later remarked that '*Sir John Herschel, for instance, when preparing his General Catalogue of nebulae, seems to have taken no notice of the errata given in his own Cape Observations.*' (Dreyer 1900). In an article of 1912 he further criticised that John Herschel had not sufficiently proved the positions of his father's objects by means of Auwers' revision: '*I assumed that Sir John Herschel had compared his positions throughout with those found by Auwers from W. Herschel's published observations. It appeared now that this comparison has not been an exhaustive one.*' (Dreyer 1912b).

7.6.4 Auwers' review

William Huggins was the first to cite Herschel's new catalogue. In his important work 'On the spectra of some of the nebulae', which appeared in the same volume of the *Philosophical Transactions* as the General

Catalogue, he already used GC-numbers (Huggins W. 1864). There Lord Rosse's publication of 1861 is mentioned too.[80]

The first review of the GC was published in 1866 – no less a person than Arthur Auwers felt a vocation for this task. It appeared in the first volume of the *Vierteljahrsschrift der Astronomischen Gesellschaft* and is eminently positive (Auwers 1866). He wrote that '*The study of nebulae has considerably boomed in the last quarter century, supported by the erection of many telescopes, having the power to reveal the proper appearances of these heavenly bodies*.' Auwers mentions the dominance of William and John Herschel: '*Over a period of nearly 60 years it was almost exclusively the tremendous task of the two Herschels to advance this large field introduced by Sir William into astronomy.*' But then he turns his view to recent observers and their remarkable contributions: Lord Rosse, Lassell, d'Arrest, Secchi, Bond and Mason – not forgetting himself. He complains that Herschel has not credited further observations, which '*however, treated only single objects, but revealed special results of high interest*'. Explicitly he mentions Lamont ('*the work published until 1863 or at least announced*'), Otto Struve, Schmidt, Laugier and Schönfeld. About the latter Auwers remarked that his '*first part [Erste Abtheilung] of the Mannheim observations appeared as early as in 1862, but became known in England only much later*' (see Section 8.5). For Auwers Herschel's late work was triggered not only by the great number of new discoveries but also by '*the recent prospect of upgrading Melbourne Observatory with a large reflector to study southern nebulae*'.[81]

Auwers writes that, concerning the Herschel objects (H, h), the GC contains '*all nebulae and star clusters observed by himself* [John Herschel] *or his father, except for one object listed in the catalogues of the latter; a diffuse nebulosity (V. 37), spread about many square degrees, which was omitted like 52 similar regions described in the Phil. Transactions of 1811*'. Indeed, John Herschel did ignore the 52 regions with 'extensive diffused nebulosity' listed by his father.[82] However, in the case of V 37 Auwers is wrong: the object in Cygnus later known as the North America Nebula (NGC 7000; see Section 2.4) is actually listed in the General Catalogue as GC 4621 (h 2096).

With satisfaction Auwers noticed that the 50 nebulae of his own list had been included. Concerning the Birr Castle objects, he remarks that '*in most cases their positions were only roughly determined*', though he is convinced about the quality of the final reductions because they were made '*under Airy's control by a Greenwich computer*'. Being a master of positional astronomy, he must add that '*the small systematic errors committed in this task are negligible in view of the uncertainty of the places*'. Auwers once again regrets that '*Miss Carolina's zone catalogue will ever remain as a mere manuscript*'. In this connection he did not miss the opportunity to mention his own results: '*A work of the same kind, however, excluding the nebulae not visible from Europe, had been made by the Reviewer as early as in 1857 for his own use and was a very helpful tool for observations*.' Auwers wrote that his own '*revision of W. Herschel's catalogues of nebulae was noticed by Sir John only when he* [Herschel] *had already filled gaps in his observational data with the aid of Miss Carolina's catalogue*'. He is proud that his own work has '*initiated numerous corrections on all sides*'.

Auwers' review contains a list of 26 corrections, sorted by GC-number. However, most of them concern his own publication: typos, wrong reference stars or wrong identifications. He mentions, for instance, GC 710 (NGC 1333), for which Herschel notes 'Schönfeld 1858'. Auwers corrects this to '1855' – actually, it was his own mistake! About the GC-number he remarks that '*A general and exclusive use could be in the interest of an upcoming generation of observers.*' But he recommends use of the '*new nomenclature not exclusively, but only in connection with the existing standard to avoid any confusion*'. Anyway, the GC-numbers were immediately accepted, mainly because Herschel had set a new standard, presenting the first reliable source for all known nebulae and star clusters.

7.6.5 Plotting the distribution of GC objects

Cleveland Abbe was the first to investigate the distribution of GC objects. On the basis of John Herschel's ideas presented in the Cape catalogue, he published a paper 'On the distribution of the nebulae in space' (Abbe 1867). He divides the sky into fields of 30^m in

---

[80] It was due to this quotation that d'Arrest became aware of Lord Rosse's work.
[81] About the Great Melbourne Telescope, see Section 9.21.
[82] They were published in Herschel W. (1811); see also Latußeck (2003a, b, 2004).

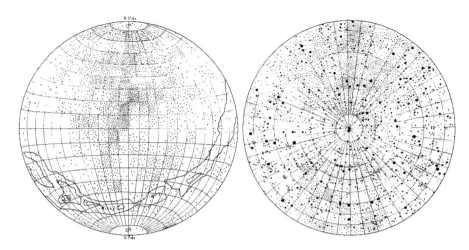

Figure 7.14. Proctor's presentation of the GC objects (Proctor 1869a). Left: the equatorial view, centred on 12^h, is dominated by the Virgo Cluster; right: the view from the northern pole with BAC stars (12^h up).

right ascension and 10° in declination (Herschel had used 1^h and 15°, respectively). Abbe distinguishes 'clusters' (637 objects), 'resolved or resolvable nebulae' (397) and 'unresolved nebulae' (4053).[83] Obviously he miscounted, because his total number of objects gives 5087 – instead of 5079. The publication presents data only for 'unresolved nebulae'. Abbe drew the following conclusions.

- *'The Clusters are members of the Via Lactea* [Milky Way], *and are nearer to us than the average of its faint stars.'*
- *'The Nebulae resolved and unresolved lie in general without the Via Lactea, which is therefore essentially stellar.'*
- *'The visible universe is composed of systems, of which the Via Lactea, the two Nubeculae, and the Nebulae, are the individuals, and which are themselves composed of stars and of gaseous bodies of both regular and irregular outlines.'*

Two years later Richard Proctor showed graphs of all GC objects in his paper 'Distribution of nebulae' (Proctor 1869a). He also referred to John Herschel, who had visualised (though only nominally) the distribution of nebulae for both hemispheres in the Cape catalogue.[84] Using Abbe's data, Proctor presents four plates (I–IV) with two charts each. The first shows distributions for the northern and southern sky, seen from the poles. The second and third plates present a side view to the equator for four different sectors (Fig. 7.14 left). The last plate shows the 'Stellar and nebulae systems' for both hemispheres in polar view (Fig. 7.14 right); here stars from the *British Association Catalogue* (BAC) and the band of the Milky Way are plotted too. On chart II.1 the Virgo Cluster is clearly visible. In a later publication, titled 'The rich nebular region in Virgo and Coma Berenices', Proctor presents this area in detail (Proctor 1872a). However, the galaxy cluster in Coma Berenices (Abell 1656) is not visible.[85] Proctor also plotted the distribution of the BD stars (Proctor 1871a, 1872b).

In 1873 Sidney Waters presented another GC plot in his paper on 'The distribution of the clusters and nebulae' (Waters 1873b). The resolution is much higher than that of Proctor's map: 4^m in right ascension and 1° in declination. Moreover, Waters separates three object-classes: 'Clusters (globular and irregular)', 'Resolvable Nebulae' and 'Irresolvable

---

[83] In his 1875 paper he also treated 'extended nebulae' (Abbe 1875).

[84] Herschel divided each hemisphere into fields, in which the numerical value of the particular frequency is entered (Herschel J. 1847: Plate X).

[85] Steinicke (2005a).

Nebulae', indicated by different symbols. His first plate presents all BAC stars and all GC objects for both hemispheres in polar view. Since both the resolved and the unresolved nebulae are not concentrated on the Milky Way, Waters concluded that the *'resolvability of a star group is no criterion of its distance'*.[86] The unresolved nebulae (i.e. galaxies) are crowded in Virgo and Coma Berenices, whereas the star clusters are clearly focused on the Milky Way, confirming earlier results.

---

[86] Waters (1873a).

8 • Dreyer's first catalogue: the supplement to Herschel's General Catalogue

Dreyer's first catalogue, titled 'A supplement to Sir John Herschel's 'General Catalogue of Nebulae and Clusters of Stars', appeared in March 1878 in the *Transactions of the Royal Irish Academy* (Dreyer 1878a). The work – henceforth abbreviated GCS – had already been received by the Academy on 26 February 1877, but afterwards was corrected by the author. Besides the new objects, it contains a large number of notes and corrections to Herschel's GC. At the same time Dreyer was working on the compilation of the Birr Castle observations made during 1848–78. These two projects influenced each other. In the GCS all objects newly found at Birr Castle were entered with calculated coordinates. The same applies for discoveries of other observers, such as Schultz and Tempel. Dreyer's supplement was the essential step towards the NGC. His first appearance as author of an important catalogue of nebulae and star clusters gives us the occasion for a detailed review of his life.

8.1 DREYER'S BIOGRAPHY

8.1.1 Copenhagen

Johan Ludvig Emil Dreyer (in English later: John Louis Emil) was born on 13 February 1852 in Copenhagen (Fig. 8.1). Many of his ancestors served in the army and navy. His father, Johan Christopher Friedrich Dreyer,[1] became Danish Minister of War and the Navy in 1864. Also three of Dreyer's three sons kept this tradition, e.g. Frederic Charles Dreyer became an Admiral of the Royal Navy. But the Dreyer with whom we are concerned obviously took a very different path, becoming an eminent astronomer. He attended school in Copenhagen and was initially interested in languages, history and science. At the age of 14 he got a book about Tycho Brahe, from which originated his lifetime

Figure 8.1. John Louis Emil Dreyer (1852–1926) during his time in Armagh.[2]

fascination for the great Danish astronomer. Now his aim was to study astronomy. Dreyer often visited the new Copenhagen Observatory at Østervold; especially Schjellerup took care of the interested pupil. Even at that time, his talent for observations and precise analyses had already appeared.

In 1869, at the age of 17, Dreyer began his study of mathematics and astronomy at Copenhagen University. There he was particularly fascinated by the lectures of Heinrich Ludwig d'Arrest, director of the observatory.

---

[1] He was married to Ida Nicoline Margrethe Dreyer (*née* Randrup); they had four sons and two daughters.

[2] See also the frontispiece and Fig. 1.2.

Throughout his life, d'Arrest was his great idol, whose influence can be traced in all his work, especially on nebulae. Just a year later Dreyer was allowed to use the telescopes. In 1872 he wrote some popular articles on Tycho Brahe, and a year later his first scientific paper was published.[3] Dreyer graduated in 1873 as M.A. with a work on observational errors at the Pistor and Martins meridian-circle of Copenhagen Observatory. An enlarged version was published in 1877, titled 'On personal errors in astronomical transit observations' (Dreyer 1877b).

8.1.2 Birr Castle

In August 1874 Dreyer, then aged 22, left Copenhagen for Irish Parsonstown (now Birr). He was employed as Astronomer at the Earl of Rosse's Observatory to assist Lawrence Parsons, the Fourth Earl of Rosse. He followed Ralph Copeland, who changed to Dunsink. The 15-years-older English astronomer was a life-long friend. In 1875 Dreyer became a Fellow of the Royal Astronomical Society.

At Birr Castle he had access to the 36" and 72" reflectors and his main task was observing and cataloguing nebulae. Here Dreyer wrote his first papers on the subject: reviews of the publications of Schultz, Schönfeld and Vogel, which appeared in the *Vierteljahrsschrift der Astronomischen Gesellschaft* (Dreyer 1875b, 1876b, c). In 1877 he published his first catalogue of non-stellar objects, the 'Supplement to John Herschel's General Catalogue' (GCS). He was also responsible for the compilation of all Birr Castle observations of nebulae, which appeared in 1880. That work contains some of his sketches.

In Parsonstown Dreyer met the woman who was later to become his wife, Katherine 'Kate' Hannah Tuthill, daughter of John Tuthill from Kilmore (County Limerick). The marriage took place on 11 November 1875 at the local church. On 24 December 1876 their son John 'Jack' Tuthill Dreyer was born, followed by Frederic Charles (8 January 1878). Dreyer stayed at Birr Castle for four years. No doubt he was the driving force and his departure initiated the decline of the well-respected observatory.

8.1.3 Dunsink Observatory

In August 1878, aged scarcely 26, Dreyer moved to Dunsink, the site of the Trinity College Observatory of Dublin University.[4] He became the assistant of Robert Stawell Ball, following Charles Burton (both former Birr Castle observers). Ralph Copeland – who had been an assistant at Dunsink during 1874–76, before changing to Lord Lindsay's Dun Echt Observatory in Scotland – might have recommended him for the job. Ball entrusted the young astronomer with observations at the meridian-circle by Pistor and Martins (well known from Copenhagen). Dreyer measured, for instance, the positions of 321 red stars (a catalogue appeared in 1882 in Dublin). His daughter Alice Beatrice Dreyer (later Shaw-Hamilton) was born in Dunsink on 18 October 1879.

Dreyer and Copeland edited the Dublin magazine *Copernicus* bearing the subtitle 'International Journal of Astronomy'. From January to July 1881 it was first titled 'Urania'. Since a publication with the same name already existed (though with astrological content), a new name was chosen. Lord Lindsay, the owner of Dun Echt Observatory, financially supported the project, but obviously the enthusiastic publishers had underestimated the international competition in the shape of the *Astronomical Journal*, *Bulletin Astronomique* and *Astronomische Nachrichten* (which was effectively remodelled by Krüger in Kiel). Thus, after a notable initial success, *Copernicus* was abandoned in 1884, after only three volumes.

Alongside the regular work at Dunsink, Dreyer was able to continue his cataloguing of non-stellar objects. During his first period, until the end of 1879, he finished the publication of the Birr Castle observations; Lawrence Parsons had left him all of the necessary records and Ball, as a former 'Rosse man', of course supported this task.

8.1.4 Armagh Observatory

With the death of Thomas Romney Robinson on 28 February 1882 the office of the Director of Armagh Observatory became vacant.[5] He was in charge for

---

[3] It contains an orbit calculation of the comet of 1870 (Dreyer 1873). Sampson's bibliography of Dreyer lists 122 publications (Sampson 1934).

[4] Dunsink lies about 100 km east of Birr; see O'Hora (1961) and McKenna (1967): 283–285.

[5] Armagh, now belonging to Northern Ireland, lies about 100 km north of Dunsink; see McKenna (1967): 285–287, Michaud (1983); McFarland (1990).

Figure 8.2. Dreyer at Armagh in about 1890 (Armagh Observatory).

59 years, a unique record. Dreyer was under discussion as a possible successor and traditionally it was up to the Royal Astronomer in Greenwich to render an official 'certification of fitness'. William Christie stated on 9 May 1882 *'I have much pleasure in stating that I consider Mr. J. L. E. Dreyer a fit and proper person for the post of Astronomer of the Armagh Observatory.*'[6] On the following day, a formal offer was sent to the 30-year-old Dreyer. After visiting Armagh, he eventually accepted on 22 May, and on 16 June was appointed as the new Director of Armagh Observatory.[7]

Owing to an urgent refurbishment of the observatory, Dreyer could not move into his new home until 31 August 1882. The site is a bit off the town centre on the wooded, 70-m-high College Hill. The main building offered a commodious flat (Fig. 8.2). There Dreyer's fourth and last child, George Villiers Dreyer, was born on 27 February 1883. Two years later the Armagh astronomer became a British citizen.

As little as six weeks after Dreyer had moved in, he had to write his first annual report. In 1882 he had been awarded his doctorate degree from Copenhagen University for his work on the constant of precession.[8]

It was particularly appreciated by the 'Pope' of celestial mechanics, Simon Newcomb, who was normally feared for his critical judgement.

Dreyer's first concern was the instrumental upgrade of the observatory. To study nebulae, he required a mid-sized refractor. The 15" Grubb reflector (which had been erected in 1835) with a metal mirror was outdated,[9] but funding the new instrument – later called the 'Robinson Equatorial' – caused difficulties. Owing to Dreyer's tenacity, the state, the church and the Royal Society eventually provided £2000 – however, under the condition that there should be no request for any further grants. Dreyer chose a 10" refractor by Grubb with a focal ratio of 1:12 (Fig. 8.3). Since his great model d'Arrest had impressively demonstrated that such a modest aperture was sufficient to observe nebulae.

Dreyer used the time while the instrument was being built for reducing Robinson's star observations, which were completed by additional measurements with the meridian-circle. The result appeared in 1886 in Dublin as the Second Armagh Catalogue, containing positions of 3300 stars for the epoch 1875. The Grubb refractor, which was equipped with two micrometer eye-pieces (one without illumination), was ready to use on 28 July 1885. The dome, which was located in

---

[6] Bennett (1990: 155).
[7] Dreyer's activity in Armagh was described in detail by Patrick Moore (Moore 1967).
[8] Dreyer (1882).

[9] Glass (1997: 17–19).

Figure 8.3. Dreyer's 10" Grubb refractor at Armagh Observatory.

the observatory park south of the main building, has a diameter of 4.9 m.

Besides observing nebulae, Dreyer carried on his catalogue work. In 1886 he offered the RAS a 'Second supplement' to the General Catalogue (see Section 9.1). Surprisingly, the Society refused it, favouring rather a completely new catalogue. Dreyer agreed and only two years later the work was published in the *Memoirs* of the RAS as the 'New general catalogue of nebulae and clusters of stars'. Afterwards, special objects (especially those supposed to be variable) were observed with the Grubb refractor. The results were published in 1894 in the *Transactions of the Royal Irish Academy*, giving positions and notes to 100 nebulae. By and by a growing number of new nebulae and star clusters were discovered – partly by photography. Thanks to the NGC, the rising interest in non-stellar objects brought a considerable amount of corrections. Dreyer, accurately following the issue, thus created two additional works: the Index Catalogues, which were published in 1895 and 1908 in the *Memoirs*. He also wrote various articles about nebulae. He was also interested in double stars, but his observations remained unpublished.

As an assistant in Dunsink, Dreyer's main duty was astrometry, but in Armagh, being now the director, he had complete freedom to do whatever he liked. After the legacy of the Second Armagh Catalogue, he focused not only on nebulae but also on another subject that would eventually become most important: the history of astronomy. As early as in 1883 Dreyer had written a 'Historical account of the Armagh Observatory', which was published in Liverpool (Dreyer 1883), but his main interest, originating from his youth, related to his home country Denmark and its great astronomer Tycho Brahe. In 1890 he published in Edinburgh a biography titled *Tycho Brahe, a Picture of Scientific Life and Work in the Sixteenth Century*. The 405-page book is still a standard work. There he described the importance of historical studies: '*Astronomers are so frequently obliged to recur to observations made during former ages for the purpose of supporting the results of the observers of the present day, that there is a special inducement for them to study the historical development of their science.*'

Though Armagh Observatory hosted a respectable library, which had been set up by Robinson, it was not sufficient for Dreyer's historical research. Thus he often visited Dun Echt Observatory in Scotland. Its 'Crawford Library', which had been amassed by Lord Lindsay, was a unique treasure of rare old books. After the closure of the site in mid 1890, the instruments and library found a new home at the Royal Observatory Edinburgh on Blackford Hill. It now owns more than 15 000 books and manuscripts on astronomy, philosophy and science, many from the fifteenth and sixteenth centuries. The library is one of the largest of its kind. Naturally, Dreyer was a frequent visitor on Blackford Hill – meeting there Copeland, who was appointed the first director in 1889 and third Astronomer Royal of Scotland. Despite health problems lasting for nearly two years, Dreyer's historical studies led to another important work: his *History of Planetary Systems from Thales to Kepler*, which was published in 1906 in Cambridge and is 432 pages long.

In early 1910 the Royal Society and the Royal Astronomical Society decided to produce an edition of the complete work of William Herschel.[10] A committee was founded and Dreyer became a member. Given his broad knowledge about observing and cataloguing nebulae (especially concerning Herschel's contribution) and history, he was the ideal person for this task. Consequently the editing of the bulky matter was up to him. The result was published in 1912 in two volumes with altogether 1441 pages: the *Scientific Papers of Sir William Herschel*.[11] Dreyer not only revised Herschel's three catalogues of non-stellar objects but also wrote an important biography, using unpublished material (the first chapter in Vol. I). For his important contributions to astronomy, Dreyer was awarded the RAS gold medal on 11 February 1916. Ralph Allan Sampson, President and fifth Astronomer Royal of Scotland, gave the laudation (Sampson 1916).

At that time, Dreyer had nearly terminated his observations; his last entry in the notebook is dated 7 November 1914, treating the transit of Mercury. Since he was exclusively concentrated on historical studies, he eventually decided to resign from his office. In his last report of 1916 one reads that '*The difficulty carrying on this work, which often requires that I should refer to rare old books, found from great libraries, and the desire to be able to devote my whole time to the work, have after much consideration decided me to resign from my appointment here and to remove to Oxford.*'[12]

An enduring issue in his 34-year office was financial problems. He always had to beg for money to operate the observatory, to maintain its instruments or for his publications. In contrast to Robinson, Dreyer had no assistants. Occasionally Charles Faris, who had already been employed by his predecessor in 1868, helped him with meteorological measurements. In Armagh Dreyer was a lone fighter, which did not suit his nature. He was very communicative, wrote and received many letters and was often visited by astronomers.[13] Among them were local figures, such as Robert Ball, Ralph Copeland and William Rambaut, but also people from the continent and even the United States. For instance, Edward Emerson Barnard and his wife Rhoda visited Armagh on 25 July 1893. Dreyer's children and grandchildren were frequent guests too.

In June 1917 Joseph Hardcastle was appointed as Dreyer's successor.[14] Unfortunately, he suddenly died on 10 November (at the early age of 49) without acceding to the office. Eventually William Ellison became the new Director of Armagh, holding the office until his death in 1936.

8.1.5 Oxford

From 30 September 1916 onwards Dreyer (now 64 years old) and his wife, Kate, lived in Oxford, at 14 Staverton Road. Above all, the famous Bodleian Library promised new material for his historical studies. Dreyer was engaged with a mammoth task: the edition of the complete work of Tycho Brahe.[15] Since 1908 he had frequently obtained manuscripts from the Royal Library in Copenhagen. The final work, funded by the Danish Carlsberg Institute and completely written in Latin, was to comprise 15 volumes; eight were published in Armagh (the first in 1913) and the remainder appeared in Oxford (the last three in 1929, after Dreyer's death).

Owing to the proximity to London, Dreyer was often present at RAS meetings. Together with Turner he edited the *History of the Royal Astronomical Society*, which appeared in 1923, covering the first 100 years (1820–1920).[16] The chapters (treating 10-year periods until 1880) were written by several authors; Dreyer contributed those on the periods 1830–40 and 1880–1920.

---

[10] Just after William Herschel's death in 1822, his son John had planned such a work, but he found on inquiry that no publisher would be willing to undertake the risk.

[11] Dreyer (1912a); in 1918 he compiled a 'Descriptive Catalogue of a Collection of William Herschel's Papers' for the RAS (Dreyer 1918).

[12] Report Armagh Observatory 1916.

[13] The 'Visitor's Book' is stored in the Armagh archive.

[14] Hardcastle was a grandson of William Herschel and member of the committee to edit his work.

[15] See e.g. Gingerich (1982).

[16] A second volume, covering 1920–80, appeared in 1987.

At the age of 72, Dreyer received another honour:[17] on 9 February 1923 he was elected President of the RAS, succeeding Arthur Eddington. Unfortunately, a month before, his wife, Kate, had died – after 47 years of marriage this was a severe loss. Dreyer's wish for an edition of the complete work of Isaac Newton, which he proposed to the RAS in 1924, could not be realised. He fell ill in 1925 and had to resign from the presidency (James Jeans followed him). John Louis Emil Dreyer did not recover from his illness and died in Oxford on 14 September 1926 at the age of 74.[18]

Dreyer was a friendly, helpful and highly educated man, always sharing his knowledge with others. He was never a professor – perhaps this explains the lack of a university connection and the fact that no estate could be localised. Anyway, there must have been a large number of letters, manuscripts and records. Up to now, the intensive efforts of the author (investigating archives, querying descendents and scholars) have brought no result. Since Dreyer's wife died first, it is possible that the estate was auctioned.

8.2 ORIGINS AND INTENTION OF THE GCS

Dreyer's interest in observing and cataloguing nebulae had two origins. First, there was the early contact with d'Arrest in Copenhagen. The admiration for his fellow countryman lasted throughout his life. Dreyer was impressed by the observations with the 11" refractor and, above all, he valued his endurance and accuracy, which eventually led in 1867 to the monumental *Siderum Nebulosorum Observationes Havnienses*. As stated in the introduction of the GCS, he judged d'Arrest's catalogue to be *'entirely free from the large accidental errors which may not seldom be found in the latter* [Herschel's GC]'.[19]

The second origin was Dreyer's appointment as Lawrence Parsons' assistant at Birr Castle in 1874. When he arrived in Ireland, he was only 22 years old (d'Arrest was already 52 at that time). The 72" reflector, which he could use without restriction, was still a powerful instrument for visual observations. However, d'Arrest's refractor had other qualities, being a true measuring machine. Dreyer remarked on this issue that *'Formerly it was only the possessors of large reflecting telescopes who thought their optical means sufficient for work on these faint objects, but when D'Arrest had shown how much could be done with a small refractor for the determination of the positions of the brighter Nebulae several astronomers turned their attention in this direction.'*[20] He mentions Schönfeld, Schultz, Vogel and Rümker. Dreyer was especially impressed by Herman Schultz' accurate observations of 500 nebulae published in 1874; actually, his first publication about nebulae was a review of this work (Dreyer 1875b).

Studying the literature was (alongside observing) a major task from the beginning. At Birr Castle, Dreyer had access to many books, astronomical magazines and observatory publications. Moreover, there was an archive with observational records, which had been opened by Lord Rosse and continued by his son Lawrence Parsons (actually the assistants had done the major work). Owing to Dreyer's accurateness and interest in current and historical data, the trend towards cataloguing nebulae was inevitable at a place like this. Of course, Lawrence Parsons, who had appointed the young astronomer, soon recognised his great talents. Thus Dreyer was charged with editing all Birr Castle records on nebulae for a comprehensive publication. The result eventually appeared in 1880 (see Section 8.18.1).

Dreyer's most important reference was John Herschel's GC of 1864, the standard catalogue of nonstellar objects at that time, but he soon recognised that the work had to be corrected and enlarged – after having existed for only 10 years. For Dreyer, as stated in the GCS, d'Arrest's *Siderum Nebulosorum* was an important supplement: "*The Copenhagen Observations [...] are in many ways quite necessary as a supplement to the 'General Catalogue'.*"[21] Confronted by the amount of new nebular data collected *'with first ranked instruments'*, he saw the need for an additional catalogue (at least for his own use): *'The necessity of arranging such a supplement for my own use soon became obvious to me, when, in 1874, I began to work at the Earl of Rosse's Observatory.'*

---

[17] In 1970 a crater on the back side of the moon with a diameter of 61 km was named after Dreyer.
[18] Obituaries: Knobel (1927), Fotheringham (1926), Anon (1926a), Anon (1962b) and Burrau (1927); see also Sampson (1934), Lindsay E. (1965), Alexander A. (1971), Ashbrook (1984), Gingerich (1988), Marriot (1992), Chapman D. (2002) and McFarland (2002).
[19] Dreyer (1878a: 382).
[20] Dreyer (1878a: 381).
[21] Dreyer (1878a: 382).

Like John Herschel before, Dreyer prepared 'working lists' for his observations, *'for all objects, which an examination of all the previous Birr observations had shown in want of being re-observed for one reason or another'*. He added that *'This occupation, as well as the reduction of the current observations, necessarily involved a careful study of the work done on the Nebulae by other observatories – especially by D' Arrest, and Dr. Schultz at Upsala'*.

Dreyer was interested in getting a complete picture about the current observations of known or new objects. He was right to assume that not all data were actually published. He therefore wrote a 'Request to astronomers' on 11 November 1876, which appeared a week later in the *Astronomische Nachrichten* (Dreyer 1876d). It reads as follows: "*As I am preparing for publication a supplement to Sir John Herschel's 'General Catalogue of nebulae and clusters of stars', I should be very much obliged to any astronomer who would be kind enough to send me unpublished observations of new nebulae or information about such important errors in the general Catalogue as he might have noticed by his own observations. It is desirable that such communication should reach me as below before December.*"

Despite the short deadline, his request yielded a remarkable reply. In the GCS, Dreyer explicitly thanked Otto Struve, Winnecke, Tempel, Stephan and Copeland. He further used lists of new nebulae published by d'Arrest (Copenhagen), Marth (Malta) and Stephan (Marseille). Additionally Lawrence Parsons allowed him to include all nebulae discovered at Birr Castle after 1861 (the date of Lord Rosse's last report). In the GCS, Dreyer for the first time mentions the upcoming publication of the Birr Castle observations: "*From these it will be seen that our attention has been of late especially directed towards finding the exact positions of all such R. [Lord Rosse] novae, for which no exact positions are given in the 'General Catalogue'.*"[22] About d'Arrest's new objects he remarked "*I have not entered such new Nebulae of D'Arrest's into this Catalogue, which I could see with certainty were identical to R. novae, but have only given their positions among the notes to the 'General Catalogue'.*" Dreyer, in the first instance, wanted to wait for a maximum number of new nebulae, but then decided to publish immediately the collected corrections and additions to the GC: '*In the first place, the list of corrections could not possibly be increased much more, if its publication was deferred; and, secondly, I have lately had proofs enough that a catalogue of new Nebulae will be useful at the present moment.*'[23]

During his work on the GCS, Dreyer published three further papers about nebulae. The first two are reviews of the 'Zweite Abtheilung' of Schönfeld's Mannheim observations and Vogel's 'Positionsbestimmungen von Nebelflecken und Sternhaufen' (Dreyer 1876b, c). The third article treats Stephan's observations in Marseille (Dreyer 1877a).

8.3 HARVARD OBJECTS

The GCS contains four Harvard objects from Bond's list, part II, which was published in 1864 (Bond G. 1864a).[24] Three were found by Bond and one by Coolidge (Table 8.1). The list was posted to John Herschel in October 1863. Although the main catalogue of the GC had already been closed, Herschel nevertheless was able to include objects from part I (the reliable cases) in the 'Supplementary list'; the doubtful cases (part II) were omitted. But Dreyer entered three of them (nos. 29, 32 and 33) in the GCS. Though he had checked only one object (GC 5354), all seemed to be sufficiently reliable for him. GC 5354 (NGC 2054) was observed on 13 January 1877 at Birr Castle: '*vF, pS, iR, at times I thought it was a v s Cl, but it is doubtful*'.[25] But in February 1898, Howe saw '*only 3 S st and no neb*' with the 20" refractor at Chamberlin Observatory, Denver (Howe 1898b); there was no trace of an 'irregular round' nebula. Bond's no. 33, listed as GC 6233 (NGC 7793) by Dreyer, is a galaxy that had already been discovered in 1826 by Dunlop in Paramatta. The remaining three (nos. 27, 30 and 31) were ignored by Dreyer, probably because Bond was unable to confirm two of them (no. 27 was checked on 9 September 1863 and no. 31 as soon as on 17 August), and Coolidge's no. 30 is described as a '*nebulous object (?)*' in Bond's list. GC 6000 originates from the Harvard zone observations (Fig. 8.4); therein the star no. 191 is marked 'planetary?' (Bond W. 1855). D'Arrest could verify on 4 August 1862 that the object is only a 9-mag star

---

[22] Dreyer (1878a: 384).
[23] Dreyer (1878a: 383).
[24] See Sections 6.7.4 and 7.2.
[25] Parsons L. (1880: 51).

Table 8.1. *Harvard objects in the GCS*

GCS	NGC	Observer	Date	No.	Type	V	Con	Remarks
5354	2054	Bond	6.10.1850	29	4 stars		Ori	
6000		Bond	1853		Star	9.0	Aqr	'Planetary?', BD +0° 4741
6092	7403	Coolidge	15.11.1859	32	Star	13.9	Psc	
6233	7793	Bond	7.11.1850	33	Gx	9.0	Scl	Dunlop 14.6.1826 (Δ 335)

Figure 8.4. Bond's 'planetary' GC 6000 in Aquarius (DSS).

(BD +0° 4741).[26] Owing to spectroscopic observations by Copeland, Dreyer later omitted GC 6000 from the NGC (see Section 8.19.1).

8.4 SCHWEIZER'S NEW 'NEBULA'

In 1860 Gottfried Schweizer saw a 'nebula', which later turned out to be a double star. The object is not listed in the GC because the discovery was not published until 1875.

8.4.1 Short biography: Kaspar Gottfried Schweizer

Gottfried Schweizer was born on 16 February 1816 in Wyla, in the canton of Zurich (Fig. 8.5 left). After studying in Zurich and Jena, he went to Königsberg Observatory in 1839 to become an assistant. From 1841 to 1845 he worked under Wilhelm Struve at Pulkovo, then moving to Moscow, where he was employed as astronomer in 1852. Four years later he was appointed Director of Moscow University Observatory. He discovered five comets and was an industrious observer, especially of sunspots. Gottfried Schweizer died on 6 July 1873 in Moscow at the age of 57. He was succeeded by his student Fedor Bredikhin.[27]

8.4.2 Discovery of NGC 7804

During his systematic study of star positions, Schweizer discovered a 'nebula' in Pisces on 22 October 1860 (Fig. 8.6). He used the 23-cm Merz refractor of Moscow University Observatory (Fig. 8.5 right), which had been installed in 1859. His note reads '*the nebula appears to be double*'. A second observation was made by his successor Bredikhin on 26 August 1875. Both records are published in Vol. 2 of the Moscow Observations (Bredikhin 1875). In the GCS Dreyer entered the object as GC 6235, giving Schweizer as its discoverer. However, this information is missing from the New General Catalogue (NGC 7804), where one reads '*Neb. (Obs. de Moscou II.)*'.

Engelhardt looked at GC 6235 in 1886 with his 30.6-cm Grubb refractor in Dresden. His four observations revealed the object to be only a double star (Engelhardt 1886). This was confirmed by Burnham in 1891, using the 36" Lick refractor. He determined NGC 7804 to be two stars of 12.5 mag and 13.0 mag at 10″ distance.[28] Dreyer mentions Engelhardt's observations in the NGC notes. On the basis of Burnham's result he added in the IC I that it was '*to be struck out*'.

---

[26] d'Arrest (1867a: 247).

[27] Obituaries: Anon (1873a) and Anon (1873b); see also Lynn (1904).

[28] Burnham S. (1892b). The two stars are GSC 594–657 and GSC 594–1044.

Figure 8.5. Gottfried Schweizer (1816–1873) and his 23-cm Merz refractor at Moscow University Observatory (Repsold J. (1908), Vol. 2, Fig. 23).

Figure 8.6. Schweizer's 'nebula' NGC 7804 in Pisces (DSS).

8.5 SCHÖNFELD'S MANNHEIM OBSERVATIONS

Schönfeld, who was already noted for his work on the Bonner Durchmusterung, used his new appointment as Court Astronomer in Mannheim for a systematic study of nebulae. From 1860 to 1868 he measured accurate positions of about 500 objects with a ring-micrometer. The results were published in two parts ('Abtheilungen'), receiving considerable attention. But the working conditions at the tower observatory were problematic, as explained in a public talk given in 1861. It also describes in some detail the basis of his programme and critically discusses the relation to the works of William Herschel and others. In the middle of the nineteenth century, Schönfeld was an outstanding astronomer. He discovered five new nebulae.

8.5.1 The public talk 'Ueber die Nebelflecke'

On 10 November 1861 Schönfeld gave a remarkable public talk at the Mannheimer Verein für Naturkunde, of which he was a member (Schönfeld 1862b). Shortly before the manuscript was published, he added two notes in February 1862. Schönfeld starts with a short historical outline, focused on the work of William Herschel and his idea about the nature and evolution of nebulae. The main intention of the talk was to describe the origin, method and results of his observing programme at Mannheim Observatory, which (the first part) had been started in November 1860 and finished in mid 1862.

Schönfeld first explains how the nebulous impression depends on the telescope and the air quality. He then reviews the changing point of view of Herschel, who at first believed that all nebulae were star clusters, but later revealed the existence of true nebulosity. This led to the popular nebular hypothesis, whereby stars and planets result from condensed nebulous matter.[29] Schönfeld sees indications for both nebulae as star clusters (disguised by distance) and true nebulosity, but states that currently the observational data are not sufficient to reveal the true nature of the nebulae.

The published drawings are judged as not really helpful in this regard: '*Owing to the large differences between the many drawings of Lord Rosse and* [William and John] *Herschel we must fear that, with the present resources, we are not at all able to detect the true shapes of most unresolved nebulae with sufficient distinctness, and thus all ideas based on these drawings are of most sensitive nature.*' Schönfeld further wrote that, '*Indeed, a simple comparison of the many drawings of William and John Herschel, Lord Rosse, and occasionally of Bond, Secchi and others, will do to demonstrate the imperfection of our descriptions and drawings presented so far.*'[30]

Schönfeld now states that '*The greatest support of Herschel's ideas would obviously result from observing real changes, indicating the concentration of nebulous matter.*' However, the existing drawings were unsuitable as means with which to prove the occurrence of such behaviour. He optimistically points to '*recently discovered variable stars in nebulae, e.g. the one in Herschel's nebula of fourth class* [IV 2 = NGC 2261] *by Julius Schmidt, a star in the cluster 3624 of John Herschel's Catalogue* [M 80] *by Auwers and Pogson, and, if they are confirmed, some stars in the Orion nebula by Otto Struve*'. In his note no. 1, added to the manuscript in February 1862, he reported that '*recently another, highly interesting matter came along: the vanishing of a complete nebula*'.[31] This is Hind's Variable Nebula in Taurus (NGC 1555).[32]

According to Schönfeld, the existence of true nebulosity is supported by comets: '*we must concede to see it* [a comet] *as a whole, i.e. a complete heavenly body and not a conglomerate of other individual bodies*'.[33] He explains the matter further: '*The fact that the great nebula near ζ Herculis* [M 13] *consists of thousands of stars obviously does not prove the same for the entirely differently formed nebula in Orion* [M 42]'.[34] He is, however, '*forced to concede that our century has deprived us of the basis of parts of Herschel's ideas about star formation*',[35] because '*Lord Rosse has investigated some nebulae with his 6-foot reflector and could mostly resolve them into stars*'.[36] Owing to the discovery of spiral structure, a '*much greater irregularity of nebular forms is present than previously thought*'. Schönfeld concludes '*But, to the degree to which the form of a nebulous mass becomes more asymmetric, the probability of nebulous matter in the sense of Herschel decreases; because, in empty space, only matter consisting of discrete and connected solid bodies can take any form; gases or fluids can permanently exist only in spherical or at least symmetrical shape.*' Concerning the distribution of nebulae it appears curious '*that nearly all nebulae in the Milky Way are resolved, i.e. true clusters of stars, whereas those at the poles* [of the Milky Way] *show not a trace of resolution*'.[37]

Schönfeld demands a clarification, which can be achieved only by systematic observations. Lamont had already tried to reveal the nature of nebulae, describing his ambitious project in a similar talk in 1837 (see Section 6.1.2), but Schönfeld lamented that '*Lamont's work appears to stagnate, as a few results were soon published but for years nothing more was heard.*'[38] For him,

---

[29] See Section 6.4.8.
[30] Schönfeld (1862c: 68).
[31] Schönfeld (1862c: 85).
[32] See Section 11.4.
[33] Schönfeld (1862c: 51).
[34] Schönfeld (1862c: 52).
[35] Schönfeld (1862c: 68).
[36] Schönfeld (1862c: 67).
[37] Schönfeld (1862c: 58).
[38] Schönfeld (1862c: 66).

any attempt to unmask the physical nature of nebulae by mere observation makes no sense. Much more promising are precise measurements of their position and possible proper motion: '*The motions of heavenly bodies, regardless of their complexity, are, however, infinitely simpler to study than their other characteristics; this is because we have* [Newton's] *celestial mechanics but no celestial physics*.'[39] For nebulae this means that, '*Once one knows a sufficient number of such motions of nebulae, the mean magnitude and direction of motion will yield results about their mean distance and thereby about their spatial distribution that will be incomparably more reliable than the mere contemplation of such, even with the best equipment, can ever grant.*'[40] Here he argues against Herschel and Lord Rosse, favouring the work of d'Arrest, Auwers and others. The determination of precise positions in order to derive possible motions requires extensive measuring campaigns, which, moreover, must be repeated at large time intervals – this is Schönfeld's own field.

Concerning positional accuracy, Schönfeld does not take the data of Messier, John Herschel and Laugier as being reliable. Therefore they are not a basis for comparisons. In contrast, he is enthusiastic about d'Arrest's measurements: '*Professor d'Arrest in Leipzig (now Director in Copenhagen) has done us the great service of starting in the year 1856 a series of observations that, with an only-six-foot* [focal length] *telescope and a ring-micrometer, surpasses in extent and methodicalness everything achieved hitherto regarding the accurate determination of the positions of the nebulae.*'[41] At the time of his talk, Schönfeld still was not aware that d'Arrest has continued his work in Copenhagen. In his second note of February 1862, however, he mentions the 'Kopenhagener Universitätsprogramm', obviously referring to d'Arrest's monograph 'Instrumentum magnum' (d'Arrest 1861). Schönfeld notes that '*we may expect the richest results from this already started work – so beautiful that some may perhaps think that, in comparison, the Mannheim observations will lose much of their value*'.[42] He, admittedly, rates the Copenhagen refractor (28.5-cm Merz) to be much better than his 16.5-cm Steinheil, but sees no reason to stop his programme: '*Certainly, astronomy would not have reached its present high standard, if the owners of small telescopes thereby were put off using them.*' Schönfeld is pleased that '*Julius Schmidt in Athens has effected a similar observing programme*'.[43] Schmidt's refractor by Plössl, having an aperture of 15.7 cm, was even smaller.

In the last part of his talk, Schönfeld describes his observing method. He emphasises the benefits of a ring-micrometer, because a nebula remains visible in the centre and no illumination is needed, which could reduce the contrast. The measurement itself is made by recording the passage times of the nebula and a reference star. Moreover, '*the data were written down in full darkness, to keep the adaptation of the eye to the dark, which could be reduced by faint external light for some period of time*'.[44] He further explains that, '*in the darkness, I often continuously count the clock ticks for more than an hour and, in the case of an exterior disturbance, which may cause a miscounting, I briefly check the clock-face in the dim light of a cigar*'. For each nebula 20 to 24 measurements were made, normally spread over five different nights. Since Schönfeld determined relative positions, the resulting coordinates depended on the quality of the reference stars too. He did not worry about it, '*because Professor Argelander in Bonn, due to his kindness and friendship and knowing about the importance of my task, has re-measured all* [reference stars]; *thus I have a uniform system of excellent star positions, on which the micrometer observations are based*'.[45]

8.5.2 'Erste Abtheilung' of the Mannheim observations

Schönfeld's first monograph, namely the *Astronomische Beobachtungen an der Großherzöglichen Sternwarte zu Mannheim. Erste Abtheilung. Beobachtung von Nebelflecken und Sternhaufen*[46] was published in 1862 by Bennsheimer, Mannheim. It contains observations from 1 December 1860 to 26 May 1862. When

---

[39] Schönfeld (1862c: 70).
[40] Schönfeld (1862c: 71).
[41] Schönfeld (1862c: 74).
[42] Schönfeld (1862c: 87).

[43] Schönfeld (1862c: 75).
[44] Schönfeld (1862c: 80).
[45] Schönfeld (1862c: 81).
[46] '*Astronomical Observations at the Mannheim Observatory of the Grand Duke. First Part. Observations of Nebulae and Star Clusters*' (Schönfeld 1862d).

he started, the comparable campaigns by Schmidt in Athens and d'Arrest in Copenhagen were not known to him. In the introduction one reads that, '*By the way, had this been the case, I should hardly had changed my plans, as, due to the expected smallness of proper motions of nebulae and the manifold uncertainties of positional measurements, any duplication of efforts in this field is still desirable.*'[47]

For selecting the targets from the catalogues of William and John Herschel, Schönfeld used the following criteria: the object should be bright enough (good visibility), concentrated (a well-defined centre that could be easily measured) and sufficiently high above the horizon ($\delta > -25°$). Only one nebula was added from a different source: Hind's globular cluster in Aquila (NGC 6760). Schönfeld eventually observed 235 objects. To complete the task in due time, a detailed plan was made. All observations were executed with the 16.5-cm refractor by Steinheil, which had been erected in 1860.[48] Schönfeld used eye-pieces with powers of 48 and 64. The telescope, which was equipped with a 46-mm finder, stood in a dome of diameter 4.5 m, located on the flat roof of the 30-m-high observatory tower (Schönfeld 1861). The dome, which was covered with tarpaulin, was pretty susceptible to wind. Not only the wind, whose noise often drowned out the clock ticks, but also the nearby chimney caused trouble. Concerning the measurements, Schönfeld was not a friend of the illuminated bar-micrometer, since he feared that '*unavoidable frequent glare badly influences the eye*'.[49] Instead, a ring-micrometer was preferred, the effective use of which he had learned from Argelander. Each object was measured three to five times. Schönfeld always observed alone; only for the data reduction was he occasionally supported by students.

The main part of the publication is taken up by the table of results (98 pages). The objects are sorted by right ascension (epoch 1865) and identified by their catalogue designation (h, H, M, L). Besides the observational data, each object is described using a German version of Herschel's code. Schönfeld remarks that '*Whether the given description of nebulae, lacking only for some early observations, is of any value, may be seen by comparing with other observers.*'[50]

There follows a list of 254 reference stars, whose positions were specifically determined by Argelander with the Bonn meridian-circle.[51] Next, Schönfeld presents a table giving the calculated positions of nebulae plus the differences to Laugier (Paris) and d'Arrest (Leipzig). The comparison with Laugier's data yielded larger deviations. Moreover, in relation to d'Arrest, Schönfeld's right ascensions were too small (for the 99 common objects), an unfortunate trend that was later confirmed by Schmidt and Schultz. The same applies in comparison with Auwers' data.

The final seven pages contain notes on some objects, concerning errors in the catalogues of d'Arrest and the two Herschels. A few other observed, though not measured, objects are mentioned here: NGC 607 (d'Arrest), NGC 1333 (Schönfeld), NGC 1555 (Hind), NGC 4383 (Schönfeld's new nebula), NGC 7005 (d'Arrest) and GC 4941 (d'Arrest).

8.5.3 The 'Zweite Abtheilung' of the Mannheim observations

In 1875 Schönfeld's 'Zweite Abtheilung' appeared, now printed by G. Braun'sche Hofbuchhandlung, Karlsruhe (Schönfeld 1875b). The preface is dated 31 May 1875. The work contains observations from 7 September 1861 to 14 June 1868. Most were made during 1862–64; in 1868 only a few gaps were filled.

The work appeared just after Schönfeld's move to Bonn and the question of the late publication arises. The Mannheim Observatory was under the authority of the government of the Grand-Duchy of Baden, which obviously had financial problems, perhaps intensified by the Franco-Prussian War of 1870–71. Moreover, from 1865 onwards Schönfeld had concentrated on variable stars and two important catalogues had appeared in 1866 and 1875 (Schönfeld 1866, 1875a).

The second catalogue lists 820 observations of 343 nebula and star clusters, among them many relatively faint ones. Altogether the two works contain nearly 500 objects (82 are common). Thus Schönfeld's result can be compared in quantity with that of Schultz (1874).

---

[47] Schönfeld (1862d: III).

[48] In 1885 the refractor was moved to Karlsruhe and was eventually erected in 1885 at Königstuhl, Heidelberg.

[49] Schönfeld (1862d: IV).

[50] Schönfeld (1862d: V).

[51] D'Arrest had failed to present such a list, which was later criticised by Auwers.

Concerning its quality, it is even superior, insofar as absolute positions are given, which makes it possible to compare directly with other results. Schönfeld gives differences relative to the results of Auwers, Laugier, Rümker, Schmidt, Vogel and Schultz. The latter are based on the 'preliminary coordinates' published in February 1875 (Schultz 1875). Concerning the right ascensions, which were found to be too small in the first campaign, Schönfeld was very careful in using the ring-micrometer – the error did not appear again.

The results are compiled on 74 pages. The objects are ordered by right ascension and now designated by GC-number (with h, H, Auwers and other labels added if possible). There is a German description in Herschel's style. To determine the coordinates (epoch 1865), Schönfeld once again could use reference stars (not listed now) measured by Argelander at the Bonn meridian-circle. The remaining seven pages contain notes on single objects, among them those targets that had not been found.

In 1876 Dreyer wrote a review applauding the publication, which will '*take an excellent place among the current works on the topic*' (Dreyer 1876b). He corrected some of the notes. The reason why Schönfeld was unable to find a few nebulae was seen to lie in the small aperture; these objects were all observed at Birr Castle.

8.5.4 Three new nebulae

Schönfeld discovered six nebulae (Table 8.2). The first two, from his observations for the Bonner Durchmusterung, are listed in the GC (see Sections 6.15.2 and 6.15.4). NGC 3185 was found on 15 January 1862; in the 'Erste Abtheilung' he wrote '*it seems that the reference star is preceded about 20s by a very faint nebula (which is not listed by h and H)*.' D'Arrest saw the object too – but the true discoverer is J. Stoney (Birr Castle). Schultz called it 'Nova Schönfeld' (observation on 24 March 1863).[52]

On 1 April 1862 Schönfeld investigated William Herschel's quartet II 568–71 in Virgo (see Table 2.20). Actually there are seven galaxies in a field of 15′ diameter (Fig. 8.7). John Herschel described six objects in the Slough catalogue, from which, due to identities, only four remain – however, they are not exactly those of his father. Schönfeld was able to resolve the case by showing that William Herschel had made an error with the reference star (second catalogue): '*the difference relative to 11 Virginis must read 0° 26′ south, instead of 0° 34′ north (Vol. 79 pag. 223)*'.[53] While observing the group, Schönfeld discovered a new nebula: NGC 4268. The correct identifications are NGC 4259 = h 1178, NGC 4270 = II 568, NGC 4273 = II 569 = h 1183 = h 1189, NGC 4277 = II 570 = h 1190 and NGC 4281 = II 571 = h 1187 = h 1194. There is another galaxy in the group: IC 3153, which was found by Kobold in Straßburg on 8 April 1894. D'Arrest first observed the field on 21 April 1862 in Copenhagen, but was unable to establish order. It was otherwise for Schultz, who came to a similar conclusion to Schönfeld, on the basis of observations during spring 1865 and 1866 in Uppsala.[54]

Also on 1 April 1862, Schönfeld saw NGC 4324 (GC 2892), a galaxy in Virgo the existence of which he published in 1862 in the 'Erste Abtheilung'. The object was discovered by d'Arrest on 4 March 1862 with the Copenhagen 11″ refractor. On 23 May 1862 Schönfeld found the galaxy NGC 4383 (GC 5644) in Coma Berenices, which is listed in both works.

The question of why NGC 4268 and NGC 4383 are not listed in the General Catalogue arises. Obviously John Herschel was not aware of the 'Erste Abtheilung', otherwise he would not have overlooked them. But the issue is even more serious, insofar as Schönfeld had published both discoveries as early as in 1862 in *Comptes Rendus*. Leverrier communicated a note, in which position, description and date are given (Schönfeld 1862c). It was not recognised by Herschel either.

To sum up, Schönfeld is the discoverer of four objects (marked in bold in Table 8.2): NGC 1333, NGC 6643, NGC 4268 and NGC 4383. All are non-stellar – an optimal rate, indicating his excellent skills.

8.6 D'ARREST'S MASTERPIECE: SIDERUM NEBULOSORUM OBSERVATIONES HAVNIENSES

Heinrich d'Arrest (Fig. 8.8) crowned his long-time work on nebulae with the monumental *Siderum Nebulosorum Observationes Havnienses* (*SN*), which inspired the

---

[52] Schultz (1864: 14).

[53] Schönfeld (1862d: 112).

[54] Schultz (1874: 88–89).

238　Dreyer's first catalogue

Table 8.2. *Schönfeld's objects in chronological order*

Date	NGC	Au	GC(S)	Type	V	Con.	Rr	Remarks
31.12.1855	**1333**	17	710	RN		Per	3"	BD +30° 548; Tuttle 5.2.1859
1858?	**6643**	40	4413	Gx	11.0	Dra	3"	BD +74° 766; Tuttle 1.9.1859
15.1.1861	3185		2054	Gx	12.0	Leo	6"	1. Abt. p. 22; J. Stoney Jan. 1850; d'Arrest 1.1.1862
1.4.1862	**4268**		5632	Gx	12.3	Vir	6"	1. Abt. p. 112
1.4.1862	4324		2892	Gx	11.6	Vir	6"	1. Abt. pp. 42 + 112; d'Arrest 4.3.1862
23.5.1862	**4383**		5644	Gx	11.8	Com	6"	1. Abt. p. 113, 2. Abt. p. 90

Figure 8.7. The galaxy group around NGC 4273 in Virgo; Schönfeld discovered NGC 4268 (DSS).

admiration of all observers – especially Dreyer. He set new standards concerning precision, skill and perseverance. The work had significant influence on both the GCS and the NGC. Unfortunately, d'Arrest could not realise his ambitious plan to revise all visible Herschel nebulae. Lately, time was lacking and his power had decreased. During the last part of his career – which was terminated by his early death – he made important contributions to the spectroscopy of nebulae.

8.6.1 An ambitious observing plan and its implementation

In Copenhagen d'Arrest planned to observe all of the nebulae of William and John Herschel's catalogues which would be visible in the 11" Merz refractor (Fig. 8.9). His aim was not only a revision but also getting reliable data for proper motions by measuring absolute positions. Obviously his ambitious task was inspired by the optical quality of his refractor, which '*shows all Herschel nebulae, even the most difficult*'

8.6 D'Arrest's masterpiece

Figure 8.8. D'Arrest during his time in Copenhagen.

Figure 8.9. D'Arrest's 11" Merz refractor in Copenhagen (d'Arrest 1861).

(d'Arrest 1862c). D'Arrest valued it *'exactly in the middle between Herschel's 20ft reflectors [18"] in their best condition, and the excellent telescope [24"] used with great*

success by Lassell in the years 1852–54'. He omitted all objects of classes VI, VII and VIII ('clusters of stars'), which he interpreted as mere accumulations of stars, essentially different from the nebulae. Moreover, he rated these classes as arbitrary and unnatural, because he had noticed that in regions of the Milky Way there are many similar objects, often near to a catalogued one, that were not considered by Herschel.

A first report, titled 'Vorläufige Mittheilung, betreffend eine auf der Kopenhagener Sternwarte begonnene Revision des Himmels in Bezug auf die Nebelflecken'[55] and dated 20 May 1862, appeared in *AN* 1366. D'Arrest had already distanced himself from his original goal of determining the absolute position of each nebula: *'As I'm observing alone, it soon became clear that the continuous alternation between finding and observing of the mostly faint objects and the microscopic reading of the circles was impracticable, harmful for both the eye and the work itself.'* He added that *'All positions are given only to the nearest arcminute in right ascension; against it, exact data for a future detection of proper motion of nebulae relative to the system of stars (a main goal of the study) must be derived from measuring distances to nearby, mostly faint stars using the ring- or bar-micrometer.'* So, d'Arrest measured precise relative positions only. He normally used a ring-micrometer eye-piece with 17′ field of view and a power of 123. The targets were observed in declination zones of 4° to 5° width. He generally used four to seven reference stars between 7 and 8 mag, which had been measured by Bessel and Argelander (a few were taken from Lalande's *Histoire Céleste*).

A further report, dated 21 October 1864, appeared in *AN* 1500 – this was probably his most important (d'Arrest 1865a). There one reads that meanwhile *'more than 2500* Gegenstände *were observed, some repeatedly'*. But d'Arrest's 'Gegenstände' cannot mean 'objects', since the final *SN* contains only 1942; it is more likely that 2500 is the number of observations. D'Arrest wrote *'I'm sure to finish the task in one or two years.'* He now was even optimistic about absolute positions: *'Since my first announcement [AN 1366] it has become possible, partly due to help for the readings, to get absolute positions more accurately.'* Obviously he was supported by Schjellerup, who determined star positions within the

---

[55] 'Preliminary report on a revision of the sky relating to nebulae started at Copenhagen Observatory' (d'Arrest 1862c).

declination zone −15° to +15° with the meridian-circle by Pistor and Martins.

Anyway, d'Arrest's plan could not be completed. After five and a half years (October 1861 to May 1867) quite a number of objects still remained unobserved. The main reasons were bad weather, lack of time and d'Arrest's deteriorating physical condition. He therefore decided to terminate the observations and to publish the results that had been achieved. D'Arrest's main work was printed in Copenhagen in August 1867, titled *Siderum Nebulosorum Observationes Havnienses Institutae in Specula Universitatis per Tubum Sedecimpedalem Merzianum ab Anno 1861 ad Annum 1867* (d'Arrest 1867a).

D'Arrest's 415-page monograph was completely written in Latin – which was hardly conventional at that time. It contains 1942 objects, among them 308 new ones (however, d'Arrest speaks about 390). The tabular data, listing about 4800 observations, cover 385 pages. Usually, each object was observed several times (up to 11, as for IV 19 = NGC 2170; see Fig. 3.5).

The table gives designations from the catalogues of William and John Herschel (H and h), calculated right ascension and declination (1861) for each observation, notes and a date. D'Arrest did not create a new designation (he had numbered the observations only for his own use, as can be seen in *AN* 1500). The notes ('Nebulosarum facies et indoles') contain the object description and, occasionally, observations by others (a few GC-numbers are given here). They also present 14 sketches (Table 8.3), made with powers of up to 356.

411 observing nights are listed in the appendix; the first on 27 October 1861 and the last on 11 March 1867. On the next 20 pages a list of all objects with positions for 1860 is given. D'Arrest calculated his errors as 0.81^s in right ascension and 17.6″ in declination; which is better than John Herschel. Finally he compares his positions with Schönfeld's (for 223 common objects); generally, d'Arrest's right ascensions are a bit larger.

8.6.2 Recognition of d'Arrest's work

Dreyer appreciates the *Siderum Nebulosorum* in the introduction of his GCS. In comparison with Schönfeld, Schultz, Vogel or Rümker he sees '*the great work of D'Arrest [...] far exceeding them all in importance*'.[56] He further wrote the following: '*This indefatigable observer – whose early death all astronomers lament – succeeded in forming a work, in which he is not surpassed by anybody as regards the extent and value of his observations, while he often surprises the reader by the sharp and critical acumen with which he analyses and explains the work of his predecessors.*' D'Arrest was not just Dreyer's teacher at Copenhagen University but his lifetime idol too. His scientific work is characterised by great clarity and quality. In contrast to Tempel, who had a disposition to fancy, ignoring new developments (e.g. spectroscopy), d'Arrest was always up to date. Owing to his university connection he not only recognised astrophysics but also played an active part in it. After his work on nebulae, he investigated gaseous nebulae by visual spectroscopy.

For Dreyer d'Arrest's monograph was recommended reading: "*The Copenhagen Observations may be supposed to be in the hands of every observer of Nebulae, and they are in many ways quite necessary as a supplement to the 'General Catalogue'. Although the probable errors of D'Arrest's observations are not much smaller than those in Sir John Herschel's positions, still the former are entirely free from the large accidental errors which may not seldom be found in the latter, and at which nobody wonders when he considers the construction of Herschel's instrument.*"[57] Dreyer praises the observing style: '*What makes the work so important is, that it alone of all similar ones, except those of the two Herschels, is founded on zone-observations (sweeps), made with a powerful instrument in order to determine and describe all the Nebulae which came into the field.*' Later, in introduction to the NGC, Dreyer wrote about the value of the *SN*: '*it is to that thesaurus, more than to the exertions of any other observer, that the credit should be given for whatever superiority as to accuracy the present work may possess in comparison with Herschel's*'.[58]

In 1868 another review was published in the *Vierteljahrsschrift der Astronomischen Gesellschaft* (Anon 1868). Mainly the content of d'Arrest's monograph is treated. For the unknown writer it is '*congenial to the work of Herschel*'.

In February 1875 d'Arrest received the RAS gold medal for his Leipzig and Copenhagen observations; the laudation was given by President John Couch

---

[56] Dreyer (1878a: 381).
[57] Dreyer (1878a: 382).
[58] Dreyer (1953: 6).

8.6 D'Arrest's masterpiece   241

Table 8.3. *D'Arrest's sketches in* Siderum Nebulosorum

NGC	H/h	Page	Date	Type	Remarks
1931	355	77	25.3.1865	OC+EN	
1999	IV 33	80	18.2.1865	EN+RN	
2024	V 28	80	18.2.1865	EN	d'Arrest: V. 28A
2261	399	86	19.2.1865	RN	Hubble's Variable Nebula
2392	450	92	19.2.1865	PN	Eskimo Nebula
3239	710	133	16.3.1865	Gx	
4291	1192	207	17.9.1866	Gx	Pair with NGC 4319
5144	IV 70	290	14.9.1866	Gx	
5907	1817	319	15.9.1866	Gx	Extreme edge-on galaxy
6720	2023	334	2.10.1861	PN	Ring Nebula M 57
6818	2047	336	1.9.1861	PN	J. Herschel: 2 'satellites'
6853	2060	338	18.8.1862	PN	Dumbbell Nebula M 27
7463	2202	360	27.8.1864	Gx	Galaxy trio[59]
7464		360	27.8.1864	Gx	Nova, galaxy trio
7465	2203	360	27.8.1864	Gx	Galaxy trio
7479	2205	362	30.9.1866	Gx	Lord Rosse: 'spiral'

Adams (Adams 1875). Like Airy in 1836 (awarding a medal to John Herschel), Adams gives a comprehensive review of the research on nebulae.[60] It reaches from Herschel's Slough catalogue to d'Arrest's spectroscopic studies, which led to the conclusion that gaseous nebulae are members of the Milky Way. Adams said '*I feel confident that a plain statement of the nature and extent of the work accomplished by Professor D'Arrest will be sufficient to convince you that he richly deserves our medal.*' He ended with the hope that '*health and strength may long be spared to him, so that he may be able to make many further contributions to the progress of Astronomy*'. Sadly, Heinrich Ludwig d'Arrest died only four months later, on 14 June 1875 in Copenhagen, of a heart attack. His death at the early age of 53 was a shock for the astronomical community (see Section 6.14.1).

8.6.3 D'Arrest's observations and new nebulae

In his second report (*AN* 1500) of 1865 d'Arrest presented a list of 215 new nebulae. Twenty-five were identified with objects of Lord Rosse, and one had already been seen by Bond (NGC 223); 117 objects were included in his earlier list of 125 new nebulae sent to John Herschel in the letter of 1 February 1863. Table 8.4 shows d'Arrest's *AN* articles related to his Copenhagen observations appearing prior to *SN*.

D'Arrest's 'Zweites Verzeichnis von neuen Nebelflecken, aufgefunden am Kopenhagener Refractor im Winter 1864/65',[61] which was published in *AN* 1537, was a follow-up of the *AN* 1500 paper. For 35 objects of the latter, new positions are given ('*given the faintness of most objects, a repeated observation is of interest*'). Moreover, the previous list is continued with objects nos. 216 to 306; for 13 objects, d'Arrest sees an identity with known ones. Meanwhile d'Arrest has got the GC, which is mentioned in the notes to some objects. In a footnote he wrote that '*Concerning this invaluable catalogue it is a pleasure to note that it is exceptionally correct (despite some critical remarks, which must be given due to the insufficient knowledge of the matter). When*

---

[59] See Tables 8.27 and 11.4; see Steinicke (2001c).
[60] Adolph Gautier published a similar report in 1863 (Gautier A. 1863). An extensive theoretical analysis on nebulae was presented by Stephen Alexander in 1852, titled 'On the origins of the forms and the present conditions of some of the clusters of stars, and several of the nebulae' (Alexander S. 1852). Therein, for the first time, the possibility of a spiral structure of the Milky Way is treated.

[61] 'Second list of new nebulae found at the Copenhagen refractor in Winter 1864/65' (d'Arrest 1865c).

Table 8.4. *D'Arrest's reports in the* AN *about the Copenhagen observations*

AN	Date	Content
1366	20.5.1862	Objects from his catalogue, double nebulae, variable nebulae, missing nebulae, proper motion
1369	2.6.1862	50 double nebulae
1500	21.10.1864	Lord Rosse's 'Nebulae not found', Coma Cluster, 215 new nebulae
1537	5.6.1865	91 new nebulae, new positions of former objects
1624	24.12.1866	Missing nebulae

continuously used and compared with the sky for nearly a year, I have found only five errors.'

In *AN* 1624 d'Arrest laments that 15 nebulae of William Herschel, which the latter had discovered on 2 April 1801 at high declinations, are missing (d'Arrest 1867b). The brightest are I 282–284, the fainter ones II 903–905 and III 963–971. He remarks that *'None is contained in Lord Rosse's publication of 1861; the northern regions were not at all investigated by the Birr Castle observers.'* Therefore d'Arrest asked Rümker, '*who had observed so many of the bright nebulae at high northern declinations*', to search for the missing objects. This action was unsuccessful too. Eventually Dreyer was able to solve the mystery, writing in the notes to NGC 2977 *"The places of all the nebulae [...] are affected by some large error. They were all compared with one star only, which was assumed to be '208 (N) Camelop. of Bode's Cat.'."*[62] On his inducement the field in Draco was photographed in 1911 with the 30" Grubb reflector at Greenwich Observatory (Christie 1911). Dreyer localised the correct reference star (BD +78° 412), thus saving the existence of all 15 nebulae.[63] The brightest are the galaxies NGC 2977 (distance to reference star about 9°), NGC 3218 and NGC 3397.

The *SN* contains 307 new nebulae; altogether d'Arrest discovered 319. Of them, 215 are catalogued in the GCS, 97 in the GC and 7 in the NGC. Table 8.5 shows their type distribution; galaxies obviously predominate, and 89% of all discoveries are non-stellar – a high success rate. D'Arrest's objects lie within the declination range −12° to +78°, thus near-pole regions were observed (later Tempel and Swift were successful there too).

Table 8.6 gives special objects and observing data. In addition to NGC 1, the galaxy NGC 7548 in Pegasus was found on 30 September 1861. After some testing, d'Arrest started his Copenhagen observations on 20 October 1861. He immediately discovered nine galaxies in Draco: NGC 6299, NGC 6307, NGC 6310, NGC 6359, NGC 6411, NGC 6512, NGC 6516, NGC 6521 and NGC 6566. Similarly to Lord Rosse's discoveries, the new nebulae were found near known ones – this was a consequence of his observing method.

Table 8.7 shows the seven objects catalogued by Dreyer in the NGC. All but one were independently found by Bigourdan in Paris. The reason why they are not in the GCS is their membership of the galaxy cluster in Coma Berenices, whose crowded appearance had caused identification problems. Earlier d'Arrest had assumed an identity with Herschel objects in four cases. Dreyer tried to resolve the puzzle in the NGC – and was not quite successful (Bigourdan's observations were still unknown at that time). NGC 4396 is not II 87. For NGC 6966 d'Arrest made a (rare) error, giving the wrong declination sign.

Table 8.8 shows d'Arrest's open clusters, emission/reflection nebulae and parts of galaxies (three belong to M 33). Both the reflection nebula NGC 2064 and the emission nebula NGC 2313 (no. 50 in d'Arrest's list of 125 objects; Fig. 8.11), were later catalogued as 'planetary nebulae' by Perek and Kohoutek (PK).

Table 8.9 lists some interesting galaxies, stars and star groups. D'Arrest was the discoverer of the first entry in the New General Catalogue: NGC 1, which is GC 1 too. He found the galaxy in Pegasus on 30 September 1861 during his tests of the new Merz refractor (see Section 6.14.7). However, he did not see the faint companion NGC 2 (14.2 mag), which was eventually discovered on 20 August 1873 by Lawrence

---

[62] Dreyer (1953: 217).
[63] Dreyer (1912a: 234–235).

Table 8.5. *Statistics of d'Arrest's discoveries, catalogued in the GC, GCS and NGC ('SN' = objects found during SN observations)*

Type	GC	GCS	NGC	Sum	SN
Galaxy	86	182	6	274	267
Galactic nebula	1	3		4	4
Open cluster		2		2	2
Part of galaxy	2	2		4	2
Star	1	6		7	7
Pair/group of stars	6	19	1	26	23
Not found	1	1		2	2
Sum	97	215	7	319	307

Table 8.6. *Special objects and observing data of d'Arrest*

Category	Object	Data	Date	Remarks
Brightest object	NGC 133	OC, 9.4 mag, Cas	4.2.1865	Fig. 8.10 left
Faintest object	NGC 7477	Gx, 15.7 mag, Psc	9.9.1866	Fig. 8.10 right
Most northern object	NGC 2591	+78°, Gx, 12.2 mag, Cam	12.8.1866	
Most southern object	NGC 7005	−12°, star group, Aqr	23.8.1855	Leipzig
First object	NGC 1	Gx, 12.8 mag, Peg	30.9.1861	
Last object	NGC 5072	Gx, 13.5 mag, Vir	26.4.1867	
Best night			6.4.1864	9 objects (Gx in Leo/Com, UMa)
Best constellation	Coma Berenices		1861–67	66 galaxies

Table 8.7. *Seven of d'Arrests objects that were eventually catalogued in the NGC*

NGC	Date	Type	V	Con.	d'Arrest	Remarks
4396	20.4.1865	Gx	12.4	Com	II 87?	
4867	10.5.1863	Gx	14.5	Com	h 1500?	Coma Cluster, Bigourdan 28.4.1885
4871	10.5.1863	Gx	14.2	Com	II 388?, h 1501?	Coma Cluster, Bigourdan 16.5.1885
4873	10.5.1863	Gx	14.1	Com	II 389?, h 1502?	Coma Cluster, Bigourdan 16.5.1885
4883	22.4.1865	Gx	14.4	Com		Coma Cluster, Bigourdan 16.5.1885
4898	6.4.1864	Gx	13.6	Com	h 1510?	Coma Cluster, Bigourdan 15.5.1885
6966	26.7.1865	2 stars		Aqr		Declination error, Bigourdan 27.7.1884

Table 8.8. *D'Arrest's open clusters, galactic nebulae and galaxy parts*

Type	NGC	GC(S)	Date	V	Con.	Remarks
Open cluster	133	5103	4.2.1865	9.4	Cas	Fig. 8.10 left
Open cluster	609	5187	9.8.1863	11.0	Cas	
Emission nebula	1973	5352	16.12.1862		Ori	North of M 42
Emission nebula	1975	5353	3.10.1864		Ori	North of M 42
Emission nebula	2313	1475	4.1.1862		Mon	Near V565 Mon (14.2–15.7 mag); PK 26−2.2
Reflection nebula	2064	5355	11.1.1864		Ori	Southwest of M 78; PK 2+2.1
Part of galaxy	588	348	2.10.1861	13.5	Tri	Au 13; H II region in M 33
Part of galaxy	592	349	2.10.1861	13.0	Tri	Au 14; H II region in M 33
Part of galaxy	595	5186	1.10.1864	13.5	Tri	H II region in M 33; Safford 1.11.1866
Part of galaxy	5471	5757	22.8.1863	14.7	UMa	H II region in M 101

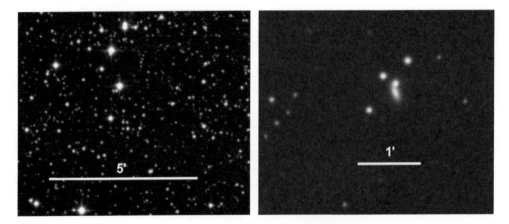

Figure 8.10. D'Arrest's brightest and faintest objects. Left: the open cluster NGC 133 in Cassiopeia; right: the galaxy NGC 7477 in Pisces (DSS).

Parsons. NGC 4064 in Coma Berenices (10.7 mag) is his brightest galaxy; the faintest is NGC 7477 (15.7 mag) in Pisces – a remarkable case. At d'Arrest's position we have the faint galaxy and four stars (Fig. 8.10 right). He correctly describes the constellation; the star north of the galaxy is said to have '17 mag' (14.2 mag is correct). On 14 February 1863 d'Arrest found NGC 1275, the brightest galaxy in the Perseus Cluster (Abell 426). The night of 9 September 1866 was quite exceptional; d'Arrest noted '*coelo omnio nitidissimo*' ['*sky perfectly brilliant*']. This explains why d'Arrest could perceive these faint objects in the 11-inch at all. Burnham later claimed (Burnham S. 1892b) that NGC 7477 = NGC 7472 (O. Struve) = NGC 7482 (Marth). The second identity is correct, but the first is not. NGC 4949 does not appear in the classic galaxy catalogues (CGCG, MCG, UGC); the object was eventually entered in the Catalog of Principal Galaxies (1987) as PGC 45161.

Of the 30 stars and star groups found by d'Arrest and seen as 'nebulae', some were taken as examples. NGC 162 is involved in the mysterious case of 'GC 80' (see Section 7.1). NGC 3932 was found while

8.6 D'Arrest's masterpiece

Table 8.9. *Special discoveries of d'Arrest: interesting galaxies, stars, star groups and the two missing objects*

Type	NGC	GC(S)	Date	V	Con.	Remarks
Galaxy	1	1	30.9.1861	12.8	Peg	First NGC object
Galaxy	1275	675	14.2.1863	11.7	Per	Brightest galaxy in Perseus cluster
Galaxy	4064	2688	29.12.1861	10.7	Com	Brightest galaxy
Galaxy	4949	5715	19.4.1865	14.9	Com	Not in classic catalogues
Galaxy	7477	6119	9.9.1866	15.7	Psc	Faintest galaxy; Fig. 8.10 right
Star	162	5107	22.8.1862	15.0	And	L. Parsons 16.10.1866; Schultz 5.9.1867
Star	3932	2593	4.12.1861	13.6	UMa	Brightest star: GSC 3452–1488
Star group	272	5119	2.8.1864		And	Around BD +35° 154 (9.4 mag)
Star group	1746	5349	9.11.1863		Tau	Near NGC 1750 and NGC 1758
Star group	7005	5983	28.2.1855		Aqr	Near M 73
Not found	3167	2039	1.5.1862		LMi	1.5.1862
Not found	3927	5586	27.4.1864		Leo	27.4.1864

Figure 8.11. The small emission nebula NGC 2313 associated with V565 Monocerotis (DSS).

observing William Herschel's II 740 (NGC 3928); the star's brightness was estimated as '17 mag' (no. 85 in his list of 125 nebulae). NGC 272 is a striking star group (reversed 'L') around BD +35° 154. The inconspicuous group NGC 1746 overlaps with the open clusters NGC 1750 (VIII 43) and NGC 1758 (VI 21) discovered by William Herschel in 1785. Both are significant objects (NGC 1750 being the brighter one); owing to their different distances, they are not physically connected. D'Arrest's group is located in the middle – a mere 'optical' mix of stars of both clusters (Leiter 2007).

Seventeen d'Arrest objects had been discovered earlier by other observers (Table 8.10). All but one are due to Marth. Between these two observers there was a 'remote duel' (albeit unnoticed) during the years 1863–65 regarding new nebulae. Using Lassell's 48" reflector on Malta, Marth discovered 587 objects between 10 June 1863 and 2 April 1865. At the same time, d'Arrest found 97 with the Copenhagen 11" refractor; there are 22 common objects. However, the two observing programmes were totally different. While d'Arrest was working on a revision of the Herschel nebulae (measuring precise positions to derive proper motions), Marth was purely focused on finding new nebulae. The planetary nebula NGC 6842 in Vulpecula is interesting; it was discovered by Marth on 28 June 1863. Whereas d'Arrest saw it on 26 August 1864 while searching for William Herschel's II 202 (NGC 6847).[64] In the case of the galaxy NGC 4908, d'Arrest did not notice the identity with III 363, which is understandable due to its position in the crowded Coma Cluster.

Nineteen objects credited to d'Arrest by Dreyer are identical with other NGC objects that had been found earlier (Table 8.11). In two cases d'Arrest discovered an object twice. The reason for the first one, NGC 4882 = NGC 4886, is obvious: the galaxy is

---
[64] See Tempel's observations in Section 6.17.3.

Table 8.10. *D'Arrest objects that had been discovered earlier by others*

NGC	GCS	Date	Type	V	Con.	Remarks
2713	5442	15.3.1866	Gx	11.9	Hya	Marth 3.3.1864 (m 135)
2716	5444	15.3.1866	Gx	12.2	Hya	Marth 3.3.1864 (m 136)
2795	5465	15.3.1866	Gx	12.9	Cnc	Marth 21.12.1863 (m 156)
2861	5475	8.2.1866	Gx	13.0	Hya	Marth 28.3.1864 (m 162)
3357	5533	22.2.1865	Gx	13.0	Leo	Marth 5.4.1864 (m 202); Tempel 18.11.1881
3391	5535	19.3.1865	Gx	13.0	Leo	Marth 1.4.1864 (m 204)
4908	5704	22.4.1865	Gx	13.2	Com	W. Herschel 11.4.1785 (III 363), Coma Cluster
6384	5865	8.4.1866	Gx	10.4	Oph	Marth 10.6.1863 (m 339); Stephan 19.7.1870
6570	5895	23.8.1865	Gx	12.7	Oph	Marth 2.6.1864 (m 363)
6842	5947	26.8.1864	PN	13.1	Vul	Marth 28.6.1863 (m 403); Safford 12.7.1866
7053	5990	8.10.1865	Gx	13.1	Peg	Marth 2.9.1863 (m 438)
7065	5993	24.8.1865	Gx	13.7	Aqr	Marth 3.8.1864 (m 440); Stephan 22.9.1876
7147	6013	15.9.1865	Gx	13.6	Peg	Marth 11.8.1863 (m 457)
7367	6082	30.8.1865	Gx	13.9	Peg	Marth 29.8.1864 (m 496)
7422	6101	29.9.1866	Gx	13.5	Psc	Marth 11.8.1864 (m 509); O. Struve 6.12.1865; Safford 27.9.1867
7608	6181	2.10.1866	Gx	14.2	Peg	Marth 25.11.1864 (m 569)
7673	6201	30.9.1866	Gx	12.5	Peg	Marth 5.9.1864 (m 577)

in the dense Coma Cluster. But the second case is strange: NGC 4078 = NGC 4107, a 13.2-mag galaxy in Virgo (Fig. 8.12). Dreyer mentions Marth (m 231) and d'Arrest as discoverers of NGC 4078; using the description of the former ('faint, very small, round'). The sequence is incorrect, in that d'Arrest's find (23 March 1865) came two days earlier than Marth's. NGC 4107 (GC 5617) was even discovered by d'Arrest on 17 April 1863, being described as a 'planetary nebula, pretty bright, small, very elongated' with a star of 10–12 mag 7′ east. Considering the form, the attribute 'planetary' is strange. Burnham's observation with the 36″ Lick refractor in 1891 is interesting (Burnham S. 1894a). He notes that NGC 4107 is not 'planetary' and has a star to the west; perhaps he meant the attached faint star. Anyway, his position of the galaxy is correct. A modern analysis shows that d'Arrest made a right ascension error of 2^m for NGC 4107 (which is about ½°).

Figure 8.12. The galaxy NGC 4078 = NGC 4107 in Virgo and its neighbours (DSS).

Table 8.11. *D'Arrest objects that are identical with other NGC objects that had been discovered earlier*

NGC	GCS	Date	Type	V	Con.	Remarks
875	5230	26.9.1865	Gx	13.1	Cet	NGC 867, W. Herschel 21.12.1783 (III 2)
3144	5522	25.9.1865	Gx	13.3	Dra	NGC 3174, W. Herschel 2.4.1801 (III 964)
3183	5523	28.9.1865	Gx	11.8	Dra	NGC 3218, W. Herschel 2.4.1801 (I 283)
3575	5552	21.2.1863	Gx	11.4	Leo	NGC 3162, W. Herschel 12.3.1784 (II 43)
3856	5580	8.5.1864	Gx	13.3	UMa	NGC 3847, J. Herschel 3.4.1831 (h 965)
3966	5592	8.5.1864	Gx	12.9	UMa	NGC 3986, J. Herschel 29.4.1827 (h 1027)
4046	5602	10.4.1863	Gx	11.9	Vir	NGC 4045, W. Herschel 20.12.1784 (II 276); Todd 2.1.1878
4078	5607	23.3.1865	Gx	13.2	Vir	Marth 25.3.1865 (m 231); = NGC 4107, d'Arrest 17.4.1863
4130	5621	15.3.1866	Gx	12.5	Vir	NGC 4129, W. Herschel 3.3.1786 (II 548)
4140	5623	10.4.1863	Gx	13.3	Vir	NGC 4077, W. Herschel 20.12.1784 (III 258); Todd 5.1.1878
4325	5638	15.4.1865	Gx	13.4	Vir	NGC 4368, W. Herschel 15.3.1784 (III 38)
4338	5640	18.5.1863	Gx	12.4	Com	NGC 4310, W. Herschel 11.4.1785 (II 378)
4702	5669	4.3.1867	Gx	12.7	Com	NGC 4692, W. Herschel 11.4.1785 (II 381)
4797	5679	21.4.1865	Gx	13.2	Com	NGC 4798, W. Herschel 11.4.1785 (II 382)
4884	5698	6.4.1864	Gx	11.5	Com	Coma Cluster; = NGC 4889, W. Herschel 11.4.1785 (II 391)
4886	5699	22.4.1865	Gx	13.9	Com	Coma Cluster; = NGC 4882, d'Arrest 22.4.1865
4960	5716	23.4.1865	Gx	13.1	Com	NGC 4961, W. Herschel 11.4.1785 (II 398)
5519	5759	26.4.1865	Gx	13.4	Boo	NGC 5570, W. Herschel 23.1.1784 (III 12)
5819		6.10.1861	Gx	13.5	UMi	d'Arrest: III 311?; = NGC 5808, W. Herschel 16.3.1785 (III 311)

The other identities are due to William Herschel (15) and John Herschel (2); among them is NGC 4889 = NGC 4884, the brightest galaxy in the Coma Cluster. NGC 5819 is not in the GCS. Obviously, for Dreyer the case was unclear, because d'Arrest supposed an identity with III 311, which was accepted in the NGC. On the other hand, we have NGC 5808 = III 311 there. Dreyer did not notice the identity with NGC 5819.

Figure 8.13 shows the brightness distribution of the d'Arrest objects. The mean is 13.3 mag, which is comparable to that for Tempel, who was using a 11" refractor too. The brightest object is the open cluster NGC 133 in Cassiopeia (9.4 mag), which was discovered on 4 February 1865.

8.6.4 Discovery of the galaxy cluster in Coma Berenices

In *AN* 1500 d'Arrest mentions two large 'groups of nebulae'. The first is located at 11^h58^m $+21°$, in the western region of Coma Berenices, near the border with Leo. Here he found six galaxies on 2 and 4 May 1864 (NGC 4086, NGC 4089, NGC 4090, NGC 4091, NGC 4092 and NGC 4093) and another six from William Herschel's catalogue were observed. The objects cover a field of 45′ diameter. John Herschel discovered four other nebulae there (which were not observed by d'Arrest) and Marth three (unknown to d'Arrest). The group is the centre of a loose cluster of diameter 1.7°, containing 555 galaxies. It was first catalogued by Zwicky as CGCG 1202.0+2028.

The second group was located by d'Arrest in the area 12^h53^m to 58^m and $+28°$ to $+29°$ (the northeastern part of Coma Berenices). This is the famous Coma

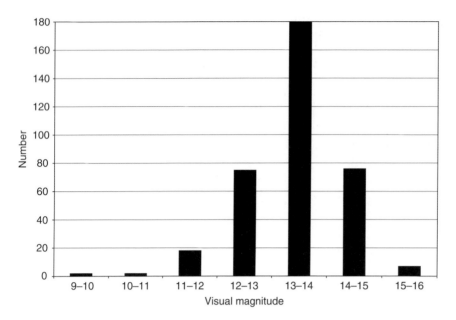

Figure 8.13. The brightness distribution of d'Arrest's objects.

Cluster, listed as no. 1656 in George Abell's catalogue of rich clusters of galaxies.[65] William Herschel discovered 23 members, including the brightest (NGC 4889, 11.5 mag). Because the cluster covers a field of 1.5° diameter, the local accumulation was not noticed. However, he detected that nebulae are conspicuously crowded together in the constellations Virgo and Coma Berenices, *'that remarkable collection of many hundreds of nebulæ which are to be seen in what I have called the nebulous stratum of Coma Berenices'*.[66] Most of these galaxies belong to the nearby (and thus extended) Virgo Cluster. Richard Proctor was the first to visualise the 'stratum' in his publication 'Distribution of nebulae' (Proctor 1869a); a more detailed map, focused on Virgo and Coma Berenices, was published three years later (Proctor 1872a).

The next to find members of the Coma Cluster, which is much more remote than the Virgo Cluster, was John Herschel. While observing objects that had been discovered by his father, he discovered three galaxies. He too did not notice the clustering. Thus d'Arrest was the first to detect it, during his Copenhagen observations. He discovered 34 members, mainly in April 1865. Among them is the second brightest, NGC 4874 (11.9 mag). He saw the brightest too, which had been catalogued as NGC 4884 by Dreyer, but neither of them noticed that it is identical with William Herschel's II 391 = NGC 4889 (see Table 8.12). Moreover, it is surprising that Herschel did not see NGC 4874, which is only 7′ east of NGC 4889 (Fig. 8.14).

On the basis of the known and new nebulae in the field, d'Arrest for the first time got a clear picture of the Coma Cluster: *'The nebulae are incredibly numerous and dense, and, though very faint, one can get a priori no idea of its manifoldness. Sometimes, in the best moments, I had the certain impression that the nebulae, having diameters of only a few arcseconds, are mixed in a crowd of larger, elongated, stellar or cometary ones, much like oysters packed in a barrel.'* (d'Arrest 1865a).

Most of d'Arrest's galaxies are of about 14 mag. It is interesting that the faintest, NGC 4858 (15.2 mag), is significantly below the observational limit of both Herschels, though their telescopes were larger. For d'Arrest, the cluster was a welcome example to demonstrate the technological superiority of his 11″ refractor: *'The observation and cataloguing of individual nebulae with the telescopes of Lord Rosse or Lassell will be, for known reasons, very difficult. In the dense crowd of objects, reliable data can only be obtained through precise measurements.'* The main problems of these large reflectors were the small field of view and the deficient mounting. Accurate micrometric measurements were impossible, which feature Otto Struve had already

---

[65] The story of the Coma Cluster discovery is told in Steinicke (2005a).
[66] Herschel W. (1785: 255–256).

Figure 8.14. The rich galaxy cluster in Coma Berenices (Abell 1656). NGC 4889 is central, NGC 4874 to the right (below the bright star BD +28° 2171) and NGC 4921 at the lower left edge (field 35′ × 35′).

criticised. Especially positions determined at Birr Castle were not reliable, which led to some confusion, particularly in crowded fields.

From Lord Rosse's 1861 publication d'Arrest did know that NGC 4864 (h 1500) was observed on 9 March 1850; the Birr Castle report mentions '*numerous nebulae around*'. But for him this 13.2-mag galaxy is located only '*at the edge of the large group of nebulae*', which is, however, not quite correct. According to the 1880 publication, the Birr Castle notes are more detailed, mentioning about 15 nebulae: '*Met 8 nebulae sweeping p the star to about 15^m and 6 or 7′ others NPD 63° 15′ ± and 15^m p last*.'[67] The text is, however, erroneous, for the second part should read '6 or 7 others NPD 61° 15′± and 15^m f last'. Since the star is BD +28° 2171 (7.2 mag), the sweep had an extension of ±3° east–west and probably something similar north–south. Even from the extended notes it must be concluded that the observers (Lord Rosse, J. Stoney?) had described neither the cluster as a whole nor individual members. Therefore d'Arrest, who could map the nebulae due to his precise measurements, must be credited as the discoverer of the Coma Cluster. Nevertheless, often Max Wolf is mentioned in the literature as having found it. Actually, he was the one to make the first photographic survey visualising the true galaxy distribution (Wolf M. 1901).

8.6.5 Other clusters of galaxy

D'Arrest discovered members of other clusters too (Table 8.12). He found 4 of the 13 NGC galaxies in Abell 194 (Cetus Cluster) which had been catalogued by Dreyer in the GCS. The first four are due to William Herschel (1 October 1785): NGC 545 (II 448, 12.3 mag), NGC 547 (II 449, 12.3 mag), NGC 560 (III 441, 12.9 mag) and NGC 564 (III 442, 12.5 mag). Lewis Swift contributed four objects on 20 November 1886 (see Table 9.19); the remaining one was found by George Searle on 2 November 1867: NGC 548 (13.7 mag). The central region of the cluster has a diameter of 1°.

---

[67] Parsons L. (1880: 123).

Table 8.12. *D'Arrest's discoveries in various galaxy clusters*

Abell	NGC	GC(S)	Date	V	Con.	Remarks
194	535	5177	31.10.1864	14.0	Cet	
194	541	5178	30.10.1864	12.2	Cet	
194	543	5179	31.10.1864	13.3	Cet	
194	558	5180	1.2.1864	14.0	Cet	
262	710	5196	12.8.1863	13.5	And	Parsons L. 18.11.1876
262	714	5197	2.12.1863	13.2	And	Parsons L. 18.11.1876
262	717	5198	16.9.1865	14.0	And	Parsons L. 18.11.1876
426	1267	668	14.2.1863	13.1	Per	Perseus Cluster
426	1268	669	14.2.1863	13.8	Per	Perseus Cluster
426	1270	671	14.2.1863	13.3	Per	Perseus Cluster
426	1272	672	14.2.1863	12.0	Per	Perseus Cluster
426	1273	673	14.2.1863	13.3	Per	Perseus Cluster
426	1275	675	14.2.1863	11.7	Per	Perseus Cluster; brightest member
	2560	1644	17.3.1862	13.7	Cnc	Cancer Cluster
	2569	1650	19.2.1862	14.3	Cnc	Cancer cluster
1367	3816	5574	9.5.1864	12.5	Leo	Leo Cluster
1367	3840	5576	8.5.1864	13.7	Leo	Leo Cluster
1367	3844	5577	8.5.1864	14.0	Leo	Leo Cluster
1367	3873	5582	8.5.1864	12.9	Leo	Leo Cluster
1367	3886	5583	9.5.1864	13.2	Leo	Leo Cluster

D'Arrest discovered three members of Abell 262 in Andromeda: NGC 710, NGC 714 and NGC 717. They were independently seen by Lawrence Parsons in 1876 and catalogued in the GCS. William Herschel was the most successful, with six galaxies. On 14 February 1863 d'Arrest discovered six members in the Perseus Cluster (Abell 426), among them the brightest (NGC 1275). He was nearly as successful as Bigourdan, who contributed seven galaxies (see Table 9.48). D'Arrest's objects had already been entered in the GC. In the Cancer Cluster, which bears no Abell number due to its loose structure, d'Arrest found two of nine NGC galaxies. The others were discovered by William Herschel (NGC 2558, NGC 2562 and NGC 2563), Marth (NGC 2553 and NGC 2556), Stephan (NGC 2557) and Copeland (NGC 2570). On 8 and 9 May 1864 d'Arrest found 5 of 21 NGC galaxies in Abell 1367 (the Leo Cluster); William and John Herschel were the most successful, with seven members each; and Stephan discovered five. In all these cases, none of the observers noticed the cluster structure.

8.6.6 D'Arrest's spectroscopic investigations

Inspired by the work of William Huggins and Lt John Herschel,[68] d'Arrest turned to spectroscopy after finishing his Copenhagen observations. A first publication, titled 'Spectroskopische Beobachtungen zweier Nebelflecke',[69] treats the planetary nebulae IV 45 (NGC 2392) and IV 37 (NGC 6543). Shortly after, a second paper on 'Gas-Nebulosen' ['gaseous nebulae'] appeared. Therein d'Arrest formulates the suggestion that *'the 31 or 32 gaseous nebulae, which have thus far become known in both hemispheres (altogether the light of only 140 nebulae has been analysed), probably belong to our own stellar system'*. The reason for this is that, *'While, as is generally known, the main part of*

---

[68] Huggins W. (1864, 1866a); see also Hearnshaw (1985, 2009). Lt Herschel, a son of John Herschel, made the first spectroscopic observation of the southern sky (Herschel Lt 1868a, b).

[69] 'Spectroscopic observations of two nebulae' (d'Arrest 1872a).

the nebulae show a distribution quite different from the stars, 24 or 25 true gaseous nebulae are located within the Milky Way, visible to the naked eye, and only 6 or 7 are outside.' (d'Arrest 1873). He concluded that '*It seems to be a remarkable fact that the distribution of the gaseous nebulae obviously matches that of the stars of our system.*'

About his spectroscopic research, d'Arrest published a monograph in Danish: *Undersogelser over de nebulose Stjerner i Henseende til deres spektralanalytiske Egenskaber.*[70] It contains the first spectral catalogue of nebulae. Thirty-eight objects with discrete spectra were treated in detail, ordered by GC-number. A subsequent table lists 115 objects with continuous spectra, of which 30 had been observed by d'Arrest with a visual spectroscope. In the appendix drawings of the Orion Nebula and its inner part (trapezium)[71] are presented – certainly d'Arrest's best. He had treated the object already in 1868: his paper 'Ueber eine Darstellung des grossen Orionnebels vom Jahre 1779'[72] draws attention to a hitherto overlooked drawing by Professor Lefèbvre in Lyon, which was published in 1783 in Paris. D'Arrest wrote that '*Only Schræter knew about its existence; but he has not seen it and, moreover, his quotation is erroneous.*'

8.7 MARTH ON MALTA: 600 NEW NEBULAE

Most of the time during his career the German Albert Marth was an astronomical 'foreign worker' in England and Ireland. Being single, his restless life was driven by the search for the perfect telescope at the optimal site. Except the few years on Malta, he could not realise his dream – a depressing situation. Marth was physically not robust and had a sensitive, reserved and sometimes anxious character. Anyway, with his Malta catalogue, presenting almost 600 new nebulae, he made an important contribution to Dreyer's GCS and NGC. Despite the deficient mounting of Lassell's 48" reflector, Marth determined astonishingly precise positions.

---

[70] 'Investigation of nebulous stars concerning their spectralanalytic features' (d'Arrest 1872b).
[71] See Section 9.6.10.
[72] 'About a drawing of the great Orion Nebula from the year 1779' (d'Arrest 1868a). See also Holden (1882b), Fig. 13.

Figure 8.15. Albert Marth (1828–1897).

8.7.1 Short biography: Albert Marth

Albert Marth was born on 5 May 1828 in Kolberg, Pomerania (Fig. 8.15). Early in his life he became an orphan. He initially studied theology at Berlin University, but his interest in mathematics and astronomy became stronger. Marth moved to Königsberg, the domain of Friedrich Wilhelm Bessel, where he was taught astronomy by Christian August Friedrich Peters. He observed comets and minor planets. At the age of 24, Marth published his first paper in the *Astronomische Nachrichten* (edited by Peters).

After that Marth travelled cross-country through England, to hire himself out as an astronomer. In 1853 he got his first employment. George Bishop, the owner of a remarkable observatory next to his feudal 'South Villa' in London's Regent's Park, sought an observer to replace Eduard Vogel (the elder brother of Hermann Vogel). Marth became the assistant of John Russell Hind and succeeded him as observatory director a year later. On 1 March 1854 he discovered the minor planet (29) Amphitrite, using the main instrument, a 7" Dollond refractor, and Hind's 'ecliptical charts'. Marth stayed only two years at Bishop's Observatory, but his name already became known.

Marth changed to Durham Observatory, Potters Bank, to succeed George Rümker as Temple Chevallier's assistant. From 1855 to 1862 he observed planets and moons with the 6.75" Fraunhofer refractor and worked on problems in mathematics and celestial mechanics. His critical analysis of the Greenwich meridian-circle observations caused some trouble. Airy, Astronomer Royal and a pretty pedantic person, was not at all amused by these comments (of a German).

In 1862 Marth was chosen by Lassell to assist him on Malta during his second sojourn. He was charged with searching for new nebulae with the 48" reflector. After the successful completion of this task he left the island in mid 1865 and returned to London; with a few breaks he lived there until 1883. Occasionally he worked for Warren de la Rue, evaluating solar observations. Quite unedifying was his activity in Gateshead, the home of Robert Sterling Newall, who had erected a 25" Cooke refractor in 1869 (Anon 1870a).[73] The heavy instrument was the prototype of a 'battleship refractor'. Astronomically it was a complete failure: confined to the smoky sky of Newcastle, only a few observations could be made over 15 years.[74] After a visit in summer 1875, Hermann Vogel lamented *'that up to now practically nothing was achieved with this fine instrument'*.[75] In 1885, Newall wrote to Denning that *'Atmosphere has an immense deal to do with definition. I have only had one fine night since 1870! I then saw what I have never seen since.'*[76]

Marth, upset by the Gateshead experience, looked for a new job. Unfortunately, his sometimes strange attitude was not helpful. He applied for employment as an assistant of Lawrence Parsons at Birr Castle (probably to follow Charles Burton). Parsons asked Airy, of all people, about Marth's qualification – of course, the failure was predictable. An application to Melbourne (Ellery), with the opportunity to observe with the 48" reflector (Great Melbourne Telescope), was unsuccessful too. In 1882 Marth joined the Venus transit expedition to the Cape of Good Hope.

A year later he eventually got a full-time job at Colonel Edward Henry Cooper's Markree Observatory near Sligo (Ireland) – it was to be his final station. He calculated ephemerides and observed planets with the aged 14" refractor which had been erected in 1831 by Col. Cooper's father Edward Joshua Cooper; the instrument was in a bad condition. Having been made of iron and permanently exposed to the Irish weather, it was corroded – reasonable observations were out of reach. After many years in England and Ireland, Albert Marth spent his last years in Germany. On 5 August 1897 he died of cancer in Heidelberg at the age of 69.[77]

8.7.2 Corrections to John Herschel's catalogues

In 1855, working as observer in Durham, Marth published his only paper on nebulae (the Malta catalogue bears the name of Lassell): a list of remarks and corrections to John Herschel's Slough and Cape catalogues. It appeared on 22 November 1855 in the *Astronomische Nachrichten*, titled 'Bemerkungen zu Sir John Herschel's Nebel-Katalogen'.[78] At the beginning Marth wrote *'Considering the not small number of errors and misprints which have crept into Sir John Herschel's catalogues of nebulae, contributions to clean them might perhaps be welcomed by some astronomers, wherefore I take the liberty of giving some relevant remarks here.'* There follows a list of 63 cases, sorted by h-number. His remarks concern typos, incorrect identifications (between William and John Herschel), wrong reference stars and positional errors. Additionally he identifies Brorsen's nebula of 1850 (NGC 2024) with William Herschel's V 28 (see Section 6.16.3).

Marth does not state whether his remarks result from his own observations or are due to a persual of the catalogues only. In Durham he had access to a humble 6.75" Merz refractor and nothing is known about any systematic observations of nebulae. However, Marth had checked some cases. For instance, on the planetary nebula h 2618 (IV 26, NGC 1535) in Eridanus he wrote

---

[73] It surpassed the short-lived 24" Craig refractor (and Porro's 20-inch in Paris) and remained the world's largest refractor until 1873, then being beaten by the 26" Clark refractor at the US Naval Observatory, Washington. In 1960 the Newall telescope was erected in Penteli near Athens; see Binnewies *et al.* (2008) and Dialetis *et al.* (1982).

[74] Some were devoted to the Merope Nebula (see Sections 11.6.9 and 11.6.10).

[75] Dick W. (2000: 115).

[76] Denning (1891b: 25).

[77] Obituaries: Dreyer (1897), Holden (1897b) and Knobel (1898); see also Steinicke (2003b).

[78] 'Remarks on John Herschel's catalogues of nebulae' (Marth 1856).

'*By looking up at the sky I have convinced myself that Sir William Herschel's place of the nebula IV. 26 is almost correct.*' Like d'Arrest and Auwers, Marth must have studied William Herschel's catalogues. Unfortunately, Marth's list was barely noticed. Only Auwers took over the corrections into his work.

A peculiar case is h 782; Marth wrote "'*or I. 118*' *and JH's remark too are to be struck out. The error seems to follow from a mistake during the reduction, because 46 Leonis Minoris instead of 46 Ursae* [Majoris] *was taken as reference star. The nebula I. 118 was not observed by J.H. a second time.*" Auwers remarked '46 L. m. = 46 Urs.', i.e. the two stars are identical.[79] In a paper of 7 May 1864 on the 'missing nebula H. I, 118', d'Arrest could finally clear the case, finding the true identity of the object (d'Arrest 1864). Therein Marth is criticised because of his '*conjecture, made without checking the sky*'. He shows that the nebula exists; the error was not due to William Herschel's reference star, but caused by a '[circle] *reading or writing error of 1° in declination*'. John Herschel was involved too, making the case even more complicated. A modern analysis gives that I 118 (h 779 = h 782, GC 2236 = GC 2239) is an 11.5-mag galaxy in Leo Minor, which was catalogued as NGC 3430 by Dreyer. To save Marth's honour, Auwers added that '*With the numerous other amendations we owe to him, he has been luckier.*'

8.7.3 The objects of the Malta catalogue

During the years 1862–65 Albert Marth was a scientific assistant of William Lassell during his second stay on Malta (see Section 7.5.1). His only duty was the search for new nebulae with the 48" reflector (Fig. 8.16). Given the difficult handling of the huge, equatorially mounted instrument, his reliable observations and precise positions were a remarkable effort. Marth fancied the idea of searching for Martian moons, but he did not dare to hamper Lassell's observations of planets.

After his visit to Malta (7 to 11 October 1863), Otto Struve gave an interesting report about the observations: '*Mr Lassell has decided to carry out a sky survey on Malta with respect to the nebulae, which, as observing in small* [declination] *zones, requires only a gradual change of the instrument and thus of the observer in each night, nearly avoiding the mentioned mechanical difficulties and,*

Figure 8.16. Lassell's 48" reflector on Malta (Repsold J. (1908), Vol. 2, Fig. 164).

on the other hand, fully effecting the great optical power of the telescope. Already the beginning of the task shows what important results we can expect. Mr Marth reported that, in an equal area, he has found twice as many nebulae as listed in the Herschel catalogues. That such a work, particularly considering such a rich harvest, even in the persistently clear sky on Malta, cannot be finished in a few months is evident. Therefore Lassell has appointed his stay on the island to spring 1865; but due to the experience hitherto made, Mr Marth guesses that, in the chosen way, even in this period and with the greatest effort, it would be hardly possible to scan even that part of the sky culminating south of the zenith. Of course, it would be very sad if the work could not be completely finished, and thus it is to be hoped that Lassell will still make the future offer to science to either extend his stay on Malta or at least leave his telescope to the able hands of Mr Marth until the observations have covered the whole Maltese sky.'[80]

The results were published by Lassell in 1867, titled 'A catalogue of new nebulae discovered at Malta with the four-foot equatorial' (Lassell 1867b). Unfortunately, he did not accede to Struve's request and thus the observations ended in April 1865. Since Marth's highest declination was +37°, which equals the zenith, the remaining

---

[79] Auwers (1862a: 52).

[80] Struve (1866a: 549–550).

Table 8.13. *Special objects and observational data of Marth*

Category	Object	Data	Date	Remarks
Brightest object	NGC 6384 (m 339)	Gx, 10.4 mag, Oph	10.6.1863	Fig. 8.17
Faintest object	NGC 4042 (m 226)	Gx, 16.4 mag, Com	18.3.1865	Faintest NGC object
Most northern object	NGC 6117 (m 309)	+37°, Gx, 13.8 mag, CrB	5.7.1864	
Most southern object	NGC 7513 (m 530)	−28°, Gx, 11.9 mag, Scl	24.9.1864	
First object	NGC 6308 (m 332)	Gx, 13.6 mag, Her	6.6.1863	Two more nebulae
Last object	NGC 2934 (m 174)	Gx, 16.0 mag, Leo	2.4.1865	
Best night			25.3.1865	26 galaxies in Vir, Ser, Leo
Best constellation	Pegasus		1863–65	87 galaxies

Figure 8.17. Marth's brightest object, the galaxy NGC 6384 in Ophiuchus (DSS).

zones towards the pole were not scanned. Obviously, due to the low latitude of Malta (36°), it was more important to search at declinations below the celestial equator (the lowest was about −38°) and off the Milky Way. In the short introduction Lassell lamented the '*extremely small portions of the heavens which have been thus surveyed – and that in a very cursory manner*'. Thus the catalogue is understood as a '*fragmentary list*' only. Anyway, for Dreyer it was a real treasure chest. All of the objects were entered into the GCS. Following the numbering of the Malta catalogue, they are designated m 1 to m 600 in an extra column – only Marth was accorded such a privilege. In the NGC, the m-number is given in the column 'Other observers'.

The quality of the observations is high. The coordinates (R.A. = right ascension, P.D. = polar distance, epoch 1860) and the short 'Summary description' in Herschel style mostly allow the clear identification of an object. Marth determined the position from known reference stars. Unfortunately, the southern sky offered not enough appropriate stars and thus not all discoveries were catalogued: '*Some 200 to 250 more Nebulae were discovered, but are not included in this list, from the want of any Catalogue or Map containing the small stars with which they were compared.*' The column 'Date first found' gives a decimal date; for instance, 1864.62 is 5 September 1864. The note 'ver.' means 'verified', i.e. another independent observation (this applies to about 10% of all objects).

Table 8.13 shows special objects and data of Marth's observations. The first nebulae (galaxies in Hercules) were found on 6 June 1863: NGC 6308 (m 325), NGC 6314 (m 326) and NGC 6315 (m 327). Malta offered '*many opportunities of clear sky*'; the best month was October 1864 with 61 discoveries. Concerning the declination, Marth found three objects at −38° in Corona Australis: the reflection nebulae NGC 6726, NGC 6727 and NGC 6729; but they had already been discovered three years earlier by Schmidt (see Section 6.18.3). Marth was the discoverer of the NGC object with the second largest distance (see Table 10.17): the 15.0-mag galaxy NGC 5535 (m 273, GC 5670) in Bootes, which he found on 8 May 1864. The object is 795 million ly away.

8.7 Marth on Malta   255

Table 8.14. *Statistics of Marth's discoveries*

Type	Number	Remarks
Galaxy	556	Brightest: NGC 6384 (Oph, 10.4 mag); last entry: NGC 7840 (Peg)
Planetary nebula	4	
Emission nebula	3	
Star	6	Brightest: NGC 7830 (Psc) = GSC 594–743 (12.7 mag)
Pair/group of stars	8	Largest group: NGC 6669 (Her) = 6 stars
Not found	5	NGC 2637, NGC 2643, NGC 6693, NGC 7028, NGC 7405
Sum	582	

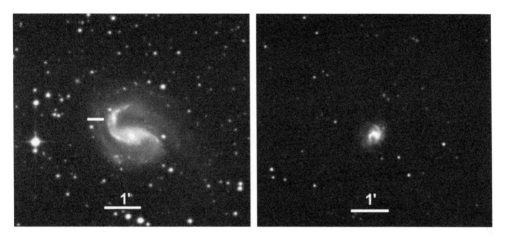

Figure 8.18. Left: NGC 6908 (marked), a background galaxy of NGC 6907 in Capricorn; right: the galaxy NGC 7422 in Pisces (DSS).

Lassell wrote that all of the objects discovered were '*new to us*', which is explained as '*not contained in Sir John Herschel's General Catalogue of Clusters and Nebulae (*Phil. Trans. *1864)*'. He further remarked that "*Also, many not included in this Catalogue were thought to be new when detected; but on the 'General Catalogue' reaching us in July 1864 were found to be inserted there, and therefore excluded from this list.*" The net result was 600 objects. As mentioned in the notes, two of them were '*found by L*[assell]': NGC 2620 (m 125) and NGC 7489 (m 523), galaxies in Cancer and Pegasus, respectively (see Table 7.16). A modern analysis has shown that 582 objects must be credited to Marth (Table 8.14). The rest had been discovered earlier by other observers, including identities with known NGC objects.

Among the many galaxies, NGC 6908 (m 411) in Capricorn is an interesting case (Fig. 8.18 left). Marth described it as '*extremely faint, very small, large elongated (close to h 2076)*'. The coordinates indicate an oval object in the northeastern spiral arm of NGC 6907 (John Herschel's h 2076), about 1′ off the centre. It first appears like an H II region, but it is actually a faint background galaxy! With NGC 1141/42 (Arp 118) in Cetus, Marth discovered a second ring galaxy; the first (NGC 4774) had been found by William Herschel. He saw the object on 5 January 1864, but did not detect any ring structure: '*vF, S, R, D neb*' (m 83/84); the components have 13.1 mag and 12.8 mag.

NGC 7422 (GC 6101, m 509) was discovered on 11 August 1864 (Fig. 8.18 right). The 13.5-mag galaxy is – apart from Swift's NGC 1360 and Voigt's NGC 6364 – the NGC object with the most independent discoveries (see Table 10.7); it was also found by Otto Struve (6 December 1865), d'Arrest (29 September 1866) and Safford (27 September 1867). Marth was responsible for the last NGC entry: NGC 7840, a member of a small galaxy group in Pisces (diameter 15′), together with NGC 7834, NGC 7835, NGC 7837,

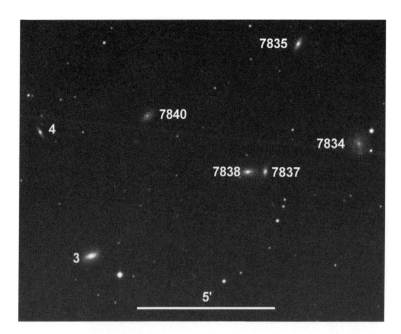

Figure 8.19. The small galaxy group in Pisces around NGC 7840, the last NGC entry (DSS).

NGC 7838, NGC 3 and NGC 4 (Fig. 8.19). The first five objects were discovered on 28 November 1864, the last two on 29 October. NGC 7837/38 is an interacting pair (Arp 246). Marth found many groups of nebulae (as did Stephan later too).

Table 8.15 shows Marth's four planetary nebulae and three emission nebulae. He discovered the second PN in Lyra: NGC 6765 – though it lies in the shadow of M 57, it is remarkable too (Fig. 8.20 right). The description reads *'faint, small, much elongated or ray'*, expressing the bipolar structure pretty well. Safford and Stephan independently found the object on 12 July 1866 and 20 July 1870, respectively.

NGC 2238 is a 'knot' of 1′ diameter in the Rosette Nebula (Monoceros), described by Marth as a *'small star in nebulosity'* (Fig. 8.20 left). The star appears in a few images only. The important parts of the Rosette Nebula (NGC 2237 and NGC 2246) were later discovered by Lewis Swift and Barnard. Perhaps Marth's field of view was too small for him to see the large complex. The emission nebulae NGC 6813 and NGC 6820 in Vulpecula, which were found on 7 August 1864, are not related. The former is small, described as a *'double star in very faint, small nebula'*. Though the latter is much larger, Marth could see only the brightest part (*'faint, small, round, brighter middle'*).

Two of the five missing objects are interesting by virtue of their location: NGC 2637 (m 130) and NGC 2643 (m 131), lying inside the open cluster M 44 (Praesepe), but nothing can be seen at the positions indicated. But the night of 30 October 1864 brought three existing objects 'into' M 44 too: the galaxies NGC 2624 (m 128), NGC 2625 (m 129) and NGC 2647 (m 129).

The 16 known objects are listed in Table 8.16. There are 11 'direct' discoveries (B. Stoney, d'Arrest, Schmidt), the others being 'indirect', i.e. the NGC contains identical objects ('NGC id'). Of the first category, NGC 5538 is remarkable, having been found by Bindon Stoney at Birr Castle and catalogued as GC 3830. Dreyer listed Marth's discovery as GC 5762, but in the NGC one eventually reads *'Ld R, m 275'* and *'3830=5762'*. The three reflection nebulae in the Corona Australis, which was discovered by Schmidt in 1861 (see Section 6.18.3), are Marth's most southern nebulae ($\delta = -36°$ 55′). In some cases, d'Arrest was earlier to observe than Marth. Dreyer always mentioned both observers in the GCS, except for NGC 7612 (GC 6183); but in the NGC one correctly reads *'m 571, d'A'*. NGC 4078 =

8.7 Marth on Malta

Table 8.15. *Marth's planetary and emission nebulae*

NGC	GCS	m	Date	Type	V	Con.	Remarks
2238	5361	99	28.2.1864	EN		Mon	Part of Rosette Nebula
6813	5944	400	7.8.1864	EN		Vul	
6820	5945	401	7.8.1864	EN		Vul	
6751	5940	397	20.7.1863	PN	11.9	Aql	Stephan 17.7.1871
6765	5941	398	28.6.1864	PN	12.9	Lyr	Safford 12.7.1866, Stephan 20.7.1870
6842	5947	403	28.6.1864	PN	13.1	Vul	d'Arrest 26.8.1864, Safford 24.8.1867
6852	5949	404	25.6.1863	PN	12.6	Aql	

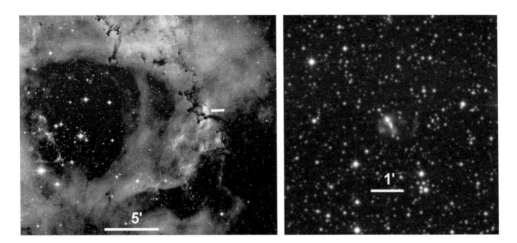

Figure 8.20. Two interesting Marth objects. Left: NGC 2238, a small nebulous area around a star in the Rosette Nebula in Monoceros; right: the bipolar planetary nebula NGC 6765 in Lyra (DSS).

NGC 4107 is an 'indirect' discovery. D'Arrest made an error of 1^h in right ascension (see Table 8.11). Marth is responsible for the case of NGC 7583 (m 555) = NGC 7605 (m 568); he found the object a second time three months later, now including the neighbouring galaxy NGC 7604 (m 567).

Concerning the brightness distribution, Marth scored lower than Stephan (Fig. 8.21), even though the latter used the better telescope. The mean value is 13.9 mag (for Stephan, 13.5 mag). In the introduction of the Malta catalogue one reads that *'the majority of objects […] are faint, or very faint, and could not be expected to be visible in* [Herschel's] *twenty-foot telescope'*. In contrast to Stephan, Marth did not underestimate the brightness (eeF was used only 11 times). The variation is large for each class, ranging, for instance, from 13.9 to 15.0 mag for eeF. On the other hand, four galaxies are fainter than 16 mag, but Marth gives only vF and eF; the 15.2-mag galaxy NGC 1111 (m 76) is even called F. As has already been noticed for other observers, the classification is thus pretty inhomogeneous.

Finally, a curious case: m 327 in Hercules (discovered on 10 August 1864), listed by Dreyer as GC 5842 and NGC 6276 (Fig. 8.22). The first Index Catalogue now gives for IC 1238 '*m 327, eF (not observed by St.)*' and, moreover, the following entry (Bigourdan's IC 1239) presents '*eF, eF* [eS] *stell N [6276?]*', assigning a possible identity with NGC 6276. Thus, do we have IC 1238 = m 327 and IC 1239 = NGC 6276? Dreyer explains the situation in the NGC notes. Two other objects are involved: NGC 6277 (called m 328 in the NGC) and William Herschel's NGC 6278 (III 124). The second to inspect the field was Stephan, on 18

Table 8.16. *Marth objects that had been discovered earlier by others*

NGC	GCS	m	Date	Type	V	Con.	Discoverer	NGC id	Discoverer
4053	5603	228	18.3.1865	Gx	13.9	Com	d'Arrest 9.5.1864		
4078	5607	231	25.3.1865	Gx	13.2	Vir	d'Arrest 23.3.1865	4107	d'Arrest 17.4.1863
5100	5734	255	22.3.1865	Gx	13.9	Vir		5106	W. Herschel 23.1.1784
5538	5762	275	8.5.1864	Gx	14.7	Boo	B. Stoney 3.6.1851 (GC 3830)		
5841	5777	286	12.4.1864	Gx	13.9	Vir		5848	d'Arrest 6.5.1862
6052	5802	302	2.7.1864	Gx	13.3	Her		6064	W. Herschel 11.6.1784
6726	5935	393	2.7.1864	RN		CrA	Schmidt 15.6.1861		
6727	5936	394	2.7.1864	RN		CrA	Schmidt 15.6.1861		
6729	5937	395	2.7.1864	EN+RN		CrA	Schmidt 15.6.1861		
6778	5942	399	25.6.1863	PN	12.3	Aql		6785	J. Herschel 21.5.1825
7464	6112	514	23.10.1864	Gx	13.3	Peg	d'Arrest 27.8.1864		Vogel 10.8.1869
7605	6180	568	29.11.1864	Gx	14.3	Psc		7583	Marth 2.9.1864 (m 555)
7612	6183	571	2.9.1864	Gx	12.7	Peg	d'Arrest 26.5.1863		
7617	6186	572	25.11.1864	Gx	13.8	Psc	d'Arrest 23.9.1864		
7679	6203	579	23.10.1864	Gx	12.5	Psc	d'Arrest 23.9.1864		
7710	6212	586	18.11.1864	Gx	13.9	Psc	d'Arrest 24.9.1862		

July 1871, finding three objects. We now have seven candidates – but only two galaxies (in an area of 5′). For Dreyer only Herschel's NGC 6278 (Stephan's third object) was confirmed – the rest was lost in confusion. He asked Marth to send his observing notes, then finding out that Stephan's first object (NGC 6276) must be identical with m 328. Therefore NGC 6277 (Stephan's second object) and m 327 are two other, different objects. Since the NGC had already been published, Dreyer saved m 327 for the IC I. Unfortunately, Bigourdan introduced another object: IC 1239. A modern analysis of the case shows that NGC 6276 is the second galaxy, identical with IC 1239, and both NGC 6277 and IC 1238 are mere star pairs.

8.8 SCHMIDT'S POSITIONAL MEASUREMENTS AND DISCOVERY OF NEW NEBULAE

During the years 1860–67 Julius Schmidt measured positions of nebulae with the 15.7-cm Plössl refractor. His favourite target was the southern sky, as visible from

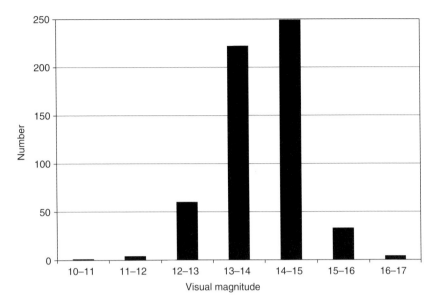

Figure 8.21. The brightness distribution of Marth's objects.

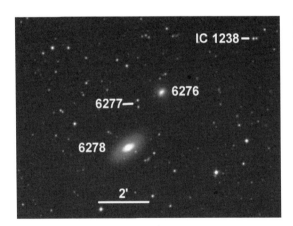

Figure 8.22. The curious field of NGC 6278 in Hercules (DSS).

Athens Observatory (latitude 38°). The observations brought some new objects. However, Schmidt's publications were often delayed, making his observations difficult to follow. Obviously, his notes were not correctly ordered at the observatory. Admittedly, Schmidt was an industrious observer, but, of his many discoveries, only one-third could be verified.

8.8.1 Positions of nebulae

As a student of Argelander, it was Schmidt's natural duty to determine nebular positions with the ring-micrometer. He measured 62 objects, which he ordered into three classes in terms of their degree of visibility (accuracy). When he got Schönfeld's work *Beobachtungen von Nebelflecken und Sternhaufen* (1862), containing data collected with a comparable refractor of aperture 16.5 cm, he immediately compared the positions with his own. The result was published in 1864 as 'Ortbestimmung der Nebelgestirne'.[81] For those nebulae which were difficult to observe (class III), the mean error was 15″.

In 1865 Schmidt published further measurements in his paper 'Ueber die Beobachtung von Nebelgestirnen'.[82] He wrote that for comparison he could use only Herschel's Cape catalogue and the observations of d'Arrest (Leipzig) and Schönfeld (Mannheim). The situation was, moreover, complicated by the lack of a catalogue of southern stars (an experience had by Lassell too). In Athens Schmidt saw himself as being apart from the action: '*all new developments in the field became only accidentally known to me, I have not seen them, and probably I have got information about very few of them only*'.

In 1868 Schmidt's main work appeared: 'Mittlere Oerter von 110 Nebeln für 1865'.[83] Therein he

---

[81] 'Determination of nebular positions' (Schmidt 1864).
[82] 'On the observation of nebulae' (Schmidt 1865a).
[83] 'Mean places of 110 nebulae for 1865' (Schmidt 1868b).

mentions his observations made over 23 years. Meanwhile he received the works of Laugier (Paris), Auwers (Königsberg), Schultz (Uppsala), Rümker (Hamburg) and Vogel (Leipzig). About Herschel's GC of 1864 he sadly remarked '*I still do not know the Gen.-Catalogue of J. Herschel.*' Schmidt compares his positions with those of the other observers. Surprisingly, the largest differences appeared with Schönfeld's data: '*it must, however, be noted that Schönfeld and I, both students of Argelander, strictly followed Argelander's instructions in using the ring-micrometer*'.

Schmidt also observed planetary nebulae, e.g. NGC 6629 and NGC 7009 (Schmidt 1868a). For NGC 6629 in Sagittarius he quoted John Herschel: '*It is not a stellar nebula, but a medium between a planetary nebula and a star cluster.*' For Schmidt it resembles a '*common core-nebula or a nebulous star*'. The Saturn Nebula NGC 7009 in Aquarius was studied during 1860–67. He noticed a faint 'satellite', which had not been mentioned by John Herschel and Lamont. The paper also contains observations of M 8 and M 80.

8.8.2 New nebulae

Between 1865 and 1867 Schmidt discovered 13 objects, mostly at southern declinations (Table 8.17). All but three are galaxies; those located in Fornax and Eridanus belong to the Fornax Cluster. NGC 1378 and NGC 1408 are missing; NGC 5877 in Libra is only a star trio. Schmidt's objects came too late for the GC and were eventually catalogued in the GCS (except for John Herschel's NGC 1375). The other Schmidt objects in the GCS – NGC 4471, NGC 6519, NGC 6726/27 and NGC 6729 – had already been discovered in 1860–61 (see Table 6.26).

While searching for comet Bruhns, Schmidt discovered an object in Virgo on 21 January 1865, which he thought to be the comet. He then noticed that it must be '*either another comet or an unknown nebula*' (Schmidt 1865b). Palisa found it again on 20 March 1884 in Vienna, mentioning Schmidt's priority (Weiss 1885). The true identity was cleared up by Tempel in 1885: '*The nebula mentioned by Dr J. Palisa in A.N. 2958, p. 287, which is identical with the object first thought to be Comet Bruhns by Schmidt on 21 Jan. 1865, is an old nebula of W. Herschel: III 312, No. 3482 in the General Catalog.*' (Tempel 1885c). This is the bright galaxy NGC 5068 (Fig. 8.23).[84]

While observing comet II 1867, Schmidt found, probably on 24 April 1867, a new nebula in Libra (NGC 5877). Differently from all his other discoveries, and probably due to the topicality of the comet, he immediately published his observation. He noted that '*Also a very difficult, small nebula, unknown to me, was found*' (Schmidt 1867b).

His most important find was made in 1865, when he saw 11 new objects while inspecting Cape catalogue nebulae. As usual, the publication was delayed – more than 10 years in this case! The discoveries are contained in his paper 'Über einige im Cape-Catalog fehlende Nebel'[85] of 1876; the new nebulae are called n^a to n^l. Schmidt wrote '*That [Herschel's] survey with the 20ft reflector has left out a faint object should not be surprising; but when in an area of $5° \times 2°$ (only $12°$ to $16°$ above the horizon) the Athens 6ft refractor shows a dozen more nebulae than listed in Herschel's catalogue, this fact should attract attention.*' Owing to the lack of a southern star catalogue, Schmidt determined the nebular positions relative to Cape catalogue objects. He correctly supposed that n^c is identical with Herschel's h 2558 (NGC 1375). All of these galaxies are located at the Eridanus–Fornax border and belong to the Fornax Cluster (Abell S373); most members were found by John Herschel during 1835–37 and four bright ones were due to Dunlop. Three of Schmidt's nebulae (NGC 1378, NGC 1389 and NGC 1408) had already been cited in his paper 'Beobachtungen der Nebelgestirne und des Faye'schen Cometen'.[86] They were found while observing NGC 1380 (h 2559 = Δ 574; Dunlop 2 September 1826) and described as '*conspicuous and easily visible*'. In contrast to what he did in his later paper in AN 2097 (Schmidt 1876a), Schmidt gives dates: 19 and 22 February 1865. Probably this applies for all other objects lying closely together. It is unknown why they are not mentioned in AN 1553 (Schmidt 1865c). Vice versa, in AN 2097 the publication AN 1553 is not mentioned.

Schmidt is known for his incomplete publications. He often referred to future reports, which should appear, for instance, in the annals of Athens

---

[84] See also Weiss (1885).
[85] 'On some missing nebulae of the Cape catalogue' (Schmidt 1876a).
[86] 'Observations of nebulae and comet Faye' (Schmidt 1865a).

Table 8.17. *Objects discovered by Schmidt during 1865–67 ('Name' from Schmidt (1876a))*

NGC	Date	Publication	Name	GC(S)	Type	V	Con.	Remarks
1318	19.1.1865	1876	n^a	5312	Gx	11.1	For	NGC 1317, Dunlop 24.11.1826
1369	19.1.1865	1876	n^b	5316	Gx	12.9	Eri	
1375	19.1.1865	1876	n^c	737	Gx	12.2	Eri	J. Herschel 25.12.1835
1378	19.1.1865	1865, 1876	n^d	5317	NF		For	
1381	19.1.1865	1876	n^e	5318	Gx	11.5	For	
1382	19.1.1865	1876	n^f	5319	Gx	12.9	For	
1386	19.1.1865	1876	n^g	5321	Gx	11.2	Eri	
1389	19.1.1865	1865, 1876	n^h	5322	Gx	11.4	Eri	
1396	19.1.1865	1876	n^i	5323	Gx	13.8	For	
1408	19.1.1865	1865, 1876	n^k	5324	NF		For	
1428	19.1.1865	1876	n^l	5326	Gx	12.8	For	
5068	21.1.1865	1865		3482	Gx	9.8	Vir	W. Herschel 10.3.1785 (II 312)
5877	24.5.1867	1867		5779	3 stars		Lib	date probable

Figure 8.23. Found by Schmidt during comet-seeking: the galaxy NGC 5068 in Virgo (DSS).

Observatory – however, nothing happened. In *AN* 1553 he wrote that '*the publication of my observations will be still a long time in coming*'; and in *AN* 2097 he put the reader off: '*All details will be found in my future collection of measurements (1842–1875).*' A visit of Charles Pritchard to the observatory in 1883 confirmed the deficient organisation: '*To me it seemed, and it still seems, a thousand pities that the invaluable results of many years of labours still remain unpublished.*' (Pritchard 1883).

How should Schmidt's discoveries be rated? The answer results from his report in *AN* 1553 (Schmidt 1865c), where altogether 69 new nebulae are listed, sorted by date (8 October 1860 to 22 January 1865). Most were found near objects from John Herschel's catalogues, especially at southern declinations. On checking the positions, a disappointing picture results: only 7 objects are new, whereas 17 are known, existing objects (mostly discovered by William Herschel). The rest (39) cannot be identified; either the field is empty or the object cannot be identified with an existing one. To sum up: nearly two-thirds of Schmidt's 'discoveries' are incomprehensible.

8.9 WINNECKE'S OBSERVATIONS IN KARLSRUHE AND STRASSBURG

In Karlsruhe, Winnecke ran a private observatory. There he discovered two nebulae, which were independently found by Eugen Block and Lewis Swift. Later he was successful in Straßburg, but his ill health was a permanent problem. Unfortunately he could use only a small telescope in a light-polluted environment. The 48.7-cm Merz refractor of the new University Observatory was still under construction.

8.9.1 August Winnecke and Eugen Block

In January 1868 Winnecke found the planetary nebula NGC 1360 in Fornax, using a 9.7-cm comet-seeker in Karlsruhe (Winnecke 1879a). Owing to the low surface brightness the object appeared *'faint and 10' large'*. Dreyer noted in the NGC that Swift had already discovered it in 1857. Apart from the date, this is correct: Lewis Swift saw NGC 1360 with his 4.5" comet-seeker in 1859 (Swift 1885c). The planetary nebula is among the three NGC objects with the most independent discoveries (see Table 10.7).

In Karlsruhe, Winnecke found the galaxy NGC 1398 in Fornax too. His observation on 17 December 1868 was made with a 12.2-cm refractor by Reinfelder & Hertel, with his noting *'Nebula I. class, 1½' long, bright'* (Winnecke 1879a). About the reason why the object (in contrast to NGC 1360) is not listed in the GCS, Winnecke wrote *'At that time I thought the nebula to be h 2574 and therefore did not communicate its place to Dreyer.'* Obviously it should read GC 2574 (NGC 1412). For identification he could use only Harding's star atlas of 1822: *'Considering the poorness of Harding's map in this area, such a mistake is easily possible.'* The true discoverer of NGC 1398 was Tempel; though, having found the object on 9 October 1861 (together with NGC 1360), it was not published until 1882 (Tempel 1882). Obviously, Dreyer overlooked Tempel's observation.

Both nebulae were independently found by Eugen Block. The observation took place on 18 December 1879 at Odessa Observatory with a 10.5-cm comet-seeker (Block 1880). Dreyer, however, mentions him only as co-discoverer of NGC 1398. Not much is known about Block. He was born on 15 April 1847 in Baldonu (Latvia), studied astronomy in Dorpat and became an assistant of Thomas Clausen at the local observatory in 1868. Three years later Block changed to Pulkovo as assistant of Otto Struve. In 1873 he moved to Odessa to become Professor of Astronomy and Director of the Maritime Observatory. During the years 1873–1910 Block was a member of the Astronomische Gesellschaft; he died in about 1920.

Block placed the new objects in Eridanus (since 1930 its southern part has belonged to Fornax). About NGC 1360 he noted that it was *'difficult to see at higher magnification, but pretty conspicuous in the comet-seeker at a power of 27'*. NGC 1398 is described as *'bright, strongly condensed to the middle [...] even visible well in moonlight'*. He wondered *'that these nebulae were not observed by Herschel, as two neighbouring nebulae are listed in the [Slough] catalogue'*. These are NGC 1371 (II 262 = h 734) and NGC 1412 (h 757), only 1° away. Block was particularly astonished, because NGC 1412 (confused by Winnecke with NGC 1398) appeared *'significantly fainter than the two new nebulae'* in the 13-cm Steinheil refractor and, moreover, NGC 1371 could not be found. Since he had seen it 1878 in the comet-seeker, he supposed that *'I have thought one of the new nebulae to be No. [h] 734, because I can see the latter [h 757] in the comet-seeker only with great effort.'* Like Winnecke, Block too confuses one of the objects. By the way, in his report on the discovery night (18 December 1879), a meteor shower radiating from Auriga is mentioned.

In Winnecke's article of 1879 (mainly treating the 'periodic variability' of NGC 3666), Block's observations are mentioned. He asked too why the two Herschels had not already found the two nebulae. Was this another case of variability? Winnecke wrote that *'The fact that occasionally new nebulae were found in sky regions often surveyed, which had escaped previous investigations, does immediately lead, in connection with the currently confirmed brightness variations of some nebulae, to the interesting question of whether these newly discovered nebulae have actually only lighted up.'* Anyway, he was unable to detect any change: *'Both nebulae were observed by me 11 years ago, and, so it seems, appeared with the same brightness.'*

8.9.2 Straßburg observations

From 1875 to 1880 Winnecke observed nebulae in Straßburg. Because the University Observatory was still under construction (it was not finished until 1881), he made do with the old Academy Observatory located in the city centre. In 1875 an 'orbit-sweeper' by Reinfelder & Hertel with an aperture of 16.3 cm was erected. It was equipped with a bar-micrometer (Fig. 8.24). The mounting by Repsold had a third axis so that one could easily follow comets or minor planets.

In his paper of 1875, titled 'Ueber die auf der Universitäts-Sternwarte Straßburg begonnenen Beobachtungen von Nebelflecken',[87] Winnecke describes his task. He lamented the bad conditions at the site of the Academy Observatory: *'the place in the city where it is located is on two sides flanked by streets with high traffic*

---

[87] 'About observations of nebulae started at the University Observatory Straßburg' (Winnecke 1875).

8.10 Tempel in Arcetri

Figure 8.24. Winnecke's 16.3-cm orbit-sweeper by Reinfelder & Hertel (Repsold J. (1908), Vol. 2, Fig. 49b).

and extremely worrying lamps'. The paper describes the orbit-sweeper in detail. Winnecke termed his observations a 'preliminary study', and a continuation with the 48.7-cm Merz refractor was planned. Unfortunately, the telescope was not ready until November 1880. So, due to his bad physical condition, Winnecke could use it for only a few observations.

The results achieved with the orbit-sweeper were not published until 1909, i.e. 12 years after Winnecke's death, in Vol. 3 of the *Straßburg Annalen*, titled 'Beobachtung von Nebelflecken, ausgeführt in den Jahren 1875–1880 am 6-zölligen Refraktor der provisorischen Sternwarte von A. Winnecke, bearbeitet von E. Becker'.[88] On the first 118 pages, 406 nebulae are listed, sorted by date (5 March 1875 to 14 July 1880) and identified by GC-number. Detailed descriptions and a few sketches are given. Winnecke was not able to reduce the data any more. The catalogue of absolute positions (starting at page 189) is the work of his successor Ernst Becker; now NGC-numbers are used.

8.9.3 New nebulae

Winnecke discovered four objects in 1876 with the orbit-sweeper; they are catalogued in the GCS. Dreyer was immediately informed by letter (except regarding NGC 1398) – obviously a reaction to his 'Request for astronomers' of 1876. Winnecke did not publish his discoveries until 1879 in *AN* 2293 (Winnecke 1879a); they also appear in the third volume of the *Straßburg Annalen* (1909). Table 8.18 shows all of the objects found in Karlsruhe and Straßburg.

The first object found in Straßburg is the galaxy NGC 4760 (GC 5674) in Virgo. Three months later Winnecke discovered the galaxy pair NGC 2276/2300 (GC 5364/70) in Cepheus (Fig. 8.25); a second observation was made a day later. The brighter (and larger) of the two nebulae, NGC 2300, was first seen by Borrelly 1871(?) in Marseille (see Table 8.25). Both objects were independently found by Tempel in 1876 and 1877 in Arcetri. There is no reference for the galaxy NGC 2146 (GC 5357) in Camelopardalis. In the GCS (and in the NGC too) Dreyer mentions Winnecke and Tempel, who must have seen it later. It is also listed in the second part of the Straßburg observations, covering 1880–1902 (Becker E. 1909b).

8.10 TEMPEL IN ARCETRI

When Tempel arrived, the Arcetri Observatory was in a bad condition, especially the two Amici refractors. Nevertheless, he was optimistic and started working. The harvest was striking. Already in the first years 1875–77 he discovered a large number of new nebulae, which were communicated to Dreyer by letter. Tempel also published many papers in the *Astronomische Nachrichten*. Despite the humble resources, he was a well-known observer, who often tended to overestimate his own capabilities.

8.10.1 The condition of the observatory

When Tempel started at Arcetri Observatory in 1875, the main instruments were two refractors: 'Amici I'

---

[88] 'Observations of nebulae made in the years 1875–1880 with the 6" refractor of the provisional observatory by A. Winnecke, edited by E. Becker' (Becker E. 1909a).

Table 8.18. *Winnecke's discoveries in Karlsruhe and Straßburg*

NGC	GCS	Date	Type	V	Con.	Ap	Remarks	Reference
1360	5315	Jan. 1868	PN	9.4	For	9.7	L. Swift 1858, Tempel 9.10.1861, Block 18.10.1879	*AN* 2293, *Straßb. Ann.* p. 105
1398		17.12.1868	Gx	9.8	For	12.2	Tempel 9.10.1861, Block 18.10.1879	*AN* 2293, *Straßb. Ann.* p. 105
2146	5357	1876	Gx	10.5	Cam	16.3	Tempel 1876	No reference
2276	5364	26.6.1876	Gx	11.3	Cep	16.3	Tempel 1876	*Straßb. Ann.* p. 47
2300	5370	26.6.1876	Gx	11.1	Cep	16.3	Borrelly 1871?, Tempel 1877	*Straßb. Ann.* p. 47
4760	5674	30.3.1876	Gx	11.6	Vir	16.3		*Straßb. Ann.* p. 38

Figure 8.25. The galaxy pair NGC 2276 and NGC 2300 in Cepheus (DSS).

with an aperture of 11″, which had been made in 1841, and the smaller 'Amici II' with 9.4″ aperture of 1844; see Fig. 8.26 (Bianchi S. 2010). Though both telescopes were not in good shape, Tempel began his nebular observations with great enthusiasm. In his first paper he wrote '*On Febr. 7* [1875] *I first tried the smaller local Amici telescope* [Amici II] *and it turned out to be excellent for nebulae.*' (Tempel 1875a). He added that '*The casters, cords and winches of the setting needed great practice and skill to adjust a certain part of the sky.*' One of his first targets was the Merope Nebula (Tempel 1875b).

In January 1877 Tempel wrote a detailed report about his early work, which appeared in *AN* 2138–39 (Tempel 1877a). Amici II, which had been erected on the observatory terrace, was criticised because the wooden mounting made observing troublesome: '*it was highly dangerous pointing the telescope to an elevation of 40° and, due to the inclined plane of the terrace, it often rolls off*'. He further wrote that '*It is to be hoped that this fine instrument of great light-gathering power will get an equatorial mounting and a dome.*' Amici I was mounted equatorially already, but '*even standing on the highest step of the ladder, only targets over 20° elevation can be viewed; also it costs many valuable hours to get, without setting circles, known nebulae into the eye-piece.*' Equipment was sparse: '*Unfortunately the two large telescopes share only one eye-piece.*' It had been made by Amici and had a focal length of 45 mm, offering powers of 113 and 75 (fields of view 20′ and 34′) for Amici I and II, respectively. Nonetheless Tempel remained

Figure 8.26. Tempel's refractors. Left: 'Amici II'; right: 'Amici I' (Arcetri Observatory).

optimistic: *'Being familiar and trained with some inconvenience, these large telescopes offered sufficient pleasure to challenge my weak forces.'*

Concerning nebulae, Tempel wrote *'Observing pretty unsystematic, I nonetheless saw some 30 new nebulae, of which but seven had been discovered earlier, as was amicably communicated to me by Dr Dreyer, which shows that his supplementary catalogue [GCS] will be of some value.'* He added that *'The present condition of the observatory does not allow any measurement, thus I rely on drawings only. My former project on the many unconfirmed nebulae and the 116 nebulae of class II still unobserved since the older Herschel, and much more, I must leave to better times too.'*

In *AN* 2138 two other objects are mentioned for the first time: NGC 2067, an appendage of the reflection nebula M 78 in Orion (see below), and the galaxy NGC 7450 in Aquarius. The latter was discovered on 19 November 1876 while observing the double galaxy NGC 7443/44. Because William Herschel had not seen the object, Tempel assumed variability. He wrote the following about this issue: "*if nebulae, due to a new hypothesis, 'consist of glowing gas', I hope to live long enough, to see some burned out or converted into other entities*". Tempel mentions another possibility for why Herschel missed NGC 7450: *'the two Amici refractors here are comparable to his reflectors concerning the light-gathering power […] actually my drawings prove that they are even better'*. Often Tempel exaggerates the power of his instruments, not even hesitating to compare them with really large ones, such as Lord Rosse's 72″ reflector.

8.10.2 Tempel's nebulae in the GCS

Twenty-five objects in the GCS are due to Tempel (Table 8.19). All of the nebulae credited to him by Dreyer were probably found in late 1876 and communicated by letter. Tempel published them, except for the Aquarius objects, in *AN* 2212 (Tempel 1878d); 56 are listed there (in the column 'Ref.'). This and other papers were cited by Dreyer in the NGC; they will be discussed in Section 9.3.1.

Most of these objects are galaxies (24). NGC 2067 is the western part of the reflection nebula M 78 (NGC 2068) in Orion; another, southwestern extension (NGC 2064) was found by d'Arrest on 11 January 1864. The objects NGC 2700/2/3/5/7 in Aquarius are faint stars or pairs of stars; there is no reference. In many cases Tempel gives no discovery date. For the planetary nebula NGC 6309 in Ophiuchus he noted '1876?'. This

Table 8.19. *Tempel's GCS objects*

NGC	GCS	Date	Type	V	Con.	Ref.	Remarks
113	5100	27.8.1876	Gx	13.1	Cet	I-1, IV-1	
195	5110	1876	Gx	13.7	Cet	I-2	
309	5127	1876	Gx	11.9	Cet	I-4	
926	5236	1876	Gx	13.2	Cet	I-9	L. Swift 3.10.1886
934	5240	1876	Gx	13.1	Cet	I-10	
976	5248	1876	Gx	12.5	Ari	I-11	
1015	5265	27.12.1875	Gx	12.2	Cet	I-13, V-1	
1530	5334	1876	Gx	11.5	Cam	I-15	
1544	5338	1876	Gx	13.6	Cep	I-16	
2067	5356	1876	RN		Ori	AN 2139, I-17	Part of M 78
2336	5372	1876	Gx	10.3	Cam	I-22	
2700	5437	1876	Star	15.2	Hya		
2702	5436	1876	Star		Hya		
2703	5438	1876	2 stars		Hya		
2705	5439	1876	Star		Hya		
2707	5440	1876	Star	14.9	Hya		
2919	5490	1.2.1877	Gx	12.9	Leo	I-24, V-4	
3580	5553	1876	Gx	14.0	Leo	I-32	
6309	5851	1876?	PN	11.5	Oph	I-46, V-31, AN 2269	Pickering 15.7.1882
7450	6108	19.11.1876	Gx	13.2	Aqr	AN 2138, I-47, IV-11	
7520	6137	1876	Gx	13.1	Aqr	I-49	
7712	6213	1876	Gx	12.8	Peg	I-53	
7717	6214	1876	Gx	12.8	Aqr	I-54	
7730	6219	1876	Gx	13.9	Aqr	I-56	

year is probable, since in a paper published in *AN* 2269, dated 15 June 1879, he wrote '*Of the new nebulae which I mainly found in 1876 and only roughly documented, one* [NGC 6309] *was observed later.*' (Tempel 1879b). On 15 June 1878 he determined a new, corrected position with the ring-micrometer ('*my value in [AN] No. 2212 is quite wrong*'). He further wrote that '*At the time of discovery I probably had Harding's atlas at hand only and not the Berlin star charts.*' NGC 6309 was found independently by Pickering on 15 July 1882 by visual spectroscopy (see Section 9.13.2).

Table 8.20 lists those of Tempel's GCS objects which had been found earlier by other observers. 'T' marks cases for which Tempel noted this in his publication of 1878 (Tempel 1878d). By that time he had acquired Dreyer's GCS. The case of the galaxy pair NGC 2276/2300 in Cepheus is interesting. Borrelly found NGC 2300 earlier (in 1871?), but, strangely, he did not notice the companion. Both galaxies were seen by Winnecke on 26 June 1876. For NGC 2276 Dreyer gives 'Tempel, Winnecke', supposing that Tempel owns the priority; but the object is assigned by Tempel as '*found earlier by others*' (Ref. I). In the case of NGC 2300 Dreyer gives 'Winnecke, Borrelly' – not mentioning Tempel. Obviously Tempel found the object later (1877) and therefore it is contained in his letter to Dreyer sent in late 1876. 'D' marks cases in which Dreyer credits Tempel in the GCS. GC 6153 (NGC 7553) is ignored here: by mistake Dreyer indicates Tempel as the discoverer (apart from Schultz). The NGC correctly states 'LdR' (B. Stoney) and Schultz in this case.

Table 8.20. *Tempel's GCS objects that had been discovered earlier by other observers*

NGC	GCS	T	D	Date	Type	V	Con.	Ref.	Remarks
316	5129			29.8.1877	Star	14.5	Psc	I-3	B. Stoney 29.11.1850
1016	5264	*	*	1876	Gx	11.6	Cet	I-12	Marth 15.1.1865 (m 68), Safford 1.11.1867
2146	5357	*	*	1876	Gx	10.5	Cam	I-18	Winnecke 1876
2268	5362	*		1877	Gx	11.4	Cam	I-19	Borrelly 1871?
2276	5364	*	*	1876	Gx	11.3	Cep	I-20	Winnecke 26.6.1876
2300	5370	*		1877	Gx	11.1	Cep	I-21	Borrelly 1871?, Winnecke 26.6.1876
2720	5445	*	*	1876	Gx	13.0	Cnc	I-23	Marth 10.3.1864 (m 137)
3353	5533			18.11.1881	Gx	13.0	Leo	V-5	Marth 5.4.1864 (m 202), d'Arrest 22.2.1865
3419	5539	*	*	15.3.1876	Gx	12.9	Leo	I-27, V-7	Marth 1.4.1864 (m 208)
3598	5555	*	*	1876	Gx	12.3	Leo	I-33	Marth 4.3.1865 (m 217)
3641	5560	*	*	1876	Gx	13.1	Leo	I-34	Marth 22.3.1865 (m 105)
4165	5625			22.3.1878	Gx	13.4	Vir	V	d'Arrest 8.4.1864
7620	6187	*		1876	Gx	12.9	Peg	I-50	Marth 5.9.1864 (m 573)
7664	6194		*	1876	Gx	12.7	Peg	I-51	Stephan 17.10.1876
7683	6204		*	1876	Gx	12.7	Peg	I-52	Ferrari 1865
7724	6217			26.11.1877	Gx	13.5	Aqr	I-55, IV-14	Stephan 23.9.1873

8.11 RÜMKER AND THE 'CIRCUMPOLAR NEBULAE'

Though George Rümker discovered only one NGC object, he was a well-respected visual observer. His most important work on the subject was the study of 'circumpolar nebulae'. The positions of these near-pole objects were precisely determined with the instruments at the old Hamburg Observatory near the Millerntor.

8.11.1 Short biography: George Friedrich Wilhelm Rümker

George Rümker was born on 31 December 1832 in Hamburg (Fig. 8.27). He was the son of Charles Rümker, the first Director of Hamburg Observatory at Millerntor and former leader of the Australian Paramatta Observatory. Through his father young George came into contact with astronomy early in life; at the age of 15 he had already observed planets, comets and star occultations. In 1851 he went to Berlin to study astronomy. While working for Encke at the University Observatory he met the young d'Arrest. Rümker calculated orbits of minor planets and comets. In 1853 he became an observer at Durham Observatory as the assistant of Temple Chevallier. He departed in 1855 (followed by Albert Marth), returning to Hamburg to support his ill father. When Charles Rümker died in 1862 in Lisbon, his 30-year-old son was told that he was too young to succeed him as director. Actually George Rümker was responsible for the office, but was not awarded the directorship of Hamburg Observatory until 1866.

Rümker's main task was the determination of nebular positions. During the years 1860–67 he carried out a first measuring campaign. When, in 1867, the 26-cm Repsold equatorial was installed in a dome, a second study was planned.[89] It ran from 1870 until 1880 and was temporarily supported by Carl Frederik Pechüle (who was employed as an assistant during the years 1870–72). Rümker was interested in comets and planets too. Additionally he was active in the fields of navigation and the time service and was responsible for the equipment of the German Venus-transit

---
[89] Since 1912 the refractor has been located in Hamburg-Bergedorf.

Figure 8.27. George Rümker (1832–1900).

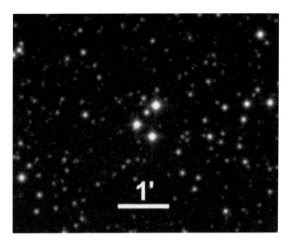

Figure 8.28. Rümker's discovery in Auriga: the small star group NGC 1724 (DSS).

expedition. A chronic illness forced him to resign as director of Hamburg Observatory in 1899 (he was succeeded by Richard Schorr). George Rümker died on 3 March 1900 in Hamburg at the age of 67.[90]

8.11.2 Observations at the old Hamburg Observatory

On 30 April 1864 Rümker discovered the star group NGC 1724 (GC 5347) in Auriga (Fig. 8.28). The observation was made with the 10-cm meridian-circle by Fraunhofer (which had been erected in 1836) of Hamburg Observatory at Millerntor.[91] It was published in 1865 in the *Astronomische Nachrichten* (Rümker 1865b). The object ('Nova') was described as '*several stars with nebulosity standing closely together*'. Further observations were made on 1, 3 and 6 May 1864. The 'nebulous' appearance was caused by the small telescope.

Rümker's publication, titled 'Beobachtungen von Circumpolarnebeln auf der Hamburger Sternwarte', was the first in a series of five.[92] Altogether 151 objects at declinations generally greater than +30° were observed (11 bright objects are more southern). Rümker selected 'core-nebulae', i.e. objects with a prominent central condensation, which are good to measure. They were chosen from the catalogues of William and John Herschel (GC-numbers were not used). Additionally Hind's nebula NGC 4125 and M 92 were observed. The work was done during the years 1860–67 with the meridian-circle by Fraunhofer, equipped with a ring-micrometer (magnification 70). Rümker apologised that the precision might not be high enough: '*Should later observers record such differences, I must regret that my time and care applied on these objects has not led to better results; the only reason lies in the insufficient power of the instrument used.*' (Rümker 1867). The results were, however, not criticised by contemporary astronomers.

The second measuring campaign (1870–80) was carried out with the new 26-cm equatorial, which was by Repsold and erected in 1867. Rümker assisted by Pechüle during the years 1870–72) published the results in 1895 as 'Positionsbestimmungen von Nebelflecken und Sternhaufen'.[93]

---

[90] Obituaries: Schorr (1900), Anon (1900), Neumayer (1901); see also Schramm (1996: 110–121).
[91] Repsold A. (1837).

[92] 'Observations of circumpolar nebulae at Hamburg Observatory'; the other publications are Rümker (1865c, d, 1866, 1867).
[93] 'Position measurements of nebulae and star clusters' (Rümker 1895).

8.12 FERRARI – IN THE SHADOW OF SECCHI

Gaspare Ferrari was an assistant of Angelo Secchi at the Collegio Romano Observatory. There he discovered 14 nebulae with the 24-cm Merz refractor in 1865–66. Unfortunately, Dreyer incorrectly credited Secchi, although the prominent Jesuit astronomer had only published the discovery – the actual observer was Ferrari.

8.12.1 Short biography: Gaspare Stanislao Ferrari

Gaspare Ferrari, who was born on 23 October 1834 in Bologna, became a Jesuit at the age of 18. From 1865 he was an assistant of Angelo Secchi at the observatory of the Collegio Romano, located on the roof of the adjacent Church of St Ignatius in Rome. The main instrument was the 24-cm Merz refractor, which had been erected in 1853. After Secchi's death (26 February 1878), Ferrari became his successor as director – but only for short time, because the Jesuit institution was appropriated by the Italian government on 1 May 1879. Under the threat of violence, Ferrari had to leave his office. The Vatican protested, but could not stop the action of the Liquidation Committee for Clerical Estates. The nationalised observatory was directed by Pietro Tacchini.[94] Ferrari was appointed Professor of Astronomy at the Università Gregoriana in Rome. From his Villa Cecchina he observed with a private 9" Merz refractor. Ferrari resigned in 1894, and died on 20 June 1903 in Paris at the age of 68.[95]

8.12.2 Fourteen new nebulae

During the unsuccessful search for comet Biela from 11 November 1865 to 18 January 1866 Ferrari found 14 nebulae (Table 8.21). Additionally, together with Secchi, positions of two of William Herschel's objects were measured: NGC 157 (II 3) and NGC 7648 (III 218), galaxies in Cetus and Pegasus (which were seen again by Stephan on 3 and 28 October 1878).

Because Secchi published all of the observations in *AN* 1571 (Secchi 1866), he is often incorrectly mentioned as the discoverer of the 14 objects. What he wrote about the Merz refractor is interesting: '*From this study we have convinced ourselves that the refractor at our observatory is at least as keen and powerful as the Herschels' telescopes.*' For the observations '*a large eye-piece which gives a 27′ field*' was used.

Dreyer listed all 14 objects in the GCS and NGC. However, in both catalogues only Secchi is mentioned. The analysis shows that only five are non-stellar: the galaxies NGC 50 (Fig. 8.29 left; Fig. 9.4), NGC 7667, NGC 7683, NGC 7738 and NGC 7739. NGC 7614 is a group of four stars (Fig. 8.29 right). All of the other objects are missing – a pretty bad rate for Ferrari.

8.13 OBSERVATIONS IN MARSEILLE: VOIGT, STEPHAN, BORRELLY AND COGGIA

The Observatoire de Marseille on Plateau Longchamps was erected in 1862. It was to host the new Foucault reflector. The revolutionary instrument, equipped with a silvered glass mirror of diameter 78.8 cm, was built in Paris and saw first light on 17 January 1862.[96] Chacornac was the first to observe nebulae with it. However, the atmospheric conditions at Paris Observatory were too bad for its great light-gathering power (focal ratio 1:5.8). Director Leverrier thus decided to move the telescope to southern France. Besides Marseille, Montpellier and Toulon were discussed as possible sites too. Marseille won the competition, and a round building with a flat roof was built to host the large reflector on its wooden mounting. After some technical difficulties, it was ready in November 1864. The first director, Auguste Voigt, and his successor, Edouard Stephan, used it for observations of nebulae.

Together with the discoveries of Marth on Malta and d'Arrest in Copenhagen, it was those of Stephan in Marseille which prompted Dreyer to put together a supplement to the General Catalogue. The combination of superior telescopes (Marth's 48" and Stephan's 80-cm) with systematic observations had shown that many objects could still be found after the two Herschels. Dreyer received a list by Stephan of all of the new nebulae discovered in Marseille by 1877. Therein were corrected errors that had lain

---

[94] Anon (1879b), Maffeo (2002).
[95] Obituary: Anon (1904).

[96] Often it is simply called the '80-cm reflector'.

Table 8.21. *Ferrari's objects discovered during 1865–66 (No. = object number in AN 1571)*

NGC	No.	GCS	Type	V	Con.	Remarks
50	13	5092	Gx	12.3	Cet	L. Swift 21.10.1886
116	14	5101	NF		Cet	
7565	2	6158	NF		Psc	
7613	3	6184	NF		Psc	
7614	4	6185	4 stars		Psc	
7663	5	6193	NF		Aqr	
7666	8	6196	NF		Aqr	
7667	9	6197	Gx	14.1	Psc	
7668	10	6198	NF		Psc	
7669	11	6199	NF		Psc	
7670	12	6200	NF		Psc	
7683	1	6204	Gx	12.7	Peg	Tempel 1876
7738	6	6222	Gx	13.2	Psc	
7739	7	6223	Gx	14.1	Psc	

Figure 8.29. Two of Ferrari's objects. Left: the galaxy NGC 50 in Cetus; right: the faint star quartet NGC 7614 in Pisces (DSS).

hidden in the published data from 1870. Alphonse Borrelly and Jérôme Coggia found some new objects, too, while searching for comets and minor planets with small instruments.

8.13.1 Short biography: Auguste Voigt

Auguste Voigt was born on 18 January 1828 in Gevrey-Chamberin, France. After having attended the Ecole Normale Supérieur he worked as a teacher for some years. Owing to his interest in astronomy, he caught the attention of Leverrier, Director of Paris Observatory, who engaged him as an assistant in October 1863. Voigt was soon sent to the old Marseille Observatory to succeed Charles Simon as Director. There he finished the move to the new site at Longchamps. In November 1864 the main instrument was ready to use: this was the 80-cm Foucault reflector, which had been transferred from Paris. Like other persons before (and later), Voigt got into trouble with Leverrier and eventually resigned in 1865, followed by Stephan. Auguste Voigt died on 12 March 1909 in Géanges at the age of 81.

8.13.2 Voigt's list of nebulae

Owing to the power of the Foucault reflector and the expected quality of the site, Voigt intended to search for nebulae; but an appropriate use of the instrument was impossible, because building up the new observatory absorbed too much of the Director's time. The only opportunity arose from March to August 1865. The result was a list of 10 new nebulae (Table 8.22). Since the objects remained in Voigt's observing book, Dreyer was not aware of the discoveries and credited them to others. The list was eventually found in 1987 by William Tobin (Tobin 1987).

One object had already been listed in the Slough catalogue: NGC 5619, which was discovered by John Herschel in 1828 and entered as h 1806 (GC 3887). Three objects are contained in the GCS, NGC and IC II, respectively. All are galaxies – thus Voigt's success rate is 100%. In 1863 Marth discovered NGC 6921. In Table 8.22 'Dreyer' shows the discoverer mentioned in the NGC/IC. The objects of Arnold Schwassmann were found photographically at Königstuhl Observatory, Heidelberg. NGC 6364 is one of the NGC objects with the most independent discoveries (together with Swift's NGC 1360 and Marth's NGC 7422; see Table 10.7).

NGC 7242 is a member of a group of seven galaxies in Lacerta (Fig. 8.30). Among them is NGC 7240, which was found by Stephan together with NGC 7242 on 24 September 1873; the faint object (14.3 mag) was obviously overlooked by Voigt. The remaining galaxies, which are listed in the Second Index Catalogue, are even fainter: IC 5191–93, which were discovered by Barnard on 5 December 1888 with the 36" Lick refractor (Barnard 1907a), and Bigourdan's IC 5195 (B 449, 16 October 1895). Detecting this 15.4-mag galaxy with the Paris 12" refractor was an observational feat.

8.13.3 Short biography: Jean Marie Edouard Stephan

Edouard Stephan was born on 31 August 1837 in St Pezenne, France (Fig. 8.31). He attended the Ecole Normale Supérieur, finishing in 1862 as the best in his class. Leverrier soon engaged him for Paris Observatory. In June 1866 Stephan moved to Marseille Observatory to succeed Voigt as director. During the first year he managed the building of the new site and had little time to observe with the Foucault reflector. He did, however, discover the minor planet Julia on 6 August 1866, and in late 1867 observed the transit of Mercury.

From 1870 to 1885 Stephan measured nebulae, discovering many new ones. Meanwhile, his assistants Borrelly and Coggia observed comets and minor planets. Inspired by Hippolyte Fizeau, Stephan was interested in interferometry too. So the tube entry of the Foucault reflector was equipped with a diaphragm with two slits 65 cm apart, which for the first time yielded an upper limit for the diameter of stars. He retired in 1907 and was succeeded by Henry Bourget. Edouard Stephan died on 31 December 1923 in Marseille at the age of 86.[97]

8.13.4 Stephan's observations with the Foucault reflector

In mid 1870 Stephan started his systematic observations of nebulae with the Foucault reflector (Fig. 8.32). He pursued two goals: the discovery of nebulae and positional measurement of old and new objects. Since nebulae had been found to be fixed objects, he wanted to use them as a reference system to determine proper motions of stars.[98]

The campaign was limited to the regions 0^h to 8^h and 15^h to 23^h in right ascension and $-12°$ to $+46°$ in declination. Here the transparency of the Marseille sky was at its best and the reflector could develop its full power. The instrument was moved along the meridian only – a sweeping method introduced by William Herschel. At low magnification Stephan first wrote down the rough position of an object. The precise measurement was done later with the bar-micrometer (normally at a power of 250). Mainly reference stars from the catalogue of the Astronomische Gesellschaft (AGK) were used for this task. However, the wooden mounting turned out to be pretty unhandy. Stephan's wish for a more stable device was never fulfilled. The observation method is described in a report of 1884 (Stephan 1884).

---

[97] Obituary: Bosler (1924); see also Tobin (1987).
[98] This is nowadays realised with quasars and remote compact galaxies.

272 Dreyer's first catalogue

Table 8.22. *Objects found by Voigt (March to August 1865); nos. 5 and 9 were earlier discoveries*

No.	Object	GC(S)	Type	V	Con.	Dreyer	Remarks
1	**NGC 4255**		Gx	12.7	Vir	Peters	Peters 1881?
2	IC 3136		Gx	14.3	Vir	Schwassmann	Schwassmann 1899
3	IC 3155		Gx	14.2	Vir	Bigourdan	Kobold 1895, Bigourdan 1907
4	IC 3268		Gx	13.5	Vir	Schwassmann	Schwassmann 1899
5	NGC 5619	3887	Gx	12.9	Vir	J. Herschel	J. Herschel 10.4.1828 (h 1806)
6	**NGC 6364**		Gx	13.1	Her	Stephan	Safford 5.9.1866, Stephan 21.7.1879, L. Swift 11.9.1885
7	**NGC 6675**	5924	Gx	12.5	Lyr	Stephan	Safford 28.9.1866, Stephan 27.7.1870
8	**NGC 6759**		Gx	14.3	Dra	Swift	L. Swift 16.10.1886
9	NGC 6921	5960	Gx	13.5	Vul	Marth	Marth 6.9.1863
10	**NGC 7242**	6035	Gx	12.9	Lac	Stephan	Stephan 24.9.1873

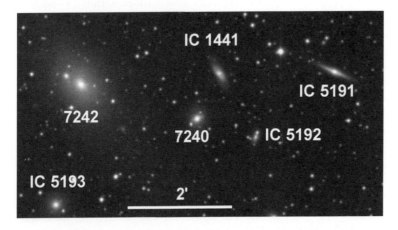

Figure 8.30. The galaxy group around NGC 7242 in Lacerta (DSS).

On 23 June 1870 Stephan discovered his first object, the galaxy NGC 6431 (GC 5873) in Hercules. The last ones were found on 2 February 1877: the galaxies NGC 2557 (GC 5414) and NGC 2572 (GC 5416) in Cancer and NGC 2538 (GC 5411) in Canis Minor. A remarkable find was made on 23 September 1876: the galaxy group in Pegasus, consisting of NGC 7317 (GC 6061), NGC 7318A/B (GC 6062), NGC 7319 (GC 6063) and NGC 7320 (GC 6064). This is the famous Stephan's Quintet (Fig. 8.33).[99] Only four nebulae are described, because Stephan could not resolve the close pair NGC 7318. It was one of 15 groups found by 1877. He was '*impressed by the high frequency of groups of nebulae that populate some regions of the sky*' (Stephan 1871a).

On 23 March 1884 Stephan discovered five members of the galaxy cluster Abell 1367 (the Leo Cluster): NGC 3857, NGC 3859, NGC 3864, NGC 3867 and NGC 3868. The first three were independently seen by Lewis Swift on 13 April 1885. D'Arrest contributed five galaxies (see Table 8.13), but most successful was William Herschel with seven. On 27 June 1870 Stephan discovered the galaxies NGC 6040–42 (GC 5799–5801) in Abell 2151; but the king of the famous Hercules Cluster

---

[99] On 20 June 1882 Stephan found NGC 6027, the brightest member of Seyfert's Sextet in Serpens (Seyfert 1951).

8.13 Observations in Marseille

Figure 8.31. Edouard Stephan (1837–1923).

Figure 8.32. The Foucault reflector with an aperture of 78.7 cm (Observatoire de Marseille).

is Lewis Swift, who contributed 11 members in 1886 (see Table 9.18). Like Stephan, he was astonished by the many pairs and groups of nebulae.

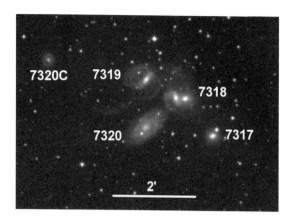

Figure 8.33. Stephan's Quintet – a compact group of galaxies in Pegasus. NGC 7318 is a double galaxy; NGC 7320C was added in the twentieth century (DSS).

By the press date of the GCS (February 1877) Stephan had published nine lists of new nebulae (Table 8.23). They appeared in parallel in the *Comptes Rendus* (*CR*), *Astronomische Nachrichten* (*AN*) and *Monthly Notices* (*MN*); the first two were combined in *MN* 32. A* for *CR* means that the first 40 nebulae appeared in no. 7 (16 July 1870), no. 12 (3 August 1870) and no. 3 (17 October 1871) of the *Bulletin Astronomique* (Paris Observatory). In the GCS (and NGC) Dreyer called Stephan's lists St I to St VIII (combining the last two). 'Num.' and 'S' show the number of objects and Stephan's discoveries. He gave positions with an accuracy of 0.01^s (AR) and $0.1''$ (NPD), using the epoch ('Ep.') of the year. For each object there is a short French description (translated into English in *MN*). The discovery date is missing; but fortunately it is mentioned, with a few exceptions, in a work published later by Emmanuel Esmiol (Esmiol 1916).[100]

Except for three cases (see the column 'x'), all of the objects were entered in the GCS. The first case (in list II) concerns the galaxy NGC 6278 in Hercules. Stephan correctly assumed that this is GC 4266 (which had been found by William Herschel as III 124). The others are the first two objects of his last list. Again he is right that these are GC 332 and GC 333, which had been observed on 27 November 1876. The galaxies in Cetus are discoveries of William Herschel (III 441 = NGC 560, III 442 = NGC 564). Dreyer later noted that John Herschel's GC positions are wrong: "*The 'General*

---

[100] See Section 9.8.2.

Catalogue' is therefore wrong with respect to these two Nebulae, while Auwers finds positions which agree better with modern determinations." Dreyer noticed the true identities of Stephan's nebulae shortly after the GCS had been published and wrote a 'Note on some of M. Stephan's new nebulae' (Dreyer 1877a).

Dreyer credits four objects to d'Arrest (d'A) and 24 to Marth (m). One identity had already been revealed by Stephan; in list VIII he wrote that *'No. 12 is probably identical with 440 Lassel* [Lassell]*'*, where Marth's m 440 (NGC 7065) is meant. This case is treated in Dreyer's note too. The object is catalogued as GC 5993 in the GCS (though without mentioning Stephan).

Two nebulae are due to Voigt (V): the galaxies NGC 6675 (GC 5924) and NGC 7242 (GC 6035). These observations were not known to Dreyer (see Table 8.23). It appears too that three objects had already been seen by William Herschel (WH): the galaxies NGC 972 (GC 5247, II 211) in Aries and NGC 6239 (GC 5832, III 727) in Hercules and the globular cluster NGC 6426 (GC 5870, II 587) in Ophiuchus. Dreyer missed these cases. He was not aware either of Safford's (Sf) observations with the 18.5″ refractor of Dearborn Observatory, Chicago (see Section 9.6.5). Safford did not publish his discoveries until 1887 and thus eight of his objects, which had been found during the years 1866–68, were credited to Stephan in the GCS.

Six of Stephan's galaxies are identical with other GC/NGC objects that had been found earlier by others (Table 8.24). One object was found twice by him: GC 6019 = GC 6020. Dreyer noticed this too late; the numbers were eventually combined to give NGC 7190. The error was due to the reference star.

With respect to these identities, 139 objects were discovered by Stephan until 1877. Among them are 127 non-stellar cases (an excellent rate of 97%): 125 galaxies, one reflection nebula (NGC 2149 = GC 5358 in Monoceros; Fig. 8.34) and one open cluster (NGC 6846 = GC 5948 in Cygnus). NGC 6414 (GC 5867) is a group of four stars in Ophiuchus; three objects are missing: NGC 952 (GC 5243), NGC 6748 (GC 5939) and NGC 7054 (GC 5991).

Stephan's values concerning brightness and size were deemed too small by Dreyer. He deduced this from the data of d'Arrest and Marth for common objects (see Table 8.24). Dreyer corrected the values thus: eeF → eF, eeS → eS, eF → vF, eS → vS. Obviously Stephan's low values were influenced by the light pollution of the city of Marseille. The average brightness of 13.5 mag is thus too high (see Fig. 9.38); even Lewis Swift in Rochester, using a smaller telescope (a 16″ refractor), had a lower mean of 13.6 mag.

8.13.5 Short biography: Alphonse Louis Nicholas Borrelly

Alphonse Borrelly was born on 8 December 1842 in Roquemaure, France (Fig. 8.35). From December 1864 he worked as an assistant at Marseille Observatory (Longchamps), first for Director Auguste Voigt and then, from 1866, for Edouard Stephan. Before his retirement in 1913, Borrelly discovered 20 minor planets (the first in 1866) and 18 comets (the first in 1873), among them the periodic 19P/Borrelly (28 December 1904). Therewith he was even more successful than his colleague Jérôme Coggia. The first observations were made with the 18.2-cm comet-seeker by Eichens (optics by Foucault), which was installed in 1866. From 1872 he used the 25.8-cm Eichens refractor with an objective by Merz. With this instrument Borrelly determined 4000 positions of minor planets and comets. He also made positional measurements of 50 000 stars with the 18.8-cm meridian-circle by Martin/Eichens, which was erected in 1876. Alphonse Borrelly died on 28 February 1919 at the age of 83 in Marseille. The minor planet (1539) bears his name.[101]

8.13.6 Borrelly's harvest: six new nebulae

With the comet-seeker of aperture 18.2 cm, Borrelly discovered six nebulae (Table 8.25), which was published on 5 April 1872 in his paper 'Nébuleuses nouvelles, découvertes et observées par Alph. Borrelly, à l'Observatoire de Marseille a l'aide du chercheur Eichens' (Borrelly 1872a).[102] No dates are given. Because the paper also contains the discovery of a variable star in November 1871 (BD −0° 62 = 'suspected variable' NSV 152 in Cetus), the nebulae were probably found in the same year (thus his list is a collection of results).

All of these objects are galaxies and were catalogued by Dreyer in the GCS. NGC 2268 (GC 5362) and NGC 2300 (GC 5370) were found independently

---

[101] Obituary: Bosler (1926).
[102] Shortly thereafter the report appeared in the *Monthly Notices* too (Borrelly 1872b).

Table 8.23. *Numbers of objects and references for Stephan in the GCS (see the text)*

St	Num.	S	Sf	H	m	D'A	V	x	Ep.	CR Vol.	CR p.	CR Year	AN No.	AN Vol.	AN p.	AN Year	MN Vol.	MN p.	MN Year
I	10	8			2				1870	*			1810	76	159	1870	32	23	1871
II	30	10			19			1	1870	*			1867	78	295	1872	"	"	"
III	20	16		1		3			1870	74	444	1872	1876	79	61	1873	32	231	1872
IV	15	14	1						1872	76	1073	1873	1939	81	303	1873	33	433	1873
V	15	13	1				1		1873	77	1365	1873	1872	83	51	1874	34	75	1873
VI	10	8				1			1874				1877	83	137	1874			
VII	23	17	2	2	1	1		1	1876	83	328	1876							
VIII	30	26	3		1				1876	84	641	1877	2129	89	263	1877	37	334	1877
''	30	27	1				2	2	1877	84	704	1877	2126	89	213	1877	37	337	1877
Sum	183	139	8	3	24	4	2	3											

Table 8.24. *Identities with NGC objects found earlier*

NGC	GCS	Date	Type	V	Con.	NGC	GC	Discoverer
211	5115	18.11.1876	Gx	14.1	Psc	203	5113	Copeland 19.12.1873
6431	5873	23.6.1870	Gx	13.4	Her	6427	5869	Marth 2.7.1864
6610	5908	13.7.1876	Gx	12.0	Her	6574	5397	Marth 9.7.1863
7111	6005	31.9.1872	Gx	14.1	Aqr	7108	6004	Marth 3.8.1864
7190	6019	23.7.1870	Gx	13.9	Peg	7190	6020	Stephan Sept? 1872
7260	6040	22.9.1876	Gx	12.9	Aqr	7257	6039	Marth 1.10.1864
7568	6160	17.10.1876	Gx	13.6	Peg	7574	6163	d'Arrest 2.10.1866

Figure 8.34. Stephan's only reflection nebula: NGC 2149 in Monoceros (DSS).

Figure 8.35. Alphonse Borrelly (1842–1919); Observatoire de Marseille.

by Tempel in 1877 (which is not mentioned by Dreyer). NGC 2300 was also seen by Winnecke in 1876. It is curious that Borrelly did not notice the bright companion NGC 2276, which was clearly visible for Winnecke and Tempel in 1876. The galaxy pair NGC 3933/34 (GC 5588/89; Fig. 8.36) was independently found in 1884 by Pechüle in Copenhagen.

8.13.7 Short biography: Jérôme Eugène Coggia

Jérôme Coggia was born on 2 February 1849 in Ajaccio, Corsica. When Edouard Stephan took over the directorship of Marseille Observatory, Coggia became an assistant on 1 October 1866 (his colleague Borrelly having been there for two years). He mainly observed with the 18.2-cm comet-seeker by Eichens; between 1867 and 1899 he found eight comets (the first in 1868; among them was the bright one of 1874) and five minor planets (the first in 1867). He received the Lalande prize in 1873. Jérôme Coggia left the observatory in 1899 and died on 15 January 1919 at the age of 69.[103]

8.13.8 Coggia's nebula

At Marseille Observatory Coggia found the 11.0-mag galaxy NGC 6952 (GC 6250) in Cepheus (Fig. 8.37).

[103] Obituaries: Anon (1919a) and Anon (1919b).

Table 8.25. *Borrelly's discoveries (date 1871 estimated)*

NGC	GCS	Date	Type	V	Con.	Remarks
2268	5362	1871	Gx	11.4	Cam	Tempel 1877
2300	5370	1871	Gx	11.1	Cep	Winnecke 28.6.1876, Tempel 1877
2715	5443	1871	Gx	11.1	Cam	
3853	5578	1871	Gx	12.4	Leo	
3933	5588	1871	Gx	13.4	Leo	Pechüle 1884
3934	5589	1871	Gx	14.0	Leo	Pechüle 1884

Figure 8.36. The galaxy pair NGC 3933/34 in Leo, found by Borrelly (DSS).

Figure 8.37. Coggia's discovery: the galaxy NGC 6952 in Cepheus (DSS).

Dreyer entered it in the 'Addenda' of the GCS, without giving a source. The late listing indicates that Coggia did not report his find until late 1877 or early 1878, perhaps in reaction to Dreyer's 'request'. It cannot be said whether the object was found during that period. Probably Coggia used the 18.2-cm comet-seeker by Eichens, because Borrelly found his objects with that instrument. The galaxy was independently discovered by Lewis Swift on 14 September 1885 with the 16" refractor of Warner Observatory. Dreyer catalogued this object as NGC 6951, without noticing the identity with Coggia's NGC 6952 (moreover, GC 6250 is not mentioned in the NGC). This identity was first stated by Denning (1892b).

8.14 VOGEL'S OBSERVATIONS IN LEIPZIG AND VIENNA

Hermann Vogel, a student of Carl Bruhns, mainly became famous by virtue of his contributions to astrophysics. As co-founder and Director of the Astrophysical Observatory in Potsdam he had access to outstanding instruments. Less known might be Vogel's visual observations of nebulae and star clusters, which were highly esteemed by Dreyer. He left significant traces in Leipzig and Vienna.

At Leipzig Observatory Vogel at first observed with the 11.7-cm refractor, which had formerly been used by d'Arrest, later changing to the new one of aperture 21.5 cm. The results were published in 1867 and 1876, presenting precise positions, descriptions and drawings (in 1876 only) of non-stellar objects. In 1883 Vogel observed with the 27" Grubb refractor in Vienna, making some excellent drawings. There he independently discovered one of Stephan's nebulae.

8.14.1 Short biography: Hermann Carl Vogel

Hermann Vogel was born on 3 April 1841 in Leipzig (Fig. 8.38). He was the son of a well-known pedagogue and came into contact early in life with the old Leipzig Observatory (Pleißenburg), where he met Director d'Arrest. Vogel attended the Polytechnic School in Dresden. In 1865 he returned to Leipzig, becoming the second assistant of Carl Bruhns at the new observatory in Johannistal. There he was awarded his doctorate for a work on nebulae. In Leipzig Vogel was in contact with Zöllner, a founder of astrophysics. After at first working on classical astronomy, Vogel later became a leading representative of stellar spectroscopy.

In 1870 Vogel became the director of the private observatory of Friedrich von Bülow in Bothkamp near Kiel, which was at that time equipped with the largest German refractor (aperture 28 cm, made by Schröder). Together with Oswald Lohse, he worked on spectroscopy. This qualified him, from 1874 in Berlin onwards, to build up the new Astrophysical Observatory in Potsdam. In 1879 he became the provisional Director, and three years later he was made the full Director of the new institution. In 1899 the new large double refractor was installed, which did not meet Vogel's high expectations. He achieved important astrophysical results, among them the discovery of spectroscopic double stars, and received many honours. Still in office, Hermann Vogel died on 13 August 1907 in Potsdam at the age of 66.[104]

8.14.2 The publication of 1867

Having been pointed towards the importance of positional measurements of nebulae by his teacher Bruhns, Vogel started an observing programme in 1865. He used first the 11.7-cm Fraunhofer refractor and later the 21.5-cm refractor by Steinheil on a Pistor and Martins mounting (Fig. 8.39). The smaller telescope, which had been built in 1830, known as the 'sechsfüssiger Refractor', was used by d'Arrest for his observations of nebulae (the 'Erste Reihe'). The larger, built in 1862, was called the 'zwölffüssiges Äquatoreal'. Both instruments were equipped with a bar-micrometer. Until October 1865 Vogel observed with the Fraunhofer, which could show only objects above +20° declination. This limitation was overcome with the new Steinheil refractor. The observations were finished in 1866, yielding the positions of 100 objects.

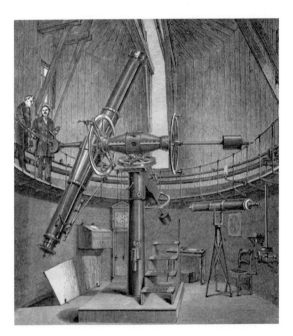

Figure 8.39. The 21.5-cm Steinheil refractor ('zwölffüssiges Äquatoreal') in Leipzig (Bruhns 1872).

Figure 8.38. Hermann Vogel (1841–1907).

---

[104] Obituaries: Lohse (1907), MacPherson (1907), Frost (1908a) and Fowler (1908).

The results were published in February 1867 by Engelmann, in Leipzig, under the title *Beobachtungen von Nebelflecken und Sternhaufen am sechsfüssigen Refractor und zwölffüssigen Äquatoreal der Leipziger Sternwarte*.[105] The work, which brought Vogel a Ph.D. from Jena University in 1868, was dedicated to Bruhns. The observations, covering 35 pages, are sorted by date (15 May 1865 to 27 October 1866); the objects are identified by their h-number. Vogel gives an (English) description. It follows a table of measured relative positions, sorted by h-number. Mostly stars of Schönfeld's Mannheim observations were used ('Erste Abtheilung', Schönfeld 1862d); the positions were measured by Argelander in Bonn. Then tables of the calculated coordinates (epoch 1865) and a list of all objects (now giving GC-numbers), with differences from Schönfeld, are presented. The final 11 pages contain remarks pertaining to individual objects.

A bit later a summary appeared in the *Astronomische Nachrichten*, dated 15 September 1867 (Vogel 1868a). Here the objects are sorted by GC-number. Vogel compares his positions with those of Schönfeld, Auwers, Schultz and Rümker. In another paper, written on 10 February 1868, differences from d'Arrest's new *Siderum Nebulosorum* are given (Vogel 1868b).

8.14.3 The publication of 1876

Vogel's observations with the 21.5-cm refractor started on 5 January 1866 and ended on 2 March 1870. The results were not published until 1876, as *Positionsbestimmungen von Nebel-flecken und Sternhaufen zw. +9° 30' und +15° 30' Declination*.[106] The objects were chosen from the GC, which lists 305 objects in this zone. Setting the brightness limit to Herschel's 'pretty faint', Vogel's sample amounted to 132 nebulae and star clusters. He expected good measurements, because d'Arrest had been successful in a comparable range with a slightly larger refractor (aperture 28.5 cm).

The object data, sorted by GC-number (h is given too), cover 19 pages. The date, reference star (BD) and a short description are presented. For five objects a GC-number is missing (Table 8.26). Vogel called NGC 63 'd'Arrest Nova', but for NGC 2194 and NGC 7464 he did not mention the Copenhagen observations. Actually NGC 2194 had been discovered by William Herschel (IV 5) on 11 February 1784; the object is listed as GC 1383 in the General Catalogue. Dreyer did not notice this, entering d'Arrest's discovery (18 September 1862) as GC 5380 in the GCS. The object was also seen in 1849 by Petersen in Altona. Besides NGC 2194, Vogel found two other 'clusters' in the Orion/Gemini region on 7 December 1869. These large, loose star groups were ignored by Dreyer in his supplement.

Vogel's publication gives a list with coordinates for all measured objects (epoch 1870). There are comparisons with d'Arrest and Schönfeld for common objects. There follow notes pertaining to individual objects; some nebulae showed 'resolvability', e.g. the galaxies M 65 and M 66 in Leo. Interesting is an appendix, in which Vogel reports that the manuscript had already been ready for printing in 1870. Obviously he delayed publication to wait for the results of Schultz and Schönfeld, which appeared in 1874 and 1875, respectively.[107] So the appendix contains a comparison with their positions. The last two pages show 15 drawings made at the 21.5-cm refractor (Table 8.27). Most of the objects are galaxies; the last three are open clusters (stars are inserted in a coordinate system). Vogel mentions no dates for his drawings (Table 8.27 gives the most probable). For some objects the designation is wrong (perhaps typos by the engraver).

In 1876 Dreyer wrote a review (Dreyer 1876c) – his third after those on the works of Schultz and Schönfeld. He discusses the precision achieved and mentions nine objects missed by Vogel. Six were observed in Birr Castle (all galaxies). The remaining ones are GC 2179 (NGC 3345), a faint star pair, GC 2436 (h 903), which is identical to h 902 (NGC 3705), and GC 4918 (NGC 7555), which is lost.

8.14.4 Observing with the great Vienna refractor

In 1883 Vogel had the opportunity to observe with the newly installed 27" Grubb refractor in Vienna (during May and June and on a few days in

---

[105] 'Observations of Nebulae and Star Clusters with the 6-ft Refractor and 12-ft Equatorial of Leipzig Observatory' (Vogel 1867a).

[106] 'Determination of Positions of Nebulae and Star Clusters between +9° 30' and +15° 30' Declination' (Vogel 1876).

[107] Schultz (1874), Schönfeld (1875b).

Table 8.26. *Vogel objects without GC-number*

NGC	GC(S)	Date	Type	V	Con.	Remarks
63	5093	16.8.1868	Gx	11.8	Psc	d'Arrest 27.8.1865, Safford 30.9.1867
		7.12.1869	Star group		Ori	'star cluster, 30–40 stars 9–11 mag'
2194	1383	7.12.1869	OC	8.5	Ori	W. Herschel 11.2.1784 (IV 5); Petersen 1849;
	5360					d'Arrest 18.9.1862
		7.12.1869	Star group		Gem	'several stars 9 mag and some fainter, altogether around 30'
7464	6112	10.8.1869	Gx	13.3	Peg	d'Arrest 27.8.1864, Marth 23.10.1864 (m 514)

Table 8.27. *Vogel's drawings made during the years 1867–69*

No.	NGC	GC	h	M	Date	Type	V	Con.	Remarks
1	2245	1425	393		16.2.1868	RN		Mon	Already drawn by B. Stoney (Table 7.1)
2	3593	2347	840		31.1.1868	Gx	11.0	Leo	'clear signs of resolvability'
3	3628	2378	859		31.1.1868	Gx	9.6	Leo	
4	3623	2373	854	65	31.1.1868	Gx	9.2	Leo	'the central part appears resolved at magn. 460'; Fig. 8.40 left
5	3627	2377	857	66	31.1.1868	Gx	8.9	Leo	'signs of resolvability'
6	4192	2786	1132	98	5.5.1869	Gx	10.1	Com	
7	4216	2806	1148		21.3.1869	Gx	10.3	Vir	'signs of resolvability'
8	4254	2838	1173	99	1.3.1868	Gx	9.7	Com	
9	4501	3049	1312	88	25.3.1868	Gx	9.4	Com	Not h 1342; 'signs of resolvability'
10	4866	3342	1498		16.5.1868	Gx	11.1	Vir	Not h 1198
11	5248	3615	1650		28.5.1867	Gx	10.1	Boo	
12	7463	4886	2202		28.8.1869	Gx	12.9	Peg	Not h 2022; trio with NGC 7464/65 (see Tables 8.3, 11.4)
13	6709	4440	2020		1.6.1867	OC	6.7	Aql	Not GC 4040
14	1662	905	332		29.2.1868	OC	6.4	Ori	
15	2169	1361	379		2.12.1868	OC	5.9	Ori	

September).[108] His report appeared a year later, titled 'Einige Beobachtungen mit dem grossen Refractor der Wiener Sternwarte'.[109] Some objects were drawn, using magnifications from 300 to 1500 (Table 8.28). Additionally there are textual descriptions of M 13, M 15 and M 57. Vogel observed stars with a visual spectroscope too.

On 3 and 4 June 1883 Vogel detected five nebulae west of BD +41° 2423 in Canes Venatici. Four could

---

[108] About Vienna University Observatory, see Holden (1891a), Firneis (1985) and Witt (2006). For one year the Grubb refractor was the world's largest. It ousted the 26-inch in Washington and was outperformed by the 30" Clark refractor at Pulkovo.

[109] 'Some observations with the great refractor of Vienna Observatory' (Vogel 1884).

Table 8.28. *Objects drawn by Vogel at the Vienna refractor*

NGC	GC	Date	Figure	Type	Remarks
5194	3572	4.6.1883	3.1	Gx	M 51
5195	3574	4.6.1883	3.1	Gx	
6210	4234	4.6.1883	Fig. 6	PN	Σ 5, sketch
6572	4390	19.9.1883	4.1	PN	Σ 6
6572	4390	19.9.1883	Fig. 7	PN	Σ 6, sketch
6853	4532	20.9.1883	3.2	PN	M 27
6905	4572	20.9.1883	4.2	PN	
7009	4628	20.9.1883	4.3	PN	Fig. 8.40 right
7662	4964	20.9.1883	4.4	PN	

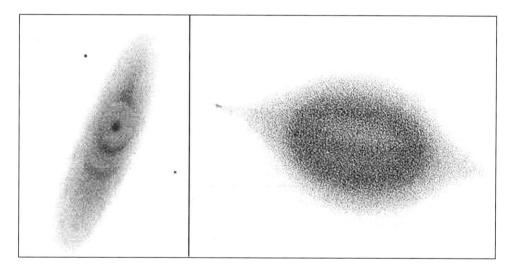

Figure 8.40. Drawings by Vogel. Left: galaxy M 65 in Leo (21.5-cm refractor, Leipzig); right: planetary nebula NGC 7009 in Aquarius (27″ refractor, Vienna).

be identified as GC 3688 (NGC 5350), GC 3693 (NGC 5353), GC 3694 (NGC 5354) and GC 3695 (NGC 5355). These galaxies had already been discovered by William Herschel on 14 January 1788. The fifth object was thought to be a 'Nova'. Actually this is the 13.7-mag galaxy NGC 5358, which had been discovered by Stephan on 23 June 1880 in Marseille.

8.15 IN THE FOOTSTEPS OF HIS FATHER: OTTO STRUVE

Otto Struve was an outstanding representative of classic astronomy, being mainly interested in precise positions. At Pulkovo Observatory he continued the study of double stars which had occupied his father, Wilhelm Struve, but the excellent 38-cm Merz refractor was used for observations of nebulae too, finding 13 new ones (which were catalogued in Dreyer's GCS). The most curious object was NGC 1554 near T Tauri, which is known as Struve's Lost Nebula. It should be noted that, due to the high northern latitude (+59° 46′, beaten only by Uppsala), visual observing was not an easy task. Severe winters with auroras caused many problems.

8.15.1 Short biography: Otto Wilhelm Struve

Otto Struve was born on 7 May 1819 in Dorpat, as this town was called by its large and culturally influential

Figure 8.41. The young Otto Struve, painted by Piazzi Smyth (Brück M. 1988).

German-speaking community (this town also went by the Russian name of Derpt and is nowadays known by its Estonian name, Tartu), as Wilhelm Struve's third child from his first marriage (Fig. 8.41). Through his father, the Director of Dorpat Observatory, he came into contact with astronomy early in life. He finished grammar school at the age of 15, but had to wait a year to enter the university. In 1837, two years before his examinations, Otto Struve became an assistant of his father in Dorpat, supporting the observations of double stars. Meanwhile Wilhelm Struve had established the new Pulkovo Observatory, and in 1839 his son became one of his first assistants. In 1845, only 26 years old, Otto Struve took over the duties of his father.[110]

His own first catalogue of double stars appeared in 1843, presenting observations with the 38-cm Merz refractor, which had been erected in 1839 (Fig. 8.42).[111] Otto Struve discovered even more objects (designated OΣ) than his father. In 1850 he was awarded the RAS gold medal for his work on the constant of precession (this was actually his thesis of 1841 at St Petersburg University). In 1862 Otto Struve succeeded his father (who died in 1864) as Director of Pulkovo Observatory. Unfortunately, due to ill health, the office was placed in the hands of his assistant August Winnecke from 1864 until the end of 1865. During the years 1867–78 Otto Struve was President of the Astronomische Gesellschaft. Another highlight was the erection of the 76-cm refractor by Clark/Repsold in 1884. Until 1889 Struve used the famous instrument (which was destroyed in the Second World War) for visual observations. Then, 50 years after the opening of Pulkovo, he resigned as Director (Fedor Bredikhin became his successor). As one of the last old-school astronomers, Otto Struve died on 14 January 1905 at the age of 85 in Karlsruhe. His ashes rest beneath a stone on which the old Merz refractor had stood at Pulkovo.[112] Like his father, Otto Struve was married twice. From the first marriage, two great astronomers emerged:[113] Karl Hermann Struve (Director of Babelsberg Observatory) and Gustav Wilhelm Ludwig Struve (Director in Karkhov, Ukraine). The latter was the father of Otto Ludwig Struve, Director of Yerkes Observatory and founder of the McDonald Observatory, Texas.

8.15.2 A 'mélange' of new nebulae

At Pulkovo Otto Struve observed nebulae with the 38-cm Merz refractor. The results were published in the observatory reports *Mélanges Mathématiques et Astronomiques de l'Académie Impériale des Sciences de St.-Pétersbourg*. In 1866 the article 'Observations de quelques nébuleuses', containing observations made during 1847–53 and 1864, was published (Struve O. 1866b). The targets were the galaxies M 31, M 49, M 51, M 94, NGC 3190/93 and NGC 4565 and the planetary nebulae NGC 2392, NGC 6543 and NGC 7662. On 28 April 1851 Struve discovered the galaxy NGC 4467 in Virgo, 5′ west of M 49 (see Fig. 6.58). It is not listed in the GC, because his publication came too late. Vice versa, Struve was not aware of Herschel's catalogue when he submitted his report (Ref. 1 in Table 8.29) on 13 April 1864. Dreyer eventually listed the object in his supplement as GC 5654.

In the same volume of the *Mélanges* there appeared the report 'Entdeckung einiger schwacher Nebel-

---

[110] On the history of Pulkovo, see Krisciunas (1978).
[111] Hartl (1987).
[112] Obituaries: Nyrén (1905a, b), Lynn (1905b), Turner H. (1906) and Anon (1906).
[113] Batten (1977, 1988).

8.15 Otto Struve

Figure 8.42. The 38-cm Merz refractor at Pulkovo Observatory (Repsold J. (1908), Vol. 2, Fig. 16).

Table 8.29. *Otto Struve's discoveries*

NGC	GCS	Date	Type	V	Con.	Remarks	Ref.
8	5082	29.9.1865	2 stars		Peg	Fig. 8.43	2
9	5083	27.9.1865	Gx	13.7	Peg	Fig. 8.43	2
107	5099	14.1.1866	Gx	15.2	Cet		2
1554	5339	14.3.1868	2 stars		Tau	Struve's Lost Nebula; d'Arrest 23.3.1868	3
3126	5521	8.4.1869	Gx	12.7	LMi	d'Arrest 30.4.1864	4
3534	5547	18.3.1869	Gx	14.5	Leo		4
3563	5550	18.3.1869	Gx	13.5	Leo		4
3609	5558	18.3.1869	Gx	13.3	Leo		4
3612	5559	16.3.1869	Gx	14.2	Leo		4
3739	5570	16.3.1869	Gx	14.5	Leo		4
3910	5584	3.3.1869	Gx	13.1	Leo		4
4005	5597	16.3.1869	Gx	13.0	Leo	NGC 4007 = III 325 (W. Herschel 6.4.1785)	4
4467	5654	28.4.1851	Gx	14.0	Vir	Near M 49; d'Arrest 27.2.1865	1
7422	6101	6.12.1865	Gx	13.5	Psc	Marth 11.8.1864 (m 509); d'Arrest 29.9.1866; Safford 27.9.1867	2
7427	6102	22.11.1865	Gx	15.1	Peg		2
7451	6109	7.12.1865	Gx	14.1	Peg		2
7472	6115	7.12.1865	Gx	13.7	Psc	NGC 7482 = m 521 (Marth 11.8.1864)	2
7688	6206	12.10.1865	Gx	14.5	Peg	Peters 13.8.1865	2

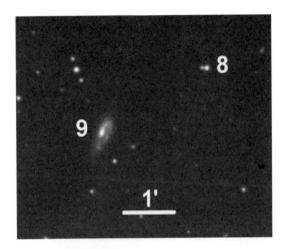

Figure 8.43. Otto Struve's discoveries in Pegasus: the double star NGC 8 and the galaxy NGC 9 (DSS).

flecke'[114] (Ref. 2), which had been submitted on 23 January 1866. The background was Struve's unsuccessful search for the comet Biela from September 1865 to January 1866 (d'Arrest had had the same experience). Herschel's GC was helpful in order to prevent '*taking nebulae, which occasionally appear in the refractor's field of view, for the comet*'. The search yielded eight objects not listed in the GC. Three of them had been discovered earlier by others. NGC 7422 is due to Marth; later the galaxy was seen by d'Arrest and Safford too. Therefore it is among the objects with the most independent discoveries (like Swift's NGC 1360 and Voigt's NGC 6364; see Table 10.7). NGC 7688 was first found by Peters, but Dreyer mentions Struve in the GCS, because Peters' publication did not appear until 1882 (see Section 9.2). We now know that NGC 7472 is identical with Marth's NGC 7482, but not, as Dreyer thought, with d'Arrest's galaxy NGC 7477 (GC 6119). Thus Dreyer's note in the IC I about NGC 7472/77 (which refers back to Burnham), '*to be struck out, both being = 7482 with errors of 2^m and 1^m in R.A. (Burnham)*', is incorrect. In his report Struve wrote the following about the nature of nebulae: '*It is known that Donati's spectral investigations too indicate an identical nature of comets and nebulous matter.*' This was, to his mind, supported by the variability of both comets and nebulae, regarding which he invoked '*d'Arrest's own observations of Hind's nebula*'.

Struve wrote another report, dated 20 April 1869 (Ref. 4), for the *Mélanges*, titled 'Wiederkehr des Winneckeschen Cometen und Entdeckung einiger neuer Nebelflecke.'[115] During his search for comet Winnecke (March–April 1869), which was in vain too, he came across many nebulae: '*In the first half of March the comet crossed an area rich in nebulae, not far from the constellation Coma Berenices. Every nebula seen in the field had to be checked to see whether this could be the comet.*' For this task, Struve used Herschel's GC, writing '*The matter became more difficult if the nebula noticed had not been listed by Herschel. Then always its fixed place had to be ascertained by micrometrical measurements against neighbouring stars.*' The search yielded eight new objects. Two turned out to be known: NGC 3126 was due to d'Arrest, as Struve revealed by consulting the *Siderum Nebulosorum*; and NGC 4005 is William Herschel's NGC 4007 (III 325), according to a modern analysis.

Table 8.29 shows Struve's discoveries. After subtracting identities, 13 objects must be credited to him (they are marked in bold); 11 are galaxies and 2 are pairs of stars (NGC 8 and NGC 1554) – a very good rate. All of these objects are catalogued in the GCS.

On 14 March 1868 Struve made a strange find in Taurus: a new object near Hind's Variable Nebula (NGC 1555). It was up to d'Arrest to publish the discovery in the *Astronomische Nachrichten* (Ref. 3), in an article titled 'Struve's Beobachtung eines neuen Nebelflecks nahe bei Hind's variablem Nebel im Taurus'.[116] The full story is told in Section 11.4.4.

8.16 NORDIC COMBINATION: SCHULTZ, DUNÉR AND PECHÜLE

Astronomers from northern Europe made great contributions to the investigation of nebulae and star clusters. It should not be forgotten that observing at these high latitudes was limited by many factors: short summer nights, auroras in winter, low temperatures and often bad weather. Outstanding is the work of d'Arrest, which is described in several sections of this book. Next in line is Herman

---

[114] 'Discovery of some faint nebulae' (Struve O. 1866c).

[115] 'Return of Winnecke's comet and discovery of some new nebulae' (Struve O. 1869).

[116] 'Struve's observation of a new nebula near Hind's variable nebula in Taurus' (d'Arrest 1868b).

Schultz. His measurements of 500 nebulae at Uppsala Observatory[117] were highly rated by Dreyer. He discovered some objects, but most of them are mere stars. Dunér, the successor of Schultz and best known for his solar research, found a galaxy. Pechüle, working in Hamburg and Copenhagen, discovered a few nebulae with d'Arrest's Merz refractor. Finally, the Dane Dreyer left only a few traces in Copenhagen – his veritable domain was Ireland and Britain.

8.16.1 Short biography: Per Magnus Herman Schultz

Herman Schultz was born on 7 July 1823 in Nygvarn, Sweden, as the son of a factory owner (Fig. 8.44). He entered Uppsala University to study astronomy. Schultz graduated in 1844 and was awarded his Ph.D. in 1855. In 1859 he became an assistant at the local observatory, which was equipped with a Steinheil refractor dating from 1850. In the same year he married Clara Charlotte Amelie von Steinheil, the daughter of the famous Munich telescope maker.

Figure 8.44. Herman Schultz (1823–1890).

With the optically good, but mechanically deficient, refractor, Schultz observed nebulae from 1863 to 1873. He amassed the considerable number of 12 000 measurements (published by 1874). Moreover, he determined star positions with the 9.6-cm meridian-circle by Repsold. In 1873 Schultz published his study of 104 stars in the cluster NGC 6885 (VIII 20). In 1878 he was appointed Professor of Astronomy and Director of Uppsala Observatory (following Gustav Svanberg). In 1882 he became a Fellow of the RAS. In 1888 he retired and was succeeded by Nils Dunér. Herman Schultz died only two years later, on 8 May 1890, in Stockholm, from influenza, at the age of 66.[118]

8.16.2 Schultz' observations in Uppsala and first discoveries

In February 1863 Schultz began to observe nebulae at Uppsala Observatory. Lying at latitude +59° 52', it is the most northern site to have contributed to the NGC.[119] From 1860 the main instrument was a 9.6" Steinheil refractor on a Repsold mounting (Fig. 8.45).

His first publication, titled *Beobachtung von Nebelflecken im Jahre 1863*,[120] appeared shortly after the GC. Schultz used a bar-micrometer, noting that '*Observations of nebulae with a bar-micrometer are among the most difficult; I have, however, the reasonable hope that much can be achieved in the field with this device*.' The first report contains 74 objects, among them the 'Novae' of Schönfeld (NGC 3185) and Hind (NGC 6760). However, the former is due to J. Stoney. Schultz' model was Schönfeld's 'Erste Abtheilung' of the Mannheim observations, using his reference stars (termed 'Sf'). The observations are sorted by date: the first is h 365 = NGC 2022 (17 February 1863) and the last h 2230 = NGC 7619 (21 October 1863). A table gives the coordinates for epoch 1865. Schultz compares them with Schönfeld, Oppolzer and Schmidt, finding differences from Schönfeld: '*At first glance one sees that my observations show generally larger right ascensions than Schönfeld*.' He proposes to check the differences with faint 'core-nebulae' (Schultz 1865a).

In June 1865 Schultz published in *AN* 1541 observations of 72 objects with detailed descriptions

---

[117] The old spelling of the city is 'Upsala'.

[118] Obituaries: Dunér (1890) and Anon (1891b).
[119] However, Pulkovo Observatory at +59° 46' is comparable.
[120] '*Observations of Nebulae in the Year 1863*' (Schultz 1864).

Figure 8.45. The 24.4-cm Steinheil refractor used by Schultz (Uppsala Observatory).

(Schultz 1865b). Three, not listed in the Herschel catalogues, are called 'Novae'. According to him, two of them are known: 'Nova I' is d'Arrest's no. 145 (NGC 4269) from *AN* 1500 (d'Arrest 1865a) and 'Hd. Nova' is Hind's NGC 6760. The third object ('Nova II'), which was observed on 23 December 1864, is, however, new, '*appearing nebulous in two nights*'. This is the pair of stars NGC 7560 in Pisces, which was discovered by Schultz on 5 October 1864 together with NGC 7561. This follows from his list of nebulae published on 30 September 1865 in *AN* 1555, '*which I have seen by chance, without knowing them from previous catalogues*' (Schultz 1865c). However, the double is only 18′ east of the galaxy pair NGC 7537/41. Though the orientation matches, Schultz' place and description clearly indicate the stars (see also Section 8.17).

Schultz' list has 11 entries. He identifies two with known objects: no. 1 = NGC 4269 (d'Arrest) and no. 4 = NGC 3222 (Winnecke); no. 5 ('*nebulosity hardly doubtful?*') and no. 11 ('*small and diffuse – nucleus $12^m 13'$*') are faint stars. These objects were catalogued by Dreyer in the GCS (see Table 8.31, column '*AN*'). Object no. 7 (GC 5859) is Schultz' reference star for the globular cluster M 92.[121] He described the object as a '*multiple star with faint nebulosity or a small planetary nebula*'. Schönfeld had used the star for his observation of M 92, without noticing any nebulosity. Owing to this uncertainty, Dreyer ignored GC 5859 in the NGC. Objects nos. 2, 3, 8, 9 and 10 are stars too. Schultz was in doubt about their character, writing '*small double star?*' (no. 2), '*hardly different from a fixed star*' (no. 3), '*probably a small double star (12^m) with faint nebulosity*' (no. 8) and '*stellar nebula*' (nos. 9 and 10). Obviously his poor yield is due to the bad observing conditions: '*When the air is not particularly good, stars acquire a nebulous appearance, even in the best instruments, thus it easily happens that a true nebulosity is thought to be unreal or vice versa.*' In winter, Schultz had trouble with auroras. Owing to its optical quality and the sharpness of its images (at the best moments), the Steinheil refractor was favoured.

8.16.3 The mystery of II 48

Schultz' next paper treats William Herschel's mysterious object II 48 (Schultz 1865d). It is a reaction to notes by Auwers and d'Arrest. John Herschel and the Birr Castle observers are involved in the matter too. Actually the field in Cancer contains three galaxies (Fig. 8.46): a close pair (Arp 167), consisting of NGC 2672 (11.6 mag) and NGC 2673 (13.0 mag), 40″ apart and the faint galaxy NGC 2677 (14.6 mag), 10′ to the southeast.

On 14 and 21 March 1784 Herschel found two 'pB, pL' objects in Datchet, which he catalogued as II 48 and II 80. The positions are 10′ apart and undoubtedly II 80 is NGC 2672. Though the separation would fit, it is, because of the description, not reasonable to identify II 48 with the faint galaxy NGC 2677 (which would certainly be of class III); thus II 48 = II 80 is more likely. Anyway, there is no hint of a double nebula in the first catalogue. Herschel had used different reference stars for II 48 and II 80: 85 Gem (12.7° west!) and δ Cnc (1.5° southwest), respectively. At another observation on 13 July 1787 he could recover II 80 – but not II 48. This indicates that his position for II 48 was wrong – which is hardly surprising, given the great distance to the reference star.

---

[121] See Butillon's observation of M 92 in July 1848 (Butillon 1848).

8.16 Nordic combination   287

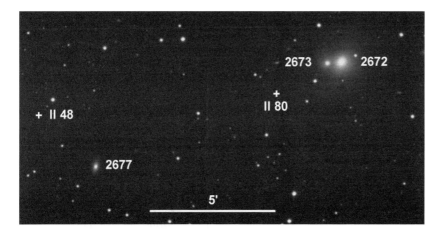

Figure 8.46. The double galaxy NGC 2672/73 and its neighbour NGC 2677 (DSS; see the text).

Things became strange, because II 80 appears in a list of 15 'Double nebulae with joint nebulosity', published in 1811.[122] However, this cannot be the 'pair' II 48/80, because Herschel must have been aware of the considerable positional difference of 10′. His list contains, for instance, the bipolar planetary nebula NGC 2371/72, with 'components' only 30″ apart. If Herschel really saw the companion NGC 2673, why is there no clear note about it?

John Herschel observed II 80 (h 526) on 25 March 1827 in Slough, noting '*Query if not bicentral*' – a first hint regarding the companion. It is now strange that he claimed to have seen 'II 48' on 17 March 1831. The object was catalogued as h 527: '*The faintest object imaginable, and discerned with the utmost difficulty.*' The position and description fit NGC 2677 well – but this can by no means be the 'pB, pL' object II 48! Obviously Herschel was confused by his father's location, which is about 3′ northeast of NGC 2677.

The next observation was made by Johnstone Stoney at Birr Castle. On 19 December 1848 he saw the companion of II 80 (h 526, NGC 2672) in the 72-inch: '*Close D neb, S star p*'.[123] Mitchell was the first to make a sketch on 9 February 1855 (Fig. 8.47): '*Very close, almost touching. [h] 526 is mbM; [h] 527 is smaller, and lbM.*'[124] The double nebula was confirmed in the great reflector, but the component (NGC 2673) was

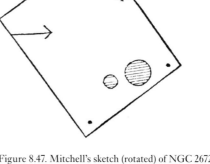

Figure 8.47. Mitchell's sketch (rotated) of NGC 2672/73 (Parsons L. 1880).

incorrectly called 'h 527'. Dreyer remarked in the 1880 publication that '*the latter is not h 527, but the close component to h 526*'.

What about II 48? The first to doubt John Herschel's claim of an identity h 527 = II 48 was Auwers: '*Since the descriptions do not match too, it is very improbable.*'[125] D'Arrest has '*searched in vain on the best winter nights in 1861/62*' for II 48 at the position of h 527 (d'Arrest 1862c). John Herschel summarised the situation in the General Catalogue: II 80 = h 526 = GC 1704 (this is NGC 2672), GC 1705 = h 526,a = 'R. nova' ('*nearly in contact with h 526*', i.e. NGC 2673) and II 48 = h 527 = GC 1707 (this is NGC 2677). All is correct – except

---

[122] Herschel W. (1811: 285); see Section 2.8.
[123] Parsons L. (1880: 69).
[124] Parsons W. (1861a: 161).

[125] Auwers (1862a: 51).

the identification of II 48! Obviously Herschel was not convinced by Auwers and d'Arrest.

On 29 January 1865 d'Arrest could see h 527 (NGC 2677) for the first time in the Copenhagen 11" refractor, as his remark in *Siderum Nebulosorum* shows.[126] There II 48 is identified with h 527 too. Schultz was not aware of this when he observed the field on 19 March 1865 with the 9.6" refractor (Schultz 1865a). At magnification 320 he saw NGC 2677 (h 527) about 8′ east of NGC 2672 (II 80). The object appeared like a 'planetary' or 'globular cluster'.[127] With respect to d'Arrest's failure (winter 1861/62), he saw his observation as *'an interesting proof of the optical power of my refractor'*. Schultz assumed II 48 *'2ᵐ east and 5′ south of h 527'* (a somewhat mysterious place), but could not find it. Because William Herschel had seen the object as being 'pretty bright', he saw *'good reasons for a probable variability'*, but, due to John Herschel's statement in the GC (II 48 = h 527), *'my unsuccessful search is no longer of interest'* – obviously, the authority of the British astronomer was too great for any doubt to be entertained. Schultz could see the companion of II 80: *'the two nuclei were perfectly separated at a power of 320'*.

Dreyer was eventually able to resolve the strange case in the GCS. In the notes to GC 1704/07 he wrote the following: *'It appears to me most likely, that II 48 = II 80 (or at least, that the descriptions belong to one nebula).'* In the New General Catalogue one reads, about NGC 2672, that *'II 48 = II 80'*. As a precaution Dreyer remarks for NGC 2677 *'II 48?'*. In the *Scientific Papers* he is a bit more detailed.[128] For him, Herschel's wrong placing of II 48 was due to a problem of the mounting: *'There was some doubt about the contraction of the rope.'*

8.16.4 Measuring 500 nebulae

In three articles, written during the years 1865–68, Schultz enlarged on his measurements of the planetary nebula NGC 2392 (IV 45), the globular cluster M 92 and the galaxy group NGC 467/70/74. For NGC 2392 George Knott had supposed a proper motion due to differences in the position angle relative to the star Baily-Lalande 145512 (Knott 1865a). According to Schultz, these could have been caused either by a wrong measurement or by motion of the star (Schultz 1865e). A subsequent note of Knott supports the former suggestion (Knott 1865b). D'Arrest, on the basis of his own long-time observations and those of Otto Struve and Schönfeld, states too that the nebula does not show any proper motion (d'Arrest 1865b). In the case of M 92, where Schultz claimed to have seen a nebulous star, a positional error was corrected (Schultz 1866a). The hint came from George Rümker. For the trio NGC 467/70/74 Schultz corrected wrong positions in John Herschel's GC (Schultz 1868). Meanwhile another paper appeared, titled 'Historische Notizen über Nebelflecke',[129] which mentions interesting sources about the first nebular discoveries.

Until 1874 no new article about nebulae was published. Schultz' time was occupied with the micrometrical measurements of 500 nebulae. The result – his main work 'Micrometrical observations of 500 nebulae' – was submitted to the *Nova Acta of the Royal Scientific Society of Upsala* on 18 April 1874 (Schultz 1874). Schultz sorted the objects – altogether 506 – by their h-number (some new ones bear GC-numbers). The work contains all his 19 discoveries (Table 8.30); some are listed in the main table, called Nova I to XII there, others are mentioned only in the 'Supplementary details of observations and remarks'. Five objects had already been listed in *AN* 1555 (Schultz 1865c). In the GCS Dreyer credited 15 objects to Schultz (13 were taken over into the NGC). Only two of them are non-stellar – a very bad rate (comparable only to that of Lohse). In the column 'Reference' 1 = Schultz (1865b) (*AN* 1541), 2 = Schultz (1865c) (*AN* 1555), 3 = Schultz (1874) (*Nova Acta*) and 4 = Schultz (1875) (*MN* 35).

Nova II (GC 5086, NGC 20), a galaxy that could have improved Schultz' rate, was discovered by Mitchell (Dreyer: 'Ld R'), but not published until 1880, by Lawrence Parsons. It was independently seen by Lewis Swift and catalogued as NGC 6 by Dreyer. The case of GC 5096 is curious. Schultz identifies the object with GC 40 (which was found in 1854 by Mitchell). Though Dreyer regards the identity as probable, he nevertheless includes Schultz' nebula in the GCS. These two

---

[126] d'Arrest (1867a: 101).
[127] Schultz (1874: 80).
[128] Dreyer (1912a: 296).

[129] 'Historical notes about nebulae' (Schultz 1866b).

in the Cape catalogue (see Section 5.4.1), which sorts the nebulae into three classes. Objects of class I ('regular nebulae') are further divided in terms of five features: magnitude, brightness, roundness, condensation and resolvability. Each has five grades (1 to 5). John Herschel coded a regular nebula by writing down the sequence of grades, e.g. 32154 (magnitude = 3, roundness = 2, ..., resolvability = 4).

Schultz revived this idea: *'This mode of description, which is very commodious and to the use of which it is very easy to accustom oneself, offers moreover many advantages, so that in my opinion it is well worth the general acceptance of nebula-observers.'* This originates from the fact that 'regular nebulae' (class I) were ideal targets to be measured with his humble refractor; anyway, he observed a few objects of class II ('irregular nebulae') too, such as M 33 and M 78. Schultz slightly modified Herschel's system by adding the brightness of the central condensation of the nebula (nucleus or star). In his catalogue a coded description is presented for each object. Contrary to his hope, the new scheme was ignored by contemporary astronomers. They were too much accustomed to William Herschel's textual abbreviations.

The central part of Schultz' work is the list of measured relative positions, covering 28 pages. The next 32 pages contain notes with further descriptions and a few new objects (here Schultz used Herschel-style descriptions). Because absolute positions (coordinates) are missing, the publication differs from those of Schönfeld, Rümker and Auwers. The reason is simple: Schultz had no precise positions for his 440 reference stars; the BD coordinates were not sufficient for the task. Despite this lack, he was unwilling to refrain from publication of his results – the observations had already taken much longer than expected.

However, in January 1875 Schultz presented a 'Preliminary catalogue of nebulae observed at Upsala' in the *Monthly Notices*, containing positions for the epoch 1865 (Schultz 1875). Reference stars, dates and notes are not given. Though the source of the star positions is not mentioned, it is quite obvious that it is the Bonner Durchmusterung. Schultz wrote that definite positions will be given not *'until after some few years [...] and through the friendly co-operation of other astronomers* [I] *shall have received reliable positions for all the stars of comparison'*. Unfortunately, this did not happen. Beside the coordinates of the observed nebulae (ordered by right ascension), the coded description is given. The paper presents a detailed explanation of the new scheme. A month later Schultz' *'most remarkable work'* was presented by Dreyer at the Annual General Meeting of the RAS (Dreyer 1875a). He stressed the small errors (according to Schultz' own calculations).

Not much later Dreyer wrote a review about Schultz' 'Micrometrical observations of 500 nebulae', which appeared in the *Vierteljahrsschrift der Astronomischen Gesellschaft* (Dreyer 1875b). After having been an assistant at Birr Castle for about a year, it was his first publication about nebulae. At the beginning he states that the work of *'Schönfeld has hitherto been the most comprehensive as containing observations of 235 nebulae; the new work of Dr. Schultz is however now superior to it, as it gives observations of about 500 nebulae.'* Actually, Schönfeld had measured about 500 objects too, but this was not known by Dreyer at that time, for Schönfeld's 'Zweite Abtheilung' appeared shortly after the review. About Schultz he wrote that he was *'the only observer, who lately has worked in this special branch'*. In a footnote Dreyer mentions the micrometrical measurements at Birr Castle. Though it was limited by the mounting of the 72-inch (an object could be traced near the celestial equator for at most 34 minutes), the great light-gathering power of the reflector was an advantage, because even faint reference stars could be seen without illumination of the micrometer. Moreover, in winter 1875/76 the 36-inch was equipped with an equatorial mounting to allow unrestricted measuring (see Section 8.18.7).

Dreyer lamented Schultz' text's lack of comparisons with results of other observers. There are, for instance, 163 common objects with Schönfeld. Schmidt had already pointed out that Schultz' right ascensions systematically differed from Schönfeld's. The Uppsala observer was able to confirm this with his preliminary positions. Dreyer remarked that *'it should be of interest to compare the results of Schönfeld and Schultz in order to see whether the very striking differences between the Mannheim- and the few earlier Upsala-observations would appear again, when all the observations of later years were employed in the comparison'*. He did not believe that this was caused by the different micrometers (Schönfeld, ring; Schultz, bar), since Schönfeld and Schmidt had used the same type. In his paper 'On systematic errors in observing right ascensions of nebulae' of 1896, Dreyer once again treated the issue, on the basis of measurements of 29 objects with the 10″ Grubb refractor in Armagh (Dreyer 1896).

In his review, Dreyer commented on Schultz' coded descriptions too: "*The Undersigned* [Dreyer] *must confess, that he does not see any reason, why the old Herschelian abbreviations, with which every astronomer is perfectly acquainted, could not be used instead of numbers, which are more difficult to remember. Is it not more convenient to read 'pL, B, R, Nucl. (star 10m) r' instead of '321.1.3 (10)', and still does it not mean just the same? Perhaps the new system does not demand so much space as the old one, but this seems to be of little importance.*" In his future catalogues Dreyer consequently used the well-tried system of William Herschel. Schultz' reply, titled 'Gewährt J. Herschel's Vorschlag, Nebelflecke mit Nummern zu beschreiben, reelle Vorteile oder nicht?',[131] is interesting. His main argument is that his new '*method to describe nebulae is independent of the observer's language*'. He explains as follows: "*In the English text of my work the appearance of abbreviations of English words is totally justified; but I must confess that, concerning the form, it seems barbaric to me to read these Herschel abbreviations in the Latin text of d'Arrest's 'Siderum Nebulosorum Observationes' or in the German text of Schönfeld's Mannheimer Beobachtungen* [Erste Abtheilung]." Interestingly, Schultz is already thinking in modern computer categories: he guessed that '*sooner or later the translation of descriptions of nebulae into the language of arithmetic will have to be done in some way*'. This would allow one to compare – '*in a certain degree of numerical account*' – the observations of different persons. In the same volume of the *Vierteljahrsschrift* Dreyer reviews Schönfeld's 'Zweite Abtheilung' – an occasion to criticise Schultz' system once again (Dreyer 1876b). He particularly commends Schönfeld for his classic descriptions of nebulae: '*the writer is right to prefer this verbose kind of description over numbers, as from the results of different observers real changes of objects can be derived only with great caution*'.

In 1915 Karl Reinmuth published his paper 'Photographische Positionsbestimmungen von 356 Schultzschen Nebelflecken'.[132] Using plates taken during the years 1900–13 with the 40-cm Bruce refractor at Königstuhl Observatory, Heidelberg, he had measured the positions of almost three-quarters of Schultz' nebulae.

Schultz' last publication on non-stellar objects appeared in 1886, under the title 'Mikrometrische Bestimmung einiger teleskopischer Sternhaufen'.[133] The work treats the open clusters NGC 1513, NGC 6341, NGC 6885 and NGC 7686. As early as in 1873 he had published his measurements of the cluster NGC 6885 around the star 20 Vulpeculae (Schultz 1873). This object was found twice by William Herschel, as VIII 20 and VIII 22 (the latter is catalogued as NGC 6882, see Table 10.8). It is no. 6 in Harding's list of nebulae too (see Section 4.1.2).

8.16.6 Short biography: Nils Christoffer Dunér

Nils Dunér was born on 21 May 1839 in Billeberga, Sweden, as the son of a priest (Fig. 8.49). From 1855 he studied at Lund University and was allowed three years later to observe at the old observatory. Shortly after his Ph.D. (with a thesis about the minor planet Panopea), in 1864 he became an assistant of Axel Møller. The new University Observatory was finished in 1867 and in August of that year the 24.5-cm Merz refractor on a Jünger mounting was ready to use. Dunér's first work at the telescope was a study of 445 double stars, which he published in 1876. He was interested in variable stars too. From 1878 he focused on spectroscopy, especially of red stars.

After 24 years in Lund, Dunér became Professor for Astronomy at Uppsala University in 1888, and succeeded Herman Schultz as director of the observatory. In 1892 the old 24.4-cm refractor, which had been used by Schultz for his nebular observations, was replaced by a double refractor by Steinheil/Repsold (apertures 36 cm and 33 cm). With it, Dunér was able to determine the rotation period of the Sun by means of the Doppler effect. He retired in 1909 and was succeeded

---

[131] 'Does J. Herschel's advice, to describe nebulae by numbers, provide real advantages or not?' (Schultz 1876).

[132] 'Photographic determination of the positions of 356 Schultz nebulae' (Reinmuth 1915).

[133] 'Micrometric analysis of some telescopic star clusters' (Schultz 1886).

8.16 Nordic combination

Figure 8.49. Nils Dunér (1839–1914).

Figure 8.50. Dunér's discovery, the bright galaxy NGC 2273 in Lynx (DSS).

by Östen Bergstrand. Nils Dunér died on 10 November 1914 in Stockholm at the age of 75.[134]

8.16.7 Dunér's late discovery of a bright galaxy

In 1871 Dunér discovered the 11.6-mag galaxy NGC 2273 (GC 5363) in Lynx with the 24.5-cm Merz refractor of Lund Observatory (Fig. 8.50). He proudly wrote in his report 'Beobachtung eines neuen Nebelflecks'[135] of 6 October 1871 that *"On September 15 of this year I discovered a nebula, which is neither listed in Sir J. Herschel's 'General Catalogue', nor was it known to the experienced observer of nebulae, Professor d'Arrest."* His description reads as follows: *'pretty bright and at least 2' in diameter, with a strong central condensation'*. Dunér added that *'Since such a bright and easily visible object was not known earlier, it should be, according to Prof. D'Arrest, of some interest.'* Indeed, it is remarkable that the bright object had been missed by all previous observers. A similar case is the discovery of the bright galaxy pair NGC 2276 (Winnecke, Tempel in 1876) and NGC 2300 (Borrelly in 1871?, Winnecke in 1876, Tempel in 1877) in Cepheus.

NGC 2273 was also observed by Schultz on 3 and 8 September 1872 in Uppsala. The object appears as 'Dunér's Nova' in his publication of 1874.[136] He thought it was *'probably a small star group'* and the nebular appearance might be an effect of the *'very bad air'*.

8.16.8 Short biography: Carl Frederik Pechüle

Frederik Pechüle was born on 8 June 1843 in Copenhagen (Fig. 8.51). After having finished school in 1865 he studied astronomy at the local university. His teacher was d'Arrest, from 1861 director of the new observatory. There Pechüle made his first observations of comets and minor planets with the 11" Merz refractor. During the years 1870–72 he was an assistant of George Rümker at Hamburg Observatory (Millerntor), observing nebulae with the 26-cm Repsold refractor. The results were published in 1895 as 'Positionsbestimmungen von Nebelflecken und Sternhaufen'.[137]

In 1873 Pechüle was awarded his M.A. at Copenhagen University. At the observatory he met the young Dreyer in 1869–70 and 1873. In 1874 Pechüle

---

[134] Obituaries: Bergstrand (1914), Ångström (1915), MacPherson (1914a), Fowler (1915) and Hasselberg (1917).
[135] 'Observation of a new nebula' (Dunér 1872).

[136] Schultz (1874: 79); see also Schultz (1875).
[137] 'Determination of the positions of nebulae and star clusters' (Rümker 1895).

Figure 8.51. Frederick Pechüle (1843–1914).

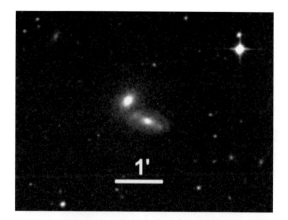

Figure 8.52. The close double galaxy NGC 78 (distance 42″) in Pisces (DSS).

joined the Venus-transit expedition to Mauritius; the second event (1885) was watched in the Caribbean. From 1875 he was permanently at Copenhagen Observatory, observing minor planets and comets, discovering three (1877, 1880 and 1886). In 1885 he succeeded Schjellerup as assistant. Frederik Pechüle died in Copenhagen on 26 May 1914 at the age of 70 after a long illness.[138]

8.16.9 Pechüle's unpublished nebula

In Dreyer's GCS Pechüle is credited as the discoverer of one object: NGC 78 (GC 5094), which is actually a double galaxy of 13.3 mag in Pisces (Fig. 8.52). Unfortunately, there is no publication. Since Pechüle and Dreyer knew each other from Copenhagen, it is probable that the find was communicated by letter, perhaps as a reaction to Dreyer's 'Request for astronomers' of late 1876. There are two options about the place (and date) of the discovery: Hamburg (1870–72) or Copenhagen (from 1875). The former is unlikely, because otherwise the object would certainly have been included in Rümker's publication of the Hamburg observations (1895). Thus the Copenhagen period, during which he was using the 11″ Merz refractor, is more probable. This limits the date of discovery to the period from 1875 to 1877 (the latter year is due to the appearance of the GCS). Perhaps Pechüle found the nebula while looking for comets or minor planets near the ecliptic (Pisces). In spring 1884 he discovered further objects; most of them were catalogued in the NGC (see Section 9.10).

8.17 HOLDEN, TUTTLE AND A POSSIBLE CASE OF IMPOSTURE

In the time before the appearance of the GCS Edward Holden worked at the US Naval Observatory, Washington.[139] According to Dreyer, he discovered one object there: GC 6166 (NGC 7581). Unfortunately, no reference is given, but the date must lie within the period 1874–76. Probably Holden wrote a letter, reacting to Dreyer's request for new nebulae. Anyway, the Washington observations covering this period were published after the GCS.

If one reviews the relevant volume, one encounters an entry 'Nebula (nova)' on 11 January 1875.[140] The observation was made with the 26″ Clark refractor (see Fig. 9.33), using a power of 392 and a field of view of 9.9′. The column 'Remarks' gives *'In looking for Encke's*

---

[138] Obituary: Strömgren (1914).

[139] Edward Singleton Holden will be introduced in Section 9.16, in connection with his work at Washburn Observatory.

[140] Newcomb (1878: 285).

8.18 FURTHER BIRR CASTLE OBSERVATIONS AND THE PUBLICATION OF 1880

Under the authority of Lawrence Parsons, Dreyer (assistant during the years 1874–78) collected the Birr Castle observations of nebulae made during the period 1848–78 for a final publication. This work appeared in 1880, having great importance for the creation of the GCS. Until 1877 Dreyer was occupied with both tasks, which meant a great synergy. However, the structure of these publications is very different. That of Birr Castle still reflects the spirit of Lord Rosse. Neglecting the demands of positional astronomy, it concentrates on the observation and description of the objects. On the other hand, the GCS appears rather sober. Here Dreyer could demonstrate for the first time his skill for systematic data reduction: the transformation of nebula discoveries from various reports into a homogeneous object catalogue.

The GCS contains 90 Birr Castle objects, among them 21 that had already been found during the years 1849–57 by J. Stoney, B. Stoney and Mitchell – which actually should have been in the GC. Not having been included in Lord Rosse's publication of 1861, they were unknown to John Herschel (see Section 7.1). Dreyer's study of the Birr Castle observing books brought them to light. The remaining objects were discovered during the years 1866–78 by Ball, Lawrence Parsons, Copeland and Dreyer. All of these new nebulae appear in the 1880 publication.

8.18.1 Lawrence Parsons' publication of the Birr Castle observations

The Birr Castle report on nebulae in the *Philosophical Transactions* of 1861 (often abbreviated '*PT* 1861') was still the work of Lord Rosse. The new one bears exclusively Dreyer's signature. In 1914 he noted that, '*When I entered Lord Rosse's observatory in 1874 it was his [Lawrence Parsons'] wish that the long series of observations of nebulae should be got ready for publication, including those previously printed, as the paper in* Phil. Trans. *1861 in many cases gave insufficient particulars, especially about the new nebulae which had been found.*' (Dreyer 1914). Dreyer further wrote that '*I therefore at once drew up a working list of nebulae which required to be re-observed, chiefly in order to fix the places of the new ones, and most of them were then looked up and*

Figure 8.53. The galaxies NGC 7541 = NGC 7581 and NGC 7537 in Pisces (DSS).

Comet, found a very faint, elongated nebula α = *23h 11m.4,* δ = *+3° 59'.*' On precessing this 1875 position to 1860, the place of GC 6166 results. Surprisingly, this note is signed by 'T', which is Horace Tuttle!

Thus not Holden, but Tuttle is the discoverer of NGC 7581. The following question arises: did Holden send false information to Dreyer? No doubt this would fit his character. He had a strong need for recognition, which caused great trouble with other astronomers, e.g. his later inferiors Burnham and Barnard at Lick Observatory (see Section 9.15.1).

Unfortunately, Tuttle's location in Pisces refers to a blank field. The nearest object is NGC 7541 (Fig. 8.53), 46' southwest. The elongated 11.9-mag galaxy was discovered by William Herschel on 30 August 1785 (II 419), together with its southwestern neighbour NGC 7537 (3' apart).[141] Probably Tuttle saw NGC 7541 in 1875, but missed its 13-mag companion (though the field of view was large enough). During comet-seeking – his major task – non-stellar objects often appeared in the eye-piece, whose positions were not measured exactly. However, a bit nearer (38' southwest) there is a pair of 'nebulae' that was found by Schultz on 5 October 1864: NGC 7560/61, which exactly matches the orientation of NGC 7537/41. But Schultz' places (relative to his reference star BD +3° 4843) and descriptions are exact: they indicate a pair of stars (NGC 7560) and a single star (NGC 7561), respectively.

---

[141] See Steinicke (2001c).

Table 8.31. *The three parts of the 1880 publication*

Part	Last Obs.	Finished	Pages	Rect.	GC	GCS	Addenda
I	31.1.1878	February 1878	7–64	$0^h\ldots8^h$	1–1625	5080–5408	6246–6248
II	27.4.1878	August 1879	65–136	$8^h\ldots14^h$	1626–3783	5409–5757	6249
III	5.5.1878	June 1880	137–178	$14^h\ldots24^h$	3784–5066	5758–6245	6251

measured.' This work, done with a micrometer at the 72-inch, established his great interest in cataloguing non-stellar objects, making him a worthy successor of John Herschel.

Lawrence Parsons published the result in the *Scientific Transactions of the Royal Dublin Society* of 1880, titled 'Observations of nebulae and clusters of stars made with the six-foot and three-foot reflectors at Birr Castle, from the year 1848 up to the year 1878' (Parsons L. 1880). In the introduction one reads that *'From time to time after the completion of the six-foot reflector at Parsonstown, my late father brought out papers on some of the nebulae and clusters of stars observed with the three-foot and six-foot instruments, the last having been published in the* Philosophical Transactions of the Royal Society of London *in 1861. Since that date no account (with the exception of the monograph on the great nebula in Orion) of the observations of nebulae, which have been, with a few interruptions, carried on up to the present time, has appeared.*' A major topic of the 1861 publication was spiral structure. New nebulae were only by-products, detected when observing Herschel objects. In the new work, discoveries play a much larger role, reflecting Dreyer's view about the subject.

The publication consists of three parts (Table 8.31). Dreyer wrote in February 1880 that, *'To facilitate the preparation of the work for publication, it was divided into three parts, the first two of which, comprising the first fourteen hours of R.A., were published last autumn.'* (Dreyer 1880a). One further reads that *'the third part, containing the last ten hours, is in press and will follow in a few months'*. In 1914 Dreyer added *'When I left in 1878 the observations were ready for press as far as 14^h* [Part I + II], *but Lord Rosse* [Lawrence Parsons] *allowed me to take the observing-books with me* [to Dunsink], *and the MS.* [manuscript] *of the remaining 10h was finished in the course of the next year.'* (Dreyer 1914). All three parts appear contiguously in Vol. II (1880) of the *Scientific Transactions*. The title page gives *'Read, February 18, 1878'*, which relates to the first part only.

On 178 pages, objects from the GC (or h) and GCS (including the 'Addenda') are presented (Table 8.32). The amount of information is great, which reduces the clarity of the work. Nevertheless, due to Dreyer it is more structured than its forerunner of 1861 (though similar in its main features). The number of observations per object – given in the text as '(IV. obs.)', for instance – varies greatly: some were viewed only once, others very often, such as M 51 (40 times) and M 1 (almost 30 times).

The observations are ordered in a continuous table with three columns (Table 8.33); the 1840 objects are sorted by GC-number (first column); sometimes a GCS-number is given (>5080). All but 129 are listed the Slough catalogue (69 are pure GC and 60 are pure GCS objects). However, the sequence is not always kept, since often several (neighbouring) objects are grouped: *'It has been found convenient, where two or more nebulae are so near as to be in the field together or nearly so, to place them in one group, even though numbered separately in the catalogues.'*[142] Birr Castle discoveries, e.g. objects not listed in the GC, are mostly termed 'Nova'. Additional new objects can be found in the third column. They are hidden in the textual description (sometimes shown in a sketch) and it is not easy to locate them.

The second column gives the classic designations (h, H, M). Concerning the Messier objects, 55 were observed at Birr Castle: M 1–3, M 5, M 9–11, M 13–15, M 17, M 27, M 29, M 30, M 33, M 35–37, M 39, M 41, M 42, M 50–52, M 56–61, M 63–66, M 72, M 73, M 75, M 76, M 78, M 80–82, M 84–86, M 90, M 92, and M 94–101.

For each observation a date is given. Though the title announces 'from the year 1848', five were made

---

[142] Parsons L. (1880: 3).

Table 8.32. *Data given in the publication*

Data	Number	Remarks
Objects	1840	h (1711), GC (69), GCS (60)
Observations	3234	1848–1878
Sketches	255	
Drawings	39	7 drawers (see Table 8.35)
Observers	11	W. + L. Parsons and assistants

Table 8.33. *The meanings of the columns*

Column	Meaning	Information
G.C.	GC-number	GC, GCS, Nova (new object)
h. & H.	Herschel designation	d'A (d'Arrest), M (Messier), R. nova (*PT* 1861)
–	Observations	Date, original description, remarks (Dreyer), measurements, sketch, number of observations

before 1848; the first two are 18 and 19 September 1843, looking at M 51 and M 27, respectively. The last observation was made by Dreyer on 5 May 1878 (NGC 5990 = GC 4135). Not all were included, as is said: '*omitting all observations inferior through weather or other causes*'.[143] Unfortunately observers are mentioned seldom.[144] It is evident that most observations (at least during the time after Lord Rosse) were made by the scientific assistants. The assignment indirectly results from the periods of their work at Birr Castle (see Table 6.4). Figure 8.54 shows the numbers of their annual observations between 1848 and 1878. It indicates that the early frequency (B. Stoney, Mitchell) was never again reached. When Mitchell left in 1859, there was no observation at all. No successor could be found, and Lord Rosse had retired from active observing. Anyway, he had enough to do with the publication of the earlier observations, which appeared in 1861. The two large reflectors were not used until Hunter's employment in February 1860.

When Lawrence Parsons finished his study at Dublin University in 1864 (at the age of 24) and Hunter left Birr Castle, Lord Rosse placed the responsibility for the telescopes on his eldest son. Dreyer notes that '*Soon afterwards he began to take part in the observations with the 6-foot and 3-foot telescopes*' (Dreyer 1909). In November 1865 Robert Ball became the first scientific assistant of Lawrence Parsons. He later wrote that, '*When I went to Parsonstown in 1865, Lord Rosse was advanced in years. He no longer took an active part in the work of observation, but he evinced a lively interest in all that went on, and was always glad to think that the telescope was being used.*'[145] When Burton, the successor of Ball, left Birr Castle due to ill-health in 1870, only two observations were made by Lawrence Parsons. It was eventually up to Copeland and Dreyer to establish a new continuity. But sadly, with Dreyer's departure in August 1878, the Birr Castle observations of nebulae ended – the 72″ reflector fell into oblivion.

The main feature of the 1880 publication is the textual description of objects. Mainly Herschel's abbreviations are used. The data correspond to the original notes from the observing books: '*almost word for word transcription of the observer's original notes*'.[146] Dreyer's notes, containing additional identifications, are given in brackets: '*All the remarks in square brackets were added by the transcriber* [Dreyer] *while arranging the observations for publication; in no case do they form a part of the*

---

[143] Parsons L. (1880: 1).
[144] It is interesting that, besides mentioning the name (e.g. 'Mr. Stoney'), occasionally the first person ('I') is given.
[145] Ball V. (1915: 64–65).
[146] Parsons L. (1880: 1).

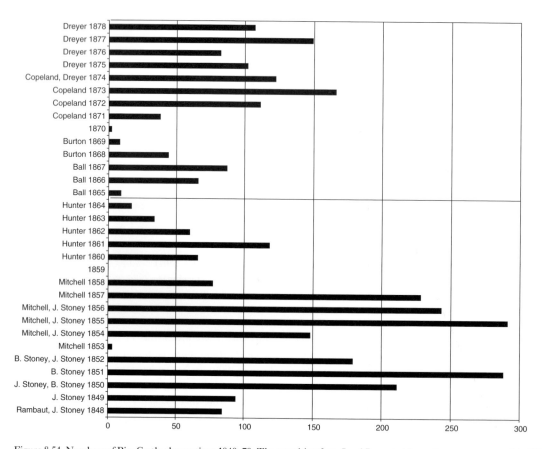

Figure 8.54. Numbers of Birr Castle observations 1848–78. The transition from Lord Rosse to his son Lawrence occurred in 1864.

original notes.[147] Their amount shows how intensively Dreyer studied the observation notes, checking many cases with the 72-inch too.

Often a 'diagram' illustrates the position or appearance of nebulae. These sketches (titled by the GC-number of the main object) are '*rough copies of those drawn at the telescope*'.[148] In some cases a new one was made for the publication, marked by a 'g' after the GC-number (e.g. '59g.' for GC 59 = NGC 125). Large nebulae are drawn with structure, small ones only as hatched ovals (sometimes marked with Greek letters). Some sketches show field stars. An arrow indicates the movement of the objects in the eye-piece, due to the rotation of the Earth (east → west): thus the arrowhead marks the westerly direction (p = preceding),

the end the easterly (f = following). North and south are perpendicular (one has to keep in mind that east is left on the sky). Figure 8.55 shows the example of GC 654–56 = NGC 1241–43 (galaxies in Eridanus). The orientation (northerly direction) of the sketches is chosen very differently.

The description also gives distances and position angles relative to nearby stars, which were determined with the micrometer. Dreyer remarks that '*a considerable number of micrometric measures were taken of groups of nebulae or of nebulae and neighbouring stars, and special pains were taken to identify all the novae which were found in the course of years*' (Dreyer 1881a). He added that, "*For obvious reasons the absolute positions of the objects examined cannot be directly obtained, and this circumstance, combined with the very condensed form of the earlier publications, made it impossible for Sir John Herschel to avoid making many mistakes in entering the 'R. Novae' in his General Catalogue. Many of these mistakes were pointed*

---

[147] Parsons L. (1880: 3).
[148] Parsons L. (1880: 4).

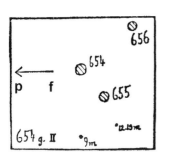 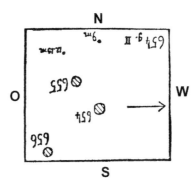

Figure 8.55. Orientation of sketches in the 1880 publication. Left: view in the eye-piece; right: sky view.

out with more or less certainty by d'Arrest." (Dreyer 1880a). From 17 November 1876 onwards Dreyer used a 'spider-line-micrometer' by Grubb (substituting an earlier one made by the same company). Occasionally a heavy Browning spectroscope was attached to the eye-piece, e.g. for the observation of the planetary nebula NGC 2438 (GC 1565) in Puppis on 15 February 1876.

Normally the 72-inch ('6 ft') was used; observations with the 36-inch ('3 ft') are explicitly mentioned. Sometimes the telescopes worked in 'front-view' ('Herschelian') mode, i.e. without a secondary mirror. Hunter's drawing of M 42 at the 72-inch (see Fig. 8.61) was partly made with the telescope in this configuration. About the observation of M 51 on 28 April 1875 one reads that '*Lord Rosse* [Lawrence Parsons] *thought the front-view showed the knots and faint parts better than he had seen it with the telescope in Newtonian mounting.*' This variation was later commented upon differently by Dreyer: '*The telescope was, in April 1875, arranged for front-view at the urgent suggestion of Dr. Robinson, but it was only used in this way on a couple of nights, as Lord Rosse* [Lawrence Parsons] *did not think that the additional light made up for the discomfort, not to say danger, to the observer.*' (Dreyer 1914). The method was dangerous, because the observer had to stay on a scaffold in front of the tube. Lawrence Parsons experimented with a 'silver glass 2nd' too, replacing the metal secondary mirror, whose surface became fragile.

Visitors are mentioned too. Romney Robinson from Armagh was still a frequent guest; an observation of M 27 (GC 4532) with J. Stoney on 29 August 1854 is documented. On 10 January 1867 Lawrence Parsons and Franz Brünnow of Dunsink Observatory observed the planetary nebula NGC 2371/72 (GC 1519/20) in Gemini. On 28 May 1875 Henry Chamberlain Russell[149] of Sydney Observatory was shown the mysterious object NGC 6049 (GC 4159) in Serpens. The discoverer, John Herschel (24 April 1830), had seen a '*nebulous atmosphere*' around a 7-mag star (SAO 121361), but on 14 May 1850 Lord Rosse noted the following: '*star looked quite sharp* […] *atmosphere which it appeared to have with a high power was illusory*'. Lawrence Parsons saw on 26 March 1875 a '*very faint surrounding atmosphere, which could be traced about 1′ from the star*'. Two days later he made the following note: '*Viewed with Mr. H. C. Russell (Sydney), who was not sure of the reality of the atmosphere, while I had no doubt of it.*' Dreyer's observation on 4 May 1877 revealed '*Nebulosity doubtful, although sometimes appearing very real*'; the NGC description reads '*star 7 [mag] in photosphere*'. Modern images show no nebula – thus Russell's doubts were justified. A similar case of a supposed nebula is the double star 2 Puppis (6 mag), 50′ east of the planetary nebula NGC 2438 (GC 1565), which belongs to the open cluster M 46. Lord Rosse noted on 28 January 1849 that it was '*very strongly nebulous*'. But wait, there is more: '*Another B neb star about 10′ n f the D star.*' This is 4 Puppis (5 mag). Dreyer was careful in the publication of 1880, speaking of '*Novae?*',[150] but in the NGC the object was ignored. The term 'Novae?' is also used for two other dubious 'nebulous stars'.[151] Obviously, the 72-inch had

---

[149] In 1872 Russell made interesting colour measurements of the open cluster NGC 4755 around κ Crucis (Russell H. C. 1872). For Sydney Observatory, see Bhathal (2009).

[150] Parsons L. (1880: 61).

[151] Parsons L. (1880: 19, 65).

Table 8.34. *Drawings in the 1880 publication*

Plate	Fig.	NGC	GC	M	Drawer	Date	Remarks
1	1	185	90		Mitchell	7.10.1855	
1	2	275, 278	156, 157		Mitchell	3.10.1856	
1	3	520	303		B. Stoney	18.12.1851	
1	4	1068	600	77	Dreyer	3.12.1874	Fig. 6 in *PT* 1861
1	5	1999	1202		Copeland	15.11.1873	
1	6	2068	1267	78	Mitchell	13.1.1858	
1	7	2170	1362		Hunter	16.2.1863	
1	8	2185	1375		Hunter	16.2.1863	
1	9	2366	1515		Copeland	9.3.1874	
2	1	2371, 2372	1519, 1520		Hunter	25.11.1862	
2	2	1952	1157	1	W. Parsons	15.1.1855	Comment by Dreyer
2	3	2024	1227		L. Parsons	18.2.1874	
3	1	2848	1829		Mitchell	15.3.1855	
3	2	3055	1964		Mitchell	15.3.1855	
3	3	3423	2234		Mitchell	29.3.1856	
3	4	3165, 3166, 3169	2037, 2038, 2041		Mitchell	19.3.1857	
3	5	3242	2102		Hunter	3.4.1861	
3	6	3440, 3445	2244, 2245		Mitchell	30.3.1856	
3	7	3521	2301		Mitchell	29.3.1856	
3	8	3628	2378		Hunter	10.2.1861	
3	9	2903	1861		B. Stoney	3.3.1851	
3	10	4214	2804		Mitchell	27.3.1856	
3	11	4501	3049	88	Mitchell	17.3.1855	
3	12	5033	3459		Hunter	19.4.1862	
4	1	5194, 5195	1622, 1623	51	Hunter	6.5.1864	Comment by Dreyer
5	1	5544, 5545	3833, 3834		Mitchell	17.3.1855	
5	2	5740	3985		Mitchell	10.5.1855	
5	3	6760	4473		Dreyer	14.8.1876	
5	4	6894	4565		Mitchell	19.8.1855	Fig. 8.56 left
5	5	6888	4561		Dreyer	20.8.1876	Fig. 8.56 centre; plus sketch
5	6	7433, 7435, 7436	4872, 4873		Dreyer	29.9.1875	Fig. 8.56 right; plus new Gx
5	7	7052	4653		Mitchell	20.9.1857	
5	8	7457	4883		Hunter	20.9.1862	
5	9	6819	4511		Copeland	4.9.1871	
5	10	7769, 7770, 7771	5020, 5021, 5022		Mitchell	8.10.1855	
5	11	7814	5046		Mitchell	3.10.1856	Fig. 42 in *PT* 1861
6	1	5907	4087		Hunter	24.4.1860	Extreme edge-on galaxy
6	2	6618	4403	17	Mitchell	19.7.1854	
6	3	6618	4403	17	Mitchell	23.7.1854	

problems in such cases: ring-like reflexes around bright stars could appear in the eye-piece (see Table 6.20).

The publication contains 39 drawings from the period 1851–76 (Table 8.34). In 1881 Dreyer remarked that '*Of the late years there was not much to be done with the pencil, as all the more interesting objects had already been drawn*' (Dreyer 1881a). The keenest drawers were Mitchell (19) and Hunter (10), followed by Copeland and Dreyer (3 each), B. Stoney (2) and William and Lawrence Parsons (1 each). As for the 1861 publication (see Table 7.1), spring remained the dominant season: 35% of all drawings were made in February/March.

NGC 5907 is a perfect edge-on galaxy. It had already been observed by Lamont and sketched by d'Arrest (Steinicke 2005b). Occasionally pairs or groups of nebulae are pictured. The drawing of the NGC 7436 group shows a fourth object, closely preceding NGC 7436 (Fig. 8.56 right). The text gives '*star or nebulous knot closely preceding*'. The object was omitted from the GCS (and NGC). Actually this is a galaxy (14.7 mag), which was first catalogued by Vorontsov-Velyamov in 1959 (VV 84b). The two drawings of the Omega Nebula (M 17) differ in contrast. Two objects (M 77 and NGC 7814) had already been shown in the 1861 publication. About M 1 and M 51, Dreyer later remarked that they were '*less successfully represented by the earlier observers*' (Dreyer 1880a). In particular, Lord Rosse's M 1 drawing of 1855 no longer shows the 'crab' (see Section 6.4.8).

8.18.2 The legacies of Johnstone Stoney, Bindon Stoney and R. J. Mitchell

Dreyer became aware of the objects found by Mitchell and the Stoney brothers in the GCS while compiling the Birr Castle observations (Table 8.35). Since some of them are hidden in the descriptions, both the page (P) and the GC-number ('GC') under which they appear are mentioned. Finally, the date indicates the precise location in the text.

The galaxy NGC 2290 is identical with William Herschel's III 897 (the NGC gives '*III 897?*'). All other objects must be credited to Mitchell and the Stoney brothers. Eleven of them had already appeared in the General Catalogue as 'R. nova' (see GC-numbers in the column 'GC id.'). Why did Dreyer revive them in the GCS? Regarding GC 5343 (NGC 1601) he noted '*probably = 867*', mentioning d'Arrest as the discoverer. In the case of GC 5200 (NGC 751) he clearly missed the identity with GC 456 (h 175,a), despite having written '*forms D neb. with II. 222*'; this is GC 455 (NGC 750), the northern component of a close double galaxy in Triangulum (Arp 166; Fig. 8.57 left). The NGC does not clear up the issue. Regarding GC 5639 (NGC 4330) one reads '*probably = 2909*'; the credit once again goes to d'Arrest. The same applies for GC 5653 (NGC 4466); no comment is given. For GC 5762 Marth is mentioned (m 275); Dreyer wrote '*must be = 3830*' – but erroneously in the following entry (GC 5763). For GC 6151 (NGC 7549) d'Arrest is mentioned and GC 6153 (NGC 7553) gives Schultz and Tempel (actually the latter was not involved). Dreyer notes here '*probably = 4913*'.

GC 5086 (NGC 20) is credited to Schultz, but the description gives '*Q y. = 6*' (possible identity with GC 6). Moreover, the galaxy is identical with Swift's NGC 6. Schultz is also mentioned for GC 5096 (NGC 90); Dreyer notes '*Q y. = 40, 41, 42*' (for this case see Section 8.16.4). The GC objects GC 5097 (NGC 91) and GC 5098 (NGC 93) are involved (Fig. 8.57 right), but here Dreyer mentions d'Arrest only. The GCS (and the NGC) erroneously credits GC 5619 (NGC 4118) to

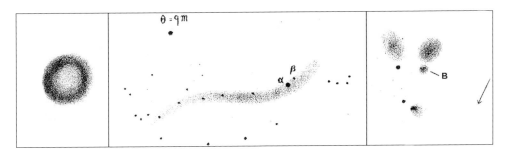

Figure 8.56. Figures 4 to 6 from Plate V (from left to right): the planetary nebula NGC 6894 in Cygnus, the Crescent Nebula NGC 6888 in Cygnus (θ = Wolf–Rayet star) and the galaxy group NGC 7433/35/36 in Pegasus (B = Dreyer's new object).

Table 8.35. *The 21 GCS objects of Johnstone Stoney, Bindon Stoney and Mitchell*

NGC	GCS	Observer	Date	'GC'	P	Type	V	Con.	GC id.	Remarks
1601	5343	J. Stoney	14.1.1849	866	42	Gx	13.8	Eri	867	GC: R. nova; GCS: d'Arrest 16.1.1865
2290	5369	J. Stoney	22.2.1849	1455	54	Gx	13.3	Gem		III 897, W. Herschel 4.2.1793; d'Arrest 19.4.1865
2321	6248	J. Stoney	18.12.1849	1481	55	Gx	13.9	Lyn		GCS Addenda
2373	5380	J. Stoney	20.2.1849	1527	58	Gx	14.0	Gem		Stephan 8.2.1878
2375	5383	J. Stoney	20.2.1849	1527	58	Gx	13.8	Gem		Stephan 8.2.1878
2694	5435	J. Stoney	9.3.1850	1720	71	Gx	14.7	UMa		
316	5129	B. Stoney	29.11.1850	173	13	Star	14.5	Psc		Tempel 29.8.1877
751	5200	B. Stoney	11.10.1850	455	24	Gx	12.2	Tri	456	GC: R. nova; pair w. NGC 750 (GC 455)
2183	5359	B. Stoney	11.12.1850	1373	52	RN		Mon		d'Arrest 11.1.1864
4109	5618	B. Stoney	21.4.1851	2723	108	Gx	14.1	CVn		
4330	5639	B. Stoney	14.4.1852	2871	113	Gx	12.4	Vir	2909	GC: R. nova; GCS: d'Arrest 15.4.1865
4466	5653	B. Stoney	26.2.1851	3020	116	Gx	14.0	Vir	3022	GC: R. nova; GCS: d'Arrest 24.4.1865
5538	5762	B. Stoney	6.3.1851	3830	138	Gx	14.7	Boo	3830	GC: R. nova; GCS: Marth 8.5.1864 (m 275)
7549	6151	B. Stoney	2.11.1850	4911	172	Gx	13.2	Peg	4912	GC: R. nova; GCS: d'Arrest 30.8.1864
7553	6153	B. Stoney	2.11.1850	4911	172	Gx	14.4	Peg	4913	GC: R. nova; GCS: Schultz 25.9.1867 (Tempel?)
20	5086	Mitchell	18.9.1857	6	7	Gx	13.1	And	6	GC: R. nova; GCS: Schultz 16.10.1866; = NGC 6 (L. Swift 1885)
90	5096	Mitchell	26.10.1854	38	8	Gx	13.8	And	40	GC: R. nova; GCS: Schultz 17.10.1866
91	5097	Mitchell	26.10.1854	38	8	Star	15.1	And	41	GC: R. nova; GCS: d'Arrest 5.10.1864
93	5098	Mitchell	26.10.1854	38	8	Gx	13.6	And	42	GC: R. nova; GCS: d'Arrest 5.10.1864
3016	5507	Mitchell	21.3.1854	1940	79	Gx	12.9	Leo		d'Arrest 31.12.1864
4118	5619	Mitchell	20.4.1857	2723	108	Gx	14.7	CVn		GCS: R_2 nova
7752	6226	Mitchell	22.11.1854	5011	176	Gx	14.0	Peg		d'Arrest 24.9.1865

8.18 Further Birr Castle observations

Figure 8.57. Left: NGC 750/51 in Triangulum (distance 24″); right: NGC 90/91/93 in Andromeda (DSS).

Lawrence Parsons. Except for this case and NGC 750, the NGC is correct.

The publication of 1880 contains 11 further objects of B. Stoney and Mitchell, which had been ignored in the GCS but were included in the NGC for unknown reasons (Table 8.36). In the cases of NGC 2689 and NGC 3460/61 Dreyer had not inspected the fields until 1878, i.e. after publishing the GCS. For NGC 3460/61 it can be said that Mitchell noted about his observation of NGC 3457 (h 793, GC 2256) on 27 March 1854 '*A S neb n of a B one, dist about 5′ or 6′.*' This correctly describes the situation: the small nebula is NGC 3461 and the bright one is NGC 3457 (which had first been seen by Francis Baily with John Herschel's 18¼-inch on 25 March 1827). Unfortunately, Dreyer was confused by his own observations of 22 March and 4 April 1878, thinking Mitchell's 'bright one' to be a new nebula, later catalogued as NGC 3460. But he was not quite sure about it, writing '*?? = h 793*' in the NGC.

To the GC score of the three observers (see Table 7.4) the following objects must be added: for J. Stoney, 5 galaxies; for B. Stoney, 9 galaxies, 1 reflection nebula and 2 stars; and for Mitchell, 11 galaxies, 1 reflection nebula (west of the Rosette Nebula) and 1 star.

8.18.3 Remains in the Index Catalogue

Traces of the early Birr Castle observers can even be found in Dreyer's Index Catalogues of 1895 and 1908 (Table 8.37). Bindon Stoney is represented in the IC I with seven galaxies in Lynx, all discovered in January 1851 (NGC 2330 and NGC 2334 are part of a complicated puzzle). IC 694, the faint northern companion of the double galaxy NGC 3690, is due to Hunter. Finally, Mitchell's galaxy involved in the case 'GC 80' (see Section 7.1.7) is listed as IC 1559 in the IC II.

8.18.4 Short biography: Robert Stawell Ball

When Hunter left Birr Castle due to ill-health, the responsibility for the observatory passed from Lord Rosse to his son Lawrence. Robert Ball became his first assistant. He stayed from November 1865 until August 1867 and discovered 11 new nebulae.

Robert Ball was born on 1 July 1840 in Dublin (Fig. 8.58). He got his scientific education at Trinity College, Dublin (1857–61). Ball was recommended by Johnstone Stoney (a life-long friend), and Lord Rosse asked him to work as the tutor for his three youngest sons.[152] Ball responded '*I would accept the post, provided that I was allowed to use the great telescope.*'[153] He was engaged in November 1865, and Lord Rosse gave him free run of the observatory. However, since all the large nebulae had already been observed and drawn, there was very little left to be done in that direction. But the micrometer had been very little used, and he found

---

[152] Randall (17 years old at that time), Richard (aged 14) and Charles (aged 11).
[153] Ball V. (1915: 62–63).

Table 8.36. *The 11 NGC objects of Bindon Stoney and Mitchell*

NGC	Observer	Date	'GC'	P	Type	V	Con.	Remarks
3179	B. Stoney	25.1.1851	2051	82	Gx	13.1	UMa	
5994	B. Stoney	9.3.1851	4139	148	Gx	14.8	Ser	
7384	B. Stoney	27.11.1850	4844	168	Star		Peg	
7390	B. Stoney	27.11.1850	4844	168	Gx	14.3	Peg	
2247	Mitchell	14.2.1857	1425	53	RN		Mon	L. Swift 24.11.1883
2689	Mitchell	11.3.1858	1714	71	Gx	16.3	UMa	Dreyer 22.5.1878
3460	Mitchell	27.3.1854	2256	91	Gx	12.3	Leo	Dreyer 4.4.1878; L. Swift 24.8.1883; = NGC 3457, Baily 25.3.1827 (h 793)
3461	Mitchell	27.3.1854	2256	91	Gx	14.5	Leo	Dreyer 4.4.1878
4607	Mitchell	24.4.1854	3143	119	Gx	13.0	Vir	
5319	Mitchell	27.3.1856	3664	133	Gx	15.5	CVn	
7402	Mitchell	2.10.1856	4853	169	Gx	14.5	Psc	

Table 8.37. *IC objects of the early Birr Castle observers*

IC	Observer	Date	Type	V	Con.	Remarks
457	B. Stoney	2.1.1851	Gx	14.8	Lyn	NGC 2330
458	B. Stoney	31.1.1851	Gx	13.5	Lyn	
459	B. Stoney	31.1.1851	Gx	14.5	Lyn	
461	B. Stoney	31.1.1851	Gx	14.9	Lyn	
463	B. Stoney	31.1.1851	Gx	15.0	Lyn	
464	B. Stoney	31.1.1851	Gx	13.8	Lyn	
465	B. Stoney	31.1.1851	Gx	13.7	Lyn	NGC 2334
694	Hunter	12.2.1860	Gx	15.0	UMa	Near NGC 3690
1559	Mitchell	18.9.1857	Gx	14.0	And	'GC 80' case

there a field worth cultivating. Additionally, visual spectroscopy was a new and promising field. Eventually Ball became Lawrence Parson's first assistant.

Ball was an industrious observer, later writing that '*I sometimes followed Herschel's strenuous example and remained observing from dusk to dawn.*'[154] However, '*I should add that the work was occasionally interrupted by little visits to the castle, where the kindness of Lord Rosse, tea and other refreshments were always available.*' Dreyer later wrote the following about Ball: "*He was an indefatigable observer, and was remembered for years after his departure by the workmen who helped to work the telescope as the man who kept them up 'terrible late' at night.*"[155] Ball did not live at Birr Castle but had rooms at Cumberland Square, about 10 minutes' walk away. It is astonishing that, alongside his astronomical interests, he was never neglectful of his duties as a tutor.

When he was offered the post as Professor of Applied Mathematics and Mechanics at the new Royal College of Science in Dublin, Ball left Birr Castle

---

[154] Ball V. (1915: 68).

[155] Cited by Ball V. (1915: 78).

8.18 Further Birr Castle observations

Figure 8.58. Robert Ball (1840–1913).

in August 1867. Seven years later he was appointed Astronomer Royal of Ireland. The title included the offices of Professor of Astronomy at Dublin University and Director of the University Observatory in Dunsink (succeeding Franz Brünnow). With the local 'South Equatorial' Ball determined star parallaxes.[156] The refractor's 11¾" Cauchoix objective originated from James South.[157] At Dunsink Ball recruited experienced observers from Birr Castle for his staff. His first assistant was Ralph Copeland, who was replaced by Charles Burton in 1876. Dreyer followed in 1878, leaving Dunsink in 1882 for Armagh.

In 1886 Ball was knighted. He stayed at Dunsink for 18 years and changed in 1892 to Cambridge University to follow John Couch Adams as Lownean Professor of Astronomy and Geometry and Director of Cambridge Observatory. In 1897–98 he was the President of the RAS.

Ball was famous for his popular books on science, among the *The Story of the Heavens* (London 1886)

and *Great Astronomers* (London 1895). In the latter he recognised the life's work of Lord Rosse, who had supported him at the beginning of his scientific career. Robert Stawell Ball stayed in his office until his death on 25 November 1913 at the age of 73.[158]

8.18.5 Ball's new nebulae

Between December 1866 and March 1867 Ball discovered 11 new objects with the 72-inch (Table 8.38). They are listed both in the GCS and in the 1880 publication. Seven of them are stars, the rest galaxies – a pretty bad rate. A peculiar case is NGC 3695 (GC 5564; Fig. 8.59). While observing John Herschel's h 899 (GC 2430, NGC 3694), Ball discovered the object on 31 March 1867, together with NGC 3700 (GC 5566). All three are galaxies in Ursa Major, correctly described as forming a triangle of extent about 10′. Dreyer inspected the field on 18 March 1876, but confused the issue. Ball's GC 5564 was interpreted as an object 4′ north of h 899: '*nnp is a pS, eeF neb. [=5565]*'; but this is a 14.8-mag star. He further noted that '*about 15′ n and a few minutes f is another eF, vS neb. [5565]*'. The 'new' object was catalogued as GC 5565 (NGC 3698) – but this is Ball's GC 5564 (NGC 3695). Dreyer did not recognise the galaxy GC 5566 (NGC 3700): the positions of this nebula and NGC 3695 are wrong in the GCS and NGC (the situation was not cleared up in the latter). The confusion might be explained by Dreyer's statement '*minima visibilia for the speculum now in use*'.

At Birr Castle, Ball and Lawrence Parsons initially used at the eye-piece a massive visual spectroscope made by Browning. Huggins had introduced this revolutionary method in 1864 (Huggins W. 1864). The 36-inch was used for this task, because it was easier to handle, especially in the 'front-view' configuration (i.e. without a secondary mirror). In 1868 Lawrence Parsons published a list of the objects that had been investigated objects.[159] M 42 (Orion Nebula) and the planetary nebulae M 97, NGC 3242 and NGC 7662 showed a '*gaseous spectrum*'; a '*continuous spectrum*' was determined for the galaxies M 31, M 65 and M 66. In the case of the galaxies M 98, NGC 3379, NGC 3384,

---

[156] Glass (1997: 29–32).
[157] In 1862 he had given the lens to Trinity College. It was part of an unfinished refractor, which had caused a hard conflict between South and the Reverend Sheepshanks (see Section 6.4.8).
[158] Obituaries: Anon (1913a), Anon (1914a), Rambaut (1914), MacPherson (1914b), Knobel (1915a) and Paterson (1916); see also Ball V. (1940).
[159] Parsons L. (1868: 71).

Table 8.38. *Ball's GCS objects*

NGC	GCS	Date	'GC'	P	Type	V	Con.	Remarks
308	5126	31.12.1866	172	13	Star		Cet	
310	5128	31.12.1866	172	13	Star	15.4	Cet	
397	5151	6.12.1866	212	15	Gx	14.8	Psc	
400	5153	30.12.1866	217	15	Star		Psc	
401	5154	30.12.1866	217	15	Star		Psc	
1742	5348	29.12.1866	965	46	Star	13.7	Ori	
2390	5385	10.12.1866	1527	58	Star	13.9	Gem	
2391	5386	10.12.1866	1527	58	Star	15.2	Gem	
3695	5564	31.3.1867	2430	97	Gx	14.0	UMa	NGC 3698, Dreyer 18.3.1876
3700	5566	31.3.1867	2430	97	Gx	14.1	UMa	Pair with NGC 3695
5601	5770	27.3.1867	3867	139	Gx	14.7	Boo	

Figure 8.59. The galaxy pair NGC 3695/3700 in Ursa Major (DSS) discovered by Ball.

NGC 3389 and NGC 3593, '*no decided spectrum seen; spectrum suspected to be continuous*'.

8.18.6 Short biography: Charles Edward Burton

Charles Burton was Ball's successor as the scientific assistant of Lawrence Parsons, staying at Birr Castle in 1868–69. The publication of 1880 contains about 50 observations by him, among them NGC 2248 (GC 1426) from the Markree catalogue.

Charles Burton was born on 16 September 1846 in Barnton, Cheshire. He was interested in astronomy from early in life. Throughout his short life he had health problems. From February 1868 to March 1869 he was an assistant of Lawrence Parsons at Birr Castle. Burton joined the 1874 Venus-transit expedition to the island of Rodrigues, analysing the data in Greenwich. In 1874 he became a Fellow of the RAS. In 1878 he was engaged by Robert Ball at Dunsink Observatory, but due to ill-health the work was often interrupted. Burton built telescopes with silvered glass mirrors of 6" to 15" diameter. He observed planets (especially the 'Mars channels') and photographed the Moon. Charles Burton died of a heart attack on 9 July 1882 in Loughlinstown near Dublin at the early age of 35.[160]

8.18.7 Short biography: Lawrence Parsons

In 1865 Lawrence Parsons took charge of the Birr Castle Observatory, becoming a keen observer and a worthy successor to his father. Lord Rosse had long since terminated active observing (there are many similarities to William and John Herschel). Lawrence Parsons mainly used the 72-inch to study nebulae. He was also interested in astrophysics and was responsible for many improvements to the telescopes.

Lawrence Parsons was born on 17 November 1840 at Birr Castle (Fig. 8.60). He was the eldest of Lord Rosse's sons (see Section 6.4.1). As such he bore the title of Lord Oxmantown, becoming the Fourth Earl of Rosse when his father died in 1867. He inevitably

---

[160] Obituaries: Erck (1882), Copeland R. (1882b) and Anon (1883a).

8.18 Further Birr Castle observations

Figure 8.60. Lawrence Parsons (1840–1908), the eldest son of Lord Rosse.

came into contact with astronomy, mechanics and optics very early (being aged five when the 72-inch was erected). Lawrence Parsons was taught at home by various tutors (J. Stoney, Mitchell, Gray and Purser). He later attended Trinity College, Dublin, graduating in 1864. On returning to Birr Castle at the age of 24, he soon started observing with both the 36- and the 72-inch. To supplement the 36-inch, he built an 18" reflector with focal length 10 ft in 1866. It was erected next to the 72-inch and used for micrometrical measurements of star fields. Its special feature was a water-powered drive constructed by Lawrence Parsons.

Lawrence Parsons' first scientific work dealt with the Orion Nebula. His first scientific assistant, Robert Ball, later wrote that *'Lord Oxmantown was an assiduous observer. Many a night did we spend together at the great telescope.'*[161] Another target was the Moon. The first studies started in 1868, assisted by Edward Burton. They succeeded in measuring the surface temperature with a thermal element at the eye-piece of the 36" reflector. Early results were published in 1870 (Parsons L. 1870). Meanwhile Lawrence Parsons made observations of nebulae with the 72-inch, first assisted by Copeland (1871–74), then by Dreyer (1874–78). He had constructed a simple drive for it, allowing more accurate micrometrical measurements: *'the micrometer has been more frequently used, and in this work the clock movement applied in the year 1868 has been found of great service'.*[162]

During the years 1874–76 Lawrence Parsons converted the 36-inch into an equatorial instrument.[163] In 1880 he published the Birr Castle observations compiled by Dreyer and covering the period from 1848 to 1878. When his keen assistant left in the latter year, Otto Boeddicker was engaged. This last scientific assistant stayed until 1916 at Birr Castle. However, no more observations were made with the 72-inch. During the periods 1884–88 improved measurements of the lunar temperature were made with the 36-inch. For this task, Lawrence Parsons later constructed a special instrument equipped with a short-focus searchlight mirror of 24" diameter. It was planned to work on the mounting of the 36-inch, but it never went into action.

In 1885 Lawrence Parsons was elected Chancellor of Dublin University. Later he became President of the Royal Dublin Society (1887–92) and the Royal Irish Academy (1895–1900). After a long illness, Lawrence Parsons died on 30 August 1908 at Birr Castle at the age of 67.[164] From the marriage with Frances Cassandra Harvey-Hawke in 1870, one daughter and two sons issued; the oldest, William Edward, became the Fifth Earl of Rosse.[165]

---

[161] Ball V. (1915: 68).

[162] Parsons L. (1880: 2).
[163] Claridge (1907), Repsold J. (1908), Vol. II, Fig. 165.
[164] Obituaries: Boeddicker (1908), Payne (1908), Anon (1908), Dreyer (1909) and Jeans (1909); Lawrence Parsons reached the same age as his father.
[165] The first names of the eldest sons (the later Earls of Rosse) traditionally alternated between William and Lawrence. After Lawrence Parsons had married, his mother, Mary Countess of Rosse, moved to London in 1870, where she died in 1885. The two women did not have good relations (Davison 1989: 54).

Figure 8.61. Hunter's drawing of the Orion Nebula (Ball R. 1886).

8.18.8 The Orion Nebula and Lawrence Parsons' discoveries

Lawrence Parsons' first object of study was the Orion Nebula. He collected all of the previous observations, including Bindon Stoney's drawing of the Huygenian region of 1852 (only an outline of the principal features is presented) and Samuel Hunter's excellent drawing of the whole nebula, which was finished in early 1864 (Fig. 8.61).[166] Robert Ball, following Hunter as assistant in 1866, wrote that it was *'an exquisite piece of work […] corrected or altered until accuracy was attained. Never before was so much pains bestowed on the drawing of a celestial object, and never again.'*[167] Lawrence Parsons contributed his own observations, which he made together with Ball. Altogether 93 new stars were found in the nebula. Even spectroscopic investigations at the 36-inch were made, using a visual spectroscope by Browning of mass about 30 kg. The result appeared in 1868 as 'An account of the observations on the great nebula in Orion, made at Birr Castle, with the 3-feet and 6-feet telescopes, between 1848 and 1867' (Parsons L. 1868).

The first to react was Lassell in April 1868. He had extensively studied M 42 on Malta with his 48″ reflector. In a letter to George Stokes, President of the Royal Society, one reads *'I have been so much interested by the perusal of Lord Oxmantown's observations and drawing of the Great Nebula in Orion […], that I venture to offer you a few remarks upon them – the more readily, as I may be supposed to be somewhat familiar with that object, though observed with less advantage of power.'* (Lassell 1868). However, many of Parsons' new stars could not be confirmed by Lassell, who added a list with his own micrometrical measurements.

The Fourth Earl and Ball also observed galaxies and planetary nebulae with the massive spectroscope. Moreover, they searched for new nebulae, mainly in the vicinity of Herschel objects or when checking earlier Birr Castle observations. In the GCS 28 objects are credited to Lawrence Parsons (Table 8.39); however, 5 of them had been discovered earlier by d'Arrest and Marth.

NGC 162, which was discovered on 13 October 1866 with the 72-inch, is a strange case involving d'Arrest and Schultz – with his mysterious object

---

[166] For the outlying portions of the Orion Nebula the 72-inch was used in 'front-view' mode. The object could be tracked for no longer than 50 minutes per night. In four years Hunter could use only five 'really good' nights and 12 'fair nights' for his observations.

[167] Ball V. (1915: 69).

Table 8.39. *The GCS objects of Lawrence Parsons*

NGC	GCS	Date	'GC'	P	Type	V	Con.	Remarks
2	6246	20.8.1873	1	7	Gx	14.2	Peg	GCS Addenda; Dreyer 29.10.1877
162	5107	16.10.1866	79	10	Star	15.0	And	d'Arrest 22.8.1862; Schultz 5.9.1867
375	5148	1.12.1874	202	14	Gx	14.5	Psc	
387	5149	10.12.1873	202	14	Gx	15.5	Psc	
399	5152	7.10.1874	217	15	Gx	13.6	Psc	
402	5155	7.10.1874	217	15	Star		Psc	
506	5171	7.11.1874	272	18	Star	15.1	Psc	
709	5195	18.11.1876	427	23	Gx	12.4	And	
710	5196	18.11.1876	427	23	Gx	13.5	And	d'Arrest 12.8.1863
714	5197	18.11.1876	427	23	Gx	13.2	And	d'Arrest 2.12.1863
717	5198	18.11.1876	427	23	Gx	14.0	And	d'Arrest 16.9.1865
1177	5296	29.11.1874	637	34	Gx	14.6	Per	
1274	5302	4.12.1875	675	35	Gx	14.1	Per	
1277	5304	4.12.1875	675	35	Gx	13.8	Per	GC 5305, Dreyer 12.12.1876
1608	5344	1.1.1876	860	42	Gx	13.8	Tau	NGC 1593, Marth 7.11.1863 (m 97)
2386	5384	1.1.1876	1527	58	2 stars		Gem	
2871	5478	7.3.1874	1845	76	Star		Leo	
2875	5479	7.3.1874	1845	76	GxP		Leo	NO-part of NGC 2874, W. Herschel 3.3.1786
3013	5505	18.3.1874	1944	79	Gx	15.2	LMi	
3950	5591	27.4.1875	2604	103	Gx	15.7	UMa	
3975	5593	21.2.1874	2627	105	Gx	15.3	UMa	
4912	5705	24.4.1865	3365	124	NF		CVn	
4913	5706	24.4.1865	3365	124	NF		CVn	
4916	5707	24.4.1865	3365	124	NF		CVn	
6002	5788	20.4.1873	4143	138	Star		CrB	
6974	5975	20.8.1873	4601	148	EN		Cyg	Middle part of Veil Nebula (with NGC 6979)
7326	6068	7.10.1874	4815	158	2 stars		Peg	
7388	6090	11.10.1873	4844	167	Star		Peg	

'GC 80' (see Section 7.1.7). NGC 709, NGC 710, NGC 714 and NGC 717 are members of the rich galaxy cluster Abell 262 in Andromeda – it is no wonder that the priorities were difficult to determine here. Expect for NGC 709 all of these galaxies were first seen by d'Arrest. NGC 1608 is identical with Marth's NGC 1593 (m 97). Both Marth and Lawrence Parsons had made position errors here.

NGC 2875 is a part of the galaxy NGC 2874, which was seen separately; NGC 2871 is a star (Fig. 8.62). GC 5304 (NGC 1277) is the same as Dreyer's GC 5305, which was found a year later. The identity was first mentioned in the NGC.

The case of NGC 4912, NGC 4913 and NGC 4916 is mysterious. In connection with his observation of the galaxy NGC 4914 (II 645, GC 3365) in Canes Venatici,

Figure 8.62. The galaxy trio NGC 2872–74 in Leo; Lawrence Parsons found NGC 2871 and NGC 2872 (DSS).

Lawrence Parsons had sketched three new objects. Unfortunately the related sky area is empty and even the broader environs shows no such constellation of four galaxies. Thus Lawrence Parsons' discoveries count as 'not found'. Probably he erroneously did not observe NGC 4914.

Concerning the GCS, 22 discoveries remain for the period 1865–76; 11 are galaxies. Interesting is the emission nebula NGC 6974 lying to the north between the two arcs of the Veil Nebula (supernova remnant) in Cygnus; William Herschel's NGC 6979 is connected with NGC 6974.

There are 16 additional objects of Lawrence Parsons, which later were catalogued in the NGC; they originate from the period 1873–78 (Table 8.40). One, the galaxy NGC 5279 in Ursa Major, is due to John Herschel. On 4 May 1831 he had observed William Herschel's II 798 (h 1665, GC 3639) in Slough. The object appeared to be a double nebula. Dreyer catalogued h 1665 as NGC 5278 and its fainter companion as NGC 5279 with the addition 'h 1665a'. On 2 May 1872 Lawrence Parsons saw a double nebula too. Therefore Dreyer also noted the Birr Castle observation in the NGC, but erroneously mentioning Lord Rosse. Altogether 27 objects must be credited to Lawrence Parsons; the non-stellar ones were 23 galaxies and 1 emission nebula. The rest are 10 stars and 3 missing objects.

8.18.9 Short biography: Ralph Copeland

From January 1871 to May 1874 Ralph Copeland was the scientific assistant of Lawrence Parsons at Birr Castle and a very active observer at the 72" reflector. His last target was the famous spiral galaxy M 51 (6 May 1874). In September 1874 Dreyer replaced him.

Ralph Copeland was born on 3 September 1837 at Moorside Farm near Woodplumpton, Lancashire (Fig. 8.63). In 1853, after finishing school in Kirkham, he moved to Victoria, Australia. He occupied himself with sheep farming, occasionally searched for gold and was impressed by the southern sky. In 1858 Copeland returned to England, hoping to study in Cambridge. It came to nothing and he worked in a locomotive factory in Manchester. A year later he married Susannah Milner, who died in 1866 (the two of them had a daughter). With some like minds he built up a small observatory in Manchester, which was equipped with a 5" Cooke refractor. Copeland's first publication, dated 1863, treats an occultation of κ Cancri by the Moon.

Copeland now tried to study astronomy in Germany – and was successful at Göttingen University. His professor was Wilhelm Klinkerfuess, Director of the University Observatory. Until 1867 he observed, together with Carl Börgen, stars between −1° and 0° declination with the meridian-circle by Reichenbach. The result (the Göttingen zone catalogue) was published in 1869. In the same year Copeland was awarded his Ph.D. for a work on the orbit of the double star α Centauri. Meanwhile he and Börgen prepared an expedition to Greenland, which started on 15 June 1869. Magnetical and meteorological measurements were planned.

Back in Germany, Copeland received an offer from Birr Castle, and in January 1871 he started his work as a scientific assistant of Lawrence Parsons. In December of that year he married Theodora Benfey, the daughter of a Professor of Oriental Studies in Göttingen.[168] At Birr Castle he measured the temperature of the Moon and observed nebulae with the 72-inch, equipped with a Grubb micrometer.

In April 1874 Copeland changed to Dunsink Observatory as an assistant of Robert Ball, but immediately was on tour again. The destination was Mauritius, where an expedition, led by Lord Lindsay, observed the transit of Venus. Copeland left Dunsink in the summer of 1876, because Lord Lindsay had engaged him as the successor of David Gill for his private Dun Echt Observatory in Scotland.[169] There he had excellent

---

[168] Three daughters and a son were born.
[169] About Dun Echt Observatory, see Brück H. (1992) and Chapman A. (1998: 133–135).

Table 8.40. *The NGC objects of Lawrence Parsons*

NGC	Date	'GC'	P	Type	V	Con.	Remarks
2846	4.4.1874	1829	75	2 stars		Hya	Dreyer 25.3.1878
2961	26.12.1873	1893	77	Gx	14.9	UMa	
3382	5.4.1874	2205	88	2 stars		LMi	Dreyer 24.3.1878
3410	1.4.1878	2222	89	Gx	14.2	UMa	
3889	1.4.1878	2555	102	Gx	14.8	UMa	
3999	25.4.1878	2641	105	Gx	14.7	Leo	
4000	25.4.1878	2641	105	Gx	14.5	Leo	
4870	1.4.1878	3365	124	Gx	14.6	CVn	
5279	2.5.1872	3637	133	Gx	14.0	UMa	h 1665a, J. Herschel 4.5.1831
5338	3.5.1877	3696	133	Gx	12.8	Vir	
5348	3.5.1877	3696	133	Gx	13.4	Vir	
5700	4.5.1877	3943	141	Gx	14.4	Boo	
5752	1.4.1878	3992	142	Gx	14.1	Boo	
5753	1.4.1878	3992	142	Gx	15.0	Boo	
5755	1.4.1878	3992	142	Gx	13.5	Boo	
7756	11.12.1873	5014	176	Star	12.8	Psc	

Figure 8.63. Ralph Copeland (1837–1905).

instruments at his disposal, e.g. the 15.1″ Grubb refractor.[170] Apart from observing, he was responsible also for building up the famous library of Lord Lindsay, which contains many rare historical books and manuscripts.[171] Additionally he edited the *Dun Echt Circular*, containing up-to-date information on astronomical events. The first circular appeared on 24 November 1879, the last (no. 179) on 29 January 1890. During the years 1881–84 Copeland was, together with his lifelong friend Dreyer, editor of the astronomical magazine *Copernicus*. Unfortunately, it had to be terminated after only three issues. At Dun Echt, a well-known centre of visual spectroscopy, Copeland often observed together with his German assistant Gerhard Lohse.

In October 1882 he travelled to Jamaica to observe the second transit of Venus. There he had the idea of high-altitude observations in the Andes. Using a 6.1″ Simms refractor from Dun Echt, equipped with a spectroscope by Vogel, he observed from 17 March to 27 June 1883 in the Peruvian highlands, mainly at Lake Titicaca (4200 m) and in Vincocaya (4800 m).[172] There Copeland discovered some planetary nebulae. On his way back to Scotland he visited major observatories in the eastern part of the United States (Copeland R. 1884b). In September 1883 he arrived in Dun Echt and started a study of bright nebulae with a new Cooke spectroscope. Copeland continued his work on the library of Lord

---

[170] Glass (1997: 69–76).
[171] On Lord Lindsay, see Anon (1914b).
[172] Holden (1892: 35).

Lindsay (now Lord Crawford), and in 1890 his 'Catalogue of the Crawford Library' was published, being particularly appreciated by Dreyer (Dreyer 1891a).

When Piazzi Smyth resigned as Astronomer Royal of Scotland in 1888, the Royal Observatory Edinburgh, located on the inner-city Calton Hill,[173] entered a state of crisis. Owing to air and light pollution the site had become worse and a relocation of the observatory was being considered. Lord Crawford intervened, offering to close Dun Echt and provide his instruments and library. The only condition was the preservation of the Royal Observatory – at a new and better place. This was accepted and Copeland was appointed as the successor of Smyth. In April 1899 he moved to Edinburgh. As the site of the new Royal Observatory, Blackford Hill, south of Edinburgh, was selected, and the relocation took place in May 1895. After further journeys Ralph Copeland's health deteriorated in 1901. Still in office, he died in Edinburgh on 27 October 1905 at the age of 68.[174]

8.18.10 Copeland's discoveries at Birr Castle

The GCS lists 18 objects discovered by Copeland (Table 8.41). NGC 203 is identical with NGC 211 of Stephan, who saw the galaxy three years later, giving a wrong position. In the 'Addenda' Dreyer notes for NGC 1062 (GC 6247) '*verified 1877*'. Anyway, no observation is listed in the publication of 1880. NGC 295 is missing. When Copeland observed NGC 296 (GC 167, II 214) he saw a new nebula 5' southwest and a 10-mag star 2' northeast. Dreyer catalogued it as GC 5123 (NGC 295). The description does not match NGC 296 and therefore Copeland must have observed a different field. A similar case – happening during the same night! – is NGC 930 (GC 5238), which is close to NGC 932 (II 489, GC 543). But here his measurements of the field stars are correct, with one exception: for the position of the new object (actually a knot in a spiral arm of NGC 932, Fig. 8.64) he noted '$3.1^s p, 42.0'' n$'. Instead of 'p' (preceding) it should read 'f' (following). Dreyer could not find the object on 13 January 1876.

The right identification of NGC 930 is due to Karl Reinmuth (Reinmuth 1926).

At Birr Castle Copeland discovered another 10 objects, which were catalogued in the NGC (Table 8.42). Except for NGC 81, all lie in the constellation Leo. The most prominent find was a compact group of interacting galaxies, known as the Copeland Septet (Arp 320). He first saw the group in spring 1874. The seven members cover an area of only 7' diameter (Fig. 8.65). Two other galaxies, NGC 3743 and NGC 3758 (which was also seen by Stephan in 1884), are 30' southwest and do not belong to the group. In connection with his observation on 13 April 1876, Dreyer explains why the group is not listed in the GCS: '*None of the above novae are included in the Suppl. to the G.C. as the star 8.9m has only a short time ago been identified by means of the star 9m 11' np.*' Copeland's reference star, described as 'reddish' by him, was mistaken by Dreyer. This led to the adoption of wrong coordinates for all of these galaxies in the NGC.

Hermann Kobold was the first to notice the error, writing about his observation with the 48.7-cm Merz refractor of Straßburg Observatory in 1894: '*After some searching, Copeland's group of nebulae was detected about 25' displaced from its NGC position.*' (Kobold 1894). He corrected the positions of the galaxies, curiously repeating Dreyer's error for the reference star! Dreyer finally clarified the matter in the same year, writing '*Setting with Lord Rosse's six foot reflector for d'Arrest's missing nebula GC. 2464 [NGC 3760], Dr. Copeland found on Febr. 9, 1874, a group of six nebulae somewhere in the neighbourhood.*' (Dreyer 1894b). For NGC 3760 d'Arrest had caused an error of 1^h in right ascension, which led to NGC 3760 = NGC 3301, as was found out by Kobold. On 18 March 1874 Copeland discovered another two galaxies nearby (NGC 3743 and NGC 3768) and determined their positions relative to his 'reddish' star (BD +22° 2380). In 1876 Dreyer observed the group. He used a 'very red star' (BD +22° 2385), thinking it was Copeland's, which actually lies 26.6' to the southwest. Therefore his coordinates were wrong by this amount. The error was eventually corrected by Dreyer in the IC I notes.[175] This, however, was carelessly overlooked by the authors of the first modern revision of the NGC, the *Revised New General*

---
[173] There he had access to a notable 21¼" refractor, which originated from Buckingham, who had erected it in 1862 in Walworth (Anon 1870b).
[174] Obituaries: Anon (1905c), Anon (1905d), Halm (1906), MacPherson (1906) and Dreyer (1906, 1907).

[175] Dreyer (1953: 282).

8.18 Further Birr Castle observations   313

Table 8.41. *Copeland's GCS objects*

NGC	GCS	Date	'GC'	P	Type	V	Con.	Remarks
85	5095	15.11.1873	38	9	Gx	14.8	And	
203	5113	19.12.1873	101	11	Gx	14.1	Psc	NGC 211, Stephan 18.11.1876
295	5123	26.10.1872	167	13	NF		Psc	
739	5199	9.1.1874	442	24	Gx	14.1	Tri	
760	5203	19.12.1873	455	24	2 stars		Tri	
930	5238	26.10.1872	543	29	GxP	14.5	Ari	Knot in NO-part of NGC 932
1062	6247	11.10.1873	599	32	Star	14.1	Tri	GCS Addenda
2363	5377	9.3.1874	1515	57	Gx	14.9	Cam	
2429	5391	10.3.1874	1554	60	Gx	13.8	Lyn	
2457	5396	10.3.1874	1577	62	Gx	15.4	Lyn	
2570	5415	20.2.1873	1650	67	Gx	14.3	Cnc	
2799	5467	9.3.1874	1788	73	Gx	13.7	Lyn	
3214	5525	9.3.1874	2082	84	Gx	14.4	UMa	
4072	5606	3.4.1872	2686	108	Gx	14.8	Com	
7486	6251	25.8.1871	4895	171	2 stars		Peg	GCS Addenda; Dreyer 3.12.1877
7766	6227	9.10.1872	5018	176	Gx	15.5	Peg	
7767	6228	9.10.1872	5018	176	Gx	13.5	Peg	
7819	6239	26.10.1872	5049	178	Gx	13.4	Peg	

Figure 8.64. Copeland's NGC 930, a knot in the galaxy NGC 932 in Leo (DSS).

Catalogue (RNGC), which was published in 1977. Consequently, Sulentic and Tifft could not localise the galaxy group for the *Palomar Observatory Sky Survey* (POSS). The embarrassing result is that all members of the Copeland Septet (except NGC 3746) are marked 'nonexistent' in the RNGC!

From Tables 8.41 and 8.42, Copeland's discoveries at Birr Castle (not counting NGC 930, this being part of NGC 932) amount to the following result: 23 objects are galaxies, 3 are stars and 1 is 'not found'. The final result must include the objects found later in Dun Echt and Peru (mostly by spectroscopy); they are listed in Section 9.14.

8.18.11 Dreyer's observations with the 72-inch

From August 1874 to August 1878 Dreyer worked at Birr Castle as the scientific assistant of Lawrence Parsons, following Copeland. He intensively used the 72-inch, equipped with a Grubb micrometer. Once a year

Table 8.42. *Copeland's NGC objects, among them the famous septet in Leo*

NGC	Date	'GC'	P	Type	V	Con.	Remarks
81	15.11.1873	38	9	Gx	15.7	And	
3743	18.3.1874	2454	99	Gx	14.4	Leo	
3745	5.4.1874	2454	99	Gx	15.2	Leo	Copeland Septet
3746	9.2.1874	2454	99	Gx	14.0	Leo	Copeland Septet
3748	5.4.1874	2454	99	Gx	14.8	Leo	Copeland Septet
3750	9.2.1874	2454	99	Gx	13.9	Leo	Copeland Septet
3751	5.4.1874	2454	99	Gx	14.3	Leo	Copeland Septet
3753	9.2.1874	2454	99	Gx	13.7	Leo	Copeland Septet
3754	5.4.1874	2454	99	Gx	14.3	Leo	Copeland Septet
3758	18.3.1874	2454	99	Gx	14.3	Leo	Stephan 18.3.1884

Figure 8.65. The Copeland Septet in Leo (DSS).

Lawrence Parsons informed the RAS about the progress made at the observatory. The reports for 1874–78 can be found in volumes 35–39 of the *Monthly Notices*.[176]

Dreyer's first observation took place on 12 September 1874; his target was the galaxy NGC 6926 (GC 4582) in Aquila. His last object, the galaxy NGC 5990 (GC 4135) in Serpens, was observed on 5 May 1878. The publication of 1880, which is a testimony to Dreyer's skill and accuracy concerning the identification and cataloguing of nebulae, lists 479 observations by him. Moreover, he wrote many comments about other objects (inserted within brackets).

In the GCS, Dreyer credits 12 objects to himself, but 4 of these had been found earlier by others (Table 8.43). NGC 372 is identical with d'Arrest's NGC 370 (GC 197, Au 5), a group of three stars found 1861. D'Arrest was the discoverer of NGC 2806 (GC 5469) too. Dreyer's GC 5305 is actually Lawrence Parsons' GC 5304, which had been found a year earlier. Later these two entries were combined as NGC 1277. NGC 2054 had already been discovered by Bond. NGC 3698 is Ball's NGC 3695; however, Dreyer's observation was 13 days earlier – it was his first discovery of a nebula. Dreyer's objects found in 1878 (Table 8.44) came too late for the GCS; they all appear in the NGC.

On 26 April 1878 Dreyer stirred up a hornet's nest: the rich galaxy group around NGC 4005, lying at the border of Leo and Coma Berenices. He discovered six new members (Fig. 8.66). The others are due to William Herschel (NGC 3987 and NGC 3993), John Herschel (NGC 3997), Mitchell (NGC 3989) and Stone (NGC 4005).

While observing NGC 1441 (GC 722) on 8 January 1877, Dreyer noticed '*perhaps a F neb. f*'. His sketch correctly shows the situation: the bright galaxy NGC 1441, d'Arrest's NGC 1449 and NGC 1451 and a trapezium of four stars (Fig. 8.67). The new object is missing; it was later catalogued by Dreyer in the NGC. Probably he was encouraged by Tempel's publication of 1882 (Tempel 1882).

---

[176] Parsons L. (1875, 1876, 1877a, 1878, 1879).

Table 8.43. *Dreyer's objects in the GCS*

NGC	GCS	Date	'GC'	P	Type	V	Con.	Remarks
372	5146	12.12.1876	202	14	3 stars		Psc	NGC 370, d'Arrest 7.10.1861
373	5147	12.12.1876	202	14	Gx	14.9	Psc	
1195	5297	8.1.1877	642	34	Gx	14.5	Eri	
1276	5303	12.12.1876	675	36	2 stars		Per	
1277	5305	12.12.1876	675	36	Gx	13.8	Per	GC 5304, L. Parsons 4.12.1875
1279	5306	12.12.1876	675	36	Gx	15.0	Per	
1281	5307	12.12.1876	675	36	Gx	13.8	Per	
2054	5354	13.1.1877	1227	51	4 stars		Ori	Bond 6.10.1850
2806	5469	22.3.1876	1792	74	Star	14.8	Cnc	d'Arrest 17.2.1862
3069	5513	15.3.1877	1975	80	Gx	14.2	Leo	
3698	5565	18.3.1876	2430	97	Gx	14.0	UMa	NGC 3695, Ball 31.3.1876
5226	6249	5.4.1877	3595	131	Gx	15.8	Vir	GCS Addenda

Table 8.44. *Dreyer's objects in the NGC*

NGC	Date	'GC'	P	Type	V	Con.	Remarks
1446	8.1.1877	772	39	Star	14.2	Eri	Tempel 1882; Fig. 8.67
4009	26.4.1878	2641	106	Star	13.4	Leo	Fig. 8.66
4011	26.4.1878	2641	106	Gx	14.7	Leo	Fig. 8.66
4015	26.4.1878	2641	106	Gx	13.5	Com	Fig. 8.66
4018	26.4.1878	2641	106	Gx	13.2	Com	Fig. 8.66
4021	26.4.1878	2641	106	Gx	14.8	Com	Fig. 8.66
4022	26.4.1878	2641	106	Gx	13.2	Com	Fig. 8.66
4023	26.4.1878	2641	106	Gx	13.7	Com	Fig. 8.66

Three known objects are described there, plus two new ones. Dreyer catalogued them as NGC 1443 and NGC 1446, identifying the former with his own find, though his position does not match Tempel's description. To sum up: of Dreyer's 20 objects catalogued in the GCS and NGC, 16 must be credited to him (13 galaxies and 3 stars).

In 1881 Dreyer wrote a brief description of the Birr Castle publication for *The Observatory* (Dreyer 1881c). There one reads that '*all the new nebulae which were found at Birr Castle in the course of years and formed the weak point of Herschel's General Catalogue, have now, by re-examination and comparison with d'Arrest's observations, been identified and their positions determined*'. Dreyer added that '*The work done during the last five or six years (1872–78) differs in many particulars from the observations taken in earlier years, and with which the paper of 1861 made the scientific world acquainted.*' This is shown by the '*micrometric measures of groups of nebulae or of nebulae and neighbouring comparison stars*'.

8.18.12 The further destiny of Birr Castle

Dreyer was succeeded by Otto Boeddicker, who worked at Birr Castle from 1880 to 1916. The 72-inch was no longer being used and fell into disrepair. Under the instruction of Lawrence Parsons, Boeddicker initially

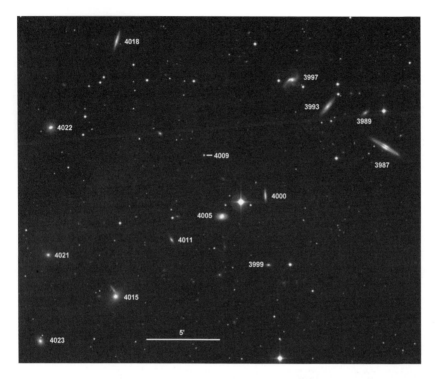

Figure 8.66. Dreyer discovered six galaxies in the NGC 4005 group in Coma Berenices (DSS).

measured the lunar temperature.[177] Then his target became the Milky Way, of which he made naked-eye drawings, which were later presented to the RAS (Boeddicker 1936).[178] In February 1916 Boedikker left Ireland to return to Germany. He died in 1937 in Berlin.

When in 1914 the great metal mirror was stored in the Science Museum, South Kensington, William Denning remarked 'So the active career of the mammoth telescope, which caused so much wonderment among men 70 years ago, has terminated in its becoming a museum curiosity!' (Denning 1914). He further wrote that 'It was supposed that its great penetrating power had resolved the nebulae, properly so called, but the spectroscope of Huggins showed this to be a misconception.' About the idleness of the telescope in Boeddicker's time, Denning remarked sarcastically that 'Its effectiveness as a working tool must have been regarded as small in the later period of its use, for Boeddicker, the observer in charge, spent several years in a naked-eye review of the Milky Way! Strange and certainly suggestive that anyone should

Figure 8.67. Dreyer's star NGC 1446 and the surrounding galaxies in Eridanus (DSS).

---

[177] Boeddicker (1886).
[178] The observations made during the years 1884–89 resulted in four large-format charts; see Boeddicker (1889), Parsons L. (1890) and Backhouse (1892). About Boeddicker's and other drawings of the Milky Way, see Ashbrook (1984: 373–379).

8.19 The structure of the GCS

Table 8.45. *Special objects and observational data at Birr Castle*

Category	Object	Discoverer	Data	Date	Remarks
Brightest object	NGC 4443	J. Stoney	Gx, 11.1 mag, Vir	13.4.1849	
Faintest object	NGC 2689	Mitchell	Gx, 16.3 mag, UMa	11.3.1858	Fig. 8.68 left
Most northerly object	NGC 2363	Copeland	+69°, Gx, 14.9 mag, Cam	5.3.1855	
Most southerly object	NGC 2847	Mitchell	−16°, GxP, Hya	9.3.1874	H II region in NGC 2848
First object	NGC 4110	W. Parsons	Gx, 13.8 mag, Com	1.4.1848	
Last object	NGC 4021	Dreyer	Gx, 14.8 mag, Com	26.4.1878	Fig. 8.68 right, Fig. 8.66
Best night		J. Stoney		13.3.1850	12 objects
Best constellation	Pisces	Various		1850–76	36 objects

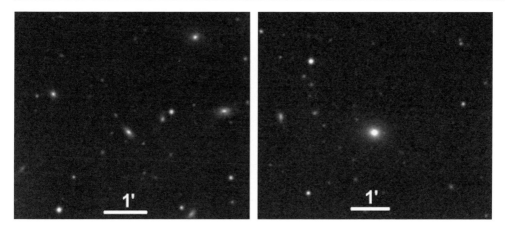

Figure 8.68. Remarkable Birr Castle objects. Left: the galaxy NGC 2689 in Ursa Major, the faintest; right: the galaxy NGC 4021 in Coma Berenices, the last to be discovered (DSS).

engage systematically in naked-eye studies with the greatest telescope in the world at his elbow!'

The final abandonment of the 72-inch in 1914 led Dreyer to write an obituary. He saw himself as *'the last survivor of the astronomers whose observations of nebulae with the six-foot reflector were published by the third and fourth Earls of Rosse'* (Dreyer 1914). In a sad mood, he wrote the following about the unused reflector: *'In 1878 [when I left Birr Castle] it was still in perfect order; and many questions, to solve which required great optical power, could have been dealt with with the 6-foot reflector, if work so long done with it – work in which several subsequently distinguished men had shared – had been continued.'*

Thanks to the efforts of the British astronomer Patrick Moore, the Sixth Earl of Rosse and local organisations, the Leviathan of Parsonstown could be completely restored in 1997. It received an aluminium mirror and electric motors for positioning. Today there is an astronomy club attending the 72-inch. Even public observations are offered – these are still a rare event, since the weather conditions have not improved since the days of Lord Rosse (Steinicke 2001d).

8.18.13 Special objects and brightness distribution

Table 8.45 shows special objects and observational data. The declination range used was pretty small; in particular, the near-pole regions were not observed due to the restrictions of the mounting. Johnstone Stoney

Table 8.46. *Galaxies of the NGC 383 group in Pisces discovered at Birr Castle*

NGC	Discoverer	Date	V	Remarks
373	Dreyer	12.12.1876	14.9	
375	L. Parsons	1.12.1874	14.5	
382	B. Stoney	4.11.1850	13.2	d'Arrest 26.8.1865
384	B. Stoney	4.11.1850	13.0	d'Arrest 12.10.1861
385	B. Stoney	4.11.1850	13.0	d'Arrest 12.10.1861
386	B. Stoney	4.11.1850	14.5	
387	L. Parsons	10.12.1873	15.5	
388	B. Stoney	4.11.1850	14.6	

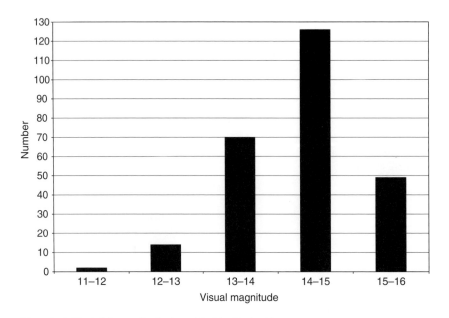

Figure 8.69. The brightness distribution of the Birr Castle objects.

is responsible for the most distant NGC object (see Table 10.17): the 15.5-mag galaxy NGC 2603 (GC 1667) in Ursa Major, which he found on 9 February 1850. In third place is Mitchell's NGC 870 (GC 514), a 15.5-mag galaxy in Aries. The objects are at distances of 799 and 795 million ly, respectively.

The Birr Castle observers discovered 11 NGC galaxies in the NGC 383 group in Pisces (Table 8.46). Most of them were found by Bindon Stoney on 4 November 1850; a few were added later by Lawrence Parsons and Dreyer. The first three were found by William Herschel on 12 August 1784: NGC 379 (II 215, 12.9 mag), NGC 380 (II 215, 12.6 mag) and NGC 383 (II 215, 12.2 mag).

The group is located 20′ northeast of the 6.4-mag star BD +31° 180 and has a diameter of 15′.

Figure 8.69 shows the brightness distribution of the Birr Castle objects. The mean is 14.3 mag, thus they are considerably fainter than those of d'Arrest, Stephan, Marth and Tempel. The 16.3-mag galaxy NGC 2689 is counted in the column '15–16'.

8.19 THE STRUCTURE OF THE GCS

8.19.1 Content and structure

The five-sided introduction treats the fundamentals and lists references. The next 14 pages contain 'Notes

8.19 The structure of the GCS

Table 8.47. *The columns of the GCS (main catalogue and 'Addenda')*

Column	Meaning
No. of Catalogue	GC-number (5079 to 6251)
No. in Marth's Catalogue	Number in Malta catalogue
Reference to other authorities	Other observers/discoverers (except Marth)
Right ascension for 1860, Jan. 0	Right ascension; accuracy 1^s
Annual precession for 1880	Precession in right ascension
North polar distance for 1860, Jan. 0	N.P.D. = angle to pole $(90° − δ)$; accuracy $0.1'$
Annual precession for 1880	Precession in N.P.D.
Summary description	Description (Herschel-style), remarks
No. of observations	Number of observations (sev = several)

and corrections to the General Catalogue of Nebulae and Clusters of Stars'. Altogether 446 GC objects are mentioned. Additionally, references for drawings of 67 objects that are not catalogued in the GC are presented. Some notes are given here, which, for instance, direct the reader to the drawings of Tempel (Arcetri) and Trouvelot (Harvard). The catalogue section covers 24 pages and contains 1172 entries. Dreyer has continued the GC-numbering, though the right ascension starts again at 0^h. The arrangement (columns) and the epoch (1860) follow the GC (Table 8.47).

The main catalogue, which lists 1166 objects (GC 5080 to 6245), is followed by 'Addenda' with six objects (GC 6246 to 6251), which were added in proof and separately sorted by right ascension. The final object (GC 6251 = NGC 7486), a pair of stars in Pegasus, was discovered by Copeland on 25 August 1871. On this occasion Dreyer added notes to a further 21 GC objects and corrected two errors in the main catalogue. He also mentions the upcoming first part (0^h to 8^h) of the 'Observations of nebulae' at Birr Castle, '*which were prepared for publication, and sent to the Royal Dublin Society after this paper* [GCS] *had been laid before the Academy, and while it was in the press (February 1878)*'.[179] According to this, the last corrections of the GCS were made in early 1878 – the catalogue appeared in March.

Marth's objects discovered on Malta are given a separate column – recognising him as the most important contributor to the GCS. Indeed, Marth ranks first with 588 objects (Table 8.48), followed by d'Arrest (232) and Stephan (147). The discoverers are given in the column 'References to other authorities'. Sometimes more than one person is mentioned there, which induces a certain priority (correctly given by Dreyer in most cases). The Birr Castle discoveries are divided into 'R nova', subsuming all observers of the 1861 publication ('*Phil. Trans.* 1861' is used too); 'R_2 nova' (Lawrence Parsons); 'R_2 nova B' (Ball); 'R_2 nova C' (Copeland); and 'R_2 nova D' (Dreyer). Explicit references are given only for Stephan (St I to St VIII, explained in the introduction) and Borrelly. For Otto Struve the discovery year is always given, for Schmidt only occasionally.

The collection in Table 8.48 is the result of a modern analysis (the sources are given in the sections on the individual observers). 'Dreyer' shows the cases (and frequency) for which the GCS gives a different discoverer. For GC 5619 (NGC 4118) Dreyer's 'R_2 nova' must read 'R nova'. Since Lassell's two objects are contained in the Malta catalogue, Dreyer mentions the m-number only. Instead of 'Schweizer' the source is presented. Italics in the column 'Period' indicate that many objects should already have been included in the General Catalogue. Their omission was mainly due to late publication (after October 1863); Marth is the best example. But, in some cases, John Herschel simply overlooked objects (e.g. NGC 4383, which had been found by 1862 by Schönfeld).

Dreyer recognised that some GCS objects had already been listed in the catalogues of William and John Herschel. For example, GC 5684 (NGC 4840) is called II 385 in the GCS; John Herschel had overlooked this galaxy in Coma Berenices. The other identities result from a modern analysis (Table 8.49).

---
[179] Dreyer (1878a: 426).

Table 8.48. *The discoverers of the GCS objects (italics in 'Period' denote discovery before the GC)*

Observer	Main Cat.	Addenda	Period	Dreyer	Remarks
Ball	11		1866–67		R$_2$ nova B
Bond	2		*1850–53*	Dreyer (1)	
Borrelly	6		1871		
Coggia	–	1	1877		No source
Coolidge	1		*1859*		
Copeland	16	2	1872–77		R$_2$ nova C
d'Arrest	232		*1855–67*	Marth (6), Schultz (1)	
Dreyer	11	1	1876–77		R$_2$ nova D
Dunér	1		1867		
Dunlop	1		*1826*	Bond (1)	
Ferrari	14		1865	Secchi (16)	
Herschel W.	10		*1784–1802*	d'Arrest (5), Stephan (3), Stoney J. (1)	
Lassell	2		*1863*		m 126, m 523
Marth	588		*1863–65*		
Mitchell	7		*1854–57*	d'Arrest (3), Schultz (2)	PT 1861
Parsons L.	22	1	1865–76		R$_2$ nova
Pechüle	1		1876?		No source
Peters	1		1865	Struve O. (1)	
Rümker	1		1864		
Schmidt	16		*1860–67*	Marth (3)	
Schönfeld	2		*1862*		
Schultz	15		*1863–70*		
Schweizer	1		1860		Obs. de Moscou
Stephan	147		1870–77		
Stoney B.	8		*1850–52*	d'Arrest (3), Marth (1), Schultz (1)	PT 1861
Stoney J.	4	1	*1849–50*		PT 1861, R nova
Struve O.	15		*1851–69*	d'Arrest (1)	
Tempel	24		1875–77		
Tuttle	1		1875	Holden (1)	See Section 8.17
Voigt	2		1865	Stephan (2)	
Winnecke	4		1868–74		

There are even two identities in the GCS itself. The first case is GC 5304 = GC 5305 (a galaxy in the Perseus Cluster). Lawrence Parsons discovered GC 5304 on 4 December 1875 and Dreyer found it again on 12 December 1876. Later he noticed the identity, combining the entries into NGC 1277. The second case is GC 6019 = GC 6020; Stephan had discovered the object twice, on 23 July 1870 and in 1872. Dreyer had already suspected the identity in the GCS ('*are these identical?*'); it is a galaxy in Pegasus (NGC 7190).

After subtracting the 21 objects in the catalogues of the two Herschels and the two identities, the GSC contains 1149 independent objects. Most of them are galaxies (1012); 2 are parts of galaxies. There are 7 emission nebulae, 6 reflection nebulae, 6 planetary

8.19 The structure of the GCS   321

Table 8.49. *GCS objects that had already been listed in the catalogues of William and John Herschel*

GCS	NGC	H	h	GC	Discoverer	Dreyer
5086	20			5	Mitchell	Schultz
5096	90			40	Mitchell	Schultz
5097	91			41	Mitchell	d'Arrest
5098	93			42	Mitchell	d'Arrest
5200	751			456	B. Stoney	
5247	972	II 211	232	560	W. Herschel	Stephan
5343	1601			867	J. Stoney	d'Arrest
5360	2194	VI 5		1383	W. Herschel	d'Arrest
5369	2290	III 897			W. Herschel	B. Stoney
5512	3066	II 334	655	1971	W. Herschel	d'Arrest
5639	4330			2909	B. Stoney	d'Arrest
5642	4352	II 64	1227	2929	W. Herschel	d'Arrest
5653	4466			3022	B. Stoney	d'Arrest
5668	4669	III 778			W. Herschel	d'Arrest
5684	4840	II 385			W. Herschel	
5704	4908	III 363			W. Herschel	d'Arrest
5762	5538			3830	B. Stoney	Marth
5832	6239	III 727		4247	W. Herschel	Stephan
5870	6426	II 587		4325	W. Herschel	Stephan
6151	7549			4912	B. Stoney	d'Arrest
6153	7553			4913	B. Stoney	Schultz

Table 8.50. *GCS objects not included in the NGC*

GCS	Discoverer	Date	Star	Con.	Remarks
5859	Schultz	30.9.1863	9.8	Her	Star near M 92; see Section 8.16.2
6000	Bond	1853	9.0	Aqr	Bond: 'planetary?'; Copeland 1879 (BD +0° 4741)
6195	Schultz	22.7.1865	8.2	And	'Nebula' around BD +41° 4780; see Section 8.16.4

nebulae and 4 open clusters; 87 cases are stars or star groups and 25 are 'not found'. Only three GCS objects (stars) were not taken over into the NGC (Table 8.50).

Bond's GC 6000, which had been discovered in 1853 and marked 'planetary?' (see Fig. 8.4), was investigated by Ralph Copeland in Dun Echt with a spectroscope. In 1879 Lord Lindsay published a note 'Observation of General Catalogue (supplement) no. 6000', writing '*At the suggestion of Mr. Dreyer, it was observed prismatically at Dun Echt on December 10* [1879] *by Ralph Copeland. The spectrum is quite continuous and offers no peculiarity;* *the object is therefore not a nebula at all. It is identical with D.M. +0° No. 4741, 8.8 mag.*'[180]

8.19.2 Copeland's review

The only review of the GCS was written by Dreyer's close friend, Ralph Copeland. After a short presentation of the catalogue structure, he wrote that '*To many the most interesting part of the work will be the notes*

---

[180] Lindsay J. (1879c); D.M. means 'Durchmusterung'.

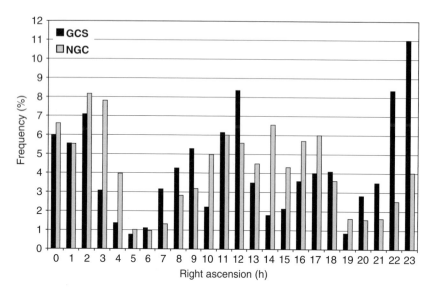

Figure 8.70. Object frequencies by right ascension for the GCS (discoveries until 1877) and NGC (later discoveries).

and corrections to the general catalogue. A comparison of these notes with Sir John Herschel's shows that out of 370 notes, some 73 are amplifications, confirmations or refutations of the notes in the General Catalogue.' (Copeland R. 1878).

Copeland had taken the trouble to sort all objects by their discoverer and right-ascension hour. He differentiates cases in which several observers are mentioned. He noticed that *'D'Arrest was most successful in finding new nebulae in 11^h and 12^h of R.A., no less than 112 having been found by him in these two hours while the region 16^h to 19^h inclusive yielded but 4 nebulae.'* The reason is *'the long twilight of the Copenhagen summer'*. On the other hand, Marth and Stephan were successful in regions neglected by d'Arrest. At Birr Castle no nebulae were found between right ascensions 5^h and 6^h. Copeland knows the reason from his own experience: when this region was in the meridian (December to February), it was *'devoted in considerable measure to the great Nebula in Orion'*. Copeland recognised the accumulations at 0^h and (opposed) 12^h, a trend already visible in the GC.

Owing to the distribution of objects in the GCS Copeland advised the following: *'it would seem to be desirable that the users of large telescopes in the low or southern latitudes should, for a time at least, give the preference to regions near 18^h of R.A. in their search for nebulae'*. As Fig. 8.70 shows, Copeland's desire was only partly realised by the NGC observers. There, an increase is visible mainly between 14^h and 17^h, whereas the interval 22^h to 23^h shows a decrease. Finally Copeland voices his great admiration for Dreyer's work: *'The whole volume evinces the most scrupulous and painstaking accuracy, all cases of doubtful identity being clearly pointed out, thus rendering the work in proportion to its extent fully as suggestive as the great catalogue it so well supplements.'*

9 · Compilation of the New General Catalogue

On coming to Dunsink Observatory in 1878, Dreyer was tasked with the completion of the Birr Castle publication. The work was finished in early 1880, whereafter he had time to focus on newly discovered nebulae. The cataloguing was done in parallel with his regular work as the assistant of Robert Ball, who was familiar with this issue, due to his own stay at Birr Castle. From 1882 Dreyer, as Director of Armagh Observatory, had complete freedom to work on a new catalogue of nebulae and star clusters.

The result was presented to the RAS in 1886 as a 'second supplement' to the General Catalogue, a follow-up of the GCS. It was surprising for Dreyer to learn that the society had refused his new work. However, the reason was honourable: the RAS favoured a 'new general catalogue' instead. It should contain all known non-stellar objects, thus being a complete new edition of Herschel's GC, which it had published in 1864. Since the bulk of the work had already been done, this was not a difficult task for Dreyer – it was carried out in Armagh within a year! The NGC contains a large number of new objects (as of December 1877) and became his most important (and most popular) work.

9.1 DREYER'S UNPUBLISHED 'SECOND SUPPLEMENT'

After the publication of the GCS (1878) and the Birr Castle observations (1880), Dreyer recurrently scanned the astronomical journals and observatory publications for new nebulae. In 1881, still in Dunsink, a first report ('Nebulae') appeared in the magazines *Copernicus* and *The Observatory* (Dreyer 1881a). The article describes observations of Burnham, Copeland, Stephan, Tempel, Winnecke and Baxendell. Moreover, Pickering's spectroscopic search for planetary nebulae is mentioned. In a subsequent paper ('Nebulae and clusters'), Dreyer treats Stephan's observations (as published in *Comptes Rendus*) and mentions new nebulae of Peters, Burnham, Block and Webb (Dreyer 1881c).

In the middle of 1885 a letter by Johann Holetschek, observer at Vienna Observatory, appeared in the *Astronomische Nachrichten* (Holetschek 1885). He points out the necessity of 'supplements' to the published standard catalogues: '*Since, due to newly discovered objects, every general catalogue of nebulae, double stars, red stars etc. loses its original completeness very soon, it happens that already a few years after the appearance of such a catalogue, not quite a new one, but at least a supplement is necessary, and the lack of it becomes more noticeable every year.*' Holetschek explicitly appreciates Dreyer's supplement to John Herschel's General Catalogue. He mentions too that '*assistant Spitaler has compiled a list of all nebulae found recently*'.

Dreyer, who was virtually the only person to catalogue nebulae and star clusters after John Herschel, immediately replied in a 'Notice of a new catalogue of nebulae' (Dreyer 1885a). He wrote '*I ought to announce that I have been for some time collecting materials for a second supplementary catalogue, which I hope to have printed before very long.*' He took this opportunity to request that unpublished discoveries be sent (a successful method, as the GCS showed). It seems that Dreyer was pressured by Holetschek's note. He was unwilling to let the priority of cataloguing nebulae out of his hands, fearing competition, e.g. by Rudolf Spitaler in Vienna.

Therefore, he worked at full speed on a new compilation. On 12 May 1886, shortly before it was finished, Dreyer wrote a letter to the Royal Astronomical Society, asking whether a second supplement to the GC ('*a catalogue of all the Nebulae discovered since 1877*') could be published in the *Memoirs* of the RAS.[1] He proposed that he attend the meeting on 14 May to discuss the issue. He was present, but, strangely, reported 'On

---

[1] RAS Letters 1886, Dreyer, 12 May.

the proper motions of 29 telescopic stars' – the minutes of the meeting do not mention a second supplement (Dreyer 1886).

However, on 8 December 1886 Dreyer sent his 'Second supplement to Sir John Herschel's General Catalogue of Nebulae and Clusters of Stars' to the RAS. In the covering letter the work is described as the '*continuation of a similar paper published nearly nine years ago* [GCS], *which I believe has been found useful by observers of Nebulae*'.[2] Again he requested publication in the *Memoirs*, 'as both Herschel's Catalogue and my former Supplement were printed in quarto form'. However, this did not happen. Dreyer's work was announced in the RAS meeting on 10 December 1886, but he was not present and there is no discussion recorded (Dreyer 1887a). Strangely, his manuscript is mentioned in the RAS files as 'No. 1073', but the archive does not contain it. Probably Dreyer picked it up at a later time.

Just because Dreyer's second supplement was not published, it was essential for the creation of the New General Catalogue. In the introduction of the NGC he wrote that '*In December 1886 I submitted to the Council of the Royal Astronomical Society a second supplementary catalogue arranged exactly like the first one. But considering the circumstance that Herschel's work is practically out of print, and that the simultaneous use of three catalogues and two copious lists of corrections would be very inconvenient, the Council proposed me to amalgamate the three catalogues into a new General Catalogue. I agreed to do so, and have adopted the following plan in compiling the present work.*'[3] Obviously, (undocumented) negotiations between Dreyer and the RAS board had taken place, as a result of which he had been persuaded to compile a completely new catalogue.

As little as half a year later, at the RAS meeting on 10 June 1887, it was announced as 'A new general catalogue of nebulae and clusters of stars, being the catalogue of the late Sir F. J. W. Herschel, revised, corrected, and enlarged' (Dreyer 1887b). This shows how far Dreyer had already got with his work in 1886 – obviously, it was only a small step from the second supplement to the NGC. Most of the NGC objects (mentioned in the following sections) must already have been contained in Dreyer's second supplement.

9.2 STAR CHARTS AND NEBULAE OF PETERS

Peters is known for his 'Celestial charts', which he made at Litchfield Observatory. Until 1882 he published 20 charts (of 182 planned). Unassisted, the huge project outgrew him. During his observations, Peters discovered some nebulae, which were mainly published in Dreyer's magazine *Copernicus*.

9.2.1 Short biography: Christian Heinrich Friedrich Peters

Peters[4] was born on 19 September 1813 in Coldenbüttel (in the Duchy of Schleswig) as a son of the local parish priest (Fig. 9.1). During the years 1825–32 he attended the secondary school in Flensburg. After that, he studied mathematics and astronomy in Berlin, receiving his Ph.D. in 1836 at the early age of 23. Peters went to Göttingen Observatory, working for Gauß. In 1838 he moved to Sicily, where he surveyed Mount Etna until 1843. Then he occasionally worked for Ernesto Capocci at Capodimonte Observatory, Naples (at that time Naples was the capital of the Kingdom of the

Figure 9.1. C. H. F. Peters (1813–1890).

---

[2] RAS Letters 1886, Dreyer, 8 December.
[3] Dreyer (1953: 2).

[4] Often called C. H. F. Peters, he should not be confused with Christian August Friedrich Peters (C. A. F. Peters) or Carl Friedrich Wilhelm Peters (C. F. W. Peters).

Two Sicilies).[5] There he discovered comet 1864 VI and observed sunspots with a 3.5″ refractor. In 1848 Sicily revolted against the tyrannical misrule of King Ferdinand II of Naples, following which Bourbon troops under General Filangieri invaded and occupied the island. Peters, who had sided with the uprising against the government, was forced to flee to France. After an unsettled time with much travelling, he eventually, on the recommendation of Alexander von Humboldt, moved to the United States in 1854.

After having been in Cambridge and Washington, Peters joined the Dudley Observatory, Albany, in 1856 to assist Benjamin Gould. There he discovered the comet 1857 IV. In 1858 Peters became the Director of Hamilton College Observatory, Clinton (NY) and Professor of Astronomy. There a new 13.5″ refractor by Spencer was erected in a 6-m dome. When the railway tycoon Litchfield took over the financing of the observatory in 1867, it was renamed 'Litchfield Observatory'. There Peters discovered 48 minor planets (being beaten in this score only by Palisa), the first on 29 May 1861 (Feronia) and the last on 25 August 1889 (Nephthys). The success was strongly connected with the making of the 'Celestial charts', showing stars of up to 11 mag along the ecliptic. Twenty charts appeared until 1882. Peters regularly recorded sunspots and did research on old star catalogues, an immense work, which unfortunately remained unpublished due to a severe conflict with his assistant Charles Borst.

Another fight was carried out with James Watson, Director of Michigan University Observatory, Ann Arbor. It concerned the supposed existence of the intramercurial planet 'Vulcan'. Watson and Lewis Swift claimed to have seen it in 1878.[6] Peters was a virulent opponent of this idea – and was proved right in the end. In 1879 he became a Fellow of the RAS. On 19 July 1890 at 7 a.m. C. H. F. Peters was found dead on his way from the observatory to his house. He had obviously died of a heart attack in the night, being 77 years old.[7]

9.2.2 First observations of nebulae and the 'Celestial charts'

A first note about Peters' observations of nebulae appeared in late 1856 in the *Astronomical Journal* (Peters C. H. F. 1856). At that time he was employed as assistant at Dudley Observatory. Seven 'new' objects, which had been observed by Peters in Naples (Capodimonte), probably with the local 17.6-cm Fraunhofer refractor, are listed. Except for one, all are located in the southern sky. From the positions and descriptions, their identity can easily be cleared up – all are known objects. The first six are contained in the catalogues of Messier, William and John Herschel: NGC 5986, M 92, NGC 6441, NGC 6553, M 69 and NGC 6723. The last object is the planetary nebula NGC 7293 in Aquarius, which had been discovered by Harding (see Section 4.1.2).

Peters' main work was the 'Celestial charts'. In about 1860 he started to observe stars near the ecliptic at Hamilton College Observatory, using the 13.5″ refractor at a power of 80 (Fig. 9.2). His model was Chacornac's charts (at the same scale), which first supported his search for minor planets. Later his own charts were used immediately after their drawing. They

Figure 9.2. The 13.5″ refractor of Litchfield Observatory (Hamilton College, Clinton).

---

[5] Peters wrote an obituary for Capocci (Peters C. H. F. 1864).
[6] Eggen (1953a), Baum and Sheehan (1997); see also Section 9.5.1.
[7] Obituaries: Krüger (1890), Knobel (1890), Porter (1890), Anon (1891c) and Anon (1891d); see also Sheehan (1998) and Ashbrook (1984: 56–66).

Table 9.1. *Objects discovered by Peters*

NGC	Date	Type	V	Con.	Remarks	Ref.
787	27.2.1865	Gx	12.9	Cet	Tempel 9.11.1879; Fig. 9.3 left	2, 3
3054	3.4.1859	Gx	11.4	Hya	Stone 14.1.1886; Fig. 9.3 right	2
3153	1880?	Gx	12.8	Leo	W. Herschel 19.3.1784 (III 53); Todd 5.2.1878	1
3328	27.3.1880	2 stars		Leo	Tempel 21.5.1879	1
3492	1880?	Gx	13.2	Leo	No reference (list I?)	1
4182	1881?	Star	11.5	Vir		1
4255	1881?	Gx	12.7	Vir	Voigt June? 1865	1
4276	1881?	Gx	12.7	Vir		1
4307	1881?	Gx	12.1	Vir	Tempel 17.3.1883	1
4309	1881?	Gx	12.8	Vir	No reference (list I?)	1
4347	5.5.1881	2 stars		Vir		1
4353	1881?	Gx	13.6	Vir		1
4411	1881?	Gx	12.3	Vir		1
4604	1881?	Gx	13.8	Vir	No reference (list I?)	
6389	7.7.1878	Gx	12.1	Her	W. Herschel 29.6.1799 (II 901)	4
6481	21.8.1859	Star group		Oph		2
6797	1860?	2 stars		Sgr		1
7134	1860?	Star group		Cap		1
7453	7.11.1860	3 stars		Aqr		1, 2
7688	13.8.1865	Gx	14.5	Peg	GC 6206, O. Struve 12.10.1865	2

contain stars from 6 to 11 mag (some of up to 14 mag); each shows an area of 20^m in right ascension and 5° in declination (the positions refer to the epoch 1860).

Using photolithography, Peters published in 1882 – at his own expense – 20 charts, titled 'Celestial charts made at the Litchfield Observatory of Hamilton College'. The set covers regions in the constellations Leo, Virgo, Libra, Capricorn, Aquarius and Pisces. Shortly thereafter a review appeared in the *Monthly Notices* (Anon 1883b). Unfortunately, no further charts (of the planned 182) appeared. Obviously the influence of photography became too strong. Peters himself was involved in the international project 'Carte du Ciel', which was started in 1887 in Paris.[8] Anyway, at least 25 additional charts must have existed, because Peters later gave the drafts to Holden, who used them for comparison with the work of Chacornac and others. He mainly was interested in determining the number and distribution of stars (Holden 1884b). On Peters' charts he counted altogether 112 338 stars (the published 20 show 43 928).

9.2.3 Peters and *Copernicus*

Peters published two lists with 'Positions of nebulae' (Refs. 1 and 2 in Table 9.1) in the first two issues of the astronomical magazine *Copernicus*, edited by Dreyer and Copeland (Peters C. H. F. 1881c, 1882). The objects presented there were a mere by-product of his observations for the 'Celestial charts' with the Spencer refractor (Table 9.1). The first list contains 131 objects, sorted by right ascension and identified by GC-number (including GCS). The positions, which were determined graphically from the charts, refer to the epoch 1860. Peters estimated the error of this method to be $1/3'$ for small nebulae. The objects discovered lie in three areas along the ecliptic: $\alpha = 1^h$, $\delta = 0°$ to $+15°$; $\alpha = 10^h$ to 12^h, $\delta = -1°$ to $+15°$; and $\alpha = 21^h$ to 22^h, $\delta = -20°$ to $-6°$. The column 'Notes' contains identifications (e.g. Marth, Stephan) and corrections to GC/GCS positions. Dates are not given (except for

---

[8] Mouchez (1885), Ashbrook (1984: 436–440), Urban and Corbin (1998), Lamy (2008).

Figure 9.3. Two of Peters' galaxies. Left: NGC 787 in Cetus; right: NGC 3054 in Hydra (DSS).

NGC 3328). Eleven objects, which were not found in the catalogues, are called 'Nova'.

Peters' second list contains 34 objects along the ecliptic. Now an observation date is given: 18 observations were made in 1859 (the first on 3 April 1859), 15 during the years 1860–65, 1 in 1878 and 8 in 1880–81 (the last on 23 May 1881). About the reason for creating two lists, Peters wrote '*Before continuing that series I have thought it well to publish a short list of nebulae determined occasionally, either by the filar micrometer or by the ring micrometer, or simply after the manner of d'Arrest by putting the nebula and comparison star into the centre of the field and reading off the circles.*' Two objects are in both lists: NGC 7393 (GC 4851) and NGC 7453; the latter is one of four termed 'Nova' in list II.

The discovery of NGC 787 had already been announced by Peters in January 1881 (Peters C. H. F. 1881b = Ref. 3). This galaxy (Fig. 9.3 left) was detected while observing NGC 731 (III 266) and NGC 755 (III 265). Another find was reported even earlier (Peters C. H. F. 1881a). While observing comet Cooper he encountered a '*faint nebulous object*' in early January 1881. A later data reduction revealed '*that it was a Herschel nebula*'. Peters identified it with GC 574 (I 102 = NGC 1022). This galaxy in Cetus is not included in Table 9.1.

There are no sources for three of Peters' discoveries (NGC 3492, NGC 4309 and NGC 4604). According to their positions, they fit well into the first list. Dreyer, who included all of these objects in the NGC, was probably informed by letter.

The discovery date of the 'Nova' NGC 4347 from list I is derived from a remark by Peters in his second publication. For 11 cases a date can only be estimated. For those from list I (including the three without a reference) the date might have been after the making of the 'Celestial charts', since their positions were determined graphically. The objects from list II, with positions measured with a bar- or ring-micrometer, originate from 1859, 1860 and 1865.

Four objects had been discovered earlier by other observers. Tempel found NGC 3328 on 21 May 1879 with the 11″ refractor in Arcetri. About NGC 3153, which was first seen by William Herschel (III 53 = h 677), Peters wrote '*AR in G. Cat. differs about 15^s, as h. 677; but H. III. 53 agrees better*'. He struggled with inexact data from the two Herschels. NGC 4255 was discovered in 1865 by Voigt in Marseille. The galaxy NGC 6389 seen by Peters while observing comet 1878a on 7 July 1878 (Peters C. H. F. 1878 = Ref. 4) is actually William Herschel's II 901. Dreyer mentions Peters as the co-discoverer of the galaxies NGC 475 (GC 5166) in Pisces and NGC 4116 (GC 2728) in Virgo. About the former, Peters noted in list II 'Marth 43'; for the latter a GC-number is given. Both objects are therefore excluded from Table 9.1. Finally, Peters must be credited with 16 objects (marked in bold); 10 are galaxies, the rest are stars or star groups.

9.3 TEMPEL'S NEW NEBULAE AND A CONTROVERSIAL TREATISE

Around 1880 Tempel had settled down in Arcetri, coming to terms with the straitened circumstances. The observatory was somewhat better equipped now and the

Table 9.2. *Tempel's references in the NGC*

T	Original title	Publication (*AN*)	Ref.	Date	E
I	Schreiben des Herrn W. Tempel and den Herausgeber	[2212] **93**, 49–62 (1878)	1878d	May 1878	56
II	Beobachtung des Brorsen'schen Cometen und eines neuen Nebelflecks	[2253] **94**, 335 (1879)	1879a	23.5.1879	
III	Schreiben des Herrn W. Tempel an den Herausgeber	[2284] **96**, 61–64 (1880)	1880b	28.10.1879	
IV	Beobachtung, angestellt auf der Sternwarte Arcetri bei Florenz	[2347] **98**, 299–304 (1881)	1881a	29.10.1880	14
V	Neue Nebelflecke, aufgefunden und beobachtet auf der Sternwarte zu Arcetri	[2439] **102**, 225–238 (1882)	1881b	May 1882	31
VI	Einige neue Nebel und ein Nebelnest	[2511] **105**, 235–238 (1883)	1883a	26.4.1883	8
VII	Beobachtung von Cometen und Nebelflecken, angestellt auf der Sternwarte Arcetri	[2522] **106**, 27–30 (1883)	1883b	10.6.1883	
VIII	Notiz über einige neue und ältere Nebel	[2527] **106**, 107–110 (1883)	1883c	11.7.1883	10
IX	Neue Nebel	[2660] **111**, 315 (1885)	1885b	25.3.1885	10
X	Notizen über verschiedene neue Nebel	[2691] **112**, 47 (1885)	1885d	23.10.1885	

productivity increased. Besides the 45-mm eye-piece, giving a power of 113 at Amici I (field of view 20′), Tempel was now using two others with focal lengths of 28 mm and 13 mm (powers 190 and 400, fields of view 12.5′ and 5.8′), which had been made in Munich (by Fraunhofer) and Florence, respectively. Moreover, Tempel applied a ring-micrometer eye-piece with a field of 20′ to measure relative positions. Observations of old and new nebulae were regularly published in the *Astronomische Nachrichten*. He also found time for writing a treatise 'Über Nebelflecken' ['On nebulae'], which was published in 1885. Dreyer, who was otherwise happy about Tempel's success, criticised this work.

9.3.1 Tempel's publications

From 1877 to 1886 Tempel discovered another 123 objects catalogued in the NGC. The new ones and a large number of those which had already been listed in the GCS were published in 10 *AN* reports (Table 9.2). Dreyer termed them T I to T X. 'Date' is the date of writing. Some reports list the new objects in tabular form ('E' gives the number of entries). Often they are only mentioned in the text, appearing in the context of descriptions of nebulae from William Herschel or the GC.

There is no reference for the galaxies NGC 47 and NGC 54 in Cetus; both were probably discovered in 1886 (Fig. 9.4 left). NGC 54 was seen independently by Lewis Swift on 21 October 1886. Dreyer credits Tempel, suggesting that he was informed by letter. Swift's observation of NGC 54 yielded another object, catalogued as NGC 58, which turned out to be identical with Tempel's NGC 47. Dreyer reflected this case in the IC II notes, writing '*probably = 47*'.

Tempel discovered altogether 149 NGC objects; their statistics are presented in Table 9.3. Columns 'GC(S)' and 'NGC' contain objects catalogued in the GC or GCS (see Table 8.20) and NGC, respectively. A curious case is the 15-mag star NGC 5467 in Bootes (Fig. 9.4 right), which was seen by Tempel probably

9.3 Tempel's new nebulae 329

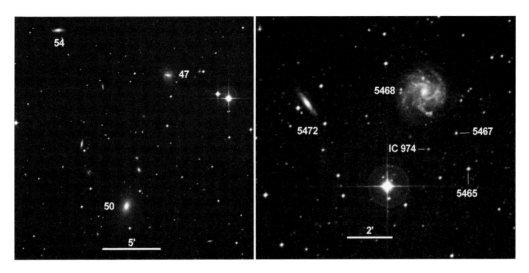

Figure 9.4. Left: the galaxy trio around NGC 47 in Cetus; right: the stars NGC 5465 = IC 973, NGC 5467 and IC 974 near the galaxy NGC 5468 in Bootes (DSS).

Table 9.3. *Statistics of Tempel's objects, divided into GC/GCS and NGC*

Type	GC(S)	NGC	Sum	Examples
Galaxy	17	90	107	Brightest: NGC 1398 (For) 9.8 mag
Planetary nebula	1		1	NGC 6309 (Oph)
Galactic nebula	3		3	NGC 1435 (RN, Tau), NGC 2067 (RN, Ori), NGC 2361 (EN, CMa)
Star	4	13	17	Brightest: NGC 7338 (Peg) = GSC 2743–2192 (12.2 mag)
Star pair or group	1	16	17	NGC 4982 (Vir): 4 stars
Not found		4	4	NGC 2630, NGC 2631, NGC 4322, NGC 4327
Sum	26	123	149	

in April 1882, together with another star, NGC 5465 (Tempel 1881b). NGC 5467 is the only single star which also bears an IC-number. It was found again by Bigourdan on 21 May 1890 (together with the star IC 974) and catalogued by Dreyer as IC 973. The identity was noticed by Karl Reinmuth. There is also a star quartet in Orion with this feature: NGC 1707 (John Herschel) = IC 2107 (Bigourdan).

Table 9.4 shows special objects and observational data. Tempel covered a wide declination range, paying attention especially to the polar region, which had been neglected by William Herschel. Only one object was discovered with the (smaller) Amici II refractor: the 12.1-mag galaxy NGC 2441 in Camelopardalis on 8 August 1882.

On 18 August 1882 Tempel discovered – after William Herschel (NGC 4774) and Marth (NGC 1141/42) – another ring galaxy: NGC 2685 (11.2 mag) in Ursa Major. It shows a 'polar ring' and is called the Helix Galaxy (Arp 336). Of course, Tempel did not recognise the peculiar structure; the object was seen only as being 'round'.

Tempel discovered eight of nine NGC galaxies in the NGC 5416 group in Bootes (Table 9.5).

Table 9.4. *Special objects and observational data of Tempel*

Category	Object	Data	Date	Remarks
Brightest object	NGC 1398 (V)	Gx, 9.8 mag, For	9.10.1861	
Faintest object	NGC 4764 (V)	Gx, 15.1 mag, Vir	March? 1881	Fig. 9.5
Most northern object	NGC 1544 (I-16)	+86°, Gx, 13.6 mag, Cep	1876	
Most southern object	NGC 1398 (V)	−26°, Gx, 9.8 mag, For	9.10.1861	
First object	NGC 1435	RN, Tau	19.10.1859	Merope Nebula
Last object	NGC 47	Gx, 13.1 mag, Cet	1886	With NGC 54, Gx, 13.7 mag, Cet
Best night			25.4.1883	7 galaxies in Bootes (see Table 9.5)
Best constellation	Virgo		1875–83	32 galaxies

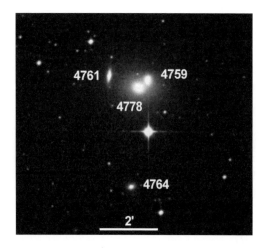

Figure 9.5. NGC 4761 and NGC 4764, two Tempel galaxies in the neighbourhood of the bright double galaxy NGC 4759/78 discovered by William and John Herschel in Virgo (DSS).

NGC 5416 had already been detected by William Herschel on 19 March 1784 (III 56). Tempel mentions five objects in his paper 'Einige neue Nebel und ein neues Nebelnest'.[9] The galaxies were found on 25 April 1883 (his best night) together with a sixth (NGC 5414), located 30′ north of the group. Three other members were first seen on 28 June 1883, among them NGC 5438. This galaxy is due to William Herschel, but was catalogued by Dreyer as NGC 5446, without

noticing the identity. Strangely, Herschel did not see the brightest objects (NGC 5423 and NGC 5424). The group covers a field of 40′ in diameter; pretty central is the 6.2-mag star BD +10° 2617.

Figure 9.6 shows the brightness distribution of Tempel's objects. The mean is 13.1 mag. The brightest object is the galaxy NGC 1398 (9.8 mag) in Fornax, which was found on 9 October 1861 with the 4″ Steinheil refractor in Marseille.

In this context, a report about a 'missing star' in the Bonner Durchmusterung is interesting (Tempel 1891b). It was first noticed by Schmidt in Athens: '*1869 Aug. 6 at $12^h 4$ I observed a star of $9^m 3$, which is not in the B.D.*' (Schmidt 1881). The place given by him is wrong, since the star is actually in the Bonn catalogue: BD +8° 215 in Pisces, located 45′ west of the bright galaxy NGC 524 (I 151). There are other, fainter galaxies in the vicinity, which had been discovered by d'Arrest and Marth. Tempel investigated the field on 1 November 1880 with the 11-inch and noted that '*The other three nebulae of d'Arrest are really the faintest ever seen with Amici I; Lord Rosse's are brighter.*' These galaxies are NGC 516, NGC 522 and NGC 525 with visual magnitudes of about 13.3 mag. Tempel's words are astonishing, insofar as he had discovered some distinctly fainter objects. On 31 January 1881 he once again looked at the place of the missing star, now writing that '*there was no star at the place, but a faint small nebulous light; its position coincides exactly with that given in the B.D. for 215 […] I have seen such a nebulous light once around a variable star. This may be a sign that +8° 215*

---

[9] 'Some new nebulae and a nest of nebulae' (Tempel 1883a).

9.3 Tempel's new nebulae

Table 9.5. *Galaxies discovered by Tempel in the NGC 5416 group in Bootes*

NGC	Ref.	Date	V	Remarks
5409	VI, VIII-5	25.4.1883	13.4	
5423	VI, VIII-7	25.4.1883	13.0	
5424	VI, VIII-8	25.4.1883	13.0	
5431	VI, VIII-9	25.4.1883	14.3	
5434	VI, VIII-10	25.4.1883	13.2	
5436	VII	28.6.1883	14.0	
5437	VII	28.6.1883	14.3	
5438	VII	28.6.1883	13.6	NGC 5446, W. Herschel 19.3.1784 (III 57)

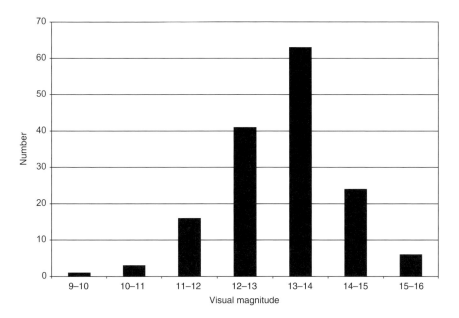

Figure 9.6. The brightness distribution of Tempel's objects in the NGC.

will become visible again.' (Tempel 1881c). To solve the mystery: the 'missing' star is AH Psc, an irregular variable (9.9 to 14 mag); but there is no nebula. Tempel's perception is difficult to assess. Sometimes yielding remarkably keen observations of very faint objects, he also reports odd things, such as 'pulsating stars' surrounded by nebulosity, that are beyond any reality.

Table 9.6 lists those of Tempel objects which had been discovered earlier by others and bear no GCS-number.[10] The case of NGC 577/80 is remarkable (see Fig. 9.27). Tempel writes that, '*Following the star Lalande 2666* [BD –2° 221], *which is near to the nebulae III 441–42* [NGC 560/64], *there are two others in +2^m 50^s* [NGC 577/80], *one has the same brightness as the preceding one of Herschel* [NGC 564].'[11] Tempel gives a precise position for only one of these nebulae. Then another nebula is mentioned: '*Class II, a bit brighter then the two preceding nebulae of H. class III: 441, 442.*'[12] A discovery date is given (14 August 1877). Because there is only one object at Tempel's place (13-mag galaxy), his first observation was erroneous; thus we have

---

[10] See the list of GCS objects in Table 8.21.
[11] Tempel (1878d), entries no. 7 and 8.
[12] Tempel (1881a), entry no. 5.

Table 9.6. *Those of Tempel's NGC objects which had been discovered earlier by others*

NGC	GC	Date	Type	V	Con.	Ref.	Remarks
114		27.9.1880	Gx	13.3	Cet	IV-2	Safford 23.9.1867
118		27.9.1880	Gx	13.6	Cet	IV-3	Safford 23.9.1867
124		27.9.1880	Gx	13.1	Cet	IV-4	Safford 23.9.1867
577		14.8.1877	Gx	13.0	Cet	I-7, IV-5	Skinner 23.10.1867; = NGC 580 Tempel
580		14.8.1877	Gx	13.0	Cet	I-8	L. Swift 20.11.1886; = NGC 577, Skinner 23.10.1867
787		9.11.1879	Gx	12.9	Cet	IV-7	Peters 27.2.1865
1446		1882	Star	14.2	Eri	V	Dreyer 8.1.1877
3425	2237	1877	Gx	13.3	Leo	I-28	W. Herschel 17.4.1784 (III 108)
4206	2795	1877	Gx	12.0	Vir	I-39	W. Herschel 17.4.1784 (II 165)
4222	2814	1877	Gx	13.2	Com	I-40	W. Herschel 8.4.1784 (II 109)
4307		17.3.1883	Gx	12.1	Vir	VII	Peters 1881?
4308		17.2.1882	Gx	13.2	Com	V-16	Safford 11.6.1868
4802		20.4.1882	Gx	11.6	Crv	V-21	NGC 4804, W. Herschel 27.3.1786 (IV 40)
5165		11.5.1883	Gx	13.9	Vir	VII, VIII-1	Burnham 5.5.1883
5171		11.5.1883	Gx	12.9	Vir	VII, VIII-2	Hough 5.5.1883, Hartwig 29.6.1883
5179		11.5.1883	Gx	14.2	Vir	VII, VIII-3	Burnham 5.5.1883, Hartwig 29.6.1883
5438		28.6.1883	Gx	13.6	Boo	VII	NGC 5446, W. Herschel 19.3.1784 (III 57)
5472	3781	April? 1882	Gx	14.3	Vir	V	Mitchell 19.4.1855
5534		29.4.1881	Gx	12.3	Vir	V-30	Stephan 17.5.1881

NGC 577 = NGC 580. Thus Aron Skinner (Dearborn Observatory) was the true discoverer.

9.3.2 The treatise 'Über Nebelflecken' and Dreyer's critical review

In December 1879 Tempel finished a manuscript titled 'Über Nebelflecken. Nach Beobachtungen angestellt in den Jahren 1876–1879 mit dem Refractor von Amici auf der Königl. Sternwarte zu Arcetri bei Florenz'.[13]

---

[13] 'About nebulae. According to observations made in the years 1876–1879 with the refractor by Amici at the Royal Observatory in Arcetri near Florence' (Tempel 1885a).

He at first could find no appropriate publisher, and there were problems with financing too. Moreover, the production of the figures (two tables) turned out to be difficult: Tempel, as a professional lithographer, was exceptionally discerning. The 28-sided treatise was eventually printed in 1885 in Prague. It was to be available in autumn 1886 in a limited edition.

The work reflects Tempel's personal, sometimes self-willed, point of view about nebulae. He focuses on fundamental questions concerning their observation and nature, mentioning special objects only as examples. In late 1886 Dreyer wrote a review, in which Tempel's opinions were critically discussed (Dreyer 1887c). At that time he was much concerned with the issue, being in the final phase of compiling the NGC.

Tempel starts with the problem of resolvability.[14] Resolvable nebulae (i.e. star clusters) are rarely questionable. For the unresolvable case he peculiarly doubts the well-established spectroscopic results: '*That the unresolvable nebulae consist of gas or a luminous fluid, as Herschel used to assume and as is again assumed nowadays, would be easier to believe if this nebulous matter were not located far above our atmosphere; how from there, at such immense distance, even beyond that of the stars, such a gas can still be visible remains inexplicable.*'[15] Tempel goes back to Mädler, who '*comes to the conclusion too that the hypothesis of gas-nebulae is inconsistent with the law of gravity*'.[16] The latter astronomer had argued in about 1840 that such objects are physically impossible. Tempel further writes that, '*had I ever believed in spectral analysis, I would never have tried even to look at a nebula or to sketch it*'.[17] He states critically that the new methods waste too much money and observing time, which is much needed for more essential problems. He laments that '*with all these colossal sacrifices there was obtained concerning the nebulae only the result: they would consist of gas!*'[18]

Tempel writes the following about the unresolvable nebulae: '*with the Amici I telescope giving such brilliant images, I did not find any nebula that does not show more or less tiny stars in the centre or in the nebulous knots, thus it is not a pure gasous mass! Occasionally the number of these pulsating stars becomes so large that I certainly believed Lord Rosse's giant telescope must have resolved it into a star cluster; but in his catalogues and drawings the faint stars are not mentioned. Certainly remarkable and very strange!*'[19] Tempel's poster child was the Orion Nebula; the appendix (Table II) shows a large-format white-on-black drawing based on observations with Amici I.

In his review Dreyer strongly attacks Tempel's ignorance: '*He appears to doubt the existence of gaseous nebulae and refers to the difficulty of understanding how a fast accumulation of gas of very irregular form could for centuries retain its shape unaltered.*'[20] He felt that Tempel's position was self-contradictory: on the one hand Tempel perceives '*transient glimpses of star-like points in such objects [...] but on referring to Lord Rosse's publications this has not been found to be the case*', but then on the other '*Mr. Tempel does not consider such objects resolvable.*' Dreyer cites reports of Edward Stone and William Huggins[21] showing that the spectra of irregular unresolvable nebulae '*not necessarily indicate that a nebula is solely composed of gas, but that a dense cluster of which the single stars were surrounded by large gaseous envelopes, if removed to a very great distance, would give a linear spectrum*'.[22] For Dreyer the spectroscopic analysis (from Huggins to Pickering) evidently shows that '*the regular-shaped disc-like planetary nebulae are probably composed of gas*'.

At the same time Lewis Swift, who was just as enthusiastic an observer of nebulae as Tempel, was convinced by the new results, without their influencing his search for nebulae in any way. In 1885 he wrote '*But spectroscopic examination settled at once the vague conjectures of a century. It tells us in language too plain to be misunderstood or doubted, that some are clusters of stars, while others are masses of gas, or nebulous matter, whatever that may be.*' (Swift 1885c).

Tempel also covers the question of distance. He believes the Merope Nebula to be 'in front of the stars', but the Andromeda Nebula[23] 'beyond' (see Section 11.6.5). Another topic is the appearance and form of nebulae. He first discusses the Herschel classes: '*This classification is of great value and one is astonished that the subsequent observers of nebulae did not follow the scheme.*'[24] Tempel states that rendering the true form is a big problem: '*It is not surprising that, for the largest and most interesting nebulae in the sky that have been sketched by the most diverse astronomers, of six illustrations of the same nebula no two concur, not even in the outer contours, in the general shape, and each drawing presents a different curious figure.*'[25] To produce evidence, Tempel presents in Table I of the appendix a set of six drawings of the Crab Nebula in Taurus (M 1, h 357). They were made by John Herschel (1827), Secchi (1852), Lassell (1854), Lord Rosse (1855), d'Arrest (1861) and

---

[14] See also Section 6.4.8.
[15] Tempel (1885a: 4).
[16] Mädler (1849: 451).
[17] Tempel (1885a: 27).
[18] Tempel (1885a: 26).
[19] Tempel (1885a: 4).
[20] Dreyer (1887c: 59–60).
[21] Stone E. (1877), Huggins W. (1877).
[22] Dreyer (1887c: 60).
[23] In Arcetri, Tempel made a fine drawing of M 31; see Radrizzani (1989).
[24] Tempel (1885a: 10).
[25] Tempel (1885a: 11).

himself (1879).²⁶ In fact, there are striking differences. Dreyer lauds the fact that Tempel did not select Lord Rosse's drawing of 1844 showing the ominous 'crab', which was *'copied in a more or less exaggerated form in most popular books on astronomy'*.²⁷ Instead, the drawing of 1855 is presented. Since it was published in 1880, Tempel must have produced his Table I after finishing the manuscript. Strangely, Secchi's drawing of M 1 looks like a 'crab' too! Dreyer comments as follows: *"It is extremely curious that the long arms of the 'crab' also appear on Secchi's drawing."* Tempel disliked Secchi's sketches, which, *'compared with all others, invited criticism all the more'*.²⁸

For Tempel it is definite that the differences are not due to the telescopes: *'The principal forms of the nebulae must appear equal in all telescopes, a large telescope only showing more details.'* He concludes that *'The reason for the incongruity of their drawings is due to the draughtsmen themselves.'*²⁹ Dreyer does not agree, stressing the difference between the 72-inch reflector at Birr Castle and all other telescopes.

Tempel also treats the question 'Are the nebulae variable?' Regarding the problems of perception of nebulae and their textual or graphic description, he is not able to find a proper answer. As an example he mentions d'Arrest: *"To convince oneself of the uncertainty present in this issue, it is sufficient to take the latest and best work of d'Arrest 'Siderum nebulosorum observationes', reading the descriptions of one nebula for different nights; nearly every nebula seems to be variable from one to the other night, which cannot be real, but must be due to the varying transparency of the atmosphere."*³⁰ Dreyer always reacted very touchily when his teacher and idol d'Arrest was attacked: *'In view of the rash assertions as to variability in the appearance of the nebula in Andromeda [M 31] which have been made during the autumn of 1886, it is pleasant to see M. Tempel express himself with the same caution which has always characterized experienced observers of nebulae.'*³¹

The last chapter contains a description of the observatory with its two Amici refractors. In the 'final remarks' Tempel laments the stagnation of research on nebulae and points out two reasons: first, the superiority of large telescopes (such as Lord Rosse's) had discouraged the users of small instruments;³² and secondly, there was the bad influence of spectroscopy. Thus Tempel appeals to future observers: *'May therefore more talented forces lay all prejudices aside, and still devote themselves to the study of nebulae; an activity that will certainly provide completely new opinions about quite a few hypotheses.'*³³ He added: *'But worse luck! After a careful study of the previous work on this issue, there are so many gaps to fill, so many errors and aberrations to correct that it might be impossible for one person to master this task, and only combined forces and systematic efforts – such as zone observations – should be able to create a gas-free basis for nebulae!'* Perhaps Tempel alludes here to d'Arrest and a little jealousy about his success and academic recognition might be present too. D'Arrest had gained a considerable reputation for his *Siderum Nebulosorum* (which actually is a zone observation³⁴) and his subsequent spectroscopic studies of nebulae. A 'gas-free' basis would be impossible for him.

In his review Dreyer explicitly supports Tempel's request for further work on nebulae: *"Though it may be doubtful whether it would be easy to unite a number of observers in an undertaking of this kind, it is very desirable indeed, that a few refractors of from 15 to 18 inch aperture erected during the years should be devoted, not to hunt for those most uninteresting 'eeF, vS, R' nebulae (as to which nobody doubts that they are next to innumerable), but to a systematic examination and, whenever possible, micrometric observations of all known nebulae not accurately observed of late. Much has been done in this direction by d'Arrest, Schönfeld, Schultz and others, but a very great deal will have to be done yet."*³⁵

The appendix contains a list of all drawings made by Tempel up to the end of 1879. There are 186 objects sorted by GC-number; 'reliable' drawings are marked. There is another list of 204 drawings, published

---

²⁶ See Section 11.2.3.
²⁷ Dreyer (1887c: 62).
²⁸ Tempel (1885a: 12).
²⁹ Tempel (1885a: 13); see also Section 11.2.
³⁰ Tempel (1885a: 15); see also his remarks on Chacornac's 'nebula' NGC 1988 (Section 7.3.2).
³¹ Dreyer (1887c: 64); about the changes in M 31, see Section 9.20.
³² Tempel (1885a: 26).
³³ Tempel (1885a: 27).
³⁴ In the introduction to the GCS (which Tempel owned), Dreyer commended d'Arrest's work, especially due to the systematic zone observations.
³⁵ Dreyer (1887c: 66).

later in *AN* 2439, which is complete up to May 1881 (Tempel 1881b). The originals are stored at the Arcetri Observatory (see Section 6.17.1).

9.4 HARVARD'S NEW GUARD: AUSTIN, LANGLEY, PEIRCE, SEARLE AND WINLOCK

During the years 1866–70 Edward Austin, Samuel Langley, Charles Peirce, George Searle and Joseph Winlock discovered some objects at Harvard College Observatory. At that time Winlock was director. His successor, Edward Pickering, later published these observations in Volume 13 of the *Harvard Annals*, which appeared in 1882. Thereby Dreyer became aware of the new objects.

9.4.1 Short biography: Edward P. Austin

Not much is known about Edward Austin. He was born in 1843 in the USA, and was employed as an assistant of Joseph Winlock, Director of Harvard College Observatory, from 1869 to 1871. In spring 1870 he observed nebulae with the 15″ Merz refractor. Later he took part in cartographic work in Arizona and Nevada. Edward Austin died in 1906.

9.4.2 Short biography: Samuel Pierpont Langley

Samuel Langley was born on 22 August 1834 in Roxbury, MA (Fig. 9.7 left). In 1851 he finished school in Boston and studied engineering and architecture. Though his great love was astronomy, he worked as an engineer for some years. In 1864–65 he travelled through Europe, visiting some major observatories. After that, he became an assistant of Joseph Winlock at Harvard College Observatory for one year. Then Langley joined the US Naval Academy, Annapolis, as an assistant professor. In 1867 he was appointed Director of the new Allegheny Observatory in Pittsburgh and Professor of Physics at Western University, Pennsylvania. Langley's main target was the Sun, which he studied using astrophysical methods. He invented the bolometer and determined the solar constant. In 1887 he left the Allegheny Observatory to work for the Smithsonian Institution, Washington. Langley founded the Smithsonian Astrophysical Observatory (SAO) and became its first director in 1890. He was well known too as a pioneer of aircraft construction on the basis of aerodynamic studies. Samuel Langley died on 27 February 1906 in Washington at the age of 71.[36]

9.4.3 Short biography: Charles Sanders Peirce

Charles Peirce (Fig. 9.7 right) was born on 10 September 1839 in Cambridge, MA, as the son of the well-known mathematician and Professor at Harvard University Benjamin Peirce. He graduated in 1859, and was employed as an assistant of Joseph Winlock at Harvard College Observatory in 1868–69 and again during the years 1872–75. From 1879 to 1884 Peirce taught at the mathematical faculty of the Johns Hopkins University. He also worked for the US Coastal Survey. Charles Peirce died on 19 April 1914 in Milford, PA, at the age of 73.[37]

9.4.4 Short biography: George Mary Searle

George Searle was born on 27 June 1839 in London (Fig. 9.8 left). He was the younger brother of the Harvard astronomer Arthur Searle (who worked at the observatory from 1869 to 1887). He graduated from Harvard University in 1857, and served in 1858–59 as an assistant at Dudley Observatory, Albany, where he

Figure 9.7. Left: Samuel Langley (1834–1906); right: Charles Peirce (1839–1914).

---

[36] Obituaries: Brashear (1906), Abbot C. (1906a, b) and Cookson (1907); see also Bailey S. (1931: 256–257), Beekman (1983a) and Eddy (1990).
[37] Bailey S. (1931: 260).

Figure 9.8. Left: George Searle (1839–1918); right: Joseph Winlock (1826–1875).

discovered the minor planet Pandora on 11 September 1858. From 1859 to 1862 Searle was engaged in the US Coastal Survey, followed by two years as an assistant professor at the US Naval Academy, Annapolis. From June 1866 to March 1868 he worked at Harvard College Observatory. Afterwards his life changed radically: Searle joint the Paulist order and became a priest in 1871. Later he taught as Professor of Mathematics at the Catholic University, Georgetown. One of his fields of interest was the habitability of planets (Searle G. 1890). George Searle retired in 1916 and died on 7 July 1918 in Washington at the age of 79.[38]

9.4.5 Short biography: Joseph Winlock

Joseph Winlock was born on 6 February 1826 in Shelbyville, KY (Fig. 9.8 right). After graduating from Shelby College in 1845 he became Professor of Mathematics and Astronomy there. In 1852 he went to Cambridge, effecting calculations for the *Nautical Almanac*. In 1857 he was appointed Professor of Mathematics at the US Naval Academy, Annapolis, and two years later he became the head of the mathematical faculty. Winlock also worked for the US Naval Observatory, Washington, where he eventually was appointed Superintendent of the Nautical Almanac Office.

In 1865 he succeeded George Phillips Bond as Director of Harvard College Observatory and Professor of Astronomy at Harvard University. In 1876 he published Etienne Trouvelot's drawings made with the 15″ Merz refractor in Vol. 8 of the *Harvard Annals*, titled 'Astronomical engravings of the Moon, planets, etc.'

(Winlock 1876). The drawings (including some non-stellar objects) are among the best ever made.[39] Joseph Winlock was still in office when he died on 11 June 1875 in Cambridge, MA, at the age of 54.[40]

9.4.6 New nebulae at Harvard College Observatory

The visual observations (among them some spectroscopic measurements by Winlock and Peirce) were made with the 15″ Merz refractor, which had been erected in 1847. Later Edward Pickering listed them in Vol. 13. of the *Annals of the Harvard Observatory* (Pickering E. 1882a). The publication contains all observations from 12 April 1866 to 24 March 1870. For the 402 objects, sorted by right ascension, the number, date, GC/GCS-number, position (epoch 1860), observer and remarks are given. Additional notes are presented. Besides the persons mentioned in Table 9.7, Arthur Searle (brother of George Searle) and Oliver Clinton Wendell were involved too. They observed only one (known) object, the galaxy NGC 1521 (GC 815, h 2612) in Eridanus (1 February 1869).

Fourteen objects bear no GC-number, thus being new to the Harvard observers (Table 9.7); 'No.' gives the number used in the publication of 1882. Later Pickering compiled all 108 new nebulae that had been found up to 1907, assigning them an H.N.-number ('Harvard nebula'; Pickering E. 1908). Pickering's publication of 1908, titled 'Nebulae detected at the Harvard College Observatory', includes all objects found during the years 1866–70, except for one: no. 325. Winlock noted about this object, which he found on 20 June 1868, '*Faint neb susp.*' Pickering added '*Place approx.; possibly G.C. 3803.*' This is NGC 5495, which had been discovered in 1834 by John Herschel (h 3561 in the Cape catalogue). All of the other objects are new (and were catalogued by Dreyer in the NGC). Most of them were found by George Searle (6); all are non-stellar. For NGC 548 and NGC 565 Dreyer notes 'Searle', otherwise 'G M Searle'.

Austin observed on only two nights. During the first (20 February 1870), the Orion Nebula was the

---

[38] Obituary: Anon (1918a); see also Bailey S. (1931: 259–260).

[39] They include M 13, M 20 (see Fig. 2.12), M 27, M 31, M 42, M 57 and M 92. Later Trouvelot published comprehensive descriptions (Trouvelot 1882).

[40] Obituary: Rogers (1875).

9.5 Warner Observatory

Table 9.7. *NGC objects discovered by Austin, Langley, Peirce, Searle and Winlock at Harvard (sorted by date)*

NGC	H.N.	No.	Discoverer	Date	Type	V	Con.	Remarks
3355	29	215	Langley	12.4.1866	NF		Hya	IC 625?
5495		325	Winlock	20.6.1868	Gx	12.8	Hya	J. Herschel 13.5.1834 (h 3561)
5872	30	329	Winlock	30.7.1867	Gx	13.8	Lib	
5883	31	331	Winlock	30.7.1867	Gx	13.6	Lib	Leavenworth 6.6.1885
570	32	27	Searle	31.10.1867	Gx	12.8	Cet	
548	33	25	Searle	2.11.1867	Gx	13.7	Cet	
565	34	26	Searle	2.11.1867	Gx	13.6	Cet	
4247	35	265	Searle	25.2.1868	Gx	14.4	Vir	
5487	36	324	Searle	22.3.1868	Gx	14.0	Boo	
4058	37	257	Searle	24.3.1868	Gx	13.2	Vir	
1170	38	47	Peirce	31.12.1869	NF		Ari	Comet tail?
3097	39	177	Austin	24.3.1870	Star	14.1	UMa	Fig. 9.9
3315	40	207	Austin	24.3.1870	Gx	13.3	Hya	In Hydra Cluster
3317	41	210	Austin	24.3.1870	2 stars		Hya	

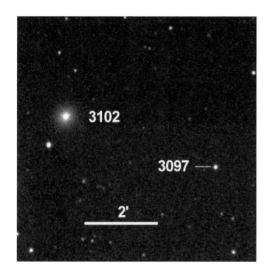

Figure 9.9. Austin's star NGC 3097 near the galaxy NGC 3102 in Ursa major (DSS).

target ('*edges less sharply defined than in Bond's drawing*'). On 24 March 1870 he scanned large parts of the constellation Hydra. He observed 15 objects, among them two new ones. NGC 3317 is only a pair of stars, 5′ north of the galaxy NGC 3316. His second find lies 20′ further north, this being the galaxy NGC 3315, a member of the Hydra Cluster Abell 1060 (which was first noticed by John Herschel in March 1835). Additionally, Austin observed eight objects in Leo and three in Ursa Major. All were already known, except NGC 3097 (Fig. 9.9), for which is noted '*G.C. 1998 s f neb; p 45°, s 2.*' Pickering added '*Place only approximate*', and the notes to no. 177 give '*Perhaps a nebulous star. It is half-way between G.C. 1998 and a star 11 magn.*' GC 1998 is the 13.3-mag galaxy NGC 3102. The data are a bit confusing and Dreyer has derived a position 2′ northwest of the galaxy – but there is no star. Possibly Austin saw a 14-mag star 4′ southwest of the galaxy.

NGC 1170, found by Peirce, probably was part of the tail of comet Tempel–Swift. He noted '*Comet 1869 III p neb 2m 31s, a little s*'; Pickering added '*approximate place results from the comparison with the comet*'. The position corresponds to a blank field. Dreyer says nothing about this strange case. Langley's object in Hydra, which he found during his search for Biela's comet, is missing. Because the place is flagged 'approx.' it could be identical to Muller's IC 625, which is, however, 45′ to the south.

9.5 WARNER OBSERVATORY: LEWIS SWIFT AND HIS SON EDWARD

Lewis Swift's contribution to the NGC is remarkable. Between 1859 and 1887 he discovered 473 objects, of which 450 are non-stellar – a very high rate of 95%.

Therefore he was called the 'American Herschel'. He used an excellent 16″ Clark refractor. Unfortunately his life was crisscrossed by all kinds of adversities, with which he coped with an iron will. Swift's observations were supported by his son Edward, who during the years 1883–87 discovered 25 objects. Together with William and John Herschel, William and Lawrence Parsons and Wilhelm and Otto Struve, this is another example of a successful astronomical connection of father and son (Fig. 9.10).

9.5.1 Short biography: Lewis Swift and Edward Swift

Lewis Swift was born on 29 February 1820 in Clarkson, NY. He was the sixth of nine children of the settler family of General Lewis Swift and his wife Anna (*née* Forbes). At the age of 13 he broke his left hip in an accident. Owing to bad medical treatment he remained handicapped and had to rely on crutches for the rest of his life. Since the young Swift could do only light farm work, he had plenty of time for reading. He was especially interested in science books, and did not hesitate to travel the arduous route to the library at Rochester. As an autodidact, he eventually discovered astronomy. A key event was the comet Halley in 1835, but it was another 20 years before Swift got a book on optics, which led him to practical astronomy. In 1850 he married Lucretia Hunt and fathered two children. He was at the time working as postmaster of Hunt's Corners, Cortland County, owning a small store. In 1857 he bought a 4.5″ refractor by Fitz and built a small dome. Swift observed comet Donati and wrote a note for the *Astronomical Journal* (Swift 1858). On 15 July 1862 he discovered his first comet (together with Horace Tuttle) – it became one of his famous: comet Swift–Tuttle, the origin of the Perseid meteor shower.

In 1862 Swift's wife died and he moved with his children to Marathon. Two years later, he married Caroline Topping. Three sons were born, among them Edward (1871). In 1865 the family settled in Rochester, where Swift opened a new shop. He organised public observations and gave talks. Having become well known by virtue of his comet discoveries, he was regarded kindly. He was offered the opportunity to install his 4.5-inch on the roof of Duffy's Cider Mill (Fig. 9.11 left). For each observation Swift brought the objective and accessories from home. After a walk of half a

Figure 9.10. The Swift family (right to left): Lewis Swift, his son Edward and his wife Caroline (Wlasuk 1996).

mile came the most difficult part: to reach the roof, he had to climb over two ladders, which, especially in winter, was difficult for a limping man. Nearly every year Swift discovered a new comet – altogether 13 are due to him. However, there was a rivalry with Edward E. Barnard in Nashville (Tennessee) and William Brooks in nearby Phelps.[41] After some trouble with priorities, the trio eventually decided to allocate certain areas of the sky to each observer.

In 1878 a curious observation happened in Denver. Swift, now a well-known astronomical authority, was invited by James Watson to erect his small refractor at the local observatory to search for 'Vulcan', the suspected innermost planet, during a total eclipse. Actually, he saw an unknown object near the Sun, which later was missing. The discovery was noted worldwide – but also criticised (e.g. by C. H. F. Peters). In 1879 Swift became a Fellow of the RAS and an honorary doctor of Rochester University. He was widely known as 'Professor Swift'.

However, an appropriate observatory was lacking. Therefore a public campaign was started. In Hulbert Harrington Warner, a successful local businessman (offering everything from steel safes to pills), a well-heeled patron was soon found. He was so impressed by Swift – who had just been awarded the gold medal of the Vienna Academy – that the necessary money to build an observatory was provided. The Warner Observatory, located in an urban park (Fig. 9.12), was finished in late 1882. It was equipped with an excellent 16″ refractor by Alvan Clark & Sons. As soon as in February 1883 Swift

---

[41] Ashbrook (1984: 92–97).

9.5 Warner Observatory   339

Figure 9.11. Swift with his 4.5″ Fitz refractor on the roof of Duffy's Cider Mill (Ashbrook 1984).

Figure 9.12. Warner Observatory in Rochester (Swift 1887b).

discovered a new comet with it. Shortly thereafter – now already 63 years old – he started to search for nebulae. The work proceeded successfully over many years. He was supported by his son Edward, who was awarded the 'Warner Gold Medal' for his discoveries of nebulae. The objects of Lewis and Edward Swift, which had been discovered in Rochester, were catalogued by Dreyer in the NGC and IC I.

Twice a week Swift offered public observations at Warner Observatory. However, the increasing light and air pollution in Rochester became a problem. In March 1889 he visited Barnard on the Californian Mt Hamilton to test the new Lick refractor – and, of course, was impressed by the telescope and the observing conditions. When in 1893 Warner's empire broke down due to the general financial crisis, Swift (now 73) thought about a move to California.

Fortunately, Swift found a new patron: Professor Thaddeus Lowe, who had erected an amusement park on Echo Mountain near Pasadena. The 1100-m-high summit, only a few miles west of Mt Wilson, offered a railway, a hotel – and an observatory. Impressed by Swift's astronomical reputation, he offered to erect the 16″ refractor on 'Mt Lowe'. The plan was accepted in Rochester and the Lowe Observatory was operational in 1894.

For the next seven years the veteran observer scanned the dark Californian sky, 9° further south than Rochester. Often the sky was clear for months. Again, Swift offered public observations. He discovered three more comets and nearly 300 nebulae, most of them in

Table 9.8. *Swift's nebular discoveries with the 4.5″ refractor*

NGC	Date	Type	V	Con.	Remarks
1360	1859	PN	9.4	For	GC 5315, Tempel 9.10.1861; Winnecke Jan. 1868; Block 18.10.1879
2237	1865	EN		Mon	II-31, part of Rosette Nebula; Barnard 29.1.1883
5707	1878	Gx	12.6	Boo	I-36, 22.7.1885; see Swift 1885d
6951	1878?	Gx	11.7	Cep	II-85, 14.9.1885; = NGC 6952, Coggia 1877?

the southern sky. As in Rochester he was supported by his son Edward, who also discovered some nebulae and a comet (1984 IV). The new objects were catalogued in the Second Index Catalogue of 1908. However, Dreyer lamented that Swift's observations had lost their earlier quality. Swift, Burnham, Barnard and Bigourdan are the only persons to be present in all three catalogues (NGC, IC I and IC II).

Swift discovered his last comet in 1899, when he was 79 years old. But his eyesight was deteriorating, and in 1901 he returned to Marathon, NY. Lewis Swift died there on January 1913 at the age of 92. Barnard wrote an obituary for his long-time friend.[42] Edward Swift died in 1945.

Swift's successor at Lowe Observatory, Edgar Larkin, had less luck, because Lowe's realm soon broke down. In 1905 a fire raged on Echo Mountain, and in 1928 a storm destroyed the observatory. Fortunately, the refractor was rescued. It was sold to Santa Clara University in 1941, and is still standing at Ricard Memorial Observatory.

9.5.2 Swift's first discoveries with the 4.5-inch

In 1857 Swift started to search for comets with his 4.5″ Fitz refractor; first in Hunt's Corners and Marathon, then, from 1872, in Rochester. His favourite eye-piece, with a focal length of 50 mm, was made by the ethnic German optician Ernst Gundlach. It was periscopic and had an inner ring with crosshairs, which were well visible in darkness. At the 4.5-inch it offered a power of 34 and a 3° field of view. During comet-

---

[42] Obituaries: Knobel (1913), Barnard (1913b) and Anon (1913b); see also Bates and McKelvey (1947), Ashbrook (1984: 71–78), Sheehan (1995), Wlasuk (1996) and Steinicke (2002c).

Figure 9.13. The large planetary nebula NGC 1360 in Fornax.

seeking Swift often detected nebulae unknown to him. They were recorded in his *Atlas Designed to Illustrate the Geography of the Heavens* by Elliah Burritt, which he purchased in 1855. Until his systematic search at Warner Observatory, which started in 1883, he saw '*some two hundred and fifty of these objects with no suspicion that a single one of them written down before 1879 was new*' (Swift 1892b). He copied them in the atlases of friendly comet-observers (among them Barnard and Brooks). Unfortunately, their early discovery is noted for a few nebulae only – mostly in connection with later observations at the 16-inch (Table 9.8). The galaxy NGC 6951 in Cepheus is identical with NGC 6952, which was discovered by Coggia in about 1877.

Swift's first object is remarkable: NGC 1360, a large planetary nebula in Fornax (Fig. 9.13). It was catalogued in Dreyer's GCS (GC 5315). There Winnecke

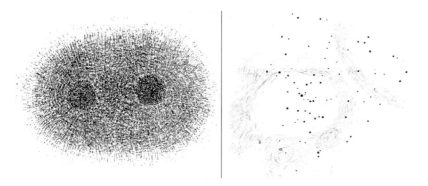

Figure 9.14. Sketches of the Rosette Nebula NGC 2237 in Monoceros by Swift (left) and Barnard (right) (Swift 1884b; Barnard 1889a).

is mentioned, who found the nebula in January 1868 in Karlsruhe with a 9.7-cm comet-seeker. About the description '*F 10*' *L*' Barnard remarked in 1885 that '*Winnecke's description is erroneous; the place is also slightly out*.' (Barnard 1885f). The NGC gives '*Swift 1857, Winnecke*'. Meanwhile Swift's discovery, which had been published in 1885 in the magazine *Sidereal Messenger*, was known to Dreyer. There Swift wrote '*In 1859 while searching in Eridanus for comets I ran across the most conspicuous nebulous star visible from this latitude – a 7th magnitude star nearly in the center of a bright nebulosity. As both were so bright I, of course, supposed they were well known. Not until five years since was I aware, that this wonderful object was not in the G.C.*'[43]

Swift also mentions the GCS entry which credits Winnecke (GC 5315) and refers to an observation by Barnard, who had sent a letter. Finally he writes '*Being ignorant of the date of Winnecke's observation, I am unable to decide to whom the honor of its discovery belongs.*' NGC 1360 was independently found by Tempel in Marseille (9 October 1861), who therefore is the veritable discoverer, and Block in Odessa (18 October 1879). Since this was unknown to Swift (see Sections 6.17.2 and 8.9.1), the PN is among the three NGC objects with the most independent discoveries (see Table 10.6).

Another interesting object is NGC 2237, a part of the Rosette Nebula in Monoceros. Swift did not publish his find until 1884. In his article, titled 'A new and remarkable nebula', he wrote that '*Some ten years ago, while searching for comets, I ran across an exceedingly large and fairly bright nebula near twelve Monocerotis [12 Mon] which I of course supposed was familiar to every astronomer.*'[44] In the 16″ refractor the object appeared '*quite sharply defined and in a shape of a perfect ellipse, having at each focus either a round and much brighter nebula, or, it has two centres of condensation, probably the latter*'. The article contains a sketch too (Fig. 9.14 left). Swift further notes that '*Not until three years ago, did I ascertain that the object was not recorded in any published catalogue of nebulae.*' He got John Herschel's GC in 1880.

Swift reports that his friend Barnard had found the nebula independently: '*About eighteen month ago Mr. Barnard […], in his sweeps for comets, picked it up and called my attention to it, thinking it might possibly be a comet.*' Barnard observed on 29 January 1883 with his private 5″ refractor in Nashville: '*I ran upon this object while seeking comets with the 5-inch refractor, and never having heard of it, thought it might be a comet, and watched it a large part of the night for motion.*' (Barnard 1885d). In an earlier note he mentions '*a remarkable object, but probably the most wonderful thing about it is that observers should have failed to see it long ago, since it is very noticeable*' (Barnard 1884f). In June 1889 he again wrote about the nebulae, adding a sketch (Barnard 1889a; Fig. 9.14 right). It was made in that year with the 12″ refractor at Lick Observatory and shows a ring of 40′ diameter, enclosing Swift's ellipse and the central cluster (VII 2 = NGC 2244).

In 1890 Swift wrote a paper about NGC 2237, titled 'A wonderful nebulous ring', in which the discovery date is mentioned: 1865 (Swift 1890b). He had

---

[43] Swift (1885b: 39).

[44] Swift (1884b); the name Rosette Nebula is a twentieth-century creation.

seen Barnard's ring during his visit to Mt Hamilton and hoped that '*the distinguished celestial photographer, Mr. Isaac Roberts, will be induced to photograph it*'. This favour was eventually done by Barnard, who in January 1894 produced an image with the 6″ Willard lens on Mt Hamilton (Barnard 1894d).[45] In an article published shortly before, Barnard noted that '*What I had seen previously and what Swift had sketched, was simply the brightish knot in a vast nebulous ring that entirely surrounded the cluster.*' (Barnard 1894c). On 27 February 1886 Swift discovered another part of the Rosette Nebula, NGC 2246, writing that it was '*Probably an offshoot of 31 of my catalogue No. 2* [NGC 2237]. *Two or three others suspected.*' (Swift 1886a).

9.5.3 Swift's observations at the Warner Observatory

In 1887 Swift published a leaflet, titled *History and Work of the Warner Observatory Rochester, N.Y., 1883–1886* (Swift 1887b). He wrote that '*The Warner Observatory is distinctively a private institution built for the purpose of original discovery rather than the ordinary routine work of most other observatories.*'[46] He added that '*The desire of its founder was that the great telescope (then the third largest in the United States) be used in such work; so, selecting as my principal field of labor, the discovery of new nebulae, which, since the death of the Herschels and of d' Arrest, had been much neglected.*' Swift was not enthusiastic about the urban site (at an elevation of only 172 m): '*The observatory is situated in the most cloudy section, saving Alaska, in the United States, and since its telescope was mounted the hindrance of work from this cloudiness has been very great.*' He further wrote that '*Much time is, of course, also lost by moonlight, and about half the observing weather is given to comet-seeking.*'

The 16″ Clark refractor, which was erected in a 9.5-m dome on top of the building, was ready to use in spring 1883 (Fig. 9.15); at that time it was the fourth-largest telescope in the United States, after those at Washington, Princeton and Chicago. The finder had an aperture of 8.3 cm. Swift's standard eye-piece was the

Figure 9.15. Swift and his 16″ Clark refractor at Warner Observatory; at bottom left the 4.5-inch can be seen (Rochester Public Library).

50-mm Gundlach with an inner ring plus crosshairs, which had already been proven at his private 4.5-inch. It now offered a power of 132 and 33′ field of view and Swift remarked that '*The power is as high as the most of the undiscovered nebulae will bear, even with a telescope of 16 inches aperture. If a much higher power be used and the seeing be poor a few small stars close together will appear nebulous when, in fact, no nebulosity exists.*'[47] The field was not illuminated: '*The faintness of a majority of those in my list of novae is inconceivable except to those who are engaged in similar work. With a luminous field, the largest telescope in the world would not reveal one of them.*'[48] Other eye-pieces had focal lengths of 33 mm, 27 mm and 25 mm, which implied powers of 200, 250 and 265, respectively. Swift also had a bar-micrometer by Clark with four eye-pieces (magnification 105 to 1250), but did not use it: '*it is very doubtful if, even with a dark field and luminous wires, micrometrical measures of positions can ever be successfully made*'.

---

[45] Max Wolf photographed the nebula in 1906 too (Wolf M. 1906).
[46] Swift (1887b: 5).

[47] Swift (1885b: 4).
[48] Swift (1885c: 43).

9.5 Warner Observatory

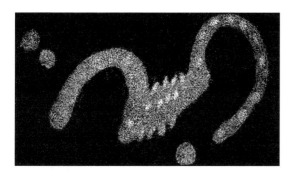

Figure 9.16. Swift's 'Swan', the emission nebula M 17 in Sagittarius (Swift 1885c).

In spring 1883 Swift started observing. As early as on 26 April he discovered the first nebulae with the 16-inch: the galaxies NGC 3588 and NGC 3522 in Leo (see Table 9.16). The former was found near the bright star δ Leonis (see Section 9.5.7).

He was interested in well-known objects too: '*As soon as the 16-inch refractor of this observatory was mounted, I began the scrutiny of several of the most conspicuous nebulae for the purpose not only of personal gratification, but also to discover, if possible, some new features overlooked by previous observers as delineated by them in pictorial illustrations.*'[49] The Omega Nebula (M 17) in Sagittarius, called the 'Swan' by him, is an example. On 4 July 1883 he recognised a '*curious appendage from the following end of the Swan nebula not represented in any known drawing*'.[50] Swift reports that Copeland, while visiting the Warner Observatory in 1883,[51] had seen the appendage too. However, it was never catalogued. Swift's paper contains a sketch. Apart from that of the Rosette Nebula it was his only published one. In contrast to such observers as Tempel and d'Arrest, he seldom drew objects (obviously owing to a lack of talent). Swift's sketch of M 17 shows a monstrous object (Fig. 9.16), much different from that in published figures by John Herschel and Tempel.

After a test phase, during which a few objects were found with the Clark refractor, Swift started his systematic search for nebulae on 9 July 1883, '*selecting for my field of observation the region between the head and the third coil of Draco, and on that short night found fourteen*'.[52] At first he was surprised about his success, finding 120 new nebulae by 1885: '*When we reflect how thoroughly, during the last hundred years, the sky has been searched over by seekers after comets, nebulae, double-stars, etc., it would seem that in the heavens, and, more especially, north of 40° south declination, there could be hardly a single undiscovered nebula as bright as Herschel's Class II.*' (Swift 1885d). Later he wrote the following about the discoveries: '*It was formerly supposed that because the heavens had been so thoroughly searched by the Herschels, D'Arrest and others, the quest for new ones would be an almost fruitless one, indeed time well nigh lost, but the numbers since discovered by Tempel, Stephan, Stone, and at this observatory, show that the nebulae, like the stars, are inexhaustible.*' (Swift 1888b).

The visual observation, which was normally done in the meridian, was Swift's favourite task, whereas he disliked measuring positions. His method was cumbersome, imprecise and, above all, fault-prone: '*When a nebula is found, it is brought roughly to the center of the field, the driving block started, the telescope clamped in R.A., the optical center of the nebula bisected with the wires* [of the 50-mm Gundlach] *in both coordinates, the telegraph sounder connected with the break circuit sidereal clock started; all without removing my eye from the telescope and without an assistant.*'[53] Still at the eye-piece, Swift memorised the basic facts: '*After noting its size, shape and brightness and configuration with some of the nearer stars in the field, and if excessively faint, and therefore probably to be re-found with difficulty, the number of seconds as counted from the sounder by which it follows or precedes the nearest bright star.*' After that, he had to step down from his observing ladder to read the setting circles – in the darkness, and, due to his handicap, this was a rather dangerous task: '*I then descend from the observing-chair to read the circles, make the records, both as to position and appearance.*' The essential tool used to determine a position was the 'automatic R. A. circle', an invention by Burnham. Swift speaks about it as an '*exceedingly useful and time-saving device where right ascensions, already reduced to any desired epoch, are read off directly from the circle*'.[54] He added that '*the places obtained are purely differential*'.

---

[49] Swift (1885b: 2).
[50] Swift (1885c: 38).
[51] Copeland R. (1884b).
[52] Swift (1885c: 38).

[53] Swift (1885c: 41).
[54] Swift (1887b: 8).

Swift's difficult measuring method influenced visual observing too: '*The dome-room is absolutely dark while I am sweeping, and the pupil of the eye is greatly expanded (an essential preparation for the observation of such faint bodies). The gas is lighted only while reading the circles and making records, but, as after its extinguishment, I am for a few minutes nebula-blind, a considerable loss of precious time is involved.*'[55] Others, such as William Herschel and Tempel, had had to grapple with this problem too. Swift was aware that his method was not exact: '*The discovery of nebulae and the getting of their approximate places is one thing, while the obtaining of their micrometrical positions with mathematical exactness, is another and very different matter.*'[56] He had no university background – unlike d'Arrest and Dreyer – and therefore was not familiar with classical astronomy, especially measuring positions with a micrometer or meridian-circle. Swift was an amateur astronomer and discoverer, comparable to the early William Herschel or Barnard. He enjoyed the liberty of a 'private' observatory, not being bound to routine work like university or government astronomers. Swift further writes that '*For the former work (their discovery) one set of appliances only can be used, viz; an eye-piece of low power and a large field for sweeping. For the latter quite another plan must be adopted, and the two cannot be combined. A micrometer eye-piece is as poorly adapted to nebula as to comet-seeking.*' For him, micrometrical work was a waste of time and he was happy '*to get positions with all accuracy possible with the means adopted for their discovery, leaving the specialists to pick them up without sweeping and to fix their places with such an exactness that those who come after us may be able to determine whether they also, like the stars, are drifting*'.[57] In contrast to classical astronomers such as d'Arrest, Schönfeld and Schultz, Swift was never interested in proper motions of nebulae. Some of his southern positions were later checked by Jermaine Porter at Cincinnati Observatory, who found large errors (Porter 1910).

9.5.4 Swift's lists of nebulae

Swift published altogether 13 lists with new nebulae in the *Astronomische Nachrichten* and *Monthly Notices* (Table 9.9). Dreyer termed them Sw I to Sw XII and another, short list Sw (X). The first six contain objects catalogued in the NGC. About Swift's list VI (*AN* 2798) Dreyer remarked in the appendix of the NGC that, '*While the present Catalogue was in press, Mr. Swift published in the* Astronomische Nachrichten, *No. 2798, his sixth list of new nebulae, containing most of the objects marked here Sw VI.*'[58] He had received the data by letter in advance. This earlier version differs from the published list VI, as Dreyer notes: '*The positions of some of the objects in the hours 12 to 17 differ slightly from those communicated to me.*'

The remaining lists were considered in the two Index Catalogues. An exception is list IX, which contains, according to Swift, 25 objects that had been found earlier but '*by an oversight are not in any of my previous lists*' (Swift 1890a). About the reason why they appear in the NGC, despite their having been published too late, Dreyer noted '[list] *VI. communicated by degrees in MS*'.[59] The 25 objects (plus 5 others not contained in list IX) are referred to as 'Sw VI' in the NGC. Each of the first nine lists shows exactly 100 objects; obviously Swift initially published his discoveries in portions (as did William Herschel). All of his lists have the same structure: object number (mostly 1 to 100), discovery date, position (various epochs), descriptions and remarks. Notes are given below for a few objects and occasionally corrections to earlier lists.

An analysis of all of the lists containing NGC objects is presented in Table 9.10. The column '(VI)' contains the five objects referred to by Dreyer as 'Sw VI' (NGC 5783, NGC 6212, NGC 6285, NGC 7759 and NGC 7780). As mentioned above, they do not appear in list IX, but were communicated in a manuscript sent by letter. Column 'VI' shows 101 objects, because Swift's entry no. 91 in list VI contains two different objects: NGC 3645 and NGC 4646. Column 'X' refers to the only object not in any of the lists: the planetary nebula NGC 1360 in Fornax, which had been discovered in 1859. Those 111 objects which had been discovered earlier by other observers (23 persons, from Baily to Voigt) cannot be presented here. Some are included in one of the subsequent tables.

---

[55] Swift (1885c: 41).
[56] Swift (1885c: 42).
[57] Swift (1885c: 42–43).

[58] Dreyer (1953: 235).
[59] Dreyer (1953: 11).

Table 9.9. *Swift's lists of nebulae (Obs. = observatory)*

Sw	Objects	Catalogue	Date	Publication	Ref.	Epoch	Obs.
I	100	NGC	Aug. 1885	*AN* [2683] **112**, 313–318 (1885)	1885e	1885	Warner
II	100	NGC	Nov. 1885	*AN* [2707] **113**, 305–310 (1885)	1886a	1885	Warner
III	100	NGC	7.6.1886	*AN* [2746] **115**, 153–158 (1886)	1886b	1885	Warner
IV	100	NGC	Sep. 1886	*AN* [2752] **115**, 257–262 (1886)	1886c	1885	Warner
V	100	NGC	2.11.1886	*AN* [2763] **116**, 33–38 (1886)	1886d	1885	Warner
VI	100	NGC	23.5.1887	*AN* [2798] **117**, 217–222 (1887)	1887a	1885	Warner
VII	100	IC I	4.8.1888	*AN* [2859] **120**, 33–38 (1888)	1888a	1890	Warner
VIII	100	IC I	29.5.1889	*AN* [2918] **122**, 241–246 (1889)	1889	1890	Warner
IX	100	NGC/IC I	July 1890	*AN* [3004] **126**, 49–54 (1890)	1890a	1890	Warner
X	60	IC I	20.2.1892	*AN* [3094] **129**, 361–364 (1892)	1892a	1890	Warner
(X)	7	IC I	Aug. 1892	*MN* **53**, 273 (1893)	1893	1890	Warner
XI	243	IC II	June 1898	*AN* [3517] **147**, 209–220 (1898)	1898a	1900	Lowe
XII	45	IC II	9.7.1899	*MN* **59**, 568–569 (1899)	1899	1900	Lowe

The entry II-30 (i.e. no. 30 in list II) is the 10.1-mag star GSC 4754–1474, which is located in the reflection nebula NGC 1788 (V 32, h 347). Swift describes it as a 'nebulous star'. When he got the NGC, he realised that the object was not listed there. Thereupon he entered it as no. 9 in list VII: '*I thought it advisable to re-insert it here.*' But Dreyer ignored it once again, writing in a note published 1889 in the *AN* that '*The object is a star of 10th magnitude and is involved in the nebulosity pf h 347 = V 32, as pointed out many years ago by d'Arrest and the observers at Birr Castle.*' (Dreyer 1889). His decision was final: '*There appeared therefore to be no reason for entering the star in the catalogue as a separate object, but it is referred to in the description of h 347* [NGC 1788].'

Swift was greatly interested in 'nebulous stars', writing in 1885 that '*Our Sun is supposed by many astronomers to be one of this class of objects, which, seen from a planet belonging to another sun, would exhibit the appearance of a nebulous star surrounded with a luminous atmosphere which we call the Zodiacal light.*' (Swift 1885d). He added that '*they should form a distinct class from those in which the star is not centrally placed, and presumably are stars that happen to be situated in our line of sight with the nebula, but probably far this side of the nebula itself, and with which it has no physical connection*'. Swift presents three examples of 'nebulous stars': William Herschel's NGC 7023 (IV 74) plus his own finds NGC 2247 and NGC 7094. The reflection nebula NGC 2247 in Monoceros was found on 24 November 1883, but the true discoverer was Mitchell at Birr Castle (14 February 1857). NGC 7094 is a planetary nebula in Pegasus, which was found on 10 October 1884 (see Table 9.15). Swift notes that '[it] *is the most wonderful of all* [nebulous stars] – *in fact it is the only instance known to me – for instead of the central star being single, it is double*'.

Three entries in Swift's lists are comets (I-19, II-29 and VI-14), and thus were ignored by Dreyer.

Table 9.10. *Assignment of objects in Swift's lists of nebulae (see the text)*

	List									Sum
	I	II	III	IV	V	VI	IX	(VI)	X	
Lewis Swift	73	73	75	78	68	76	20	3	1	467
Edward Swift	9	3	2		1	5	5			25
Baily	1									1
Barnard			3							3
Bond						1				1
Burnham					1					1
Coggia		1								1
d'Arrest		3	5	2	1	1				12
Ferrari					1					1
Hartwig	1									1
J. Herschel	1	1	1	1		1				5
W. Herschel	3	2	3	6	5	5				24
Holden			1							1
Hunter						1				1
Leavenworth					2	1		1		4
Marth	3	1	1	2						7
Mitchell	1	2								3
Muller						1				1
Palisa			1							1
Safford		1	1	2		3				7
Skinner						1				1
Stephan	3	5	4	3	9	1		1		26
Stone					1					1
Tempel		1	1	1	3					6
Voigt		1			1					2
Star		1								1
Comet	1	1				1				3
List identity		2	1	2	5					10
NGC identity	4	4	1	2	2	1				14
Catalogued in IC					1	2				3
Number of objects	100	100	100	100	100	101	25	5	1	632

About I-19 Swift notes in the introduction of his second list that '*It was found 1885 Apr. 6, while searching for Tempel's comet. It was of course a Comet, but I thought at the time that the Decl. was too great to be Tempel's.*' For II-29 one reads '*Resembles a Comet.*'[60]

Ten objects appear twice in the lists (i.e. Swift later found them a second time). These 'list identities' are shown in Table 9.11. An interesting case is the close double galaxy NGC 6621/22 in Draco; the fainter component NGC 6622 lies only 50″ southeast of NGC 6621 (see Section 9.5.6). The latter was first seen by Edward Swift on 2 June 1885 (I-95). His father inspected the field and detected the fainter object (I-96). A second observation was made on 11 August 1885 by Lewis Swift alone, who noted that the fainter nebula was '*Very difficult to separate with power of 265.*' He did not realise that the double nebula was already contained in list I. This is not surprising, given the many objects

---

[60] About VI-14, see Section 9.5.5.

9.5 Warner Observatory 347

Table 9.11. *Objects appearing twice in the lists (except for I-95, all of the objects were found by Lewis Swift)*

NGC	List	Date	List	Date	Type	V	Con.	Remarks
6214	I-47	2.8.1884	IV-39	3.8.1886	Gx	13.5	Dra	
6621	I-95	2.6.1885	II-66	11.8.1885	Gx	13.2	Dra	I-95: E. Swift, II-66: L. Swift; pair w. NGC 6622
6622	I-96	2.6.1885	II-65	11.8.1885	Gx	15.0	Dra	I-96, II-65: L. Swift; pair w. NGC 6621
6869	II-83	26.8.1884	IV-79	6.9.1886	Gx	12.1	Dra	
1038	III-16	17.10.1885	V-34	3.8.1886	Gx	13.5	Cet	
6677	I-98	8.6.1885	III-100	25.10.1885	Gx	13.1	Dra	Pair w. NGC 6679 (IX-90), 24.6.1887 (see Section 6.5.6)
6488	IV-59	1.9.1886	V-81	29.9.1886	Gx	14.5	Dra	
7066	IV-80	31.8.1886	V-92	31.8.1886	Gx	14.2	Peg	
7511	IV-92	6.9.1886	V-95	25.9.1886	Gx	14.0	Peg	
6690	V-86	16.8.1884	V-85	31.10.1886	Gx	12.3	Dra	NGC 6689, d'Arrest 22.8.1863

Table 9.12. *Entries catalogued twice in the NGC ('NGC identity')*

NGC	List	Date	NGC	List	Date	Type	V	Con.	Remarks
859	V-23	3.10.1886	856	V-22	31.10.1886	Gx	13.3	Cet	
866	V-24	3.10.1886	885	V-27	31.10.1886	Gx	13.0	Cet	NGC 863, W. Herschel 6.1.1785 (III 260)
5502	I-29	9.5.1885	5503	I-30	11.5.1885	Gx	15.3	UMa	NGC 5502: E. Swift, NGC 5503: L. Swift
5785	VI-71	21.4.1887	5783	VI	1887?	Gx	12.9	Boo	
5826	I-39	9.6.1885	5870	I-41	11.6.1885	Gx	14.4	Dra	
6053	III-86	8.6.1886	6057	III-89	8.6.1886	Gx	14.7	Her	
6127	IV-29	6.7.1886	6128	IV-30	28.7.1886	Gx	12.0	Dra	NGC 6125, W. Herschel 24.4.1789 (II 810)
6170	IV-35	9.7.1886	6176	V-70	1.10.1886	Gx	13.8	Dra	
6189	II-41	3.8.1885	6191	IV-36	6.7.1886	Gx	12.6	Dra	
6297	II-49	8.7.1885	6298	II-50	1.8.1885	Gx	13.6	Dra	
6497	I-80	16.9.1884	6498	I-81	26.9.1885	Gx	13.4	Dra	
6511	I-83	9.10.1884	6510	IV-61	30.5.1886	Gx	13.7	Dra	
6667	II-69	11.9.1883	6678	I-99	8.6.1885	Gx	13.0	Dra	NGC 6668, L. Swift 31.7.1886 (IV-70)
6763	II-76	30.8.1883	6762	II-75	30.4.1884	Gx	13.4	Dra	

348 Compilation of the NGC

Table 9.13. *Double entries catalogued in the NGC and IC*

NGC	List	Date	IC	List	Date	Type	V	Con.	Remarks
223	VI-5	21.11.1886	44	X-1	12.11.1890	Gx	13.4	Cet	Bond 5.1.1853 (GC 119), d'Arrest 1.1.1862
3979	III-61	27.4.1886	2976	XI-129	23.5.1897	Gx	12.7	Vir	Holden 23.4.1881
5876	I-43	11.6.1885	1111	VIII-86	27.8.1888	Gx	12.9	Boo	
6050	IV-26	27.6.1886	1179	VII-71	3.6.1888	Gx	14.1	Her	Hercules galaxy cluster
6056	III-88	8.6.1886	1176	VII-69	8.6.1888	Gx	14.1	Her	Hercules galaxy cluster
6798	II-80	5.8.1885	1300	X-47	2.10.1891	Gx	13.2	Cyg	
7649	VI-96	25.9.1886	1487	IX-99	15.10.1887	Gx	13.9	Peg	

Table 9.14. *Objects from lists I–VI that appear in the IC*

IC	List	Date	Type	V	Con.	Remarks
884	VI-53	24.4.1887	NF		Vir	VII-30
887	VI-55	24.4.1887	NF		Vir	VII-31
1712	III-6	30.11.1885	Gx	10.5	Cet	Barnard; = NGC 584, W. Herschel 10.9.1785 (I 100)
1761	V-18	2.10.1886	Gx	14.4	Cet	Javelle 18.12.1897

discovered in Draco (see Table 9.16). The double entry NGC 6690 was first seen by d'Arrest in 1863, though it was catalogued as NGC 6689 by Dreyer.

'NGC identity' refers to 14 cases in which Swift found an object twice or even thrice (NGC 6667), but Dreyer, not having resolved the identity, catalogued it as different objects in the NGC (Table 9.12). For two of them (NGC 866 and NGC 6127) William Herschel was the true discoverer. Additionally, there are seven cases in which double entries appear both in the NGC and in the IC (Table 9.13). Two objects (NGC 223 and NGC 3939) must be credited to Bond and Holden, respectively.

Table 9.14 shows the objects from lists I to VI (i.e. the published prior to the NGC) which appear in the two Index Catalogues. Only for IC 884 and IC 887 (both in IC I) does Dreyer mention Swift. They were entered again in list VII, where Swift wrote that '*These are numbers 53 and 55 of List VI, which inadvertently were not sent to Dr. Dreyer, and, therefore, are not incorporated in the body of the NGC, but are in the appendix and, being liable to be overlooked, are inserted here.*' Dreyer notes in the appendix of the NGC that '*The list [VI] contains two objects, Nos. 53 and 55, which accidentally had not been communicated to me, and are therefore not included in the present Catalogue.*'[61]

9.5.5 Lewis and Edward Swift's discoveries

Altogether 466 objects must be credited to Lewis Swift – too many for a presentation. His son Edward discovered another 25 (see Table 9.21). Table 9.15 shows the numbers according to type; 443 and 23, respectively, are non-stellar objects – very high rates.

---
[61] Dreyer (1953: 235); see also Swift's remarks in Swift (1888b).

Table 9.15. *Swift's objects, divided by type*

Type	L. Swift	E. Swift	Objects
Galaxy	438	23	Brightest (L. Swift): NGC 2663 = III-40 (Pyx), 10.9 mag, 8.2.1886
			Brightest (E. Swift): NGC 2523 = II-32 (Cam), 11.8 mag, 7.9.1885
Planetary nebula	3		NGC 1360 (For), 9.4 mag, 1859
			NGC 2242 = VI-27 (Aur), 15.0 mag, 24.11.1886; Fig. 9.17 left
			NGC 7094 = II-88 (Peg), 14.4 mag, 10.10.1884; Fig. 9.17 right
Emission nebula	2		NGC 2237 = II-31 (Mon), 1865, part of Rosette Nebula
			NGC 2246 = III-36 (Mon), 27.2.1886, part of Rosette Nebula
Reflection nebula	1		NGC 2296 = VI-28 (CMa), 3.11.1187
Star	2		
Star pair or group	4		
Not found	17	2	E. Swift: NGC 5309 (Vir), NGC 6666 (Lyr)
Sum	467	25	

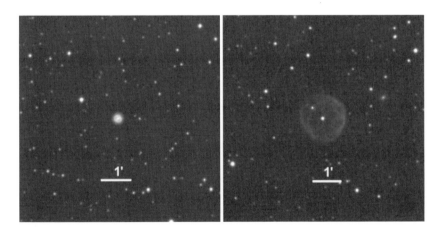

Figure 9.17. Two planetary nebulae of Swift. Left: NGC 2242 in Auriga; right: NGC 7094 in Pegasus (DSS).

Lewis Swift's emission, reflection and planetary nebulae are given in detail in the column 'Objects'; for the other types, important examples are presented. NGC 2242 actually holds two PN records in the NGC: it is both the faintest and the last to have been discovered (it was catalogued as 'galaxy' CGCG 204–5 too). NGC 7094 is an equally remarkable NGC PN, being the second last one to have been discovered and the only one in Pegasus. NGC 2237 and NGC 2246 are parts of the Rosette Nebula in Monoceros (see Section 9.5.2).

Table 9.16 shows special objects and observational data. Lewis Swift found the greatest number of new nebulae (13) on 22 October 1886. The most fruitful constellation (for both father and son) was Draco. William Herschel and other observers had not seriously scanned this near-pole region, which fact was known

Table 9.16. *Special objects and observational data of Lewis and Edward Swift*

Category	Object	Data	Date	Remarks
Brightest object	NGC 1360	PN, 9.4 mag, For	1859	See Table 10.7
Faintest object	NGC 5919 = VI-77	Gx, 15.9 mag, Ser	30.3.1887	
Most northerly object	NGC 6414 = III-94	+74°, Gx, 15.3 mag, Dra	30.5.1886	NGC 2760 (NF) at +76°
Most southerly object	NGC 2663 = III-40	−33°, Gx, 10.9 mag, Pxy	8.2.1886	NGC 1392 (NF) at −37°
First object	NGC 1360	PN, 9.4 mag, For	1859	See Table 10.7
First object (Warner)	NGC 3588 = I-12	Gx, 14.4 mag, Leo	26.4.1883	Plus NGC 3522 (III-59), Gx, 131 mag, Leo
Last object (Warner)	NGC 6916 = IX-94	Gx, 14.1 mag, Cyg	26.6.1887	Last-discovered NGC object
Best night			22.10.1886	13 objects
Best constellation	Draco			154 objects (10 by E. Swift)

to Swift. As the '*first neb. discovered at this Observatory [Warner]*' Swift mentions I-28 (galaxy NGC 5439 in Canes Venatici), which was discovered on 9 July 1883, but list I gives for its no. 2 (NGC 3588) the date 26 April 1883. Swift explained the discrepancy by invoking the fact that the start of his systematic observations with the 16-inch was meant.

It is interesting that both the most northerly and the most southerly object are missing. The position of NGC 1392 (VI-15) lies about 2° south of the Fornax Cluster. Perhaps Swift saw a galaxy, but noted a wrong position. During the same night (13 February 1887) he found another object, which he placed 4' to the south: VI-14 (not catalogued by Dreyer). Swift wrote in the following a footnote in list VI: '*Regarding this object I have reasons for supposing with a considerable degree of probability that it was the great comet 1887 I.*' He further wrote that '*No. 15 was by estimation exactly 4' north of it.*' This explains the rough position of NGC 1392.

Figure 9.18 shows the brightness distribution of Swift's objects, giving a mean of 13.6 mag. Thus he is in this regard comparable to Stephan (13.5 mag), though the French observer used a 31-inch with a silvered glass mirror. About the comparison with objects of William Herschel, Swift remarked in February 1887 that '*The greater part of Herschel's class III* [very faint nebulae] *are bright objects compared with most of those discovered at this observatory. Nearly one-half of the 200 novae can be seen only by averted vision and by the most persistent and long-continued effort.*'[62] Averted vision is a useful technique to glimpse faint objects; the eye does not look centrally at the object, but to the side. The light then enters the more sensitive rods.[63]

Table 9.17 lists objects without date (an estimate is given). Three were referred in the NGC to '*List VI*'. However, they are not included there, but were communicated by letter.

9.5.6 The strange case of NGC 6677/79 in Draco

On 8 June 1885 Swift for the first time observed a field in the central region of Draco. The number '3' is magical in this case: there are three galaxies, three entries in Swift's lists (I-98, III-100 and IX-90), three objects in Dreyer's catalogues (NGC 6677, NGC 6679 and IC 4763)[64] and three visual observers involved (Lewis Swift, Guillaume Bigourdan and Herbert Howe). This looks like perfect harmony – but really was far from it: the case is among the strangest in the NGC/IC.

NGC 6677 and NGC 6679 are two of the many nebulae Swift discovered in Draco. They are connected with the entries I-98, III-100 and IX-90. The field in

---

[62] Swift (1885c: 40).
[63] Steinicke (2004a: 166).
[64] See Steinicke (2001b).

Table 9.17. *Swift's objects without date*

NGC	Date	Type	V	Con.	Remarks
6123	1885?	Gx	13.7	Dra	II-40, no date given
6285	1886?	Gx	13.6	Dra	VI
6612	1886?	Gx	14.5	Lyr	VI
5783	1887?	Gx	12.9	Boo	VI; = NGC 5785 (VI-71), L. Swift 21.4.1887

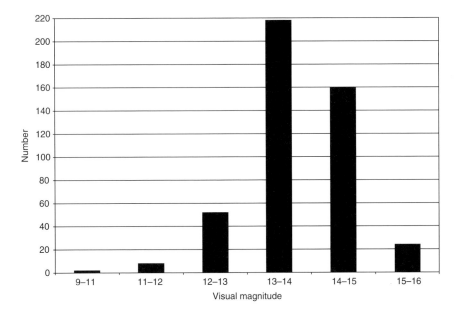

Figure 9.18. The brightness distribution of Swift's objects.

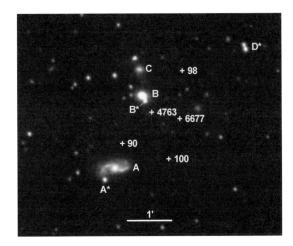

Figure 9.19. The galaxy trio NGC 6677/79/IC 4763 in Draco and Swift's objects I-98, III-100 and IX-90 (DSS; see the text).

question exhibits three galaxies (Fig. 9.19), which are first marked as A, B and C. Object B is the central and brightest (13.1 mag). The galaxy is small and compact with a 14.5-mag star (B*) 7″ to the southwest. Galaxy A (of type Sbc) is more extended and a bit fainter (13.6 mag), being located 1.7′ south-southwest of B. The 17″ southeast of A is a 14.7-mag star (A*). At 15.1 mag the small galaxy C, 0.6′ north of B, is the faintest. Then trio is also known as VV 672 in the *Catalogue of Interacting Galaxies* by Vorontsov-Velyaminov.

Swift's position and description of I-98 '*pF, pS, R*' ('pretty faint, pretty small, round') in the first instance fits the central galaxy B pretty well; its place is 1.2′ to the northwest (marked '98'). The observation was made on 8 June 1885. Did he overlook the companion A, only 1.7′ south following? To settle this question, the second observation (III-100), made 4½ months later (25 October

1885) must be taken into account. The position (marked '100') lies 1′ west of A and 1.6′ south of B, respectively. This indicates galaxy A, especially since the line 100–98 is nearly parallel to A–B and of the same length. This is supported by Swift's description '*eF, eS, bet. a star v close and a vF D star*' ('excessively faint, excessively small, between a very close star and a very faint double star'). Let's start with brightness. Assuming that I-98 is galaxy B (13.1 mag) and III-100 is A (13.6 mag), Swift's data 'vF' and 'eF' fit very well. But the sizes ('pS' and 'vS') nominally do not: the compact object B (0.4′ × 0.3′) is smaller than the elongated spiral A (0.9′ × 0.5′). However, the spiral arms are difficult to see; the core of A actually appears smaller than B.

Do the mentioned stars near A exist? The first choice for Swift's 'star v close' is A*, 17″ to the southeast. Because of the word 'between', the double star ('vF D star') must be northwest of A. There are two candidates. About 4′ distant, i.e. pretty far away, we find a double star (marked 'D*') with components of 14 mag (separation 7″). In this case A is located very asymmetrically between D* and A* (ratio 14:1). The alternative depends on the question of whether Swift saw B during his second observation (anyway, nothing is noted about it). Perhaps he did indeed see B: as the 'very faint double star'. In fact, A is located between A* and B, but less asymmetrically (6:1). Moreover, B has a close companion, the star B* (7″ to the southwest). With a power of 132, as was used by Swift, the pair B (compact galaxy) and B* (star) could look like a close 'double star'. No matter which choice is correct, the position and description initially lead to the conclusion that III-100 = galaxy A. Does this automatically imply that I-98 = B? To clear this up, the third observation (IX-90) is needed.

Owing to his entries I-98 and III-100, Swift obviously was forced to revisit the field, which took place on 24 June 1887. The observed object IX-90 is one of the 25 which erroneously are not included in list VI, which was published in 1887. The missing data appeared in 1890 in list IX, i.e. two years after the NGC. However, Dreyer was aware of them, because Swift had sent a complete version of list VI as a manuscript in due time (this communication is called 'VIm' here). Compared with this manuscript, list IX gives more information about IX-90: '*An e close D with* [NGC] *6679; suspected with* [power] *132, confirmed with 200; perfectly separated with 250.*' Obviously, the term '6679' was added.

At this point an important question must be treated: how did Dreyer interpret and catalogue Swift's observations? As a reminder: the NGC was a mere desk-top product made without knowing the real sky. Dreyer had only Swift's data from lists I, III and VIm. To begin with NGC 6677, Dreyer refers to '*Swift I & III*', which indicates that he has identified I-98 and III-100. Actually, he has mixed up Swift's data. The position of NGC 6677 (marked '6677') is 1.6′ to the southwest of B. This is neither the place of I-98 nor that of III-100 – but it is the mean position! He also has averaged the descriptions: '*vF, vS, bet star v close & vF D star*'. From 'pF, pS' (I-98) and 'eF, eS' (III-100) he derived 'vF, vS'; 'very' (v) is between 'pretty' (p) and 'excessively' (e). The rest originates from the description of III-100. Except for 'R' ('round'), Dreyer has not omitted information. Since III-100 had already been identified with galaxy A, we now also have NGC 6677 = A. Dreyer interpreted I-98 as a mere re-observation of this object. Swift's positional differences relative to A are 2.8′ (I-98) and 0.8′ (III-100), which is not unusual. He was not satisfied with the precision of his measurements, based solely on reading the setting circles. In list III he remarks that '[I] *hope that improved methods of measurement would give more exact positions*'. In March 1888 Swift was able to use improved mechanics and optics. Thus Swift's positions for the three entries (determined between 1885 and 1887) can hardly solve the problem. One must rely on the descriptions.

Dreyer refers NGC 6679 to '*Swift VI*', giving the description '*eF, close double*'. This shows, too, what Swift's original description of IX-90 was in his manuscript VIm. Dreyer omitted the information regarding magnification from the NGC (perhaps to save space). Let's look at Swift's probable original note (VIm): it first mentions a double object. In this observation (as noted in list IX) higher powers were used – finding a 'close double'. At first glance, Swift's place of IX-90 and 'eF' fit A well. Is IX-90 therefore another observation of galaxy A? The key to the solution must lie in the term 'close double': Swift could have meant either the galaxy pair A+B or the galaxy–star pair B+B*.

The possibility A+B is supported by Swift's later addition 'with 6679' (IX). If IX-90 = A, then the second nebula is the one Dreyer has called NGC 6679 – this then must be B. The only problem is that Dreyer has called exactly this object a 'close double'! The first alternative thus leads to a contradiction. The

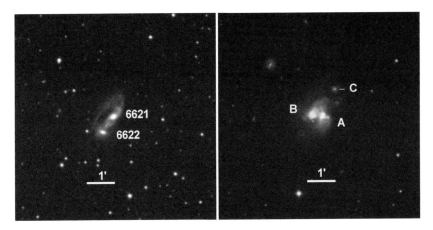

Figure 9.20. Left: the double galaxy NGC 6621/22 in Draco. Right: the double galaxy NGC 3690A/B in Ursa Major; the faint companion C is IC 694 (DSS).

error must be due to Swift incorrectly expanding his original description. Obviously he has revised the issue in his paper 'Nebulae and clusters', dated 29 January 1894 (Swift 1894). About 'NGC 6679' he remarks that, '*Like many stars, some of the nebulae are double and triple. One of my own, New General Catalogue, No. 6679, is a close double, suspected with a power of 132, confirmed with 200, and well separated with 250.*' It is unambiguously stated that 'NGC 6679' is the double nebula. Thus we have IX-90 = NGC 6679 = B+B*. Of course, Swift's estimate 'eF' fits too.

It is interesting to investigate similar cases generated by Swift. On 2 June 1883 he discovered (assisted by his son) the double galaxy NGC 6621/22 (Arp 81; Fig. 9.20 left) in Draco.[65] It is located 2.3° to the northwest of NGC 6679. The components are separated by 42″. In the Gundlach eye-piece, giving a power of 132 (field of view 33′), young Edward could see only the northeastern component NGC 6621 (I-95) of 13.2 mag. When Lewis Swift checked the field, he detected the 15.0-mag galaxy NGC 6622 (I-96), perhaps using a higher magnification. Though the distance is smaller in the case of B+B*, the 'components' (compact galaxy plus star) are better defined. Anyway, both doubles are difficult in their own ways. This was confirmed in a second observation (II-66) of NGC 6621 on 11 August 1885; Swift noted '*forms a close double with the preceding [NGC 6621 = II-65]. Very difficult to separate with a power of 265. Well seen.*'

Another comparable case is the double galaxy NGC 3690 (Arp 299) in Ursa Major, which was discovered by William Herschel on 18 March 1790 (I 247).[66] Neither he nor his son John (who catalogued the object as h 896 and GC 2425) noticed the double structure. The two components (A: 11.2 mag, B: 10.9 mag) are only 22″ apart. The first to separate them was Bindon Stoney on 27 January 1852, observing with the 72″ reflector at Birr Castle: '*Neb. div. into two parts [NGC 3690 A/B], F appendage np about one diam. distant.*'[67] The 'appendage' is a new object: a small 15.0-mag galaxy C, located 1′ to the northwest. Swift observed NGC 3690 in mid 1883: '*mentioned by all observers as very little elongated. Chancing to run across it with a power of 132, I immediately suspected it to be a close double, which suspicion a power of 200 confirmed. It is probably the closest double nebula known. I shall reexamine it at the first opportunity.*'[68] In this case he had not consulted the Birr Castle observations (normally this was mentioned). Nine years later, on 18 April 1892, a second observation was made, which appears in list (X): '*vs, close D with 3690 = H 247.1 suspected with 132, ver. with 200.*' Only after this confirmation did he deem it necessary to list his find. Again Swift uses the term 'with', which is misleading. Dreyer, obviously not knowing of Swift's first observation, catalogued the galaxy C as IC 694. It is credited to '*Ld R. Sw (X)*',

---

[65] See Steinicke (2001b).

[66] See Steinicke (2001a).
[67] Parsons W. (1861a: 722), Parsons L. (1880: 97).
[68] Swift (1885c: 39).

using Swift's description: '*vS, forms D neb with I 247*'. However, this interpretation is wrong. Evidently, Swift meant the second component of the double galaxy NGC 3690 – not IC 694! Curiously, Swift mentions a possible proper motion. Because the two Herschels did not see a double nebula and '*while now the components are very easily separated*' he believes that '*If they have proper motion in opposite directions it is not unreasonable to conclude that in ninety-six years* [since W. Herschel] *the separation ought to be noticeable.*' (Swift 1886d). He recommends micrometrical measurements.

Regarding these examples, the conclusion for NGC 6677/79 is that I-98 = III-100 = NGC 6677 = A and IX-90 = NGC 6679 = B+B*. Dreyer's NGC 6677 position is the mean of Swift's data. Since Swift's IX-90 position lies nominally about 2′ southeast of that given in the NGC, one might expect the same for NGC 6679. This would place NGC 6679 (B) below NGC 6677 (A), which is not the case. Curiously Dreyer's NGC 6679 position is above NGC 6677 – even 8.5′ farther north and thus far outside the field!

The first to inspect the objects (after Swift) was Bigourdan. The visual observation was made on 5 September 1891 with the 12″ refractor at Paris Observatory. He saw two nebulae. Following Dreyer's data, the southern one was interpreted as NGC 6677. The position determined with the micrometer exactly matches galaxy A. Now Bigourdan searched for the 'double nebula' NGC 6679, which he expected to be 8.5′ farther north – of course, in vain. Surprisingly, he detected a 'new' nebula. The measured position is that of galaxy B. He wrote '*No doubt this object of 13.3 mag is the double star mentioned for NGC 6677; it cannot, however, be resolved with powers 159 and 344; with the latter it even appeared nebulous.*'[69] This interpretation is interesting, since it supports the suggestion that Swift saw the pair B+B* (compact galaxy plus star) while observing III-100. A second observation of Bigourdan on 25 September 1897 confirmed his first result: '*The object is a mix of nebula and star; consisting of a pretty stellar core of 13.3 mag, surrounded by an oval nebula of 10″ diameter.*' Bigourdan entered the object as no. 333 in his list of new nebulae.

Further important results are due to Herbert Howe, Director of Chamberlin Observatory, Denver (Howe 1900). On the basis of the NGC data he visually inspected the field in 1900 with the 20″ Clark refractor,

certainly using a sufficiently high magnification. His micrometrical positions of NGC 6677 and NGC 6679 are very good. NGC 6679 (galaxy B) is described as follows: '*This is a nebulous double star of mags. 12.5, distance 5″, and angle 60°.*' This is the pair B+B*. About NGC 6677 (galaxy A) he notes that '*I did not notice the very faint double star; the other one is of mag. 12, and follows the nebula 2^s, 10″ south.*' Howe also saw A*, the star southeast of A, although he had been expecting a double star in the opposite direction. He obviously did not account for his 'nebulous double star' (B+B*) as the proper candidate. For Swift, who was the first observer, using too little magnification, the situation of course looked different.

In contrast to Bigourdan, Howe (not knowing of the Paris observations) corrected Dreyer's drastic position error of NGC 6679: '*The N.G.C. place is 8.5′ out in declination.*' In the notes of the Second Index Catalogue Dreyer refers to Howe's observation. The work lists Bigourdan's new nebula as IC 4763. The description '*vF, ? neb star; 6677 nr*' is a mix of Bigourdan's and Howe's data ('nebulous double star') given for the northern object (B), which had been measured by both observers. Dreyer, however, does not dare to state an identity with NGC 6679, voicing his doubt by including a question mark. That NGC 6677 is nearby ('6677 nr') is unquestionable.

The discovery night of I-98 (8 June 1885) brought yet another object: I-99. Swift's place is 11′ west of NGC 6677/79. This case is odd too, concerning the entries NGC 6667, NGC 6668, NGC 6678 and IC 4762. The first three are due to Swift – but they are identical (see Table 9.12). This is one of five cases in the NGC with three different numbers (however, it is the only one with a single observer). Lately, the three 'objects' have been considered as corresponding to a spiral galaxy of 13.0 mag, 54′ northwest of NGC 6677/79. Swift's places deviate by about 1° from the true position – another proof of his deficient measuring method. IC 4762 was discovered by Bigourdan (no. 332 in his list of new nebulae); it is a double star, 13′ southeast of the galaxy. There is no doubt that Draco is an exceptional constellation!

9.5.7 Galaxy clusters, galaxies near bright stars and a 'superthin galaxy'

Swift shared Stephan's opinion about pairs/groups of nebulae: '*Nebulae are often associated in pairs and sets,*

---

[69] Bigourdan (1907a: E 53).

Table 9.18. *Members of the Hercules Cluster (Abell 2151) that were discovered by Lewis Swift*

NGC	List	Date	V	Remarks
6034	IV-20	19.6.1886	13.6	
6039	IV-21	27.6.1886	13.7	NGC 6042, Stephan 27.6.1870
6043	IV-22	27.6.1886	14.4	
6044	IV-23	27.6.1886	14.3	
6045	IV-24	27.6.1886	14.1	
6047	IV-25	27.6.1886	13.7	
6050	IV-26	27.6.1886	14.1	IC 1179 (VI-71), L. Swift 3.6.1888
6054	IV-27	27.6.1886	14.3	
6055	III-87	8.6.1886	14.1	
6056	III-88	8.6.1886	14.1	IC 1176 (VI-69), L. Swift 8.6.1888
6057	III-89	8.6.1886	14.7	NGC 6053 (III-86), L. Swift 8.6.1886
6061	III-90	8.6.1886	13.6	

and so frequently has this been observed that on the discovery of one, search in its immediate vicinity is invariably made for a mate and often as the result of chance, one is found apparently connected by physical relations like the binary system of double stars.'[70] In June 1886 he discovered 11 NGC galaxies of the Hercules Cluster Abell 2151 (Table 9.18), also finding Stephan's NGC 6039 = NGC 6042 (Stephan is credited for NGC 6040 and NGC 6041 too). Later he discovered further members, which were entered in the IC. But two of them had already appeared in the NGC, without their being noticed by Dreyer due to the complex situation: NGC 6050 = IC 1179 and NGC 6056 = IC 1176.

Five galaxies are published in list III (including NGC 6053). Swift notes at the beginning that, '*In some cases where many nebulae are in one field or nearly so, identification from imperfect or erroneous descriptions becomes difficult and sometimes impossible. As one instance among many, M. Stephan has three nebulae, [GC] 5799, 5800, and 5801 [NGC 6040–42] in one field, wherein I see six, but I have been able to identify with certainty only one of them as, when the positions agree the descriptions are at variance.*' Eight galaxies are in list IV, where one reads that '*Three of the ten or more nebulae in this interesting group are Mr. Stephan's presumably G.C. 5799, and certainly 5800 and 5801. Two or 3 more suspected. They are very difficult to see and especially to measure, atmospheric*

Table 9.19. *Members of the galaxy cluster Abell 194 in Cetus that were discovered by Lewis Swift*

NGC	List	Date	V
519	VI-8	20.11.1886	14.4
530	VI-9	20.11.1886	13.1
538	VI-10	20.11.1886	13.8
557	VI-11	20.11.1886	13.7

conditions seldom allowing them to be seen at all except Stephan's last two, which are quite interesting objects, but those he describes as faint and small and very small, I call pretty large.' In list VII Swift remarks that, '*After the completion of my improved method of reading right ascensions directly from the circle, I remeasured the members of this interesting group, and am enabled to identify M. Stephan's 3, which are NGC 6040, 1, 2. His places are in close agreement with mine.*' It follows an account with positions (epoch 1890) for 12 objects.

On 20 November 1886 Lewis Swift discovered 4 of 13 NGC galaxies in the cluster Abell 194 (Cetus Cluster), which has a central part of 1° diameter (Table 9.19). The rest were found by William Herschel, d'Arrest (see Table 8.12) and Searle.

In 1884 Swift had the idea of searching for nebulae near bright stars: '*On the night of April 26, 1884, while engaged in nebula-seeking, there occurred to me the thought, that perhaps there are undiscovered nebulae in close*

---

[70] Swift (1885c: 40).

Figure 9.21. The galaxy trio NGC 1618/22/25 near ν Eridani (DSS).

proximity to some of the brighter stars.'[71] The most prominent case is due to William Herschel, who discovered the 10.3-mag galaxy NGC 404 (II 224) 6.6′ northwest of Mirach (β And, 2.1 mag) on 13 September 1784.[72] A similar find was made by d'Arrest on 2 December 1861: the galaxy NGC 772 (13.6 mag), 4.7′ south of β Ari (2.6 mag).

A curious case is the galaxy trio NGC 1618/22/25 near the 3.9-mag star ν Eridani (Fig. 9.21). The objects of 12.7 mag, 13.1 mag and 12.3 mag form a chain of 17′ length in the northwest–southeast direction, only 10′ north of the star. One could assume that a single observer would have found the spectacular trio – of course, William Herschel is the first choice. He did indeed discover NGC 1618 (II 524) on 1 February 1786, but mentioned only '*p 2 S st*' ('preceding two faint stars'); ν Eri cannot be one of them. Nothing is said about the other objects – they were even found by different observers! John Herschel saw NGC 1625 on 24 November 1827 (h 322), mentioning at least ν Eri. He was, however, not aware of his father's object (catalogued as h 320 at the correct position): '*No description; observation marked as doubtful.*' The first to recognise all three galaxies was Johnstone Stoney on 16 January 1850 at Birr Castle, using the 72-inch (a sketch shows the situation). The case was further complicated by an observation of d'Arrest on 1 January 1862, when he discovered another object, which was catalogued by John Herschel as GC 878 = h 320,a ('d'Arrest 44'). Stoney's nebula is listed as GC 881 = h 322,a ('R. nova') – with a declination error of 3° into the bargain! In the Birr Castle publication of 1880 Dreyer, on the basis of his own observation of 1 December 1874, states that these two objects are identical (NGC 1622). This example shows how difficult the perception could be in the vicinity of a bright star.

Swift's very first attempt to find a galaxy near a bright star was successful: '*I turned the telescope upon Delta Leonis, and immediately there was a nebulous object which proved to be new.*' He had found the galaxy NGC 3588 (14.4 mag), located 8.2′ south of the 2.5-mag star Zosma (Fig. 9.22 left). Additionally, this was Swift's first object at Warner Observatory. Encouraged by this event, he made a find near Algol too (14 October 1884). On 29 May 1887 his son Edward discovered three galaxies near Vega.[73] All of these cases are listed in Table 9.20.

On 10 November 1885 Swift discovered the remarkable galaxy NGC 100 (III-1), noting that it was '*very elongated*' (Fig. 9.22 right). With an axis ratio of 9.6:1 it is the flattest NGC galaxy (13.6 mag). It belongs among the rare cases of a 'superthin galaxy'.[74] NGC 100 was also observed on 7 September 1891 by Bigourdan, who wrote that its '*form and extension are incredible*' (Bigourdan 1904c). The second place is held by NGC 5023 in Canes Venatici (12.1 mag), showing a ratio of 9.3:1. The galaxy was discovered by William Herschel on 9 April 1787 (II 664) and described as '*much elongated, 5′ long, ¾′ broad*'. With a ratio of 8.9:1 the 12.3-mag galaxy IC 2233 in Lynx, which was found photographically by Roberts on 25 March 1894, is an extreme case too (Roberts 1903c).

9.5.8 Objects of Edward Swift

Edward Swift discovered 25 objects, mostly in Draco (Table 9.21). All bar two are galaxies. NGC 5502 (I-29) was again found by his father, three days later, who put both 'objects' in his first list, without noticing the identity. The same is true for Dreyer, who catalogued I-30 as NGC 5503. The positions are 2.5′ apart, with Lewis Swift's as the better one. The objects near Vega and the double galaxy NGC 6621/22 have already been treated above.

---

[71] Swift (1885c: 41).
[72] See Steinicke (2001e).
[73] See also Swift's remarks in Swift (1888b).
[74] See Steinicke (2000b).

Table 9.20. *The galaxies near bright stars which were found by Lewis and Edward Swift*

NGC	List	Discoverer	Date	Type	V	Con.	Star	Position
1212	I-5	L. Swift	18.10.1884	Gx	14.6	Per	Algol (β Per)	18′ SE
1213	I-6	L. Swift	14.10.1884	Gx	14.7	Per	Algol (β Per)	2.3° S
1224	II-28	L. Swift	20.8.1885	Gx	14.0	Per	Algol (β Per)	42′ SE
3588	I-12	L. Swift	26.4.1883	Gx	14.4	Leo	Zosma (δ Leo)	8.2′ S
6663	IX-88	E. Swift	29.5.1887	Gx	13.9	Lyr	Vega (α Lyr)	1.4° NW
6685	IX-91	E. Swift	29.5.1887	Gx	13.4	Lyr	Vega (α Lyr)	1.3° NO
6686	IX-92	E. Swift	29.5.1887	Gx	14.7	Lyr	Vega (α Lyr)	1.5° NW

Figure 9.22. Left: The galaxy NGC 3588 near Zosma (δ Leonis); right: the 'superthin galaxy' NGC 100 in Pegasus (DSS).

9.6 DEARBORN OBSERVATORY: SAFFORD, SKINNER, BURNHAM AND HOUGH

From 1866 to 1883 Safford, Skinner, Burnham and Hough discovered nebulae at Dearborn Observatory, Chicago. Safford had already been successful at Harvard College Observatory. Burnham ranked as the leading double-star observer of his time. This was proved not only in Chicago, but also later at Lick and Yerkes Observatories. Hough was well known for his studies of Jupiter.

9.6.1 Short biography: Truman Henry Safford

Truman Henry Safford was born on 6 January 1836 in Royalton, VT (Fig. 9.23 left). Already during his childhood a remarkable talent appeared: he was a master of mental arithmetic, able to perform complicated calculations in a very short time. At the age of 9 he published a table with planet ephemerides, and he calculated the orbit of comet 1849 at the age of 14. After studying at Harvard University (1852–54) he worked as an assistant at the observatory. He was entrusted with observations and reductions of star positions and celestial mechanics. After the death of George Phillips Bond (in 1865) Safford was provisional Director of Harvard College Observatory for a year. Then he became Professor of Astronomy at the University of Chicago and in 1866 Director of Dearborn Observatory, which was being run by the Chicago Astronomical Society (located at that time in the urban Douglas Park). His assistants were Aaron Skinner and Ormond Stone.

During the years 1866–68 Safford discovered 47 nebulae with the 18.5″ Clark refractor of Dearborn Observatory. Later he measured star positions with the meridian-circle. On 9 October 1871 a fire destroyed large parts of Chicago; the observatory was affected too. Because there was no money for its reconstruction,

Table 9.21. *NGC objects discovered by Edward Swift*

NGC	List	Date	Type	V	Con.	Remarks
851	III-10	30.11.1885	Gx	13.7	Cet	
1009	III-15	1.1.1886	Gx	14.5	Cet	
2128	VI-25	27.12.1886	Gx	12.6	Cam	
2523	II-32	7.9.1885	Gx	11.8	Cam	Arp 9
4544	VI-45	27.4.1887	Gx	13.1	Vir	
4633	VI-46	27.4.1887	Gx	13.3	Com	
4969	VI-52	27.4.1887	Gx	13.9	Vir	
5309	VI-60	27.4.1887	NF		Vir	
5502	I-29	9.5.1885	Gx	15.3	UMa	NGC 5503 (I-30), L. Swift 11.5.1885
6091	II-39	8.7.1885	Gx	14.2	UMi	
6288	I-52	19.8.1884	Gx	14.5	Dra	
6289	I-53	19.8.1884	Gx	14.5	Dra	
6382	I-60	2.6.1883	Gx	14.6	Dra	
6395	I-63	18.9.1884	Gx	12.2	Dra	
6418	I-68	4.5.1885	Gx	14.4	Dra	
6457	I-71	8.6.1885	Gx	14.3	Dra	
6532	V-83	19.9.1886	Gx	14.0	Dra	
6536	I-84	18.8.1884	Gx	13.4	Dra	
6585	IX-87	25.5.1887	Gx	12.9	Her	
6621	I-95	2.6.1885	Gx	13.2	Dra	II-66 (Fig. 9.20 left)
6663	IX-88	29.5.1887	Gx	13.9	Lyr	1.4° NW of Vega
6666	IX-89	25.5.1887	NF		Lyr	
6685	IX-91	29.5.1887	Gx	13.4	Lyr	1.3° NE of Vega
6686	IX-92	29.5.1887	Gx	14.7	Lyr	1.5° NE of Vega
6825	II-82	18.9.1884	Gx	14.3	Dra	

Safford abandoned his office. The successor was Elias Colbert, who had been its Vice-director since 1870. Until 1876 Safford worked on cartography for the government, then he became Professor of Astronomy at Williams College, Williamstown (MA). There he observed with the 4.5″ meridian-circle by Repsold and published a catalogue of near-pole stars in 1888. Truman Henry Safford died of a heart attack on 13 June 1901 in Newark, NJ, at the age of 64.[75]

9.6.2 Short biography: Aron Nichols Skinner

Aron Skinner was born on 8 August 1845 in Boston, MA. During the years 1866–71 he worked as Safford's assistant at Dearborn Observatory. Later he became Professor of Mathematics at the US Naval Academy, Annapolis, and an assistant at the US Naval Observatory, Washington. There he measured star positions for the catalogue of the Astronomische Gesellschaft (AGK). Aaron Skinner died on 14 August 1918 in Framingham, MA, at the age of 73.[76]

9.6.3 Short biography: Sherburne Wesley Burnham

Sherburne Wesley Burnham was born on 12 December 1838 in Thetford, VT (Fig. 9.23 centre). After professional training as a stenographer he went to New York in 1858 and later was in charge in New Orleans for the Confederate Army in the American Civil War. From

---

[75] Obituaries: Hollis (1901), Knobel (1901), Jacoby (1901) and Searle A. (1902); see also Fox (1913), Bailey (1931: 257–258) and Ashbrook (1976).

[76] Obituary: Anon (1918b).

9.6 Dearborn Observatory

Figure 9.23. Observers at Dearborn Observatory (left to right): Truman Henry Safford (1836–1901), Sherburne Wesley Burnham (1838–1921) and George Washington Hough (1836–1909).

1866 Burnham worked as a court reporter in Chicago. There he discovered his interest in astronomy. Having been inspired by Webb's *Celestial Objects for Common Telescopes*, he began to observe double stars with a 3.75″ Fitz refractor. He lived near Dearborn Observatory, which was located at that time in Douglas Park. Director Safford allowed him to use the 18.5″ Clark refractor. When Safford left the site in 1871, Burnham backed away too.

In 1868 Burnham married Mary Cleland; they had three sons and three daughters. In 1870 he purchased a fine 6″ Clark refractor to be erected in the garden of his house near Ellis Park. There Burnham discovered his first double star on 27 May 1870. The find was a surprise, because it had been assumed that, with the work of John Herschel, James South and Wilhelm and Otto Struve, all of the double stars that would be visible with existing telescopes had already been found. In 1873 Burnham published his first catalogue, with 81 new objects; another followed. During the years 1872–77 he discovered altogether 451 pairs with his 6-inch. Because he had no micrometer at the beginning, the famous Milan double-star observer Baron Dembowski was asked for measurements with his 19-cm Merz refractor. This contact was the origin of a friendship lasting until Dembowski's death in 1881. Not only the Baron but also Otto Struve, a proven expert on double stars, was impressed by Burnham's results. On Webb's suggestion Burnham became a Fellow of the RAS in 1874. In autumn 1874 he visited the US Naval Observatory in Washington. Newcomb allowed him to observe with the 26″ Clark refractor. There he met Edward Holden, his later superior on Mt Hamilton. After that he visited Charles Young at Dartmouth Observatory, Hanover, using the local 9.4″ refractor.

Thanks to Safford's successor Colbert, the great refractor at Dearborn Observatory was again ready for action in mid 1875. His dramatic appeal of May 1873 eventually found an attentive ear: '*It seems not very creditable to the Observatory that the great telescope should stand unused, given over to rust and dust, while novice Burnham is publishing from Chicago lists of new double stars discovered with his 6-inch telescope.*'[77] From September 1876 to April 1877 Burnham was the temporary Director of Dearborn Observatory. He used the 18.5-inch to search for and measure double stars, finding 413 new ones. Volume 1 of the *Annals of the Dearborn Observatory*, published in 1915, is dedicated to him: '*To Sherburne Wesley Burnham, whose labours brought new life to double star astronomy, who spared not himself in his own heroic vigils, whose personal encouragement has been the direct inspiration for these observations, this volume is gratefully inscribed.*'

In 1879 Burnham erected his 6-inch on Mt Hamilton to test the site for the planned Lick Observatory (Burnham S. 1880). In 1881 he visited Washburn Observatory, where Holden was at that time director, to use the 15.1″ Clark refractor. Then he took a five-year break to work again as a court reporter in Chicago. His refractor was sold to Madison (Fig. 9.24). Holden, who had meanwhile become Director of Lick Observatory, was eventually able to hire him for Mt Hamilton in 1888. Burnham was offered two nights per week for observing double stars. With the 12″ and 36″ refractors Burnham found 198

[77] Fox (1915: 4).

Figure 9.24. Burnham's old 6″ Clark refractor at Washburn Observatory, Madison.

new objects. As by-products some new nebulae were seen. At Lick Observatory a life-long friendship with Barnard began. They shared not only their enthusiasm for visual observing, but also a growing dislike for the authoritarian director Holden. In 1892 Burnham left Mt Hamilton to return to Chicago; Barnard was to follow him three years later, when the crisis reached its ugly climax.

During the years 1892–1902 Burnham worked as an amanuensis at the US Circuit Court. In 1894 he was awarded the RAS gold medal for his double-star observations; the laudation was given by President William Abney (Abney 1894). In the same year he was appointed Professor of Practical Astronomy at the University of Chicago, operating the new Yerkes Observatory in Williams Bay (120 km northwest of Chicago). Hale's attractive offer included the use of the 40″ Clark refractor.[78] Of course, Burnham could not resist and from 1895 to 1914 he was employed on a freelance basis at the observatory. During the years 1897–1902 he used the 40-inch at weekends; later, only sporadically (the last observation is dated 13 May 1914). Often about 100 measurements were made per night. Thirty-four new double stars were found.

In 1906 Burnham published his famous *General Catalogue of Double Stars*, containing 13 665 objects (his own 1300 discoveries are termed 'β'). He was the typical American self-made man. As an astronomical autodidact he always refused university teaching; his favourite task was visual observing and micrometrical measuring. He was an excellent photographer too, but concentrated on terrestrial subjects. Astrophotography (and astrophysics) was not his concern. Sherburne Wesley Burnham died on 11 March 1921 in Chicago, IL, at the age of 82.[79]

9.6.4 Short biography: George Washington Hough

George Washington Hough was born on 24 October 1836 in Tribes Hill, NY (Fig. 9.23 right). He studied at Union College, Schenectady, and graduated in 1859 from Harvard University. Then he became the assistant of Ormsby Mitchel at Cincinnati Observatory. Both of them moved to Dudley Observatory (Albany) in 1850, where in the same year a 13″ Fitz refractor was installed.[80] Hough was appointed director in 1862. From 1874 he was charged with business affairs, but could exploit his own inventions. Following Colbert and Burnham, in 1879 he became Director of Dearborn Observatory, which at that time was still located in Douglas Park. Hough and Burnham observed double stars with the 18.5″ Clark refractor. A critical response caused them to publish their paper on Tempel's Merope Nebula, which could not be seen with the large refractor.

However, Hough's primary target was Jupiter. His numerous micrometrical measurements brought him the nickname 'Jupiter'. In 1888 the building of the new Dearborn Observatory located on the campus of Northwestern University, Evanston, north of Chicago, was started. The opening took place on 9 June 1889. Now Hough became Professor of Astronomy, keeping all his offices until his death. George Washington

---

[78] It surpassed even the 36-inch at Lick Observatory and has been the world's largest (existing) refractor from 1897 onwards.

[79] Obituaries: Barnard (1921), Frost (1921a, b), Aitken (1921), See (1921), Anon (1921) and Jackson (1921, 1922); see also Frazer (1889), Eggen (1953b), Ashbrook (1984: 84–92) and Sheehan (1995).

[80] Weddle (1986).

Figure 9.25. The 18.5″ Clark refractor (left) of Dearborn Observatory (right), which was initially located in Chicago's Douglas Park; the telescope is in the cylindrical dome.

Hough died on 1 January 1909 in Evanston, IL, at the age of 72.[81]

9.6.5 New nebulae of Safford and Skinner

From May 1864 onwards the Dearborn Observatory (in Chicago's Douglas Park) possessed an excellent 18.5″ Clark refractor (Fig. 9.25). It was erected in a cylindrical dome on top of a 30-m-high tower. During the test phase in Cambridgeport (MA), Alvan Graham Clark, the son of the famous telescope maker, discovered the companion of Sirius on 31 January 1862.[82] The 18.5-inch was the largest American telescope until 1873, when it was surpassed by the 26″ Clark refractor of the US Naval Observatory, Washington.

During the years 1866–68 Safford discovered, together with his assistant Skinner, altogether 108 nebulae with the refractor. The list was not published until autumn 1887 (at that time Safford was Director of Hopkins Observatory, Williamstown). Titled 'Nebulae found at the Dearborn Observatory, 1866–68' it appeared in the *Annual Reports of the Dearborn Observatory* (Safford 1887). The basic data were taken from Safford's observing books. In the introduction he wrote that '*It was drawn up at the request of Mr. Colbert, in order to complete the record of observed nebulae.*' This must have happened before Colbert retired in May 1879. In Williamsburg Safford's resources were limited: '*As I have no access to the original journals, and here at Williamsburg have no copies of the principal catalogues of nebulae, I must simply rely upon my remembrance that the positions were carefully determined.*' He added '*I remember rightly, to assure myself that the nebulae in question were not in Herschel's general catalogue.*' About his motivation Safford remarks that '[I] *took up this subject of the nebulae mainly to gain a practical acquaintance with these very interesting bodies*'.

The objects in Safford's list are sorted by discovery date (3 May 1866 to 12 June 1868) and numbered from 1 to 108. Therefore Safford is occasionally mentioned as the discoverer of 108 nebulae – but nos. 1, 4 and 17 are missing and one object was inserted (without a number) between no. 37 and no. 38. The positions given are for epoch 1870. Objects from no. 1 to no. 83 are briefly described. About the missing descriptions (no. 84 onwards), Safford remarks that '*From this point on the descriptions have not been copied from the observing book into the book sent me.*'

For some objects footnotes, mostly written by Safford, are given. One must distinguish between those originating from the observation period and others added later. There are insertions in brackets, which are probably due to Colbert, who edited the list. For instance, the note to no. 83 (NGC 6941 = GC 5966) shows all of these variants. First one reads '*Position doubtful [5966].*', followed by '*I suppose these numbers [5966] refer to a continuation of the G.C. not within my reach. They were inserted by some one since I left Chicago.*

---

[81] Obituaries: Curtis (1909) and Anon (1910).
[82] On Sirius B, see Ashbrook (1984: 155–160); in 1888, while testing the 36″ Lick refractor on Mt Hamilton, Clark also discovered the star 'G' in the trapezium of the Orion Nebula (see Section 9.6.5).

The numbers of G.C. end with 5070, and contain all nebulae then (1866) known to me.' Someone has added numbers from Dreyer's GCS (1878). The most probable candidate is Colbert, who was again Director of Dearbon Observatory (1878–79) after Burnham's departure.

Owing to its late appearance in a barely known series, Dreyer at first missed Safford's list. Therefore the objects arise neither in the GCS nor in the NGC. They were eventually presented in the 'Appendix' of the NGC, where Dreyer remarked that *'Professor Safford's list* [is] *published in a place where many observers might overlook it'*. In a table 47 objects with Safford's numbers and description are given, sorted by right ascension (1860), *'in order that this volume may contain a record (I hope a complete one) of all nebulae of which the places have been published up to December 1887'*. About the remaining 59 Dreyer wrote *'Fifty-nine of these were found to occur in the present catalogue, having been recorded independently by other observers.'* The 47 new objects were entered in the first Index Catalogue of 1895, as Dreyer explained: *'I have inserted them here (though found before 1888), as very few people ever think of referring to an appendix.'*

Of the 106 objects, Safford found 103 and Skinner 5. According to Dreyer, 59 of them had already been catalogued in the NGC. This is incorrect, since four others listed in the IC I are duplicates of NGC objects: IC 1026 = NGC 5653 (William Herschel), IC 1030 = NGC 5672 (William Herschel), IC 1148 = NGC 6020 (Stephan) and IC 1253 = NGC 6347 (Stephan).

Table 9.22. *The 27 NGC objects from Safford's list ('No.') which were discovered by himself*

NGC	GCS	No.	Date	Type	V	Con.	Remarks
114		90	23.9.1867	Gx	13.3	Cet	Tempel 27.9.1880
118		91	23.9.1867	Gx	13.6	Cet	Tempel 27.9.1880
124		92	23.9.1867	Gx	13.1	Cet	Tempel 27.9.1880
140		60	8.10.1866	Gx	13.3	And	Stephan 5.11.1882
183		65	5.11.1866	Gx	12.6	And	Stephan 6.10.1883
237		94	27.9.1867	Gx	12.9	Cet	L. Swift 21.11.1886
425		62	29.10.1866	Gx	12.7	And	Stephan 11.10.1879
591		61	10.10.1866	Gx	13.0	And	L. Swift 30.11.1885
769		68	9.11.1866	Gx	12.9	Tri	Stephan 5.11.1882
778	5205	64	5.11.1866	Gx	13.2	Tri	Stephan 17.11.1876
1211		102	31.10.1867	Gx	12.4	Cet	Stephan 27.11.1880
2410	5388	74	5.2.1867	Gx	13.0	Gem	Stephan 2.2.1877
4308		106	11.6.1868	Gx	13.2	Com	Tempel 17.2.1882
5657		16	14.5.1866	Gx	13.4	Boo	Stephan 5.6.1880
6020	5794	10	9.5.1866	Gx	12.8	Ser	Stephan 27.6.1876; IC 1148
6098		76	24.4.1867	Gx	13.3	Her	L. Swift 3.4.1887
6099		76	24.4.1867	Gx	14.2	Her	L. Swift 3.4.1887
6347		29	6.6.1866	Gx	13.7	Her	Stephan 6.7.1880; IC 1253
6484	5883	41	11.7.1866	Gx	12.5	Her	Stephan 27.6.1876
6527		46	1.8.1866	Gx	14.2	Her	L. Swift 6.6.1886
6575		33	7.6.1866	Gx	13.0	Her	Stephan 1.7.1880
6589		81	28.8.1867	RN		Sgr	L. Swift 12.7.1885; Fig. 9.26
6641	5916	47	9.8.1866	Gx	13.5	Her	Stephan 20.8.1873
6941	5966	83	29.8.1867	Gx	12.8	Aql	Stephan 1.9.1872; [5966]
7291	6055	56	1.10.1866	Gx	13.5	Peg	Stephan 21.9.1876
7343	6072	53	14.9.1866	Gx	13.5	Peg	Stephan 26.9.1876
7375		57	1.10.1866	Gx	13.7	Peg	L. Swift 2.9.1886

9.6 Dearborn Observatory 363

Figure 9.26. Reflection nebulae in Sagittarius: Safford's NGC 6589 and Swift's NGC 6590 (DSS).

The last two were actually first seen by Safford. Thus 63 objects from Safford's list were already in the NGC (some can even be traced back to the GCS). Safford and Skinner must be credited with 61 and 2, respectively. Seven of Safford's 61 objects cannot be identified – the fields are blank. Of the remaining 54, Safford was the discoverer of 27 (Table 9.22). Most were credited by Dreyer to Stephan (16), followed by Swift (7) and Tempel (4). Eight objects were already in the GCS, having been known to Dreyer prior to 1877 (see the 'Remarks'). All bar one are galaxies – a remarkable rate. About the reflection nebulae NGC 6589 (Fig. 9.26) Safford notes '*star 10^m pretty faint nebulosity*'. The double galaxy NGC 6098/99 in Hercules is given a single number (no. 76): '*double nebula, pretty faint, dist 40″*', which correctly describes the situation.

Also 27 of Safford's objects had been discovered earlier by others (Table 9.23); in these cases the priority given in the NGC is correct. Except for one (NGC 6364), all had already been catalogued in the GC or GCS. A few remarks on interesting objects follow. NGC 595 is an H II region in M 33, which was discovered by d'Arrest. Safford notes that it was '*pretty faint, very small, probably a well-known outlier of M 33*'. Both of the planetary nebulae (NGC 6765 and NGC 6842) were found by Marth in a single night. The former is the second PN in Lyra (after M 57); it was later seen by Stephan too.

The galaxy NGC 6674 in Hercules appears twice in the list (nos. 32 and 34), as had already been supposed by Safford, who noted regarding no. 34 '*Same as 32?*'. For NGC 1016 Colbert (?) had seen an identity with GC 5266 (NGC 1020) and Safford wrote '*My own note is that this nebula possibly = G.C. 581*'; this is NGC 1032. But, due to the positions given, both assignments are wrong. However, in the case of NGC 5088 Safford's suggestion '*G.C. 3489?*' has been confirmed. Finally, the galaxy NGC 7422 is remarkable: it is among the NGC objects with the most independent discoveries (together with Swift's NGC 1360 and Voigt's NGC 6364; see Table 10.7); between 1864 and 1867 it was found by Marth, Otto Struve, d'Arrest and Safford.

Skinner discovered the galaxy NGC 577 in Cetus (Fig. 9.27). Dreyer mentions Tempel, who saw two nebulae on 14 August 1877; but the second, NGC 580, is identical with NGC 577. In the case of NGC 7416 Marth was faster than Skinner. Table 9.24 shows Skinner's IC objects too.

9.6.6 New nebulae of Burnham and Hough

Five of the NGC objects discovered in Chicago are due to Burnham and four to Hough (Table 9.25). Except for the small planetary nebula NGC 7026, which was first seen by Burnham with his private 6-inch, all were found with the 18.5″ Clark refractor.

9.6.7 By-products of Burnham's double-star observations

On 6 July 1873 Burnham discovered the planetary nebula NGC 7026 in Cygnus with his 6″ Clark refractor (Fig. 9.28). The telescope was erected in 1870 in his small private observatory in Chicago and used to observe double stars. The nebula, which he first considered to be a component of a double star, is mentioned in his 'Third catalogue of new double stars' (Burnham S. 1873 = Ref. 1): '*A very remarkable and curious double, or elongated planetary (?) nebula. It is close to a 9.3 m. star, Arg. (+47°) 3298.*'[83] About the identification he remarks that '*This may have been noted before, but it is not in Herschel's General Catalogue, or Lassell's Catalogue of new Nebulae.*'

---

[83] Burnham S. (1873: 71).

Table 9.23. *The 27 NGC objects listed by Safford which had been discovered earlier by others*

NGC	GC(S)	No.	Date	Type	V	Con.	Remarks
63	5093	96	30.9.1867	Gx	11.8	Psc	d'Arrest 27.8.1865
595	5186	63	1.11.1866	GxP	13.5	Tri	d'Arrest 1.10.1864; in M 33
1016	5264	103	1.11.1867	Gx	11.6	Cet	Marth 15.1.1865, Tempel 1876; [5266]; 'G.C. 381'
4012	5598	108	12.6.1868	Gx	13.5	Vir	Marth 25.3.1865 (m 225); [5589]
4314	2881	20	16.5.1866	Gx	10.5	Com	W. Herschel 13.3.1785 (I 76)
4585	5664	21	16.5.1866	Gx	14.1	Com	d'Arrest 21.4.1865
4614	5666	11	11.5.1866	Gx	13.4	Com	d'Arrest 9.5.1864
4615	5667	12	11.5.1866	Gx	13.1	Com	d'Arrest 9.5.1864
5088	3489	104	20.5.1868	Gx	12.6	Vir	Mitchell 18.4.1855, d'Arrest 26.4.1867; 'G.C. 3489?'
5653	3914	13	11.5.1866	Gx	12.1	Boo	W. Herschel 13.3.1785; IC 1026
5672	3930	6	4.5.1866	Gx	13.4	Boo	W. Herschel 13.3.1785; IC 1030
5936	4105	24	1.6.1866	Gx	12.3	Ser	W. Herschel 12.4.1784 (II 130)
5996	4139	25	1.6.1866	Gx	12.5	Ser	W. Herschel 21.3.1784 (II 97)
6364		49	5.9.1866	Gx	13.1	Her	Voigt June? 1865, Stephan 21.7.1879, L. Swift 11.9.1885
6371	5861	45	1.8.1866	Gx	14.3	Her	Marth 24.6.1864 (m 336)
6577	5899	30	6.6.1866	Gx	12.7	Her	Marth 7.8.1864 (m 367)
6580	5901	31	6.6.1866	Gx	13.1	Her	Marth 7.8.1864 (m 369)
6674	5923	32, 34	6.6.1866	Gx	12.2	Her	Marth 6.6.1864, Stephan 18.7.1871; present twice in list
6675	5924	54	28.9.1866	Gx	12.5	Lyr	Voigt July? 1865, Stephan 27.7.1870; 1.3° N of Vega
6688	5926	35	9.6.1866	Gx	12.7	Lyr	Marth 3.8.1864, Stephan 25.7.1870
6765	5941	42	12.7.1866	PN	12.9	Lyr	Marth 28.6.1864, Stephan 20.7.1870
6842	5947	43	12.7.1866	PN	13.1	Vul	Marth 28.6.1864, d'Arrest 26.8.1864
6897	5952	80	24.8.1867	Gx	14.0	Cap	Marth 28.6.1864 (m 406)
6898	5953	79	24.8.1867	Gx	13.6	Cap	Marth 28.6.1864 (m 407)
7288	6053	84	19.9.1867	Gx	13.0	Aqr	Marth 1.10.1864 (m 482); [6053]
7422	6101	93	27.9.1867	Gx	13.5	Psc	Marth 11.8.1864, O. Struve 6.12.1865, d'Arrest 29.9.1866
7783	6230	99	23.10.1867	Gx	13.0	Psc	Marth 9.9.1864 (m 591); [6230]

Copeland independently found the PN with his visual spectroscope on 18 November 1880 (Copeland R. 1881). Dreyer, knowing about this observation, overlooked Burnham's publication; otherwise the nebula would have been catalogued in the GCS. Obviously Dreyer, when compiling lists of new nebulae, did not scan papers on double stars (later his searching was cleverer). Burnham read Copeland's discovery note, which appeared in early 1881 in *Copernicus* – and was forced to define the priority in an article titled 'The planetary nebula in Cygnus' (Burnham S. 1881). Like his later friend Barnard, he always paid attention to correct acknowledgement, especially concerning his own results. In his article dated 8 March 1881 Burnham wrote that '*Astronomers interested in nebulae seem to have overlooked the fact*

contains four measurements of NGC 7026 made in 1878. For all observations the Dearborn 18.5-inch was used.

9.6.8 Nebulae discovered by Burnham and Hough in 1883

Hough's publication 'New nebulae' lists six objects (Hough 1883 = Ref. 3). They were discovered by Burnham and him while they were searching for comet d'Arrest in April and May 1883 with the 18.5-inch. All are galaxies in Sextans, Virgo and Bootes. Dreyer credits NGC 3047 (Sextans) to 'Burnham and Hough'. However, Hough's report clearly states that both observers 'examined' the galaxy on 5 May 1883, but it had already been found by Hough on 24 April. The Virgo objects discovered on 5 May lie in a field of diameter 45′. Three were independently found by Tempel on 11 May (i.e. only six days later) with the Amici I refractor in Arcetri. He was on the track of comet d'Arrest too. Tempel proudly wrote on 11 June (knowing of Hough's publication) that '*It is a great pleasure for me to see my earlier statement that the 10½-inch Amici I shows the nebulae as well as an 18½-inch confirmed.*' (Tempel 1883c). He missed Hough's NGC 5191, not examining the object until 5 July. However, he discovered a new brighter one (the galaxy NGC 5178), commenting that '*my nebula was not noticed in Chicago*'. It is unknown why Hough calls NGC 5171 a '*double*'. Like the Dearborn observers, Ernst Hartwig in Straßburg searched for comet d'Arrest with the 18″ Merz refractor. On 29 June he independently found NGC 5171 and NGC 5179. He also noticed two other galaxies (NGC 5176 and NGC 5177). All four form a small group of diameter 5′ (see Fig. 9.71 left).

Hough's final object, NGC 5511, causes a problem. The field in Bootes shows an interacting pair of galaxies (VV 299). The components (A and B) are only 1′ apart (Fig. 9.29); A is to the southeast (14.0 mag) and B to the northwest (14.5 mag). B is fainter and smaller, presenting a somewhat higher surface brightness. So, which object was seen by Hough? His description is not helpful: '*Small, very faint. star 10^m, star 10^m preceding.*' There is no 10-mag star to the west – let alone two! Probably a typo happened. Tempel noticed only '*a star 12^m with very little nebulosity*'. Checking the field with a modern telescope

Figure 9.27. Skinner's galaxy NGC 577 in Cetus, which is identical with Tempel's NGC 580 (DSS).

that the Planetary Nebula found by Dr. Copeland at the Dun Echt Observatory last November was discovered a good many years ago.' He also reports about observations of the PN with the 18.5″ Clark refractor made during the years 1878–80. Already on first seeing the nebula he had suggested that '*With a large aperture it would probably be a very interesting object.*' In 1891 he was able to study NGC 7026 with the 36″ Lick refractor (Burnham S. 1891).

In a paper of 1881 Burnham compared his micrometrical measurements of the 'pair' (BD +47° 3298 plus NGC 7026) with those of Copeland, finding good agreement. Being a 'classical' visual observer, he made snide remarks about Copeland's spectroscopic search method: '*Whatever the merits or advantages of the spectroscopic plan for the discovery of nebulae may be generally, certainly no new method is required for this particular object; for any good refractor of four or five inches aperture is quite sufficient to detect it at once with an ordinary eyepiece.*'

Two other nebulae discovered by Burnham in 1877–78, are also by-products of double-star observations: the galaxies NGC 1363 and NGC 4997. They appear in his catalogue of 1879 (Burnham S. 1879 = Ref. 2). Again, this was noticed by Dreyer too late; not so for Tempel, who observed NGC 1363 in about 1880 (Tempel 1882). Entry no. 424 of Burnham's catalogue

Table 9.24. *Objects from Safford's list that had been discovered by Skinner (sorted by date)*

Object	No.	Date	Type	V	Con.	Remarks
NGC 7416	86	21.9.1867	Gx	12.4	Aqr	GC 6099, Marth 25.8.1864 (m 507)
IC 1528	88	23.9.1867	Gx	12.7	Cet	
IC 138	95	27.9.1867	Gx	14.0	Cet	
NGC 577	100	23.10.1867	Gx	13.0	Cet	= NGC 580, Tempel 14.8.1877; Fig. 9.27
IC 210	101	23.10.1867	Gx	13.1	Cet	

Table 9.25. *The NGC objects found by Burnham and Hough (for 'Ref.' entries see the following section).*

NGC	Discoverer	Date	Type	V	Con.	Ref.	Remarks
1363	Burnham	31.12.1877	Gx	13.1	Eri	2	L. Swift 21.10.1886
3047	Hough	24.4.1883	Gx	14.1	Sex	3 (1)	Dreyer: 'Burnham and Hough'
4997	Burnham	28.3.1878	Gx	12.9	Vir	2	Holden 17.5.1881
5165	Burnham	5.5.1883	Gx	13.9	Vir	3 (2)	Tempel 11.5.1883
5171	Hough	5.5.1883	Gx	12.9	Vir	3 (3)	'double', Tempel 11.5.1883; Hartwig 29.6.1883
5179	Burnham	5.5.1883	Gx	14.2	Vir	3 (4)	Tempel 11.5.1883; Hartwig 29.6.1883
5191	Hough	5.5.1883	Gx	14.1	Vir	3 (5)	Tempel 5.7.1883
5511	Hough	10.5.1883	Gx	14.5	Boo	3 (6)	
7026	Burnham	6.7.1873	PN	10.9	Cyg	1	Copeland 18.11.1880

Figure 9.28. Burnham's first find, the small planetary nebula NGC 7026 in Cygnus (DSS).

revealed that B is the better candidate, since A looks too diffuse.

9.6.9 Burnham at Lick and Yerkes Observatories

Burnham's astronomical career reached its highest points when he was working at these two great observatories. However, the stay at Mt Hamilton was overshadowed by the enduring conflict with its authoritarian director Holden (Barnard was affected even more). There Burnham used the 36-inch for micrometrical measurements of NGC objects. The amateur astronomer often welcomed Barnard in the large dome, since Holden had refused him a weekly observation night at this telescope (this did not change until 1892). Barnard described his friend as someone '*whose excellent custom it is to measure everything he comes across that is measurable*' (Barnard 1895b).

Figure 9.29. The double galaxy NGC 5511A/B (VV 299) in Bootes; probably Hough saw component B (DSS).

A first report with observations of 27 planetary nebulae appeared in November 1891 (Burnham S. 1891). A second followed in April 1892, containing 32 NGC objects (Burnham S. 1892b). Four new nebulae are listed, which were catalogued in the IC. Later both reports appeared unchanged in the second volume of the *Publications of the Lick Observatory* (Burnham S. 1894a, b).

At Yerkes Observatory Burnham continued his observations of NGC/IC objects with the 40″ refractor; he discovered some nebulae too.[84] The results are listed in the first two volumes of the *Publications of the Yerkes Observatory*, which were published in 1900 and 1904. Burnham is among the four observers (together with Barnard, Bigourdan and Swift) who are represented in the NGC and both ICs.

9.6.10 Burnham's favourite target: the trapezium in the Orion Nebula

For the well-versed double-star observer – with a weakness for nebulae – the famous quadruple star θ Orionis in the centre of M 42 was, of course, a very special object.[85] The four stars are called the 'trapezium', due to the geometrical form. Burnham had observed the field with a variety of refractors; for instance, his private 6-inch, the 18.5-inch at Dearborn Observatory and the 36-inch on Mt Hamilton. His measurements were published in 1889 (Burnham S. 1889). He wrote that '*Perhaps no object in the sidereal heavens has received more attention from astronomers than the multiple star θ Orionis.*' He was mainly interested in the faint stars located in the nebular area: '*Certainly no equal area in any part of the sky has furnished room for the location of so many purely imaginary stars.*' Again and again there appeared reports about supposed stars that were visible even with a 4-inch. The centre of the Orion Nebula had something mystical about it, exciting the fantasy of observers. In contrast, Burnham was a realist: '*all the large and most perfect modern refractors, directed by the most experienced double-star observers, have utterly failed to show, under the best atmospheric conditions, the least trace of a single one of the dozen or more supposed stars*'. The sketch of the θ Orionis region presented in his article gives the measure of things (Fig. 9.30).

The first three stars were detected by Cristiaan Huygens in 1656. His impression was that no telescope but his (chromatic) refractor of focal length 23 ft would show it – a curious example of self-deception.[86] The fourth star was added on 20 March 1673 by Jean Picard in Paris.[87] William Herschel sketched the trapezium (including M 42) in 1773. Later, on 19 February 1787, the Orion Nebula was the first object to be observed in his famous 40-ft reflector, which was still under construction. Because '*the nebula was of extreme brightness*', only four stars were seen.[88] Therefore the addition of a fifth (E) by Wilhelm Struve in Dorpat was a surprise (since the group had been studied so frequently with telescopes fully able to show it). Having examined the nebula several times during a period of two years with the 24.4-cm Fraunhofer refractor, he discovered the 10.3-mag star on 11 November 1826.[89] It is listed in his double-star catalogue *Mensurae Micrometricae* of 1837 (in which he introduced the common nomenclature A–F).

---

[84] In the IC, Burnham is represented with 23 objects.
[85] See for instance Smyth W. (1844: 130). The quadruple star is also called θ¹ Orionis, because there is a star trio called θ² Orionis nearby.

[86] For the history of the trapezium, see Webb (1864b: 259–265), Holden (1882b: 35–36) and Burnham R. (1966, Vol. 3: 1327–1331).
[87] The visual brightnesses of the four stars are A, 6.6 mag; B, 7.5 mag; C, 5.4 mag (variable, BM Ori); and D, 6.4 mag. The diameter of the quartet is about 20″.
[88] Dreyer (1912a, Vol. 1: xlvii).
[89] Struve F. G. W. (1828: 86–87).

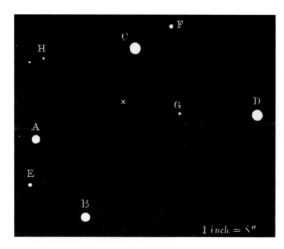

Figure 9.30. Burnham's sketch of the trapezium region in M 42 (N below; field size 25″ × 15″).

It is surprising that the fifth star was not noticed by John Herschel during his M 42 observation in 1824 and again on 13 March 1826 (together with John Ramage). Wilhelm Struve wrote in 1828 that, '*since I have communicated the discovery to Mr. Herschel junior, he has also distinguished it with his reflecting telescope of 20 feet. Is this little star of the number of the changeable* [variable] *stars, or does it still exist?*' Shortly later Herschel noted that the star '*was so bright as not to be overlooked with the most ordinary degree of attention*', considering it therefore '*as a new star, at least as a variable one of very singular character*'.[90]

As a delayed gratification, John Herschel eventually found the sixth star (F, 10.2 mag) on 13 February 1830, using the 11¾″ refractor of his friend James South.[91] Interestingly, Struve never saw this star in Dorpat. In the Cape catalogue, John Herschel mentions other observations.[92] It was, for instance, visible in his 18¼″ reflector, even when it was stopped to 12″. Stars E and F were seen at Birr Castle too.[93] It is interesting that Lassell independently found F with his 9″ reflector in Liverpool, which, however, is not shown in his (unpublished) M 42 drawing of 1862.[94]

Other 'stars' were reported, but later turned out to have been illusions. For instance, Lord Rosse was 'successful' at Birr Castle on 10 and 20 February 1846, using the 7″ reflector. Robinson reported that '*on the same evening* [the 20th] *an eighth star was found in the trapezium, a seventh being discovered on the 10th; the first near Herschel's α, and in the opposite direction from the sixth one discovered by Sir James South's large achromatic*'.[95] Another supposed star was reported by Lassell (Lassell 1862c).

The next real star must be credited to the Italian optician and astronomer Ignazio Porro. In 1856 he had erected an azimuthally mounted 52-cm refractor in Paris (de Senarmont 1857). After the failure of the Craig telescope in 1852 with its defective lens of diameter 62 cm, it was the largest in the world.[96] While testing it in Paris, Porro discovered a 12-mag star in the trapezium region, using a power of 1200. He published his find in the next year and made a sketch (Fig. 9.31 left), in which the new star is marked 'P' (Porro 1857). Leverrier, Director of Paris Observatory and in conflict with Porro (as with many others), questioned its existence. Secchi reported that he had seen the star on 10 February 1857 with the 24-cm Merz refractor in Rome (d'Abbadie 1856b). This is doubtful, because he also 'confirmed' two stars that had been claimed to have been seen by Dumouchel (a former Director of the Collegio Romano Observatory) with the 15.9-cm Cauchoix refractor. These objects, marked D' and D″ in Porro's sketch, do not exist.[97] Porro credited Herschel's star ('H') to South, placing it wrongly, moreover (d'Abbadie 1856a).[98]

On 6 January 1866 Huggins observed the field with his 8″ Clark refractor in Tulse Hill (Huggins W. 1866b). He made a sketch, using powers of 60, 135 and 220 (Fig. 9.31 right).[99] He perhaps saw Porro's star (no. 7?), but was, however, sceptical: '*This star can scarcely be the same as the one which M. Porro saw in 1857.*' Star no. 8 is definitely nonexistent, but possibly no. 9 is real, insofar as Huggins wrote '*rather feebler than 7*' and '*I*

---

[90] Brewster (1828: 93).
[91] Mayall (1962: 194).
[92] Herschel J. (1847: 29–30).
[93] Robinson (1840: 9). The star E was easily visible in the 36-inch, even when reducing the aperture to 18 inches.
[94] Herschel J. (1849), Lassell (1856); for the drawing, see Holden (1882b: 76).

[95] Robinson (1848: 7).
[96] It was surpassed in 1869 by Newall's 63-cm refractor in Gateshead.
[97] The 'Cauchoix' was known to have a bad habit of pointing out the existence of things that were not really there.
[98] By the way, Porro is not mentioned in Holden's monograph on the Orion Nebula (Holden 1882b).
[99] The sketch was given later in Huggins W. (1867).

9.7 Todd and the trans-Neptunian planet

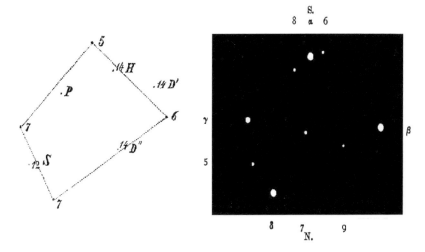

Figure 9.31. Sketches of the trapezium by Porro (left) and Huggins (right); see the text.

was able to see it only by occasional glimpses'. Actually, there is a star of 14.5 mag at this place ('G' in Burnham's sketch), but the matter is questionable. In his report Burnham wrote that Alvan Graham Clark had found it in 1888 while testing the new 36″ Clark refractor on Mt Hamilton.[100] Because 'G' obviously needed an aperture of 36″ (or at least Porro's 20″), Huggins' discovery with an 8″ is very doubtful – particularly since his sketch shows stars that clearly do not exist!

As so often, Barnard closed the issue. At one of the first observations with the new Lick refractor he demonstrated his exceptional vision: in October 1888 he discovered 'H', a double star with components of 14.5 mag and 15.5 mag, only 1.3″ apart. Perceiving this object in front of the brightened background at the centre of M 42 is astonishing. Burnham, one of the best visual observers, applauded this in his report: '[the star] *had been missed by all who had examined* [the trapezium] *with this telescope*'. He further wrote that Barnard '*also saw that this exceedingly faint star was itself a double, but I was unable to see the two components on any night when it was measured from the bright stars of the trapezium. Later, and on a remarkable perfect night, I saw the minute pair well, though with great difficulty, and obtained a fairly good measure.*'

But Barnard saw another star (marked 'x' by Burnham): '*Barnard also discovered a second star within

the trapezium, more centrally placed than the other, of the existence of which he has no question, having seen it on two occasions.*' Burnham had to pass on that '*I have not satisfactorily seen this star, and cannot speak of it from personal knowledge, but, from Mr. Barnard's great experience in studying faint objects and his remarkable acuteness of vision, I am confident this new star will be found where he saw it nearly on the diagonal from B to C [...] It is certainly very much fainter than the other stars.*' This 15-mag star is now called 'I' – obviously it is Porro's! In addition to the trapezium (A–D = θ Ori), we finally have five other stars (E–I) at the centre of the Orion Nebula (counting Barnard's double 'H' as one star).

9.7 TODD AND THE SEARCH FOR THE TRANS-NEPTUNIAN PLANET

David Todd was a controversial person with many interests. During his unsuccessful search for the trans-Neptunian planet with the 26″ Clark refractor at the US Naval Observatory, Washington, in 1877–78 he discovered a few galaxies. Dreyer did not examine Todd's descriptions carefully enough, otherwise some more objects would have appeared in his catalogues.

9.7.1 Short biography: David Peck Todd

David Todd was born on 19 March 1855 in Lake Rich, NY (Fig. 9.32). From 1874 he attended the Columbia College in New York for two years. Then he changed to Amherst College (MA), where he

---
[100] While testing the new 18.5″ refractor for Dearborn Observatory, Clark had already discovered the faint companion of Sirius on 31 January 1862.

Figure 9.32. David Todd (1855–1939).

finished at the age of 20. Its Lawrence Observatory with a 7.25″ Clark refractor fascinated him. He used the telescope to observe Jupiter and its moons, analysing their occultations. This attracted the interest of Simon Newcomb, Director of the US Naval Observatory in Washington.[101] During the period 1873–83 the institution held the largest refractor in the world, a 26-inch by Clark.[102] In 1875 Newcomb offered Todd a job as an assistant. Todd hesitated because his plan was to become an organ player, but eventually accepted and was lucky enough to witness Asaph Hall's discovery of the Martian moons in 1877.[103] He worked on the solar parallax and orbit perturbations of Uranus, from which originated his interest for the hypothetical trans-Neptunian planet. In 1879 he married Mabel Loomis, the daughter of the Washington astronomer Eben Jenks Loomis. Todd's wife was interested in astronomy too, writing many popular articles.

In 1881 Todd became the Director of Amherst College Observatory. He stayed there as Professor of Astronomy until his retirement in 1920. In 1903 the observatory was equipped with an 18″ Clark refractor. Todd's interest in American telescopes was great, as his comprehensive report of 1888 shows (Todd 1888).

Solar eclipses fascinated him and his wife, both making many voyages to observe them; but Todd was even more attracted by the 'Martian channels', especially Percival Lowell's observations in Flagstaff, Arizona. In 1907 Lowell funded an expedition to Chile to prove the phenomenon by photography. Todd boxed the Amherst 18-inch and during five weeks more than 13 000 plates were exposed – yielding 'clear evidence'.[104] Lowell was enthusiastic about the results. However, Lowell and Todd parted in acrimony due to a conflict about the copyright of the images.

Todd was a keen balloonist too, trying to observe comet Halley in 1910 from an altitude of 3000 m. Three years later he even reached 6600 m with an army balloon – in an attempt to communicate with the Martians! In another attempt in 1924 he used a radio receiver for this task. Of course, the noise detected was not due to Mars.

David Peck Todd's eventful life ended on 1 June 1939 in Lynchburg, VA, at the age of 84.[105] Already in life he and his wife Mabel Loomis-Todd had been eternalised by the naming of the minor planets (510) Mabella and (511) Davida, both of which were discovered in 1903 in Heidelberg.

9.7.2 Galaxies as candidates for a new planet

The search for the trans-Neptunian planet was executed between 3 November 1877 and 6 March 1878. Todd had calculated an orbital period of 375 years and a mean distance of 52 AU; a small disc of 13 mag was expected. With the 26″ Clark refractor (Fig. 9.33) ecliptic regions in Leo and Virgo were scanned at powers of 400 to 600. The yield was altogether 32 non-stellar objects – unfortunately there was no new planet among them. Todd did not publish the results until October 1885, i.e. seven years after his observations (Todd 1885b). At first he was not willing to report about a failure, but his mind was changed in June 1885 by Dreyer's request for new nebulae for his *'second supplementary catalogue'* (Dreyer 1885a). Todd announced his report in a note of August 1885 in the *Astronomische Nachrichten*: '*Dreyer's note in the late number of the* Astron. Nachr. *calls for my saying here that the observations which I made seven years*

---

[101] The history of the observatory is described in Dick S. (1980, 2003).
[102] It surpassed Newall's 25-inch and was beaten by the 27″ Grubb refractor in Vienna.
[103] Steinicke (2003a).

[104] Fischer (1990).
[105] Obituaries: Hudson (1939), Anon (1939a) and Anon (1939b) see also Steinicke (2002d).

9.7 Todd and the trans-Neptunian planet

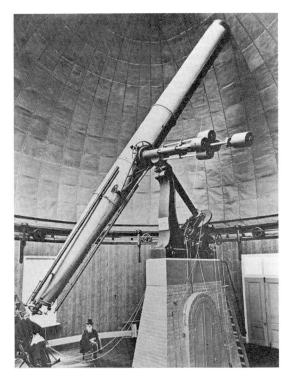

Figure 9.33. The 26″ Clark refractor at the US Naval Observatory, Washington.

 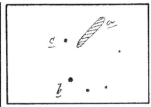

Figure 9.34. Todd's sketch of object no. 30 = NGC 3279, a galaxy in Leo (DSS).

ago, in search for the Trans-Neptunian planet are soon to be published [...]. They include observations of many nebulous stars of the type O eeF.' (Todd 1885a). What is meant by 'type O' is unknown.

Todd fixed his finds in 30 sketches. They are pretty exact and were used to check for a possible motion of the particular object: '*On the succeeding night of observation they were re-observed; and, at an interval of several weeks thereafter, the observation was again verified.*' By means of the field stars drawn, all non-stellar objects can be easily identified (south is up on all sketches; see Fig. 9.34 for an example).[106] Twenty are catalogued in the NGC (Table 9.26). However, Dreyer credited only eight of them to Todd (* in the column 'D'). In the introduction to the NGC one reads that '*A number of nebulous looking objects were found by Professor D. P. Todd during his search for an ultra-Neptunian planet, but I have inserted only eight of them. Of the rest, some were near the place of nebulae already catalogued, while the nebular character of others seemed very doubtful.*'[107] For NGC 4202 'Tod' is a typo. Todd was the discoverer of eight NGC objects (marked in bold). All are galaxies, with NGC 3279 as the brightest (13.3 mag).

Four of Todd's nebulae were independently found later by Javelle, Lewis Swift and Bigourdan (catalogued in the IC I). For NGC 4063 Dreyer did not notice that Todd was the true discoverer (marked (*) in the table); he credited it to Stephan ('St XI'). Vice versa, NGC 4355 must be taken away from Todd: the galaxy is identical with William Herschel's III 492 (NGC 4418), which fact was overlooked by Dreyer. Of the remaining 11 galaxies, 8 are due to William Herschel and 1 each to John Herschel, d'Arrest and Tempel. Two of William Herschel's objects (NGC 4045 and NGC 4077, only 30′ apart) were found independently by d'Arrest and catalogued a second time by Dreyer (NGC 4046 and NGC 4140).

NGC 3342 (Todd 24) is remarkable. The galaxy in Leo was found twice by William Herschel: as III 5 on 18 January 1784 and again as I 272 (NGC 3332) on 4 March 1796 while observing the moons of Uranus (the planet served as a 'reference star'). Schönfeld observed NGC 3332 on 26 April 1862, noting '*I. 272?*'.[108] Vogel (in 1867) in Leipzig and Tempel in Arcetri (21 May 1879) saw the galaxy too. Dreyer did not notice the identity of NGC 3342 and NGC 3332. Another double entry of William Herschel is NGC 3611 = NGC 3604. NGC 4179 was correctly identified by Todd as GC 2776 (I 9). For NGC 3427 Dreyer mentions Tempel only. In the cited publication of May 1878 no date is given (Tempel 1878d). It probably lies before 11 November 1877, by which time Tempel had collected his discoveries; thus Todd could be the veritable discoverer.

---

[106] See also Wenzel (2005).

[107] Dreyer (1953: 11).
[108] Schönfeld (1862d: 22).

Table 9.26. *Todd's NGC objects, sorted by his number, which correlates with the discovery date (see the text)*

Todd	NGC	D	Date	Type	V	Con.	Remarks
1	3611		3.11.1877	Gx	11.9	Leo	W. Herschel 27.1.1786 (II 521); = NGC 3604, W. Herschel 30.12.1786 (II 626)
4	3427		11.11.1877	Gx	13.2	Leo	Tempel 1877
5	3462		13.11.1877	Gx	12.3	Leo	W. Herschel 13.1.1784 (II 16)
6	3436	*	30.11.1877	Gx	13.9	Leo	
9	3685	*	11.12.1877	Gx	14.1	Leo	
10	3849	*	11.2.1878	Gx	13.7	Vir	IC 730 (Javelle)
11	4075		27.12.1877	Gx	13.7	Vir	J. Herschel 14.4.1828 (h 1074)
12a	4073		2.1.1878	Gx	11.4	Vir	W. Herschel 20.12.1784 (II 277)
12b	4063	(*)	2.1.1878	Gx	13.9	Vir	Stephan 3.5.1881
13a	4045		2.1.1878	Gx	11.9	Vir	W. Herschel 20.12.1784 (II 276); = NGC 4046, d'Arrest 10.4.1863
14	4077		5.1.1878	Gx	13.3	Vir	W. Herschel 20.12.1784 (III 258); = NGC 4140, d'Arrest 10.4.1863
15	4179		5.1.1878	Gx	10.9	Vir	W. Herschel 24.1.1784 (I 9)
16	4139		6.1.1878	Gx	13.7	Vir	d'Arrest 10.8.1863; = IC 2989, Bigourdan 28.3.1895
17	4355	*	5.2.1878	Gx	13.2	Vir	= NGC 4418, W. Herschel 1.1.1786 (III 492)
18	4202	*	6.2.1878	Gx	13.7	Vir	Dreyer: 'Tod'
20	3153		5.2.1878	Gx	12.8	Leo	W. Herschel 19.3.1784 (III 53); Peters 1880?
21	3134	*	6.2.1878	Gx	13.7	Leo	
24	3342		26.2.1878	Gx	12.5	Leo	W. Herschel 18.1.1784 (III 5); = NGC 3332, W. Herschel 4.3.1796 (I 272)
29	3217	*	4.3.1878	Gx	14.6	Leo	IC 606 (Javelle)
30	3279	*	5.3.1878	Gx	13.3	Leo	IC 622 (L. Swift); Fig. 9.34

Twelve of Todd's objects are not listed in the NGC (Table 9.27). However, two appear in Dreyer's IC I, where Javelle is mentioned as the discoverer. IC 591 was found again by Palisa in Vienna. The rest are galaxies from modern catalogues or stars from the Guide Star Catalogue (GSC).

not show its full power. The effect was that Stephan estimated his objects to be fainter than did other observers, such as Swift. Thus Dreyer had to correct his brightness and size data. Altogether, Stephan might have observed about 6000 objects within 15 years, 420 of which were new.

9.8 STEPHAN'S NEBULAE IN THE NGC

Edouard Stephan was still very active, producing lists of new objects for the NGC. Unfortunately, the observing conditions in Marseille deteriorated. Owing to light and air pollution the 80-cm Foucault reflector could

9.8.1 Nine new lists from Marseille

During the years 1878–85 Stephan published nine further lists in the *Comptes Rendus* (*CR*); see Table 9.28. Again, the epoch ('Ep.') is given for the observation year; there is a short (French) description for

9.8 Stephan's nebulae in the NGC   373

Table 9.27. *The remaining Todd objects (see the text)*

Todd	Object	Date	Type	V	Con.	Remarks (Todd)
2	GSC 838–889	11.11.1877	Star	10.8	Leo	'object a star'
3	GSC 848–1219	11.11.1877	Star	11.0	Leo	'object a star'
7	GSC 268–113	3.12.1877	Star	14.2	Leo	'not like a star, though I cannot see a disk well'
8	IC 669	3.12.1877	Gx	13.4	Leo	Javelle 7.4.1893
13b	PGC 38075	2.1.1878	Gx	14.3	Vir	
19	GSC 836–339	5.2.1878	Star	11.0	Leo	'object suspected'
22	IC 591	6.2.1878	Gx	13.2	Leo	Javelle 31.3.1892, Palisa 13.3.1899
23	PGC 29879	6.2.1878	Gx	15.3	Leo	
25	UGC 5864	16.2.1878	Gx	15.0	Leo	
26	PGC 31610	16.2.1878	Gx	14.9	Leo	
27	GSC 841–1088	28.2.1878	Star	11.0	Leo	'bright and very star-like – disk slightly suspected'
28	PGC 31589	28.2.1878	Gx	15.0	Leo	

each object. Some notes are added, containing identifications with nebulae from the GC. The papers in the *Astronomische Nachrichten* (*AN*) appeared a bit later (combining two or three lists). In the NGC, Dreyer called Stephan's publications St IX to St XIII (continuing the GCS numbering); he used the *AN* versions. The second column gives the number of objects; altogether there are 335 ('S' is Stephan's fraction). Together with the GCS lists (see Table 8.24) the total number of entries is 518.

Twelve of the 15 objects that had been found earlier by William Herschel (WH) had already been identified by Stephan (as noted in the lists). The rest are the galaxies NGC 2774 (III 61), NGC 7648 (III 218) and NGC 407 (II 219). Because they are in the GC (which was used as a source together with the GCS), he could have noticed the identity. For NGC 407 (no. 9 in list XIII) he noted that it '*is not identical with 5156 and 5157 of Dreyer–Schultz*'. These objects are NGC 408 (GC 5156) and NGC 414 (GC 5157) in Pisces, which had been discovered by Schultz on 22 October 1867. The former is a faint star (14.3 mag), preceding the latter, a 13.5-mag galaxy. Then, 15′ to the southwest, is Herschel's NGC 407 – and exactly this object is described by Stephan in his observation of 2 October 1883. He also missed the identity of NGC 2379 (GC 1527), the 11th entry in list XIII, which is John Herschel's (JH) h 446. This galaxy in Gemini is the third in a trio, following NGC 2373 (GC 5380) and NGC 2375 (GC 5383), both of which had been discovered by Johnstone Stoney (JS) at Birr Castle on 20 February 1849. Stephan found them too, during the night of 8 February 1878, but again did not notice the GCS identity. He found yet a fourth object: NGC 2378, 2.5′ northwest of NGC 2379, which is only a pair of stars. He could have identified Bindon Stoney's (BS) NGC 318 (GC 177) too, a galaxy in Pisces that was observed on 6 November 1882.

Mitchell's (Mi) galaxy NGC 3165 (GC 2037) in Sextans, no. 56 in list XIII, is correctly identified by Stephan (18 March 1884). The same is true for Marth's objects (m) from list XII (nos. 39 and 90): NGC 3362 (GC 5534 = m 203) and NGC 6751 (GC 5940 = m 397). The former is a galaxy in Leo (18 March 1882). The latter is a bright planetary nebula in Aquila, which had already been found on 17 July 1871 ('*bright; small; round; central condensation appears resolvable*'). It is unknown why the object is not present in an earlier list (second observation 3 August 1881).

Stephan could not have known anything about the other cases from Table 9.28. NGC 3758 in Leo, which had been discovered by Copeland (Co) on 18 March 1874, was found by him exactly 10 years later. In 1877 Tempel (T) discovered NGC 338 (Pisces) and NGC 4360 (Virgo); Stephan saw them on 6 November 1882 and 1 April 1884, respectively. Stephan observed NGC 3375 (Sextans) and NGC 4316 (Virgo) on 23 April 1881 and 1 April 1884, respectively, but Tempel had seen

Table 9.28. *Number of entries and references for Stephan's list in the NGC (see the text)*

St	Num	S	Co	JH	WH	m	Mi	BS	JS	Sf	T	To	V	Ep.	CR Vol.	CR P.	CR Year	AN No.	AN Vol.	AN P.	AN Year
IX	39	31		1	4				2	1				1878	87	869	1878				
X	40	38								1				1880	90	837	1880				
XI	20	18									1	1		1880	92	1128	1881	2390	100	209	1881
"	20	19								1				1880	92	1183	1881	"	"	"	"
"	20	19								1				1880	92	1260	1881	"	"	"	"
XII	50	41			4	1		1		2	1			1880	96	546	1883	2502	105	81	1882
"	46	39			4	1				1	1			1880	96	609	1883	"	"	"	"
XIII	50	48			1		1			1				1885	100	1043	1885	2661	111	321	1885
"	50	44	1		2						2			1885	100	1107	1885	"	"	"	"
Sum	335	297	1	1	15	2	1	1	2	8	5	1	1								

374

9.8 Stephan's nebulae in the NGC

Table 9.29. *Identities with NGC objects that had been discovered earlier*

NGC	Date	Type	V	Con.	NGC	Discoverer	Remarks
1002	14.12.1881	Gx	13.2	Tri	983	Stephan 13.12.1871	GC 5249, list III; Fig. 9.35 left
3110	17.3.1884	Gx	12.7	Sex	3122	W. Herschel 5.3.1785 (II 305)	NGC 3518, Stone 1885
5706	12.5.1883	Gx	14.8	Boo	5699	W. Herschel 16.5.1784 (III 127)	
5709	12.5.1883	Gx	13.6	Boo	5703	W. Herschel 16.5.1784 (II 128)	
6363	24.7.1879	Gx	13.5	Her	6138	Stephan September? 1872	GC 5816, list II; Fig. 9.35 centre
6550	19.7.1882	Gx	14.0	Her	6549	Marth 27.7.1864 (m 361)	
6599	27.7.1880	Gx	12.8	Her	6600	Marth 6.6.1864 (m 374)	

Figure 9.35. Objects of Stephan (left to right): the galaxies NGC 1002 and NGC 6363 (Triangulum, Hercules) and the bipolar reflection nebula NGC 2163 in Orion (DSS).

them earlier (21 February 1878 and 17 March 1882). Finally, Tempel discovered NGC 5534 in Virgo on 29 April 1881 and Stephan saw it on 17 May 1881. The remaining objects, NGC 4063 in Virgo (Todd (To), 2 January 1878) and NGC 6364 in Hercules (Voigt (V), summer 1865) were found by Stephan on 3 May 1881 and 21 July 1879, respectively. NGC 6364 was independently seen by Safford (5 September 1866) and Lewis Swift (11 September 1885); this galaxy is among the objects with the most independent discoveries (together with Swift's NGC 1360 and Marth's NGC 7422; see Table 10.7).

Because Dreyer did not know about Safford's (Sf) discoveries with the 18.5″ refractor at Dearborn Observatory, Chicago, eight were credited to Stephan. Safford's observations were made during the years 1866–68, but did not appear until 1887 (see Section 9.6.5).

In 1883 some objects from list XII were published by an unknown writer (Dreyer?) and identified with GC and GCS objects (Anon 1883d); eight are due to William Herschel and two to Marth. The paper refers to Stephan's results; thus it contains no news.

Seven of Stephan's galaxies are identical with GCS or NGC objects that had been discovered earlier (Table 9.29). In two cases the second find was due to himself: NGC 1002 = NGC 983 and NGC 6363 = NGC 6138 (Fig. 9.35 left and centre). For NGC 6138 Stephan made a right-ascension error of 1^h (this happened twice for d'Arrest). In the work of Esmiol, described in the

Table 9.30. *Stephan's 'anonymous' objects from Esmiol's compilation*

Object	Date	Type	V	Con.	Remarks
MCG 6–2–16	17.10.1876	Gx	12.6	And	UGC 480; Fig. 9.36
MCG 3–19–10	6.2.1874	Gx	13.0	Gem	UGC 3840
CGCG 85–33	6.2.1874	Gx	14.5	Gem	PGC 20152
	31.8.1872	NF		Vul	NW of NGC 7080

following section, an identity NGC 6138 = NGC 6262 is mentioned, which must be a typo. NGC 3110 = NGC 3122 = NGC 3518 is curious: the galaxy in Sextans was discovered three times (there are eight such cases in the NGC). For NGC 3518 Ormond Stone made a 1^h error. NGC 6550 is a strange case too, involving NGC 6549 and William Herschel's NGC 6548 (III 555).[109]

After subtraction of identities, Stephan must be credited for 290 objects: 280 galaxies, 2 planetary nebulae (NGC 7027 and NGC 7048), 2 reflection nebulae (NGC 2163 and NGC 6914), 1 emission nebula (NGC 2174) and 1 open cluster (NGC 7827). NGC 453, NGC 1330, NGC 2378, NGC 5948 and NGC 6672 are mere star pairs or groups. NGC 7027 in Cygnus is interesting, being the only object found by the Reverend Webb (see Section 9.9.2). The bipolar reflection nebula NGC 2163 in Orion had already been discovered by Stephan on 6 February 1874 (Fig. 9.35 right); it is unknown, why it appears in list IX rather than in an earlier one (second observation: 6 January 1878). The nebula was lost and was found again by Sven Cederblad (no. 62 in his catalogue; Cederblad 1946) – a curious story.[110] Together with NGC 2644, a galaxy in Hydra, NGC 2174 is the first object in Stephan's list IX (i.e. after the GCS). His observation took place on 6 February 1877. The last discovery was NGC 3007 (galaxy in Sextans) on 16 March 1885.

9.8.2 Esmiol's publication and special objects

At Bigourdan's suggestion, Emmanuel Esmiol, an assistant at Marseille Observatory, made a comprehensive reduction of Stephan's observations. Reference stars from the catalogue of the Astronomische Gesellschaft (AGK) were used. The result was published in 1917

---

[109] Thomson (1991).
[110] Steinicke (2001f).

Figure 9.36. Found by Stephan but not catalogued in the NGC/IC: MCG 6–2–16 (A) in Andromeda; the fainter companion MCG 6–2–17 (B) was not noticed (DSS).

as 'Réduction des observations de nébuleuses découvertes par M. Stephan' (Esmiol 1916). It contains 546 objects, presented in four tables. The first shows the observational data for each object (date, relative position and reference star), sorted by NGC-number. Since Stephan's publications did not give any discovery dates, Esmiol's work is the only source. The second table lists the calculated absolute positions for the epoch 1900. The third gives Stephan's descriptions and the last (in the appendix) shows positional differences relative to Bigourdan's measurements (from the Paris *Annales*).

Interesting are four discoveries of Stephan, called 'Anonyme' by Esmiol (Table 9.30). Since they are not contained in any of the lists, Dreyer was not aware of them. Three objects can be identified with galaxies that were catalogued in the 1960s. The field of the fourth is blank.

Table 9.31 shows special objects and observational data of Stephan. The small declination range

Table 9.31. *Special objects and observational data of Stephan*

Category	Object	Data	Date	Remarks
Brightest object	NGC 7027 (IX-27)	PN, 8.5 mag, Cyg	Oct.? 1878	
Faintest object	NGC 903 (XIIIa-17)	Gx, 15.7 mag, Ari	13.12.1884	Fig. 9.37 left
Most northerly object	NGC 7048 (IX-28)	+46° (Cyg), PN, 12.1 mag	19.10.1878	
Most southerly object	NGC 62 (XIIIa-2)	−13° (Cet), Gx, 13.5 mag	8.10.1883	
First object	NGC 6431 (I-7)	Gx, 13.4 mag, Her	23.6.1870	
Last object	NGC 2581 (XIIIa-38)	Gx, 13.5 mag, Cnc	7.3.1885	With NGC 2657
Best night			30.11.1878	7 galaxies (And)
Best constellation	Hercules		1870–84	74 objects

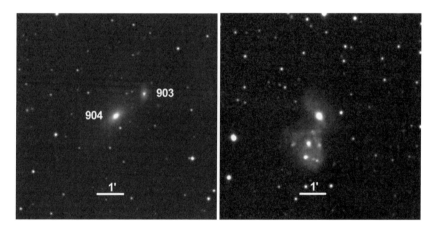

Figure 9.37. Left: the galaxy NGC 903 in Aries; right: the ring galaxy NGC 2444/45 (Arp 143) in Lynx (DSS).

and the large brightness difference are remarkable. Owing to the light-polluted sky above Marseille, he could survey only small areas. The great light-gathering power of the Foucault reflector was a disadvantage at that site. Only at high magnifications was the contrast acceptable.

On 18 January 1877 Stephan discovered the ring galaxy NGC 2444/45 (Arp 143, 12.9 mag) in Lynx (Fig. 9.37 right); however, the peculiar structure was not noticed. Earlier, William Herschel, Marth and Tempel had found examples of this rare type. In one night (30 October 1878), Stephan discovered seven of nine NGC galaxies in the cluster Abell 347 (Table 9.32). It is located in Andromeda, only 40′ southeast of the prominent edge-on galaxy NGC 891 (V 19), which was found by William Herschel on 6 October 1784.[111] He was also the discoverer of the two other NGC galaxies: NGC 898 (III 570) and NGC 910 (III 571), both seen on 17 October 1786.

Just as he had done earlier in the GCS, Dreyer once again increased Stephan's magnitude and size classes in the NGC by one level. Referring to galaxies in the Hercules Cluster, Swift remarked about Stephan's data that '*those* [objects] *he describes as faint and small and very small, I call pretty large*' (Swift 1886b). The brightness distribution is given in Fig. 9.38; the mean is 13.5 mag. Compared with Marth's observations (see Fig. 8.21), the value is 0.4 mag brighter (0.1 mag for

---
[111] Steinicke (2006).

Table 9.32. *Stephan's discoveries in the rich galaxy cluster Abell 347*

NGC	List	Date	V
906	X-5	30.10.1878	13.0
909	X-6	30.10.1878	13.7
911	X-7	30.10.1878	12.8
912	X-8	30.11.1878	14.1
913	X-9	30.11.1878	15.0
914	X-10	30.11.1878	13.0
923	X-11	30.10.1878	13.7

all, Webb was famous for his book *Celestial Objects for Common Telescopes*, which went through many editions. It stands out for its many objects and coherent descriptions and is esteemed by visual observers to this day. Probably the book inspired Burnham to observe double stars.

9.9.1 Short biography: Thomas William Webb

Thomas Webb was born on 14 December 1806 in Tretire, Herefordshire (Fig. 9.39).[112] His mother died early and his father, a vicar of the Church of England,

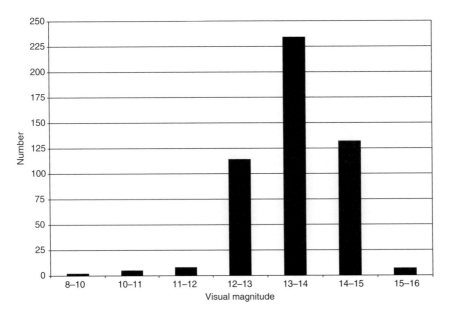

Figure 9.38. The brightness distribution of Stephan's objects.

Swift). Owing to the sky brightness, as mentioned earlier, Stephan estimated the objects to be fainter.

9.9 THE REVEREND WEBB AND HIS PLANETARY NEBULA NGC 7027

Webb's discovery of the tiny planetary nebula NGC 7027 in Cygnus on 14 November 1879 is an example of a quick international cooperation to reveal its nature. Additionally the story shows the successful application of spectroscopy. Some of the leading astronomers and observatories of the time were involved. Unfortunately – due to Dreyer's research – Webb had to concede the priority of his find to Stephan. First of

was responsible for his education. Already in his early years Webb showed an interest in science and mathematics. From 1826 he studied at Magdalen College in Oxford, and in 1829 graduated in mathematics. He went to Hereford and was ordained as a priest in 1831. Until 1841 he served at Gloucester Cathedral. Back in Tretire he married Henrietta Wyatt in 1843.

Webb's astronomical observations started in 1834 with a 3.7″ refractor by Tully. In 1852 he became Reverend Webb in Hardwicke, and he became a Fellow of the RAS in the same year. The double burden of

---

[112] The often-given year of birth 1807 is wrong.

9.9 The Reverend Webb

Figure 9.39. The Reverend Thomas Webb (1806–1885) and his wife Henrietta.

ministry and astronomical observing was mastered with the aid of his diligent wife. Over the years observational notes accumulated and Webb decided to publish them in a book. His model was the Bedford Catalogue of William Henry Smyth, which had appeared in 1844.[113] However, this book had been out of print since the late 1850s and to some extent had become out of date. Webb's *Celestial Objects for Common Telescopes* appeared in 1859. The bestseller published in many editions became the 'bible' of visual observers until the twentieth century. In 1859 Webb bought a fine 5.5″ Clark refractor from his friend the Reverend William Rutter Dawes, who was one of the best double-star observers.

During the following years Webb was a keen writer for various magazines. Especially his column 'Clusters and nebulae' in the *Intellectual Observer* (1863–67) was popular. In 1864 he purchased his first reflector from George With, a well-known telescope maker in Hereford. The 8-inch was equipped with a silvered glass mirror. In 1868 his father, who was to die a year later, made a generous gift: a 9.38″ With reflector. This, Webb's largest telescope, had a mounting and dome by Edward Berthon. With it he discovered the planetary nebula NGC 7027 in Cygnus in 1879. Until 1884 he observed in Hardwicke (the observation book contains altogether 3463 entries). In 1884 Henrietta passed away. Shocked by this loss, Thomas William Webb outlived her by only seven months, dying in Hardwicke on 19 May 1885 at the age of 78.[114]

9.9.2 Webb's planetary nebula

In the late evening of 14 November 1879 (perhaps at about 11 p.m.) Webb scanned the constellation Cygnus with his 9.38″ With reflector (Fig. 9.40 left).[115] At a power of 50 he encountered an unknown non-stellar object, about 2° south of ξ Cygni and not far from the (then still undiscovered) North America Nebula. Entry no. 3216 in his observing book is titled '*New Comet or Neb: Cyg*'.[116] About the atmospheric conditions one reads that it was a '*transparent night, thin haze however forming*'. The object is described as a '*blueish 9m star, which appeared rather large and much resembling a pair too close to be separated with that power*'. Higher powers (212, 375 and 450) showed '*no other result, but the proportional enlargement of a confused disc of about 4″, brighter apparently in the centre, but without any stellar point and surrounded by a little faint glow*'. Webb wrote that '*It reminded me somewhat of the appearance of Uranus on an extremely bad night.*' A sketch was added (Fig. 9.40 centre). After checking the GC, he concluded that '*There is no cluster or nebula near the spot in Gen. Cat.*'

---

[113] As mentioned in Section 6.5, Smyth's book was not the first popular presentation of deep-sky objects; that was one due to Joseph v. Littrow (Littrow J. 1835). Later his son Karl published the *Atlas des gestirnten Himmels*, describing the best deep-sky objects and adding 52 drawings (Littrow K. 1866).

[114] Obituaries: Ranyard (1885, 1886) and Chambers (1885); see also Mee (1897, 1905), Chapman A. (1998: 225–228), Carver (2006) and Robinson and Robinson (2006). On Webb's deep-sky discoveries, see Steinicke (2008).

[115] See also Steinicke (2005d).

[116] It is now in the RAS archive: RAS MS Webb 4; see also Robinson and Robinson (2006: 148).

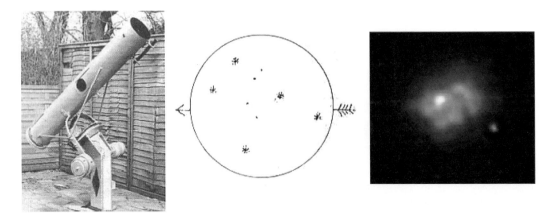

Figure 9.40. Left: a 9″ With reflector, similar to Webb's (Webb Deep-Sky Society); centre: the discovery sketch of NGC 7027 (south up, field 1°); right: a modern image of the PN.

On about 18 November 1879, Webb informed Lord Lindsay (Dun Echt Observatory, Aberdeenshire) by letter and on the 22nd he wrote to his friend George Knott in Knowles Lodge, Cuckfield. In Dun Echt, Copeland attended to the issue. His first observation, using the 15.1″ Grubb refractor, is dated 22 November. In Lindsay's 'Note on the Rev. T. W. Webb's new nebula' (*Monthly Notices*) one reads that '*The nebula under a power of 307 shows a sharp nucleus near the north-preceding edge, while the opposite side fades away like a very short tail or wisp.*' (Lindsay J. 1879b.) Copeland and his assistant Lohse also made the first spectroscopic observation of the nebula: '*At 6 p.m. on November 22 Rev. Webb's planetary nebula was seen approximately monochromatic with a prism in front of the micrometer eyepiece.*' It is further reported that '*This observation was confirmed on the following evening, and late in the night Mr. Lohse saw a spectrum of three bright lines with the Grubb spectroscope.*'

This interesting case caused the writing of the first Dun Echt Circular, which was sent out by Lord Lindsay on 24 November 1879. It states that '*On Nov. 14, 1879, the Rev. T. W. Webb discovered a very small nebula or nebulous star in Cygnus. It is identical with D.M. +41° 45′.3″*'. (Lindsay J. 1879a). What attracts attention is that obviously Copeland had identified the 9-mag star in the Bonner Durchmusterung; the terminology, however, is uncommon – not because of 'D.M.' (meaning 'Durchmusterung'), but due to the odd coordinate designation.

On 25 November Knott saw the nebula in his 7.3″ Clark refractor, which was equipped with a 'McClean star spectroscope'.[117] In his report 'Note on a gaseous nebula in Cygnus' (*Monthly Notices*) he wrote that '*it is Zone +41° No. 4004 and with a small aperture and low magnifying power has the appearance of a hazy star of the 8.5 magnitude, which is the magnitude assigned by Argelander.*' (Knott 1879). Knott added '*With the full aperture it presents the appearance of a bright bluish white nebulosity slightly elongated n.p., s.f., and I have the impression that its brightest part is at its north preceding extremity.*' In the visual spectroscope he noticed that the '*spectrum consists of one bright line*'. Since Knott refers to the BD star (with the correct designation), he obviously had received not only Webb's letter but also the Dun Echt Circular. On 26 November Copeland and Lohse again observed with the Grubb spectroscope, now detecting three lines: '*wave-lengths 500.1, 495.6 [O III], 486.0 nm [Hβ]*' (Lindsay J. 1879b).

Meanwhile, Webb had not been idle. On 29 November he published his discovery in the *Astronomische Nachrichten* (*AN* 2292) and *Astronomical Register*, and on 12 December it appeared also in the *Monthly Notices* and *The Observatory*.[118] Similar notes appeared also in the popular magazines *Nature* (12 December), *English Mechanic* (5 December) and

---

[117] On Knott see e.g. Anon (1895).
[118] Webb (1879a, 1880a, 1879b, c).

Sirius.[119] For a short period Webb's nebula was a top astronomical issue.

Webb received the Dun Echt Circular too, since he speaks in *AN* 2292 about '*No. 4004 in Argelander, + 41*' (Webb 1879a). He also was aware of Copeland's spectroscopic results. In the *Astronomical Register* Webb adds an interesting detail: he compares the new nebula with NGC 6210 in Hercules (Wilhelm Struve's no. 5; see Table 4.2), writing that '*The planetary nebula Σ5 greatly resembles it in character, though on a much smaller scale.*' (Webb 1880a). Amateur observers are warned as follows: '*Any of your readers who may feel disposed to search for it must of course employ a sufficient power to bring out the minute soft disk.*' Webb mentions Knott's observation too.

On the continent Webb's discovery was popularised by the Dun Echt Circular. Winnecke led the way with his observation on 28 November in Straßburg, which he reported in his *MN* paper on 'The nebula in Cygnus' (Winnecke 1879b). He used the 6″ orbit-sweeper by Reinfelder & Hertel and a Merz spectroscope. The observation was made at full moon. At a power of 260 he saw the object as a '*star of 8th magnitude out of focus* [...] *oblong in the direction 136.1° with a lucid point like a star 10–11 mag. in the preceding part; greatest diameter 5.5″*'. About the spectrum Winnecke wrote that it '*appeared to be continuous, but rather knotty (perhaps bright lines?)* [...] *quite different from the nearly monochromatic gaseous spectrum of H. IV 18* [NGC 7662] *which was looked at immediately afterwards*'. A rather strange find. However, a further observation on 2 December ('*the night was very foggy*') led to the expected result: '*I was very much surprised to find the light nearly monochromatic!*'. This was confirmed on 6 December ('*very clear*'), when Winnecke saw a gaseous spectrum: '*I have therefore no doubt that the new nebula is a gas nebula; still I can by no means understand the observation of Nov. 28 (the nebula was not visible that day in the finder).*' Owing to bright moonlight and bad air, he obviously had observed the wrong target.

It is interesting that the above-mentioned *Monthly Notices* reports by Webb, Knott, Lord Lindsay (Copeland) and Winnecke appear successively in the December 1879 issue (Vol. 40). Because they partly relate to each other (e.g. using the same data) it is evident that both the Dun Echt Circular and private communications played an important role in this issue.

In parallel with Winnecke, Vogel investigated Webb's object in Potsdam. On 30 November he saw it in the 12″ Schröder refractor as a '*round nebula with diffuse border of around 3″ to 4″ diameter*' (Vogel 1880a). At first the spectroscope did not show a definite image; the spectrum appeared '*to consist of a single line, at a further observation a very faint continuous spectrum was suggested to both sides of the bright line*'. Another observation by Gustav Müller, made under better conditions on 9 December revealed that '*The spectrum consists of three lines of very different intensity and a faint continuous spectrum*' (Vogel 1880b). The Potsdam result thus confirmed the '*identity of the spectrum with that of a planetary nebula*'.

Having been informed by the Dun Echt Circular, Julius Schmidt in Athens piped up too. His observation of the nebula on 6 December with the 6.2″ Plössl refractor revealed that '*With a power of 300 and even more it appears elliptical, elongated W–E, very small, oddly glowing, very condensed, stellar to the middle.*' (Schmidt 1880). However – but rather typically for Schmidt – his final remark sounds a bit strange: '*It is to some degree similar to the nebula near ν Piscium.*' Actually, there is no nebula near this star; perhaps NGC 7009 (the Saturn Nebula) near ν Aquarii was meant.

9.9.3 Dreyer's note and further observations

In his note on 'Mr. Webb's planetary nebula' of March 1880, Dreyer surprisingly clarified that Stephan was the true discoverer: "*the nebula is identical with No. 27 of Mr. Stéphan's list in the* '*Comptes Rendus*' *for Dec. 2, 1878.*" (Dreyer 1880b). Working in Dunsink, he had just compiled the Birr Castle observations and was now collecting new nebulae for his 'Second supplement to the General Catalogue'. The New General Catalogue gives for NGC 7027 '*St IX, Webb*'.

The nebula is no. 27 in Stephan's list IX and was found in autumn (October?) 1878 with the 80-cm Foucault reflector at Marseille Observatory (Stephan 1878). A bit more detail is given in Stephan's report of 1884: '*4004 B. D. +41°. Beautiful blue planetary nebula, irregular round, very small. The star is classified as 7th*

---
[119] Webb (1879d, e, 1880b).

magnitude.' (Stephan 1884). The nebula is the third-last-discovered planetary nebula in the NGC (followed by Lewis Swift's NGC 7094 and NGC 2242 of 1884 and 1886).

In December 1880 Copeland published another paper about the issue (Copeland R. 1881). It was caused by the spectroscopic discovery of the planetary nebula NGC 7026 on 7 November 1880. Both NGC 7026 and NGC 7027 (the objects are only 5° apart) were studied with the visual spectroscope on 6 December 1880. Copeland reported that, '*Although far fainter, this body [NGC 7026] has so much in common with its not distant neighbour, the Stephan–Webb nebula.*' The spectra appeared to be nearly identical: in addition to three bright lines a very faint fourth was present.

After the publication of the New General Catalogue, Bigourdan observed NGC 7027 with the 12″ refractor at Paris Observatory (on 16 August 1889 and 10 October 1895).[120] On 29 July 1891 Burnham was the first to see the object in a large refractor, namely the 36-inch of Lick Observatory, writing the following: '*Discovered by Webb. It has two nuclei, the following one of which is fairly well defined, but the brighter is too large and diffused for reliable measures of distance. There is nothing planetary about the appearance of this nebula. PA 131°.*' (Burnham S. 1891).

The first systematic spectroscopic investigation of planetary nebulae was made by James Keeler, using the 36″ refractor at Lick Observatory, equipped with a visual spectroscope by Brashear. NGC 7027 was observed several times between 15 August and 10 October 1890 (Keeler 1894). Johannes Hartmann was the first to photograph the spectrum (on 31 October 1901) with the 80-cm refractor at the Astrophysical Observatory of Potsdam (Hartmann 1902). He detected a radial velocity of 4.9 km/s. Francis Pease and Heber Curtis produced the first images of NGC 7027. Pease exposed '*Webb's bright planetary*' on 15 August 1912 with the 60″ reflector on Mt Wilson (Pease 1917b). Curtis, using the Crossley reflector, noted '*Binuclear planetary; long exposure shows no outer nebulosity*' (Curtis 1912a). Johannes Wilsing's attempt in summer 1892 appears a bit curious: he wanted to measure the parallax of '*the planetary nebula found by Webb*' with the 32-cm Steinheil refractor in Potsdam (Wilsing 1893). Of course, the result was negative. These examples show, too, that it was not commonly known that Stephan was the discoverer of NGC 7027.

9.9.4 Webb's *Celestial Objects for Common Telescopes* and five new open clusters

Webb's book appeared on 20 August 1859 in London (Webb 1859). The basic observations were made during the years 1847–56 with the 3.7″ Tully refractor, which had been purchased in 1834 (i.e. this is the 'common telescope'). Webb observed, for instance, all of the objects from Admiral Smyth's Bedford Catalogue. The book impresses by its large number of double stars, nebulae and star clusters and the clear descriptions. In contrast to Smyth's florid style of writing, Webb (following Dawes) omits every exaggeration. The non-stellar objects were identified by reference to the catalogues of Messier and William Herschel. The book contains practical hints and historical information too.

Webb received very good reviews and international success came soon. The next editions appeared in 1868, 1873 and 1881. The fifth, which was published in 1893 by the Reverend Thomas Espin, a keen visual observer too,[121] was the first to appear in two volumes (Vol. I: The Solar System, Vol. II: The Stars). The introduction contains a short biography of Webb, called the '*father of all amateur astronomers*'. This edition gives NGC-numbers for the first time. In 1917 Espin was responsible for the sixth edition too. In 1962 an American reprint, edited by Margaret Mayall, appeared showing modern images (Mayall 1962).[122]

The first edition of 1859 contains 60 nebulae and 50 star clusters. Recent research has revealed that five 'anonymous' open clusters that had been discovered during his observations are hidden in Webb's descriptions.[123] Since this remained unnoticed, the objects bear modern designations. One is IC 4756 in Serpens. In the IC II Dreyer mentions Solon Bailey, who found the open cluster on plates taken in 1896 in Arequipa. Actually Webb, in noting '*Group. Very large,*

---

[120] Bigourdan (1899b: G59).

[121] Between 1893 and 1907 he discovered altogether 20 IC objects.

[122] About the editions and reviews of Webb's book, see Robinson and Robinson (2006: 215–234).

[123] See also Fritz (2007).

Table 9.33. *Objects discovered by Pechüle*

GCS	NGC	Date	Type	V	Con.	Remarks
5094	78	1876?	Gx	13.3	Psc	
5588	3933	Spring 1884	Gx	13.4	Leo	Borrelly 1871?
5589	3934	Spring 1884	Gx	14.0	Leo	Borrelly 1871?
	4239	Spring 1884	Gx	12.6	Com	
		Spring 1884	Gx	14.2	Com	MCG 3–31–25, CGCG 98–33
		Spring 1884	Gx	13.5	Com	MCG 3–31–27, CGCG 98–35

subdivided, chiefly 9 and 10 mag', was half a century earlier! The other cases are similar. Two objects were found by Jürgen Stock in 1954: Stock 2 in Cassiopeia and Stock 23 in Camelopardalis. Trümpler 2 in Perseus was noticed in 1930 by Robert Trümpler and the 'δ Lyrae Cluster' (Stephenson 1) was found by Bruce Stephenson in 1959. In all cases Webb's descriptions clearly point to these objects.

The tradition of Smyth and Webb was soon continued by William Arthur Darby, who presented *The Astronomical Observer. A Handbook to The Observatory and The Common Telescope* (Darby 1864). The guide contains a fine collection of deep-sky objects. Another remarkable book was Leo Brenner's *Beobachtungsobjekte für Amateur-Astronomen*.[124] It contains 391 non-stellar objects (sorted by NGC-number) and gives many observational hints. Brenner was the editor of the magazine *Astronomische Rundschau*, which appeared during the years 1899–1909. Webb's spirit was still present in the twentieth century, as was demonstrated by the three-volume *Burnham's Celestial Handbook* written by Robert Burnham.[125] Finally, the British 'Webb Deep-Sky Society' founded in 1967 bears his name. The society published the valuable *Deep-Sky Observer's Handbook* in seven volumes (1979–87).

9.10 NEW NEBULAE DISCOVERED BY PECHÜLE

After his first object, NGC 78 (which he found in about 1876), Pechüle made further discoveries in

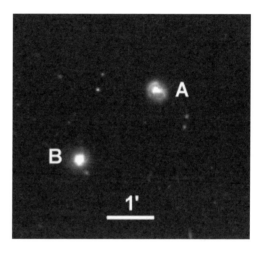

Figure 9.41. Found by Pechüle, but not catalogued in the NGC: the galaxy pair MCG 3–31–25 (A) and MGC 3–31–27 (B) in Coma Berenices (DSS).

spring 1884, while searching in vain for comet Tempel (Table 9.33). They were published in 1886 in his report 'Beobachtungen angestellt mit dem grossen Kopenhagener Refractor'.[126] The short list contains four objects with rough positions for 1855 (based on the BD). The first two are the galaxy pair NGC 3933/34 in Leo, described as '*faint*' and '*fainter*', respectively. However, both objects were discoveries of Borrelly (in about 1871) and had already been listed in the GCS (GC 5588/89). The third ('*faint*') must be credited to Pechüle: it is the galaxy NGC 4239 in Coma Berenices. The last entry gives '*two exceedingly small nebulae, which perhaps are but mere stars*'. Dreyer, despite accepting all other objects, was sceptical in this case; thus the pair

---

[124] '*Observational Objects for Amateur Astronomers*' (Brenner 1902). The author (aka Spiridon Gopčević) was a controversial person; see Ashbrook (1984: 103–111).
[125] Burnham R. (1966).

[126] 'Observations made with the great Copenhagen refractor' (Pechüle 1886).

was ignored in the NGC (Fig. 9.41). Actually, two faint galaxies are present, and were eventually catalogued in the 1960s by Vorontsov-Velyaminov and Zwicky in the MCG and CGCG, respectively. During the years 1888–92 Pechüle discovered three more nebulae (two galaxies and an open cluster) with the 11″ Merz refractor, which were entered by Dreyer in the first Index Catalogue as IC 4, IC 503 and IC 1369. Pechüle's rate is perfect: all his objects are non-stellar.

9.11 BAXENDELL'S 'UNPHOTOGRAPHABLE NEBULA'

Joseph Baxendell is well known through a curious find: NGC 7088 near the globular cluster M 2 in Aquarius – his only discovery and perhaps one of the most controversial NGC objects. Some, including Dreyer, believed that they had seen the 'nebula', but others were convinced of its nonexistence. The problem is that NGC 7088 does not appear on any photographic image. Today it is thought to be an illusion, though there are still some positive votes in favour of its existence.

9.11.1 Short biography: Joseph Baxendell

Joseph Baxendell was born on 19 April 1815 in Bank Top, Manchester (Fig. 9.42). His mother encouraged his interest in astronomy. He travelled to Central and South America (where he watched the Leonid meteor shower in 1833). Back in Manchester, he was able to use the 13″ and 5″ refractors of Arthur Worthington's private Crumpsall Observatory. Baxendell's favourites were variable stars (he found 18); his first report of 1837 treats the variation of Rigel. In 1857 Baxendell became a Fellow of the RAS and two years later he was appointed Astronomer to the Corporation of Manchester. In 1865 he married Mary Anne, a sister of Norman Pogson, another variable-star specialist.

In his later years Baxendell lived in Southport, where he first worked as the Superintendent at John Fernley's Hesketh Park Observatory, which was equipped with a 6.5″ Cooke refractor. In 1871 his own observatory (Birkdale) was erected, housing a 6″ refractor by Thomas Cooke & Sons. There his son Joseph Jr assisted him. Baxendell was a diligent writer, working on planets, comets, meteors and the Sun; he was the first to describe the 'green flash' at sunset. In

Figure 9.42. Joseph Baxendell (1815–1887).

1884 he became a Fellow of the Royal Society. Joseph Baxendell died on 7 October 1887 in Birkdale at the age of 72.[127]

9.11.2 The mystery of NGC 7088

On 28 September 1880, using the 6″ Cooke refractor in Birkdale, Baxendell saw a large faint nebula 30′ north of the globular cluster M 2 in Aquarius. In the same year a short report appeared in the *Monthly Notices*, titled 'A new nebula' (Baxendell 1880). The object is described as being of '*irregular oval form, its longer axis lying in a nearly east and west direction*' (size 75′ × 52′). Baxendell further wrote that '*It seems to be similar in character to the large nebula near the Pleiades* [Merope Nebula], *but is slightly less bright. I have, however, seen it on several nights, and have no doubt of its existence.*' But exactly the question of its 'existence' is the very problem!

The object could be visually confirmed with refractors of various apertures. Dreyer himself saw the nebula in 1885 with the 10-inch in Armagh. He thus catalogued it as NGC 7088, writing in notes '*seen without difficulty* [...] *it seems to extend about 35′ northwards from the parallel of the star 10′ north following M 2, and to be about 45′ in length*'. This followed positive observations by Bigourdan in 1897 (with a 12″refractor), Hagen in 1915 and 1917

---

[127] Obituaries: Espin (1887), Anon (1887a), Anon (1888a) and Anon (1888b).

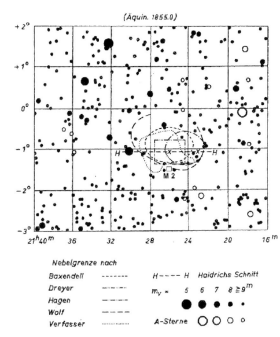

Figure 9.43. A chart of the NGC 7088 region (Strohmeier and Güttler 1952).

(16″), Wolf in 1927 (6″), O'Connor in 1929 (15″), F. Becker in 1930 (12″) and G. Lehner in 1930 (4″).[128]

But the object could not be detected on any photographic image – whence the popular name 'Baxendell's Unphotographable Nebula'! Many attempts were made, e.g. by Max Wolf, Baade and Shapley. They used various emulsions and filters – with no success. Later Strohmeier and Güttler comprehensively investigated the case (Fig. 9.43), including by carrying out their own observations (Strohmeier and Güttler 1952). Their conclusion was that the nebula does not exist! Obviously, the visual sightings were the result of optical illusions, perhaps due to reflections caused by the bright globular cluster M 2.

Other examples of such illusions might be the obscure 'Hagen clouds' (Hagen 1921).[129] In 1929 Johann Georg Hagen wrote a paper 'On the significance of Baxendell's nebulosity', defending the idea of its existence (Hagen 1930). He also mentions peculiar observations of Edward O'Connor in Stoneyhurst: '*After finding Baxendell's nebula without difficulty in the Stoneyhurst 15-inch refractor, Father O'Connor discovered another extensive nebula about one degree further south* [of M 2], *and even a third one, half a degree in a preceding direction from the second.*' All these objects are nonexistent. However, NGC 7088 was entered by Sven Cederblad as Ced 193 in his catalogue of galactic nebulae (Cederblad 1946).

In 1887 Dreyer called attention to a similar object: '*Another still larger one is north following IV. 3 (G.C. 1425), first seen at Birr Castle in 1850 and reobserved by Mr. Swift (A.N. 2683).*' (Dreyer 1887c). IV 3 is the reflection nebula NGC 2245 in Monoceros, which had been discovered by William Herschel and drawn by B. Stoney and Vogel (see Tables 7.1 and 8.27). Then, 15′ to the northeast, there is another one, NGC 2247, which was found independently by Mitchell (on 14 February 1857) and Lewis Swift (on 24 November 1883). In contrast to Baxendell's NGC 7088, this object exists.

9.12 COMMON'S DISCOVERIES WITH THE 36-INCH REFLECTOR

Andrew Common was among the most important constructors of telescopes. He marked the climax of a long tradition of technically gifted amateur astronomers working in nineteenth-century Britain and Ireland. William Herschel's 18.7″ and 48″ reflectors were followed by those of Lord Rosse of sizes 36 and 72 inches. Then Lassell developed equatorially mounted 24″ and 48″ reflectors (still with manual tracking). Eventually Common constructed clock-driven reflectors with silvered glass mirrors of diameter 36″ and 60″. All these telescopes made history. Common first used his instruments for visual observations of nebulae and star clusters, discovering some galaxies in 1880. In the same year a controversial sketch of the Pleiades nebulae was made. He was one of the first to use a large reflector for astrophotography; his image of the Orion Nebula made with 36-inch in 1883 is famous.

9.12.1 Short biography: Andrew Ainslie Common

Andrew Common was born on 7 August 1841 in Newcastle-on-Tyne (Fig. 9.44). By the age of 10 he was already interested in astronomy. His father died young and young Andrew had to earn money. He

---

[128] References are given in Strohmeier and Güttler (1952).
[129] See Latußeck (2009) and the remarks in Section 10.3.2.

Figure 9.44. Andrew Ainslee Common (1841–1903).

entered the company of his uncle in London and married in 1867. In 1874 the first astrophotographic attempts were made with a 5.5″ refractor and at the same time his interest for telescope making grew. Two years later he became a Fellow of the RAS. His first large telescope was erected in 1877 at his house in the London suburb of Ealing. The 18″ reflector had a silvered glass mirror made by George Calver; the mounting was Common's work. Since the observation of the Martian moons, which had just been discovered by Asaph Hall, failed, he immediately ordered a 36″ mirror with which to construct a new telescope. It was ready in 1880, and on 31 January 1883 Common managed to obtain a spectacular image of the Orion Nebula. For his achievements concerning telescopes and astrophotography he received the RAS gold medal in 1884; the laudation was given by President Edward Stone (Stone E. 1884). In 1885 Common became a Fellow of the Royal Society.

In 1885 Common sold the 36-inch to Edward Crossley to clear a space for his newest project: the construction of a 60″ Cassegrain reflector, which now included the making of the glass mirror. After several setbacks the great telescope was finished in 1891. Impressive photographs of the Pleiades, the Orion Nebula and the Dumbbell Nebula were taken. Common built reflectors for private persons and observatories too. He was President of the RAS in 1895–96. Andrew Ainslie Common died unexpectedly on 2 June 1903 in Ealing at the age of 61.[130]

9.12.2 New nebulae

With his 36″ reflector in Ealing, Common visually discovered 42 NGC objects (Table 9.34). His list of 'New nebulae' was published 1881 in Dreyer's magazine *Copernicus*[131] (Common 1881a). Eleven objects (marked with * in column 'S') were presented in *Sirius* too (Common 1881b).

Although Common's list contains only 32 entries, some numbers ('No.') cover more than one object. In the notes other nebulae in the field are mentioned. The objects – all galaxies – are sorted by right ascension. He gives only '*approximate places*' for 1880: '*The positions are obtained from the circles of the instrument, but may be near enough for identification. Almost in every case the objects have been seen on more than one night.*' Some objects '*seem worth re-examination, which might be done now, as they are conveniently placed*'.

Common gives no precise dates, noting only that the objects were '*found in 1880*'. Nevertheless, some objects can be dated. In August 1880 he tracked the periodic comet Faye, which was then in Pegasus, with the 36-inch. On the 21st he wrote a note on 'Faye's comet' for *The Observatory* (Common 1880d). There one reads that '*On the 4th (the night not being very fine) it* [comet Faye] *had moved from this place, and a nebula s f some few minutes was mistaken for it.*' This is NGC 7601 (no. 26), discovered on 4 August. In his list Common remarks '*Found in looking for Faye's comet.*' NGC 7630, NGC 7638 and NGC 7639 were seen four days later (Fig. 9.45): '*On the 8th, the night being fine, this last nebula* [NGC 7601] *was seen, as also another rather large and faint, about 5′ following the place of the comet on the 2nd, and much fainter […] and there was another nebula very like, and about 15′ preceding it.*' The three objects are described in no. 32. Common finally wrote that '*During the search several nebulae not in Herschel's* [GC] *or Dreyer's Catalogues* [GCS] *were found, besides those already mentioned as near the comet.*'

In 12 cases Common was beaten by other observers; sometimes only 'indirectly' via identities in the NGC. Five objects are due to William Herschel: Common's NGC 3388 and NGC 3402 are identical with NGC 3425 (III 108) and NGC 3411 (III 522); the galaxies NGC 3636, NGC 3637 and NGC 3732 in Crater had been discovered on 4 March 1786 by Herschel

---

[130] Obituaries: Turner H. (1903a–c, 1904), Greig (1903), Lockyer (1903) and Dyson (1904a).

[131] At that time it was still named *Urania*.

Table 9.34. *Common's discoveries (sorted by list number)*

NGC	No.	S	Date	Type	V	Con.	Remarks
3143	1	*	1880	Gx	14.3	Hya	
3280	2		1880	Gx	14.7	Hya	NGC 3295, Leavenworth 26.2.1886
3322	3		1880	Gx	13.5	Sex	NGC 3321, Leavenworth 3.1.1887
3360	4		1880	Gx	13.7	Sex	
3361	4		1880	Gx	12.8	Sex	
3388	5		1880	Gx	13.3	Leo	NGC 3425, W. Herschel 17.4.1784 (III 108)
3402	6		1880	Gx	11.9	Hya	NGC 3411, W. Herschel 25.3.1786 (III 522)
3404	7	*	1880	Gx	13.2	Hya	
3421	8		1880	Gx	13.7	Hya	
3422	8		1880	Gx	14.1	Crt	
3429	9		1880	Gx	13.2	Leo	NGC 3428, Marth 25.3.1865 (m 208)
3452	10	*	1880	Gx	14.1	Crt	
3480	11		1880	Gx	14.0	Leo	NGC 3476, Marth 25.3.1865 (m 213)
3490	12		1880	Gx	14.0	Leo	
3537	13	*	1880	Gx	12.8	Crt	Tempel 7.2.1878 and 21.2.1878, double nebula
3541	14	*	1880	Gx	14.5	Crt	Tempel 7.2.1878
3636	15	*	1880	Gx	12.4	Crt	W. Herschel 4.3.1786 (II 550)
3637	15	*	1880	Gx	12.7	Crt	W. Herschel 4.3.1786 (II 551)
3663	16		1880	Gx	12.5	Crt	
3688	17	*	1880	Gx	14.3	Crt	Leavenworth 1886
3704	18	*	1880	Gx	12.9	Crt	Tempel 23.2.1878
3707	18	*	1880	Gx	14.7	Crt	Tempel 23.2.1878
3723	19		1880	Gx	13.3	Crt	
3722	20		1880	Gx	14.1	Crt	Leavenworth 1886
3724	20		1880	Gx	14.1	Crt	Leavenworth 1886
3732	20		1880	Gx	12.5	Crt	W. Herschel 4.3.1786 (II 552)
	20		1880	Gx	14.0	Crt	MCG –1–30–8
3763	21	*	1880	Gx	12.8	Crt	
3775	22	*	1880	Gx	13.8	Crt	
3779	22		1880	Gx	13.8	Crt	
3865	23		1880	Gx	12.2	Crt	NGC 3854, Leavenworth 1886
3866	23		1880	Gx	13.2	Crt	NGC 3858, Leavenworth 1886
3905	24		1880	Gx	12.8	Crt	Stone 1886
4240	25		1880	Gx	12.7	Vir	Tempel 20.5.1875 (Dreyer); = NGC 4243, L. Swift 27.4.1886
7528	26		Aug. 1880	Gx	15.1	Peg	
7594	27		Aug. 1880	Gx	13.8	Peg	
	27		Aug. 1880	Gx	14.8	Peg	IC 5306, Kobold 26.10.1897
7595	28	*	Aug. 1880	Gx	14.9	Peg	
7601	29		4.8.1880	Gx	13.8	Peg	
7610	30		Aug. 1880	Gx	12.7	Peg	NGC 7616, Common
7616	31		Aug. 1880	Gx	12.7	Peg	NGC 7610, Common
7630	32		8.8.1880	Gx	14.4	Peg	
7638	32		8.8.1880	Gx	14.9	Peg	Fig. 9.45
7639	32		8.8.1880	Gx	14.6	Peg	Fig. 9.45

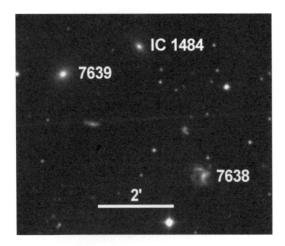

Figure 9.45. Common's galaxies NGC 7638/39 in Pegasus; IC 1484 was found by Javelle (DSS).

Figure 9.46. Discovered by Common: a chain of galaxies in Crater with NGC 3722, NGC 3724 and MCG -1-30-8 (B); the faint member MCG -1-30-5 (A) was missed (DSS).

(II 550–552). The first two are listed as no. 15, for which he noted '*2 Planetary nebulae*'. Common's NGC 3429 and NGC 3480 are equal to Marth's NGC 3428 (m 208) and NGC 3476 (m 213).

Five galaxies were first discovered by Tempel: NGC 3537, NGC 3541, NGC 3704, NGC 3707 and NGC 4240. NGC 3537 is a double nebula, described as '*2 stars involved in haze, 160°*' by Common. Tempel too had seen two objects (on different nights). NGC 3704/07 is no. 18: '*2, F, R, on the parallel*'. For NGC 4240 (no. 25) Dreyer – as editor of *Copernicus* – added a footnote: '*Found by Tempel in 1876, A.N. 2212.*' In *AN* 2212 Tempel gives 16 March 1876 as the discovery date (Tempel 1878d), but later this became 20 May 1875 (Tempel 1881b). The galaxy NGC 4240 was also seen in 1886 by Lewis Swift and catalogued as NGC 4243. NGC 7610 was found twice by Common, which was not noticed by Dreyer, who listed the second discovery as NGC 7616.

Two objects are not in the NGC. For no. 20 (NGC 3732) Common noted: '*F, R, a cluster of 3 similar ones 15' n.*' The three 'similar ones' are NGC 3722 (MCG -1-30-5), NGC 3724 (MCG -1-30-7) and MCG -1-30-8, forming a chain of length 4' in the southwest–northeast direction (Fig. 9.46). Between the NGC objects there is another galaxy, MCG -1-30-5, which was too faint for Common to see. Moreover, he discovered the galaxy IC 5306 in Pegasus, while observing comet Faye. About no. 27 (NGC 7594) one reads '*F, R, f 3 stars in a line 90° pointing to another fainter neb s.*' Thus Common had beaten Kobold at Straßburg Observatory to it.

Common must be credited for 29 NGC objects (marked in bold); all are galaxies – the maximum possible rate. However, Dreyer assigned not all of them to him: for NGC 3688 (no. 17) and NGC 3722/24 (no. 20) he mentions only Leavenworth.

9.12.3 Common's large reflectors

Common's first reflector was an 18-inch, built in 1877, with a silvered glass mirror by George Calver; the equatorial mounting was self-made. The same cooperation applied for the 36-inch of 1879 (Common 1879). The reflector – called the '3-foot' – had a truss-tube held by a fork, erected in a wooden shelter with a movable roof (Fig. 9.47 left).

In 1885 Common sold it to the amateur astronomer Edward Crossley, who ran a private observatory in Bermerside, near Halifax. Crossley built an iron dome of diameter 11.7 m. Because the weather was constantly bad, he eventually offered the telescope, mounting and dome to the Lick Observatory. Though the instrument was transferred in 1893, interpersonal problems on Mt Hamilton, due to the director, Edward Holden, delayed its erection. By the way, Holden wanted Common to participate in a '*generous co-operation between the great telescopes of the world*' (Common 1889), but the British astronomer, not agreeing about the benefits, remained aloof. It was Holden's successor

9.13 Pickering

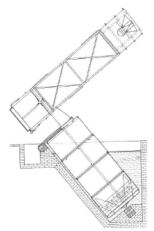

Figure 9.47. Common's reflectors in Ealing with 36″ (left) and 60″ apertures (right).

James Keeler who was to use the 'Crossley reflector' in 1898, but, after his sudden death on 12 August 1900, the 36-inch was redesigned by Charles Perrine. From 1905 onwards the telescope had a closed tube on an English axis mounting. Heber Curtis continued Keeler's work and observed planetary nebulae (Curtis 1918b). In 1933 the main mirror was the first to receive an aluminium coating.

Common's largest reflector, the 60-inch ('5-foot'), was completely self-made (Fig. 9.47 right). To make the glass mirror he constructed a grinding machine similar to Lord Rosse's. The optical design was a Cassegrain (the blank had a central hole of diameter 25 cm); the tube and mounting were similar to those of the 36-inch. First light was in 1889, but unfortunately the optical quality was not satisfying and Common made a second mirror for a Newtonian system. The new 60-inch was ready to use in 1891. For visual observations, he had to climb a high ladder to reach the eye-piece. Because one time he had nearly fallen down, he decided to shift the focus near to the main mirror with the aid of a tertiary mirror.[132] Common wrote that '*the power of the 5-foot over the 18-inch and 3-foot is proportionate to the size. On nebulae this is seen to great advantage, both visually and photographically.*' (Common 1892). After his death in 1903 Edward Pickering asked his descendants for the 60-inch. Having been inspired by the success of the Crossley reflector on Mt Hamilton, Pickering wanted it for Harvard College Observatory. Just a year later the telescope was erected at its new site (Pickering E. 1904). It now had a Coudé-like fixed focus at the polar axis; the observer was placed in an adjoining cabin.[133] However, the site was not suitable for such a powerful instrument.

9.13 PICKERING'S SPECTROSCOPIC SEARCH FOR PLANETARY NEBULAE

Edward Pickering was the first to perform a systematic search for new planetary nebulae by visual spectroscopy. With the old 15″ refractor at Harvard College Observatory he discovered some examples that were barely distinguishable from stars. Of the 22 NGC objects found by spectroscopy, Pickering contributed 15; the rest are due to Copeland.

9.13.1 Short biography: Edward Charles Pickering

Edward Pickering was born on 19 July 1846 in Boston, MA (Fig. 9.48). He built his first telescope at the age of 11. When he had finished school in Boston, he studied at Harvard University, Cambridge, graduating in 1865. Until 1867 he was an assistant in the mathematical

---

[132] A similar optical design was later used by George Ritchey for the 60″ reflector on Mt Wilson.

[133] On Common's telescopes, see Stone R. (1979), Ashbrook (1984: 168–176) and Chapman A. (1998: 136–137).

Figure 9.48. Edward Charles Pickering (1846–1919).

faculty and then became Professor of Physics at the Massachusetts Institute of Technology (MIT) in Boston.

In 1877, now 42 years old, he succeeded Joseph Winlock as Professor of Astronomy at Harvard University and Director of its observatory. Pickering first applied himself to photometry. Assisted by Arthur Searle and Oliver Clinton Wendell, he determined the brightness of stars down to 6 mag north of −30° declination ('Harvard photometry' 1879–82).[134] Later this was extended to the whole sky down to 6.5 mag ('Revised Harvard photometry'). To observe the southern sky, a station in Arequipa (Peru) at −16° latitude and located at an elevation of 2457 m was installed in 1891.[135] First an 8″ astrograph (Bache telescope) was erected, followed by the famous 24″ Bruce refractor in 1896.

During the surveys, nearly 3500 new variable stars were detected (called 'Harvard variables' HVs). Pickering was excellent in theory too, originating the physics of eclipsing binaries. Additionally, he was an expert in astrophotography and spectroscopy. With financial support from the 'Henry Draper Memorial' he directed a comprehensive spectral investigation of the whole sky. Starting in 1884, the spectral classes of about 222 000 stars could be determined, using objective-prism plates; the main work was done by Annie Cannon. A staff of 'women computers' crowded the observatory, among them Williamina Fleming, Henrietta Leavitt and Antonia Maury.[136] Pickering's classification scheme was eventually accepted worldwide. In 1901 he was awarded the RAS gold medal for his work on variable stars; in 1905 he became President of the American Astronomical Society.

Edward Charles Pickering died on 3 February 1919 in Cambridge, MA, at the age of 72.[137] He was Director of Harvard College Observatory for 42 years and must be considered as the leading collector of astronomical data of his time. His younger brother, William Henry Pickering, was a well-known astronomer too. He discovered the large supernova remnant in Orion known as Barnard's Loop (see Section 11.6.15).

9.13.2 Successful searches with a visual spectroscope

In 1880 Pickering first reported in the *American Journal of Science* about his new method (Pickering E. 1880). The article appeared a year later in *The Observatory* too (Pickering E. 1881a = Ref. 1). There one reads that '*a direct-vision prism was placed between the eyepiece and objective of the telescope, thus forming a spectroscope without a slit [...] to detect any minute planetary nebulae*'. Pickering used the aged 15″ Merz refractor of Harvard College Observatory, observing with a power of 140 (field of view 12′). He made systematic 'sweeps': '*The Telescope is clamped in right ascension and moved through 5° in declination. This is repeated so frequently that the successive sweeps shall overlap, the region continually varying by the diurnal motion.*' The search was focused on Milky Way regions (e.g. in Sagittarius, Aquila, Cygnus and Sagitta), because many planetary nebulae had already been found there.

Success came promptly: '*The first sweep was made on July 13* [1880]*, and revealed in a few minutes a bright point of light wholly unlike the lines formed by the stars. This proved to be a new planetary.*' This is NGC 6644

---

[134] By the way, Wendell is mentioned in the introduction of the NGC (see 'Harvard College'), but does not appear in any entry.

[135] See Holden's book *Mountain Observatories in America and Europe* (Holden 1896).

[136] Fleming later discovered 46 IC objects (mostly planetary and emission nebulae) by photographic spectroscopy.

[137] Obituaries: Kobold (1919), Strömgren (1919), King (1919), Cannon (1919), Bailey (1919), Campbell (1919) and Turner H. (1920); see also Plotkin (1990).

Table 9.35. *Pickering's planetary nebulae (see the text)*

NGC	HN	Date	V	Con.	Ref.	Remarks
6309		15.7.1882	11.5	Oph	3	Tempel 1876? (GC 5851)
6439	48	18.8.1882	12.6	Sgr	4, 6	
6537	45	15.7.1882	11.6	Sgr	3, 6	
6565	42	14.7.1880	11.6	Sgr	1, 6	
6567	49	18.8.1882	11.0	Sgr	4, 6	AR 1^m too small
6578	47	18.8.1882	12.9	Sgr	4, 6	AR 1.5^m too small, declination $10'$ too small
6620	43	3.9.1880	12.7	Sgr	1, 6	AR 1^m too small; Fig. 9.49 centre
6644		13.7.1880	10.7	Sgr	1, 6	Fig. 9.49 left
6741	50	19.8.1882	11.5	Aql	4, 6	
6766	53	8.5.1883	10.9	Cyg	7	AR 1^h too small!; = NGC 6884, Copeland 20.9.1884
6790	46	16.7.1882	10.5	Aql	3, 6	Near BD $+1°$ 3979
6778		Sept.? 1882	12.3	Aql	4	Marth 25.6.1863 (m 399); = NGC 6785, J. Herschel 21.5.1825 (h 2038)
6803	52	17.9.1882	11.4	Aql	5	Called no. 12 in Ref. 5 (too late for list Ref. 6)
6807	51	4.9.1882	12.0	Aql	4, 6	
6833	54	8.5.1883	12.1	Cyg	7	
6879	55	8.5.1883	12.5	Sge		Copeland 9.9.1884; Fig. 9.49 right
6881	44	25.11.1881	13.9	Cyg	2, 2a, 6	

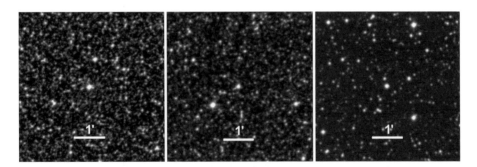

Figure 9.49. Three spectroscopic PN of Pickering (left to right, at the centre of each image): NGC 6644, NGC 6620 and NGC 6879 (DSS).

in Sagittarius (Fig. 9.49 left); Pickering noted that '*Its disk is so small that it can scarcely be distinguished from a star and would not probably have been detected with an ordinary eyepiece even if brought into the field of view.*' The next day he discovered NGC 6565, describing it as '*somewhat fainter, but with a larger disk*'. The article contains another planetary nebula in Sagittarius: NGC 6620, '*the smallest planetary nebula known*', which he found on 3 September 1880, but giving a wrong position (Fig. 9.49 centre). A fourth object, which had been discovered as soon as on 28 August is the O-star BD $-21°$ 4864, which is striking by virtue of its extraordinary spectrum (type 'WC'). All of these planetary nebulae are listed in Table 9.35. Some of the published positions are, however, wrong.

Pickering's next find was NGC 6881 in Cygnus on 25 November 1881 (Fig. 9.49 right), which he published in *Copernicus* (Pickering E. 1881b = Ref. 2) and *The Observatory* (Pickering E. 1882d = Ref. 2a). He wrote that '*The nebula is indistinguishable from a star of the fourteenth magnitude.*' Another short paper lists three PN: NGC 6309, NGC 6537 and NGC 6790 (Pickering

E. 1882b = Ref. 3). The first is correctly identified with Tempel's GC 5851, which had perhaps been found in 1876 (Tempel 1882). Pickering justly criticised his description, quoted as '*between two stars*'.

The next publication contains another six objects from August/September 1892 (Pickering 1882c = Ref. 4). All were new except for NGC 6778; although this PN had been found in 1863 by Marth on Malta, it is, however, identical with NGC 6785 (h 2038) of John Herschel, who gave only a rough position. On 17 September 1882 Pickering discovered NGC 6803 in Aquila (Pickering E. 1882f = Ref. 5). In this month the paper 'Small planetary nebulae discovered at the Harvard College Observatory' appeared, presenting a list of 11 objects sorted by date (Pickering E. 1882e = Ref. 6). Pickering mentions having seen GC 4333 (NGC 6445), a PN in Sagittarius, which had been found in 1786 by William Herschel (II 586). Dreyer used Pickering's list for the NGC (also mentioning a letter of July 1885).

The last three planetary nebulae were found on 8 May 1883. Two of them, NGC 6766 and NGC 6833, are published in Ref. 7 (Pickering E. 1883). It is unknown why NGC 6879 does not appear; the object was independently found in 1884 by Copeland. For NGC 6766 Pickering made an error of 1^h in right ascension. Dreyer did not notice it; thus the object appears again in the New General Catalogue as Copeland's NGC 6884 (at the correct position).

Pickering's spectroscopic search from 1880 to 1883 yielded 15 new planetary nebulae (marked in bold). Most of the objects from Table 9.35 also appear in his publication 'Nebulae discovered at the Harvard College Observatory' of 1908, in which he used the abbreviation 'H.N.' for 'Harvard nebula' (Pickering E. 1908). However, NGC 6644 is missing, but, vice versa, NGC 6879 is listed here only.

9.14 COPELAND: ON PICKERING'S TRAIL

In 1877 Ralph Copeland was the first to detect a nebula surrounding Nova Cygni, which had flared up a year earlier. In Dun Echt he used a spectroscope at the eyepiece of Lord Lindsay's 15.1″ Grubb refractor (Fig. 9.50 left). This nebula, catalogued by Dreyer as NGC 7114, was the first non-stellar object to be discovered by spectroscopy. In the following years Copeland – inspired by Pickering's success at Harvard – scanned the Milky Way regions for stellar objects with a 'gaseous spectrum', finding some new planetary nebulae.

9.14.1 The 'planetary' nebula around Nova Cygni 1876

On 24 November 1876 Julius Schmidt in Athens discovered a nova in Cygnus, which appeared as a star of about 3 mag (Schmidt 1877). On 2 January 1877 Copeland

Figure 9.50. Left: The 15.1″ Grubb refractor (here at its final site in Edinburgh); right: Copeland at the visual spectroscope.

Table 9.36. *Copeland's discoveries*

NGC	Date	Rr	Type	V	Con.	Ref.	Remarks
3736	1885?	6.1″	Gx	14.9	Dra		No reference; no spectroscope; Fig. 9.52 right
5315	4.5.1883	6.1″	PN	9.8	Cir	3	Peru
5873	2.5.1883	6.1″	PN	11.0	Lup	3	Peru; first South American/mountain object; Fig. 9.52 left
6153	27.5.1883	6.1″	PN	10.9	Sco	3	Peru
6879	9.9.1884	6.1″	PN	12.5	Sge	4	Pickering 8.5.1883
6884	20.9.1884	6.1″	PN	10.9	Cyg	4	NGC 6766, Pickering 8.5.1883
6886	17.9.1884	6.1″	PN	11.4	Sge	4	
6891	22.9.1884	6.1″	PN	10.5	Del	4	
7026	18.11.1880	15.1″	PN	10.9	Cyg	2	Burnham 6.7.1873
7114	2.9.1877	15.1″	(EN)		Cyg	1	Nova Cygni 1876; Lohse 3.10.1885

observed it with a visual spectroscope at the 5.1″ Grubb refractor in Dun Echt. He detected a continuous spectrum with five bright lines, looking like that of a Wolf–Rayet star (Copeland R. 1877a, b). Eight months later (2 September 1877) the situation had changed: '*Viewing through a low power eye-piece and a powerful direct vision prism, held between the eye & the eye-piece, the light of the star was found to be absolutely monochromatic.*' (Copeland R. 1877c = Ref. 1). He concluded that the '*star has changed into a planetary nebula of small angular diameter [...] no astronomer discovering the object in its present state would, after viewing it through a prism, hesitate to pronounce as to its present nebulous character*'. His assistant Gerhard Lohse took part in the observations. Worrying that the nova might become too faint, both measured the relative positions to 112 neighbouring stars (Copeland R. 1882a). Copeland mentions that the star showed a '*small disc with a soft margin*'. On 3 October 1885 Lohse was able to confirm the emission nebula with the new 15.5″ Cooke refractor at Wigglesworth Observatory, Scarborough (Lohse G. 1887b).

The next to see the nebula was Burnham in 1891, using the 36″ refractor at Lick Observatory (Burnham S. 1892b). In a later publication he praised the precise work of Copeland and Lohse: '*The arrangement of the catalogue is perfect in every respect, and the chart showing the relative positions of the stars could not be improved. Both should serve as models for all works of this kind.*' (Burnham S. 1894b). Burnham's opinion was

Figure 9.51. The faint stellar remnant of the Nova Cygni 1876; NGC 7114 has long since faded away (DSS).

important, since he was one of the leading measuring observers. About the nebula around what had meanwhile become a 13.5-mag faint star (25 September 1891), he wrote that '*At times the new star did not seem to have a perfectly stellar appearance under moderately high powers, but rather to resemble an exceedingly minute nebula.*' In 1902 Barnard investigated the case in a comprehensive study (Barnard 1902). Dreyer, cataloguing the Nova Cygni nebula as NGC 7114 (Fig. 9.51), mentions only Lohse's observations of 1885 in the notes.

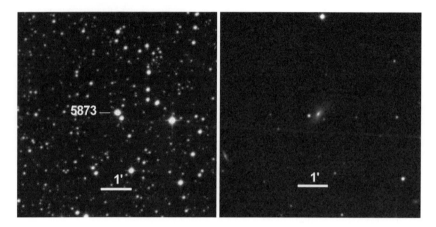

Figure 9.52. Left: the planetary nebula NGC 5873 in Lupus, the first 'South American' NGC object; right: the galaxy NGC 3736 in Draco (DSS).

9.14.2 Nine planetary nebulae and a galaxy

Seven of the 22 NGC objects discovered by visual spectroscopy are due to Copeland (Table 9.36). The search method was developed by Edward Pickering at Harvard College Observatory, who found altogether 15 nebulae from 1880 onwards (see Section 9.13.2).

After having been delayed by bad weather, Copeland started his search on 18 November 1880 – and was immediately successful. In the discovery note, Pickering is mentioned expicitly: '*A new planetary nebula has been discovered by Dr. Copeland at Dunecht, using Prof. Pickering's method of search.*' (Copeland R. 1880 = Ref. 2). The new object was NGC 7026 in Cygnus. No date is given; it can be found in Dun Echt Circular no. 13 (Lindsay J. 1880). In a later report, appearing in 1881 in *Copernicus*, Copeland tells the story of its discovery: '*By placing a direct-vision prism between the eye-piece and the object glass the images of the stars are drawn out into long spectra, while those of the planetary nebulae remain nearly undistorted.*' (Copeland R. 1881). One further reads that '*On the suggestion of Lord Lindsay, on November 7* [1880], *a prism was combined on this plan, with a Huyghenian eye-piece in the 15.06 in. Dunecht refractor.*' When the search started during the night of 11 November Copeland noticed that '*a round disc, altogether different from the spurious images* [...] *entered the field*'. Later NGC 7026 was observed visually too: "On November 21, Mr. Lohse, Mr. Carpenter and I all saw that the nebula had two imperfectly defined nuclei, giving it very much the appearance of the older drawings of the 'Dumb bell' nebula [M 27]." Copeland was not aware of Burnham's observation of 6 July 1873. The American observer had discovered the nebula in a 'classic' manner, while searching for double stars with his private 6" Clark refractor in Chicago (see Section 9.6.7). Copeland, just like later Dreyer too, did not notice Burnham's publication (Burnham S. 1873).

Copeland discovered the next three planetary nebulae on his expedition to Peru in May 1883. He used a 6.1" Simms refractor with a spectroscope by Vogel. His report 'Experiments in the Andes' appeared in *Copernicus* (Copeland R. 1884a = Ref. 3).[138] A fourth nebula with a brightness of 9.5 mag (found on 24 May) was unfortunately lost: '*thunder clouds advancing across Lake Titicaca prevented further observations*'.

His last four objects were discovered with the Simms refractor after Copeland's return to Dun Echt in 1884. For the first time he used an objective-prism, speaking of a '*Secchi prism in front of the object glass*' (Copeland R. 1884c = Ref. 4). He further wrote that '*In examining the richer part of the milky way about Cygnus four nebulae and one star with a spectrum of bright lines, which seem not to have been previously noticed, were found in the earlier sweeps.*' The mentioned star is BD +37° 3821, located in the centre of the Crescent Nebula NGC 6888 in Cygnus (see Fig. 8.56 centre), which had been found by William Herschel on 15 September

---

[138] About Copeland's expedition, see Holden (1892: 35).

1792 (IV 72). The prism did not show the nebula, but revealed the central star of the rare 'Wolf–Rayet' type. Two of the four nebulae found in Dun Echt must be credited to Pickering (8 May 1883): NGC 6879 in Sagitta and NGC 6884 in Cygnus. For the former Dreyer correctly mentions Copeland and Pickering, but for NGC 6884 the identity with Pickering's NGC 6766 remained unnoticed. The reason might have been a typo by Pickering: in his report the right ascension should read '20h 06m 32s' instead of '19h 06m 32s' (Pickering E. 1883). Copeland's position (Ref. 4) is, however, correct.

Only the galaxy NGC 3736 in Draco is left (Fig. 9.52 right). There is no source; probably Dreyer was informed by letter. Anyway, Draco is not the place to look for planetary nebulae, thus the usage of a spectroscope or objective-prism is unlikely. Given that there is no *Copernicus* note plus the fact that publication of the magazine was abandoned in 1884, a discovery date between 1885 and 1887 is probable.

9.15 BARNARD: THE BEST VISUAL OBSERVER

Edward E. Barnard is among the four observers (together with Bigourdan, Burnham and Swift) who contributed to all of Dreyer's catalogues (the NGC, IC I and IC II). Because of his exceptional vision and the high quality of his observations – which initially was achieved with rather small telescopes – Barnard rates as the best visual observer of the nineteenth century. Especially Swift and Burnham admired his work. Important objects bear his name, such as Barnard's Galaxy (NGC 6822), a Local Group member in Sagittarius. Concerning the priority of his discoveries Barnard appeared quite fussy. He started his unusual career in his home town of Nashville, first becoming known for his frequent comet discoveries. Later he was able to use the largest refractors: the 36-inch at Lick Observatory and the 40-inch at Yerkes Observatory.

9.15.1 Short biography: Edward Emerson Barnard

Barnard was born on 16 December 1857 in Nashville, TN (Fig. 9.53). His father died prior to his birth. Young Edward grew up in poor circumstances and was educated by his mother, attending school for only two months. At the age of nine he got a job with a local photographer. Barnard had to operate the monstrous 'Jupiter camera', to follow the Sun for blowups – a task perfectly mastered by Barnard. He stayed for 17 years, supporting his family and gaining a comprehensive knowledge of optics and photography. Of course, his interest in astronomy originated there.

Barnard's first telescope was a self-made refractor of aperture 2.4″. In 1877 he bought a 5″ refractor by Byrne (Fig. 9.53 right). In the same year the American

Figure 9.53. Edward Emerson Barnard (1857–1923) and his 5″ Byrne refractor (Sheehan 1995).

Astronomical Association held its meeting in Nashville and Barnard plucked up the courage to address Simon Newcomb, one of the leading American astronomers. He was advised to search for comets and to master mathematics. The first task was no problem – he became a very successful comet discoverer. However, he found mathematics pretty difficult.

In 1881 Barnard married Rhoda Calvert, who originated from England and was the sister of his employer. Both of them had to live in his mother's house. He started to search for comets, discovering his first on 17 September 1881. Two years later he was offered a job at the Vanderbilt University Observatory in Nashville. Now Barnard had access to a fine 6″ Cooke refractor, which he mainly used for measurements. He stayed for four years and found eight more comets – always in competition with William Brooks in Phelps and Lewis Swift in Rochester.[139] His fame increased and the discoveries yielded prize money. In 1887 he graduated at Vanderbilt University and was soon engaged by Edward Holden for the new Lick Observatory in California.

Barnard and his wife moved to Mt Hamilton. Director Holden allocated him the 12″ Clark refractor. He usually observed throughout the whole night, additionally taking plates of Milky Way fields with the 6″ Willard lens, which were exposed during the day – sleep was a rare matter. However, the main instrument, the 36″ refractor, was reserved for Holden and Burnham. Not until 1892 was Barnard allowed to use it for one night a week – this was the result of a bitter conflict with the authoritarian director. At the drop of a hat he discovered the fifth moon of Jupiter (the first since Galileo), which enormously increased his fame. In 1893 Barnard and his wife Rhoda travelled to Europe, visiting some of the most important observatories. On 25 July he met Dreyer in Armagh. Back on Mt Hamilton his position against Holden became stronger, resulting in even more altercations. Barnard was supported by his friend Burnham, who eventually fled from Lick Observatory in 1892 – Barnard followed suit three years later. His Milky Way photographs made with the Willard lens did not appear until 1913 (Barnard 1913c).

The new destination was Chicago. When Barnard arrived in 1895 the Yerkes Observatory in Williams Bay with its huge 40″ Clark refractor was still under construction. He was appointed Professor at the University of Chicago, but was exempted from teaching. At the observatory, which was finished two years later, he again met his friend Burnham. Unfortunately, the observing conditions at the low-level site were pretty poor – not comparable to the excellent Mt Hamilton.

In 1905 Barnard went to Mt Wilson, photographing Milky Way fields with the 10″ Bruce astrograph (lent by Yerkes Observatory). A first catalogue of 182 dark nebulae appeared in 1919 (Barnard 1919). In 1921 his wife Rhoda died. The target of Barnard's last visual observation with the 40-inch, dated 21 December 1922, was the Nova Persei of 1901 near which he had discovered a faint nebulosity on 16 December 1916 (Barnard 1916c). Edward Emerson Barnard died on 6 February 1923 in Williams Bay, WI, at the age of 65.[140]

The Bruce photographs were not published until 1927, i.e. four years after his death, by his niece and former clerk Mary Calvert. The work is titled *A Photographic Atlas of Selected Regions of the Milky Way* (Barnard 1927). It also contains the final catalogue of 346 dark nebulae, now designated 'Barnard' (B).[141] The delay was caused by Barnard's desire for perfection in reproducing the images – much similar to Tempel.

Barnard's name lives on in many popular objects. Barnard's Galaxy (NGC 6822) was found in 1884 at Vanderbilt Observatory. Barnard's Loop, a large supernova remnant in Orion, was detected on 13 October 1894 on a plate taken with the Willard lens (Barnard 1894e). However, William Pickering had photographed the object already in 1889 on Mt Wilson (Pickering W. 1890). Barnard's Arrow Star in Ophiuchus, showing the largest known proper motion of all stars, was discovered in 1916 (Barnard 1916a). Moreover, he discovered 16 comets. An odd story was told about

---

[139] Ashbrook (1984: 92–97).

[140] Obituaries: Aitken (1923), Boss (1923), Frost (1923), Fox (1923), Mitchell S. (1923), Parkhurst (1923), Paterson (1923), See (1923), Wolf M. (1923) and Dyson (1924); see also Burnham S. (1893, 1894c), Hardie (1964), Ashbrook (1979), Mumford (1987) and Sheehan (1995).

[141] Verschuur (1989), Kerner H. (1995). A large fraction of Barnard's dark nebulae was later observed visually by William Franks (Franks 1930). A similar study was made by Thomas Espin (Espin 1922).

Table 9.37. *NGC objects discovered by Barnard (see the text)*

NGC	Date	Rr	Type	V	Con.	Remarks	Ref.
206	1883	5″	GxP		And	Star cloud in M 31, W. Herschel 17.10.1786 (V 36)	12, 14
281	16.11.1881	5″	EN		Cas	IC 11, Barnard	4, 7
661	11.10.1882	5″	Gx	12.2	Tri	W. Herschel, 26.10.1786 (II 610)	3
1255	30.8.1883	6″	Gx	10.7	For		4, 7, 8
1292	Nov. 1885	6″	Gx	12.0	For		17
1297	Feb.? 1885	5″	Gx	11.8	Eri		13
1302	Feb.? 1885	5″	Gx	10.7	For		13
1499	3.11.1885	6″	EN		Per	California Nebula, Archenhold 27.10.1891	18, 19, 22
1721	10.11.1885	6″	Gx	12.8	Eri	L. Swift 2.12.1885	15, 17
1723	10.11.1885	6″	Gx	11.7	Eri	Tempel 12.1.1882	15, 17
1725	10.11.1885	6″	Gx	12.8	Eri	L. Swift 2.12.1885	15, 17
1728	10.11.1885	6″	Gx	13.9	Eri	L. Swift 2.12.1885	15, 17
1798	Nov. 1885	6″	OC	10.0	Aur		17
2141			OC	9.4	Ori	No reference	
2226			OC		Mon	No reference; core of NGC 2225 (OC)	
2282	3.3.1886	6″	RN		Mon	IC 2172, Barnard	16, 20
2568	1881	5″	OC	10.7	Pup		7, 8, 13
2835	Feb.? 1885	6″	Gx	10.3	Hya	Tempel 13.4.1884	13
5584	27.7.1881	5″	Gx	11.5	Vir	2° northwest of φ Vir	1, 2, 3
5824	1882	5″	GC	9.1	Lup	NGC 5834, Dunlop 4.5.1826 (Δ 611)	16, 21
6302	1880	5″	PN	9.6	Sco	First object; Dunlop 5.6.1826 (Δ 567); Fig. 9.54	7, 23
6352			GC	7.8	Ara	No reference; Dunlop 14.5.1826 (Δ 411)	
6354	1884	6″	3 stars		Sco		5, 9
	July 1883	5″	DN		Sgr	B 86; W. Herschel; Secchi	6, 7
6822	17.8.1884	6″	Gx	8.7	Sgr	Barnard's Galaxy; IC 4895, M. Wolf 16.7.1906	10, 11, 19
7513	1883	5″	Gx	11.9	Scl	GC 6131, Marth 27.9.1864 (m 530)	4

the ostensible 'Barnard's automatic comet seeker'.[142] Barnard, a member of many astronomical societies, was often awarded medals and honoured. Having been a Fellow of the RAS from 1888, he received their gold medal in 1897; the laudation was given by President Andrew Common (Common 1897).

9.15.2 New nebulae in Nashville

During the years 1881–86 Barnard found 26 objects in Nashville (Table 9.37). All bar one are catalogued in the NGC. Seventeen must be credited to him (marked in bold): nine galaxies, four open clusters, two emission nebulae, one reflection nebula and a star quartet. For three objects among them the globular cluster NGC 6352, which was first seen by Dunlop, information on the source is missing.

---

[142] About this remarkable hoax, see also Anon (1955).

Table 9.38. *Barnard's publications (sorted by date)*

Ref	Title	Publication	Ref.
1		*Sid. Mess.* **1**, 135 (1883)	1883c
2	A new nebula near φ Virginis	*AN* [2490] **104**, 285 (1883)	1883b
3		*Sid. Mess.* **1**, 238 (1884)	1883e
4	New nebulae	*Sid. Mess.* **2**, 226 (1883–84)	1884c
5	A new faint nebula	*Obs.* **7**, 269 (1884)	1884b
6		*Sid. Mess.* **2**, 259 (1883–84)	1884d
7	New nebulae – small black hole in the Milky-Way	*AN* [2588] **108**, 369–372 (1884)	1884a
8		*Sid. Mess.* **3**, 60 (1884)	1884k
9		*Sid. Mess.* **3**, 184 (1884)	1884g
10		*Sid. Mess.* **3**, 254 (1884)	1884i
11	New nebula near General Catalogue No. 4510	*AN* [2624] **110**, 125 (1885)	1885a
12	Small nebulae near great Andromeda Nebula	*AN* [2687] **112**, 391–392 (1885)	1885b
13	New nebulae	*Obs.* **8**, 123 (1885)	1885e
14		*Obs.* **8**, 386 (1885)	1885g
15	A correction to Dr. Swift's list of new nebulae in A.N. 2746	*AN* [2755] **115**, 315–316 (1886)	1886c
16	Ring-micrometer observations of comets and nebulae	*AN* [2756] **115**, 323–328 (1886)	1886d
17	New nebulae	*Sid. Mess.* **5**, 25 (1886)	1886e
18	An excessively faint nebula	*Sid. Mess.* **5**, 27 (1886)	1886f
19	Large nebula in field with General Catalogue 4510	*Sid. Mess.* **5**, 31 (1886)	1886g
20	New nebulous star	*Sid. Mess.* **5**, 154 (1886)	1886h
21	Note on the nebula G.C. 4036 = N.G.C. 5834	*AN* [2995] **125**, 315 (1890)	1890
22	Photograph of the nebula N.G.C. 1499 near the star ξ Persei	*ApJ* **2**, 350 (1895)	1895e
23	The nebula NGC. 6302	*AN* [4136] **173**, 123 (1907)	1907b

Barnard initially used his private 5″ Byrne refractor, which he had purchased in 1877. From 1883 he could observe with the 6″ Cooke refractor at Vanderbilt Observatory. The nebulae appeared during comet seeking and Barnard announced them in the new magazine *Sidereal Messenger*.[143] He published a lot – but often just not in time. Sometimes he was forced into publication by others who presented an object that had already been discovered by him. Like Burnham, he was sensitive about priorities. This was later criticised by Denning on the occasion of his discovery of the galaxy IC 356 in Camelopardalis (Denning 1892a). Barnard had found the object in 1889, but did not publish his discovery until 1892. Denning, a bit angry about the issue, wrote that '*Mr. Barnard claims to have discovered [IC 356] in August 1889 whereas I did not pick it up until November 1890. While admitting this claim I would venture to remark that any one who makes a discovery ought to be prompt in announcing it as a delay of several years is very likely to cause misconception and unnecessary trouble to others.*' Table 9.38 lists the 23 Barnard papers, giving information on new nebulae (those without title are mere notes). He was the NGC observer with the largest number of publications (see Fig. 10.5).

Barnard discovered his first non-stellar object, the bright planetary nebula NGC 6302 in Scorpius, in 1880. The find was published four years later (object 'e' in Ref. 7): '*A small flickering nebula slightly elongated (e. and w.) with 5 inch refractor. Prof. Swift, with his 16 inch refractor finds it to be triple and elongated.*' In 1892

---

[143] About American astronomical journals, see Holden (1895).

Barnard made detailed observations with the 36″ Lick refractor (Ref. 22) and created a name for the peculiar object: "*From its singular appearance, I have called it the 'Bug Nebula'.*" The report contains a sketch (Fig. 9.54). Dreyer correctly suggested in the notes of the IC I that the nebula had already been discovered by Dunlop (Δ 567, 5 June 1826).

In 1881 Barnard discovered NGC 281, NGC 2568 and NGC 5584 with his 5-inch. NGC 281 is an emission nebula in Cassiopeia, described as a '*large faint nebula, very diffuse, not less than 10′ diameter*' (Ref. 4). It is curious that Dreyer catalogued it a second time as IC 11 with a nearly identical description. Though Barnard is mentioned, there is no source. NGC 2568 is an open cluster in Puppis. The first note (Ref. 8) was followed by another (object 'c' in Ref. 7), giving a precise position and the description '*very faint nebulosity of moderate extension; pretty even in light*'. Obviously Barnard could not resolve the cluster. For NGC 5584, a galaxy in Virgo, he first wrote a note (Ref. 1), giving '*July, 1881*' as the discovery date. In his report 'A new nebula near φ Virginis' (Ref. 2), dated 8 January 1883, he was more precise: 27 July 1881. NGC 5584 is probably Barnard's first original discovery. He communicated it to Harvard College Observatory and Oliver Clinton Wendell determined a precise position with the 15″ refractor, which was published separately (Ref. 3). In his first note Barnard wrote that '*Professor Swift has examined it, and says that it appeared mottled, and as though more light and power would resolve it.*' But this was corrected shortly after: '*that remark was intended to apply to a different nebula*' (Barnard 1883d). Another nebula, which was found on 23 August 1883, is mentioned here, but '*Herschel's III 245 occupies about that place, but the description does not answer*'. This is the galaxy NGC 1187 in Eridanus, which had been discovered by William Herschel on 9 December 1784.

In 1882 Barnard saw the galaxy NGC 661, which he described as a '*minute speck of a nebula […] which I suppose to be new*' (Ref. 3). He added that '*By a direct vision it is quite faint, but by averted vision it is pretty distinct.*' However, the object is William Herschel's II 610 (discovered on 20 October 1786).

NGC 206 and NGC 1255 were found in 1883. On 14 September 1885 Barnard wrote the following about the former (Ref. 12): '*About two years ago, I found with my 5 inch refractor, a moderate size nebula involved in the extreme p. end of the Great Nebula of Andromeda. I*

Figure 9.54. Barnard's drawing of the Bug Nebula NGC 6302 in Scorpius (Barnard 1895e).

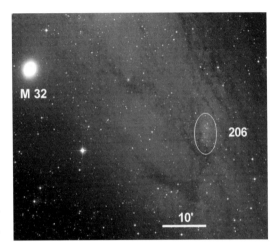

Figure 9.55. The conspicuous star cloud NGC 206 in the Andromeda Nebula (M 31).

have now confirmed the observation with the 6 inch Cooke Equatorial and as I can find no record of such nebula I suppose it is new.' NGC 206 is a conspicuous star cloud in M 31 (Fig. 9.55), which was first noticed by William Herschel on 17 October 1786 and described as '*vF, vL, E*' (V 36). After recognising this fact, Barnard wrote in *AN* 2691 that it '*is nothing else but G.C. 106 (V 36), and therefore not new*' (Barnard 1886b). By the way, on the same *AN* page Tempel mentions an observation of '*Barnard's new nebula in the southern spindle of the Andromeda nebula*' (Tempel 1885d). In Barnard's correction another object is mentioned: '*The new nebula in the f. end [A.N. 2690] has been seen repeatedly.*' This refers to his report 'A small faint nebula near the n. f. end of the great nebula of Andromeda' (Barnard 1886a). On 8 October 1885 he had found a bright H II region in the northeastern corner of M 31. This object was not catalogued by Dreyer, but appeared in 1964 as BA 602 in a list of Baade and Arp.

Figure 9.56. Barnard's Galaxy NGC 6822 in Sagittarius, a dwarf member of the Local Group (DSS).

After a preliminary note (Ref. 4), the galaxy NGC 1255, which had been found in 1883, was described a year later (object b in Ref. 7) as a '*Faint nebula, not large, pretty even in light.*' Barnard's second find in 1883 was the galaxy NGC 7513 (GC 6131). However, Marth was the first to see it, on 27 September 1864, which was stated by Barnard in another note: '[it] *proves to be number 6131 of Dreyer's Supplement*' (Barnard 1884e).

In July 1883 Barnard discovered another remarkable object: the dark nebula B 86 in Sagittarius.[144] He wrote (Ref. 6) that '*It is a small triangular hole in the Milky Way, as black as midnight. It is some 2′ diameter, and resembles a jet black nebula. There are one or two faint stars in the following part of it with a small cluster f. A small bright orange star is close n.p., on the border of the opening. Numerous larger dark openings are in its neighbourhood but none is as small and decided as this.*' The cluster mentioned is NGC 6520, only 9.5′ south of B 86. Barnard's declination sign is incorrect (+27° instead of −27°), but was soon corrected (Barnard 1884e). The dark nebula had been noticed earlier by Secchi; however, a report by William Herschel was wrongly interpreted as pertaining to this object too (see Section 6.11.3).

In 1884 Barnard discovered the first objects to be found with the 6″ Cooke refractor of Vanderbilt Observatory: NGC 6354 and NGC 6822. The former is a mere star trio in Scorpius (Ref. 5), but the latter is truly spectacular: this is Barnard's Galaxy in Sagittarius, which he found on 17 August 1884 (Fig. 9.56). NGC 6822 is a dwarf galaxy in the Local Group and thus of great astrophysical importance. As before, Barnard first published a discovery note in the *Sidereal Messenger* (Ref. 10): '*It lies in a low power field, with the small bright planetary No. 4510 of the General Catalogue, and is south of that object.*' The nearby PN is William Herschel's NGC 6818 (IV 51), 42′ to the northwest. Barnard's report in *AN* 2624 (Ref. 11) is a bit more detailed. About this '*excessively faint nebula*' he wrote that '*It is some 2′ diameter, and very diffuse and even in its light. With 6 inch Equatorial it is very difficult to see, with 5 inch and a power of 30± (field about 1¼°) it is quite distinct. This should be borne in mind in looking for it.*' The warning is justified, since the object is indeed difficult to observe: a small aperture giving a large field of view is recommended. The find demonstrates Barnard's high ability as a visual observer. Another publication of 1886 is even more detailed (Ref. 19): '*it certainly seems to be much larger and brighter than last year and I certainly think it has increased in density and size since that time*'. Barnard estimates a diameter of 10′ to 15′, giving a strange assumption: '*Probably this is a variable nebula.*' The increase in size and brightness is, however, not real, but due to a more favourable observation with a 'comet eye-piece' at the Cooke refractor.

Max Wolf was the first to photograph NGC 6822, on 16 July 1906 with the 72-cm Waltz reflector at Heidelberg/Königstuhl. His report is titled 'Ein Nebelfleckhaufen mit Nebelreichtum im Sagittarius'.[145] Another image of 8 August 1907 showed that '*the nebula appears to consist of numerous single nebulous patches*'. These are bright H II regions; two of them were discovered 1887 by Leavenworth (an assistant of Stone). Wolf wrote that '*At the northern border of the cluster there are two nebulae, appearing bright for the reflector: IC 1308 and NGC 6822. The latter was found by Barnard, the former by Stone. Barnard's nebula precedes Stone's by

---

[144] Such dark objects were also studied by Max Wolf, who called them 'cave nebulae' (Wolf M. 1917, 1919). A remarkable case is IC 5146 in Cepheus (Wolf 1908); see also Steinicke and Stoyan (2005b).

[145] 'A rich cluster of nebulae in Sagittarius' (Wolf M. 1907).

Figure 9.57. The galaxy trio NGC 1721/25/28 in Eridanus (DSS).

12s.' These misleading descriptions caused serious confusion: Dreyer catalogued Wolf's 'cluster of nebulae' as IC 4895, noting '*Group of neb, 25′ diam.*'[146] Actually, the object is identical with NGC 6822!

In 1885 Barnard was particularly successful. His report 'New nebulae' (Ref. 17) lists six objects found in November with the 6-inch. Except for NGC 1798, an open cluster in Auriga, all are galaxies in the Eridanus/Fornax region. NGC 1292 is described as '*rather faint, moderate size, elongated nearly north and south*'. The trio NGC 1721/25/28 with a diameter of only 3′ (Fig. 9.57) is interesting; a fourth galaxy, NGC 1723, is 8′ to the north. Barnard discovered the objects on 10 November 1885. The first three are in Swift's list III too, where a date, 2 December 1885, is given (Swift 1886a). When this list appeared in *AN* 2746, Barnard reacted to state his priority. In *AN* 2755 (Ref. 15) he wrote that Swift's nebulae '*have evidently crept into his catalogue by mistake*'. It proved that Swift had certainly not discovered the objects; in fact, Barnard communicated his find in a letter. Swift replied that '*I made a search for your trio near Rigel and saw all three [...] I don't see how you ever saw them.*' The observation entered Swift's list by mistake. Barnard wrote that '*Of course their being in Dr. Swift's catalogue as new on Dec. 2nd is purely accidental.*' He added that '*They were in field with a nebula, that Dr. Swift afterwards informed me was discovered by Tempel.*' This is NGC 1723. Once again, this case shows Barnard's exceptional vision: where it was difficult for Swift to see even known objects in his 16″ refractor, Barnard was able to discover them in his 6-inch (certainly under similar observing conditions). Barnard ended by commenting placatively that '*The great number of new nebulae discovered at the Warner Observatory shows well the industry of its keen-sighted Director.*' In 1891 Burnham measured the trio with the 36″ Lick refractor (Burnham S. 1892b).

Tempel was faster not only with NGC 1723 but also with the galaxy NGC 2835 in Hydra (Ref. 13). The priority was accepted. This was, however, not the case for the California Nebula NGC 1499 in Perseus (Fig. 9.58 right).[147] In 1891 Simon Archenhold published a paper in *AN* 3082, titled 'Ein ausgedehnter Nebel bei ξ Persei'.[148] The object appeared on a plate taken on 27 October 1891 with a 3.1″ portrait lens in Halensee near Berlin. Instead of the photograph only a sketch is given (Fig. 9.58 left). Archenhold wrote that '*Dreyer's new catalogue of nebulae lists under No. 1499 the following nebula, discovered by Barnard*'. Owing to the apparent size of nearly 2°, he obviously believed that Barnard's NGC 1499 was not, or was at best partly, identical with his object – even though the NGC gives '*vF, vL, Ens, dif*'! Another plate was taken in 1892 by Julius Scheiner in Potsdam. However, the title of his publication, 'Ueber den grossen Nebel bei ξ Persei (NGC. 1499)',[149] implies that he assumed an identity.

Not until 1894 did Barnard pipe up, mentioning that the nebula was seen on a plate taken with a 'lantern lens' on Mt Hamilton (Barnard 1894). The objective has only 1.5′ diameter and 3.5′ focal length. Since his paper (treating the independent discovery of Barnard's Loop in Orion) is dated 27 October, the image was probably produced in the autumn. It was focused on the Pleiades, but, due to its large field of about 30°, it showed NGC 1499 too. However, the nebula is erroneously called '*N. G. C. 1497*'.[150] Concerning the discovery, Barnard wrote '*This nebula, which was discovered by me with the 6-inch Cook equatorial of the Vanderbilt University, Nashville, Tenn., on Nov. 3, 1885, was photographed by Dr. Archenhold in October, 1891.*'

---

[146] Steinicke (2005c).

[147] Probably this name is due to Barnard, but there is no source.

[148] 'An extended nebula near ξ Persei' (Archenhold 1891).

[149] 'On the large nebula near ξ Persei (NGC. 1499)' (Scheiner 1893).

[150] NGC 1497 is a galaxy in Taurus, which was discovered by Stephan in 1876.

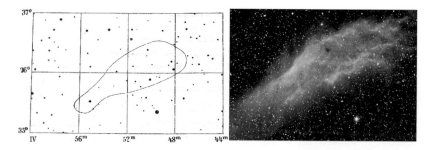

Figure 9.58. The California Nebula NGC 1499 in Perseus. Left: Archenhold's sketch; right: a DSS image (the bright star below to the right is ξ Persei).

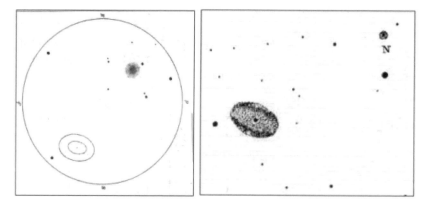

Figure 9.59. M 57 with its 'companion galaxy' IC 1296. Left: Barnard's sketch (Barnard 1894a); right: a sketch of von Gothard based on his photography of 1888 (Gothard 1894).

He added '*Dr. Archenhold is wrong in speaking of this object as having been discovered by photography. Visually, on account of its diffused nature, it is a very faint object in any telescope.*' Surprisingly, Barnard does not mention his discovery note of 1886 (Ref. 18). In November 1895 he wrote that, in the meantime, altogether five plates had been taken with the 6″ Willard lens at Lick Observatory (Ref. 21). A picture of the nebula, dated 21 September 1895, was added. Once again, Archenhold is criticised for his claim. The object is described as '*excessively faint, but rather large* [...] *probably about ½° long*'. It was difficult to see: '*This requires the lowest power and cannot be seen by direct vision. It is only by directing the vision slightly to one side of its place that it is possible to see it, it then flashes out feebly.*' This publication caused Dreyer's NGC entry. Certainly Barnard saw the brightest part of NGC 1499 – but there is no doubt that he was the discoverer, not Archenhold.[151] In 1904 the California Nebula was photographed by Max Wolf too (Wolf. M. 1903).[152]

In 1886 Barnard discovered another two objects at Vanderbilt Observatory: NGC 2282 and NGC 5824 (Ref. 16). The reflection nebula NGC 2282 in Monoceros is called a 'nebulous star'; it surrounds the 10.3-mag star HD 289120 and has a diameter of 3′. The case of Barnard's NGC 5824 is remarkable. Though he supposed the identity with John Herschel's h 2000 (GC 4036, NGC 5834), Dreyer catalogued the nebula as a new object, mentioning Barnard as the discoverer, because both the position and the description of h 2000 are wrong in the Slough catalogue. Barnard wrote in 1885 "*Not 'eeF'* [...] *it is visible in a 1¼ in. finder* [...] *the place does not quite agree*" (Barnard 1885f). The difference is 31′. He describes his find as '*small and very bright with a decided nucleus*', while Herschel gives

---

[151] Steinicke (2004d).

[152] Further studies of NGC 1499 were made by Curtis (1912a), Hubble (1920) and de Kerolyr (1937).

'*A very strongly suspected nebula; but I can not be quite sure (from low position) it is not a star.*' No doubt the observation of the object in Lupus ($\delta = -32°$) was critical at Slough. In 1890 Barnard finally stated the identity (Ref. 21).[153] However, the veritable discoverer of the globular cluster is Dunlop (Δ 611, 4 May 1826).

9.15.3 Further publications and objects

Barnard was an industrious writer: the number of his publications exceeds 500. They cover a wide range of astronomical themes and objects. Concerning nebulae, he is the undisputed number one (Fig. 10.5). After the NGC had appeared in 1888, Barnard checked many objects by visual observations and measurements. Dreyer included Barnard's corrections in the notes of both Index Catalogues. In 1892 Barnard investigated the complicated case of NGC 6589, NGC 6590 and NGC 6595 in Sagittarius (Barnard 1892b). This is a complex of reflection and emission nebulae connected with an open cluster, which were discovered by Safford, Swift and John Herschel.[154] In the same year he detected a nebula around Nova Aurigae (Barnard 1892c), which was not catalogued by Dreyer (in contrast to Copeland's Nova Cygni nebula NGC 7114). The 'new star' was first seen on 1 February 1892 by Thomas Anderson in Edinburgh.[155] In 1895 Barnard corrected and extended Dreyer's data for the galaxy NGC 532 in Pisces (Barnard 1895d).

Barnard proved his fantastic vision at one of his first observations with the 36″ Lick refractor. In October 1888 he discovered the star pair 'H' in the trapezium region of M 42 (see Section 9.6.10). On 2 October 1893 he discovered the faint galaxy IC 1296 (14.3 mag), 4′ northwest of the Ring Nebula M 57, with the 36-inch. In his report 'On a small nebula close to M. 57' in *AN* 3217, which includes a sketch (Fig. 9.59 left), one reads that '*This Nebula is about ½′ diameter. Not round. A little brighter in the middle. About 14th magnitude.*' (Barnard 1894a). However, the nebula is shown as a tiny spot on a plate taken by Eugen von Gothard on 13 June 1888 with his 26-cm Browning reflector in Herény, Hungary (Fig. 9.59 right). Although it had initially been overlooked, it was detected after reading Barnard's note: '*The small nebula is, however, so tiny that it is impossible to distinguish it from the numerous stars in the vicinity.*' (Gothard 1894). In the IC I Dreyer correctly assigns Barnard the priority for IC 1296.

Barnard was the discoverer of 129 IC objects, which mainly appear on photographs of Milky Way regions taken at Lick and Yerkes Observatory.[156] He found even more nebulae, but these were first seen by others. An example is the galaxy NGC 3596 in Leo (William Herschel's II 102), which was detected by Barnard in 1906 of a 10″ Bruce astrograph plate. It shows a '*fine planetary nebula which does not appear to be in the catalogues*' (Barnard 1906a). In the years 1906–8 he published detailed studies of the planetary nebulae M 57, M 97 and NGC 7662 (Barnard 1906a, 1907c, 1908b).

9.16 HOLDEN AT WASHBURN OBSERVATORY

Holden was one of the most controversial American astronomers in the nineteenth century. He must be blamed for this, since he was overambitious, bureaucratic and authoritarian. This caused conflicts and even enmity, e.g. with Barnard and Burnham. Holden showed his unpleasant character especially as Director of Lick Observatory, where he ended ingloriously. However, we owe to him a number of distinguished works made with great diligence. Examples are his monograph on the Orion Nebula and the compilation of publications on nebulae. In his time as Director of Washburn Observatory, Madison, Holden discovered some nebulae. He was a skilful visual observer, having access to the largest refractors.

9.16.1 Short biography: Edward Singleton Holden

Edward Holden was born on 5 November 1846 in St Louis, MO (Fig. 9.60). From 1862 he studied at Washington University, St Louis, graduating in 1866. At that time he visited his cousin George Phillips Bond, Director of Harvard College Observatory, where he came into contact with astronomy. During the years

---

[153] See also Barnard (1884h, 1886i).
[154] NGC 6595 was also photographed in 1895 in India with a 16.5″ reflector (Naegamvala 1895).
[155] Holden (1892); the nebula was catalogued in 1947 as Ced 50 by Sven Cederblad.

[156] See the 'Historic IC' by the author (www.klima-luft.de/steinicke).

Figure 9.60. Edward Singleton Holden (1846–1914).

1866–70 Holden was a cadet at the Military Academy at West Point (NY), after which he took over the scientific–technical education there. In 1871 he married Mary, the daughter of William Chauvenet, Professor of Mathematics and Astronomy at his former university.

In 1873 Holden became Professor of Mathematics at the US Naval Academy, Annapolis, and an assistant of Simon Newcomb at the US Naval Observatory, Washington. At first he worked there as a librarian. His major task was the compilation of a library catalogue. Newcomb, impressed by Holden's industry, allowed him to use the new 26″ Clark refractor for visual observations. When the planning for a new Californian observatory, to be funded by James Lick, became known in 1874, Newcomb suggested Holden as its future director. The matter became protracted and Holden stayed in Washington until 1881, eventually following James Watson as Director of Washburn Observatory in Madison, Wisconsin. There he observed double stars and nebulae with the 15.6″ Clark refractor erected in 1881. Holden stayed in office until 1885.

Meanwhile the construction of the Lick Observatory on Mt Hamilton had advanced. Holden first became President of the University of California and then on 1 June 1888 took up his new office as observatory director.[157] Later the same year the 36″ Clark refractor was ready to use – it was the largest in the world.[158] Holden did not take his wife to live with him in California, which was socially uncommon; she lived alone in St Louis. As assistants Holden engaged James Keeler, Sherburne W. Burnham, Edward E. Barnard, Martin Schaeberle and Wallace W. Campbell – an impressive staff.

While Holden concentrated on administration, his employees provided spectacular results. However, he reserved the right to publish them – and received the honour. Such issues, plus his authoritarian leadership, created tensions, mainly concerning Barnard and Burnham. Barnard, for instance, who was the most successful visual observer of his time, was not allowed to use the great Lick refractor; he had to put up with the 12-inch. On the other hand, Holden often allowed his allocated time at the 36-inch to lapse or took images of the Moon, which were criticised by Barnard as incondite and useless. The crisis escalated and even the press reported about the issue. The first to draw the conclusions was Burnham, who left the observatory in 1892. Barnard followed three years later. The two of them met again at the University of Chicago's new Yerkes Observatory. A palace revolution of staff members eventually caused Holden to abandon his position on 31 December 1897.[159] He was followed by Keeler. Holden first went to New York and in 1901 he became a librarian at the US Military Academy, West Point, NY. Edward Singleton Holden died there on 16 March 1914 at the age of 67.[160]

9.16.2 New nebulae

Between 17 March 1881 and 14 May 1882 Holden made zone observations of nebulae at the Washburn Observatory in Madison, Wisconsin. He used the newly erected 15.6″ Clark refractor (Fig. 9.61), equipped with a bar-micrometer by Kahler (power 145, field of view 25.5′).

The results appeared in the first two volumes of the *Publications of the Washburn Observatory*. Holden's first report is titled 'List of new nebulae and clusters discovered in the zone-observations at the Washburn Observatory, from 23 April to 30 September 1881' (Holden 1882a = Ref. 1).[161] It contains 23 objects, sorted by right ascension (epoch 1860). Usually several

---

[157] Neubauer (1950).

[158] It surpassed the 30-inch at Pulkovo of 1884 and was beaten in 1897 by the 40″ Yerkes refractor.

[159] Holden (1897).

[160] Obituaries: Campbell (1914, 1916) and Knobel (1915b); see also Osterbrock (1984) and Sheehan (1995).

[161] See also Holden's Washburn Observatory report for 1882 (Holden 1883).

9.16 Holden at Washburn Observatory

Table 9.39. *Holden's NGC objects*

NGC	No.	Date	Type	V	Con.	Ref.	Remarks
3200	1	10.4.1882	Gx	12.1	Hya	1	
3441	2	6.4.1882	Gx	13.5	Leo	1	
3531	3	27.4.1881	Gx	13.1	Leo	1	NGC 3526, Marth 25.3.1865 (m 215)
3792	4	27.4.1881	2 stars		Vir	1	
3843	5	27.4.1881	Gx	13.4	Vir	1	
3979	6	23.4.1881	Gx	12.7	Vir	1	L. Swift 27.4.1886
4576	7	27.4.1881	Gx	13.5	Vir	1	
4592	8	23.4.1881	Gx	11.6	Vir	1	W. Herschel 23.2.1784 (II 31)
4997	9	17.5.1881	Gx	12.9	Vir	1	Burnham 28.3.1878
4581	24	20.4.1882	Gx	12.3	Vir	2	
5030	10	17.3.1881	Gx	12.7	Vir	1	
5031	11	17.3.1881	Gx	13.5	Vir	1	
5035	12	17.3.1881	Gx	12.8	Vir	1	
5038	13	28.5.1881	Gx	13.5	Vir	1	
5046	14	17.5.1881	Gx	13.5	Vir	1	
5080	15	27.4.1881	Gx	13.6	Vir	1	
5496	16	23.4.1881	Gx	12.2	Vir	1	
5769	17	27.4.1881	Gx	14.4	Boo	1	
5858	25	14.5.1882	Gx	12.8	Lib	2	
6912	21	14.8.1881	Gx	13.6	Cap	1	
7226	22	20.6.1881	OC	9.6	Cep	1	Fig. 9.62 left
7490	23	21.6.1881	Gx	12.5	Peg	1	Stephan 11.10.1879

Figure 9.61. The 15.6″ Clark refractor of Washburn Observatory, Madison.

observations were made. For some objects a short description is given. The second publication lists only 'Two new nebulae' (Holden 1884a = Ref. 2). The numbers are continued (giving nos. 24 and 25). The 22 objects catalogued in the NGC are listed in Table 9.39. Eighteen of them must be credited to Holden (marked bold). Except for two objects, all are galaxies; most of them are located in Virgo. NGC 7226 is an open cluster (Fig. 9.62 left) and NGC 3792 is only a pair of stars. Holden's rate is thus fine.

In the case of NGC 3531 (no. 3) Holden suggested an identity with Marth's m 215 (GC 5546). This object was independently catalogued as NGC 3526 by Dreyer, who, however, added '? = 5546'. Finally Spitaler proved the identity, as mentioned by Dreyer in the notes to the IC I. That no. 8 (NGC 4592) is William Herschel's II 31 was not noticed by Holden, who is not mentioned in the NGC. When Burnham visited Washburn Observatory in spring 1881 to measure double stars with the new refractor, he probably observed NGC 4997 (no. 9). Holden notes that '*This nebula was first discovered by S. W. Burnham at the Dearborn Observatory, 1878, March 28.*' For NGC 7490 Dreyer notes '*St X, Holden*', since

Table 9.40. *The remaining objects from Holden's list (Ref. 1)*

Object	No.	Date	Type	Con.	Remarks
NGC 6416?	18	6.5.1881	NF	Sco	OC: 1° NW
B 92	19	2.5.1881	DN	Sgr	Barnard; Fig. 9.62 right
Trümpler 33	20	20.5.1881	OC	Sgr	

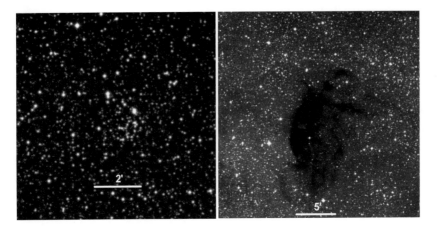

Figure 9.62. Left: the open cluster NGC 7226 in Cepheus; right: the dark nebula B 92 in Sagittarius (DSS).

this galaxy was first seen by Stephan in Marseille (it is contained in his list X). NGC 3979 was found independently by Lewis Swift in 1886.

Three objects from Holden's list were ignored by Dreyer in the NGC (Table 9.40). Object no. 18, described as a '*coarse cluster of 30–40 stars; several 8 mag*', could be NGC 6416. But this open cluster, which was discovered by Dunlop (Δ 612), lies 1° to the northwest and does not contain many bright stars. There is no doubt about no. 19, since Holden writes that '*This is a black circular hole in the Milky Way (10′ ± in diameter). The stars about it are excessively crowded, and inside there is but two stars; one tolerably bright 10^m, and the other very small.*' This is the dark nebula B 92 (Fig. 9.62 right) discovered by Barnard, who wrote in 1913 '*Known to me in my early days of comet-seeking, this object has always been of the deepest interest*' (Barnard 1913a). Holden's no. 20 ('*Coarse cluster detached from the Milky Way. It contains a reddish star.*') is the open cluster Trümpler 33. Since Robert Trümpler did not catalogue the object until 1930, Holden is the veritable discoverer. Holden's first report (Ref. 1) contains a compilation of a further 75 observed GC objects; notes are given for 20 of them.

9.16.3 Holden's publications

Holden is known as an industrious worker, compiler of astronomical data and writer of many publications.[162] The jobs as librarian at the US Naval Observatory and (after the Lick disaster) at West Point were ideal for him. Scientific questions were rarely treated; an exception is his work on the three-dimensional shape of nebulae.[163] Holden was not an explorer or scientist, literally, but had a somewhat bureaucratic mind, which originated from his military education at West Point. Some objects particularly fascinated him: the Orion Nebula (M 42), the Ring Nebula in Lyra (M 57) and the Omega Nebula (M 17) and Trifid Nebula (M 20) in Sagittarius. He observed them continually with various refractors.

The article 'Drawing of the Ring Nebula in Lyra' (Holden 1875) originates from Holden's time at the US Naval Observatory. It collects all published reports and drawings of various observers, reaching

---

[162] For a bibliography see Campbell (1916).
[163] See 'On the helical nebulae' (Holden 1889b) and 'What is the real shape of the spiral nebulae?' (Holden 1890).

back to Schroeter and Harding in 1797. In another one, titled 'The Ring Nebula in Lyra', he updated the existing knowledge on the basis of observation with the 36″ refractor at Lick Observatory, in which he was assisted by Martin Schaeberle (Holden 1888). As director he tried to present the benefits of the new observatory: '*In order to give or to obtain an idea of the power of the 36-inch Clark refractor of the Lick Observatory, it is necessary to compare its performance with that of other telescopes on familiar objects.*' He further wrote that '*One of the first objects examined here was the ring nebula in Lyra; and I wish to communicate a few particulars regarding its aspect in this telescope, not so much for their strictly scientific value, as to enable others (and myself) to estimate the gain of power by the use of the 36-inch lens.*' In 1888 Holden and Schaeberle also observed NGC 6543 (GC 4373) in Draco and the Saturn Nebula NGC 7009 (GC 4628) in Aquarius (Holden and Schaeberle 1888). A report titled 'Die ersten Beobachtungen von Nebelflecken am grossen Refraktor der Lick Sternwarte in Californien' appeared in 1889 in the German magazine *Sirius*.[164]

As early as in 1876 Holden had already treated the Omega Nebula (M 17). His work on historical and new observations appeared in the *American Journal of Science* as 'On supposed changes in the nebula M. 17 = h. 2008 = G.C. 4403' (Holden 1876). He analysed the drawings of John Herschel, Lamont, Mason, Lassell and Trouvelot and the spectroscopic studies of Huggins. Finally the results achieved with the Washington refractor, including his own observations, are mentioned. Holden sees signs for a proper motion (which is actually not real): '*On the whole, then, these drawings show that the western end of this nebula has moved relatively to its contained stars.*'

In his report 'On the proper motion of the Trifid Nebula', Holden treated similar questions (Holden 1877b). It analyses the observations and drawings of Lassell, William and John Herschel and Mason and Smith. Holden presents results from Harvard College Observatory and the US Naval Observatory too (the latter from him and Burnham). He concludes that '*The evidence as recorded with regard to this nebula indicates marked changes of position or brilliancy, or both, during the period 1784–1877.*' A summary appeared as 'Der dreifache Nebel im Schützen' in *Sirius*.[165]

No doubt Holden's masterpiece was the 'Monograph of the central parts of the nebula of Orion', which was published in the appendix of the *Washington Astronomical Observations for 1878* (Holden 1882b). On 230 pages the history of the Orion Nebula observations is presented in detail, containing the contributions of 44 observers (from Cysat in 1619 to Draper in 1880). Thirty-eight drawings are reproduced and numerous measurements, descriptions and references are given. This is Holden's world, as a collector and editor of astronomical data.[166] As the reference system for stars in and near M 42, he used the catalogue of George Phillips Bond, reduced to the epoch 1877. Holden created a new system to designate the various parts of nebulae. A large fraction of the monograph is reserved for his own visual observations made during the years 1874–80 with the 26″ Clark refractor at the US Naval Observatory. Holden comes to the conclusion that brightness changes occurred for some parts of the nebula. But the form seemed to be constant. Since it was published at the advent of photography, Holden's monograph was one of the last works on nebulae to be based exclusively on visual observing.[167] This qualified its practical importance for the astronomy of the then-ending nineteenth century. Nevertheless, the work is of historical interest.

In parallel to it, Holden edited a comprehensive compilation of the literature on nebulae and star clusters up to 1876, titled *Index Catalogue of Books and Memoirs Relating to Nebulae and Clusters, etc.* (Holden 1877a). A year earlier, his friend Edward Knobel had published the 'Reference catalogue of astronomical papers and researches, 4. Nebulae and clusters' (Knobel 1876). Since the two of them regularly communicated by letter, Holden was obviously inspired by this work. The 'Index catalogue' gives, spread over 111 pages, a pretty complete review of articles, books and drawings about non-stellar objects. Section I is a list of all publications, sorted by author. However, John Herschel's 'Cape observations' are missed out. Section II gives references for the Orion Nebula (not

---

[164] 'The first observations of nebulae at the Lick Observatory in California' (Holden 1889a).

[165] 'The threefold nebula in Sagittarius' (Holden 1878).

[166] A review appeared 1883 in *Monthly Notices* (Anon 1883c).

[167] However, at the end the first plate of the Orion Nebula, which was taken by Henry Draper, is reproduced.

as complete as in Holden's monograph). Section III treats the literature on variable nebulae. Section IV lists the sources of published drawings – an extension of John Herschel's list in the General Catalogue. Section V contains cross references between William Herschel's catalogues (objects sorted by class) and the GC. This is presented analogously for the Messier catalogue in Section VI. A review of Holden's catalogue appeared a year later in the *Monthly Notices* (Anon 1878).

In 1896 Holden published a book on *Mountain Observatories in America and Europe* (Holden 1896), presenting the development of 21 sites (after all, Mt Hamilton was the first permanent one). He also wrote biographies: *Sir William Herschel: His Life and Works* and *Memorials of William Cranch Bond and of His Son George Phillips Bond* (Holden 1881, 1897a).

9.17 HARRINGTON'S GALAXY

Harrington contributed one object to the NGC – an accidental find. He also worked on the globular cluster M 13 in Hercules.

Mark Walrod Harrington was born on 18 August 1848 in Sycamore, Illinois (Fig. 9.63). He graduated from the University of Michigan, Ann Arbor, in 1871 and then taught astronomy at Louisiana State University in Baton Rouge. After visiting Germany (Leipzig), China and Alaska he returned to Ann Arbor. There Harrington became Professor of Astronomy and Director of the University Observatory in 1879, succeeding James Watson. His assistants were Martin Schaeberle and William W. Campbell (both later made a career at Lick Observatory). In 1884 Harrington founded the *American Meteorological Journal*. In 1899 he had to abandon his offices due to ill-health. Mark Walrod Harrington died on 5 January 1926 in Newark, New Jersey, at the age of 77.

On 18 August 1882 Harrington discovered the 14.0-mag galaxy NGC 7040 in Equuleus (Fig. 9.64 left). He used the 12.5″ Fitz refractor which had been erected in 1853 at the University of Michigan Observatory in Ann Arbor. In his note of 31 October he wrote that '*It is so faint that I can only see it after resting my eyes in the dark a few moments.*' (Harrington 1883). The small constellation Equuleus hosts (together with Chamaeleon) the least NGC objects: only four.

Figure 9.63. Mark Harrington (1848–1926).

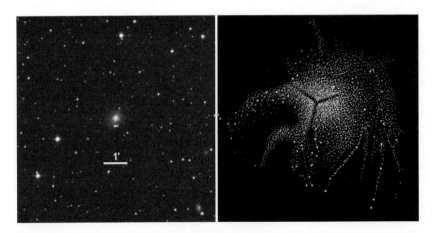

Figure 9.64. Left: Harrington's galaxy NGC 7040 in Equuleus (DSS); right: a drawing of M 13 by Bindon Stoney (Parsons W. 1861a).

In 1887 Harrington wrote a paper titled 'On the structure of 13 M Herculis' (Harrington 1887). He was interested in the symmetrical 'rifts', three radial dark lines that had been seen by the Birr Castle observers in the bright globular cluster. John Herschel had reported '*hairy-looking curvilinear branches*'. Lord Rosse even supposed a spiral structure of M 13 in his 1850 publication; he was supported by George Phillips Bond, who noticed an '*unquestionable curvilinear sweep in the disposition of its exterior stars*' (Bond G. 1860). Harrington presented Bindon Stoney's drawing of M 13, which had been made on 26 May 1851 (Fig. 9.64 right). He observed the globular cluster for a month, assisted by Schaeberle and Markham. The phenomenon of 'rifts' could barely be seen in the Fitz refractor, using a power of 500: '*they are so elusive that I sometimes almost doubted their existence*'. On 13 August 1887 Markham made a drawing, which, however, was not published.

9.18 HALL, THE MARTIAN MOONS AND A GALAXY

Asaph Hall is famous for his discovery of the two Martian moons, Phobos and Deimos, on 12 August 1877. He mainly observed planets, moons and comets with the 26″ Clark refractor at the US Naval Observatory in Washington, which had been erected in 1873. In doing so, he discovered NGC 7693, a faint galaxy in Pisces.

9.18.1 Short biography: Asaph Hall

Asaph Hall was born on 15 October 1829 in Goshun, CT, as the son of a clock maker (Fig. 9.65). He became interested in mathematics and astronomy at an early age. After his marriage he moved to Ann Arbor in 1856 to study astronomy at the University of Michigan; his teacher was Franz Brünnow. In the following year, being short of money, he applied for a job at Harvard College Observatory in Cambridge. He was fortunate, and the director, William Cranch Bond, engaged him. Under his successor, George Phillips Bond, Hall calculated orbits. A beneficiary was comet-observer Horace Tuttle, who started working at the observatory in 1857. In 1862 Hall changed to the US Naval Observatory in Washington. As little as a year later he became Professor of Mathematics. He worked mainly on celestial mechanics and could only infrequently use the 26″ Clark refractor, which was installed in 1873. It was reserved for the director, Simon Newcomb, and his ambitious (and not particularly popular) assistant Edward Singleton Holden.

Hall's favourite was Mars and the question of moons. At the perihelion opposition in August 1877 the opportunity for a search with the 26-inch appeared – Holden was away visiting Henry Draper in Dobb's Ferry (NY). On 8 August Hall discovered two moons, as was confirmed by Newcomb and Todd a few days later. Abruptly, the jealous Holden claimed to have found three other moons even earlier. However, their orbits proved to be physically impossible – an embarrassing issue for him.

In 1891 Hall retired at the age of 62, leaving the US Naval Observatory for Connecticut. Five years later he became honorary Professor for Astronomy at Harvard. Asaph Hall died on 22 November 1907 in Annapolis, MD, at the age of 78.[168]

9.18.2 NGC 7693, M 57 and the supernova in M 31

While observing comet Faye on 1 December 1882, Hall detected a faint nebula in Pisces (Fig. 9.66). He used the 26″ Clark refractor at a power of 383 (see Fig. 9.33). His short report gives the following comment: '*On Dec. 1 a small nebula or nebulous star, was found which at first was taken for the comet.*' (Hall A. 1881). A day later he checked the field and determined the position; about the brightness he noted '*Its magnitude I estimate as 14.*'

Figure 9.65. Asaph Hall (1829–1907).

---

[168] Obituaries: Anon (1907), Lewis (1908), Pritchett (1908) and Eichelberger (1908); see also Bailey (1931: 254–255), Ashbrook (1984: 290–297), Dick S. (1988) and Steinicke (2003a).

Figure 9.66. Hall's discovery: the galaxy NGC 7693 in Pisces (DSS).

Figure 9.67. Left: Johann Palisa (1848–1925); right: Samuel Oppenheim (1857–1928).

Dreyer catalogued the 13.6-mag galaxy as NGC 7693. In 1891 Burnham measured it with the 36″ Lick refractor (Burnham S. 1892b).

Hall was interested in the Ring Nebula M 57 too, particularly the question of its proper motion relative to the surrounding stars.[169] Tempel had given an account of 12 stars outside and a large number inside the nebula, which he had seen with his 11″ refractor in Arcetri (Tempel 1877a). Hall checked this with the 26-inch in the summer of 1877, but could detect only nine stars of 12 to 15 mag in the vicinity, which were measured (Hall A. 1878). In Tempel's reply (Tempel 1878c), he now speaks about even more stars. Instead of 12, his sketch now shows '*16 faint stars and a small star group*', one lying outside Hall's field. A bit arrogantly, he wrote '*I have no doubt that Prof. Hall, when considering my data, will eventually perceive the remaining 6 stars near the ring and perhaps too – using a very low power and large field of view – both the large preceding faint nebulous shine with the small star group and the following short, but brighter beard-nebula.*' Many a time did Tempel give his fancy full scope. Moreover, he showed little respect for superior telescopes.

In 1888 Holden commented on the issue in his publication on M 57 (Holden 1888). He had observed with the 36″ Lick refractor, whose great optical power made things more complex, however: '*On one of the nights of observations Mr. Schaeberle and myself were mapping new stars in and about the nebula, and we finally gave up our task not because it was ended, but because it seems to be endless.*' In his comprehensive study 'On the proper motion of the annular nebula in Lyra (M 57) and the peculiarities in the focus for the planetary nebulae and their nuclei' Barnard treated the case too (Barnard 1900a). He was able to measure all of Hall's stars with the 40″ Yerkes refractor and even some more (including the famous central star; see Section 2.7.2). No proper motion could be detected: '*I would suggest that the question of motion in this nebula is one particularly suited for the photographic plate to decide.*' This was eventually done by Leavenworth, using the 10.5″ refractor of the University of Minnesota, Minneapolis (Leavenworth 1900) – however, in vain too. In 1906 Barnard published another paper on M 57 and its stars (Barnard 1906b).

By the way, Hall was the last visual observer to see the fading supernova S And (in 1885).[170] On 7 February 1886 the now 16-mag star near the core of the Andromeda Nebula was '*at the limit of visibility to the naked eye to that in a 26-inch telescope*' (Hall A. 1886b).

9.19 PALISA, OPPENHEIM AND THE NEW VIENNA UNIVERSITY OBSERVATORY

At the time of their discoveries of nebulae, Johann Palisa and Samuel Oppenheim were assistants at the new University Observatory in Vienna. Both mainly

---

[169] See Section 2.7.2.

[170] About S And, see Section 9.20.2.

9.19 Palisa and Oppenheim   411

Table 9.41. *NGC objects found by Palisa and Oppenheim ('Ref.': see the text)*

NGC	Observer	Date	Rr	Type	V	Con.	Ref. (Obj.)	Remarks
927	Palisa	18.1.1885	27″	Gx	13.5	Ari	3 (1)	L. Swift 2.12.1885; Fig. 9.69 left
2819	Palisa	2.4.1886	27″	Gx	13.0	Cnc	4 (1)	GC 5472, Marth 21.10.1863 (m 160)
2926	Palisa	27.3.1886	12″	Gx	13.6	Leo	4 (2)	
2944	Palisa	27.3.1886	12″	Gx	14.1	Leo	4 (3)	
2981	Oppenheim	27.3.1886	12″	Gx	14.0	Leo	4 (4)	Fig. 9.69 right
3071	Palisa	10.3.1886	12″	Gx	14.5	Leo	4 (5)	
3094	Palisa	31.12.1885	12″	Gx	12.5	Leo	3 (2)	
3116	Palisa	10.3.1886	12″	Gx	14.5	LMi	4 (6)	
3976	Palisa	26.3.1886	12″	Gx	11.7	Vir	4 (7)	W. Herschel 13.4.1784
4587	Palisa	17.4.1882	12″	Gx	13.3	Vir	1	
6654	Palisa	20.9.1883	12″	Gx	11.9	Dra	2	L. Swift 11.9.1883, Lamp 23.9.1883

observed comets and minor planets with the 12″ and 27″ refractors. The new nebulae were published by director Edmund Weiss in the annual reports.

9.19.1 Short biography: Johann Palisa

Johann Palisa was born on 6 December 1848 in Troppau, Silesia (Fig. 9.67 left). After studying mathematics and astronomy at Vienna University (1866–70) he worked as an assistant of Karl von Littrow at the old Academy Observatory for one year. Then he got a job at the Naval Observatory in Pola, east of Trieste, where he became Director in 1873. There Palisa discovered 28 minor planets with the 15-cm refractor; his first was Austria on 18 March 1874. For his achievements (including the discovery of a comet) he was awarded the French Lalande prize for 1879. A year later Palisa returned to Vienna to work at the new University Observatory on the 'Türkenschanze', which had been finished in 1878.[171] The reason why he exchanged his directorship in Pola for a job as an 'Adjunkt' (night assistant) might have been the much better equipment in Vienna.

At the new site, Palisa found another 93 minor planets – his total of 121 visual discoveries is unique. In 1906 he received the French Valz prize. He was interested in astrophotography too. From 1900 to 1909 he edited the star charts of Max Wolf in Heidelberg. The Wolf–Palisa charts contain 123 large-format sheets with graticules, showing stars of up to about 15 mag. Already in 1902 he had published a star dictionary with data and identifications of catalogued stars between −1° and +19° declination. In 1909 Palisa was appointed vice-director of the University Observatory. He retired in 1919, but could still use the telescopes. Johann Palisa died on 2 April 1925 in Vienna at the age of 76.[172]

9.19.2 Short biography: Samuel Oppenheim

Samuel Oppenheim was born on 19 November 1857 in Braunsberg, Moravia (Fig. 9.67 right). In 1876 he began to study mathematics, physics and astronomy in Vienna. In 1882 he was appointed as a guest observer at the new University Observatory, where he eventually became an assistant in 1884. In the same year he was awarded his doctorate. In 1888 Oppenheim changed to the private Kuffner Observatory in Vienna-Ottakring, where he observed until 1896. Then he worked as a teacher at the secondary high school in Arnau (in Bohemia). In 1902 he became a professor at Prague University and later,

---

[171] The history of Vienna's observatories is described by Witt (2006).

[172] Obituaries: Hepperger (1925) and Anon (1925).

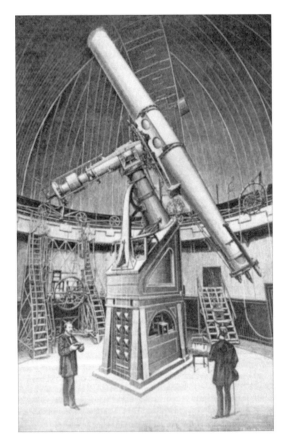

Figure 9.68. The 27″ Grubb refractor of 1883 at the University Observatory, Vienna.

in 1911, at Vienna University. His work was focused on celestial mechanics. Samuel Oppenheim died on 15 August 1928 in Vienna at the age of 70.[173]

9.19.3 New nebulae discovered by Palisa and Oppenheim

Between 1882 and 1886 Palisa discovered 10 objects; 7 of them were new. Oppenheim is credited for one discovery in 1886 (Table 9.41). All are galaxies; they were catalogued by Dreyer. The main instrument used was the 12″ Clark refractor in the west dome of the Vienna University Observatory. Two of Palisa's nebulae (NGC 927 and NGC 2819) were found with the 27″ Grubb refractor (Fig. 9.68) installed in 1883. However, since NGC 2819 is due to Marth, NGC 927 (Fig. 9.69

---

[173] Obituaries: Rheden (1928) and Wirtz (1929).

left) is the only NGC object to have been discovered with the great Vienna refractor.

Palisa's first galaxy, NGC 4587 in Virgo, was found on 17 April 1882; it was published by Edmund Weiss in 'Planeten- und Cometenbeobachtungen auf der neuen Wiener Sternwarte' (Weiss 1883 = Ref. 1). NGC 6654, which was discovered on 20 September 1883, is mentioned in a note in the *AN* (Palisa 1883 = Ref. 2). It is followed by another one by Ernst Lamp (Kiel Observatory), who observed the object three days later (Lamp 1883). Both had searched for comet Swift. This was Lewis Swift's task too, which led to his discovering the galaxy in Draco nine days earlier than Palisa did. Vice versa, Palisa was faster in the case of NGC 927, beating Swift by 10 months. The find was published by Weiss in 1886, together with NGC 3094 (Weiss 1886b = Ref. 3). The object number in this publication is given in the table ('Obj.').

The remaining objects were published in 1887 (Weiss 1887 = Ref. 4). Among them is Oppenheim's NGC 2981, which was found on 27 March 1886 (Fig. 9.69 right). However, the galaxy in Leo was credited to Palisa by Dreyer, though Oppenheim is clearly mentioned in Weiss' text. During the same night, Palisa discovered the galaxies NGC 2926 and NGC 2944 in Leo. Both observers used the 12″ refractor.

Two other objects had been found earlier (which was not noticed by Palisa): William Herschel saw NGC 3976 on 13 April 1784, and NGC 2819 is Marth's m 160, catalogued by Dreyer as GC 5472. The NGC correctly gives '*m 160, Palisa*'. Later Palisa discovered two more galaxies, which were entered in the Second Index Catalogue as IC 1748 and IC 5290.

9.20 HARTWIG'S OBSERVATIONS IN DORPAT AND STRASSBURG

Ernst Hartwig is primarily known for the discovery of the epochal supernova in the Andromeda Nebula (in 1885). He also found – during his time at Straßburg Observatory – some galaxies. They were noticed while searching for the comets of d'Arrest and Swift.

9.20.1 Short biography: Carl Ernst Albrecht Hartwig

Ernst Hartwig was born on 14 January 1851 in Frankfurt as the son of a post-office clerk (Fig. 9.70).

9.20 Hartwig 413

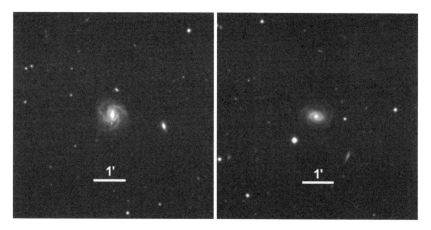

Figure 9.69. Left: Palisa's galaxy NGC 927 in Aries; right: the galaxy NGC 2981 in Leo, which was discovered by Oppenheim (DSS).

Figure 9.70. Ernst Hartwig (1851–1923).

He attended the secondary high schools in Frankfurt and Nürnberg. Afterwards he studied in Erlangen, Leipzig, Göttingen and Munich, where he passed his examination for a lectureship in 1873. A year later he moved to Straßburg Observatory to assist Winnecke. Hartwig travelled a lot, watching, for instance, the transit of Venus in 1874 in Brasilia. In 1880 he was awarded his doctorate at Straßburg University. He changed to Dorpat Observatory in 1884, staying there for two years as an observer. There, during the evening of 20 August 1885, Hartwig noticed the supernova S And in M 31, using the 24.4-cm Fraunhofer refractor. In 1886 he became the Director of Remeis Observatory in Bamberg, which was equipped with a 27.3-cm refractor by Schröder. He mainly observed variable stars. Together with Gustav Müller, he wrote the standard work *Geschichte und Literatur des Lichtwechsels der Veränderlichen Sterne*,[174] which appeared in 1918. After 37 years he retired from the directorship. Ernst Hartwig died on 3 May 1923 in Bamberg at the age of 72.[175]

9.20.2 New nebulae and the supernova in M 31

In 1883 Hartwig found seven new nebulae in Straßburg (Table 9.42). His first discovery, the galaxy NGC 5405 in Bootes, was made with a 16.3-cm orbit-sweeper by Reinfelder & Hertel, which had been installed in 1875. In his note 'Auffindung eines neuen Nebels' ['Discovery of a new nebula'] Hartwig wrote that the nebula was seen accidentally on 3 April 1883, while searching for comet d'Arrest (Hartwig 1883a = Ref. 1). He at first thought that it was the comet: '*As, moreover, this nebula appeared more conspicuous that night than the nearby nebula h. 1731 (3747 in the General catalogue)* [NGC 5418], *in whose vicinity the comet had to be that night, I supposed that I was seeing d'Arrest's comet and immediately measured its position with the 18-inch refractor.*' All of the

---

[174] '*History and Literature of the Light Variation of Variable Stars*'; for a later edition see Hartwig and Müller (1923).
[175] Obituary: Hoffmeister (1923).

414 Compilation of the NGC

Table 9.42. *Hartwig's NGC objects ('Ref.': see the text)*

NGC	Date	Type	V	Con.	Ref.	Remarks
5171	29.6.1883	Gx	12.9	Vir	3 (2)	Hough 5.5.1883; Tempel 11.5.1883
5176	29.6.1883	Gx	14.4	Vir	3 (4)	Fig. 9.71 left
5177	29.6.1883	Gx	14.6	Vir	3 (5)	Fig. 9.71 left
5179	29.6.1883	Gx	14.2	Vir	3 (3)	Burnham 5.5.1883; Tempel 11.5.1883
5186	29.6.1883	Gx	14.8	Vir	3 (1)	
5405	3.3.1883	Gx	13.8	Boo	1	Found with orbit-sweeper
6508	19.9.1883	Gx	13.3	Dra	2	L. Swift 17.6.1884

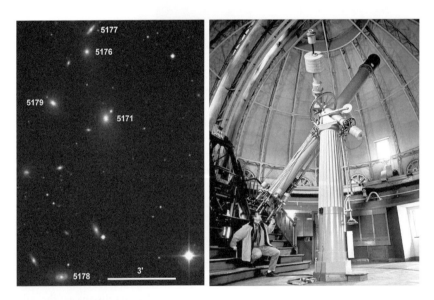

Figure 9.71. Left: the galaxy group around NGC 5171 in Virgo (DSS); right: the author at the 18″ Merz refractor of Straßburg Observatory (picture: S. Binnewies).

other objects were found with the 18″ Merz refractor erected in 1880 (Fig. 9.71 right).

The continuing search for comet d'Arrest brought another five galaxies on 29 June 1883, all in Virgo. Hartwig did not report the find until 17 September 1885, meanwhile working in Dorpat (Hartwig 1885 = Ref. 3). Possibly he had not had enough time in Straßburg. The delayed note was triggered by '*Mr Dreyer's request (Astr. Nachr. 2667)*'. Dreyer collected all of these discoveries for his 'second supplementary catalogue' (Dreyer 1885a). The galaxies of the quartet NGC 5171/76/77/79 form a close group of 5′ diameter (Fig. 9.71 left). Two of its members had already been discovered in early May by Hough, Burnham and Tempel, who were searching for comet d'Arrest too (see Section 9.6.8).

The galaxy NGC 6508 in Draco was found on 19 September 1883. This time, Hartwig was looking for comet Swift (Hartwig 1883b = Ref. 2). His note in the *Astronomische Nachrichten* is followed by two of Johann Palisa and Ernst Lamp, concerning the galaxy NGC 6654 in Draco; both had been searching for the comet too (see Section 9.19.3).

Hartwig's discovery of S And in M 31 on 20 August 1885 was at first accompanied by trouble: his report got lost in the post. It eventually appeared on 1 September 1885 in Dun Echt Circular no. 97 (Copeland R. 1885). The supernova – called 'Nova

Andromedae' at that time – was an astronomical event of first rank.[176] All of the major visual observers recorded its appearance, among them Barnard on 1 September (Barnard 1885c, h). Today it is debated whether Hartwig was the first to see the supernova. There is, for instance, a report by the Welsh Isaac Ward about a sighting on 19 August (Ward 1885) – one day earlier than Hartwig.

In October 1886 the Andromeda Nebula caused a stir again, when the Hungarian observer Radó von Kövesligethy reported 'Ueber wahrscheinliche neue Veränderungen im grossen Andromeda-Nebel'.[177] At first is was not clear whether a new 'flare up' of the supernova had happened or whether the core of M 31 was changing. The issue could not be decided for the time being (Krüger 1886).[178] Hartwig piped up in 1898, pointing out that all such observations were *mostly due to confusion of the place of the Nova 1885 with a neighbouring star of 11 mag [...] or with the optical centre of the nebula itself* (Hartwig 1898a). However, another 'supernova' did not appear.

9.21 ELLERY, LE SUEUR, MACGEORGE, TUNER, BARACCHI AND THE GREAT MELBOURNE TELESCOPE

The Great Melbourne Telescope (GMT) erected in 1869 in Australia was the great hope for southern-sky astronomy. For some time the 48″ reflector was the largest telescope in that hemisphere. It was the last major instrument with a metal mirror too.[179] In Europe silvered glass mirrors of diameter up to 80 cm (the Foucault reflector in Marseille) were already in use. Not only the extremely heavy instrument but also the bad site conditions near Melbourne caused problems. Thus the scientific value, especially concerning visual observing, photography and spectroscopy of non-stellar objects, was limited. However, it was a significant step for Australian astronomy.[180]

The GMT was used between 1869 and 1885 – a short time for such an elaborate instrument. It was installed under the directorship of Robert Ellery. His assistants, working successively at Melbourne Observatory, were the main observers: Albert Le Sueur, Farie MacGeorge, Joseph Turner and Pietro Baracchi (Table 9.43). The latter became Ellery's successor in 1895. It should be mentioned that Melboure was the most southern site contributing to the NGC. Its latitude of −38° even beats that of Paramatta and Feldhausen (both −34°).

9.21.1 Short biography: Robert Lewis John Ellery

Robert Ellery was born on 14 July 1827 in the English town of Cranleigh, Surrey (Fig. 9.72 left). Visits to the Royal Observatory Greenwich led him to astronomy. In 1852 he settled in Williamstown near Melbourne. A year later an observatory was built there. Ellery became its director and was appointed Victorian Government Astronomer. In 1863 it moved to a better site: the Flagstaff Hill near Melbourne. In 1868 the Great Melbourne Telescope with its 48″ mirror was erected there. Ellery's duties were positional measurements, meteorology and the time service, thus there was little time to use the large reflector. During the years 1866–84 he was President of the Royal Society of Victoria. Robert Ellery retired in 1895 and died on 14 January 1908 in Melbourne at the age of 80.[181]

9.21.2 Short biography: Adolphus Albert Le Sueur

Not much is known about Albert Le Sueur. He was born on 8 December 1849. He studied in Cambridge until 1863. He got his astronomical education from John Couch Adams at the local observatory. In 1866 Le Sueur moved to Dublin and supervised the construction of the Great Melbourne Telescope at the Grubb factory. In 1868 he escorted the transportation of the telescope to Melbourne and was responsible for its erection. He was not only technically talented but also a keen visual observer and drawer. After disputes with director Ellery, he returned to England in July 1870.

---

[176] Glyn Jones (1976), Ashbrook (1984: 404–407), Beesley (1985), de Vaucouleurs (1985), Kippenhahn (1985).
[177] 'About possible changes in the great Andromeda Nebula' (Kövesligethy 1886a).
[178] See also Konkoly (1886) and Kövesligethy (1886b).
[179] Perdrix (1992).
[180] Haynes et al. (1996).

[181] Obituaries: Lynn (1909b) and Turner H. (1909).

Table 9.43. *Ellery's assistants and GMT observers*

Observer	Life data	Observing period
Albert Le Sueur	8.12.1849 – 25.4.1906	Apr. 1869 – July 1870
Farie MacGeorge		Aug. 1870 – Sep. 1872
Joseph Turner	2.7.1825 – 25.8.1883	Feb. 1873 – Aug. 1883
Pietro Baracchi	25.2.1851 – 23.7.1926	Sep. 1883 – 1886

Figure 9.72. Astronomers at Melbourne Observatory (from left to right): Robert Ellery (1827–1908), Pietro Baracchi (1851–1926) and Joseph Turner (1825–1883).

He married in 1875 in London. Albert Le Sueur died there on 25 April 1906 at the age of 58.

9.21.3 Short biography: Farie MacGeorge

Farie MacGeorge is an almost anonymous person. He was an employee at the Geodetic Survey of Victoria. Ellery's job advertisement in a newspaper in August 1870 caught his attention and he became the successor of Le Sueur at Melbourne Observatory. MacGeorge observed for two years with the GMT and was followed by Joseph Turner.

9.21.4 Short biography: Joseph Turner

Joseph Turner was born on 2 July 1825 in Kirkconnel, Dumfriesshire, Scotland (Fig. 9.72 right). In 1852 he went to Melbourne. In 1856 he opened a photo studio in Geelong, which was destroyed by a fire in 1869. In 1873 he followed MacGeorge as Ellery's assistant at Melbourne Observatory. For 10 years Turner observed and sketched nebulae with the Great Melbourne Telescope. He succeeded in obtaining good photographs of the Moon at the prime focus. Whilst still in office, Joseph Turner died unexpectedly on 26 August 1883 in East Hill, Victoria, at the age of 58.

9.21.5 Short biography: Pietro Paolo Giovanni Ernesto Baracchi

Pietro Baracchi was born on 25 February 1851 in Florence, Italy (Fig. 9.72 centre). After an education as an engineer he moved to New Zealand in 1876. Being unable to find work there, he went to Melbourne Observatory (Williamstown) later in the same year, but stayed only briefly. During the years 1877–83 Baracchi worked as a surveyor. When Turner died in August 1883, Baracchi followed him as the assistant of Ellery. He was the last GMT observer. A year later he became a Fellow of the RAS. After Ellery's retirement in 1895 he was appointed Director of Melbourne Observatory and Victorian Government Astronomer (both first in an acting capacity and from 1900 official). In 1908–9 he was the President of the Royal Society of Victoria. In 1911 he founded the Mt Stromlo Observatory near Canberra. He retired in 1915. Pietro Baracchi died on 23 July 1926 in Melbourne at the age of 75.

9.21.6 The Great Melbourne Telescope (GMT)

In the middle of the nineteenth century, Great Britain and Ireland had acquired the greatest experience with large reflectors. Airy proudly remarked that *'the reflecting telescope is exclusively a British instrument in its invention and improvement, and almost exclusively in its use'* (Airy 1849). The builders were wealthy amateur astronomers, such as William and John Herschel, Lord Rosse and Lassell. To demonstrate this dominance in the southern hemisphere too, a plan to install a great reflector there was conceived. The result was the Great Melbourne Telescope (GMT). It was problematic from the beginning, and contributed very little to the prestige of the Empire.[182]

In 1850 Romney Robinson, Director of Armagh Observatory, formulated the idea of a large telescope for the southern hemisphere. His plan was *'a minute re-examination of all brighter nebulae [...] embodied in drawings so correct that each of them may be referred to as an authentic record'* (Robinson 1851). The instrument should be designed for 'hand-and-eye drawing', a common task for a reflecting telescope at that time. The objective was to repeat John Herschel's observations of southern nebulae and star clusters to detect possible changes. On the basis of a suggestion of Lord Rosse, the possibility of erecting the telescope at the Cape Observatory in South Africa (Warner 1982) was discovered.

In 1852 the Royal Society appointed a 'Southern Telescope Committee', which should direct *'the establishment in the Southern Hemisphere of a Telescope not inferior in power to a three-foot reflector'*.[183] By this Lord Rosse's 36-inch was meant. Among others, Romney Robinson (as speaker), Lord Rosse, John Herschel, William Lassell and Warren de la Rue were on the committee. Because of Robinson and Lord Rosse, the project had a strong Irish influence. Consequently Thomas Grubb in Dublin was consulted, and he suggested a 48″ reflector with a metal mirror.

After a promising start the issue was put off in 1853.[184] Not until 1862 was it revived, for a simple reason: the British colony in Victoria was willing to fund a large reflector in Melbourne. Meanwhile telescopes with silvered glass mirrors had been built, such as the Foucault reflectors with apertures of 40 cm (1859) and 80 cm (1862), these being superior to the old 'speculum' type.[185] Of course, the committee was aware of this development, but believed that the (British) experiences were not sufficient to build a glass mirror of 48 inches. So it was decided in favour of a classical metal one, whose costs were calculable. This turned out to be a fatal error. Lord Rosse advocated that *'someone should accompany the telescope as a mechanical assistant – he having been the principal operator in grinding and polishing the specula'*.[186] The problems were known: high weight, fragility, tarnishing, sensitivity to dew and thermal expansion (which was, however, less than for the first glass mirrors).

Lassell made the enticing offer of making his 48″ reflector available for use after the Malta mission, *'if it should be found suitable in all its points'*.[187] In its report the committee refused it. The issue elicited a critical reply from Lassell. Back in England he stripped down the 48-inch – it was never again assembled.

Thomas Grubb was charged with the construction of the reflector.[188] A Cassegrain design was chosen, allowing better access to the focus. In 1866 Grubb's 21-year-old son Howard produced two metal mirrors (with central hole) of diameter 1.22 m and focal length 9.3 m. Albert Le Sueur was appointed as the 'principal operator' supervising the whole process at the Grubb factory in Dublin. The final telescope was tested and shown to be of good optical and mechanical quality.

Le Sueur supervised the dismantling of the instrument, joined the transport by ship and finally assembled it in Melbourne. In June 1869 the GMT – the new landmark on Flagstaff Hill – was ready to use. The Cassegrain system had an enormous focal length of 51 m and a moving mass of 8.5 t (Fig. 9.73). The combined truss/closed tube was 12 m long, resting on a massive English axis mounting. The GMT was sheltered by a rectangular barracks with sliding roof. The construction allowed observations at above 10° elevation.

When Le Sueur installed the first mirror, a severe problem occurred, probably caused by him. Since

---

[182] Hogg (1959), Hyde (1987), Gascoigne (1996).
[183] Hyde (1987: 227).
[184] Probably as a consequence of the Crimean War.
[185] Ashbrook (1984: 148–154). For Foucault's 40-cm reflector see Foucault (1859) and Tobin (2003: 215–218).
[186] Hyde (1987: 228).
[187] Hogg (1959: 108).
[188] Glass (1997: 39–61).

Figure 9.73. The Great Melbourne Telescope, a Cassegrain reflector with 48″ metal mirror.

it was not usable, it had to be re-polished in July. Unfortunately, the Australian winter was bad and the observations could not be started until September (Sabine 1869). Le Sueur has described the first period in a report, titled 'An account of the Great Melbourne Telescope from April 1868 to its commencement of operations in Australia in 1869' (Le Sueur 1870a).

Soon the disadvantages of telescope and site came to light. Only 17% of all nights were usable. Since there was no dome due to the length of the tube, the telescope was exposed to the wind and tended to vibrate. Moreover, dust attacked the mirror, which soon fogged up (additionally to tarnishing). Finally, large temperature variations influenced the optical image. Ellery noted that '*I never once obtained with it what I would call moderately good definition.*'[189]

Despite the adverse conditions, Le Sueur made some interesting observations, which he published in the *Proceedings of the Royal Society*.[190] The objects described are the Orion Nebula (M 42), M 17, Eta Carinae, 30 Doradus, the planetary nebula NGC 5189, the reflection nebula NGC 2316/17 and Jupiter (Le Sueur 1870b, c, 1871). In 1870 he produced a fine drawing of the nebula around Eta Carinae (called η Argo). The visual spectroscope was used too, though the great focal length and the resulting high magnification were disadvantageous (the regular field of view was only 12.5′). Thus an object could hardly be focused on the slit, especially under windy conditions. The first attempts to take photographs at the prime focus showed that only bright objects could be imaged, allowing a short exposure time. Since he was isolated, the bad conditions more and more frustrated Le Sueur. Melbourne Observatory was simply too far away from the major astronomical centres. Stress with Ellery, exacerbated by the situation, was the consequence. As a representative of classical astronomy (positional measurements with a meridian-circle or refractor), the director had no experience with large reflectors. Furthermore, the GMT was the first one to be erected at a non-private observatory.[191] In July 1870, Albert Le Sueur left Melbourne Observatory and returned to England. His successor was Farie MacGeorge, who stayed for two years. In 1874 he continued the spectroscopic observations; his main target was Eta Carinae (MacGeorge 1874).

The GMT had its best time under Joseph Turner, who used it for 10 years. However, there were some flops. For instance, in 1877, when he tried to observe the newly discovered Martian moons, they were really drowned out in the stray light of the metal mirror, whose reflectivity had noticeably decreased.[192] When it became known that the two moons were not visible in the GMT – the even better appearance of the planet in the southern sky notwithstanding – the end of the telescope, in which such great hopes had been placed, was nigh. At times the GMT distinguished itself. For instance, Turner succeeded in taking a fine photograph of the Moon in the prime focus, and his follower Baracchi made the first images of southern nebulae in 1883. Though the exposure times for M 42 and Eta Carinae were only 5 minutes, the images were, due to vibrations, not very sharp. Given that Common, at the same time, produced an award-winning photo of M 42 with his 36″ silver-on-glass reflector in Ealing, this demonstrates the full dilemma of the GMT.

---

[189] Hyde (1987: 229).
[190] In some cases the aperture of the GMT was reduced to 18 inches.
[191] Forty of the 48 private, university or governmental observatories in Britain were equipped with refractors.
[192] In contrast, Phobos and Deimos could be seen that year at Birr Castle with the 72″ metal mirror reflector.

Table 9.44. *A comparison of the GMT and the Foucault reflector*

	Great Melbourne Telescope	Foucault reflector
Mirror	Metal	Glass (silvered)
Reflectivity	65%	95%
Diameter	122 cm	78.8 cm
Focal length (system)	9.3 m (primary), 51 m (Cassegrain)	5.8 m (Newton)
Relative light-gathering power	1	2
Mirror mass	1000 kg	300 kg
Moving mass	8600 kg	1500 kg
Producer	Grubb	Foucault/Eichens
Site	Melbourne	Marseille
Erection	June 1869	January 1862
End of operation	1886	1965

The origin of the problem was the design of the telescope, which had been built for visual observations. It had not been realised that the 'hand-and-eye' method was an obsolescent model due to the advent of photography. For the GMT there was nothing meaningful to do – thus it eventually became an astronomical curiosity. Even Robinson's objective of a re-examination of John Herschel's survey was not achieved. Although many observations and drawings were made by Le Sueur, MacGeorge, Turner and Baracchi, a general continuity and systematic approach were never present. For a long time, Ellery could not find an able engraver; moreover, he was absorbed by other duties. In 1884 an 8″ meridian-circle was erected and Ellery concentrated on positional astronomy. However, in 1885 a modest selection of GMT observations appeared (see the following section). But the publication attracted little interest – it simply came too late. In May 1886 Ellery wrote two short reports for *The Observatory* (Ellery 1886a, b).

In 1886 the work with the GMT came to an end; lately Baracchi had used it only sporadically. The reflectivity of the metal mirror was poor. However, Ellery refurbished the instrument in 1890, but this generated no further observations and the telescope was eventually mothballed in 1893. George Ritchey later remarked that the GMT – due to the problems that had been made public – had put back the development of reflectors by nearly 30 years.[193] No doubt the late decision to adopt a metal mirror, when one should already have been aware of the benefits of silvered glass mirrors, was wrong and serious. It is informative to compare the specifications of the GMT and the 80-cm Foucault reflector (Table 9.44). However, Charles Wolf was a critic of large telescopes in general, asking: '*of what good are these enormous instruments, so costly to construct, so inconvenient to manage, if it be true that a telescope of 16 inches aperture, and 95 inches long, can supply their place?*'[194] His favourite was a 40-cm Foucault reflector with silvered glass mirror that had been erected at Paris Observatory.

When Melbourne Observatory closed in March 1944, the GMT was transferred to Mt Stromlo near Canberra. It was modified to receive a glass mirror of 50″, a closed tube and a new English axis mounting. The telescope was refurbished once more in 1990, this time receiving a truss-tube. Unfortunately the new GMT was destroyed in a disastrous bush fire on 18 January 2003.

9.21.7 Observations of nebulae and star clusters with the GMT

The visual observations of Le Sueur, MacGeorge, Turner and Baracchi were (partly) published by Ellery in 1885. The report *Observations of the Southern Nebulae Made with the Great Melbourne Telescope from 1869 to 1885* covers 17 pages (Ellery 1885). It describes 24 fields, some containing several objects.

There are drawings of the objects observed (Table 9.45). All bar one were made by Turner; they

---

[193] Hyde (1987: 230).

[194] Wolf C. (1886b: 227).

Table 9.45. *Objects observed and drawn at the GMT by Turner and Le Sueur (see the text)*

Pl.	Fig.	Drawer	Date	Objects (NGC)	Type	Con.	Remarks
I	1	Turner	Nov. 1875	134	Gx	Scl	Fig. 9.74 right
I	2	Turner	29.10.1875	55	Gx	Scl	Sculptor group; Fig. 9.74 left
I	3	Turner	Dec. 1875	300	Gx	Scl	Sculptor group
I	4	Turner	Dec. 1875	346	EN+OC	Tuc	SMC
I	5	Turner	Nov. 1876	808	Gx	Cet	
I	6	Turner	Nov. 1876	823	Gx	For	
I	7	Turner	Nov. 1876	833, 835, 838–39	Gx	Cet	Arp 318
I	8	Turner	Nov. 1875	986	Gx	For	
III	20	Turner	Nov. 1876	1735	OC	Dor	LMC
III	21	Turner	Nov. 1876	1736	EN	Dor	LMC
III	22	Turner	Nov. 1876	1737, 1743, 1745, 1748	EN	Dor	LMC; NGC 1745: EN+OC
III	23	Turner	Dec. 1876	1744	Gx	Lep	
III	24	Turner	Dec. 1876	1760	EN	Dor	LMC
III	25	Turner	19.12.1876	1779	Gx	Eri	
III	26	Turner	19.12.1876	1808	Gx	Col	
III	27	Turner	Dec. 1875	1847	OC	Dor	LMC
III	28	Turner	4.1.1877	1888–89	Gx	Lep	Arp 123
III	29	Turner	21.12.1875	1950, 1958–59, 1969, 1971–72	OC	Dor	LMC; NGC 1958: GC, NGC 1959: Men
III	30	Turner	Dec. 1875	1962, 1965–66, 1970	EN+OC	Dor	LMC
III	31	Turner	Dec. 1876	2068	RN	Ori	M 78
IV	32	Le Sueur	7.2.1870	2046–47, 2057–59, 2065–66	OC	Men	LMC
IV	33	Turner	26.4.1876	2046–47, 2057–59, 2065–66	OC	Men	LMC; Baracchi: NGC 2043, NGC 2072
IV	34	Turner	Jan. 1876	2736	EN	Vel	Herschel's Ray[195]
IV	35	Turner	19.5.1876	3576, 3579, 3581–82, 3584, 3586	EN	Car	

are positive (white on black background) and collected on three plates. Altogether 44 different objects are presented, among them 15 galaxies. Nine fields are located in the Large Magellanic Cloud (LMC), one lies in the Small Magellanic Cloud (SMC). The galaxy quartet around NGC 835 (Arp 318), the interacting pair NGC 1888/89 (Arp 123) and the nearby galaxies NGC 55 (Fig. 9.74 left) and NGC 300, both members of the Sculptor group, are interesting.

9.21.8 Poor yield: two new nebulae

When observing the group of open clusters around NGC 2065 in the LMC, Baracchi found two new objects: the open clusters NGC 2043 and NGC 2072 (Fig. 9.75). He noted that '*Preceding H. 1259* [NGC

---

[195] This name originates from Herschel's description in the Cape catalogue (h 3145).

Figure 9.74. Turner's drawings at the GMT: the edge-on galaxies NGC 55 (left) and NGC 134 (right) in Sculptor.

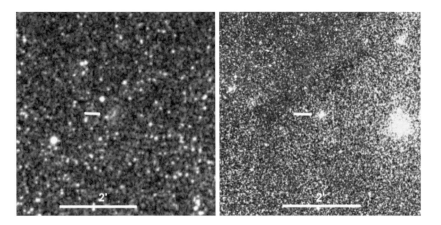

Figure 9.75. The two inconspicuous open clusters in Mensa found with the GMT: NGC 2043 (left) and NGC 2072 (right), to an identical scale (DSS).

2058] *by 79.5^s and 4' 30" north is a small elongated group of pf minute stars in very thin nebula* [NGC 2043], *and following H. 1265* [NGC 2065] *by 47^s and 40" is a very faint, small, indistinct patch* [NGC 2072]. *Neither of these appear to have been hitherto observed.*'[196] The group had already been observed and sketched on 7 February 1870 by Le Sueur and on 26 April 1876 by Turner; neither had noticed the new objects.

Baracchi discovered NGC 2072 on 18 December 1884, but had to terminate the observation because of the '*night becoming hazy*'.[197] On 20 December he tried again and this time was able to confirm the object. On that occasion he found NGC 2043 too. In the NGC column 'Other observers' Dreyer notes merely 'Melbourne obs.'; and in the introduction one reads

that '*In the first part of the Melbourne observations of southern nebulae there are a few novae.*' Sometimes he gave pretty poor references. Contemporary visual observers might have been familiar with such information, but today it is not always obvious what is meant.

From the observing book, now archived at Mt Stromlo Observatory, it is evident that many more nebulae were observed and sketched with the GMT – among them there might be some new ones.[198] It remains to be investigated which additional objects were found and why they were not published. Since Dreyer speaks about a 'first part' of the Melbourne observations, it can be assumed that Ellery had planned a continuation. Owing to the progress in astrophotography, he obviously came to believe the project to be

---

[196] Ellery (1885: 23); 'H' refers to John Herschel's Cape catalogue (instead of 'h').
[197] Ellery (1885: 22).

[198] Gascoigne (1996).

needless. Anyway, Ellery's publication was one of the last of its kind.

9.22 BIGOURDAN, MASTER OF THE NGC

Bigourdan's achievement was the visual observation of 6380 NGC objects with the 12″ refractor at Paris Observatory. During this enormous task, many new nebulae were found. The result appeared in 1919, as a monumental five-volume work. Unfortunately, in terms of astrophysics it came much too late. Of course, Dreyer was a great admirer of Bigourdan, cataloguing 76 objects found by him in the NGC and 322 in the IC. Many were found to be mere stars, perhaps as a result of the poor observing conditions in Paris around the turn of the century.

9.22.1 Short biography: Guillaume Bigourdan

Guillaume Bigourdan was born on 6 April 1851 in the French town of Sistels, Tarn-et-Garonne (Fig. 9.76). He was educated at a private school in Valence d'Agen and began a study of mathematics and physics at Toulouse University in 1873. After graduating three years later he became the assistant of Félix Tisserand at the local observatory, where he mainly observed at the meridian-circle.

Figure 9.76. Guillaume Bigourdan (1851–1932).

In 1879 Bigourdan changed to Paris Observatory, where in 1883 he married (at the age of 34) Sophie Mouchez, the daughter of the director. He observed with the 12.4″ refractor by Secrétan in the west dome. First he made measurements of comets, minor planets and double stars. In 1884 he concentrated on a new field: nebulae. Bigourdan observed for nearly 30 years, using every clear night to determine relative positions with the micrometer. His plan was a comprehensive study of a large fraction of the NGC/IC objects. The precise positions were to form a basis for future determinations of proper motion. During his observations, Bigourdan discovered many new objects. For his monumental work he received the RAS gold medal in 1919.

In 1903 Bigourdan joined the Bureau de Longitudes to work on time measurements and signal transfer. In 1912 he became the first director of the newly founded Bureau de l'Heure. In 1923 he was elected Vice-President of the Académie des Sciences. After his retirement in 1928 he wrote a history of the Bureau de Longitudes, which was published in 1932. Guillaume Bigourdan died on 28 February 1932 in Paris at the age of 80.[199]

9.22.2 New nebulae

In May 1884 Bigourdan started his observations of nebulae with the 12.4″ Secrétan refractor, which had been erected in 1858 in the west dome of Paris Observatory (Fig. 9.77). For his measurements of relative positions he used a bar-micrometer without illumination. His standard eye-piece had a field of view of 16′ and a power of 169 (sometimes a power of up to 953 was applied). His targets were known objects from the GC and GCS. He started with NGC 5976, NGC 5981 and NGC 5982 in Draco, forming a nice galaxy chain (the first two were found by J. Stoney, the third by J. Herschel).[200] By June 1887 Bigourdan had discovered many new objects, which were catalogued in the NGC.

Bigourdan published his first discoveries in two papers, titled 'Nébuleuses nouvelles, découvertes à l'observatoire de Paris', which appeared in the *Comptes Rendus* of 1887. The first contains 50 objects (Bigourdan 1887a = Ref. 1), the second another 52 (Bigourdan 1887b = Ref. 2). The 102 discoveries are sorted by right

---

[199] Obituaries: Dyson (1932, 1933) and Anon (1933).
[200] See Steinicke (2001b).

9.22 Bigourdan, master of the NGC

Figure 9.77. The 12.4″ Secrétan refractor of Paris Observatory.

Table 9.46. *The contents of Bigourdan's lists*

	List 1	List 2	Sum
Bigourdan (NGC)	36	34	70
Bigourdan (non-NGC)	3	5	8
d'Arrest		7	7
W. Herschel	2	3	5
Marth		1	1
Stephan		1	1
B. Stoney	8	1	9
Sum	50	52	102

ascension for epoch 1860 (GC) and numbered continuously. For every object a short description is given. At the end of the second paper, notes to entries of the first list and 15 further GC objects are presented. Unfortunately, discovery dates are missing. These did not appear until 1917 in volume 56 of the *Annales de l'Observatoire de Paris*. This publication also includes a list of all of the objects discovered and further important data.[201]

The new objects were found near groups of nebulae from the GC and GCS: '*they generally verge on groups of known nebulae, and in most cases the study of these groups is as complete as possible with my instrument*'. Considering the modest telescope at the inappropriate urban site, the yield is astonishing. However, among Bigourdan's objects there are many stars and star groups. Table 9.46 shows the content of both lists (Refs. 1 and 2). Except for eight objects, all were catalogued in the NGC. Altogether 71 of these must be credited to Bigourdan. The rest had been discovered earlier by other observers, either directly or indirectly (i.e. via a duplicate NGC entry).

Three of the eight objects ignored by Dreyer are stars: 1–8, 2–58 and 2–93 (list, number). Four entries indicate an empty field: 1–47, 1–48, 2–94 and 2–100. However, the entry 2–75 is interesting, because there is a galaxy at the place in Virgo. Bigourdan noted that it was '*round, stellar centre, forms a very close companion to 4045 G.C.* [NGC 5846]'. This compact 12.8-mag galaxy is located 19″ south of NGC 5846 (Fig. 9.78 right); it was listed in 1927 by Eric Holmberg as no. 694b in his catalogue of double and multiple galaxies (other designations are MCG 0–38–26 and NGC 5846A). It is not known why Dreyer omitted this object.

In Table 9.47 the 70 Bigourdan objects are divided by type (Dreyer mentions him in 76 cases). Among them are 38 independent non-stellar objects (all galaxies), which gives a moderate success rate of 56%. Of the remaining objects, many are stars or star groups. NGC 2361 is a small knot in the nebulous complex NGC 2359 (V 31, found by William Herschel on 31 January 1785). In the case of NGC 2404 (Fig. 9.78 left) Bigourdan speaks about a 'companion of GC 1541' (the 8.2-mag galaxy NGC 2403 in Camelopardalis).

In October and November 1884, Bigourdan discovered 7 of 20 NGC galaxies in the Perseus Cluster (Abell 426), all of which feature in his first list (Table 9.48). D'Arrest ranks second with six members. Bigourdan's 'stellar-appearing' object NGC 1257 (1–16), which he found on 19 October 1884 near the cluster, is only a double star (12.5 and 13.6 mag, distance 3″).

On his best night (9 March 1886) Bigourdan found six objects in Lynx, which had all been seen earlier by

---
[201] Bigourdan (1917d: E344–E356).

Figure 9.78. From left to right: NGC 2404, an H II region in the bright galaxy NGC 2403 in Camelopardalis; the galaxy NGC 76 in Andromeda, Bigourdan's first object (A = MCG 5–1–73 was not seen); and NGC 5846A, a compact galaxy near NGC 5846 in Virgo (DSS).

Table 9.47. *Bigourdan's discoveries by type*

Type	Number	Remarks
Galaxy	38	
Part of galaxy	1	NGC 2404 (I–28), 2.2.1886, H II region in NGC 2403 (Cam)
Star	16	Brightest: NGC 3948 (I–50), 23.6.1886, 13.4 mag (Leo)
Star pair/group	10	
Not found	5	NGC 1173, NGC 2529, NGC 2531, NGC 4160, NGC 7133

Table 9.48. *Bigourdan's galaxies in the Perseus Cluster*

NGC	List	Date	V
1259	1–17	21.10.1884	14.7
1260	1–18	19.10.1884	13.3
1264	1–19	19.10.1884	14.3
1265	1–20	14.11.1884	12.2
1271	1–21	14.11.1884	14.2
1282	1–22	23.10.1884	12.7
1283	1–23	23.10.1884	13.8

Bindon Stoney at Birr Castle (Table 9.49). Obviously the declination range was fairly small, perhaps as a result of the viewing conditions in Paris.

Table 9.50 shows 23 objects that had been found earlier by other observers. Dreyer gives most of the priorities correctly. For five cases there is an identity with another NGC object. The galaxies NGC 2330 and NGC 2334 in Lynx, which were discovered by Bindon Stoney, are interesting. Dreyer catalogued them once again in the IC I (see Section 8.14.3). The galaxies NGC 4867, NGC 4871, NGC 4873, NGC 4883 and NGC 4898 are members of the Coma Cluster (Abell 1656), which was first noticed by d'Arrest (see Section 8.6.4).

Bigourdan discovered pretty faint objects, with a mean of 14.1 mag (Fig. 9.79). Given his observing conditions, this is even better than Marth on Malta (13.9 mag) and comparable to Birr Castle (14.3 mag). This shows Bigourdan's remarkable power of observation.

9.22.3 Bigourdan's main work: 'Observations de nébuleuses et d'amas stellaires'

Bigourdan's observations and micrometrical measurements lasted from 20 May 1884 until 27 April 1911. Initially he observed objects from the GC and GCS, but, upon the appearance of the NGC in 1888, he focused on the new catalogue. Bigourdan's action was systematic, scanning sky areas of 4° × 4°. The targets and reference stars (mostly from the Bonner Durchmusterung) were defined in advance. At the end, Bigourdan had observed 6380 objects from the NGC and IC (a significant proportion of those visible from Paris). His list of new nebulae has 559 entries (322 were catalogued in the IC). Apart from Barnard, Burnham and Swift, he was the

Table 9.49. *Special objects and observational data of Bigourdan*

Category	Object	Data	Date	Remarks
Brightest object	NGC 1265 = 1–20	12.2 mag, Gx, Per	14.11.1884	Perseus Cluster
Faintest object	NGC 7577 = 2–97	15.7 mag, Gx, Psc	7.10.1885	
Most northerly object	NGC 2404 = 1–28	+66°, GxP, Cam	2.2.1886	Part of NGC 2403
Most southerly object	NGC 2361 = 1–27	−13°, EN, CMa	25.2.1887	
First object	NGC 76 = 1–1	Gx, 13.0 mag, And	22.9.1884	Fig. 9.78 centre
Last object	NGC 5509 = 2–71	Gx, 14.1 mag, Boo	10.6.1887	
Best night			9.3.1886	6 objects (Lynx)
Best constellation	Coma Berenices		1885–86	15 objects

Table 9.50. *Those of Bigourdan's objects which had been discovered earlier by others*

NGC	List	Date	Type	V	Con.	Discovery	Dreyer
2330	1–25	16.11.1885	Gx	14.8	Lyn	B. Stoney 2.1.1851; = IC 457, B. Stoney	Ld R?, Bigourdan
2334	1–26	16.11.1885	Gx	13.7	Lyn	B. Stoney 2.1.1851; = IC 465, B. Stoney	Ld R?, Bigourdan
2458	1–29	9.3.1886	Gx	14.5	Lyn	B. Stoney 20.2.1851	Ld R, Bigourdan
2461	1–30	9.3.1886	Star		Lyn	B. Stoney 20.2.1851	Ld R, Bigourdan
2462	1–31	9.3.1886	Gx	13.3	Lyn	B. Stoney 20.2.1851	Ld R, Bigourdan
2464	1–32	9.3.1886	3 stars		Lyn	B. Stoney 20.2.1851	Ld R, Bigourdan
2465	1–33	9.3.1886	Star		Lyn	B. Stoney 20.2.1851	Ld R, Bigourdan
2471	1–34	9.3.1886	2 stars		Lyn	B. Stoney 20.2.1851	Ld R, Bigourdan
2661	1–38	8.3.1886	Gx	13.0	Cnc	W. Herschel 13.9.1784 (III 50)	W. Herschel, Bigourdan
2863	1–39	15.2.1887	Gx	12.9	Hya	W. Herschel 25.3.1786 (III 520); = NGC 2869, Muller 1886	W. Herschel, Bigourdan
4357	2–52	8.3.1886	Gx	12.4	CVn	NGC 4381, W. Herschel 9.2.1788 (III 743)	Bigourdan
4427	2–53	22.4.1886	2 stars		Com	NGC 4426, d'Arrest 21.4.1865	Bigourdan
4738	2–57	9.3.1885	Gx	14.0	Com	B. Stoney 1.3.1851	Ld R, Bigourdan
4867	2–62	28.4.1885	Gx	14.5	Com	d'Arrest 10.5.1863	d'A, Bigourdan
4871	2–63	16.5.1885	Gx	14.2	Com	d'Arrest 10.5.1863	d'A, Bigourdan
4873	2–64	16.5.1885	Gx	14.1	Com	d'Arrest 10.5.1863	d'A, Bigourdan
4883	2–67	16.5.1885	Gx	14.4	Com	d'Arrest 22.4.1865	d'A, Bigourdan

Table 9.50. (Cont.)

NGC	List	Date	Type	V	Con.	Discovery	Dreyer
4898	2–69	15.5.1885	Gx	13.6	Com	d'Arrest 6.4.1864	d'A, Bigourdan
5648	2–74	23.5.1887	Gx	13.3	Boo	NGC 5649, W. Herschel 19.3.1787 (III 645)	Bigourdan
6028	2–76	4.5.1886	Gx	13.5	Her	NGC 6046, W. Herschel 14.3.1784 (III 33)	Bigourdan
6240	2–80	2.7.1886	Gx	12.8	Oph	Stephan, 12.7.1871; = IC 4625, Barnard	St II, Bigourdan
6966	2–86	27.7.1884	2 stars		Aqr	d'Arrest 26.7.1865	d'A, Bigourdan
6975	2–88	23.9.1886	Gx	14.0	Aqr	NGC 6976, Marth 12.7.1864 (m 427)	Bigourdan; = m 427?

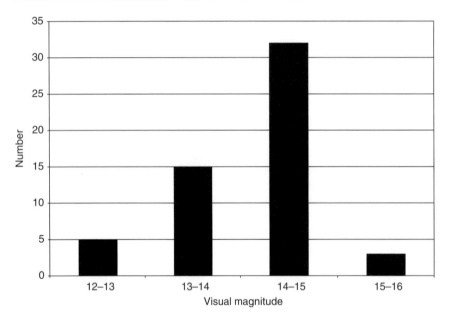

Figure 9.79. The brightness distribution of Bigourdan's objects.

only observer who contributed to all three of Dreyer's catalogues. Concerning quantity, quality, methodicalness and homogeneity, Bigourdan's work exceeded everything else done in visual observations of nebulae – both prior to him and up to the present day.

From 1892 to 1917 he published the results as 'Observations de nébuleuses et d'amas stellaires' in the *Annales de l'Observatoire de Paris*. The observations are sorted by right-ascension hour and fill 17 volumes with 2600 pages (Table 9.51). In Table 9.51 'Ref.' refers to the entries on Bigourdan in the references.

For each object (usually observed several times) Bigourdan gives a multitude of data: catalogue designations, earlier observers, dates, magnification, estimated visual brightness, reference star (position for 1900), relative position, position angle, description and notes. Since there were no data reductions, absolute positions are missing. Obviously, Bigourdan's time and power were exhausted. When the first volumes appeared in 1892, Dreyer could use many of the observations for his notes and corrections to NGC objects in the first Index Catalogue.

Table 9.51. *Bigourdan's publications in the* Paris Annales, *sorted by right-ascension hour*

Rect.	Vol.	Pages	Ref.	Rect.	Vol.	Pages	Ref.
0	49	F1–F141	1904c	12	55	E1–E243	1912
1	51	F1–F137	1906	13	56	E1–E119	1917a
2	50	F61–F153	1905b	14	48	F1–F139	1904a
3	49	F143–F207	1904d	15	36	G6–G103	1892b
4	50	F1–F59	1905a	16	42	D49–D151	1898a
5	47	G101–G131	1902b	17	42	D153–D241	1898b
6	48	F141–F187	1904b	18	43	E1–E73	1907a
7	52	F1–F75	1907c	19	43	E75–E111	1907b
8	53	F1–F89	1908	20	46	G1–G53	1899a
9	44	D1–D119	1910	21	46	G55–G115	1899b
10	45	E1–E117	1911a	22	47	G1–G99	1902a
11	54	E1–E173	1911b	23	40	D9–D107	1896

Volume 56 of the *Paris Annales* (published in 1917) contains an 'Introduction', an 'Appendix' and 'Mesures complémentaires' (additional measurements).[202] In the introduction, Bigourdan gives a comprehensive overview about observations of nebulae, starting in ancient times. The objects that had been discovered by 1700 (including by Flamsteed) are compiled in a table. Of special interest is the list of objects discovered by de Chéseaux, which was barely known before. Bigourdan gives a great number of references, which were useful for this work, though he found them quite late.

Because the presentation of observations – spread over 20 volumes of the *Paris Annales* – was hardly user-friendly, Bigourdan decided on a reprint in five volumes. It was published in 1919 and fills altogether 3000 pages. His well-deserved reward was the RAS gold medal, which was presented in February 1919. In his laudation President Percy MacMahon emphasised Bigourdan's achievements and gave an overview (following the tradition of Airy and Adams) about the history and cataloguing of nebulae (MacMahon 1919).

Unfortunately, Bigourdan's observations came too late, since astrophysics had already become the primary field of interest. Especially astrophotography, as done by Max Wolf in Heidelberg, James Keeler and Charles Perrine at Lick Observatory and others, yielded a large number of new objects and positions. A similar destiny was experienced by another elaborate work: the visual observations of NGC objects made by Johann Georg Hagen at the Vatican Observatory in the early twentieth century (see Section 10.3.2).

9.23 YOUNG'S DISCOVERY IN PRINCETON

Young contributed only one object to the New General Catalogue, namely the galaxy NGC 6187 in Draco. It was probably an accidental find, for which no source can be given.

Charles Augustus Young was born on 15 December 1834 in Hanover, New Hampshire (Fig. 9.80 left). His grandfather and father were Professors of Natural Philosophy at the local Dartmouth College, where Young too studied during the years 1849–53. In 1856 he became Professor of Mathematics and Astronomy at Western Reserve College (Ohio). In 1866 he succeeded his father in Hanover and changed eventually to Princeton University in 1877. Young was a pioneer of solar physics and spectroscopy. In 1869 he detected a new element in the corona ('Coronium'). A year later he managed to take the first photograph of a protuberance and discovered the flash spectrum. In Princeton he established the Halsted Observatory, becoming its first director (Fig. 9.80 right).

In 1882 a 23″ Clark refractor was installed and 10 years later it was equipped with the best spectroscope of its time, which was built by Brashear.[203] Young

---

[202] Bigourdan (1917b–d).

[203] Gegen (1988).

Figure 9.80. Left: Charles Young (1834–1908); right: Halsted Observatory, Princeton.

Figure 9.81. Left: Young's only discovery, the galaxy NGC 6187 in Draco (DSS); right: the 23″ Clark refractor (Princeton University).

retired in 1905 and returned to Hanover. He was followed at Halsted by the famous astrophysicist Henry Norris Russell. Charles Young died on 3 January 1908 in Hanover at the age of 73.[204] When Halsted Observatory closed in 1932, the refractor was dismantled and refurbished. Until 1964 it was located at the new Princeton University Observatory. Since 1986 the 23-inch has been the main instrument at the Charles E. Daniel Observatory in South Carolina.

In about 1885 Young discovered the 14.6-mag galaxy NGC 6187 in Draco (Fig. 9.81 left). Probably Dreyer was informed by letter. The object might have been found with the 23-inch of Halsted Observatory (Fig. 9.81 right), possibly while observing a comet.

9.24 LOHSE AT THE WIGGLESWORTH OBSERVATORY

Gerhard Lohse started his astronomical career as Ralph Copeland's assistant in Dun Echt, later changing to the new Wigglesworth Observatory in Scarborough. There he discovered 17 nebulae with the 15.5″ Cooke refractor. Unfortunately his list, which was sent by letter to Dreyer, is lost. Lohse's result looks pretty bad: only three objects are non-stellar.

---

[204] Obituaries: Frost (1908b, 1909), MacPherson (1908), Poor (1908), Russell H. N. (1909) and Towley (1909).

9.24 Lohse

Figure 9.82. Gerhard Lohse (1851–1941); from the 'portrait gallery' of the Astronomische Gesellschaft.

9.24.1 Short biography: J. Gerhard Lohse

Gerhard Lohse was born on 10 January 1851 in the German village of Fünfhausen near Oldenburg (Fig. 9.82).[205] After studying astronomy at Göttingen University he went to Scotland in 1877. Recommended by Klinkerfuess, Ralph Copeland's former teacher, he became the second assistant at Lord Lindsay's Dun Echt Observatory, Aberdeenshire (Copeland was the first assistant). There he discovered the returning comet Tempel–Swift on 8 November 1880. In 1884 Lohse changed to the new Wigglesworth Observatory in Scarborough, Yorkshire.

The observatory was the private property of James Wigglesworth (born in 1815), the owner of a soap factory in Scarborough and an avid amateur astronomer. For 14 years a 6″ refractor, which had been built in 1853 by Thomas Cooke of York, was used. In 1879 Wigglesworth bought the optical branch of Cooke's company and became the partner of the Cooke sons. In 1885 he became an affiliate of the RAS, and in the same year his new observatory in Scarborough was finished. It was equipped with a 15.5″ Cooke refractor in a 9-m dome[206] (Lohse G. 1887a). On 1 September 1885 Lohse and Wigglesworth started visual observing. One

Figure 9.83. The 15.5″ Cooke refractor of Wigglesworth Observatory, Scarborough.

of their first targets was the supernova S Andromedae in M 31 (Lohse G. 1885). James Wigglesworth died in 1889 and his son continued the business of the Cooke company.[207] The observatory was closed and the great refractor was eventually sold to the Italian astronomer Vincenzo Cerulli, who erected it in his private Collurania Observatory near Teramo.

Lohse was fortunate in getting a job at the new Royal Observatory Edinburgh, which at that time was being directed by the Astronomer Royal of Scotland, Ralph Copeland. After his active career had ended, Lohse returned to Germany. Gerhard Lohse died on 2 January 1941 in Oldenburg at the age of 90.[208]

9.24.2 Observations of nebulae and a nova

In 1885 Lohse observed the Nova Cygni 1876 with the 15.5″ Cooke refractor (Fig. 9.83). The 'new star' was

---

[205] He should not be confused with the German astronomer Oswald Lohse; there was no relation between them.
[206] Lohse G. (1887a), McConnell (1992: 57–59), Emery and Hawkridge (2007).
[207] Obituary: Anon (1889b).
[208] Burke-Gaffney (1968).

Table 9.52. *Lohse's discoveries (the date 1886 is estimated)*

NGC	Date	Type	V	Con.	Remarks
793	1886	2 stars		Tri	
1456	1886	2 stars		Tau	
1655	1886	2 stars		Tau	
1674	1886	Star group		Tau	
1675	1886	Star group		Tau	Possibly identical with NGC 1674
2195	1886	2 stars		Ori	
2412	1886	2 stars		CMi	
2518	1886	Gx	13.6	Lyn	
2519	1886	Star	14.3	Lyn	
2565	1886	Gx	12.6	Cnc	
4345	1886	Gx	12.0	Dra	NGC 4319, W. Herschel 10.12.1797 (I 276)
5884	1886	2 stars		Boo	
6344	1886	2 stars		Her	
6353	1886	3 stars		Her	
6731	1886	2 stars		Lyr	
6767	1886	2 stars		Lyr	
6792	1886	Gx	12.5	Lyr	
7114	3.10.1885	(Star)		Cyg	Nebula around Nova Cygni 1876; Copeland 2.9.1877

noticed on 24 November 1876 by Schmidt in Athens as a 3-mag object (Schmidt 1877). On 3 October 1885 Lohse saw it, with its having in the meantime decreased to 15 mag, as '*certainly surrounded by nebulosity*' (Lohse G. 1887b). This was confirmed on 7 and 26 October. Dreyer catalogued the new nebula as NGC 7114, giving no discoverer. However, in the NGC notes he wrote that '*Mr. Lohse asserts, that this star is surrounded by nebulosity.*' The Nova Cygni nebula must be credited to Copeland, who detected its line spectrum visually on 2 September 1877 in Dun Echt, using the 15.1″ refractor (Copeland R. 1877c).

During the years 1885–87 Lohse determined the positions of 17 new nebulae at Wigglesworth Observatory (Table 9.52). He used a Merz micrometer lent by Lord Lindsay (Dun Echt). Dreyer in Armagh was informed about the discoveries by letter. In the introduction of the NGC he speaks about a '*list of about 20 new nebulae [...] kindly communicated by letters*'. Unfortunately it is lost.

NGC 4345 is identical to William Herschel's I 296 (NGC 4319). This is a galaxy in Draco, well known by virtue of its proximity to the quasar Mrk 205, which is only 40″ to the south (Fig. 9.84). Having detected a 'light bridge', Halton Arp claimed that there was a physical connection between the two objects. However, the extremely different redshifts cause a severe problem for this idea. Arp thus doubted the common cosmological interpretation, in that he put both objects at the same distance. This has been disproved: the quasar is a background object (as it should be). Lohse's position of NGC 4319 is 1^m too far east, but the description fits; Dreyer did not notice the error. By the way, Lord Rosse did not observe the object (h 1210), perhaps due to the high declination (+76°). The 'star' (quasar) south of NGC 4319 was first noticed by Karl Reinmuth on a plate; he described it as a '*star 15.5 mag 0.8′ south of nucleus*'.[209] The quasar itself was detected as a UV source in 1969 by Beniamin Markarian (Mrk).

Lohse's balance is – like that of Schultz – quite bad. Subtracting NGC 4345 and NGC 7114, 16 new objects remain (marked in bold in Table 9.52). Only three of them are non-stellar: the galaxies NGC 2518, NGC 2565 and NGC 6792. The rest are stars or star groups (13). For NGC 1655 (probably a star pair)

---
[209] Reinmuth (1926: 58).

Figure 9.84. The galaxy NGC 4319 in Draco and its close 'companion', the quasar Mrk 205 (DSS).

Dreyer noted '*pB, R, gbM, star 10 s*'. Delisle Stewart searched for the nebula on a Harvard plate, detecting nothing but a '*hazy star p 1ᵐ, same decl*'.[210] Obviously Stewart's object was just a plate flaw.

9.25 THE LEANDER MCCORMICK OBSERVATORY: STONE, LEAVENWORTH AND MULLER

From 1885 the Leander McCormick Observatory in Charlottesville, Virginia, was for a few years a place of active nebular observations. Ormond Stone and his assistants Francis Leavenworth and Frank Muller searched for new objects, concentrating on southern declinations (the latitude is 38°). Using the 26″ Clark refractor, they discovered 400 nebulae that were catalogued by Dreyer in the NGC. Though the positions are not very precise, the identifications are easy in most cases. For the first time, estimated values of visual magnitudes and sizes are given.

9.25.1 Short biography: Ormond Stone

Ormond Stone was born on 11 January 1847 as the son of a Methodist preacher in Pekin, IL (Fig. 9.85 left). Though he could attend school only sporadically due

Figure 9.85. Left: Ormond Stone (1847–1933); right: Francis Leavenworth (1858–1928).

to the Civil War, Stone was able to become a student at the University of Chicago in 1866. During his studies, in which he mainly concentrated on mathematics, he occasionally worked as a school teacher. After graduating in 1870, Stone was engaged by the US Naval Observatory, Washington. Because of Stone's mathematical knowledge, Simon Newcomb entrusted him with astronomical calculations. On the recommendation of Newcomb, Stone became Director of Cincinnati Observatory in 1875. With the local 11″ Merz refractor he discovered southern double stars.

In 1870 the Chicago industrial magnate Leander McCormick had the idea of funding the largest refractor in the world. It should outclass the 26-inch at the US Naval Observatory in Washington. Not until 1877 did the University of Virginia in Charlottesville get the money for a new observatory on the 260-m-high Mt Jefferson, which lies about 1.5 km away from the campus. Stone was appointed director in 1882 and supervised the construction, which was finished three years later.[211] On 13 April 1885 the new refractor by Alvan Clark & Sons (see Fig. 9.86) was inaugurated in his 13.7-m dome. However, the lens was only ¼ inch larger than that in Washington. Unfortunately, at that time both instruments had already been surpassed by the new refractors in Vienna and Pulkovo. Stone's first target was the Orion Nebula. Then he started the search for new southern nebulae, in which task he was assisted by Francis Leavenworth and Frank Muller.

In his time at the University of Virginia Stone trained many well-known astronomers, e.g. Heber

---

[210] Dreyer (1953: 369).

[211] Sayre (1892).

Figure 9.86. The 26″ Clark refractor of the Leander McCormick Observatory, Charlottesville.

Curtis. He retired from his office in 1912 and settled at a farm in Centreville, VA. Six days after his 86th birthday, namely on 17 January 1933, Ormond Stone died near his house – while walking he was knocked down by a car.[212]

9.25.2 Short biography: Francis Preserved Leavenworth

Francis Leavenworth was born on 3 September 1858 in Mount Vernon, IN (Fig. 9.85 right). After finishing school he studied at the University of Indiana. He graduated in 1880 and became an assistant of Ormond Stone at Cincinnati Observatory (Herbert Howe and Herbert C. Wilson were colleagues there). In 1881 he joined Stone at the new Leander McCormick Observatory in Charlottesville. There he mainly searched for nebulae with the 26″ Clark refractor (Fig. 9.86).

In 1887 Leavenworth became Director of Haverford College Observatory, which possessed a 10″ Clark refractor. After five years he changed to Minneapolis, where he was appointed Professor of Astronomy at the University of Minnesota and Director of its observatory, which was equipped with a 10.5″ Brashear refractor. Leavenworth observed double stars and carried out astrophotography. Occasionally he used the 40″ refractor at Yerkes Observatory. In 1922 Leavenworth became a Fellow of the RAS. He stayed for 35 years as the observatory director in Minneapolis. One year after his retirement, Francis Leavenworth died on 12 November 1928 in St Paul, MN, at the age of 70.[213]

9.25.3 Short biography: Frank Muller

Not much is known about Frank Muller. He was born on 10 September 1862 in Virginia. He probably started his work at the Leander McCormick Observatory in 1885. Initially he observed double stars with the 26″ Clark refractor, as can be seen from the reports in Vol. I of the *Publications of the Leander McCormick Observatory*. In 1886 Muller was engaged in the search for southern nebulae. In 1888 he graduated from the University of Virginia (as the first in the faculty of astronomy) with a work on the orbit of comet 1887 IV. Afterwards he was awarded a Vanderbilt grant for a year. Frank Muller died on 19 April 1917 in Virginia at the age of 54.

9.25.4 Two lists of new nebulae

During the years 1885–87 Leavenworth, Stone and Muller searched for new southern nebulae with the 26″ Clark refractor at the Leander McCormick Observatory of the University of Virginia, Charlottesville. The telescope, which had been in operation since January 1885, was first used by Stone and Leavenworth to observe the Orion Nebula until the end of March (Stone O. 1893a). The regular magnification was 163.[214] The results, including sketches, were published in Vol. I of the *Publications of the Leander McCormick Observatory* (Stone O. 1893a).

Leavenworth was the one who spent the longest amount of time searching for nebulae: from June 1885 until the end of January 1887. Stone observed between July 1885 and the end of 1886; Muller was involved only in 1886. With a few exceptions, the search concentrated on declinations between −2° and −32°, scanning

---

[212] Obituary: Oliver (1933).

[213] Obituaries: Beal (1929), Anon (1928) and Anon (1929).

[214] The later paper 'Southern nebulae' mentions a power of 175 and the use of a filar-micrometer (Stone O. 1893b).

9.25 The Leander McCormick Observatory

Table 9.53. *Numbers of objects in the published lists (NGC = catalogued objects)*

	List 1	List 2	Sum	NGC
Leavenworth (L)	159	113	272	269
Stone (S)	105	10	115	114
Muller (M)	9	80	89	89
Sum	273	203	476	472

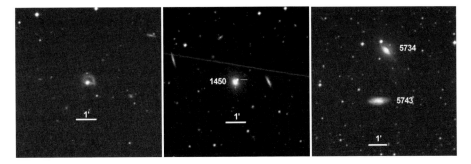

Figure 9.87. Leander McCormick's objects (from left to right): the ring galaxy NGC 985 in Cetus; NGC 1450 in Eridanus with its compact companion (marked); and the galaxy pair NGC 5734/43 in Libra (DSS).

the constellations Aquarius, Capricorn, Cetus, Crater, Eridanus, Hydra, Fornax and Lepus. With such a large instrument, success was guaranteed.

In 1886–87 Stone published two lists of new nebulae in the *Astronomical Journal* (Table 9.53). The first, titled 'List of nebulas observed at the Leander McCormick Observatory, and supposed to be new', contains 273 objects (Stone O. 1886 = Ref. 1). The 'Second list of nebulas observed at the Leander McCormick Observatory, and supposed to be new' gives another 203 objects (Stone O. 1887 = Ref. 2). The numbering of the lists is continued: they run from 1 to 273 and from 274 to 476, respectively. For NGC 167 = 2–286 (list, number) Dreyer gives Muller as the discoverer, but this should be Leavenworth.

Both lists are separately ordered by right ascension. The positions (epoch 1890) are not very precise. Additionally, visual brightness, size, form and concentration are given for each object. About the size Stone remarked that '*Still later the custom was instituted of estimating diameters of the nebulas observed in fractions of the diameter of the field, and from these deducing their dimensions in minutes of arc.*'[215] There are two columns, giving the number of observations (1 or 2), the observer (L, M or S) and notes. Additional remarks are presented at the end of each publication. Unfortunately, no discovery date is given. This can be derived (in most cases) from the observing books archived at the Leander McCormick Observatory.[216] Sketches exist for a large number of objects (mentioned in Ref. 1): '*Sketches have been made of the larger portion of the nebulas contained in this list.*' Fifteen galaxies discovered by Muller appear also in Stone's publication 'Southern nebulae' of 1893. It lists observations of 890 non-stellar objects made during the years 1887–93 (Stone O. 1893b).

Leavenworth discovered the largest number of objects, followed by Stone and Muller. All were entered in the NGC, except four. Leavenworth's no. 71 is missing (blank field) and three objects appear twice. The first case is no. 113 = no. 114 = NGC 1450 (S); Stone correctly notes '*D, P 310° Δ 0.4″*', which means a double nebula with separation 0.4′ (Fig. 9.87 centre). The other cases are no. 141 = no. 142 = NGC 1730 (L) and no. 161 = no. 162 = NGC 3058 (L); for no. 162 Stone noted '*same as (161)?*'. Of the 472 NGC objects,

---
[215] Stone O. (1886: 14).

[216] They were inspected by Harold Corwin (private communication).

Table 9.54. *Numbers of NGC objects discovered*

Type	L	S	M	Sum
Galaxy	245	83	64	392
Single star	1	1	3	5
Star pairs/trios		1	8	9
Not found	5	4	7	16
Sum	251	89	82	422

Table 9.55. *Special objects and observational data of the discoverers at Leander McCormick Observatory*

Category	Object	Data	Disc.	Date	Remarks
Brightest object	NGC 1640 = 1–136	11.6 mag, Gx, Eri	S	11.12.1885	
Faintest object	NGC 344 = 2–298	16.1 mag, Gx, Cet	M	1886	
Most northerly object	NGC 7164 = 2–465	+1°, Gx, 14.7 mag, Aqr	L	1886	NGC 2901 (S, NF): +31°
Most southerly object	NGC 7187 = 1–246	–32°, PsA, Gx	L	1886	
First object	NGC 5734 = 1–206	Gx, 12.8 mag, Lib	L	3.6.1885	Fig. 9.87 right
Last object	NGC 5039 = 2–459	Gx, 15.4 mag, Vir	L	25.1.1887	
Best night			S, L	31.12.1885	S: 10 objects, L: 6 objects
Best constellation	Cetus		S, L, M	1885–86	141 objects

422 must be credited to the three observers (Table 9.54); 393 of them are galaxies, of which Leavenworth found 245. The rate of non-stellar objects is high for Leavenworth and Stone: 99% and 94%, respectively. Muller's rate is lower (78%); 29 of his objects are stars or 'not found' (probably due to wrong positions).

Table 9.55 shows special objects and observational data. Only two are located at northern declinations: NGC 7164 (galaxy, Aquarius) and NGC 2901 (not found, Leo). One object lies in Sagittarius: the missing NGC 6551 (1–230, Leavenworth). Two exotic examples of the many non-stellar objects are the ring galaxies NGC 985 (VV 285, 13.5 mag; Fig. 9.87 left) and NGC 7828/29 (Arp 144, 13.9 mag) in Cetus. Leavenworth discovered them in 1886, not noticing their peculiar structure. Altogether there are six ring galaxies in the NGC; the others were found by William Herschel, Marth, Stephan and Tempel.

Fifty objects had been discovered earlier by other observers, as was often noticed by Stone himself. There are three double entries in the lists. Leavenworth's 1–30 (NGC 563) is identical with 1–28 (NGC 539), which he had found earlier. The same applies to Stone's 1–87 (NGC 1205) and 1–64 (NGC 1182). Moreover, 1–180 (NGC 3479), found by Stone, is the same as Leavenworth's 1–181 (NGC 3502); the priority is open (there are other such cases, due to missing data). Twenty-five objects appear twice in the NGC. Those of Leavenworth's objects which had been found earlier by others are listed in Table 9.56.

Table 9.57 shows those of Stone's objects which had been found earlier by others. NGC 1794 = NGC 1781 (GC 998) was noticed by him (see Ref. 2). NGC 3518 is curious: the galaxy is identical with NGC 3122 (William Herschel) and NGC 3110 (Stephan). Stone's right ascension is too large by 1^h.

Finally, Table 9.58 shows those of Muller's objects which had been found earlier by others. In Ref. 2 Stone remarks that 1–6 (NGC 155) is the same as Swift's object no. 2 in his fourth list.

Table 9.56. *NGC objects of Leavenworth that had been found earlier by others*

NGC	List	Date	Type	V	Con.	NGC	Discoverer	Remarks
62	1–2	1886	Gx	13.5	Cet		Stephan 8.10.1883	
210	1–11	2.10.1886	Gx	10.9	Cet		W. Herschel 3.10.1785 (II 452)	Stone: 'G.C. 107?' [NGC 210]
481	2–303	1886	Gx	13.5	Cet		L. Swift 20.11.1886	Priority?
563	1–30	1886	Gx	13.5	Cet	539	Leavenworth 31.10.1885 (1–28)	
1148	2–352	21.12.1886	Gx	14.9	Eri		L. Swift 10.11.1885	
1307	2–366	1886	Gx	13.5	Eri	1304	W. Herschel 5.10.1785 (III 444)	
1455	2–386	1886	Gx	11.9	Eri	1452	W. Herschel 6.10.1785 (II 458)	
1458	2–387	1886	Gx	11.6	Eri	1440	W. Herschel 6.10.1785 (II 459)	NGC 1442, W. Herschel 20.9.1786 (II 594)
3007	1–158	23.2.1886	Gx	13.6	Sex		Stephan 16.3.1885	
3295	1–173	26.2.1886	Gx	14.7	Hya	3280	Stephan 17.3.1884	Galaxy trio NGC 3280A–C (A = IC 617)
3321	2–423	3.1.1887	Gx	13.5	Sex	3322	Common 1880	
3502	1–181	1886	Gx	13.0	Crt	3479	Stone 1886 (1–180)	Priority?
3688	2–437	1886	Gx	14.3	Crt		Common 1880	
3722	2–442	1886	Gx	14.1	Crt		Common 1880	
3724	2–443	1886	Gx	14.1	Crt		Common 1880	
3854	2–449	1886	Gx	12.2	Crt	3865	Common 1880	
3858	2–450	1886	Gx	13.2	Crt	3866	Common 1880	
5796	1–214	31.5.1886	Gx	11.6	Lib		Tempel 23.5.1884	

Table 9.57. *NGC objects found by Stone that had been found earlier by others*

NGC	List	Date	Type	V	Con.	NGC	Discoverer	Remarks
207	1–9	3.11.1885	Gx	13.7	Cet		Mitchell 7.12.1857	
757	1–43	1886	Gx	12.1	Cet	731	W. Herschel 10.1.1785 (III 266)	
763	1–44	1886	Gx	12.6	Cet	755	W. Herschel 10.1.1785 (III 265)	
787	1–46	1885?	Gx	12.9	Cet		Peters 27.2.1865	Tempel 9.11.1879
948	1–56	1886	Gx	13.7	Cet		L. Swift 1.11.1886	Priority?
961	2–338	1886	Gx	12.8	Cet	1051	Stephan 27.11.1880	
1017	1–61	1886	Gx	14.4	Cet		L. Swift 29.9.1886	Priority?

Table 9.57. (*Cont.*)

NGC	List	Date	Type	V	Con.	NGC	Discoverer	Remarks
1205	1–87	1886	Gx	14.7	Eri	1182	Stone 1886	
1214	1–94	1886	Gx	14.1	Eri		L. Swift 21.10.1886	Priority?
1215	1–95	1886	Gx	14.1	Eri		L. Swift 21.10.1886	Priority?
1255	1–98	1886	Gx	10.7	For		Barnard 30.8.1883	
1367	1–106	1886	Gx	10.7	For	1371	W. Herschel 17.11.1784 (II 262)	
1450	1–113	31.12.1886	Gx	14.1	Eri		L. Swift 22.10.1886	
1509	1–122	1886	Gx	13.7	Eri		L. Swift 22.10.1886	Priority?
1524	1–123	31.12.1885	Gx	11.8	Eri	1516	W. Herschel 30.1.1786 (III 499)	
1794	1–146	11.12.1885	Gx	12.7	Lep	1781	W. Herschel 6.2.1785 (III 268)	
2652	2–406	1886	Gx	10.9	Sex	2974	W. Herschel 25.3.1786 (III 520)	Bigourdan 15.2.1887
2876	2–413	1886	Gx	13.1	Hya		Stephan 5.3.1880	
3054	1–160	14.1.1886	Gx	11.4	Hya		Peters 3.4.1859	
3518	1–182	31.12.1885	Gx	12.7	Sex	3122	W. Herschel 5.3.1785 (II 305)	NGC 3110, Stephan 17.3.1884
3525	1–183	1886	Gx	11.6	Crt	3497	W. Herschel 8.3.1790 (III 824)	NGC 3528, J. Herschel 22.3.1835 (h 3316)
3544	1–184	7.1.1886	Gx	12.2	Crt	3571	W. Herschel 8.3.1790 (II 819)	Stone: 'G.C. 2330?' [NGC 3571]
3905	1–192	1886	Gx	12.8	Crt		Common 1880	
5022	1–196	1886	Gx	12.9	Vir		Tempel 31.3.1881	
5069	1–197	1886	Gx	13.5	Vir	5066	Marth 30.5.1864 (m 251)	

Table 9.58. *NGC objects found by Muller that had been found earlier by others*

NGC	List	Date	Type	V	Con.	NGC	Discoverer	Remarks
35	2–277	1886	Gx	14.2	Cet		L. Swift 21.11.1886	Priority?
155	1–6	1886	Gx	13.3	Cet		L. Swift 1.9.1886	Priority?
599	2–310	1886	Gx	13.5	Cet		W. Herschel 27.11.1785 (II 473)	
1575	2–395	1886	Gx	12.7	Eri	1577	L. Swift 10.11.1885	
2869	2–412	1886	Gx	12.9	Hya	2863	W. Herschel 6.1.1785 (I 61)	
3050	2–418	1886	Gx	12.7	Sex	2979	W. Herschel 25.3.1786 (III 521)	
7254	2–467	1886	Gx	13.2	Aqr	7256	Marth 27.9.1864 (m 472)	

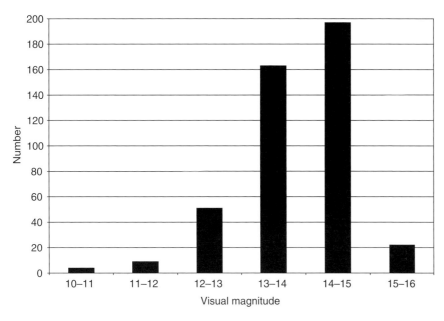

Figure 9.88. The brightness distribution of the Leander McCormick objects.

About visual magnitude Stone wrote that '*Afterwards numerical magnitudes were employed to indicate brightness, assuming that the faintest nebula visible in the 66^{cm} refractor, with power 167, is 16.3, that being a theoretical limit for stars. The magnitudes refer to the nucleus, or, in case there is no nucleus, to the brightest part.*'[217] The magnitudes given in the lists (normally rounded to the nearest 0.5 mag) are not compatible with modern values; there are differences of up to ±2 mag. For NGC 210 (1–11) '15.5' is given (perhaps a typo); actually this galaxy, which had already been discovered by William Herschel, has a brightness of 10.9 mag. The mean visual magnitude is 13.8 mag (Fig. 9.88); lying between that of Swift (13.6 mag, 16″ refractor) and Marth (13.9 mag, 48″ reflector).

9.26 THE FIRST PHOTOGRAPHIC DISCOVERY: NGC 1432

The first discovery of a non-stellar object by photography was that of the nebula around Maia in the Pleiades in 1885. The credit goes to the brothers Henry at Paris Observatory. They constructed a famous 33-cm double astrograph (Fig. 9.89), which was the model for all of the instruments used in the international project 'Carte du Ciel' (Astrographic Catalogue) started in 1887.[218] Later Paul and Proper Henry built many large telescopes, among them the 32″ photographic refractor, 24″ visual refractor and 40″ reflector at Meudon Observatory, the 24″ equatorial coudé of Paris Observatory, the 30″ refractor of Nice Observatory and the 31.5″ reflector of Toulouse Observatory.

9.26.1 Short biographies: Paul Pierre and Prosper Mathieu Henry

Paul Henry was born on 21 August 1848 in Nancy. He was the older brother of Prosper Henry. After having begun his working life as an optician in his home town, he came to Paris in 1864. In 1868 he became an assistant astronomer at the observatory, where he completed Chacornac's charts of the ecliptic. For this task the brothers Henry tried photography to remove the many stars in Milky Way regions. The results, which were presented by Admiral Mouchez in August 1884, were so satisfactory that they were commenced to construct an astrograph of aperture 33 cm. Giving a field of 3°

---

[217] Stone O. (1886: 14).

[218] See Mouchez (1885), Ashbrook (1984: 436–440), Urban and Corbin (1998) and Lamy (2008).

Figure 9.89. The brothers Henry working at the 33-cm astrograph in Paris.

diameter, it was used for the 'Carte du Ciel' project. The brothers discovered 14 minor planets between 1872 and 1882. In 1899 Paul Pierre Henry was elected an Associate of the RAS. He died on 4 January 1905 in Grand-Montrouge near Paris of a heart attack at the age of 56.[219]

Prosper Henry was born on 10 December 1849 in Nancy. He was the younger brother of Paul Henry. Since they worked together throughout their lives, their careers were nearly identical. Prosper Henry started as an optician in his home town. On 1 May 1865 he entered Paris Observatory to work for Leverrier, being initially responsible for the meteorological service. The brothers (both were assistant astronomers from 1868), completed Chacornac's 'Atlas ecliptique' with the aid of photography. When the international 'Carte du Ciel' project was initiated, the brothers Henry naturally were put in charge of the work at Paris Observatory. This included all optical and mechanical arrangements, especially the construction of the standard 33-cm double astrograph. Prosper Mathieu Henry was elected an Associate of the RAS in 1889. As the consequence of an accident he died on 25 July 1903 in Pralognan (in the Savoy mountains) at the age of 53 only.[220]

9.26.2 The brothers Henry and the Maia Nebula

On 11 November 1885 Paul and Prosper Henry took the first plate of the Pleiades in Taurus. For this task the famous double refractor with objectives of diameters 33 cm (photographic) and 25 cm (visual) and a focal length of 3.44 m was used. The brothers had made the optics; the mounting was due to Gauthier. Besides the known Merope Nebula (NGC 1435), which had already been discovered visually in 1859 by Tempel, the images showed a new one around Maia. It was catalogued by Dreyer as NGC 1432. The whole story is told in Section 11.6.10.

---

[219] Obituaries: Kreutz (1905), Anon (1905a) and Anon (1905b).
[220] Obituaries: Callandreau (1903a,b), Dyson (1904b), Anon (1903).

10 • The New General Catalogue: publication, analysis and effects

After having presented all of the information which contributed to the making of the NGC, this section treats Dreyer's publication of 1888 and analyses the structure of the catalogue. It appeared in Vol. 49 of the *Memoirs of the Royal Astronomical Society* and is titled 'A new catalogue of nebulae and of clusters of stars, being the catalogue of the late Sir John F. W. Herschel, Bart., revised, corrected, and enlarged' (Dreyer 1888b). The main topics of the following Sections 10.1 and 10.2 are:

- the temporal development of discoveries (1800–1887)
- the number of independent NGC objects
- discoverers (success rate, nationality)
- telescopes (aperture, type, site, nationality)
- the distribution according to modern object types
- the presentation of special objects
- the distribution of visual brightness
- missing data

After 1888, Dreyer and other astronomers collected corrections and supplementary data regarding NGC objects. Moreover, many new non-stellar objects were found, mainly by photography. To keep track, Dreyer published two Index Catalogues, in 1895 and 1908, containing the new data (Section 10.3). NGC-numbers are mentioned in nearly all modern catalogues (e.g. of galaxies, galactic nebulae and open clusters), albeit not always correctly. This is mainly due to the fact that the original NGC is not user-friendly and thus difficult to compare with modern data. Moreover, it contains errors, which were simply reproduced. Therefore a revised New General Catalogue, connecting historical and modern data, was urgently wanted. The first attempts were made in 1977 and 1988 (the centenary of the NGC), but neither was satisfactory. Fortunately, on the basis of thorough historical investigations, a reliable revision of the NGC/IC could be presented recently (Section 10.4).

10.1 DREYER'S PUBLICATION OF 1888

10.1.1 Structure of the work

In contrast to the three catalogues of William Herschel and John Herschel's Slough and Cape catalogues, Dreyer's NGC does not rely on his own observations. It is a compilation of all non-stellar objects known until the end of the year 1887. The model was the General Catalogue published by John Herschel in 1864. Though the structures are very similar, the NGC is not a continuation of Herschel's GCS, but a completely new catalogue.

Actually, Dreyer would have preferred to catalogue only nebulae – omitting all kinds of star clusters (which would still involve about 700 objects). In a letter to Hagen of 3 December 1914 he wrote '*I was very much tempted to leave them* [star clusters] *out in the NGC, for they have nothing to do with nebulae, at least not those of classes VII and VIII.*'[1] However, to maintain continuity vis-à-vis the earlier catalogues, he refrained from implementing his inclination towards including only nebulae.

Table 10.1 shows the structure of Dreyer's publication. In the introduction he describes the case history, especially stressing the importance of William Herschel's and John Herschel's work. He also mentions the remarkable results of Caroline Herschel and Auwers in revising the catalogues of William Herschel, ignoring the insufficient attempts of Bode and Jahn. As a professional astronomer, Dreyer emphasises the importance of precise positions, mentioning the basic measurement campaigns of Laugier, d'Arrest, Schönfeld and Schultz and the smaller series of Auwers, Rümker, Vogel, Schmidt and Engelhardt (see Section 11.1). As one of the key works, Dreyer presents the *Siderum Nebulosorum Observationes Havnienses* of his former teacher d'Arrest, which contains many valuable positions and

---

[1] Private communication by Arndt Latußeck, who recently found Dreyer's letters to Hagen at the Vatican Observatory.

Table 10.1. *The structure of Dreyer's publication*

Part	Pages	Explanation
Introduction	1–4	Historical basics
Arrangement of the Catalogue	4–13	Meaning of the columns (see Table 10.2); references
Catalogue	14–211	Table of 7840 NGC objects
Notes	212–225	Notes (* in column 11)
Index to Published Figures of Nebulae and Clusters	226–234	List of published drawings until 1887 († in column 11)
Appendix	235–237	Additional objects; list of Safford's nebulae; errata

Table 10.2. *The meanings of the NGC columns*

Column	Designation	Explanation
1	No.	NGC-number (1 to 7840)
2	G. C.	GC-number (John Herschel's General Catalogue)
3	J. H.	h-number (John Herschel's Slough and Cape catalogues)
4	W. H.	H-designation (William Herschel's three catalogues)
5	Other Observers	Discoverer or other observers; other designations
6	Right Ascension, 1860.0	Right ascension for 1860
7	Annual Precession, 1880.0	Annual precession in right ascension for 1880
8	North Polar Distance, 1860.0	NPD = $90° - \delta$ for 1860
9	Annual Precession, 1880.0	Annual precession in NPD for 1880
10	Summary Description	Short description (Herschel-style), remarks
11	Notes	* = sign for note, † = sign for drawing

descriptions, albeit confined to the northern sky: '*Of course very many positions had been left unaltered, chiefly those of objects situated in the southern hemisphere which is still waiting for its d'Arrest.*'

In addition to positional measurements, the search for nebulae was an essential task in the nineteenth century. Dreyer mentions the observations of d'Arrest, Marth, Stephan, Tempel, Swift and Stone. Concerning the identification of new objects, he remarks that, '*With regard to the very numerous new nebulae recorded of late years, it was frequently a matter of some difficulty to decide about the identity of objects announced independently by several observers, and differing little as regards place, but often much as to description.*'[2]

After this introduction, Dreyer explains the structure of the catalogue, which is very similar to that of John Herschel's GC, in order to guarantee continuity and comparability. Consequently, the 'North Polar Distance' (instead of declination) and the epoch 1860 are used, with however, the addition of the precession for 1880. The meanings of the NGC columns are given in Table 10.2.

Like Herschel, Dreyer introduces a new numbering – a necessary action due to the large number of new objects. Now astronomers, in addition the classical H-, h- and GC designations, were given yet another numbering. Dreyer, being aware of the possibility of confusion, recommended that the common numbers should be kept as far as possible: '*It was with much regret that I found it necessary to introduce new numbers, and it is greatly to be hoped that these will be quoted as little as possible, but that old nebulae, as hitherto, will be chiefly mentioned by their h-numbers, or failing such by their H class and number.*'[3] Fortunately, this advice was not

---

[2] Dreyer (1953: 4).

[3] Dreyer (1953: 4–5).

followed and the new NGC-numbers were quickly accepted – to date they remain unchallenged by any alternative.[4]

From 1874 Dreyer had treated many identification problems for new objects appearing in the GCS and the published Birr Castle observations. This was his special field, in which he was not surpassed by anyone. As Director of Armagh Observatory he had enough time to continue the work (after managing to sort out some problems concerning the condition of the buildings and the acquisition of new instruments). The conditions for the creation of a new catalogue of non-stellar objects were excellent and thus it took not much more than a year from the rejection of his second supplement to the GC by the RAS to the publication of the NGC. However, in order to identify its objects correctly with H, h and GC labels, Dreyer once again had to delve deeply into the previous data. Fortunately he could use the valuable work of Auwers and John Herschel.

About the column 'Other observers' Dreyer remarks that '*It was not possible to indicate the source of each position in the catalogue, nor was it necessary, as a catalogue of this kind can only be a work of reference or an index, but not a systematic catalogue of final positions representing the observations of this or that observer. But whenever considerable alterations, the authority (or the principal one if there were several) has always been quoted in the column.*'[5] The column also tells the reader something about the quality of the Herschel positions: '*Whenever the name of an observer later than the two Herschels is given at an object observed by h or H, it means that the place given in the general Catalogue was considerably in error, and has been corrected by means of the observations of the astronomer mentioned in this column.*'[6] But the column serves yet another purpose for newly found objects, namely to mention the discoverer (or independent discoverers). In these cases the columns H, h and GC are, of course, blank. In contrast to John Herschel, Dreyer also gives the names of discoverers for objects found prior to Messier. This shows his interest in the history of astronomy: '*For the sake of the historical interest attached to early observations of nebulae and clusters, I have inserted the names of observers before Messier (Hipparch, Sûfi, Cysat, Flamsteed, Méchain, &C.).*' In some cases a year is given.

Concerning the discoverers after the two Herschels, Dreyer collects the references used on pages 6 to 12. However, the list is incomplete, with some gaps. Sometimes the source is given very roughly (e.g. 'Melbourne') or is even missing. This could concern a private communication (letter, manuscript). Since there is no known estate, this valuable information is lacking. Examples are Lohse's objects found at Wigglesworth Observatory and some of Barnard's. Anyway, except in a few cases, the required sources could be deduced.

An example of an unexplained entry in the column 'Other observers' is the strange phrase 'Greenwich IX yr C' for NGC 2392. It was not easy to locate the appropriate source; it is the 'Nine-year catalogue of 2263 stars: deduced from observations extending from 1868 to 1876. Made at the Royal Observatory, Greenwich. Reduced to the epoch 1872. Appendix I to the Greenwich Observations, 1876'. Or take 'Engelhardt', as is mentioned for some objects such as NGC 227 and NGC 1491[7] (see Section 11.1). For NGC 2194 Dreyer noted 'DM 1066', meaning the star BD +12° 1066.[8] Sometimes stars from Lalande's *Histoire Céleste* are mentioned by noting 'LL' (NGC 2045) or 'Lal' (NGC 3242). The designation varies for other names too, such as 'Bond' (NGC 223) and 'G. P. Bond' (NGC 219).

The descriptions follow the well-known scheme of William Herschel. Dreyer had in mind not only the desirability of comparability with previous catalogues; he was also convinced of the effectiveness of textual abbreviations. Revisions, such as the number codes suggested by John Herschel and Herman Schultz, were strictly refused (see Section 8.16.5).

What could not be included in the column 'Summary description' was placed in the notes. There, 387 objects are mentioned, sorted by NGC-number. Dreyer writes 'N.G.C.' to show how the new designation should be used. Therewith he created the popular name New General Catalogue – following the tradition

---

[4] In a publication of 1896 Dreyer still used GC numbers (Dreyer 1896).

[5] Dreyer (1953: 3).

[6] Dreyer (1953: 5).

[7] The emission nebula NGC 1491 in Perseus (which was discovered by William Herschel in 1790) was later classified as a 'barred spiral' by Curtis 1918a. It is also catalogued as 'galactic nebula' Ced 25.

[8] Instead of BD often DM ('Durchmusterung') was used in the literature.

of John Herschel's General Catalogue. Though the designation and new numbering was soon accepted, there was no common notation at first. Dreyer, for instance, wrote 'N.G.C. 1234', but also '1234 N.G.C.', 'New G.C. 1234' and 'Dreyer 1234' were used by other authors.[9]

Dreyer followed John Herschel concerning references for published drawings. The number increased from 298 (GC) to 464. To allow a comparison, the list is sorted by GC-number. A few early sources that had been missed by Herschel were added. It is interesting that Dreyer (who was otherwise very familiar with the literature) does not mention Holden's *Index Catalogue of Books and Memoirs relating to Nebulae and Clusters* (Holden 1877a). It contains a compilation of 451 drawings up to 1877; some of Holden's references are missing from the NGC.

Shortly before the printing, Dreyer wrote an appendix. There he mentions Swift's newly published list VI of objects found at Warner Observatory (see Section 9.5.4). Dreyer detected some differences relative to a manuscript sent to him in advance by letter (which unfortunately is lost). He also mentions discoveries made by Safford in 1866–67 at Dearborn Observatory, but not published until autumn 1887 (see Section 9.6.5). Dreyer lists 47 objects, sorted by right ascension (1860).

10.1.2 The review of Bauschinger

In 1889 a review of the NGC appeared in the *Vierteljahrsschrift der Astronomischen Gesellschaft*, written by Julius Bauschinger, Director of the Astronomisches Recheninstitut in Berlin, and titled 'J. L. E. Dreyer, A New General Catalogue' (Bauschinger 1889). First he points out the necessity of the catalogue, since John Herschel's General Catalogue had been rendered outdated by the large number of new discoveries and corrections. Moreover, the mishmash of various circulating discovery lists is seen as an unbearable state for all observers of nebulae. Therefore Dreyer's work is urgently needed: '*This up-to-date adaption of a new General Catalogue, achieved by Dr Dreyer on the request of the Royal Astronomical Society, is now available for all astronomers in the above-mentioned volume and will become a valuable tool both for telescopic observations and for work on nebulae in general, yielding considerably more benefit than the older catalogue.*'

Bauschinger criticised Dreyer's suggestion that one should keep the old GC designations: '*For the objects of the G.C. this might be expedient, but for the others, which are compiled in lists that are not always readily available, it is not feasible, and here certainly the New G.C. will be cited, if one doesn't prefer simply to give the position.*'[10] He would have appreciated it if (in the column 'Other observers') '*other existing observations, namely the micrometrical by Schönfeld, Schultz, d'Arrest and others, had been given completely*', rather than just when '*their results have led to corrections of the G.C.*' On the other hand, he considers the '*indication of observers prior to Messier, like Hipparch, Sufi, Cysat and others, as obsolete*', because '*the researcher will a priori take other works for his historical studies*'.

Finally, Bauschinger presents the result of a statistical analysis of the NGC: the distribution of objects on the sphere according to 'type'. He follows Copeland, who had given something similar in his GCS review (see Section 8.19.2). Bauschinger divides the sky into areas of 1^h in right ascension and $15°$ in declination. He distinguishes among 'faint nebulae' (pF to eF, including objects with 'no description'), 'bright nebulae' (pB to eB), 'planetary nebulae' and 'star clusters' (including globular clusters). The division into faint and bright nebulae should show '*whether the known fact that the nebulae generally avoid the Milky Way could be due to the brightness of the Milky Way affecting the visibility of faint objects; in this case, the bright nebulae should obviously show a different distribution*'.

From the NGC data, Bauschinger concludes that '*The faint nebulae avoid the Milky Way; the largest accumulations are located near the poles of the Milky Way.*' He added that '*The bright nebulae show exactly the same behaviour as the faint ones, which proves that the general brightness of the Milky Way alone does not cause the characteristic distribution.*' Vice versa, planetary nebulae and star clusters are located '*in or near the Milky Way, with a few exceptions only*'.

---

[9] By the way, this is a galaxy in Eridanus, which was discovered in 1886 by Leavenworth.

[10] This advice was not followed until the late twentieth century; today many catalogues use coordinate abbreviations to designate celestial objects (e.g. IRAS 12323+6348, an infrared source).

10.2 CONTENT OF THE NGC AND STATISTICAL ANALYSIS

10.2.1 Finding methods and number of discoveries

Figure 10.1 shows the number of discoveries from 1800 until the end of 1887 (the press date of the NGC) in intervals of 10 years. The period 1800–1810 brought William Herschel's last objects in Slough. Afterwards the discoveries stopped for a decade until John Herschel started his observations at the same place. The objects found during the years 1820–30 are mainly due to him. The next decade was dominated by his work at Feldhausen. Then another 10 years of idleness followed, due to the opinion that nothing more could be discovered.

The revival originated from the Birr Castle activities starting in the mid 1840s. The next decade was characterised by observations made by the local astronomers, mainly Mitchell and the Stoney brothers. The period 1860–70 was dominated by the work of d'Arrest (Copenhagen) and Marth (Malta). The decade 1870–80 was the high time of Stephan (Marseille) and Tempel (Arcetri), which continued into the final period lasting until 1887. Also Swift (Warner Observatory) and the observers at Leander McCormick Observatory (Stone, Leavenworth and Muller) contributed to it. The last seven years brought the largest number of discoveries – the crucial reason for Dreyer to create the NGC.

Concerning the kind of discovery, the bulk of the objects had been found by visual observing (Table 10.3). This fact makes the NGC a very homogeneous catalogue (compared with the later Index Catalogues). A few objects were found by visual spectroscopy and only one by photography.

10.2.2 Independent NGC objects

The next question that arises is that of how many independent objects are catalogued in the NGC. Since there are 254 identities between two or more objects, the total number of entries is reduced to 7586 independent objects (Table 10.4).

Table 10.5 shows the objects with triple identities. For the galaxy NGC 3497 in Crater there is even an additional identity with an IC object: Lewis Swift's IC 2624. NGC 6667 is unique too: all three discoveries were made by one observer – Lewis Swift.

10.2.3 Discoverers of NGC objects and their success rates

The scores of the most successful discoverers are shown in Fig. 10.3. As expected, William and John Herschel dominate, but it should be pointed out that 3480 of the

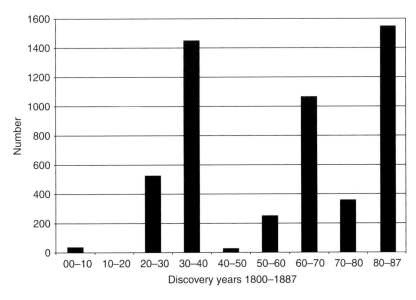

Figure 10.1. Numbers of discoveries per decade (1800–1887).

Table 10.3. *Object numbers according to discovery methods*

Discovery method	Number	Discoverer	Example
Visual observation	7817	Various	NGC 6916 (last), 26.6.1887
Visual spectroscopy	22	Pickering (15), Copeland (7)	NGC 7114 (first), 2.9.1877
Photography	1	Paul and Prosper Henry	NGC 1432 (Maia Nebula), 16.11.1885

Table 10.4. *The number of independent NGC objects*

	Number
Entries	7840
Catalogue identities (2 objects)	238
Catalogue identities (3 objects)	16
Independent objects	7586

independent NGC objects (7586 in total) were found by other observers. Thus the NGC is by no means a 'Herschel catalogue' (as is occasionally claimed in the literature). Marth is placed third, followed by Lewis Swift and Stephan.

The NGC objects were found by 100 persons; 26 of them made their discoveries prior to William Herschel (Table 10.6). The brothers Paul and Prosper Henry, who discovered the Maia Nebula NGC 1432 by photography, are counted as a single entry. The name Kirch stands for Gottfried and Maria Kirch; while Gottfried Kirch found the open cluster M 11 in Scutum, his wife was concerned in the discovery of the globular cluster M 5 in Serpens.

Column 'D' gives the numbers of objects which were credited by Dreyer to the individual observers in the NGC, whereas 'NGC' shows the true rates, i.e. considering independent objects only. The large differences for the Birr Castle observers (W. Parsons, Mitchell, B. Stoney and J. Stoney) are due to the fact that Dreyer assigned 'Ld R' (William Parsons) for all of their discoveries. Our knowledge of the true discoverers results from a careful analysis of the observations published in 1861 and 1880 (see Sections 7.1 and 8.18). For Dunlop the difference is explained by the difficult interpretation of his observations. Dreyer knew only the *Philosophical Transactions* publication of 1828, whereas a modern analysis uses Dunlop's original notes (see Section 4.4.2). Safford is mentioned by Dreyer only in connection with the objects found

Figure 10.2. NGC 3497, a galaxy in Crater with many identities (DSS).

at Harvard College Observatory. Some of those discovered later at Dearborn Observatory are credited to other observers in the NGC (due to lack of knowledge). The differences for William and John Herschel, Lewis Swift and Stephan are explained by the large number of objects, which naturally caused identification problems.

Dreyer does not mention 13 discoverers: Aratos, Baily, Baracchi, Ferrari, Hunter, Oppenheim, Peiresc, Pigott, Ptolemy, Skinner, Edward Swift, Vespucci and Voigt. In some cases he was not aware of the original notes or publications (e.g. Pigott, Vespucci and Voigt),

Table 10.5. *The NGC objects with triple identities*

NGC	Discoverer	NGC	Discoverer	NGC	Discoverer	Type	V	Con.	Remarks
614	W. Herschel	618	J. Herschel	627	J. Herschel	Gx	12.7	Tri	
863	W. Herschel	866	L. Swift	885	L. Swift	Gx	13.0	Cet	
1440	W. Herschel	1442	W. Herschel	1458	Leavenworth	Gx	11.6	Eri	
3122	W. Herschel	3110	Stephan	3518	Stone	Gx	12.7	Sex	
3497	W. Herschel	3528	J. Herschel	3525	Stone	Gx	11.6	Crt	IC 2624, L. Swift; Fig. 10.2
4664	W. Herschel	4665	W. Herschel	4624	J. Herschel	Gx	10.3	Vir	
6125	W. Herschel	6127	L. Swift	6128	L. Swift	Gx	12.0	Dra	
6667	L. Swift	6668	L. Swift	6678	L. Swift	Gx	13.0	Dra	

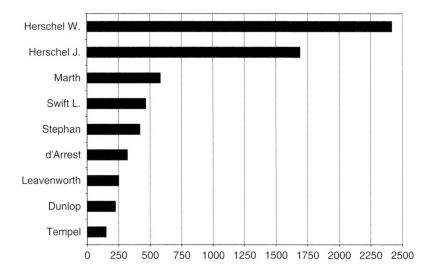

Figure 10.3. Discoverers of more than 100 NGC objects.

made wrong interpretations (e.g. Baily, Baracchi, Ferrari, Hunter, Oppenheim and Skinner) or simply omitted the observer, as was the case for Edward Swift. Vice versa, Dreyer mentions discoverers to whom no priority can be assigned: Cysat, Ellery, Magellan, Secchi and Tuttle. In the case of Secchi, all of the objects were found by his assistant Ferrari (which, however, should have been known to Dreyer); the cases of Ellery and Barrachi are similar. It is known today that Peiresc saw the Orion Nebula one year earlier than Cysat (see Section 2.1).

The column 'D' refers to Dreyer's credits (note that the sum is less than 7840). The column 'NS' shows the fraction of non-stellar objects (see the definition in Section 1.1) and '%' gives the percentage (relative to 'NGC') – this is the observer's success rate! Altogether 93% of the NGC objects are non-stellar; the rest are parts of galaxies, stars or star groups or missing objects ('not found'). Among the observers, who discovered more than 15 independent objects, Messier, Méchain, Common and Pickering reached the maximum possible success rate (100%). Excellent scores were obtained by Lewis Swift, Marth, Leavenworth, William Herschel, Stephan and Dunlop, 95%–99% and large numbers of objects. Especially Dunlop is surprising, because his observations were badly rated in the Cape catalogue. Its author, John Herschel, achieved a very good result too (93% and a large number of objects); the same is

Table 10.6. *The discoverers of the 7586 independent NGC objects and the telescopes used (italics = prior to William Herschel; E = naked eye; bold numbers are totals for column 'D'; 'NGC' and 'NS' and the average rate on the column '%')*

Discoverer	Telescope	D	NGC	NS	%	Discoverer	Telescope	D	NGC	NS	%
Aratos	E		1	1	100	*Kirch*	Rr	2	2	2	100
Aristotle	E	1	2	2	100	Koehler	Rr	2	3	3	100
As-Sufi	E	2	1	1	100	Lacaille	0.5 Rr	27	23	21	91
Austin	15.0 Rr	3	3	1	33	Langley	15.0 Rr	1	1	0	0
Auwers	2.6 Rr; 6.2 Rr	2	2	2	100	Lassell	24.0 Rl; 48.0 Rl	3	4	4	100
Baily	18.3 Rl		1	1	100	Leavenworth	26.3 Rr	260	251	245	98
Ball	72.0 Rl	11	11	4	36	*Legentil*	Rr	4	2	2	100
Baracchi	48.0 Rl		2	2	100	Lohse	15.5 Rr	18	16	3	19
Barnard	5.0 Rr; 6.0 Rr	20	17	16	94	*Mairan*	Rr	1	1	1	100
Baxendell	6 Rr	1	1	0	0	*Maraldi*	4.0 Rr	2	2	2	100
Bevis	3.0 Rr	1	1	1	100	Marth	48.0 Rl	594	582	563	97
Bigourdan	12.4 Rr	76	70	38	56	Méchain	3.0 Rr	26	26	26	100
Bode	Rr	5	4	4	100	Messier	3.3 Rr	33	40	39	98
Bond	4.0 Rr; 15.0 Rr	15	14	5	36	Mitchell	72.0 Rl		89	83	93
Borrelly	7.2 Rr	5	6	6	100	Muller	26.3 Rr	87	82	64	78
Brorsen	9.6 Rr	1	1	1	100	Oppenheim	12.0 Rr		1	1	100
Bruhns	Rr	1	1	1	100	Oriani	3.6 Rr	3	1	1	100
Burnham	6.0 Rr; 18.5 Rr	4	5	5	100	Palisa	12.0 Rr; 27.0 Rr	8	7	7	100
Cacciatore	3.0 Rr	1	1	1	100	Parsons L.	72.0 Rl	44	38	24	63
Cassini	5.0 Rr	1	1	1	100	Parsons W.	72.0 Rl	225	1	1	100
Chacornac	10.0 Rr	1	1	0	0	Pechüle	11.0 Rr	2	2	2	100
Coggia	7.2 Rr	1	1	1	100	Peirce	15.0 Rr	1	1	0	0
Common	36.0 Rl	33	29	29	100	*Peiresc*	Rr		1	1	100
Coolidge	15.0 Rr	9	9	0	0	Peters	13.5 Rr	16	16	10	63
Cooper	13.5 Rr	7	7	0	0	Pickering	15.0 Rr(s)	15	15	15	100
Copeland	6.1 Rr(s); 15.1 Rr(s); 72.0 Rl	35	35	29	83	*Pigott*	5.0 Rr		1	1	100

446

Name	Telescope				
Darquier	3.5 Rr			1	100
d'Arrest	4.6 Rr; 11.0 Rr	344	319	280	88
de Chéseaux	Rr	4	6	6	100
Dreyer	72.0 Rl	19	15	12	80
Dunér	9.6 Rr	1	1	1	100
Dunlop	9.0 Rl	116	225	223	99
Ferrari	9.5 Rr		14	5	36
Flamsteed	9.7 Rr	1	1	1	100
Hall	26.0 Rr	1	1	1	100
Halley	Rr	4	2	2	100
Harding	8.5 Rl	2	1	1	100
Harrington	12.5 Rr	1	1	1	100
Hartwig	6.8 Rr; 18.0 Rr	4	5	5	100
Henry	13.0 Rr(p)	1	1	1	100
Herschel C.	4.2 Rl; Rr	13	10	10	100
Herschel J.	5.0 Rr; 18.3 Rl	1866	1691	1569	93
Herschel W.	12.0 Rl; 18.7 Rl	2439	2415	2363	98
Hind	7.0 Rr	4	4	4	100
Hipparch	E	3	2	2	100
Hodierna	Rr	8	11	11	100
Holden	15.6 Rr	20	18	17	94
Hough	18.5 Rr	2	4	4	100
Hunter	72.0 Rl		1	1	100
Ihle	Rr	1	1	1	100
Ptolemy	E		1	1	100
Rümker	4.0 Rr		1	0	0
Safford	15.0 Rr; 18.5 Rr	2	29	27	93
Schmidt	6.2 Rr	14	16	10	63
Schönfeld	3.0 Rr; 6.0 Rr	3	4	4	100
Schultz	9.6 Rr	14	12	1	8
Schweizer	9.0 Rr	1	1	0	0
Searle	15.0 Rr	6	6	6	100
Skinner	18.5 Rr		1	1	100
Stephan	31.0 Rl	457	420	411	98
Stone	26.3 Rl	99	89	83	94
Stoney B.	72.0 Rl		86	64	74
Stoney J.	72.0 Rl		49	45	92
Struve O.	15.0 Rl	16	13	11	85
Struve W.	9.6 Rr	7	7	4	57
Swift E.	16.0 Rr		25	23	92
Swift L.	4.5 Rr; 16.0 Rr	560	467	443	95
Tempel	4.0 Rr; 9.4 Rr; 11.0 Rr	156	149	111	74
Todd	26.0 Rr	8	8	8	100
Vespucci	E		1	1	100
Voigt	31.0 Rl		5	5	100
Winlock	15.0 Rr	2	2	2	100
Winnecke	3.0 Rr; 6.5 Rr; 9.6 Rr	9	8	7	88
Young	23.0 Rr	1	1	1	100
		7821	7586	7064	93

Table 10.7. *The NGC objects which were found four times*

NGC	GCS	Type	V	Con	Discoverer 1	Discoverer 2	Discoverer 3	Discoverer 4
1360	5315	PN	9.4	For	L. Swift 1859	Tempel 9.10.1861	Winnecke Jan. 1868	Block 18.10.1879
6364		Gx	13.1	Her	Voigt June? 1865	Safford 5.9.1866	Stephan 21.7.1879	L. Swift 11.9.1885
7422	6101	Gx	13.5	Psc	Marth 11.8.1864	O. Struve 6.12.1865	d'Arrest 29.9.1866	Safford 27.9.1867

Table 10.8. *Identical NGC objects with small time gaps between their discoveries*

NGC	Type	V	Con.	Discoverer	Date	H/h	D	NGC	H/h
4208	Gx	11.1	Com	W. Herschel	8.4.1784	II 107	0	4212	II 108
4358	Gx	14.4	UMa	W. Herschel	17.4.1789	III 799	0	4362	III 800
4888	Gx	13.4	Vir	W. Herschel	23.3.1789	II 778	0	4879	III 759
3284	Gx	13.6	UMa	W. Herschel	8.4.1793	III 912	1	3286	III 917
6885	OC	8.1	Vul	W. Herschel	9.9.1784	VIII 20	1	6882	VIII 22
7140	Gx	11.7	Ind	J. Herschel	4.10.1834	3892	1	7141	3893

true for Lacaille, Edward Swift, Johnstone Stoney, Safford, Mitchell, Barnard, Holden and Stone.

Tempel (74%) and Bigourdan (56%) attained poorer results; many of their objects are stars. The ratings of Schultz and Lohse are really bad, but Coolidge is the worst case: none of his discoveries is non-stellar! Thirty-three observers contributed only one find; in some cases this unfortunately turned out to be a star or star group, or is even missing (NF): Chacornac (NGC 1988, a star), Peirce (NGC 1170, NF), Rümker (NGC 1724, a star group), Langley (NGC 3355, NF) and Schweizer (NGC 7804, a pair of stars).

There are NGC objects that were found twice, three times or even four times (Table 10.7). However, Dreyer does not mention all of the discoverers in these cases: 'Swift 1857, Winnecke' (NGC 1360), 'Stephan' (NGC 6364) and 'Marth, Struve, d'Arrest' (NGC 7422).[11]

Table 10.8 gives the reversed view, showing identical objects that were found twice during the same night (i.e. $D = 0$ days) or once again in the following ($D = 1$). As expected, the first two cases are curious.

The explanations were given by Dreyer in the *Scientific Papers*.[12]

After William Herschel had observed II 107 (NGC 4208) he looked up the reference star 6 Comae, then he '*again moved the telescope 1° south and took the nebula a second time without noticing that it was the same object*'. He later catalogued the second find as II 108 (NGC 4212). Dreyer had not noticed the identity when compiling the NGC of 1888. III 799 is the galaxy NGC 4358 (which is identical to III 800 = NGC 4362), located 2' northwest of NGC 4364 (III 801). For III 800/801 Herschel had noted '*Two, both cF. cS. R.*' Dreyer explained this as follows: "*Very probably the word 'two' refers to III. 799 and III. 800, as nobody seems to have seen three nebulae in the pla*ce." Indeed, these are the only non-stellar objects in the field. The same thing applies to III 758 (NGC 4878) and III 759 (NGC 4879), which is the same as II 778 (NGC 4888). The identity of VIII 20 (NGC 6885) and VIII 22 (NGC 6882), the cluster around the star 20 Vulpeculae, was not noticed by Dreyer.[13] It is no. 6 in Harding's list of

---

[11] More information is given in Sections 9.5.2, 8.13.2 and 8.7.3.

[12] Dreyer (1912a: Vol. 1, 297 and Vol. 2, 236).

Table 10.9. *Identical NGC objects with large time gaps between their discoveries*

NGC	Type	V	Con.	Discoverer 1	NGC	Discoverer 2	Years	Remarks
2244	OC	4.8	Mon	Flamsteed 17.2.1690	2239	J. Herschel March 1830	140	12 Mon
2422	OC	4.4	Pup	Hodierna 1654	2478	Messier 19.2.1771	117	M 47
5174	Gx	12.5	Vir	W. Herschel 15.3.1784	5162	L. Swift 19.4.1887	103	III 45

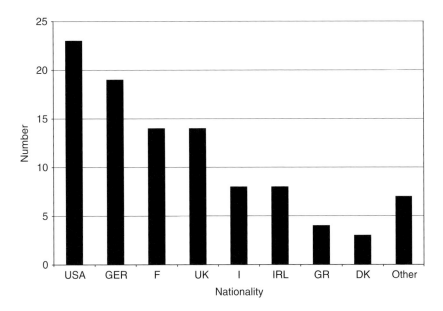

Figure 10.4. The nationalities of the discoverers (GER, German; F, French; I, Italian; IRL, Irish; GR, Greek; and DK, Danish).

nebulae (see Section 4.1.2) and was later studied by Schultz in Uppsala (Schultz 1873). On h 3892/93 John Herschel wrote in the Cape catalogue that an identity '*is not improbable* [...] *one or other being mistaken 1° in P.D.*' There is only one galaxy in the field, located in Indus. Other cases that are different parts of a single nebula are not included here. These are the planetary nebula NGC 650/51 = M 76 in Perseus (Méchain, 5 September 1780), the galaxy NGC 2242/43 in Volans (John Herschel, 23 December 1834) and the planetary nebula NGC 2371/72 in Gemini (William Herschel, 5 September 1780).

Another point concerns the gap (in years) between two independent discoveries of the same object bearing different NGC-numbers. Table 10.9 gives the three record holders.

Figure 10.4 shows the nationalities of the discoverers; the USA clearly leads. 'Other' contains Austria, Switzerland and Sweden with two persons each and Persia with one (As-Sufi).

It is interesting to differentiate the 7064 independent non-stellar NGC objects by hemisphere. However, it would not be fair to take the celestial equator as the division line. Thus the 'northern sky' is defined by the region accessible to William Herschel (Slough). Since his most southerly object was the galaxy NGC 3621 in Hydra at $\delta = -32°\ 48'$ (see Table 2.7), all objects south of it are allocated to the 'southern sky'. Tables 10.10 and 10.11 present the discoverers of northern and southern objects, in

---

[13] See Archinal and Hynes (2003: 187).

Table 10.10. *Discoverers of the 5880 northern NGC objects (see the text)*

Number	Discoverers
2363	W. Herschel
645	*J. Herschel*
563	Marth
442	*L. Swift*
411	Stephan
280	d'Arrest
245	Leavenworth
111	Tempel
83	Mitchel, Stone
64	Muller, B. Stoney
45	J. Stoney
39	Messier
38	Bigourdan
29	Common
27	Safford
26	*Copeland*, Méchain
24	L. Parsons
23	E. Swift
17	Holden
15	Barnard, Pickering
12	Dreyer
11	O. Struve
10	C. Herschel, *Hodierna*, Peters
8	Todd
7	Palisa, Winnecke
6	Borrelly, de Chéseaux, Searle
5	Bond, Burnham, Ferrari, Hartwig, Voigt
4	Ball, Bode, *Dunlop*, Hind, Hough, Lassell, Schönfeld, W. Struve
3	Koehler, Lohse
2	Aristotle, Auwers, Hipparch, Kirch, *Lacaille*, Legentil, Maraldi, Pechüle, Winlock
1	As-Sufi, Aratos, Austin, Baily, Bevis, Brorsen, Bruhns, Cassini, Coggia, Darquier, Dunér, Flamsteed, Hall, *Halley*, Harding, Harrington, Henry, Hunter, Ihle, Mairan, Oppenheim, Oriani, W. Parsons, Peiresc, Pigott, Schultz, Skinner, Young

Table 10.11. *Discoverers of the 1184 southern NGC objects (see the text)*

Number	Discoverers
924	*J. Herschel*
219	*Dunlop*
19	*Lacaille*
10	Schmidt
3	*Copeland*
2	Baracchi
1	*Barnard*, Cacciatore, *Halley*, *Hodierna*, Ptolemy, *L. Swift*, Vespucci

this sense. However, some persons (marked in italics) contributed to both regions, because they were located in southern Europe, made trips to the southern hemisphere (e.g. Copeland's Andes observations) or discovered northern objects from the south.

About 40% of these persons might be called professional astronomers, qualified by a scientific education and working at a government or university observatory. The rest are amateurs: owners of private observaories and their assistants.[14] The only female discoverers were Caroline Herschel and Maria Kirch; the latter found the globular cluster M 2 together with her husband Gottfried. More frequent among observers were relatives: brothers and sisters (Paul and Prosper Henry, Bindon and Johnstone Stoney, Caroline and William Herschel), father and son (William and John Herschel, William and Lawrence Parsons, Wilhelm and Otto Struve, Lewis and Edward Swift) and married couples (Gottfried and Maria Kirch, Margaret and William Huggins). William Lassell's daughter Caroline was astronomically active too; she made an excellent drawing of the Orion Nebula based on observations made by her father.

Figure 10.5 shows the observers with the largest number of publications on non-stellar objects appearing in the nineteenth century. Barnard is top, followed by Lewis Swift, who contributed many lists of nebulae (like Stephan and Tempel).

---

[14] Chapman A. (1998).

10.2 Content and statistical analysis 451

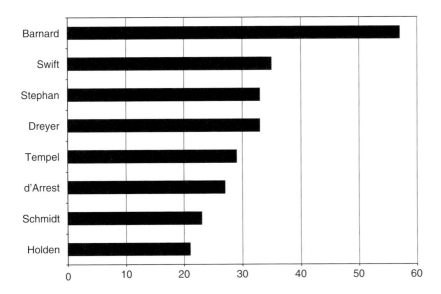

Figure 10.5. The most active publishers (1800–1900).

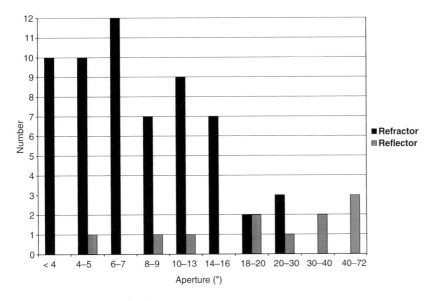

Figure 10.6. Telescopes and their apertures.

10.2.4 Statistics of telescopes

Figure 10.6 shows the number and apertures of refractors and reflectors used for observations of non-stellar objects in the nineteenth century. Altogether we have 71 instruments: 60 refractors (85%), mainly of small and moderate size, and 11 reflectors (15%), dominating the large apertures. Most of the refractors had lenses of 6–7" diameter; the smallest were the 3" telescopes of Schönfeld (Bonner Durchmusterung) and Winnecke (private). The smallest reflector was Caroline Herschel's 4.2-inch, which was built by her brother. The largest telescopes were all reflectors: Lassell's 48-inch (Malta), the Great Melbourne Telescope of the same aperture and Lord Rosse's

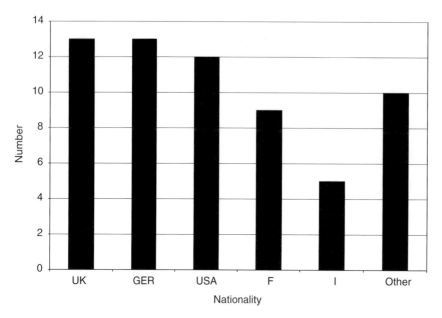

Figure 10.7. Nationalities of observatories (sites) (GER, Germany; F, France; I, Italy).

72-inch at Birr Castle. The leading refractor was the 27" Grubb in Vienna, with contributed only one NGC object (Palisa's galaxy NGC 927). The first non-stellar object to be discovered with a reflector was NGC 7009, the Saturn Nebula in Aquarius (William Herschel, 7 September 1782, Datchet). The first to be found with a refractor – of course, this was the first telescopic discovery too – was NGC 224, the Andromeda Nebula M 31 (Simon Marius, 15 December 1612, Gunzenhausen).

Figure 10.7 shows the nationalities of the observatories. Great Britain and Germany are in front. 'Other' includes Ireland (Birr Castle, Markree). All of the observatories were located in flat country, except for Copeland's expedition site in the Andes, where he found NGC 5315, NGC 5873 and NGC 6153. The first objects found at mountain observatories (e.g. Mt Hamilton, Mont Gros near Nice, Königstuhl near Heidelberg) are catalogued in the Index Catalogue.[15] The sites with extreme latitudes were Uppsala and Pulkovo (both +60°), Melbourne (−38°), Paramatta and Feldhausen (both −34°).

---

[15] The Chamberlin Observatory, located in the city of Denver, has an elevation of 1644 m, but cannot be termed a 'mountain observatory'. The same applies for Arequipa (Peru).

10.2.5 Types of objects

Table 10.12 shows the distribution of the 7586 independent NGC objects according to type (by definition, the first five categories are the non-stellar objects). The term 'galactic nebula' generally covers emission nebulae (EN), reflection nebulae (RN), dark nebulae (DN) and supernova remnants (SNRs), including mixed forms. No real dark nebulae are in the NGC, but there are four SNRs (see below). The column 'Maximum' shows who found the most objects of each type. As expected, William and John Herschel dominate. Despite the high rate, Lewis Swift is the observer with the most missing objects – a result of his bad positions. Strangely, Dreyer calls them 'very good' in the NGC; in the IC I he was, however, a bit more critical: *'The positions [...] are generally reliable within one or two minutes of arc, but larger errors occur occasionally.'* In the IC II, Swift is eventually called a *'veteran observer* [whose] *places are not as good as those formerly found by him at Rochester'*.

Table 10.13 shows the four supernova remnants listed in the NGC. However, until 1933 the most prominent one, M 1 in Taurus, was classified as a 'planetary nebula' (see e.g. Curtis (1918b)); later it became evident that it is the remnant of the supernova of 1054 (see e.g. Baade (1942)). A fifth case, NGC

10.2 Content and statistical analysis   453

Table 10.12. *Types of the 7586 independent NGC objects*

Type	Number	%	Brightest object	Maximum
Galaxy	6028	79.5	NGC 292 = SMC (2.2 mag, Tuc)	W. Herschel (2108)
Open cluster	683	9.0	NGC 6231 (2.6 mag, Sco)	J. Herschel (338)
Globular cluster	114	1.5	NGC 104, (4.0 mag, Tuc)	W. Herschel (37)
Galactic nebula	147	2.0		J. Herschel (50)
Planetary nebula	93	1.2	NGC 7293 (7.3 mag, Aqr)	W. Herschel (37)
Part of galaxy	28	0.4	NGC 604 (12.0 mag, Tri); H II region in M 33	B. Stoney (8)
Single star	120	1.6	NGC 771 = 50 Cas (4.0 mag)	Tempel (18)
Star pair, trio, quartet	163	2.1		J. Herschel (24)
Star group, cloud	116	1.5		J. Herschel (74)
Not found	94	1.2		L. Swift (17)

Table 10.13. *The four supernova remnants (SNRs) in the NGC*

NGC	h	V	Con.	Discoverer	Remarks
1918	1125		Dor	Dunlop 27.9.1826	In the LMC
1952	1157	8.4	Tau	Bevis 1731	M 1, Crab Nebula (SN 1054)
2060	1261	9.6	Dor	J. Herschel	In the LMC
6960	4614	13.1	Ind	J. Herschel 9.7.1834	Veil Nebula with NGC 6974/79/95

2020 in Dorado (LMC), is sometimes called a 'candidate'. Actually this shell-like object, which was discovered by Herschel on 30 December 1836 (h 2903), is a Wolf–Rayet nebula.

Table 10.14 lists all of the single stars in the NGC brighter than 10th magnitude. The phenomenon of supposed nebulous appearance has already been discussed in connection with 'nebulous stars' (see Tables 2.17 and 6.20). Most of these objects were found by John Herschel. The original description is given. The case of NGC 405 is symptomatic. In the Cape catalogue (h 2380) John Herschel wrote *'After a long and obstinate examination with all powers and apertures I cannot bring it to a sharp disc and leave it, in doubt whether it be a star or not. The star B 137* [β Phe, 3.4 mag] *immediately preceding offered no such difficulty, giving a good disc with 320.'*

Figure 10.8 shows the brightness distribution of the single stars. The mean is 13.1 mag. The objects '<8' are those of Table 10.14. Obviously the 'nebulous stars' are concentrated in this class. All others are fainter than 10 mag (starting with Chacornac's NGC 1988 in Taurus of 11.0 mag; see Section 7.3.2). Here the appearance was different: the faint star itself was confused with a small nebula. This explains the remarkable gap. The faintest single star is NGC 6973 in Aquarius, which was found by Bigourdan on 5 July 1886.

Finally, it may be interesting to learn when the first and last object of the types listed in Table 10.12 were found. However, for galaxies and open clusters this is not quite straightforward, since the brightest examples (e.g. the LMC and M 44) had already been seen in ancient times. As usual, Table 10.15 gives the 'published' discoveries (see the following section). Only single stars are presented; additionally, double discoveries are included.

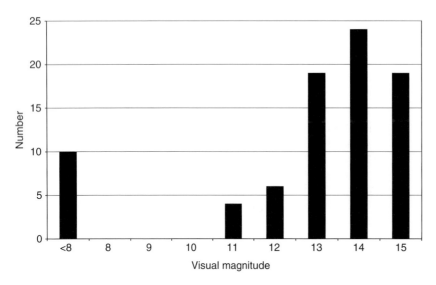

Figure 10.8. The brightness distribution of the single stars in the NGC.

Table 10.14. *The brightest stars in the NGC (sorted by visual magnitude)*

NGC	h	Star	V	Con.	Discoverer	Appearance of star
771	179	50 Cas	4.0	Cas	J. Herschel 29.10.1831	Nebulous
4530	1332	β CVn	4.3	CVn	J. Herschel May 1828	Nebulous atmosphere (see Fig. 6.45)
2542	3115	19 Pup	4.7	Pup	J. Herschel 11.12.1836	Nebulous (CC)
2142	373	3 Mon	5.0	Mon	J. Herschel 6.1.1831	F neb atmosphere of diameter 2′ or 3′
5856	1904	BD +19° 2924	6.0	Boo	W. Herschel (IV 41) 24.5.1791	Enveloped in extensive milky nebulosity
6049	1945	BD +8° 3134	6.3	Ser	J. Herschel 24.4.1830	Nebulous atmosphere of diameter about 2′
2045	367	BD +12° 884	6.5	Tau	J. Herschel 23.1.1832	F neb
3148	675	BD +51° 1585	6.6	UMa	J. Herschel 17.2.1831	Photosphere of diameter 2′ or 3′
405	2380	CD −47° 333	7.1	Phe	J. Herschel 6.9.1834	Stellar nebula (CC)
7748	2266	BD +68° 1393	7.2	Cep	J. Herschel 16.11.1829	Affected with nebulosity

10.2.6 Special objects

Table 10.16 shows special objects. No 'first object' is specified, since that information cannot be given. There are documented sightings of the open cluster NGC 2287 (Canis Major) and NGC 7092 (Cygnus) by Hipparch, dated to about 325 BC, but undoubtedly other NGC objects might have been seen even earlier – for which, however, there is no known source.

For instance, the SMC (NGC 292) was seen by the Australian aborigines many thousands of years ago. The same applies for Praesepe (NGC 2362, M 44) and other naked-eye objects. However, a few 'first objects' relating to special equipment are given.

Which is the constellation with the largest number of NGC objects? The question is, however, of limited value, due to the large size differences. Virgo

Table 10.15. *The first and last to be discovered, objects according to type*

Type	NGC	Discoverer	Date	V	Con.	Other
Gx	224	As-Sufi	905	3.5	And	M 31, Andromeda Nebula, Marius 15.12.1612
Gx	6916	L. Swift	26.6.1887	14.1	Cyg	
EN	1976	Peiresc	1610	3.7	Ori	M 42, Orion Nebula, Cysat 1611
EN	2246	L. Swift	27.2.1886		Mon	
RN	2068	Méchain	March 1780	8.0	Ori	M 78
RN	2296	L. Swift	11.3.1887		CMa	IC 452
PN	6853	Messier	12.7.1764	7.4	Vul	M 27, Dumbbell Nebula
PN	2242	L. Swift	24.11.1886	15.0	Aur	
OC	2287	Aristotle	325 BC	4.5	CMa	M 41, Hodierna 1654?
OC	1798	Barnard	Nov. 1885	10.0	Aur	
GC	6656	Ihle	26.8.1665	5.2	Sgr	M 22, Halley 1715
GC	6366	Winnecke	12.4.1860	9.5	Oph	Au 36
GxP	604	W. Herschel	11.9.1784	12.0	Tri	III 150, in M 33
GxP	2404	Bigourdan	2.2.1886		Cam	In NGC 2404
Star	5175	W. Herschel	15.3.1784	14.4	Vir	III 46
Star	7493	Bigourdan	28.10.1886	14.9	Psc	
NF	3401	W. Herschel	13.4.1784		Sex	III 88
NF	6666	E. Swift	25.5.1887		Lyr	
Double	2478	Messier	19.2.1771	4.4	Pup	OC, M 47, NGC 2422, Hodierna 1654?
Double	5648	Bigourdan	23.5.1887	13.3	Boo	Gx, NGC 5649, W. Herschel 19.3.1787 (III 645)

Table 10.16. *Special objects in the NGC*

Object	NGC	Discoverer	Instrument	Date	Type	V	Con.	Remarks
Brightest	292	Vespucci	Naked eye	1501	Gx	2.2	Tuc	SMC
Faintest	4042	Marth	48" Rl	18.3.1865	Gx	16.4	Com	Fig. 10.9 left
Most northerly	3172	J. Herschel	18.25" Rl	4.10.1831	Gx	14.4	UMi	+89°, Polarissima Borealis
Most southerly	2573	J. Herschel	18.25" Rl	29.3.1837	Gx	13.4	Oct	−89°, Polarissima Australis
First telescopic	224	Marius	Rr	15.12.1612	Gx	3.5	And	Andromeda Nebula (M 31)
First with reflector	7009	W. Herschel	12" Rl	7.9.1782	PN	8.0	Aqr	Saturn Nebula
First spectroscopic	7114	Copeland	15.1" Rr	2.9.1877	Nova		Cyg	Nova Cygni 1876
First photographic	1432	Henry	13" Rr	16.11.1885	RN		Tau	Maia Nebula
Last discovered	6916	L. Swift	16" Rr	26.6.1887	Gx	14.1	Cyg	Fig. 10.9 right

Figure 10.9. Special NGC objects (from left to right): the faintest, the galaxy NGC 4042 in Coma Berenices; the faintest found with a refractor, the galaxy NGC 344 in Cetus (16.1 mag); and the last to be discovered, the galaxy NGC 6916 in Cygnus (DSS).

Table 10.17. *Nearest and farthest objects in the NGC*

NGC	Distance (ly)	Discoverer	Type	V	Con.	Remarks
1435	430	Tempel 19.10.1859	RN		Tau	Merope Nebula; in M 45 (with NGC 1432)
2632	610	Aratos 260 BC	OC	3.1	Cnc	Praesepe, M 44
6475	980	Ptolemy 138 BC	OC	3.3	Sco	Ptolemy's Cluster, M 7
870	795×10^6	Mitchell 22.11.1854	Gx	15.5	Ari	
5535	795×10^6	Marth 8.5.1864	Gx	15.0	Boo	
2603	799×10^6	J. Stoney 9.2.1850	Gx	15.5	UMa	

dominates, with 677 objects (355 were found by William Herschel); most of them are Virgo Cluster galaxies. Both Chamaeleon and Equuleus host only four NGC objects. Whereas all those in Chamaeleon (in the southern sky) were found by John Herschel, each of those in Equuleus (in the northern sky) had its own discoverer: William and John Herschel, Harrington and Stephan.

Finally, which are the nearest and farthest objects? Table 10.17 gives the three record holders in each category. The nearest objects are open clusters or nebulae associated with them; in the case of M 45 (Pleiades) there are two associated NGC objects: NGC 1435 (the Merope Nebula) and NGC 1432 (the Maia Nebula), both of which are reflection nebulae (a third one is IC 349 near Merope; see Section 11.6). Of course, the most remote NGC objects are galaxies. Their distances have been calculated from their redshifts using Hubble's law with a Hubble parameter of $H_0 = 70$ km/s per Mpc.

10.2.7 The visual brightness of NGC objects

Figure 10.10 shows the distribution of visual brightness for all NGC objects. The mean is 12.6 mag and the minimum 16.4 mag. Owing to the advent of photography, both the mean and the minimum are lower in the following Index Catalogues. The mean of the IC I is 14.0 mag (minimum 17.0 mag) and that of the IC II is 14.2 mag (minimum 17.9 mag).

As expected, there is a weak correlation between the limiting magnitude (faintest discovered object) and the aperture used, both for refractors and for reflectors (Fig. 10.11). This, of course, depends on the experience and eyesight of the observer, the quality of the telescope and the site. The graph shows that, even with small refractors, surprisingly faint objects were found. Examples are the 15.7-mag galaxies NGC 7477 and NGC 7577 in Pisces, which were discovered by d'Arrest and Bigourdan with refractors of 11″ and 12″ aperture, respectively. That Dunlop saw the 14.5-mag open cluster NGC 1751 in

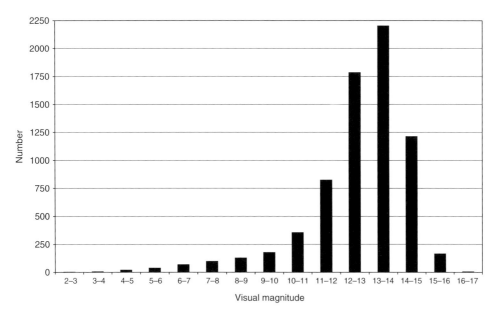

Figure 10.10. The brightness distribution of NGC objects.

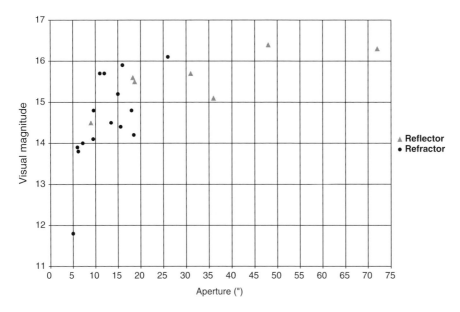

Figure 10.11. The correlation of limiting magnitude and aperture.

Dorado in his 9" reflector is astonishing too. The faintest object to be discovered with a refractor is the 16.1-mag galaxy NGC 344 in Cetus (Fig. 10.9 centre). The observer was Muller in 1886, using the 26-inch of the Leander McCormick Observatory. In the case of reflectors the winner is Marth, who discovered the 16.4-mag galaxy NGC 4042 in Coma Berenices on 18 March 1865 with the 48-inch on Malta (Fig. 10.9 left).

Table 10.18 shows the ranges of visual magnitude for observers with many discoveries at different sites (sorted in order of the mean value). In the cases of

Table 10.18. *Mean, maximum and minimum visual magnitudes (V) for the most successful discoverers (Ap = aperture)*

Observer	Site	Ap (")	Tel.	V mean	V max	V min
Dunlop	Paramatta	9.0	Rl	9.8	4.2	14.5
J. Herschel	Feldhausen	18.3	Rl	11.1	2.8	15.5
W. Herschel	Datchet, Clay Hall, Slough	18.7	Rl	12.0	3.8	15.5
J. Herschel	Slough	18.3	Rl	13.0	4.2	15.6
Tempel	Arcetri	11.0	Rr	13.1	9.8	15.1
d'Arrest	Copenhagen	11.0	Rr	13.3	9.4	15.7
Stephan	Marseille	31.0	Rl	13.5	8.5	15.7
Swift	Rochester (Warner)	16.0	Rr	13.6	9.4	15.9
Various	Charlottesville	26.0	Rr	13.9	11.6	16.1
Marth	Malta	48.0	Rl	13.9	10.4	16.4
Bigourdan	Paris	12.0	Rr	14.1	12.2	15.7
Various	Birr Castle	72.0	Rl	14.3	11.1	16.3

Charlottesville (the Leander McCormick Observatory) and Birr Castle, results for different observers are collected together.

William Herschel developed a scheme of brightness classes to describe the visual appearance. The main levels are B (bright) and F (faint), with the qualifiers e (excessively), v (very), c (certainly) and p (pretty); their doubling marks extreme values, e.g. eeB or eeF (Stephan even used eeeF). This was the common scale for most observers in the nineteenth century (see Section 10.3.2).[16] For fixed instruments (e.g. John Herschel's 18¼" reflector in Feldhausen), differences were caused by atmospheric conditions and the condition of the observer. It is much more difficult to compare different persons, instruments and sites. Even for such an experienced observer as Dreyer this was an unfeasible task. There was no unique and calibrated system for the brightness of non-stellar objects. Terms like integrated magnitude and surface brightness were defined much later.[17]

In some cases an object was classified by different observers and Dreyer could compare their brightness estimates. For instance, he found out that Stephan gave much lower values than Marth. Thus Dreyer incremented Stephan's data by one step: e.g. from eF to vF and from pB to cB (he raised his size estimates too). It is interesting to compare the brightness classes with modern visual magnitudes. Figure 10.12 shows, as expected, a rough correlation. However, the standard deviations are particularly large, leading to significant overlaps between classes. So, the faintest NGC object, the 16.4-mag galaxy NGC 4042, was given an estimated brightness of 'vF' by its discoverer Marth, though it might deserve 'eeeF'. This shows the influences of aperture and site: 48" on Malta.

A similar graph about the size class versus angular diameter cannot be given, since the visual estimate depends too much on the atmospheric conditions. Additionally, in most cases only a small part of an extended object could be seen, namely the region of greatest surface brightness. Only photography can give a reliable size. Examples of such discrepancies are the California Nebula (NGC 1499) and the North America Nebula (NGC 7000), whose brightest parts were first seen visually by William Herschel and Barnard, respectively. For visually small galaxies one might see only the central portion ('bulge'), since spiral arms are difficult to see. This results in a significant underestimation of the true size – especially in the discovery situation. For example, the spiral structure of M 51 was not detected until 1845 with the Birr Castle 72-inch (see Section 11.3.1). Because we now know about spiral

---

[16] Occasionally John Herschel used star magnitudes that strongly deviate from modern values; see Hearnshaw (1992) and the comparison between the scales of J. Herschel and Argelander (W. Struve) in Bigourdan (1917b: E144).

[17] These data are found in the author's 'Revised NGC/IC'.

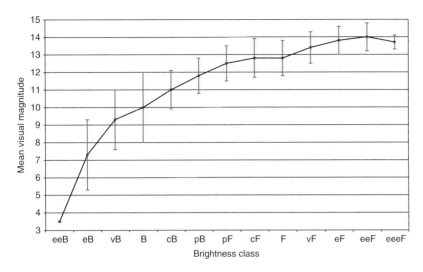

Figure 10.12. Correlation of brightness class and mean of visual magnitude.

arms, they are visible today in a 12″ reflector. Planetary nebulae are in most cases much larger than their visual image; there are, for instance, extended haloes around M 27, M 57 and NGC 6543.

10.2.8 Missing data

For a few objects there are no data about the discovery date (Table 10.19), telescope size and publication. For 34 NGC objects a reference is missing (Table 10.20); in these cases Dreyer was probably informed by letter. As mentioned in the introduction to the NGC, this is the case for Lohse (leading the list with 17 objects), Barnard, Tempel and Winnecke. Unfortunately, Dreyer's estate could not be located.

For 19 NGC objects the discovery date is missing (for many others the year was estimated). Most of them were found by John Herschel at the Cape of Good Hope (during the years 1834–38) and are in the LMC (about their designation see Section 5.1). Barnard's two open clusters were found in Nashville (they are listed again in Table 10.20).

In some cases the telescope aperture is unknown, which concerns refractors only. Moreover, this mainly applies to observers prior to William Herschel: Bode, de Chéseaux, Halley, Hodierna, Ihle, Kirch, Koehler, Legentil, Mairan and Peiresc. As usual at that time, only the focal length was given (normally in feet). However, for a few later observers information regarding the aperture is lacking too: this concerns Caroline Herschel (NGC 2349 and NGC 2360), Bruhns (NGC 2175), Copeland (NGC 3736) and Barnard (NGC 2141 and NGC 2226).

10.3 CORRECTIONS AND ADDITIONS TO THE NGC

10.3.1 Dreyer's Index Catalogues

As has already been mentioned, the period after William and John Herschel brought a state of idleness, because the astronomers did not believe that they could discover any new objects. This was different in the case of the New General Catalogue in 1888. In the following years there was a continual output of corrections, additions and new objects. So Dreyer's work was a great source of inspiration. Essential contributions were due to Bigourdan, Burnham, Denning, Kobold, Stone and Spitaler. Dreyer noted them in his special edition of the NGC, which is now on display at Armagh Observatory.[18] There one reads *'In this volume are entered all corrections known to me.'*

Here are some examples of corrected objects, starting with the observations of Rudolph Spitaler. He investigated the galaxy NGC 163 in Cetus, which had been

---

[18] This is a common edition of all three publications (NGC, IC I + II) from the *Memoirs* of 1888, 1895 and 1908 (Butler, Hoskin 1987). It must have contained the NGC first, later completed by the Index Catalogues.

Table 10.19. *NGC objects with unknown discovery date*

NGC	GC	h	Discoverer	Type	V	Con.
1767	5062	(123)	J. Herschel	OC	10.6	Dor
1785	1002	(147)	J. Herschel	Star group		Dor
1825	1036	(199)	J. Herschel	OC	12.0	Dor
1913	1120	(356)	J. Herschel	OC	11.1	Dor
1922	5063	(374)	J. Herschel	OC	11.5	Dor
1967	1172	(456)	J. Herschel	OC	10.8	Dor
2037	1240	(593)	J. Herschel	OC	10.3	Dor
2044	1246	(608)	J. Herschel	OC	10.6	Dor
2050	1252	2928	J. Herschel	OC	9.3	Dor
2052	1254	2929	J. Herschel	EN		Dor
2060	1261	(642)	J. Herschel	SNR	9.6	Dor
2074	1272	2942	J. Herschel	EN		Dor
2081	1279	2951	J. Herschel	OC		Dor
2091	1289	2957	J. Herschel	OC	12.1	Dor
2092	1290	2962	J. Herschel	OC		Dor
2096	1294	(725)	J. Herschel	OC	11.3	Dor
2102	1300	(730)	J. Herschel	OC	11.4	Dor
2141			Barnard	OC	9.4	Ori
2226			Barnard	OC		Mon

discovered by William Herschel and was found again by Lewis Swift on 9 August 1886 (Spitaler 1892). Another such object was the galaxy NGC 1186 in Perseus (Spitaler 1891b).[19] The Vienna observations of 1890–92 made with the 27" Grubb refractor appeared in 1896 and contain a large number of notes about NGC objects (Spitaler 1896). Two years later Friedrich Bidschof published a second compilation; another followed in 1900.[20] In 1893 Hermann Kobold published his corrections, titled 'Berichtigungen zum New General Catalogue of Nebulae'.[21] He had observed the galaxies NGC 2332 and NGC 2340 in Gemini and its vicinity with the 18" Merz refractor in Straßburg. Another report followed a year later (Kobold 1894). It contains the solution of the puzzle about the Copeland Septet in Leo (see Section 8.18.10). Vincenzo Cerulli, the owner of an observatory in Teramo, Italy, reported an 'Erratum in Dreyer NGC. of nebulae', concerning the galaxies NGC 5898 and NGC 5903 in Libra (Cerulli 1896). He had purchased the 15.5" refractor of Wigglesworth Observatory, Scarborough.

Dreyer especially watched the publication of new nebulae and star clusters with great attention. Interesting new objects were still being found in the region of the northern pole. For instance, Denning discovered 10 objects at high declinations while comet-seeking with his 10" reflector in 1889 and 1890.[22] Altogether the output was so large that as early as in 1895 a first additional listing with 1529 new objects was published by Dreyer. It is titled 'Index catalogue of nebulae found in the years 1888 to 1894, with notes and corrections to the New General Catalogue', here abbreviated as IC I (Dreyer 1895). Most of the new objects had been discovered by Stéphane Javelle (Nice, 794), Lewis Swift (Echo Mountain, 315) and Guillaume Bigourdan (Paris, 134). Dreyer used the opportunity to include known corrections and additional data for NGC objects (notes to 134 cases are given).

---

[19] On 26 May 1884 Spitaler found the galaxies NGC 6446 and NGC 6447; it was soon noticed that they had first been seen by Marth (Weiss 1884).

[20] See Bidschof (1898, 1900); Bidschof discovered two IC objects: IC 1872 and IC 2951.

[21] 'Corrections to the New General Catalogue of nebulae' (Kobold 1893).

[22] Denning (1890, 1891b: 342).

10.3 Corrections and additions 461

Table 10.20. *NGC objects without source (sorted by discoverer)*

Discoverer	NGC	GCS	Year?	Type	V	Con.	Remarks
Barnard	2141			OC	9.4	Ori	
Barnard	2226			OC		Mon	
Coggia	6952	6250	1877	Gx	11.0	Cep	NGC 6951, L. Swift 14.9.1885
Copeland	3736		1885	Gx	14.9	Dra	
Lohse	793		1886	2 stars		Tri	
Lohse	1456		1886	2 stars		Tau	
Lohse	1655		1886	2 stars		Tau	
Lohse	1674		1886	NF		Tau	
Lohse	1675		1886	NF		Tau	
Lohse	2195		1886	2 stars		Ori	
Lohse	2412		1886	2 stars		CMi	
Lohse	2518		1886	Gx	13.6	Lyn	
Lohse	2519		1886	Star	14.3	Lyn	
Lohse	2565		1886	Gx	12.6	Cnc	
Lohse	4345		1886	Gx	12.0	Dra	NGC 4319, W. Herschel 10.12.1797 (I 276)
Lohse	5884		1886	2 stars		Boo	
Lohse	6344		1886	2 stars		Her	
Lohse	6353		1886	3 stars		Her	
Lohse	6731		1886	Star group		Lyr	
Lohse	6767		1886	2 stars		Lyr	
Lohse	6792		1886	Gx	12.5	Lyr	
Pechüle	78	5094	1876	Gx	13.3	Psc	
Peters	3492		1880	Gx	13.2	Leo	
Peters	4309		1881	Gx	12.8	Vir	
Peters	4604		1883	Gx	13.8	Vir	
Tempel	47		1886	Gx	13.1	Cet	NGC 58, L. Swift 21.10.1886
Tempel	54		1886	Gx	13.7	Cet	L. Swift 21.10.1886
Tempel	2700	5437	1877	Star	15.2	Hya	
Tempel	2702	5436	1876	Star		Hya	
Tempel	2703	5438	1876	2 stars		Hya	
Tempel	2705	5439	1876	Star		Hya	
Tempel	2707	5440	1876	Star	14.9	Hya	
Winnecke	2146	5357	1876	Gx	10.5	Cam	Tempel 1876
Young	6187		1885	Gx	14.6	Dra	

In 1908 Dreyer's 'Second index catalogue of nebulae and clusters of stars containing objects found in the years 1895 to 1907, with notes and corrections to the New General Catalogue and to the Index Catalogue for 1888–94' appeared (Dreyer 1908).[23] This work (called IC II here) contains 3857 new objects, most of them

---
[23] The occasionally given publication year of 1910 is wrong.

Table 10.21. *The two parts of the Index Catalogue*

Catalogue	Date	Objects	NGC notes	IC notes
IC I	10.1.1895	1529	134	
IC II	4.5.1908	3857	542	57
Sum		5386	676	57

found by Max Wolf (Heidelberg, 1117), Delisle Stewart (Arequipa, 672), Stéphane Javelle (Nice, 637), Royal Frost (Arequipa, 454), Lewis Swift (Echo Mountain, 270) and Guillaume Bigourdan (Paris, 188). The IC II gives notes to 542 NGC objects, which are mainly based on micrometrical observations by Herbert Howe at Chamberlin Observatory in Denver. Barnard, Bigourdan, Frost, Stewart and Max Wolf contributed too. Table 10.21 shows the content of both Index Catalogues (the 'sum' is commonly called the IC). Dreyer held the epoch of the NGC (1860) constant, but gives the differences relative to 1900. Initially the designation of IC objects was not unique; e.g. 'NGC II' was used.[24]

A large fraction of the IC objects was found by photography. It is interesting that the IC contains bright objects that were discovered visually after 1888. Table 10.22 shows examples of those 82 which were brighter than 12 mag (there are even 2270 brighter than Herschel's magnitude limit of 15.5 mag). Obviously the sky still afforded some surprises! Among them are galaxies, e.g. IC 342 in Camelopardalis, which was found by Denning (Fig. 10.13 right), but planetary nebulae too, like those of Robert Aitken, Walter Gale, Thomas Espin and Edward E. Barnard. Aitken contributed one IC object; Espin found 20. Gale's PN (his only IC object) bears two IC-numbers: IC 5148 and IC 5150 (Fig. 10.13 left).[25] Robert Innes[26] discovered 13 objects, Joseph Lunt 7, Herbert Howe 60, William Denning 18 and Rudolph Spitaler 64. Finally, Barnard was the most successful, with 155 discoveries.

Additionally there are 206 IC objects discovered before the end of 1887, the press date of the NGC (Table 10.23). The reasons are as follows: (a) for Dreyer their publication came too late or did not come to his notice in due time, e.g. discoveries by Bigourdan and Safford; and (b) he had not studied matters sufficiently carefully, overlooking objects or assigning priorities erroneously; e.g. in the cases of Dunlop and Webb. This had already happened in the publication of the Birr Castle observations (1880) and concerns objects of Hunter, Mitchell and B. Stoney.

The most interesting case is IC 4703, the Eagle Nebula in the star cluster M 16 (NGC 6611), which was detected by Trouvelot in 1876. Often the cluster is identified with the nebula in the literature, but both of the discoverers of M 16 (de Chéseaux in about 1745 and Messier on 3 June 1764) speak only about an agglomeration of stars. Though the latter mentions that the object '*appears nebulous in a small telescope*' and '*enmeshed in a faint glow*', this is due to the low resolution – a common issue in nebular history. Otherwise William Herschel would have noticed the nebula in his 12" reflector at Datchet on 20 July 1783. His description reads '*Large stars with small ones among them; within a small compass I counted more than 50, and there must be at least 100 without taking in a number of straggling ones, everywhere dispersed in the neighbourhood.*' This is the origin of John Herschel's description in the General Catalogue (GC 4400): '*Cl; at least 100 st L & S*'. His own observation of the cluster is dated 31 August 1826, albeit giving '*no description*' (h 2206). The GC note was copied by Dreyer for NGC 6611 (= M 16).

The nebula in which M 16 is embedded was catalogued as IC 4703 by Dreyer ('*B, eL, Cl M. 16 inv*'). Isaac Roberts is mentioned as the discoverer. His first plate was made in 1894 with the 20" reflector (often a date of 'August 1875' is erroneously given), but the true discoverer – and, moreover, visually – was Etienne Trouvelot. He noticed the nebula in 1876 with the 26" refractor of the US Naval Observatory, Washington. The observation is described in his manual on 'Astronomical drawings'. There one reads the following: '*The star cluster, No. 4,400, in Scutum Sobieskii, which is described by Sir J. Herschel as a loose cluster of at least 100 stars, I have*

---

[24] For instance, IC 1613 was called 'NGC II 1613' by Baade (Baade 1928).

[25] Hoffmeister (1961).

[26] In 1913 Innes discovered Proxima Centauri, the nearest star.

Table 10.22. *Bright IC objects that were discovered visually*

IC	Discoverer	Date	Telescope	Type	V	Con.
342	Denning	19.8.1892	10" Rl	Gx	8.4	Cam
356	Barnard	23.8.1889	12" Rr	Gx	10.6	Cam
529	Denning	1890?	10" Rl	Gx	11.3	Cam
334	Denning	30.9.1891	10" Rl	Gx	11.9	Cam
361	Denning	11.2.1893	10" Rl	OC	11.7	Cam
3568	Aitken	31.8.1900	12" Rr	PN	10.6	Cam
166	Denning	1890?	10" Rl	OC	11.7	Cas
4329	Howe	21.6.1900	20" Rr	Gx	11.1	Cen
4291	Innes	1901	7" Rr	OC	9.7	Cen
2163	Howe	11.2.1898	20" Rr	Gx	11.7	CMa
1369	Pechüle	27.4.1891	11" Rr	OC	8.8	Cyg
2157	Espin	11.1.1899	17.3" Rl	OC	8.4	Gem
5201	Lunt	1900	18" Rr	Gx	10.8	Gru
5181	Lunt	1900	18" Rr	Gx	11.5	Gru
5148	Gale	4.6.1894	8.5" Rr	PN	11.0	Gru
1954	Innes	1898	7" Rr	Gx	11.4	Hor
2035	Innes	1898	7" Rr	Gx	11.5	Hor
2482	Howe	14.3.1899	20" Rr	Gx	11.5	Hya
4351	Innes	1898?	7" Rr	Gx	11.8	Hya
1434	Espin	1893?	17.3" Rl	OC	9.0	Lac
1442	Espin	1893?	17.3" Rl	OC	9.1	Lac
4662	Innes	1901	7" Rr	Gx	11.1	Pav
2003	Espin	18.1.1907	17.3" Rl	PN	11.4	Per
351	Barnard	5.12.1890	36" Rr	PN	11.9	Per
2311	Howe	16.2.1898	20" Rr	Gx	11.5	Pup
2056	Innes	1898	7" Rr	Gx	11.7	Ret
750	Spitaler	22.4.1892	27" Rr	Gx	12.0	UMa

Figure 10.13. Two bright IC objects. Left: Gale's planetary nebula IC 5148 in Grus; right: Denning's galaxy IC 342 in Camelopardalis (DSS).

Table 10.23. *The IC objects discovered before 1888*

Discoverer	Objects	Discovery	Publication	Remarks
Bigourdan	67	1884–87	1892–1912	*Ann. Obs. Paris*
de Chéseaux	2	1745?	1917	IC 4665, OC (Oph); IC 4725 = M 25, OC (Sgr)
Dunlop	10	1826	1826	Paramatta
Engelhardt	1	1886	1890	IC 1463, 2 stars (Aqr); Fig. 10.14 left
Finlay	5	1883–87	1898	Cape Observatory
Herschel J.	1	1827	1833	IC 344, Gx (Eri)
Herschel W.	1	1786	1789	IC 434, EN (Ori)
Lacaille	3	1751	1761	IC 2391, IC 2395, IC 2488 (all: OC in Vel)
Leavenworth	4	1887	1886	Leander McCormick Observatory
Marth	1	1864	1867	IC 1238, 2 stars (Her)
Messier	1	1764	1771	IC 4715 = M 24, Sagittarius Star Cloud (Sgr)
Mitchell	1	1857	1880	IC 1559, Gx (And), near NGC 169
Muller	10	1887	1890	Leander McCormick Observatory
Safford	44	1867	1887	Dearborn Observatory
Skinner	3	1867	1887	IC 138, IC 210, IC 1528 (all: Gx in Cet)
Stone	5	1887	1890	Leander McCormick Observatory
Stoney B.	8	1851	1880	Birr Castle
Swift E.	2	1885–86	1888	IC 1129, IC 1145 (both: Gx in UMi); Warner Obs.
Swift L.	26	1886–87	1886–90	Warner Observatory
Tempel	1	1877		IC 1271, EN (Sgr); sketch only
Thome	2	1887?	1892	IC 229, NF (For); IC 1290, star group (Sgr), Fig. 10.14 right
Todd	3	1877	1885	IC 591, IC 622, IC 669 (all: Gx in Leo)
Trouvelot	1	1876	1882	IC 4703 (Eagle Nebula in M 16)
Voigt	3	1865	1987	IC 3136, IC 3155, IC 3268 (all: Gx in Vir)
Webb	1	<1859	1859	IC 4756, OC (Ser)

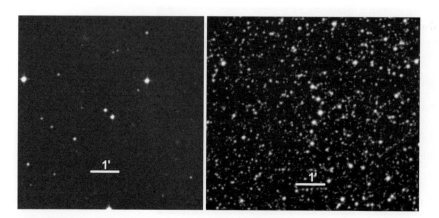

Figure 10.14. Left: von Engelhardt's star pair IC 1463 in Aquarius; right: the star chain IC 1290 in Sagittarius found by Thome (DSS).

found to be involved in an extensive, although not very bright, nebula, which seems to have escaped his scrutiny. In a study and drawing of this nebula made in 1876, its general form is that of an open fan, with the exception that the handle is wanting, with deeply intended branches on the preceding side, where the brightest stars of the cluster are grouped. From this peculiar form, this object might appropriately be called the Fan nebula.[27] Obviously this name of the nebula was not generally accepted. It is now known as the Star-Queen Nebula (after Robert Burnham Jr.) or Eagle Nebula (a name of unknown provenance).

10.3.2 Later publications and photographic surveys

The most important work is probably Bigourdan's 'Observations de nébuleuses et d'amas stellaires', which was first published between 1892 and 1917 in the *Annales de l'Observatoire de Paris* (see Section 9.22.3).[28] It contains observations of 6380 NGC objects. As they became available, Dreyer used the results for the Index Catalogues.

Jermaine Porter, Director of Cincinnati Observatory, published two series of micrometrical observations of nebulae. The first contains positions measured during the years 1884–86 with the old 11" refractor[29] (Porter 1891). The sketches of 96 objects presented in the appendix are of interest. During the years 1905–10 a second campaign was carried out (Porter 1910), this time using the new 16" Clark refractor which had been erected in spring 1904 on Mt Lookout. 669 objects were observed, among them 31 by Swift (who detected significant position errors) and 19 'novae'. However, 16 of the newly found objects were already in the NGC, one is listed in the first Index Catalogue (Burnham's IC 168 in Cetus), one is a faint double star and only one is actually new. This is UGC 492, a galaxy in Cetus, which was probably the first object (of only a few) to be found visually after the Second Index Catalogue.

In 1911 Carl Wirtz published the Straßburg observations as 'Generalkatalog der am großen Refraktor in Straßburg beobachteten Nebelflecke 1881–1910'.[30] This completed the study of nebulae that had been started by Winnecke. The catalogue contains the positions of 1257 objects for the epoch 1900, among them all of those discovered in Straßburg, mainly with the 48.7-cm Merz refractor. Wirtz compares the results with other works, listing corrections too (Wirtz 1911c).

Dreyer's last contribution appeared in 1912. The RAS had ordered an edition of the complete work of William Herschel. The result was published in 1912 as *Scientific Papers of Sir William Herschel* in two volumes (Dreyer 1912a). It contains a revision of all three Herschel catalogues. During the making of his edition Dreyer noticed a number of errors in the NGC, which were also published as 'Corrections to the New General Catalogue resulting from the revision of Sir William Herschel's three catalogues of nebulae' (Dreyer 1912b). The list appeared in November 1912 and contains notes for 117 NGC objects. He also used plates taken in 1911 with the 30" Grubb reflector in Greenwich. Dreyer had asked the Astronomer Royal William Christie for the photographs to verify Herschel's observations of 2 April 1801 (Christie 1911). They show 40 nebulae, among them a few new ones.

Even after Bigourdan and despite the dominance of photography, systematic visual observations of NGC objects were made. The last representative of this tradition was the Jesuit Johann Georg Hagen, Director of the Vatican Observatory in Castelgandolfo. He planned the observation of a large fraction of NGC objects with the 16" Zeiss refractor in order to develop a numerical visual brightness scale for non-stellar objects similar to that for stars. The study was also intended to lead to a revision of the NGC by creating a uniform database. Hagen started in 1911 and observed about 5800 objects.

The result was published 1922 as 'A preparatory catalogue for a Durchmusterung of nebulae. The zone catalogue' (Hagen 1922). In the introduction Hagen wrote that *'The revision of the NGC will, at the same time, answer a desideratum of Dr. Dreyer, who is better qualified than any other to state the needs in this branch of astronomy.'* However, the *'discovery of new nebulae [was] strictly excluded from the programme as hindrance to its*

---

[27] Trouvelot (1882: 151).

[28] In 1919 Bigourdan published a five-volume version.

[29] The former 12" lens (Merz & Mahler) was replaced by an 11" Clark objective.

[30] 'General catalogue of nebulae observed with the great refractor in Straßburg 1881–1910' (Wirtz 1911b). For the observations made during the years 1902–11 see Wirtz (1911a).

completion'. He added *'However, the principal result of the present Catalogue is the establishment of the much needed numerical scales, one for the light of the bright nebulae and another for the density of the obscure cosmic clouds. Both scales may, in the course of time, be perfected by photometric results, but one of their immediate efforts will be the abolition of verbal descriptions that have been in use for over a century.'*

The NGC objects are arranged in declination zones of width 10°, thus the name 'Zone catalogue'. No positions are given (or measured). The catalogue columns are NGC-number, discoverer, size, visual brightness, description and remarks. For each NGC object there is a parallel entry on the table 'Obscure nebulae'. These are 'nebulous clouds' at the object's place, as noticed by Hagen. Their 'density' is given on a five-level scale, from I ('thinnest veil, almost imperceptible') to V ('densest cloud, almost obscuring all stars'). In 1921 Hagen first reported about 'Die dunklen kosmischen Nebel'.[31] Carl Wirtz later noted that *'There are doubts about which subject is meant by the observations of the Vatican Durchmusterungen.'* (Wirtz 1925). He interpreted Hagen's density scale as *'a measure of the starlight, i.e. the apparent brightness of the dark sky background'*. The existence of these 'Hagen clouds' is, however, very doubtful.[32]

Later Friedrich Becker converted Hagen's zone catalogue into a new form. Using the NGC positions (regarding known corrections), he sorted the objects by right ascension for the epoch 1925. The result was published in 1928 as 'A preparatory catalogue for a Durchmusterung of nebulae. The general catalogue' (Becker F. 1928).

Besides Hagen's visual magnitude scale, there were others. As early as in 1907 Johann Holetschek published his 'Beobachtungen über den Helligkeitseindruck von Nebelflecken und Sternhaufen'.[33] The visual observations were made during the years 1896–1902 with the 6" Fraunhofer refractor of the University Observatory Vienna. Holetschek had selected objects of William Herschel's class I and estimated their brightness relative to the stars (mainly from the Bonner Durchmusterung). Altogether 300 nebulae and star clusters were observed under uniform conditions. A similar campaign was 1911–13 by Wirtz at the 48.7-cm Merz refractor of Straßburg University Observatory, but using a photometer. The result appeared in 1923 under the title 'Flächenhelligkeiten von 566 Nebelflecken und Sternhaufen'.[34] This and other works were discussed by Becker in his 1928 publication.

The first photographic investigation of NGC/IC objects was made by Solon Bailey in 1896. Using a Cooke lens of diameter only 1" he photographed the sky in Arequipa, at the Peruvian station of Harvard College Observatory. The result appeared in 1908 as 'A catalogue of bright clusters and nebulae' (Bailey 1908). On the images, 263 nebulae and star clusters could be identified, including 252 NGC objects and 6 IC objects (among them the open clusters IC 4665 and IC 4756).[35] Two of the remaining five are the Pleiades (M 45) and the Hyades.

In 1914 Joseph Hardcastle published a study based on the 'Franklin-Adams charts' (Hardcastle 1914).[36] The 206 plates, covering the whole sky, had been taken during the years 1902–10 by the British amateur astronomer John Franklin-Adams with a 10" astrograph. Hardcastle, supported by Grace Cook, identified 785 NGC/IC objects, describing them as 'spiral', 'elongated', 'diffused' or 'small'. It was Arthur Hinks who had the idea for the project. On the basis of the NGC/IC data and information in known studies (such as Bailey's), he had already analysed the distribution of nebulae and star clusters (Hinks 1911a, b). With this project Hinks continued the work of Abbe, Proctor and Waters (see Section 7.6.5).

The first to study non-stellar objects with high photographic resolution was Heber Curtis, using plates taken during the years 1898–1918 with the 36" Crossley reflector at Lick Observatory. He published the results

---

[31] 'The dark cosmic clouds' (Hagen 1921).

[32] See Becker and Meurers (1956), Latußeck (2009) and also Section 11.6.14.

[33] 'Observations on the apparent brightnesses of nebulae and star clusters' (Holetschek 1907). Later Kasimir Graff studied his objects by photometry, using the 60-cm refractor in Hamburg-Bergedorf (Graff 1948).

[34] 'Surface brightnesses of 566 nebulae and star clusters' (Wirtz 1923).

[35] IC 4665 (Oph) was found by Caroline Herschel on 31 July 1783, and IC 4756 (Ser) was discovered in 1859 by Webb.

[36] Hardcastle was designated as Dreyer's successor at Armagh Observatory; unfortunately, he died shortly before assumption of office (see Section 8.1.4).

in 1918 as 'Descriptions of 762 nebulae and clusters photographed with the Crossley reflector' and 'The planetary nebulae' (Curtis 1918a, b). Most of the objects were selected from the NGC.

In 1912 Max Wolf charged his assistant Adam Massinger with carrying out a photographic study of GC and GCS objects north of −20° declination. Unfortunately, before the work had even been started, Massinger died in 1914 near Ypres (Belgium) in the First World War. It was Karl Reinmuth who eventually took over the task in 1919, using plates taken with the 40-cm Bruce refractor, the 71-cm Waltz reflector and Max Wolf's private 16-cm refractor. The final catalogue, titled 'Die Herschel-Nebel', appeared seven years later.[37] It contains 4445 NGC objects, sorted by right ascension (epoch 1875). The following data are given: NGC-number, position (taken from the NGC), galactic longitude and latitude, type (according to Wolf's scheme), size ($a \times b$), position angle and description (based on the classic Herschel abbreviations). Reinmuth's work brought great progress, due to its homogeneity and completeness.

A first attempt to extend the NGC/IC was made by Knut Lundmark. On the basis of Mt Wilson images, a catalogue of 16 000 objects, mainly galaxies, was derived. For each object a file card was created; about 35 000 were planned. In 1930 Lundmark described the huge project in his report 'A new general catalogue of nebulae' (Lundmark 1930). He wanted to give the photographic magnitude, size, position angle, concentration and type for each object. Lundmark's aim was a 'General catalogue of nebulae of the Observatory of Lund' (L.G.C.), consisting of two parts: an index with data and a compilation of observations and literature. Unfortunately, the ambitious project was soon terminated; apart from some file cards, nothing remains of it.

In the early 1930s the Harvard College Observatory contributed important efforts on the identifications of NGC/IC objects. In 1930 Adelaide Ames published 'A catalogue of 2778 nebulae including the Coma–Virgo group' (Ames 1930). The work identifies 214 NGC and 342 IC objects in the area of the Virgo Cluster. Two years later she published 'A survey of external galaxies brighter than the thirteenth magnitude' in collaboration with director Harlow Shapley (Shapley and Ames 1932). The important Shapley–Ames Catalogue contains 1249 galaxies of the whole sky; 1189 are in the NGC and 48 in the IC.

In the late 1930s Edwin Hubble charged his assistant Dorothy Carlson with the revision of the NGC/IC on the basis of plates taken with the 60" and 100" reflectors at Mt Wilson Observatory. About 2000 NGC and 700 IC objects were investigated. Carlson also used the publications of Reinmuth, Curtis and Ames. The result appeared as 'Some corrections to Dreyer's catalogues of nebulae and clusters' (Carlson 1940). Alongside the publications of Dreyer and Bigourdan, Carlson's compilation was for a long time the most important source on corrections to the NGC/IC. She gives notes for 717 objects, divided into four categories (Table 10.24), and suggests the omission of many entries. The reasons were (a) identification with stars, (b) identity with another object and (c) the object was not found at the given position. However, 98 objects termed 'missing' by Dreyer in the NGC/IC notes were not validated by Carlson.

10.4 REVISIONS OF THE NGC

Ever since the appearance of Dreyer's catalogues, non-stellar objects have mainly been named by their NGC- or IC-number. It is, however, a rather elaborate undertaking to consult the original work published in the *Memoirs*. Fortunately the RAS created (albeit pretty late) a collected edition in 1953.[38] It was re-published, without any changes, in 1962 and 1971.

Authors of many modern catalogues have tried to cross-identify their objects with respect to the NGC/IC. The main works are those of Boris Vorontsov-Velyaminov, Fritz Zwicky, Peter Nilson and Andris Lauberts, all of which are related to inspections of the *Palomar Observatory Sky Survey* (POSS) and its southern extension.

No doubt these catalogues made NGC objects much more popular, but it became necessary too, to prove their identity reliably for a correct identification. The problem was particularly evident in digital versions of the modern catalogues and their computer visualisation (later in the Internet). Owing to growing confusion, 'clusters of identical objects' appeared

---

[37] 'The Herschel nebulae' (Reinmuth 1926).

[38] This is the main reference for NGC/IC citations (here termed 'Dreyer (1953)').

Table 10.24. *Dorothy Carlson's corrections to the NGC/IC*

Category	NGC	IC
Objects to be struck out from the NGC	331	
Objects to be struck out from the IC		314
Corrections to Dreyer's positions	43	6
Nebulae that had reported not found, but which appear on Mt Wilson plates in Catalogue positions	22	1
Sum	396	321

on the screen – a result of lacking or incorrect identifications. At least the NGC (as the basic catalogue) should be 'clean' – a satisfying solution was achieved only recently.

10.4.1 Modern catalogues of non-stellar objects

The trail-blazers were catalogues of galaxies.[39] This is not surprising, since 77% of all NGC objects are galaxies. In 1964 Gérard de Vaucouleurs published the *Reference Catalogue of Bright Galaxies* (RC1). The subtitle ('being the Harvard survey of galaxies brighter than the 13th magnitude of H. Shapley and A. Ames, revised, corrected, and enlarged, with notes, bibliography and appendices') already demonstrates the ambition of this revolutionary compilation. The work on the RC1 started in 1949. The final catalogue contains 2599 objects, which is twice the number of the original. In 1981 Allan Sandage and Gustav Tammann published a *Revised Shapley–Ames Catalogue* with modern data on 1249 galaxies (including about 90 images of NGC/IC objects).

Between 1962 and 1974 Boris Vorontsov-Velyaminov and his co-workers published the data of 31 917 galaxies north of −45° declination in the monumental *Morphological Catalogue of Galaxies* (MCG). A similar study was presented by Fritz Zwicky and his co-workers, listing between 1963 and 1968 altogether 29 378 galaxies and 9133 galaxy clusters north of −3.5° declination in the *Catalogue of Galaxies and of Clusters of Galaxies* (CGCG).

Peter Nilson's *Uppsala General Catalogue of Galaxies* (UGC), which appeared in 1973, followed with 12 940 galaxies north of −2.5° declination. The first complete Durchmusterung of the southern sky was finished in 1982. Andris Lauberts and his co-workers catalogued 18 438 non-stellar objects south of −17.5° declination. The resulting *ESO/Uppsala Survey of the ESO(B) Atlas* contains 16 019 galaxies.

However, these catalogues were not intended to give historically correct identifications of NGC/IC objects. They mainly refer to the data given by Dreyer. Of course, this caused errors, which were continuously copied. Only Nilson corrected a few of those appearing in the MCG and CGCG. In the appendix of the UGC he gives an interesting review, titled 'Catalogues of galaxies – a short survey of their history'.[40] Another positive exception is the *Second Reference Catalogue of Bright Galaxies* (RC2) published by de Vaucouleurs in 1976, in which, for the first time, an identity check for bright galaxies was performed. The RC2 also introduced additional NGC objects in the vicinity of known ones, designated by suffices (A, B, C, …). An example is NGC 337A, a companion of NGC 337 in Cetus (Fig. 10.15). Unfortunately, this method – especially when imitated by other authors – led to some confusion.

In 1980 Dixon and Sonneborn published their bulky *Master List of Nonstellar Optical Astronomical Objects* (MOL). Actually, this is a computer print-out of a mix of 270 catalogues, sorted by right ascension (1950). The MOL contains all non-stellar objects known up to 1980; thus the work is something like a modern version of the NGC. The 'Master list' has almost 180 000 entries. For this task, the original NGC

---

[39] References to modern catalogues can be found in Steinicke and Jakiel (2006). For individual objects see the NASA Extragalactic Database (NED), http://nedwww.ipac.caltech.edu.

[40] Nilson (1973: 449–455).

Figure 10.15. NGC 337A, a companion of NGC 337 in Cetus.

and IC were punched in for the first time. Since there are no cross-identifications, many objects appear several times and – due to slightly differing coordinates – not always at adjacent list positions. Anyway, due to its completeness, the MOL is an important tool with which to search for and identify objects – however, a keen eye is needed at the desktop.

In 1989 George Paturel and his co-workers published the *Catalog of Principal Galaxies* (PGC), which is a collection of all known galaxy catalogues. Unfortunately, the cross-identifications to NGC/IC objects are not always reliable.[41] The PGC was later used for de Vaucouleur's *Third Reference Catalogue of Bright Galaxies* (RC3), which appeared in 1991. This important work gives precise data and references for 23 022 galaxies.

10.4.2 RNGC and NGC 2000.0

The first catalogue with the aim of updating the original NGC was the *Revised New General Catalogue* (RNGC), presented by Jack Sulentic and William Tifft in 1977 (Sulentic and Tifft 1977). However, the attempt to improve Dreyer's work with modern data, especially positions derived from the POSS, was a failure. Obviously due to time pressure, the authors not only ignored published corrections but also introduced a considerable number of new errors. An example, demonstrating the confusion, is the case of the Copeland Septet (see Section 8.18.10). The well-known compact galaxy group in Leo appears as 'nonexistent' in the RNGC.[42] Moreover, at places where no object could be found (mainly due to wrong positions), often an RNGC-number was assigned to the nearest 'anonymous' candidate – without consulting the original data or published corrections. Now some plate flaws on the POSS bear RNGC-numbers! The catalogue of Sulentic and Tifft contains additional objects too, marked by a suffix (most of them were taken from the RC2). The limited value of the RNGC might also be caused by the infelicity of the chosen epoch 1975 – obviously a compromise between 1950 and 2000.

In 1988 Roger Sinnott published the first revision of the complete NGC/IC. His *NGC 2000.0* appeared just in time for the centennial of the New General Catalogue (Sinnott 1988). Obviously, this caused time-pressure too. Unfortunately, the original observations and published corrections were ignored in favour of modern data taken from the CGCG and MCG, for instance. However, due to the fact that these modern catalogues are not reliable sources for the NGC/IC, many historical errors entered the NGC 2000.0. The old Dreyer positions were used without verification and precessed to the epoch 2000. Sinnott ordered the 13 226 objects by right ascension – an awkward presentation. Since many of the positions are not accurate or even are doubtful, this does not make any sense. Any correction would lead to a different order. Moreover, the natural

---

[41] At present, the database contains over 3 million objects (http://leda.univ-lyon1.fr).

[42] There is even a treatise on *The 'Non-existent' Star Clusters of the RNGC* (Archinal 1993).

sequence of objects is broken and the NGC and IC are now mixed, which makes any search difficult (use of the index given in the appendix is needed). In spite of these defects, especially the lack of diligence, Sinnott's NGC 2000.0 was more successful than the RNGC. To sum up: both of these publications are not reliable sources for NGC objects.

10.4.3 The 'Revised New General and Index Catalogue' and the 'NGC/IC Project'

Since the 1970s the author has been occupying himself with data of non-stellar objects. Early results are the 'Catalogue of galaxy groups' and the 'Catalogue of bright quasars and BL Lacertae objects'. During this work again and again errors or gaps in the catalogues used were noticed, especially in the NGC and IC. The RNGC of 1977 was no real help and the IC remained 'terra incognita'. On the basis of the MOL by Dixon and Sonneborn, a first 'Revised Index Catalogue' was compiled in 1982, still in analogue form. The digital age started in 1987, when magnetic tapes of the most important catalogues were ordered from the Centre du Données Stellaires (CDS) in Strasbourg.[43] The data were analysed and, after extensive cross-checking, a database with about 150 000 non-stellar objects north of $-30°$ declination has been created, accompanied by self-made visualisation software. This was very helpful for solving NGC/IC identification problems. The result was the first version of the 'Revised NGC/IC' (RNI) in 1997, containing all available data and identifications for 13 775 objects with positions for the epoch 2000. The additional 549 objects bear a suffix (A, B, C, ...). Since then a new version has been produced every year.

With the appearance of the Digitized Sky Survey (DSS), the digital version of the POSS and its southern extension, the 'real sky' was available on the computer.[44] At first only on CD-ROM and later via the Internet, this offered the opportunity to check all NGC/IC objects. For the 1998 version of the RNI about 12 600 positions were newly determined with a precision of 1″. Moreover, the photographic magnitudes, sizes and position angles of a large number of galaxies were revised; many gaps in the data could be filled. The work also revealed a considerable number of errors in modern galaxy catalogues; for instance, more than 800 wrong identifications were detected in the PGC. Vice versa, there are about 700 NGC/IC galaxies that are not listed there.

However, the DSS is not always helpful, especially concerning brightness: most galaxies are shown fainter than they appear visually. This is due to their manifesting different intensities in the photographic and visual region of the spectrum. Therefore it was necessary to check critical cases by visual observation. For this task reflectors of 8″, 14″ and 20″ aperture were used – reproducing the nineteenth-century observations. Since 2002 the RNI has contained visual and surface brightness of galaxies calculated from photographic data with due consideration of type and angular size.

The latest version (2009) of the 'Revised New General and Index Catalogue' has 14 002 entries and presents 42 413 cross-identifications to 79 catalogues.[45] 316 objects are listed as 'not found' (2.3%). This catalogue has been used in many desktop planetarium programs and printed sky atlases. An exception is the *Uranometria 2000.0* and its database, the 'Deep sky field guide' (DSFG). As expected, a thorough analysis has revealed a large number of errors for the NGC/IC objects included (Steinicke 2002e).

In the mid 1990s the international 'NGC/IC Project' was founded. Its goal is the complete revision of Dreyer's original catalogues. It is headed by Harold Corwin (California Institute of Technology, Pasadena).[46] For the author it was exciting to compare his own results with those of the project, which was first done in 2000.[47] Apart from a few differences concerning identification problems, they match very well. Indeed, such 'puzzles' allow different solutions or even cannot be solved due to incomplete historical data. An example is Barnard's IC 919 group in Ursa Major. Fortunately, in many cases a final and satisfying state has obviously been reached.

---

[43] This database is now part of Simbad: http://simbad.u-strasbg.fr/simbad.

[44] http://archive.eso.org/dss/dss.

[45] See the author's website: http://www.klima-luft.de/steinicke.

[46] He is, for instance, co-author of the RC3 (Third Reference Catalogue), the SGC (Southern Galaxy Catalogue) and of a recent book about de Vaucouleurs' classification of galaxies (Buta et al. 2007).

[47] The author has been a core team member since that time; see also Steinicke (2000a).

Besides the author's data, the website of the NGC/IC Project[48] offers three important – and independent – contributions: (a) Harold Corwin's 'Precise position list' and 'NGC/IC puzzle solutions', (b) the 'Identity survey of IC galaxies' by Malcolm Thomson and (c) Steve Gottlieb's 'Corrections to the RNGC', which are based on visual observations of a large fraction of the NGC objects with a 17.5" reflector.

Recently a 'photographic catalogue' of all 13 226 NGC/IC objects has appeared; it is based on the author's revised data (Numazawa and Wakiya 2009). It shows DSS images of all objects with the relevant information in a standardised format. The large-format book is the first printed illustrated version of Dreyer's nebulae and star clusters.

10.4.4 The 'Historic NGC'

Since 2000 (and in parallel with the RNI) the author has occupied himself with the historical sources of the NGC objects. The aim is to determine the publication, discoverer, discovery date, instrument and site for each object. For many this is quite easy (e.g. those from the Herschel catalogue), but often problems occur, due to incomplete data given by Dreyer. At present, only 34 NGC objects are without source (see Table 10.20). The result of the work was first presented in early 2006 as 'Historic NGC' – the only listing of its kind.[49] It includes a compilation of all of the contributing observers: the 'Observers list' cites 172 persons with their biographical data and portraits and pictures of their instruments and observatories.[50] Like the 'Revised NGC/IC' mentioned above, the 'Historic NGC' is subjected to continuous updating. Meanwhile, the 'Historic IC', which first appeared in 2008 on the author's website, has become available.

---

[48] See: www.ngcicdetectives.org.

[49] There still is a 'Historically corrected NGC' in the Internet (www.ngcicproject.org), which is merely a copy of an early version of the 'Historic NGC'. It is presented as an independent work, giving no credit to the author. However, there is overwhelming evidence for the identity. For instance, all errors were copied.

[50] These data can also be found on the author's website: www.klima-luft.de/steinicke.

11 • Special topics

The last chapter covers themes, spreading over a large fraction of the nineteenth century and involving many eminent visual observers. Thus it was impossible to treat them in closed form in the above sections. This concerns observation methods, measurements and presentations of some key objects. The topic 'resolvability of nebulae' is excluded here, because it is strongly related to William Herschel's ideas about the nature and evolution of nebulae and therefore cannot be separated from the previous text.[1] Section 11.1 treats positional measurements, often made by professional astronomers using refractors with micrometers or meridian-circles. Inspired by the success for stars, it was hoped that it would be possible to detect proper motions of nebulae to determine their distances and spatial distribution. The result was negative, since most objects are too far away and the accuracy of the measuring method was too low. Another, much more debated topic is drawings. Section 11.2 describes the main problem: the subjectivity in perception and presentation of non-stellar objects. Prior to photography, textual descriptions or sketches were the only way to capture the appearance. An important goal was the detection of change concerning form or brightness. This attempt failed as well – the differences due to the observers were too large. A special kind of object was discovered by Lord Rosse in the year 1845: the spiral nebula. The first was M 51, a bright galaxy in Canes Venatici. The controversial discussion of the reality of the spiral appearance is presented in Section 11.3. The main protagonists were Tempel and Dreyer. However, there was a remarkable case showing real change: Hind's Variable Nebula in Taurus, which was later catalogued as NGC 1555 by Dreyer. It is associated with the variable star T Tauri. Its history – and that of the neighbouring object NGC 1554 – is discussed in detail in Section 11.4, where all of the important publications, drawings and images are presented. The general theme 'variable nebulae' is treated in Section 11.5. D'Arrest, Dreyer and others contributed important investigations. Many objects were observed, but, as has already been mentioned, the common problem was the subjectivity of visual observing. This can be demonstrated in much detail for another prominent target, the Merope Nebula NGC 1435 in the Pleiades. This was discovered by Tempel in 1859, and became one of the key objects in the nineteenth century. Section 11.6 describes the discussion about its existence, appearance and supposed variability, which lasted for more than half a century, continually being fuelled by new discoveries in and around the open cluster. Finally, photography became an essential element in resolving this issue.

11.1 POSITIONAL MEASUREMENTS

Positional measurements were a central field of work in the nineteenth century. Before the advent of astrophysics this was the veritable domain of professional astronomers. For some, such as Bessel, Airy and Auwers, it was even the main purpose of astronomy. First, stars, minor planets and comets were measured, but in the second half of the century the campaigns were extended to nebulae. However, the first results were undistinguished.

The main instruments were the meridian-circle and the equatorially mounted refractor (favouring the German design). Both had pros and cons. The former yielded accurate absolute positions, but the small refractor restricted the measurement to bright nebulae. With the latter, fainter objects could be observed, but only positions relative to known reference stars were determined using the micrometer eye-piece. Thus, to achieve a reasonable precision in this case, accurate star positions were needed from meridian-circle measurements or perhaps could be taken from published star

---
[1] See Section 6.4.3.

catalogues. Though there was a considerable number, they normally covered only small areas.[2] The publications of the time reflect this problem: even when the number of measurements was small (e.g. when a comet was observed), different catalogues had to be consulted – an elaborate and error-prone undertaking. The situation changed with the Bonner Durchmusterung, whose precision was sufficient for positions of nebulae. From the (given) star position and the relative position of the nebula determined with the micrometer, an absolute position was calculated. This 'reduction' yields object coordinates (right ascension and declination) for a fixed epoch.

The goal of all these campaigns was the detection of proper motions. The astronomers were aware about the period and accuracy needed to notice positional change. William and John Herschel had already done important work in this field. Considering their deficient equipment, the results are astonishingly precise (especially in the case of John Herschel). However, significant progress was not achieved until Laugier and d'Arrest.

The main problem for all observers was the diffuse character of the objects (in contrast to stars), which significantly reduced the accuracy.[3] 'Core-nebulae' were preferred: objects showing a sharply defined centre (ideal targets were planetary nebulae with central stars). Normally, the positional error due to the non-stellar shape was much larger than that caused by the position of the reference star (except when an inaccurate catalogue was used).

Though the precision of nebular measurements increased (especially due to the work of Auwers, Schönfeld and Vogel) and the time interval became longer, proper motions could not be confirmed. According to today's view, the reason is simple: the objects are too far away. Even for a stellar-appearing nebula and maximum precision there is no detectable proper motion after many decades. Fortunately, astrophysics developed new methods – otherwise astronomy would be buried in the past.

In the NGC Dreyer refers to positional measurements in the column 'Other observers', mentioning Vogel, Schönfeld, d'Arrest, Schultz and Engelhardt.[4] Table 11.1 shows the campaigns which were performed

or at least started in the nineteenth century (thus William Herschel is not listed). Additionally, there is a large number of single measurements of nebulae and star clusters.

John Herschel, despite his not being a classical representative of positional astronomy, is listed. He was the first to give precise coordinates for all known nebulae and star clusters in the Slough and Cape catalogues. Except for Lamont and Bigourdan (the leader by quantity), all observers derived absolute positions.

Observers treated in a special section, where their campaigns are described in detail, are marked in bold. In the column 'Type' we have Rr = refractor, Rl = reflector and H = heliometer and M = meridian-circle. The column 'Micro.' gives the tools used: ring-, bar- or lamella-micrometer. From the columns 'Observer' and 'Publ.' the reference can be derived, e.g. Becker L. (1902). Specially marked instances are (a) see Table 9.51; (b) = 1886, 1890 and 1895; (c) = 1865a–d, 1866 and 1867; (d) = 1874 and 1875; (e) Esmiol (1916); and (f) 1886 and 1890.

11.2 DRAWINGS OF NEBULAE: FACTS AND FICTION

Prior to photography, textual descriptions and drawings were the only ways to document the features of non-stellar objects.[5] The problems connected with them have already been mentioned in several sections. Here, fundamental questions are treated, going back to John Herschel.

First, the observation is influenced by exterior factors: telescope, site and atmospheric conditions. Moreover, what the eye sees in the case of a faint object cannot be adequately processed by the brain. Thus the reproduction in the form of a sketch is necessarily subjective. Talent plays a significant role too. A subsequent problem was engraving. What finally appears in the publication often does not bear much resemblance to the impression at the eye-piece. At the end of the sequence, the quality depends on the ability of the engraver.[6] Since good drawers and artists were rare, the

---

[2] See the list of Chambers (1890: 487–495).
[3] Of course, comets are diffuse objects too.
[4] At his observatory in Dresden, Baron von Engelhardt used a 30-cm Grubb refractor.

[5] The role of visual representations is treated by Hentschel (2000); on astronomical drawing see Schaffer (1998b). See also Nasim (2010b).
[6] William and John Herschel (and later Lord Rosse and his son) had the best experiences with the Basire family, the official engravers of the Royal Society; see Hentschel (2000).

Table 11.1. *Positional measurements of nebulae and star clusters in the nineteenth century*

Observer	Observatory	Tel. (cm)	Type	Objects	Period	Micro.	Epoch	Publ.	Ref.
Auwers	Königsberg	15.8	H	40	1859–62	Ring	1860	1862	1862c
Becker L.	Dun Echt	21.5	M	217	1886–91	Bar	1890	1902	1902
Bidschof	Vienna	38.0	Rr	31	1897–98	Bar	1897/98	1899	1899
Bigourdan	Paris	31.6	Rr	6380	1884–1911	Bar	–	1892–1917	(a)
d'Arrest	Leipzig	11.7	Rr	230	1852–58	Ring	1850	1856	1856a
d'Arrest	Copenhagen	28.5	Rr	1942	1861–67	Ring	1861	1867	1867a
Dreyer	Armagh	25.4	Rr	100	1887–93	Bar	1890	1894	1894a
Engelhardt	Dresden	30.6	Rr	390	1883–94	Bar	1885	1886–95	(b)
Engelmann	Leipzig	16.2	M	124	1866–74	Bar	1870	1883	1883
Ginzel	Vienna	18.9	Rr	54	1884–86	Ring	1890	1888	1888
Herschel J.	Slough	47.5	Rl	2306	1823–32	Bar	1830	1833	1833a
Herschel J.	Feldhausen	47.5	Rl	1714	1834–38	Bar	1830	1847	1847
Kempf	Potsdam	20.7	Rr	82	1880–84	Lamella	1875	1892	1892
Lamont	Munich	26.7	Rr	32	1836–37	Bar	–	1869	1869
Laugier	Paris	9.7	Rr	53	1848–49	Bar	1850	1853	1853
Merecki	Warsaw	16.2	Rr	72	1900–02	Bar	1900	1902	1902
Mönnichmeyer	Bonn	15.9	Rr	236	1891–93	Ring	1892	1895	1895
Oppolzer	Vienna	18.9	Rr	13	1863–65	Ring	1865	1868	1868
Porter	Cincinnati	30.5	Rr	105	1884–86	Bar	1890	1891	1891
Rümker	Hamburg	10.0	Rr	151	1860–67	Ring	1865	1865–67	(c)
Rümker	Hamburg	26.0	Rr	105	1871–80	Bar	1875	1895	1895
Schmidt	Athens	15.7	Rr	110	1860–67	Ring	1865	1868	1868b
Schönfeld	Mannheim	16.5	Rr	235	1860–62	Ring	1865	1862	1862d
Schönfeld	Mannheim	16.5	Rr	343	1862–64	Ring	1865	1875	1875b
Schultz	Uppsala	24.4	Rr	500	1863–74	Bar	1865	1874–75	(d)
Spitaler	Vienna	68.6	Rr	136	1891–92	Ring	1891–92	1893	1893
Stephan	Marseille	78.8	Rl	546	1870–85	Bar	1870–85	1916	(e)
Stone	Charlottesville	66.0	Rr	310	1887–91	Bar	1890	1893	1893b
Vogel	Leipzig	11.7/21.5	Rr	100	1865–66	Bar	1865	1867	1867
Vogel	Leipzig	21.5	Rr	132	1866–70	Bar	1870	1876	1876
Weinek	Prague	16.3	Rr	44	1884–87	Ring	1884/7	1886/90	(f)

products were inevitably hardly comparable, especially concerning faint details.

11.2.1 From the beginnings to scientific drawings

The first presentations of nebulae and star clusters can be found in early star charts. Often they were treated as 'exotic stars', as was the Andromeda Nebula (M 31) in the atlas of the Persian astronomer As-Sufi. The first individual sketch of a nebula was made by Hodierna in the year 1654, showing the Orion Nebula (M 42). A certain structure is visible, which, however, does not have much to do with reality. The same object was rendered much better by Christiaan Huygens in 1656 (Fig. 11.1 left) and by Jean Jacques de Mairan in 1731. The first sketch of a star cluster, showing M 11 in Scutum, was due to Gottfried Kirch in 1682; however, the dominant star is exaggerated (Fig. 11.1 right). Johann Elert Bode added some sketches of nebulae in his atlas *Vorstellung der Gestirne*, among them M 31.[7] Messier has left only two drawings, which are pretty realistic. The first, of 1771, shows the Orion Nebula and the second, of 1806, the Andromeda Nebula with its companions M 32 and NGC 205 (M 110).[8]

William Herschel published some drawings too, which are, however, not very detailed. His preference was for textual description, recording the form, brightness and place – essential data for identification.[9] Additionally, due to the large number of objects, his standard scheme allowed a subsequent analysis. Herschel's aim was to reveal the physical connection of the detected forms of non-stellar objects for an evolutionary scenario. Therefore, his sketches were mainly intended to illustrate his ideas. The individual object was a mere example for a certain state. All his sketches of nebulae were made at his standard instrument, the 18.7-inch reflector. It is surprising that he did not use the 48-inch, his biggest telescope, for this task. Certainly, the large aperture would have led to important results about the nature of nebulae – in particular,

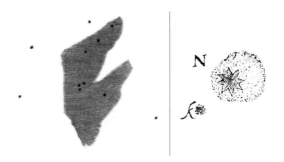

Figure 11.1. Left: Huygens' drawing of the Orion Nebula; right: the Wild Duck Cluster[10] M 11, sketched by Kirch.

he should have noticed the spiral form of M 51 prior to Lord Rosse.[11]

William Herschel's publication on the 'Construction of the heavens' contains altogether 74 sketches: 15 in the paper of 1784, 40 in that of 1811 (two are earlier sketches of M 31) and 17 in the 1814 paper.[12] Furthermore, a few rough presentations can be found in his observing books, which are stored in the RAS archive.

It was John Herschel who gave drawings a new meaning: they should show individual objects. Realistic representations rather than merely class features were demanded. The qualitative, abstracting sketch should be transformed into a detailed image of reality. The final drawing should show the form, structure, resolution (if any) and brightness distribution. Surrounding stars must be included for orientation and separation. The need originates from the important issue of the possibility of changes concerning shape, brightness and position (proper motions relative to the stars). Herschel also required the specification of important observational data (e.g. date, air quality and instrument). In 1826 he wrote that '*The same nebula viewed on the same night with different telescopes, presents very different degrees of light, distinctness and magnifying power employed; – so different, indeed, as to be scarcely recognizable for the same object: and if viewed on different nights with the same telescope, atmospherical circumstances will make hardly less variation in it; and the presence or absence of the moon is equally or more efficacious, in changing entirely the shape and magnitude of the visible outline.*

---

[7] 'Presentation of the stars' (Bode 1782).

[8] The history of the M 31 companions is told in Steinicke (2004c).

[9] In addition to the arduous handling, the image quality of the reflector was quite bad; see Section 2.3.

[10] Dreyer (1912a: xl).

[11] The name is due to Smyth, who wrote that the cluster '*resembles a flight of wild ducks in shape*' (Smyth 1844: 431).

[12] Herschel W. (1784, 1811, 1814).

If we add to this the extreme difficulty of representing such appearances on paper, and the hardly inferior one in getting them faithfully engraved; – if we add too, that astronomers are seldom draftsmen, and have hitherto evidently (with one honourable exception) contented themselves with very general and hasty sketches, it will be no matter of surprise that the published engravings of these objects present a mass of concentrations, and for the most part offer as little resemblance to the objects themselves as to each other.[13]

Drawing and sketch should be treated as different subjects; the former shows the physical appearance and the latter is used for identification. A drawing requires not only the will to make it ('*astronomers are seldom draftsmen*') but also artistic talent. Herschel provided excellent examples in the Slough and Cape catalogues. His drawings of the Orion Nebula and the Lagoon Nebula (M 42 and M 8) set new benchmarks concerning quality and accuracy (Fig. 11.2). Olbers found them '*as felicitous as one could ever imagine*' (Olbers 1834). Airy wrote in 1836 that '*few observers possess the delicacy of hand of Sir John Herschel; yet it were to be wished that this example might be imitated by many, and that careful drawings, the best that circumstances admit of, might frequently be made of the same nebula.*'[14] The young American astronomer Mason was an outstanding 'imitator', who broached new fields. In his drawings (see Section 6.2.2), he used a graticule and produced the first chart showing lines of equal brightness (the target was the Trifid Nebula M 20 in Sagittarius). Herschel was impressed by Mason's scientific treatment of nebulae.

However, such masterpieces were copied in various textbooks to attract the public. Lord Rosse's drawing of the spiral nebula M 51 was a popular example. Astronomers treated their pictures as immediate representations of the heavens and regarded themselves as uniquely skilled representatives.[15] Sometimes their drawings were even weapons in a struggle between oppositional ideas; the nebular hypothesis, claiming the existence of true nebulosity to form stars and clusters, is a good example (see Section 6.4.8). True nebulosity

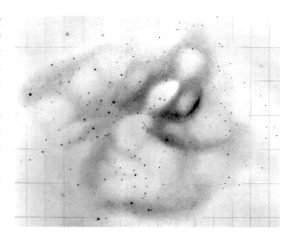

Figure 11.2. John Herschel's remarkable drawing of the Lagoon Nebula M 8 in Sagittarius (Herschel J. (1847), Plate I, Fig. 1).

was to be recognised through changes of shape, so pictures acquired an unusually high status.

11.2.2 Methods and tools

Charles Piazzi Smyth, the son of William Henry Smyth and something of an amateur draftsman, described the scientific foundations of astronomical drawing in a text written in 1841 at the Cape Observatory. Two years later it was read at a meeting of the RAS.[16] Smyth was not sparing with his critique, stating that John Herschel's '*recommendations [...] have never been followed up*'. Drawings of nebulae, if reliably made, were the only means of getting at changes in their form. He stressed the need for '*faithful imitation, for want of which none of the so-called high finishing can ever atone, and which can only be accomplished by correctness of eye, facility of hand, and a due appreciation for the subjects*'. The principles formulated by Smyth (summarised below) are still valid today. Basically there are two styles of presentations:

- positive (white on black background, simulating the visual appearance)
- negative (black on white background)

Smyth favoured the former, whereas the latter '*is extremely likely to puzzle and to create misconception*'. Of

---

[13] Herschel J. (1826a: 488–489); by 'one honorable exception' obviously William Herschel is meant. This is a typical example of Herschel's long, convoluted sentences.
[14] Airy (1836: 311).
[15] Schaffer (1998a: 204).

[16] A summary appeared in the meeting report as 'Memoir on astronomical drawing' (Smyth P. 1843). The full text was later published in the RAS *Memoirs*, under the title 'On astronomical drawing' (Smyth P. 1846); see also Hentschel (2000).

course, both styles have their pros and cons. The positive one is problematic, because it is difficult to render different degrees of brightness correctly. With a white pen, it is hardly possible to reproduce intensities (grey levels). Thus positive images are, concerning contrast, not very realistic. An early example is Messier's drawing of the Orion Nebula published in 1774. Much more common is the negative image, which, however, '*gives the objects a better definition than they really possess*'. Though this style allows a fine graduation of intensity (e.g. with a pencil), many drawers tend to exaggerate contrast. A practical reason is that extremely faint parts of nebulae are difficult to reproduce in a publication. The first realistic negative image was John Herschel's drawing of the Orion Nebula made in 1825 and presented in the RAS *Memoirs* (Herschel J. 1826a).

At the telescope, sketches were made with the aid of a tray and faint light, often with red filters to retain the adaptation of the eye to low light levels. Some people used templates, which showed the principal stars in order to position the object correctly. The drawing itself was not made while observing. It was desktop work. Cardboard was a favoured surface, using Indian ink, chalk or a carbon graphite to draw. With a wiper or brush areas were smoothed out. Of course, the drawing was created from memory with the aid of the sketch in the observing book. However, this harbours the danger that some information could be lost or that some features that are not present might be added. To counter the problem, the result was verified in a subsequent observation – and perhaps corrected.

Another problem is orientation, which can be very different at the telescope. In a refractor the view is upside down, in a Newtonian reflector it can be turned arbitrarily and in 'front-view' mode (which was sometimes used by William and John Herschel and by Lord Rosse) it is reversed. The reason for the last case is that in the case of a direct view, there is only one reflection (with a secondary mirror there are two). Thus, comparing drawings can be complicated if there are no hints about orientation. Therefore, a proper image should show the north and east directions.

A final point concerns the reproduction of drawings. Common techniques were engraving, use of woodcuts and lithography. None of these guaranteed a true copy of the original. Especially Tempel, who was an astronomer, draughtsman and skilled lithographer, was critical about the issue. However, photography later faced similar problems. Barnard, for instance, struggled for many years to obtain faithful reproductions of his famous Milky Way images.

11.2.3 Tempel's criticism: drawings and reality

Is a drawing a representation of reality? This was a fundamental question in the nineteenth century, which led to many debates and conflicts. Since a drawing was the only way to image an observation, the observer bore great responsibility. Basically, as had already been stated by John Herschel and was repeated by Smyth, a drawing is influenced by the following factors:

- instrument: system, aperture, magnification, field of view, mechanical stability
- site: location, elevation
- atmospheric conditions: view, transparency, air pollution, wind, humidity
- external light: Moon, aurora, light pollution
- observer: perception, physical condition, experience, drawing talent

Therefore, scientific statements about nebulae depend on exterior (objective) and personal (subjective) factors. These made it difficult to compare different drawings (even when they had been made by the same person). A standard result was that an object was claimed to exhibit changes (e.g. form, brightness) that were not real. But who was able to decide? Photography improved the situation. However, modern image processing again raised the question of objectivity (see the next section).

Drawings were criticised at all times. Secchi, for instance, called the results of John Herschel, Lord Rosse and Lassell 'monstrosities'. Tempel agreed, but remarked that Secchi's drawings would '*provoke criticism even more*'.[17] In 1877 he drew the Lagoon Nebula M 8 in Sagittarius, comparing his result with John Herschel's.[18] The question of variability was central. Even though Tempel appreciated Herschel's drawing of M 8 as the best, the comparison with his own drawing did not lead to definitive results. A bit desperately he asked himself '*Shouldn't we be able in the future to gain securer basis with exacter drawings?*'

---

[17] Tempel (1885a: 12).
[18] Tempel (1877b); the drawing is reproduced in Clausnitzer (1989: 59).

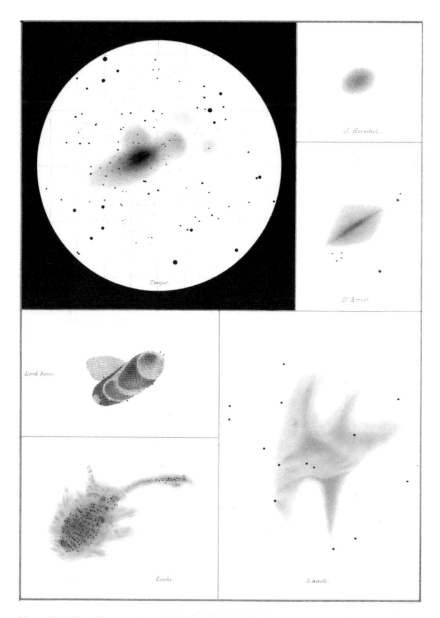

Figure 11.3. Tempel's comparison of M 1 drawings (see the text).

Giving himself the answer: '*I doubt it; because the receptivity (sensitivity) of the eye for such fine nebulous structures is too different for each observer for one to hope for an exact match.*'

Tempel's comparison of drawings of the Crab Nebula (M 1) presented in his treatise 'Über Nebelflecken' (1885) is spectacular. His Table I (Fig. 11.3) shows, besides his own result, those of John Herschel (1827), Secchi (1852), Lord Rosse (1844 and 1855), d'Arrest (1861) and Lassell (1864). Indeed, the differences in appearance are striking. Tempel noted that, '*If the large differences of the same object are due only to the various telescopes used – as is commonly accepted –, one should outlaw them.*'[19] According to him,

---

[19] Tempel (1885a: 29).

11.2 Drawings of nebulae

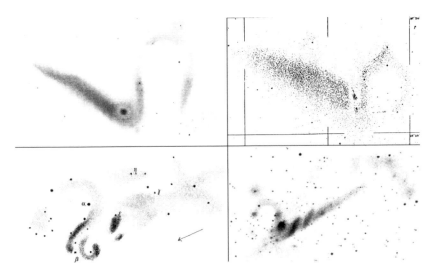

Figure 11.4. Drawings of M 17: upper left, John Herschel in 1828; upper right, Mason in 1835; lower left, Mitchell in 1854; and lower right, Tempel in 1878 (see the text).

they are caused neither by the telescopes nor by the atmospheric conditions, since large telescopes are comparable and, moreover, drawings were normally made under the best conditions (*'One does not make a drawing in a single night and under a dull sky.'*[20]). His conclusion is simple: *'The reason why drawings do not match is due to the draughtsman itself.'* According to Tempel, talent, perception and experience are the essential factors.

As a result of his own methods, he pleads for a standardisation of drawings, including a common scale. If this could be realised, he is convinced that drawings would be superior to textual descriptions: *'True exact drawings of nebulae say much more than many observations. Even the fleeting sketches which Lord Rosse has added in his two catalogues of nebulae are of greater account than his descriptions; however, they should have been copied a bit better.'* (Tempel 1882). Sometimes Tempel's words – due to his lithographic education – appear pretty schoolmasterly. This elicited some critical remarks, e.g. by Dreyer in the hot debate about the reality of spiral structure (see Section 11.3.3).

Similar differences were noticed for many other objects too. Take, for instance, the drawings of the Omega Nebula M 17 by John Herschel (in the Slough and Cape catalogues), Lamont, Mason, Mitchell, Swift,

William Smyth and Tempel (Fig. 11.4).[21] A further complication is due to fact that John Herschel's drawing is reversed (because he observed the nebula using the 'front-view' mode). The same applies for Mason, who might have mirrored the image in print to make it comparable with Herschel's. Also drawings of the Merope Nebula (NGC 1435) in the Pleiades, Hind's Variable Nebula (NGC 1555) in Taurus and M 57 (Ring Nebula in Lyra) were subjected to enduring discussion.

11.2.4 Drawing versus photography

With the advent of astrophotography, many questions were cleared up. Concerning drawings, Isaac Roberts was the first to point out the following: *'But all drawings alike fail to present to the eye proportions, details, and outlines as they are shown on photographs'* (Roberts 1889b). The basis for this statement was his plate of M 51 taken on 28 April 1889, which he compared with drawings by John Herschel, Lord Rosse and Lassell.

James Keeler continued the discussion in his paper 'Note on a case of differences between drawings and photographs of nebulae' (Keeler 1895). He wrote that *'A comparison of the best drawings and photographs of nebulae reveals at once the existence of considerable discrepancies between the forms depicted by methods so widely*

---

[20] Tempel (1885a: 12).

[21] See also the sketches of William Smyth and Lewis Swift: Figs. 6.17 and 9.17.

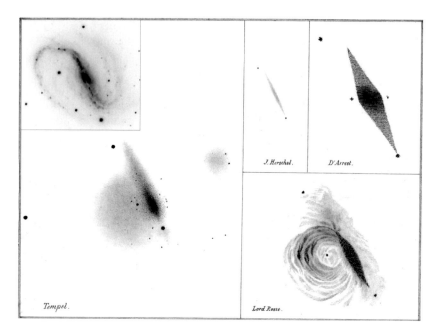

Figure 11.5. Drawings of the galaxy NGC 7479 in Pegasus, as collected by Tempel (that on the left is his own drawing); and, in the upper left corner, a modern image, rotated and flipped to give the corresponding orientation (see the text).

different.' Keeler was the first to mention that the different spectral sensitivities of eye and photographic plate could cause the differences. For instance, the hydrogen lines of the Orion Nebula appear different for these two 'detectors'. Since the associated emission regions have different structures and distributions, their visual and photographic appearances do not match.

In 1889 Wilhelm Foerster wrote an article for the German magazine *Himmel und Erde*, titled 'Ueber die Verschiedenheiten der Wahrnehmung und Darstellung von Nebelflecken'.[22] Several drawings of the spiral nebula NGC 7479 (I 55, GC 4892) in Pegasus are presented (Fig. 11.5).

The collection was made by Tempel shortly before his death, containing drawings by John Herschel (1825), Lord Rosse (1849), d'Arrest (1866) and himself (1878?). The first two show an elongated nebula, whereas Lord Rosse has appended a spiral, which appears as a diffuse oval structure in Tempel's drawing. Regarding the latter difference, Foerster supports Tempel, who was very critical about spiral nebulae: '*That this is not caused by the inferior light-gathering power of Tempel's instrument is obvious from the fact that Mr Tempel has seen a faint nebula near the larger one, which is not present on any of the other images.*' He refers here to the patch to the right of the galaxy.

Keeler treated the case in his paper 'The spiral nebula H I, 55 Pegasi', which this time includes photographic results (Keeler 1900b). In August 1899 he had taken two plates of NGC 7479 with the 36" Crossley reflector at Lick Observatory. Foerster's statement about Tempel was refuted: the argument '*loses all its force when it is shown that Tempel's supposed nebula certainly does not exist*'. He further wrote that '*There is no doubt that Lord Rosse saw the nebula to much greater advantage than Tempel.*' Keeler stated that '*A comparison of the photograph with the figures in the plate illustrates in a very interesting and instructive manner, the personality of the draftsman which is commented upon by Tempel, and from which his own drawing is by no means exempt.*' One further reads that '*There is also a natural tendency in drawing to emphasize the details caught by the eye, while others, which are missed, may be nearly or quite as prominent. Thus the drawing and the photograph differ.*' Thus, the question 'facts or fiction?' arises – and was answered by Keeler's advocacy of photography. Nevertheless, the

---

[22] 'On the discrepancies concerning perception and presentation of nebulae' (Foerster 1889). See also the remarks of Margaret Huggins (Huggins M. 1882).

issue was still irritating in the early twentieth century. This was demonstrated by the debate about the Martian channels: photography – especially the 'evidence' produced by David Todd – could not decide the matter (see Section 9.7.1).

Keeler also used the Crossley reflector to photograph John Herschel's double nebulae, which had been sketched in the Slough catalogue.[23] He did not find any correspondence: '*The actual nebulae, as photographed, have almost no resemblance to the figures. They are, in fact, spirals, sometimes of very beautiful and complex structure.*' (Keeler 1900c).

Visual observers of today, keeping the tradition of drawing, share the old problem of subjectivity. Of course, the technical equipment has improved, but paper, pencil and red light are still in use. However, two other problems have been added. The first is related to the computer. The drawing can be scanned and, therefore, is subject to digital manipulation, such as inversion, rotation or contrast enhancement. Of course, the temptation to change the content of the drawing is great; the result can even be made to look like photography. Fortunately, drawings have completely lost their scientific weight. The second problem concerns the influence of visual observing by photography: one already knows what should be seen. Having been biased by 'pretty pictures', the eye is influenced by the brain. Compared with this situation, the (uninfluenced) nineteenth-century observer was actually more 'objective'.

Early drawings of the spiral nebula M 51 were recently studied (Nasim 2010a) and compared with modern images made with the Hubble Space Telescope (Kessler 2007). This offered an opportunity to examine shifts in the object's representation within a given period, as well as over the long history of observing it. This demonstrates the consistent interest of astronomy in structure and improved resolution, as well as the subjective treatment of light and colour.

11.2.5 Statistics of published drawings

In the nineteenth century three lists of published drawings giving object and source appeared (Table 11.2). The first was compiled by John Herschel in the General Catalogue; Holden and Dreyer followed. The list presented in the NGC is an updated version of that in the GC. Surprisingly, Holden's comprehensive compilation is not mentioned, though it appeared 11 years earlier. Although Dreyer presents new drawings, some of those mentioned by Holden are missing from Dreyer's publication.

Apart from his own drawings, John Herschel mentions only the classic sources (Lord Rosse, Bond, Mason, d'Arrest, Lamont, Lassell) plus Messier and de Vico. William Herschel's sketches are not listed. Dunlop is omitted, but for another reason: '*The figures annexed to Mr. Dunlop's catalogue are not included, as for the main part they offer no resemblance to the objects figured (when identifiable), and would serve only to mislead.*'[24] This is wrong – the targets of his drawings can be identified very well.

Holden takes over the GC nomenclature (abbreviations of the sources) and especially adds drawings of the Orion Nebula, the theme of his comprehensive monograph (see Section 9.16.3). Dreyer too uses the designations introduced by John Herschel. All three authors do not distinguish among the Birr Castle drawings, which were generally credited to 'Lord Rosse'; they are divided only by publication (1844, 1850, 1861).

An analysis of the drawings made until 1900 yielded a number of 720 (from 55 observers). This covers those mentioned in the references. Generally the drawing is counted, not the imaged objects. Of course, some sources still might have been overlooked. Figure 11.6 shows the distribution. The drawings prior to 1800 are mainly due to William Herschel and Bode. During the subsequent two decades, there were only a few drawings until John Herschel entered the scene.

As Fig. 11.7 shows, Tempel has left 190 drawings, which are stored at Arcetri Observatory. Second best is John Herschel (125). Together they produced about 40% of the total number. William Herschel is in the third place (72), followed by Lassell (51). Bode's presentations, shown in Table XXX of his *Vorstellung der Gestirne*, are mainly sketches (Bode 1782). Those given in Smyth's Bedford Catalogue are of low quality too (Smyth W. 1844). Like William Herschel, Barnard published only sketches. The other observers have left drawings of good quality; some even show great artistic talent. An example is Caroline Lassell, who made a fine drawing of the Orion Nebula based on her father's observations with the 48" reflector on Malta.

---

[23] Herschel J. (1833a: Figs. 68–79 (Plate XV)).

[24] Herschel J. (1864: 40–41).

Special topics

Table 11.2. *Lists of published drawings*

Author	Publication	Year	Drawings	Drawers
J. Herschel	General Catalogue	1864	298	9
Holden	Index Catalogue of Books and Memoirs	1877	451	38
Dreyer	New General Catalogue	1888	464	40

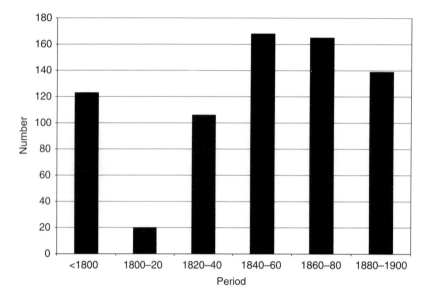

Figure 11.6. Numbers of drawings per decade until 1900.

Figure 11.8 shows that the Orion Nebula (M 42) was the most portrayed target. Its 29 published drawings originate from 20 different observers. Nearly all are presented in Holden's monograph (Holden 1882b). Other popular subjects were the Crab Nebula (M 1) and Dumbbell Nebula (M 27).

11.3 M 51 AND THE SPIRAL STRUCTURE OF NEBULAE

Lord Rosse was the first to notice the spiral structure of nebulae.[25] The key object was the 8.1-mag galaxy M 51 (h 1622, GC 3572, NGC 5194) in Canes Venatici, which had been discovered by Messier on 13 October 1773 (and found independently by Bode on 5 January 1774). It was first seen as a 'spiral nebula' in April 1845 with the brand-new 72" reflector at Birr Castle.[26] The circumstances of the discovery led to some interesting questions. Another point is the meaning of this unexpected feature for the celebrated nebular hypothesis, according to which the spinning of true nebulosity led to the formation of stars and planets. Otherwise, a 'resolved' spiral nebula like M 51, consisting of rotating masses of stars, was the prototype of an 'island universe'. However, the question of 'external' galaxies could not be treated scientifically at that time.[27]

---

[25] See also Sections 6.4.10 and 7.1.3.

[26] Lord Rosse was unable to see the spiral form in his 36-inch. However, William Herschel could have been successful with his 48-inch, but he never looked at M 51 with the famous '40ft'.

[27] Fernie (1970), Wolfschmidt (1995).

11.3 M 51 and the spiral structure

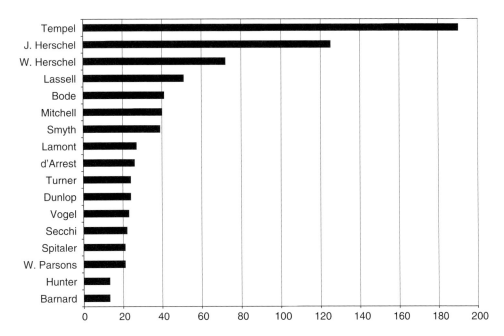

Figure 11.7. The most industrious drawers.

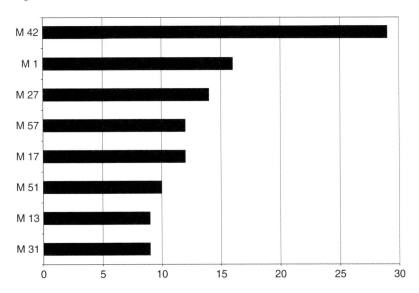

Figure 11.8. Numbers of drawings for popular objects.

Later, spiral structure could be confirmed by other contemporary observers. But there were critics too. Tempel (contra) and Dreyer (pro) conducted a heated debate. In time the controversy was resolved by photography – of course, in favour of the spiral nebulae. The crucial image was obtained by Eugen von Gothard in 1888, clearly showing the spiral arms of M 51.

11.3.1 Lord Rosse's discovery of spiral nebulae

Between 4 and 13 March 1845 the 72" reflector at Birr Castle was pointed at nebulae for the first time. Lord Rosse had invited James South and Romney Robinson to join his observations (see Section 6.4.9). Shortly after, Robinson reported the results at a meeting of the

Royal Irish Academy in Dublin. In the published version one reads that *'most of the lucid interval from the 4th to the 13th of March was devoted to nebulae, and after that it again became cloudy [...] forty were examined by Dr. Robinson and also by Sir James South'*.[28] Robinson was focused on the resolvability of nebulae. Given the popularity of the nebular hypothesis, this was perhaps the dominant issue of the time (see Section 6.4.8). This theory, proclaimed by the Glasgow astronomer John Pringle Nichol, implied the existence of true nebulosity ('luminous fluid'), which forms stars and clusters. Robinson was its strongest opponent, enthusiastically claiming that all nebulae must be resolvable, i.e. mere clusters of stars, disguised only by distance. A sufficiently large telescope – like the Leviathan of Parsonstown – should be able to verify his view.

The bright nebula M 51 in Canes Venatici, which in spring is high in the sky, was a primary target. Robinson reported that *'the central nebula is a globe of large stars; as indeed had been previously discovered with the three-feet telescope [36-inch]: but it is also seen with [magnification] 560 that the exterior stars, instead of being uniformly distributed as in the preceding instances, are condensed in a ring, although many are also spread over its interior.'*[29] He added that, *'Were the centre absent, we should have a ring nebula; and were the line of vision near the plane of this ring, it would become one of those rays with a bright nucleus and parallel band or satellite nebulae which occur so frequently in the catalogue.'* South wrote his own report, which appeared in *The Times* of 4 April 1845 and was reprinted in the *Astronomische Nachrichten* the following year (South 1846). He remarked that *'The most popularly known nebulae observed this night [March 5] were the ring nebula in the Canes Venatici, or the 51 of Messier's catalogue, which was resolved into stars with a magnifying power of 548; and the 94th of Messier, which is in the same constellation, and which was resolved into a large globular cluster of stars, not much unlike the well-known cluster in Hercules, called also 13th Messier.'* Though South did not always share Robinson's radical views, he agreed with him in the case of M 51.

The term 'ring nebula' refers to John Herschel's popular drawing in the Slough catalogue of 1833 (Fig. 11.9 left). He had observed M 51 six times with his 18¼" reflector.[30] The drawing was made on 4 April 1830 (during the third observation) under favourable conditions. It shows the object as a spherical centre surrounded by a ring with a single chord (spiral arm?); the companion NGC 5195 (h 1623) appears as a separate nebula.[31] Herschel made the following note in his observing book on M 51: *'It is a vB neb 1' in diameter of a resolvable kind of light with a double ring or rather 1½ ring like an armillary sphere.'* A similar drawing was later presented by Smyth (Fig. 11.9 right).[32] Interestingly, Herschel's ring was not seen with the 36-inch at Birr Castle on 18 September 1843.[33]

William Herschel observed M 51 four times.[34] On 29 September 1783, using his 'small 20ft' reflector in Datchet, he claimed to have resolved the nebula.[35] But his description on 12 May 1787 (with the 'large 20ft' in Slough) sounds similar to that of his son, though without noticing the subdivision in the ring: *'nebulosity in the center with a nucleus surrounded by a detached nebulosity in form of a circle'*. However, John Herschel added an essential detail: a visionary interpretation. In his observation book he noted that *'It is like our Milky Way & perhaps this is our Brother System'*.[36] In the Slough catalogue he explained this conjecture as follows: *'Supposing it [M 51] to consist of stars, the appearance it would present to a spectator placed on a planet attendant on one of them eccentrically situated towards the north preceding quarter of the central mass, would be exactly similar to that of our Milky Way.'*[37] His father had interpreted the Milky Way as the optical effect of a

---

[28] Robinson T. (1845: 12).
[29] Robinson T. (1845: 15–16).
[30] The dates are 17.3.1828 (sweep 138), 20.3.1828 (140), 26.4.1830 (225), 27.4.1830 (226), 13.5.1830 (257) and 7.3.1831 (329). During the second and third observations Herschel made a sketch in his notebook, see Hoskin (1987), Figs. 4 and 5.
[31] Herschel J. (1833a), Fig. 25. NGC 5195 was discovered by Méchain on 21 March 1781 and independently found by William Herschel on 12 May 1787 in Slough (I 186).
[32] Chambers (1867: Plate XXVII); Glyn Jones (1967b: 182).
[33] Parsons W. (1850a: 509–510).
[34] Dreyer (1912a, Vol. 2: 657). The observations were made on 17.9.1783 (6.2"), 20.9.1783 (12"), 12.5.1787 (18.7", sweep 734) and 29.4.1788 (18.7", sweep 836). On the last date Herschel wrote *'vB. L., surrounded with a beautiful glory of milky nebulosity with here and there small interruptions that seem to throw the glory at a distance'*.
[35] Herschel W. (1784: 440).
[36] Hoskin (1987: 12).
[37] Herschel J. (1833a: 496–497).

11.3 M 51 and the spiral structure

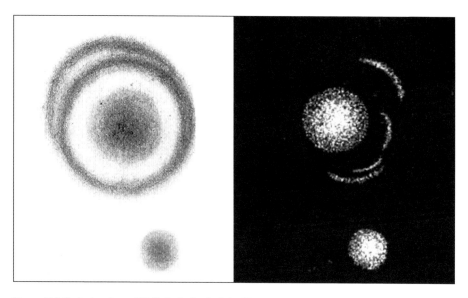

Figure 11.9. Early drawings of M 51. Left: that by John Herschel in 1830; right: that by Smyth in 1836; south is up.

continuous, though bifurcated layer ('stratum of stars'), consisting of (nearby) bright stars and (more distant) faint stars. But John Herschel, inspired by the remarkable appearance of M 51, drew a modified picture of the Milky Way with two separated 'strata of stars': a central cluster and a surrounding ring of faint stars, divided by a void.[38] Concerning the genesis, John Herschel conjectured as follows: '*Were it not for the subdivision of the Ring, the most obvious analogy would be that of the system of Saturn, and the ideas of Laplace respecting the formation of that system would be powerfully recalled for this object.*'[39] Here he cites Laplace's cosmogony, which formed the basis of the nebular hypothesis.

In April 1845 Lord Rosse detected the spiral structure of M 51. The approximate date of this revolutionary find can be derived from his report of 1850: '*The spiral arrangement of Messier 51 was detected in the spring of 1845.*'[40] But the most precise hint appears in the Birr Castle observations, which were compiled by Dreyer and published by Lawrence Parsons in 1880: '*1845, Apr. During this month M. 51 was for the first time examined with the 6 foot and its spiral character immediately noticed, but no record is left of these early observations.*'[41] The limits of the discovery date can be set by two factors. New moon was on 6 April, half moon on the 14th. Lord Rosse noted that the spring sky was optimal, i.e. dark enough and of good transparency, at about 11 p.m., whereas during the second half of these nights it became '*luminous and the faintest nebulae disappear.*'[42] Therefore it is likely that the observation took place on about 6 April at approximately 1 a.m., when M 51 stood near the meridian at an elevation of 85°.[43] At that position the object could be tracked for some time for a thorough study.

We now return to the reports of Robinson and South about their March observations with the 72-inch. Concerning M 51, only known features are mentioned: the ring structure (which had been confirming John Herschel) and the resolution of the central 'globular cluster' (which had already been seen with the 36-inch).

---

[38] Hoskin (1987: 13).

[39] Herschel J. (1833a: 497). This analogy was the motive for Smyth to describe M 51 as a '*ghost of Saturn, with his ring in vertical position*' (Smyth W. 1844: 302).

[40] Parsons W. (1850a: 505).

[41] Parsons L. (1880: 127). However, the statement '*M 51 was for the first time examined with the 6 foot*' is not correct, since it has already been observed in March by Lord Rosse and his guests.

[42] Parsons W. (1861a: 125).

[43] Observations near the zenith caused no problems for the 72-inch, since it could even be tilted 12° beyond it. However, the celestial pole (at 53° above the horizon) was out of sight; see Dreyer (1914).

Now, given Lord Rosse's result of April, an important question arises, which was first treated in 2005 by Mark Bailey (the present Director of Armagh Observatory), John McFarland and John Butler:[44] why had Lord Rosse, Robinson and South not already noticed the spiral structure of M 51 in March? The authors conclude that *'It seems likely that Rosse, Robinson and South could have seen the spiral arrangement [...] though there is no evidence that they noticed it.'*

However, the question must be attacked from different directions. If one assumes that Robinson and South actually saw M 51 as a spiral nebula, why was nothing reported?[45] A spectacular discovery like this would immediately have been exploited by them. Since nothing happened, we may conclude that it was not perceived by the observers (including Lord Rosse). But this, vice versa, automatically leads to the following crucial question: why did the spiral structure immediately become obvious for Lord Rosse in April, when he was observing alone with his large reflector?[46]

Were there any differences (meteorological or instrumental) between the observations in March and April? The first factor is the metal mirror. Since it had been polished on 3 March, it probably was still in good condition in April (a worse mirror would even make the problem harder). This is supported by Lord Rosse, who wrote in his publication of 1861 that *'In the early observations* [March] *with the 6-feet telescope we had the advantage of a very fine speculum [...] and many nebulae were resolved. Very soon after, the spiral form of arrangement was detected.'*[47] Concerning the weather, it is questionable whether April was as clear as early March (again making the problem harder). According to South's report, *'the night of the 5th of March was, I think, one of the finest I ever saw in Ireland. Many nebulae were observed by Lord Rosse, Dr. Robinson and myself'* (South 1846). Finally, the eye-pieces must be taken into account (see Table 6.5). Perhaps different ones were used for the observations. With a regular eye-piece (higher magnification) only parts of M 51 appeared in the field of view: the central 'globular cluster' or the outer 'ring'. But in the finding eye-pieces the 11′-large galaxy should have been visible as a whole: the 76-mm eye-piece offered a field of view of 31′ (however, some light was lost due to the large exit pupil of diameter 8.5 mm) and the 46-mm eye-piece still gave 13.7′. It is reasonable to assume that these tools were used on every observation. The conclusion is that, since no great exterior differences can be found (two factors even favour March), it is barely imaginable that the spiral structure of M 51 could have been overlooked by Robinson, Lord Rosse and South – on a very good night with a new polished mirror in a wide-field eye-piece. Thus the non-detection remains a miracle – all the more so insofar as spiral arms were so obvious later!

To solve the problem, one therefore must concentrate on internal (psychological and psychosomatic) factors, i.e. mind and perception. Was there any difference between March and April in this regard? The crucial terms might be 'resolvability of nebulae' and 'nebular hypothesis' – and the different views of the protagonists, namely Robinson and Lord Rosse, about them. It seems to be possible that merely ideological reasons led them to ignore the spiral structure of M 51 in March 1845.

Robinson was an ardent – and thus uncritical – advocate of the resolvability of nebulae, regardless of their type. The result that nebulae appeared as clusters of stars became the main argument against the nebular hypothesis. However, these classes of objects were merely proxies in the underlying battle between atheism and religion. The protagonists were Robinson, as the representative of the Church of Ireland, and the materialist and free-thinking Scot Nichol. Robinson's observational result, proving the stellar nature of nebulae, had already damaged the nebular hypothesis.

No doubt Robinson was completely focused on this issue, deleting all distracting and unwanted structures from his perception. The consequences are straightforward: a brain biased in this way forces the eye to 'see' a mottled structure everywhere. Consequently, if an observer is concentrated on the 'resolution' of a nebula (supported by former positive experience), the mind is not prepared for unexpected features (such as spiral arms) – and ignores them!

What about the two critical minds, Lord Rosse and South? One may think that Lord Rosse had left the 72-inch to his guests, when observing M 51.[48] But

---

[44] Bailey et al. (2005); see also Ashbrook (1984: 397–404).
[45] See Section 6.4.9.
[46] There was no scientific assistant until 1848, when Rambaut was engaged.
[47] Parsons W. (1861a: 147).

[48] In 1970 it was guessed that *'Lord Rosse himself was much occupied at that time by an active political career, and, as a major*

South wrote (as quoted above) that the owner was present on the best night (5 March). Anyway, it is probable that Robinson took command, which would be in accord with his character. Thus the trio was focused exclusively on his concern – and saw nothing but resolved nebulae, which is obvious from their independent reports. Lord Rosse later conceded that, *'When certain phenomena can only be seen with great difficulty, the eye may imperceptibly be in some degree influenced by the mind.'*[49] If his mind had been free and unbiased – as was the case later, in April, when he could observe M 51 alone – he probably would have noticed the spiral structure. But under the strong influence of Robinson none of the observers was prepared for a new structure. Everybody saw what he wanted to see.

The situation was rather different in April. Lord Rosse, now using the 72-inch without ideological constraints, was open to new features such as spiral pattern. Once something has been detected, the information is stored in the brain and it becomes easily visible in later observations and, moreover, can be noticed for other objects too, even if the structure is less obvious. The galaxies M 74 (h 142, NGC 628) in Pisces and M 101 (h 1744, NGC 5457; see Fig. 7.2) in Ursa Major are good examples. Owing to their lower surface brightness (compared with M 51), the spiral pattern is more difficult to see. For M 74 Johnstone Stoney noticed it on 13 December 1848; his rough sketch shows two spiral arms.[50] Concerning 'known facts', Bindon Stoney's note about M 74 on 24 October 1851 is significant: *'spirality would have been overlooked, had it not been previously seen'*. Already on 1 May he had described M 101 as *'large spiral faintish'*. In 1861 Hunter made a fantastic drawing, which shows several arms and knots.[51]

Later the spiral form of M 51, which is apparent in the drawings (see next section), became obvious, and Lord Rosse wrote in 1850 that *'This nebula has been seen by a great many visitors, and its general resemblance to the sketch at once recognised even by unpractised eyes.'*[52] Another point is important: with the detection of spiral nebulae, the focus changed at Birr Castle. Lord Rosse wrote that after *'the spiral form of arrangement was detected […] our attention was then directed to the form of the nebulae, the question of resolvability being a secondary object.'*[53] Again, it was Robinson who overshot the mark – now detecting spiral structure everywhere.

What was the impact of spiral structure on the tarnished nebular hypothesis? Obviously such a feature was able to reanimate it: a spiral nebula could be interpreted as visible testimony for the formation of cosmic bodies out of a spinning luminous fluid under the action of gravity. Such an evolutionary scenario – based on pure materialism and proclaimed by atheists – might have been unacceptable for Robinson. Surprisingly, he later had no ideological problems with spiral nebulae, because there was an elegant way to save his view: he interpreted the rotating mass as an ensemble of bodies (stars). Obviously, these stars were seen in his observations – true nebulosity was obsolete. Of course, his opponent Nichol was fascinated by the spiral nebulae too. Obviously, a rotating self-luminous fluid – a star in the making – leaves no room for individual stars within it. Thus both Robinson and Nichol saw evidence for their own views. As we know today, both views have true aspects, insofar as spiral nebulae are a complex mix of stars, gas and dust.

11.3.2 Lord Rosse's first drawing of M 51

During his 72-inch observations in April, Lord Rosse made a drawing of M 51, which was presented on 19 June 1845 at the 15th annual meeting of the British Association in Cambridge.[54] In the 'Address of the President', John Herschel praised Lord Rosse's new reflector as *'an achievement of such magnitude […] that I want words to express my admiration of it'* (Herschel J. 1845). For him a new period of astronomy has begun

---

landlord, also by the great potato famine then raging in Ireland, so that many of the more important discoveries were made by others working with the telescope in his absence.' (Fernie 1970: 1196–1197). Though this is true for many occasions (mostly later), there is no evidence for early March 1845.

[49] Parsons W. (1850a: 503).
[50] Parsons L. (1880: 21).
[51] Parsons W. (1861, Fig. 35).

[52] Parsons W. (1850: 504). On page 509 he added that *'This object has been observed twenty-eight times with the 6-feet instrument; it had been repeatedly previously observed with the 3-feet instrument.'* However, the 36-inch observations did not reveal the spiral pattern (see Section 6.4.8).
[53] Parsons W. (1861a: 147).
[54] In Kessler (2007, Fig. 2) a sketch from Lord Rosse's observing book, which has been archived at Birr Castle, is presented. It captures the nebula's characteristic form and its companion. However, the statement that it preceded his first drawing of

with the giant reflector, especially concerning the question of the resolvability of nebulae, a matter in which he and before him his father had been greatly interested. Herschel must have known of South's report in *The Times* of 4 April. His key object was the Orion Nebula, which had not been observed with the 72-inch. If it could be resolved at Birr Castle, this would mean that such objects are *'only optically not physically nebulous'*. Given the immense potential of the telescope, Herschel had to explain away his own drawings as illusory.[55]

The Cambridge meeting brought nothing new on the question of 'resolution'. However, an unexpected new matter arose: the spiral structure of M 51. In his opening speech, Herschel said *'I have not myself been so fortunate as to have witnessed its [72-inch] performance, but from what its noble constructor has himself informed me of its effects on one particular nebula, with whose appearance in powerful telescopes I am familiar, I am prepared for any statement which may be made of its optical capacity.'* Obviously he was aware about the discovery, though the M 51 drawing was still unknown to him. In the 'Notes and abstracts of communications' it is not mentioned.

The drawing must have been handed round to the participants during Lord Rosse's talk 'On the nebula 25 Herschel, or 61 of Messier's catalogue'; the content is summarised in the report on the meeting (Parsons W. 1845). The designation '25 Herschel' refers to Fig. 25 in the appendix of the Slough catalogue, which shows the nebula, and '61' is obviously a typo. The text describes a *'working plan of this nebula'*. According to this, Lord Rosse initially used his smallest telescope,[56] which was equatorially mounted, to render *'the great features of the nebula'*. The sketch was then filled in with details, as were revealed by the 72-inch. One further reads that, *'as the equatorial mounting of the latter was not yet complete he [Lord Rosse] could not lay these smaller portions down with rigorous accuracy; yet as he had repeatedly gone over them, and verified them with much care, though by estimation, he did not think the drawing would be found to need much future correction'*.

The original of Lord Rosse's drawing, showing M 51 negative (black on white), is stored in the Birr Castle archive (Fig. 11.10).[57] On the double page one reads in the lower right corner *'Fig. 25 Herschell [Herschel] 51 Messier, sketched April 1845, carefully compared with original on different nights, but no micrometer employed. Handed round the Section at the Cambridge meeting'* (Hoskin 1982a). The upper left corner gives the SC-number (h 1622), the right ascension (13^h 26^m) and north-pole distance (42° 0′).

At about the time of Christmas 1845 John Pringle Nichol visited Lord Rosse at Birr Castle. Both observed M 51, M 42 and other objects. It is probable that Nichol was given a copy of the drawing on that occasion. In the following year the Director of Glasgow Observatory presented a double-sided reproduction in his new book *Thoughts on Some Important Points Relating to the System of the World*, which features the nebular hypothesis. In the introduction one reads that *'The remarkable Spiral Nebula is now published for the first time, through his [Lord Rosse's] kindness; and I am glad to state, that – aided by willing and ingenious artists – my rather venturous attempt to represent the masses of stars in the light in which they appear – viz: white on a dark background, has been considered by his Lordship to be successful.'*[58] (Fig. 11.11 left). Nichol wrote that *'The nebula formerly referred to [John Herschel's drawing of M 51], situated in the Dog's Ear [near η Ursae Majoris], seemed a portrait astonishingly close, of what we might conceive our own galaxy to be; and, although unresolved, it was by common consent considered a mighty cluster.'*[59] One further reads that *'This plate, which is copied from the sketch shown by his lordship at the recent meeting of the British Association, (1845) is an eye sketch only – unverified by micrometrical measurement; though these, when made, will alter no essential feature.'* Another drawing, made by Rambaut with chalk on black cardboard in spring 1848, was never published and is now stored in the cellar of Armagh Observatory (Fig. 11.11 right). It looks remarkably similar to Lord Rosse's.

An impressive new drawing by Lord Rosse, made on 31 March 1848, was presented as Fig. 1 in his publication of 1850 (Fig. 11.12 left). He wrote that John Herschel's 'divided ring' has proved to be *'bright convolutions of the spiral'*[60] and, moreover, *'with each*

M 51 is wrong. The paper compares early representations of the spiral nebula with Hubble Space Telescope images. See also Nasim (2010a).
[55] Schaffer (1989: 205).
[56] Probably the 15″ reflector, which was built in 1830.

[57] Bennett and Hoskin (1981).
[58] Nichol (1846: x); Lord Rosse's notes are omitted.
[59] Nichol (1846: 23).
[60] Parsons W. (1850a: 504).

11.3 M 51 and the spiral structure 489

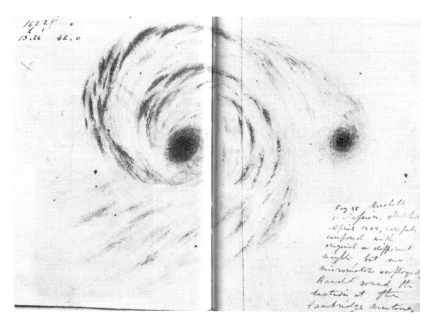

Figure 11.10. Lord Rosse's first drawing of M 51, made in April 1845 (Hoskin 1982a).

Figure 11.11. Left: Lord Rosse's drawing as published by Nichol (Nichol (1846), Plate VI, here rotated 90° clockwise); right: the chalk drawing by Rambaut, made in spring 1848.

successive increase of optical power, the structure has become more and more complicated and more unlike anything which we could picture to ourselves as the result of any form of dynamical law, of which we find a counterpart in our system [the Milky Way]'. However, proving this law by the determination of motion with the micrometer was seen critically: '*Measurements of the points of maximum brightness in the mottling of the*

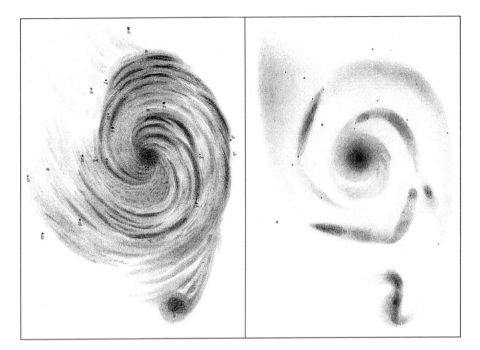

Figure 11.12. Further Birr Castle drawings by M 51. Left: that by Lord Rosse in 1848 (Parsons W. (1850), Fig. 1); right: that by Hunter in 1864 (Parsons L. (1880), Plate IV).

different convolutions must necessarily be very loose.' Lord Rosse, following Robinson, believed that he had resolved the spiral arms into stars sometimes: '*on the finest nights we see them breaking up into stars, the exceedingly minute stars cannot be seen steadily, and to identify one in each case should be impossible with our present means*'. Because both features – spiral structure and resolution – were now fixed in Lord Rosse's mind, they were perceived together.

Hunter observed M 51 six times with the 72" reflector, starting (as his debut at Birr Castle) on 13 April 1860. Five days later he noticed that the companion (NGC 5195) '*is shaped like an S*'. The supposed spirality of NGC 5195 (which actually is an irregular galaxy) had already been mentioned by Lord Rosse, who in 1850: '*saw also spiral arrangement in the smaller nucleus.*'[61] Hunter was responsible for the last drawing of M 51, which was finished on 6 March 1864 (Fig. 11.12 right). Compared with the earlier ones, here the spiral nebula looks more abstract; only the principal features are shown. On 6 May 1874 Copeland contributed a final sketch (shortly before leaving Birr Castle), which shows his measured stars and knots.[62]

11.3.3 Observations of Struve, Bond, Lassell and Chacornac

Otto Struve wrote in a letter to Lord Rosse, dated 2 June 1851 that the '*spiral structure* [of M 51] *is very distinctly seen*' in the Pulkovo refractor.[63] The observations took place in April and May of that year. George Phillips Bond was able to confirm this with the 15" Merz refractor at Harvard College Observatory, which was comparable to Struve's. In 1860 he wrote that '*the spirality is seen with perfect distinctness in a refractor of 15 inches aperture*' (Bond G. 1860). It interesting to read his analysis of the M 51 observations made by William and John Herschel and Lord Rosse – it goes to the core of the matter. Concerning the spiral form he remarks that '[it] *must be within the reach of the 20-foot Herschelian reflectors*', particularly insofar as the object '*had been subjected to a careful examination and description by*

---

[61] Parsons W. (1850a: 510).

[62] Parsons L. (1880: 130).

[63] Parsons W. (1861a: 741–743).

both the Herschels; but neither their drawings nor descriptions furnished the slightest intimation of a spiral structure'. For Bond their drawings are correct, but he added that 'They were simply made at a great disadvantage in the absence of a clear conception of the general plan of structure presented in the object.' According to Bond, William and John Herschel were simply not prepared to see something radically new – the same applies to Robinson, South and Lord Rosse in March 1845. He further wrote that 'Some of the details indispensable to its recognition, being only faintly presented, were overlooked, or, appearing by mere suggestions and glimpses of vision, they conveyed an erroneous impression: in this way the mutual relation of the various parts came to be entirely misconceived.' What then caused Lord Rosse's success in April 1845, which was unprepared too? Bond gives the following answer: 'The missing links were supplied by the large optical power of Lord Rosse's telescope, too plainly not to insure notice; and the nebula then presented itself under a totally different aspect.'

Lassell first reported in May 1862 about his observations of M 51 (Lassell 1862b). The spiral arms of the nebula (and others) were clearly seen in his 48″ reflector on Malta: 'I have indeed carefully observed some of Lord Rosse's nebulae, and in at least two or three instances can fully confirm the spiral structure attributed to them by his Lordship – not, I think, when the objects are well seen, to be overlooked, even when the mind is not previously possessed with the idea.' The result is also mentioned in a letter to Herschel of 23 May 1862, to which drawings of the spiral nebulae M 83, M 99, M 100 and NGC 5247 were appended.[64] Lassell drew M 51 on 27 June and 2 July 1862 (Fig. 11.13); the images appear in his publication of 1867 as Figs. 27 and 27[A], respectively (Lassell 1867a).

In spring 1862 Chacornac observed the nebula with the new 80-cm Foucault reflector, which had been erected at Paris Observatory in January (see Section 8.13). With it he could easily see the spiral structure, as noted in a report communicated by Leverrier (Chacornac 1862b). There a drawing is mentioned. Fortunately, it was recently located at Paris Observatory (Fig. 11.14).[65] The manuscripts found specify the date: 25 April 1862. For the drawing the low-magnification eye-piece 'no. 1'

was used, which probably offered a power of 90. In his report Chacornac wrote 'The appearance of a vortex of small stars surrounding a central star that does not possess the planetary character indicated by Lord Rosse. These stars, of which those near the centre appear as if though a haze of mist, are not the only new ones; up to nine are distributed along the spiral arms of the great nebula and are not to be found on Lord Rosse's drawing. Apart from these objects, the number of which I hope to raise by later observations, I shall locate various crossing branches of this spiral nebulae.' Thus Chacornac noticed certain differences from what Lord Rosse had observed.

Surprisingly, Chacornac supports Herschel's impression of a ring with a chord: 'The configuration of the brightest convolutions, as shown in my drawing, confirms the drawing of Sir John Herschel.' He also describes the companion NGC 5195 and its connection to M 51: 'The arm connecting the small nebula with the large one cuts the two main arms near the [marked] place, or these arms cross each other, so that the entanglement of arms appears like a spherical triangle.' Chacornac adds that 'The nebulous companion is a spiral too and appears like a smooth atmosphere.' This matches the earlier observations of Lord Rosse and Hunter.

Chacornac's report, which was reprinted in the magazine *Intellectual Observer* (number VI), was criticised by William Darby in a letter to the *London Review*, dated 18 October 1862.[66] The main point was Chacornac's review of Lord Rosse's observations. However, Darby (not knowing of his drawing) concludes that the French observer was not aware of the Birr Castle drawing in *PT* 1850. Thus his arguments, favouring his one observation against those made by Lord Rosse, were not well founded. Darby was, for instance, astonished about Chacornac's preference for Herschel's representation of M 51. Later another drawing was made at Paris Observatory by Charles Wolf. His observation took place in 1878, using the new equatorial 47.2-inch reflector by Martin/Eichens.

In Table 11.3 the drawings of M 51 are collected. The column 'S' gives the style: p = positive (white on black background), or n = negative. Rambaut's chalk drawing is stored at Armagh Observatory; that of Tempel was not published and is archived at Arcetri

---

[64] RAS Herschel J. 12/1.7.
[65] Tobin and Holberg (2008).

[66] The whole issue can be found in Darby (1864: viii–x) (in a lengthy footnote).

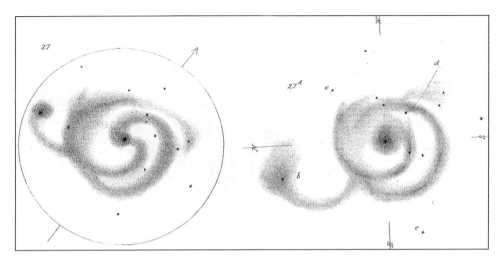

Figure 11.13. Lassell's drawings of M 51, made in 1862 (Lassell 1867a): Fig. 27 (left) and Fig. 27^A (right).

Figure 11.14. Chacornac's drawing of M 51, which was recently found at Paris Observatory (Tobin and Holberg 2008).

Observatory (see Section 6.11.1); Widt used for his drawing the first photograph of M 51 by von Gothard, which was taken on 11 April 1888 in Herény (see Section 11.3.5).

Could the spiral pattern be seen with smaller telescopes? Lord Rosse wrote that '*the discovery of these strange forms may be calculated to excite our curiosity* […] *we are sometimes enabled to recognize a faint object with a small instrument, having had our attention previously directed to it by an instrument of greater power.*'[67] He added that '*The 6-feet aperture so strikingly brings out the characteristic features of 51 Messier, that I think considerably less power would suffice, on a very fine night, to bring out the principal convolutions.*'[68] However, many observers, mainly using smaller telescopes, failed to see spiral structure – and doubted its existence. Ball later wrote that, '*When the extraordinary structure of these objects was first announced, the discovery was received with some degree of incredulity. Other astronomers looked at the same objects, and when they failed to discern – and they frequently did fail to discern – the spiral structure which Lord Rosse had indicated, they drew the conclusion that this spiral structure did not exist. They thought it must be due to some instrumental defect or to the imagination*

---

[67] Parsons W. (1850a: 503).
[68] Parsons W. (1850a: 504). William Henry Pickering remarked in 1917 that '[M 51] *is always very disappointing to those who have seen only its photograph. This is because it is so faint. Nevertheless the spiral structure can without much difficulty be distinguished in the telescope* [11"], *and the two condensations are conspicuous.*' (Pickering W. 1917: 79). Today the spiral structure of M 51 can be noticed with reflectors of 10" aperture (Lord Rosse's 72-inch might be regarded as comparable to a modern 30-inch).

11.3 M 51 and the spiral structure   493

Table 11.3. *Drawings of M 51, sorted by date*

Drawer	Date	Tel.	Site	Reference	S	Figure	Remarks
Bode	1782	Rr	Berlin	Bode (1782), Fig. 15	n		Sketch
J. Herschel	26.4.1830	18.25" Rl	Slough	Herschel J. (1833a), Fig. 25	n	11.9 left	
Smyth	1836	5.9" Rr	Bedford	Glyn Jones (1967b), 182	p	11.9 right	
W. Parsons	April 1845	72" Rl	Birr	Hoskin (1981)	n	11.10	
W. Parsons	April 1845	72" Rl	Birr	Nichol (1846), Plate VI	p	11.11 left	
W. Parsons	31.3.1848	72" Rl	Birr	Parsons W. (1850a), Fig. 1	n	11.12 left	
Rambaut	1848	72" Rl	Birr		p	11.11 right	At Armagh
Chacornac	25.4.1862	31" Rl	Paris	Chacornac (1862b)	n	11.14	Unpublished
Lassell	27.6.1862	48" Rl	Malta	Lassell (1867a), Fig. 27	n	11.13 left	
Lassell	2.7.1862	48" Rl	Malta	Lassell (1867a), Fig. 27^A	n	11.13 right	
Hunter	6.5.1864	72" Rl	Birr	Parsons L. (1880), Fig. 4.1	n	11.12 right	
Tempel	1877?	11" Rr	Arcetri		n		Unpublished; at Arcetri
C. Wolf	1878	47.2" Rl	Paris				Unpublished
Vogel	4.6.1883	27" Rr	Vienna	Vogel (1884), Fig. 1	n	11.15 left	
Widt	11.4.1888	10" Rl		Vogel (1888)	n	11.15 right	From photo. (von Gothard)

of the observer. It will hardly be possible for any one who was both willing and competent to examine the evidence, to doubt the reality of Lord Rosse's discoveries.'[69] In 1864 Darby quoted a very critical voice: '*Spiral! hem! rather say, coil-tracings left on the face of the speculum by the grinder, or the polisher!*'[70] More factual was the leading critic among astronomers, Wilhelm Tempel. Despite his being an experienced observer, he was unable to see M 51 as a spiral nebula in his 11" refractor at Arcetri. His published doubts about the existence of spiral structure eventually led to a bitter conflict with Dreyer (see the next section).

Nevertheless, some critical statements notwithstanding, the existence of spiral nebulae was widely accepted. Most astronomers were quite modest, admitting the inferiority of their equipment. For instance, on 6 October 1861 d'Arrest described M 51 as '*very large and bright with double annulus.*'[71] Obviously his 11" refractor in Copenhagen showed what John Herschel had perceived earlier. Bigourdan, using a 12" refractor at Paris Observatory, noted on 1 May 1897: '*One does not see any detail indicating a spiral nebula.*' (Bigourdan 1917a). Similar impressions were reported by Auwers, Secchi, Schönfeld, Schmidt, Winnecke, Rümker and

---

[69] Ball R. (1895: 286).
[70] Darby (1864: viii).

[71] d'Arrest (1867a: 291). He had first observed M 51 on 1 June 1856 at Leipzig with an 11.7-cm refractor (d'Arrest 1856a: 337).

Schultz. Schönfeld, for instance, noted in his public talk of 1861 about Lord Rosse's discovery that, '*Without taking all these forms for real – since the heavy weight of the* [metal] *mirror can change its form, causing catacaustic lines in the image – one generally must take the drawings of Lord Rosse as being more lifelike than those of Herschel.*'[72]

11.3.4 The conflict between Tempel and Dreyer about spiral nebulae

Tempel was bothered by the popular view that some nebulae are spirals. The published pictures, mainly originating from observations by Lord Rosse and Lassell, could not be confirmed with his 11" Amici refractor in Arcetri. He first formulated his critical view in 1877 in the *Astronomische Nachrichten* (*AN* 2139): '*one cannot fend off the thought that these forms and shapes are only figments of the imagination, even that their description and drawing can be recognised as an endeavour to assign this form to all nebulae* [...] *This statement sounds hard, but why should I withhold it, since it expresses my firm conviction that I'm able to prove.*'[73] Tempel justifies his bold statement by invoking the experience that drawings of one nebula made by different observers (such as Lord Rosse, Lassell and himself) often show great differences: '*If, despite all these large instruments, we are still in the dark regarding the outer shape of the nebulae, how can one claim that these hypotheses built on such a weak basis are more valuable? Harsh criticism in this field of research has long been needed and, even if it should seemingly offend, the exact sciences demand a free opinion.*'[74]

Tempel, being an experienced visual observer, excellent draughtsman and lithographer, is convinced of his right to take the liberty of making this judgement. However, his critical statements about the existence of spiral nebulae did not remain unchallenged. They led to a bitter dispute with Dreyer, who was at that time the assistant astronomer of Lawrence Parsons at Birr Castle. The dispute took place in *The Observatory*.

The story starts in 1878 with a report on Tempel's work written by the editor William Christie (Christie 1878), which was based on the paper in *AN* 2139. It led to a strong reply by Dreyer, who was especially disturbed by the sentence: '*Mr. Tempel further disputes the reality of the spiral form represented in so many of Lord Rosse's drawings.*' Since he was undoubtedly already aware of the German original, he might have been angry about the fact that Tempel's provoking assertions were now being popularised on English territory.

In an open letter Dreyer wrote on 19 January 1878 that '*I cannot agree with M. Tempel in his opinion that the remarkable features represented on many drawings made with Lord Rosse's telescopes are mere optical illusions arising from a sort of pulsation of light, which M. Tempel says he has remarked in some nebulae.*' (Dreyer 1878b). One further reads that '*M. Tempel supposes that the spiral shapes are only creations of fantasy, in which a desire of giving all nebulae this shape is perceptible.*' Dreyer is convinced that the German astronomer is trying to discredit the Birr Castle observers: '*This does not look as if M. Tempel believes much in the good faith of the observers with the 6-foot telescope.*' He feels Tempel's statements to be arrogant, because he '*has never seen any nebula through this powerful telescope*'. Temple's not having seen the spiral structure proves only, "*that his 11-inch is inferior not only to the 6-foot and to Mr. Lassell's 4-foot, but also to the Pulkovo refractor, in which the spiral form 'is very distinctively seen'.*" For Dreyer, spiral nebulae are by no means illusions created in the observer's mind: '*this spiral* [M 51] *and many others exist as such,* in rerum naturae, *and not only in the eyes of a good many observers*' (Dreyer 1878b). He characterises Tempel as a pig-headed, egocentric person: '*any thing he cannot see does not exist*'. There follows an attack on his qualification and competence by comparing him with his great master d'Arrest: '*D'Arrest had a good deal more experience in observing nebulae than M. Tempel has, and he never doubted the reality of the spiral forms, but was always willing to acknowledge the superiority of large reflectors with respect to the bringing out of faint details of nebulae.*'

Tempel replied on 13 March 1878 (Tempel 1878b), cleverly calling attention to a letter he had sent to Dreyer in 1877, in which he announced his intention "*to publish some observations on nebulae in the 'Astronomische Nachrichten'* [*AN* 2139], *containing a criticism of Lord Rosse's drawings of those objects*" to unmask '*the unreflecting transcriptions of mere imitators, whether in pen or pencil,* [rather than] *the pardonable exaggeration of enthusiasm in a man like Lord Rosse*'. Tempel quotes

---

[72] Schönfeld (1862b: 67).
[73] Tempel (1877a: 38).
[74] Tempel (1877a: 39).

a sentence of Dreyer's answer: '*I am very glad that you* [Tempel] *should criticise the drawings made here* [Birr Castle], *of which some are exaggerated, as, for example, Messier 51, which has been far better drawn by Lassell.* [...] *I hope you will prove to the world that* [spiral] *nebulae may be drawn with refractors.*' Obviously Tempel felt betrayed by Dreyer's open letter of 19 January 1878: '*Had Dr. Dreyer informed me of his doubts as to the reliability of my data before the publication* [...], *I should have been able to send him immediately a series of my best drawings* [...] *and I am convinced that his disparaging estimate of me would have been greatly modified, and would have given way to calm reflection and consideration.*'

For Tempel it was all Lord Rosse's fault, because he had not limited his descriptions of nebulae to mere text, but had published drawings encroaching on Tempel's own subject. Thus '*it is my right and opinion as an astronomer and practised lithographer, to express my opinion of them freely*'. Dreyer's remark that he had not seen any nebula through a large telescope was countered thus: '*the increase in size from four to eleven inches gave me data for logically estimating what, and how much more, might be visible with telescopes of 15, 26, or 72 inches*'. Tempel's first telescope was his private 4" Steinheil refractor (1858); his last and largest was the 11" Amici at Arcetri (1875). It would seem very speculative to venture to extrapolate to 72" on the basis of this experience.

The published drawings were mainly criticised due to their large differences: '[they] *cannot have their cause in the different instruments employed, but in the art of seeing and in the power of copying faithfully*'. According to Tempel, not only the skill of the observer is crucial, but also time and diligence: '*Who has the time to gaze at a small nebula for hours altogether, in order to discern its form? The practical astronomer has no time, and the amateur has too many other things to do.*'

Concerning M 51, Tempel writes that it is not mentioned in his *AN* 2139 paper: '*I made no allusion to the nebula* [...] *Dr. Dreyer's remark is uncalled for*'. Next he discusses Lassell's drawings, published as Figs. 27 and 27^A in 1867 (Lassell 1867a), and regarded by Dreyer as particularly accurate. Tempel remarks that they differ distinctly: '*the first* [Fig. 27] *has two spiral curves* [...] *the second has but one*'. He now asks '*Does Dr. Dreyer see nothing remarkable in this discrepancy? and which of the two drawings does he take to be the more faithful transcript?*'

Finally Tempel mentions d'Arrest, whose name had been introduced by Dreyer in a provoking manner: '*Dr. Dreyer can testify with what esteem and veneration I have written of D'Arrest; and this appreciation grows in me day by day from the use of his admirable catalogue* [the *Siderum nebulosorum*].' He now states that d'Arrest wrote '*not a single word to show that he saw them of that* [spiral] *or any similar shape*' and claims that '*Were D'Arrest alive today and could I show him my drawings, with a few observations appended, I am certain that the great master in the realm of nebulae would accord to me greater appreciation than does his younger disciple* [Dreyer].' Eventually Tempel suggests a competition: '*Let Dr. Dreyer draw with Lord Rosse's telescope any nebula not already published, and I will draw the same nebula with Amici I., and we will submit our drawings to a commission to decide on their respective merits.*'

Dreyer replies in an article written on 5 April 1878 (Dreyer 1879). He defends his letter to Tempel, where he asked him to study the spiral nebulae: '*When I wrote that letter I could not expect, that M. Tempel would go so far as to deny the existence of spiral nebulae altogether, apparently on no grounds than his inability to recognize them in his 11-inch refractor.*' He had felt it legitimate to mention M 51 therein, since this is '*the most striking instance of a spiral*'. One further reads '*as M. Tempel, in some notes on nebulae which he kindly sent me for my supplement of the 'General Catalogue', expressly had stated that he did not see M. 51 as a spiral*'.

Dreyer can easily explain the discrepancy between Lassell's two drawings (calling Fig. 27 the better one): '*no one acquainted with the appearance of this difficult object in a large reflector will think it remarkable*'. The faint first arm, as shown in Fig. 27, is difficult to perceive: it is '*not well brought out, except with a fresh speculum and on a good night*'. Concerning d'Arrest, Dreyer states in the first instance that '[Tempel] *says with great triumph that there is not a single word in d'Arrest's work to prove his ever having seen a spiral*'. He then tries to disabuse Tempel: "*It is a pity, that M. Tempel did not take the trouble to read my letter in 'The Observatory' more carefully, as my next remarks on d'Arrest having acknowledged the superiority of large reflectors in this respect would not then have escaped his notice, and he would have written his answer with that calm reflection which he regrets not to have found in my letter.*"

Concerning the difference between an 11″ refractor and the 72″ reflector, Dreyer's trump is a letter from d'Arrest. The Copenhagen astronomer wrote on 13 November 1874 that '*My splendid telescope* [11″ Merz refractor] *is, however, immensely inferior to Lord Rosse's, so that I, although I certainly have tried to make the utmost out of it, cannot at all bear witness against what you see with the 6- or the 3-foot mirrors* [...] *I am very glad that Lord Rosse* [Lawrence Parsons] *and you seem more and more to return to the nebulae. Here every single observation made with the Rosse telescopes is a step forward which nobody else is able to make now-a-days. You have there no competition at all to contend with*'. With this clear statement, Dreyer hoped to convince Tempel – ending with some propitiatory words: '*That many nebulae may be drawn well with a good 11-inch refractor is still my opinion, and nobody will be more delighted to see M. Tempel's drawings published than I will.*'

The issue was not finished, since Tempel repeated his provoking statements in his treatise 'Über Nebelflecken'. The work was completed in December 1879, thus the gap between it and Dreyer's previous writing was not large; the publication, however, appeared after much delay in 1885. The essential statements can be found in the chapter on 'appearance and forms of nebulae'. Tempel ignores the history of the controversy, apart from mentioning his article in *AN* 2139. As a result of careful observations he believed that the '*spiral forms do not exist in the sky and that one can easily deduce from Lord Rosse's remarks about this nebula his addiction to fitting and imposing these spiral forms on most nebulae*'.[75] Tempel feels himself duty bound to point out this issue, because he '*can present sure evidence with true drawings*'.[76] He rates Lord Rosse's addiction for spirals as '*human and thus excusable*'.

Tempel finds similar cases among artists and scholars. As an example he recalls the criticism by his patron Schiaparelli in Milan on the Mars drawings made by Frederik Kaiser in Leiden (completely ignoring the problem of Schiaparelli's '*canali*'). For Tempel it is obvious that '*Neither the telescope nor the air is the cause of these spirals seen from one side only* [...], *but rather the observer's perception alone is to be blamed.*' This finally led to the fact that the drawings do not match; in other words, '*the reason for the differences is due to the draughtsmen*'.[77]

However, Tempel qualifies Lord Rosse's spiral nebulae as a '*new, very surprising idea* [...] *that supports many cosmic conceptions and thus has stimulated and satisfied many a fantasy*'.[78] But, however nice and alluring this idea may be, it holds many dangers: '*it will not easily be exercised, for it has already been too widely disseminated*'. He regrets that some people have already been 'infected' – but castigates persons for misguiding these people in their publications: '*Severe but legitimate censure is deserved, however, for the obsession of many astronomical authors, who, without thinking, without investigating or observing, carelessly and swiftly copy, reprint, ignorantly exaggerate not just these nonexistent spiral forms but many things, possible and impossible hypotheses, thereby forgetting the high standard of exact science and revealing their ignorance, through which a reprehensible literature of truly ridiculous books, leaflets, and deplorable copyings of heavenly bodies has penetrated the beautiful field of our astronomy in a rampant and confusing way.*'

In his review of Tempel's treatise 'Über Nebelflecken', Dreyer does not mention M 51 explicitly, but turns massively against Tempel's view that the instrument does not play a crucial role when drawing a nebula (Dreyer 1887c). He points out that the differences in the Birr Castle drawings (e.g. of M 1) are easily explicable in terms of the variable quality of the metal mirror. Moreover, Dreyer criticised many of Tempel's other views about nebulae (see Section 9.3.2).

11.3.5 Von Gothard and Vogel: first photographs of spiral nebulae

The matter was eventually decided – in favour of spiral nebulae – when Eugen von Gothard photographed M 51 on 11 April 1888 with his 26-cm Browning reflector in Herény (Hungary). After an exposure time of about 2½ hours the spiral form was evident. Hermann Vogel, a friend of von Gothard, presented it (and four other images) in an article of June 1888, titled 'Ueber die Bedeutung der Photographie zur Beobachtung von Nebelflecken'.[79] Since he had already made a drawing

---

[75] Tempel (1885a: 22).
[76] Tempel (1885a: 23).
[77] Tempel (1885a: 12).
[78] Tempel (1885a: 23).
[79] 'About the importance of photography for the observation of nebulae' (Vogel 1888).

11.3 M 51 and the spiral structure

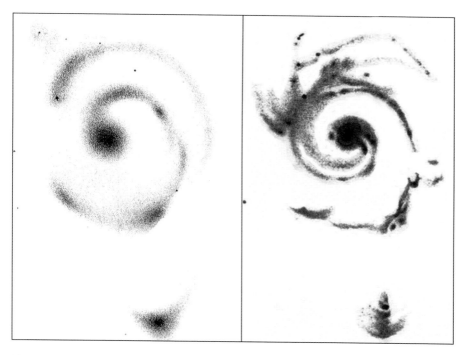

Figure 11.15. Left: Vogel's drawing of M 51 made in 1883 at the Vienna 27" refractor (Vogel (1884), Fig. 1); right: a drawing by Widt, based on a plate taken by von Gothard (Vogel 1888).

of M 51 in 1883 at the Vienna 27" refractor (Fig. 11.15 left),[80] clearly showing the spiral form, he was particularly interested in clearing up the issue.

It is interesting that the figures are drawn copies of the plates (Fig. 11.15 right) made by Seweryn Widt of Potsdam Astrophysical Observatory. This was done because of the very small plate scale resulting from the short focal length of the reflector; e.g. M 51 covers only 4 mm and '*the rich detail could only be rendered by a drawing*'. Vogel wrote that '*The spiral arrangement of the nebulous mass, which was not noticed by J. Herschel, is distinctly seen on the photograph.*' Vogel characterises Lord Rosse's drawing of 1850 '*as least matching the photography*'. He remarked that '*The drawer has rendered the character of the nebula only along general lines, not considering any details.*' Lassell's drawings (Figs. 27 and 27^A) were compared and Vogel regards '*Fig. 27 as decidedly better in agreement with the photography than 27^A.*' His own drawing made at the 27-inch was finally judged as '*most alike to the photograph*'. The other figures in Vogel's publication show the spiral nebulae M 64, M 65, M 66 and M 99. The respective plates were taken by von Gothard in April 1888 too. Obviously Tempel could not watch the new development due to his retirement and advancing illness; he died on 16 March 1889 in Florence.

Common, being an experienced astrophotographer, commented on von Gothard's result in his paper 'Photographs of nebulae' (Common 1888b). In the introduction he wrote that '*It is not many years since the idea of photographing such objects as nebulae would have been considered impossible, more particularly by those who know most about the subject.*' Here John Herschel is meant, who is quoted with the words '*photography might be used for getting the positions of stars near nebulae, in order to assist their delineation by hand*'. Common remarks that '*even he* [Herschel], *whose knowledge of the nebulae and of the powers of photography was so great, does not seem to have indulged in the hope that they would ever photographed directly*'. About von Gothard he wrote that '*The work of Herr von Gothard is a solid contribution to our practical knowledge of nebular photography, and goes a long way towards settling the capabilities of a certain aperture.*' He was particularly impressed by the image of M 51: '*This, the spiral nebula*

---

[80] See Section 8.14.4.

in Canes Venatici (G. Catalogue 3572), is a nebula that is most peculiarly fitted to deceive the eye by the shape of the outlying parts; hence the great difference in the various drawings.' Common's own photograph of 1883, taken with the 36" reflector and an exposure time of 30 min, showed 'less detail'.

11.4 HIND'S VARIABLE NEBULA (NGC 1555) AND ITS VICINITY

NGC 1555 in Taurus, which was discovered in 1852 by John Russell Hind, was the prototype of a variable nebula in the nineteenth century. This object class fascinated the observers, who were trying to confirm William Herschel's idea of cosmic change (evolution). The story of Hind's nebula covers a whole epoch of astronomical activities, from about 1850 until the middle of the twentieth century. It exemplifies the development from visual observing and drawing to photography and modern astrophysics. The object caused misunderstandings and controversial debates, often with surprising turns. Nearly all of the prominent observers are involved. The complete story of Hind's Variable Nebula[81] and its surroundings, which includes Struve's Lost Nebula (NGC 1554), is told here for the first time.

11.4.1 Hind's discovery and early observations

Hind discovered the prominent nebula, which is located about 2.5° north of the Hyades, on 11 October 1852 with the 7" Dollond refractor of Bishop's Observatory, London. In his report, written on the following day and published in the *Astronomische Nachrichten* (*AN* 839), one read 'Last night (October 11) I noticed a very small nebulous-looking object in A.R. $4^h 11^m 50^s$ $\delta = 19°\ 9'$ for 1825.0, the epoch of our ecliptic charts; it was south-preceding a star of 10th mag. which, to my surprise, has escaped insertion on the map for 4h R.A. recently published – possibly it may be variable. The sky at the time was remarkably clear but the object appeared very faint: it preceded the star 1^s2 and was $0'7$ south of it. I suppose it will prove a new nebula, none of our Catalogues having anything in the above position. Its diameter did not exceed 30'.' (Hind 1853a).

Hind refers to his 'ecliptical charts', which he made in order to search for minor planets (see Section 6.6.2).

The nebula was found during a visual check of the chart for right ascension 4^h.[82] Obviously his positions are contradictory (as was noticed by Wilhelm Tempel too): the coordinates (position 1 in Fig. 11.16) and the position relative to the star (2) correspond to two different places, northeast and southwest of the star (the latter fits much better).

The star (a) seen by Hind near the nebula, which is missing on his 'ecliptical chart', is the variable star T Tauri = BD +19° 706. Hind was the discoverer of this object too.[83] However, the designation 'T Tauri' is due to George Chambers (Chambers 1865). Astrophysical studies have shown that T Tauri and Hind's Variable Nebula (NGC 1555, GC 839), located 40" to the west, are physically connected. The star is in an early stage of its life, showing irregular brightness variations between 9.3 and 13.5 mag over a long period.[84] It illuminates a part of the protostellar dust-shell, thus NGC 1555 is a reflection nebula. The variability of T Tauri, which is the prototype of a class of young stars, is transferred to the nebula.

The first to confirm the existence of Hind's nebula was Chacornac. His observations of 1854 in Marseille were communicated eight years later by Leverrier in the *Comptes Rendus* (Leverrier 1862b). Probably he used a 4" Steinheil refractor. Chacornac entered the nebula and the nearby star, estimated as being of 10 mag, in chart no. 13 of his 'Atlas écliptique' (see Section 7.3.1).

In summer 1855 the nebula in Taurus was independently found by James Breen with the 12" Northumberland refractor of Cambridge Observatory. He published a note in *AN* 1024, writing: *'In searching for de Vico's comet [...] I found a small nebula (AR $4^h 10^m$ N.P.D. 70°50') which is not contained in Sir J. Herschel's Catalogue. It is close to and south-preceding a star of the 10th mag. and has a cometary aspect.'* (Breen 1856). In addition to the fact that the right ascension (AR) is too small by 3^m (as was later noticed by Auwers), Breen's position is inaccurate, but from his description it is clear that he definitely saw Hind's nebula.

On 3 November 1855 d'Arrest first observed the nebula in Leipzig, using the 11.7-cm Fraunhofer refractor. He described it thus: *'A pretty bright nebula, 4' diameter; star 10 at its northern end, which follows the centre by 2^s20*

---

[81] The name was probably first used by Darby (1864: 89).

[82] See Hind's letter to J. Herschel of 4 November 1862: RAS Herschel J. 12/1.6.

[83] Hartwig and Müller (1923: 108).

[84] See e.g. Bertout (1980).

11.4 Hind's Variable Nebula

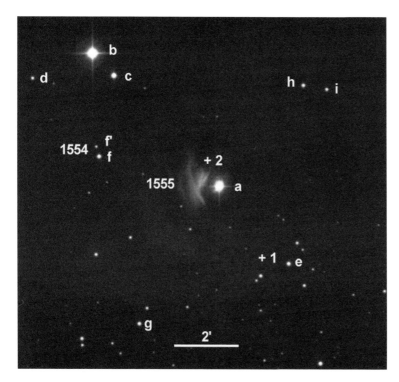

Figure 11.16. The region of NGC 1555 and NGC 1554; south is up (DSS). 1, 2 = Hind's coordinates/relative position; stars (mag): a = T Tauri = BD +19° 706 (10.9 var), b = BD +19° 704 (6.4), c (12.1), d (13.7), e (13.8), f (14.0), f′ (15.0), g (14.3), h (14.4) and i (14.6).

and is about 35″ north of it.[85] The position is near Hind's (2 in Fig. 11.16). D'Arrest saw the object three more times (28 August 1855, 7 and 12 January 1856) without noticing any difference. During the last observation, five days after new moon, the nebula was '*clearly visible*'.

Auwers observed the nebula five times between 7 January and 3 March 1858 with the 11-cm Fraunhofer refractor at Göttingen Observatory, using a power of 45 (Auwers 1862b, d). On all these dates the nebula was '*visible quite easily and without difficulty, but much fainter than it must have been appearing* [to d'Arrest] *in 1855 and 1856*'. However, because there was no micrometer to hand, Auwers could not make any measurements. About the last observation he noted that the object was '*Faint, round; ¾′ diameter; perhaps brighter at the northern edge, star 10^m, 1′ distant, PA 40–45°*.'

11.4.2 The disappearance of Hind's nebula

In his publication Auwers mentions observations by Chacornac of March 1858 stating that the nebula was invisible (Auwers 1862b). They were made with the 25-cm Lerebours refractor of Paris Observatory, which had been erected in 1823.[86] Auwers refers to a report by Leverrier dated 28 January 1862 (Leverrier 1862a). Chacornac had examined the region to correct an error in his chart no. 13. He was astonished that Hind's nebula, shown in his chart, was missing (Leverrier 1862b). Hermann Goldschmidt too could not find the object on 22 November 1859 at Paris Observatory, using a small refractor of aperture 10.5 cm. Possibly these were the first indications of its disappearance. However, due to his own positive observations of 1858, Auwers considered that Chacornac's result had been disproved.[87]

On 3 and 4 February 1861 Schönfeld could 'not see the nebula with certainty' under good conditions, considering it as being only 'possibly present'.[88] He further noted that '*Also in February of the current year 1862 I believed from time to time that I had seen traces*

---

[85] d'Arrest (1856a: 315).

[86] The refractor was briefly described by South (1826: 229).
[87] Auwers (1865: 227).
[88] Schönfeld (1862d: 110).

of the nebula, without being able to be certain about it.' He observed with the 16.5-cm Steinheil refractor at Mannheim Observatory. Hind's nebula is mentioned too in Schönfeld's first note to his Mannheim talk 'Ueber die Nebelflecke', which he gave on 10 November 1861. In this text, added in February, one reads *'Whether the change lies in the nebula itself or is caused by processes that are in space nearer to us and only by chance lie in the line of sight [...] cannot be answered with equal certainty. Is the nebula missing from the catalogues of the two Herschels because it was then invisible, or was it merely overlooked? Is the star actually variable or is its present faintness the result of a solitary occurrence?'*[89]

In 1863 another variable star only 19′ north of T Tauri was discovered by Baxendell: U Tauri = BD +19° 705 (Baxendell 1863). In his report, the light variation of T Tauri between March 1862 and January 1863 is described too; obviously the nebula was invisible in his 6″ Cooke refractor.

At that time, a comparable instrument was being used by Auwers, who was now in Königsberg. His six attempts between January and September 1861 to find the nebula with the 15-cm Fraunhofer heliometer, albeit with the handicap of often misty air, were unsuccessful. But even on the good night of 14 September 1861 only *'traces of the nebula [were] suggested'* (Auwers 1862b), and the exceptional night of 3 November brought nothing too. In a later publication of 1865 Auwers added that *'I very carefully searched for Hind's nebula in Taurus, at Breen's [incorrect] place, but completely in vain. The air was of an exceptional transparency; the nebula has disappeared. The nearby companion [T Tauri] is not 10ᵐ, as seen earlier, but hardly brighter than 11–12ᵐ. It must be variable.'*[90]

D'Arrest observed in Copenhagen at the same time. On 3 October 1861 the nebula was invisible also in the superior 11″ Merz refractor.[91] He tried again on the following night, this time assisted by Schjellerup, but the object had disappeared. D'Arrest's observations in January and February 1862 were unsuccessful too. Fascinated by the case, he wrote in *AN* 1341 *'If I'm not wrong, this is the first recorded example of a variable nebula; my observations in several cases already give me reason to suspect changes in other nebulae. It seems that there will soon be life and movement in the hitherto dead regions of the nebulae.'* (d'Arrest 1862b). Such an object was unique in the sky and thus caused much excitement – at last a promising candidate for William Herschel's idea of the evolution of nebulae had been found.

As early as in December 1861 d'Arrest had sent a letter about the disappearance of Hind's nebula to Otto Struve in Pulkovo. Struve and Winnecke used the first clear night (29 December) to examine the object with the 15″ Merz refractor (Winnecke 1862) – and were successful! In his report 'On the missing nebula in Taurus' Struve wrote that *'in the first moment [we] thought that its faintness exceeded even the power of our instrument; but, after some minutes, when the eyes had sufficiently adapted themselves to the darkness, we distinctly recognised some traces of nebulosity to the south of a star of 11th magnitude [T Tauri]'* (Struve O. 1862). Not until 22 March 1862 was the weather fine again; Struve now saw the nebula even more distinctly, and made a sketch (Fig. 11.17): *'At first glimpse the nebula appeared considerably brighter than it did in December, so that it even bore a feeble illumination of the wires. [...] The difficulty of the first observation made upon me the impression that the brightness of the nebula had considerably increased; but this impression is probably in great part due to the extraordinary transparency of the atmosphere we enjoyed in the month of March.'* The nebula was subsequently monitored by Winnecke with the 15-inch. In 1863 he wrote that *'The strange variable nebulae of Hind in Taurus appeared in Oct. 1863 no brighter than in Dec. 1861 and March 1862, but was undoubtedly visible.'* (Winnecke 1866). Interestingly this successful period

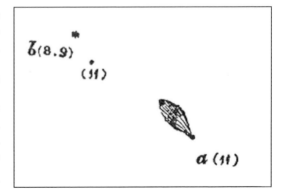

Figure 11.17. Struve's sketch of Hind's nebula; south is up (Struve O. 1862).

---

[89] Schönfeld (1862b: 86).
[90] Auwers (1865: 227).
[91] d'Arrest (1867a: 66–67).

in Pulkovo is peppered with failures in Paris, Rome, Athens, Copenhagen and on Malta.

The Paris observations were triggered by a letter of Hind to Director Leverrier, dated 2 January 1862. Hind calls attention to the strange behaviour of the object. Leverrier quoted some phrases in his *Bulletin* of January 29: '*I [Hind] suspected at the time that the star which almost touched the n.f. edge of the nebula must be variable from the fact of it having escaped notice during the frequent examination of that part of the heavens on previous dates, but the idea of a variable nebula never occurred to me.*' (Leverrier 1862a). Confronted with this issue, Hind appears a bit helpless: '*What now becomes of our theories as to the stellar constitution of these objects?*' He is supported by John Herschel, who esteemed the nebula (according to Hind) as '*one of the strangest facts in astronomy*'.

Owing to bad weather and moonlight, Leverrier and Chacornac could not start their observations until 26 January 1862. In the 31.6-cm Secrétan refractor of Paris Observatory the nebula was not visible. Fortunately there was a superior alternative: the new 80-cm reflector by Foucault, which had been in operation since 17 January. It was the first large telescope with a silvered glass mirror. Surprisingly, even this telescope yielded nothing. Further attempts with the refractor, when the weather was fine again between 14 and 18 February, didn't change anything. The disappearance of the nebula was a topic at the meeting of the Académie des Sciences of 17 February (Leverrier 1862b). Leverrier's report is nearly identical with his *Bulletin* (29 January 1862).

Secchi in Rome too tried to find the nebula on 27 January 1862 with the 24-cm Merz refractor. The negative result was communicated in a letter to Hind, who mentions it in a report of 3 February in the *American Journal of Science* (Hind 1862a). Therein Hind writes about the attempts of Chacornac and Leverrier in Paris too. Obviously, the successful Pulkovo observation of 29 December 1861 was unknown to him. He assumed that the nebula had disappeared either in 1856 or early in 1857 (about this issue he had written a letter to *The Times*, which appeared on 4 February 1862). Hind was the first to conjecture about the possible causes of the disappearance, coming pretty near to the truth: '*How the variability of a nebula and a star closely adjacent is to be explained, it is not easy to say in the actual state of our knowledge of the constitution of the sidereal universe. A dense but invisible body of immense extent interposing between the Earth and them might produce effects which would accord with those observed; yet it appears more natural to conclude that there is some intimate connexion between the star and the nebula upon which alternations of visibility and invisibility of the latter may depend. If it be allowable to suppose that a nebula can shine by light reflected from a star, then the waning of the latter might account for the apparent extinction of the former; but in this case it is hardly possible to conceive that the nebula can have a stellar constitution.*'

Julius Schmidt, Director of Athens Observatory, again played a curious role, which was typical for him. Being interested in variable nebulae, he had read d'Arrest's note in *AN* 1341, which, however, gives no position. In *AN* 1360 Schmidt wrote that he had got the information about "*which nebula was meant [...] from Argelander's description in the* 'Kölnische Zeitung'" (Schmidt 1862b). He searched for the object from 19 to 25 March 1862 with his 15.7-cm Plössl refractor; however, '*no trace of the nebula was seen*'. Schmidt further wrote that '*On March 26 and 27, after the lenses of the refractor had been cleaned again, I once again scanned the field, but became convinced that the nebula could not be seen, even if the star of 9^m, which is 1' north, was masked by the ring of the micrometer.*' The experienced observer was desperate: '*There is no earlier observation in my journal, though the nebula was positioned in vain several times in 1860.*' The reason for his failure was later explained by Auwers: the wrong star was observed! He wrote that '*an indication by Schmidt, A.N. 1360, 9^m, is due to a mix-up with a Bessel star*' (Auwers 1862b). Actually Schmidt had considered BD +19° 704 to be T Tauri, which is almost 2 mag brighter and located to the southwest (b in Fig. 11.16); but, even using the correct reference star, there was no chance of seeing the faint nebula in his small refractor at that time.

D'Arrest summarised the chronology in an article of 20 May 1862 (d'Arrest 1862c). In the section 'Variabilität bei Nebelflecken' ['Variability of nebulae'] he describes the observations of Hind, Chacornac, Breen and Otto Struve. For Lassell's attempt with the 48" reflector on Malta he refers to 'Leverrier's Bulletin of 17 April 1862', which is actually a letter of Lassell to Leverrier of April 5 (Leverrier 1862c). Therein he first apologises for the late observation: '*Although I generally receive the London* Times *Newspaper, an accident prevented my seeing that copy which contained Mr. Hind's letter on the subject of the lost nebula in Taurus until long after its publication.*' He continues: '*I immediately pointed my large Equatorial Reflector [...] to the place where the nebula formerly existed, but I was not able to detect certainly any nebula there at all.*' Despite the moderate observing conditions (due to zodiacal light) Lassell concludes that '*there can be no doubt of the fact of the nebula having disappeared – strange and unaccountable as such a phenomenon may be.*'

Since Lassell's observations were unsuccessful (as was later mentioned by Struve too), d'Arrest states in his review that the '*Pulkovo sighting [by Struve and Winnecke] of the exceedingly faint object is perhaps the greatest triumph of the local refractor*'. Unfortunately his collection of observations of Hind's nebula contains a few errors. For instance, James Breen is confused with his brother Hugh.[92] Auwers later lamented that his own observations had been '*incorrectly cited*' (Auwers 1862b).

In his own report, Auwers compiled his observations in Göttingen (1858) and Königsberg (1861). He concludes that Hind's nebula '*did not become visible for telescopes of medium aperture until 1852, reached its maximum brightness in 1856 and vanished again in about 1860*' (Auwers 1862b). A similar statement appears in his list of new nebulae (1862),[93] showing the object as no. 20. Auwers' remarks given in the publication of 1865 are comparable.[94]

In a letter to John Herschel of 4 November 1862,[95] Hind mentions the wrong positions of himself (*AN* 839) and Breen (*AN* 1024); he is happy that in the meantime Breen has confessed his error in the *London Review*, which error must have happened either during the observation or in the data reduction. Challis noted in the same journal that Breen's notes about the observation were not to be found at Cambridge Observatory. About his own imprecise position (1 in Fig. 11.16), Hind wrote that it had been '*estimated upon our eclipical chart for 4^h*'. He precessed the coordinates to 1862 and compared them with those, derived by John Herschel using d'Arrest's observations; the result '*agrees nearly enough with d'Arrest's for an estimated place*'. This is quite a favourable formulation, because Hind's position is northeast of T Tauri, whereas d'Arrest's (at the right place 2) is southwest of it, i.e. in the diametrically opposite direction.

On 12 December 1863 Hind re-examined the field. At Meadowbank Observatory in Twickenham (the successor to Bishop's Observatory) he was again able to use the old 7" Dollond refractor. The night was as clear as it has been at the time of discovery, when the nebula '*could not be overlooked with a low power on his [Bishop's] 7-inch refractor*' (Hind 1864). Hind explained the delayed observation thus: '*I've long been anxious, since the variability was detected, to examine the vicinity under the like favourable conditions but, from one cause or another, have never succeeded until the above mentioned date [12.12.1863].*' Moreover, Hind was strongly involved in the making of the *Nautical Almanac* from 1853 onwards and was not in the best of health either. The result of his new observation was unambiguous: '*On applying the same power, with which I saw it readily in 1852, and repeatedly distinguished it afterwards, when the sky was not so clear, I was unable to perceive the least trace of it [...] there was an entire absence of nebulosity*'. This was confirmed by his assistant George Talmage, '*whose sight is remarkably acute*'. Chambers summarised the observations of Hind's nebula in the 1877 edition of his book *Descriptive Astronomy*.[96]

11.4.3 The nebula becomes temporarily visible again

In 1863 Otto Struve visited Lassell on Malta to inspect the 48" reflector (see Section 7.5.2). On 10 October Hind's nebula was observed. In Struve's report of 9 December one reads the following: '*Concerning the variable nebula of Hind [...] Lassell, as is generally known, has recounted that he was not able to detect it in spring 1862, while it was continuously visible under somewhat favourable conditions in the Pulkovo refractor. I guessed then that possibly the position of the nebula was not accurately enough known to Mr Lassell, and this conjecture now seems to be confirmed, since we recognised the nebula very well on Oct. 10.*'[97] Lassell had already doubted his vision, as Struve reported: '*Now Lassell saw it and was thereby, so it seems, somewhat cured of the idea that his former acuity of the eye, which had led to so*

---

[92] Hugh Breen was an assistant of Airy in Greenwich; about the Breen family of astronomers see Brück M. (1999).
[93] Auwers (1862a: 74).
[94] Auwers (1865: 227).
[95] RAS Herschel J. 12/1.6.

[96] Chambers (1877: 543–545).
[97] Struve O. (1866a: 538).

11.4 Hind's Variable Nebula

many discoveries, had been lost.' Struve now describes the differences between the Pulkovo and Malta observations: *'Although not having the impression that the nebula here was more easily visible than it appeared to us at Pulkovo in March, I, however, must admit that we could perceive here particularities of the object that our refractor was unable to show similarly. Whereas in Pulkovo it had the appearance of two nebulous streaks merging at an acute angle, here we could clearly discern it as consisting of three or four individual masses separated from each other by black sky background, or at most linked to one another by exceedingly faint nebulous traces.'*

It is owing to Thomas Webb that *'Hind's wonderful nebula in Taurus'*[98] became known to a general public, at least in the English-speaking regions. He wrote the column 'Cluster and nebulae' in the popular magazine *Intellectual Observer*. There he warned visual observers that *'The student, though he would now look in vain for this mysterious nebula, may be glad to see its data, as given by d'Arrest and Auwers.'*[99] Webb's friend George Knott could not see the nebula with certainty in his 7.3" Clark refractor (Knowles Lodge, Cuckfield): *'I have never seen any trace of Hind's nebula, but on two occasions in 1877 suspected a minute glimpse star near the place.'* (Knott 1891). In Germany the Vienna astronomer Karl von Littrow popularised Hind's nebula in his *Atlas des gestirnten Himmels* of 1866.[100] It is mentioned as an example of a 'variable heavenly body' (together with Tempel's Merope Nebula and Schönfeld's NGC 1333).

John Herschel, who had never observed the object, entered it in his General Catalogue as GC 839, using d'Arrest's Leipzig position (which agrees with Auwers'). The history of Hind's nebula is mentioned in the 'Notes on the catalogue'.[101]

In 1865 and 1866 Vogel was able to see the object five times at the new observatory in Leipzig-Johannisthal. For the first four observations the 'sechsfüssiger Refractor' by Fraunhofer (aperture 11.7 cm) was used. On 1 August 1865 he saw Hind's nebula under very good conditions ('air 1–2') as *'very faint, irregular round'*;[102] then, on 26 August, as *'faint pretty large, irregular round'* ('air 2'); however, a day later under similar conditions it appeared *'bright large, irregular round'*, but on the 28th the nebula was *'extremely faint'* ('air 2–3'). On 18 August 1866 Vogel observed with the 'zwölffüssiges Aequatoreal' by Steinheil (aperture 21.5 cm), finding the nebula *'pretty bright, irregular round, diameter 130", partly resolved'* ('air 1–2').[103] All in all quite different results, which he might have attributed to variability (there is no note about it). Vogel abstained from measuring the position with a micrometer. According to his notes (using the designation GC 839 for the first time), he could *'not see anything nebulous'* on 13, 14 and 20 November 1865, probably with the 21.5-cm refractor.[104]

Schönfeld, who had been interested in variable stars ever since his work for Argelander in Bonn, listed T Tauri in his first catalogue (Schönfeld 1865).[105] In the second, of 1866, he wrote that *'Its variability was soon after d'Arrest's discovery noticed simultaneously by Auwers, Chacornac and Hind. Owing to the incomplete publications, it was impossible for me to derive elements [period, amplitude], but the variability itself is confirmed by my local observations.'* (Schönfeld 1866). In *AN* 1648, which appeared in 1867, he reported about his observations with the Steinheil refractor in Mannheim, which covered the following periods: 2 January to 20 March 1865, 13 December 1865 to 15 April 1866 and 28 November 1866 to 13 April 1867. He wrote about the nebula that *'The neighbouring Hind–d'Arrest nebula was invisible at all times in the refractor.'* (Schönfeld 1867). However, for 13 April 1867 one reads the following: *'Indeed, it seems to be a faint object of doubtful nature preceding T Tauri that I have not seen previously.'* This and other observations of T Tauri (until 1873) were later published by Wilhelm Valentiner, Schönfeld's successor at Mannheim Observatory (Valentiner 1900). The nebula remained invisible.

Winnecke looked for the object in vain too. About his observation of 2 November 1875 with the 16.3-cm comet-seeker in Straßburg he wrote *'Air very transparent. – No clear trace of the nebula (near T Tauri); perhaps it is very faintly visible in its associated place.'*[106] An added sketch shows it as an oval spot at d'Arrest's

---

[98] Webb (1864a: 61).
[99] Webb (1864d: 450).
[100] *Atlas of the Starry Heavens*; (Littrow K. 1866: 58).
[101] Herschel J. (1864: 18).
[102] Vogel (1867: 36–37).
[103] Vogel (1867: 59).
[104] Vogel (1867: 83).
[105] Concerning the publication of the first catalogue of variable stars, he was beaten by Chambers (Chambers 1865).
[106] Becker E. (1909a: 26).

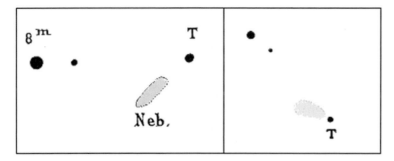

Figure 11.18. Winnecke's sketches of Hind's nebula, made in 1875 (left) and 1877 (right); south is up (Becker E. 1909a).

position. A second observation took place on 15 January 1876; it too was in vain. Exactly a year later, the situation had changed; Winnecke now saw a '*nebulous mass, sticking out from T Tauri nearly as far as the 11m star from that 8ᵐ*'. The line T Tauri–nebula draws an angle of 20° to that of T Tauri–star 8ᵐ'. Again a sketch was made, which differs from the first (Fig. 11.18).

11.4.4 Struve's Lost Nebula (NGC 1554) and Tempel's drawing

In 1868 a new surprising development arose, which was first reported by d'Arrest on 26 March in *AN* 1689 (d'Arrest 1868b). He wrote that near to T Tauri '*very strange changes have appeared*'. He refers to an (obviously private) communication by Otto Struve about an observation on 14 March 1868 with the 15" Merz refractor, which reads '*Now Hind's nebula has completely disappeared in the Pulkovo instrument too. However, 4 arcminutes apart from the position of Hind's nebula, at a place which I had so often investigated with the most powerful telescopes in Paris, Petersburg, Valetta [Malta] and Copenhagen, a new nebula has become visible.*' D'Arrest was able to confirm it in several observations (on 23 to 25 March): '*Struve's nebula is pretty small, nearly round, with an eccentric core like a star of 14th magnitude […] approximately of Herschel's second class. […] on the other hand, in the year 1856 Hind's variable nebula was larger and much brighter, as I remember weakly of first class.*' The new nebula (NGC 1554) was later termed Struve's Lost Nebula – a name characterising the object's fate.[107]

Lawrence Parsons looked for Hind's nebula in vain three times with the 72" reflector at Birr Castle. In his 1880 publication one reads[108] '*not found*' (11 October 1872), '*nothing decisive, sev e F patches suspected*' (12 December 1876) and '*No nebulosity seen with certainty near the star. […] Some object in Pos. 257°.5±, Dist. 231"±, I am almost sure it was only a star 14–15m., probably the star Struve's Nova from 1868 was then attached*' (9 January 1877). Lawrence Parsons' assistant Dreyer confirmed the negative result; the note on GC 839 in the GSC reads '*At present [1877] there is no nebulosity distinctly visible, neither round this star, nor near the well known variable star.*'[109] Dreyer mentions that Ralph Copeland was unsuccessful too, using the 15.1" Grubb refractor in Dun Echt: '*I am in perfect accordance with Dr. Copeland, observing with the large Dunecht refractor.*'

The data of d'Arrest and Lawrence Parsons clearly point to the star pair f–f', 3' north of BD +19° 704 (b, see Fig. 11.16). The components have magnitudes of 14.0 and 15.0 and are 20" distant. This faint double can look nebulous in a large telescope. Struve's nebula was first catalogued in the GCS as GC 5339 at d'Arrest's position and with the description '*!!! var. S, R, Nu = star 13*' ('variable, small, round, nucleus = star 13 mag'). In his note on GC 839 Dreyer mentions a private communication by Struve: '*M. Otto Struve informs me, that he, from time to time, has observed the variable Nebula* [Hind], *but that he avoids reducing and comparing his observations for fear of being preoccupied with respect to this minimum visible. He does not consider the nova from 1868* [Struve] *as separate nebula.*'

---

[107] The name Struve's Lost Nebula was probably introduced in 1946 by Sven Cederblad.

[108] Parsons L. (1880: 41).
[109] Dreyer (1878a: 389).

11.4 Hind's Variable Nebula

This surprising change of mind is explained by quoting Struve: '*What I* [Struve] *see is certainly the variable Nebula itself, only in altered brightness and spread over a larger space. Some traces of nebulosity are still to be seen exactly on the spot, where Hind and d'Arrest placed the variable Nebula.*' This contradicts d'Arrest's observation, which, however, dates back 10 years. Struve's latest result is entirely incompatible with an observation made in 1877 by Wilhelm Tempel.

On 8 November 1877 Tempel saw Struve's nebula – at the right place – with his 11" refractor (Amici I) in Arcetri. He discovered, as described in *AN* 2212, '*a nebulous glow of 1½' diameter, in which a faint star appeared to the north. This faint nebula was about 1' south and 15^s preceding the variable star* [T Tauri].' (Tempel 1878d). Using d'Arrest's description, he was convinced that the object is '*identical with Struve's nebula*'. On the 12th '*the nebula was not seen, but two faint stars instead, the northern being the same as the one I already saw on Nov. 8 in the nebula and the southern follows at about 40'''*. He added that '*the brighter star lies eccentrically in the nebula, but d'Arrest mentions the south following star, and my observation of the northern star gives only 16^m, while d'Arrest gives 14^m, which certainly indicates a variability, since d'Arrest's 14th class is equal to my 12th class*'. Obviously Tempel had misunderstood d'Arrest, who does not speak about stars at all, but merely mentions a round nebula with an eccentric nucleus (as the brightest point), which appears like a star of 14 mag. The very centre (perhaps interpreted by Tempel as the second star) is located south of it. Did Tempel really see the pair f–f′ (see Fig. 11.16)? That would be surprising, since it was later extremely difficult for Barnard to perceive the 15-mag star f′ in the great refractors of Lick and Yerkes Observatories.

Concerning the direct vicinity of T Tauri, Tempel had written in a previous article that '*quite near to the variable star a nebulous glow was perceived*' (Tempel 1877a), but about his observation on 5 November 1877 he noted that '*the small 11^m, star described as variable no longer has anything nebulous about it*' (Tempel 1878d). Furthermore, for 12 December one reads that '*while the sky was clear I saw two faint stars quite near to the variable star, but no longer any nebulous glow, which I had seen so often before*'. Tempel's conclusion was that '*both nebulae, that around the variable star and Struve's, had emerged as stars*', but with the proviso that '*The nature of a nebulous appearance must apply only for certain stars, since faint stars show no nebulosity.*' He refers to '*two small star groups*', which he claimed to have detected a few arcminutes to the north and northeast of T Tauri, which '*were not seen as nebulae*'. The northeastern object might have been the small group around the 13.8-mag star e (see Fig. 11.16); the northern one was perhaps identical with star g (14.3 mag).

Tempel was a gifted drawer, who always pleaded that one should fix observations graphically. In 1877 he made a map of the T Tauri region (Fig. 11.19). His own results are marked a to d. The numbers 1 to 5 refer to the data of Hind, d'Arrest, Auwers and Struve. However, Auwers' and d'Arrest's positions (*SN*) of NGC 1555 are incorrectly plotted. According to Tempel's legend we have Var = T Tauri, a = 'nebulous glow' near T Tauri, b = Struve's nebula (NGC 1554); c and d = 'small star clusters', 1 = Hind's position of NGC 1555 (according to his coordinates), 2 = Hind's position of NGC 1555 (according to his description), 3 = Auwers' position of NGC 1555 (GC), 4 = d'Arrest's position of NGC 1555 (*SN*), and 5 = the Struve–d'Arrest position of NGC 1554. Concerning Tempel's 'nebulous glow' near T Tauri, which he later saw as 'two faint stars' (a on his map), it is doubtful that it is identical with NGC 1555. At that time Dreyer was unable to see Hind's nebula with the 72-inch. However, the pair of stars labelled (a) does not exist. If the other pair, labelled (b), is identical with f–f′, the orientation is wrong, anyway.

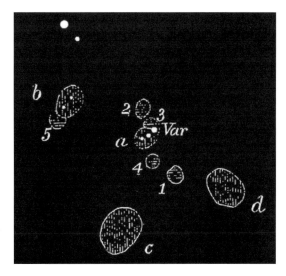

Figure 11.19. Tempel's map of the T Tauri region, made in 1877; south is up (Tempel 1878d).

Engelhardt in Dresden looked up Hind's nebula with his 30.6-cm Grubb refractor. About his observation of 31 December 1883 he wrote '*I see no trace of a nebula around this star.*'[110] Two further attempts (on 16 December 1884 and 8 January 1885) were unsuccessful too. Since no position of NGC 1555 (GC 839) could be determined, only T Tauri was measured.

Dreyer entered the nebulae of Hind and Struve as NGC 1555 and NGC 1554, respectively, in the New General Catalogue, using the positions of d'Arrest. In the notes the status of 1887 is summarised thus: '*The latter [NGC 1555] is the well-known nebula found by Hind, Oct. 11. 1852 (Astron. Nachr. 839), observed by d'A[rrest] at Leipzig four times in 1855–56 as a pB or pF neb, about a minute in diameter, and found missing by d'A in Oct. 1861. G.C. 5339 [NGC 1554] is a neb, S. with an eccentric nucleus = star 14 mag, which was found early in 1868 by O. Struve, and was also observed by d'A (Astr. Nachr. 1689), who was sure that no nebulosity had formerly existed in that place (4′ p Hind's nebula). This object must also have disappeared since, as I was unable to perceive any nebulosity near the place with Lord Rosse's 6-foot Reflector in 1877. The place has also been examined of the late years by Tempel (A.N. 2212) with a similar result. The place for G.C. 839 [NGC 1555] in the catalogue is that resulting from d'A's four obs. at Leipzig.*'[111]

On 7 January 1886 Bigourdan tried to observe both objects with the 12.4″ Secrétan refractor in Paris.[112] He described NGC 1554 as a: '*Faint star of 13.3–13.4 mag, possibly accompanied by an extremely faint nebula.*' The faint star with its supposed, extremely faint nebula is the one which had already been described by Struve, d'Arrest and Tempel (f in Fig. 11.16). A second observation was made on 12 December 1890. Once again he thought that he had seen traces of the nebula 4′ north of the star BD +19° 704 (b): '*One cannot see the nebula with certainty, however, the star could be accompanied by a very faint nebula. At the NGC-position nothing was seen, i.e. at BD +19° 704 (0m0s, +4′).*' Thus Bigourdan was the last to have seen Struve's nebula – if it ever existed.

However, Hind's nebula was '*Not seen. According to the NGC-coordinates this nebula must be at the place of BD +19° 706, this is the variable star T Tauri, whose brightness is at present 12.5–13 mag. Engelhardt could no longer see the nebula.*' But Bigourdan found a new 'nebula', which he catalogued as no. 144 in his discovery list (Bigourdan 1891b). He wrote '*Star 13.4–13.5 mag, which could be surrounded by nebulosity. The existence of this object is certain.*' The further story of this 'nebula' is quickly told: it does not appear on any photograph and was never mentioned again. The object (wrongly called '143 Big.') is the 14.4-mag star h (Fig. 11.16), 4′ southeast of T Tauri.

11.4.5 Observations of Burnham and Barnard

Already prior to Bigourdan's find, the number of 'new nebulae' in the field of T Tauri had increased in a curious manner. When the 36″ Clark refractor at Lick Observatory, which was the largest telescope in the world, was ready to use in 1888, it was only a question of time before someone would point it at Hind's nebula. This was up to Burnham on 11 October 1890, shifting the issue from Europe to America. However, the observation was accidental, being occasioned by his favourite target: double stars. Near to T Tauri he discovered, under optimal atmospheric conditions, a very close pair: '*Not long since I found a new double star in the vicinity of Hind's supposed variable nebula, and took the occasion when the measures of the new pair were finished to examine the place of the nebula.*' (Burnham S. 1890).

Burnham's observation, made almost in passing, led to astonishing entanglements. The reason was detected much later: an error in the Bonner Durchmusterung. Therein the star BD +19° 706 (T Tauri) is catalogued with wrong coordinates. The difference is small, but accidentally identical with that between T Tauri and Hind's nebula! Thus, according to the Bonn catalogue, the star is located exactly at the position of the nebula given by d'Arrest – an error with dramatic consequences!

Burnham was characterised by his friend Barnard as someone '*whose excellent custom it is to measure everything he comes across that is measurable*' (Barnard 1895b). The experienced observer had great respect for all those who were committed to exact measurements. Since, of course, Argelander and Schönfeld were among them,

---

[110] Engelhardt (1890: 135).
[111] Dreyer (1953: 214–215); it is curious that the number 839 appears for the *AN* and the GC.
[112] Bigourdan (1905a: F8–F11).

11.4 Hind's Variable Nebula

Burnham did not doubt for a second that the BD data were correct. He stated that '*The place of the nebula, as given by Dreyer* [NGC], *on the authority of d'Arrest, is identical with that of D.M. +19°, 706, the magnitude of which was estimated by Argelander as 9.4; and this is T Tauri of the variable star catalogues.*'[113] (Burnham S. 1890). Misguided in this way, Burnham searched for NGC 1555 with the 36-inch at the wrong position: that of T Tauri – and surprisingly was successful! Glad to have recovered Hind's Variable Nebula after its long period of invisibility, he noted that '*This small star, if it is a star, is placed in a very small condensed nebula.*' However, due to the location of the object, the slightly oval shape and a measured diameter of only 4.4″, he raised slight doubts about its identity: '*this description of the nebula does not correspond with that in the early observations, where it was noted, when it was seen at all, as about 1′ in diameter*'.

Burnham immediately asked Barnard for help, due to his '*great experience in work of this kind, and remarkable acuteness of vision in detecting extremely faint, diffused objects, which would escape the ordinary observer*'. On 15 October 1890 both of them observed T Tauri with the Lick refractor. Barnard confirmed the object as '*conspicuous and definite, not a nebulous glow*'. Was it identical with the 'nebulous glow' Tempel claimed to have seen in 1877 close to the star? Taking into account the small size and the inferior telescope, this is most unlikely.

Barnard's extraordinary vision yielded yet another surprise: '*an excessively faint, round nebula, about ¾′ from the one previously described, in the estimated direction of 185°. This faint nebulosity was about 40′ or 50′ in diameter, and apparently not connected with the variable, and was of the last degree of faintness for the light-power of the large instrument. It is too faint for any other telescope.*' Burnham had to admit '*I should not have seen it independently.*' In 1899 Barnard published a sketch of his observation, showing the small nebula around T Tauri and the large one to the south of it (Fig. 11.20). Obviously, the large 'Barnard nebula' is the real 'Hind nebula'! Because the experienced observers had not realised the BD error, they did not know about the identity at that time. Moreover, they obviously were convinced that NGC 1555 had vanished. An attempted

---

[113] Instead of BD, often DM (Durchmusterung) is used.

Figure 11.20. Barnard's observation of T Tauri made on 15 October 1890 (Barnard 1899).

observation of Struve's NGC 1554 was unsuccessful. They inspected the field with the 12″ Clark refractor too, which revealed no object at all. Burnham's nebula '*was completely lost with the smaller instrument*' and T Tauri appeared '*precisely like any ordinary star*'.

On 1 November 1890 Keeler entered the scene, investigating the tiny Burnham nebula around T Tauri with a visual spectroscope at the 36-inch. Detecting a weak line at a wavelength of 5005 Å, he supposed the presence of a gaseous nebula. This is the prominent O III line of doubly ionised oxygen, arising at high temperature (due to the exciting star) and low gas density. In contrast to NGC 1555 (reflection nebula), Burnham's object is an emission nebula.

Motivated by Burnham's report, Isaac Roberts tried to image the field. Since October 1890 his new 'Starfield' Observatory in Crowborough had been ready.[114] Using the 20″ Grubb reflector, equipped with a silvered glass mirror, the first image of T Tauri was made on 9 December 1890. However, the result of the three-hour exposure was bleak: '*The photograph does not show any nebulosity or nebula or nebulous star anywhere about the region.*' (Roberts 1891). The self-confident Roberts claimed that the failure was not due to his equipment but caused by '*rapid changes [...] in the object between October 15, 1890, when observed at Lick, and 9 December when the photograph was taken*'. That was, of course, too much variability – the objects around T Tauri were simply too small to appear on Roberts' plate.

Burnham was able to observe Barnard's faint nebula near T Tauri – still not identified as Hind's – several times in September and October 1891 with the 36-inch: '*At first it was thought to be a little brighter than in 1890, but subsequent examinations made this doubtful.*' (Burnham S. 1892b).

For the next three years no observation of the field was made until Johann Georg Hagen used the 12″ Clacy refractor at Georgetown College Observatory. Obviously, he had no problem with detecting Hind's nebula at the correct position: '*–2s, –0.7′ from T Tauri* [the nebula] *was seen between Dec. 1894 and Jan. 1895*'.[115] The object seemed to have reappeared. Anyway, a critical assessment of Hagen's observation will be given below.

Dreyer summarised the status of 1894 in the notes to his IC I.[116] Unfortunately his report starts with an error, which has led to some confusion: '*N.G.C. 1554 Hind's variable nebula*'. Of course, it should read 'NGC 1555'. Struve's nebula was clearly not meant, as the subsequent text shows: '*2s p and 40″ south of the variable star T Tauri. Barnard in 1890 found an e F neb in Pos. 185°, dist ¾′ from T, which agrees well with Hind's and d'Arrest's observations.*' Thus, Dreyer was the first to point out a possible identity of Barnard's nebula with NGC 1555. About the former he noted that '*Barnard and Burnham also saw T Tauri within a very small condensed nebula (often seen by Tempel).*' The invocation of Tempel is, however, somewhat bold. Obviously, Dreyer identifies Tempel's 'nebulous glow' near T Tauri with Burnham's object around the star, but the latter could by no means be seen with the 11″ refractor! About Bigourdan's new nebula Dreyer notes that '*Bigourdan's No. 144 (star 13 nebulous?) [...] was apparently not seen at the Lick Observatory.*' With good cause, he had refrained from cataloguing the dubious object.

11.4.6 Barnard corrects a fatal error in the Bonner Durchmusterung

Concerning the further observations, the matter was left completely to Barnard. In 1895 he wrote a comprehensive report, titled 'On the variable nebulae of Hind (N.G.C. 1555) and Struve (N.G.C. 1554) in Taurus, and on the nebulous conditions of the variable star T Tauri' (Barnard 1895b). Therein he stressed that '*The subject of the variability of the light of a nebula is of the highest importance.*' Barnard is sceptical, arguing similarly to Lamont 60 years earlier, who claimed that supposed changes of nebulae are mainly due to the observer (especially his experience), the quality of the instrument and the seeing (see Section 6.1.3).

In his report, Barnard describes the observations of Hind, d'Arrest, Struve, Tempel, Dreyer and Burnham, mentioning his own role as 'eyewitness' too. He presents his latest results, which were based on observations from 25 February to 23 September 1895. Already on the first date, Barnard was surprised that the faint object near T Tauri was seen '*very easily; it was faint, but not extremely so*' in the 36-inch (the following night it was even visible in the 12-inch). He described it as '*round, quite definite, and very feebly brighter in the middle*'. After measuring the position relative to T Tauri (BD +19° 706) using the star BD +19° 704 (b) to the southwest, he was astonished to learn that the BD coordinates of T Tauri were wrong! He immediately realised that this error had misled Burnham (and himself) in 1890.

The correction brought an unexpected result: using the true position of T Tauri, Barnard's object lands exactly on NGC 1555! After checking Hind's discovery note (*AN* 839) Barnard realised that his nebula '*by no means was identical with T Tauri, and that he* [Hind] *had observed both objects at that time*'. He now officially stated (confirming Dreyer's assumption) that '*this faint nebula, seen by Burnham and myself in 1890, was nothing else but Hind's nebula of 1852, which was supposed to have vanished from the face of the heavens!*' A simple

---

[114] The name 'Starfield' had earlier been used by Lassell for his observatory in Liverpool.

[115] Hagen (1922: 208).

[116] Dreyer (1953: 281).

error had led to major confusion about the objects of Burnham, Barnard and Hind. The identity of the last two meant the reappearance of a long-missing nebula (NGC 1555). Unfortunately, the new result did not get through to Burnham, who had already left Lick Observatory in 1892 (due to the conflict with Holden). On 26 February 1895 Barnard examined the field with the 12-inch and could see Hind's nebula; however, only by masking T Tauri with the micrometer bar.

What about Burnham's tiny nebula around T Tauri? It caused another surprise to Barnard: '*the small nebula in which it shone in* 1890 *had absolutely disappeared. The star seemed perfectly stellar*'. However, on comparing the star with another one nearby, he detected a '*very feeble indefinite nebulous glow for a few seconds of arc*'. This appearance could be confirmed on 4 March with the 36-inch at a power of 520: T Tauri was surrounded by an extremely faint nebula of '*several seconds in diameter*'. This was, as noted on 24 March, not identical with the tiny compact nebula discovered by Burnham in 1890, which '*certainly does not now exist*'. Barnard tried to explain the changes: '*At the time of Mr. Burnham's observations this variable star was at its minimum brightness, and at the present observations it was bright though at its maximum. Is it possible that in its fainter phases this star becomes essentially a very small nebula, or the nucleus of such, and at its maximum the nebulosity is absorbed into the star, or otherwise disposed of?*'

Barnard searched also for the faint stars described by Tempel near T Tauri and at the position of Struve's nebula. Nothing could be seen in the vicinity of the star and at the location of NGC 1554 only d'Arrest's 14-mag star was visible. About Tempel's map of 1877 (reproduced in the report) he remarked that '*Tempel's work seems to have been carefully done – his stars are missing.*' Barnard produced a new sketch of the field (Fig. 11.21).

The observations of 15, 22 and 23 September 1895 ('*under the very finest conditions*') brought another turn about: '*To my surprise no trace of Hind's nebula now exists – it seems to have entirely vanished!*' (Barnard 1895c). Barnard used all the tricks available to him, such as star masking and averted vision. However, he discovered – '*at the very limit of the 36-inch*' – a second, extremely faint star at the position of Struve's 'lost' nebula: '*I could not be certain, that this faint object was really stellar; sometimes it looked nebulous.*' This was probably the 15-mag star f' (see Fig. 11.16), which

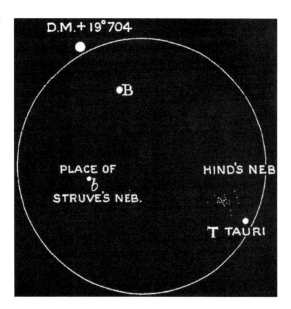

Figure 11.21. Barnard's sketch of the T Tauri region (Barnard 1895b).

Tempel claimed to have seen in the 11-inch (which must be doubted).

Barnard had to wait two years, then he got the ultimate chance to clarify the matter: the new 40" Yerkes refractor was ready to use. The issue had a high priority: '*One of the first things examined here* [Yerkes] *with the 40-inch was T Tauri and the region of Hind's and Struve's nebulae.*' (Barnard 1899). Barnard monitored the field from 20 September 1897 to 10 December 1898. Though the seeing was good, no nebula was visible on the first date. Two days later things started similarly, but then he detected a new nebula near T Tauri: '*At first the star appeared to be double, but the higher powers showed the appearance to be due to a very small nebula or nebulous patch very close south following the star, perhaps from* 1″ *to* 2″ *distant.*' He made a sketch (Fig. 11.22).

On 26 September Barnard and Burnham (who was working at Yerkes Observatory too) observed the field. The small nebula was invisible, probably due to bad visibility. The night of 28 September was much better: '*Hind's nebula could be very faintly seen, but with the utmost difficulty, occupying the position of 1890 and 1895.*' The new 'nebulous patch', appended to the southeast of T Tauri, could be seen too. Perhaps this was Burnham's nebula of 1890 – now in a new shape? At the place of Struve's nebula the second star (f') was visible '*at the

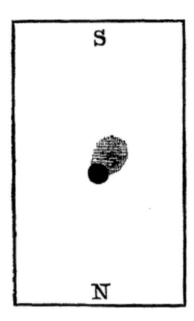

Figure 11.22. T Tauri, sketched by Barnard on 22 September 1897 at the 40" Yerkes refractor (Barnard 1899).

limit of vision'. NGC 1554 itself could not be found in the 40-inch. The further observations were similar, varying depending on the air quality. Tempel's pair of stars closely northwest of T Tauri (see Fig. 11.16) was never seen.

At the end of 1898 Barnard summarised the situation thus: '*we have evidence that Hind's nebula still exists, but in a most excessively faint condition in the most powerful telescopes under the very best conditions*'. He added a sketch that he had already made on 15 October 1890 with the 36" Lick refractor (see Fig. 11.20), showing Hind's nebula (which had not been identified at that time) and the tiny, bright Burnham nebula around T Tauri, which was not visible in 1895. As has already been mentioned, the disappearance of a nebula contradicted William Herschel's idea of a condensation into a star. Therefore it was interesting to learn that an object could lose its light and become 'dark'. Perhaps this was another reason for Barnard to search systematically for 'dark nebulae', which he had already imaged in Milky Way fields (Barnard 1916b).

11.4.7 Images of Keeler and Reinmuth, Dreyer's error and last visual observations

After Roberts' failure in 1890, Keeler successfully photographed the field on 6 December 1899 using the 36" Crossley reflector of Lick Observatory (Keeler 1900a). After an exposure time of almost four hours Hind's nebula appeared as a '*faint and irregular*' object (the plate limit was 16–17 mag). Keeler wrote that '*Three patches, of which the middle one is the brightest, and which are not sharply bounded on any side, are connected by still fainter nebulosity. […] From sketches made by Professor Barnard and other observers, it appears that the brightest patch, or the two patches nearly in line with D.M. +19° 704* [T Tauri] *are the parts of the nebula which have been observed visually.*' About Struve's nebula he noted that '*The photograph shows no nebulosity whatever.*' Another plate, taken on 27 December confirmed the results.[117] Agnes Clerke summarised the history up to this time in her book *A Popular History of Astronomy during the Nineteenth Century*.[118]

In view of Keeler's photographs, which confirmed Barnard, the positive sightings of Johann Georg Hagen (December 1894 to January 1895) with the 12" refractor in Georgetown appear very doubtful. The assessment gets even worse for Hagen if one takes into account his observation of 1900. About Struve's nebula NGC 1554, which had been lost from 1877, he notes '*–17s, –0.9' from T Tauri was certainly in the 12" refractor at Georgetown, 1900 Dec. 14*'.[119] Given the big problems Burnham and Barnard – the two leading observers with a superior telescope – faced at that time when it came to seeing the nebulae near T Tauri, Hagen's results must be strongly questioned. He was as controversial in visual astronomy as Roberts was in astrophotography.[120] On 17 February 1912 Hagen unsuccessfully observed the T Tauri region with the 16" refractor at the Vatican Observatory, Castelgandolfo. Highly doubtful too is his observation of 1919: Hagen saw Struve's nebula as '*elongated preceding-following* [east–west]', not matching the north–south-oriented star pair f–f'.[121]

At Keeler's request Robert Aitken and Charles Perrine made another visual observation on 20 January 1899 ('*a fine night*') with the 36" refractor of Lick

---

[117] Keeler's images are not contained in his publication; only a sketch of the region is given.
[118] Clerke (1902: 403).
[119] Hagen (1922: 208).
[120] By virtue of the obscure issue of the 'Hagen clouds' these two fields later became connected; see Latußeck (2009) and Section 10.3.2.
[121] Hagen (1922: 208).

Observatory to check his photography (Keeler 1900a). Both saw Hind's nebula '*with difficulty, at the very limit of visibility*'. Struve's star and its faint companion (f and f′) were visible too, and they incorrectly noted that '*the latter was not seen by Barnard*'. However, they noticed that Tempel's star '*does not agree in position with the star shown by photograph*'. Aitken and Perrine estimated the brightnesses of the components to be 13 mag and 15.5–16 mag, respectively. For Keeler, the comparison between photography and visual observations (especially those made in the past) was a cause for criticism: '*With respect to some of the details of the various observations of these nebulae, which relate to objects at the very limit of vision, there is, I think, room for considerable doubt.*' One further reads that '*So skilful an observer as Tempel has, as I have shown in an article which will be printed in another place, drawn stars and nebulae where none exist, while stars and nebulae that certainly do exist escaped his notice. The fallibility of the observer must not be lost of sight.*' In the announced paper Keeler contrasts visual and photographic results for NGC 7479, a bright spiral nebula in Pegasus (see Section 11.2.4).

Leo Brenner (aka Spiridon Gopčević) summarised the results on NGC 1554/55 in his popular book *Beobachtungsobjekte für Amateur-Astronomen*.[122] However, the chapter on Hind's and Struve's nebula is incorrectly titled 'NGC 1988 (h 1191)' – this is Chacornac's nebulae in Taurus! The given position refers to this object too, but the text undoubtedly concerns the objects of Hind and Struve. Brenner wrote '*From all this it almost certainly follows that not only Hind's nebula is variable but several stars in the vicinity too.*' Brenner observed with a 17.8-cm refractor by Reinfelder & Hertel in his Manora Observatory (Mali Lošinj), which was erected in 1895.[123]

As one of the last visual observers, Wirtz explored the region around T Tauri with the 48.7-cm Merz refractor in Straßburg. His observations of 16 and 20 February 1906 were negative: '*both nebulae could not be seen, the same applies for the nebulous nature of the variable* [T Tauri]'.[124]

In early January 1907 Burnham, who was looking for double stars with the 40″ Yerkes refractor, engaged with the field for the last time. About Hind's Variable Nebula he noted that it is '*still invisible*' (Burnham S. 1907). Had he not come across Barnard's results on the identity of Hind's nebula? Anyhow, his note sounds curious, given the successful rediscovery of NGC 1555. Burnham further wrote that '*In 1890 the nebula was very plane [...] with a length of 4″ or 5″ in the direction of 150°.*' This implies that he thought the nebula (which had been accidentally discovered by him around T Tauri and later disappeared) to be Hind's! Obviously he did not follow up the case after his departure from Lick Observatory in 1892.[125] Thus he interpreted Barnard's faint nebula (the true Hind nebula), to the southwest of T Tauri, as an individual object. Indeed, he wrote that '*the faint nebula in the field sp, photographed by Keeler [...] was not seen. This is too faint to be visible in the largest apertures except under the most favorable conditions.*'

Once again Dreyer reviewed the case (up to 1907) in the notes of the Second Index Catalogue.[126] His error that had appeared in the IC I remained uncorrected; only the designation is slightly improved: '*N.G.C. 1554–55. Hind's variable nebula.*' Dreyer further wrote that '*Barnard in February and March 1895 found (with the 36-inch refr.) that T Tauri was not, as in 1890, the nucleus of a pB, S neb; the star was perfectly stellar; but involved in a nebulous glow; Struve's neb was not seen, though there was possibly a slight haziness there [...] In September 1895, on three nights, no trace of Hind's neb was seen with the 36-inch [...] Keeler on two photos taken in December 1899 found three vF, irregular patches, connected by still fainter nebulosity, sp and p T Tauri, but clear of the star; no trace of Struve's neb [...] Not visible to Burnham around 1907.*'

The first to fall for Dreyer's confusing designation were Max Wolf and his assistant Karl Reinmuth in Heidelberg. The latter describes in his publication 'Die Herschel-Nebel' of 1926 a plate taken by Wolf on 11 November 1906 with the 16″ Bruce refractor at Königstuhl Observatory. Reinmuth noted the following about NGC 1554: '*Hind's variable nebula; eeeF pL neb E 40° sp var star BD +19° 706*'.[127] This undoubtedly refers to Struve's nebula. However, even more interesting than the wrong catalogue designation is the

---

[122] '*Observational Objects for Amateur Astronomers*' (Brenner 1902).
[123] Ashbrook (1984: 103–111).
[124] Wirtz (1911b: 26).

[125] Burnham thereafter again worked as a court reporter in Chicago.
[126] Dreyer (1953: 369).
[127] Reinmuth (1926: 19).

512   Special topics

information that the missing object should be visible on the plate! The descriptions 'eeeF' and 'pL' indicate a plate flaw, appearing as a faint extended haze. About NGC 1555 one reads '= *T Tauri = var star BD +19° 706 = star 11 (1906 November 11), nf vnr of var neb, not inv*'. The identification of the star with NGC 1555 looks strange; obviously, Reinmuth is referring to Burnham's nebula, insofar as he describes a separate 'variable nebula', closely following the star ('nf vnr') and not surrounding it ('not inv'). No doubt Barnard's results of 1895 did not find their way to the Königstuhl Observatory.

11.4.8 New studies of the T Tauri region

Heber Curtis too used a wrong designation. In his article 'Note on Hind's variable nebula' one reads that '*This nebula (N.G.C. 1554), close to the irregularly variable star T Tauri, is an object of special interest.*' (Curtis 1915). Curtis' photographic study of the T Tauri region with the 36″ Crossley reflector at Lick Observatory continues that of Keeler, who took two plates in December 1899. The new exposures were made in March and November 1914. Curtis wrote that '*The nebula is of the utmost faintness on all plates, but is apparently brighter in 1914.*' He believed that the apparent changes were due to the bad condition of the early plates. The new ones show a structure 4″ south of the star: '*T Tauri itself has a very interesting nebulous wing which is either variable or must rotate so as to be concealed at times. This appendage is a small cone shaped projection.*' Giving too much credence to Tempel's observations, Curtis remarks that '*It was often seen by Tempel, and was fairly conspicuous when examined by Barnard and Burnham in 1890, but entirely invisible when Barnard looked for it in February 1895.*'

This work was continued by Francis Pease at Mt Wilson Observatory. During the year 1911–17 he made seven images with the 60″ reflector (Fig. 11.23). In a footnote to his publication, he clears up the designation error, mentioning Dreyer as the cause: '*N.G.C. 1554, which precedes No. 1555 [...] is sometimes referred to as Hind's variable nebula, probably because of the note on p. 225 of the First Index Catalogue.*' (Pease 1917a). In two sketches – based on the photographs – Pease describes the complex variable structures in the region of T Tauri (Fig. 11.24). The first shows the close vicinity of the star, the other one the regional nebulosities and dark zones. With reference to this presentation the reason for the many identification problems can be imagined.

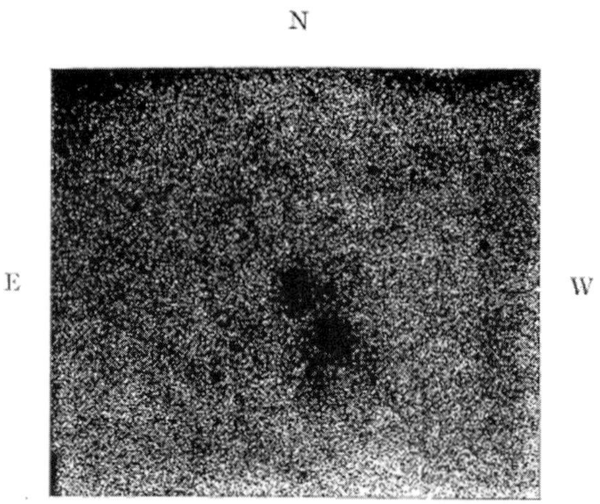

Figure 11.23. Pease's exposure of 13 February 1913, made with the 60″ reflector on Mt Wilson (Pease 1917a).

Another, even larger study was presented by Carl Lampland of Lowell Observatory, Flagstaff (Lampland 1936). In 1935 NGC 1555 was well visible in the 24″ Clark refractor. Moreover, those who had thought that in the time of astrophysics professional visual observing was over were surprised by two eminent astronomers: Edwin Hubble and Walter Baade. According to George Herbig, both ventured a glimpse at T Tauri with the 100″ Hooker reflector on Mt Wilson in 1935 (Herbig 1953). Five years later Baade made an exposure of the region with the same telescope.

Today there are known to be two nebulae in the T Tauri region that are variable in form and brightness.[128] The first is NGC 1555, the reflection nebula to the west of T Tauri, which was discovered by Hind in 1852. It is also catalogued as vdB 28 and Ced 32b.[129] At the moment the object is easily visible.[130] The second is the emission nebula close to the star, which was discovered by Burnham in 1890 and is now called Sh2–238 and Ced 32c. NGC 1554 (Ced 32a), Struve's Lost Nebula, is indeed missing. Since it is pretty far from T Tauri and

---

[128] See also Herbig (1949, 1953), Bertout (1980, 1981), Schwartz (1974) and Lorre (1975).

[129] The first designation is due to Sidney van den Bergh and the second to Sven Cederblad (Cederblad 1946).

[130] This was confirmed by the author on 11 December 2004 with a 20″ reflector.

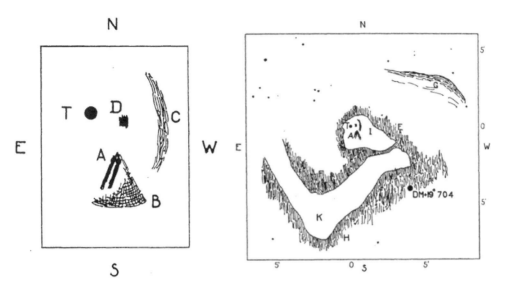

Figure 11.24. Sketches made by Pease, on the basis of photographs of 1911–16 taken at Mt Wilson (Pease 1917a). Left: the region around T Tauri; right: regional nebulosities.

there is no other variable star in its vicinity, it probably never existed. Obviously, the nebulous appearance was caused by the faint pair of stars f–f′. Tempel's various objects ('nebulous glows') and Bigourdan's no. 144 were mere illusions too.

11.5 D'ARREST, DREYER AND THE VARIABLE NEBULAE

The term 'variable nebula' has already been mentioned in connection with several objects. It threads through nineteenth-century astronomy and was a subject of controversy. The essence of the problem is the subjectivity of visual observing. Owing to the lack of reliable measurements, sustainable data and thus real progress in the physical description of nebulae could not be achieved.

The uncertainty caused many reports about changes in nebulae. Each new candidate brought new excitement, but in the end the observations demonstrated only the helplessness of astronomers in the face of a phenomenon that could not be revealed by classic methods. Already Lamont, doubting the reality of the reported changes, had stressed the difference between the 'subjective' research on nebulae and the 'objective' study of double stars. The reasons for the problem were seen to lie in the telescope, the atmospheric conditions and the individual perception of the observer. Barnard summarised the situation in 1892: '*There are many of these objects that have been reported variable; but so differently do many of the nebulae appear in different telescopes and with different magnifying powers, that no sort of confidence can be placed in most of these statements, and a positive proof of an actual variable nebula, except in one case* [Hind's nebula], *is sadly wanting.* [...] *It is quite reasonable to believe that some of the nebulae do vary their light, but, from the peculiarities above mentioned, it will take decidedly strong proof to convince astronomers who are familiar with these objects that some deception has not misled the observer.*' (Barnard 1892c).

11.5.1 Early cases and d'Arrest's papers of 1862

The first speculations about variable nebulae were due to Legentil (M 31) and William and John Herschel (M 42). For a long time the Orion Nebula was the prototype: various observers believed that they had noticed changes in form and brightness of particular parts.[131] Other objects were given descriptions different from those by William Herschel. For instance, Friedrich von Hahn noted that the shape and position of the planetary nebula NGC 3242 had changed since Herschel

---

[131] Parsons L. (1868).

(Hahn 1800).[132] In 1837 Lamont reported differences for NGC 2695, NGC 6905, NGC 7009 and NGC 7662,[133] mainly concerning position (see Section 6.1.3). Schmidt wrote in 1861 a paper 'Über einen neuen veränderlichen Nebelstern'.[134] This is the variable star R Mon, illuminating the reflection nebula NGC 2261 – a case of real variability, which was investigated by Hubble in the twentieth century (Hubble's Variable Nebula).[135]

As early as in 1854, during his Leipzig observations, d'Arrest had noticed some nebulae that appeared brighter or fainter than William Herschel's descriptions (see Table 6.18). In 1862 he treated the theme in his publication in *AN* 1366 (d'Arrest 1862c); a chapter is titled 'Variabilität von Nebelflecken' ['Variability of nebulae']. Four cases are mentioned: the Orion Nebula (M 42), Hind's Variable Nebula NGC 1555 in Taurus,[136] NGC 4473/77 in Coma Berenices[137] and NGC 3662 in Leo[138] (an object studied by Schmidt).

The topic suddenly gained momentum. When in the same year new cases emerged, d'Arrest wrote two further papers. The first, titled 'Auffindung eines zweiten variablen Nebelflecks im Stier',[139] treats the Merope Nebula NGC 1435 in the Pleiades. Schönfeld's discovery of NGC 1333 in Perseus was the reason to report the 'Auffindung eines dritten variablen Nebelflecks'.[140]

In 1863 Tempel published an interesting note in the French magazine *Le Monde* (Tempel 1863c). He explicitly doubted that variability was occurring at all, claiming that the effect was due to the 'variable atmosphere'. He criticised Chacornac for his 'variable nebula' NGC 1988 in Taurus and discussed the Merope Nebula too.[141] Tempel wrote that '*my experience has led to the conviction that the nebulae are invariant, because lately large instruments have resolved them into stars and it would be unlikely that so many stars are variable simultaneously*'. But, he concedes, '*I think that true variability is possible for very small and simply structured nebulae, which, however, do not deserve closer attention. But the new variable nebula announced by Mr Chacornac ranks among the larger ones and I really doubt it to be variable.*' To reinforce his thesis, he gives an interesting example: the Virgo galaxy cluster. '*The atlas, attached to Mädler's* 'Populäre Astronomie', *contains a chart of the main group of nebulae in Virgo. If one draws in all objects, as seen with my 4-inch, in the Berlin chart*[142] *and then marks the Herschel-objects and those found by d'Arrest, one gets an idea about the changes in this small area of nebulae. Simultaneously you will agree that we are far from making any statement about variability of nebulae.*' He added that '*The different appearance is caused not only by different telescopes and eyes, but also by the variable clearness of the atmosphere.*'

Variable nebulae soon entered the astronomical textbooks. For instance, Chambers included a section in his *Descriptive Astronomy* of 1867; because additional cases had arisen it was even enlarged in the 1877 edition.[143] The following objects are treated there: Hind's nebula NGC 1555, the Nova T Scorpii in the globular cluster M 80,[144] NGC 6643 in Draco, which was discovered by Schönfeld and Tuttle,[145] the Merope Nebula NGC 1435, Chacornac's 'nebula' NGC 1988 in Taurus,[146] and the η Carinae Nebula NGC 3372 (see Section 11.15.3). Agnes Clerke too describes 'variable nebulae' in her book *A Popular History of Astronomy during the Nineteenth Century*, mentioning NGC 1555, M 42 and M 20.[147]

The reported 'changes' of objects were manifold and concerned their

- shape and structure
- brightness (variability)
- appearance ('Nova' in a star cluster, nebula around 'Nova')

---

[132] See Section 2.7.2.
[133] In February 1908 Barnard reported a '*variability of the nucleus of NGC 7662*', which probably concerns the 13.6-mag central star of the planetary nebula (Barnard 1908b).
[134] 'About a new variable nebulous star' (Schmidt 1861b).
[135] See Sections 2.6.2 and 6.18.3.
[136] See Section 11.4.
[137] See Section 6.14.4.
[138] See Section 6.15.3.
[139] 'Detection of a second variable nebula in Taurus' (d'Arrest 1862e) (see Section 11.6).
[140] 'Detection of a third variable nebula' (d'Arrest 1862f) (see Section 6.15.2).
[141] See also Sections 7.3.2 and 11.6.1.
[142] *Berliner Akademische Sternkarten*, Hora XII, made by Karl August von Steinheil in 1834; see Bessel (1834).
[143] Chambers (1877: 543–547).
[144] See Section 6.13.2.
[145] See Section 6.7.4.
[146] See Section 7.3.2.
[147] Clerke (1902: 403–404).

- disappearance and reappearance
- position, i.e. proper motion or orbital motion (double nebulae)

Some observers were rather critical, but others let their imagination run wild. Until 1887 many cases circulated, without there being any clear idea about their nature and the supposed changes. Dreyer was among the critical minds trying to shed light on the dubious matter, which he attempted in a report published in 1887.

11.5.2 Dreyer's report of 1887

In May 1877 Dreyer wrote a report 'On some nebulae hitherto suspected of variability or proper motion' (Dreyer 1887d).[148] It starts with the statement '*The discrepancies met with in comparing observations of nebulae by different observers are frequently so great that more or less positive assertions have naturally from time to time been made as to variability or changes in the objects.*' Seventeen objects, ordered by their SC-numbers, for which several observers had indicated changes, are treated (Table 11.4). For Dreyer, brightness variability is the key feature: '*It seems that the only well authenticated cases of change in nebulae are changes of brightness only.*' He sees only two cases of '*nebulae having disappeared*' that he considers '*quite certain*': those of Hind (NGC 1555) and Chacornac (NGC 1988) – both located in Taurus.

All other claims about changes were rated by Dreyer as spurious. Most of them could be traced back to deficient observation or documentation and poor atmospheric or instrumental conditions. He is critical too about positional changes and shape variations: '*we so far do not possess any clear evidence of change of the form or change of place*'. Despite extensive measurements, no sign of proper motion was detected – given the great distances of the nebulae, this is not surprising.

About supposedly 'missing' William Herschel objects, Dreyer remarked that '*It is true that some of William Herschel's nebulae cannot now be found, but these may either have been comets, or, more probably, some error of observation has vitiated the position he gives for the object in question.*' For instance, d'Arrest had indicated the absence of 15 nebulae, all of which had been discovered by Herschel on 4 April 1801 (d'Arrest 1867b). Dreyer wrote that, '*As not one of these objects can now be found, it is evident that he* [Herschel] *made a mistake either in identifying the star or in making or recording the observation for it.*' The matter kept bothering him, and he was later able to locate the correct reference star, leading to the identification of all of the objects concerned (see Section 8.6.3). The three brightest are the galaxies NGC 2977, NGC 3218 and NGC 3397 in Draco. D'Arrest also pointed out the case of Herschel's I 26 (GC 2179, NGC 3345), which he treated in the chapter about 'missing nebulae' in his *AN* 1366 paper (d'Arrest 1862c). This is the prominent galaxy M 95 in Leo.

Dreyer describes still another variant of supposed changes: Schönfeld's NGC 1333 (BD +30° 548) and Tempel's NGC 1435 (the Merope Nebula); both were invisible for some observers. He wrote that '*This difficulty is, however, now universally understood to arise from the use of too high power with consequent smallness of field; and nobody now suspects these two nebulae of variability.*'

Most objects in Dreyer's list are double nebulae. The collection was based on a publication by Camille Flammarion (Flammarion 1879a). But the true origin was d'Arrest's micrometrical work on double nebulae, which appeared in *AN* 1369 (d'Arrest 1862d). A relative motion was assumed, because the positions appeared different from those given by William or John Herschel. Dreyer unambiguously stated that '*Sir W. Herschel never employed a micrometer, but merely estimated the position-angles and distances of neighbouring nebulae, and that Sir John Herschel did the same from commencement of his observations, and up to July 5, 1828.*' After that date John Herschel used a bar-micrometer. However, only position angles could be determined and often he abstained from doing so, merely estimating the angle. Dreyer claimed not to use such weak data for such critical issues as positional changes of nebulae. Therefore the cited cases were treated by him as doubtful. He had measured NGC 2371/72 and NGC 7463/65 with the 10" Grubb refractor in Armagh. The former is a bipolar planetary nebula, which had been described as a double nebula by most visual observers. Only at Birr Castle was a slight connection seen. However, the 'orbital motion' supposed by d'Arrest could not be confirmed.

Ismaël Boulliau (Boulliau 1667), who coined the name Andromeda Nebula, was the first to suppose a

---

[148] This was in the final phase of the making of the NGC.

Table 11.4. *Supposed cases of variable nebulae investigated by Dreyer*

NGC	H	h	Discoverer	Type	V	Con.	Remarks
224		50	As-Sufi	Gx	3.5	And	Andromeda Nebula (M 31)
1044	III 228	251	W. Herschel	Gx	13.4	Cet	
1045	III 229	252	W. Herschel	Gx	13.8	Cet	
1293	III 574	294	W. Herschel	Gx	13.5	Per	
1294	III 575	295	W. Herschel	Gx	13.4	Per	
1587	II 8	316	W. Herschel	Gx	11.7	Tau	
1588	II 9	317	W. Herschel	Gx	13.2	Tau	
1976		360	Peiresc	EN		Ori	Orion Nebula (M 42)
2327	IV 25	428	W. Herschel	EN		CMa	
2371	II 316	444	W. Herschel	PN	11.2	Gem	Dreyer
2372	II 317	445	W. Herschel	PN	11.2	Gem	Dreyer
3230		705	J. Herschel	Gx	13.3	Leo	
3372		3295	Lacaille	EN		Car	Eta Carinae Nebula
3894	I 248	983	W. Herschel	Gx	11.6	UMa	
3895	II 832	984	W. Herschel	Gx	13.1	UMa	
4061	III 394	1065	W. Herschel	Gx	13.2	Com	
4065	III 395	1067	W. Herschel	Gx	12.7	Com	
5857	II 751	1905	W. Herschel	Gx	13.0	Boo	Drawing by Hunter
5859	II 752	1905	W. Herschel	Gx	12.5	Boo	Drawing by Hunter
6514	V 10–12	1991	Messier	EN		Sgr	Trifid Nebula (M 20)
6618		2008	de Chéseaux	EN		Sgr	Omega Nebula (M 17)
6962	II 426	2087	W. Herschel	Gx	12.2	Aqr	In a group with five other NGC galaxies[149]
6964	II 427	2089	W. Herschel	Gx	12.9	Aqr	See NGC 6962
7463	III 210	2202	W. Herschel	Gx	12.9	Peg	Dreyer, sketch by d'Arrest, drawing by Vogel[150]
7465	III 211	2203	W. Herschel	Gx	12.3	Peg	Dreyer, sketch by d'Arrest, drawing by Vogel
7805	III 855	2294	W. Herschel	Gx	13.2	Peg	
7806	III 856	2295	W. Herschel	Gx	13.5	Peg	

variability of M 31. Later Kirch (in 1676) and Legentil (in 1759) followed. Eventually George Phillips Bond was able to prove that no change had happened (Bond G. 1867). However, observers had described the nucleus differently: it had been described '*by some as a starlike, by others at the very same time as a very soft and gradual condensation*'.[151] Copeland's tests with various eyepieces, showing the influence of magnification on the appearance of the nucleus, were interesting: '*the lower powers make it more star-like, the higher ones more soft-looking and extensive*' (Copeland R. 1886).

In the case of M 42, d'Arrest stressed that '*the observed changes in this vast mass of gas seem exclusively*

---

[149] See Steinicke (2001b).

[150] In a galaxy trio with NGC 7464 and NGC 7465 (see Tables 8.3 and 8.27); see Steinicke (2001c).

[151] In 1885 a real change appeared in the nucleus of M 31: the supernova S And (see Section 9.20.2).

11.5 D'Arrest, Dreyer and variable nebulae

Figure 11.25. NGC 3230 in Leo (DSS).

to turn out to be temporary fluctuations of brightness' (d'Arrest 1872b). Finally, Holden concluded that '*the figure of the nebula in Orion has remained the same from 1758 till now* [1880], *but that in the brightness of its parts undoubted variations have taken place, and that such changes are even now going on*' (Holden 1882b).

The case of NGC 3230 (GC 2091) is due to Herbert Sadler, who claimed to have detected a proper motion (Sadler 1885). This 13.3-mag galaxy in Leo (Fig. 11.25), located 3.8° to the east of α Leonis, was found by John Herschel on 24 March 1830 (h 705). D'Arrest observed it in 1864–65 in Copenhagen; later Copeland and Burnham made contributions. According to Sadler, the data implied a motion relative to a nearby double star ('*irregular but certain motion seems well established*'). Dreyer investigated the case and concluded that '*When the observations of this object are examined with care, it appears, however, that there is no certain evidence of proper motion.*' (Dreyer 1885b). This was confirmed by Barnard. His measurements at the Lick refractor made on 5 March 1889 together with Burnham showed that the available data '*do not indicate any change*' (Barnard 1889b). Swift thought the object remarkable too; he had studied the similar case of NGC 3690 (Swift 1886d).[152]

For NGC 5857/59 the orientation of the galaxies was discussed; the Birr Castle observations were

---

[152] See Section 9.5.6.

confusing (see Section 7.1.4). Dreyer measured the pair in Armagh and was able to clear up the confusion. The Trifid Nebula (M 20) appeared to have moved relative to the triple star in its southeastern part. Holden had studied the issue in his work 'On the proper motion of the Trifid Nebula' (Holden 1877b). As for the similar case of M 17, which had been treated by Holden earlier (Holden 1876), Dreyer doubted the motion.

Finally, Dreyer wrote '*I have spared no trouble in going through these cases one by one, although in some evidence was of such a character as hardly to deserve a refutation.*' He added that '*In making micrometric observations of these interesting objects we must be content to work for unborn generations, or at least not to expect immediate and startling results, which would look well in popular books.*' A summary of his report appeared in the magazine *Naturwissenschaftliche Rundschau* (Anon 1887b).

11.5.3 Abbott, Lieutenant Herschel and the changes in the nebula around η Carinae

The 'Great nebula around η Argûs' is only briefly mentioned in Dreyer's report. Anyway, the case is worth presenting in detail. The object is a large nebulous complex centred on the star η Carinae (formerly called η Argus), which was catalogued by Dreyer as NGC 3372. Some changes were reported.[153] The case became well known through the contributions of eminent astronomers: John Herschel and his son Lt John Herschel, Lassell, Proctor and, last but not least, the Astronomer Royal Airy. However, it eventually became an awkward matter for the main originator, Francis Abbott, an amateur astronomer and the owner of a private observatory in Hobart, Tasmania.

On the basis of John Herschel's results, Abbott first claimed in 1863 to have detected a decrease in brightness and size of the nebula and, moreover, a 'displacement' relative to the star (Abbott 1863). His observations, made with a 4" Dallmeyer refractor, had started in 1861.[154] Without question, the star η Car had passed through an extreme phase between 1827

---

[153] The centre of NGC 3372, the Homunculus Nebula around η Carinae, is actually variable (see Table 11.5). The complex contains the Keyhole Nebula, a striking dark nebula.

[154] Abbott had already published a study about the open cluster NGC 4755 around κ Crucis (Abbott 1862).

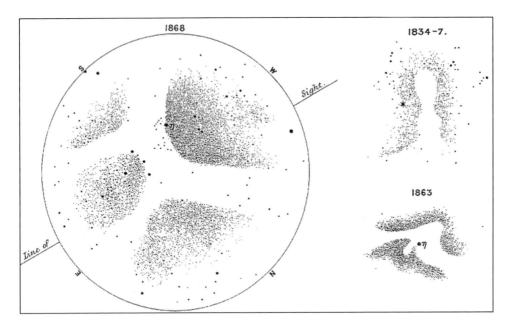

Figure 11.26. Abbott's drawing of the Eta Carinae Nebula NGC 3372 (left), compared with earlier presentations (Abbott 1868).

and 1856, during which the brightness increased from 4 to even −0.8 mag (1843); by 1863 it had fallen to 6 mag.[155] Abbott, who was publishing further papers and drawings of the object (Fig. 11.26),[156] was supported by Burton Powell at Madras Observatory (Powell 1863). In contrast, Lt John Herschel[157] in Bangalore could not approve Abbott's result: "*I do not recognize such a total 'subversion of all the greatest and most striking features &c.', as Mr. Abbott's diagrams and statements appear to imply.*" (Herschel Lt 1869). He was supported by his father, who kindly explained his earlier drawings (Herschel J. 1871).

Airy piped up too, feeling obliged to defend the honour of the two Herschels. He traced the different opinions about the shape and position of the nebula back to Abbott's poor sketches: '*In points of geometry, therefore, Mr. Abbott is a most inaccurate man. It is impossible for us to publish maps in this state.*' (Airy 1871). For Lt Herschel the problem was caused by the incompatible fields of view of his father's 18¼" reflector and Abbott's 4" refractor: '*the truth seems to be that a low power drawing has been looked at as if it were a small portion full of detail*' (Herschel Lt. 1871). Indeed, confronted by the complex structure of the nebula, discrepant interpretations could easily happen. In his reply to Airy, Abbott defended his drawings: '*In making comparisons it would be desirable to refer to the original drawings, as in the lithographs, which are on a reduced scale, some trifling inaccuracies occur*' (Abbott 1871).

In July 1871 Lassell published a critical report on the annoying matter (Lassell 1871). Therein he also mentions observations of Le Sueur and MacGeorge at the Great Melbourne Telescope, who did not notice any displacement between star and nebula.[158] Lassell concludes that '*no proof has yet been given of any change in the nebula at all*'. Proctor confirmed this in the same year. Obviously angry about Abbott's wrong maps, he remarked: '*It is most unfortunate that by this mistake Mr. Abbott has caused so much valuable time to be wasted by Sir J. Herschel,*

---

[155] In 1871 John Tebbutt from Windsor (New South Wales) published his brightness observations (Tebbutt 1871). At present the star is of about 5 mag.

[156] Abbott (1868).

[157] The son of John Herschel also made spectroscopic studies of southern nebulae from the Cape catalogue (Herschel Lt 1868a, b).

[158] However, in 1870 Le Sueur made a fine drawing of the nebula that differs from Herschel's – perhaps indicating true changes of its form.

11.5 D'Arrest, Dreyer and variable nebulae

the Astronomer Royal [Airy], Mr. Lassell, and others.' (Proctor 1871b). This closed the case and Abbott's reputation was completely ruined. Dreyer concluded in his report that "*There has ever since been perfect unanimity among astronomers that the changes were 'altogether imaginary'.*"

11.5.4 Further publications of Dreyer, Denning and Bigourdan

In 1891 Dreyer published a 'Note on some apparently variable nebulae' (Dreyer 1891b). The first case treats the galaxies NGC 1417, NGC 1418 and IC 344 in Eridanus. The two NGC objects were discovered by William Herschel on 5 October 1785. Lewis Swift found IC 344 (no. 13 in his list IX) on 23 December 1889, noting '*I strongly suggest it to have been a comet, as at two subsequent examinations it could not be found. It was in line with N.G.C. 1417 and 1418, and all three were seen simultaneously.*' Dreyer too could not clear up the matter: '*Observers of nebulae are accustomed occasionally to find a particular object very difficult to see owing to atmospheric causes, but the present case is certainly a very suspicious one.*' Although Swift's object is pretty faint (14.2 mag), it does exist at the measured position. Later Dreyer realised (see notes in IC II) that the object had already been discovered by John Herschel (17 October 1827) and appears as h 305 in the Slough catalogue; it is his only IC object (see Section 3.2).

The next objects presented by Dreyer are NGC 955 and NGC 3666 (Fig. 11.27). The former, a 12.1-mag galaxy in Cetus, had already been mentioned by Winnecke in his paper 'On the evidence of periodic variability of the nebula H. II. 278' (Winnecke 1878a). It was found on 6 January 1785 by William Herschel (II 278) and later observed by his son (h 229) and Mitchell (Birr Castle). In 1861 the nebula could not be seen by Schönfeld, but it was visible in 1863 and 1864, which was confirmed by d'Arrest. In 1865 it disappeared once more, as was noticed by Vogel, but it was seen again in 1868 by Schönfeld. Therefore Winnecke concluded that it exhibited a 'periodic variability', which is, of course, not real. Dreyer wrote that '*The evidence of Variability does not seem very strong in this case.*'

However, for NGC 3666 (h 882), which was treated in Winnecke's paper 'Über die periodische Veränderlichkeit in der Helligkeit des Nebelflecks h 882 = H I 20',[159] '*the variability seems scarcely better established*'. William Herschel found the galaxy in Leo on 15 March 1784 and put it in class I ('bright nebulae'). In 1830 John Herschel saw the object as '*eF, 2rd or 3rd class*', but for d'Arrest (in 1863) and Winnecke (in 1878–79) it appeared 'pretty bright' again. When Dreyer first observed NGC 3666 in 1887 with the 10" Grubb refractor in Armagh, it was barely visible, but in 1891 the object appeared bright. He wrote the following in his paper: '*It should be remembered that this is a diffused nebula with very slight central condensation, and, as far as my experience goes, the appearance of objects of this kind is far more influenced by the state of our atmosphere than that of nebulae with a distinct condensation generally is.*'[160]

Finally, Dreyer mentions the cases of NGC 4731 and NGC 4471. The galaxy NGC 4731 (I 41) in Virgo was seen with differing brightnesses by its observers; John Herschel had already supposed its variability in the General Catalogue (GC 3254). NGC 4471 is Schmidt's nebula near M 49 (see Section 6.18.3), which was not seen by d'Arrest. Dreyer warned others not '*to draw important conclusions from very uncertain observations*' in such cases and referred to the hypothetical intramercurial planet 'Vulcan' and the supposed 'satellite' of Venus.

In 1891 Denning published a report on 'Variations in nebulae'.[161] He claimed that '*in observations of nebulae great differences are induced by atmospheric variations*'. As an example the discovery of a nebula with his 10" Browning reflector (Fig. 11.28) is presented. This is the galaxy IC 512 in Camelopardalis, which he found on 23 August 1890. Denning wrote '*I felt inclined to believe that real variations occurred in the inherent light of the nebula, but further observations proved the changes were due to the varying transparency of the atmosphere.*' As further examples he mentions NGC 6015 (a galaxy in Draco), NGC 1469 (a galaxy in Camolopardalis) and NGC 6760 (Hind's globular cluster in Aquila). In another paper, titled 'Supposed variable nebula' (Denning 1892a), three other cases

---

[159] 'About the periodic variability in brightness of the nebula h 882 = H I 20' (Winnecke 1879a).

[160] NGC 3666 is of Hubble-type SBc, i.e. wide open spiral arms wind around a small bulge.

[161] Denning (1891a); see also the chapter on variable nebulae in his textbook *Telescopic Work* (Denning 1891b).

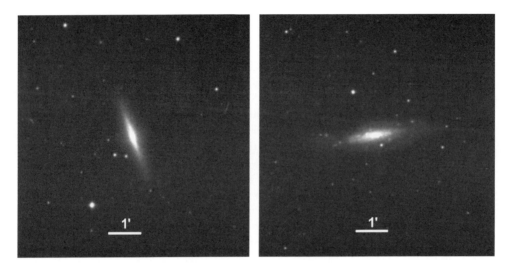

Figure 11.27. The 'variable' galaxies NGC 955 in Cetus and NGC 3666 in Leo (DSS).

Figure 11.28. Denning at his 10" Browning reflector in Bristol.

are presented: Tuttle's NGC 6643, IC 356 (which was found independently by Barnard and Denning) and IC 334 (Denning). The two IC objects are galaxies in Camelopardalis. In March 1892 Barnard investigated IC 356 (Fig. 11.29 left) and another nebula found by himself: IC 48 ('*I am positive of a light change.*').[162]

---

[162] Barnard (1892a).

Figure 11.29. The 'variable' galaxies IC 356 in Camelopardalis (left) and NGC 1186 in Perseus (DSS).

Bigourdan too announced a 'variable' nebula (Bigourdan 1891a): William Herschel's galaxy NGC 1186 in Perseus (Fig. 11.29 right), located only 2° northwest of Algol. Swift found it independently (this find was catalogued by Dreyer as NGC 1174). Herschel classified the object as 'planetary nebula' IV 43. John Herschel saw it too (h 281), but Mitchell and Hunter failed at Birr Castle[163] and d'Arrest too. For Bigourdan NGC 1186 was visible. Perhaps the problems of d'Arrest and the Birr Castle observers are due to the low surface brightness, making observation difficult with a long-focus instrument. Spitaler in Vienna published a paper 'Ueber den Nebel NGC. 1186', in which he wrote *'This nebula, thought to be variable by Bigourdan, was looked up in March 12 of this year* [1891] *with the great refractor in very clean and transparent air.'* (Spitaler 1891b). Since he could not find Swift's nebula NGC 1174 with the 27" refractor, he correctly supposed an identity with NGC 1186.

Even today the number of confirmed variable non-stellar objects is low – there are only a few reflection and emission nebulae (Table 11.5). The main features are variations of brightness and shape caused by a nearby irregularly variable star (illumination, excitation). The column 'Year' gives the date of discovery; in most cases the variability was detected much later (e.g.

in 1916 for NGC 2261). Supernova remnants such as M 1 in Taurus, which show slowly expanding nebulosities, are omitted.

11.6 THE PLEIADES NEBULAE

The open cluster of the Pleiades (M 45) is one of the most popular deep-sky objects. Being visible with the naked eye, it has been known since ancient times; however, it bears no NGC-number. The object is not a mere cluster – it contains many faint patches of nebulosity (Fig. 11.30). The brightest one, starting at the 4.2-mag star Merope, was not found until 1859, when it was discovered by Tempel in Venice. The object was catalogued by Dreyer as NGC 1435. With a distance of about 430 ly the Pleiades are among the nearest open clusters. Therefore the Merope Nebula is the nearest NGC object (together with the Maia Nebula NGC 1432; see Table 10.17).

However, for a long time the appearance (and even the existence) of NGC 1435 was a subject of controversial discussion because while the faint nebula, which was outshone by the bright star Merope, was visible in small telescopes, many large ones failed to reveal it. Thus the visual observers were split into two camps: believers and doubters. Of course, its variability of brightness and shape were debated too. Photography was tried in order to clear up the issue – but the result was new conflict! Not only observational data but also

---

[163] This object is one of Lord Rosse's 'nebulae not found' (see Section 7.1.5).

522   Special topics

Table 11.5. *The known variable nebulae (sorted by discovery date)*

Object	Type	Con.	Star	Discoverer	Year	Remarks
NGC 2261	RN	Mon	R Mon	Herschel	1783	Hubble's Variable Nebula; LBN 920
NGC 1555	RN	Tau	T Tau	Hind	1852	Hind's Variable Nebula; Ced 32b
NGC 6729	RN	CrA	R CrA	Schmidt	1861	LBN 561
	EN	Car	η Car	Gaviola	1944	Homunculus Nebula; in NGC 3372
Ced 59	RN	Ori	FU Ori	Cederblad	1946	LBN 879
GM 1–29	RN	Cep	PV Cep	Gyulbudaghian, Magakian	1977	Gyulbudaghian's Nebula
PP 2	RN	Cas	V633 Cas	Parsamyan, Petrosian	1979	LkHα 198
	RN	Ori	V1647 Ori	McNeil	2004	McNeil's Nebula; near M 78

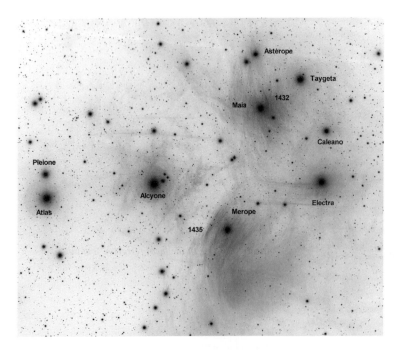

Figure 11.30. The nebulae in the Pleiades (north is up).

the honour of astronomers played a role: a Vanity Fair. In addition to the nebula near Merope, those connected with other stars were debated, among them the Maia Nebula (NGC 1432), the only NGC object to have been discovered by photography (by Paul and Prosper Henry, in 1885). There were even visual sightings of 'exterior nebulosities' surrounding the open cluster, which were later seen on long-exposure plates. The history of all these objects will be presented in detail for the first time.

11.6.1 Discovery of the Merope Nebula (NGC 1435) and early observations

The reflection nebula close to the star Merope was discovered by Wilhelm Tempel on 19 October 1859 in Venice. At that time he was using a Steinheil refractor of aperture 10.5 cm, observing from the staircase ('escalier Lombard') of the Palazzo Contarini del Bovolo. He used an eye-piece with a power of 45 and 2° field of view. As he later wrote, the nebula was seen during the next night too (Tempel 1863). Tempel did not publish his discovery until the end of 1860, by which time he was in Marseille. In his letter to the editor of the *Astronomische Nachrichten*, Christian August Friedrich Peters, dated 23 December 1860, one reads *'Perhaps it is of interest that yesteryear in Venice, after not observing the Pleiades for half a year, I discovered a large bright nebula near Merope on Oct. 19th, which was first thought to be a large comet, but on the following evening, the 20th, I was convinced that it is fixed.'* (Tempel 1861).

At Marseille Observatory the director Benjamin Valz and others were able to confirm the nebula in Tempel's small telescope. About the inner structure Tempel noted the following: *'Not so long ago I clearly saw individual stars pulsating in this nebula and it is considerably brighter on one side.'* He often talks about such dubious 'pulsating stars'; the phenomenon is perhaps related to his perception. In a footnote to Tempel's letter Peters mentions his own observation, which he made together with Carl-Ferdinand Pape at the 7" Repsold refractor of Altona Observatory: *'1860 Dec. 31 this nebula was seen under fairly good conditions with the local 6ft equatorial; however, it was only narrowly perceived by me and Dr. Pape. It does not appear in any catalogue.'* Now the Merope Nebula was officially confirmed. In Marseille Tempel made a drawing, which was later published in Milan (Fig. 11.31). Obviously the striking 'comet tail' emerging from Merope was added later (by Tempel?). Despite this feature, the drawing is nearly identical to that published in 1880 in Arcetri (see the cover and Fig. 11.37).

In 1862 Auwers entered the object as no. 18 in his list of new nebulae, noting *'That the nebula was not seen before 1859, might be sufficiently explained by the proximity of many bright stars; it can easily be overlooked.'*[164]

Previously Bessel had studied the Pleiades during the years 1838–40 with the Königsberg heliometer, measuring the positions of 53 stars – without noticing the nebula.[165] Auwers first saw it on 14 January 1861 and reported that it was *'faint and large under slightly hazy air. At one place the glow is a bit brighter in the middle'.*[166] Two observations on 19 and 21 February 1862 followed. About the first one reads *'I saw Tempel's nebula in the Pleiades of triangular shape, around 15′ large and pretty bright, visible in the finder* [aperture 4.6 cm] *of the heliometer [...] several faint stars placed inside.'* During the second observation, Auwers determined the positions of the stars limiting the nebula; the object was *'seen only vaguely in hazy air'*.

D'Arrest, who had for some time been much interested in nebulae, first tried to observe the object on 20 August 1862 with the 11" Merz refractor of Copenhagen Observatory. Nothing was seen by him and his assistant Schjellerup; two days later the result was negative too. Being convinced of the outstanding quality of his telescope, d'Arrest concluded that the nebula must be variable in brightness! After Hind's nebula (NGC 1555), which he had studied in detail, he subsequently announced on 23 August 1862 the 'Auffindung eines zweiten variablen Nebelflecks im Stier'.[167] In his report he wrote *'Earlier than expected, I today can announce the localization of another variable nebula, again in Taurus, only 9° distant from Hind's variable nebula of yesteryear.'* D'Arrest's confirmation is based on his own observations and those of Tempel, Peters and Pape: *'The bright large nebula, discovered by Tempel in Venice on 19 October 1859 [...], which Prof. Peters and Dr. Pape could see just barely in December 1860 under fairly good air in the 6ft Altona equatorial, is at present entirely invisible in the local refractor.'* He added *'I have searched for it on several nights; Dr. Schjellerup too tried in vain to find it on the refractor.'* However, the nights in August were very good, since some very faint nebulae could be seen. D'Arrest reported *'that H. III. 166, the companion of I. 53, described as extremely faint by W. H. and excessively by J. H., could not only be seen easily but actually was perceived as a double nebula'*.

---

[164] Auwers (1862a: 74).

[165] Engelmann (1876: 299–305).
[166] Auwers (1865: 277).
[167] 'Location of a second variable nebula in Taurus' (d'Arrest 1862e).

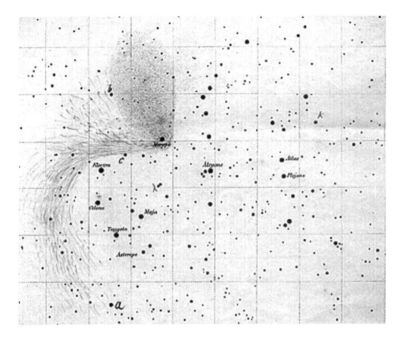

Figure 11.31. Tempel's first drawing of the Merope Nebula (with an added 'comet tail'), which was made in Marseille and published in Milan; south is up (Tempel 1874).

This is the galaxy NGC 7331 (9.5 mag) in Pegasus with its faint companion NGC 7335 (13.8 mag). Since the Merope Nebula had been seen by Tempel (using a much smaller refractor) as resembling a bright comet, d'Arrest had *'no doubt that this was another case of a variable nebula'*.

This report initiated a discussion that lasted more than 30 years; nearly all of the notable visual observers were involved. As early as on 1 September 1862 d'Arrest announced the location of a third variable nebula – another piece of evidence to support his thesis (d'Arrest 1862f). This is Schönfeld's NGC 1333 in Perseus. He stressed the remarkable fact that all three objects *'are located only a few degrees from each other'*.

On 20 September 1862 Schmidt piped up (Schmidt 1862c). He was an expert on the matter, as shown by his paper 'Ueber veränderliche Nebelgestirne'.[168] He indirectly supports d'Arrest's claim about the variability of the Merope Nebula, writing *'if the nebula is not variable, it should have been discovered much earlier'*. He states that *'it is impossible that it would have escaped me, since I have frequently observed and drawn the Pleiades since 1861, not cursorily but carefully'*. From 1841 onwards (working in Hamburg, Düsseldorf-Bilk, Bonn and Olmütz) Schmidt had studied the Pleiades with various telescopes, determining the positions and brightnesses of stars – without noticing the nebula. Now knowing its place, it was easy for him to locate in the 15.7-cm Plössl refractor at Athens Observatory: *'On 5 Febr. 1861 I first saw this nebula in the refractor through pretty clear, steady air. It was very large, very pale and quite shapeless.'* Apart from the better telescope, Schmidt mentions another reason why he was successful in February 1861: the use of low-power eyepieces, *'which are suitable for the detection of medium bright or even faint nebulae'*. In almost all observations made until 26 March 1862 the *'large triangular nebula in the Pleiades was easily visible'*.

In a letter of 29 September 1862, Auwers dealt critically with d'Arrest's 'variable nebulae', writing as follows about the Merope Nebula and NGC 1333: *'Moreover d'Arrest recently was unable to see Tempel's nebula [...] with the 10½-inch Copenhagen refractor, and the one found in the Bonner Durchmusterung by Schönfeld in 1858 appeared very faint, therefore both were announced*

---

[168] 'About variable nebulae' (Schmidt 1862b).

11.6 The Pleiades nebulae

as variable. But, according to my own observations, both nebulae are unchanged.' (Auwers 1862b). He explains *'that large, blurred, faint objects are much more easily visible in small instruments than in large ones'*. Auwers concludes that *'d'Arrest was unable to perceive* [the Merope Nebula] *because he probably used a pretty high magnification, implying a small field of view, completely filled by the 15' large nebula, whereas it is visible without difficulty in a 2ft comet-seeker. With such an instrument I saw it on Sep. 23 and 24* [1862], *when the Pleiades stood only 16° high, looking much like in earlier observations (1861 Jan. 14, 1862 Febr. 19 and 21), which agree with that in Altona.'* However, in contrast to Tempel ('large bright nebula'), Auwers saw the objects as being *'always faint, at most fairly bright'*. Because he considered Tempel's description too vague, the difference was not significant enough that d'Arrest's *'assumption of a variability of the nebulae could be confirmed'*.

Auwers' result was supported by Chacornac at Paris Observatory.[169] He also noticed that the nebula was difficult to see at higher power: *'On September 16* [1862], *shortly before moonrise, I could perceive the nebula in the Pleiades which Mr. d'Arrest claims has vanished. In 1860 it appeared large, diffuse, somewhat enclosing the star Merope and of even brightness, in instruments with higher magnification it is faint and difficult to see.'* (Chacornac 1862d). Obviously, Chacornac had already observed the object in 1860, i.e. shortly after Tempel's publication.

On the basis of observations made with the 16.5-cm Steinheil refractor in Mannheim, Schönfeld took a firm stand against d'Arrest too. In his article of 3 October 1862 he wrote *'Even concerning the high air quality, which d'Arrest reported for these August nights, doubts should be allowed, because Tempel's nebula in the Pleiades, missing in his 15ft-telescope, was since then not only seen by Chacornac in its brightness of 1860, but also instantly stuck out in the local telescope on Sept. 20, 1862 when I pointed it freely towards Merope, without knowing the exact place, looking like a blurred nebula with the shape and size described by Auwers.'* (Schönfeld 1862a). In 1875 Schönfeld added that the nebula had been observed *'very clearly, immediately conspicuous, without accurate knowledge of the position'*.[170]

On 12 November 1862 d'Arrest reacted to the criticism. In *AN* 1393 he first of all somewhat ruefully reported that *'after a long effort I actually set eyes on Tempel's Nebula to the south of Merope, whose invisibility I had announced in August of this year'* (d'Arrest 1863a). He, adds however, "*This is the faintest object which I remember ever having seen in the refractor, even fainter than the nebula h. 2084 = H. IV. 76* [NGC 6946], *where Sir John Herschel says: 'Requires the eye to be well prepared for seeing it', and which I have observed on 19 February 1862 without much difficulty.*"[171] To his critics d'Arrest replies *'That I was unable to see this Pleiades–nebula in late August at a place where a large and bright nebula was expected is comprehensible, because many others, who had previously studied the Pleiades constantly and carefully, had not noticed it either.'* He uses Schmidt's arguments and proceeds to defend his thesis (preventatively shifting any responsibility onto Tempel): *'But I have certainly not seen an object for which a place and contour can be given in any way […] thus, to this very day, I am still convinced that the nebula is variable; otherwise the discovery report must be seen as highly exaggerated.'* Obviously, this was a tangential criticism of Tempel. D'Arrest added that, *'Even if I had taken Tempel's description of the Merope nebula more literally than it probably was meant to be taken, it is nonetheless good to have directed the attention of a few sharp-eyed astronomers to the Pleiades region.'* D'Arrest fails to reply to Schönfeld's and Auwers' statements about his observation method (too high a power, too small a field of view). He only mentions *'Auwers' convenient catalogue of the Herschel nebulae'*, which in the meantime had reached Copenhagen.

For d'Arrest the idea of variability was proved by the detection that "*on Jeaurat's chart of the Pleiades […] a star located about 16' north of Pleione is marked 'Nébuleuse'*". In 1778–79 Edme-Sébastien Jeaurat had observed the Pleiades, measuring the positions of 64 stars (Jeaurat 1782). Indeed, two stars on his chart (listed as nos. 46 and 48; later called 'Anonyma' nos. 31 and 32 by Bessel) are marked as being 'nebulous'. The close pair can produce a 'nebulous' appearance. This was, however, not confirmed by Charles Wolf in 1877 (Wolf C. 1877).

---

[169] Probably he used the 25-cm Lerebours refractor, which had been erected in 1823.
[170] Schönfeld (1875b: 88).

[171] The galaxy in Cygnus was observed by Auwers in 1854 too (see Section 6.19.1).

In November 1862 Winnecke, who was at that time working at Pulkovo, entered the discussion. First of all he remarks with a didactic undertone *'About the difficult visibility of Tempel's nebula near Merope, which, however, had to be noticed from the text of the report* [by Tempel], *I own several letters of Auwers, Krüger, Pape from the first months of 1861.'* (Winnecke 1863). He now comes to the crucial factor which contradicts the thesis of variability: the telescope size. In March 1862 he observed the nebula with an *'excellent telescope of 48'''* aperture [10.5 cm] *from the factory of Michael Baader in Munich* [...]. *Using the lowest power it appeared large, very blurred but easily visible.'* He asked Otto Struve to examine the object with the 38-cm Merz refractor, *'convinced that it would be difficult to see the large nebula with it'*. On 22 March 1862, after observing Hind's nebula, both of them observed the object. Winnecke noted that *'Indeed, we were not convinced about its existence until the telescope was moved quickly back and forth at a power of 150.'* Thus a technique called 'field sweeping' was needed to reveal the Merope Nebula. At low light levels, the eye perceives moving objects better than it does static ones.[172] However, no tricks had to be used at the 11.8-cm Steinheil refractor (observation on 29 September 1862); the object appeared as a *'milky, very easily visible nebula, in which some stars twinkle from time to time'*. Winnecke made a sketch, which was not published. In the 7.4-cm comet-seeker *'the nebula was conspicuous* [...] *but, due to the many bright stars, it could easily be overlooked'*. He summarised the situation as follows: *'According to all available data, I hold that there is no reason to assume any brightness variation.'* About d'Arrest's object of comparison, NGC 6946, which supposedly demonstrated the difficult visibility of the Merope Nebula, Winnecke mockingly remarked that he had *'found it on 21 Jan. 1854 during comet-seeking with a tube of 34'''* aperture [7.4 cm; in Göttingen] *and it was seen later many times'*. He was surprised about the *'quite unexpected visibility, even conspicuousness, of such a faint nebula in the comet-seeker'*.[173] With Winnecke's report, the matter was obviously finished for d'Arrest – he never again joined the discussion on the Merope Nebula.

Hind, the discoverer of NGC 1555 in Taurus, had never observed the object. However, in a letter to Herschel, dated 8 December 1862, he wrote *'I have never seen this nebula but I have repeatedly fancied, without satisfying myself, that several of the others of the Pleiades which fall within our Ecliptical Chart for 3^h R.A. were involved in a faint diffuse nebulosity. Chacornac mentioned in a late number of the Paris Bulletin* [18.9.1862] *that Merope itself is involved in the nebula.'*[174]

In a letter to Leverrier in Paris of 21 September 1863, Hermann Goldschmidt reported an observation of the Pleiades (Goldschmidt 1863b). Since this observation was made in Fontainbleau, he probably used a 10.5-cm refractor with a power of 36.[175] Goldschmidt not only saw the Merope Nebula, but, moreover, also noticed that the Pleiades are completely surrounded by diffuse nebulosity! A sketch published shortly thereafter in the Paris Observatory's *Bulletin* (Goldschmidt 1863c) shows the structure; see Fig. 11.32.

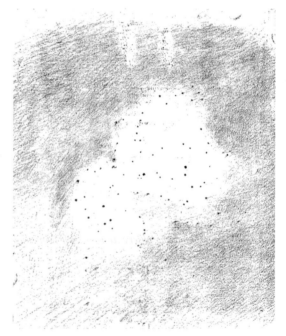

Figure 11.32. Goldschmidt's sketch showing the nebulosity surrounding the Pleiades star cluster; south is up (Goldschmidt 1863c).

---

[172] Steinicke (2004a: 174).

[173] Barnard too could easily see NGC 6946 on 28 May 1886 with his 5" Byrne refractor in Nashville (Barnard 1886k). He also mentions the nearby open cluster NGC 6939.

[174] RAS Herschel J. 12/1.6. However, Chacornac speaks only about a 'diffuse nebula' around other Pleiades stars.

[175] The telescope was probably made by Gambey.

In another report he affirmed his observation: '*While drawing the Pleiades I have just this moment made the discovery that they are surrounded on all sides by a nebula. This appearance is relatively easy to confirm, but the details demand attentiveness.*' (Goldschmidt 1864). He added '*Of course, one may ask why this nebula was not noticed by Tempel; perhaps a sign of variability. Though faint, it can hardly be missed by an observer with a good telescope.*' The faint exterior nebulae later appeared on plates, but it is doubtful whether Goldschmidt actually saw them. There were, however, similar observations by Charles Wolf, Schiaparelli and Spitaler (see Section 11.5.7). The positions and dimensions of these structures do not agree very well. In this connection d'Arrest's note about '*changing or variable light condensations in a hitherto unobserved, extended nebulosity of the Pleiades*' is interesting (d'Arrest 1863a).

In 1863, Tempel, working at Marseille Observatory, wrote a letter to the French magazine *Le Monde* (Tempel 1863). Primarily, it was a reaction to Chacornac's announcement of the 'variable nebula' NGC 1988 in Taurus. But for Tempel this was a chance to attack the idea of variable nebulae. He wrote '*Concerning this claim I want to remark that my experience has led to the conviction that the nebulae are unchanging, because lately large instruments have resolved them into stars and it would be unlikely that so many stars are variable simultaneously.*' About his key object, the Merope Nebula, he reported that it had already been seen in a 2" seeker in Marseille (which was even witnessed by the concierge of the observatory). On the other hand, '*d'Arrest was unable to see it at the same time with his large refractor in Copenhagen*'. Tempel concluded that '*All these facts show that the atmosphere is not always equal and the observation of such faint and vague objects needs some experience.*' He was forced to add '*My statement that the nebula in the Pleiades looked like a comet on 19 and 20 October 1859, and that I saw no trace of it six months earlier when I sketched the cluster, has been criticised. But these different observations do not indicate a variability of the nebula but are caused by the cleanness of the air. The clear air on 19 and 20 October 1859 is rarer than one might think, even for Marseille.*' He quotes a letter by Peters, written on 27 December 1860, which reads '*Your observation of a nebula in the Pleiades looks very extraordinary. If one could ascertain variability, this would be of great importance.*' Tempel remarks '*I answered Mr. Peters that I don't believe in the variability of nebulae and it more and more became clear that the variability of the atmosphere is the only reason for the phenomenon.*'

On 6 October 1863 the Reverend Webb in Hardwicke observed the Merope Nebula with his 5.5" Clark refractor. In an article for the magazine *Intellectual Observer* of 1864 he mentions earlier observations too (Webb 1864d): '*Schönfeld at Mannheim doubts the fact of variation [...] and he thinks that this and other suspected nebulae, being very feeble, large, and diffused, are influenced in visibility by magnifying power, varying transparency of the air, and practice of the eye, so that aperture is less concerned in their case than in that of minute stars. Auwers, at Göttingen, argues on the same side.*' He further remarks about Auwers that '*It has often, this observer says, been remarked – Encke's comet being an instance of it – that large, ill-defined, faint objects are best seen with small instruments, and that probably this nebula, having 15' of extent, filled d'Arrest's field under a considerable magnifier, and so became inconspicuous; he found it an easy object in a comet finder of 2 feet focus, and saw it repeatedly.*'

Concerning his own observation, Webb writes '*On turning the telescope upon the group with powers 29 and 64, though I probably should not have it discovered unknown, I found it with ease, as a very ill-defined, but on the whole egg-shaped haze, encompassing a brilliant star with its smaller but rather brighter end.*' The comparison with a chart provided no definite identification, but another observation on 10 November brought clarity: '*the star was Merope, and the glow connected with it the nebula in question*'. This marks the beginning of an observation series covering almost 20 years. In his book *Celestial Objects for Common Telescopes* Webb summarises the results: '*found readily with 5½-in.*'; '*very feeble*' (25 September 1865); '*a mere glow when star out of field, 9-in. spec.*' (4 March 1872); and '*in presence of star, 9⅓-in. spec.*' (15 January 1876).[176] The telescopes mentioned are the 5.5" Clark refractor and the 9.38" With reflector. By the way, 'spec.' does not mean a metal mirror, called 'speculum' by Herschel, Lord Rosse and others. Later the term was used in its original sense, meaning simply 'mirror'. In Webb's case it is a silvered

---

[176] This report first appeared in the fourth edition of 1881. The quotation is from the latest one, edited by Margaret Mayall; see Mayall (1962: 237).

glass mirror, termed a 'silvered speculum', 'glass speculum' or 'silvered reflector' too. In Britain they were produced at that time by George With (Hereford), John Browning (London) and Herbert Cooper Key (Stretton Sugwas).

In Germany, the Vienna astronomer Karl von Littrow popularised Tempel's Merope Nebula in his *Atlas des gestirnten Himmels* of 1866.[177] It is presented as an example of a 'variable heavenly body' (together with Hind's nebula NGC 1555 and Schönfeld's NGC 1333).

W. Matthews of Hill House, Gorleston, observed the Merope Nebula in autumn 1864, using an 8" With reflector with a silvered glass mirror.[178] Perhaps he had read Webb's article in the *Intellectual Observer*. He reported *'I had some excellent views of the nebula in the Pleiades with my 8-inch speculum; the nebula was something the shape of a fan.'* (Matthews 1866). Matthews was the only one who had the impression of colour: *'I was most struck with was the colour of the nebula, which has a dull dingy appearance, very unlike the bright white mist of Orion* [Nebula].' One further reads that *'its extent is immense and its colour remarkable'*.[179] Without mentioning a reason, he claims that *'The nebula must be variable.'*

John Herschel, who had never observed the Merope Nebula, entered it as GC 768 in the General Catalogue of 1864. The description reads: '*!!!B; vL; iF; VAR*'. After correcting Herschel's typo '!!!B' to '!!!; B' we get 'extremely remarkable (instead of extremely bright!), bright, very large, irregular formed, variable'.

This GC entry and the ongoing discussion about variability caused Lawrence Parsons and his assistant Ralph Copeland to examine the Merope Nebula with the 36" reflector at Birr Castle. Five observations are documented – unfortunately, all in vain![180] About the first, made on 10 February 1871 it is noted that there was *'No nebulosity detected about any of the Pleiades.'* Again ten days later, under *'very favourable circumstances,'* nothing was seen; 7 August 1872 yielded the same result: *'Not a trace of nebulosity visible.'* Another two unsuccessful attempts (now using the 72-inch) were made on 11 October 1872 and 28 September 1873.

In 1872 Bruhns' *Atlas der Astronomie* [*Atlas of Astronomy*] appeared. Therein the Director of Leipzig Observatory wrote about Tempel's nebula that *'it can be stated with certainty that some* [nebulae] *have changed their brightness* [...] *also a nebula in the Pleiades (AR = 3h 38m 45s, δ = +23° 24') can only narrowly be found, whereas it was visible in 1859 in a small telescope.'*[181]

In the same year a quarrel broke out between two amateur astronomers, which was publicly settled in the magazine *Astronomical Register*. The origin was a note in the June issue, stating that the Merope Nebula could not be seen in an 8" reflector, followed by the conjecture that it had vanished between the Pleiades stars: *'There is something peculiar about all the brighter stars of this group, which for months past have appeared to me as if surrounded by nebulous lights.'* On 10 June 1872 Charles Grover of Richmond replied in a letter to the editor: *'I think it should be known, that since December, 1869, this object has always been seen without difficulty in the 12¼ speculum.'* (Grover 1872a). While observing with his 12.25" Browning reflector, a 'comet eye-piece' with a power of 45 was used. The telescope was located at John Browning's home in Clapham, London. At a power of 150 (without moving the telescope) the nebula was invisible. Grover remarked about the problems of other observers that *'with higher powers it is seen with difficulty, and, in fact, might be easily missed altogether, and this is especially the case with an equatorial telescope accurately driven by clockwork'*. He concluded that *'we have a very probable explanation of many of the supposed changes which this nebula has been supposed to undergo'*. Especially his last sentence was the stumbling block: *'we see it now exactly as it has existed for many hundreds and, perhaps, thousands of years'*.

However, the fight was not carried out in the open: the opponent mysteriously signed himself as 'Facts Rather Than Fancy'. He invoked the authority of John Herschel and his characterisation 'VAR' in the General Catalogue: "*Even without the known facts which led Sir J. Herschel to include this nebula among variable objects* [...] *one can scarcely see how the observation of a nebula for 2½ years should make us altogether confident as to its condition 'many hundreds and perhaps*

---

[177] *Atlas of the Starry Heavens* (Littrow K. 1866: 58).
[178] Matthews (1865).
[179] In 1965 the colour was measured at McDonald Observatory: the nebula is bluer near Merope than the star and becomes increasingly redder at greater distances (O'Dell 1965). It is doubtful that Matthews saw this feature.
[180] Parsons L. (1880: 38).

[181] Bruhns (1872: 9).

thousands of years ago'." (Anon 1872a). Grover replied on 12 August 1872: *'The note on this subject in your last issue is very unsatisfactory, for two reasons. – 1st, it contains but a very imperfect statement of previous observations; and, 2nd, the writer totally fails to understand the meaning of the concluding sentences.'* (Grover 1872b). Before responding to the anonymous writer and defending his own line, he briefly reviewed the history, mentioning Tempel, d'Arrest, Otto Struve and Webb. Grover's main argument was that the appearance of the nebula strongly depends on the telescope.

'Facts Rather Than Fancy' promptly returned to the fray: *'It would appear that Mr. Grover considers that this nebula has probably not changed for thousands of years, because it was singularly variable a few years ago. One fails to see the force for this reasoning.'* (Anon 1872b). He further wrote that *'How a nebula can be at one time an easy object, and at another, a very difficult one, and yet not be variable, is the problem Mr. Grover has to solve.'* Thereupon Grover remarked that the reason for the change was not variability but the instrument used: *'it is only one example of the well-known fact that many celestial objects, and faint diffused nebula in particular, are best seen with low powers and abundant illumination'* (Grover 1873). He cites also John Herschel, who had presented a similar statement in the GC. However, when it came to the crunch, the unknown critic did not want to comprehend what it was really all about, whereas Grover had grasped the true nature of the problem. In his last letter the anonymous author wrote: *'I will dispute this point no longer with you; though I must confess, you are a truly moderate and polite arguer, for almost every third word you say is on my side of the question.'* (Anon 1873c). He was visibly annoyed that Grover had invoked 'his' witness John Herschel against him.

11.6.2 Tempel's observations in Arcetri

Owing to personal problems Tempel could not follow the ongoing debate, let alone make any observations himself (see Section 6.17.1). Except for his note in *Le Monde* of 1864 there was nothing from him on the issue. Not until 1874 did he publish, in the meantime having moved to Milan (Brera Observatory), a drawing of the Pleiades and the Merope Nebula (Tempel 1874). In the text he explains the circumstances of the discovery in Venice: *'When starting in 1859 to observe the sky with my telescope of 5ft focal length and 4 inches aperture, I immediately made a sketch of the group […] When I compared this lithographic sketch with the sky 6 months later, on 23 October 1859, during a very clear night, I was astonished to see a large and bright nebula near Merope, which I thought to be a comet at that time. On the next evening I was able to check that the nebula remained exactly at the same place. At that time I had no catalogue at my disposal, to see whether this nebula was already known or not.'*

Tempel gives an incorrect date, since his discovery was actually made on October 19. About the delayed publication of his find he remarked *'But a year later, after convincing myself and the others in Marseille of its constant visibility, I wrote a report […] as a result, the nebula could be well seen by other astronomers, however, it generally appeared better and brighter in small telescopes than in large ones.'* The drawing published in Milan had already been made when Tempel was in Marseille (Fig. 11.31). Tempel first mapped the stars on the basis of positions from the Bonner Durchmusterung and then inserted the nebula as it *'exactly appears in the telescope at a power of 24'*. Apart from his 10.5-cm Steinheil refractor he could use only a Plössl of comparable size at Brera Observatory; Schiaparelli's 22.2-cm Merz refractor was not erected before his departure.

When Tempel moved to Arcetri in 1875, for the first time substantial observations could be made. Two refractors with apertures of 28.3 cm and 23.8 cm (called Amici I and II) were available. Tempel's first report is dated 2 May 1875; this was a letter to the editor of the *Astronomische Nachrichten* (C. A. F. Peters), which appeared in *AN* 2045 (Tempel 1875b). In the introduction one reads *'I owe the first confirmation of my announcement about a nebula near Merope in 1859 to you [Peters] and Mr Pape.'* He mentions the performance of the refractors by Steinheil and Plössl: *'the difference in visibility of nebulae or faint comets was sometimes incomprehensible: in the Plössl stars and planets were beautifully clear images, but there was often no trace of nebulae and comets, which were seen so well in the Steinheil'*. In the Plössl the Merope Nebula was barely silhouetted against the sky background; however, *'it showed many small stars twinkling in this nebular region, which I could not see in the Steinheil'*. The two Amici refractors differed optically too. The smaller (Amici II) showed the Merope Nebula considerably better; Tempel was *'highly surprised to see this nebula so clearly, large and (at least on one side) sharply bounded'*. On the other hand, *'Amici I performed like the Plössl'*. He asked Schiaparelli to test the new 22.2-cm Merz refractor on the Merope Nebula. In a letter, dated 7 March 1875, the Milan astronomer wrote back about his observation of 25 February that *'many persons have observed the Pleiades without noticing this large nebula, which is so easily seen when the sky is clear'*. Chambers later commented in the third edition of his *Descriptive Astronomy* (1877) that *'Schiaparelli, at Milan, trying a new telescope on 25 February 1875, saw this nebula very clearly, and was much surprised by its size. He noted it to extend from the star Merope, beyond Electra and as far as Celæno.'*[182]

In the last section of his report, Tempel treats d'Arrest's *AN* 1393 article, in which it was critically stated that his first description was 'greatly exaggerated'. He now replies that *'After 10 years of patience it is quite a satisfaction for me that I now can answer this uncomplimentary judgement about my first announcement with the above statements, but also once again to draw the main attention to the actual variability of this nebula of such an important extent or to the disappointing and uncertain knowledge about the power of telescopes.'* There was no reply from d'Arrest – he died on 14 June 1875. However, the common interest in the Merope Nebula was undiminished, shifting now from variability to the question of its very existence.

11.6.3 Reactions to Tempel's observations

In 1875 Charles Wolf of Paris Observatory reported about a comprehensive study of the Pleiades, including micrometrical measurements of star positions. He mentions an interesting observation of the Merope Nebula with the 31.6-cm Sécretan refractor by Benjamin Baillaud and Charles André: *'On 7 March 1874 the Merope nebula consisted of two nuclei: one nearly concentric around the star and spreading a bit to the east; the other, brighter, is located 7ˢ west of the star at the same declination with a diameter of about 1ˢ.'* (Wolf C. 1875).

In spite of good weather, Wolf was unable to see the Merope Nebula from November 1874 to February 1875. This was confirmed by Stephan, using the 80-cm Foucault reflector in Marseille. Wolf concluded that *'This nebula is truly variable and its period seems to be rather short.'* His result appeared in the *Monthly Notices* too (Anon 1876b). The author commented on the opposed positions about variability held by Wolf (pro) and Tempel (contra): if the latter were right it would be *'difficult to explain the apparently decisive observations of M. Wolf, who has examined the spot with the same telescope under equally good atmospheric conditions, when the nebula has been visible at certain times and invisible at others'*.

In 1875 Holden observed the Merope Nebula several times with the 26" Clark refractor at the US Naval Observatory, Washington.[183] He wrote the following about his observation of 25 October 1875: *'Merope is a nebulous star, with a double space on its preceding north and following sides. To the south it is joined by nebulosity to the main body of nebulosity above.'* At a power of 158 he saw a nebula of 6′ to 7′ diameter, which was lost at a power of 400, except for a central part around Merope. This proved – contrary to Tempel's opinion – that a large telescope was able to show the nebula. In the comet-seeker it appeared even larger than on Tempel's Milan drawing: *'I am sure it is further extended to the south and west than Temple* [Tempel] *has it. No nebulosity seen north of Merope.'* About the Paris observation, Holden wrote (on the basis of a subsequent observation on the 28th) *'I see nothing of Wolf's 2nd nucleus 7ˢ following Merope.'*

Dreyer, who since August 1874 had been employed as an observer at Birr Castle, used Tempel's report as an occasion for another attempt with the 72" reflector.[184] About his observation of 24 December 1875 he wrote *'examined Merope & environs in consequence of Mr. Tempel's letter in the Astr. Nachr. No. 2045. No nebulosity seen, only the same false light as around the other bright stars. Sky misty.'* Again a Birr Castle observer had failed to see the Merope Nebula – this, fortunately, would change.

11.6.4 Birr Castle observations and the drawing of Maxwell Hall

In a note appearing in the magazine *Nature* on 11 January 1877, Maxwell Hall reported his observation of the

---

[182] Chambers (1877: 546); in 1889 a note about the issue appeared in the *Sidereal Messenger* (Vol. 8) too.

[183] Newcomb (1878: 319–320).

[184] Parsons L. (1880: 38).

Merope Nebula made on 20 October 1876 with his 4" Cooke refractor at a power of 55 (Hall M. 1877). From 1872 he ran the private Kempshot Observatory, located in the mountains of Jamaica. Hall wrote that "*the nebula was 'bright', according to Sir John Herschel's scale, and extended in a parabolic form at least 40' from Merope*". Concerning the difficulty of seeing the faint object near the bright star, especially in a large telescope, he quoted Dreyer: '*The Merope-nebula is never perceived with Lord Rosse's telescopes.*' Since the Birr Castle observer did not publish his report until March 1877 (see the following section), and the complete results appeared even later, in 1880, the following question arises: what was the source for Dreyer's remark? It can be found in his review of Vogel's 'Positionsbestimmungen von Nebelflecken' (published in 1876); in the note to GC 1440 (NGC 2264, the bright cluster around 15 Mon) one reads '*As is generally known, it is always difficult to see faint nebulosity around bright stars in a large reflector (the Merope nebula is never perceived with Lord Rosse's telescopes).*'[185]

In January 1877 Tempel published another report on the Merope Nebula, which appeared in *AN* 2139 (Tempel 1877a). About the inconsistency of its visibility he remarked '*what is seen with my small telescope, however, should certainly be seen with a large one too*'. He concluded that '*the invisibility of the Merope Nebula in a large telescope is due to the eye-piece and its field of view*'. This was clearly aimed at d'Arrest: '*if d'Arrest had used an eye-piece of lower power than that of magnification 95, giving a field of view of 20 to 25': he would have seen the nebula very easily, not leaving room for any hypothesis about a variability*'. Tempel's refractors offered powers of 133 (Amici I) and 155 (Amici II) with fields of view of 20' and 34', respectively. The same argument is quoted against Lord Rosse: '*That the nebula was not seen with L. Rosse's big telescopes is only caused by the eye-pieces used*' (obviously Dreyer's statement of 1876 was unknown to him). Tempel took this opportunity to make a general attack on the Birr Castle observations. On the basis of the drawings published by Lord Rosse, he claimed that the 36" and 72" reflectors did not show more stars than his 11" refractor. With a provocative tone, he suggested that the issue should be clarified by star counts: '*I cannot understand why the happy owners of these giant telescopes do not prompt their astronomers to carry out such studies.*'

In the March issue of *Nature* Lawrence Parsons replied to Hall's report (Parsons L. 1877b). This was the first 'official' response from Birr Castle. Judging Dreyer's words '*a little too strong*', he could well believe that Dreyer had claimed about the Merope Nebula that it could 'never' be seen with Birr Castle reflectors – he had to challenge this! He specified five observations between 1871 and 1875, which had indeed been carried out in vain. Lawrence Parsons attributed this to the metal mirror: '*dust and other opaque substances will interfere more with the action of a speculum than an object-glass in searching for faint nebulosity near a bright star*'. He concluded that '*It may therefore still be possible that under peculiarly favourable atmospheric conditions, and with a speculum just repolished, we may still be able to detect the nebulosity, but it appears far more probable that we must look for an explanation of the difficulty of seeing the nebulosity to the comparative smallness of field of so large an instrument.*' He mentions his observations of faint extensions of the Orion Nebula, using a '*finding eye-piece of 26' field*'. Lawrence Parsons cites d'Arrest's problems and Tempel's success too, which he ascribes to the '*more southern position*'.

On 6 March 1877 Hall drew the Merope Nebula at his 4" refractor with a power of 100 and moving of the field (Hall M. 1879) (see Fig. 11.33). He was still astonished that it was invisible in large telescopes, such as Lord Rosse's and Newall's 25" Cooke refractor.[186] Obviously, the common optical principle – '*the ratio of the light of the star to that of the nebula remains the same whatever the aperture may be*' – was invalid in this case. Hall also worried about Rudolph Engelmann's illustration (Fig. 11.34), which was published in 1876 in his edition of Bessel's work.[187] It was based on observations of the Pleiades made in Leipzig with a 12.2-cm comet-seeker by Reinfelder & Hertel at a power of 40. Hall wrote that the nebula '*is represented as a very small circular patch of light in no way connected with Merope, and situated about 11' from the star*'. This did not match Tempel's description at all. No doubt Engelmann just wanted to mark the position of the nebula ('neb') north

---

[185] Dreyer (1876c: 280).

[186] Hall had visited Robert Newall in Gateshead (see Section 11.6.9).

[187] Engelmann (1876: 304).

532   Special topics

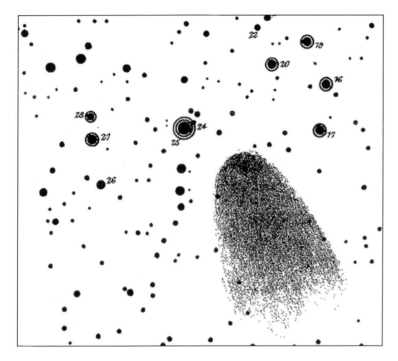

Figure 11.33. Maxwell Hall's drawing of the Merope Nebula; north is up (Hall. M 1879).

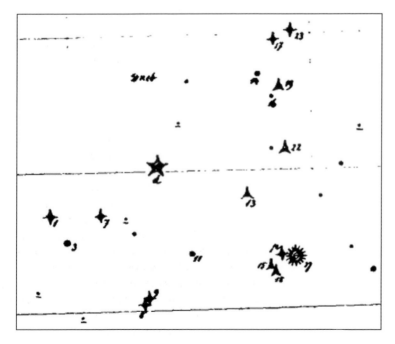

Figure 11.34. Engelmann's 'sketch' of the Merope Nebula; north is up (Engelmann 1876).

11.6 The Pleiades nebulae 533

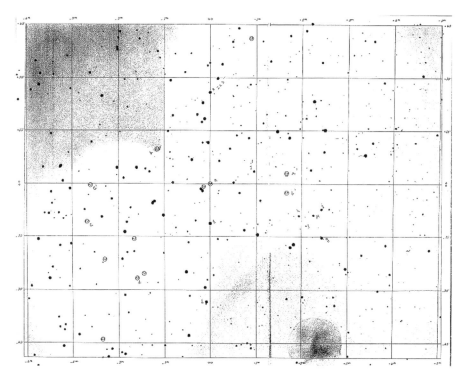

Figure 11.35. Charles Wolf's drawing of the Merope Nebula; south is up (Wolf C. 1877).

of Merope ('d'). The reason was simple: due to the bad night, he was unable to see the object.

Hall was astonished about Schiaparelli's observation of 1875 too, in which the nebula was seen '*extending in a direction at right angles to the axis of my figure*'. He hoped that '*further observations will be made of this most interesting nebula by possessors of telescopes with both large and small apertures*'. Concerning the history of the nebula, he emphatically refers to the '*excellent summary given by Mr. Webb, in the* Intellectual Observer, *vol. iv., p. 440*'. This was Webb's merit, pointing out remarkable objects, such as the Merope Nebula and Hind's nebula, in his column 'Clusters and nebulae'. This challenged many visual observers. Hall's article and drawing, which were published in 1879 in the *Monthly Notices*, triggered an intensive debate involving Charles Wolf, Common, Tempel, Hough, Burnham, Lewis Swift, Hunt, Barnard and Backhouse.

11.6.5 Charles Wolf, Dreyer and Tempel

In 1877 Wolf published his (already announced) work on the Pleiades, titled 'Description du Groupe des Pléiades'.[188] It first outlines the current research on the Pleiades. Concerning the Merope Nebula, Wolf mentions the results of various observers, especially about its possible variability. After his first success of 7 March 1874 the nebula remained invisible until it reappeared on 22 November 1875.

Wolf wrote that '*On 22 November* 1875, *when the sky was extremely transparent, I immediately saw the triangular nebula near Merope, this large comet of Tempel; Merope seems to be its nucleus and the wide tail reaches the edge, which is clearly defined in the east but smeared to the north.*' He made a sketch (Fig. 11.35). In a note of 11 January 1880, Charles Wolf refers to his work of 1877 (Wolf C. 1880a). He mentions Goldschmidt's observations of nebulous masses surrounding the Pleiades (see Fig. 11.32): '*As my map of the Pleiades shows, a nebula, first seen by Goldschmidt, completely surrounds the star group, shining on the black sky background; it looks like a hole drilled into the centre of the nebula.*' Like Schiaparelli, Wolf sees the nebula extending to Electra and Celaeno. He attributed the different results to the

---

[188] 'Description of the Pleiades group' (Wolf C. 1877).

quality of the air and the telescope. The method of observing is important too: Wolf had masked the bright star with the micrometer bar. Variability of the Merope Nebula was ruled out – at least after 1864.

On 10 December 1877 Dreyer tried again to see the nebula in the 72-inch, obviously having been motivated to do so by Lawrence Parsons. He now used an eye-piece with 26′ field of view, writing '*I saw first, that f a line through 2 stars sf the ground is blacker than p that line; when I pointed this to Lord R.* [Lawrence Parsons] *he saw it also, and could distinguish the p limit too, forming an angle of about 40° with the faint limit*.'[189] This could be confirmed on 27 December : '*I was struck with it at first sight when the star had been placed in the middle of the field.* [...] *I think the brilliancy of the B star* [Merope] *is the principal reason why it is not seen with a high power.*' The successful observation with the wide-angle eye-piece was important enough to be mentioned in the report of the 'Earl of Rosse Observatory' for 1877 (Parsons L. 1878).[190] There the magnification is given as 130.

William Christie, editor of *The Observatory*, made Tempel's publication (*AN* 2139) available to English-language astronomers (Christie 1878). The article triggered an open exchange of blows between Dreyer and Tempel. Although the argument was focused on the 'spiral form of nebulae' (see Section 11.3.4), the Merope Nebula was an important piece of evidence. In Dreyer's reply, his recent success is mentioned: '*M. Tempel is right in considering that its visibility depends on the use of a large field and a low power; in fact, our own recent experience with the 6-foot-reflector proves this perfectly*' (Dreyer 1878b). Thereupon Tempel asks with satisfaction '*Is it not singular that it should now be seen with all instruments, while at the time of its discovery I was overwhelmed with reproaches by D'Arrest, as he was unable to see it with his 11-inch telescope, and Dr. Dreyer failed to find it with Lord Rosse's too. Is it possibly variable, as some astronomers claim?*' (Tempel 1878b). He gives the following answer: '*I cannot myself admit the probability of this idea for an object of such vast magnitude, and have almost abandoned the belief in the variability of such numbers of heavenly bodies, since I have come to see that nothing under the sun is so variable as poor mortals like us.*'

In December 1879 Tempel wrote his treatise 'Über Nebelflecken' ['On nebulae']; it did not appear until 1885, when it was published in a limited edition. It contains a chapter on the Merope Nebula, which treats variability and its location relative to the Pleiades stars.[191] Tempel mentions a private communication by Secchi, quoting '*that he had never seen the nebula*'. He once again responds to d'Arrest: "*I even was criticised by d'Arrest about my exaggerated statement that it has been bright like a comet*". Tempel explained that the nebula even appeared much brighter than his first comet, which he had discovered half a year earlier (2 April 1859) with his 4-inch in Venice. He wrote, in a conciliatory mood, '*Nineteen years have passed since its discovery, and also the reports on its variability have quietened down, for it certainly seems unbelievable that such a great nebulous mass as this Merope nebula contains should be variable.*' He correctly predicts that '*In 50 and more years' time there will certainly be discovered in the heavens many things that we have at present overlooked.*'

Finally, Tempel speculates about the nature of the nebula. For him its situation is already remarkable, in that '*no other nebula presents itself in this manner*'. What makes the Pleiades unique in the northern sky is that the cluster encloses a considerable number of bright stars within a small area. Tempel wrote '*and this mass of light has not been able to prevent (indeed has even facilitated) the seeing of the large, albeit faint, nebula right in the middle of it*'. Since the nebula obviously is not dimmed by the stars, he suggests that the nebula is located in front of the stars: '*Just as through a bright chandelier, the objects on the opposite side are hard to make out or not visible at all, whereas those in front of it, in our line of sight, are seen well.*' If this is true, '*observers should easily measure its faster motion against the* [background] *stars*'. But Tempel sees a problem: '*unfortunately such measurements are quite impossible for this weakly bounded nebula, without a core or stellar centre*'.

However, in the case of the Andromeda Nebula, Tempel states that '*this nebula must be located far beyond the stars*'.[192] His conclusion is based on an unpublished drawing made with Amici I, showing '*200 fine small stars, sharply contrasting with the nebulous background*

---

[189] Parsons L. (1880: 38).
[190] The reports hardly exceed half a page.

[191] Tempel (1885a: 6–8).
[192] Tempel (1885a: 9–10).

11.6 The Pleiades nebulae

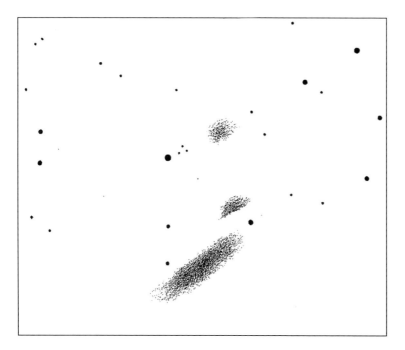

Figure 11.36. Common's drawing of the Pleiades nebulae; north is up (Common 1880b).

in their different magnitudes'. This would be impossible, '*if the stars were located in the background, sending their light to us through the gaseous nebula*'. Tempel also considers a possible compromise ('*in the middle between the two hypotheses*'), namely '*the matter visible as the nebula is associated with certain little stars, i.e. the nebula belongs to the star or the stars belong to the nebula [...] at the same distance as the stars and physically closely connected to them*'.[193] Dreyer comments on this opinion in his review of Tempel's treatise: '*There does not however, seem to be any reason why the nebula should not belong to the Pleiades and be physically connected to the star Merope, an idea that derives some support from the number of stars with cometary tails or similar nebulous appendages which we know are found in the heavens.*'[194]

11.6.6 Common's controversial drawing

On 8 February 1880 Common visually observed the Pleiades with his 36″ reflector in Ealing. The results were reported in the RAS meeting on 9 April 1880 (Common 1880a). He presented a remarkable drawing, which was published the same year (Common 1880b); see Fig. 11.36. It considerably differs from Maxwell Hall's and caused a controversy. The star positions were taken from short-exposure plates. The drawing shows three distinct nebulae: a long one to the southeast of Merope (at the bottom), above it a smaller one and a third, oval, to the northwest of Alcyone.

The first to react was James McCance, of Putney Hill, who wrote the following in a letter of 16 April 1880: '*I looked for this nebula on many nights during the past winter, using my 10-inch (Calver) mirror, and I cannot say that I ever saw any thing very like what either Mr. Common or Mr. Hall has drawn; I merely caught at times a very faint glow close to Merope, chiefly north and south of it, but could not make out any thing more definite, and did not see a tail of light stretching in any direction.*' (McCance 1880a.) Nonetheless, he respected Common's telescope: '*Of course Mr. Common's great aperture has enormous advantage in this kind of work, and we cannot hope to see all he sees or compete with him in any way.*' McCance refers also to Charles Wolf, who '*sees nebulous matter almost entirely surrounding the group*' with his 12″ refractor. Obviously, there are doubts: '*I see many small stars which are not on Mons. Wolf's map, though I understand that he inserts all stars*

---

[193] Tempel (1885a: 10).
[194] Dreyer (1887c: 61).

therein he sees with at least 12 inches aperture.' This opened the debate.

As early as on 8 May 1880 Wolf defended his opinion, reporting observations of the Pleiades, made in November 1878 with the new 1.2-m reflector at Paris Observatory (Wolf C. 1880a). He is astonished that not much more was seen than in the 31.6-cm refractor: '*I expected to see – thanks to the great power of this magnificent instrument – a much larger number of stars than the reflector* [refractor] *of 0.31 metre had shown me. My expectation was completely disappointed.*' He concludes that '*the space-penetrating power of an objective glass of 0.31 metre is sufficient to attain the limit of the visible universe in that region of the sky – a conclusion which is further supported by the fact that the background of the sky appears there perfectly black in the middle of the nebulosity of the Pleiades*'.[195] Later Solon Bailey and Vsevolod Stratonov found out that the Pleiades region is an area with very few faint stars.[196] Actually, Wolf had not reached the 'limit of the visible universe', but rather experienced interstellar absorption. On Common's observation, he remarked that '*I have never seen the nebulosity in the neighbourhood of Alcyone of which Mr. Common speaks.*' His results were also presented in the magazine *Sirius* (Anon 1880d). The article erroneously speaks about a '*Foukault'schen Spiegelteleskop von 4 Fuss Durchmesser*' ['Foucault reflector of diameter 4 ft]; actually, the Paris reflector, which was equipped with a silvered glass mirror (as had been introduced by Foucault in France), had been made by Martin/Eichens.

On 10 May 1880, George Knott, of Cuckfield, wrote the following: '*Tempel's Merope Nebula I believe I caught with my 7 1/3-inch refractor on January 5* [1864], *it was very faint, stretching in a s p direction from Merope.*' (Knott 1880). Despite confirming Hall's drawing, he was, however, '*a little puzzled by its direction*'. Concerning the exterior nebulosities, Knott refers to the drawing of Goldschmidt in Leverrier's *Bulletin*: '*the keen-eyed observer and indefatigable discoverer of minor planets, observed an extensive nebulosity surrounding the Pleiades group*'.

---

[195] Wolf later explained in detail why a 15.7-inch telescope is sufficient for all nebular observations (Wolf C. 1886b).

[196] Herbert Couper Wilson made a similar study (Wilson H. 1907).

11.6.7 Tempel's drawing of 1880

In a letter of 22 May 1880 to Lord Lindsay, who was at that time the Foreign Secretary of the RAS, Tempel gave his view on these papers (Tempel 1880a): '*it is but natural that I should be much interested in anything written or observed in reference to this nebula, which I was the first to discover*'. Attached is a new drawing (actually a modification of the one published in 1874 in Milan): '*all the minute stars and the exact form of the nebula have been added with my Amici No. I, with a magnifying power of 113*'. It does not show the curved streamer to the northwest of Merope, pointing in the direction of Electra (Fig. 11.37).

Tempel took the opportunity to list the advocates of the opposite opinions about the existence of the Merope Nebula. In addition to himself, '*various illustrious astronomers, such as Schmidt, Winnecke, Auwers, Schönfeld*', had confirmed the object. The opponents were '*D'Arrest, Padre A. Secchi, and the observers at Parsonstown, who failed to see the nebula with their great telescopes, and consequently doubted its existence.*' The Birr Castle observers were immediately exonerated by Tempel, because they had followed his urgent suggestion: on '*fitting the large telescope with eye-pieces of a low magnification the nebula becomes distinctly visible*'.

Tempel's concluding remarks reflect a certain gratification: '*It is now ascertained beyond question that the nebula exists […] and anyone publishing statements about its non-existence merely uses vain words, and proclaims himself wanting in knowledge of the history of nebulae and of the management of telescopes.*' He clearly refused to accept Goldschmidt's observation of 1864 that the Pleiades are surrounded by vast nebulous masses: '*this is an optical illusion*'. Tempel compares his new drawing with those of Hall and Common, noticing a perfect agreement with the former. Concerning Common, he believed that his sketch must have '*evidently been executed with a telescope of insufficient power to show the Merope nebula*' – a fatal misunderstanding.

Common, the owner of a 36" reflector, was alarmed that Tempel speaks about 'insufficient power'. In a letter of 11 November 1880 he wrote that by the '*three feet telescope*' mentioned in his report, the reflector of '*three feet aperture*' is meant (Common 1880c). Obviously, Tempel had assumed that this meant a focal length of three feet! About the place of the oval nebula to the northwest of Alcyone, Common noted that '*By the aid*

11.6 The Pleiades nebulae

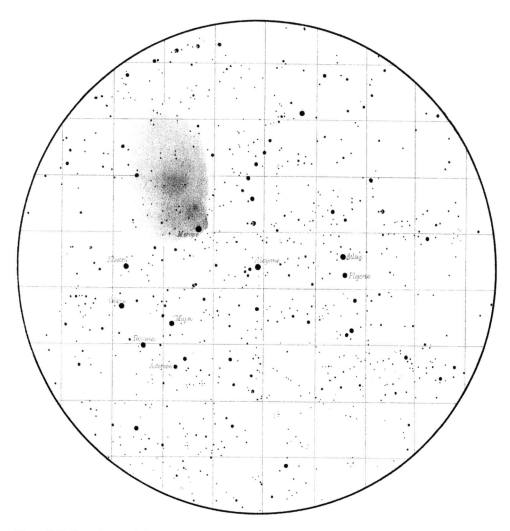

Figure 11.37. Tempel's second drawing of the Merope Nebula; south is up (Tempel 1880a).

of a sketch which includes the small stars near, I was enabled to locate the Nebula on the map that Prof. Tempel gives.' (The title 'Professor' is incorrect.)

It is interesting that on the same page of the *Monthly Notices*, immediately following Common's report, there is Baxendell's discovery note of his 'unphotographable nebula' (NGC 7088) in Aquarius. The name is due to the fact that it was impossible to verify it by photography (see Section 9.11.2). The discussion about the obscure object shows some parallels with the case of the Merope Nebula. Indeed, Baxendell wrote in his note that *'It seems to be similar in character to the large Nebula near the Pleiades, but is slightly less bright.'* (Baxendell 1880).

McCance piped up again on 10 June 1880 (McCance 1880b). Though not knowing Tempel's last publication, he was aware of his drawing, which had been sent in advance to the RAS. McCance termed it *'a charming little map or drawing'*, writing that *'The drawing shows the nebula much as Mr. Hall has drawn it.'* He mentions the English translation of Tempel's description, in which the German astronomer *'considers the nebulosity surrounding the group, which some observers [Goldschmidt, Wolf] have thought they saw, an optical illusion'*. About Wolf's *'nuclei or nodules'* to the south of Merope, McCance remarked that they *'are but little more luminous than the rest of the nebula'*.

On 13 December 1880 Hall wrote another report (Hall M. 1881). Therein he describes an observation of 21 November 1880 on Jamaica (a drawing made there is not included). Surprisingly, the form of the nebula appeared different: '*a change in the preceding boundary of the Nebula was noticed; it was found to extend as far as Electra, and the parabolic form of the Nebula, as seen 1877, was destroyed*'. This could be confirmed on subsequent nights. His comments on Common's observation of 2 December 1880 and Tempel's latest drawing are interesting. About the former Hall wrote: '*I failed to make out any of its features*'; and about the latter, '*I was very pleased with Prof. Tempel's beautiful drawing [...] his remarks are very valuable.*' He added that '*the form of the Nebula, as drawn by Prof. Tempel, is exactly the mean of the two forms drawn by me in 1877 and 1880*'. However, Hall asked why Tempel had overlooked the '*extension of the Nebula in the direction of Electra*'. Finally, he remarked '*I am anxious that the present indications of systematic change should be confirmed by Prof. Tempel and other observers.*'

Henry Pratt observed the Pleiades in Brighton with an 8.15" reflector by With/Browning. He thought himself to hold an intermediate position between Tempel/Hall and Common. In a letter of 11 January 1881 he first admits that he wanted to enclose a drawing made in October 1876, but backed down when confronted with '*Signor Tempel's beautiful map*' (Pratt 1881). He largely supports Tempel's opinion, '*with one slight exception*', however: '*I always see the Nebula extend to some little distance beyond Electra in a fainter condition than the portion drawn by Tempel.*' Pratt used Kellner eye-pieces with powers of 45 and 84, giving fields of view of 85' and 50', respectively. It is surprising that he could confirm one of Common's nebulae between Merope and Alcyone with the higher power: '*I am able to see just a trace of nebulosity where Common has drawn it.*'

11.6.8 The awkward failure of Hough and Burnham

The camp of Tempel's opponents was unexpectedly augmented by two distinguished visual observers: Hough and Burnham of Dearborn Observatory, Chicago. On 29 March 1881 they published the results of their Merope Nebula observations, comparing them with previous ones by other astronomers (Hough and Burnham 1881). In the introduction one reads that '*The observations already made in regard to Tempel's Nebula in the Pleiades, although very numerous, are nevertheless so discordant that additional facts may be of interest.*' Both of them had already tried to observe the nebula: '*The casual examination of Merope and its immediate neighbourhood on a number of nights during 1879 and 1880 with the 18½-inch Clark Refractor, with powers 120 and upwards, failed to show any nebulosity whatever.*' But now, confronted with Tempel's '*recent very positive statement [...] in which the nebula is so sharply delineated*', Hough and Burnham were possessed with ambition. Between 29 November 1880 and 22 March 1881 intensive observations were made on '*first-class observing nights*'. Various eye-pieces, including those with large focal length (low power, large field of view), were used. Sometimes the refractor was stopped down to 12"; one had tried to mask Merope too. All their efforts were in vain. Hough and Burnham were frustrated, but saw a solution.

They compared the positions of the nebula as given by Tempel, Hall, Engelmann, Wolf, Schiaparelli, Common and Holden. The surprising result was that the position angles relative to Merope greatly differed. Doubts about the existence of the nebulae arose: '*In view of such great discrepancies by observers with first-class telescopes, we think it may with great propriety be asked whether these various appearances are due to real matter, or are simply an illusion.*' Given their own failure, it was understandable that they favoured the second alternative – and directly provided an explanation: '*We are strongly inclined to the opinion that the phenomena in question are due to the glow proceeding from Merope and the neighbouring stars.*' Concerning the supposed visibility of the nebula in various instruments, Hough and Burnham remarked that '*With a small telescope and low power we should have to contend with the light from numerous stars in different parts of the field. As the size of the telescope is increased, the field will be diminished, and a less number of stars will be included, which might materially modify the phenomena.*' What was the effect of these statements of two eminent astronomers? They met with opposition!

Dreyer was the first to give his opinion, in the new astronomical magazine *Copernicus* (Dreyer 1881b). He cited his own successful observations of the Merope Nebula made with the 72" reflector in Birr Castle between 10 and 27 December 1877. Then, for the first time, he brought into play the optical ratio of entrance and exit pupil: *'the pupil of the observer's eye did not receive light from the whole of the large speculum'*. A brief calculation shows that Dreyer was right. With a mirror of diameter 1830 mm and an assumed power of 200, an exit pupil of 1830 mm/200 = 9.15 mm results.[197] This is considerably larger than the maximum entrance pupil of the eye of 7 mm (when fully adapted to darkness). Owing to the ratio $(7/9.15)^2 = 0.58$, the eye loses more than 40% of the light leaving the eye-piece (vignetting). Nevertheless, the apparent surface brightness of the nebula remains constant. The nebula does not fill the field of view and appears by virtue of its contrast to the dark background. In 1875 Dreyer was not successful when using a power of 400: the apparent surface brightness is halved in value and, moreover, the nebula covers the whole field of view, leaving no contrast. As early as in 1878 Dreyer had mentioned this fact: *'a higher power which receives light from the whole surface of the speculum does not show it'*.

Dreyer sees his observations as being *'in good accordance with the drawing of M. Tempel'* made in 1874 in Milan. Since he had not been aware of them before autumn 1878, his own observation was *'entirely free from any bias'*. Dreyer stressed both the agreement with Hall and Pratt and the significant differences from Schiaparelli, Wolf and Common. He advised that one should ignore the *'tiny speck'* in Engelmann's map of the Pleiades which had caused some confusion. He quoted Engelmann's own words: *'the sky was not clear and the Merope nebula not visible on the night when the map was made'*.

After this brief historical outline, Dreyer turns to Hough and Burnham: *'The above accounts do not appear to be quite as irreconcilable as supposed by Messrs. Hough and Burnham.'* He does not see any contradiction between the observations of Tempel, Lord Rosse, Dreyer, Hall, Pratt and Holden and those of Wolf and Schiaparelli. According to Dreyer, they differ only by the apparent size of the nebula (in the direction of Electra) and the position angle (differences of about 100°): *'I have myself seen many cases of a large and diffused nebulosity on different nights appearing to extend more or less in different directions.'* It is interesting that Dreyer ignores Common's drawing.

The idea of Hough and Burnham that the nebula is an optical reflex caused by Merope was rejected: *'That what the observers [...] have seen, should be nothing but an illusion, is not likely to be conceded by one of them, and it seems difficult to imagine how the glare of the bright star could produce an appearance of nebulosity at one side only.'* Dreyer further wrote that *'There are many cases, in which an observer has believed a bright star to be nebulous, while all subsequent observers have failed to see any nebulosity, but in all such cases the star has by the first observer been described as being surrounded by an atmosphere.'* Here he refers to the cases mentioned by John Herschel in the Slough catalogue.[198] Since Hough and Burnham also quote his remark in *The Observatory*, in which the benefit of a low power was pointed out,[199] Dreyer cannot deny himself the final comment that, *'As any statement made by the two exceptional observers [...] cannot fail to carry certain weight with it, and as they quoted my remark [...] about the visibility of the object in question, the above description of what I have seen may not be entirely uncalled for.'*

Lewis Swift, obviously irritated by Hough and Burnham, wrote on 2 December 1881 that *'this nebula [...] is par excellence the greatest enigma in observational astronomy'* (Swift 1882).[200] He reported his own experiences, dating from 1874: *'while searching for comets, I ran upon it, and having never heard of a nebula in the Pleiades, strongly suspected that it was a new comet, which illusion the following night quickly dispelled, as the object was stationary'*. Therefore Swift was an independent discoverer of the object! At that time he was observing with his 4.5" refractor from the flat roof of 'Duffy's Cider Mill' in Rochester. He informed Burnham about his find, who in return enlightened him on Tempel's discovery in 1859 and the supposed variability of the nebula. Later Swift looked at the Merope Nebula whenever possible, without, however, detecting any change.

In his article, Swift critically treated the following question: *'Why will not a large telescope be as effectual as a small one in revealing faint nebulae close to bright stars?'* Like Hall he doubted that the brightness ratio between star and nebula should be independent of aperture. He concluded that this could apply to the internal

---

[197] Steinicke (2004a: 167–169); see also Section 6.4.10.

[198] Herschel J. (1833a: 499–400).
[199] Here Dreyer (1878b) is meant.
[200] This was Swift's first publication on nebulae.

comparison between nebulae and stars only, because a point-like object (star) could not be magnified, in contrast to an extended one (nebula). Swift wrote that *'Then, too, with large telescopes, eyepieces of higher power are generally used and the field correspondingly contracted, and so, of course, the opportunity for contrasting the faint light of the nebula with the surrounding dark sky is diminished if not entirely lost.'* This is thought to be the essential reason for the failure of Hough and Burnham. He stated that *'It is quite natural for those trained observers, with such an instrument, to call into question the reality of a nebula which they are unable to see.'* If Merope, according to the Chicago astronomers, would create such an illusion, then it must be *'certainly very strange that of all the stars in the heavens, Merope alone should show nebulosity immediately following it!'* Swift could see the nebula even with an aperture of 2" (using a diaphragm) and a magnification of 25 (the same as Tempel's minimum). He recommended the following: *'If Messrs. Hough and Burnham will contract the aperture of their telescope to from 4 to 6 inches, and use a power of about 30, they will see as a reality what they now believe to be a myth.'* Hough and Burnham had tested a reduction to 12" – but obviously this was not enough! Their reaction to Swift's article is not known.

The next reply was due to George Hunt of Hopefield, West Dulwich, who had purchased the old 8" Cooke refractor of the Reverend Dawes, who had been living nearby since 1857. In a letter dated 17 December 1881, he reported his observation of 17 November (using a Kellner eye-piece with magnification 42 and field of view 55'): *'at once [I was] struck with the glow of the nebula near Merope'.* About Burnham Hunt wrote *'Notwithstanding the dictum of the distinguished and conscientious observer Mr. Burnham (and I wish to express myself with all becoming deference to so great an authority), I cannot resist the conviction that the nebula in the Pleiades is not a mere subjective illusion.'* (Hunt 1882).

11.6.9 Further observations and an important RAS meeting

In 1883 Barnard observed and drew the Merope Nebula in Nashville with his private 5" Byrne refractor (Barnard 1883a).[201] His interest was awakened by Swift, who had mentioned the object in one of his first letters. Barnard criticised Common's drawing. Concerning the positions of stars and nebula, his own drawing (Fig. 11.38) agrees with Tempel's.

Barnard's description reads *'It is plainly visible in my five-inch refractor, it has been seen with a 2½-inch telescope, in the presence of a quarter-full moon.'* He is in agreement with Swift too: *'Prof. Swift sees it as I do, even with two inches of aperture.'* Finally, he promises that *'observations will be continued to detect variability'*.[202]

Barnard mentions observations made by Etienne Trouvelot at the US Naval Observatory, Washington: '[Trouvelot] *reports that it is variable, and that it has become very faint since, and that it can now be seen only by those acquainted with its former appearance'*. Though no source is given, there exists an unpublished manuscript by Trouvelot, titled 'On the Tempel nebula in the Pleiades', which was presented at the meeting of the American Academy of Arts and Sciences on 14 December 1875 (Anon 1876c). Fortunately, Trouvelot mentions his observations in his manual on 'Astronomical drawings'. There one reads *'In 1875 I made a long study of this object, and drew it carefully a dozen times, but I was not able to see any changes in it within the two or three months during which my observations were continued. But on Nov. 24, 1876, it was found of a different color, being purplish and very faint. On Dec. 23, 1880, it was found just as bright and visible as when I drew it in 1875, and on Oct. 20, 1881, it appeared faint and purplish again, as in 1876. On this last night, and on those which followed it, it was impossible for me to trace the nebulosity as far as in 1875. I consider this as due to a variation in the light of this object, which in 1875 was bright enough to be well seen while the Moon after her First Quarter was within ten or fifteen degrees from the Pleiades.'*[203]

On 25 April 1882, Thomas Backhouse commented on the controversial case (Backhouse 1882). He owned a 4.5" Cooke refractor at West Hendon House, Sunderland. On 13 February 1882 the nebula was observed carefully with a power of 38: *'In order that I might be unbiased, I avoided looking at the descriptions of it in the* Monthly Notices *previous to looking for it. I did not even know which star was Merope.'* Backhouse was

---

[201] This was Barnard's first publication on nebulae.

[202] See also the note in *The Observatory* (Anon 1884).
[203] Trouvelot (1882: 155).

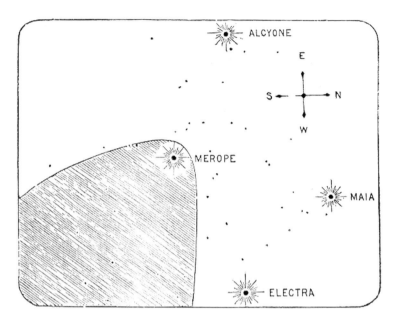

Figure 11.38. Barnard's sketch of the Merope Nebula (Barnard 1883a).

successful: '*I soon found the nebula, though to see it well it was necessary to have Merope out of the field. It was very faint, and the only definite feature was the following edge, which however, was quite plain.*' Like Barnard, he noticed a complete accordance with Tempel's drawing – but not with Common's: '*Mr. Common's drawing appears to be very rough, or perhaps he has confounded two stars [...] the sky is perfectly dark where he draws it.*' At places where Common had seen nebulosity the sky appeared completely black to Backhouse.

The Merope Nebula was a topic at the RAS meeting on 12 May 1882 (Stone E. 1882). First a report by Backhouse (probably identical to that published in the *Monthly Notices*) was read by the Secretary James Glaisher. A controversial debate between Newall and Common developed. The former sketched the nebula on the blackboard, as he had seen it with his 25" refractor in Gateshead. He obviously was generally in accord with Backhouse, Tempel and most other observers. Common replied '*I have never seen the nebula like that; indeed the drawing appears to me upside down [...] did you use an inverting eye-piece?*' Newall said no and Common briefly countered with '*it is a very interesting object [...] it demands further scrutiny*'. Edward Stone, being in the chair, now entered the discussion. He suggested taking John Brett, a well-known draughtsman, as referee: '*I think we shall be obliged to get Mr. Common to show Mr. Brett this nebula, and get a drawing from him, for at present I cannot see that these are drawings of the same object at all.*' Stone now focused on the weak point of the issue, the subjectivity of visual observing: '*I think that the great dissimilarity of these drawings is a point of very considerable importance, because we are every now and then discussing supposed changes that are taking place in nebulae; and if we cannot get two or three observers to agree better than this in representing something that they are supposed to have seen, when we begin to infer changes from the drawings we are walking upon very treacherous ground.*'

Owing to the ongoing discussion, Denning felt the need to publish his observations of the Merope Nebula made in autumn 1881 in Bristol (Denning 1882). In a letter of 9 June 1882, he wrote that he had seen the object clearly with his 10" Browning reflector and a 'comet eye-piece': '*the nebula is an easy object; its position and brightness are as evident as they are unvarying*'. His drawing matched Tempel's. Denning could not detect any nebulosity around other stars, thus concluding '*I cannot agree with the theory of Messrs. Hough and Burnham that this nebula is simply due to the glow proceeding from Merope and neighbouring stars.*' He held that such a claim plus sketches like Common's had caused damage: *'the discrepancies [...] have originated the scepticism with which the existence of this very interesting object is regarded'*. Denning further wrote that *'differences such as these have also given the erroneous idea, that the nebula is variable'*. He clearly was on the side of Tempel, Hall, Barnard and Swift.

In October 1885 the Netherlands astronomer Jean Oudemans visually observed the Merope Nebula, using a Steinheil refractor stopped down to 7" (Oudemans 1888). He was impressed by the unusual darkness of the sky background and *'the Merope nebula as beautifully visible as one could want'*. John Hooper noted *'the faint light in the region about Merope'* on 17 August 1885, using a 5" Clark refractor with magnification 45 and 'field sweeping'. He wrote that *'The general outlines* [of the nebula] *can be made out by slowly moving the telescope about, when the contrast with the surrounding darker sky becomes very evident.'* (Hooper 1885). Another observation, made with an 8" Grubb refractor in Leuven, Belgium, was published by François Terby on 1 November 1886 (Terby 1886). He confirmed the triangular shape described by Tempel.

Georges Rayet reported an observation made with the 14" refractor of Bordeaux Observatory: *'Concerning the Merope nebula, it always was easily visible with the equatorial 14-inch (power 140) when the nights were moonless and the clear sky showed stars of 13 mag.'* (Rayet 1886). He detected faint extensions to the west and south. It is interesting that Rayet confirms Goldschmidt's observation *'In Bordeaux the form of the nebula appeared very similar to that described by Goldschmidt.'* It is not known whether this impression included the nebulosities surrounding the Pleiades.

11.6.10 Photographic discovery of the Maia Nebula (NGC 1432) and other Pleiades nebulae

The honour of obtaining the first (acceptable) photographic image of Pleiades nebulae is due to the brothers Paul and Prosper Henry (see Section 9.26). The plate was taken on 16 November 1885 at Paris Observatory, using the 33-cm astrograph (see Fig. 9.89); two other exposures followed on 8 and 9 December. In a note of 22 December 1885, titled 'Découverte d'une nébuleuse par la photographie', the brothers Henry published their success in the *Astronomische Nachrichten*.[204]

However, the subject under discussion was not the Merope Nebula but a new one around the star Maia (NGC 1432). The brothers Henry wrote *'This nebula appears very clearly and has spiral form [...] it was impossible to see it in our telescopes.'* Unfortunately the first image was not published; a later one appeared in 1888 (Fig. 11.39). The Maia Nebula is the only one in the NGC to have been discovered by photography.

Edward Pickering was the first to react. In his letter of 21 January 1886 he points out that he already had photographed the Pleiades on 3 November 1885 with an 8" lens (Pickering 1886). The 65-minute exposure shows only *'certain irregularities [...] due merely to defects in the photographic process'*. However, a comparison with the image of the brothers Henry revealed

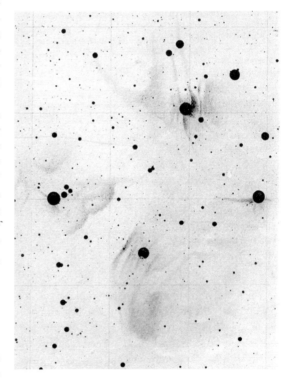

Figure 11.39. The image of the Pleiades taken by the brothers Henry of 1888; north is up (Henry and Henry 1888).

---

[204] 'Discovery of a nebula by photography' (Henry and Henry 1886a). The image is not given there. At the same time a note appeared in the *Monthly Notices* (Henry and Henry 1886b).

that these are indeed the nebulae near Merope and Maia! Krüger, editor of the *Astronomische Nachrichten*, remarked in a footnote to Pickering's report that the '*reproduction of the original negative clearly shows a nebulous rudiment west of Maia, pointing to the north, and near Merope a faint diffuse remnant can be traced to the south*'. However, it was not possible to reproduce Pickering's image in the *AN*. Of course, the priority of the discovery of the Maia Nebula should remain with the brothers Henry. Pickering had intially interpreted the nebulae as plate flaws, not realising their nature until he inspected the French image.

In his footnote, Krüger also mentions the first visual observation of the Maia Nebula by Otto Struve. The Pulkovo astronomer had informed Amédée Mouchez, Director of Paris Observatory about his success, whereupon the later immediately published a note in Vol. 102 of *Comptes Rendus* (Mouchez 1886).[205] The observation was made on 5 February 1886 with the brand-new 76-cm refractor by Clark/Repsold. Struve wrote that the nebula '*was not difficult to see*' (Struve O. 1886a), but with a simple '*Durchmusterung it would probably barely have been noticed, in that on the one hand weak nebular light mixes with the stray light surrounding the brighter stars in all telescopes, on the other hand the human eye is made less sensitive to the slight differences in light in the immediate vicinity of the bright star by its glare*'. Another observation followed on 23 February this time inspecting the '*nebula with more leisure and making the appended sketch, though the temperature was only −16.5° C.*' Struve's sketch shows a curved nebula, stretching from Maia to the east (Fig. 11.40 left). For orientation he used stars from Charles Wolf's list of 1877. A day later the object was seen in the old 15" Merz refractor too. Another observation (with the large refractor) was made on 3 March assisted by Struve's son Hermann. The Merope Nebula is not mentioned in the paper. In May 1886 a summary appeared in *The Observatory* (Struve O. 1886b).

Shortly afterwards, Charles Wolf published a comparison between visual observations of the Merope Nebula (including his drawing of 1874) and the image of the brothers Henry (Wolf C. 1886a). Significant changes could be detected: compared with the repeated sightings of Wolf and others, the Merope Nebula now appeared different. Moreover, the nebula '*20′ north of Atlas seems to have vanished; it was discovered by Jeaurat, observed by Goldschmidt and Mr Wolf has plotted it in his chart*'. Wolf defends his results and refers to the latest, successful observations of the Merope Nebula by Rayet. He claims that nebulae that are invisible on photographs can nevertheless exist for the eye and that visual differences are not automatically due to erroneous observations. Wolf concludes that '*Photography shows an aspect of the sky that can be very different from direct observation.*'[206] He proposed that the two kinds of observations should be considered separately, because photography will never be able to substitute for the eye.

In 1886 the brothers Henry treated the Merope Nebula (Henry and Henry 1886c). They wrote that '*Photography has likewise allowed the determination of the form of the Merope nebula, whose appearance has been seen differently by all observers and whose existence has been disputed by several astronomers.*' However, their result was not the expected one, as it supported Common instead of Tempel: '*In our image the object looks very similar to a drawing made by Common.*' The work of Goldschmidt and Wolf could not be confirmed: '*however, the image is completely different from earlier drawings by Goldschmidt and Wolf*'. The brothers Henry also criticised Wolf's statements on photography and visual observation: '*photographic mapping shows the relative brightness of stars much better than by direct observing*'. In addition to the Maia Nebula, another one was detected near Electra.

In a subsequent article the brothers Henry explicitly stated that the Merope Nebula looks completely different from what had been drawn by Tempel. On the contrary, the match with Common's sketch was striking: '*We see on all images of the Pleiades faint traces of nebulosity near Merope, but [...] they do not resemble the nebula Tempel has seen. On the contrary, they have much similarity to what Common saw.*' (Henry and Henry 1886d). Of course, the surprising result was gratifying for Common – particularly by virtue of the fact that it was published in the British *Monthly Notices*!

Common reacted promptly – he was visibly proud that his work had been vindicated. About the Merope

---

[205] This volume contains five articles on the Pleiades nebulae.

[206] Later Keeler advanced a similar view, pointing out the different spectral sensitivities of eye and plate (see Section 11.2.4).

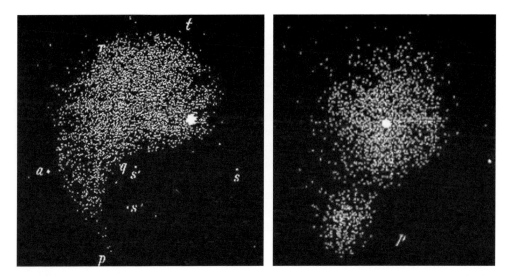

Figure 11.40. The Maia Nebula (NGC 1432), drawn by Otto Struve (left) and Rudolph Spitaler (right); south is up (Struve O. 1886a; Weiss 1886a).

Nebula, as imaged by the brothers Henry, he wrote '*this being the same one near Merope that I found with the 3-foot Reflector in 1880*' (Common 1886). He quoted his former notes: '*There is strong suspicion in my mind of more nebulae there.*' Indeed, the French image showed three nebulae – in agreement with his sketch! Thus, Common was convinced that he had seen the Maia Nebula in 1880: '*This I believe alluded to an observation of the Maia nebula which will be doubtless easily seen in the 3-foot telescope.*' However, on comparing the two pictures, the identity is pretty vague: Common's 'Maia nebula' is much closer to Alcyone than to Maia.

The images of the brothers Henry and Pickering prompted Edmund Weiss, Johann Palisa and Rudolph Spitaler in Vienna to search the Pleiades for nebulae with the 27" Grubb refractor. However, it was not until 26 February 1886 ('*the first clear night*') that the observations were started (Weiss 1886a). The Maia Nebula appeared as a '*small flaky nebulosity, completely separated from Maia, well bounded to the right* [west], *but to the left gradually fading in to the background*'. On March 3 the air was optimal and the part seen on 26 February was seen to be '*only the brightest knot of an extended nebulosity, completely covering Maia*'. Spitaler, '*possessing a very sensitive eye for faint nebulae*', made a sketch (Fig. 11.40 right).

Motivated by this success, the Vienna astronomers now studied Jeaurat's 'nebulous stars' of 1779, mentioned in 1863 by d'Arrest. Spitaler noticed that '*both stars are placed within faint nebulous envelopes*'. He also saw an isolated nebula to the northeast of Taygeta on several nights, but bad weather made further observations impossible. Given the photographs and the observations of Common and Schiaparelli ('*supported by a particularly keen eye and the clear sky of Italy*'), Spitaler wrote '*one can hardly refrain from thinking that at least the whole Pleiades region west and north of Alcyone is covered by an extended nebulosity, of which all previously perceived, apparently isolated nebulae are merely bright knots of light.*'

In a paper of 5 March 1886 Henri Perrotin, Director of Nice Observatory, reported observations of the Maia Nebula with the 38-cm refractor by Henry/Gautier (Perrotin 1886). They were made between 28 February and 4 March, assisted by Louis Thollon and Auguste Charlois. Like Otto Struve and the Vienna astronomers, they could confirm the Maia Nebula visually. However, it was visible to them only because they already knew about its existence from photography. Thollon made a sketch, which was not published.

Asaph Hall too tried to see the Maia Nebula at the US Naval Observatory, Washington, with the 26" Clark refractor. However, comparing the star with others at a power of 400 yielded nothing that '*I should call a nebula without other information*' (Hall A. 1886a). But he detected something different: '*there is a bluish halo preceding this star*'; Merope showed the same phenomenon, but fainter. Hall wrote that '*the appearances around Maia and Merope are peculiar*'. The reason was never established – perhaps there was a reflection inside the optical system.

On the suggestion of Raoul Gautier, Director of Geneva Observatory, his assistant Arthur Kammermann tried to observe the Maia Nebula with the 10″ Merz refractor. Professor Thury of Geneva University decreed a 'safety regulation' that seems stange from today's standpoint, stipulating '*visualization of the photographic rays, for it is possible that the nebula in question emits chemical rays for the most part*' (Kammermann 1886). For this purpose the 'Soret process' should be applied, i.e. a '*blade of uranium-glass or a compound of Aesculin*' had to be inserted into the eye-piece. Thus 'protected', Kammermann was able to perceive the nebula on 30 March 1886 at a power of 80 (on 2 April he used a power of 40). In the eye-piece Maia was masked by an '*obturator of about 1 mm width*' (the uranium-glass being still in the light path). Kammermann wrote that '*These measures allowed me to see the Maia nebula clearly.*'

On 23 October 1886 Isaac Roberts started to photograph the Pleiades in Mughull, publishing the results in three *Monthly Notices* papers. However, his first attempt with the 20″ Grubb reflector was disturbed by clouds (Roberts 1886). During the following night all went well, and he noted about his three-hour exposure that '[the plate] *showed that not only are the stars* [Alcyone, Maia, Electra, Merope] *surrounded by nebulae, but that the nebulosity extends in streamers and fleecy masses, till it seems almost to fill the spaces between the stars, and to extend far beyond them*'. It appeared that the Pleiades '*are involved either directly or else in sight alignment with one vast nebula*'. On 29 December Roberts took another plate: '*the evidence of their reality should be so clear that it cannot reasonably be doubted*' (Roberts 1887). This and further images (made until the end of 1888) were not published.

Roberts showed his images at the RAS meeting on 12 November 1886 (Anon 1886). Common remarked '*I am extremely gratified to see what Mr. Roberts has done, not only because it has been done by a person in England, but because it has been done with a reflector.*' He reported that '*the images, which a reflector gives, are smaller*'. Roberts' results were again presented by Edward Knobel at the RAS meeting on 14 January 1887 (Knobel 1887). The only, albeit pretty long, statement was due to Common, once again quoting his visual observation of 1880: '*seven years ago I described some nebulae I had discovered, or thought I had discovered, in the Pleiades with the 3-feet reflector*'. After this unusually self-critical opening, he immediately went into attack mode. Once again, his target was Tempel: '*when I brought this to the notice of the Society, I was rather roughly handled by Prof. Tempel, who thought I had not used a sufficiently large telescope*' (which obviously was due to a misunderstanding). About the nebula to the northwest of Alcyone Common remarked '*I have been anxiously waiting to see if any photographs would show* [it]'. With gratification he now stated that '*In two of Mr. Roberts' photographs [...] these very objects of light are unmistakable.*'

In the New General Catalogue of 1888, the Maia Nebula and the Merope Nebula are entered as NGC 1432 and NGC 1435, respectively. The former is described as '*eF, vL, diff (Maja Plejadum)*', referring to the brothers Henry, Otto Struve, Spitaler and Roberts. For NGC 1435 (GC 768) one reads '*vF, vL, diff (Merope)*'; Dreyer mentions the drawings of Tempel, Charles Wolf, Maxwell Hall and Common and Roberts' image.

In 1888 the brothers Henry piped up for the last time (Henry and Henry 1888). To detect possible changes, plates with exposure times of four hours had been taken every year (Fig. 11.39 shows the 1888 version). Each image shows nearly 2000 stars down to 18 mag. Like Roberts, the French astronomers could identify many individual nebulae and filaments. They guessed that most of them could be observed visually only with the largest telescopes, such as that at Pulkovo.

Common commented on the latest image of the brothers Henry in the RAS meeting on 8 June 1888 (Common 1888a). Again he felt confirmed in his 1880 observation. His opening is a strike against Tempel: '*Everyone knows the Pleiades and the Nebula which I think Tempel discovered some years ago.*' He then moves on to his pet subject: '*In 1880, using a 3-feet telescope, I gave two or three hours to this object and found that in addition to the Merope Nebula there was, between Alcyone and two stars near, another nebula; and that nearly on Merope there was also another.*' Aiming to Tempel, he continues: "*It was the publication of this sketch in 1880 that brought the remarks from Professor Tempel that are published in Vol. xl. of the*

Figure 11.41. Isaac Roberts' Pleiades image of 8 December 1888 (Roberts (1893), Plate 11).

'Monthly Notices' [Tempel 1880a]". As evidence that his own sketch matches the new image of the brothers Henry, Common explains that *'I immediately compared my sketches with it and found that every star I had seen, except one, was there, and, of course, in their proper places.'* In the minutes of the meeting one reads that *'Mr. Common showed on the blackboard the position of the stars as observed and photographed.'*

Robert Newall attended the meeting and, of course, the old controversy with Common (of 12 May 1882) flared up again. First he summarised the history of their dispute: *'I made a drawing of what I saw of the nebulae; Mr. Common made a drawing of his: he said he could not understand my drawing; I said I could not understand his. I do not know whether the Fellows understand either.'* About the present state of affairs, he added *'I have seen it [the Merope Nebula] several times since, and I am quite certain what I saw was there, but it differs from what Mr. Common saw a little time ago. It is a nebula very like an oval comet, Merope being in the focus of the nebulous matter surrounding it. I think I remarked at that time that I had observed throughout the whole constellation of the Pleiades a large amount of nebulous light. [...] I have not seen the patches of light which Mr. Common has on the blackboard.'* Newall's witness is Maxwell Hall: *'I have made a drawing on Argelander's map [the BD] of what I saw, and I gave Mr. Hall a copy of it when he visited me, and this compared exactly with his drawing.'* Thereupon Common remained calm. However, in a paper on 'Photographs of nebulae', which appeared in the same year, he defended his position: *'no one had been able to localize any of the brighter parts that have since been shown by photography, except perhaps an observation made with a 3-foot in 1880, that showed two or three of the brighter portions that have since been shown beautifully on the photographs of the Brothers Henry'* (Common 1888b).

In his third paper, Roberts presents another image, which was made on 8 December 1888 with an exposure time of four hours (Fig. 11.41). It shows a large number of individual nebulae, and he notes that *'there is a remarkable absence of symmetry between these nebulae, that leads us to infer that we are looking at a number of separate nebulae one behind the other in the line of sight'* (Roberts 1889a). The accordance with the image made by the brothers Henry of that year is remarkable.[207]

Maxwell Hall's last paper is dated 6 December 1889 (Hall M. 1889). It was caused by an action of the RAS. The society had allocated Hall a 9" Browning reflector with a 'glass speculum' for the testing of its performance in the humid climate of Jamaica.

---

[207] Roberts' images are stored in the RAS archive.

11.6 The Pleiades nebulae 547

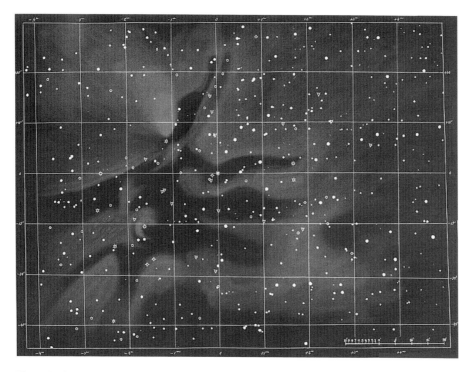

Figure 11.42. Spitaler's map of the Pleiades region. The brightest (cometary) nebulosity, on the left above the centre, is the Merope Nebula; south is up (Spitaler 1891a).

Alongside the reflector, which was '*found to be excellent*', he used his old 4″ Cooke refractor ('*its definition is perfect*'). Of course, the target of comparison was the Merope Nebula; the result was unambiguous: '*a glance through the refractor showed the well-known nebula projected against the dark background or field of view; but in the reflector there was so much light scattered around the field of view that the nebula was invisible*'. Hall further wrote that '*It was not the light of Merope which overpowered the nebula; it was scattered light which filled the field of view to whatever part of the midnight sky the telescope was turned.*' This explained the positive sightings in smaller telescopes (especially refractors) and, vice versa, the failures with large ones, such as the 72″ reflector at Birr Castle (referring to Lawrence Parsons' report of 1877 in *Nature*).

11.6.11 Spitaler's treatise 'Nebel in den Plejaden'

In 1891 Rudolph Spitaler reviewed the major observations of the Pleiades nebulae in a 20-page treatise on 'Nebel in den Plejaden', which he published in the *Annalen der K.K. Universitäts-Sternwarte Wien*.[208] His motivation was '*the photographic discovery of a nebula close to the star Maia by the brothers Henry at Paris Observatory, and the fact that certain nebulous objects were seen earlier at some places in the Pleiades group by various observers*'. He created a map of the Pleiades region, covering 2° × 1.5° (Fig. 11.42). It shows extended nebulosities, qualitatively matching the observations of Goldschmidt, Wolf and Schiaparelli.

Spitaler concludes that either the Pleiades group is '*surrounded by an extended pale nebulosity*' or '*it is located at a dark part of the sky, similar to some holes in the Milky Way*'.[209] Alluding to the 'illusion theory' of Hough and Burnham, he again stresses the importance of a sufficient field of view when one is observing faint extended nebulae: '*When only a particular place in the extended nebula is fixed upon with a large telescope, the commensurately small field of view is illuminated evenly and, although one is gazing at the nebula, one does not*

---

[208] 'Nebulae in the Pleiades' (Spitaler 1891a: 185).
[209] Spitaler (1891a: 192).

notice it. Therefore the nebula was not seen at all, and its existence was doubted, by many observers with large telescopes.'

According to Spitaler, the question of the variability of the Pleiades nebulae, especially of the Merope Nebula, is '*still not decided, for neither the grounds for nor those against are free of objections*'.[210] However, '*due to my own observations and in the face of the above mentioned history, I consider the nebula to be invariable*'. For Spitaler it was essential to clarify the meaning of Tempel's first description, which marks the beginning of the whole debate. In a letter he had asked Tempel '*to intimate his memory about the discovery of the Merope nebula*'. In his reply Tempel referred to his treatise 'Über Nebelflecken' of 1885. But the statements presented there were not really enlightening for Spitaler, causing even more confusion. He therefore substantiated his critique of Tempel. So, in his discovery note there was talk about a '*large bright nebula […] which I [Tempel] thought to be a large comet in the first instance*'. However, in 1885 Tempel compared the nebula with the comet 1859 I, which appeared '*very faint and small compared with the Merope nebula*'. According to Spitaler, Tempel had eventually '*made another mistake by his comparison with comet 1859 I*'.[211] Given the fact that the comet 1859 I was far from being faint (as Spitaler could show), this would imply that the Merope Nebula was actually very bright at the time of its discovery: '*If this comparison [with the comet 1859 I] is correct, then the question of variability is posed beyond any doubt […] this must have been realised by Tempel in the first place.*'

Spitaler now explains the making of his map of the Pleiades in 1886–87. When observing with the 27" Grubb refractor in Vienna, mainly the eye-piece of lowest power (field of view 16′) was used. The map was based on Charles Wolf's star list. Spitaler wrote that '*Mapping the nebulae onto the chart was very difficult, since, after each view on the illuminated paper, it took some time for the eye to reach its former state of sensibility to perceive the faint objects.*'[212]

For him it was essential to compare his map critically with the images of Roberts (obtained using a 50-cm reflector) and the brothers Henry (obtained with a 33-cm refractor), which he attested were of great value: '*Since the photographs […] made with two very different instruments match so significantly, no doubts remain about the reality of the imaged nebulae.*'[213] Spitaler's description of the one around Alcyone, as seen on Roberts' photograph of 1888, is interesting: '*Alcyone lies in a dense spiral nebula*' (see Fig. 11.41).

Spitaler continues: '*If one compares the images with my map, they hardly seem to be in accordance at first sight*'; but he adds that the '*main structure […] were seen correctly and generally only the boundaries vary*'.[214] One further reads that '*The nebulous bridge connecting the Merope and Alcyone nebulae visible on Roberts' image is completely missing on the Paris chart, while it appears on my drawing as the northwestern edge of a large pale nebulosity, extending vastly beyond the Pleiades to the southeast.*'[215] On the significance of visual observations relative to photography, Spitaler argues like Wolf: '*Insofar as the other tremendously pale and diffuse nebulae on my chart are concerned, it would not surprise me if photography were not able to depict them, for they are not much more than a slightly brighter sky background.*' Finally, Spitaler states '*thus here the possibility that the eye draws to one's attention where on the photograph images of very faint nebulae are to be found is not out of the question*'.

During the year 1888–91 the Merope Nebula was again observed in Vienna. Using the 6" Fraunhofer refractor, Johann Holetschek carried out his work, which he published in 1907 under the title 'Beobachtungen über den Helligkeitseindruck von Nebelflecken und Sternhaufen'.[216] The nebula could '*almost always easily be seen when the air was clear*'. Three observations were made: on 18 January 1888 (when the Moon was 5.5 days old), in September 1891 (when comet Wolf passed the Pleiades) and on 23 February 1892. Holetschek estimated the brightness of the nebula as 9.5 to 11 mag.

11.6.12 Barnard's 'new Merope nebula' (IC 349) and the dispute with Pritchard

In late 1890 Barnard observed the Pleiades with the 36" refractor at Lick Observatory (Barnard 1891a). He was

---

[210] Spitaler (1891a: 193).
[211] Spitaler (1891a: 194).
[212] Spitaler (1891a: 195).
[213] Spitaler (1891a: 199).
[214] Spitaler (1891a: 199–200).
[215] Spitaler (1891a: 200).
[216] 'Observations on the apparent brightness of nebula and star clusters' (Holetschek 1907).

11.6 The Pleiades nebulae

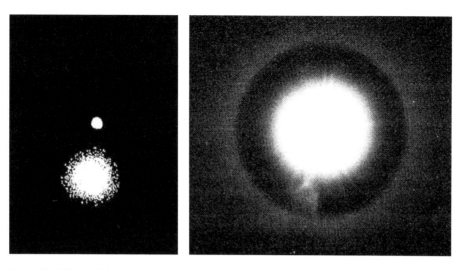

Figure 11.43. Barnard's 'new Merope nebula' IC 349. Left: his own sketch (Barnard 1891a); right: the Allegheny Observatory image of 1981 with overexposed star (Reiland 1982); north is up.

able to confirm the nebulae near Merope, Maia and Electra. Whereas the first two caused no problems, he wrote about the last '*This curved streak emanating from the place of Electra is extremely difficult.*'

However, with his observation of 14 November 1890, Barnard added a new fascinating detail: very close to Merope he discovered a tiny object, which he termed the 'new Merope nebula' and described as '*comparatively bright round cometary nebula close south and following Merope. [...] It is about 30" in diameter, of the 13^m, gradually brighter in the middle, and very cometary in appearance.*' Barnard used powers of 300, 520 and 1500 and made a sketch (Fig. 11.43 left). The distance to Merope was measured to be only 36". In the 12" Clark refractor the new nebula was seen '*with some difficulty [...] by occulting Merope with a wire in the eyepiece*'. Dreyer listed it as IC 349 in the first Index Catalogue.

Though IC 349 is the brightest Pleiades nebula, it remained hidden in the glare of Merope – until the best visual observer, Barnard, entered the scene. He could easily explain why it does not appear on any photograph: '*A sufficiently easy* [long] *exposure to secure an impression of the nebula, would so overexpose Merope, that its light would coalesce with that of the nebula.*' However, Barnard's statement caused an exciting dispute.

On 23 February 1893, Charles Pritchard of Oxford University Observatory wrote a note titled 'A newly discovered Merope nebula' (Pritchard 1891a). He was an approved expert on star imaging, being proficient at measuring their positions and brightness on plates (Pritchard 1891b). Therefore, Oxford took part in the international 'Carte du Ciel' project. Pritchard had a 13" Grubb astrograph (in the standard Henry design) and two photographic 15" reflectors with lenses of $f/5.3$ and $f/8$, respectively, at his disposal. Test plates were taken, e.g. of the Pleiades (Pritchard 1888). In 1893 Agnes Clerke remarked about Pritchard's attempts to measure the parallax of the Pleiades by photography that '*Professor Pritchard possesses an instinct of selection much needed by the sidereal astronomer. The problems he attacks are always of high interest and importance.*' (Clerke 1893).

Pritchard replied to Barnard's statement that the 'new Merope nebula' was unphotographable: '*I think that these latter remarks are scarcely borne out of the facts of the case. I will not venture to say they are distinctly misleading.*' The nebula was '*plainly impressed on a photographic plate taken at Oxford, 1889 Jan. 29, after an exposure of 120 minutes. [...] I regarded this apparently insignificant fleck, simply as the brightest portion of this widely distributed nebulous matter.*' Four other plates showed the nebula too; Pritchard wrote that '*The separation from Merope is in all cases too saliently marked to be overlooked, and with the exception of some nebulous matter round Maia is the only nebulous matter which has hitherto been impressed on the Oxford photographs.*' On a short-exposure plate, the Merope image was '*not one half of that implied by Prof. Barnard's description*'. Anyway, Pritchard had not published his result: '*I* should not have considered it as possessing any great interest or importance.' However, Barnard's article changed his mind. Moreover, he felt a *'natural jealousy'* against the opulent equipment of the Lick Observatory.

Barnard reacted in a letter of 7 April 1891 (Barnard 1891b). First he apologised to Pritchard, admitting his mistake: *'The statement that this nebula had never been photographed was erroneous.'* His remarks on the chance of being able to photograph the nebula had been based on images taken by the brothers Henry and Roberts with exposure times of 3 and 4 hours, respectively, showing Merope as a blob of about 2′ diameter. About Pritchard's work Barnard had to concede the following: *'I was not aware that an impression of the nebula could be obtained with such a short exposure.'* Indeed, the object appeared on a plate, exposed for only 21 minutes that was taken afterwards by Martin Schaeberle and Edward Holden with the 36″ refractor.

After so much repentance, Barnard started his counterattack: *'I must deprecate the tone in which Prof. Pritchard criticizes the remainder of my paper.'* He considers himself as the discoverer of the nebula – though Pritchard had imaged it two and a half months earlier: *'no statement of the existence of this nebula had ever appeared in print […] it has so long escaped detection in a region so thoroughly studied, visually and photographically'.* Concerning priority of discoveries, especially his own, Barnard was very sensitive.[217] However, he had several times failed to announce a discovery in due time, which was later criticised by Denning.[218] Barnard complained about Pritchard's disparaging qualification of the new nebula as an *'insignificant fleck'*, which contradicted his further statement that it is the *'brightest portion'* of the Pleiades nebulae. For him the object was of great scientific importance: *'a remarkable object in this well known cluster […] by far visually the closest known nebula to a naked-eye star'.*

Barnard speculated about the object's nature: *'it is so distinctly different in aspect from all these* [nebulous matter in the Pleiades] *that from appearance alone one would say that its presence in the group was merely an accident of projection, and that it has no real connection with the Pleiades'.* However, Barnard qualified his view: *'its extreme closeness to Merope […] would rather suggest some physical connection with that star'.* He added that *'Prof. Pritchard is wrong in speaking of it as simply a condensation in the nebulosities of the Pleiades. If appearances are to be trusted, it is too distinct in form and individuality to countenance any such suggestion.'* Barnard's visionary interpretation has been confirmed by modern astrophysical studies!

Barnard afforded himself a last dig at Pritchard and his views about research: *'It is to be regretted that Prof. Pritchard, if he was aware of this singular object at the time the photographs he refers to were made, did not call attention to it at that time so that visual observations might have been made as was done when the Maia Nebula was first photographed by the Henry Brothers a few years ago.'* This primarily shows Barnard's anger that, due to Pritchard's public criticism, the professional world could doubt his discovery priority. Obviously, he had intially undervalued the reputation of the Oxford astronomer. There was no (public) response from Pritchard, who died two years later.

Burnham observed Barnard's new Merope nebula in August 1891, calling it *'one of the most singular objects in the heavens'* (Burnham S. 1892a). He used the Lick refractor at a power of 500 to measure the distance and position angle, obtaining 36.1″ and 166.3°, respectively.[219] Barnard's diameter was confirmed too. Burnham saw the importance of these data, *'to ascertain whether the new nebula is drifting in space with Merope and the other stars of this famous group'.* Barnard made further measurements in autumn 1895 too, confirming the previous ones (Barnard 1896a).

11.6.13 Photography of the exterior Pleiades nebulae

Concerning the Pleiades, Barnard's motto was that: *'anything new about it is of the highest interest'* (Barnard 1894b). In late 1893 he used for the first time the 6″ Willard lens of the Lick Observatory, which gave a field of 15°. The first plate (1 December), taken with an exposure time of 4 hours, showed the region north of the cluster (Fig. 11.44).

Another plate, now centred on the Pleiades, was taken in two parts. The task was started on 6 December

---

[217] See e.g. the case of the California Nebula NGC 1499 (Section 9.15.2).
[218] See the case of the galaxy IC 356 (Section 9.15.2).

[219] Burnham S. (1894b: 173–174).

11.6 The Pleiades nebulae

Figure 11.44. Barnard's first image of the Pleiades, made 1 December 1894; south is up (Clerke 1895).

with an exposure time of 5 hours, and was finished two days later (meanwhile the camera was carefully covered); the total exposure time was 10¼ hours. Barnard explained that *'For many years, during my comet-seeking I have known of a vast and extensive, but very diffused, nebulosity north of the Pleiades.'* The result confirmed his visual observations, showing *'exterior nebulosities of the Pleiades'*. His article of 25 July 1894 contains just a sketch (Fig. 11.45). The long-exposure image itself was not published until 1913.[220] Dreyer listed the detected nebulae as IC 336, IC 341, IC 353, IC 354 and IC 360 in the first Index Catalogue.

In January 1895 Max Wolf took up Barnard's discovery in his 'Notiz über die Plejaden-Nebel'.[221] First, he revealed that he had *'often photographed this nebulous region too'*, referring to his earlier note in the magazine *Sirius* (Wolf M. 1891b). He mainly stressed *'an interesting feature of these nebulae […], appearing after a longer exposure than Prof. Barnard achieved'*. According to Barnard's sketch *'one could guess that the nebulae emanate from the Pleiades as the centre of the structure, then trailing away in space to all sides […] But this would be an erroneous idea.'* According to Wolf the Pleiades rather lie *'in a relatively nebulosity-free region […] in the middle of an irregularly shaped cave'*.[222] Thus the Pleiades cluster, including their interior nebulae (such as those near Maia and Merope), are located in an *'oasis'*. It is interesting that this matches Goldschmidt's view of 1863 (see Fig. 11.32).

In 1895 Agnes M. Clerke reported about the discoveries of Barnard and Wolf in the December issue of the magazine *Knowledge* (Clerke 1895).[223] Here Barnard's images of 1 and 6–8 December 1893 and his sketch of the exterior Pleiades nebulae are reproduced. Clerke mentions that – between the Pleiades and the Milky Way to the north there lies one of William Herschel's 52 regions with 'extensive diffused nebulosity'.[224] She wrote that *'In obtaining pictures of them, telescopes are left nowhere by ordinary portrait lenses.'* The object is no. 13, which was described by Herschel as 'much affected'. As early as in 1891 Barnard had treated these regions, because he often encountered some of them while comet-seeking. He designed a project for their photographic exploration with the 6" Willard lens (Barnard 1892d). In early December 1893 he exposed the area north of the Pleiades – and was indeed able to detect faint vast nebulosities.

Immediately following Clerke's article in *Knowledge* there is one by Barnard, treating his Pleiades images (Barnard 1895a). He had wanted to take even more, but was stopped by the weather: *'It was my intention to repeat this exposure in the past winter* [1894/95] *with greater duration, but the winter proved the most unfortunate for observation.'* Barnard mentions a letter by Herbert Cooper Wilson (Goodsell Observatory, Northfield, Minnesota), dated 26 June 1895. He quotes from as follows: *'I* [Wilson] *have an eleven hour exposure on the Pleiades region taken on October 23rd, 24th, 26th, 1894, which shows most of the great nebula sketched by you from your photographs. It is so delicate that I did not see it until I made a positive which considerably increased the contrasts.'* For Barnard, this finally confirmed the exterior nebula. However, Wolf's work is not mentioned. As early as on 30 January, 1894 Wilson had made a four-hour exposure of the inner Pleiades nebulae with an 8" Clark lens (Fig. 11.46). This was published in the new

---

[220] Barnard (1908: 13c), Plate 15.
[221] 'Note on the Pleiades nebulae' (Wolf M. 1895).
[222] Wolf later coined the term 'Höhlennebel' ['cave nebula'], which accords with Barnard's 'dark cloud'.

[223] She confuses the Merope Nebula with the Maia Nebula: *'Tempel's discovery, in 1859, of the great Maia nebula gave the first hint of this state of things.'*
[224] They were published in 1811 (Herschel W. 1811).

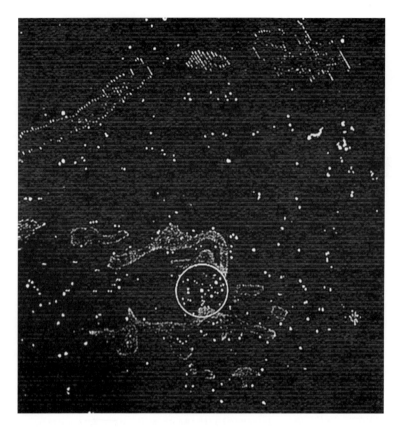

Figure 11.45. Barnard's sketch of the exterior Pleiades nebulae; north is up (Barnard 1894b).

Figure 11.46. Wilson's first Pleiades image of 30 January 1894; north is up (Payne 1894).

magazine *Popular Astronomy*[225] (Payne 1894). Wilson wrote that '*The connection of the nebula with the brighter stars of the Pleiades is so obvious that one could hardly doubt it after inspecting the photograph.*'

Vsevolod Stratonov added another exterior nebula: IC 1990, which he discovered in May 1895 with the 33-cm astrograph by Repsold/Henry at Tashkent Observatory (Stratonov 1896).

11.6.14 The conflict between Roberts and Barnard

Barnard photographed large sky areas with the small, short-focus Willard lens (6", $f/5.2$), which had originally been constructed for taking portraits. The main

---

[225] *Popular Astronomy* was one of three journals edited by William Wallace Payne (Goodsell Observatory). It appeared from 1894 until 1951. The others were the *Sidereal Messenger*

11.6 The Pleiades nebulae 553

Figure 11.47. Isaac Roberts and his 'Starfield' Observatory, Crowborough.

critic of this method was Isaac Roberts, who was following a quite different path. He used a 20" Grubb reflector with a silvered glass mirror (f/4.9), which he had erected in his new 'Starfield' Observatory in the English town of Crowborough (Fig. 11.47). Roberts felt confident about the superiority of his equipment. So, the Pleiades nebulae were an ideal target with which to demonstrate this. The results led to a public altercation with Barnard.

The first step was Roberts' publication 'On the relative effectiveness of a reflector and of portrait lenses for the delineation of celestial objects' (Roberts 1896a). He compared his 20-inch, giving a field of 2.5°, with his two portrait lenses by Dallmeyer (3.5", f/2.7) and Cooke (5", f/4.8). Using all three instruments, the Pleiades were imaged on 4 February 1896 (with exposure times from 50 minutes to 4 hours). He counted the stars and analysed the quality of the images of stars and nebulae: '*The results [...] prove conclusively the greater efficiency of the reflector form of instrument over the portrait-lens or refractor form, for certain work in celestial photography.*' He claimed that Barnard's exposure of 10¼ hours does not show more stars than his own of 4 hours: '*This fact throws some doubt upon the reality of the distant nebulosity which is shown on the plate; and when we consider that the whole patch, that covers the group of the Pleiades, is due to halation, and not to nebulosity, the doubt is further strengthened.*' By 'halation' he meant the light-halo around a star which is caused by reflection from the plate background. It can be prevented by 'coating' ('backing').

Thereupon Barnard wrote a paper 'On the comparison of reflector and portrait lens photography' (Barnard 1896b), dated 19 July 1896.[226] He opened with the remark that '*Dr. Roberts takes occasion to criticise several of my star photographs.*' To him Roberts' experiments – with results to the detriment of the portrait lens – sounded very dubious. The reason for taking the Willard lens was '*its advantage, on account of small scale and wide field, in showing very large and faint nebulous areas, the phenomena of the tails of comets, and the structural forms of the Milky Way [...] in these special fields it is supreme*'. Roberts' method – comparing the image quality in terms of star counts – was bashed by Barnard: '*it is very unfair to the short focus lens*'. With

---

(1882–91) and *Astronomy and Astro-Physics* (1892–95), the progenitor of the distinguished *Astrophysical Journal*.

[226] At that time he was visiting George Ellery Hale at Kenwood Observatory near Chicago.

a small focal length and long exposure time, many stars overlap; and, moreover, '*a very slight difference in the actual magnitudes of the smallest stars shown on the two plates may double or treble the number of stars revealed by the larger instrument*'. Barnard could not resist checking Roberts' star counts – finding many more stars on his own image: '*After seeing his article, I was curious to know how many stars were really shown on my plate.*' Moreover, his own star images appeared distinctly smaller.

Concerning the exterior nebulae, Barnard cited a remark made by Roberts during the meeting of the British Astronomical Association on 29 April 1896, according to which Barnard's objects '*were not nebulosity, and he [Roberts] made the assertion without fear of anyone proving the contrary*'. Barnard now states '*That these nebulosities exist as shown on my photographs there can be no doubt whatever. Indeed, it was the knowledge of their existence by visual observations alone that led me to make the photographs.*' In his counterattack, he criticises a number of Roberts' 20" reflector images:[227] '*Notwithstanding his condemnation of the faint nebulosities shown by portrait lenses to exist in certain parts of the sky, Dr. Roberts claims that his photographs of the dense globular clusters show essentially all these objects to be nebulous.*' For him this was '*a photographic effect only, due to the crowding of the bright stars, combined with an unsteady atmosphere, and a lack of exact focus, and further increased by the great enlargement from the originals*'. This hits Roberts' most sensitive point – his qualification as an astrophotographer. But Barnard had even more trump cards to play: the optimal conditions on Mt Hamilton and his unchallenged reputation as a visual observer: '*Under first-class conditions, however, when the air was very steady, the impression was that these clusters were free from true nebulosity.*' Roberts was not able to counter this situation. Concerning Barnard's images presented to the RAS, Roberts advanced the view that '*an unprejudiced inspection of these will show their value to the science of astronomy*'. Moreover, he

Figure 11.48. Roberts' image of the Pleiades made with the 20" reflector on 22 December 1897; north is up (Roberts (1899), Plate XXV).

claimed that '*It must be understood, that they are to be looked at as pictures of the objects they represent, and not to be examined with the microscope for details which in the nature of things it is impossible for them to show.*'

Roberts felt that he had been challenged – however, his answer took until May 1898 (Roberts 1898). He first describes new Pleiades plates, taken with the 20" reflector and the 5" Cooke lens (1 October to 25 December 1897).[228] The reflector showed the known nebulae near Merope, Maia and Alcyone surprisingly well (Fig. 11.48). Because of the appearance of those images Roberts supposed that stars and nebulae must be physically connected (as had been pointed out earlier by Tempel): '*It appears to me that the evidence, taken as a whole, points strongly to the probability that the Pleiades consist of a group of stars seen by us either behind or else in front of a group of nebulae.*'

With the Cooke lens the Pleiades appeared '*completely obscured by the photographic effects of atmospheric glare*'. Roberts now focused on Barnard's images, whose 'quality' can be seen by everyone: '*Those who will take the trouble to examine these photographs will be struck*

---

[227] Roberts had published them in 1893 in his book *A Selection of Photographs of Stars, Star-Clusters and Nebulae* (Roberts 1893); a second volume appeared in 1899 (Roberts 1899). Later Dreyer used Roberts' image of M 33 to determine the positions of 431 objects in this galaxy with the aid of a plate-measurement device (Dreyer 1904).

[228] The photograph taken on 22 December 1897 was published as in Plate XXV Roberts (1899).

by the very coarse character of the film [...] *with the distorted images of the stars; with the halation circles round the brighter stars, caused by the omission of backing the plate; and with the large patch of something (stated to be nebulosity) covering the group of the Pleiades.*' Now he strikes the key blow: '*it is upon this one defective photograph that Professor Barnard based his unqualified assertion that there is no doubt whatever that these nebulosities exist as shown on his photograph*'. Barnard's visual observations were brushed off too: '*The vague suggestion of visibility in a telescope is quite inadmissible so long as the bright nebulosities known to exist in the group cannot be seen.*' It is not clear what Roberts is referring to here, since Barnard could see the inner nebulae very well (perhaps his own attempts are meant?). About his new images, which were '*taken with proper care*', he remarks that they '*entirely discredit the existence of Professor Barnard's nebulosity*'. He emphatically refuses Barnard's criticism about star counts, existence of nebulae and image quality: '*he misquotes from my work – misconstrues my statements*'. Roberts now treats the reasons for the 'spurious nebulosity' seen on his plates. On the basis of results from experiments, he admits that there can be problems during exposure and processing. Finally, he wrote that '*The spurious nebulosity is shown in cloud-like patches, and simulates the real nebulosity so closely that its true character could not be decided with certainty, without careful correlation of two or more photographs taken at different times of the same area in the sky.*' Such artefacts were a permanent problem of astrophotography on plates or film.

In autumn 1898 Barnard sent a copy of the Pleiades image made by Herbert Cooper Wilson (Goodsell Observatory) with a 6" Brashear lens, to the RAS. In his report in *The Observatory* one reads that '*This picture shows well the exterior nebulosities of the group, just as shown in my photographs, and even stronger.*' (Barnard 1898). The editor, Thomas Lewis, explained that '*It will be remembered that the above has reference to a controversy between Dr. Roberts and Prof. Barnard some years ago.*' The direct comparison of Barnard's and Wilson's results yielded the conclusion that '*the coincidence in position of the wisp-like markings on two independent photographs affords strong evidence as to the reality of their existence*'.

In the same issue of *The Observatory* another note on the matter appeared (Lewis 1898). Therein it is mentioned that Solon Bailey had shown a photograph of the Pleiades to Barnard (who wrote about it in a letter), while he was visiting Harvard College Observatory. It was taken on 29 October 1897 with the 8" Bache reflector in Arequipa and an exposure time of 5 hours. One reads that '[it] *distinctly showed the nebulous streams, the existence of which is a matter of argument between Prof. Barnard and Dr. Roberts*'. Bailey had found out (as communicated by Pickering) that the Pleiades region is compared with other clusters, lacking in faint stars (Pickering E. 1897). A possible cause is given: '*This absorption of faint stars is probably due to the nebulosity surrounding this group.*' This could be confirmed by Stratonov in Tashkent too (Stratonov 1897).

The Pleiades nebulae were again treated in the RAS meeting of 11 November 1898 with Robert Ball in the chair (Ball R. 1898). Barnard had sent his photographs of the Milky Way taken with the Willard lens; they were presented by Newall. Roberts was among the participants and explained that the extended nebulosities appearing on Barnard's images are not real, but rather '*the effects of starlight and atmospheric glare concentrated by a small lens*'. He claimed that '*The real nebulosity [...] is limited to relatively small areas, and generally it is not difficult, to distinguish it from the large areas of spurious nebulosity.*' Concerning the exterior Pleiades nebulosities, he remarked that '*the Meeting is referred to two photographs presented to the Society years ago* [Barnard], *and to statements made that two gentlemen in America* [Wilson and Bailey] *had confirmed the existence of the nebulosity, but no evidence whatever is presented to show that there is reliable data available*'. Roberts added '*If the nebulosity really exists why are the proofs not presented?*' About the images sent by Barnard, he said '*I have many times examined the two photographs referred to by Prof. Barnard, and have not succeeded in finding upon them any nebulosity such as he describes.*' Quite the converse: '*I have shown before the Society my photographs upon which the non-existence of nebulosity of a real character could be detected in the region.*'

The Astronomer Royal, William Christie, was not convinced by Roberts' arguments: '*I had very great difficulty in following Dr. Roberts's criticism of these photographs.*' For him it was not the nebulous regions in Barnard's images which are essential, but rather '*the remarkable rifts or dark spaces which are shown in the Milky Way, and in those rifts we have an utter absence of nebulosity and an almost complete absence of stars*'. He added *'The few stars that are shown are shown so sharply that I think that fact negatives the idea Dr. Roberts is trying to put forward – that this nebulosity is due to diffused starlight or defects of the lens.'* Christie asked Roberts *'If so, why is it not in those rifts, and why is not the diffused starlight shown all over the plate?'*, who replied as follows: *'The rifts appear because of the absence of stars. There are no stars to give light to fill them up, as is the case in the surrounding neighbourhood. That is the case of the remarkable rifts in the light. The other parts are crowded with stars, which give light to illuminate the atmosphere and so cause apparent nebulosity.'* Christie was not convinced: *'If you get these stars very sharply defined without diffused light, it is not fair to say the diffused light comes from the faults of the lens. The point is, in these rifts we have an absence of diffused light and star-images sharply defined. If the stars are sharply defined there they ought to be elsewhere [...] we have evidence here that the so-called nebulosity is not due to defects of the lens.'* This terminated the discussion. Roberts, however, ensured that a footnote was added to the published minutes: *'these vacant areas – reflecting little if any light – give rise to the rifts, tortuous and otherwise, which are seen on photographs in the midst of the crowded star-areas. In this way is produced the appearance of nebulosity which is spurious; it is simply the effect of contrast between light and dark areas.'*

A certain aversion against Americans and their achievements can be discerned in Roberts' words. Obviously, these 'upstarts', equipped with the best and most expensive instruments, were a threat to British observational astronomy, which had for a long time maintained an unrivalled domination of the terrain. The February 1899 issue of *Popular Astronomy* contains a comprehensive report by Herbert Cooper Wilson (Wilson H. 1899). Some of his results had already been announced by Barnard. The review article contains Wilson's exposure of the Pleiades of 30 January 1894. To clarify the set of problems to the reader, an extract of Roberts' response from the RAS meeting was published in the same issue (Anon 1899).

In 1894 Wilson tried to confirm Barnard's photographic results with the 6" Brashear lens ($f/5.2$), giving a sharp image to the edge. The plate was taken on 23/24 October 1894, with a total exposure time of 11¼ hours.[229] One reads that *'The photograph obtained with this camera showed all the nebulous areas sketched by Professor Barnard and other fainter details between them.'* In the summer of 1898 Wilson visited Barnard at Yerkes Observatory: *'I compared copies of these negatives with those obtained by him and we found that they agreed in almost every detail of the distant nebulous areas.'* He further wrote that *'It was therefore with considerable surprise, that I read Dr. Roberts' article in* Monthly Notices *June, 1898* [Roberts 1898], *received just when we were comparing the photographs.'* He possessed an image taken by Roberts dating from December 1897. About these unsuccessful attempts with the 20" reflector and the 5" portrait lens Wilson made the following remark: *'All of which proves nothing, being only negative testimony.'* Of course, being an experienced photographer,[230] he was familiar with the plate defects mentioned by Roberts.

In late 1898, Wilson was again able to confirm the exterior nebulae, using a 2.5" lens too. The image made on 15 November 1898 made with the standard 6" Brashear lens (exposed to 5.5 hours) is reproduced in the article. The plate was *'backed with a paste of lampblack, oil of cloves and terpentine, to prevent halation, so that there are no rings around bright stars'*. Such rings on Roberts' images were criticised by Barnard. Wilson supposed that some plates show the 'Gegenschein' (scattered interplanetary light). Nevertheless, all of Barnard's features could be confirmed. Wilson also mentions that on one plate taken with the 2.5" lens a large bright nebula north of ξ Persei is visible. This is NGC 1499, which had been discovered visually by Barnard on 3 November 1885 and photographed independently by Archenhold in 1891 (see Section 9.15.2).

Finally Wilson compares his images with one made by Roberts', who had taken it with his 5" lens (exposed for 10 hours): *'I can see traces of the brighter nebular streams on his own photograph, where he* [Roberts] *says is nothing but spurious nebulosity.'* He was aware why Roberts *'has failed to get more than the merest traces of the disputed regions'*, namely because *'The atmospheric glare around his star images is something tremendous when compared with that on our plates. The Pleiades region is completely blotted out.'* Wilson spots the main reason as Roberts' bad site: *'This shows that our atmosphere must be much clearer than that in which Dr. Roberts*

---

[229] These images are reproduced in the article.

[230] He certifies this for Barnard too, Barnard *'having been a photographer from his boyhood'*.

Figure 11.49. Calvert's drawing of the Pleiades region (from Bailey's image of 1897); north is up (Brenner 1902).

works, for he states that the photographs in question were taken on very clear nights.'

In 1900 Barnard, now at Yerkes Observatory, wrote another paper on the 'Exterior nebulosities of the Pleiades' (Barnard 1900b). He first mentions the images of Wilson (15 November 1898) and Bailey (29 October 1897). Using the Harvard plate, Ebenezer Calvert[231] had produced a fine drawing, which was fitted by Barnard with coordinate lines (Fig. 11.49). It deliberately waives the inner nebulae.[232] The brightest part of the exterior nebulae, surrounding the 5.2-mag star BD +25° 624 north of the cluster, is described by Barnard; the object was later catalogued as IC 1995 by Dreyer.

Barnard summarises as follows: '*Every detail has been perfectly and satisfactorily verified, leaving no question, in the mind of an unprejudiced person, of the existence of those singular nebulosities exterior to, but connected with the Pleiades.*'

In his paper Barnard mentions an excellent plate taken by Max Wolf '*in which the exterior nebulosities are distinctly traced*'. He had seen it during his visit to the Königstuhl in early 1900. Barnard describes visual observations with the 40" Yerkes refractor: '*The nebulosities in the Region of Merope are perhaps the most striking. […] The original nebulosity discovered by Tempel, extending south-westerly from Merope, and which is specially suited for small telescopes, comes out very strong in the 40-inch.*' His new Merope nebula appeared as '*a decidedly conspicuous object […] there is an excessively faint point of light on the edge of the nebulosity towards Merope*'.

---

[231] Calvert was Barnard's brother-in-law and an excellent drawer.

[232] For a reproduction see Brenner (1902: 118).

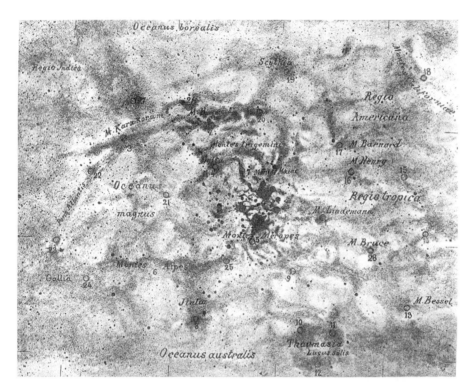

Figure 11.50. Max Wolf's map of the Pleiades, labelled with his own designations (Wolf M. 1900).

In April 1900 Max Wolf published a treatise, titled 'Die Aussen-Nebel der Plejaden'.[233] The reason was that *'just now Barnard has published a sketch of the exterior nebulae'*. He mentions that they were already present on an exposure made by him on 9 October 1890 with a Steinheil achromat. Later plates, exposed in autumn 1891 for up to 7.75 hours, confirmed this. Thus he had beaten Barnard. In 1894 Wolf had sent a drawing to some astronomers, but an even better one based on three plates of 1894–98, taken with a 15.9-cm portrait lens by Voigtländer ($f/5$). In his treatise Wolf presented a positive and a negative version of this drawing (Fig. 11.50). Conspicuous nebular structures were named, mainly inspired by geography; for instance 'Karakorum', 'Himalaya', 'Alpes', 'Montes Meropes' and the 'Regio Americana' with 'Mons Barnard'. Wolf analyses the different brightness rates and draws a comparison with prominent objects such as the Orion Nebula and the North America Nebula. Given the richness of detail, he wrote *'I want to spare the reader further description, for the image gives a better presentation of the multifariousness of these mysterious objects than words can.'*

11.6.15 Barnard, Roberts and the 52 nebulous regions of William Herschel

The conflict between Barnard and Roberts about the Pleiades nebulae afforded a last – but particularly juicy – situation. The issue was caused by Roberts' latest task: a photographic study of William Herschel's 52 regions with 'extensive diffused nebulosity',[234] using the established combination of a 20" Grubb reflector and a 5" Cooke lens. In 1902 Roberts sent the result to the distinguished *Astrophysical Journal*. Its editor, George Ellery Hale, asked (of all people) Barnard for a peer review of the manuscript. Barnard was not convinced by Roberts' work and demanded from Hale the opportunity to state his point of view in a separate paper.

The result was that the both articles – Roberts' 'Herschel's nebulous regions' and Barnard's 'Diffused

---

[233] 'The exterior nebulae of the Pleiades' (Wolf 1900).

[234] Herschel W. (1811: 275–276); see also Section 2.4.

nebulosities in the heavens' – appeared back-to-back in Vol. 17 (1903).[235] Given the issue's turbulent history, Roberts was probably not amused. Barnard remarked in his paper that he had already pointed out the importance of Herschel's 52 regions in January 1892 (Barnard 1892d).[236] At the beginning of Roberts' paper, one reads in a footnote by the editor that '*The manuscript of this article was accepted by the editors on the supposition that it was not to be published in other current journals.*' However, Roberts did not obey the injunction: nearly identical versions appeared in the *Monthly Notices* – as soon as in the November issue of 1902 (Roberts 1902) – and in the *Astronomische Nachrichten* (Roberts 1903a). In an article on 'The large nebulous areas of the sky' Wilson later wrote about Roberts' results that '*Dr. Roberts' work was not done under the most favourable conditions*' (Wilson H. 1904).

Barnard was disappointed that Roberts had confirmed only four regions. This – and, at the same time, his admiration of Herschel – is expressed by the statement '*It is a little unreasonable to suppose that Herschel, who made so few blunders compared with the wonderful and varied work that he accomplished, should be so palpably mistaken in forty-eight of fifty-two observations of this kind.*' He incidentally clarifies that the four detected regions were already known[237] and that '*one of these very objects shown to be free from nebulosity, is really the brightest portion of one of the most extraordinary nebulae in the sky*'. Here no. 27 in Orion is meant. Barnard had detected this nebula on 3 October 1894 with his 1.5" 'magic lantern lens' of focal ratio *f*/3 (Barnard 1894e); later he was successful with the 6" Willard lens too. Though the object had already been discovered in 1889 by William Henry Pickering, using a 2.6" portrait lens by Voigtländer on Mt Wilson, it is now called Barnard's Loop.[238]

Later Johann Georg Hagen showed much interest in Herschel's nebulous regions and Roberts' images, investigating the similar 'dark cosmic clouds'.[239] In 1925 Hagen imaged some regions with the 33-cm astrograph at the Vatican Observatory, Castelgandolfo. Moreover, four could be observed visually by Friedrich Becker with the 16" refractor by Zeiss (magnification 122, field of view 25.4′) – interestingly, nos. 1 to 4 (Hagen 1926). Hagen claimed that '*Herschel saw exactly that, which we see at the Vatican Observatory, which we call dark cosmic clouds. To Herschel belongs the credit of discovery.*' Especially the last sentence forced Dreyer to write a critical reply (Dreyer 1926).[240] He saw a significant difference between Hagen's 'dark cosmic clouds' and Herschel's nebulous regions, writing '*Before acknowledging that W. Herschel was the discoverer of dark cosmic clouds, it will be well to bear in mind that he does not anywhere make any distinction between the general appearance of the objects examined by Father Hagen and the rest of the fifty-two objects.*' Dreyer added that '*He certainly saw, or believed that he saw [...] luminous nebulosities. Considering his vast experience it is difficult to believe that he saw something totally different in the four places examined by Father Hagen, without realising this and drawing special attention to them.*' Thus he draws a clear distinction between Herschel's 'bright' regions and Hagen's 'dark' clouds.

Finally, in his *Astrophysical Journal* paper Barnard cannot refrain from mentioning once again the issue of the 'exterior Pleiades nebulae' – and the unfortunate role played by Roberts: '*It was with the same instruments described in this paper that Dr. Roberts failed to get any traces of the exterior nebulosities of the Pleiades, which have been shown by four observers with four different instruments not only to exist, but to be not at all difficult objects.*' The long-running battle between Barnard and Roberts was finished now – obviously, with a victory of the former.

11.6.16 Dreyer's cataloguing of the Pleiades nebulae and new observations

Dreyer entered Barnard's nebulae in the Index Catalogue of 1895. Concerning the new Merope nebula IC 349 he noted '*eF, vS, Pos. 165°, Dist. 36″ from Merope*'. The exterior Pleiades nebulae IC 336, IC 341, IC 353, IC 354 and IC 360 are collectively described as '*vF, eeL, v dif*'. For IC 360 (representing all of the others) Dreyer added in the IC II notes "*Compare M.N., lvii. p. 12, picture of the exterior nebulosities*

---

[235] Roberts (1903b), Barnard (1903).

[236] It is juicy that the table which Roberts presents in his paper was taken from this publication.

[237] Regions nos. 44 and 46 are parts of the North America Nebula NGC 7000 in Cygnus (Fig. 2.17).

[238] Pickering W. (1890); see Section 9.15.1.

[239] Also called 'Hagen clouds' (see Section 10.3.2).

[240] This was Dreyer's last publication too.

around the Pleiades [Barnard 1896b], *and M. Wolf: 'Die Aussen-Nebel der Plejaden', Abh. d. K. Bayer. Akad., 1900, 4°* [Wolf M. 1900]." Stratonov's nebula was catalogued in the IC II as IC 1990 with the description '*vL, mE pf, 15′ l*'. Also the brightest nebula in Calvert's drawing IC 1995, was included ('*star 6* [mag] *in eF, eeL neb*'); Dreyer wrongly mentions Barnard as the discoverer. The American astronomer had only described the object in his drawing – it first appeared on Bailey's plate taken in 1897. Probably the nebula was noticed then, but the find was not published.

The Merope Nebula became a popular target for amateur astronomers on account of Webb's column in the *Intellectual Observer* and several notes in the *Astronomical Register*. At the beginning of the twentieth century, there were presentations in the books of Agnes Clerke and Leo Brenner.[241] Both authors focused on the history too. Brenner's *Beobachtungsobjekte für Amateur-Astronomen* contains two drawings, both based on plates: Calvert's (using Bailey's images) and Max Wolf's made at Königstuhl. It is interesting to read Brenner's version of the unsuccessful observations of Hough and Burnham. For instance, about the 18.5-inch of Dearborn Observatory he blasphemed: '[it] *is broadcasted with pure American arrogance as incomparable; but it does not impress me, since I noticed the miserable Jupiter drawing made with it by Hough* [...] *The success was a disgrace too*.' Brenner could see the Maia Nebula (NGC 1432) with his 7" refractor (however, observations of the Merope Nebula are not reported). The brightest part '*shows a singular smooth, almost bluish light, while the remaining part appears to consist of glimmering, extremely faint and densely packed points of light*'. Similar doubtful impressions are known from Tempel's descriptions.

In late 1898 Keeler reported about photographs of Barnard's new nebula close to Merope (IC 349), which he had taken with the 36" Crossley reflector at Lick Observatory (Keeler 1898a, b). After some precautions had been taken ("*The plates were coated on the back to prevent 'halation'*."), the tiny object clearly appeared and was best seen on the image with the shortest exposure time of 15 minutes. He wrote '*My attention was, in fact, attracted by it before I recalled the note by Professor Barnard* [Barnard 1891a].' Keeler described the form of the nebula as '*roughly pentagonal in shape,* the most salient angle pointing directly to Merope. From opposite sides, symmetrically placed with respect to the line joining the nebula and the star, two wisps of nebulosity stream away, and join the other nebulous wisps which are characteristic of the region*.' For him the object is '*not a cometary or planetary nebula*'. He photographed the Pleiades again on 28 December 1899 with the Crossley reflector (Keeler 1908). After 4 hours, the inner Pleiades nebulae appeared; however, the image field was pretty small.

Vesto Slipher of Lowell Observatory, Flagstaff, was the first to take a spectrum of the Pleiades nebulae. He used the 24" Clark refractor, equipped with a Brashear spectrograph. The plates of 7 to 13 December, 1911 (exposed for up to 21 hours) showed a continuous spectrum with strong hydrogen lines. Slipher wrote that '*It contains no traces of any bright lines found in the spectra of gaseous nebula.* [...] *the whole spectrum is a true copy of that of the brighter stars of the Pleiades.*' (Slipher 1911). He at first believed that there had been an instrumental reflection: '*The nature of the spectrum at once raised the question whether it might not be due to light from Merope scattered and reflected by the 24-inch objective*.' But this was ruled out through tests: '*there seems to be support for the conclusion that the Pleiades nebula shines by reflected light*'. Slipher's conclusion was that the Pleiades nebulae are reflection nebulae. However, he believed that the Andromeda Nebula and spiral nebulae are of this type too, insofar as they '*might consist of a central star enveloped and beclouded by fragmentary and disintegrated matter which shines by light supplied by the central sun*'.

In 1915–16 Robert Trümpler took plates of Barnard's new Merope nebula with the 30" Thaw refractor of Allegheny Observatory, Pittsburgh (Trümpler 1922). About the one taken on 7 December 1916 (exposed for 1 hour) he noted that '*the nebula is well visible and shows a much more complicated structure than Barnard's drawing indicates*'. He compared it with others made during the years 1890 to 1921; most of them had been exposed too briefly or for too long or showed too little detail. Even allowing for the poor quality of the images, he had the impression that '*the nebula has changed in appearance during the last thirty years*'. Tom Reiland repeated Trümpler's work with the same telescope in September 1981 (see Fig. 11.43 right); unfortunately he could not find the original plate of 1916 (Reiland 1982).

---

[241] Clerke (1902: 410–411), Brenner (1902: 104–120).

11.6 The Pleiades nebulae

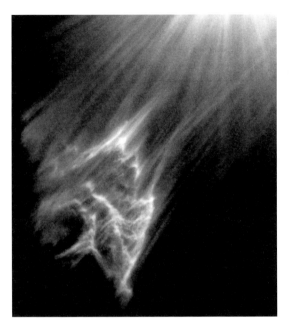

Figure 11.51. IC 349 as seen by the Hubble Space Telescope (Hubble Heritage Project, 12/2000).

In 1995 George Herbig managed to take an image of IC 349 with the 2.2-m reflector on Hawaii (Herbig 1995). About the Pleiades nebulae he wrote that '*The familiar nebulosity illuminated by the Pleiades stars is known to be the consequence of a chance encounter of the cluster with a molecular cloud.*' Barnard's IC 349 is '*a dense condensation […] in the cloud that is moving almost directly towards the star […] being dissipated in the radiation field of the star*'. The spectacular HST image shows this too (Fig. 11.51).[242] Thus the Pleiades nebulae are not remnants of interstellar matter, which once formed the cluster stars, but passing dust clouds. Exactly this had already been conjectured in 1891 by Barnard, who was no doubt the greatest visual observer of the nineteenth century.

[242] See http://hubblesite.org/newscenter/newsdesk/archive/releases/2000/36.

12 • Summary

The subject 'nebulae and star clusters', centred on the nineteenth century, has hitherto not historically been treated in detail. At best, a few aspects were studied and occasionally results were published. Particularly the 'history of the NGC' is new territory and the present work is the first of its kind. The following comprehensive summary concentrates on the following issues (see also Chapter 1):

(1) the subject and line of questioning
(2) the importance of the New General Catalogue and the motivation of the work
(3) objects, observers and methods
(4) milestones of the cataloguing of nebulae and star clusters (non-stellar objects)
(5) statistical analysis and the way ahead

12.1 THE SUBJECT AND LINE OF QUESTIONING

This work deals with the discovery, visual observation, description and cataloguing of nebulae and star clusters, focusing on the nineteenth century. Astronomers, sites and a large number of objects are presented. The climax of the development, starting with William and John Herschel was the publication of the New General Catalogue (NGC) by John Louis Emil Dreyer in 1888. The date also marks the transition from classical astronomy based on visual observations and positional measurements to modern astrophysics represented by spectroscopy, photography and photometry.

The following questions are central.

- Why did astronomers search for nebulae and star clusters and what were the scientific goals?
- Who were the discoverers and what can be said about their motivation and success rate?
- What were the circumstances and instruments (discovery date, site, telescope)?
- How were the results communicated or published and what was the reaction of other astronomers?
- What can be said about the successes and problems associated with visually observing non-stellar objects (identification, textual description, drawing)?
- What was the influence of visual observations of non-stellar objects on astronomical methods (measurements) and techniques (instrumental development)?
- What ideas were formulated concerning the nature and evolution of these objects?
- Did visual observation, classification and cataloguing bring forth any development concerning their physical nature and cosmic evolution?
- What affect did catalogues of non-stellar objects have on the developing field of astrophysics and what is the significance of the NGC today?

Answers do not result straightforwardly from the NGC itself. A comprehensive research, which must include all relevant publications, is required, thus the list of references contains more than 1600 sources.

12.2 THE IMPORTANCE OF THE NEW GENERAL CATALOGUE AND THE MOTIVATION OF THE WORK

Nebulae and star clusters considerably differ from stellar objects (stars, double stars). They form a unique category, and were mysterious from their first discovery onwards. Some appear resolved (star clusters), whereas others, such as the Orion Nebula and the Andromeda Nebula, remain nebulous even in the largest telescopes. Nineteenth-century observations were carried out in the attempt to reveal their physical nature, evolution and spatial distribution, but reliable answers had to wait until the 1920s.

Visually the objects offered only a few features: position, brightness, structure and form. Thus the observations yielded an inhomogeneous mixture of more or less precise measurements and subjective descriptions – the ideal background for conflicts. William Herschel provided a way out of this impasse, by concentrating not on individual objects but on the maximum number of objects of all detected types. Therefore, he established systematic description, classification and cataloguing.

Dreyer brought Herschel's concept to a convincing completion in his New General Catalogue. The genesis of this work shows what was achievable by the means of the time in terms of understanding the nature of non-stellar objects. Actually, he put together the last 'visual' catalogue containing all kinds of non-stellar objects of the whole sky. Thus it is both of great historical and of great astrophysical importance.

The latter is demonstrated by the fact that NGC objects are still primary targets of research. This is due to the following factors. First there is the homogeneity of the catalogue, in terms of the selection, brightness and size of the objects. The NGC contains the most important types: galaxies, open and globular clusters, galactic and planetary nebulae. Their number is large, but still manageable, presenting sufficient material for observations. The objects' moderate magnitudes and sizes allow astrophysical studies even with small telescopes, but finer details appear with the largest. Therefore NGC-numbers are still part of the astronomer's ordinary work. There is no doubt that the catalogue is the one that is most often used in observational astronomy.

A profound historical analysis of the New General Catalogue thus appears to be important. However, in contrast to the popular Messier catalogue with its 103 entries, this was new territory. The task necessarily involved consideration of all 7840 NGC objects with reference to their original and modern data. For each entry the discoverer, date, instrument, site and publication had to be investigated. Substantial research was needed to reveal the background. Except in a few cases, all of the relevant information could be compiled (only for 34 objects is a source lacking). It is, however, regrettable that Dreyer's estate could not be located. This eminent treasure (which must have been large) seems to have been lost.

12.3 OBJECTS, OBSERVERS AND METHODS

The main issues about nebulae and star clusters were their form (e.g. spiral structure), resolvability, variability and proper motion. A central problem was the subjectivity of visual observation, description and drawing (later in its relation to photography). Non-stellar objects were connected with other fields of classical astronomy in the nineteenth century too. Examples are observations of comets, minor planets and double stars, the making of star charts and the search for a 'trans-Neptune' planet.

Peculiar cases, which have not yet been studied comprehensively, are presented. This includes the galaxy M 51, whose spiral structure was detected by Lord Rosse in 1845, Hind's Variable Nebula (NGC 1555), which shows real changes, and Tempel's Merope Nebula (NGC 1435) in the Pleiades. The published observations and drawings were subjected to a long-lasting debate in the nineteenth century, involving many eminent astronomers. The history of the NGC is thus a psychogram of its observers too. Since they were lone fighters in most cases, such negative attributes as vanity, prejudice, jealousy, arrogance and deception appeared.

The list of astronomers who contributed to the NGC is long, containing great names, for instance, John Herschel, Wilhelm and Otto Struve, d'Arrest, Auwers, Lord Rosse, Lassell, Secchi, Schmidt, Schultz, Schönfeld, Winnecke, Webb, Hall, Pickering, Vogel, Swift, Tempel, Rümker, Peters, Common, Stephan, Burnham and Barnard. Most of them were known as the discoverers of comets, minor planets, moons and double stars too; some were famous for the construction of large telescopes or became pioneers of astrophysics.

What can be said about the motivation to search for nebulae and star clusters? Cataloguing them to avoid confusion with comets (as in the case of Messier) no longer played a significant role in the nineteenth century. The objects were ascribed an individual meaning and were recognised as steps of the cosmic evolution described by Newton's theory of gravity and celestial mechanics. William and John Herschel tried to substantiate this idea with statistical arguments. Therefore, a representative sample of objects from which to develop a standardised description and classification was needed.

The discovery of a large number of nebulae and star clusters, or certain peculiar objects, brought not only new insights but also recognition and fame. As for discoveries of comets and minor planets, prizes were awarded, such as the gold medal of the Royal

Astronomical Society. Some objects are named after their discoverers, e.g. Hind's Variable Nebula and Barnard's Galaxy. However, less fame was won in a case like Struve's Lost Nebula (NGC 1554).

Investigating the structure of non-stellar objects with respect to their resolvability, spiral form or variability, and also the search for new ones, needed large telescopes. Impressive reflectors were constructed, with mirrors first of metal and later of silvered glass. This led to essential innovations, bringing their constructors fame (and prizes). Examples are Lord Rosse, Lassell, Nasmyth and Common. The situation for refractors was different: the reason for larger instruments was not the observation of nebulae and star clusters but solar-system objects. Nearly all telescopes shared a common problem: the increasing light and air pollution at urban sites, causing bad seeing and transparency. The tendency towards their location at remote places, especially on high mountains, was still low. Early exceptions were Lassell's expeditions to Malta with his 24" and 48" reflectors and the Lick Observatory on Mt Hamilton with its famous 36" refractor erected in 1888.

Naturally, nebulae and star clusters show a larger spectrum of forms than do stellar objects. However, there are related phenomena. William Herschel suggested a similarity between double stars and double nebulae. A milestone was the detection of (hot) gas in the cosmos, confirming the unity of physics. As shown by Huggins in 1864, spectral lines of gaseous elements, which had been measured in the laboratory, appeared in planetary nebulae and the Orion Nebula too. This offered the chance to search for such objects with a visual spectroscope. With this revolutionary method, Pickering and Copeland detected cases that optically could not be distinguished from stars. On the other hand, faint, large nebulae were a challenge for astrophotography, needing long exposure times. Not only astrographs and reflectors were used, but also small portrait lenses. However, photometry of nebulae was a difficult task. Reliable integrated and surface magnitudes could not be determined until the twentieth century.

A complete disaster was associated with the attempted detection of proper motions of nebulae to get information about their relative distances. Extensive campaigns were performed to measure positions with refractors, absorbing a considerable part of observing time in the nineteenth century. The attempt inevitably failed due to the extreme distance of the objects; even the 'nearby' Orion Nebula is more than 1300 light-years away.

Did classical methods (visual observation and position measurements) bring any progress concerning elucidation of the physical nature and evolution of non-stellar objects? The answer must be no. The old ideas of William Herschel, which he formulated in the celebrated 'nebular hypothesis', remained nearly unchanged until 1864, when Huggins introduced spectroscopy. Up to this epochal event, neither large telescopes with their great light-gathering power nor extensive measurement campaigns to determine precise positions were able to terminate the various speculations. The research was confined between 'natural history', treating cosmic objects like plants, and 'astrometry', dealing with mere celestial positions.

12.4 MILESTONES OF THE CATALOGUING OF NEBULAE AND STAR CLUSTERS

The discovery of nebulae and star clusters visible with the naked eye, such as the Andromeda Nebula, reaches far back. Although they were first mentioned only sporadically, Messier eventually set a standard with his first catalogue of 1781. Starting in the late eighteenth century, William Herschel greatly influenced all subsequent work. His investigations were unique insofar as quality and quantity were concerned, raising the number of known objects from about 100 (in Messier's catalogue) to 2500! His son John continued the observations, first in Slough and then at the Cape of Good Hope, discovering the bulk of southern non-stellar objects.

There followed a time of stagnation due to the common opinion that new nebulae and star clusters scould barely be found – the completed work of the two Herschels was considered all-powerful. Thus most astronomers concentrated on astrometry, measuring positions with refractors and meridian-circles. The main targets were single/double stars, minor planets and comets, with new nebulae and star clusters, which were occasionally found during the observations, being mere by-products.

Many publications show that the discoverers were not focused on the correct identification of new objects. Those taking the trouble to prove whether an object was already known normally consulted John Herschel's Slough catalogue of 1833. Anything not listed there had to be new! Actually, it was overlooked that a considerable number of objects appear in William Herschel's catalogues only, namely those not re-observed by his son. Thus many objects found later were not really new, but due to the old master. In the nineteenth century, we have, apart from John Herschel, only a few astronomers (e.g. d'Arrest, Auwers, Winnecke and finally Dreyer) who concerned themselves with the following standard questions:

- Is the object already known (identity)?
- Who is the true discoverer (priority)?
- Are the published data unambiguous and sufficient (identification)?

Many errors and misunderstandings could be cleared up in the past; especially Dreyer was very successful at this task. But, as new studies have shown, there remained a considerable number of unsolved puzzles. The present work offers many examples.

The dominance of Messier, William and John Herschel concerning the historical treatment of the matter has led to the fact that many other important persons were more or less ignored in the literature. An example is Arthur Auwers and his remarkable contributions about nebulae. He was well known as the leading representative of fundamental astronomy, having started with accurate observations of non-stellar objects. We owe to him the first reliable published revision of William Herschel's three catalogues. He also edited the works of Messier and Lacaille, making them popular too. Moreover he published the first list of new nebulae after John Herschel. Herschel profited much from Auwers' work, which appeared just in time for his General Catalogue of 1864 – this was the new standard, listing all known objects of the whole sky.

Another important person was Heinrich Ludwig d'Arrest, Dreyer's teacher and life-long idol. Starting in Leipzig, he developed the precise measurement of nebular positions. The highlight of his career was the *Siderum Nebulosorum*, the result of an immense observation programme carried out with the 11" refractor in Copenhagen. A by-product was a systematic study of double nebulae. In d'Arrest's last years he concentrated on spectroscopy – a new field of research, participation in which was strictly refused by another eminent visual observer, Wilhelm Tempel.

For Dreyer, the work of d'Arrest was the vital motivation to focus on nebulae and star clusters. No doubt the practical experience concerning observation and cataloguing was gained during his stay at Birr Castle as the assistant of Lawrence Parsons, Lord Rosse's oldest son. There he could use the world's largest telescope, the 72" reflector, which was widely known as the Leviathan of Parsonstown. Major results were Dreyer's supplement to the General Catalogue and the publication of the Birr Castle observations from 1848 to 1878. From there it was only a small step to the NGC. Dreyer, who had in the meantime been appointed Director of Armagh Observatory, collected all reports about discoveries of nebulae, carefully checking their consistency. He was the unchallenged expert in the field – a second Dreyer never appeared.

12.5 STATISTICAL ANALYSIS AND THE WAY AHEAD

A central point of the present work is the statistical analysis of the original and modern data of NGC objects, involving observers, methods, instruments, sites and publications. Altogether 2144 NGC and 107 IC objects are mentioned.

Concerning the numbers of discoveries from 1800 to 1887 (just before the publication of the NGC), there were considerable fluctuations. John Herschel's Slough and Cape catalogues marked a first boom. Astonishingly, a period of idleness followed. It was felt among astronomers that nothing new could be discovered. Fortunately, from 1848 onwards things changed due to the work of the Birr Castle astronomers. A new climax appeared due to the discoveries of d'Arrest (Copenhagen) and Albert Marth (Malta); similar numbers of new objects were not reached by Edouard Stephan (Marseille) and Wilhelm Tempel (Arcetri). However, the period from 1880 to 1887 brought a new rise, mainly through the work of Lewis Swift and the observers at the Leander McCormick Observatory.

The statistics of the NGC observers give their numbers of discoveries and success rates (percentage of non-stellar objects). As expected, William and John Herschel dominate, followed by Marth and Swift. The worst success rate is that of Coolidge (Harvard College Observatory): his nine objects are either stars or 'not found'. Most discoverers came from the USA, followed by Germany. The most industrious writer was Edward

E. Barnard, who had published 60 articles on nebulae and star clusters by 1900. His results established him as the best visual observer.

It is interesting to divide the 7586 independent NGC objects (i.e. subtracting identities) by their modern types: galaxies dominate (79.5%), with open clusters (9.0%) far behind. The visual magnitudes range from 2.2 mag (NGC 292, the Small Magellanic Cloud) to 16.4 mag (NGC 4042). The average is 12.6 mag, reflecting the visual discovery of most objects; only 22 were found by spectroscopy and 1 by photography (NGC 1432, the Maia Nebula in the Pleiades).

Seventy-one different telescopes were used: 60 refractors (85%) and 11 reflectors. The apertures range from 3" (refractor) to 72" (reflector). Most instruments were located in the UK and Germany. It is significant that reflectors were mainly used by rich amateurs at their private observatories. On the other hand, a refractor was the standard telescope of professional astronomers at university or government observatories.

The time after the publication of the NGC, concerning observations, supplements, corrections and revisions, is considered too. The first contribution was due to Dreyer himself: the two Index Catalogues published in 1895 and 1908 with a total of 5386 objects (mostly discovered by photography). They show considerably lower mean brightness values (IC I, 14.0 mag; IC II, 14.2 mag).

Extensive visual observations were made by Bigourdan and Hagen, both covering a major part of the NGC objects. A comprehensive photographic investigation of 'Herschel nebulae' was presented by Reinmuth. This, together with the important work of Carlson, led to many corrections. Unfortunately, most of the modern catalogues (e.g. of galaxies) have ignored these important results. The same applies for the two revisions of the NGC published in 1977 and 1988, which therefore are not reliable sources. A comprehensive modern revision is the 'Revised New General and Index Catalog' of the author, which was compiled in cooperation with the international 'NGC/IC Project'. It is followed by of another catalogue, the 'Historic NGC', which forms an essential basis this work.

Finally, subsequent historical research should first concentrate on Dreyer's Index Catalogues. The relevant database (the author's 'Historic IC') is already available. Most interesting in this context is the transition from visual observing (which was still central in the IC I) to spectroscopy and photography (which dominate the IC II). The catalogued nebulae and star clusters are key objects both for classical astronomy and for modern astrophysics. Further study of the Index Catalogues might lead to new insights, especially for the history of astrophysics. Relevant persons are, for instance, Williamina Fleming, James Keeler and Max Wolf. Further interesting fields of work are given in Section 10.3.2: the work on the surface brightness of nebulae by Carl Wirtz, Solon Bailey's photographic survey of the southern sky, the detailed images of nebulae made by Heber Curtis, Knut Lundmark's unfinished Lund General Catalogue and the famous Shapley–Ames catalogue of galaxies.

Appendix

TIMELINE

The following table gives 152 major events in the history of the New General Catalogue, sorted by date. It covers observations, discoveries, publications (marked in italic, with the title occasionally shortened), erection of telescopes at various sites and a few personal events. The persons and sites involved are presented too. It is instructive to compare this with the general astronomical timeline given by Chambers, covering the period 720 BC to 1889 (Chambers 1890: 468–486).

Date	Person	Event	Site
1781	Charles Messier	*Catalogue des nébuleuses et des amas d'étoiles* (Messier's catalogue)	Paris
7.9.1782	William Herschel	First discovery: NGC 7009 (Saturn Nebula)	Datchet
26.2.1783	Caroline Herschel	First discovery: NGC 2360	Datchet
23.10.1783	William Herschel	18.7" reflector ('large 20ft') operational	Datchet
28.10.1783	William Herschel	First sweep	Datchet
27.4.1786	William Herschel	*Catalogue of One Thousand New Nebulae and Clusters of Stars*	Slough
24.10.1786	William Herschel	Discovery of NGC 7000 (North America Nebula)	Slough
11.6.1789	William Herschel	*Catalogue of a Second Thousand New Nebulae and Clusters of Stars*	Slough
1789	William Herschel	48" reflector operational	Slough
17.2.1790	William Herschel	Discovery of most southerly Herschel object: NGC 3621	Slough
13.11.1790	William Herschel	Discovery of NGC 1514	Slough
1795	Friedrich v. Hahn	Discovery of the M 57 central star	Remplin
1.7.1802	William Herschel	*Catalogue of 500 New Nebulae, Nebulous Stars, Planetary Nebulae, and Clusters of Stars*	Slough
30.9.1802	William Herschel	Last sweep	Slough
1820	John Herschel	Erection of 18¼" reflector	Slough
29.3.1821	John Herschel	First sweep	Slough
6.8.1823	John Herschel	First discovery: NGC 7010	Slough
Sep.? 1823	Karl Ludwig Harding	Discovery of NGC 7293 (Helix Nebula)	Göttingen
1824	John Herschel	Drawing of M 42 (Orion Nebula)	Slough
1826	William Parsons	Constructed his first telescope (6" reflector)	Birr Castle
29.4.1826	James Dunlop	First discovery: NGC 5128	Paramatta
11.11.1826	Wilhelm Struve	Discovery of the fifth star in the trapezium (M 42)	Dorpat

567

Continued

Date	Person	Event	Site
1827	Caroline Herschel	Zone catalogue of W. Herschel's observations	Hannover
1828	James Dunlop	*A Catalogue of Nebulae and Clusters of Stars in the Southern Hemisphere Observed at Paramatta*	Paramatta
1831	Edward J. Cooper	Erection of 14" refractor	Markree Castle
4.10.1831	John Herschel	Discovery of Polarissima Borealis (NGC 3172)	Slough
1833	John Herschel	*Observations of Nebulae and Clusters of Stars, Made at Slough* (SC)	Slough
5.3.1834	John Herschel	First southern sweep	Feldhausen
8.3.1834	John Herschel	First southern object: NGC 2887	Feldhausen
3.4.1834	John Herschel	Discovery of NGC 3918 ('Uranus')	Feldhausen
7.10.1836	Johann v. Lamont	First observation (Saturn Nebula NGC 7009)	Munich
29.3.1837	John Herschel	Discovery of Polarissima Australis (NGC 2573)	Slough
27.6.1837	John Herschel	Drawing of M 8 (Lagoon Nebula)	Feldhausen
25.8.1837	Johann v. Lamont	Public lecture *Ueber die Nebelflecken*	Munich
1837	John P. Nichol	*The Architecture of the Heavens*	Glasgow
1839	Wilhelm Struve	15" Merz refractor operational	Pulkovo
Sep. 1839	William Parsons	36" reflector operational	Birr Castle
17.4.1840	Ebeneezer P. Mason	*Observations on Nebulae with a Fourteen Feet Reflector*	Yale
29.10.1840	William Parsons	Begins 36" reflector observations	Birr Castle
1844	William H. Smyth	*Cycle of Celestial Objects* (Bedford Catalogue)	Bedford
1844	William Parsons	Drawing of M 1 (Crab Nebula)	Birr Castle
11.2.1845	William Parsons	First light of 72" reflector	Birr Castle
4.3.1845	William Parsons	Begins 72" reflector observations	Birr Castle
Apr. 1845	William Parsons	Detection of spiral structure of M 51 (Whirlpool Nebula)	Birr Castle
1845	William Lassell	Equatorial 24" reflector operational	Liverpool
1847	John Herschel	*Results of Astronomical Observations Made at the Cape of Good Hope* (CC)	Feldhausen
June 1847	William C. Bond	15" Merz refractor operational at Harvard College Observatory	Harvard
Jan. 1848	William Rambaut	First assistant of Lord Rosse	Birr Castle
10.2.1848	George P. Bond	First American discovery: NGC 7150	Harvard
31.3.1848	William Parsons	Second drawing of M 51 (Whirlpool Nebula)	Birr Castle
1.4.1848	William Parsons	First discovery: NGC 4110	Birr Castle
Sep. 1852	William Lassell	First observation campaign	Malta
11.10.1852	John R. Hind	Discovery of Hind's Variable Nebula NGC 1555	London
Dec. 1853	Ernest Laugier	Catalogue of 53 nebulae	Paris
23.8.1855	Heinrich d'Arrest	First discoveries: NGC 607, NGC 7005	Copenhagen
1855	Heinrich d'Arrest	*Resultate aus Beobachtungen der Nebelflecke und Sternhaufen – Erste Reihe*	Leipzig
19.10.1855	Jean Chacornac	Discovery of NGC 1988	Paris

Date	Person	Event	Site
31.12.1855	Eduard Schönfeld	Discovery of NGC 1333 (Bonner Durchmusterung)	Mannheim
1859	Thomas W. Webb	*Celestial Objects for Common Telescopes*	Hardwicke
1859	Lewis Swift	First discovery: NGC 1360	Rochester
19.10.1859	Wilhelm Tempel	Discovery of NGC 1435 (Merope Nebula)	Venice
Jan. 1860	William Lassell	First light of 48" reflector	Liverpool
Mar. 1860	Julius Schmidt	Beginning of observations at Athens Observatory	Athens
1.12.1860	Eduard Schönfeld	Beginning of observations at Mannheim Observatory	Mannheim
29.4.1861	Samuel Hunter	Drawing of M 101 (Pinwheel Galaxy)	Birr Castle
5.6.1861	William Parsons	*On the Construction of Specula of Six-feet Aperture*	Birr Castle
Sep. 1861	William Lassell	Second observation campaign	Malta
26.9.1861	William Lassell	Erection of 48" reflector	Malta
30.9.1861	Heinrich d'Arrest	First discovery with 11" refractor: NGC 1	Copenhagen
10.11.1861	Eduard Schönfeld	Public lecture *Ueber die Nebelflecke*	Mannheim
15.11.1861	Arthur Auwers	*William Herschel's Verzeichnisse von Nebelflecken und Sternhaufen*	Königsberg
17.1.1862	Jean Chacornac	First light of 80-cm Foucault reflector at Paris Observatory	Paris
May 1862	Arthur Auwers	List of 50 new nebulae	Königsberg
1862	Eduard Schönfeld	*Beobachtungen von Nebelflecken und Sternhaufen; Erste Abtheilung*	Mannheim
Feb. 1863	Herman Schultz	Beginning of observations at Uppsala Observatory	Uppsala
8.5.1863	Heinrich d'Arrest	List of 125 new nebulae	Copenhagen
6.6.1863	Albert Marth	First discovery: NGC 6308	Malta
23.6.1863	John Herschel	Editorial deadline of General Catalogue	Collingwood
Apr. 1864	Samuel Hunter	Drawing of M 42 (Orion Nebula)	Birr Castle
2.5.1864	Heinrich d'Arrest	Detection of Coma Cluster of galaxies	Copenhagen
6.5.1864	Samuel Hunter	Drawing of M 51 (Whirlpool Nebula)	Birr Castle
1864	John Herschel	*Catalogue of Nebulae and Clusters of Stars* (GC)	Collingwood
Mar. 1865	Auguste Voigt	First observations with 80-cm Foucault reflector	Marseille
18.3.1865	Albert Marth	Discovery of faintest object: NGC 4042	Malta
Apr. 1865	Albert Marth	End of observations on Malta	Malta
15.5.1865	Hermann Vogel	Beginning of observations at Leipzig Observatory	Leipzig
1865	Lawrence Parsons	Succession of Lord Rosse	Birr Castle
Nov. 1865	Robert S. Ball	First assistant of Lawrence Parsons	Birr Castle
3.3.1866	Truman H. Safford	Beginning of observations at Dearborn Observatory	Chicago
May 1867	Heinrich d'Arrest	End of observations at Copenhagen Observatory	Copenhagen
May 1867	Julius Schmidt	End of observations at Athens Observatory	Athens
1867	Heinrich d'Arrest	*Siderum Nebulosorum Observationes Havnienses* (*SN*)	Copenhagen
14.3.1868	Otto Struve	Discovery of NGC 1554 (Struve's Lost Nebula)	Pulkovo

Continued

Date	Person	Event	Site
12.6.1868	Truman H. Safford	End of observations at Dearborn Observatory	Chicago
14.6.1868	Eduard Schönfeld	End of observations at Mannheim Observatory	Mannheim
June 1869	Robert Ellery	Great Melbourne Telescope (GMT) operational	Melbourne
2.3.1870	Hermann Vogel	End of observations at Leipzig observatory	Leipzig
23.6.1870	Edouard Stephan	First discovery: NGC 6431	Marseille
Jan. 1871	Ralph Copeland	Assistant at Birr Castle	Birr Castle
6.7.1873	Sherburne W. Burnham	First discovery: NGC 7026	Chicago
5.4.1874	Ralph Copeland	Discovery of Copeland Septet	Birr Castle
Aug. 1874	John L. E. Dreyer	Assistant at Birr Castle	Birr Castle
1874	Wilhelm Tempel	Drawing of NGC 1435 (Merope Nebula)	Mailand
1874	Herman Schultz	*Micrometrical Observations of 500 Nebulae*	Uppsala
1875	Wilhelm Tempel	Beginning of work at Arcetri Observatory	Arcetri
1875	Eduard Schönfeld	*Beobachtungen von Nebelflecken und Sternhaufen; Zweite Abtheilung*	Mannheim
18.3.1876	John L. E. Dreyer	First discovery: NGC 3698	Birr Castle
23.9.1876	Edouard Stephan	Discovery of Stephan's Quintet	Marseille
1877	Edward S. Holden	*Index Catalogue of Books and Memoirs Relating to Nebulae and Clusters*	Washington
2.9.1877	Ralph Copeland	First spectroscopic object: NGC 7114	Dun Echt
5.5.1878	John L. E. Dreyer	Last 72" observation (end of use)	Birr Castle
Aug. 1878	John L. E. Dreyer	End of work at Birr Castle	Birr Castle
1878	John L. E. Dreyer	*Supplement to Sir John Herschel's General Catalogue* (GCS)	Birr Castle
1879	Andrew Common	Erection of 36" reflector	Ealing
14.11.1879	Thomas W. Webb	Discovery of NGC 7027	Hardwicke
1880	Edward E. Barnard	First discovery: NGC 6302 (Bug Nebula)	Nashville
June 1880	Lawrence Parsons	*Observations of Nebulae and Clusters of Stars Made with the Six-Foot and Three-Foot Reflectors (1848–78)*	Birr Castle
13.7.1880	Edward Pickering	First discovery in spectroscopic search: NGC 6644	Harvard
28.9.1880	Joseph Baxendell	Discovery of NGC 7088 (Baxendell's Unphotographable Nebula)	Birkdale
17.3.1881	Edward S. Holden	Beginning of observations at Washburn Observatory	Madison
1882	Edward S. Holden	*Monograph on the Central Parts of the Nebula of Orion*	Madison
31.8.1882	John L. E. Dreyer	Beginning of directorship at Armagh Observatory	Armagh
Nov. 1882	Lewis Swift	Warner Observatory operational	Rochester
16.4.1883	Lewis Swift	First object at Warner Observatory: NGC 3588	Rochester
1883	Hermann Vogel	Observations with 27" Clark refractor	Vienna
20.5.1884	Guillaume Bigourdan	Beginning of observations at Paris observatory	Paris
17.8.1884	Edward E. Barnard	Discovery of NGC 6822 (Barnard's Galaxy)	Nashville

Date	Person	Event	Site
22.9.1884	Guillaume Bigourdan	First discovery: NGC 76	Paris
18.12.1884	Pietro Baracchi	First GMT discovery: NGC 2072	Melbourne
7.3.1885	Edouard Stephan	Last discovery at Marseille Observatory: NGC 2581	Marseille
1885	Wilhelm Tempel	*Über Nebelflecken*	Arcetri
3.6.1885	Francis Leavenworth	First object at Leander McCormick Observatory: NGC 5734	Charlottesville
20.8.1885	Ernst Hartwig	Discovery of supernova in M 31	Dorpat
3.11.1885	Edward E. Barnard	Discovery of NGC 1499 (California Nebula)	Nashville
16.11.1885	Paul & Prosper Henry	First photographic object: NGC 1432 (Maia Nebula)	Paris
1886	Wilhelm Tempel	Last observation at Arcetri Observatory	Arcetri
8.12.1886	John L. E. Dreyer	Second Supplement to the General Catalogue	Armagh
25.1.1887	Francis Leavenworth	Last discovery at Leander McCormick Observatory: NGC 5039	Charlottesville
26.6.1887	Lewis Swift	Last NGC discovery: NGC 6916	Rochester
Dec. 1887	John L. E. Dreyer	Editorial deadline of New General Catalogue	Armagh
1888	John L. E. Dreyer	*New General Catalogue of Nebulae and Clusters of Stars* (NGC)	Armagh
1888	Edward S. Holden	36" refractor at Lick Observatory	Mt Hamilton
1889	Andrew Common	Erection of 60" reflector	Ealing
1.6.1891	Max Wolf	Image of NGC 7000 (North America Nebula)	Heidelberg
1894	Lewis Swift	Lowe Observatory operational	Echo Mountain
10.1.1895	John L. E. Dreyer	*Index Catalogue of Nebulae* (IC I)	Armagh
1897	Sherburne W. Burnham	Beginning of observations with 40" Yerkes refractor	Williams Bay
4.5.1908	John L. E. Dreyer	*Second Index Catalogue of Nebulae and Clusters of Stars* (IC II)	Armagh
27.4.1911	Guillaume Bigourdan	End of observations at Paris Observatory	Paris
1912	John L. E. Dreyer	*Scientific Papers of Sir William Herschel*	Armagh
30.9.1916	John L. E. Dreyer	End of directorship at Armagh Observatory	Armagh
1919	Guillaume Bigourdan	Publication of *Observations de nébuleuses et d'amas stellaires*	Paris

ABBREVIATIONS AND UNITS

The following abbreviations are used in the text and the tables. One must distinguish among between historical catalogues and lists (see Table 1.2), object types (in all tables), object descriptions (Herschel-style) and other designations, among them conventional units (see also the section 'Telescope data'). Not given are the official abbreviations of constellations, which can be found in many textbooks.

Historic catalogues, lists and designations

Au	Auwers
B	Barnard
BD (DM)	Bonner Durchmusterung
Big	Bigourdan
CC	Cape catalogue
D	Dreyer
dA (d'A)	d'Arrest
GC	General Catalogue

GCS	GC Supplement		S, L	small, large
H, h	William Herschel's catalogues, John Herschel's catalogues (SC, CC)		R, E, r	round, extended (elongated), resolvable
IC	Index Catalogue (both)		n, s, p, f	north, south, preceding (west), following (east)
IC I, IC II	Index Catalogue, Second Index Catalogue		np, sf	north preceding (northwest), south following (southeast)
JH	John Herschel			
L	Laugier		v, c, p, e	very, considerably, pretty, excessively (extremely)
Ld R	Lord Rosse (William Parsons)			
M	Messier			
NGC	New General Catalogue			
NGC/IC	New General Catalogue and both Index Catalogues			

Abbreviations and units

AR, Rect., α	Right ascension
Decl., δ	Declination
SC	Slough catalogue
SN	*Siderum Nebulosorum*
WH	William Herschel
DSS	Digitized Sky Survey
GMT	Great Melbourne Telescope
HST	Hubble Space Telescope
LMC, SMC	Large/Small Magellanic cloud ('Nubecula Major', 'Nubecula Minor')

Object type

DN	Dark nebula
EN	Emission nebula
GC	Globular cluster (not to be confused with 'General Catalogue')
Gx, GxP	Galaxy, part of galaxy (e.g. H II region)
NF	Not found
OC	Open cluster
PN	Planetary nebula
RN	Reflection nebula
SNR	Supernova remnant

mag	Magnitude
NPD, SPD	North/South Pole Distance
POSS	Palomar Observatory Sky Survey
RAS	Royal Astronomical Society
Rl, Rr	Reflector, refractor
Con	Constellation
V	Visual magnitude
h m s	Hour, minute, second (α or time)
° ′ ″	Degree, arcminute, arcsecond (δ, NPD/SPD or angular distance)

Object descriptions (Herschel, Dreyer)

cl, neb	cluster, nebula
F, B	faint, bright

The following table shows the metric and non-metric units used in the text and the tables; they mainly concern the aperture and focal length of telescopes.

Country	Unit	Name	Equivalent	cm
Germany	1‴	Linie (L)		0.218
	1 z	Zoll	12 L	2.615
	1 Fuß	Fuß	12 z	31.385
UK	1″	inch		2.540
	1 ft	foot	12″	30.480
France	1 Par. L	Paris line (PL)		0.226
	1 Par. Zoll	Paris inch	12 PL	2.712

TELESCOPE DATA

Here the data of 226 historical telescopes mentioned in the text are given; they are sorted by site and year (see also the 'Site index'). The columns are explained below.

Site (Observatory)	Type	Maker	D	F	F/D	Name	Year	E	L	Observer
Albany, Dudley Obs.	Rr	Fitz	33.0	4.10	12.4	13"	1860	52	43	Hough
Altona	Rr	Repsold	18.0	1.90	10.6	7"/6 ft		31	54	Petersen
Amherst Coll. Obs.	Rr	Clark	45.7	7.80	17.1	18"	1905	110	42	Todd
Amherst, Lawrence Obs.	Rr	Clark	18.4			7.25"	1854	100	42	Todd
Ann Arbor	Rl*		30.5	50.8	1.5	12"	1902	110	42	Schaeberle
Ann Arbor, Univ. of Michigan	Rr	Fitz	31.8	5.20	16.4	12.5"	1853	110	42	Harrington
Arcetri (Florence)	Rr	Amici (I)	28.3	5.37	19.0	11"	1841	73	44	Tempel
Arcetri (Florence)	Rr	Amici (II)	23.8	3.18	13.4	9.3"	1844	73	44	Tempel
Arequipa, Harvard Coll. Obs.	AC	Clark	20.0	1.15	5.8	8"	1885	2451	−16	Pickering
Arequipa, Harvard Coll. Obs.	Rr	Clark	61.0	3.40	5.6	24"	1896	2451	−16	Pickering
Armagh	Rl	Grubb	38.0	2.74	7.2	15"	1835	70	55	Robinson
Armagh	Rr	Grubb	25.4	3.05	12.0	10"	1885	70	55	Dreyer
Athens	Rr	Plössl	15.7	1.83	11.7	6"	1861	110	38	Schmidt
Aylesbury, Hartwell Obs.	Rr	Tully	15.0	2.65	17.7	5.9"	1830	50	52	Smyth, Lee
Bamberg, Remeis Obs.	Rr	Schröder	26.4	3.87	14.7		1889	288	50	Hartwig
Bath	Rl	W. Herschel	15.7	2.10	13.4	6.2"/7 ft	1778	10	51	W. Herschel
Bedford	Rr	Tully	15.0	2.65	17.7	5.9"	1830	50	52	Smyth
Berlin	Rr	Merz	24.4	4.30	17.6	9"	1835	47	53	Winnecke
Birr Castle	Rl	Parsons	15.2	6.10	6.0	6"	1828	56	53	W. Parsons
Birr Castle	Rl	Parsons	38.0	3.70	9.6	15"	1830	56	53	W. Parsons
Birr Castle	Rl	Parsons	61.0			24"	1831	56	53	W. Parsons
Birr Castle	Rl	Parsons	91.4	8.23	9.0	36"/27 ft	1839	56	53	W. + L. Parsons etc.
Birr Castle	Rl	Parsons	182.9	16.5	9.0	72"/54 ft	1845	56	53	W. + L. Parsons etc.
Birr Castle	Rl	Parsons	45.7	5.50	12.0	18"		56	53	W. Parsons
Bonn	HE	Merz & Mahler	15.7	2.50	15.9	6"	1845	62	51	Winnecke
Bonn	Rr	Schröder	15.9	1.93	12.1	6"	1874	62	51	Schönfeld, Mönnichmeyer

Appendix

Continued

Site (Observatory)	Type	Maker	D	F	F/D	Name	Year	E	L	Observer
Bonn	CS	Fraunhofer	7.8	0.63	8.1	2 ft	1851	62	51	Schönfeld, Winnecke
Bordeaux	Rr		35.6			14"	1878		42	Rayet
Bothkamp	Rr	Schröder	29.3	4.90	16.7		1870	32	54	Vogel
Breslau	Rr		9.7	1.50	15.5	43 L/5 ft		147	51	Boguslawski
Brighton	Rl*	With	20.7			8.15"		20	51	Pratt
Bristol	Rl*	Browning	25.4	1.65	6.5	10"	1871	10	51	Denning
Cambridge, Harvard Coll. Obs.	CS	Bowditch	10.0	0.80	8.0	4"	1845	24	42	Bond, Tuttle
Cambridge, Harvard Coll. Obs.	CS	Quincy	6.4	0.71	11.1	2.5"	1845	24	42	Bond, Tuttle
Cambridge, Harvard Coll. Obs.	Rr	Merz & Söhne	38.1	6.83	17.9	15"	1847	24	42	W. + G. Bond etc.
Cambridge, University Obs.	Rr	Cauchoix	29.5	5.90	20.0	12"	1838	24	52	Breen
Castelgandolfo, Vatican Obs.	Rr	Zeiss	40.6	6.00	14.8	16"	1906	450	42	Hagen
Castelgandolfo, Vatican Obs.	AC	Henry	33.0	3.40	10.3		1912	450	42	Hagen
Charlottesville, Leander McCormick Obs.	Rr	Clark	66.8	9.93	14.9	26"	1885	259	38	Stone, Leavenworth, Muller
Chicago	Rr	Clark	15.2	2.30	15.1	6"	1870	170	42	Burnham
Cincinnati	Rr	Merz & Mahler	30.5	5.30	17.4	12"	1845	247	39	Porter, Stone
Cincinnati	Rr	Clark	40.6	6.24	15.4	16"	1904	247	39	Porter
Clapham	Rl*	Browning	31.1			12.25"			51	Grover
Clinton, Litchfield Obs.	Rr	Spencer	34.3	4.90	14.3	13.5"	1858	276	43	Peters
Copenhagen	Rr	Merz	28.5	4.90	17.2	11"	1861	10	56	d'Arrest, Pechüle
Copenhagen	Rr	Merz & Söhne		1.50				14	56	d'Arrest
Crowborough, Starfield	Rl*	Grubb	50.8	2.49	4.9	20"	1886	238	51	Roberts
Crowborough, Starfield	AC	Cooke	12.7	0.61	4.8	5"		238	51	Roberts
Crowborough, Starfield	AC	Dallmeyer	8.8	0.24	2.7	3.5"		238	51	Roberts
Crumpsall Hall	Rr		12.7			5"		20	53	Baxendell
Crumpsall Hall	Rr		33.0			13"		20	53	Baxendell

Appendix 575

Site (Observatory)	Type	Maker	D	F	F/D	Name	Year	E	L	Observer
Cuckfield	Rr	Clark	18.6	2.74	14.7	7.3"	1859	113	51	Knott
Datchet/Clay Hall	Rl	W. Herschel	30.0	6.00	20.0	12"/20 ft	1776	50	51	W. Herschel
Datchet/Clay Hall	Rl	W. Herschel	11.4	0.69	6.1	4.5"	1783	50	51	C. Herschel
Denver, Chamberlin Obs.	Rr	Clark	50.8	7.80	15.4	20"	1891	1644	40	Howe
Dorpat	Rr	Fraunhofer	24.4	4.10	16.8		1824	67	58	W. Struve
Dresden	Rr	Grubb	30.6	3.85	12.6		1879	124	51	Engelhardt
Dun Echt	Rr	Grubb	38.4	4.60	12.0	15.1"	1873	141	57	Copeland, Lohse
Dun Echt	MC	Troughton	21.5	2.60	12.1		1873	141	57	L. Becker
Dun Echt	Rr	Simms	15.5			6.1"		141	57	Copeland
Dunsink, Trinity Coll. Obs.	Rr	Grubb	30.0	5.80	19.3	11.75"	1868	91	53	Brünnow
Durham	Rr	Merz & Mahler	17.2	2.50	14.5	6.75"	1841	108	55	Marth
Ealing	Rl*	Calver	45.7			18"	1877	20	51	Common
Ealing	Rl*	Calver	91.4	5.35	5.9	36"	1879	20	51	Common
Ealing	Rl*	Common	152.4	8.70	5.7	60"	1891	20	51	Common
Echo Mountain, Lowe Obs.	Rr	Clark	40.6	6.70	16.5	16"	1894	1100	34	L. + E. Swift
Evanston, Dearborn Obs.	Rr	Clark	47.0	8.20	17.4	18.5"	1863	175	42	Hough, Burnham
Gateshead	Rr	Cooke	63.0	9.10	14.4	25"	1869	20	55	Newall, Marth
Geneva	Rr	Merz & Söhne	25.0			10"		407	46	Kammermann
Georgetown	Rr	Clacy	30.5	4.60	15.1	12"	1895	62	39	Hagen
Gorleston	Rl*	With	20.0	1.75	8.8	8"	1864	10	53	Matthews
Gotha	Rr	Steinheil	11.3	1.93	17.1	6 ft	1863	360	51	Auwers
Göttingen	Rr	Fraunhofer	6.3			29 L		150	52	Auwers
Göttingen	CS	Merz	7.4			34 L		150	52	Winnecke
Göttingen, University Obs.	Rl	Herschel	21.7	3.00	13.8		1786	161	52	Harding
Göttingen, University Obs.	MC	Reichenbach	10.1	1.64	16.2	4"	1819	161	52	Harding
Göttingen, University Obs.	Rr	Fraunhofer	11.0	1.90	17.3			161	52	Auwers
Greenwich, Royal Obs.	Rl	Ramage	38.1	7.60	19.9	15"/25 ft	1820	47	51	Pond
Greenwich, Royal Obs.	Rl*	Grubb	71.1	8.46	11.9	30"	1896	47	51	Christie

Continued

Site (Observatory)	Type	Maker	D	F	F/D	Name	Year	E	L	Observer
Hamburg-Bergedorf	Rr	Steinheil/Repsold	60.0	9.00	15.0		1914	50	54	Graff
Hamburg, Millerntor	Rr	Merz	10.0	1.50	15.0	5 ft	1836	0	54	Rümker
Hamburg, Millerntor	Rr	Repsold	26.0	3.00	11.5	10"	1867	0	54	Rümker
Hanover, Dartmouth Coll. Obs.	Rr		23.9			9.4"		160	44	Young
Hardwicke	Rr	Tully	9.4	1.52	16.2	3.7"/5 ft	1834		52	Webb
Hardwicke	Rr	Clark	14.0	2.13	15.2	5.5"/7 ft	1859		52	Webb
Hardwicke	Rl*	With	20.3	1.83	9.0	8"/6 ft	1864		52	Webb
Hardwicke	Rl*	With	23.8	1.90	8.0	9.38"	1866		52	Webb
Haverford College Obs.	Rr	Clark	25.4			10"		116	40	Leavenworth
Heidelberg	Rr	Kranz	13.4	0.77	5.7	5"		126	49	M. Wolf
Heidelberg, Königstuhl	Rr	Reinfelder & Hertel	16.0	0.81	5.1	6"	1885	570	49	M. Wolf, Reinmuth
Heidelberg, Königstuhl	AC	Voigtländer	15.9	0.80	5.0		1896	570	49	M. Wolf
Heidelberg, Königstuhl	AC	Grubb	40.6	2.08	5.1	16"	1900	570	49	M. Wolf, Reinmuth
Heidelberg, Königstuhl	Rl*	Zeiss	72.0	2.88	4.0		1906	570	49	M. Wolf, Reinmuth
Hereford	Rl*	Key	45.7	3.00	6.6	18"	1865		52	Key
Herény	Rl*	Browning	26.0	2.00	7.7	10.25"	1881	229	47	v. Gothard
Hobart	Rr	Dallmeyer	10.2	1.50	14.7	4"		−43		Abbott
Hopefield	Rr	Cooke	20.3			8"	1869		51	Hunt
Jamaica, Kempshot Obs.	Rr	Cooke	10.0			4"	1872	540	18	M. Hall
Jamaica, Kempshot Obs.	Rl*	Browning	22.9			9"	1889	540	18	M. Hall
Karlsruhe	Rr	Reinfelder & Hertel	12.2			54 L		110	49	Winnecke
Karlsruhe	CS		9.7			43 L		110	49	Winnecke
Königsberg	HE	Merz	15.0	2.40	16.0	6"/8 ft	1829	10	55	Auwers
Königsberg	Rr	Merz & Mahler	15.7	1.80	11.5	6 ft	1841	10	55	Auwers
Leiden	Rr	Merz	16.6	2.40	14.5		1838	0	52	Kaiser
Leipzig	CS	Reinfelder & Hertel	12.2	1.15	9.4	54 L/44"		120	51	Engelmann
Leipzig, Johannisthal	Rr	Steinheil	21.5	3.87	18.0	12 ft	1862	119	51	Vogel
Leipzig, Johannisthal	MC	Pistor/Martins	16.2	2.44	15.1		1866	119	51	Engelmann

Site (Observatory)	Type	Maker	D	F	F/D	Name	Year	E	L	Observer
Leipzig, Pleißenburg	Rr	Merz	11.7	1.96	16.8	52 L/6 ft	1830	155	51	d'Arrest, Vogel
Liverpool	Rl	Lassell	17.8			7"	1830	50	53	Lassell
Liverpool	Rl	Lassell	22.8	2.84	12.5	9"	1840	50	53	Lassell
Liverpool	Rl	Lassell	61.0	6.10	10.0	24" (2 ft)	1845	50	53	Lassell
London, Bishops Obs.	Rr	Dollond	17.8	3.28	18.4	7"	1836	20	51	Hind, Marth etc.
Lund	Rr	Merz & Söhne	24.5			9.6"	1867	34	56	Dunér
Madison, Washburn Obs.	Rr	Clark	39.6	6.20	15.7	15.6"	1881	292	43	Holden
Maidenhead	Rl	Lassell	61.0	6.10	10.0	24" (2 ft)	1867		52	Lassell
Milan	Rr	Merz & Söhne	19.1				1862	120	45	Dembowski
Milan (Brera)	Rr	Merz & Söhne	22.0			8.5"	1874	120	45	Schiaparelli
Milan (Brera)	Rr	Plössl	10.5			4"		120	45	Tempel
Mali Lošinj, Manora Obs.	Rr	Reinfelder & Hertel	17.5			7"	1895		45	Brenner
Malta	Rl	Lassell	122.0	11.2	9.2	48" (4 ft)	1860	100	36	Marth, Lassell
Manchester	Rr	Cooke	12.7			5"		20	53	Copeland
Mannheim	Rr	Steinheil	16.5	2.60	15.8	73 PL	1860	98	49	Schönfeld
Markree Castle	Rr	Cauchoix	35.5	7.60	21.4	14"	1831	45	54	Cooper, Doberck
Markree Castle	CS	Ertel	7.6	0.76	10.0	3"	1842	45	54	Graham
Marseille	Rr	Steinheil	10.0			4"			43	Chacornac
Marseille, Longchamps	Rl*	Foucault	78.8	4.54	5.8		1862	75	43	Voigt, Stephan
Marseille, Longchamps	CS	Foucault	18.2	2.10	11.5		1866	75	43	Coggia, Borrelly
Marseille, Longchamps	Rr	Merz & Söhne	25.8	3.10	12.0		1872	75	43	Coggia, Borrelly
Marseille, Longchamps	MC	Martin/ Eichens	18.8	2.25	12.0		1876	75	43	Borrelly
Melbourne	Rl	Grubb	122.0	51.0	41.8		1869	100	−38	Baracchi, Turner etc.
Minneapolis, University Obs.	Rr	Brashear	26.7			10.5"		260	45	Leavenworth
Modena	MC	Reichenbach		1.50		5 ft		30	45	Bianchi
Moscow, University Obs.	Rr	Merz & Söhne	23.0	4.50	19.6	9"/15 ft	1859	166	56	Schweizer
Mt Hamilton, Lick Obs.	Rr	Clark	30.5	3.65	12.0	12"	1881	1283	37	Barnard
Mt Hamilton, Lick Obs.	Rr	Clark/ Warner & Swasey	91.0	17.6	19.3	36"	1888	1283	37	Barnard, Burnham

578 Appendix

Continued

Site (Observatory)	Type	Maker	D	F	F/D	Name	Year	E	L	Observer
Mt Hamilton, Lick Obs.	Rr		15.0	0.78	5.2	6"	1889	1283	37	Barnard
Mt Hamilton, Lick Obs.	Rl*	Common	91.0	5.35	5.9	36"	1896	1283	37	Keeler
Mt Wilson	Rl*	Ritchey	1.5	7.60	506.7	60"	1908	1742	34	Pease
Mughull	Rl*	Grubb	50.8	2.49	4.9	20"	1886		54	Roberts
Munich, Bogenhausen	Rr	Merz	28.5	4.90	17.2	10.5"	1835	500	48	Lamont
Naples, Capodimonte	Rr	Fraunhofer	17.6	3.02	17.2		1812	150	41	Capocci
Naples, Capodimonte	Rr	Ertel	8.9			3.5"		150	41	C.H.F. Peters
Nashville	Rr	Byrne	12.7	1.60	12.6	5"	1877	180	36	Barnard
Nashville, Vanderbilt Obs.	Rr	Cooke	15.2			6"	1876	174	36	Barnard
New Haven, Yale Coll. Obs.	Rr	Dollond	12.7	3.05	24.0	5"/10 ft	1830	21	41	Mason
New Haven, Yale Coll. Obs.	Rl	Mason/Smith	30.5	4.27	14.0	12"/14 ft	1838	21	41	Mason
Nice (Mont Gros)	Rr	Henry	38.0	6.90	18.2		1884	372	44	Perrotin
Nice (Mont Gros)	Rr	Henry	74.0	17.9	24.2		1887	372	44	Javelle
Northfield, Goodsell Obs.	AC	Brashear	15.2	0.79	5.2	6"		290	44	Wilson
Odessa	Rr	Steinheil	13.0			5"		60	46	Block
Odessa	CS		10.5			4"		60	46	Block
Olmütz	Rr	Merz	13.6	1.55	11.4	5"/5 ft		200	50	Schmidt
Oxford, Radcliffe Obs.	AC	Grubb	31.0	4.50	14.5	13"	1875	64	52	Pritchard
Oxford, Radcliffe Obs.	Rl*		38.1	2.02	5.3	15"		64	52	Pritchard
Oxford, Radcliffe Obs.	Rl*		38.1	3.05	8.0	15"		64	52	Pritchard
Palermo	Rr	Ramsden	7.5	1.50	20.0		1790	72	38	Cacciatore
Paramatta	TT	Troughton	8.9	1.67	18.8		1822	50	−34	Dunlop
Paramatta	Rl	Dunlop	22.8	2.70	11.8		1824	50	−34	Dunlop
Paris	Rr	Porro	52.0	15.00	28.8		1856	60	49	Porro
Paris Obs.	Rr	Lerebours	25.0	3.15	12.6	9.5"	1823	67	49	Chacornac
Paris Obs.	Rr	Gambey	9.7	1.65	17.0		1826	67	49	Laugier
Paris Obs.	Rr	Sécretan	31.6	5.15	16.3	12.4"	1858	67	49	Bigourdan, Chacornac
Paris Obs.	Rl*	Foucault	40.0	2.50	6.25	15.7"	1859	67	49	Chacornac, C. Wolf

Site (Observatory)	Type	Maker	D	F	F/D	Name	Year	E	L	Observer
Paris Obs.	Rl*	Martin/ Eichens	120.0	7.20	6.0	47.2"	1877	67	49	C. Wolf
Paris Obs.	AC	Henry	33.0	3.44	10.4		1885	67	49	P. + P. Henry
Paris Obs.	Rr		10.5			4" (48 L)		67	49	Goldschmidt
Pittsburgh, Allegheny Obs.	Rr	Brashear	76.2	14.4	18.9	30"	1912	370	40	Trümpler
Pola	Rr	Steinheil	15.0				1872		45	Palisa
Potsdam, Astrophysikal. Obs.	Rr	Grubb	20.7	3.20	15.5		1877	100	53	Kempf
Potsdam, Astrophysikal. Obs.	Rr	Schröder	29.8	5.20	17.4	12"	1878	100	53	Vogel
Potsdam, Astrophysikal. Obs.	Rr	Steinheil	32.0	3.40	10.6		1889	100	53	Wilsing
Prague, University Obs.	Rr		16.3	2.27	13.9			267	50	Weineck
Princeton, Halsted Obs.	Rr	Clark	58.0	9.80	16.9	23"	1882	75	40	Young
Putney Hill	Rl*	Calver	25.4			10"			51	McCance
Remplin	Rl	Herschel/ Hahn	20.3	2.10	10.3	8"/7 ft	1793	10	54	Hahn
Remplin	Rl	Herschel/ Hahn	30.5	6.10	20.0	12"/20 ft	1794	10	54	Hahn
Remplin	Rl	Herschel/ Hahn	47.5	6.10	12.8	18.7"/20 ft	1700	10	54	Hahn
Remplin	Rl	Cary	5.1	0.84	16.5	2"		10	54	Hahn
Rochester	Rr	Fitz	11.4	1.14	10.0	4.5"	1857	170	43	Swift
Rochester, Warner Obs.	Rr	Clark	40.6	6.70	16.5	16"	1883	172	43	L. + E. Swift
Rome	Rr	Merz & Söhne	22.9			9"		50	42	Ferrari
Rome, Collegio Romano	Rr	Cauchoix	15.9	2.30	14.5	6"/7 ft	1825	51	42	de Vico, Secchi
Rome, Collegio Romano	Rr	Merz & Söhne	24.0	4.35	18.1	9.5"	1853	51	42	Secchi, Ferrari
Scarborough	Rr	Cooke	39.4	5.90	15.0	15.5"	1885	45	54	Lohse
Senftenberg (Žamberk)	CS	Merz	9.4	1.25	13.3	43 L	1842	150	50	Brorsen
Slough	Rl	W. Herschel	47.5	6.10	12.8	18.7"/20 ft	1783	50	52	W. Herschel
Slough	Rl	W. Herschel	122.0	12.2	10.0	40 ft	1789	50	52	W. Herschel
Slough	Rl	W. Herschel	23.4	1.60	6.8	9.2"	1791	50	52	C. Herschel
Slough	Rl	W. Herschel	61.0	7.60	12.5	25 ft	1797	50	52	W. Herschel
Slough	Rl	W. Herschel	61.0	3.05	5.0	X-feet	1799	50	52	W. Herschel

Continued

Site (Observatory)	Type	Maker	D	F	F/D	Name	Year	E	L	Observer
Slough	Rl	J. Herschel	46.4	6.10	13.1	18¼"/20 ft	1820	50	52	J. Herschel
Slough	Rr	Tully	12.7	2.10	16.5	7 ft		50	52	J. Herschel
Feldhausen	Rl	J. Herschel	46.4	6.10	13.1	18¼"/20 ft	1820	20	−34	J. Herschel
Feldhausen	Rr	Tully	12.7	2.10	16.5	7 ft		20	−34	J. Herschel
Southport, Birkdale	Rr	Cooke	15.2			6"	1871	12	54	Baxendell
Southport, Hesketh Park Obs.	Rr	Cooke	16.5			6.5"	1869	10	54	Baxendell
Stoneyhurst	Rr	Grubb	38.1	3.80	10.0	15"	1894	117	60	O'Connor
St Petersburg, Pulkovo	Rr	Baader	10.5			48'''	1839	75	60	Winnecke
St Petersburg, Pulkovo	Rr	Merz & Mahler	38.0	6.90	18.2	15"	1839	75	60	O. Struve
St Petersburg, Pulkovo	HE	Merz & Mahler	15.7	3.27	20.8	72 L (7")	1839	75	60	Winnecke
St Petersburg, Pulkovo	CS	Plössl	7.4			34 L	1842	75	60	Winnecke
St Petersburg, Pulkovo	Rr	Clark	76.0	13.7	18.0	30"	1884	75	60	O. Struve
St Petersburg, Pulkovo	Rr	Steinheil	11.8			54 L		75	60	Winnecke
Straßburg, Academy Obs.	OS	Reinfelder & Hertel	16.3	2.63	16.1	6"	1875	140	49	Winnecke
Straßburg, University Obs.	Rr	Merz & Söhne	48.7	6.92	14.2	18"	1880	144	49	Winnecke, Kobold
Sunderland	Rr	Cooke	11.4					49	55	Backhouse
Taschkent	AC	Henry Repsold	33.0	3.44	10.4			477	41	Stratonov
Teramo, Collurania Obs.	Rr	Cooke	39.4	5.90	15.0	15.5"	1890	388	43	Cerulli
Toulouse	Rr	Dollond	9.5	1.14	12.0	3.5"	1770	195	44	Darquier
Tulse Hill	Rr	Clark/Cooke	20.0	2.00	10.0	8"	1858	50	52	Huggins
Twickenham, Meadowbank	Rr	Dollond	17.8	3.28	18.4	7"	1863	20	51	Hind, Talmage
Uppsala	Rr	Steinheil	24.4	4.00	16.4	13 ft	1860	50	60	Schultz
Uppsala	Rr	Steinheil	36.0	5.40	15.0		1892	50	60	Dunér
Uppsala	MC	Repsold	9.6					50	60	Schultz
Venice	Rr	Steinheil	10.8	1.62	15.0	4"	1858	10	45	Tempel
Walworth Common	Rr	Wray	54.0	8.70	16.1	21¼"	1862	10	52	Buckingham
Wandsworth	Rr	Chance/Gravatt	61.0	11.00	18.0	24"	1852	10	51	Craig

Site (Observatory)	Type	Maker	D	F	F/D	Name	Year	E	L	Observer
Warsaw	Rr	Merz & Söhne	16.2				1859	121	52	Merecki
Washington, US Naval Obs.	Rr	Clark	65.5	9.90	15.1	26"	1873	38	39	Holden, Hall, Tuttle
Vienna	Rr	Plössl	18.9	2.44	12.9		1863	186	48	Oppolzer, Ginzel
Vienna, University Obs.	Rr	Merz	15.0			6"	1832	240	48	Holetschek
Vienna, University Obs.	Rr	Clark	29.8	5.20	17.4	12"	1880	240	48	Palisa
Vienna, University Obs.	Rr	Grubb	68.6	10.5	15.3	27"	1883	240	48	Palisa, Spitaler, Vogel
Vienna, University Obs.	Rr*	Gautier	38.0	9.27	24.4		1890	240	48	Bidschof
Williams Bay, Yerkes Obs.	Rr	Clark	102.0	19.4	19.0	40"	1897	334	43	Barnard, Burnham
Williamstown, Hopkins Obs.	MC	Repsold	11.4					213	43	Safford

'Type' gives the telescope type. All, except for the reflectors (Rl), are achromatic-lens systems. The abbreviations are

AC   Astrocamera (astrograph, portrait lens)
CS   Comet-seeker
HE   Heliometer
MC   Meridian-circle
OS   Orbit-sweeper (triaxial comet-seeker)
Rl   Reflector (* = silvered glass mirror)
Rr   Refractor (* = Coudé system)
TT   Transit-telescope

D is the optical diameter (aperture in cm) and F the focal length (in m); F/D is the inverse focal ratio (also called the 'aperture number'). The column 'Name' contains the common telescope name; the text often includes the maker, e.g. 11" Merz refractor (in the case of 'Fraunhofer', the refractors can have been made by the following: Merz, Merz & Mahler and Merz & Söhne – see Mädler (1873: 86–87)). However, the table always gives the correct company. In the eighteenth century a telescope was often designated by its focal length, e.g. Herschel's '40ft'. Depending on the context (common use, publication), both metric and non-metric units are used in the text (see 'Abbreviations and units'). The column 'Year' gives the date of first use, E is the site elevation (in m) and L its (rounded) latitude (negative = southern). The column 'Observer' lists the primary users (as mentioned in the text). The following list shows some of the general literature on observatories and instruments not contained in the reference list.

L. Ambronn. *Handbuch der Astronomie und Instrumentenkunde*, vol. 2 (Berlin: Springer Verlag 1899).
G. Anderson. *The Telescope. Its History, Technology, and Future* (Princeton: Princeton University Press 2007).
D. Andrews. Cyclopaedia of telescope makers. *Irish Astronomical Journal*, **20** (1992) 102–183, **21** (1993–94)

1–82, 167–249, **22** (1994) 43–96, **23** (1995) 57–117, 215–242, **24** (1997) 125–192.

Anon. Die astronomischen Observatorien der Erde. *Sirius*, **18** (1890) 1–4, 25–29, 84–90, 204–209, 229–234, 252–256, 280–283, **19** (1891) 39–43, 88–89.

The Large Refractors of the World. *The Observatory*, **21** (1898) 239–271.

S. Binnewies, W. Steinicke, J. Moser. *Sternwarten – 95 astronomische Observatorien in aller Welt* (Erlangen: Oculum Verlag 2008).

S. Brunier, A.-M. Lagrange. *Great Observatories of the World* (Richmond Hill: Firefly Books 2005).

G. Chambers. *Astronomy, Vol. 2. Instruments and Practical Astronomy.* 4th edn. (Oxford: Clarendon Press 1890).

T. N. Clarke, A. D. Morrison-Low, A. D. C. Simpson. *Brass and Glass. Scientific Instrument Making Workshops in Scotland as Illustrated by Instruments from the Fink Collection of the Royal Museum of Scotland* (Edinburgh: National Museums of Scotland 1989).

J. Classen. Sternwarten vor 100 Jahren. *Die Sterne*, **47** (1971) 29–31, 85–89, 147–150, **48** (1972) 177–181.

I. Glass. *Victorian Telescope Makers. The Lives and Letters of Thomas and Howard Grubb* (Bristol: Institute of Physics Publishing 1997).

B. Gould. Latitudes and longitudes of the principal observatories. *Astronomical Journal*, **3** (1853) 17–23.

H. Grubb. *An Illustrated Catalogue of Astronomical Instruments, Observatories etc.* (Dublin: Astronomical Instrument Works 1903).

J. Hamel, I. Keil (eds.). *Der Meister und die Fernrohre. Das Wechselspiel zwischen Astronomie und Optik in der Geschichte.* Acta Historiae Astronomiae, vol. 33 (Frankfurt: Harri Deutsch 2007).

E. S. Holden. *Mountain Observatories in America and Europe* (Washington: Smithonian Institution 1896).

H. Hollis. The large telescopes. *The Observatory*, **37** (1914) 245–252.

D. Howse. The Greenwich list of observatories. A world list of astronomical observatories, instruments and clocks, 1670–1850. *Journal for the History of Astronomy*, **17** (1986) part 4, **25** (1994) 207–218.

R. Hutchins. *British University Observatories 1772–1939* (Farnham: Ashgate Publishing 2008).

G. Jahn. *Geschichte der Astronomie*, vol. 2 (Leipzig: Verlag Heinrich Hunger 1844).

K. King. *The History of the Telescope* (New York: Dover Publications 1979).

R. Learner. *Astronomy through the Telescope* (London: Evans Brothers 1982).

E. Loomis. Astronomical observatories in the United States. *Harpers New Monthly Magazine*, **13** (1856) 25–52, **48** (1874) 526–518, **49** (1875) 518–531.

S. Marx, S. Pfau. *Sternwarten der Welt* (Freiburg: Herder 1980).

A. McConnell. *Instrument Makers to the World. A History of Cooke, Troughton & Simms* (York: William Sessions 1992).

A. Mee. *The Story of the Telescope* (Cardiff: Lianishen 1909).

P. Moore. *Eyes on the Universe. The Story of the Telescope* (London: Springer 1997).

P. Müller. *Sternwarten. Architektur und Geschichte der astronomischen Observatorien* (Bern: Peter Lang 1978).

Sternwarten in Bildern (Berlin: Springer Verlag 1992).

D. Northrop. *A Selection of Astronomical Observatories Owned by Amateur Astronomers from the End of the 18th Century to the Present Day* (Kent: Iredale Press 1997).

J. Repsold. *Zur Geschichte der astronomischen Messwerkzeuge*, 2 vols. (Leipzig: Engelmann 1908).

R. Riekher. *Fernrohre und ihre Meister* (Leipzig: VEB Verlag Technik 1990).

A. Schweiger-Lerchenfeld. *Atlas der Himmelskunde auf Grundlage der coelestischen Photographie* (Vienna: Hartleben 1898).

F. Watson. *Stargazer. The Life and Times of the Telescope* (Cambridge: Da Capo Press 2005).

V. Witt. *Astronomische Reiseziele für unterwegs. Sternwarten, Museen und Schauplätze der Astronomie* (Heidelberg: Spektrum Verlag 2004).

References

Altogether 1628 references are given. The publications mostly originate from astronomical journals. They are referred by author/editor and year. If there is more than one article by an author, they are distiguished by an appended letter a, b, c etc. For authors with identical names the first letter of their forename is added to citations in the text (e.g. Wolf C. 1880a). 'Anon' refers to anonymous articles. No distinction among primary and secondary literature and obituaries has been made for citations in the text; otherwise they would have had to be marked differently in the text in order to present them in different lists (due to the large number of references, this would have been an impossible task). Articles from the *Astronomische Nachrichten* (*AN*) are occasionally referred by their message number (e.g. *AN* 1500). It should be noticed that *AN* counts columns instead of pages; therefore, first and last columns are given here (if an article covers only one page, only the starting column is given). Obituaries and other biographical sources are mentioned in the footnotes. Secondary literature can be identified by the year, which normally is later than 1900. A primary source is the RAS archive; citations follow their standard (Bennett 1978). General literature about observatories and telescopes can be found in the section 'Telescope data'. The following abbreviations are used for frequently appearing journals:

A&G	*Astronomy & Geophysics* (RAS publication)
AJ	*Astronomical Journal*
AN	*Astronomische Nachrichten* (German; message number within [ ])
AHO	*Annals of the Harvard College Observatory* (*Harvard Annals*)
AOP	*Annales de l'Observatoire de Paris* (French; *Paris Annales*)
ApJ	*Astrophysical Journal*
AR	*Astronomical Register*
Cop	*Copernicus*
CR	*Comptes Rendus Hebdomadaires des Séances de l'Académie des Sciences* (French; *Comptes Rendus*)
DSO	*Deep-Sky Observer*
IAJ	*Irish Astronomical Journal*
IS	*Interstellarum* (German)
JBAA	*Journal of the British Astronomical Association*
JHA	*Journal for the History of Astronomy*
JRASC	*Journal of the Royal Astronomical Society of Canada*
LASP	*Leaflets of the Astronomical Society of the Pacific*
MRAS	*Memoirs of the Royal Astronomical Society* (*Memoirs*)
MN	*Monthly Notices of the Royal Astronomical Society* (*Monthly Notices*)
Obs	*The Observatory*
PA	*Popular Astronomy*
PASP	*Publications of the Astronomical Society of the Pacific*
PLO	*Publications of the Lick Observatory*
PRS	*Proceedings of the Royal Society* (*Proc. Roy. Soc.*)
PT	*Philosophical Transactions of the Royal Society* (*Philosophical Transactions*)
QJRAS	*Quarterly Journal of the Royal Astronomical Society*
RBA	*Reports of the British Association for the Advancement of Science*
S&T	*Sky & Telescope*
SM	*Sidereal Messenger*
SuW	*Sterne und Weltraum* (German)
VA	*Vistas in Astronomy*
VdSJ	*Journal für Astronomie der Vereinigung der Sternfreunde* (German)
VJS	*Vierteljahrsschrift der Astronomischen Gesellschaft* (German; *Vierteljahrsschrift*)

Abbe, C. (1867). On the distribution of nebulae in space. *MN*, **27**, 257–264.

(1875). The extended nebulae of Sir J. Herschel's catalogue. *MN*, **35**, 236.

(1906a). Samuel Pierpont Langley. *AN*, [4086] **171**, 91–96.

(1906b). Samuel Pierpont Langley. *ApJ*, **23**, 271–283.

Abbott, F. (1862). On the cluster κ Crucis. *MN*, **23**, 32–33.

(1863). Note on Eta Argûs. *MN*, **24**, 2–6.

(1868). On the variability of Eta Argûs and surrounding nebula. *MN*, **28**, 200–202.

(1871). Reply to the notes and queries made by the Astronomer Royal on the 'observations of Eta Argûs and its nebula'. *MN*, **32**, 61–62.

Abietti, G. (1960). Father Angelo Secchi – a noble pioneer in astrophysics. *LASP*, **8**, 135–142.

Abney, W. (1894). Gold medal to Prof. S. W. Burnham. *MN*, **54**, 277–283.

Adams, J. C. (1853). Gold Medal to John Russell Hind. *MN*, **13**, 93.

(1875). Address delivered by the President, Prof. Adams, on presenting the gold medal of the Society to Professor Heinrich d'Arrest. *MN*, **35**, 265–276.

Airy, G. B. (1836). Honorary medal to Sir J. F. W. Herschel. *MRAS*, **9**, 303–312.

(1845). Honorary medal to William Henry Smyth. *MN*, **6**, 193–199.

(1849). Substance of a lecture delivered by the Astronomer Royal on the large reflecting telescopes of the Earl of Rosse and Mr. Lassell. *MN*, **9**, 110–121.

(1851). Report of the Council to the thirty-first Annual General Meeting. *MN*, **11**, 104.

(1871). *MN*, **31**, 233–234.

Aitken, R. (1899). Bruce medal to Arthur Auwers. *PASP*, **11**, 61–70.

(1921). Sherburne Wesley Burnham, 1838–1921. *PASP*, **33**, 85–90.

(1923). Edward Emerson Barnard, 1857–1923. *PASP*, **35**, 87–94.

Alexander, A. F. (1971). Dreyer, Johann Louis Emil. *Dictionary of Scientific Biography*, **4**, 185–186.

Alexander, S. (1852). On the origins of the forms and the present conditions of some of the clusters of stars, and several of the nebulae. *AJ*, **2**, 95–96, 97–103, 105–111, 113–115, 126–128, 140–142, 148–152, 158–160.

Ames, A. (1930). A catalogue of 2778 nebulae including the Coma–Virgo Group. *AHO*, **88**, 1–40.

Ångström, A. (1915). Nils Christofer Dunér. *ApJ*, **41**, 81–85.

Anon (1835). Karl Ludwig Harding. *MN*, **3**, 86.

(1836). Mr. John Ramage. *MN*, **4**, 37–38.

(1844a). Niccolò Cacciatore. *MN*, **6**, 29–31.

(1844b). Francis Baily. *MN*, **6**, 90–128.

(1845). Ebenezer Porter Mason. *The New Englander*, **3** (no. 11), 313–324.

(1849). Francis de Vico. *MN*, **9**, 65–66.

(1856). D'Arrest. Resultate aus Beobachtungen der Nebelflecken und Sternhaufen. *MN*, **17**, 48.

(1860). Sir Thomas Macdougall Brisbane. *PRS*, **11**, iii–vii.

(1862a). George Bishop. *MN*, **22**, 104–106.

(1862b). George Bishop. *PRS*, **12**, iii.

(1863a). Edward Joshua Cooper. *PRS*, **13**, i–iii.

(1863b). William Herschel's Verzeichnisse von Nebelflecken und Sternhaufen. *MN*, **24**, 49–47.

(1864). The Hartwell Observatory. *AR*, **2**, 230–233.

(1865a). The Hartwell Observatory. *AR*, **3**, 35–37.

(1865b). Admiral Smyth. *AR*, **3**, 241–243.

(1865c). George Phillips Bond. *Proceedings of the American Academy of Arts and Science*, **6**, 499–500.

(1866a). William Henry Smyth. *PRS*, **15**, xliii–xliv.

(1866b). Returns of private observatories. *AR*, **4**, 91.

(1868). Siderum Nebulosorum Observationes Havnienses. *VJS*, **3**, 94–99.

(1870a). The great Newall telescope. *MN*, **30**, 112–114.

(1870b). Mr. Buckingham's telescope. *MN*, **30**, 114–115.

(1872a). The nebula in the Pleiades. *AR*, **10**, 186.

(1872b). The nebula in the Pleiades. *AR*, **10**, 264.

(1873a). Caspar Gottfried Schweizer. *VJS*, **8**, 163–169.

(1873b). Todes-Anzeige. *AN*, [1951] **82**, 97.

(1873c). The nebula in the Pleiades. *AR*, **11**, 181.

(1876a). Heinrich Ludwig d'Arrest. *MN*, **36**, 155–158.

(1876b). The Pleiades group. *MN*, **36**, 196–197.

(1876c). *Proceedings of the American Academy of Arts and Science*, **11**, 322.

(1878). The literature of nebulae and clusters. *MN*, **38**, 213–214.

(1879a). Angelo Secchi. *MN*, **39**, 238–241.

(1879b). Das Observatorium des Collegio Romano. *Sirius*, **12**, 165–166.

(1880a). Johann von Lamont. *Sirius*, **13**, 191–196.

(1880b). The old Paramatta Observatory. *Obs*, **3**, 614–616.

(1880c). Henry Cooper Key. *MN*, **40**, 199–200.

(1880d). Die Gruppe der Plejaden. *Sirius*, **13**, 150.

(1880e). William Lassell. *Sirius*, **13**, 245–246.

(1880f). The variable nebula near Zeta Tauri. *Nature*, **22**, 231.

(1881a). Carl Christian Bruhns. *Cop*, **1**, 165–166.

(1881b). Carl Christian Bruhns. *AN*, [2385] 100, 129.
(1881c). Karl Bruhns. *Sirius*, **14**, 190.
(1882). Carl Christian Bruhns. *MN*, **42**, 147–148.
(1883a). Charles E. Burton. *MN*, **43**, 159–160.
(1883b). The celestial charts of Prof. C. H. F. Peters. *MN*, **43**, 224–226.
(1883c). Professor Holden's monograph on the nebula of Orion. *MN*, **43**, 227–229.
(1883d). New nebulae. *Obs*, **6**, 162.
(1884). Nebulae. *Obs*, **7**, 340.
(1886). RAS meeting (November 12, 1886). *AR*, **24**, 297–302.
(1887a). The late Joseph Baxendell, F.R.S. *Obs*, **10**, 399–400.
(1887b). J. L. E. Dreyer. Ueber einige Nebelflecke, die bisher für veränderlich oder mit Eigenbewegung behaftet gehalten werden. *Naturwissenschaftliche Rundschau*, **2**, 282.
(1888a). Joseph Baxendell. *MN*, **48**, 157–160.
(1888b). Joseph Baxendell. *PRS*, **43**, iv–vi.
(1889a). Wilhelm Tempel. *Sirius*, **17**, 97–98.
(1889b). James Wigglesworth. *MN*, **49**, 169.
(1891a). Edward Schönfeld. *PASP*, **3**, 255.
(1891b). Herman Schultz. *MN*, **51**, 205–207.
(1891c). Christian Heinrich Friedrich Peters. *Himmel und Erde*, **3**, 173–175.
(1891d). Death of Dr. C. H. F. Peters. *Obs*, **13**, 311–312.
(1895). George Knott. *MN*, **55**, 195–197.
(1896). John Russell Hind. *MN*, **56**, 200–205.
(1899). Nebulosities exterior to the Pleiades. *PA*, **7**, 109–110.
(1900). Dr. G. F. W. Rümker. *Obs*, **23**, 183–184.
(1903). Prosper Henry. *Obs*, **26**, 396–397.
(1904). Gaspare Stanislao Ferrari. *Atti della Pontificia Accademia Romana dei Nuovi Lincei*, **57**, 61–67.
(1905a). Paul Henry. *Obs*, **28**, 110.
(1905b). Paul Henry. *MN*, **65**, 349.
(1905c). Ralph Copeland. *Obs*, **28**, 472–473.
(1905d). The late Astronomer-Royal for Scotland. *PASP*, **17**, 204–206.
(1906). Otto Wilhelm von Struve. *MN*, **66**, 179.
(1907). Asaph Hall. *PASP*, **19**, 264.
(1908). The Earl of Rosse. *PASP*, **20**, 272.
(1909). Andrew Graham. *JBAA*, **19**, 33.
(1910). George Washington Hough. *MN*, **70**, 302–304.
(1912). George Johnstone Stoney. *MN*, **72**, 253–255.
(1913a). Death of Sir Robert Ball. *JRASC*, **7**, 474–475.
(1913b). Lewis Swift. *PASP*, **25**, 51.
(1914a). Sir Robert Ball. *PASP*, **26**, 63–64.
(1914b). James Ludovic Lindsay. *MN*, **74**, 271–273.
(1918a). George Mary Searle. *Nature*, **101**, 430.
(1918b). A. N. Skinner. *Obs*, **41**, 453.
(1919a). Jérôme Coggia. *Journal des Observateurs*, **2**, 142.
(1919b). Jérôme Coggia. *Obs*, **42**, 138.
(1921). From an Oxford note-book. *Obs*, **44**, 163–164.
(1925). Death of Johann Palisa. *PASP*, **37**, 174.
(1926a). Dr. J. L. E. Dreyer. *Obs*, **49**, 293–294.
(1926b). Dr. J. L. E. Dreyer. *PASP*, **38**, 400–401.
(1928). Professor Frances P. Leavenworth. *PASP*, **40**, 413–414.
(1929). Francis P. Leavenworth. *MN*, **89**, 312–313.
(1933). Guillaume Bigourdan. *PASP*, **44**, 133.
(1939a). David Todd. *JRASC*, **33**, 262.
(1939b). Dr. David Todd. *PASP*, **51**, 247.
(1955). Barnard's automatic comet seeker. *LASP*, **7**, 81–87.
Arago, F. (1842). *Annuaire du Bureau des Longitudes 1842*, 409–411.
Archenhold, F. (1891). Ein ausgedehnter Nebel bei ξ Persei. *AN*, [3082] **129**, 153–158.
Archinal, B. (1993). *The 'Non-existent' Star Clusters of the RNGC*. Webb Society Monograph no. 1, Portsmouth.
Archinal, B., Hynes, S. (2003). *Star Clusters*. Richmond, VA: Willmann-Bell.
Argelander, F. (1856). Aus einem Schreiben des Herrn Professors Argelander an den Herausgeber. *AN*, [1045] **44**, 203.
(1868a). Ueber die Instrumente, deren sich Messier bei seinen Cometenbeobachtungen bedient hat. *VJS*, **3**, 10–28.
(1868b). Schreiben des Herrn Prof. Dr. Argelander, Director der Sternwarte in Bonn, an den Herausgeber. *AN*, [1698] **71**, 287.
Ashbrook, J. (1976). Truman Henry Safford. *S&T*, **52** (no. 5), 346.
(1977). Julius Schmidt and his book about lunar eclipses. *S&T*, **53** (no. 3), 173–174.
(1979). E. E. Barnard and the globular clusters. *S&T*, **58** (no. 6), 523–524.
(1984). *The Astronomical Scrapbook. Skywatchers, Pioneers, and Seekers in Astronomy*. Cambridge, MA: Sky Publishing Corporation.
Ashworth, W. (2003). Faujas-de-Saint-Fond visits the Herschels at Datchet. *JHA*, **34**, 321–324.
Auwers, A. (1862a). William Herschel's Verzeichnisse von Nebelflecken und Sternhaufen. *Astronomische Beobachtungen auf der Königlichen Universitäts-Sternwarte zu Königsberg*, **34**, 155–229.

(1862b). Aus einem Schreiben des Herrn Dr. Auwers. *AN*, [1391] 58, 361–364.
(1862c). Verzeichnis der Örter von vierzig Nebelflecken, aus Beobachtungen am Königsberger Heliometer abgeleitet. *AN*, [1392] 58, 369–378.
(1862d). Extract of a letter from Mr. A. Auwers to the Rev. R. Main. *MN*, 22, 148–150.
(1863). Aus einem Schreiben des Herrn Dr. Auwers an den Herausgeber. *AN*, [1409] 59, 271.
(1865). Beobachtung von Nebelflecken. *Astronomische Beobachtungen auf der Königlichen Universitäts-Sternwarte zu Königsberg*, 35, 193–239.
(1866). John Herschel, Catalogue of nebulae and clusters of stars. *VJS*, 1, 176–184.
(1898). August Friedrich Theodor Winnecke. *AN*, [3467] 145, 161–166.
Baade, W. (1928). Der Nebel NGC II 1613. *AN*, [5612] 234, 407.
(1942). The Crab Nebula. *ApJ*, 96, 188–198.
Backhouse, T. (1863). Lord Rosse on the nebulae. *AR*, 1, 33–35, 49–51.
(1882). On the nebula near Merope. *MN*, 42, 358–359.
(1892). Boeddicker's drawings of the Milky Way. *Obs*, 15, 193–194.
Bagdasarian, N. (1985). Sir James South and the Cauchoix objective. *S&T*, 70 (no. 6), 525–526.
Bailey, M., Butler, J., McFarland, J. (2005). Unwinding the discovery of spiral nebulae. *A&G*, 46, 2.26–2.28.
Bailey, S. I. (1908). A catalogue of bright clusters and nebulae. *AHO*, 60, 199–229.
(1919). Edward Charles Pickering, 1846–1919. *ApJ*, 50, 233–244.
(1931). *The History and Work of the Harvard Observatory, 1839 to 1927*. New York, NY: McGraw-Hill.
Ball, R. (1911). George Johnstone Stoney. *Obs*, 34, 287–290.
Ball, R. S. (1886). *The Story of the Heavens*. London: Cassell.
(1895). *Great Astronomers*. London: Isaac Pitman.
(1898). RAS meeting (November 11, 1898). *Obs*, 21, 430–433.
Ball, V. ed. (1915). *Reminiscences and Letters of Sir Robert Ball*. London: Cassell.
ed. (1940). Extracts from the diary of Sir Robert Ball. *Obs*, 63, 199–206.
Barnard, E. E. (1883a). *SM*, 1, 33–35.
(1883b). A new nebula near Phi Virginis. *AN*, [2490] 104, 285.
(1883c). *SM*, 1, 135.
(1883d). *SM*, 1, 168.
(1883e). *SM*, 1, 238.
(1884a). New nebulae – small black hole in the Milky-Way. *AN*, [2588] 108, 369–372.
(1884b). A new faint nebula. *Obs*, 7, 269.
(1884c). New nebula. *SM*, 2, 226.
(1884d). *SM*, 2, 259.
(1884e). *SM*, 2, 287.
(1884f). *SM*, 3, 91.
(1884g). A new and faint nebula. *SM*, 3, 184.
(1884h). Erroneous description of a nebula. *SM*, 3, 189.
(1884i). New nebula. *SM*, 3, 254.
(1884k). *SM*, 3, 60.
(1885a). New nebula near General Catalogue no. 4510. *AN*, [2624] 110, 125.
(1885b). Small nebula near great Andromeda Nebula. *AN*, [2687] 112, 391–392.
(1885c). The new star in the nebula of Andromeda. *SM*, 4, 240–243.
(1885d). The great nebula in Monoceros. *SM*, 4, 313–314.
(1885e). New nebulae. *Obs*, 8, 123.
(1885f). Nebulae. *Obs*, 8, 210.
(1885g). *Obs*, 8, 386.
(1885h). Ueber den neuen Stern im grossen Andromeda-Nebel. *AN*, [2688] 112, 403.
(1886a). A small faint Nebula near the n. f. end of the great nebula of Andromeda. *AN*, [2690] 113, 31.
(1886b). Berichtigung zu 'Small nebula near great Andromeda Nebula' in *A.N.* 2687. *AN*, [2691] 113, 47.
(1886c). A correction to Dr. Swift's list of new nebulae in *A.N.* 2746. *AN*, [2755] 115, 315–316.
(1886d). Ring-micrometer observations of comets and nebulae. *AN*, [2756] 115, 323–328.
(1886e). New nebulae. *SM*, 5, 25.
(1886f). *SM*, 5, 27.
(1886g). Large nebula in field with General Catalogue 4510. *SM*, 5, 31.
(1886h). New nebulous star. *SM*, 5, 154.
(1886i). On the nebula 4036 of Herschel's General Catalogue. *SM*, 5, 255.
(1886k). General Catalogue no. 4594. *SM*, 5, 286.
(1889a). The cluster G.C. 1420 and the nebula N.G.C. 2237. *AN*, [2918] 122, 253.
(1889b). The nebula G.C. 2091. *MN*, 49, 418–419.
(1890). Note on the nebula G.C. 4036 = N.G.C. 5834. *AN*, [2995] 125, 315–318.
(1891a). On the nebulosities of the Pleiades and on a new Merope nebula. *AN*, [3018] 126, 293.
(1891b). The new Merope nebula. *AN*, [3032] 127, 135.
(1892a). Two probably variable nebulae. *AN*, [3097] 130, 7.

(1892b). A new nebulous star, and corrections to Dreyer's NGC. *AN*, [3101] **130**, 77–80.

(1892c). Nova Auriga a nebula. *AN*, [3118] **130**, 407.

(1892d). The great nebulous areas of the sky. *Knowledge*, **15**, 14–16.

(1894a). On a small nebula close to M. 57. *AN*, [3200] **134**, 129.

(1894b). On the exterior nebulosities of the Pleiades. *AN*, [3253] **136**, 193.

(1894c). Photographic nebulosities and star clusters connected with the Milky Way. *Astronomy and Astro-Physics*, **13**, 177–182.

(1894d). Photograph of Swift's nebula in Monoceros N.G.C. 2237. *Astronomy and Astro-Physics*, **13**, 642–644.

(1894e). The great photographic nebula of Orion, encircling the belt and Theta nebula. *Astronomy and Astro-Physics*, **13**, 811–814 (reprint in *PA*, **2**, 151–154).

(1895a). On the exterior nebulosities of the Pleiades. *Knowledge*, **18**, 282.

(1895b). On the variable nebulae of Hind, J. R. (N.G.C. 1555) and Struve (N.G.C. 1554) in Taurus, and the nebulous condition of the variable star T Tauri. *MN*, **55**, 442–452.

(1895c). Invisibility of Hind's Variable Nebula (N.G.C. 1555). *MN*, **56**, 66–67.

(1895d). Note on the nebula Dreyer N.G.C. no. 532. *AJ*, **15**, 24.

(1895e). Photograph of the nebula N.G.C. 1499 near the star ξ Persei. *ApJ*, **2**, 350.

(1896a). On the new Merope nebula. *AN*, [3315] **139**, 41.

(1896b). On the comparison of reflector and portrait lens photographs. *MN*, **57**, 10–16.

(1898). The exterior nebulosities of the Pleiades. *Obs*, **21**, 351.

(1899). Observations of Hind's Variable Nebula in Taurus (N.G.C. 1555), made with the 40-inch refractor of the Yerkes Observatory. *MN*, **59**, 372–376.

(1900a). On the proper motion of the annular nebula in Lyra (M 57) and the peculiarities in the focus for the planetary nebulae and their nuclei. *MN*, **60**, 245–257.

(1900b). The exterior nebulosities of the Pleiades, with a drawing from the different photographs and on the appearance of the involved nebulosities of the cluster with the 40-inch refractor. *MN*, **60**, 258–261.

(1902). Micrometrical and visual observations for Nova Cygni (1876) made with the 40-inch refractor of the Yerkes Observatory. *MN*, **62**, 405–419.

(1903). Diffuse nebulosities in the heavens. *ApJ*, **17**, 77–80.

(1906a). On a planetary nebula. *AN*, [4112] **172**, 123.

(1906b). The annular nebula in Lyra (M 57). *MN*, **66**, 104–113.

(1907a). Groups of small nebulae. *AN*, [4136] **173**, 117–121.

(1907b). The nebula NGC. 6302. *AN*, [4136] **173**, 123.

(1907c). The 'owl' nebula Messier 97 = NGC 3587. *MN*, **67**, 543–550 (plus Pl. 3).

(1908a). Some notes on nebulae and nebulosities. *AN*, [4239] **177**, 231–236.

(1908b). The variability of the nucleus of the planetary nebula NGC 7662. *MN*, **68**, 465–480.

(1913a). Dark regions in the sky suggesting an obscuration of light. *ApJ*, **38**, 496–501.

(1913b). Lewis Swift. *AN*, [4639] **194**, 133–136.

(1913c). Photographs of the Milky Way and of comets made with the six-inch Willard lens and Crocker telescope during the years 1892 to 1895. *PLO*, **11**.

(1915). A great nebulous region near Omicron Persei. *ApJ*, **41**, 253–258.

(1916a). A small star with large proper motion. *AJ*, **29**, 181–183.

(1916b). Some of the dark markings on the sky and what they suggest. *ApJ*, **43**, 1–8.

(1916c). Nova Persei. *AJ*, **30**, 68.

(1919). On the dark markings of the sky with a catalogue of 182 such objects. *ApJ*, **49**, 1–23.

(1921). Sherburne Wesley Burnham. *PA*, **29**, 309–324.

(1927). *A Photographic Atlas of Selected Regions of the Milky Way*. Washington D.C.: Carnegie Institution.

Bates, R., McKelvey, B. (1947). Lewis Swift. The Rochester Astronomer. *Rochester History*, **9**, 1–20.

Batten, A. (1977). The Struves of Pulkowo – a family of astronomers. *JRASC*, **71**, 345–372.

Batten, A. H. (1988). *Resolute and Undertaking Characters: The Lives of Wilhelm and Otto Struve*. Dordrecht: Reidel Publishing.

Baum, R., Sheehan, W. (1997). *In Search of Planet Vulcan*. New York, NY: Plenum Trade.

Bauschinger, J. (1889). J. L. E. Dreyer, a new General Catalogue. *VJS*, **24**, 43–51.

Baxendell, J. (1863). Variable stars. *AR*, **1**, 39–40.

(1880). A new nebula. *MN*, **41**, 48.

Beal, W. (1929). Francis Preserved Leavenworth. *PA*, **37**, 117–119.

Becker, E. (1909a). Beobachtung von Nebelflecken, ausgeführt in den Jahren 1875–1880 am 6-zölligen Refraktor der provisorischen Sternwarte von A. Winnecke, bearbeitet von E. Becker. *Annalen der Kaiserlichen Universitäts-Sternwarte Strassburg*, **3** (part 1).

(1909b). Beobachtung von Nebelflecken, ausgeführt in den Jahren 1880–1902 am 18-zölligen Refraktor der neuen Sternwarte von H. Kobold, A. Winnecke und W. Schur, bearbeitet von C. W. Wirtz. *Annalen der Kaiserlichen Universitäts-Sternwarte Strassburg*, 3 (part 2).

Becker, F. (1928). A preparatory catalogue for a Durchmusterung of nebulae. The general catalogue. *Specola Astronomica Vaticana*, **6B**.

Becker, F., Meurers, J. (1956). The problem of Hagen's clouds. *VA*, **2**, 1069–1073.

Becker, L. (1902). Observations of 217 nebulae made with the transit circle at Dun Echt Observatory. *Annals of the Edinburgh Observatory*, **1**, 1–46.

Beekman, G. (1983a). Der Astronom, der fliegen wollte. *SuW*, **22**, 18–19.

(1983b). The long thread of Danish astronomy. *S&T*, **65** (no. 6), 487–491.

Beesley, D. E. (1985). Isaac Ward and S Andromedae. *IAJ*, **17**, 98–102.

Bennett, J. (1976a). 'On the power of penetrating into space'. The telescopes of William Herschel. *JHA*, **7**, 75–108.

(1976b). William Herschel's Large Twenty-Foot Telescope. *QJRAS*, **17**, 303–305.

(1978). Catalogue of the archives and manuscripts of the Royal Astronomical Society. *MRAS*, **85**, 1–90.

(1981). A surviving flat from William Lassell's four-foot equatorial Newtonian telescope. *JHA*, **12**, 195–197.

(1990). *Church, State and Astronomy in Ireland. 200 Years of Armagh Observatory*. Armagh: Armagh Observatory.

Bennett, J., Hoskin, M. (1981). The Rosse papers and instruments. *JHA*, **12**, 216–229.

Bergstrand, Ö. (1914). N. C. Dunér. *AN*, [4775] **199**, 391.

Bertout, C. (1980). T Tauri – ein Portrait. Erster Teil. T Tauri und seine Umgebung. *SuW*, **19**, 205–209.

(1981). T Tauri – ein Portrait. Zweiter Teil. Physikalische Eigenschaften. *SuW*, **20**, 164–168.

Bessel, F. W. (1823). *Astronomische Beobachtungen auf der Königlichen Universitäts-Sternwarte zu Königsberg*, **8**, 107–108.

(1834). Bericht über den Fortgang des Unternehmens der akademischen Sternkarten. *AN*, [243] **11**, 33–40.

(1835). *Astronomische Beobachtungen auf der Königlichen Universitäts-Sternwarte zu Königsberg*, **17**, 95–96.

Bhathal, R. (2009). 150 Years of Sydney Observatory. *A&G*, **50**, 3.26–3.30.

Bianchi, G. (1839). Schreiben des Herrn Bianchi, Director der Sternwarte zu Modena, an den Herausgeber. *AN*, [383] **16**, 371–376.

Bianchi, S., Gasperini, A., Galli, D., Palla, F. (2010). Wilhelm Tempel and his private telescope. *Journal of Astronomical History and Heritage* in press.

Bidschof, F. (1898). Catalog der auf der k. k. Sternwarte zu Wien beobachteten Nebelflecken. *AN*, [3520] **147**, 257–264.

(1899). Beobachtungen an dem Equatoreal Coudé der k.k. Sternwarte zu Wien. *AN*, [3582] **150**, 81–94.

(1900). Nebelfleck-Beobachtungen. *Annalen der K.K. Universitäts-Sternwarte Wien*, **14**, 128–134.

Biela, W. (1827). Auszug aus einem Schreiben des Herrn Hauptmanns und Ritters v. Biela an den Herausgeber. *AN*, [120] **5**, 425–428.

Bigourdan, G. (1887a). Nébuleuses nouvelles, découvertes à l'Observatoire de Paris. *CR*, **105**, 926–929.

(1887b). Nébuleuses nouvelles, découvertes à l'Observatoire de Paris. *CR*, **105**, 1116–1119.

(1891a). Sur une nébuleuse variable. *CR*, **112**, 471–474.

(1891b). Nébuleuses nouvelles, découvertes à l'Observatoire de Paris. *CR*, **112**, 647–650.

(1892a). Observations de nébuleuses et d'amas stellaires (introduction). *AOP*, **36**, G1–G15.

(1892b). Observations de nébuleuses et d'amas stellaires (XV heures). *AOP*, **36**, G6–G103.

(1896). Observations de nébuleuses et d'amas stellaires (XXIII heures). *AOP*, **40**, D9–D107.

(1898a). Observations de nébuleuses et d'amas stellaires (XVI heures). *AOP*, **42**, D49–D151.

(1898b). Observations de nébuleuses et d'amas stellaires (XVII heures). *AOP*, **42**, D153–D241.

(1899a). Observations de nébuleuses et d'amas stellaires (XX heures). *AOP*, **46**, G1–G53.

(1899b). Observations de nébuleuses et d'amas stellaires (XXI heures). *AOP*, **46**, G55–G115.

(1902a). Observations de nébuleuses et d'amas stellaires (XXII heures). *AOP*, **47**, G1–G99.

(1902b). Observations de nébuleuses et d'amas stellaires (V heures). *AOP*, **47**, G101–G131.

(1904a). Observations de nébuleuses et d'amas stellaires (XIV heures). *AOP*, **48**, F1–F139.

(1904b). Observations de nébuleuses et d'amas stellaires (VI heures). *AOP*, **48**, F141–F187.

(1904c). Observations de nébuleuses et d'amas stellaires (0 heures). *AOP*, **49**, F1–F141.

(1904d). Observations de nébuleuses et d'amas stellaires (III heures). *AOP*, **49**, F143–F207.

(1905a). Observations de nébuleuses et d'amas stellaires (IV heures). *AOP*, **50**, F1–G59.

(1905b). Observations de nébuleuses et d'amas stellaires (II heures). *AOP*, **50**, F61–F153.

(1906). Observations de nébuleuses et d'amas stellaires (I heures). *AOP*, **51**, F1–F137.

(1907a). Observations de nébuleuses et d'amas stellaires (XVIII heures). *AOP*, **43**, E1–F73.

(1907b). Observations de nébuleuses et d'amas stellaires (XIX heures). *AOP*, **43**, E75–E111.

(1907c). Observations de nébuleuses et d'amas stellaires (VII heures). *AOP*, **52**, F1–F75.

(1908). Observations de nébuleuses et d'amas stellaires (VIII heures). *AOP*, **53**, F1–F89.

(1910). Observations de nébuleuses et d'amas stellaires (IX heures). *AOP*, **44**, D1–D119.

(1911a). Observations de nébuleuses et d'amas stellaires (X heures). *AOP*, **45**, E1–E117.

(1911b). Observations de nébuleuses et d'amas stellaires (XI heures). *AOP*, **54**, E1–E173.

(1912). Observations de nébuleuses et d'amas stellaires (XII heures). *AOP*, **55**, E1–E243.

(1916). La découverte de la Nébuleuse d'Orion (N.G.C. 1976) par Peiresc. *CR*, **162**, 489–490.

(1917a). Observations de nébuleuses et d'amas stellaires (XIII heures). *AOP*, **56**, E1–E119.

(1917b). Observations de nébuleuses et d'amas stellaires (introduction). *AOP*, **56**, E121–E194.

(1917c). Observations de nébuleuses et d'amas stellaires (appendices). *AOP*, **56**, E265–E422.

(1917d). Observations de nébuleuses et d'amas stellaires (esures complémentaires). *AOP*, **56**, E357–E397.

Binnewies, S., Steinicke, W., Moser, J. (2008). *Sternwarten – 95 astronomische Observatorien in aller Welt*. Erlangen: Oculum.

Birmingham, J. (1879). Schmidt's lunar maps. *Obs*, **3**, 10–17.

Bishop, G. (1852). *Astronomical Observations made at the Observatory, South Villa, Regent's Park, London, 1839–1851*. London: Taylor, Walton and Maberly.

Block, E. (1880). Schreiben des Herrn E. Block an den Herausgeber. *AN*, [2287] **96**, 110–112.

Bode, J. E. (1777). Ueber einige neuentdeckte Nebelsterne und einem vollständigen Verzeichnisse der bisher bekannten. *Berliner Jahrbuch für 1779*, **65**–71.

(1782). *Vorstellung der Gestirne auf XXXIV Tafeln*. Berlin.

(1785a). Ein Sternring oder ein Nebelfleck mit einer Öffnung. *Berliner Jahrbuch für 1788*, 242.

(1785b). Planetenähnliche Nebelflecke. *Berliner Jahrbuch für 1788*, 242–245.

(1788). Verzeichnis von tausend neuen Nebelflecken und Sternhaufen, vom 7. Sept. 1782 bis 26. April 1785 entdeckt. *Berliner Jahrbuch für 1791*, 157–173.

(1791). Verzeichnis von tausend neuen Nebelflecken und Sternhaufen, vom 26. April 1785 bis 26. November 1788 entdeckt. *Berliner Jahrbuch für 1794*, 151–167.

(1794). Verzeichnis der vorzüglichen in dem astronomischen Salon des Herrn Erblandmarschal von Hahn zu Remplin befindlichen Instrumente. *Berliner Jahrbuch für 1797*, 240–244.

(1804a). Bemerkungen über den Bau des Himmels. *Berliner Jahrbuch für 1807*, 113–128.

(1804b). Verzeichnis von fünfhundert neuen Nebelflecken und Sternhaufen, vom 3. Dec. 1788 bis 26. Sept. 1802 entdeckt. *Berliner Jahrbuch für 1807*, 129–137.

(1806). Friedrich v. Hahn. *Berliner Jahrbuch für 1809*, 272.

(1808). *Berliner Jahrbuch für 1811*, 204–205.

(1813). Chronologisches Verzeichnis der berühmtesten Astronomen, seit dem dreizehnten Jahrhundert, ihrer Verdienste, Schriften und Entdeckungen. *Berliner Jahrbuch für 1816*, 92–124.

Boeddicker, O. (1886). Ueber die Änderungen der Wärmestrahlung des Mondes während einer totalen Sonnenfinsternis am 4. Oktober 1884. *Naturwissenschaftliche Rundschau*, **1**, 193–194.

(1889). Note to accompany a drawing of the Milky Way. *MN*, **50**, 12–15.

(1908). The Earl of Rosse, K.P., L.L.D., D.C.L., F.R.S. *Obs*, **31**, 374–376.

(1936). Description of chart of the Milky Way deposited with the Society. *MN*, **96**, 641–642.

Bond, G. P. (1848). An account of the nebula in Andromeda. *Memoirs of the American Academy of Arts and Sciences, New Series*, **3**, 75–86.

(1859). Letter from Mr. Bond to Mr. Carrington. *MN*, **19**, 224.

(1860). On the spiral structure of the great nebula of Orion. *MN*, **21**, 203–207.

(1862). Schreiben des Herrn Prof. Bond, Directors der Sternwarte in Cambridge, an den Herausgeber. *AN*, [1337] **56**, 269–272.

(1864a). List of new nebulae seen at the observatory of Harvard College. *AN*, [1453] **61**, 193–198.

(1864b). Aus einem Schreiben von Prof. G. P. Bond, Director der Sternwarte in Cambridge Mass., an den Herausgeber. *AN*, [1456] **61**, 255.

(1866). A list of new nebulae seen at the observatory of Harvard College, 1847–1863. *Proceedings of the American Academy of Arts and Sciences*, **6**, 177–182.

(1867). On the great nebula in Orion. *AHO*, **5**.

Bond, W. C. (1847). Schreiben des Herrn Bond an den Präsidenten Everett. *AN*, [611] **26**, 167–172.

(1848). Description of the nebula about the star Theta Orionis. *Memoirs of the American Academy of Arts and Sciences, New Series*, **3**, 87–96.

(1855). Zone Catalogue of 5500 stars situated between the equator and 0° 20′ north declination observed in the years 1852–53. *AHO*, **1** (part 2).

Borrelly, A. (1872a). Nébuleuses nouvelles, découvertes et observées par Alph. Borrelly, à l'Observatoire de Marseille à l'aide du chercheur Eichens. *AN*, [1885] **79**, 205–208.

(1872b). New nebulae discovered and observed by Alph. Borrelly at the Marseille Observatory with the Eichens searcher. *MN*, **32**, 248.

(1913). Changements observés dans la nébuleuse de Tuttle N.G.C. 6643, à l'Observatoire de Marseille, au chercheur de comètes. *CR*, **157**, 1377.

Bosler, J. (1924). Edouard Stephan. Directeur honoraire de l'Observatoire de Marseille (1837–1923). *Journal des Observateurs*, **7**, 9–10.

(1926). Alphonse Borrelly (1842–1926). *Journal des Observateurs*, **9**, 169–170.

Boss, B. (1923). Edward Emerson Barnard. *AJ*, **35**, 25–26.

Boulliau, I. (1667). Observations of the star, called nebulosa, in the girdle of Andromeda. *PT*, **2**, 459–460.

Brachner, A. (1988). Die Sternwarten Münchens (Teil 2). *SuW*, **27**, 80–85.

Brashear, J. (1906). Samuel Pierpont Langley. *PA*, **14**, 257–264.

Bredikhin, F. (1875). Observations des nébuleuses. *Annales de l'Observatoire de Moscou*, **3**, 113–134.

Breen, J. (1856). Observations of Circe. *AN*, [1024] **43**, 241–246.

Brenner. L. (1902). *Beobachtungsobjekte für Amateur-Astronomen*. Leipzig: Mayer.

Brewster, D. (1828). Notice on 'The third series of observations with the twenty-feet reflecting telescope, containing a catalogue of 384 new double and multiple stars, completing a first thousand of these objects detected in sweeps with that instrument. By J. F. W. Herschel'. *Edinburgh Journal of Science*, **9** (no. 17), 90–93.

Brorsen, T. (1851). Auszüge aus Briefen des Herrn Observators Th. Borsen an den Herausgeber. *AN*, [751] **32**, 105–110.

(1856). Entdeckung eines neuen Nebelflecks. *Unterhaltungen im Gebiete der Astronomie, Geographie und Meteorologie*, **10**, 292.

Brück, H. (1979). P. Angelo Secchi, S. J., 1818–1878. *IAJ*, **14**, 9–13.

(1992). Lord Crawford's observatory at Dun Echt 1872–1892. *VA*, **35**, 81–138.

Brück, M. (1988). The Piazzi Smyth collection of sketches, photographs and manuscripts at the Royal Observatory, Edinburgh. *VA*, **32**, 371–408.

(1999). A family of astronomers. The Breens of Armagh. *IAJ*, **26**, 121–127.

Bruhns, C. (1862). Über die neue Sternwarte in Leipzig. *AN*, [1342] **56**, 337–346.

(1872). *Atlas der Astronomie*. Leipzig: Brockhaus.

Brünnow, F. (1872). Dublin (Dunsink) Observatory. *MN*, **32**, 151–152.

Budde, K. (2006). *Sternwarte Mannheim. Die Geschichte der Mannheimer Sternwarte 1772–1880*. Mannheim: Verlag Regionalkultur.

Burke-Gaffney, M. (1968). Copeland and Lohse and the comet, 1882 II. *JRASC*, **62**, 49–51.

Burnham, R. (1966). *Burnham's Celestial Handbook*, 3 vols. New York, NY: Dover Publications.

Burnham, S. W. (1873). A third catalogue of 76 new double stars, discovered with a 6-inch Alven Clark-refractor. *MN*, **34**, 59–71.

(1879). Double star observations made in 1877–8 at Chicago with the 18.5-inch refractor of the Dearborn Observatory. *MRAS*, **44**, 169, 216.

(1880). Location of Lick Observatory on Mount Hamilton, California. *American Journal of Science*, **20**, 338–339.

(1881). The planetary nebula in Cygnus. *MN*, **41**, 409–410.

(1889). The trapezium of Orion. *MN*, **49**, 352–358.

(1890). Note on Hind's Variable Nebula in Taurus. *MN*, **51**, 94–95.

(1891). Measures of planetary nebulae with the 36-inch equatorial of the Lick Observatory. *MN*, **52**, 31–46.

(1892a). Measures of Barnard's new Merope nebula. *AN*, [3074] **129**, 17.

(1892b). Observations of nebulae with the 36-inch equatorial of the Lick Observatory. *MN*, **52**, 440–461.

(1893). E. E. Barnard's work at the Lick Observatory. *SM*, **1**, 441–447.

(1894a). Measures of planetary nebulae with the 36-inch equatorial of the Lick Observatory. *PLO*, **2**, 159–167.

(1894b). Observations of nebulae with the 36-inch equatorial of the Lick Observatory. *PLO*, **2**, 168–181.

(1894c). Early life of E. E. Barnard. *PA*, **1**, 193–195.

(1907). Double star measures. *AN*, [4209] **176**, 129–147.

Burrau, C. (1927). J. L. E. Dreyer. *PA*, **35**, 325–327.

Burton, E. (1872a). Refractors and reflectors compared. *AR*, **10**, 281–282.

(1872b). Reflectors versus refractors. *AR*, **11**, 289–290.

Buta, R., Corwin, H., Odewahn, S. (2007). *The de Vaucouleurs Atlas of Galaxies*. Cambridge: Cambridge University Press.

Butillon (1848). Sur une nébuleuse et une étoile qui paraissent devoir fixer l'attention des astronomes. *CR*, **27**, 112–113.

Butler, J., Hoskin, M. (1987). Archives of the Armagh Observatory, 1790–1916. *JHA*, **18**, 295–307.

Buttmann, G. (1961). *William Herschel. Leben und Werk*. Stuttgart: Wissenschaftliche Verlagsgesellschaft.

(1970). *The Shadow of the Telescope*. New York, NY: Charles Scribner's Sons.

Cacciatore, N. (1827). Neuer Nebelfleck. *AN*, [113] **5**, 281.

Callandreau, O. (1903a). Prosper Henry. *AN*, [3912] **163**, 381–384.

(1903b). Prosper Henry. *PA*, **11**, 558–560.

Campbell, W. W. (1914). Edward Singleton Holden, Sc.D., LL.D., LITT.D. *PASP*, **26**, 77–87.

(1916). Edward Singleton Holden, 1846–1914. *Biographical Memoirs of the National Academy of Science*, **8**, 347–372.

(1919). Edward Charles Pickering, 1846–1919. *PASP*, **31**, 73–76.

Cannon, A. (1919). Edward Charles Pickering, 1846–1919. *PA*, **27**, 177–182.

Capocci, E. (1827). Posizioni di alcune nebulose osservate nel Giugno del 1826 da Capocci. *AN*, [120] **5**, 427.

Carlson, D. (1940). Some corrections to Dreyer's catalogues of nebulae and clusters. *ApJ*, **91**, 350–359.

Carver, R. (2006). Thomas William Webb. *DSO*, **141**, 10–18.

Cederblad, S. (1946). Catalog of bright diffuse galactic nebulae. *Meddelanden Från Lunds Astronomiska Observatorium, Ser. II*, **119**, 1–166.

Cerulli, V. (1896). Erratum in Dreyer NGC. of nebulae. *AN*, [3315] **139**, 47.

Chacornac, J. (1855). On several stars which have disappeared from his ecliptical charts. *MN*, **15**, 199–202.

(1862a). Schreiben des Herrn Chacornac an den Herausgeber. *AN*, [1368] **57**, 373.

(1862b). *CR*, **54**, 888–889.

(1862c). On the missing nebula in Coma Berenices. *MN*, **22**, 277.

(1862d). Nébuleuse des Pléiades. *Bulletin quotidien de l'Observatoire impérial de Paris* (18.9.1862).

(1863a). Note de M. Chacornac sur une nébuleuse variable. *Bulletin quotidien de l'Observatoire impérial de Paris* (28.4.1863).

(1863b). Note de M. Chacornac sur une nébuleuse variable. *Bulletin quotidien de l'Observatoire impérial de Paris* (29.4.1863).

(1863c). Nébuleuse variable de zeta Taureau. *CR*, **56**, 637–639.

Challis, J. (1869). William Parsons, Earl of Rosse. *MN*, **29**, 123–130.

Chambers, G. (1865). A catalogue of variable stars. *AN*, [1496] **63**, 117–124.

(1867). *Descriptive Astronomy*. Oxford: Clarendon Press.

(1877). *A Handbook of Descriptive Astronomy*. Oxford: Clarendon Press.

(1881). *A Cycle of Celestial Objects*, 2nd edn. Oxford: Clarendon Press.

(1885). The Rev. T. W. Webb. *Nature*, **32**, 130.

(1890). *A Handbook of Descriptive and Practical Astronomy, Vol. 2. Instruments and Practical Astronomy*, 4th edn. Oxford: Clarendon Press.

Chant, C. (1931). Sir Charles Parsons. *JRASC*, **25**, 185–191.

Chapman, A. (1988). William Lassell, W. (1799–1880). Practitioner, patron and 'grand amateur' of Victorian astronomy. *VA*, **32**, 341–370.

(1989). William Herschel and the measurement of space. *QJRAS*, **30**, 309–418.

(1993). An occupation for an independent gentleman. Astronomy in the life of John Herschel. *VA*, **36**, 71–116.

(1998). *The Victorian Amateur Astronomer. Independent Astronomical Research in Britain 1820–1920*. Chicester: John Wiley and Sons.

Chapman, D. (2002). J. L. E. Dreyer. *JRASC*, **96**, 20–21.

Christie, W. (1878). The Arcetri Observatory. *Obs*, **1**, 292–294.

(1911). Observations of nebulae at the Royal Observatory Greenwich. *MN*, **71**, 509–511.
Claridge, J. T. W. (1907). Lord Rosse's Smaller Telescope. *Knowledge*, **30**, 5–7.
Clausnitzer, L. (1989). Wilhelm Tempel und seine kosmischen Entdeckungen. *Vorträge und Schriften der Archenhold-Sternwarte, Berlin-Treptow*, no. 70.
Clerke, A. M. (1891). New nebulae in Cygnus. *Obs*, **14**, 301–303.
(1893). The distance of the Pleiades. *Obs*, **16**, 198–199.
(1895). The exterior nebulosities of the Pleiades. *Knowledge*, **18**, 280–282.
(1902). *A Popular History of Astronomy during the Nineteenth Century*, 4th edn. London: A. and C. Black.
(1905). *Modern Cosmogonies*. London: A. and C. Black.
Coe, S. (2006). *Nebulae and How to Observe Them*. London: Springer.
Coffey, P. A. (1998). Recently discovered photograph of Edward Joshua Cooper of Markree (1798–1863). *IAJ*, **25**, 47–48.
Common, A. A. (1879). Description of the three-feet telescope. *Obs*, **3**, 167–169.
(1880a). RAS meeting (April 9, 1880). Paper on 'The nebula in the Pleiades'. *Obs*, **3**, 405.
(1880b). The nebula in the Pleiades. *MN*, **40**, 376–377.
(1880c). Note on the nebula near Merope. *MN*, **41**, 48.
(1880d). Faye's Comet. *Obs*, **3**, 575.
(1881a). New nebulae. *Cop*, **1**, 50.
(1881b). Neue Nebelflecke. *Sirius*, **14**, 117.
(1886). The nebulae in the Pleiades. *MN*, **46**, 34.
(1888a). *Obs*, **11**, 282–284.
(1888b). Photographs of nebulae. *Obs*, **11**, 390–394.
(1889). Great telescopes. *Obs*, **12**, 138–140.
(1892). On the construction of a five-foot equatorial reflecting telescope. *MRAS*, **50**, 113–204.
(1897). Gold medal to Prof. E. E. Barnard. *MN*, **57**, 321–328.
Cookson, B. (1907). Samuel Pierpont Langley. *MN*, **67**, 239–241.
Cooper, E. J. (1848). Extract of a letter from E. J. Cooper, Esq. Markree Castle. *MN*, **8**, 221.
(1851–56). *Catalogue of Stars Near the Ecliptic Observed at Markree*, 4 vols. Dublin: Alex Thom.
(1852). *MN*, **12**, 100.
Copeland, L. (1955). Adventuring in the Virgo cloud. *S&T*, **14** (no. 4), 147–151.

(1960). An amateur tour of planetary nebulae. *S&T*, **19** (no. 2), 214–217.
Copeland, R. (1877a). Schreiben an den Herausgeber. *AN*, [2116] **89**, 61.
(1877b). The new star in Swan. *AN*, [2117] **89**, 79.
(1877c). On Schmidt's Nova Cygni. *AN*, [2158] **90**, 351.
(1878). J. L. E. Dreyer, a supplement to Sir John Herschel's 'General Catalogue of nebulae and clusters of stars'. *VJS*, **13**, 274–278.
(1880). A new planetary nebula. *Obs*, **3**, 663.
(1881). A new planetary nebula. *Cop*, **1**, 2–3.
(1882a). Spectroscopic and other observations of Schmidt's Nova Cygni at Dunecht Observatory. *Cop*, **2**, 101–120.
(1882b). Charles Edward Burton. *Cop*, **2**, 158.
(1884a). Experiments in the Andes. *Cop*, **3**, 206–231.
(1884b). Notes on a recent visit to some North American observatories. *AR*, **22**, 210–215, 235–243.
(1884c). Spectroscopic observations made at the Earl of Crawford's Observatory, Dun Echt, Aberdeen. *MN*, **45**, 90–91.
(1885). Dun Echt Circular, no. 97. *AR*, **23**, 248.
(1886). On Hartwig's Nova Andromedae. *MN*, **47**, 49–61.
Cozens, G., White, G. (2001). James Dunlop. Messier of the southern sky. *S&T*, **101** (no. 6), 112–116.
Cragin, M., Bonanno, E. (2001). *Uranometria 2000.0, Vol. 3. Deep Sky Field Guide*, 2nd edn. Richmond, VA: Willmann-Bell.
Crowe, D. (1971). Thomas Romney Robinson (1792–1882), Director of Armagh Observatory (1832–1882). *IAJ*, **10**, 93–101.
Curtis, H. (1909). George Washington Hough. *ApJ*, **30**, 68–69.
(1912a). Descriptions of 132 nebulae and clusters photographed with the Crossley reflector. *Lick Observatory Bulletin*, no. 219.
(1912b). Three interesting spiral nebulae. *PASP*, **24**, 227–228.
(1915). Note on Hind's Variable Nebula. *PASP*, **27**, 242–243.
(1918a). Descriptions of 762 nebulae and clusters photographed with the Crossley reflector. *PLO*, **13**, 11–42.
(1918b). The planetary nebulae. *PLO*, **13**, 57–74.
d'Abbadie, A. (1856a). Discovery of a new star in the nebula of Orion. *MN*, **17**, 245.
(1856b). Note on the star recently discovered in the trapezium of the nebula of Orion. *MN*, **17**, 266.

Darby. W. A. (1864). *The Astronomical Observer. A Hand-Book to the Observatory and the Common Telescope.* London: Robert Hardwicke.

d'Arrest, H. L. (1852a). Beobachtungen auf der Leipziger Sternwarte. *AN*, [809] **34**, 269.

(1852b). From letters of Professor d'Arrest to the editor. *AJ*, **2**, 130–131.

(1855). Einige Verbesserungen zu Sir John Herschel's Nebel-Catalogen. *AN*, [972] **41**, 191.

(1856a). Resultate aus Beobachtungen der Nebelflecke und Sternhaufen. Erste Reihe. *Abhandlungen der Königlich Sächsischen Gesellschaft der Wissenschaften*, **3**, 293.

(1856b). Verzeichnis von fünfzig Messier'schen und Herschel'schen Nebelflecken, aus Beobachtungen auf der Leipziger Sternwarte hergeleitet. *AN*, [997] **42**, 193–200.

(1856c). Aus einem Schreiben des Herrn Prof. d'Arrest an den Herausgeber. *AN*, [1039] **44**, 109.

(1861). *Instrumentum magnum aequatoreum in Specula Universitatis Havniensis super erectum.* Copenhagen.

(1862a). Literarische Anzeige. Instrumentum magnum aequatoreum in specula Universitatis Havniensis super erectum. *AN*, [1337] **56**, 271.

(1862b). Beobachtung des Merkurdurchgangs nebst Anzeige von der Vollendung der neuen Kopenhagener Sternwarte/Bemerkungen über den neuen Hind'schen Nebel. *AN*, [1341] **56**, 327.

(1862c). Vorläufige Mittheilungen, betreffend eine auf der Kopenhagener Sternwarte begonnene Revision des Himmels in Bezug auf die Nebelflecken. *AN*, [1366] **57**, 337–348.

(1862d). Verzeichnis von 50, theilweise neuen, Doppelnebeln für den Anfang des Jahres 1861. *AN*, [1369] **58**, 1–6.

(1862e). Auffindung eines zweiten variablen Nebelflecks im Stier. *AN*, [1378] **58**, 155.

(1862f). Auffindung eines dritten variablen Nebelflecks. *AN*, [1379] **58**, 175.

(1863a). Über den Nebel von Merope und einen zweiten Nebel in den Plejaden, sammt Beobachtungen eines neuentdeckten Planeten. *AN*, [1393] **59**, 13–16.

(1863b). Aus einem Schreiben des Herrn Prof. d'Arrest in Kopenhagen an den Herausgeber. *AN*, [1407] **59**, 231–236.

(1863c). Über einen angeblich von Maskelyne beobachteten, gegenwärtig unsichtbaren Nebelfleck. *AN*, [1440] **60**, 377–380.

(1864). Über den vermissten Nebelfleck H.I, 118. *AN*, [1477] **62**, 197–200.

(1865a). Über einige am Kopenhagener Refractor beobachtete Objecte aus Lord Rosse's 'List of nebulae not found'. *AN*, [1500] **63**, 177–190.

(1865b). Aus einem Schreiben des Herrn Prof. d'Arrest, Director der königl. Sternwarte in Kopenhagen, an den Herausgeber. *AN*, [1520] **64**, 125–128.

(1865c). Zweites Verzeichnis von neuen Nebelflecken, aufgefunden am Kopenhagener Refractor im Winter 1864/65. *AN*, [1537] **65**, 1–8.

(1867a). *Siderum Nebulosorum Observationes Havnienses.* Copenhagen.

(1867b). Über drei am Himmel fehlende Nebelflecken erster Classe, nebst einigen nachträglichen Bemerkungen den Biela'schen Cometen betreffend. *AN*, [1624] **68**, 251–256.

(1868a). Ueber eine Darstellung des grossen Orionnebels vom Jahre 1779. *AN*, [1678] **70**, 337–342.

(1868b). Struve's Beobachtung eines neuen Nebelflecks nahe bei Hind's variablem Nebel im Taurus. *AN*, [1689] **71**, 143.

(1872a). Spectroskopische Beobachtungen zweier Nebelflecke. *AN*, [1885] **79**, 193–196.

(1872b). *Undersøgelser over de nebulose Stjerner i Henseende til deres spektralanalytiske Egenskaber.* Copenhagen.

(1873). Aus einem Schreiben des Herrn Prof. d'Arrest an den Herausgeber, betreffend die Gas-Nebulosen. *AN*, [1908] **80**, 189.

Davison, D. (1989). *Impressions of an Irish Countess. The Photographs of Mary Countess of Rosse.* Birr: Birr Scientific Heritage Foundation.

Dawes, W. (1865). The Observations at South Villa. *AR*, **3**, 43–44.

de Kerolyr, M. (1937). Photographs of two nebulae. *ApJ*, **85**, 340.

Dekker, E. (1990). The light and the dark. A reassessment of the discovery of the Coalsack nebula, the Magellanic Clouds and the Southern Cross. *Annals of Science*, **47**, 529–560.

de la Rue, W. (1865). Gold medal for George Philips Bond. *MN*, **25**, 125–137.

Denning, W. (1882). The nebula in the Pleiades. *Obs*, **5**, 199–200.

(1890). New nebulae. *MN*, **51**, 96.

(1891a). Variations of nebulae. *Obs*, **14**, 196–197.

(1891b). *Telescopic Work for Starlight Evenings*. London: Taylor and Francis.

(1892a). Supposed variable nebulae. *AN*, [3111] **130**, 231.

(1892b). Notes on new and old nebulae. *Obs*, **15**, 104–106.

(1914). Lord Rosse's telescope. *Obs*, **37**, 347–348.

de Quincey, T. (1846). System of the heavens as revealed by Lord Rosse's telescopes. *Tait's Edinburgh Magazine*, **13**, 566–579.

de Senarmont, H. H. (1857). *CR*, **44**, 1294–1295.

de Vaucouleurs, G. (1985). The supernova of 1885 in Messier 31. *S&T*, **70** (no. 2), 115–118.

(1987). Discovering M 31's spiral structure. *S&T*, **74** (no. 6), 595–598.

de Vico, F. (1839). La nebulosa d'Orione. *Memoria intorno a parecchie osservazioni dell'Universita Gregoriana in Collegio Romano*. 1839, 34–37.

(1841a). Nébuleuse d'Orion. *CR*, **13**, 449–450.

(1841b). Nebulose. *Memoria intorno ad alcune osservazioni fatte alla specula del Collegio Romano*. 1841, 22–29.

(1843). Nebulose. *Memoria intorno ad alcune osservazioni fatte alla specula del Collegio Romano*. 1843, 27–28.

Dewhirst, D., Hoskin, M. (1991). The Rosse spirals. *JHA*, **22**, 257–266.

Dialetis, D., Matsopoulos, N. and Prokakis, T. (1982). The wanderings of a 25-inch refractor. *S&T*, **64** (no. 2), 136–137.

Dick, S. J. (1980). How the U.S. Naval Observatory began, 1830–1865. *S&T*, **60** (no. 6), 466–471.

(1988). Discovering the moons of Mars. *S&T*, **76** (no. 3), 242–243.

(2003). *Sea and Ocean Joined*. Cambridge: Cambridge University Press.

Dick, T. (1845). *The Practical Astronomer*. London: Seeley, Burnside, and Seeley.

Dick, W. R. (2000). Hermann Carl Vogels Bericht über eine Reise nach England, Schottland und Irland im Jahr 1875. *Acta Historica Astronomiae*, **11**, 97–126.

Dick, W. R., Brüggenthies, W. (2005). *Biographischer Index der Astronomie* (Acta Historica Astronomiae, vol. 26). Frankfurt: Harri Deutsch.

Dien, C. (1855). Nébulosité observée dans le voisinage de l'étoile O du Rameau. *CR*, **40**, 775.

Doberck, W. (1876). Colonel Cooper's Markree Observatory. *MN*, **36**, 171.

(1884). Markree Observatory. *Obs*, **7**, 283–288, 329–332.

Drews, J., Schwier, H. eds. (1984). *Lilienthal oder die Astronomen*. Munich: edition text + kritik.

Dreyer, J. L. E. (1873). Sur l'orbite de la première comète de l'an 1870. *AN*, [1910] **80**, 219.

(1875a). Schultz's micrometrical measures of 500 nebulae. *MN*, **35**, 233–245.

(1875b). Dr. Herman Schulz. Micrometrical observations of 500 nebulae. *VJS*, **10**, 64–73.

(1876a). Heinrich Louis d'Arrest. *VJS*, **11**, 1–14.

(1876b). E. Schönfeld, Astronomische Beobachtungen auf der Grossherzöglichen Sternwarte zu Mannheim. *VJS*, **11**, 269–276.

(1876c). Vogel. Positionsbestimmungen von Nebelflecken und Sternhaufen zw. +9° 30' und +15° 30' Declination. *VJS*, **11**, 276–280.

(1876d). Request to astronomers. *AN*, [2111] **88**, 359.

(1877a). Note on some of M. Stephan's new nebulae. *MN*, **37**, 427–428.

(1877b). On personal errors in astronomical transit observations. *Proceedings of the Royal Irish Academy*, **2**, 484–528.

(1878a). Supplement to Sir John Herschel's 'General Catalogue of nebulae and clusters of stars'. *Transactions of the Royal Irish Academy*, **26**, 381–426.

(1878b). Spiral form of nebulae. *Obs*, **1**, 370–371.

(1879). Spiral form of nebulae. *Obs*, **2**, 22–23.

(1880a). Lord Rosse's observations of nebulae and clusters. *MN*, **40**, 252–254.

(1880b). Mr. Webb's planetary nebula. *Obs*, **3**, 360.

(1881a). Nebulae. *Cop*, **1**, 77–78.

(1881b). Note on the nebula near Merope. *Cop*, **1**, 156–157.

(1881c). Nebulae and clusters. *Obs*, **4**, 61–62.

(1882). A new determination of the constant of precession. *Cop*, **2**, 135–155.

(1883). *An Historical Account of the Armagh Observatory*. Liverpool: H. Greenwood.

(1885a). Notice of a new catalogue of nebulae. *AN*, [2667] **112**, 41.

(1885b). On the supposed proper motion of the nebula h 705. *Obs*, **8**, 175–176.

(1886). On the proper motions of 29 telescopic stars. *AR*, **24**, 134–136.

(1887a). Papers announced to the RAS. Second supplement to Sir John Herschel's general catalogue of nebulae and clusters of stars. *MN*, **47**, 196.

(1887b). Papers announced to the RAS. A new general catalogue of nebulae and clusters, being the catalogue of the late Sir F. J. W. Herschel, revised, corrected, and enlarged. *Obs*, **10**, 241.

(1887c). Wilhelm Tempel, Ueber Nebelflecken. *VJS*, **22**, 59–66.

(1887d). On some nebulae hitherto suspected of variability or proper motion. *MN*, **47**, 412–421.

(1888a). Hans Carl Frederick Christian Schjellerup. *MN*, **48**, 171–174.

(1888b). New General Catalogue of nebulae and clusters of stars. *MRAS*, **49**, 1–237.

(1889). Note on the nebula h 347. *AN*, [2863] **120**, 111.

(1890). Ernst Wilhelm Leberecht Tempel. *MN*, **50**, 179–182.

(1891a). Astronomical bibliography. *Obs*, **14**, 322–325.

(1891b). Note on some apparently variable nebulae. *MN*, **52**, 100–103.

(1894a). Micrometric observations of nebulae made at the Armagh Observatory. *Transactions of the Royal Irish Academy*, **30**, 513–558.

(1894b). Note on a group of nebulae. NGC. 3743–58. *AN*, [3246] **136**, 93.

(1895). Index catalogue of nebulae found in the years 1888 to 1894. *MRAS*, **51**, 185–228.

(1896). On systematic errors in observing right ascensions of nebulae. *MN*, **57**, 44–50.

(1897). Albert Marth. *AN*, [3446] **144**, 223.

(1900). Corrections to the Armagh catalogue for 1840. *MN*, **61**, 10–12.

(1904). A survey of the spiral nebula Messier 33 by means of photographs taken by Dr. Issac Roberts. *Proceedings of the Royal Irish Academy, sect. B*, **25**, 3–30.

(1906). Ralph Copeland. *MN*, **66**, 164–174.

(1907). Ralph Copeland. *VJS*, **42**, 230–238.

(1908). Second index catalogue of nebulae and clusters of stars containing objects found in the years 1895 to 1907. *MRAS*, **59**, 105–198.

(1909). Lawrence Parsons, Fourth Earl of Rosse. *MN*, **69**, 250–253.

(1912a). *The Scientific Papers of Sir William Herschel*, 2 vols. London: Royal Society.

(1912b). Corrections to the New General Catalogue resulting from the revision of Sir William Herschel's three catalogues of nebulae. *MN*, **73**, 37–40.

(1914). Lord Rosse's six-foot reflector. *Obs*, **37**, 399–400.

(1918). Descriptive catalogue of a collection of William Herschel's papers presented to the RAS by the late Sir W. J. Herschel. *MN*, **78**, 547–554.

(1926). Note on the preceding paper. *MN*, **86**, 146.

(1953). *New General Catalogue, Index Catalogue, Second Index Catalogue*. London: Royal Astronomical Society.

Dreyer, J. L. E., Turner, H. H. eds. (1923). *History of the Royal Astronomical Society*, vol. 1. London: Royal Astronomical Society.

Duncan, J. (1921). Bright and dark nebulae near ζ Orionis photographed with the 100-inch Hooker telescope. *ApJ*, **53**, 392–396.

(1949). Messier's nebulae and star clusters. *LASP*, **5**, 328–335.

Dunér, N. (1872). Beobachtung eines Nebelflecks. *AN*, [1864] **78**, 251.

(1890). Professor Dr. Herman Schultz. *AN*, [2968] **124**, 271.

Dunkin, E. (1873). Paul Auguste Ernest Laugier. *MN*, **33**, 211–214.

(1880). Johann von Lamont. *MN*, **40**, 208–212.

(1885). Johann Friedrich Julius Schmidt. *MN*, **45**, 211.

Dunlop, J. (1828). A catalogue of nebulae and clusters of stars in the southern hemisphere observed at Paramatta in New South Wales. *PT*, **118**, 113–152.

(1829a). Approximate places of double stars in the southern hemisphere. *MRAS*, **3**, 257–275.

(1829b). Verzeichnis von Doppelsternen. *AN*, [149] **7**, 113.

Dyson, F. W. (1904a). Andrew Ainslie Common. *MN*, **64**, 274–278.

(1904b). Mathieu-Prosper Henry. *MN*, **64**, 296–298.

(1915). Arthur Auwers 1838–1915. *Obs*, **38**, 177–181.

(1924). Edward Emerson Barnard. *MN*, **84**, 221–224.

(1932). Guillaume Bigourdan. *Obs*, **55**, 116–117.

(1933). Guillaume Bigourdan. *MN*, **93**, 233–234.

Eddington, A. (1916). Georg Friedrich Julius Arthur Auwers. *MN*, **76**, 284–289.

Eddy, J. (1990). Founding the Astrophysical Observatory. The Langley years. *JHA*, **21**, 111–120.

Eggen, O. (1953a). Vulcan. *LASP*, **6**, 291–298.

(1953b). Sherburne Wesley Burnham and his double star catalogue. *LASP*, **6**, 354–360.

Eichelberger, W. (1908). Asaph Hall. *AN*, [4232] **177**, 127.

Eichhorn, E. (1963). Ernst Wilhelm Leberecht Tempel. *Die Sterne*, **39**, 113–116.

Ellery, R. L. J. (1885). *Observations of the Southern nebulae made with the Great Melbourne Telescope from 1869 to 1885*. Melbourne: Melbourne Observatory.

(1886a). The great Melbourne reflector. *Obs*, **9**, 204–205.

(1886b). Melbourne Observatory. *Obs*, **9**, 267–268.

Emery, R., Hawkridge, D. (2007). James Wigglesworth and the Scarborough Telescope. *JBAA*, **117**, 118–125.

Engelhardt, B. (1886). *Observations astronomiques, première partie*. Dresden: G. Beansch.

(1890). *Observations astronomiques, deuxième partie.* Dresden: G. Beansch.

(1895). *Observations astronomiques, troisième partie.* Dresden: G. Beansch.

Engelmann, F. (1876). *Abhandlungen von F. W. Bessel*, vol. 2. Leipzig: Engelmann.

(1883). Meridianbeobachtungen von Nebelflecken. *AN*, [2485] **104**, 193–208.

Erck, W. (1882). The late C. E. Burton. *AR*, **20**, 173–174.

Esmiol, E. (1916). Réduction des observations de nébuleuses découvertes par M. Stephan. *Traveaux de l'Observatoire de Marseille*, **4**, 1–67.

Espin, T. H. E. (1887). Joseph Baxendell, F.R.S., F.R.A.S. *AN*, [2819] **118**, 175.

(1922). Dark structures in the Milky Way (second contribution) with catalogue and description. *JRASC*, **16**, 218–229.

Everett, E. (1850). Observations on the belts and satellites of Jupiter, and on certain nebulae by the Messrs. Bond. *AN*, [702] **30**, 93–96.

Fernie, J. (1970). The historical quest for the nature of the spiral nebulae. *PASP*, **82**, 1189–1230.

Ferris, F. (1982). *Galaxies.* Cambridge, MA: Harvard University Press.

Firneis, M. (1985). Construction and scientific activities of the new Vienna Observatory 1870–1900. *VA*, **28**, 401–404.

Fischer, D. (1990). Der 18-Zoll-Refraktor des Amherst College. *SuW*, **29**, 396–387.

Fitzsimons, J. (1980). *S&T*, **59** (no. 2), 108.

(1982). *Notes on the History of Markree Castle Observatory.* Sligo: Markree Castle.

Flammarion, C. (1879a). Nébuleuses doubles en mouvement. *CR*, **88**, 27–30.

(1879b). In Bewegung begriffene Doppelnebel. *Sirius*, **12**, 93.

(1917). Nébuleuses et amas d'étoiles de Messier. *L'Astronomie*, **31**, 385–399.

Fletcher, I. (1866). Admiral William Henry Smyth. *MN*, **26**, 121–129.

Foerster, W. (1881). Carl Christian Bruhns. *VJS*, **18**, 1–5.

(1889). Ueber die Verschiedenheiten der Wahrnehmung und Darstellung von Nebelflecken. *Himmel und Erde*, **1**, 179–181.

(1891). Eduard Schönfeld. *Himmel und Erde*, **3**, 418–420.

Fotheringham, J. (1926). Dr. J. L. E. Dreyer. *Nature*, **118**, 4.

Foucault, L. (1859). Essai d'une nouveau télescope parabolique en verre argenté. *CR*, **49**, 85–87.

Fowler, A. (1908). Hermann Carl Vogel. *MN*, **68**, 254–257.

(1915). Nils Christoffer Dunér. *MN*, **75**, 256–258.

Fox, P. (1913). The celebration of the semi-centennial of the Chicago Astronomical Society and the dedication of a tablet to the memory of Truman Henry Safford. *PA*, **21**, 473–479.

(1915). General account of the Dearborn Observatory. *Annals of the Dearborn Observatory*, **1**, 1–20.

(1923). Edward Emerson Barnard. *PA*, **31**, 195–200.

Fraissinet, A. (1873). Chacornac. *Le Nature*, **1**, 358–360.

Franks, W. (1930). Visual observations of dark nebulae. *MN*, **90**, 326–331.

Fraser, J. (1889). An American amateur astronomer. *Century Magazine*, **38**, 300–307.

Fraunhofer, J. (1826). Ueber die Construction des so eben vollendeten großen Refractors. *AN*, [74] **4**, 17–24.

Fritz, M. (2007). Webb's clusters. *DSO*, **143**, 25–27.

Frommert, H. (2006). Messier 102. Status der Identifizierung dieses Messier-Objekts. *VdSJ*, **19**, 69–71.

Frost, E. B. (1908a). Hermann Carl Vogel. *ApJ*, **27**, 1–11.

(1908b). Charles August Young. *JRASC*, **2**, 27–31.

(1909). Charles August Young. *ApJ*, **30**, 323–338.

(1921a). Sherburne Wesley Burnham, 1838–1921. *ApJ*, **54**, 1–8.

(1921b). Sherburne Wesley Burnham, 1838–1921. *JRASC*, **15**, 269–275.

(1923). Edward Emerson Barnard. *ApJ*, **58**, 1–35.

Fürst, D., Hamel, J. (1983). Friedrich von Hahn und die Sternwarte in Remplin/Mecklenburg. *Die Sterne*, **59**, 88–99.

(1999). Graf Friedrich von Hahn auf Remplin. *VdSJ*, **3**, 79–81.

Gärtner, H. (1996). *Er durchbrach die Schranken des Himmels. Das Leben des Friedrich Wilhelm Herschel.* Leipzig: Edition Leipzig.

Gascoigne, S. (1996). The Great Melbourne Telescope and other 19th-century reflectors. *QJRAS*, **37**, 101–125.

Gautier, A. (1863). Recent research on nebulae. *American Journal of Science and Arts*, **35**, 101–110.

Gautier, E. (1889). The late W. Tempel. *Obs*, **12**, 442.

Gegen, D. (1988). Rebirth of a glass giant. *S&T*, **75** (no. 5), 486–488.

Gingerich, O. (1953a). Messier and his catalogue I. *S&T*, **12** (no. 8), 255–258.

(1953b). Messier and his catalogue II. *S&T*, **12** (no. 9), 288–291.

(1954). Observing the Messier catalogue. *S&T*, **13** (no. 3), 157–159.

(1960). The missing Messier objects. *S&T*, **20** (no. 4), 196–199.

(1967). Messier's clusters and nebulae. *LASP*, **10**, 73–79.

(1982). Dreyer and Tycho's world system. *S&T*, **64** (no. 2), 138–140.

(1987). The mysterious nebulae, 1610–1924. *JRASC*, **81**, 113–127.

(1988). J. L. E. Dreyer and his NGC. *S&T*, **76** (no. 6), 621–623.

Ginzel, F. (1888). Beobachtungen von Nebelflecken. *AN*, [2829–30] **128**, 321–344.

Glaisher, J. (1888). Gold medal of the Society to Dr. Arthur Auwers. *MN*, **48**, 236–251.

Glass, I. (1997). *Victorian Telescope Makers. The Lives and Letters of Thomas and Howard Grubb*. Bristol: Institute of Physics Publishing.

Glyn Jones, K. (1967a). Some new notes on Messier's catalogue. *S&T*, **33** (no. 3), 156–158.

(1967b). Messier's clusters and nebulae. *LASP*, **10**, 73–79.

(1975). *The Search for Nebulae*. Chalfont St. Giles, Bucks.

(1976). S Andromedae, 1885. An analysis of contemporary reports and a reconstruction. *JHA*, **7**, 27–40.

ed. (1981). *Webb Society Deep-Sky Observer's Handbook, Vol. 4. Galaxies*. Hillside, NJ: Enslow Publishers.

(1991). *Messier's Nebulae and Clusters*. Cambridge: Faber & Faber.

Goldschmidt, H. (1863a). Aus einem Schreiben von Herrn Goldschmidt an den Herausgeber. *AN*, [1394] **59**, 31.

(1863b). Lettre de M. Goldschmidt. *Bulletin quotidien de l'Observatoire impérial de Paris* (6.10.1863).

(1863c). Carte des pleiades de Ms. Goldschmidt. *Bulletin quotidien de l'Observatoire impérial de Paris* (4.12.1863).

(1864). Etude du groupe des Pléiades. *CR*, **58**, 72–74.

Gore, J. (1902). Messier's nebulae. *Obs*, **25**, 264–269, 288–293, 321–326.

Gothard, E. (1886a). Photographische Aufnahmen. *AN*, [2749] **115**, 221.

(1886b). Bemerkung zu A.N. 2749 betr. den Ringnebel in der Leyer. *AN*, [2754] **115**, 303.

(1894). Der kleine Barnard'sche Nebel bei M. 57. *AN*, [3217] **135**, 11.

Gould, B. (1853). New works. *AJ*, **3**, 8.

(1854). Professor A. C. Petersen. *AJ*, **3**, 160.

Graney, G. (2007). On the accuracy of Galileo's observations. *Baltic Astronomy*, **16**, 443–449.

Greig, A. (1903). Andrew Ainslie Common. *PASP*, **15**, 209.

Grosser, H. (1998). *Historische Gegenstände an der Universitätssternwarte Göttingen*. Göttingen: Akademie der Wissenschaften zu Göttingen.

Grove-Hills, E. (1923). The decade 1850–1860, in *History of the Royal Astronomical Society*, vol. 1, eds. J. L. E. Dreyer, H. H. Turner. London: Royal Astronomical Society, pp. 110–128.

Grover, C. (1872a). The nebula in the Pleiades. *AR*, **10**, 173–174.

(1872b). The nebula in the Pleiades. *AR*, **10**, 218–219.

(1873). The nebula in the Pleiades. *AR*, **11**, 159–160.

Graff, K. (1948). Visuelle Gesamthelligkeiten von hellen Nebeln, Sternhaufen und Milchstraßenwolken. *Mitteilungen der Wiener Sternwarte*, **4** (no. 2), 57–68.

Grubb, H. (1910). Bindon Blood Stoney, 1828–1909. *PRS*, **85**, viii–x.

Haffner, H. (1963). 100 Jahre Astronomische Gesellschaft 1863–1963. *SuW*, **2**, 220–223.

Häfner, R. (1990). Die Zeit Johann von Lamonts an der Königlichen Sternwarte München-Bogenhausen. *SuW*, **29**, 13–18.

Hagen, J. (1912). Wilhelm Tempel. In *The Catholic Encyclopedia*, vol. 14. New York, NY: Robert Appleton.

(1921). Die dunklen kosmischen Nebel. *AN*, [5110] **231**, 351–354.

(1922). A preparatory catalogue for a Durchmusterung of nebulae. The zone catalogue. *Specola Astronomica Vaticana*, vol. **10**.

(1926). First note on W. Herschel's observations of extended nebulous fields. *MN*, **86**, 144–146.

(1930). On the significance of Baxendell's nebulosity. *MN*, **90**, 331–333.

Hahn, F. (1796). Gedanken über den Nebelfleck im Orion. *Berliner Jahrbuch für 1799*, 235–238.

(1799). Ueber den planetarischen Nebel bey μ Wasserschlange. *Berliner Jahrbuch für 1802*, 231–233.

(1800). Einige Beobachtungen über Mira Ceti. Über Nebelflecken in der Leyer und der Hydra. *Berliner Jahrbuch für 1803*, 106–107.

Hall, A. (1878). Observations of stars around the Ring Nebula in Lyra. *AN*, [2186] **92**, 27.

(1881). Observations of comets. *AN*, [2394] **100**, 273–278.

(1886a). Nebulae in the Pleiades. *AN*, [2723] **114**, 167.

(1886b). Parallax of Nova Andromedae. *Obs*, **9**, 204.

Hall, M. (1877). The nebula in the Pleiades. *Nature*, **15**, 244.
(1879). The nebula in the Pleiades. *MN*, **40**, 89–90.
(1881). Further notes on the nebula in the Pleiades. *MN*, **41**, 315–316.
(1889). The nebula in the Pleiades. *Obs*, **12**, 443–444.
Halm, J. (1906). Ralph Copeland. *AN*, [4058] **170**, 29–32.
Hamel, J. (1988). *Friedrich Wilhelm Herschel*. Leipzig: Teubner.
(1989). Bessels Projekt der Akademischen Sternkarten. *Die Sterne*, **65**, 11–19.
Hardcastle, J. (1914). Nebulae seen on the Franklin-Adams' plates. *MN*, **74**, 699–706.
Hardie, R. (1964). The early life of E. E. Barnard. *LASP*, **9**, 113–128.
Harding, K. L. (1824). Neue Nebelflecke. *Berliner Jahrbuch für 1827*, 131–134.
Harrington, M. W. (1883). New nebula by M. W. Harrington, Director of the Observatory Ann Arbor, Mich. *AN*, [2479] **104**, 111.
(1887). On the structure of 13 M Herculis. *AJ*, **7**, 156–157.
Harrison, T. (1984). The Orion Nebula. Where in history is it? *QJRAS*, **25**, 65–79.
Hartl, G. (1987). Der Refraktor der Sternwarte Pulkowa. *SuW*, **26**, 397–404.
Hartmann, J. (1902). Spectrographic measures of velocities of gaseous nebulae. *ApJ*, **15**, 287–295.
Hartwig, E. (1883a). Auffindung eines neuen Nebels. *AN*, [2507] **105**, 173.
(1883b). Beobachtung von neuen Nebeln. *AN*, [2544] **106**, 381.
(1885). Neue Nebel. *AN*, [2688] **112**, 407.
(1898a). Ueber den grossen Andromedanebel. *AN*, [3529] **148**, 11.
(1898b). Friedrich August Theodor Winnecke. *VJS*, **33**, 5–13.
(1898c). Friedrich August Theodor Winnecke. *MN*, **58**, 155–159.
Hartwig, E., Müller, G. (1923). *Geschichte und Literatur der Veränderlichen Sterne*, vol. 1. Berlin: Ferdinand Dümmler.
Hasselberg, B. (1917). Nils Christoffer Dunér. *VJS*, **52**, 2–31.
Haynes, R., Haynes, R. D., Malin, D., McGee R. (1996). *Explorers of the Southern Sky. A History of Australian Astronomy*. Cambridge: Cambridge University Press.
Hearnshaw, J. (1985). William Huggins und die Anfänge der Spectroscopie. *SuW*, **24**, 140–142.
(1992). Origin of the stellar magnitude scale. *S&T*, **84** (no. 5), 494–499.
(2009). *Astronomical Spectrographs and Their History*. Cambridge: Cambridge University Press.
Henry, P., Henry, P. (1886a). Découverte d'une nébuleuse par la photographie. *AN*, [2702] **113**, 239.
(1886b). On photographs of a new nebula in the Pleiades, and of Saturn. *MN*, **46**, 98.
(1886c). Sur une carte photographique du groupe des Pléiades. *CR*, **102**, 848–851.
(1886d). The photographic nebulae in the Pleiades. *MN*, **46**, 281.
(1887). Photographie de la nébuleuse 1180 du Catalogue générale d'Herschel. *CR*, **104**, 394–396.
(1888). Nouvelles nébuleuses remarquables, découvertes, a l'aide de la photographie, dans les Pléiades. *CR*, **106**, 912–914.
Hentschel, K. (2000). Drawing, engraving, photographing, printing. Historical studies of visual representations, esp. in astronomy, in *The Role of Visual Representation in Astronomy. History and Research Practice*, eds. K. Hentschel, A. Wittmann. Frankfurt: Harri Deutsch, pp. 11–43.
Hepperger, J. (1925). Johann Palisa. *AN*, [5383] **225**, 125–128.
Herbig, G. (1949). The spectrum of the nebulosity surrounding T Tauri. *ApJ*, **111**, 11–14.
(1953). T Tauri and Hind's Variable Nebula. *LASP*, **6**, 338–345.
(1995). IC 349. Barnard's Merope nebula. *JRASC*, **89**, 133–134.
Herczog, T. (1998). The Orion Nebula. A chapter of early nebular studies, in *The Message of the Angles. Astrometry from 1798 to 1998*, eds. P. Brosche, W. R. Dick, O. Schwarz, R. Wielen. Frankfurt: Harri Deutsch, pp. 246–258.
Herschel, C. (1827). A catalogue of the nebulae which have been observed by William Herschel in a series of sweeps. Brought into zones of N.P. Distance and order of R.A., for the years 1800, by applying to the determining stars the variations given in Wollaston's of Bode's catalogues. Unpublished, Hannover.
Herschel, J. (1826a). An account of the actual state of the great nebula in Orion, compared with those of former astronomers. *MRAS*, **2**, 487–495.
(1826b). Of the nebula in the girdle of Andromeda. *MRAS*, **2**, 495–497.

(1826c). Schreiben des Herrn J. F. W. Herschel F.R.S. an den Herausgeber. *AN*, [85] **4**, 231–234.

(1829). Observations with the 20 feet reflecting telescope. Third series. *MRAS*, **3**, 177–213.

(1830). Gold medal to Thomas Brisbane and James Dunlop. *MN*, **1**, 56–62.

(1833a). Observations of nebulae and clusters of stars, made at Slough, with a twenty-feet reflector, between the years 1825 and 1833. *PT*, **123**, 359–509.

(1833b). *A Treatise on Astronomy*. London: Longman, Rees, Orme, Brown, Green and Longman, and John Taylor.

(1835). Schreiben von Sir John Herschel Ritter des Bath-Ordens an den Herausgeber. *AN*, [281] **12**, 273–276.

(1845). Address of the President. *RBA*, **15**, xxxvii–xxxviii.

(1847). *Results of Astronomical Observations Made during the Years 1834, 5, 6, 7, 8 at the Cape of Good Hope*. London: Smith, Elder and Co.

(1848). Testimonial to John Russell Hind and George Bishop. *MN*, **8**, 104–107.

(1849). Honorary medal of the Society to William Lassell. *MN*, **9**, 87–92.

(1862). Letter from Sir John Herschel to Mr. Hind on the disappearance of a nebula in Coma Berenices. *MN*, **22**, 248–250.

(1864). Catalogue of nebulae and clusters of stars. *PT*, **154**, 1–137.

(1869). *Outlines of Astronomy*, 10th edn. 2 vols., New York, NY: Cambridge University Press (1st edn. 1849).

(1871). Remarks on Abbott's foregoing paper on Eta Argus. *MN*, **31**, 228–230.

Herschel, J., South, J. (1824). Observations of 380 stars. *PT*, **114** (part 3), 1–412.

Herschel, J. Lt (1868a). Observations of spectra of some southern nebulae. *PRS*, **16**, 417–418.

(1868b). Results of examination of southern nebulae with the spectroscope. *PRS*, **16**, 451–455.

(1869). The great nebula round Eta Argus. *MN*, **29**, 82–88.

(1871). On the nebula of Eta Argus. *MN*, **31**, 235.

Herschel, W. (1783). On the diameter of the Georgium Sidus. With a description of the dark and lucid disk and periphery micrometers. *PT*, **73**, 4–14.

(1784). Account of some observation tending to investigate the construction of the heavens. *PT*, **74**, 437–451.

(1785). On the construction of the heavens. *PT*, **75**, 213–266.

(1786). Catalogue of one thousand new nebulae and clusters of stars. *PT*, **76**, 457–499.

(1789). Catalogue of a second thousand new nebulae and clusters of stars. *PT*, **79**, 212–255.

(1791). On nebulous stars, properly so called. *PT*, **81**, 71–88.

(1802). Catalogue of 500 new nebulae, nebulous stars, planetary nebulae, and clusters of stars. *PT*, **92**, 477–528.

(1811). Astronomical observations relating to the construction of the heavens. *PT*, **101**, 269–336.

(1814). Astronomical observations relating to the sidereal part of the heavens, and its connection with the nebulous part. *PT*, **104**, 248–284.

Hetherington, N. S. (1974). Edwin Hubble's examination of internal motions of spiral nebulae. *QJRAS*, **15**, 392–418.

Hind, J. R. (1845). Schreiben des Herrn J. R. Hind an den Herausgeber. *AN*, [549] **23**, 351–356.

(1850a). Auszug aus einem Briefe des Herrn Hind. *AN*, [712] **30**, 257.

(1850b). Auszug aus einem Schreiben des Herrn Hind an den Herausgeber. *AN*, [713] **30**, 275.

(1850c). Extrait d'une lettre de M. Hind. *CR*, **30**, 358–359.

(1850d). *MN*, **10**, 141.

(1852). New nebula. *MN*, **12**, 208.

(1853a). Auszug aus einem Schreiben des Herrn Hind an die Redaktion. *AN*, [839] **35**, 371.

(1853b). Nachricht über 4 bei der Anfertigung von Mr. Bishop's ecliptic charts bemerkte und später nicht wieder gefundene kleine Sterne. *AN*, [849] **36**, 147.

(1853c). On four small stars which had been observed during the preparation of Mr. Bishop's ecliptical charts, and have since disappeared. *MN*, **13**, 168.

(1862a). Letter from the eminent astronomer, J. R. Hind of London, announcing the disappearance of a nebula. *American Journal of Science and Arts*, **33**, 436–437.

(1864). Note on the variable nebula in Taurus. *MN*, **24**, 65.

(1876). Chacornac's variable nebula near Zeta Tauri. *Nature*, **14**, 545.

Hinks, A. R. (1911a). The galactic distribution of spiral nebulae. *MN*, **71**, 588–595.

(1911b). On the galactic distribution of gaseous nebulae and of star clusters. *MN*, **71**, 693–701.

Hirsch, F. (1932). *100 Jahre Bauen und Schauen*. Karlsruhe: Badenia.

Hirshfeld, A. (2001). *Parallax – The Race to Measure the Cosmos*. New York, NY: Freeman.

Hoffmeister, C. (1923). Carl Ernst Albrecht Hartwig. *AN*, [5243] **219**, 185–188.

(1961). Der Ringnebel IC 5148, 5150. *Die Sterne*, **37**, 204–205.

Hogg, A. (1959). The last of the specula. *LASP*, **8**, 105–112.

Holden, E. S. (1875). Drawing of the Ring Nebula in Lyra. *MN*, **36**, 61–69.

(1876). On supposed changes in the nebula M. 17 = h. 2008 = G.C. 4403. *American Journal of Science*, **11**, 341–361.

(1877a). *Index Catalogue of Books and Memoirs Relating to Nebulae and Clusters, etc.* Smithsonian Miscellaneous Collection 311, VIII. Washington D.C.: Smithsonian Institution.

(1877b). On the proper motion of the Trifid Nebula. *American Journal of Science*, **14**, 433–458.

(1878). Der dreifache Nebel im Schützen. *Sirius*, **11**, 158–159.

(1881). *Sir William Herschel. His Life and Works.* New York, NY: Charles Scribner's Sons.

(1882a). List of new nebulae and clusters discovered in the zone observations at the Washburn Observatory, from April 23 to September 30, 1881. *Publications of the Washburn Observatory*, **1**, 73–76.

(1882b). Monograph on the central parts of the nebula of Orion. *Washington Astronomical Observations for 1878*, Appendix I.

(1883). Washburn Observatory. *Obs*, **6**, 280–281.

(1884a). Two new nebulae. *Publications of the Washburn Observatory*, **2**, 101.

(1884b). Statistics of stellar distribution derived from star-gauges and from celestial charts of Peters, Watson, Chacornac, and Palisa. *Obs*, **7**, 249–256.

(1888). The Ring Nebula in Lyra. *MN*, **48**, 383–388.

(1889a). Die ersten Beobachtungen von Nebelflecken am grossen Refraktor der Lick-Sternwarte in Californien. *Sirius*, **17**, 14–17, 40–44.

(1889b). On the helical nebulae. *PASP*, **1**, 25–31.

(1890). What is the real shape of the spiral nebulae? *Century Magazine*, **39**, 456–459.

(1891a). The Imperial Observatory of Vienna. *PASP*, **3**, 243–247.

(1891b). The University Observatory of Strassburg. *PASP*, **3**, 279–282.

(1892). The new star in Auriga, February, 1892. *PASP*, **4**, 84–85.

(1895). American astronomical journals. *PASP*, **7**, 59–61.

(1896). *Mountain Observatories in America and Europe.* Washington D.C.: Smithsonian Institution.

(1897a). *Memorials of William Cranch Bond and his Son George Phillips Bond.* New York, NY: Charles Scribner's Sons.

(1897b). Albert Marth. Born 1828, died 1897. *PASP*, **9**, 202–203.

(1897c). Resignation of Professor E. S. Holden as Director of the Lick Observatory. *PASP*, **9**, 235–238.

Holden, E. S., Schaeberle, M. (1888). Observations of nebulae at the Lick Observatory. *MN*, **48**, 388–392.

Holetschek, J. (1885). Schreiben des Herrn Dr. J. Holetschek an den Herausgeber. *AN*, [2664] **111**, 399.

(1907). Beobachtungen über den Helligkeitseindruck von Nebelflecken und Sternhaufen. *Annalen der K.K. Universitäts-Sternwarte Wien*, **20**, 39–120.

Hollis, H. (1901). Hollis, Henry. Truman Henry Safford. *Obs*, **24**, 307–309.

Hooper, J. R. (1885). Merope Nebula. *SM*, **4**, 311.

Hoskin, M. (1959). *William Herschel, Pioneer of Sidereal Astronomy.* London: Sheed and Ward.

(1963). *William Herschel and the Construction of the Heavens.* London: Oldbourne.

(1979). William Herschel's early investigations of nebulae. A reassessment. *JHA*, **10**, 165–176.

(1982a). The first drawing of a spiral nebula. *JHA*, **13**, 97–101.

(1982b). Archives of Dunsink and Markree Observatories. *JHA*, **13**, 146–152.

(1982c). *Stellar Astronomy. Historical Studies.* Cambridge: Science History Publications.

(1984). Astronomical observations of William Rowan Hamilton. *JHA*, **15**, 69–73.

(1987). John Herschel's cosmology. *JHA*, **18**, 1–34.

(1989). Astronomers at war. South v. Sheepshanks. *JHA*, **20**, 175–212.

(1990). Rosse, Robinson, and the resolution of the nebulae. *JHA*, **21**, 331–344.

(2002a). The Leviathan of Parsonstown. Ambitions and achievements. *JHA*, **33**, 57–70.

(2002b). Caroline Herschel. Assistant astronomer or astronomical assistant? *History of Science*, **40**, 425–444.

(2003a). *Caroline Herschel's Autobiographies.* Cambridge: Science History Publications.

(2003b). *The Herschel Partnership.* Cambridge: Science History Publications.

(2003c). Herschel's 40ft reflector. Funding and function. *JHA*, **34**, 1–32.

(2005a). Caroline Herschel's 'small' sweeper. *JHA*, **36**, 28–30.

(2005b). Caroline Herschel as an observer. *JHA*, **36**, 373–406.

(2005c). William Herschel's sweeps for nebulae. *History of Science*, **43**, 305–320.

(2007). *The Herschels of Hannover*. Cambridge: Science History Publications.

Hoskin, M., Warner, B. (1981). Caroline Herschel's comet sweepers. *JHA*, **12**, 27–34.

Hough, G. W. (1883). New nebulae discovered at the Dearborn Observatory, Chicago, U.S.A., with 18.5-inch refractor. *AN*, [2524] **106**, 63.

Hough, G. W., Burnham, S. W. (1881). The nebula near Merope. *MN*, **41**, 410–413.

Houghton, H. (1942). Sir William Herschel's 'Hole in the sky'. *Monthly Notices of the Astronomical Society of South Africa*, **1**, 107–108.

Howard-Duff, I. (1985). George Bishop (1785–1861) and his South Villa Observatory in Regents Park. *JBAA*, **96**, 20–26.

Howe, H. (1898a). Observations of nebulae. *MN*, **58**, 356–361.

(1898b). Observations of nebulae. *MN*, **58**, 515–522.

(1900). Observations of nebulae made at the Chamberlin Observatory, University Park, Colorado. *MN*, **61**, 29–51.

Hubble, E. P. (1916). The variable nebula N.G.C. 2261. *ApJ*, **44**, 190–197.

(1920). The spectrum of N. G. C. 1499. *PASP*, **32**, 155–156.

Hudson, C. (1939). David Todd 1855–1939. *PA*, **47**, 472–477.

Huggins, M. (1880). The late Mr. William Lassell. *Obs*, **3**, 587–590.

(1882). Astronomical drawing. *Obs*, **5**, 358–362.

Huggins, W. (1864). On the spectra of some nebulae. *PT*, **154**, 437–444.

(1866a). Further observations on the spectra of some nebulae, with a mode of determining the brightness of their bodies. *PT*, **156**, 381–397.

(1866b). On the stars within the trapezium of the nebula of Orion. *MN*, **26**, 72–73.

(1867). The trapezium of Orion. *AR*, **5**, 54–55.

(1877). On the inference to be drawn from the appearance of bright lines in the spectra of irresolvable nebulae. *PRS*, **26**, 179–181.

(1881). William Lassell. *MN*, **41**, 188.

Humboldt, A. (1850). *Kosmos. Entwurf einer physikalischen Weltbeschreibung*, vol. 3. Stuttgart: Gotta'scher Verlag.

Hunt, G. (1882). The 'Merope' nebula. *Obs*, **5**, 58–59.

Hutchins, R. (2008). *British University Observatories 1772–1939*. Farnham: Ashgate Publishing.

Huth, H. (1804). An Nebelflecken. *Berliner Jahrbuch für 1807*, 192–194.

Hyde, L. (1987). The calamity of the Great Melbourne Telescope. *Proceedings of the Astronomical Society of Australia*, **7**, 227–230.

Hynes, S. (1991). *Planetary Nebulae*. Richmond, VA: Willmann-Bell.

Ilgauds, H.-J., Münzel, G. (1995). *Die Leipziger Universitätssternwarten auf der Pleißenburg und im Johannistal*. Beucha: Sax-Verlag.

Jackson, J. (1921). Sherburne Wesley Burnham. *Obs*, **44**, 154–158.

(1922). Sherburne Wesley Burnham. *MN*, **82**, 258–263.

Jacoby, H. (1901). Truman Henry Safford. *PASP*, **13**, 181–182.

Jahn, G. A. (1844). *Geschichte der Astronomie vom Anfange des neunzehnten Jahrhunderts bis zum Ende des Jahres 1842*. Leipzig: Verlag Heinrich Hunger.

Jeans, J. (1909). Lord Rosse, 1840–1908. *PRS*, **83**, xv–xix.

(1911). George Johnstone Stoney, 1826–1911. *PRS*, **86**, xx–xxxv.

Jeaurat, E.-S. (1782). Détermination de la position de soixante-quatre étoiles des Pléiades. *Histoire de l'Académie des Sciences 1779*, 505–520.

Jones, B., Boyd, L. (1971). *The Harvard College Observatory. The Four Directorships, 1839–1919*. Cambridge, MA: Harvard University Press.

Kaiser, F. (1839). Schreiben des Herrn Professors Kaiser, Directors der Sternwarte zu Leiden, an den Herausgeber. *AN*, [391] **17**, 97–100.

Kammermann, A. (1886). Ueber den Majanebel. *AN*, [2730] **114**, 313.

Kanipe, J., Webb, D. (2006). *The Arp Atlas of Peculiar Galaxies*. Richmond, VA: Willmann-Bell.

Keeler, J. (1892). On the central star of the Ring Nebula in Lyra. *AN*, [3111] **130**, 227–230.

(1894). Spectroscopic observations of nebulae, made at Mount Hamilton, California, with the 36-inch telescope of the Lick Observatory. *PLO*, **3**, 161–230.

(1895). Note on a case of differences between drawings and photographs of nebulae. *PASP*, **7**, 279–282.

(1898a). The small bright nebula near Merope. *JBAA*, **9**, 133–134.

(1898b). The small bright nebula near Merope. *PASP*, **10**, 245–246.

(1900a). Photographic observations of Hind's Variable Nebula in Taurus, made with the Crossley-reflector of the Lick Observatory. *MN*, **60**, 424–427.

(1900b). The spiral nebula H I, 55 Pegasi. *ApJ*, **11**, 1–5.

(1900c). The Crossley reflector of the Lick Observatory. *ApJ*, **11**, 325–349.

(1908). List of nebulae and clusters photographed. *PLO*, **8**, 30–46.

Kempf, P. (1892). Beobachtungen von Nebelflecken und Sternhaufen mit einem Lamellenmikrometer. *AN*, [3086] **129**, 233–239.

ed. (1911). *Newcomb-Engelmann. Populäre Astronomie*, 4th edn. Leipzig: Engelmann.

Kemps, T. (1955). Caroline Herschel. *Scripta Mathematica*, **21**, 237–251.

Kerner, C . ed. (2004). *Sternenflug und Sonnenfeuer. Drei Astronominnen und ihre Lebensgeschichten*. Weinheim: Beltz & Gelberg.

Kerner, H. (1995). Edward E. Barnard und die Dunkelwolken der Milchstraße. *SuW*, **34**, 844–845.

Kessler, E. (2007). Resolving the nebulae. The science and art of representing M 51. *Studies in History and Philosophy of Science*, **38**, 477–491.

Key, H. C. (1863). On the mode of figuring glass specula for the Newtonian telescope. *MN*, **23**, 199–202.

(1868a). Mr. Lockyer on telescopes. *AR*, **6**, 182.

(1868b). Telescopes. Reflectors and refractors. *AR*, **6**, 226–228.

(1868c). On the planetary nebula 45 H IV. Geminorum. *MN*, **28**, 154–156.

King, E. (1919). Edward Charles Pickering. *JRASC*, **13**, 165–173.

King-Hele, D . ed. (1992). *John Herschel 1792–1871. A Bicentennial Commemoration*. London: Royal Society.

Kippenhahn, R. (1985). Die Supernova im Andromedanebel. *SuW*, **24**, 432–435.

Klare, G. (1970). Ein Jahrhundert wechselvoller Geschichte der Mannheimer Sternwarte 1783–1883. *SuW*, **9**, 148–150.

Knobel, E. B. (1876). Reference catalogue of astronomical papers and researches, 4. Nebulae and clusters. *MN*, **36**, 377–381.

(1887). RAS meeting (January 14, 1887). Pleiades. *Obs*, **10**, 87–88.

(1890). Christian Heinrich Friedrich Peters. *MN*, **51**, 199–203.

(1898). Albert Marth. *MN*, **58**, 139.

(1901). The late Professor Safford. *Obs*, **24**, 349–351.

(1913). Lewis Swift. *MN*, **73**, 217–219.

(1915a). Robert Stawell Ball. *MN*, **75**, 230–236.

(1915b). Edward Singleton Holden. *MN*, **75**, 264–268.

(1927). John Louis Emil Dreyer. *MN*, **87**, 251–257.

Knott, G. (1865a). On the nebulous star 45 H IV. Geminorum. *MN*, **25**, 62–63.

(1865b). Note on the nebulous star 45 H IV. Geminorum. *MN*, **25**, 191–192.

(1879). Note on the gaseous nebula in Cygnus. *MN*, **40**, 91.

(1880). *Obs*, **3**, 452–453.

(1891). U and T Tauri. *Obs*, **14**, 97–98.

Kobold, H. (1893). Berichtigungen zum New General Catalogue of nebulae. *AN*, [3184] **133**, 269–272.

(1894). Berichtigungen zu Dreyer New General Catalogue of nebulae. *AN*, [3241] **136**, 11.

(1919). Edward Charles Pickering. *VJS*, **54**, 274–278.

Konkoly, N. T. (1886). Beobachtungen Andromeda-Nebels. *AN*, [2751] **115**, 251–254.

Kopff, A. (1902). Die Vertheilung der Fixsterne um den grossen Orion-Nebel und den America-Nebel. *Publikationen des Astrophysikalischen Observatoriums Königstuhl-Heidelberg*, **1**, 177–184.

Kövesligethy, R. (1886a). Ueber wahrscheinliche Veränderungen im grossen Andromeda-Nebel. *AN*, [2750] **115**, 231.

(1886b). Beobachtungen Andromeda-Nebels. *AN*, [2755] **115**, 305–310.

Kreutz, H. (1905). Todes-Anzeige Paul Henry. *AN*, [3997] **167**, 223.

Krisciunas, K. (1978). A short history of Pulkovo Observatory. *VA*, **22**, 27–37.

Krüger, A. (1884). Johann Friedrich Julius Schmidt. *AN*, [2577] **108**, 129.

(1886). Beobachtungen Andromeda-Nebels. *AN*, [2752] **115**, 265–268.

(1890). Christian Heinrich Friedrich Peters. *AN*, [2984] **125**, 127.

(1891a). E. Schönfeld. *VJS*, **26**, 173–185.

(1891b). Eduard Schönfeld. *AN*, [3033] **127**, 151.

Lamont, J. (1837). *Über die Nebelflecken*. Munich: K. B. Akademie der Wissenschaften.

(1838). Die königliche Sternwarte bei München. *Jahrbuch der Königlichen Sternwarte bei München für 1838*, **1**, 141–169.

(1839). Jahresbericht der königlichen Sternwarte 1837. *Jahrbuch der Königlichen Sternwarte bei München für 1839*, **2**, 174–188.

(1843). Stellarum in Nebula Orionis. *Observationes Astronomicae in Specula Regia Monachiensi*, **11**, 17–22.

(1869). Refractor-Beobachtungen angestellt an der königl. Sternwarte bei München, B. Messungen an verschiedenen Nebelflecken. *Annalen der Königlichen Sternwarte München*, **17**, 305–316.

Lamp, E. (1883). Beobachtung von neuen Nebeln am Refractor der Kieler Sternwarte. *AN*, [2544] **106**, 381.

Lampland, C. (1936). Hind's Variable Nebula (NGC 1555). *PASP*, **48**, 318–320.

Lamy, J. (2008). *La Carte du Ciel. Histoire et actualité d'un point scientifique international*. Les Ulis: EDP Sciences.

Lankford, J. (1997). *American Astronomy. Community, Careers, and Power, 1859–1940*. Chicago, IL: University of Chicago Press.

Lassell, W. (1841). Description of the observatory erected at Starfield. *MN*, **5**, 107–109.

(1842). Description of an observatory erected at Starfield near Liverpool. *MRAS*, **12**, 265–272.

(1848). Schreiben des Herrn Professors Lassell an den Herausgeber. *AN*, [635] **27**, 171–176.

(1854a). Observations of the nebula in Orion, made at Valletta, with the twenty-foot equatorial. *MRAS*, **23**, 53–57.

(1854b). Miscellaneous observations, chiefly of clusters and nebulae. *MRAS*, **23**, 59–62.

(1854c). Miscellaneous observations, chiefly of clusters and nebulae. *MN*, **14**, 76–77.

(1856). Trapezium in the nebula of Orion. *MN*, **17**, 68.

(1862a). Lettre de M. Lassell. *CR*, **55**, 606–607.

(1862b). Letter to the President from Mr. William Lassell, F.R.S. *PRS*, **12**, 108–110.

(1862c). Extract from a letter from Mr. Lassell. *MN*, **22**, 162–164.

(1863). Letter to Professor Stokes, Sec. R.S., containing observations made at Malta on a planetary nebula. *PRS*, **12**, 269–270.

(1865a). Schreiben des Herrn William Lassell an den Herausgeber. *AN*, [1512] **63**, 369–376.

(1865b). Messier no. 20. *MRAS*, **33**, 121.

(1867a). Miscellaneous observations with the four-foot equatorial at Malta. *MRAS*, **36**, 33–51.

(1867b). A catalogue of new nebulae discovered at Malta with the four-foot equatorial. *MRAS*, **36**, 53–75.

(1868). Remarks on the great nebula in Orion. *PRS*, **16**, 322–329.

(1871). Remarks on the evidence brought forward on the question of a supposed remarkable change in the great nebula near Eta Argus. *MN*, **31**, 249–254.

(1877). On the space-penetrating power of Mr. Lassell's 2-foot and 4-foot reflectors. *MN*, **37**, 124–126.

Latußeck, A. (2003a). Die Nebelschleier des Sir William Herschel. *VdSJ*, **11**, 120–123.

(2003b). Die Nebelschleier des Sir William Herschel. *VdSJ*, **12**, 58–60.

(2004). Die Nebelschleier des Sir William Herschel. *VdSJ*, **13**, 62–65.

(2009). Via Nubilia – Am Grund des Himmels. Johann Georg Hagen und die Kosmischen Wolken. Hamburg: tredition.

Laugier, E. (1847). Sur le mouvement propre de trois amas d'étoiles du catalogue de Messier. *CR*, **24**, 1021–1022.

(1849). Sur l'utilité d'un Catalogue de nébuleuses. *CR*, **28**, 573–576.

(1853). Sur un nouveau Catalogue de nébuleuses observées à l'Observatoire de Paris. *CR*, **37**, 874–879.

Leavenworth, F. (1900). Photographic measures of the Ring Nebula in Lyra and of the neighbouring faint stars. *MN*, **61**, 25–29.

Leiter, F. (2007). Ein Trio von Offenen Sternhaufen? NGC 1746, NGC 1750 und NGC 1758 im Sternbild Taurus. *VdSJ*, **22**, 52–54.

Le Sueur, A. (1870a). An account of the Great Melbourne Telescope from April 1868 to its commencement of operations in Australia in 1869. *PRS*, **18**, 216–222.

(1870b). Spectroscopic observations of the nebula in Orion, and of Jupiter, made with the Great Melbourne Telescope. *PRS*, **18**, 242–245.

(1870c). On the nebulae in Argo and Orion, and on the spectrum of Jupiter. *PRS*, **18**, 245–250.

(1871). Observations with the Great Melbourne Telescope. *PRS*, **19**, 18–19.

Leverington, D. (1995). *A History of Astronomy from 1890 to the Present*. London: Springer.

Leverrier. J. J. (1862a). Nébuleuse de Hind. *Bulletin quotidien de l'Observatoire impérial de Paris* (29.1.1862).

(1862b). Nébuleuse de Hind. *CR*, **54**, 299–301.

(1862c). Lettre de Mr. Wm. Lassell. *Bulletin quotidien de l'Observatoire impérial de Paris* (17.4.1862).

Lewis, T. (1898). *Obs*, **21**, 386–387.

(1908). Asaph Hall. *MN*, **68**, 243.

Lindsay, E. (1965). J. L. E. Dreyer and his New General Catalogue of nebulae and clusters of stars. *LASP*, **9**, 289–296.

Lindsay, J. (1879a). Note on the Rev. T. W. Webb's new nebula. *Dun Echt Circular*, no. **1**.

(1879b). Note on the Rev. T. W. Webb's new nebula. *MN*, **40**, 91–92.

(1879c). Observation of General Catalogue (supplement) no. 6000. *MN*, **40**, 93.

(1880). Dun Echt Circular no. 13. *AR*, **18**, 322.

Littrow, J. J. (1835). *Sterngruppen und Nebelmassen des Himmels*. Vienna: Beck's Universitäts-Buchhandlung.

Littrow, K. L . ed. (1866). *Atlas des gestirnten Himmels für Freunde der Astronomie*. Stuttgart: Gustav Weise.

(1869). Zählung der nördlichen Sterne im Bonner Verzeichnis nach Größen. *Sitzungsberichte der Kaiserlichen Akademie der Wissenschaften, Wien*, **59** (II. Abtheilung), 569–596.

Lockyer, N. (1903). Andrew Ainslie Common. *Nature*, **68**, 132.

Lohse, G. (1885). Observations of the new star in Andromeda, made at Mr. Wigglesworth's Observatory with the 15.5-inch Cooke-refractor. *MN*, **46**, 299–302.

(1887a). Mr. Wigglesworth's Observatory, Scarborough. *MN*, **47**, 162–164.

(1887b). Observations of Nova Cygni, of some of the planets, and of comet Barnard, made at Mr. Wigglesworth's Observatory with the 15.5-inch Cooke equatorial. *MN*, **47**, 494–498.

(1907). Hermann Carl Vogel. *AN*, [4199] **175**, 373–378.

Lorre, J. (1975). Analysis of the nebulosities near T Tauri using digital computer image processing. *ApJ*, **202**, 696–717.

Lubbock, C. (1933). *Herschel Chronicle. The Life-story of Sir William Herschel and His Sister Caroline.* Cambridge: Cambridge University Press.

Lundmark, K. (1930). A new general catalogue of nebulae. *PASP*, **42**, 31–33.

Luther, E. (1860). Aus einem Schreiben des Herrn Prof. Luther der Sternwarte in Königsberg, an den Herausgeber. *AN*, [1267] **53**, 293.

Lynn, W. T. (1884). Dr. Julius Schmidt. *Obs*, 7, 118–199.

(1886a). Discovery of the star-cluster 22 Messier in Sagittarius. *Obs*, **9**, 163–164.

(1886b). The discovery of the great Andromeda Nebula. *Obs*, **9**, 192–194.

(1887). First discovery of the great nebula in Orion. *Obs*, **10**, 232–233.

(1904). Kaspar Gottfried Schweizer. *Obs*, **27**, 314–315.

(1905a). The cluster in Cancer. *Obs*, **28**, 105–106.

(1905b). Otto Struve. *Obs*, **28**, 251–252.

(1909a). H. L. d'Arrest. *Obs*, **32**, 211–212.

(1909b). Robert Lewis John Ellery. *MN*, **69**, 245.

MacGeorge, F. (1874). *Proceedings of the Royal Society of Victoria*, **10**, 106.

MacKeown, K. (2009). William Doberck – double star astronomer. *Journal of Astronomical History and Heritage*, **10**, 49–64.

MacMahon, P. (1919). Gold medal of the Society to M. Guillaume Bigourdan. *MN*, **79**, 306–314.

MacPherson, H. (1906). Ralph Copeland. *PA*, **14**, 1–3.

(1907). Hermann Carl Vogel. *Obs*, **30**, 403–405.

(1908). Charles August Young. *Obs*, **31**, 122–125.

(1914a). Nils Christoffer Dunér. *Obs*, **37**, 446–448.

(1914b). Sir Robert Ball. *PA*, **22**, 28–30.

(1919). *Herschel*. London: Society for Christian Knowledge.

Mädler, J. H. (1849). *Populäre Astronomie*, 4th edn. Berlin: Heymann.

(1873). *Geschichte der Himmelskunde*, vol. 2. Braunschweig: George Westermann.

Maffeo, S. (2002). *The Vatican Observatory. In the Service of Nine Popes*. Paris: University of Notre Dame Press.

Mallas, J. (1966). Letter to Sky & Telescope. *S&T*, **32** (no. 2), 83.

Marriott, B. (1992). J. L. E. Dreyer and the NGC. *Popular Astronomy*, 10/1992, 20–21.

Marth, A. (1856). Bemerkungen zu Sir John Herschel's Nebel-Katalogen. *AN*, [995] **42**, 169–172.

Mason, E. P., Smith, H. (1841). Observations on nebulae with a fourteen feet reflector. *Transactions of the American Philosophical Society*, **7**, 165–213.

Mathews, W. (1865). Silvered reflectors. *AR*, **3**, 75.

(1866). The trapezium, 32 Orionis, nebula in Pleiades. *AR*, **4**, 167–168.

Maurer, A. (1971). William Herschel's astronomical telescopes. *JBAA*, **81**, 284–291.

(1996). *A Compendium of All Known William Herschel Telescopes*. Dahlonega, GA: Antique Telescope Society.

Mayall, M. ed. (1962). *Celestial Objects for Common Telescopes, Vol. 2. The Stars*. New York, NY: Dover Publications.

McCance, J. (1880a). The nebula in the Pleiades. *Obs*, **3**, 418.

(1880b). The nebula in the Pleiades. *Obs*, **3**, 484.

McConnell, A. (1992). *Instrument Makers to the World. A History of Cooke, Troughton & Simms*. York: William Sessions.

(1994). Astronomers at war. The viewpoint of Troughton & Simms. *JHA*, **25**, 219–235.

McFarland, J. (1990). The historical instruments of Armagh Observatory. *VA*, **33**, 149–210.

(2002). Dreyer's sesquicentennial. *A&G*, **43**, 1.22.

McKenna, S. (1967). Astronomy in Ireland from 1780. *VA*, **9**, 238–296.

McKenna-Lawlor, S., Hoskin, M. (1984). Correspondence of the Markree Observatory. *JHA*, **15**, 64–68.

Mee, A. (1897). *Observational Astronomy. A Practical Book for Amateurs*. Cardiff: Llanishen.

(1905). Thomas William Webb. *PA*, **13**, 138–140.

Merecki, R. (1902). Observations micrométriques de nébuleuses. *Observatoire Astronomique Jedrzejewicz à Varsovie*, I. Partie.

Messier, C. (1774). Catalogue des nébuleuses et des amas d'étoiles. *Histoire de l'Académie de Sciences 1771*, 435–461.

(1780). Catalogue des nébuleuses et des amas d'étoiles. *Connaissance des Temps pour l'année 1783*, 225–254, 408.

(1781). Catalogue des nébuleuses et des amas d'étoiles. *Connaissance des Temps pour l'année 1784*, 227–272.

Michaelis, J. (1856). Chacornac's Sterncharten. *Unterhaltungen im Gebiete der Astronomie, Geographie und Meteorologie*, **10**, 301–305.

Michaud, M. (1983). Astronomy at Armagh. *S&T*, **65** (no. 1), 17–19.

Mitchell, O. (1851). *The Orbs of the Heaven*. London: Office of the National Illustrated Library.

Mitchell, S. (1923). Edward Emerson Barnard, 1857–1923. *Obs*, **46**, 158–164.

Möbius, A. (1848). Schreiben des Herrn Prof. A. F. Möbius an den Herausgeber. *AN*, [641] **27**, 265–268.

Mönnichmeyer, C. (1895). Beobachtung von Nebelflecken angestellt am sechszölligen Refractor der Bonner Sternwarte. *Veröffentlichungen der Königlichen Sternwarte Bonn*, no. **1**.

Moore, P. (1967). *Armagh Observatory 1790–1967*. Armagh: Armagh Observatory.

(1971). *The Astronomy of Birr Castle*. Birr: Tribune Printing & Publishing.

(1995). Beyond Messier. The Caldwell Catalog. *S&T*, **90** (no. 6), 38–43.

Mouchez, A. (1885). Carte photographique du ciel a l'aide des nouveaux objectives de MM. P. et Pr. Henry. *CR*, **100**, 1177–1181.

(1886). Photographie céleste. *CR*, **102**, 289–290.

Mumford, G. (1987). The legacy of E. E. Barnard. *S&T*, **74** (no. 1), 30–34.

Münzel, G. (2000). Gustav Adolph Jahn, ein Leipziger Astronom des 19. Jahrhunderts. *Beiträge zur Astronomiegeschichte*, **3**, 101–119.

Murphy, F. (1965). The Parsons of Parsonstown. *IAJ*, **7**. 53–58.

Naegamvala, K. (1895). Nebula no. 6595 of the New General Catalogue. *Obs*, **18**, 310.

Nasim, O. W. (2010a). Observation, working images and procedure: the 'Great Spiral' in Lord Rosse's astronomical record books and beyond. *British Journal for the History of Science*, **44**, in press.

(2010b). The 'landmark' and 'groundwork' of stars: John Herschel, photography and the drawing of nebulae. *Studies in History and Philosophy of Science*, in press.

Nasmyth, J. (1855). Suggestions respecting the origin of the rotatory movements of the celestial bodies and the spiral forms of the nebulae as seen in Lord Rosse's telescopes. *MN*, **15**, 220–221.

Neubauer, F. J. (1950). A short history of the Lick Observatory. *PA*, **58**, 201–222, 318–334, 369–388.

Neumayer, G. (1901). George Friedrich Wilhelm Rümker. *VJS*, **36**, 2–5.

Newcomb, S. (1878). *Astronomical and Meteorological Observations Made during the Year 1875 at the U. S. Naval Observatory*, vol. 15.

(1879). *Astronomical and Meteorological Observations Made during the Year 1876 at the U. S. Naval Observatory*, vol. 16.

Nichol, J. P. (1837). *The Architecture of the Heavens*. London: John Parker.

(1846). *Thoughts on Some Important Points Relating to the System of the World*. Edinburgh: William Tait.

Nilson, P. (1973). *Uppsala General Catalogues of Galaxies*. Uppsala: Royal Society of Sciences.

Nobel, W. (1886). *Hours with a Three-inch Telescope*. London: Longmans, Green & Co.

Nugent, R. (2002). The nature of the double star M 40. *JRASC*, **96**, 63–65.

Numazawa, S., Wakiya, N. (2009). *NGC-IC Photographic Catalogue*. Tokyo: Seibundo Shinkosha Publishing.

Nyrén, M. (1905a). Otto Wilhelm Struve. *AN*, [4013] **168**, 78.

(1905b). Otto Wilhelm Struve. *VJS*, **40**, 286–303.

O'Dell, R. (1965). Photoelectric spectrophotometry of gaseous nebulae. *ApJ*, **142**, 604–608.

Oestmann, G. (2007). Zur frühen Geschichte der Dorpater Sternwarte. *Acta Historica Astronomiae*, **33**, 315–331.

O'Hora, N. (1961). The Dunsink Observatory. *Obs*, **81**, 189–195.

Olbers, W. (1802). Neuer Nebel. *Monatliche Correspondenz zur Beförderung der Erd- und Himmelskunde*, **5**, 500.

(1827). Auszug aus einem Schreiben des Herrn Doctors und Ritters Olbers, an den Herausgeber. *AN*, [104] **5**, 121.

(1828). Auszug aus einem Schreiben des Herrn Doctors und Ritters Olbers an den Herausgeber. *AN*, [148] **7**, 61–64.

(1834). Anzeige des Herrn Doctors und Ritters Olbers, der Observations of Nebulae and Clusters of Stars made at Slough. *AN*, [261] **11**, 373–378.

(1836). Schreiben des Herrn Dr. Olbers an den Herausgeber. *AN*, [309] **13**, 337–340.

Oliver, C. (1933). Ormond Stone. *PA*, **41**, 295–198.

Olmstedt, D. (1842). *Life and Writings of Ebenezer Porter Mason. Interspersed with Hints to Parents and Instructors on the Training and Education of a Child of Genius*. New York, NY: Dayton & Newman.

O'Meara, S. (2006). M 102. Mystery solved. *S&T*, **109** (no. 3), 78–79.

(2007). *Herschel 400 Observing Guide*. Cambridge: Cambridge University Press.

O'Meara, S., Green, D. (2003). The mystery of the Double Cluster. *S&T*, **105** (no. 2), 115–119.

Oppolzer, E. (1868). Beobachtung einiger Nebelflecke. *AN*, [1666] **70**, 155–158.

Osterbrock, D. (1984). The rise and fall of Edward S. Holden. *JHA*, **15**, 81–127, 151–176.

Oudemans, J. (1888). Über die Sichtbarkeit von Nebelflecken und Protuberanzen. *Sirius*, **16**, 166.

Palisa, J. (1883). Beobachtung von neuen Nebeln am 12zöll. Refractor der Wiener Sternwarte. *AN*, [2544] **106**, 381.

Pannekoek, A. (1923). *Annals van de Sterrewacht te Leiden*, **14** (no. 3), 5–8.

Parkhurst, J. (1923). Edward Emerson Barnard, 1857–1923. *JRASC*, **17**, 97–103.

Parsons, C. (1926). *The Scientific Papers of William Parsons, Third Earl of Rosse, 1800–1867*. London: Percy Lund, Humphries and Co.

Parsons, L. (1868). An account of the observations on the great nebula in Orion, made at Birr Castle, with the 3-feet and 6-feet telescopes, between 1848 and 1867. *PT*, **158**, 57–73.

(1870). Lunar temperature. *MN*, **30**, 107–108.

(1875). The Earl of Rosse's Observatory, Birr Castle, Parsonstown. *MN*, **35**, 194.

(1876). The Earl of Rosse's Observatory, Birr Castle, Parsonstown. *MN*, **36**, 172.

(1877a). The Earl of Rosse's Observatory, Birr Castle, Parsonstown. *MN*, **37**, 175.

(1877b). Nebulous star in the Pleiades. *Nature*, **15**, 397.

(1878). The Earl of Rosse's Observatory, Birr Castle. *MN*, **38**, 185–186.

(1879). Observatory, Birr Castle. *MN*, **39**, 252–253.

(1880). Observations of nebulae and clusters of stars made with the six-foot and three-foot reflectors at Birr Castle, from the year 1848 up to the year 1878. *Scientific Transactions of the Royal Dublin Society*, **2**, 1–178.

(1890). The Earl of Rosse's Observatory, Birr Castle. *MN*, **50**, 211.

Parsons, W. (1828). Account of a new reflecting telescope. *Edinburgh Journal of Science*, **9** (no. 17), 25–30.

(1840). An account of experiments on the reflecting telescope. *PT*, **130**, 503–527.

(1844a). Observations on some of the nebulae. *PT*, **134**, 321–324.

(1844b). Observations on some of the nebulae. *PRS*, **5**, 513–541.

(1845). On the nebula 25 Herschel, or 61 of Messier's Catalogue. *RBA*, **15**, 4.

(1850a). Observations on the nebulae. *PT*, **140**, 499–514.

(1850b). Observations on the nebulae. *PRS*, **5**, 962–966.

(1861a). On the construction of specula of six-feet aperture and a selection from the observations of nebulae made with them. *PT*, **151**, 681–745.

(1861b). Further observations upon the nebulae, with practical details relating to the construction of large telescopes. *PRS*, **11**, 375–376.

Paterson, J. (1916). Sir Robert Ball. The astronomer, the mathematician and the man. *JRASC*, **10**, 42–63.

(1923). Edward Emerson Barnard. His life and work. *JRASC*, **18**, 309–318.

Payne, W. W. (1894). The Pleiades. *PA*, **1**, 456–460.

(1908). Earl of Rosse. *PA*, **16**, 570–572.

Pease, F. (1917a). Hind's Variable Nebula N.G.C. 1555. *ApJ*, **45**, 89–92.

(1917b). Photographs of nebulae with the 60-inch reflector, 1911–1916. *ApJ*, **46**, 24–55.

Pechüle, C. F. (1886). Beobachtungen angestellt mit dem grossen Kopenhagener Refractor. *AN*, [2710] **113**, 361–364.

Perdrix, J. L. (1992). The last great speculum. The 48-inch Great Melbourne Telescope. *Australian Journal of Astronomy*, **4**, 149–163.

Perrotin, H. (1886). Observation de la nébuleuse de Maia. *CR*, **102**, 544–545.

Peters, C. F. W. (1881). William Lassell. *AN*, [2341] **98**, 207.

Peters, C. H. F. (1856). New Works. *AJ*, **5**, 16.

(1864). Ernesto Capocci. *AN*, [1461] **61**, 321.

(1868). Beobachtung und Elemente eines neuen Planeten (98) Ianthe. *AN*, [1695] **71**, 239.

(1878). Verschiedene Beobachtungen von der Sternwarte des Hamilton College. *AN*, [2258] **95**, 19–22.

(1881a). Schreiben des Herrn Prof. C. H. F. Peters an den Herausgeber. *AN*, [2361] **99**, 141.

(1881b). Beobachtungen einiger Nebelflecke und des Cometen III. 1864. *AN*, [2365] **99**, 203.

(1881c). Positions of nebulae. Series I. *Cop*, **1**, 51–54.

(1882). Positions of nebulae. Series II. *Cop*, **2**, 54–55.

Petersen, A. C. (1850). Letter from Dr. A. C. Petersen, of the Altona Observatory, to the editor. *AJ*, **1**, 47–48.

(1853). A letter of Professor Petersen to the editor. *AJ*, **3**, 70–71.

Pfaff, J. W. (1826). *William Herschel's sämtliche Schriften. Erster Band. Ueber den Bau des Himmels.* Dresden and Leipzig: Arnold.

Pickering, E. C. (1880). New planetary nebulae. *American Journal of Science*, **20**, 303–305.

(1881a). New planetary nebulae. *Obs*, **4**, 81–83.

(1881b). New planetary nebula. *Cop*, **1**, 242.

(1882a). Nebulae. *AHO*, **13**, 62–85.

(1882b). New planetary nebulae. *AN*, [2454] **103**, 95.

(1882c). New planetary nebulae. *AN*, [2459] **103**, 165.

(1882d). New planetary nebula. *Obs*, **5**, 26–27.

(1882e). Small planetary nebulae discovered at the Harvard College Observatory. *Obs*, **5**, 294–295.

(1882f). Small planetary nebula. *Obs*, **5**, 342–343.

(1883). New planetary nebulae. *AN*, [2517] **105**, 335.

(1895). A new star in Centaurus. *Harvard Circular*, **4**, 1–2.

(1886). On the new nebula discovered in the Pleiades by MM. Henry. *AN*, [2712] **113**, 399.

(1897). Distribution of stars in clusters. *AN*, [3422] **143**, 229.

(1904). Common's 60-inch telescope. *PA*, **12**, 660–662.

(1908). Nebulae discovered at the Harvard College Observatory. *AHO*, **60**, 147–194.

Pickering, W. H. (1890). The great nebula in Orion. *SM*, **9**, 1–2.

(1917). The sixty finest objects in the sky. *PA*, **25**, 75–86.

Pigott, E. (1781). Account of a nebula in Coma Berenices. *PT*, **71**, 82–83.

Plotkin. H. (1990). Edward Charles Pickering. *JHA*, **21**, 47–58.

Plummer, W. (1896). Dr. John Russel Hind, F.R.S. *AN*, [3328] **139**, 255.

Pogson, N. (1860). Remarkable changes observed in the cluster 80 Messier. *MN*, **21**, 32–33.

Poor, J. (1908). Charles August Young. *PA*, **16**, 218–230.

Porro, I. (1857). Schreiben des Herrn Porro an den Herausgeber, betreffend die Entdeckung eines neuen Sterns im Trapez des Orion. *AN*, [1091] **46**, 171–174.

Porter, J. G. (1884). Note on a southern nebula. *SM*, **4**, 314–315.

(1890). Christian Heinrich Friedrich Peters. *SM*, **9**, 138–139.

(1891). Charts and micrometrical measures of nebulae, made in the years 1884, 1885, and 1886. *Publications of the Cincinnati Observatory*, vol. **11**.

(1910). Micrometrical measures of nebulae, 1905 to 1910. *Publications of the Cincinnati Observatory*, vol. **17**.

Powell, B. (1863). Notes on Alpha Centauri and other southern binaries, and on the nebula about Eta Argus. *MN*, **24**, 170–172.

Pratt, H. (1881). The nebula in the Pleiades. *MN*, **41**, 316.

Pritchard, C. (1864). Wilhelm Struve. *MN*, **25**, 83–99.

(1883). Meeting of the RAS (May 11, 1883). *Obs*, **6**, 171.

(1888). Report on the capacities, in respect of light and photographic action, of two silver on glass mirrors of different focal lengths. *PRS*, **44**, 168–182.

(1891a). A newly discovered Merope nebula. *AN*, [3024] **126**, 397–400.

(1891b). Further experiments regarding the magnitude of stars as obtained by photography in the Oxford University Observatory. *MN*, **51**, 430–435.

Pritchett, H. (1908). Asaph Hall. *PA*, **16**, 67–70.

Proctor, R. (1869a). Distribution of nebulae. *MN*, **29**, 337–344.

(1869b). The Rosse telescope set to new work. *Fraser's Magazine for Town and Country*, **80**, 754–760.

(1871a). Note on the construction of the heavens, explanatory chart of 324,198 stars. *MN*, **32**, 1–7.

(1871b). Note on Mr. Abbott's imagined discovery of great changes in the Argo nebula. *MN*, **32**, 62.

(1872a). The rich nebular regions in Virgo and Coma Berenices. *MN*, **33**, 14–17.

(1872b). Statements of views respecting the sidereal universe. *MN*, **33**, 539–552.

Radrizzani, A. (1989). Tempel a Brera. Astronomo e artista. *L'Astronomia*, **86** (no. 3), 25–31.

Ragona, D. (1868). Giuseppe Bianchi. *VJS*, **3**, 167–169.

Rambaut, A. A. (1885). The Rev. Prebendary Webb. *AR*, **23**, 148–149.

(1914). Sir Robert Stawell Ball. *Obs*, **37**, 36–41.

Ranyard, A. (1886). Thomas William Webb. *MN*, **46**, 198–201.

Rayet, G. (1886). Position d'étoiles téléscopiques de la constellation des Pléiades. *CR*, **102**, 489–492.

Reiland, T. (1982). Rare plate whereabouts? *S&T*, **64** (no. 1), 5.

Reinmuth, K. (1915). Photographische Positionsbestimmungen von 356 Schultzschen Nebelflecken. *Veröffentlichungen der Grossherzöglichen Sternwarte Heidelberg*, **7**, 141–168.

(1926). Die Herschel-Nebel. Abhandlungen der Heidelberger Akademie der Wissenschaften, 13. Abhandlung.

Repsold, A. (1837). Meridiankreis von A. und G. Repsold, aufgestellt in der Hamburger Sternwarte im Frühjahr 1836. *AN*, [349] **15**, 225–228.

Repsold, J. (1908). *Zur Geschichte der astronomischen Messwerkzeuge*, 2 vols. Leipzig: Engelmann.

(1918). Arthur v. Auwers. *VJS*, **53**, 15–23.

Reynolds, J. H. (1916). The variable nebula in Corona Australis (N.G.C. 6729). *MN*, **76**, 645–646.

Rheden, R. (1928). Samuel Oppenheim. *AN*, [5585] **233**, 295–296.

Rigge, W. (1918). Father Angelo Secchi. *PA*, **26**, 589–598.

Ring, F. (1992). John Herschel and his heritage, in *John Herschel 1792–1871. A Bicentennial Commemoration*, ed. D. King-Hele. London: Royal Society, pp. 3–15.

Roberts, I. (1886). Note on two photographs of the nebula in the Pleiades taken in October 1886. *MN*, **47**, 24.

(1887). Photographs of nebulae in Orion and in the Pleiades. *MN*, **47**, 89–91.

(1889a). Photographs of nebulae in the Pleiades and in Andromeda. *MN*, **49**, 120–121.

(1889b). Photographs of nebula M 51 Canis Venaticorum. *MN*, **49**, 389–390.

(1891). Photograph of the region of Hind's Variable Nebula in Taurus. *MN*, **51**, 440–441.

(1893). *A Selection of Photographs of Stars, Star-clusters and Nebulae*. London: Universal Press.

(1896a). On the relative efficiency of a reflector and of portrait lenses for the delineation of celestial objects. *MN*, **56**, 372–378.

(1896b). Photograph of the 'owl' nebula M 97 and of the nebula H V 46 Ursa Majoris. *MN*, **56**, 378–379.

(1898). Photographs of the nebulae in the Pleiades, of stars in the surrounding regions and of spurious nebulosity. *MN*, **58**, 392–397.

(1899). *Photographs of Stars, Star-clusters and Nebulae*. London: Universal Press.

(1902). William Herschel's observed nebulous regions, 52 in number, compared with Isaac Roberts' photographs of the same regions, taken simultaneously with the 20-inch reflector and the 5-inch Cooke lens. *MN*, **63**, 26–34.

(1903a). William Herschel's observed nebulous regions. *AN*, [3836] **160**, 337–344.

(1903b). Herschel's nebulous regions. *ApJ*, **17**, 72–76.

(1903c). On a region in Lynx rich in nebulae. *AN*, [3857] **161**, 301.

Robinson, J., Robinson, M. eds. (2006). *The Stargazer of Hardwicke*. Leominster: Gracewing.

Robinson, T. R. (1840). Account of a large reflecting telescope. *Proceedings of the Royal Irish Academy*, **2**, 2–12.

(1845). An account of the Earl of Rosse's great telescope. *Proceedings of the Royal Irish Academy*, **3**, 113–133.

(1848). An account of the present condition of the Earl of Rosse's great telescope. *Proceedings of the Royal Irish Academy*, **4**, 119–128.

(1851). Copy of the memorial to Lord John Russell. *RBA*, **20**, xvii–xix.

(1853). Drawings to illustrate recent observations on nebulae. *RBA*, **22**, 22–23.

(1867a). Sir James South. *PRS*, **16**, xliv–xlvii.

(1867b). William Parsons, Third Earl of Rosse. *PRS*, **16**, xxxvi–xlii.

(1876). On the relative power of achromatic and reflecting telescopes. *MN*, **36**, 305–309.

Rogers, W. (1875). Death of Professor Winlock, Director of Harvard College Observatory. *AN*, [2048] **86**, 113–118.

Rothenberg, M. (1990). Patronage of Harvard College Observatory, 1839–1851. *JHA*, **21**, 37–46.

Roy, A. (1993). Glasgow and the heavens. *VA*, **36**, 389–407.

Rümker, G. (1865a). Schreiben des Herrn Dr. G. Rümker an den Herausgeber. *AN*, [1542] **65**, 93.

(1865b). Beobachtungen von Circumpolarnebeln auf der Hamburger Sternwarte. *AN*, [1508] **63**, 305–318.

(1865c). Beobachtungen von Circumpolarnebeln auf der Hamburger Sternwarte. *AN*, [1531] **64**, 289–298.

(1865d). Beobachtungen von Circumpolarnebeln auf der Hamburger Sternwarte. *AN*, [1566] **66**, 81–92.

(1866). Beobachtungen von Circumpolarnebeln auf der Hamburger Sternwarte. *AN*, [1599] **67**, 225–238.

(1867). Beobachtungen von Circumpolarnebeln auf der Hamburger Sternwarte. *AN*, [1631] **68**, 353–364.

(1895). Positionsbestimmungen von Nebelflecken und Sternhaufen. *Mitteilungen der Hamburger Sternwarte*, no. 1.

Russell, H. C. (1872). The colored cluster about κ Crucis. *MN*, **33**, 66–70.

Russell, H. N. (1909). Charles August Young. *MN*, **69**, 257–260.

Sabine, E. (1853). Royal medal for Lord Rosse. *PRS*, **6**, 113–114.

(1869). Great Melbourne Telescope. *PRS*, **18**, 104–105.

Sadler, H. (1879). Note on the late Admiral Smyth's 'Cycle of Celestial Objects', volume the second, commonly known as the 'Bedford Catalogue'. *MN*, **39**, 183–195.

(1885). Note on a possible case of proper motion in a nebula. *Obs*, **8**, 127–128.

Safford, T. H. (1887). Nebulae found at the Dearborn Observatory, 1866–68. *Annual Report of the Board of Directors of the Chicago Astronomical Society, together with the Report of the Director of the Dearborn Observatory for 1885 and 1886*, 37–41.

Sampson, R. A. (1916). Gold medal to Dr. J. L. E. Dreyer. *MN*, **76**, 368–375.

(1923). The decade 1840–1850, in *History of the Royal Astronomical Society*, vol. 1, eds. J. L. E. Dreyer, H. H. Turner. London: Royal Astronomical Society, pp. 82–109.

(1934). John Louis Emil Dreyer, J. L. E. (1852–1926). *Isis*, **21**, 131–144.

Sandage, A. (1961). *The Hubble Atlas of Galaxies*. Washington D.C.: Carnegie Institution.

Sandage, A., Bedke, J. (1994). *The Carnegie Atlas of Galaxies*. Washington D.C.: Carnegie Institution.

Sawerthal, H. (1889). Notes on visits to some continental observatories. *Obs*, **12**, 344–349.

Sayre, H. A. (1892). The Leander McCormick Observatory. *PASP*, **4**, 112–114.

Sawyer-Hogg, H. (1947a). Halley's List of Nebulous Objects. *JRASC*, **41**, 69–71.

(1947b). Derham's catalogue of nebulous objects from Hevelius' Prodomus. *JRASC*, **41**, 233–237.

(1947c). Catalogues of nebulous objects in the eighteenth century. *JRASC*, **41**, 265–273.

Schaeberle, M. (1903a). On the photographic efficiency of a 13-inch reflector of 20-inches focus. *AJ*, **23**, 109–113.

(1903b). The Ring Nebula in Lyra and the Dumb-bell Nebula in Vulpecula as great spirals. *AJ*, **23**, 181–182.

Schaffer, S. (1980). 'The great laboratories of the universe'. William Herschel on matter theory and planetary life. *JHA*, **11**, 81–110.

(1981). Uranus and the establishment of Herschel's astronomy. *JHA*, **12**, 11–26.

(1989). The nebular hypothesis and the science of progress, in *History, Humanity & Evolution*, ed. J. R. Moore. Cambridge: Cambridge University Press, pp. 131–164.

(1998a). The Leviathan of Parsonstown. Literary technology and scientic representation, in *Inscribing Science. Scientific Texts and the Materiality of Communication*, ed. T. Lenoir. Oxford: Oxford University Press, pp. 182–222.

(1998b). On astronomical drawing, in *Picturing Science, Producing Art*, eds. C. Jones, P. Galison. New York: Routledge, pp. 441–474.

Scheiner, J. (1892). Ueber die Planetarischen Nebel h 2098 und h 2241. *AN*, [3086] **129**, 239.

(1893). Ueber den grossen Nebel bei ξ Persei (NGC. 1499). *AN*, [3157] **132**, 203–206.

Schiaparelli, G. (1889a). Guglielmo Ernesto Tempel. *AN*, [2886] **121**, 95.

(1889b). Wilhelm Ernst Tempel. *Himmel und Erde*, **1**, 486–488.

Schjellerup, H. (1867). Verschiedene Bemerkungen und Zusätze, mitgetheilt von Herrn Prof. Schjellerup. *AN*, [1613] **68**, 65–68.

Schmidt, J. (1854). Bericht über die neue Sternwarte des Herrn E. Ritter von Unkrechtsberg zu Ölmütz in Mähren. *AN*, [869] **37**, 73–78.

(1859). Nachrichten über die Sternwarte zu Athen, von dem Direktor derselben, Herrn J. J. Julius Schmidt. *AN*, [1193] **50**, 267–270.

(1861a). Beobachtungen auf der Sternwarte zu Athen im Jahre 1860, von Herrn J. F. Julius Schmidt. *AN*, [1293] **54**, 321–330.

(1861b). Über einen neuen veränderlichen Nebelstern. *AN*, [1302] **55**, 91–94.

(1862a). Beobachtungen auf der Sternwarte zu Athen. *AN*, [1355] **57**, 157.

(1862b). Über veränderliche Nebelgestirne, von Herrn J. F. Jul. Schmidt, Director der Sternwarte zu Athen. *AN*, [1360] **57**, 243–246.

(1862c). Beobachtungen auf der Sternwarte zu Athen. *AN*, [1391] **58**, 353.

(1864). Ueber die Ortsbestimmung der Nebelgestirne. *AN*, [1463] **61**, 365–368.

(1865a). Ueber die Beobachtung der Nebelgestirne. *AN*, [1513] **64**, 1–8.

(1865b). Beobachtungen auf der Sternwarte zu Athen. *AN*, [1529] **64**, 271.

(1865c). Beobachtungen der Nebelgestirne und des Faye'schen Cometen. *AN*, [1553] **65**, 261–268.

(1867a). Beobachtungen auf der Sternwarte zu Athen. *AN*, [1613] **68**, 69.

(1867b). Beobachtung des Cometen II. 1867 auf der Sternwarte zu Athen. *AN*, [1651] **69**, 301.

(1868a). Bemerkungen über Nebel und veränderliche Sterne. *AN*, [1672] **70**, 245–250.

(1868b). Mittlere Örter von 110 Nebeln für 1865. *AN*, [1678] **70**, 343–352.

(1868c). Beobachtungen auf der Sternwarte zu Athen im Jahre 1868. *AN*, [1745] **73**, 259–272.

(1873). Beobachtungen auf der Sternwarte zu Athen. *AN*, [1932] **81**, 177–188.

(1876a). Ueber einige im Cape-Catalog fehlende Nebel. *AN*, [2097] **88**, 137–140.

(1876b). Ueber die Lage eines neuen veränderlichen Sterns in der südlichen Krone. *AN*, [2106] **88**, 283.

(1877). Ueber den neuen Stern im Schwan. *AN*, [2113] **89**, 9–14.

(1880). Webb's Nebula im Schwan. *AN*, [2309] **97**, 69.

(1881). Anmerkung zur Bonner Durchmusterung. *AN*, [2339] **98**, 175.

(1884). Veränderliche Sterne 1883, beobachtet zu Athen. *AN*, [2577] **108**, 131–148.

Schmidt-Kaler, T. (1983). Deutsche Astronomen um 1880 – private Eindrücke und Erlebnisse. *Die Sterne*, **59**, 228–235, 290–294.

Schönfeld, E. (1861). Schreiben des Herrn Prof. Schönfeld, Director der Sternwarte Mannheim, an den Herausgeber. *AN*, [1310] **55**, 211–214.

(1862a). Über den Nebelfleck Zone +30° No. 548 des Bonner Sternverzeichnisses, mit einigen Bemerkungen über die Nebelbeobachtungen in der Bonner Sternwarte überhaupt. *AN*, [1391] **58**, 355–360.

(1862b). Ueber die Nebelflecke. *28. Jahresbericht des Mannheimer Vereins für Naturkunde*, 46–87.

(1862c). Premier cahier d'observations des nébuleuses. *CR*, **55**, 792.

(1862d). *Beobachtungen von Nebelflecken und Sternhaufen. Erste Abtheilung*. Mannheim: Bensheimer.

(1865). Schreiben des Herrn Professor Schönfeld, Director der Sternwarte in Mannheim, an den Herausgeber. *AN*, [1523] **64**, 161–176.

(1866). Catalog von veränderlichen Sternen mit Einschluss der neuen Sterne. *32. Jahresbericht des Mannheimer Vereins für Naturkunde*, 59–109.

(1867). *AN*, [1648] **69**, 246.

(1875a). Catalog von veränderlichen Sternen mit Einschluss der neuen Sterne. *40. Jahresbericht des Mannheimer Vereins für Naturkunde*, 49–120.

(1875b). *Beobachtungen von Nebelflecken und Sternhaufen. Zweite Abtheilung*. Karslruhe: Braun'sche Hofbuchhandlung.

Schorr, R. (1900). George Friedrich Wilhelm Rümker. *AN*, [3632] **152**, 127.

Schramm, J. (1996). *Sterne über Hamburg. Die Geschichte der Astronomie in Hamburg*. Hamburg: Kultur- und Geschichtskontor.

Schultz, H. (1864). *Beobachtung von Nebelflecken im Jahre 1863. Refractor-Beobachtungen der K. Universitäts-Sternwarte in Upsala*. Uppsala: Edquist & Berglund.

(1865a). Aus einem Schreiben des Herrn Schultz an den Herausgeber. *AN*, [1504] **63**, 243–246.

(1865b). Beobachtung von Nebelflecken. *AN*, [1541] **65**, 65–78.

(1865c). Beobachtung von Nebelflecken. *AN*, [1555] **65**, 297–300.

(1865d). Schreiben des Herrn Dr. H. Schultz an den Herausgeber. *AN*, [1556] **65**, 315–318.

(1865e). On the star Baily Lalande 14512 and nebula H IV. 45. *MN*, **25**, 189–191.

(1866a). Schreiben des Herrn Dr. Herman Schultz an den Herausgeber. *AN*, [1563] **66**, 47.

(1866b). Historische Notizen über Nebelflecke. *AN*, [1585] **67**, 1–6.

(1868). Schreiben des Herrn Dr. H. Schultz in Upsala an den Herausgeber. *AN*, [1665] **70**, 135.

(1873). Der Sternhaufen 20 Vulpeculae = H, VIII. 20. *AN*, [1898] **80**, 21–28.

(1874). Micrometrical observations of 500 nebulae. *Nova Acta Regiae Societatis Scientiarum Upsaliensis, ser. 3*, vol. **9**.

(1875). Preliminary catalogue of nebulae observed at Upsala. *MN*, **35**, 135–152.

(1876). Gewährt J. Herschel's Vorschlag, Nebelflecke mit Nummern zu beschreiben, reelle Vortheile oder nicht? *VJS*, **11**, 73–77.

(1886). Mikrometrische Bestimmung einiger teleskopischer Sternhaufen. *Bihang till Kongliga Svenska Vetenskaps-akademiens Handlingar*, vol. **12**.

Schwartz, R. (1974). The T Tauri emission nebula. *ApJ*, **191**, 419–432.

Searle, A. (1902). Truman Henry Safford. *AN*, [3749] **157**, 95.

Searle, G. M. (1890). Are the planets habitable? *PASP*, **2**, 165–177.

Secchi, A. (1853). Entdeckung eines neuen Nebelflecks. *AN*, [855] **36**, 243.

(1856a). Schreiben des Herrn Professors Secchi, Director der Sternwarte Coll. Rom., an den Herausgeber. *AN*, [1018] **43**, 157–160.

(1856b). Osservazioni delle nebulose. *Memorie dell'Osservatorio del Collegio Romano 1852–55*, 80–95.

(1857a). Mesures de Saturne et de ses anneaux. *AN*, [1060] **45**, 53–60.

(1857b). Notes on the nebula of Orion and other astronomical subjects. *MN*, **18**, 8–11.

(1865). Schreiben des Herrn Professors Secchi, Director der Sternwarte des Collegio Romano, an den Herausgeber. *AN*, [1553] **65**, 257–262.

(1866). Schreiben des Herrn P. A. Secchi an den Herausgeber. *AN*, [1571] **66**, 161.

(1867). Spectral studies on some of the planetary nebulae. *AR*, **5**, 40.

See, T. (1921). Biographical notice to Professor S. W. Burnham of Chicago. *AN*, [5097] **213**, 141–144.

(1923). Brief biographical note of Professor E. E. Barnard. *AN*, [5224] **218**, 247–252.

Seeliger, H. (1915). Arthur v. Auwers. *AN*, [4788] **200**, 187–190.

Seyfert, C. (1951). A dense group of galaxies in Serpens. *PASP*, **63**, 72–75.

Shapley, H., Ames, A. (1932). A survey of external galaxies brighter than the thirteenth magnitude. *AHO*, **88**, 41–75.

Shapley, H., Davis, H. (1917). Messier's catalog of nebulae and clusters. *PASP*, **29**, 177–179.

Sheehan, W. (1995). *The Immortal Fire Within. The Life and Work of Edward Emerson Barnard*. Cambridge: Cambridge University Press.

(1998). Christian Heinrich Friedrich Peters, September 19, 1813-July 18, 1890. *Biographical Memoirs of the National Academy of Sciences*, **76**, 3–26.

Sheehan, W., O'Meara, S. (1998). Phillip Sidney Coolidge. Harvard's romantic explorer of the skies. *S&T*, **95** (no. 4), 71–75.

Sinnott, R. W. (1988). *NGC 2000.0*. Cambridge, MA: Sky Publishing Corporation.

Skiff, B. (1996). What is IC 2120? *Webb Society Quarterly Journal*, no. **76**, 1–9.

Slipher, V. (1911). On the spectrum of the nebula in the Pleiades. *Bulletin of the Lowell Observatory*, **2**, 26–27.

Smyth, C. P. (1843). Memoir on astronomical drawing. *MN*, **5**, 277–279.

(1846). On astronomical drawing. *MRAS*, **15**, 71–82.

Smyth, W. H. (1830). An account of a private observatory, recently erected at Bedford. *MN*, **1**, 197–200.

(1844). *A Cycle of Celestial Objects*, Vol. 2. The Bedford Catalogue. London: J. W. Parker.

South, J. (1826). Auszug aus einem Schreiben des Herrn J. South F.R.S. an den Herausgeber. *AN*, [85] **4**, 227–232.

(1830). Gold medal for Caroline Herschel. *MN*, **1**, 62–64.

(1846). Auszug aus einem Berichte über Lord Rosse's großes Telescop, den Sir James South in The Times no. 18899, 1845 April 16 bekannt gemacht hat. *AN*, [536] **23**, 113–118.

Spitaler, R. (1887). Ueber den Ringnebel in der Leyer. *AN*, [2800] **117**, 261.

(1891a). Zeichnungen und Photographien am Grubb'schen Refractor in den Jahren 1885 bis 1890, Teil I. Nebel in den plejaden. *Annalen der K.K. Universitäts-Sternwarte Wien*, **7**, 185–205.

(1891b). Ueber den Nebel NGC. 1186. *AN*, [3030] **127**, 91.

(1892). Ueber den Nebel NGC. 163. *AN*, [3100] **130**, 57.

(1893). Beobachtungen von Nebelflecken am 27 zöll. Grubb'schen Refractor der k. k. Sternwarte in Wien. *AN*, [3167–68] **132**, 369–394.

(1896). Beobachtungen von Nebelflecken am Grubb'schen Refractor von 86 cm Öffnung in den Jahren 1890, 1891 und 1892. *Annalen der K.K. Universitäts-Sternwarte Wien*, **11**, 81–130.

Staubermann, K. (2007). *Astronomers at Work. A Study of the Replicability of 19th Century Astronomical Practice*. Frankfurt: Harri Deutsch.

Steavenson, W. (1924). The Herschel instruments at Slough. *Obs*, **47**, 262–308.

Steel, D. (1982). The Craig Telescope of 1852. *S&T*, **64** (no. 1), 12–13.

Steiner, G. (1990). *Eduard Schönfeld*. Hildburghausen: Frankenschwelle.

Steinicke, W. (2000a). Digitale Deep-Sky Daten, visuelle Beobachtung und das NGC/IC-Projekt. *VdSJ*, **4**, 49–56.

(2000b). 'Superthin Galaxies' – Objekte scharf wie eine Rasierklinge. *VdSJ*, **5**, 71–73.

(2001a). Beobachtung von Galaxiengruppen, Teil 1. *IS*, **17**, 29–31.

(2001b). Beobachtung von Galaxiengruppen, Teil 2. *IS*, **18**, 35–37.

(2001c). Beobachtung von Galaxiengruppen, Teil 3. *IS*, **19**, 46–47.

(2001d). Besuch in Birr Castle. *IS*, **19**, 58–60.

(2001e). Galaxien bei hellen Sternen. *Magellan* (no. 4), 35.

(2001f). Verloren und wiedergefunden. Der bipolare Reflexionsnebel NGC 2163. *SuW*, **40**, 182–183.
(2002a). Die kosmische A-Klasse – Beobachtung historischer Radioquellen. *IS*, **20**, 60–63.
(2002b). Auf den Spuren einer Bärentatze – NGC 2537 und ihre Nachbarn. *IS*, **21**, 74–75.
(2002c). Der NGC und seine Beobachter – Teil 1. Lewis Swift. *IS*, **22**, 56–58.
(2002d). Der NGC und seine Beobachter – Teil 2. David Peck Todd. *IS*, **24**, 58–60.
(2002e). Deep-Sky Kataloge, die neue Uranometria und andere Geschichten. *VdSJ*, **9**, 102–104.
(2003a). Asaph Hall und die Entdeckung der Marsmonde. *VdSJ*, **10**, 88–91.
(2003b). Der NGC und seine Beobachter – Teil 3. Albert Marth. *IS*, **26**, 51–55.
(2003c). NGC 4567–8 (die Siamesischen Zwillinge). *IS*, **27**, 74.
(2003d). NGC 4038–9 (die Antennen). *IS*, **27**, 76–77.
(2003e). NGC 7006. *IS*, **29**, 74–75.
(2003f). Der NGC und seine Beobachter – Teil 4. Heinrich Ludwig d'Arrest. *IS*, **30**, 50–52.
ed. (2004a). *Praxishandbuch Deep Sky*. Stuttgart: Kosmos Verlag.
(2004b). M 99/M 100. *IS*, **33**, 21–22.
(2004c). M 32/M 110. *IS*, **36**, 20–21.
(2004d). NGC 1499 (Californianebel). *IS*, **37**, 24–25.
(2005a). Die Entdeckung des Coma Berenices Galaxienhaufens. *VdSJ*, **16**, 14–17.
(2005b). NGC 5907. *IS*, **40**, 25.
(2005c). NGC 6822 (Barnard's Galaxy). *IS*, **41**, 22–23.
(2005d). Object of the season. NGC 7027. *DSO*, **138**, 17–18.
(2006). Object of the season. NGC 891. *DSO*, **141**, 17–21.
(2007a). Ein Albtraum der Familie Herschel. *SuW*, **46**, 61–65.
(2007b). Herschel, Uranus und die Planetarischen Nebel. *VdSJ*, **22**, 8–12.
(2008). Webb's deep sky discoveries, *DSO*, **145**, 14–16.
(2009). Nebel und Sternhaufen. Geschichte ihrer Entdeckung, Beobachtung und Katalogisierung – von Herschel bis zu Dreyers 'New General Catalogue'. Norderstedt: Books on Demand.
Steinicke, W., Jakiel, R. (2006). *Galaxies and How to Observe Them*. New York, NY: Springer.
Steinicke, W., Stoyan, R. (2005a). NGC 6818. *IS*, **41**, 26–27.
(2005b). IC 5146. *IS*, **42**, 22–23.
Stephan, E. (1870). Positions moyennes pour 1870, de nébuleuses nouvelles découvertes et observées par M. Stephan a l'Observatoire de Marseille. *AN*, [1810] **76**, 159.
(1871a). Sur les nébuleuses découvertes et observées par M. Stephan à l'Observatoire de Marseille. *CR*, **73**, 825–826.
(1871b). Nebulae discovered and observed at the Observatory of Marseilles with the Foucault Telescope of 0.80 m. *MN*, **32**, 23–25.
(1872a). Nébuleuses nouvelles découvertes et observées à l'Observatoire de Marseille, à l'aide du télescope Foucault de 0.80 m. *AN*, [1867] **78**, 295–298.
(1872b). Nébuleuses découvertes et observées à l'Observatoire de Marseille. *CR*, **74**, 444–445.
(1872c). Nebulae discovered and observed at the Marseilles Observatory. *MN*, **32**, 231–232.
(1872d). Nébuleuses découvertes et observées à l'Observatoire de Marseille par M. Stephan. *AN*, [1876] **79**, 61.
(1873a). Nébuleuses découvertes et observées à l'Observatoire de Marseille par M. Stephan. *AN*, [1939] **81**, 303.
(1873b). Nébuleuses découvertes et observées à l'Observatoire de Marseille. *CR*, **76**, 1073–1074.
(1873c). Nébuleuses découvertes et observées à l'Observatoire de Marseille. *CR*, **77**, 1365–1366.
(1873d). New nebulae discovered at the Marseilles Observatory. *MN*, **33**, 433–434.
(1873e). Nebulae discovered and observed at Observatory of Marseilles. *MN*, **34**, 75–76.
(1874a). Nébuleuses découvertes et observées à l'Observatoire de Marseille par M. Stephan. *AN*, [1972] **83**, 51–54.
(1874b). Nébuleuses découvertes et observées à l'Observatoire de Marseille par M. Stephan. *AN*, [1977] **83**, 137.
(1876). Nébuleuses découvertes et observées à l'Observatoire de Marseille. *CR*, **83**, 328–330.
(1877a). Nébuleuses nouvelles découvertes et observées à l'Observatoire de Marseille. *MN*, **37**, 334–339.
(1877b). Nébuleuses nouvelles découvertes et observées à l'Observatoire de Marseille par M. Stephan. *AN*, [2126] **89**, 213–216.
(1877c). Nébuleuses nouvelles découvertes et observées à l'Observatoire de Marseille par M. Stephan. *AN*, [2129] **89**, 263–266.
(1877d). Nébuleuses découvertes et observées à l'Observatoire de Marseille. *CR*, **84**, 641–642.
(1877e). Nébuleuses découvertes et observées à l'Observatoire de Marseille. *CR*, **84**, 704–706.

(1878). Nébuleuses découvertes et observées à l'Observatoire de Marseille. *CR*, **87**, 869–871.

(1880). Nébuleuses découvertes et observées à l'Observatoire de Marseille. *CR*, **90**, 837–839.

(1881a). Nébuleuses découvertes et observées à l'Observatoire de Marseille. *AN*, [2390] **100**, 209–216.

(1881b). Nébuleuses découvertes et observées à l'Observatoire de Marseille. *CR*, **92**, 1128–1129.

(1881c). Nébuleuses découvertes et observées à l'Observatoire de Marseille. *CR*, **92**, 1183–184.

(1881d). Nébuleuses découvertes et observées à l'Observatoire de Marseille. *CR*, **92**, 1260–1262.

(1883a). Nébuleuses découvertes et observées à l'Observatoire de Marseille. *CR*, **96**, 546–549.

(1883b). Nébuleuses découvertes et observées à l'Observatoire de Marseille. *CR*, **96**, 609–612.

(1883c). Nébuleuses découvertes et observées par M. E. Stephan à l'Observatoire de Marseille. *AN*, [2502] **105**, 81–90.

(1884). Note sur les nébuleuses découvertes à l'Observatoire de Marseille. *Bulletin Astronomique*, **1**, 286–290.

(1885a). Nébuleuses découvertes et observées par M. E. Stephan à l'Observatoire de Marseille. *AN*, [2661] **111**, 321–330.

(1885b). Nébuleuses découvertes et observées à l'Observatoire de Marseille. *CR*, **100**, 1043–1046.

(1885c). Nébuleuses découvertes et observées à l'Observatoire de Marseille. *CR*, **100**, 1107–1110.

Stephens, C. (1990). Astronomy as public utility. The Bond years at the Harvard College Observatory. *JHA*, **21**, 21–35.

Stone, E. (1877). On the cause for the appearance of bright lines in the spectra of irresolvable star clusters. *PRS*, **26**, 156–157, 517–519.

(1882). Meeting of the Royal Astronomical Society (Friday, May 12, 1882). *Obs*, **5**, 157–165.

(1884). Gold medal to Mr. Common. *MN*, **44**, 221–223.

Stone, O. (1886). List of nebulas observed at the Leander McCormick Observatory, and supposed to be new. *AJ*, **7**, 9–14.

(1887). Second list of nebulas observed at the Leander McCormick Observatory, and supposed to be new. *AJ*, **7**, 57–61.

(1893a). Nebula of Orion, 1885. *Publications of the Leander McCormick Observatory*, **1**, 31–43.

(1893b). Southern nebulae. *Publications of the Leander McCormick Observatory*, **1** (part 6), 173–241.

Stone, R. (1979). The Crossley-reflector. Centennial review. *S&T*, **58** (no. 4), 307–311.

Stoyan, R., Binnewies, S., Friedrich, S., Schroeder, K.-P. (2008). *Atlas of Messier Objects*. Cambridge: Cambridge University Press.

Stratonov, V. (1896). Nouvelles nébuleuses dans les Pléiades. *AN*, [3366] **141**, 103.

(1897). Note sur les Pléiades. *AN*, [3441] **144**, 137–140.

Strohmeier, W., Güttler, A. (1952). Zur Frage der Existenz des Baxendell-Nebels NGC 7088. *Veröffentlichungen der Sternwarte München*, **4** (no. 8).

Strömgren, E. (1914). Carl Frederik Pechüle. *AN*, [4749] **198**, 407.

(1919). E. C. Pickering. *AN*, [4977] **208**, 133–136.

Struve, F. G. W. (1826a). Nachricht von der Ankunft und Aufstellung des Refractors von Fraunhofer auf der Sternwarte der Kaiserlichen Universität zu Dorpat. *AN*, [75] **4**, 37–44.

(1826b). Nachricht von der Ankunft und Aufstellung des Refractors von Fraunhofer auf der Sternwarte der Kaiserlichen Universität zu Dorpat (Beschluß). *AN*, [76] **4**, 49–52.

(1826c). Nachricht von einer auf der Dorpater Sternwarte angefangenen neuen Durchmusterung des Himmels in Bezug auf Doppelsterne. *AN*, [76] **4**, 61–64.

(1826d). Nachricht von einer auf der Dorpatar Sternwarte angefangenen neuen Durchmusterung des Himmels in Bezug auf Doppelsterne (Beschluß). *AN*, [75] **4**, 65–72.

(1827). *Catalogus Novus Generalis Stellarum Duplicium et Multiplicium*. Dorpat: Schünmann.

(1828). Report on double stars. *Edinburgh Journal of Science*, **9** (no. 17), 79–90.

(1847). *Etudes d'astronomie stellaire*. St Petersburg: L'Académie Impériale des Sciences.

Struve, O. (1862). On the missing nebula in Taurus. *MN*, **22**, 242–244.

(1866a). Über das von Herr W. Lassell in Malta aufgestellte Spiegelteleskop. *Mélanges Mathématiques et Astronomiques*, **3**, 517–550.

(1866b). Observations de quelques nébuleuses. *Mélanges Mathématiques et Astronomiques*, **3**, 569–587.

(1866c). Entdeckung einiger schwacher Nebelflecke. *Mélanges Mathématiques et Astronomiques*, **3**, 689–694.

(1869). Wiederkehr des Winneckeschen Cometen und Entdeckung einiger neuer Nebelflecke. *Mélanges Mathematiques et Astronomiques*, **4**, 392–398.

(1877). Note on the relative space-penetrating power of the Pulkowa 15-inch refractor and Mr. Lassell's 4-foot reflector. *MN*, **37**, 89.

(1886a). Ueber den Majanebel. *AN*, [2719] **114**, 97–100.

(1886b). The Maia Nebula. *Obs*, **9**, 201–202.

Sugden, K. (1982). An electic astronomer. *S&T*, **63** (no. 1), 27–29.

Sulentic, J., Tifft, W. (1977). *The Revised New General Catalogue of Nonstellar Astronomical Objects.* Tucson, AZ: University of Arizona Press.

Swift, L. (1858). Appearance of the great comet of 1858. *AJ*, **5**, 176.

(1882). The Merope Nebula. *MN*, **42**, 107.

(1884a). *SM*, **3**, 26.

(1884b). A new and remarkable nebula. *SM*, **3**, 57–58.

(1885a). Catalogue no. 2 of nebulae discovered at the Warner Observatory. *AN*, [2707] **113**, 305–310.

(1885b). The nebulae. *SM*, **4**, 1–4.

(1885c). The nebulae. *SM*, **4**, 36–43.

(1885d). Curious, difficult and remarkable nebulae discovered at Warner Observatory. *SM*, **4**, 174–177.

(1885e). Catalogue no. 1 of nebulae discovered at the Warner Observatory. *AN*, [2683] **112**, 313–318.

(1886a). Catalogue no. 3 of nebulae discovered at the Warner Observatory. *AN*, [2746] **115**, 153–158.

(1886b). Catalogue no. 4 of nebulae discovered at the Warner Observatory. *AN*, [2752] **115**, 257–262.

(1886c). Catalogue no. 5 of nebulae discovered at the Warner Observatory. *AN*, [2763] **116**, 33–38.

(1886d). Suspected proper motion in a nebula. *SM*, **5**, 60.

(1887a). Catalogue no. 6 of nebulae discovered at the Warner Observatory. *AN*, [2798] **117**, 217–222.

(1887b). *History and Work of the Warner Observatory 1883–1886.* Rochester, NY: Warner Observatory.

(1888a). Catalogue no. 7 of nebulae discovered at the Warner Observatory. *AN*, [2859] **120**, 33–38.

(1888b). New nebulae at the Warner Observatory. *SM*, **7**, 38–40.

(1889). Catalogue no. 8 of nebulae discovered at the Warner Observatory. *AN*, [2918] **122**, 241–246.

(1890a). Catalogue no. 9 of nebulae discovered at the Warner Observatory. *AN*, [3004] **126**, 49–54.

(1890b). A wonderful nebulous ring. *SM*, **9**, 47.

(1892a). Catalogue no. 10 of nebulae discovered at the Warner Observatory. *AN*, [3094] **129**, 361–364.

(1892b). Notes on new and old nebulae. *Astronomy and Astro-Physics*, **11**, 566–567.

(1893). New nebulae discovered by Lewis Swift at Warner Observatory, Rochester, New York. *MN*, **53**, 273.

(1894). Nebulae and clusters. *PA*, **1**, 369–371.

(1896). Catalogue no. 1 for 1900.0 of nebulas discovered at the Lowe Observatory, California. *AJ*, **17**, 27–28.

(1897a). List no. 5 of nebulas discovered at the Lowe Observatory. *AJ*, **18**, 111.

(1897b). Catalogue no. 2 of nebulas discovered at the Lowe Observatory, California. *MN*, **57**, 629–631.

(1897c). Catalogue no. 3 of nebulas discovered at the Lowe Observatory, California. *MN*, **57**, 631–632.

(1897d). List no. 4 for 1900.0 of nebulas discovered at the Lowe Observatory, California. *MN*, **58**, 18–19.

(1897e). Catalogue no. II of nebulae discovered at the Lowe Observatory, Echo Mountain, California. *PASP*, **9**, 186–188.

(1897f). Catalogues nos. III and IV of nebulae discovered at the Lowe Observatory, Echo Mountain, California. *PASP*, **9**, 223–226.

(1898a). Catalogue no. 11 of nebulae. *AN*, [3517] **147**, 209–220.

(1898b). List no. 6 of nebulae discovered at the Lowe Observatory, Echo Mountain, California. *MN*, **58**, 331.

(1898c). List no. 7 of nebulae discovered at the Lowe Observatory, Echo Mountain, California. *MN*, **58**, 332.

(1898d). List no. 8 of nebulae discovered at the Lowe Observatory, Echo Mountain, California, for 1900.0. *MN*, **58**, 333–334.

(1899). List no. 12 of nebulae discovered at the Lowe Observatory, Echo Mountain, California, for 1900.0. *MN*, **59**, 568–569.

Tebbutt, J. (1871). On Eta Argus. *MN*, **31**, 210–211.

Tempel, W. (1861). Schreiben des Herrn Wilh. Tempel an den Herausgeber. *AN*, [1290] **54**, 285.

(1862). *AN*, [1383] **58**, 239.

(1863). Nébuleuses. *Le Monde*, **3**, 178.

(1864). Schreiben des Herrn Tempel an den Herausgeber. *AN*, [1472] **62**, 119–122.

(1871). Entdeckung eines neuen Cometen. *AN*, [1848] **77**, 375.

(1873). Aus einem Schreiben des Herrn Wilh. Tempel an den Herausgeber. *AN*, [1898] **80**, 27–30.

(1874). Osservazioni astronomiche diverse fatte nella specola di Milano. *Publicazioni del Reale Osservatorio di Brera in Milano*, no. V.

(1875a). Schreiben des Herrn Wilh. Tempel an den Herausgeber. *AN*, [2029] **85**, 203.

(1875b). Schreiben des Herrn Tempel, Astronom der Sternwarte in Florenz, an den Herausgeber. *AN*, [2045] **86**, 67–70.

(1877a). Schreiben des Herrn Tempel, Astronom der Königl. Sternwarte zu Arcetri an den Herausgeber. *AN*, [2138–39] **90**, 27–42.

(1877b). Über den scheinbaren Unterschied der zwei vorhandenen Zeichnungen vom grossen Nebelflecken. Messier 8, von John Herschel und Wilhelm Tempel. *AN*, [2159] **90**, 355–358.

(1878b). Spiral form of nebulae. *Obs*, **1**, 403–405.

(1878c). Schreiben des Herrn W. Tempel an den Herausgeber. *AN*, [2190] **92**, 95.

(1878d). Schreiben des Herrn W. Tempel an den Herausgeber. *AN*, [2212] **93**, 49–62.

(1878e). Der Nebelfleck im Orion. *Sirius*, **11**, 23.

(1879a). Beobachtung des Brorsen'schen Cometen und eines neuen Nebelflecks. *AN*, [2253] **94**, 335.

(1879b). Wiederauffindung und Beobachtung des Cometen II 1867, angestellt auf der Sternwarte Arcetri. *AN*, [2269] **95**, 199–202.

(1880a). Note on the nebula near Merope. *MN*, **40**, 622–623.

(1880b). Schreiben des Herrn W. Tempel an den Herausgeber. *AN*, [2284] **96**, 61–64.

(1881a). Beobachtung, angestellt auf der Sternwarte Arcetri bei Florenz. *AN*, [2347] **98**, 299–304.

(1881b). Schreiben des Herrn W. Tempel an den Herausgeber. *AN*, [2349] **98**, 335.

(1881c). Beobachtungen angestellt auf der Sternwarte Arcetri bei Florenz. *AN*, [2365] **99**, 197.

(1882). Neue Nebelflecke, aufgefunden und beobachtet auf der Sternwarte zu Arcetri. *AN*, [2439] **102**, 225–238.

(1883a). Einige neue Nebel und ein Nebelnest. *AN*, [2511] **105**, 235–238.

(1883b). Beobachtung von Cometen und Nebelflecken, angestellt auf der Sternwarte Arcetri. *AN*, [2522] **106**, 27–30.

(1883c). Notiz über einige neue und ältere Nebel. *AN*, [2527] **106**, 107–110.

(1885a). Über Nebelflecken, nach Beobachtungen angestellt in den Jahren 1876–1879 mit dem Refraktor von Amici auf der Königl. Sterwarte zu Arceti bei Florenz. *Abhandlungen der Königlich Böhmischen Gesellschaft der Wissenschaften*, section VII, vol. **1**.

(1885b). Neue Nebel. *AN*, [2660] **111**, 315.

(1885c). Notizen über verschiedene Nebel. *AN*, [2668] **112**, 61.

(1885d). Notizen über verschiedene neue Nebel. *AN*, [2691] **113**, 47.

Terby, F. (1886). The Pleiades and Eps. Lyrae. *Obs*, **9**, 369–370.

Thiele, T. N. (1888). Hans Carl Frederik Christian Schjellerup. *AN*, [2814] **118**, 95.

Thomson, M. (1991). Three cases of identity errors of NGC galaxies as published in major astronomical catalogues. *QJRAS*, **32**, 17–24.

(2001). Revealing the Rosse spirals. *A&G*, **42**, 4.8–4.10.

Tirion, W., Rappaport, B., Lovi, G. (1987). *Uranometria 2000.0*, 1st edn, vol. I. Richmond, VA: Willmann-Bell.

Tobin, W. (1987). Foucault's invention of the silvered-glass reflecting telescope and the history of his 80 cm reflector at the Observatoire de Marseille. *VA*, **30**, 153–184.

(2003). *The Life and Science of Léon Foucault*. Cambridge: Cambridge University Press.

(2008). Full-text search capability. A new tool for researching the development of scientific language. The 'Whirlpool Nebula' as a case study. *Note and Record of the Royal Society*, **62**, 187–196.

Tobin, W., Holberg, J. (2008). A newly-discovered and accurate early drawing of M 51, the Whirlpool Nebula. *Journal of Astronomical History and Heritage*, **11**, 107–115.

Todd, D. P. (1885a). Schreiben des Herrn David P. Todd an den Herausgeber. *AN*, [2682] **112**, 297.

(1885b). Telescopic search for the trans-Neptunian planet. *AN*, [2698] **113**, 153–160.

(1888). American telescopes. *Obs*, **11**, 394–398.

Towley, S. (1909). Charles A. Young. *PASP*, **20**, 46–47.

Treadwell, T. (1943). Notes on a forgotten episode. *PA*, **51**, 497–500.

Tromholdt, S. (1875). Heinrich Ludwig d'Arrest. *Wochenschrift für Astronomie, Meteorologie und Geographie*, **29**, 217–228.

Trouvelot, E. (1882). *The Trouvelot Astronomical Drawings Manual*. New York, NY: Charles Scribner's Sons.

(1884). Drawing on p. 421, in Flammarion, C. (1884). Les vides dans le ciel. *L'Astronomie*, **3**, 419–421.

Trümpler, R. (1922). The structure of Barnard's Merope nebula. *PASP*, **34**, 165.

Turner, A. (1988). Portraits of William Herschel. *VA*, **32**, 65–94.

Turner, H. H. (1903a). Andrew Ainslie Common. *AN*, [3886] **162**, 353–356.

(1903b). Andrew Ainslie Common. *Obs*, **26**, 304–308.
(1903c). Andrew Ainslie Common. *PA*, **11**, 367–377.
(1904). Andrew Ainslie Common. *PRS*, **75**, 313–318.
(1906). Otto Struve 1819–1905. *PRS*, **78**, liv–lix.
(1909). R. L. J. Ellery, 1828–1908. *PRS*, **82**, vi–x.
(1910). Letters from the Rev. W. R. Dawes to Mr. George Knott. *Obs*, **33**, 383–398.
(1920). Edward Charles Pickering. *MN*, **80**, 360–365.
Urban, S., Corbin, T. (1998). The astrographic catalogue. A century of work pays off. *S&T*, **95** (no. 6), 41–44.
Valentiner, W. (1900). Beobachtungen veränderlicher Sterne. *Veröffentlichen der Grossherzöglichen Sternwarte Heidelberg*, **1**, 209–210.
Vehrenberg, H. (1983). *Atlas of Deep-Sky Splendors*. Cambridge, MA: Sky Publishing Corporation.
Verschuur, G. (1989). Barnard's 'dark' dilemma. *Astronomy*, **17** (no. 2), 30–38.
Vogel, H. C. (1867). *Beobachtungen von Nebelflecken und Sternhaufen am sechsfüssigen Refractor und zwölffüssigen Äquatoreal der Leipziger Sternwarte*. Leipzig: Engelmann.
(1868a). Verzeichnis von hundert Nebelflecken, abgeleitet aus Beobachtungen am sechsfüssigen Refractor und zwölffüssigen Aequatoreal der Leipziger Sternwarte. *AN*, [1667] **70**, 161–172.
(1868b). Vergleich der von d'Arrest gegebenen Nebelpositionen mit den in No 1667 der *Astronomischen Nachrichten* mitgetheilten. *AN*, [1683] **71**, 45.
(1876). *Positionsbestimmungen von Nebelflecken und Sternhaufen zw. +9° 30' und +15° 30' Declination*. Leipzig: Engelmann.
(1880a). Ueber einen neuen Nebel im Sternbilde des Schwans. *AN*, [2289] **96**, 143.
(1880b). Ueber das Spectrum des von Webb entdeckten Nebels im Schwan und eines neuen, von Baxendell aufgefundenen Sternes im kleinen Hund. *AN*, [2298] **96**, 287.
(1884). Einige Beobachtungen mit dem grossen Refractor der Wiener Sternwarte. *Publikationen des Astrophysikalischen Observatoriums Potsdam*, **4** (no. 14), 1–38.
(1888). Ueber die Bedeutung der Photographie zur Beobachtung von Nebelflecken. *AN*, [2854] **119**, 337–342.
Ward, I. (1885). New star in Andromeda. *AR*, **23**, 242.
Warner, B. (1979). Sir John Herschel's description of his 20-feet reflector. *VA*, **23**, 75–107.
(1982). The large southern telescope. Cape or Melbourne? *QJRAS*, **23**, 505–514.
(1992). The years at the Cape of Good Hope, in *John Herschel 1792–1871. A Bicentennial Commemoration*, ed. D. King-Hele. London: Royal Society, pp. 51–66.
Waters, S. (1873a). Note on the distribution of resolvable and irresolvable nebulae. *MN*, **33**, 406–407.
(1873b). The distribution of the clusters and nebulae. *MN*, **33**, 558–559.
Webb, T. W. (1859). *Celestial Objects for Common Telescopes*. London: Longman, Green, Longmann, and Roberts.
(1864a). Clusters of stars and nebulae. *Intellectual Observer*, **4**, 56–62.
(1864b). Clusters and nebulae. *Intellectual Observer*, **4**, 257–266.
(1864c). Clusters, nebulae, and occultations. *Intellectual Observer*, **4**, 346–352.
(1864d). The nebula in the Pleiades. *Intellectual Observer*, **4**, 448–451.
(1869). Reflectors and refractors. *AR*, **7**, 21–23.
(1879a). New gaseous nebula. *AN*, [2292] **96**, 191.
(1879b). Discovery of a gaseous nebula in Cygnus. *MN*, **40**, 90–91.
(1879c). Nebulous star. *Obs*, **3**, 257.
(1879d). New gaseous nebula. *English Mechanic*, **30**, 308.
(1879e). New gaseous nebula. *Nature*, **21**, 111.
(1880a). New gaseous nebula. *AR*, **18**, 18–19.
(1880b). Entdeckung und Beobachtung eines neuen Gas-Nebels. *Sirius*, **13**, 25–26.
Weddle, K. (1986). Old stars. Ormsby Mitchel. *S&T*, **71** (no. 1), 14–16.
Weinek, L. (1886). Beobachtung von Nebelflecken. *Astronomische Beobachtungen der K.K. Sternwarte Prag 1884*, 35–40.
(1890). Beobachtung von Nebelflecken. *Astronomische Beobachtungen der K.K. Sternwarte Prag 1885, 1886 u. 1887*, 34–37.
Weiss, E. (1883). Planeten- und Cometenbeobachtungen auf der neuen Wiener Sternwarte. *AN*, [2520] **105**, 369–386.
(1884). Verschwundener Nebel. *AN*, [2601] **109**, 143.
(1885). Beobachtungen von Planeten und Cometen angestellt auf der Wiener Sternwarte während des Jahres 1884. *AN*, [2658–59] **111**, 273–296.
(1886a). Über die Nebel in den Plejaden. *AN*, [2726] **114**, 209–212.
(1886b). Beobachtungen von Planeten und Cometen angestellt auf der Wiener Sternwarte während des Jahres 1885. *AN*, [2732–33] **114**, 337–354.

(1887). Beobachtungen von Planeten und Cometen angestellt auf der Wiener Sternwarte während des Jahres 1886. *AN*, [2782–83] **116**, 337–354.

Weitzenhoffer, K. (1992). General Thomas Brisbane's astronomical adventures. *S&T*, **84** (no. 6), 620–622.

Wenzel, K. (2004). Die Geschichte um den Nebel von 55 Andromedae. *VdSJ*, **15**, 75–76.

(2005). Die Entdeckungen des David P. Todd. *VdSJ*, **17**, 61–65, **18**, 109–112.

Weyer, G. (1895). Theodor Brorsen. *AN*, [3285] **137**, 367.

Whitrow, J. (1970). Some prominent personalities and events in the early history of the Royal Astronomical Society. *QJRAS*, **11**, 89–104.

Williams, F. S. (1852). *The Wonders of the Heavens*. London: John Casell.

Wilsing, J. (1893). Ueber die Parallaxe des planetarischen Nebels BD. +41° 4004. *AN*, [3190] **133**, 353–358.

(1894). Ueber die Parallaxe des Nebels h 2241. *AN*, [3261] **136**, 349–352.

Wilson, B. (2007). Caroline Herschel. No ordinary eighteenth-century woman, in *Deep-Sky Companions: Hidden Treasures*, ed. S. O'Meara. Cambridge: Cambridge University Press, pp. 545–561.

Wilson, H. C. (1894). A photograph of the Pleiades and two asteroids. *PA*, **1**, 322–324.

(1899). The exterior nebulosities of the Pleiades. *PA*, **7**, 57–62.

(1904). The large nebulous areas of the sky. *PA*, **12**, 401–405.

(1907). The number and distribution of the stars in the vicinity of the Pleiades. *PA*, **15**, 193–204.

Winlock, J. (1872). Zone catalogue of 6100 stars situated between the 0° 40′ and 1° 0′ north declination, observed in the years 1859–60. *AHO*, **6** (part 2).

(1876). Astronomical engravings. *AHO*, **8**, 120–125.

Winnecke, A. (1857). Notiz über Nebelflecken. *AN*, [1072] **45**, 247–250.

(1862). Beobachtung von Cometen auf der Pulkowaer Sternwarte. *AN*, [1357] **57**, 203–208.

(1863). Schreiben des Herrn Dr. Winnecke an den Herausgeber. *AN*, [1397] **59**, 65–68.

(1865). Friedrich Wilhelm Georg Struve. *AN*, [1504] **63**, 241.

(1866). Über den Nebelfleck im Orion. *Mélanges Mathématiques et Astronomiques*, **3**, 499–502.

(1869). Doppelsternmessungen. *AN*, [1738] **73**, 145–160.

(1875). Ueber die auf der Universitäts-Sternwarte Strassburg begonnenen Beobachtungen von Nebelflecken. *VJS*, **10**, 297–304.

(1878). On the evidence of periodic variability of the nebula H. II. 278. *MN*, **38**, 104–106.

(1879). Über die periodische Veränderlichkeit in der Helligkeit des Nebelflecks h 882 = H I 20. *AN*, [2293] **96**, 201–206.

(1879b). The nebula in Cygnus. *MN*, **40**, 92–93.

Wirtz, C. (1911a). Beobachtung von Nebelflecken am 49 cm Refraktor III. Teil 1902 April bis 1910 März. *Annalen der Kaiserlichen Universitäts-Sternwarte Strassburg*, **4** (part 1), 1–78.

(1911b). Generalkatalog der am großen Refraktor in Straßburg beobachteten Nebelflecke 1881–1910. *Annalen der Kaiserlichen Universitäts-Sternwarte Strassburg*, **4** (part 1), 79–112.

(1911c). Vergleichung des Generalkataloges mit anderen Nebelverzeichnissen. *Annalen der Kaiserlichen Universitäts-Sternwarte Strassburg*, **4** (part 1), 113–176.

(1923). Flächenhelligkeiten von 566 Nebelflecken und Sternhaufen. *Meddelanden Från Lunds Astronomiska Observatorium*, ser. II (no. 29), 1–63.

(1925). Der Gegenstand der Vatikanischen Durchmusterungen dunkler kosmischer Nebel. *AN*, [5368] **224**, 267–270.

(1929). Samuel Oppenheim. *VJS*, **64**, 20–29.

Witt, V. (1999). Der Fraunhofer-Refraktor in München-Bogenhausen. *SuW*, **38**, 378–379.

(2006). Die Universitätssternwarten in Wien. *SuW*, **45**, 76–80.

(2007a). Erinnerungen an die Sternwarte Lilienthal. *SuW*, **46**, 84–89.

(2007b). Guiseppe Piazzi und die Entdeckung der Ceres. *SuW*, **46**, 80–85.

(2007c). Auf den Spuren Ole Roemers in Kopenhagen. *SuW*, **46**, 88–94.

Wittmann, A. (2004). Messung und Festlegung des Meridians der Göttinger Sternwarte durch Karl-Ludwig Harding (1803). *Mitteilungen der Gauß-Gesellschaft*, **41**, 91–101.

Wlasuk, P. (1996). 'So much for fame!' The story of Lewis Swift. *QJRAS*, **37**, 683–707.

Wolf, C. (1875). Description du groupe des Pléiades et mesures micrométriques des positions des principales étoiles qui le composent. *CR*, **81**, 29–32.

(1877). Description du groupe des Pléiades. *AOP*, **14**, A1–A81.

(1880a). Note on Mr. Maxwell Hall's paper on the nebula in the Pleiades. *MN*, **40**, 293.

(1880b). The nebula in the Pleiades. *Obs*, **3**, 451–452.

(1886a). Sur la comparaison des résultats de l'observation astronomique directe avec ceux de l'inspection photographique. *CR*, **102**, 467–467.

(1886b). The part taken by large instruments in astronomical observations. *AR*, **24**, 197–205, 225–232.

Wolf, M. (1891a). Ueber den grossen Nebel um ζ Orionis. *AN*, [3027] **127**, 39–42.

(1891b). Über die Verwendung gewöhnlicher photographischer Objektive bei der Himmelsphotographie. *Sirius*, **19**, 106–109.

(1891c). Ueber grosse Nebelmassen im Sternbild des Schwans. *AN*, [3048] **127**, 427.

(1895). Notiz über die Plejaden-Nebel. *AN*, [3275] **137**, 175.

(1900). Die Aussen-Nebel der Plejaden. *Abhandlungen der Bayerischen Akademie der Wissenschaften*, **20**, 615–627.

(1901). Ein merkwürdiger Haufen von Nebelflecken. *AN*, [3704] **155**, 127.

(1903). Über eine Eigenschaft der großen Nebel. *AN*, [3848] **161**, 129–132.

(1906). Sternverteilung um die großen Nebel bei ξ Persei und bei 12 Monocerotis. *Publikationen des Astrophysikalischen Instituts Königstuhl-Heidelberg*, **2** (no. 11).

(1907). Ein Nebelfleckhaufen und Nebelreichtum in Sagittarius. *AN*, [4207] **176**, 109–112.

(1908). A new 'cave-nebula' in Cepheus. *MN*, **69**, 117.

(1917). Über das Spektrum der Höhlennebel. *AN*, [4875] **204**, 41–44.

(1919). Über zwei Höhlennebel in Cepheus und einige andere Nebel dieser Gegend. *AN*, [5011] **209**, 303.

(1923). Edward Emerson Barnard. *AN*, [5224] **218**, 241–248.

(1925). Variables (? Nova) 94.1925 Virgins. *AN*, [5405] **226**, 75–78.

Wolf, R. (1854). Ueber einen Nebelfleck im Orion. *AN*, [895] **38**, 109.

(1890). *Handbuch der Astronomie*, 2 vols. Zurich: Schulthess.

Woods, T. (1844). *The Monster Telescopes Erected by the Earl of Rosse*. Parsonstown: Sheilds and Son.

Wolfschmidt, G. (1995). *Milchstraßen, Nebel, Galaxien*. Munich: Oldenbourg.

Wollaston, F. (1789). *A Specimen of a General Astronomical Catalogue*. London: G. and T. Wilkie.

Yeomans, D. (1991). *Comets. A Chronological History of Observations, Science, Myth, and Folklore*. New York, NY: John Wiley and Sons.

Zach, F. X. (1826). Retour de la brillante comète du Taureau. *Correspondance Astronomique*, **14**, 409–411.

Zwicky, F. (1957). *Morphological Astronomy*. Berlin: Springer Verlag.

Internet and image sources

Many articles – but by far not all – are accessible via the Internet. The main portal is the Astronomical Data System (ADS) of Harvard University (Cambridge, MA). CR, PT, PRS and other series can be found on the French website 'Gallica'. Many of the books cited have already been scanned by Google; these and others are available in the 'Internet Archive' too. The following URLs were active at the time of writing.

Historic images are mostly free of copyright. Most of them are from the author's archive. For others a source is given. Nearly all object images are from the Digitized Sky Survey (DSS). Others are from the author's archive or a source is given. Figure 11.51 is taken from the Hubble Heritage Project. The DSS and HST data are copyrighted by the Space Telescope Science Institute (STScI).

Astronomical Data System (ADS)	http://adsabs.harvard.edu/journals_service.html
Digitized Sky Survey (DSS)	http://archive.eso.org/dss/dss
Gallica	http://gallica.bnf.fr
Google Booksearch	http://books.google.com
Historical Scans of the ADS	http://adsabs.harvard.edu/historical.html
Hubble Space Telescope	http://hubblesite.org
Internet Archive	http://www.archive.org
NASA Extragalactic Database (NED)	http://nedwww.ipac.caltech.edu
NGC/IC Project	http://www.ngcicproject.org
Simbad	http://simbad.u-strasbg.fr/simbad
Wolfgang Steinicke	http://www.klima-luft.de/steinicke

Name index

The dates of birth and death, if available, for all 545 people (astronomers, telescope makers etc.) listed here are given. The data are mainly taken from the standard work *Biographischer Index der Astronomie* (Dick, Brüggenthies 2005). Some information has been added by the author (this especially concerns living twentieth-century astronomers). Members of the families of Dreyer, Lord Rosse and other astronomers (as mentioned in the text) are not listed. For obituaries see the references; compare also the compilations presented by *Newcomb–Engelmann* (Kempf 1911), Mädler (1873), Bode (1813) and Rudolf Wolf (1890). Markings: bold = portrait; underline = short biography.

Abbe, Cleveland (1838–1916), 222–23, 466
Abell, George Ogden (1927–1983), 47, 248
Abbott, Francis Preserved (1799–1883), 517–19
Abney, William (1843–1920), 360
Adams, John Couch (1819–1892), 122, 150, 241, 305, 415, 427
Airy, George Biddell (1801–1892), 9, 52, 99, 106, 112, 120–21, 132, 135, 146, 217, 222, 241, 252, 417, 427, 472, 502, 517–18
Aitken, Robert Grant (1864–1951), 462–63, 510–11
Alexander, Stephen (1806–1883), 83, 241
Ames, Adelaide (1900–1932), 467–68
Amici, Giovan Battista (1786–1863), 96, 171, 216, 263–64, 328–29, 334, 365, 495, 505, 529–31, 534, 536
Anderson, Thomas David (1853–1932), 403
André, Charles Louis Françoise (1842–1912), 530
Arago, Dominique Françoise Jean (1786–1853), 43, 134
Aratos (-315– -245), 16, 18, 444, 446, 450, 456
Archenhold, Friedrich Simon (1861–1939), 397, 401–02, 556
Argelander, Friedrich Wilhelm August (1799–1875), 58, 124, 144, 150–51, 158, 161–64, 166, 174–75, 181, 206, 235–37, 239, 259–60, 279, 380, 458, 501, 503, 506–07, 546
Aristotle (-384– -322), 16, 18, 446, 450, 455
Arp, Halton Christian (1927–), 44, 399, 430

As-Sufi, Abd-al-Rahman (903–986), 15–16, 167, 441–42, 446, 449–50, 455, 475, 516
Austin, Edward P. (1843–1906), 6, 82, **335**, 337, 446, 450
Auwers, Georg Friedrich Julius Arthur v. (1838–1915), 7, 10, 12, 14–15, 26–27, 47, 50–51, 61, 65, 68–69, 88, 92–93, 97, 123, 127, 130, 134–36, 139, 144–45, **146**, *147*, 148–49, 151, 154–56, 158, 160–64, 166–67, 169–70, 172, 176, 178–83, 186–87, 198, 204–05, 210, 218, 220–22, 234–37, 253, 258, 260, 279, 286–88, 291, 439, 441, 446, 450, 472–74, 493, 498, 500–03, 505, 523–27, 536

Baade, Wilhelm Heinrich Walter (1893–1960), 385, 399, 462, 512
Baader, Michael (1810–1866), 526
Backhouse, Thomas William (1842–1920), 188, 533, 540–41
Baeker, Carl Wilhelm (1819–1882), 144
Bailey, Mark (1952–), 486
Bailey, Solon Irving (1854–1931), 17, 29, 382, 466, 536, 555, 557, 560
Baillaud, Edouard Benjamin (1848–1934), 530
Baily, Francis (1774–1844), 53, 55, *56*, 303–04, 344, 346, 444–46, 450
Ball, Robert Stawell (1840–1913), 101–04, 139, 171, 189, 226, 229, 295, 297–98, **303**, 304, *305*, 306–08, 310, 314–15, 319–20, 323, 446, 450, 492, 555
Baracchi, Pietro Paolo Giovanni Ernesto (1851–1926), 415, *416*, **416**, 418–21, 444–46, 450
Barnard, Edward Emerson (1857–1923), 5, 8, 17, 33, 74, 97, 116–17, 142, 160, 164, 183, 229, 256, 271, 295, 338–42, 344, 346, 348, 360, 364, 367, 369, 393, *395*, **395**, 396–404, 406, 410, 415, 423–24, 436, 441, 446, 448, 450, 455, 458–59, 461–63, 470, 477, 481, 483, 505–11, 513–14, 517, 520, 526, 533, 540–42, 548–61

Barnard, Mary (née Calvert) (?–1921), 396
Bauschinger, Julius (1860–1934), 442
Baxendell, Joseph (1815–1887), 323, **384**, *384*, 384–85, 446, 500, 537
Bayer, Johann (1572–1625), 16
Becker, Ernst Emil Hugo (1843–1912), 263
Becker, Friedrich Eberhard (1900–1985), 385, 466, 559
Becker, Ludwig Wilhelm Emil Ernst (1860–1947), 474
Benzenberg, Johann Friedrich (1777–1846), 174
Bergstrand, Carl Östen Emanuel (1873–1948), 293
Berthon, Edward Lyon (1813–1899), 118, 379
Bessel, Franz Friedrich Wilhelm (1784–1846), 66, 68, 70, 122, 150, 152, 157, 186, 221, 239, 251, 472, 501, 523, 525, 531
Bevis, John (1695–1771), 16, 167, 446, 450, 453
Bianchi, Giuseppe (1791–1866), **96**, 97, 186
Bidschof, Friedrich (1864–1915), 460, 474
Biela, Wilhelm v. (1782–1856), 71, 269, 284, 337
Bigourdan, Guillaume (1851–1932), 8, 12, 18, 33, 62, 203, 242–43, 250, 257–58, 271–72, 290, 329, 340, 350,

354, 356, 367, 371–72, 376, 382, 384, 395, **422**, *422*, 423–27, 436, 446, 448, 450, 453, 455–56, 458–60, 462, 464–65, 467, 473–74, 493, 506, 508, 513, 521

Birmingham, John (1829–1884), 175
Bishop, George (1785–1861), 121, *122*, **122**, 123, 131, 251
Blagden, Charles (1748–1820), 59–60
Block, Eugen (1847–1912), 172–73, 261–62, 264, 323, 340–41, 448
Bode, Johann Elert (1747–1826), 15–16, 25, 35, 37, 42–43, 48–50, 64, 68, 157–58, 167, 179–80, 182, 186, 221, 242, 439, 446, 450, 459, 475, 481–83, 493
Boeddicker, Otto (1853–1937), 101, 307, 315–16
Boguslawski, Heinrich v. (1789–1851), 165
Bond, George Phillips (1825–1865), 88, 124, **125**, 126–31, 162, 183, 188, 195, 203–05, 215, 217, 221–22, 231–32, 234, 241, 314–15, 320–21, 336–37, 346, 348, 357, 403, 407–09, 441, 446, 450, 481, 490–91, 516
Bond, William Cranch (1789–1859), 111, 124, *125*, 132, 408–09
Börgen, Carl Nicolai Jensen (1843–1909), 310
Borrelly, Alphonse Louis Nicholas (1842–1926), 127, 130, 263–64, 266–67, 270–71, **274**, *276*, 277, 293, 319–20, 383, 446, 450
Borst, Charles Augustus Rufus (1851–1918), 325
Boulliau, Ismaël (1605–1694), 125, 515
Bourget, Carl Etienne Henry (1864–1921), 271
Bouris, Georg Constantin (1802–1860), 174
Bowditch, Nathaniel (1773–1838), 127
Brahe, Tycho (1546–1601), 15, 149, 225–26, 228–29
Brashear, John Alfred (1840–1920), 382, 427, 560
Bredikhin, Fedor Alexandrowitsch (1831–1904), 232, 282
Breen, Hugh (1824–1891), 502
Breen, James (1826–1866), 125, 183, 498, 500–02
Bremiker, Carl (1804–1877), 122, 149
Brenner, Leo [Spiridon Gopčević] (1855–1936), 383, 511, 560
Brett, John (1831–1902), 541
Brisbane, Thomas Makdougall (1773–1860), 72–73, 79
Brooks, William Robert (1844–1921), 338, 340, 396

Brorsen, Theodor Johann Christian Ambders (1819–1895), 88, **166**, *169*, 170, 183, 186, 221, 252, 446, 450
Browning, John (1835–1925), 299, 305, 308, 528, 538
Brück, Hermann Alexander (1905–2000), 142
Brück, Mary Teresa (1925–2008), xvii
Brünnow, Franz Friedrich Ernst (1821–1891), 106, 152, 299, 305, 409
Bruce, Catherine Wolfe (1816–1900), 177, 396
Bruhns, Carl Christian (1830–1881), 88, 149, 161, 166, *169*, **169**, 170, 172, 183, 221, 260, 277–79, 446, 450, 459, 528
Buckingham, James, 10, 312
Bülow, Friedrich Gustav v. (1817–1893), 278
Burnham, Robert jr. (1931–1993), 383, 465
Burnham, Sherburne Wesley (1838–1921), 121, 139, 154, 198, 207, 232, 244, 246, 284, 295, 323, 332, 340, 343, 346, 357, **358**, *359*, 360, 362–69, 378, 382, 393–96, 398, 403–05, 407, 414, 424, 446, 450, 459, 465, 506–12, 517, 533, 538–41, 547, 550, 560
Burritt, Eliah Hinsdale (1794–1838), 340
Burton, Charles Edward (1846–1882), 101, 139, 215, 226, 252, 297–98, 305, **306**, 307
Buta, Ronald James (1952–), 195
Butillon, 286
Butler, John (1940–), 486
Byrne, John, 395

Cacciatore, Gaetano (1814–1889), 69, 186
Cacciatore, Niccolò (1780–1841), 12, 63, **69**, *70*, 71, 80, 446, 450
Calver, George (1834–1927), 386, 388, 535
Calvert, Ebenezer (1850–1924), 557, 560
Calvert, Mary Rhoda (1885–1973), 396
Campbell, William Wallace (1862–1938), 404, 408
Cannon, Annie Jump (1863–1941), 390
Capocci, Ernesto (1798–1864), 64–66, 71, 183, 186, 324
Carlson, Dorothy, 44, 467–68
Carpenter, Henry James (1849/50–1899), 394
Carrington, Richard Christopher (1826–1875), 131, 142, 162
Cary, William (1759–1825), 41–42
Cassini, Giovanni Domenico (1625–1712), 15–16, 446, 450
Cassini, Jean Dominique (1748–1845), 60

Cauchoix, Robert-Aglae (1776–1845), 106, 137, 140–41, 368
Cederblad, Sven Aldo (1914–1991), 162, 376, 385, 403, 504, 512, 522
Cerulli, Vincenzo (1859–1927), 429, 460
Chacornac, Jean (1823–1873), 9, 122, 138, 156, 182, 188, **205**, 206–07, 221, 269, 325–26, 334, 438, 446, 448, 453, 491–93, 498–99, 501, 503, 511, 514–15, 525–27
Challis, James (1803–1882), 111, 502
Chambers, George Frederick (1841–1915), 59, 78, 120–21, 142, 498, 502–03, 514, 530
Chance, Robert Lucas (1782–1865), 10
Chapman, Allan (1946–), 9
Charlois, Auguste Honoré Pierre (1864–1910), 544
Chauvenet, William (1820–1870), 404
Chevallier, Temple (1794–1873), 252, 267
Christie, William Henry Mahoney (1845–1922), 227, 465, 494, 534, 555–56
Clacy, John, 508
Clark, Alvan Graham (1832–1897), 173, 282, 338, 342–43, 361, 369–70, 431, 543
Clausen, Thomas (1801–1885), 262
Clerke, Agnes Mary (1842–1907), 6, 510, 514, 551, 560
Coggia, Jérôme Eugene (1849–1919), 270–71, 274, **276**, 277, 320, 340, 346, 446, 450, 461
Colbert, Elias (1829–1921), 358–63
Common, Andrew Ainslie (1841–1903), **385**, *386*, 387–88, 397, 418, 435–36, 445–46, 450, 497–98, 533, 535–46
Cook, A. Grace (?–ca 1958), 466
Cooke, Charles Frederick (1836–1898), 10, 311, 466, 553–54
Cooke, Thomas (1807–1868), 217, 384, 400, 429
Coolidge, Phillip Sydney (1830–1863), *125*, **125**, 126–28, 131, 188, 203–04, 221, 231–32, 320, 446, 448
Cooper, Edward Henry (1827–1902), 137, 252
Cooper, Edward Joshua (1798–1863), 9, 12, 88, 106, 122, 135, **137**, *138*, 139–40, 157, 183, 205, 221, 252, 327, 446
Copeland, Leland Stanford (1886–1973), 16, 36, 197, 209
Copeland, Ralph (1837–1905), 8, 11, 47, 101, 170, 226, 229, 231–32, 250, 276, 295, 297–98, 300–01, 305, 307, **310**, *311*, 312–14, 317, 319–21, 323, 326, 343, 364–66, 373, 380–82, 389, 391, *392*, 391–95, 403, 428–30, 442, 444, 446, 450, 452, 455, 459, 461, 490, 504, 516–17, 528

Corwin, Harold Glenn (1943–), 433, 470–71
Craig, John (1805–1877), 10, 124, 252, 368
Crossley, Edward (1841–1905), 386, 388
Curtis, Heber Doust (1872–1942), 65, 382, 389, 402, 432, 466–67, 512
Cysat, Johann Baptist (1587–1657), 16, 18, 407, 441–42, 445, 455

Dallmeyer, Johann Heinrich (1830–1883), 553
Darby, William Arthur (1809–1879), 116, 383, 491, 493
Darquier, Antoine de Pellepoix (1718–1802), 16, 35, 43, 447, 450
d'Arrest, Heinrich Louis (1822–1875), 7–8, 10–14, 23, 35, 37, 46–47, 49–51, 65–66, 83, 88–89, 92, 98, 119, 123, 125, 128, 130–31, 134–36, 141, 146, 148, *149*, 150–60, 162–65, 169–70, 175, 177–78, 180–83, 186–88, 196–99, 201–04, 206–10, 217–22, 225–27, 230–31, 235–238, *239*, 240–51, 253, 255–57, 259, 265, 267, 269, 272, 274, 276–80, 283–90, 292–93, 297, 301–02, 308–09, 312, 314–15, 318–21, 327, 330, 333–34, 342–48, 355–56, 363–65, 371–72, 375, 412–14, 423–25, 439–40, 442–43, 447–48, 450, 456, 458, 472–74, 478, 480–81, 483, 493–95, 498–509, 514–17, 519, 521, 523–27, 529–31, 534, 536, 544
Davis, Helen, 15
Dawes, William Rutter (1799–1868), 122, 132–33, 379, 382, 540
de Chéseaux, Jean Phillippe Loys (1718–1751), 16–18, 28–29, 31, 167, 221, 427, 447, 450, 459, 462, 464, 516
de Kerolyr, Marcel (1873–1969), 402
de la Rue, Warren (1815–1889), 125, 252, 417
Dembowski, Ercole (1812–1881), 359
Denning, William Frederick (1848–1931), 252, 277, 316, 398, 459–60, 462–63, 519, *520*, 541–42, 550
de Quincey, Thomas (1785–1859), 99
de Vaucouleurs, Gérard Henry (1918–1995), 468
de Vico, Francesco (1805–1848), 139–40, **141**, 142, 481, 498
Dewhirst, David (1926–), 192
Dien, Charles (1809–1870), 186
Dixon, Robert S., 468, 470
Doberck, August William (1852–1941), 137, *138*
Döllen, Johann Heinrich Wilhelm (1820–1897), 144
Dollond, John (1709–1761), 67, 95, 123
Donati, Giovan Battista (1826–1873), 125, 171, 284, 338
Draper, Henry (1837–1882), 407, 409
Dreyer, Alice Beatrice (1879–?), 226
Dreyer, Frederic Charles (1878–1956), 225–26
Dreyer, George Villiers (1883–?), 227
Dreyer, Ida Nicoline Margrethe (née Randrup) (1812–?), 225
Dreyer, Johan Christopher Friedrich (1814–1898), 225
Dreyer, John Louis Emil (1852–1926), 2–3, 7, 17, 18, 26–27, 30, 44, 47–51, 55, 71, 79, 84, 86, 87, 96, 101, 108, 113, 126–129, 131, 133, 137, 139, 148–149, 159–160, 170, 172–174, 177–178, 183, 196, 198, 200–203, 205–206, 215–217, 220–221, **225**, *225*, 226, *227*, 228–32, 237, 240, 242, 245, 247, 249, 253, 256–58, 262–63, 265–66, 269, 273–74, 277, 279, 281–82, 284–88, 290–301, 303–07, 309–21, 323–24, 326–30, 332–37, 339–41, 344–45, 348, 350, 352, 354–56, 362–66, 369–73, 375, 377–78, 381, 384–88, 392–96, 399–403, 405–06, 412, 414, 421–24, 426, 428, 430–31, 439–45, 447–48, 450, 452, 458–62, 465–69, 471–74, 479, 481–83, 485, 493–96, 504–08, 511, 515–17, 519, 521, 531, 534–35, 538–39, 545, 549, 551, 554, 557, 559
Dreyer, John Tuthill (1876–?), 226
Dreyer, Katherine Hannah (née Tuthill) (1858–1923), 226, 230
Dumouchel, Étienne Stefano (1773–1840), 141, 368
Dunér, Nils Christoffer (1839–1914), 285, **292**, *293*, 320, 447, 450
Dunlop, James (1793–1848), 5, 8, 12, 56, 61, 63, 71, **72**, *72*, 73–81, 83, 85–86, 128, 131, 176, 181–82, 186, 188, 212, 231–32, 260–61, 320, 397, 399, 403, 406, 444–45, 447, 450, 453, 456, 458, 462, 464, 481, 483

Eddington, Arthur Stanley (1882–1944), 230
Edmondson, Neil, 102
Eichens, Friedrich Wilhelm (1820–1884), 274, 276, 419, 491, 536
Ellery, Robert Lewis John (1827–1908), 252, **415**, *416*, 418–19, 421, 445
Ellison, William Frederick Archdall (1864–1936), 229
Encke, Johann Franz (1791–1865), 62, 72, 127, 144, 147–49, 169, 267, 294, 527
Engelhardt, Basilius v. (1828–1915), 232, 439, 441, 464, 473–74, 506
Engelmann, Friedrich Wilhelm Rudolf (1841–1888), 474, 531–32, 538–39
Ertel, Traugott Lebrecht (1778–1858), 67, 138, 149
Esmiol, Emmanuel (1853–?), 273, 375–76, 473
Espin, Thomas Henry Espinall Compton (1858–1934), 382, 396, 462–63
Everett, Edward (1794–1865), 126

Faris, Charles (1847–1924), 102, 229
Faye, Hervé Auguste Etienne Alban (1814–1902), 176, 260, 386, 388, 409
Fernley, John (1796–1873), 384
Ferrari, Gaspare Stanislao (1834–1903), 143, 267, **269**, 270, 320, 346, 444–45, 447, 450
Finlay, William Henry (1849–1924), 464
Fitz, Henry (1808–1863), 409
Fizeau, Armand Hippolyte Louis (1819–1896), 271
Flammarion, Nicolas Camille (1842–1925), 17–18, 209, 515
Flamsteed, John (1646–1719), 16, 18, 25, 27–28, 31, 34, 50, 55–56, 152, 159, 427, 441, 447, 449–50
Fleming, Williamina Paton Stevens (1857–1911), 23, 390
Foerster, Wilhelm Julius (1832–1921), 161, 480
Foucault, Jean Bernard Léon (1819–1868), 110, 131, 205, 269, 271, 273–74, 377, 415, 417, 419, 501, 536
Franklin-Adams, John (1843–1912), 466
Franks, William Sadler (1851–1935), 396
Fraunhofer, Joseph v. (1787–1826), 9, 67, 90, 99, 147, 149–50, 159, 162, 186, 216–17, 268, 278, 328, 503
Frost, Royal Harwood (1879–1950), 462

Gale, Walter Frederick (1865–1945), 462–63
Galilei, Galileo (1564–1642), 15, 18, 396
Galle, Johann Gottfried (1812–1910), 10, 89, 145, 149
Gambey, Henri Prudence (1787–1847), 135, 526
Gauß, Carl Friedrich (1777–1855), 63, 144, 163, 324
Gautier, Adolph Raoul (1854–1931), 241, 545
Gautier, Ferdinand Paul (1842–1909), 544
Gaviola, Ramon Enrique (1900–1990), 522
Gill, David (1843–1914), 310

Name index 623

Gingerich, Owen Jay (1930–), 18
Ginzel, Friedrich Karl (1850–1926), 474
Glaisher, James Whitbread Lee
 (1848–1928), 541
Glenn, John Herschel (1921–), 52
Glyn Jones, Kenneth (1915–1995), 18
Goldschmidt, Hermann (1802–1866),
 165, 173, 499, 526–27, 533, 536–37,
 542–43, 547, 551
Gopčević, Spiridon; *see* Brenner, Leo
Gore, John Ellard (1845–1910), 15
Gothard, Eugen v. (1857–1909), 43, 402,
 403, 483, 492, 496–97
Gottlieb, Steve (1949–), 471
Gould, Benjamin Apthorp (1824–1896),
 66, 325
Graff, Kasimir Romuald (1878–1950), 466
Graham, Andrew (1815–1907), 138
Gravatt, William (1806–1866), 10
Gray, T., 103, 307
Groombridge, Stephen (1755–1832), 146
Grove-Hills, Rdmond Herbert
 (1864–1922), 100
Grover, Charles (1842–1921), 528–29
Grubb, Howard (1844–1931), 280, 417
Grubb, Thomas (1801–1878), 100, 106,
 132–33, 137, 227–28, 299, 310, 313,
 380, 415, 417, 419
Güttler, Adalbert Max Ferdinand Adolf
 (1912–1964), 385
Gundlach, Ernst (1834–1908), 340, 353
Gyulbudaghian, Armen L. (1948–), 522

Hagen, Johann Georg (1847–1930), 8,
 12, 384–85, 427, 439, 465–66, 508,
 510, 559
Hahn, Friedrich v. (1742–1805), 14,
 41–43, 513
Hale, George Ellery (1868–1938), 131,
 360, 553, 558
Hall, Asaph (1829–1907), 126, 133, 370,
 386, **409**, *409*, 410, 447, 450, 544
Hall, Maxwell (1845–1920), 530–33,
 535–39, 542, 545–46
Halley, Edmond (1656–1742), 16, 18,
 71–72, 79–80, 167, 338, 370, 447, 450,
 455, 459
Hamilton, James Archibald (1767–1815),
 137
Hamilton, William Rowan (1805–1865),
 38, 106
Hansen, Peter Andreas (1794–1874),
 147
Hardcastle, Joseph Alfred (1868–1917),
 229, 466
Harding, Karl Ludwig (1765–1834), 5,
 9, 29, **63**, *63*, 64, 66, 71, 97, 164, 166,

168, 183, 186, 221, 262, 266, 292, 325,
 407, 447–48, 450
Harrington, Mark Walrod (1848–1926),
 195, *408*, **408**, 409, 447, 450, 456
Hartmann, Johannes Franz (1865–1936),
 382
Hartwig, Carl Ernst Albrecht
 (1851–1923), 332, 346, 365–66, **412**,
 413, 414–15, 447, 450
Henderson, Thomas (1798–1844), 66,
 181
Henkel, Frederick William (1869–1916),
 137
Henry brothers (Paul Henry & Prosper
 Henry), 8, 86, 205, *438*, 447, 450, 455,
 543–50, 552
Henry, Mathieu-Prosper (1849–1903),
 438, 444, 450, 522, 542
Henry, Paul-Pierre (1848–1905) **438**,
 438, 444, 450, 522, 542
Herbig, George Howard (1920–), 512,
 561
Herschel, Caroline Lucretia (1750–1848),
 5, 8, 17, 19–23, 26, *28*, 29–31, 34, 36,
 47, 50, 53, 55–56, 64–65, 77, 80, 166–
 68, 173, 179–80, 182, 186, 188, 217–19,
 222, 439, 447, 450–51, 459, 466
Herschel, Friedrich Wilhelm [William]
 (1738–1822), 2, 5, 7, 8, *14*, 17–30,
 32–44, 46–52, 54, 56, 59–65, 68,
 77–81, 83–85, 88–89, 92–93, 95–100,
 104–06, 118–21, 127–31, 134–35, 139,
 141–42, 145, 148, 150–54, 156–57,
 159–60, 164, 166–67, 169–70, 172–83,
 186, 195, 198, 200–01, 204–06, 208–11,
 213, 216, 218–20, 222, 229, 233–37,
 239–40, 242, 245–50, 252–53, 255,
 257–58, 260–61, 265, 268–69, 272–74,
 279–81, 283–84, 286–88, 291–92, 295,
 301–02, 306, 309–10, 314, 318–21,
 325–32, 338, 344–50, 353, 355–56, 362,
 364, 367, 371–73, 375–77, 382, 385–87,
 392, 394, 399–400, 407–08, 411–12,
 417, 423, 425, 430, 434–37, 439–41,
 443–50, 452–56, 458–62, 464–67,
 471–73, 475–77, 481–84, 490–91, 498,
 500, 504, 510, 513–16, 519, 521–22,
 527, 551, 558–59
Herschel, John Frederick William
 (1792–1871), 2, 7–8, 21–22, 24,
 26–29, 33, 43–44, 47–51, *52*, 53–69,
 71, 73–89, 94–97, 100, 104–08, 111,
 115–16, 118–19, 122, 124, 127, 130–33,
 139, 141–43, 150–60, 162, 164, 170,
 175–84, 186, 188–89, 192, 196–98,
 200–13, 216–23, 226, 229–31, 234–37,
 240–41, 247–50, 252–53, 255,

258–61, 268, 271–73, 286–88, 290–91,
 293, 295–96, 298–99, 303, 305–06,
 310–11, 314–15, 319, 321, 323–25, 329,
 333, 336–38, 341, 343, 346, 356, 359,
 361, 363, 368, 371–73, 386, 391–92,
 397, 402–03, 405, 407–09, 417, 420–22,
 436, 439–45, 447–50, 452–56, 458–60,
 462, 464, 473–85, 487–88, 490–91, 493,
 497–98, 501–03, 513–19, 521, 525–26,
 528–29, 539
Herschel, John Lt. (1837–1921), 250,
 517–18
Hertel, Wilhelm (1837–1893), 144–45,
 262–63, 381, 413, 511, 531
Hevelius, Johannes (1611–1687), 146
Hickson, Paul (1950–), 44
Hind, John Russell (1823–1895), 5, 9,
 13, 69, 88–89, *121*, **122**, 123–25, 138,
 154–56, 163–64, 166, 168, 183, 205–07,
 221, 236, 251, 268, 284–86, 447, 450,
 498–506, 508–11, 513–15, 519, 522–23,
 526, 528, 533
Hinks, Arthur Robert (1873–1945),
 466
Hipparch (-190– -125), 5, 16, 18, 28, 31,
 56, 441–42, 447, 450, 454
Hodierna, Giovanni Battista (1597–1660),
 16–18, 28–29, 31, 56, 80–81, 167, 221,
 447, 449–50, 455, 459, 475
Holcombe, Amasa (1787–1875), 94
Holden, Edward Singleton (1846–1914),
 15, 190, 207, 211, 213, 294–95, 320,
 326, 346, 348, 359–60, 366, 368, 388,
 390, 396, **403**, *404*, 405–06, 408–10,
 442, 447–48, 450, 481–82, 509, 517,
 530, 538–39, 550
Holetschek, Johann (1846–1923), 323,
 466, 548
Holmberg, Eric Bertil (1908–2000), 44,
 423
Hooper, John R., 542
Hoskin, Michael (1930–), 12, 15, 37, 80,
 86, 192
Hough, George Washington (1836–1909),
 332, 357, *359*, **360**, 361, 363, 365–67,
 414, 447, 450, 533, 538–41, 547, 560
Howe, Herbert Alonzo (1858–1926), 216,
 231, 350, 354, 432, 462–63
Hubble, Edwin Powell (1889–1953), 40,
 118, 176, 402, 467, 512, 514
Huggins, Margaret Lindsay Murray
 (1848–1915), 133, 450, 480
Huggins, William (1824–1910), 11, 93,
 112, 144, 159, 196, 221, 250, 305, 333,
 368–69, 407, 450
Humboldt, Friedrich Willhelm Heinrich
 Alexander v. (1769–1859), 15, 18, 325

Hunt, George (1823–1896), 533, 540
Hunter, Samuel, 101, **103**, 112, 190–92, 195–96, 198, 201, 297–301, 303–04, 308, 346, 444–45, 447, 450, 462, 483, 487, 490–91, 493, 516, 521
Huth, Johann Sigismund Gottfried (1763–1818), 66
Huygens, Christiaan (1629–1695), 367, 475

Ihle, Johann Abraham (1627–1699), 15–16, 447, 450, 455, 459
Innes, Robert Thorburn Ayton (1861–1933), 177, 462–63

Jahn, Gustav Adolph (1804–1857), 64, 169, 439
Javelle, Stéphane (1864–1917), 160, 183, 203, 348, 371–73, 460, 462
Jeans, James Hopwood (1877–1946), 230
Jeaurat, Edme-Sébastien (1724–1803), 525, 543–44
Jünger, Emil, 149, 292

Kahler, William, 404
Kaiser, Frederick (1808–1872), 97, 496
Kammermann, Arthur (1861–1897), 545
Keeler, James Edward (1857–1900), 43, 117, 208, 382, 389, 404, 427, 479–81, 507, 510–12, 543, 560
Kellner, Carl (1826–1855), 162, 538, 540
Kempf, Paul Friedrich Ferdinand (1856–1920), 474
Key, Henry Cooper (1819–1879), 118–19, 215, 528
Kirch, Gottfried (1639–1710), 16, 125, 167, 444, 446, 450, 459, 475, 516
Kirch, Maria Margaretha (1670–1720), 444, 450
Klein, Hermann Joseph (1844–1914), 173
Klinkerfuess, Ernst Friedrich Wilhelm (1827–1884), 310, 429
Knobel, Edward Ball (1841–1930), 121, 407, 545
Knott, George (1835–1894), 122, 288, 380–81, 503, 536
Kobold, Hermann Albert (1858–1942), 145, 237, 272, 312, 387–88, 459–60
Koehler, Johann Gottfried (1745–1801), 16, 18, 167, 446, 450, 459
Kövesligethy, Radó v. (1862–1934), 415
Kohoutek, Lubos (1935–1951), 242
Kopff, August Adalbert (1882–1960), 33
Krüger, Karl Nikolaus Adalbert (1832–1896), 161–64, 174, 226, 526, 543

Lacaille, Nicolas-Louis de (1713–1762), 5, 16, 18, 25, 70–71, 73, 75–77, 80, 151, 181–82, 220–21, 446, 448, 450, 464, 516
Lalande, Joseph-Jerome le Francois de (1732–1807), 55, 60, 68, 97, 122, 152, 154, 159, 170, 239, 441
Lamont, Johann v. (1805–1879), 19, 88, *89*, **89**, 90–93, 95–97, 124, 151, 153, 187, 222, 234, 260, 301, 407, 473–74, 479, 481, 483, 508, 513–14
Lamp, Ernst August (1850–1901), 411–12, 414
Lampland, Carl Otto (1873–1951), 512
Langley, Samuel Pierpont (1834–1906), 6, **335**, *335*, 337, 446, 448
Laplace, Pierre-Simon de (1749–1827), 105, 118, 485
Larkin, Edgar Lucien (1847–1924), 340
Lassell, Caroline, 133, 211, 450, 481
Lassell, William (1799–1880), 9, 12, 30, 34, 88, 106, 118–19, 125, **131**, *132*, 133–35, 146, 156–58, 183, 188, 207, 211–17, 221–22, 239, 245, 248, 251–55, 259, 274, 308, 320, 333, 363, 368, 385, 407, 417, 446, 450–51, 477–79, 481, 483, 491–95, 497, 501–02, 508, 517–19
Lauberts, Andris (1942–), 467–68
Laugier, Paul August (1812–1872), 10, 88, **134**, 135–36, 150–52, 181, 187, 222, 235–37, 260, 439, 473–74
Leavenworth, Francis Preserved (1858–1928), 47, 337, 346, 387–88, 400, 410, 431–35, 442–43, 445–46, 450, 464
Leavitt, Henrietta Swan (1868–1921), 390
Lee, John (1783–1866), 120
Lefèbvre, Pierre (1726–1806), 251
Legentil, Guillaume Hyazinthe (1725–1792), 16, 18, 68, 125, 167, 186, 446, 450, 459, 513, 516
Lehner, Georg (1873 –1947), 385
Lerebours, Noël-Jean (1761–1840), 135
Le Sueur, Adolphus Albert (1849–1906), **415**, 416–21, 518
Leverrier, Urban Jean Joseph (1811–1877), 122, 134, 149, 171, 211, 237, 269–71, 368, 491, 498–99, 501, 526, 536
Lewis, Thomas (1856–1927), 555
Lick, James (1796–1876), 404
Lindsay, James Ludovic [Lord Lindsay] (1847–1913), 226, 228, 310–11, 321, 380–81, 392, 394, 429–30, 536
Littrow, Joseph Johann v. (1771–1840), 119, 379

Littrow, Karl Ludwig v. (1811–1877), 166, 379, 411, 503, 528
Lockyer, Joseph Norman (1836–1920), 215
Lohse, J. Gerhard (1851–1941), 6, 288, 311, 380, 393–94, 428, *429*, **429**, 430, 441, 446, 448, 450, 459, 461
Lohse, Wilhelm Oswald (1845–1915), 278, 429
Loomis, Eben Jenks (1828–1912), 370
Loomis-Todd, Mabel (1856–1932), 370
Lord Lindsay; *see* Lindsay, James Ludovic
Lord Rosse; *see* Parsons, William
Lowe, Thaddeus Sobieski Constantine (1831–1913), 339
Lowell, Percival (1855–1916), 370
Lundmark, Knut Emil (1889–1958), 467
Lunt, Joseph (1866–1940), 462–63
Luther, Eduard (1816–1887), 147–48, 180, 182

MacGeorge, E. Farie, 415, **416**, 418–19, 518
Maclear, George William Herschel (1836–1895), 52
Maclear, Thomas (1794–1879), 38, 52
MacMahon, Percy Alexander (1854–1929), 427
Mädler, Johann Heinrich v. (1794–1874), 61, 66, 93, 333, 514
Magakian, Tigran Yuri (1953–), 522
Magellan, Fernando de (1480–1521), 15–16, 445
Mahler, Eduard (1857–1945), 465
Mairan, Jean Jacques d'Ortous de (1678–1771), 16, 28, 446, 450, 459, 475
Mallas, John H. (1927–1975), 146
Maraldi, Jean-Dominique (1709–1788), 16, 68, 168, 186, 446, 450
Marius, Simon (1573–1624), 15–16, 452, 455
Markarian, Beniamin Egishevich (1913–1985), 430
Markham, H. C., 409
Marth, Albert (1828–1897), 8, 47, 122, 127, 133–34, 137, 166, 169, 174, 177, 179, 181, 212–13, 217, 231, 244–47, 250, *251*, **251**, 252–59, 267, 269, 271–72, 274, 276, 280, 283–84, 289–90, 301–02, 309, 318–21, 326, 329–30, 346, 363–64, 366, 373, 375, 377, 387–88, 391–92, 397, 400, 405, 411–12, 423–24, 434, 436–37, 440, 443–46, 448, 450, 455–58, 460, 464
Martin, A., 274, 491, 536

Martins, Carl Otto Albrecht (1816–1871), 226, 240, 278
Maskelyne, Nevil (1732–1811), 182, 184, 186
Mason, Ebenezer Porter (1819–1840), 27, 88, **94**, 95–96, 217, 222, 407, 476, 479, 481
Massinger, Adam (1888–1914), 467
Matthews, W., 528
Maury, Antonia Caetana de Paiva Pereira (1866–1952), 390
Mauvais, Félix Victor (1809–1854), 133
Mayall, Margaret Walton (1902–1995), 382, 527
McCance, James Law (1856–1903), 535, 537
McClean, Frank (1837–1904), 380
McCormick, Leander James (1819–1900), 431
McFarland, John (1948–), 486
McNeil, Jay (1971–), 522
Méchain, Pierre Francois Andre (1744–1804), 16–18, 28, 31, 44–45, 55–56, 64–65, 68, 79–80, 121, 151, 167, 181–82, 186, 221, 441, 445–46, 449–50, 455, 484
Merecki, Franciszek Romuald Pawel (1860–1922), 474
Merz, Georg (1793–1867), 66, 90, 144–45, 150, 161, 169, 179, 217, 242, 269, 274, 282, 285, 381, 430, 465
Messier, Charles-Joseph (1730–1817), 2–3, 7, 14–18, 24, 26–29, 31, 35, 45, 53–56, 60, 64, 71, 77–80, 104, 119–20, 134–35, 146, 151, 154, 164, 166–67, 169, 179, 181–82, 186, 211, 219–21, 235, 296–97, 325, 382, 408, 441–42, 445–46, 449–50, 455, 462, 464, 475, 477, 481–82, 488, 516
Mitchel, Ormsby MacKnight (1809–1862), 360
Mitchell, Maria (1818–1889), 94
Mitchell, R. J., 46–47, 101, **103**, 190–92, 196, 198–203, 221, 287–89, 295, 297–98, 300–04, 307, 314, 317–18, 320–21, 332, 345–46, 364, 373, 385, 435, 443–44, 446, 448, 450, 456, 462, 464, 479, 483, 519, 521
Mitchell, William (1791–1869), 94, 111
Möbius, August Ferdinand (1790–1868), 149, 169
Møller, Didrik Magnus Axel (1830–1896), 292
Mönnichmeyer, Carl (1860–1942), 474
Moore, Patrick Alfred (1923–), 2, 8, 102, 227, 317

Mouchez, Amédée Ernest Barthélémy (1821–1892), 438, 543
Müller, Carl Hermann Gustav (1851–1925), 381, 413
Muller, Frank (1862–1917), 337, 346, 425, 431, **432**, 433–34, 436, 443, 446, 450, 457, 464

Nasmyth, James (1808–1890), 118, 132, 217
Newall, Robert Sterling (1812–1888), 10, 252, 368, 370, 531, 541, 546, 555
Newcomb, Simon (1835–1909), 147, 227, 370, 396, 404, 409, 431
Newton, Isaac (1643–1727), 14, 62, 109, 118, 230, 235
Nichol, John Pringle (1804–1859), 87, 105–06, 111, 115–16, 118, 484, 486–89
Nilson, Peter (1937–1998), 467–68
Nobel, William (1828–1904), 38

O'Connor, Edward Dominic (1874–1954), 385
Olbers, Heinrich Wilhelm Matthias (1758–1840), 61–62, 70–71, 73, 186, 476
Olmstedt, Denison (1791–1859), 94
Olufsen, Christian Friis Rottbøl (1802–1855), 149
Oppenheim, Samuel (1857–1928), *410*, **411**, 412–13, 444–46, 450
Oppolzer, Theodor v. (1841–1886), 285, 474
Oriani, Barnaba (1752–1832), 16, 18, 28, 31, 446, 450
Oudemans, Jean Abraham Chrétien (1827–1906), 542

Palisa, Johann (1848–1925), 260, 325, 346, 372, *410*, **411**, 412–14, 446, 450, 452, 544
Pannekoek, Antonie (1873–1960), 175
Pape, Carl-Ferdinand (1834–1862), 144, 523, 526, 529
Parish, John (1774–1858), 166, 169
Parsamyan, Elma S. (1929–), 522
Parsons, Charles Algernon (1854–1931), 99, 101, 103, 303
Parsons, Lawrence (1840–1908), 9, 46, 98–99, 101, 103, 112, 157, 160, 170, 190, 192, 201–03, 226, 230–31, 244–45, 250, 252, 288–89, 295–301, 303–305, **306**, *307*, 308–11, 313–15, 318–20, 338, 446, 450, 485, 494, 496, 504, 528, 531, 534, 547
Parsons, William [Lord Rosse] (1800–1867), 7, 9, 12–13, 16, 34, 36, 38, 40, 43, 50, 87, **98**, *98*, 99, *100*, 101–18, 131–35, 142, 157, 159–60, 179, 183, 188–90, 192, 194–99, 201, 206, 208, 211, 216–18, 220–22, 230–31, 234–35, 241–42, 248–49, 265, 295, 297–301, 303, 305–07, 312, 317, 330, 333–34, 338, 368, 385, 389, 409, 417, 430, 444, 446, 450–51, 472–73, 475–83, 485–97, 506, 521, 527, 531, 534, 539
Paturel, Georges (1946–), 469
Payne, William Wallace (1837–1928), 552
Pease, Francis Gladheim (1881–1938), 382, 512–13
Pechüle, Carl Frederick (1843–1914), 267–68, 276–77, 285, **293**, *294*, 320, 383–84, 446, 448, 450, 461, 463
Peirce, Benjamin (1809–1880), 335
Peirce, Charles Sanders (1839–1914), 6, 62, 119, 335–37, 446
Peiresc, Nicolas-Claude Fabri de (1580–1637), 16, 18, 444–46, 450, 455, 459, 516
Perek, Lubos (1919–), 242
Perrine, Charles Dillon (1867–1951), 389, 427, 510–11
Perrotin, Henri Joseph Athanase (1845–1904), 544
Peters, Carl Friedrich Wilhelm (1822–1894), 324
Peters, Christian August Friedrich (1806–1880), 251, 324, 523, 527, 529
Peters, Christian Heinrich Friedrich (1813–1890), 9, 64, 66, 122, 138, 164, 172–73, 183, 186, 272, 283–84, 320, 323, *324*, **324**, 325–27, 332, 338, 372, 435–36, 446, 450, 461
Petersen, Adolph Cornelius (1804–1854), 124, 141, 174, 182, 186, 279–80
Petrosyan, Vahé M. (1938–), 522
Pfaff, Johann Wilhelm Andreas (1774–1835), 50, 179–81
Piazzi, Giuseppe (1746–1826), 55, 69–71, 119, 121, 157–59
Picard, Jean (1620–1682), 367
Pickering, Edward Charles (1846–1919), 8, 11, 131, 266, 323, 333, 335–37, **389**, *390*, 391–95, 444–46, 450, 542–44
Pickering, William Henry (1858–1938), 390, 396, 492, 559
Pigott, Edward Charles (1753–1825), 15–16, 167, 444, 446, 450
Pistor, Carl Phillip Heinrich (1778–1847), 226, 240, 278
Plössl, Georg Simon (1794–1868), 175, 178, 235, 529–30
Pogson, Norman Robert (1829–1891), 122, 148, 234, 384

Porro, Ignazio (1801–1875), 10, 252, 368–69
Porter, Jermain Gildersleve (1852–1933), 66, 344, 465, 474
Powell, Eyre Burton (1819–1904), 518
Pratt, Henry (1839–1891), 538–39
Preuß, Ernst Wilhelm (1796–1839), 67
Pritchard, Charles (1808–1893), 261, 549–50
Proctor, Richard Anthony (1837–1888), 30, 109, 223, 248, 466, 517–18
Ptolemy, Claudius (87–165), 16, 18, 444, 447, 450, 456
Purser, John (1835–1903), 103, 307

Quincy, Josiah (1772–1864), 127, 162

Ramage, John (1784–1835), 9–10, 94, 99, 104, 106, 368
Rambaut, Arthur Allcock (1859–1923), 102
Rambaut, William Hautenville (1822–1911), **101**, *102*, 112, 116–17, 190, 229, 298, 486, 488–89, 491, 493
Ramsden, Jesse(1735–1800), 69
Rayet, Georges Antoine Pons (1839–1906), 542–43
Reichenbach, Georg Friedrich (1771–1826), 64–65, 67, 96, 187, 310
Reiland, Tom, 560
Reinfelder, Gottlieb (1836–1898), 144–45, 262–63, 381, 413, 511, 531
Reinmuth, Karl Wilhelm (1892–1979), 292, 312, 329, 430, 467, 511–12
Repsold, Johann Adolf (1838–1919), 144–45, 147, 262, 268, 282, 285, 292, 358, 543, 552
Ritchey, George Willis (1864–1945), 9, 389, 419
Roberts, Isaac (1829–1904), 15, 103, 116, 342, 356, 462, 479, 507–08, 510, 545–46, 548, 550, *553*, 554–56, 558–59
Robinson, Thomas Romney (1792–1882), 98, 102, 105, *106*, 107–118, 132–133, 137, 215, 226–229, 368, 417, 419, 483, 484–87, 490–91
Rondoni, Francesco, 141–42
Rümker, Christian Ludwig Carl [Charles] (1788–1862), 72–73, 174, 182, 184, 186, 252, 267
Rümker, George Friedrich Wilhelm (1832–1900), 10, 51, 72, 182, 184, 230, 237, 240, 242, 260, **267**, *268*, 279, 288, 291, 293–94, 320, 439, 447–48, 474, 493
Russell, Henry Chamberlain (1836–1907), 299
Russell, Henry Norris (1877–1957), 428

Sabine, Edward (1788–1883), 100, 106, 211
Sadler, Herbert (1856–1898), 121, 517
Safford, Truman Henry (1836–1901), 125–27, 129, 131, 160, 188, 203–04, 221, 244, 246, 255–57, 267, 272, 274, 280, 283–84, 332, 346, **357**, 358, *359*, 361–64, 366, 375, 403, 440, 442, 444, 447–48, 450, 462, 464
Sampson, Ralph Allen (1866–1939), 226, 229
Sandage, Allan Rex (1926–), 468
Sawyer-Hogg, Helen Battles (1905–1993), 18
Schaeberle, Johann Martin (1853–1924), 117, 404, 407–10, 550
Scheiner, Julius (1858–1913), 43, 401
Schiaparelli, Giovanni Virginio (1835–1910) , 171, 496, 527, 529–30, 533, 538–39, 544, 547
Schjellerup, Hans Carl Frederik Christian (1827–1887), 149, 159, 225, 239, 294, 500, 523
Schmidt, Johann Friedrich Julius (1825–1884), 8, 10, 40, 68, 70, 75–76, 81, 148, 157, 161–162, 164, **174**, *174*, 175–179, 183, 208, 221–222, 234–237, 256, 258–61, 285, 290, 319–20, 330, 381, 392, 430, 439, 447, 450, 474, 493, 501, 514, 519, 522, 524–25, 536
Schönfeld, Eduard (1828–1891), 1, 8, 10, 46, 51, 88, 93, 127–29, 131, 134, 151, 155–56, **161**, *161*, 162–64, 166–67, 174, 178, 183, 199–200, 204, 221–22, 226, 230–31, 233–38, 240, 259–60, 279, 285, 288, 290–92, 319–20, 334, 344, 371, 439, 442, 447, 450–51, 473–74, 493–94, 499–500, 503, 506, 514–15, 519, 524–25, 527–28, 536
Schorr, Richard Reinhard Emil (1867–1951), 268
Schrader, Johann Gottlieb Friedrich (1763–1832), 41
Schröder, Heinrich Ludwig Hugo (1834–1902), 278, 413
Schroeter, Johann Hyronimus (1745–1816), 41, 63, 174, 251, 407
Schultz, Per Magnus Herman (1823–1890), 8, 10, 26, 43, 51, 83, 146, 151, 158, 194, 202–03, 225–26, 230–31, 236–37, 240, 245, 260, 266, 279, **285**, *285*, 286, 288–93, 295, 301–02, 308–09, 320–21, 334, 344, 373, 430, 439, 441–42, 447–50, 473–74, 494
Schumacher, Heinrich Christian (1780–1850), 70–71, 96, 123, 166, 182

Schwassmann, Friedrich Carl Arnold (1870–1964), 271–72
Schweizer, Kaspar Gottfried (1816–1873), **232**, *233*, 319–20, 447–48
Searle, Arthur (1837–1920), 335–36, 390
Searle, George Mary (1839–1918), 6, 249, **335**, *336*, 337, 355, 447, 450
Secchi, Pietro Angelo (1818–1878), 43, 88, 108, 119, **139**, *140*, 141–44, 150, 176, 182, 186, 208, 216–17, 222, 234, 269, 320, 333–34, 368, 394, 397, 400, 445, 477–78, 483, 493, 501, 534, 536
Sécretan, Marc François Louis (1804–1867), 422
Shapley, Harlow (1885–1972), 15, 385, 467–68
Sheepshanks, Richard (1794–1855), 106, 120, 305
Sievers, Johann Joachim (1805–1882), 187
Simms, William jr. (1817–1907), 394
Simon, Charles Marie Etienne Theophile (1825–1880), 270
Sinnott, Roger W. (1944–), 469–70
Skinner, Aaron Nichols (1845–1918), 332, 346, **357–58**, 361–63, 365–66, 444–45, 447, 450, 464
Slipher, Vesto Melvin (1875–1969), 560
Smith, Hamilton Landphere (1819–1903), 94, 407
Smyth, Charles Piazzi (1819–1900), 100, 106, 119–20, 133, 282, 312, 476–77
Smyth, William Henry (1788–1865), 15, 38, 43, 51, 53, 65, **119**, *120*, 121, 157–58, 379, 382–83, 475–76, 479, 481, 483–85, 493
Soldner, Johann Georg v. (1776–1833), 89
Sonneborn, George, 468, 470
South, James (1785–1867), 57, 66, 67, *106*, 107, 109, 110, 123, 205, 305, 359, 368, 483, 484–88, 491
Spencer, Charles Achilles (1813–1881), 325–26
Spitaler, Rudolf Ferdinand (1859–1946), 43–44, 62, 323, 405, 459–60, 462–63, 474, 483, 521, 527, 544–45, 547–48
Steinheil, Hugo Adolph (1832–1893), 161, 163, 171–72, 236, 278, 285–86, 292, 503, 523, 529, 542, 558
Steinheil, Karl August (1801–1870), 514
Stephan, Jean Marie Edouard (1837–1923), 8, 11, 13, 47, 231, 246, 250, 256–57, 267, 269, **271**, 272, *273*, 274–77, 281, 302, 312–14, 318–21, 323, 326, 332, 343, 346, 350, 354–55, 362–64, 371–78, 381–82, 401, 405–06, 423, 434–36, 440, 443–45, 447–48, 450, 456, 458, 474, 530

Name index

Stephenson, Charles Bruce (1929–2001), 383
Stewart, Delisle (1870–1941), 177, 431, 462
Stock, Jürgen (1923–2004), 383
Stokes, George Gabriel (1819–1903), 106, 211, 308
Stone, Edward James (1831–1897), 333, 386, 541
Stone, John, 77
Stone, Ormond (1847–1933), 314, 326, 343, 346, 357, 375–76, 387, 400, **431**, *431*, 432–37, 440, 443, 445, 447–48, 450, 459, 464, 474
Stoney, Bindon Blood (1829–1909), 47, 101, *102*, **103**, 112, 190–92, 195, 198–201, 221, 256, 258, 266–67, 280, 289, 295, 297–98, 300–04, 308, 320–21, 353, 373, 385, 408–09, 423–25, 444, 447, 450, 453, 462, 464, 487
Stoney, George Johnstone (1826–1911), 45, 101, *102*, **102**, 103, 110, 112, 114–15, 118–19, 190–91, 195–96, 198–201, 221, 237–38, 249, 285, 287, 295, 298–99, 302–03, 307, 317, 320–21, 356, 373, 422, 444, 447–48, 450, 456, 487
Stratonov, Vsevolod Viktorovich (1869–1938), 536, 552, 555, 560
Strohmeier, Wolfgang Paul (1913–2004), 385
Struve, Friedrich Georg Wilhelm v. (1793–1864), 10, 12, 38, 53–56, 58, 63, **66**, *67*, 68–69, 88, 90, 119, 121, 143–45, 164, 166–67, 181, 183, 186, 221, 232, 281–82, 338, 359, 368, 381, 447, 450
Struve, Gustav Wilhelm Ludwig (1858–1920), 282
Struve, Karl Hermann (1854–1920), 282, 543
Struve, Otto Ludwig (1897–1963), 282
Struve, Otto Wilhelm v. (1819–1905), 51, 106, 112, 132, 144, 177, 195, 211, 213–15, 222, 231, 234, 244, 246, 248, 253, 262, **281**, *282*, 283–84, 288, 319–20, 326, 338, 359, 363–64, 367, 447–48, 450, 490, 500, 502–11, 526, 529, 543–45
Sulentic, Jack William (1947–), 313, 469
Svanberg, Gustav (1802–1882), 285
Swift, Edward D. T. (1871–1945), *338*, 339–40, 346–50, 353, 356–58, 444–45, 447–48, 450, 455, 464
Swift, Lewis A. (1820–1913), 5, 8, 55, 62, 66, 95, 126, 128, 131, 173, 183, 201, 242, 249, 255–56, 261–62, 264, 266,
270–74, 277, 284, 288–89, 301–02, 304, 325, 328, 332–33, 337, **338**, *338*, *339*, 340–341, *342*, 343–58, 362–64, 366–67, 371–72, 375, 377, 382, 385, 387–88, 395–99, 401, 403, 405–06, 411–12, 414, 424, 429, 434–37, 440, 442–45, 447–50, 452–53, 455, 458, 460–62, 464–65, 479, 517, 519, 521, 533, 539–40, 542

Tacchini, Pietro (1839–1905), 269
Talmage, Charles George (1840–1886), 122, 502
Tammann, Gustav Andreas (1932–), 468
Tebbutt, John (1834–1916), 518
Tempel, Ernst Wilhelm Leberecht (1821–1889), 5, 8, 12–13, 35, 47, 51, 58, 88, 112, 117, 126, 139, 158, 163, 169–70, **171**, *171*, 172–74, 183, 206, 215, 221, 225, 231, 240, 242, 246–47, 260, 262–67, 270, 276–77, 289, 293, 301–02, 314–15, 318–20, 323, 326–34, 337, 340–41, 343–44, 346, 360, 362–66, 371–73, 375, 377, 383, 387–88, 391–92, 396–97, 399, 401, 410, 414, 429, 434–36, 440, 443, 447–48, 450, 453, 456, 458–59, 461, 464, 472, 477–80, 483, 491, 493–98, 503, 505–15, 521, 523–31, 533–43, 545, 548, 554, 560
Terby, François Joseph Charles (1846–1911), 542
Thiele, Thorvald Nicolai (1838–1910), 150
Thollon, Louis (1829–1887), 544
Thome, John Macon (1843–1908), 464
Thomson, Malcolm (1933–), 192, 471
Thormann, Friedrich (1831–1882), 162, 174
Thury, Marc (1822–1905), 545
Tifft, William (1932–), 313, 469
Tisserand, François Félix (1845–1896), 422
Tobin, William (1953–), 115, 271
Todd, David Peck (1855–1939), 247, 326, **369**, *370*, 371–73, 375, 409, 447, 450, 464, 481
Troughton, Edward (1753–1835), 72
Trouvelot, Etienne Leopold (1827–1895), 27, 143, 319, 336, 407, 462, 464, 540
Trümpler, Robert Julius (1886–1956), 383, 406, 560
Tully [Tulley], Charles (1761–1830), 378
Turner, Herbert Hall (1861–1930), 229
Turner, Joseph (1825–1883), 415, **416**, *416*, 418–21, 483
Tuttle, Charles Wesley (1829–1881), 126–27
Tuttle, Horace Parnell (1837–1923), *125*, **126**, 127–31, 145–46, 162–63, 171, 183, 204, 221, 238, 295, 320, 338, 409, 445, 514, 520

Unkhrechtsberg, Eduard v. (1790–1870), 174

Valentiner, Karl Wilhelm Friedrich Johann (1845–1931), 503
Valz, Jean Elix Benjamin (1787–1867), 171, 206, 523
van den Bergh, Sidney (1929–), 512
Vespucci, Amerigo (1451–1512), 5, 15–16, 80, 444, 447, 450, 455
Vogel, Eduard (1829–1856), 122, 251
Vogel, Hermann Carl (1841–1907), 10, 19, 43, 150–51, 213, 226, 230–31, 237, 240, 251–52, 258, 260, 277, **278**, *278*, 279–81, 311, 371, 381, 385, 394, 439, 473–74, 483, 493, 496–97, 503, 516, 519, 531
Voigt, Auguste (1828–1909), 255, 269–72, 274, 284, 320, 326–27, 344, 346, 363–64, 375, 444, 447–48, 450, 464
Voigtländer, Johann Christoph (1732–1797), 558–59
Vorontsov-Velyaminov, Boris Alexandrowitch (1904–1994), 44, 301, 351, 384, 467–68

Ward, Isaac William (1833–1905), 415
Warner, Hulbert Harrington (1842–1923), 338
Waters, Sidney (1853–1896), 223, 466
Watson, James Craig (1838–1880), 325, 338, 404, 408
Watson, William (1744–1825), 19
Webb, Thomas William (1806–1885), 18, 46, 118–19, 142, 215, 323, 359, 376, **378**, *379*, 380–83, 462, 464, 466, 503, 527–29, 533, 560
Weinek, Ladislaus (1848–1913), 474
Weiss, Edmund (1837–1917), 411–12, 544
Weisse, Maximilian v. (1798–1863), 187
Wendell, Oliver Clinton (1850–1912), 6, 336, 390, 399
Whewell, William (1794–1866), 105
Widt, Seweryn (1862–1912), 492–93, 497
Wigglesworth, James (1815–1889), 6, 429, 441
Wilsing, Johannes Moritz Daniel (1856–1905), 152, 382
Wilson, Herbert Couper (1858–1940), 432, 536, 551–52, 555–57, 559
Winlock, Joseph (1826–1875), 6, 119, 335, *336*, **336**, 337, 390, 447, 450

Winnecke, Friedrich August Theodor
    (1835–1897), 8, 17, 51, 63, 65–66, 88,
    97, 127, 129–30, **144**, *144*, 145–48,
    161, 163, 172–73, 178–80, 183, 206,
    215, 221, 231, 261–64, 266–67,
    276–77, 282, 284, 286, 293, 320, 323,
    340–41, 381, 413, 447–48, 450–51,
    455, 459, 461, 465, 493, 500, 502–04,
    519, 526, 536
Wirtz, Carl Wilhelm (1876–1939), 145,
    465–66, 511
With, George Henry (1827–1904), 379,
    528, 538
Wolf, Charles Joseph Etienne (1827–1918),
    109, 419, 491, 493, 525, 527, 530,
    533–39, 543, 545, 547–48
Wolf, Johann Rudolf (1816–1893), 18
Wolf, Maximilian Franz Josef Cornelius
    (1863–1932), 3, 23, 33, 249, 342, 385,
    397, 400–02, 411, 427, 462, 467, 511,
    551, 557–58, 560
Wollaston, Francis (1731–1815), 25, 42,
    49, 220–21
Woods, Thomas (1815–1905), 100
Worthington, Arthur Mason
    (1852–1916), 384

Wray, William (1829–1885), 10

Young, Charles Augustus (1834–1908),
    43, 359, **427**, *428*, 447, 450, 461

Zach, Franz Xaver v. (1754–1832), 70
Zeiss, Carl Friedrich (1816–1888), 557
Zöllner, Johann Karl Friedrich
    (1834–1882), 278
Zwicky, Fritz (1898–1974), 14, 46–47,
    247, 384, 467–68

Site index

Here all sites or observatories mentioned in the text are given (see also the section 'Telescope data' above). A bold page number refers to a figure.

Aberdeen, 10
Albany, 66, 325, 335, 360
Altona, 124, 141, 166, 174, 182, 279, 523, 525
Amherst, 369–70
Ann Arbor, 117, 325, 408–09
Arcetri, 112, 139, 171–73, 215, 263, 319, 327, 335, 365, 371, 410, 443, 458, 481, 491, 493–95, 505, 523, 529
Arequipa, 17, 23, 29, 177, 390, 452, 462, 466, 555
Armagh, 12, 98, 102, 105, 132, 137, 196, 215, 226–29, 292, 305, 323, 384, 396, 417, 430, 441, 459, 466, 474, 486, 488, 491, 493, 515, 517, 519
Athens, 68, 76, 81, 148, 157, 164–65, 174–75, 178, 235–36, 259–60, 330, 381, 392, 430, 474, 501, 524
Aylesbury, 120

Bamberg, 413
Bangalore, 518
Bath, 18, 38, 104
Bedford, 119–21, 159, 493
Berlin, 10, 42, 64, 88–89, 122, 144–48, 169–70, 180, 267, 493, 514
Bermerside, 388
Birkdale, 384
Birr Castle (Parsonstown), 7–8, 12–13, 46–47, 98–100, 102–03, 105–07, 109–12, 116–17, 119, 132, 139, 157, 159, 161, 170–71, 188–90, 192, 195, 198–203, 218–20, 222, 225–26, 230–31, 237, 242, 249, 252, 256, 279, 286–87, 291, 295–98, 301, 303–04, 306–08, 310, 312–13, 315, 317–19, 322–23, 334, 345, 353, 356, 368, 373, 381, 385, 409, 418, 424, 441, 443–44, 452, 458, 462, 464, 481–85, 487–88, 490–91, 493–96, 504, 515, 517, 519, 521, 528, 530–31, 536, 538, 547
Bonn, 65, 88, 144, 161–66, 174–75, 235–37, 279, 474, 503, 506, 524

Bordeaux, 542
Bothkamp, 278
Breslau, 165
Brighton, 538
Bristol, 520, 541

Cambridge (MA), 13, 124, 325, 336, 389, 409
Cambridge (UK), 105, 121, 138, 305, 415, 487, 498, 502
Cambridgeport, 361
Cape of Good Hope, 7, 38, 54, 62–63, 73, 77, 83, 105, 133, 459
Castelgandolfo, 465, 510, 559
Charlottesville, 13, 431–32, 458, 474
Chicago, 13, 121, 274, 342, 357, 361, 363, 365, 375, 394, 396, 404, 538, 540
Cincinnati, 66, 344, 360, 431–32, 465, 474
Clapham, 528
Clay Hall (Old Windsor), 19–20, 26–27, 30–31, 41, 169, 458
Clinton, 138, 165, 325
Copenhagen, 119, 146, 148–49, 156, 159–60, 163, 170, 184, 188, 196, 199, 206–08, 225–27, 229–31, 235–42, 245, 248, 250, 269, 276, 279, 285, 288, 294, 322, 443, 458, 474, 493, 496, 500–01, 504, 517, 523–25, 527
Crowborough, 507, 553
Crumpsall Hall, 384
Cuckfield, 380, 503, 536

Datchet, 19–23, 26–31, 35, 41, 46, 48, 104, 286, 452, 458, 462, 484
Denver, 216, 231, 338, 354, 452, 462
Dobb's Ferry, 409
Dorpat (Tartu), 10, 66–67, 90, 93, 145, 216, 262, 282, 367–68, 413–14
Dresden, 232, 473–74, 506
Düsseldorf, 174, 524
Dun Echt (Dunecht), 226, 228, 310–11, 313, 321, 380, 392–95, 428–30, 474, 504

Dunsink, 12, 102, 152, 226, 228, 296, 299, 305–06, 310, 323, 381
Durham, 212, 252, 267

Ealing, 386, 389, 418, 535
Echo Mountain (Mt Lowe), 339–40, 460, 462
Edinburgh, 120, 228, 312, 392, 403, 429
Evanston, 360

Feldhausen, 7, 38, 71–72, 77–78, 80, 85, 105, 415, 443, 452, 458, 474
Flagstaff, 370, 512, 560
Flagstaff Hill, 415, 417

Gateshead, 10, 252, 368, 531, 541
Genva, 545
Georgetown, 140–41, 336, 508, 510
Glasgow, 87, 105, 111, 484, 488
Gorleston, 528
Gotha, 147, 186
Göttingen, 63–65, 88, 144–47, 178–80, 310, 324, 413, 429, 499, 502, 526–27
Greenwich, 6, 10, 99, 121, 133, 146, 184, 217, 222, 227, 242, 252, 306, 415, 441, 465, 502
Gunzenhausen, 452

Halensee, 401
Hamburg, 72, 174, 182, 260, 267–68, 285, 293–94, 466, 474, 524
Hanover, 359, 427
Hardwicke, 378–79, 527
Harvard, 6, 27, 29, 66, 88, 111, 119, 124–28, 131–33, 162, 188, 195, 204, 215, 231–32, 319, 335–37, 357, 389–90, 392, 394, 399, 403, 407, 409, 431, 444, 466–67, 490, 555, 557
Haverford, 432
Hawaii, 561
Heidelberg, 23, 33, 236, 271, 292, 370, 400, 411, 427, 452, 462, 511
Hereford, 528

Herény, 43, 403, 492, 496
Hobart, 517
Hongkong, 137
Hopefield, 540

Jamaica, 530, 538, 546

Karlsruhe, 144, 236, 261–64, 341
Kensington, 106
Kiel, 226, 412
Königsberg, 68, 147–48, 158, 164, 178, 180, 182, 186–87, 232, 251, 260, 474, 500, 502
Krakau, 187

Leiden, 97, 175, 496
Leipzig, 65–66, 88, 123, 148–50, 152, 154–57, 159, 169, 180, 207–08, 210, 217, 235–36, 240, 243, 259–60, 277–78, 280, 371, 408, 413, 474, 493, 498, 503, 506, 514, 528, 531
Leuven, 542
Lilienthal, 63
Liverpool, 131–34, 213, 216, 368, 508
London (Bishop's Obs.), 88, 121–23, 138, 206, 251, 498
Lund, 292–93, 467

Madison, 359–60, 403–05
Madras, 518
Maidenhead, 133
Mali Lošinj, 511
Malta, 119, 131, 133–34, 156–57, 188, 211–16, 231, 245, 251–54, 269, 308, 319, 392, 417, 424, 443, 451, 457–58, 481, 491, 493, 501–04
Manchester, 118, 132, 310, 384
Mannheim, 1, 156, 161, 163–64, 222, 231, 233–36, 259, 279, 285, 474, 500, 503, 525, 527
Markree Castle, 12, 88, 106, 122, 135, 138, 157, 205, 252, 452
Marseille, 9, 11, 13, 170–73, 205–06, 231, 263, 269–71, 274, 276, 281, 327, 330, 341, 372, 376–77, 381, 406, 415, 419, 443, 458, 474, 498, 523–24, 527, 529–30
Melbourne, 6, 9, 13, 133, 222, 252, 415–19, 421, 441, 452
Milan, 170–71, 173, 359, 496, 523–24, 529–30, 536, 539

Minneapolis, 410, 432
Modena, 96
Moscow, 232–33
Mt Hamilton, 339, 342, 359–61, 366–67, 369, 388–89, 396, 401, 404, 408, 452, 554
Mt Palomar, 131
Mt Stromlo, 416, 419, 421
Mt Wilson, 9, 131, 170, 339, 382, 389, 396, 467–68, 512–13, 559
Mughull, 545
Munich, 88–90, 95, 151, 413, 474

Naples, 65–66, 69, 123, 137, 324–25
Nashville, 338, 341, 395–97, 401, 459, 526, 540
New Haven, 94
Nice, 437, 452, 460, 462, 544
Northfield, 551

Odessa, 262, 341
Olmütz, 161, 174, 524
Oxford, 121, 229–30, 549–50

Palermo, 69–70, 119
Paramatta, 63, 71–73, 77, 81, 231, 267, 415, 452, 458, 464
Paris, 10, 33, 43, 88, 122, 134–35, 137–38, 148, 156, 165, 171, 173, 188, 205–06, 211, 236, 242, 252, 260, 269–71, 273, 326, 354, 367–68, 382, 419, 422–24, 437–38, 458, 460, 462, 474, 491–93, 499, 501, 504, 506, 525–26, 530, 536, 542–43, 547–48
Pittsburgh, 335, 560
Pola, 411
Potsdam, 152, 277–78, 381–82, 401, 474, 497
Prague, 474
Princeton, 43, 342, 427–28
Pulkovo, 66–67, 88, 90, 112, 124, 132, 137, 144, 146, 163, 195, 206, 211, 213, 215, 232, 262, 280–83, 285, 404, 431, 452, 490, 494, 500–04, 526, 543, 545
Putney Hill, 535

Remplin, 41–42
Richmond, 528
Rochester, 66, 274, 338–40, 396, 452, 458, 539
Rome, 119, 140–41, 269, 368, 501

Scarborough, 393, 428–29, 460
Senftenberg (Žamberk), 166, 169
Slough, 7, 19–20, 22–23, 26–29, 31–32, 35, 41, 46, 48, 50–55, 57, 59–60, 63, 66, 69, 77–78, 80–81, 104–05, 180, 287, 310, 403, 443, 449, 458, 474, 484, 493
Southport, 384
Southwick, 94
St Petersburg (Pulkovo), 66, 282, 504
Stoneyhurst, 140, 385
Straßburg, 144–45, 237, 261–64, 312, 365, 381, 388, 412–14, 460, 465–66, 470, 503, 511
Stretton Sugwas, 118–19, 528
Sunderland, 540
Sydney, 73, 299

Taschkent, 552, 555
Tenerife, 133
Teramo, 429, 460
Toulouse, 43, 437
Tulse Hill, 368
Twickenham, 122–23, 502

Uppsala (Upsala), 13, 158, 202, 231, 237, 260, 281, 285, 291–93, 449, 452, 474

Vatican, 12, 427, 439, 465–66, 510, 559
Venice, 13, 170–72, 521, 523, 529, 534
Vienna, 13, 43–44, 119, 169, 260, 277, 279–81, 323, 370, 372, 410–12, 431, 452, 460, 466, 474, 493, 497, 503, 521, 528, 544, 548

Walworth Common, 10, 312
Wandsworth, 10, 124
Warsaw, 474
Washington, 126, 133, 143, 207, 252, 280, 294, 325, 336, 342, 358–59, 361, 369–71, 404, 407, 409, 431, 462, 530, 540, 544
Williams Bay, 360, 396
Williamstown, 358, 361
Windsor (NSW), 518

Yale, 88, 94–95

Object index

Here all NGC /IC objects mentioned in the text and the tables are presented. Additionally the list contains all objects shown in figures (marked italic) and common names (total number: 2521).

3 Mon 57, 157, 454
15 Mon 53, 531
19 Pup 80–81, 157, 454
30 Dor 418
47 Tuc 5
50 Cas 53–55, 58, 157, 453–54
55 And 9, 53–56, 157, *158*, 159, 197

Abell 194 249–50, 355
Abell 262 46, 250, 309
Abell 347 377–78
Abell 426 244, 250, 423
Abell 1060 81, 83, 337
Abell 1367 46–47, 250, 272
Abell 1656 46, 223, 248, *249*, 424
Abell 2151 272, 355
Abell 3526 81, 83
Abell S373 76, 81–82, 260
Andromeda Nebula 1–2, 11, 16, 18, 52, 69, 93, 103–04, 124–25, 141–42, 159, 333, 398, *399*, 410, 412, 415, 452, 455, 475, 515–16, 534, 560

B 86 *143*, 397, 400
B 92 *406*
Barnard's Galaxy 5, 395–97, *400*
Barnard's Loop 390, 396, 401, 559
Baxendell's Unphotographable Nebula 385, 537
Bear Paw Galaxy 195
Bessel's star 157–58, 184
Beta Canes Venatici (β CVn) 55, 157, *158*, 212, 454
Black Eye 16, 59–60
Blinking Planetary 143
Blue Planetary Nebula 51
Blue Snowball 5, 36
Bode's Nebulae 15
Bubble Nebula 41
Bug Nebula 74, *399*

California Nebula 5, 397, 401, *402*, 458, 550
Cancer Cluster 250

Cat's Eye Nebula 5, 98
Centaurus A 5, 76
Centaurus Cluster 81, 83
Cetus Cluster 249, 355
Christmas Tree Cluster 197
Coma Cluster 9, 46, 223, 242–43, 245–48, *249*, 424
Conus Nebula 35, 197
Copeland Septet 312–13, *314*, 460, 469
Crab Nebula 16, 61, 107, *108*, 143, 206, 333, 453, 478, 482
Crescent Nebula 5, *301*, 394

Double Cluster 5, 16, 31
Dumbbell Nebula 28, 36, *39*, 60, 103–04, *108*, 143, 195, 241, 386, 394, 455, 482

Eagle Nebula 462, 464–65
Eskimo Nebula 5, 40–41, 118, *119*, 241
Eta Carinae Nebula 5, 84, 86, 418, 514, 516, *518*

Flame Nebula 32–33, 166, 170
Fornax Cluster 75–76, 81–82, 117, 260, 350

GC 80 9, 198–99, 201–03, 244, 303–04, 309
GC 6000 231, *232*, 321
Ghost of Jupiter 5, 38, 40, 42, 152
Gyulbudaghian's Nebula 522

Helix Galaxy 329
Helix Nebula 5, 64, *65*
Hercules Cluster 272, 348, 355, 377
Herschel's Ray 420
Hind's Variable Nebula 5, 121, 124, 164, 176, 215, 234, 284, 472, 479, 498, 507, 511, 514, 522
Hind's Crimson Star 124
Homunculus Nebula 517, 522
Hubble's Variable Nebula 5, 9, 40, 61, *176*, 241, 514, 522

Hyades 17, 466, 498
Hydra Cluster 81–83, 337

IC 4 384
IC 11 397, 399
IC 44 131, 183, 348
IC 48 520
IC 138 366, 464
IC 166 463
IC 168 465
IC 210 366, 464
IC 229 464
IC 334 463, 520
IC 336 551, 559
IC 341 551, 559
IC 342 462, *463*
IC 344 55–56, *57*, 464, 519
IC 349 456, *549*, 559–60, *561*
IC 351 463
IC 353 551, 559
IC 354 551, 559
IC 356 398, 463, 520, *521*, 550
IC 360 551, 559
IC 361 463
IC 434 30, 464
IC 448 33
IC 452 455
IC 457 304, 425
IC 458 304
IC 459 304
IC 461 304
IC 463 304
IC 464 304
IC 465 304, 425
IC 503 384
IC 512 519
IC 529 463
IC 591 372–73, 464
IC 606 372
IC 617 435
IC 622 372, 464
IC 625 337
IC 669 373, 464
IC 694 303–04, *353*, 554

631

632 Object index

IC 730   372
IC 750   463
IC 819   44
IC 820   44
IC 884   348
IC 887   348
IC 919   470
IC 973   *412*
IC 974   *329*
IC 1026   362, 364
IC 1030   362, 364
IC 1111   348
IC 1129   464
IC 1145   464
IC 1148   362
IC 1176   348, 355
IC 1179   348, 355
IC 1238   257–58, *259*, 464
IC 1239   257–58
IC 1253   362
IC 1271   464
IC 1290   *464*
IC 1296   117, *402*, 403
IC 1300   348
IC 1308   400
IC 1369   384, 463
IC 1434   463
IC 1441   *272*
IC 1442   463
IC 1463   *464*
IC 1484   *388*
IC 1487   348
IC 1528   366, 464
IC 1559   201, *202*, *203*, 303–04, 464
IC 1613   462
IC 1653   160, 183
IC 1656   160, 183
IC 1712   348
IC 1748   412
IC 1761   348
IC 1954   463
IC 1990   552, 560
IC 1995   557
IC 2003   463
IC 2035   463
IC 2056   463
IC 2107   329
IC 2120   62
IC 2157   463
IC 2163   463
IC 2172   397
IC 2233   356
IC 2311   463
IC 2391   464
IC 2395   464
IC 2482   463
IC 2488   464

IC 2624   443, 445
IC 2976   348
IC 2989   372
IC 2991   *246*
IC 3136   272, 464
IC 3153   237, *238*
IC 3155   272, 464
IC 3268   272, 464
IC 3568   463
IC 4291   463
IC 4329   463
IC 4351   463
IC 4625   426
IC 4662   463
IC 4665   29, 464, 466
IC 4677   *97*
IC 4703   462, 464
IC 4715   17, 57, 220, 464
IC 4725   15, 17, 220, 464
IC 4756   382, 464, 466
IC 4762   354
IC 4763   350, *351*, 354
IC 4812   177
IC 4895   397, 401
IC 5070   2
IC 5146   400
IC 5148   462, *463*
IC 5150   462
IC 5181   463
IC 5191   271, *272*
IC 5192   271, *272*
IC 5193   271, *272*
IC 5195   271
IC 5201   463
IC 5290   412
IC 5306   387–88
Intergalactic Wanderer   30

Jewel Box   5, 86

Keyhole Nebula   517
Kidney Been Galaxy   47

Lagoon Nebula   16, *476*, 477
Large Magellanic Cloud (LMC)   3, 15, 74–81, 84–86, 219–20, 420, 453, 459
Leo Cluster   46–47, 250, 272
Little Dumbbell Nebula   16
Local Group   395, 400

M 1   4, 16, 61, 104, 107, *108*, *120*, 134–36, 143, 161, 167, 206, 213–14, 296, 301, 333–34, 452–53, *478*, 482–83, 496, 521
M 2   16, 30, 61, 109, 136, 168, 296, 384–85, 450
M 3   107, 135–36, 167, 296

M 4   16, 18, 79, 221
M 5   16, 30, 61, 107, 115, 136, 167, 296, 444
M 6   16, 18
M 7   15–16, 18, 456
M 8   16, 18, 84, 86, 144, 260, *476*, 477
M 9   143, 167, 296
M 10   115, 136, 143, 167, 296
M 11   16, 89, 135–36, 143, 296, 444, *475*
M 12   167, 194
M 13   *2*, *4*, 16, 18, 61, 84, 107, 109, 115, 136, 141, 143, 167, 175, 191, 195, 234, 280, 296, 336, *408*, 409, 483–84
M 14   18, 64–65, 168, 186, 296
M 15   16, 68–69, 108, 136, 168, 186, 280, 296
M 16   16, 462, 464
M 17   16, 39–40, 60, 86, 89, 95, 104, *120*, 142, 168, 209, 214, 296, 301, *343*, 406–07, 418, *479*, 483, 516–17
M 20   26, *27*, 31, 40, 61, 86, 95, *96*, 209, 213–14, 336, 406, 476, 514, 516–17
M 22   15–16, 151, 455
M 24   16–17, 57, 79, 220, 464
M 25   15–17, 79, 220, 464
M 27   28, 36, *39*, 40, 44–45, 53, 60, 65, 76, 84, 91, 103–04, 106, *108*, 111, 114, 117, 135–36, 143, 161, 168, 192, 195, 214, 241, 281, 296–97, 299, 336, 394, 455, 459, 482–83
M 28   135–36
M 29   296
M 30   39, 61, 143, 175, 296
M 31   2, 4, 15–16, 18, 23, 30, 37, 53, 91, 93, 104, 106–09, 120, 124–26, 136, 166–67, 175, 282, 306, 333–34, 336, *397*, *399*, 413–15, 429, 452, 455, 475, 483, 513, 516
M 32   16, 68–69, 107, 126, 136, 167, 175, 186, 475
M 33   16, 30–31, 104, 114–15, 160, 167, 183, 187, 191, 193, 242, 244, 291, 296, 363–64, 453, 455, 554
M 34   16
M 35   16, 221, 296
M 36   16, 18, 296
M 37   16, 136, 296
M 38   16, 79, 221
M 39   16, 120, 296
M 40   16–17, 79, 146, 220
M 41   16, 18, 296, 455
M 42   2, 4, 9, 16, 18, 30, 39–40, 77, 86, 90, 102–05, 108–09, 111–12, 124, 126, 141, 143–44, 173, 190, 195, 215, 234, 244, 296, 299, 305, 308, 336, 367–69, 403, 406–07, 418, 455, 475–76, 482–83, 488, 513–14, 516
M 43   16, 31, 79

Object index   633

M 44 2, 15–16, 18, 30, 256, 453–54, 456
M 45 16–17, 30, 39, 79, 220, 456, 466, 521
M 46 114, 299
M 47 15–17, 31, 219–21, 449, 455
M 48 16–17, 29, 31, 219
M 49 18, 31, 151, 154, 168, 177, *178*, 282–83, 519
M 50 16, 296
M 51 9, 13, 31, 44–45, 51, 60, 76, 98, 100, 102–03, 110, 112–16, 118, 142, 167, 175, 194–95, 205, 209, 214, 281–82, 296–97, 299, 301, 310, 458, 472, 475–76, 479, 481–484, *485*, 486–88, *489*, *490*, 491, *492*, 493–496, *497*
M 52 168, 296
M 53 16, 58, 68–69, 167, 186
M 55 16, 151
M 56 168, 296
M 57 4, 16, 35, *37*, 42–43, *44*, 60, 84, 91, 104, 106–08, 111, 116–17, 124, 136, 142, 144, 161, 205, 215, 241, 256, 280, 296, 336, 363, *402*, 403, 406, 410, 459, 479, 483
M 58 116, 168, 195, 296
M 59 16, 167, 214, 296
M 60 16, 61, 167, 196, 209, 296
M 61 16, 30–31, 61, 191, 194, 296
M 63 116, 194, 296
M 64 15–16, 59–60, 102, 167, 214, 296, 497
M 65 61, 91, *92*, 104, 110, 114, 117, 136, 164, 168, 187, 193, 213–14, 279, *281*, 296, 306, 497
M 66 54, 61, 91, 102, 104, 136, 168, 187, 191–92, 213–14, 279, 296, 306, 497
M 67 16, 18, 167
M 69 186, 220–21, 325
M 71 16, 18, 168
M 72 30, 168, 296
M 73 79, 155–56, 221, 245, 296
M 74 30, 104, 116, 145, 167, 193, 487
M 75 168, 296
M 76 15–17, 31, 44–45, 79, 136, 221, 296, 449
M 77 116, 136, 167, 175, 191, 193–94, 214, 301
M 78 41, 135–36, 167, 193, 209, 244, 265–66, 291, 296, 420, 455, 522
M 79 79, 136, 221
M 80 16, 148, 167, 176–77, 234, 260, 296, 514
M 81 16, 104, 135–36, 141, 167, 220–21, 290, 296
M 82 16, 48–49, 79, 104, 135–37, 141, 152, 167, 220, 290, 296
M 83 16, 211–12, 491

M 84 47, 168, 296
M 85 296
M 86 47, 148, 168, *186*, 296
M 87 168
M 88 116, 194, 214
M 90 79, 168, 221, 296
M 91 15, 17, 30–31, 55–57, 219
M 92 79, 107, 167, 186, 221, 268, 286, 288–89, 296, 321, 325, 336
M 93 16
M 94 136, 167, 194, 214, 282, 296, 484
M 95 136, 167, 193, 296, 515
M 96 110, 116, 136, 167, 193, 296
M 97 102, 114, 116, *117*, 136, 168, 190, 296, 305, 403
M 98 168, 296, 306
M 99 102, 112–14, *115*, 116, 193, 211–12, 214, 296, 491
M 100 116, 194, 211–12, 214, 296, 491
M 101 18, 104, 167, 191, 194, *195*, 244, 296, 487
M 102 17–18, 30–31, 114, 117–18, 121, 167, 181, 219
M 103 55, 68, 181
M 104 17, 31, 56, 60–61, 91, 168, 214
M 105 17, 31, 56, 136
M 106 17, 30–31, 56, 61, 168
M 107 16–18, 31, 55, 64–65, 71, 79–80, 167, 186
M 108 17, 31, 56
M 109 17, 31, 40, 56, 136
M 110 17, 29–31, 56, 126, 167, 193, 475
Maia Nebula 8, 438, 444, 455–56, 521–22, 542–43, *544*, 545, 550–51, 560
Markarian's Chain 47, 209
MCG -1–30–5 *388*
MCG -1–30–8 *388*
MCG 3–31–25 *383*
MCG 3–31–27 *383*
MCG 5–1–73 *424*
MCG 6–2–16 *376*
MCG 6–2–17 *376*
MCG 15–1–10 *58*
McNeil's Nebula 522
Merope Nebula 5, 9, 13, 88, 163, 171–73, 176, 209–10, 252, 264, 330, 333, 360, 384, 438, 456, 472, 479, 503, 514–15, 521, *524*, 525–31, *532*, *533*, 534, 536, *537*, 538–40, *541*, 542–43, 545–46, *547*, 548, 551, 560
Milky Way 1, 3, 39, 74, 84–86, 142, 150, 174, 223, 234, 239, 241, 251, 254, 316, 390, 392, 396, 398, 400, 403, 406, 438, 442, 477, 484–85, 489, 510, 547, 551, 553, 555
Mrk 205 *430–31*

NGC 1 4, *160*, 242–43, 245
NGC 2 4, *160*, 242, 309
NGC 3 4, *256*
NGC 4 4, *256*
NGC 5 4
NGC 6 4, 288–89, 301–02
NGC 7 4
NGC 8 4, 283, *284*
NGC 9 4, 283, *284*
NGC 10 4
NGC 12 53
NGC 16 54, 160
NGC 18 289
NGC 20 288–89, 301–02, 321
NGC 23 121
NGC 29 58
NGC 32 178–79, 182–83
NGC 35 436
NGC 40 41
NGC 46 139, 183
NGC 47 328, *329*, 330, 461
NGC 50 269, *270*, 329
NGC 54 328–30, 461
NGC 55 420, *421*
NGC 58 328, 461
NGC 62 377, 435
NGC 63 279–80, 364
NGC 67 191
NGC 68 191
NGC 69 191
NGC 70 190–91
NGC 71 191
NGC 72 191
NGC 74 191
NGC 76 424, *425*
NGC 78 *294*, 383, 461
NGC 81 312, 314
NGC 85 313
NGC 90 289, 301–02, *303*, 321
NGC 91 301–02, *303*, 321
NGC 93 301–02, *303*, 321
NGC 100 356, *357*
NGC 104 5, 453
NGC 107 283
NGC 108 193
NGC 109 160, 183
NGC 110 120
NGC 113 266
NGC 114 332, 362
NGC 116 270
NGC 118 332, 362
NGC 124 332, 362
NGC 125 298
NGC 133 243, *244*, 247
NGC 134 420, *421*
NGC 140 362
NGC 146 120

NGC 155 434, 436
NGC 157 167, 269
NGC 160 201–03
NGC 162 201, *202*, 203, 244–45, 289, 308–09
NGC 163 459
NGC 167 433
NGC 169 200–201, *202*, *203*, 464
NGC 183 362
NGC 185 300
NGC 189 29
NGC 192 127
NGC 194 127, 130
NGC 195 266
NGC 196 127, 130
NGC 197 127
NGC 201 127, 130
NGC 203 276, 312–13
NGC 205 17, 29–31, 107, 167, 193–94, 475
NGC 206 126, 397, *399*
NGC 207 435
NGC 210 193, 435, 438
NGC 211 276, 312–13
NGC 219 127–29, *130*, 131, 204–05
NGC 221 68, 135–36, 167, 186
NGC 223 127–28, *130*, 131, 183, 204, 241, 348
NGC 224 2, 91, 135–36, 167, 452, 455, 516
NGC 225 29
NGC 227 441
NGC 237 362
NGC 246 41
NGC 247 197
NGC 253 5, 20, 29, *30*, 80, 213, *214*
NGC 259 54, 198
NGC 272 245
NGC 274 188
NGC 275 188, 300
NGC 278 193, 300
NGC 281 397, 399
NGC 292 5, 16, 80, 86, 453–55
NGC 295 312–13
NGC 296 312
NGC 300 420
NGC 308 306
NGC 309 266
NGC 310 306
NGC 316 200, 267, 302
NGC 318 373
NGC 337 193, 468, *469*
NGC 337A 468, *469*
NGC 338 373
NGC 344 434, *456*, 457
NGC 346 420
NGC 370 160–61, 183, 314–15

NGC 372 160–61, 183, 314–15
NGC 373 315, 318
NGC 374 160, 183
NGC 375 309, 318
NGC 379 318
NGC 380 318
NGC 381 29, 161
NGC 382 318
NGC 383 318
NGC 384 160, 183, 201, 318
NGC 385 160–61, 183, 201, 318
NGC 386 318
NGC 387 309, 318
NGC 388 318
NGC 391 128, 183, 204
NGC 394 200
NGC 397 306
NGC 399 309
NGC 400 306
NGC 401 306
NGC 402 309
NGC 404 356
NGC 405 453–54
NGC 407 373
NGC 408 289, 373
NGC 414 289, 373
NGC 422 78
NGC 425 362
NGC 443 160, 183
NGC 447 160, 183
NGC 453 376
NGC 467 288
NGC 470 288
NGC 474 288
NGC 475 327
NGC 481 435
NGC 495 47
NGC 496 47
NGC 504 160–61, 183
NGC 506 309
NGC 507 47
NGC 508 47
NGC 510 289
NGC 516 330
NGC 519 355
NGC 520 300
NGC 522 330
NGC 523 210
NGC 524 330
NGC 525 330
NGC 530 355
NGC 532 403
NGC 535 250
NGC 537 210
NGC 538 355
NGC 539 434–35
NGC 541 250

NGC 543 250
NGC 545 249
NGC 547 249
NGC 548 249, 336–37
NGC 557 355
NGC 558 250
NGC 559 120
NGC 560 249, 273, 331
NGC 563 434–35
NGC 564 249, 273, 331
NGC 565 336–37
NGC 570 337
NGC 577 331–32, 363, *365*, 366
NGC 580 331–32, 363, *365*, 366
NGC 581 68
NGC 584 25, 348
NGC 586 25
NGC 588 160–61, 183, 244
NGC 591 362
NGC 592 160–61, 183, 244
NGC 595 161, 244, 363–64
NGC 596 151, 154
NGC 598 31, 114, 167, 191, 193
NGC 599 436
NGC 602 75
NGC 604 453, 455
NGC 607 154–56, 183, 210, 236
NGC 609 244
NGC 614 55, 198, 445
NGC 615 151, 154–55
NGC 618 55, 196–98, 445
NGC 627 55, 198, 445
NGC 628 116, 167, 193, 487
NGC 629 68–69, 183
NGC 650 16–17, 45, 136, 449
NGC 651 16–17, 31, 45, 136, 449
NGC 657 120
NGC 659 29
NGC 661 397, 399
NGC 674 210
NGC 676 159
NGC 679 46
NGC 687 46
NGC 693 191
NGC 697 210
NGC 701 153
NGC 703 46
NGC 704 46
NGC 705 46
NGC 708 46
NGC 709 309
NGC 710 250, 309
NGC 714 250, 309
NGC 717 250, 309
NGC 727 79
NGC 729 79
NGC 731 327, 435

NGC 733 199–200
NGC 739 313
NGC 741 45
NGC 742 45
NGC 750 301–02, *303*
NGC 751 301–02, *303*, 321
NGC 752 29, 31, 56
NGC 755 327, 435
NGC 757 435
NGC 760 313
NGC 763 435
NGC 769 362
NGC 770 200
NGC 771 55, 58, 157–58, 453–54
NGC 772 193, 356
NGC 778 362
NGC 780 197
NGC 787 326, *327*, 332, 435
NGC 793 430, 461
NGC 808 420
NGC 821 152–53, 209
NGC 823 58, 80, 420
NGC 833 46, 420
NGC 835 46, 420
NGC 838 46, 420
NGC 839 46, 420
NGC 851 358
NGC 856 347
NGC 859 347
NGC 863 347, 445
NGC 864 197
NGC 866 347–48, 445
NGC 867 247
NGC 869 5, 16, 31
NGC 870 318, 456
NGC 875 247
NGC 884 5, 16, 31
NGC 885 347, 445
NGC 891 5, 29, *30*, 60, 377
NGC 898 377
NGC 903 *377*
NGC 906 378
NGC 909 378
NGC 910 377
NGC 911 378
NGC 912 378
NGC 913 378
NGC 914 378
NGC 923 378
NGC 926 266
NGC 927 411–12, *413*, 452
NGC 930 312, *313*
NGC 932 312, *313*
NGC 934 266
NGC 936 161
NGC 948 435
NGC 952 274

NGC 955 519, *520*
NGC 957 120
NGC 961 435
NGC 972 191, 274, 321
NGC 976 266
NGC 983 375
NGC 985 *433*, 434
NGC 986 420
NGC 988 159
NGC 1002 *375*
NGC 1009 358
NGC 1012 191
NGC 1015 266
NGC 1016 267, 363–64
NGC 1017 435
NGC 1020 363
NGC 1022 153, 327
NGC 1023 135–36, 191
NGC 1032 21, 26, 193, 363
NGC 1038 347
NGC 1044 516
NGC 1045 516
NGC 1051 435
NGC 1052 136
NGC 1055 151, 153
NGC 1062 312–13
NGC 1068 116, 136, 167, 191, 193–94, 214, 300
NGC 1084 214
NGC 1111 257
NGC 1135 81
NGC 1141 255, 329
NGC 1142 255, 329
NGC 1148 435
NGC 1170 62, 337, 448
NGC 1173 424
NGC 1174 521
NGC 1177 309
NGC 1182 434, 436
NGC 1184 26
NGC 1186 197, 460, *521*
NGC 1187 399
NGC 1195 315
NGC 1205 434, 436
NGC 1207 197
NGC 1211 362
NGC 1212 357
NGC 1213 357
NGC 1214 436
NGC 1215 436
NGC 1224 357
NGC 1234 442
NGC 1241 298
NGC 1242 298
NGC 1243 298
NGC 1251 129, 200
NGC 1255 397, 399–400, 436

NGC 1257 423
NGC 1259 424
NGC 1260 424
NGC 1264 424
NGC 1265 117, 424–25
NGC 1267 250
NGC 1268 250
NGC 1270 250
NGC 1271 424
NGC 1272 250
NGC 1273 250
NGC 1274 309
NGC 1275 244–45, 250
NGC 1276 315
NGC 1277 309, 314–15, 320
NGC 1279 315
NGC 1281 315
NGC 1282 424
NGC 1283 424
NGC 1292 397, 401
NGC 1293 516
NGC 1294 516
NGC 1297 397
NGC 1302 397
NGC 1304 435
NGC 1307 435
NGC 1310 82
NGC 1312 129, 204
NGC 1315 82
NGC 1316 75–76, 81
NGC 1317 75, 261
NGC 1318 75, 261
NGC 1322 80
NGC 1325 62
NGC 1326 82
NGC 1330 376
NGC 1333 9, 65, 127–28, *162*, 164, 166–67, 183, 205, 209, 222, 236–38, 503, 514–15, 524, 528
NGC 1336 82
NGC 1341 82
NGC 1350 75
NGC 1351 82
NGC 1360 172–73, 255, 262, 264, 271, 284, *340*, 344, 349–50, 363, 375, 448
NGC 1363 365–66
NGC 1365 75–76
NGC 1367 436
NGC 1369 261
NGC 1371 172, 262, 436
NGC 1373 82
NGC 1374 82
NGC 1375 82, 260–61
NGC 1378 260–61
NGC 1379 82
NGC 1380 75, 260
NGC 1381 261

NGC 1382 261
NGC 1386 261
NGC 1387 82
NGC 1389 260–61, 329
NGC 1392 350
NGC 1396 261
NGC 1398 *172*, 172–173, 262–64, 330
NGC 1399 75
NGC 1400 167
NGC 1407 167
NGC 1408 260–61
NGC 1412 262
NGC 1417 55, *57*, 519
NGC 1418 55, *57*, 519
NGC 1419 82
NGC 1427 82
NGC 1428 261
NGC 1432 8, 438, 444, 455–56, 521–22, 542, *544*, 545, 560
NGC 1435 5, 9, 13, 163, 172–73, 183, 208, 329–30, 438, 456, 472, 479, 514–15, 521, 545
NGC 1436 81–82
NGC 1437 81–82
NGC 1440 435, 445
NGC 1441 314, *316*
NGC 1442 435, 445
NGC 1443 315, *316*
NGC 1446 315, *316*, 332
NGC 1449 314, *316*
NGC 1450 *433*, 436
NGC 1451 314, *316*
NGC 1452 435
NGC 1453 *316*
NGC 1455 435
NGC 1456 430, 461
NGC 1458 435, 445
NGC 1460 82
NGC 1469 519
NGC 1488 139, *140*, 183
NGC 1491 441
NGC 1497 401
NGC 1499 5, 397–98, 401, *402*, 458, 550, 556
NGC 1509 436
NGC 1513 292
NGC 1514 39, *40*, 41, 57, 104, 121, 153, 157, 191, 193–94, *195*
NGC 1516 436
NGC 1521 336
NGC 1524 436
NGC 1530 266
NGC 1535 37, 134, 161, 167, 214, 252
NGC 1544 266, 330
NGC 1550 210
NGC 1551 210
NGC 1553 81

NGC 1554 124, 281, 283–84, 472, 498, *499*, 504–12
NGC 1555 5, 9, 13, 121, 124–25, 163–64, 176, 183, 206, 215, 234, 236, 284, 472, 479, 498, *499*, 505–09, 511–12, 514–15, 522–23, 526, 528
NGC 1575 436
NGC 1576 197
NGC 1577 436
NGC 1579 191
NGC 1587 516
NGC 1588 516
NGC 1593 309
NGC 1601 301–02, 321
NGC 1608 309
NGC 1618 *356*
NGC 1622 199–200, *356*
NGC 1624 128, 204–05
NGC 1625 207, *356*
NGC 1637 116, 191, 193
NGC 1640 434
NGC 1655 430, 461
NGC 1662 280
NGC 1665 197
NGC 1674 430, 461
NGC 1675 430, 461
NGC 1699 201
NGC 1707 329
NGC 1721 397, *401*
NGC 1723 397, 401
NGC 1724 *268*, 448
NGC 1725 397, *401*
NGC 1726 210
NGC 1728 397, *401*
NGC 1730 433
NGC 1735 420
NGC 1736 420
NGC 1737 420
NGC 1742 306
NGC 1743 420
NGC 1744 420
NGC 1745 420
NGC 1746 245
NGC 1748 420
NGC 1750 245
NGC 1751 75–76, 456
NGC 1757 197
NGC 1758 245
NGC 1760 420
NGC 1767 460
NGC 1778 51, 120
NGC 1779 420
NGC 1781 434, 436
NGC 1785 78, 460
NGC 1788 345
NGC 1794 434, 436
NGC 1798 397, 401, 455

NGC 1808 420
NGC 1824 79
NGC 1825 460
NGC 1838 72, 79
NGC 1847 420
NGC 1855 79
NGC 1869 84–85
NGC 1871 84–85
NGC 1873 84–85
NGC 1888 420
NGC 1889 420
NGC 1904 136
NGC 1905 86
NGC 1913 460
NGC 1918 74, 453
NGC 1922 460
NGC 1924 127, 129, 204
NGC 1927 197
NGC 1929 84
NGC 1931 136, 241
NGC 1934 84
NGC 1937 84
NGC 1950 420
NGC 1952 107, 134, 136, 143, 161, 167, 214, 300, 453
NGC 1958 420
NGC 1959 420
NGC 1960 18
NGC 1961 22
NGC 1962 420
NGC 1965 420
NGC 1966 85, 420
NGC 1967 460
NGC 1969 420
NGC 1970 420
NGC 1971 420
NGC 1972 420
NGC 1973 244
NGC 1975 244
NGC 1976 18, 86, 143, 455, 516
NGC 1977 86
NGC 1980 114, 118, 134
NGC 1981 58
NGC 1982 31
NGC 1988 9, 182, 205, *206*, 207, 334, 448, 453, 511, 514–15, 527
NGC 1990 157
NGC 1999 241, 300
NGC 2020 452
NGC 2022 134, 143, 161, 214, 285
NGC 2023 161
NGC 2024 32–33, *142*, 166, 169, *170*, 186, 241, 252, 300
NGC 2037 460
NGC 2043 420, *421*
NGC 2044 460
NGC 2045 454

NGC 2046 420
NGC 2047 420
NGC 2050 460
NGC 2052 460
NGC 2054 128, 231–32, 314–15
NGC 2057 420
NGC 2058 420
NGC 2059 420
NGC 2060 453, 460
NGC 2064 242, 244, 265
NGC 2065 420–21
NGC 2066 420
NGC 2067 265–66, 329
NGC 2068 136, 167, 193–94, 265, 300, 420
NGC 2070 86
NGC 2071 41, 209–10
NGC 2072 420, *421*
NGC 2074 460
NGC 2081 460
NGC 2091 460
NGC 2092 460
NGC 2096 460
NGC 2099 136
NGC 2102 460
NGC 2128 358
NGC 2141 397, 459–61
NGC 2142 57, 157, 454
NGC 2146 263–64, 267, 461
NGC 2149 30, 274, *276*
NGC 2163 *375*, 376
NGC 2167 40–41, 55–56, *57*
NGC 2169 280
NGC 2170 40–41, *57*, 240, 300
NGC 2174 376
NGC 2175 *170*, 183, 459
NGC 2182 41
NGC 2183 302
NGC 2185 167, 300
NGC 2189 127, 129, 204
NGC 2194 167, 182, 186, 279–80, 321, 441
NGC 2195 430, 461
NGC 2198 127, 129, 204
NGC 2202 68
NGC 2204 29
NGC 2218 139, 183
NGC 2225 397
NGC 2226 397, 459–61
NGC 2232 35
NGC 2237 5, 256, 340, *341*, 342, 349
NGC 2238 5, 256, *257*
NGC 2239 5, 55–56, 449
NGC 2240 152
NGC 2242 *349*, 382, 449, 455
NGC 2243 75, 449
NGC 2244 16, 31, 55–56, 449

NGC 2245 191, 280, 385
NGC 2246 5, 256, 342, 349, 455
NGC 2247 304, 345, 385
NGC 2248 139, *140*, 183, 306
NGC 2261 5, 40, 61, 84, 114, 134, 136, 143, *176*, 234, 241, 514, 521–22
NGC 2264 35, 196–97, 531
NGC 2268 267, 274, 277, 455
NGC 2273 *293*
NGC 2276 263, *264*, 266–67, 276, 293
NGC 2282 397, 402
NGC 2287 18, 454–55
NGC 2290 302, 321
NGC 2296 349, 455
NGC 2300 263, *264*, 266–67, 274, 276–77, 293
NGC 2301 46
NGC 2304 61
NGC 2313 242, 244, *245*
NGC 2316 191, 418
NGC 2317 191, 418
NGC 2319 47–48
NGC 2321 201, 302
NGC 2327 41, 516
NGC 2330 303–04, 424–25
NGC 2332 460
NGC 2334 303–04, 424–25
NGC 2336 266, 350
NGC 2340 460
NGC 2346 41
NGC 2349 29, 459
NGC 2355 64, 186
NGC 2359 80, 214, 423
NGC 2360 29, 80, 167, 459
NGC 2361 329, 423, 425
NGC 2362 31, 35, 56, 80, 454
NGC 2363 313, 317
NGC 2366 22, 300
NGC 2371 44–45, *46*, 61, 114, 134, 143, 208–09, 214, 287, 299–300, 449, 515–16
NGC 2372 44–45, *46*, 61, 114, 143, 208–09, 287, 299–300, 449, 515–16
NGC 2373 200, 302, 373
NGC 2375 200, 302, 373
NGC 2378 373, 376
NGC 2379 373
NGC 2386 309
NGC 2390 306
NGC 2391 306
NGC 2392 5–6, 40–41, 84, 114, 118, *119*, *121*, *134*, 143, 157, 214, 241, 250, 282, 288, 441
NGC 2399 128, 130, 183, 204
NGC 2400 128, 130, 183, 204
NGC 2403 30, 167, 423, *424*, 425
NGC 2404 423, *424*, 425, 455

NGC 2410 362
NGC 2412 430, 461
NGC 2415 91
NGC 2422 16–17, 31, 220, 449, 455
NGC 2429 313
NGC 2438 114, 116, 134, 143, 167, 299
NGC 2440 134, 143, 167
NGC 2441 329
NGC 2444 *377*
NGC 2445 *377*
NGC 2451 80–81
NGC 2457 313
NGC 2458 425
NGC 2459 197
NGC 2461 425
NGC 2462 425
NGC 2464 425
NGC 2465 425
NGC 2467 37
NGC 2471 425
NGC 2478 16–17, 31, 220, 449, 455
NGC 2493 *199*
NGC 2495 *199*
NGC 2506 64, 115, 186
NGC 2509 47–48
NGC 2511 200, 208
NGC 2513 199, 208
NGC 2515 128, 131, 204
NGC 2518 430, 461
NGC 2519 430, 461
NGC 2523 349, 358
NGC 2529 424
NGC 2531 424
NGC 2537 193, 195
NGC 2538 272
NGC 2542 79, 81, 157, 454
NGC 2548 16–17, 29, 31, 219
NGC 2553 250
NGC 2556 250
NGC 2557 250, 272
NGC 2558 250
NGC 2560 250
NGC 2562 250
NGC 2563 250
NGC 2565 430, 461
NGC 2568 397, 399
NGC 2569 250
NGC 2570 250, 313
NGC 2572 272
NGC 2573 57, 81, *82*, 455
NGC 2581 377
NGC 2591 243
NGC 2603 318, 456
NGC 2608 193
NGC 2620 216–17, 255
NGC 2623 18
NGC 2624 256

NGC 2625 256
NGC 2629 48
NGC 2630 329
NGC 2631 329
NGC 2632 456
NGC 2637 255–56
NGC 2641 48
NGC 2643 255–56
NGC 2644 376
NGC 2647 256
NGC 2650 48
NGC 2652 436
NGC 2655 21, 127, 129, 136–37, 204
NGC 2657 377
NGC 2658 85–86
NGC 2661 425
NGC 2663 349
NGC 2672 25, 44–45, 208, 286, *287*, 288
NGC 2673 44–45, 286, *287*, 288
NGC 2677 286, *287*, 288
NGC 2681 91
NGC 2682 18, 167
NGC 2683 91, 136
NGC 2685 329
NGC 2689 303–04, *317*, 318
NGC 2694 200, 302
NGC 2695 91, *92*, 514
NGC 2698 *208*
NGC 2699 *208*
NGC 2700 265–66, 461
NGC 2702 265–66, 461
NGC 2703 265–66, 461
NGC 2705 265–66, 461
NGC 2707 265–66, 461
NGC 2713 246
NGC 2715 277
NGC 2716 246
NGC 2720 267
NGC 2727 197
NGC 2736 420
NGC 2760 350
NGC 2774 373
NGC 2775 91
NGC 2776 193
NGC 2795 246
NGC 2799 313
NGC 2804 197
NGC 2806 314–15
NGC 2807 *209*
NGC 2807A *209*
NGC 2808 80
NGC 2809 197, *209*
NGC 2818 *74*
NGC 2818A 74
NGC 2819 411–12
NGC 2826 201
NGC 2830 36

NGC 2835 397, 401
NGC 2841 136, 167
NGC 2843 35–36
NGC 2846 311
NGC 2847 317
NGC 2848 300, 317
NGC 2849 80–81, *82*
NGC 2852 208
NGC 2853 197, 208
NGC 2859 136
NGC 2861 246
NGC 2863 425, 436
NGC 2867 38
NGC 2869 425, 436
NGC 2871 309, *310*
NGC 2872 26, *310*
NGC 2873 *310*
NGC 2874 309, *310*
NGC 2875 309, *310*
NGC 2876 436
NGC 2887 80–81
NGC 2901 434
NGC 2903 44–45, 61, 91, 112–14, 116, 136, 193, 214, 300
NGC 2905 44–45, 61, 114, 214
NGC 2908 21
NGC 2912 289
NGC 2919 266
NGC 2926 411–12
NGC 2934 254
NGC 2944 411–12
NGC 2953 80
NGC 2954 80
NGC 2961 311
NGC 2973 86
NGC 2974 436
NGC 2977 242, 515
NGC 2979 436
NGC 2981 411–12, *413*
NGC 2997 30
NGC 3007 376, 435
NGC 3013 309
NGC 3016 200, 302
NGC 3031 136, 167
NGC 3034 48, 136–37, 167
NGC 3047 365–66
NGC 3050 436
NGC 3054 326, *327*, 436
NGC 3055 33, 300
NGC 3057 20–22
NGC 3058 433
NGC 3063 22, *23*, 48
NGC 3065 22, *23*
NGC 3066 22, *23*, 321
NGC 3069 315
NGC 3071 411
NGC 3075 80

NGC 3077 68, 186, 290
NGC 3080 23
NGC 3094 411–12
NGC 3097 *337*
NGC 3102 *337*
NGC 3107 23
NGC 3110 375–76, 434, 436, 445
NGC 3114 75
NGC 3115 167, 290
NGC 3116 411
NGC 3119 *134*
NGC 3120 81
NGC 3121 133, *134*, 146, 183, 216
NGC 3122 375–76, 434, 436, 445
NGC 3123 129, 204
NGC 3126 283–84
NGC 3129 197
NGC 3132 134, 143, 214
NGC 3134 372
NGC 3135 197, 238
NGC 3143 387
NGC 3144 247
NGC 3148 454
NGC 3153 326–27, 372
NGC 3162 193, 211, 247
NGC 3165 300, 373
NGC 3166 300
NGC 3167 245
NGC 3169 300
NGC 3172 24, 57, *58*, 81, 455
NGC 3174 247
NGC 3179 304
NGC 3183 247
NGC 3184 191, 193
NGC 3185 *199*, 237, 285
NGC 3190 191, 282
NGC 3193 282
NGC 3198 116, 193
NGC 3200 405
NGC 3210 48, 80
NGC 3212 48
NGC 3214 313
NGC 3215 48, 372
NGC 3218 242, 247, 515
NGC 3222 145–46, 183, 286
NGC 3226 141, 146, 150, 182, 186, 208
NGC 3227 141, 146, 150, 182, 186, 208
NGC 3229 129, 204
NGC 3230 516, *517*
NGC 3234 197, 210
NGC 3235 210
NGC 3239 241
NGC 3242 5, 37–38, 40, 42–43, 134, 143, 152, 167, 207, 214, 300, 305, 513
NGC 3244 86
NGC 3272 289
NGC 3279 *371*, 372

Object index 639

NGC 3280 387, 435
NGC 3280A 435
NGC 3280B 435
NGC 3280C 435
NGC 3284 448
NGC 3285 83
NGC 3286 448
NGC 3295 387, 435
NGC 3301 210, *211*, 312
NGC 3305 83
NGC 3307 82–83
NGC 3308 83
NGC 3309 82–83
NGC 3310 114, 190, 193
NGC 3311 82–83
NGC 3312 83
NGC 3314 83
NGC 3315 337
NGC 3316 83, 337
NGC 3317 337
NGC 3321 387, 435
NGC 3322 387, 435
NGC 3328 326–27
NGC 3332 23, 371–72
NGC 3336 83
NGC 3342 23, 371–72
NGC 3344 193
NGC 3345 208, 279, 515
NGC 3351 136, 167, 193
NGC 3353 267
NGC 3355 337, 448
NGC 3357 246
NGC 3359 197
NGC 3360 387
NGC 3361 387
NGC 3362 373
NGC 3367 193
NGC 3368 116, 136, 167, 193
NGC 3372 5, 86, 514, 516–17, *518*, 522
NGC 3375 373
NGC 3377 136, 153
NGC 3379 17, 31, 136, 306
NGC 3382 311
NGC 3384 136, 306
NGC 3388 386–87
NGC 3389 306
NGC 3391 246
NGC 3395 191, 193
NGC 3396 191
NGC 3397 242, 515
NGC 3401 455
NGC 3402 386–87
NGC 3404 387
NGC 3410 311
NGC 3411 386
NGC 3413 210
NGC 3414 136

NGC 3419 267
NGC 3421 387
NGC 3422 387
NGC 3423 91, 300
NGC 3425 332, 386–87
NGC 3427 372
NGC 3428 387–88
NGC 3429 387–88
NGC 3430 253
NGC 3436 372
NGC 3440 300
NGC 3441 387, 405
NGC 3445 300
NGC 3447 80–81
NGC 3452 387
NGC 3457 55, *56*, 303–04
NGC 3460 303–04
NGC 3461 303–04
NGC 3462 372
NGC 3476 387–88
NGC 3479 434–35
NGC 3480 387–88
NGC 3485 193
NGC 3486 136
NGC 3489 136, 153
NGC 3490 387
NGC 3492 326–27, 461
NGC 3495 81
NGC 3497 436, 443, *444*, 445
NGC 3502 434–35
NGC 3504 136, 193
NGC 3505 81
NGC 3508 81
NGC 3510 193
NGC 3518 375–76, 434, 436, 445
NGC 3521 136, 168, 300
NGC 3522 343, 350
NGC 3525 436, 445
NGC 3526 405
NGC 3528 81, 436, 445
NGC 3529 *444*
NGC 3531 405
NGC 3534 283
NGC 3537 387–88
NGC 3541 387–88
NGC 3544 436
NGC 3547 197
NGC 3556 17, 31
NGC 3563 283
NGC 3571 436
NGC 3575 211, 247
NGC 3576 420
NGC 3579 420
NGC 3580 266
NGC 3581 420
NGC 3582 420
NGC 3584 420

NGC 3586 420
NGC 3587 114, 136, 168
NGC 3588 343, 350, 356, *357*
NGC 3593 280, 306
NGC 3596 193, 403
NGC 3598 267
NGC 3604 27, 371–72
NGC 3609 283
NGC 3611 27, 371–72
NGC 3612 283
NGC 3621 23, *26*, 449
NGC 3623 91, 114, 136, 168, 193, 280
NGC 3627 91, 136, 168, 191, 214, 280
NGC 3628 110, 280, 300
NGC 3630 206
NGC 3631 193
NGC 3636 386–87
NGC 3637 386–87
NGC 3640 168, 206
NGC 3641 267
NGC 3645 206, 344
NGC 3646 193
NGC 3662 *164*, 166, 168, 176, 208, 514
NGC 3663 387
NGC 3665 136
NGC 3666 262, 519, *520*
NGC 3672 153
NGC 3675 136
NGC 3685 372
NGC 3688 387–88, 435
NGC 3690 303–04, 353–54, 517
NGC 3690A *353*
NGC 3690B *353*
NGC 3694 305
NGC 3695 305, *306*, 314–15
NGC 3698 305, *306*, 314–15
NGC 3700 305, *306*
NGC 3704 387–88
NGC 3705 279
NGC 3707 387–88
NGC 3722 387, *388*, 435
NGC 3723 387
NGC 3724 387, *388*, 435
NGC 3726 116, 193
NGC 3732 386–88
NGC 3736 393, *394*, 395, 459, 461
NGC 3739 283
NGC 3743 312, 314
NGC 3745 *314*
NGC 3746 313, *314*
NGC 3748 *314*
NGC 3750 *314*
NGC 3751 *314*
NGC 3753 *314*
NGC 3754 *314*
NGC 3758 312, 314, 373
NGC 3760 210, *211*, 312

NGC 3763 387
NGC 3766 75–76
NGC 3768 312
NGC 3775 387
NGC 3779 387
NGC 3792 405
NGC 3799 61
NGC 3800 61
NGC 3810 168, 193
NGC 3816 250
NGC 3821 22, 47
NGC 3837 22, 47
NGC 3840 250
NGC 3842 46–47
NGC 3843 405
NGC 3844 250
NGC 3847 247
NGC 3849 372
NGC 3853 277
NGC 3854 387, 435
NGC 3856 247
NGC 3857 272
NGC 3858 387, 435
NGC 3859 272
NGC 3860 47
NGC 3862 47
NGC 3864 272
NGC 3865 387, 435
NGC 3866 387, 435
NGC 3867 272
NGC 3868 272
NGC 3873 250
NGC 3875 47
NGC 3884 47
NGC 3886 250
NGC 3889 311
NGC 3893 168, 193
NGC 3894 45, 516
NGC 3895 45, 516
NGC 3905 387, 436
NGC 3910 283
NGC 3918 *38*, 80–81
NGC 3927 245
NGC 3928 208
NGC 3932 208, 245
NGC 3933 276, *277*, 383
NGC 3934 276, *277*, 383
NGC 3938 116, 193
NGC 3941 136
NGC 3948 424
NGC 3950 309
NGC 3953 191
NGC 3966 247
NGC 3975 309
NGC 3976 411–12
NGC 3979 348, 405–06
NGC 3986 247

NGC 3987 314, *316*
NGC 3989 314, *316*
NGC 3992 17, 31, 136
NGC 3993 314, *316*
NGC 3996 58
NGC 3997 314, *316*
NGC 3999 311, *316*
NGC 4000 311, *316*
NGC 4005 283–84, 314, *316*
NGC 4007 283–84
NGC 4008 55–56
NGC 4009 315, *316*
NGC 4011 315, *316*
NGC 4012 364
NGC 4015 315, *316*
NGC 4017 193
NGC 4018 315, *316*
NGC 4019 58
NGC 4021 315, *316*, *317*
NGC 4022 315, *316*
NGC 4023 315, *316*
NGC 4030 168
NGC 4038 5, 40, 45, 47–48, *49*, 191, 193
NGC 4039 5, 40, 45, 47–48, *49*, 191, 193
NGC 4042 254, 455, *456*, 457–58
NGC 4045 247, 371–72
NGC 4046 247, 371–72
NGC 4051 191, 193
NGC 4053 258
NGC 4058 337
NGC 4061 46, 516
NGC 4063 371–72, 375
NGC 4064 244–45
NGC 4065 46, 516
NGC 4069 46
NGC 4070 46
NGC 4072 313
NGC 4073 372
NGC 4074 46
NGC 4075 372
NGC 4076 46
NGC 4077 247, 371–72
NGC 4078 *246*, 247, 256, 258
NGC 4082 *246*
NGC 4083 *246*
NGC 4086 247
NGC 4089 247
NGC 4090 247
NGC 4091 247
NGC 4092 247
NGC 4093 247
NGC 4102 193
NGC 4107 *246*, 247, 257–58
NGC 4109 200, 302
NGC 4110 317
NGC 4111 136
NGC 4116 327

NGC 4118 201, 302–03, 319
NGC 4119 28
NGC 4123 193
NGC 4124 28
NGC 4125 124–25, 168, 183, 268
NGC 4129 247
NGC 4130 247
NGC 4131 46
NGC 4132 46
NGC 4134 46
NGC 4139 372
NGC 4140 247, 371–72
NGC 4151 191, 193
NGC 4156 191
NGC 4160 424
NGC 4165 267
NGC 4169 46, *46*
NGC 4170 *46*
NGC 4171 *46*
NGC 4173 46, *46*
NGC 4174 46, *46*
NGC 4175 46, *46*
NGC 4179 371–72
NGC 4182 326
NGC 4192 168, 280
NGC 4202 371–72
NGC 4203 135–37
NGC 4205 160, 183
NGC 4206 332
NGC 4208 448
NGC 4209 23
NGC 4212 448
NGC 4214 61, 300
NGC 4216 136, 280
NGC 4222 332
NGC 4228 61
NGC 4239 383
NGC 4240 387–88
NGC 4243 387–88
NGC 4247 337
NGC 4254 114, 116, 193, 214, 280
NGC 4255 272, 326–27
NGC 4258 17, 31, 168
NGC 4259 237
NGC 4268 237, *238*
NGC 4269 209–10, 286
NGC 4270 20, 46, 209–10, 237, *238*
NGC 4273 46, 237, *238*
NGC 4276 326
NGC 4277 46, 237, *238*
NGC 4281 46, 237, *238*
NGC 4291 241
NGC 4303 31, 61, 191, 194
NGC 4307 326, 332
NGC 4308 332, 362
NGC 4309 326–27, 461
NGC 4310 247

Object index   641

NGC 4314 364
NGC 4316 373
NGC 4319 241, 430, *431*, 461
NGC 4321 116, 194, 214
NGC 4322 329
NGC 4324 209, 237–38
NGC 4325 247
NGC 4327 329
NGC 4330 301–02, 321
NGC 4338 247
NGC 4345 430, 461
NGC 4347 326–27
NGC 4352 321
NGC 4353 326
NGC 4355 371–72
NGC 4357 425
NGC 4358 448
NGC 4360 373
NGC 4361 40, 136
NGC 4362 448
NGC 4364 448
NGC 4368 247
NGC 4374 168
NGC 4381 425
NGC 4383 236–38, 319
NGC 4389 191
NGC 4395 48, *49*, 61
NGC 4396 242–43
NGC 4399 47, *49*
NGC 4400 47, *49*
NGC 4401 47–48, *49*, 61
NGC 4402 148–49, 182–83, *186*
NGC 4406 168, 186
NGC 4411 326
NGC 4414 194
NGC 4418 371–72
NGC 4424 23
NGC 4426 425
NGC 4427 425
NGC 4435 47–48, 208
NGC 4438 47–48, 208
NGC 4443 317
NGC 4449 44, *46*
NGC 4452 153
NGC 4453 36
NGC 4458 208
NGC 4461 208
NGC 4464 177, *178*
NGC 4466 301–02, 321
NGC 4467 177, *178*, 282–83
NGC 4470 28, 177, *178*
NGC 4471 177, *178*, 179, 260, 519
NGC 4472 18, 31, 168
NGC 4473 155–57, 186, 514
NGC 4476 214
NGC 4477 155, *156*, 186, 212, 214, 514
NGC 4478 214

NGC 4479 155, *156*, 157, 212, 214
NGC 4485 191, 194, *195*
NGC 4486 168
NGC 4490 191, *195*
NGC 4492 177, *178*
NGC 4494 197
NGC 4501 116, 194, 214, 280, 300
NGC 4526 28, 154
NGC 4530 55, 157, *158*, 454
NGC 4536 191, 194
NGC 4544 358
NGC 4548 17, 31, 55, 219
NGC 4549 35
NGC 4559 168
NGC 4560 28, 153
NGC 4565 5, 30, 60, 91, 110, 168, 190, 213–14, 282
NGC 4567 45, 55, 61, 209
NGC 4568 45, 55, 61, 209
NGC 4569 168
NGC 4571 17, 55–56
NGC 4575 83
NGC 4576 405
NGC 4578 195
NGC 4579 116, 168
NGC 4581 405
NGC 4582 129, 204
NGC 4585 364
NGC 4587 411–12
NGC 4589 23
NGC 4592 405
NGC 4594 17, 31, 60, 91, 168, 214
NGC 4601 83
NGC 4603 83
NGC 4604 326–27, 461
NGC 4605 168
NGC 4606 214
NGC 4607 304
NGC 4610 28
NGC 4614 364
NGC 4615 364
NGC 4616 83
NGC 4618 45, 191, 194
NGC 4621 167, 214
NGC 4622 83
NGC 4624 28, 445
NGC 4625 194
NGC 4627 61, 114
NGC 4631 30, 61, 114, 117
NGC 4633 358
NGC 4634 214
NGC 4639 194
NGC 4644 48–49
NGC 4645 83
NGC 4646 48–49, 344
NGC 4647 61, 153, 196, 209
NGC 4649 61, 167

NGC 4650 83
NGC 4656 45, 61, 191
NGC 4657 45, 61, 191
NGC 4661 83
NGC 4664 27, 445
NGC 4665 27, 445
NGC 4669 321
NGC 4672 83
NGC 4676 44
NGC 4677 83
NGC 4683 83
NGC 4686 49
NGC 4689 194
NGC 4692 247
NGC 4695 48–49
NGC 4696 83
NGC 4697 167
NGC 4702 247
NGC 4706 83
NGC 4710 191
NGC 4725 116, 167, 194
NGC 4729 83
NGC 4730 83
NGC 4731 519
NGC 4736 136, 167, 194, 214
NGC 4738 425
NGC 4743 83
NGC 4744 83
NGC 4755 5, 86, 299, 517
NGC 4759 81, *330*
NGC 4760 263–64
NGC 4761 *330*
NGC 4762 91
NGC 4764 *330*
NGC 4774 47, 255, 329
NGC 4776 81
NGC 4778 *330*
NGC 4797 247
NGC 4798 247
NGC 4800 166–67
NGC 4802 332
NGC 4804 332
NGC 4826 60, 167, 214
NGC 4827 197
NGC 4840 319, 321
NGC 4858 248
NGC 4864 249
NGC 4866 280
NGC 4867 243, 424–25
NGC 4870 311
NGC 4871 243, 424–25
NGC 4873 243, 424–25
NGC 4874 248, *249*
NGC 4878 448
NGC 4879 448
NGC 4882 246–47
NGC 4883 243, 424–25

NGC 4884 247–48
NGC 4886 246–47
NGC 4888 448
NGC 4889 247–48, *249*, 424
NGC 4898 243, 426
NGC 4900 91
NGC 4908 245–46, 321
NGC 4912 309
NGC 4913 309
NGC 4914 309–10
NGC 4916 309
NGC 4921 *249*
NGC 4949 244–45
NGC 4960 247
NGC 4961 247
NGC 4969 358
NGC 4978 197
NGC 4982 329
NGC 4989 139, 182
NGC 4993 81
NGC 4994 81
NGC 4997 365–66, 405
NGC 5020 194
NGC 5022 436
NGC 5023 356
NGC 5024 68, 167, 186
NGC 5030 405
NGC 5031 405
NGC 5033 91, 300
NGC 5035 405
NGC 5038 405
NGC 5039 434
NGC 5046 405
NGC 5055 116, 194
NGC 5066 436
NGC 5068 260, *261*
NGC 5069 436
NGC 5072 243
NGC 5080 405
NGC 5088 363–64
NGC 5100 258
NGC 5101 20
NGC 5106 258
NGC 5112 191, 194
NGC 5128 5, *75*, 76
NGC 5139 80
NGC 5144 241
NGC 5162 449
NGC 5165 332, 366
NGC 5171 332, 365–66, *414*
NGC 5174 44–45, 449
NGC 5175 45, 455
NGC 5176 365, *414*
NGC 5177 365, *414*
NGC 5178 365, *414*
NGC 5179 332, 366, *414*
NGC 5186 414
NGC 5189 418

NGC 5191 365–66
NGC 5194 44–45, 60, 114, 116, 167, 194, 214, 281, 300, 482
NGC 5195 31, 44–45, 60, 114–16, 194, 209, 214, 281, 300, 484, 490–91
NGC 5200 129, 204
NGC 5226 315
NGC 5236 214
NGC 5247 211–12, 214, 491
NGC 5248 191, 194, 280
NGC 5253 23, 26
NGC 5268 139, 182
NGC 5272 136, 167
NGC 5278 310
NGC 5279 310–11
NGC 5309 349, 358
NGC 5310 129, 204
NGC 5315 393, 452
NGC 5319 304
NGC 5322 167
NGC 5338 311
NGC 5348 311
NGC 5350 281
NGC 5353 281
NGC 5354 281
NGC 5355 281
NGC 5358 281
NGC 5366 128, 204
NGC 5367 81
NGC 5377 91
NGC 5378 191
NGC 5404 129, 204
NGC 5405 413–14
NGC 5409 331
NGC 5414 330
NGC 5416 329–31
NGC 5418 413
NGC 5422 59
NGC 5423 330–31
NGC 5424 330–31
NGC 5426 45
NGC 5427 45, 194
NGC 5431 331
NGC 5434 331
NGC 5436 331
NGC 5437 331
NGC 5438 330–32
NGC 5439 350
NGC 5441 58
NGC 5443 59
NGC 5446 330–32
NGC 5457 167, 191, 194, 487
NGC 5465 *329*
NGC 5466 167
NGC 5467 328, *329*
NGC 5468 194, *329*
NGC 5471 244
NGC 5472 *329*, 332

NGC 5473 59
NGC 5487 337
NGC 5495 336–37
NGC 5496 405
NGC 5502 347, 356, 358
NGC 5503 347, 356, 358
NGC 5509 425
NGC 5511 365–66
NGC 5511A *367*
NGC 5511B *367*
NGC 5519 247
NGC 5522 44–45
NGC 5534 332, 375
NGC 5535 254, 456
NGC 5538 256, 258, 302, 321
NGC 5544 300
NGC 5545 300
NGC 5557 115–16, 194
NGC 5566 207
NGC 5570 247
NGC 5584 397, 399
NGC 5601 306
NGC 5619 271–72
NGC 5632 128, *130*, 182, 204
NGC 5634 167
NGC 5648 426, 455
NGC 5649 426, 455
NGC 5651 128, *130*, 182, 204
NGC 5653 362, 364
NGC 5657 362
NGC 5658 128, *130*, 182, 204
NGC 5660 197
NGC 5672 362, 364
NGC 5694 151
NGC 5699 375
NGC 5700 311
NGC 5703 375
NGC 5706 375
NGC 5707 340
NGC 5709 375
NGC 5713 194
NGC 5734 *433*, 434
NGC 5740 300
NGC 5743 *433*
NGC 5752 311
NGC 5753 311
NGC 5755 311
NGC 5765 *200*
NGC 5765B *200*
NGC 5769 405
NGC 5775 91, 190
NGC 5783 344, 347, 351
NGC 5785 347, 351
NGC 5796 435
NGC 5808 160–61, 247
NGC 5813 91–92, 208
NGC 5814 208
NGC 5819 160, 247

NGC 5824 397, 402
NGC 5826 347
NGC 5834 56, 397–98
NGC 5838 91, *92*
NGC 5841 258
NGC 5846 423, *424*
NGC 5846A 423, *424*
NGC 5848 258
NGC 5856 34, 41, 454
NGC 5857 61, 191, *196*, 516–17
NGC 5858 405
NGC 5859 61, 191, *196*, 516–17
NGC 5866 17, 31, 114, 117, 167, 181, 219
NGC 5865 210
NGC 5868 210
NGC 5870 347
NGC 5872 337
NGC 5873 393, *394*, 452
NGC 5876 348
NGC 5877 260–61
NGC 5879 121
NGC 5883 337
NGC 5884 430, 461
NGC 5897 167
NGC 5898 460
NGC 5903 460
NGC 5904 136, 167
NGC 5907 30, 91, 241, 300–01
NGC 5919 350
NGC 5921 194
NGC 5928 17–18
NGC 5936 364
NGC 5948 376
NGC 5964 61, 107, 110
NGC 5976 422
NGC 5981 422
NGC 5982 422
NGC 5985 192, 194
NGC 5986 186, 325
NGC 5990 20, 297, 314
NGC 5994 304
NGC 5996 364
NGC 6002 309
NGC 6015 519
NGC 6020 362
NGC 6027 272
NGC 6028 426
NGC 6034 355
NGC 6039 355
NGC 6040 272, 355
NGC 6041 272, 355
NGC 6042 272, 355
NGC 6043 355
NGC 6044 355
NGC 6045 355
NGC 6046 426
NGC 6047 355

NGC 6049 299, 454
NGC 6050 348, 355
NGC 6052 258
NGC 6053 347, 355
NGC 6054 355
NGC 6055 355
NGC 6056 348, 355
NGC 6057 347, 355
NGC 6058 191
NGC 6061 355
NGC 6064 258
NGC 6073 197
NGC 6091 358
NGC 6093 167
NGC 6098 362–63
NGC 6099 362–63
NGC 6115 175
NGC 6117 254
NGC 6121 18
NGC 6123 351
NGC 6125 347, 445
NGC 6127 347–48, 445
NGC 6128 347, 445
NGC 6138 375–76
NGC 6153 393, 452
NGC 6170 347
NGC 6171 16–17, 31, 64–65, 80, 167, 186
NGC 6176 347
NGC 6187 427, *428*, 461
NGC 6189 74, 347
NGC 6191 347
NGC 6205 136, 143, 167, 191
NGC 6210 38, 68, 91, 121, 152, 175, 186, 281, 381
NGC 6212 344
NGC 6214 347
NGC 6218 167, 194
NGC 6229 40, *97*, 167, 186
NGC 6231 453
NGC 6235 167
NGC 6239 274, 321
NGC 6240 426
NGC 6241 35
NGC 6251 26
NGC 6252 26
NGC 6254 136, 143, 167
NGC 6257 197
NGC 6262 376
NGC 6276 257–58, *259*
NGC 6277 257–58, *259*
NGC 6278 257–58, *259*, 273
NGC 6285 344, 351
NGC 6288 358
NGC 6289 358
NGC 6297 347
NGC 6298 347
NGC 6299 242

NGC 6301 41
NGC 6302 74, 397–98, *399*
NGC 6307 242
NGC 6308 254
NGC 6309 265–66, 329, 391
NGC 6310 242
NGC 6314 254
NGC 6315 254
NGC 6316 74
NGC 6326 74
NGC 6333 143, 167
NGC 6337 214
NGC 6341 167, 186, 292
NGC 6344 430, 461
NGC 6347 362
NGC 6352 397
NGC 6353 430, 461
NGC 6354 397, 400
NGC 6356 167
NGC 6359 242
NGC 6363 *375*
NGC 6364 255, 271–72, 284, 363–64, 375, 448
NGC 6366 146, 182, 455
NGC 6371 364
NGC 6382 358
NGC 6384 246, *254*, 255
NGC 6389 326–27
NGC 6395 358
NGC 6401 36
NGC 6402 64, 168, 186
NGC 6404 86
NGC 6405 18
NGC 6411 242
NGC 6412 33
NGC 6414 274, 350
NGC 6415 74, 85
NGC 6416 406
NGC 6418 358
NGC 6421 85
NGC 6426 274, 321
NGC 6427 276
NGC 6428 214
NGC 6431 272, 276, 377
NGC 6434 148
NGC 6439 391
NGC 6441 186, 325
NGC 6445 392
NGC 6446 460
NGC 6447 460
NGC 6457 358
NGC 6475 456
NGC 6480 84–85
NGC 6481 326
NGC 6484 362
NGC 6488 347
NGC 6497 347
NGC 6498 347

644 Object index

NGC 6503 *147*, 148–49, 168, 182
NGC 6508 414
NGC 6510 347
NGC 6511 347
NGC 6512 242
NGC 6514 26, 31, 86, 95, 214, 516
NGC 6516 242
NGC 6519 176, 179, 260
NGC 6520 142, *143*, 400
NGC 6521 242
NGC 6522 176
NGC 6523 18, 86
NGC 6527 362
NGC 6529 74
NGC 6532 358
NGC 6535 124–25, 182
NGC 6536 358
NGC 6537 391
NGC 6539 169–70, 182
NGC 6541 69, *71*, 80, 186
NGC 6543 5, *97*, 98, 152, 166, 168, 186, 250, 282, 407, 459
NGC 6544 36, 390
NGC 6548 376
NGC 6549 375–76
NGC 6550 375–76
NGC 6551 434
NGC 6553 186, 325
NGC 6558 81
NGC 6563 74
NGC 6565 391
NGC 6566 242
NGC 6567 391
NGC 6569 23, 26
NGC 6570 246
NGC 6572 37, 68, 91, 121, 143–44, 168, 175, 186, 281
NGC 6574 276
NGC 6575 272, 362
NGC 6577 364
NGC 6578 391
NGC 6580 364
NGC 6585 358
NGC 6589 362, *363*, 403
NGC 6590 *363*, 403
NGC 6595 403
NGC 6599 375
NGC 6600 375
NGC 6603 57, 220
NGC 6610 276
NGC 6611 462
NGC 6612 351
NGC 6618 60, 86, 95, 168, 214, 516
NGC 6620 *391*
NGC 6621 346–47, *353*, 356, 358
NGC 6622 346–47, *353*, 356
NGC 6626 136
NGC 6629 260

NGC 6633 29, 31
NGC 6634 220
NGC 6637 186, 220
NGC 6641 362
NGC 6643 127, 129–30, 168, 182, 204, 237–38, 514, 520
NGC 6644 *391*, 392
NGC 6648 68–69, *69*, 182
NGC 6654 411–12, 414
NGC 6655 146, 182
NGC 6656 455
NGC 6663 357–58
NGC 6666 349, 358, 455
NGC 6667 347–48, 354, 443, 445
NGC 6668 347, 354, 445
NGC 6669 255
NGC 6672 376
NGC 6674 363–64
NGC 6675 274, 364
NGC 6677 9, 347, 350, *351*, 352, 354
NGC 6678 347, 354, 445
NGC 6679 9, 347, 350, *351*, 352–54
NGC 6685 357–58
NGC 6686 357–58
NGC 6688 364
NGC 6689 347–48
NGC 6690 347–48
NGC 6693 255
NGC 6704 145–46, 182
NGC 6705 136, 143
NGC 6708 86
NGC 6709 280
NGC 6712 168
NGC 6717 36
NGC 6720 43, 60, 91, 136, 142, 161, 241
NGC 6723 176, 182, 325
NGC 6724 57–58
NGC 6726 *176*, 177, 179, 254, 258, 260
NGC 6727 *176*, 177, 179, 254, 258, 260
NGC 6729 *176*, 177, 179, 254, 258, 260, 522
NGC 6731 430, 461
NGC 6741 391
NGC 6748 274
NGC 6749 58
NGC 6751 257, 373
NGC 6759 272
NGC 6760 *123*, 124–25, 154, 156, 182, 236, 285–86, 300, 519
NGC 6762 347
NGC 6763 347
NGC 6765 256, *257*, 363–64
NGC 6766 391–93, 395
NGC 6767 430, 461
NGC 6775 120
NGC 6777 220
NGC 6778 258, 391–92
NGC 6779 168

NGC 6781 91, 153, 194, 214
NGC 6785 58, 258, 391–92
NGC 6790 391
NGC 6791 127, 129–30, *145*, 146, 182, 205
NGC 6792 430, 461
NGC 6797 326
NGC 6798 348
NGC 6803 391–92
NGC 6804 41
NGC 6807 391
NGC 6813 256–57
NGC 6818 91, 116, 143–44, *153*, 154, 159, 161, 168, 175, 241, 300, 400
NGC 6819 29, 64–65, 186, 300
NGC 6820 256–57
NGC 6822 5, 395–97, *400*, 401
NGC 6823 20
NGC 6825 358
NGC 6826 40, 60, 143–44
NGC 6833 391–92
NGC 6838 168
NGC 6842 245–46, 257, 363–64
NGC 6844 391
NGC 6846 274
NGC 6847 174, 245
NGC 6852 174, 257
NGC 6853 45, 60, 91, 114, 136, 143, 161, 168, 192, 214, 241, 281, 455
NGC 6857 197
NGC 6859 128, 130, 182, 204
NGC 6864 168
NGC 6866 29, 64, 186
NGC 6869 347
NGC 6871 68
NGC 6873 68
NGC 6879 *391*, 392–93, 395
NGC 6881 391
NGC 6882 64–65, 186, 292, 448
NGC 6884 392–93, 395
NGC 6885 64–65, 186, 285, 292, 448
NGC 6886 37, 393
NGC 6888 5, 300, *301*, 394
NGC 6891 393
NGC 6894 37, 116, 300, *301*
NGC 6897 364
NGC 6898 364
NGC 6903 197
NGC 6905 60, 91, *92*, 108, 116, 142, 151, 153–54, 191, 194, 214, 281, 514
NGC 6907 *255*
NGC 6908 *255*
NGC 6912 405
NGC 6914 376
NGC 6916 350, 444, 455, *456*
NGC 6921 271–72
NGC 6926 314
NGC 6933 289

Object index

NGC 6939 526
NGC 6941 361–62
NGC 6943 142
NGC 6946 179, 192, 194, 525–26
NGC 6951 277, 340, 461
NGC 6952 276, *277*, 340, 461
NGC 6960 5, 60, 96, 453
NGC 6962 516
NGC 6964 516
NGC 6966 242–43, 426
NGC 6973 453
NGC 6974 309–10, 453
NGC 6975 426
NGC 6976 426
NGC 6979 309–10, 453
NGC 6981 168
NGC 6992 5, 30, 95
NGC 6995 5, 61, 95, 453
NGC 6999 4–5
NGC 7000 2, 4–5, 30, *32*, 33, 222, 458, 559
NGC 7001 4–5, *6*
NGC 7005 154, *155*, 156, 186, 243, 245
NGC 7006 30
NGC 7008 41, 192
NGC 7009 18, *19*, 20, 22, 36–38, 43, 60, 68, 91, *92*, 114, *135*, 136, 143, 151–52, 159, 161, 168, 175, 186, 211, 214–15, 260, *281*, 381, 407, 452, 455, 514
NGC 7010 57–58
NGC 7023 41, 345
NGC 7026 363, 365, *366*, 382, 393–94
NGC 7027 9, 376–379, *380*, 382
NGC 7028 255
NGC 7040 *408*
NGC 7048 376–77
NGC 7051 197
NGC 7052 300
NGC 7053 246
NGC 7054 274
NGC 7065 246, 274
NGC 7066 347
NGC 7078 68, 136, 168, 186
NGC 7080 376
NGC 7088 *384–85*, 537
NGC 7089 136, 168
NGC 7092 454
NGC 7094 345, *349*, 382
NGC 7099 143
NGC 7108 276
NGC 7111 276
NGC 7114 392, *393*, 403, 430, 444, 455
NGC 7122 139, *140*, 182
NGC 7133 424
NGC 7134 326
NGC 7135 212
NGC 7140 448
NGC 7141 448

NGC 7143 197
NGC 7147 246
NGC 7150 124, 128, 204
NGC 7152 212
NGC 7153 212
NGC 7154 212
NGC 7164 434
NGC 7165 197
NGC 7167 212
NGC 7172 212
NGC 7173 212
NGC 7174 212
NGC 7177 192
NGC 7184 20, *21*, 65
NGC 7187 434
NGC 7190 274, 276, 320
NGC 7210 197
NGC 7226 *405–06*
NGC 7240 *271–72*
NGC 7242 *271–72*, 274
NGC 7254 436
NGC 7256 436
NGC 7257 276
NGC 7258 212
NGC 7259 212
NGC 7260 276
NGC 7284 212, *216–17*
NGC 7285 212, *216–17*
NGC 7288 364
NGC 7291 362
NGC 7293 5, 64–66, 71, 163, 168, 182, 325, 453
NGC 7317 *272–73*
NGC 7818 *273*
NGC 7318A 272
NGC 7318B 272
NGC 7319 *272–73*
NGC 7320 *272–73*
NGC 7320C *273*
NGC 7322 79
NGC 7325 289
NGC 7326 309
NGC 7331 5, 91, 116, 136, 168, 192, 195, 209, 524
NGC 7332 91, 209
NGC 7333 289
NGC 7334 79
NGC 7335 209, 524
NGC 7338 329
NGC 7339 209
NGC 7343 362
NGC 7347 159
NGC 7367 246
NGC 7375 362
NGC 7380 29
NGC 7383 199–200
NGC 7384 304
NGC 7387 201

NGC 7388 309
NGC 7390 304
NGC 7393 327
NGC 7402 304
NGC 7403 129, 232
NGC 7405 255
NGC 7416 363, 366
NGC 7422 246, *255*, 271, 283–84, 363–64, 448
NGC 7427 283
NGC 7433 *300–01*
NGC 7435 *300–01*
NGC 7436 300–01
NGC 7443 61, 265
NGC 7444 61, 265
NGC 7447 139, 182
NGC 7450 265–66
NGC 7451 283
NGC 7453 326–27
NGC 7457 300
NGC 7463 209, 241, 280, 515–16
NGC 7464 209, 241, 258, 279–80, 516
NGC 7465 209, 241, 280, 515–16
NGC 7472 244, 283–84
NGC 7477 243, *244*, 245, 284, 456
NGC 7479 91, 114, 117, 153, 159, 161, 241, *480*, 511
NGC 7482 244, 283–84
NGC 7486 313, 319
NGC 7489 216, 255
NGC 7490 405
NGC 7493 455
NGC 7507 212
NGC 7511 347
NGC 7513 254, 397, 400
NGC 7520 266
NGC 7528 387
NGC 7537 286, *295*
NGC 7541 286, *295*
NGC 7547 289
NGC 7548 160, 182, 242
NGC 7549 289–90, 301–02, 321
NGC 7550 289–90
NGC 7553 266, 289–90, 301–02, 321
NGC 7555 279
NGC 7559 159
NGC 7560 286, 289, 295
NGC 7561 286, 289, 295
NGC 7563 159
NGC 7565 270
NGC 7568 276
NGC 7571 289, *290*
NGC 7574 276
NGC 7577 425, 456
NGC 7581 131, 294, *295*
NGC 7582 212
NGC 7583 257–58
NGC 7588 *290*

NGC 7594 387–88
NGC 7595 387
NGC 7597 289, *290*
NGC 7598 *290*
NGC 7601 386–87
NGC 7602 *290*
NGC 7604 257
NGC 7605 257–58
NGC 7606 153
NGC 7608 246
NGC 7610 387–88
NGC 7612 256, 258
NGC 7613 270
NGC 7614 269, *270*
NGC 7616 387–88
NGC 7617 258
NGC 7619 47, 285
NGC 7620 267
NGC 7623 47
NGC 7626 47
NGC 7630 386–87
NGC 7631 200
NGC 7635 41
NGC 7638 386–87, *388*
NGC 7639 386–87, *388*
NGC 7648 269, 373
NGC 7649 348
NGC 7654 168
NGC 7662 5, 36, *37*, 38, 43, 60, 91, 114, 116, 143–44, 151–53, 168, 192, 194, 212, 214, 281–82, 290, 305, 381, 403, 514
NGC 7663 270
NGC 7664 267
NGC 7666 270
NGC 7667 269–70
NGC 7668 270
NGC 7669 270
NGC 7670 270
NGC 7673 246
NGC 7678 192, 194
NGC 7679 258
NGC 7683 267, 269–70
NGC 7686 292
NGC 7688 283–84, 326
NGC 7691 197
NGC 7692 128, 204
NGC 7693 409, *410*
NGC 7710 258
NGC 7712 266
NGC 7717 266
NGC 7724 267
NGC 7730 266
NGC 7738 269–70
NGC 7739 269–70
NGC 7748 454
NGC 7752 200, 302

NGC 7756 311
NGC 7759 344
NGC 7766 313
NGC 7767 313
NGC 7769 300
NGC 7770 300
NGC 7771 300
NGC 7780 344
NGC 7783 364
NGC 7789 5, 29, 168
NGC 7793 128, 131, 231–32
NGC 7804 232, *233*, 448
NGC 7805 516
NGC 7806 516
NGC 7810 48
NGC 7814 192, 300–01
NGC 7815 289
NGC 7819 313
NGC 7822 197
NGC 7827 53, 376
NGC 7828 434
NGC 7829 434
NGC 7830 255
NGC 7834 255, *256*
NGC 7835 255, *256*
NGC 7837 255, *256*
NGC 7838 *256*
NGC 7840 255–56
North America Nebula 2, 4–5, 30, *32*, 33, 222, 379, 458, 558–59
Nova Aurigae 1892 403
Nova Cygni 1876 392–93, 403, 429–30, 455
Nova Persei 1901 396
Nova T Scorpii 148, 176, 514
Nubecula Major 77, 85–86
Nubecula Minor 77, 85–86

Omega Centauri (ω Cen) 16, 79–80
Omega Nebula 16, 39, 60, 86, 89, 95, 104, 111, 142, 144, 301, 343, 406–07, 479, 516
Orion Nebula *1–2*, 9, 11, 14, 16, 18, 33, 38–39, 41, 52–53, 84, 86, 88–90, 103–05, 107, 112, 118, 124–25, 132–34, 138, *142*, 143, 150, 173, 190, 192, 195, 211, 215, 234, 251, 305, 307, *308*, 321, 333, 336, 361, 367–69, 385–86, 403, 406–07, 418, 431–32, 445, 450, 455, *475*, 476–77, 480–82, 488, 513–14, 516, 528, 531, 558
Owl Nebula 116, *117*, 190

Pegasus I Cluster 47
Pelican Nebula 2
Perseus Cluster 244–45, 250, 320, 423–25

Pleiades 1, 8, 13, 17, 30, 39, 171–73, 220, 384–86, 401, 438, 456, 466, 472, 479, 514, 521, *522*, 523–25, *526*, 527–31, 533–34, *535*, 536–40, *542*, 543–45, *546–47*, 548–550, *551–52*, 553, *554*, 555–556, *557–58*, 559–61
Polarissima Australis 57, 81, *82*, 455
Polarissima Borealis 24, 57, *58*, 81, 455
Praesepe 1, *2*, 15–16, 18, 30, 256, 454, 456
Ptolemy's Cluster 15–16, 456

Ring Nebula 16, 35–36, 42–43, 60, 107, 111, 116–17, 124, 205, 215, 241, 349, 403, 406–07, 410, 479
Rosette Nebula 5, 256, *257*, 303, 340, *341*, 342–43

Sagittarius Star Cloud 17, 57, 220, 464
S And 23, 410, 413–14, 429, 516
Saturn Nebula 5, *19*, 36, 68, 92, 151, 211, 215, 260, 381, 407, 452, 455
Sculptor group 420
Seyfert's Sextet 272
Siamese Twins 55, 61, 209
Silver Dollar Galaxy 5
Small Magellanic Cloud (SMC) 5, 15–16, 74, 77–78, 80, 84–86, 219–20, 420, 453–55
Sombrero Galaxy 31
Spindle Galaxy 290
Stephan's Quintet 272, *273*
Struve's Lost Nebula 124, 281, 283, 498, 504, 512

Tarantula Nebula 28, 86
The Antennae 5, 40, 45, 47–48, *49*
The Box 44, *46*
The Mice 44
Theta Orionis (θ Ori) 53, 86, 367, 369
Trapezium 9, 53, 77, 111, 132, 251, 361, 367, *368–69*, 403
Trifid Nebula 26, *27*, 31, 40, 86, *95*, *96*, 213, 406–07, 476, 516–17

UGC 492 465

vdB 69 *57*
Veil Nebula 5, 30, 60, 95, 309–10, 453
Virgo Cluster 3, 46–47, 86, 148–49, 155, 209, 223, 248, 456, 467, 514

Whirlpool Nebula 60, 115
Wild Duck Cluster 16, *475*
Winnecke 4 146

Subject index

The compilation excludes names of persons, telescopes, sites (observatories) and designated objects; see the indices for these categories.

annular nebula, 106–07, 114–17, 410
asterism, 1
Astronomical Society Catalogue (ASC), 78–79
Atlas écliptique, 205, 438, 498

Bedford Catalogue, 15, 119–21, 157, 159, 379, 382, 481
Berliner Akademische Sternkarten, 122, 138, 149, 174, 184, 205
binocular, 21
bipolar nebula, 16, 45, 61, 208, 287, 375–76, 515
Bonner Durchmusterung (BD), 9, 88, 131, 158, 161–63, 165–66, 173–74, 184, 214, 233, 237, 291, 330, 380, 424, 451, 466, 473, 506, 524, 529
British Association Catalogue (BAC), 55, 135, 154, 180, 223
Brisbane catalogue, 72, 78
Bruce medal, 147

Carte du Ciel, 326, 437–38, 549
catalogue identity, 11, 27, 79
Catalogus Britannicus, 159
cave nebula, 400, 551
Celestial charts, 324–27
central star, 39–40, 42–43, 104, 116, 118, 157, 345, 395, 473
classification, 8, 22, 44, 83, 290, 333
colour, 35–38, 528
comet, 2, 17, 62, 70–71, 88, 93, 126, 131, 158, 173, 178, 187, 217, 234, 260, 284, 337, 341, 345–46, 379, 515, 523
cometary nebula, 35, 40, 61, 176, 549, 560
comet eye-piece, 400, 528, 541
comet-seeker, 65–66, 77, 88, 127, 137–38, 144–46, 148, 161–66, 169–70, 173, 179, 262, 274, 341, 397, 503, 525–26
compact galaxy, 352–54, 424
contrast, 33, 43
core-nebula, 177, 260, 268, 285, 473

Crossley reflector, 208, 382, 389, 467, 481, 560

dark nebula, 1, 142, 396, 400, 406, 452, 510, 517
Digitized Sky Survey (DSS), 5, 12, 470
double galaxy, 40, 47, 61, 160, 196, 209, 216, 265, 273, 294, 301, 303, 330, 346, 353, 356, 363, 367, 423
double nebula, 8–10, 44, 46–47, 54, 60–62, 83, 148, 154, 179, 182, 187, 202–03, 208–09, 242, 286–87, 310, 346, 353–54, 387–88, 481, 515
dwarf galaxy, 400

ecliptical charts, 88, 122, 138, 205, 251, 498, 526
edge-on galaxy, 5, 12, 29–30, 60, 111, 159, 213, 241, 300–01, 377, 421
elliptical galaxy, 82, 84, 115, 194
emission nebula, 1, 12, 30, 40, 58, 74, 81, 242, 255–57, 320, 349, 390, 397, 452, 521
entrance pupil, 113, 538–39
exit pupil, 113, 486, 539
extensive diffused nebulosity, 32, 181, 551, 558

face-on galaxy, 30
field of view, 25–26, 28, 33, 41, 95, 113, 172, 212, 239, 264, 294, 308, 328, 340, 342, 390, 404, 418, 486, 523, 531, 534, 538, 540, 547, 559
field sweeping, 66, 526, 542
filter, 33
finding eye-piece, 110, 113, 116
Franklin-Adams charts, 466
front-view, 20, 103, 112, 118, 138, 299, 305, 422, 477, 479

galactic nebula, 1, 209, 213, 243, 329, 385, 452–53
galaxy, 1, 23, 30, 40, 55, 58, 61, 74, 81, 91, 104, 135, 178, 187, 213, 224, 243, 255, 265, 271, 274, 279, 282, 303, 308, 310, 315, 318, 320, 327, 329, 349, 355, 365, 376–77, 385–86, 388, 397, 401, 424, 434, 453
galaxy chain, 46, 55, 356, 388, 422
galaxy cluster, 46
galaxy group, 12, 190, 255, 271–72, 288, 290, 301, 313–14, 365, 469
galaxy trio, 82, 241, 356, 435, 516
gaseous nebula, 112, 119, 150, 250–51, 380, 507, 535
Gegenschein, 166, 169, 556
glass mirror, 34, 99, 110, 118, 131, 205, 215, 306, 388, 415, 417, 419, 501, 528
glass secondary, 299
globular cluster, 1, 30, 39, 42, 58, 61, 69, 74, 81, 111, 113, 115, 135, 187, 194, 442, 453
gold medal (RAS), 50, 52, 66, 73, 100, 121–22, 125, 132, 147, 150, 240, 282, 360, 386, 390, 397, 422, 427
gold medal (Warner), 339
grand amateur, 9
Great Melbourne Telescope (GMT), 72, 77, 100, 133, 222, 252, 415–21, 451, 518
group of nebulae, 249, 256, 273, 312, 354, 423
Guide Star Catalogue (GSC), 79, 372

HII region, 44–45, 47–49, 61, 160–61, 174, 183, 244, 255, 317, 363, 399–400, 424, 453
Hagen clouds, 385, 466, 510, 559
halation, 553, 555, 560
helical nebula, 65, 190, 213
heliometer (Bonn), 97, 144, 152, 162–63, 166
heliometer (Königsberg), 65, 69, 92, 123–24, 127, 130, 134, 139, 145–48, 154–55, 158, 163, 169–70, 182, 184, 186–87, 500, 523

Henry Draper Catalogue (HD), 9
Herschel omitted nebula (HON), 47–48, 59, 78, 85
Herschel Space Observatory, 15
Herschelian, 20, 170, 299
Histoire Céleste, 55, 68, 97, 122, 152, 154, 159, 170, 239, 441
Hubble classification, 1
Hubble Space Telescope (HST), 4, 79, 481, 487, 561
Hubble's law, 456

illumination, 42, 70, 113, 115, 138, 195, 227, 235, 290–91, 342, 422
integrated magnitude, 9, 12
irregular galaxy, 194
isophotic chart, 95–96

Jupiter, 35, 37–38, 94, 357, 360, 370, 396, 418, 560

Leviathan of Parsonstown, 100, 109, 317, 484
Lalande price, 126, 134, 276, 411
light pollution, 131, 133, 261, 274, 312, 339, 372, 377, 477
luminous fluid, 9, 40, 104–05, 484, 487
lunar temperature, 307, 310

Malta catalogue, 213, 217, 251, 254, 257, 319
Markree catalogue, 137–38, 206, 306
Mars, 133, 149, 306, 370, 409, 496
Mercury, 229, 271
micrometer, 10, 25, 42, 65, 67, 70, 90–91, 95, 102, 113, 115, 127, 135, 138, 141, 150–51, 165, 174–75, 182, 184, 187, 189, 195–96, 207, 215, 227, 235–36, 239, 248, 259, 262, 266, 268, 271, 278, 284–85, 288, 290–91, 296, 298–99, 307, 310, 313, 327–28, 342, 344, 354, 380, 398, 404, 422, 430, 432, 462, 465, 472–73, 499, 501, 503, 515, 517, 534

nebular hypothesis, 9, 11, 98, 105, 107, 110–12, 115, 118, 234, 476, 482, 484–87
nebulous star, 40–41, 43, 55, 57, 60, 62, 80, 118, 121, 139, 154, 157–59, 163–64, 299, 345, 354, 398, 402, 453, 544
Neptune, 10, 89, 122, 124, 132, 145, 148, 215
NGC/IC Project, 470
NGC identity, 11, 27, 55, 79, 346–47

North Pole Distance (NPD), 4, 8, 24, 54, 95, 219, 319, 440, 488

objective prism, 140, 390, 394–95
open cluster, 1, 33, 58, 74, 81, 84, 135, 187, 242–43, 292, 321, 382, 397, 420, 453, 459, 466
orbit-sweeper, 145, 262–63, 413–14

Palomar Observatory Sky Survey (POSS), 313, 467, 469–70
parallax, 10, 152, 154, 382
peculiar galaxy, 75–76, 195
photography, 4, 8, 11, 15, 43, 103, 125, 205, 249, 360, 382, 385, 390, 396, 398, 400, 402, 407, 411, 415, 418, 421, 427, 432, 437, 443–44, 458, 465–66, 472–73, 477, 479, 481, 492, 497–98, 510–11, 521–22, 542–46, 548–51, 553–57
photometry, 4, 8, 11, 23, 390
planetary nebula, 1, 9, 11, 14, 22, 26, 30, 32, 35, 37–40, 42, 47, 58, 60, 62, 64, 68, 74, 81, 89–91, 93, 97, 104, 108, 114, 116–17, 120, 135, 142, 144, 148, 152–54, 157, 161, 175, 180, 187, 194, 209, 213, 242, 255–57, 260, 282, 305, 308, 311, 320, 323, 333, 349, 363, 376, 389–92, 395, 403, 410, 442, 453, 459, 462, 473
portrait lens, 401, 553–54, 556, 558–59
proper motion, 8, 14, 91, 98, 134–35, 151–52, 154, 180, 187, 207, 238, 242, 288, 407, 410, 472–73, 475, 515, 517

quasar, 1, 271, 430–31

reduction, 10, 49, 473
reflection nebula, 1, 12, 40, 43, 81, 135, 176, 194, 242, 320, 349, 376, 452, 521
reflectivity, 34, 38, 110
resolvability, 8–9, 83, 86, 89, 93, 98, 104, 106–08, 110–12, 118, 190, 223–24, 279–80, 291, 333, 373, 472, 484, 486, 488
ring galaxy, 12, 47, 255, 329, 377, 433–34
Royal Medal, 100

satellite, 8, 11, 60–62, 148, 152–53, 241, 260
Saturn, 38, 94, 110, 125–26, 132, 174, 211, 215, 485
seeing, 175
Seyfert galaxy, 116, 194
silvered (glass) speculum, 188, 528, 546

South Pole Distance (SPD), 8, 74
spectrograph, 560
spectroscope, 144, 251, 280, 305, 308, 311, 321, 364, 380–82, 390, 392, 395, 418, 427, 507
spectroscopy, 1, 4, 8, 11, 98, 119, 140, 150, 250–51, 278, 292, 304, 313, 323, 333–34, 365, 389–92, 407, 415, 427, 443–44, 518
speculum, 34, 99, 110, 132, 190, 417, 495, 527, 531, 539
spiral galaxy, 110, 115, 195
spiral nebula, 9, 65, 112, 114–18, 195, 205, 472, 476, 480–82, 486–88, 493–94, 496–97, 560
star cloud, 397, 399
supernova, 23, 410, 412–13, 415, 429, 452, 516
supernova remnant (SNR), 1, 74, 95, 104, 108, 135, 161, 310, 390, 452–53, 521
superthin galaxy, 356–57
surface brightness, 9, 35, 37, 43, 64, 539

tarnishing, 30, 34, 51, 417–18
trans-Neptunian planet, 369–71
true nebulosity, 14, 39–40, 57, 62, 83, 89, 93, 98, 104–05, 109, 112, 118, 234, 476, 482

Uranus, 18, 23, 25, 35, 38, 80, 132, 215, 370–71, 379

Valz price, 411
variable nebula, 148, 162, 176, 205–07, 242, 472, 501, 512–14, 519, 521–22, 524, 527
Venus, 252, 267, 294, 306, 310–11, 413, 519
Very Large Telescope (VLT), 4
vignetting, 113, 539
Vulcan, 325, 338, 519

Webb Deep-Sky Society, 383
white nebula, 11
Willard lens, 396, 553
Wolf-Palisa charts, 411
Wolf-Rayet star/nebula, 301, 393, 395, 453
working list, 33, 50, 53, 73, 179, 189, 217, 231

zodiacal light, 345
zone catalogue, 7, 47, 50, 53, 179–80, 217, 222, 465–66